The Stability of Matter: From Atoms to Stars

Selecta of Elliott H. Lieb

Springer
Berlin
Heidelberg
New York
Barcelona
Hong Kong
London
Milan
Paris
Tokyo

Elliott H. Lieb

The Stability of Matter: From Atoms to Stars

Selecta of Elliott H. Lieb

Edited by W. Thirring

With a Preface by F. Dyson

Third Edition

Springer

Professor Elliott H. Lieb
Departments of Mathematics and Physics
Jadwin Hall
Princeton University
P.O. Box 708
Princeton, New Jersey 08544-0708, USA

Professor Walter Thirring
Institut für Theoretische Physik, Universität Wien
Boltzmanngasse 5, 1090 Wien, Austria

Professor Freeman Dyson
The Institute for Advanced Study
School of Natural Sciences
Princeton, New Jersey 08540, USA

Library of Congress Cataloging-in-Publication Data
Lieb, Elliott H. The stability of matter: from atoms to stars: selecta of Elliott H. Lieb /
edited by W. Thirring; with a preface by F. Dyson. – 3rd ed.
p. cm. Includes bibliographical references.
ISBN 3540420835 (acid-free paper)
1. Thomas-Fermi theory. 2. Matter-Properties. 3. Functional analysis. 4. Mathematical physics.
I. Thirring, Walter E., 1927– II. Title.
QC173.4.T48 L54 2001 539'.1–dc21 2001041096

ISBN 3-540-42083-5 Third Edition
Springer-Verlag Berlin Heidelberg New York

ISBN 3-540-61565-2 Second Edition
Springer-Verlag Berlin Heidelberg New York

Springer-Verlag Berlin Heidelberg New York
a member of BertelsmannSpringer Science+Business Media GmbH

http://www.springer.de

Printed on acid-free paper SPIN 10836013 55/3141/xo - 5 4 3 2 1 0

Preface to the Third Edition

The second edition of this "selecta" of my work on the stability of matter was sold out and this presented an opportunity to add some newer work on the quantum-mechanical many-body problem. In order to do so, and still keep the volume within manageable limits, it was necessary to delete a few papers that appeared in the previous editions. This was done without sacrificing content, however, since the material contained in the deleted papers still appears, in abbreviated form, at least, in other papers reprinted here.

Sections VII and VIII are new. The former is on *quantum electrodynamics* (QED), to which I was led by consideration of stability of the non-relativistic many-body Coulomb problem, as contained in the first and second editions. In particular, the fragility of stability of matter with classical magnetic fields, which requires a bound on the fine-structure constant even in the non-relativistic case (item V.4), leads to the question of stability in a theory with quantized fields. There are many unresolved problems of QED if one attempts to develop a non-perturbative theory – as everyone knows. A non-perturbative theory is essential, however, if one is going to understand the stability of the many-body problem, which is the stability of ordinary matter. Some physicists will say that a non-perturbative QED does not exist – and this might be true in the absence of cutoffs – but an effective theory with cutoffs of a few Mev *must* exist since matter exists.

At the present time physicists believe fully in the non-relativistic Schrödinger equation and even write philosophical tracts about it. This equation is the basis of condensed matter physics and, while it is understood that relativistic effects exist, they are not large and can be treated as mild perturbations. The original Schrödinger picture is alive and well as a very good approximation to reality, and shows no signs of internal inconsistency. For this reason, the 'stability of matter' question, which takes the equation literally and seriously without approximation, is accepted as a legitimate question about physical reality.

The radiation field, on the other hand, especially the quantized radiation field, is almost always treated perturbatively in one way or another. The Schrödinger equation, including coupling to the quantized radiation field treated perturbatively, describes ordinary matter with great accuracy, for it is the basis of chemistry and solid state physics.

One can ask if there is a theory, analogous to the Schrödinger theory, that incorporates particles and electromagnetic fields together without having to resort to perturbation theory. If it were really impossible to have a self-consistent, non-perturbative theory of matter and the low frequency radiation field it would not be unreasonable to call this a crisis in physical theory. It has always been assumed that at each level of physical reality there is a self consistent theory that describes phenomena well at that level. For example, we have chemistry, which is well described at the level of molecules, phase transitions and thermodynamics, which is presumably well described by statistical mechanics, etc. At the present time we have the beginnings of a consistent theory of matter and radiation at the nonrelativistic level, but a corresponding relativistic theory, is less developed.

Section VII describe some primitive attempts to address some of the questions that have to be resolved. One problem concerns renormalization. Starting from a relativistic Dirac theory, one finds that the radiation field forces a renormalization of the electron mass. This is well known, but what is not so well known is that the nonrelativistic theory forces an even bigger renormalization. Thus, if one starts from a relativistic theory, performs the renormalization, and then asks for the nonrelativistic limit of such a theory, it appears that one has to make further renormalizations – which it should not be necessary to do.

The new Section VIII is devoted to a subject that was a favorite of mine from the beginning of my career: *many-boson systems.* Some of the papers reprinted in the first and second editions on the stability of charged boson systems have been moved to this section. A very brief summary of all this material, as well as a list of my earliest work on this topic is given in an introduction VIII.1. Three new papers (VIII.2, 3, 4) are about the solution to a very old problem, namely the calculation of the ground state energy of a system of Bose particles with short-range forces at low density. The stability question is trivial, but the calculation of physical quantities in this intensely quantum-mechanical regime requires another sort of physical insight that goes back half a century but which was never rigorously validated.

Currently, the study of low density, low temperature Bose gases in "traps" is an active area of experimental physics. The extension of the homogeneous gas results to inhomogeneous gases in traps is in VIII.3. Similar subtleties arise for charged bosons in a neutralizing background ("jellium"), and this was resolved recently in VIII.7. The new material is summarized in VIII.4.

Certainly, much more remains to be understood about the ground states of bosonic systems. The two-component charged Bose gas, which is related to the jellium problem (see VIII.6) should be better understood, quantitatively. And, of course, there is the ancient quest for a proof of the existence of Bose condensation in interacting systems.

Princeton, March 2001 *Elliott Lieb*

Preface to the Second Edition

The first edition of "The Stability of Matter: From Atoms to Stars" was sold out after a time unusually short for a selecta collection and we thought it appropriate not just to make a reprinting but to include eight new contributions. They demonstrate that this field is still lively and keeps revealing unexpected features. Of course, we restricted ourselves to developments in which Elliott Lieb participated and thus the heroic struggle in Thomas-Fermi theory where the accuracy has been pushed from $Z^{7/3}$ to $Z^{5/3}$ is not included. A rich landscape opened up after Jakob Yngvason's observation that atoms in magnetic fields also are described in suitable limits by a Thomas-Fermi-type theory. Together with Elliott Lieb and Jan Philip Solovej it was eventually worked out that one has to distinguish 5 regions. If one takes as a dimensionless measure of the magnetic field strength B the ratio Larmor radius/Bohr radius one can compare it with $N \sim Z$ and for each of the domains

(i) $B \ll N^{4/3}$,
(ii) $B \sim N^{4/3}$,
(iii) $N^{4/3} \ll B \ll N^3$,
(iv) $B \sim N^3$,
(v) $B \gg N^3$

a different version of magnetic Thomas-Fermi theory becomes exact in the limit $N \to \infty$. In two dimensions and a confining potential ("quantum dots") the situation is somewhat simpler, one has to distinguish only

(i) $B \ll N$,
(ii) $B \sim N$,
(iii) $B \gg N$

and thus there are three semiclassical theories asymptotically exact. These fine distinctions make it clear how careful one has to be when people claim to have derived results valid for high magnetic fields. There is plenty of room for confusion if they pertain to different regions.

The question of stability of matter in in a magnetic field B has also been further cleaned up. Already the partial results in V.3 and V.4 of the first edition showed where the problem resides. Whereas diamagnetism poses no problem to stability since there a magnetic field pushes the energy up, the paramagnetism of the electron's spin magnetic moment can lower the energy arbitrarily much with increasing B. Only if one includes the field energy $+\frac{1}{8\pi}\int B^2$ one gets a bound uniform in B. This has now been shown to be true for arbitrarily many electrons and nuclei provided α and $Z\alpha^2$ are small enough. Though the sharp constants have not been determined, for $\alpha = 1/137$ and $Z \leq 1050$ (nonrelativistic) stability is guaranteed. Two recent works pertaining to this question have now been included:

1. A transparent and relatively simple proof of stability for Coulomb systems with relativistic kinetic energy is now available.
2. The problem has been studied with the Dirac Hamiltonians for the individual electrons in an external magnetic field. Here the question of how to fill the Dirac sea arises and it turns out that the correct way is to use the Hamiltonian including the magnetic field. Once this is done the Coulomb interaction does not introduce an instability provided the charges are below certain limits. On the contrary if one fills the Dirac sea of the free electrons and then introduces the magnetic field then one gets instability. Amusingly the relativistic theory also allows the above nonrelativistic limit on Z to be improved to $Z \leq 2265$.

Unfortunately QED has not yet matured mathematically to a state where these questions can be answered in a full-fledged relativistic quantum field theory and only in patchworks of relativistic corrections some answers are obtained. Thus this field is by no means exhausted and will keep challenging future generations.

Vienna, November 1996 *Walter Thirring*

Preface to the First Edition

With this book, Elliott Lieb joins his peers Hermann Weyl and Chen Ning Yang. Weyl's *Selecta* was published in 1956, Yang's *Selected Papers* in 1983. Lieb's "Selecta", like its predecessors, gives us the essence of a great mathematical physicist concentrated into one convenient volume. Weyl, Yang and Lieb have much more in common than the accident of this manner of publication. They have in common a style and a tradition. Each of them is master of a formidable mathematical technique. Each of them uses hard mathematical analysis to reach an understanding of physical laws. Each of them enriches both physics and mathematics by finding new mathematical depths in the description of familiar physical processes.

The central theme of Weyl's work in mathematical physics was the idea of symmetry, linking physical invariance-principles with the mathematics of group-theory. One of Yang's central themes is the idea of a gauge field, linking physical interactions with the mathematics of fibre-bundles. The central theme of Lieb's papers collected in this book is the classical Thomas-Fermi model of an atom, linking the physical stability of matter with the mathematics of functional analysis. In all three cases, a rather simple physical idea provided the starting-point for building a grand and beautiful mathematical structure. Weyl, Yang and Lieb were not content with merely solving a problem. Each of them was concerned with understanding the deep mathematical roots out of which physical phenomena grow.

The historical development of Lieb's thinking is explained in the review articles in this volume, items 65, 92 and 136 in Lieb's publication list. I do not need to add explanatory remarks to Lieb's lucid narrative. I suppose the reason I was asked to write this preface is because Lenard and I found a proof of the stability of matter in 1967. Our proof was so complicated and so unilluminating that it stimulated Lieb and Thirring to find the first decent proof, included here (paper 85 in the publication list). Why was our proof so bad and why was theirs so good? The reason is simple. Lenard and I began with mathematical tricks and hacked our way through a forest of inequalities without any physical understanding. Lieb and Thirring began with physical understanding and went on to find the appropriate mathematical language to make their understanding rigorous. Our proof was a dead end. Theirs was a gateway to the new world of ideas collected in this book.

Princeton, March 1990 *Freeman Dyson*

Contents

Part IV. Thomas–Fermi and Related Theories

Part V. Stability of Matter

Part VI. The Thermodynamic Limit for Real Matter with Coulomb Forces

Part VII. Quantum Electrodynamics

Part VIII. Bosonic Systems

Introduction

1

"Once we have the fundamental equation (*Urgleichung*[1]) we have the theory of everything" is the creed of some physicists. They go on to say that "then physics is complete and we have to seek other employment". Fortunately, other scientists do not subscribe to this credo, for they believe that it is not the few Greek letters of the *Urgleichung* that are the essential physics but rather that physics consists of all the consequences of the basic laws that have to be unearthed by hard analysis. In fact, sometimes it is the case that physics is not so much determined by the specific form of the fundamental laws but rather by more general mathematical relations. For instance, the KAM-theorem that determines the stability of planetary orbits does not depend on the exact $1/r$ law of the gravitational potential but it has a number theoretic origin. Thus a proper understanding of physics requires following several different roads: One analyzes the general structure of equations and the new concepts emerging from them; one solves simplified models which one hopes render typical features; one tries to prove general theorems which bring some systematics into the gross features of classes of systems and so on. Elliott Lieb has followed these roads and made landmark contributions to all of them. Thus it was a difficult assignment when Professor Beiglböck of Springer-Verlag asked me to prepare selecta on one subject from Lieb's rich publication list[2]. When I finally chose the papers around the theme "stability of matter" I not only followed my own preference but I also wanted to bring the following points to the fore:

(a) It is sometimes felt that mathematical physics deals with epsilontics irrelevant to physics. Quite on the contrary, here one sees the dominant features of real matter emerging from deeper mathematical analysis.

(b) The *Urgleichung* seems to be an ever receding mirage which leaves in its wake laws which describe certain more or less broad classes of phenomena. Perhaps the widest class is that associated with the Schrödinger equation with $1/r$-potentials, which appears to be relevant from atoms and molecules to bulk matter and even cosmic bodies. Thus the papers reproduced here do not deal with mathematical games but with the very physics necessary for our life.

(c) In mathematics we see a never ending struggle for predominance between geometry and analysis, the fashions swinging between extremes. Not too long ago the intuitive geometrical way in which most physicists think was scorned by

[1] *Urgleichung* (which is a word coined by Heisenberg) has now been replaced by the TOE (theory of everything) of the advocates of superstring theory.

[2] Henceforth quoted as [Lieb, ...].

mathematicians and only results from abstract analysis were accepted. Today the pendulum has swung the other way and the admired heroes are people with geometric vision, whereas great analysts like J. von Neumann tend to be thought of as degenerate logicians. As a physicist one should remain neutral with regard to internal affairs of mathematics, but it is always worthwhile to steer against the trend. Stability of matter illustrates beautifully that the great masters have forged for us the very analytical tools which we need to extract the physics from the fundamental equations.

In the selected papers the deeper results are carefully derived and what is presupposed on the reader's part are standard mathematical relations like convexity inequalities. Thus, it is to be hoped that this volume is not only of historical interest but also a useful source for workers in quantum mechanics.

2 Extensivity

We all take for granted that two liters of gasoline contain twice as much energy as one liter. In thermodynamics this fact has been dogmatized by the statement that the energy is an extensive quantity. This means that if $E(N)$ is the energy of the N-particle system then $\lim_{N\to\infty} E(N)/N$ is supposed to exist, or that the energy per particle approaches a limit in the many-body problem. This is what is meant by "stability of matter" and it was tacitly assumed by people working in this field even though at first sight it seems rather improbable for a system of particles interacting via two-body potentials. The Hamiltonian

$$H_N = \sum_{i=1}^{N} \frac{p_i^2}{2m_i} + \sum_{i>j} v(x_i - x_j) \tag{1}$$

contains a double sum and one might expect $E(N) \sim N^2$ rather than $E(N) \sim N$. Historically the existence of $\lim_{N\to\infty} \frac{1}{N} E(N)$ was first seriously studied in the framework of classical statistical mechanics by Van Hove [1] where he had to assume a potential with a hard core. However, real matter, which consists of electrons and nuclei, obeys the laws of quantum mechanics and the dominant interactions are electrostatic or, for cosmic bodies, gravitational. In the latter case the N-dependence of the quantum mechanical ground state energy E_N was first established by Levy-Leblond [2] who proved that in the gravitational case, $v(x_i - x_j) = -\kappa m_i m_j / |x_i - x_j|$, $E_N \sim N^3$ for bosons and $E_N \sim N^{7/3}$ for fermions. Consequently, as was pointed out by Fisher and Ruelle [3], the situation for purely electrostatic interactions, $v(x_i - x_j) = e_i e_j / |x_i - x_j|$, was not so clear. A new chapter in science was opened up when this question was finally settled by Dyson and Lenard [4] who derived in a seminal paper the fundamental result that in this case actually $E_N \sim N$ if all particles of one sign of charge are fermions. If this is not the case and there are bosons of both signs of charge then they could show that E_N is somewhere between $N^{5/3}$ and $N^{7/5}$. It is fairly obvious that there is no chance for stability if the potential is all attractive, but it was rather shocking that electrostatics with its screening property might lead to instability. One might be inclined to consider this as a pathology of the Coulomb potential $1/r$ with its long range and its singularity and that things would be stable once these troublemakers are removed. That this is not the case is illustrated by the following [5].

Proposition (2)

For the Hamiltonian

$$H_N = \sum_{i=1}^{N} \frac{p_N^2}{2m} + \sum_{i>j} e_i e_j v(|x_i - x_j|), \qquad e_i = \pm e, \qquad \sum_i e_i = 0,$$

the quantum mechanical ground state energy E_N satisfies

(i) $E_N \sim N$ for fermions, $E_N \sim N^{7/5}$ for bosons if

$$v(r) = \frac{1}{r},$$

(ii) $E_N \sim N$ for fermions and bosons if

$$v(r) = \frac{1 - e^{-\mu r}}{r},$$

(iii) $E_N \sim N^2$ for fermions and bosons if

$$v(r) = r e^{-\mu r}.$$

Thus it becomes clear that $E_N \sim N$ happens only under exceptional circumstances and one has to inquire what goes wrong in the case that $E_N \sim N^\gamma$, $\gamma > 1$. It is sometimes said that in this case the thermodynamic limit does not exist and therefore thermodynamics no longer applies. Though there is some truth in this statement, the argument is somewhat superficial and a more detailed analysis is warranted. One might object that in reality one deals with finite systems and the existence of $\lim_{N \to \infty}$ is only of mathematical interest. This is not quite true because the existence of the limit simply says that for a large but finite system the corresponding quantities are already close to their limiting value and the system can be described by the idealized situation $N \to \infty$. But for this to be true the kind of limit that is taken is irrelevant. For instance, in the gravitational case one can show the existence of $\lim_{N\to\infty} N^{-7/3} E_N$ for fermions (or $\lim_{N\to\infty} N^{-3} E_N$ for bosons). For stars ($N \sim 10^{57}$) one is certainly close to the limit, and their energy is predicted equally well by this kind of $N^{7/3}$ limit as in the stable situation. However I shall now discuss three dominating features which set the case $E_N \sim N$ apart from $E_N \sim N^\gamma$, $\gamma > 1$, and justify the terminology "stability of matter".

2.1 Relativistic Collapse

The relativistic expression for the kinetic energy $\sqrt{p^2 + m^2}$ weakens the zero point pressure to the extent that the "relativistic" quantum mechanical Hamiltonian

$$H_N^r = \sum_i \sqrt{p_i^2 + m_i^2} + \sum_{i>j} v(x_i - x_j) \tag{3}$$

might cease to be bounded from below if N is sufficiently big. Although this fact was recognized at the beginning of quantum mechanics [6], its dramatic consequence that a star of more than about twice the solar mass will collapse was doubted by leading astrophysicists. Only the experimental discovery of pulsars in the sixties provided overwhelming evidence for this prediction of quantum mechanics and therefore its relevance for cosmic bodies. Thus, there seems to be an essential difference between the relativistic and the nonrelativistic Hamiltonian. Whereas (1) with $1/r$-potentials is always bounded from below – and instability only implies that its lower bound E_N is not proportional to N – relativistically E_N may be $-\infty$ for all $N > N_c$ = some critical particle number ("relativistic collapse"). Related to this is the fact that nonrelativistically the size of the ground state may shrink with N but it always stays finite, whereas relativistically for $N > N_c$ the system keeps contracting until effects not contained in (3) take over. The two phenomena are related by the following fact which illustrates why instability is catastrophic [7].

Proposition (4)
If in (1) $v(x)$ scales like $v(\lambda x) = \lambda^{-1} v(x)$ then nonrelativistic instability implies relativistic collapse.

The converse statement is false since nonrelativistically matter is stable for arbitrary values of e, whereas relativistically the two-body Hamiltonian already becomes unbounded from below for $\alpha > \pi/2 = \alpha_c(2)$. Once it was found by Daubechies and Lieb [Lieb, 150] that these stability limits for α become more severe in the many-body situation, the question of stability of Coulomb matter with relativistic kinetic energy appeared more serious because $\lim_{N\to\infty} \alpha_c(N)$ could conceivably be zero. That this is not so for fermions was first shown by Conlon [8] who proved for nuclear charge $Z = 1$ $\alpha_c(N) > 10^{-200}$ for all N, a value which was subsequently improved by Fefferman and de la Llave [9]. Finally, Lieb and Yau [Lieb, 186] obtained the optimal result which is stability if $\alpha Z \leq 2/\pi$ and $\alpha \leq 1/94$. By Proposition (4) in the other cases (Coulomb matter with positive and negative bosons or with gravitation) the relativistic collapse is unavoidable.

2.2 Thermodynamic Stability

Shortly after $H_N \geq -N \cdot const$ was established, Lieb and Lebowitz showed the existence of the thermodynamic functions in the limit $N \to \infty$ for Coulomb matter [Lieb, 43]. This does not follow automatically from stability because the long range of the Coulomb potential poses additional problems. They had to demonstrate that this is sufficiently screened so that separated portions of matter are sufficiently isolated. They not only generalized the work of Van Hove et al. [1,3] to the realistic situation but also considered the microcanonical ensemble and showed that the specific heat was positive. Van Hove had shown that classically the limiting system had positive compressibility which Lieb and Lebowitz also verified for the quantum Coulomb case. These conditions for thermodynamic stability are concisely expressed by a convexity property. Denote by $H_{N,V}$ the N-particle Hamiltonian (1) in a volume V, by $\mathcal{H}_S$ an e^S-dimensional subspace of the Hilbertspace in which $H_{N,V}$ acts, and $\mathrm{Tr}_{\mathcal{H}_S}$ the trace in this subspace. Then for a finite system the energy as function of particle number, entropy and volume is

$$E(N,S,V) = \inf_{\mathcal{H}_S} \mathrm{Tr}_{\mathcal{H}_S} H_{N,V}. \tag{5}$$

In the thermodynamic limit one considers

$$\varepsilon(N,S,V) = \lim_{\lambda\to\infty} \lambda^{-1} H(\lambda N, \lambda S, \lambda V) \tag{6}$$

and thermodynamic stability requires that ε is jointly convex in its arguments. Lest the reader might think that one gets thermodynamic stability for free one has to note that it was shown by Hertel and Thirring [10] that in the gravitational case the temperature-dependent Thomas-Fermi theory becomes exact in a certain limit such that in this case $\varepsilon(N,S,V)$ not only exists but is calculable. Since this system is not stable, ε has to be defined by

$$\varepsilon(N,S,V) = \lim_{\lambda\to\infty} \lambda^{-7/3} H(\lambda N, \lambda S, \lambda^{-1} V). \tag{7}$$

It turned out that this system was also thermodynamically unstable; it showed a region of negative specific heat [11,12,13]. Such a phenomenon had been previously discovered in some models [14,15,16] and was always suspected by astrophysicists. That it was no accident that it occurred for the nonextensive Hamiltonian was clarified by Landsberg [17] by means of the following.

Theorem (8)

Let $x \to f(x)$ be a map from a convex set of $\mathbf{R}^n$ into $\mathbf{R}$. Then any two of the conditions

Homogeneity:	$H : f(\lambda x) = \lambda f(x), \quad \lambda \in \mathbf{R}^+$
Subadditivity:	$S : f(x_1 + x_2) \le f(x_1) + f(x_2)$
Convexity:	$C : f(\lambda x_1 + (1-\lambda) x_2) \le \lambda f(x_1) + (1-\lambda) f(x_2)$

imply the third.

If we take for f the function $\varepsilon(N,S,V)$ then H is obviously satisfied in the stable situation (6) but not in the unstable case (7). Thus H is a condition for stability against implosion. On the other hand, the subadditivity S means a stability against explosion; one gains energy by putting two parts together. Finally, C is thermodynamic stability. The theorem says that any two of these stability notions imply the third or, if one fails to hold, the others cannot both be true. In the cases we considered, S holds. This is clear for attractive interactions like gravity whereas for repulsive potentials (e.g. all charges having the same sign) the system would be explosive and S is violated. However, for electrically neutral systems there is always a van der Waals attraction [Lieb, 166] so that S is satisfied, which means that H and C become equivalent. Thus it turns out that our stability condition in terms of extensitivity is equivalent to thermodynamic stability provided the system is not explosive.

2.3 The Existence of Quantum Field Theory

Systems with negative specific heat cannot coexist with other systems. They heat up and give off energy until they reach a state of positive specific heat. This is what thermodynamic considerations tell us [15] but it was doubted [18] that in these cases thermodynamics is applicable and reflects the dynamics of the system. However, recent computer studies [19,20] of the dynamics of unstable systems have revealed that they behave exactly the way one expected by determining the dominant feature in phase space. In these studies one solved the classical equations of motion for 400 particles on a torus with

$$v(x_i - x_j) = -\exp[-|x_i - x_j|^2/b^2]$$

or

$$v(x_i - x_j) = e_i e_j (x_i - x_j)^2 \exp[-|x_i - x_j|^2/b^2].$$

It turned out that irrespective of the initial state, if the total energy is sufficiently small a hot cluster of size b with N_c, say 150 particles, developed, the rest being a homogeneous atmosphere. The temperature increased proportional to N_c. Thus there is no hope for a limiting dynamics for $N \to \infty$. In particular in quantum field theory, where N is not restricted, one cannot expect that a Hamiltonian H with a smooth potential that, however, leads to an unstable system leads to a dynamics in the Heisenberg representation via the usual formula

$$a(t) = e^{iHt} a e^{-iHt}.$$

Since states with unlimited N will lead to unlimited temperatures and thus unlimited velocities there cannot be a state independent dynamics. These ideas about the relevance of stability for the existence of quantum field theory were brought forward first by Dyson [21] and have been substantiated by proofs only in the converse direction. If one achieves stability by a momentum cut-off then there exists a dynamics in the Heisenberg representation for the quantum field theory [22].

Thus, if one wants a many-body system to behave in the way we are used to, the first question to be answered is its stability. Instability seems to be the rule rather than the exception and actually we owe our lives to the gravitational instability. Not only did Boltzmann's heat death not take place but, on the contrary, the universe, which originally is supposed to have been in an equilibrium state, developed hot clusters, the stars. Thus we can enjoy the sunshine which is rich in energy and lean in entropy, the kind of diet we need.

3

We have seen how a seemingly inconspicuous inequality turned into a key notion for understanding many-body physics. One of the great contributors to this development was Elliott Lieb. He usually did not say the first word on any of these issues but, with various collaborators, the last. His work (with B. Simon) [Lieb, 97] which put Thomas-Fermi (T.F.) theory on a firm basis opened the way for a better understanding of stability of matter. Since they proved that there is no chemical binding in T.F. theory, stability of matter becomes obvious once one knows that the T.F. energies are lower bounds for the corresponding quantum problem. That this is so is still a conjecture but it was shown by E. Lieb (with W. Thirring) [Lieb, 85] that this is true if one changes the T.F.-energies by some (N-independent) numerical factor.

If the electrons were bosons there remained the question whether the lower bound $\sim -N^{5/3}$ or the upper bound $\sim -N^{7/5}$ given by Dyson and Lenard [4] reflects the true behavior. This question was settled in an unexpected way. Lieb showed that if the masses of the nuclei are ∞ then $-N^{5/3}$ is correct [Lieb, 118] whereas if they are finite he supplied (with J. Conlon and H.-T. Yau) [Lieb, 188] a lower bound $\sim -N^{7/5}$, thereby showing the essential difference between these two cases. This is all the more surprising since in the justification of the Born-Oppenheimer approximation, which says that there is not much difference between big and infinite nuclear mass, usually no reference is made to the fermionic nature of the electrons.

Once one has established the gross features of Coulombic matter reflected by T.F. theory one can worry about finer details not described by it. Of these I will single out two to which Lieb has significantly contributed. One is the question of negative ions (which do not exist in pure T.F. theory but exist if the von Weizsäcker correction to the kinetic energy is added [Lieb, 130]). Ruskai [24] was the first to show that $N(Z)$, the maximal number of electrons bound by a nucleus (in the real quantum theory), is finite. Lieb gave an elegant proof that $N(Z) < 2Z + 1$ [Lieb, 157]. Furthermore he showed (with I. Sigal, B. Simon and W. Thirring) [Lieb, 185] that if the electrons are fermions then $\lim_{Z\to\infty} N(Z)/Z = 1$. (If the electrons were bosons this limit would be 1.21 [Lieb, 160], [23].) The latter result on real electrons has recently been improved by Fefferman and Seco but the proof of the conjecture $N(Z) < 2 + Z$ is still a challenge for the future.

A typical feature of quantum mechanics which is not included in T.F. theory or any single electron theory (or density functional theory for that matter) is the attraction between neutral objects. There Lieb (with W. Thirring) [Lieb, 166] showed that the Schrödinger equation predicts a potential which is below $-c/(R + R_0)^6$ where R is the distance between their centers and R_0 a length related to their size.

From Proposition (4) it follows that the relativistic Hamiltonian with gravitation becomes unbounded from below if $N > N_c$ = some critical particle number. It remained to be demonstrated that it stays bounded for $N < N_c$ and to determine N_c. Lieb (with W. Thirring) [Lieb, 158] showed that this was actually the case and gave some bounds for N_c which show the expected behavior $\kappa^{-3/2}$ for fermions and κ^{-1} for bosons (κ = gravitational constants in units $\hbar = c = m = 1$). Later he sharpened (with H.-T. Yau) [Lieb, 177] these results and showed that, for $\kappa \to 0$, N_c for fermions is exactly the Chandrasekhar limit.

The papers selected here show how powerful modern functional analysis is in determining the gross features of real matter from the basic quantum mechanical equations. Whereas a century's effort of the greatest mathematicians on the classical N-body problem with $1/r$ potentials only produced results for the simplest special cases, in quantum mechanics the contours of the general picture have emerged with much greater clarity.

The papers are grouped in various subheadings, the first paper being a recent review which will serve as a convenient introduction.

Vienna, November 1990 *Walter Thirring*

References

[1] L. Van Hove, Physica *15*, 137 (1950).

[2] J.M. Lèvy-Leblond, J. Math. Phys. *10*, 806 (1969).

[3] M. Fisher, D. Ruelle, J. Math. Phys. *7*, 260 (1966).

[4] F.J. Dyson, A. Lenard, J. Math. Phys. *8*, 423 (1967).

[5] W. Thirring, *A Course in Mathematical Physics*, Vol IV, Springer, Wien, New York, 1980.

[6] S. Chandrasekhar, Month. Not. Roy. Ast. Soc. *91*, 456 (1931).

[7] W. Thirring, Naturwissenschaften *73*, 605 (1986).

[8] J. Conlon, Commun. Math. Phys. *94*, 439 (1984).

[9] Ch. Fefferman, R. de la Llave, I. Rev. Math. Iberoamerican *2*, 119 (1986).

[10] P. Hertel, W. Thirring, J. Math. Phys. *10*, 1123 (1969).

[11] P. Hertel, W. Thirring, in *Quanten und Felder*, ed. H.P. Dürr, Vieweg, Braunschweig, 1971.

[12] A. Pflug, Commun. Math. Phys. *22*, 83 (1980).

[13] J. Messer, J. Math. Phys. *22*, 2910 (1981).

[14] D. Lynden-Bell, R. Wood, Month. Not. *138*, 495 (1968).

[15] W. Thirring, Z. f. Physik *235*, 339 (1970).

[16] P. Hertel, W. Thirring, Ann. of Phys. *63*, 520 (1971).

[17] P. Landsberg, J. Stat. Phys. *35*, 159 (1984).

[18] H. Jensen, private communications to W. Thirring.

[19] A. Compagner, C. Bruin, A. Roelse, Phys. Rev. *A 39*, 5989 (1989).

[20] H. Posch, H. Narnhofer, W. Thirring, in *Simulations of Complex Flows*, NATO A.S.I., ed. M. Mareschal (1990) and "Dynamics of Unstable Systems", UWThPh-1990-2, submitted to Phys. Rev. A

[21] F.J. Dyson, Phys. Rev. *85*, 613 (1952).

[22] H. Narnhofer, W. Thirring, "On Quantum Field Theories with Galilei-Invariant Interactions", Phys. Rev. Lett. *54*, 1863 (1990).

[23] B. Baumgartner, J. Phys. *A 17*, 1593 (1984).

[24] M.B. Ruskai, Commun. Math. Phys. *82*, 457 (1982).

Part I
Review

THE STABILITY OF MATTER: FROM ATOMS TO STARS

ELLIOTT H. LIEB

Why is ordinary matter (e.g., atoms, molecules, people, planets, stars) as stable as it is? Why is it the case, if an atom is thought to be a miniature solar system, that bringing very large numbers of atoms together (say 10^{30}) does not produce a violent explosion? Sometimes explosions do occur, as when stars collapse to form supernovae, but normally matter is well behaved. In short, what is the peculiar mechanics of the elementary particles (electrons and nuclei) that constitute ordinary matter so that the material world can have both rich variety and stability?

The law of motion that governs these particles is the quantum (or wave) mechanics discovered by Schrödinger [SE] in 1926 (with precursors by Bohr, Heisenberg, Sommerfeld and others). Everything we can sense in the material world is governed by this theory and some of its consequences are quite dramatic, e.g., lasers, transistors, computer chips, DNA. (DNA may not appear to be very quantum mechanical, but notice that it consists of a very long, thin, complex structure whose overall length scale is huge compared to the only available characteristic length, namely the size of an atom, and yet it is stable.) But we also see the effects of quantum mechanics, without realizing it, in such mundane facts about stability as that a stone is solid and has a volume which is proportional to its mass, and that bringing two stones together produces nothing more exciting than a bigger stone.

The mathematical proof that quantum mechanics gives rise to the observed stability is not easy because of the strong electric forces among the elementary constituents (electrons and nuclei) of matter. The big breakthrough came in the mid sixties when Dyson and Lenard [DL] showed, by a complicated proof, that stability

Received by the editors August 15, 1989.

1980 *Mathematics Subject Classification* (1985 *Revision*). Primary 81H99, 81M05, 85A15; Secondary 81C99, 82A15.

This paper was presented as the Sixty-Second Josiah Willard Gibbs Lecture on January 11, 1989 at the 95th annual meeting of the American Mathematical Society in Phoenix, Arizona.

Work partially supported by U.S. National Science Foundation grant PHY-85-15288-A03.

is, indeed, a consequence of quantum mechanics. (Part of their motivation came from earlier work by Van Hove, Lee and Yang, van Kampen, Wils, Mazur, van der Linden, Griffiths, Dobrushin, and especially Fisher and Ruelle who formulated the problem and showed how to handle certain well chosen, but unrealistic forces.) This was a milestone but there was room left for improvement since their results had certain drawbacks and did not cover all possible cases; for instance, it turns out that quantum mechanics, which was originally conceived to understand atoms, is also crucial for understanding why stars do not collapse. Another problem was that they proved what is called here stability of the second kind while the existence of the thermodynamic limit (Theorem 3 below), which is also essential for stability, required further work [LL]. The full story has now, two decades later, mostly been sorted out, and that is the subject of this lecture. The answer contains a few surprises, some of which are not even discussed in today's physics textbooks.

No physics background will be assumed of the reader, so Part I reviews some basic facts. Part II contains a synopsis of the aspects of quantum mechanics needed here. Part III treats the simplest system—the hydrogen atom, and Part IV introduces the strange Pauli exclusion principle for many electrons and extends the discussion to large atoms. Part V deals with the basic issue of the stability of matter (without relativistic effects) while Part VI treats hypothetical, but interesting, matter composed of bosons. Part VII treats the problems introduced by the special theory of relativity. Finally, Part VIII applies the results of Part VII to the structure of stars.

PART I. THE PHYSICAL FACTS AND THEIR PREQUANTUM INTERPRETATION

While it is certainly possible to present the whole story in a purely mathematical setting, it is helpful to begin with a brief discussion of the physical situation.

The first elementary constituent of matter to be discovered was the **electron** (J.J. Thomson, 1897). This particle has a **negative electric charge** (denoted by $-e$) and a **mass**, m. It is easy to produce a beam of electrons (e.g., in a television tube) and use it to measure the ratio e/m quite accurately. The measurement of e alone is much trickier (Millikan 1913). The electron can be considered to be a point, i.e., it has no presently discernible geometric structure. Since matter is normally electrically neutral (otherwise we would feel electric fields everywhere), there must also be another constituent with positive charge. One of the early ideas about this positive object was that it is a positively charged ball of a radius about equal to the radius of an atom, which is approximately 10^{-8} cm. (This atomic radius is known, e.g., by dividing the volume of a solid, which is the most highly compressed form of matter, by the number of atoms in the solid.) The electrons were thought to be stuck in this charged ball like raisins in a cake; such a structure would have the virtue of being quite stable, almost by fiat. This nice picture was destroyed, however, by Rutherford's classic 1903 experiment which showed that the positive entities were also essentially points. (He did this by scattering positively charged helium nuclei through thin metal foils and by showing that the distribution of scattering angles was the same as for the Kepler problem in which the trajectories are hyperbolas; in other words, the scatterers were effectively points—not extended objects.)

The picture that finally emerged was the following. Ordinary matter is composed of two kinds of particles: the point electrons and positively charged **nuclei**. There are many kinds of nuclei, each of which is composed of positively charged **protons** and chargeless **neutrons**. While each nucleus has a positive radius, this radius (about 10^{-13} cm) is so small compared to any length we shall be considering that it can be taken to be zero for our purposes. The simplest nucleus is the single proton (the nucleus of hydrogen) and it has charge $+e$. The number of protons in a nucleus is denoted by z and the values $z = 1, 2, \dots, 92$, except for $z = 43, 61, 85$, are found in nature. Some of these nuclei, e.g., all $84 \le z \le 92$, are unstable (i.e., they eventually break apart spontaneously) and we see this instability as naturally occurring radioactivity, e.g., radium. Nuclei with the missing z values 43, 61, 85, as well as those with $92 < z \le 109$ have all been produced artificially, but they decay more or less quickly [AM]. Thus, the charge of a naturally occurring nucleus can be $+e$ up to $+92e$ (except for 43, 61, 85), but, as mathematicians often do, it is interesting to ask questions about "the asymptotics as $z \to \infty$" of some problems. Moreover, in almost all cases we shall consider here, the physical constraint that z is an integer need not and will not be imposed. The other constituent—the neutrons—will be of no importance to us until we come to stars. They merely add to the mass of the nucleus, for they are electrically neutral. For each given z several possible neutron numbers actually occur in nature; these different nuclei with a common z are called isotopes of each other. For example, when $z = 1$ we have the **hydrogen nucleus** (1 proton) and the deuterium nucleus (1 proton and 1 neutron) which occur naturally, and the tritium nucleus (1 proton and 2 neutrons) which is artificial and decays spontaneously in about 12 years into a helium nucleus and an electron, but which is important for hydrogen bombs. Isolated neutrons are also not seen naturally, for they decay in about 13 minutes into a proton and an electron.

Finally, the **nuclear mass**, M, has to be mentioned. It satisfies $zM_p \le M \le 3zM_p$ where $M_p = 1837m$ is the mass of a proton. Since the nuclear mass is huge compared to the electron mass, m, it can be considered to be infinite for most purposes, i.e., the nuclei can be regarded as fixed points in $\mathbf{R}^3$, although the location of these points will eventually be determined by the requirement that the total energy of the electron-nucleus system is minimized. A similar approximation is usually made when one considers the solar system; to calculate the motion of the planets the sun can be regarded as fixed.

The forces between these constituents of matter (electrons and nuclei) is given by **Coulomb's inverse square law** of electrostatics: If two particles have charges q_1 and q_2 and locations x_1 and x_2 in $\mathbf{R}^3$ then F_1—the force on the first due to the second—is minus F_2—the force on the second due to the first—and is given by

$$-F_2 = F_1 = q_1 q_2 \frac{(x_1 - x_2)}{|x_1 - x_2|^3}. \tag{1.1}$$

(Later on, when stars are discussed, the gravitational force will have to be introduced.) If $q_1 q_2 < 0$ then the force is **attractive**; otherwise it is **repulsive**. This force can also be written as minus the gradient (denoted by ∇) of a **potential energy function**

$$W(x_1, x_2) = q_1 q_2 \frac{1}{|x_1 - x_2|}, \tag{1.2}$$

that is

$$F_1 = -\nabla_1 W \quad \text{and} \quad F_2 = -\nabla_2 W. \tag{1.3}$$

If there are N electrons located at $\underline{X} = (x_1, \dots, x_N)$ with $x_i \in \mathbf{R}^3$, and k nuclei with positive charges $\underline{Z} = (z_1, \dots, z_k)$ and located at $\underline{R} = (R_1, \dots, R_k)$ with $R_i \in \mathbf{R}^3$, the **total-potential energy function** is then

$$W(\underline{X}) = -A(\underline{X}) + B(\underline{X}) + U \tag{1.4}$$

with

$$A(\underline{X}) = e^2 \sum_{i=1}^{N} V(x_i) \tag{1.5}$$

$$V(x) = \sum_{j=1}^{k} z_j |x - R_j|^{-1} \tag{1.6}$$

$$B(\underline{X}) = e^2 \sum_{1 \le i < j \le N} |x_i - x_j|^{-1} \tag{1.7}$$

$$U = e^2 \sum_{1 \le i < j \le k} z_i z_j |R_i - R_j|^{-1}. \tag{1.8}$$

The A term is the electron-nucleus attractive potential energy, with $eV(x)$ being the electric potential of the nuclei. B is the electron-electron repulsive energy and U is the repulsive energy of the nuclei. A, B, U and V depend on $\underline{R}$ and $\underline{Z}$, which are fixed and therefore do not appear explicitly in the notation. It is then the case that the force on the ith particle is

$$F_i = -\nabla_i W. \tag{1.9}$$

In the case of an **atom**, $k = 1$ by definition. The case $k > 1$ will be called the **molecular** case, but it includes not only the molecules of the chemist but also solids, which are really only huge molecules.

So far this is just classical electrostatics and we turn next to classical dynamics. Newton's law of motion is (with a dot denoting $\frac{d}{dt}$, where t is the time)

$$m\ddot{x}_i = F_i. \tag{1.10}$$

This law of motion, which is a system of second order differential equations, is equivalent to the following system of first order equations. Introduce the **Hamiltonian function** which is the function on the **phase space** $\mathbf{R}^{6N} = (\mathbf{R}^3 \times \mathbf{R}^3)^N$ given by

$$H(\underline{P}, \underline{X}) = \frac{1}{2m} \sum_{i=1}^{N} p_i^2 + W(\underline{X}). \tag{1.11}$$

The notation $\underline{P} = (p_1, \dots, p_N)$ with p_i in $\mathbf{R}^3$ is used, and the quantity

$$T = \frac{1}{2m} \sum_{i=1}^{N} p_i^2 \tag{1.12}$$

is called the **kinetic energy**. The equations of motion (1.10) are equivalent to the following first order system in $\mathbf{R}^{6N}$

$$v_i \equiv \dot{x}_i = \frac{\partial H}{\partial p_i} \qquad (1.13)$$
$$\dot{p}_i = -\frac{\partial H}{\partial x_i}.$$

The **velocity** of the ith electron is v_i and p_i is called its **momentum**: $p_i = mv_i$ by the first equation in (1.13).

From (1.13) it will be seen that $H(\underline{P}, \underline{X})$ is constant throughout the motion, i.e., $dH(\underline{P}(t), \underline{X}(t))/dt = 0$. This fixed number is called the **energy** and is denoted by E; it depends, of course, on the trajectory, and it is important to note that it can take all values in $(-\infty, \infty)$.

Another interesting fact about the flow defined by (1.13), but one which will not be important for us, is that it preserves Lebesgue measure $dx_1 \cdots dx_N dp_1 \cdots dp_N$ on $\mathbf{R}^{6N}$; this is Liouville's theorem and it follows from the fact that the vector field that defines the flow, $(\partial H/\partial p_1, \dots, \partial H/\partial p_N, -\partial H/\partial x_1, \dots, -\partial H/\partial x_N)$, is divergence free. This theorem is one important reason for introducing the Hamiltonian formalism, for it permits a geometric interpretation of classical mechanics and is crucial for ergodic theory and statistical mechanics. The analogue in quantum mechanics turns out to be that quantum mechanical time evolution is given by a one parameter *unitary* group in Hilbert space (see (2.18))—but time evolution will not concern us here.

Consider the simplest possible case, neutral *hydrogen*, with $z = 1$ (a proton) and one electron ($N = 1$ and $k = 1$). With the proton fixed at the origin (i.e., $R_1 = 0$) the Hamiltonian is $p^2/2m - Ze^2|x|^{-1}$ and classical bound orbits (i.e., orbits which do not escape to infinity) of the electron are well known to be the ellipses of Kepler with the origin as a focus. These can pass as close as we please to the proton. Indeed, in the degenerate case the orbit is a radial line segment and in such an orbit the electron passes through the nucleus. One measure of average closeness of the electron to the nucleus in an orbit is the energy E, which is always negative for a bound orbit. Moreover E can be arbitrarily negative because the electron can be arbitrarily close to the nucleus and also have arbitrarily small kinetic energy T. A consequence of this fact is that the hydrogen atom would be physically unstable; in a gas of many atoms another particle or atom could collide with our atom and absorb energy from it. After many such collisions our electron could find itself in a tiny orbit around the nucleus and our atom would no longer be recognizable as an object whose radius is supposed to be 10^{-8} cm. Each atom would be an infinite source of energy which could be transmitted to other atoms or to radiation of electromagnetic waves.

The problem was nicely summarized by Jeans [J] in his 1915 textbook.

> "There would be a very real difficulty in supposing that the (force) law $1/r^2$ held down to zero values of r. For the force between two charges at zero distance would be infinite; we should have charges of opposite sign continually rushing together and, when once together, no force would be adequate to separate them... Thus the matter in the universe would tend to shrink into nothing or to diminish indefinitely in size."

The inability to account for stable atoms in terms of classical trajectories of pointlike charged particles was the major problem of prequantum physics. Since the existence of atoms and molecules was largely inferential in those days (nowadays we can actually "see" atoms with the tunneling electron microscope), the inability to account for their structure even led some serious people to question their existence—or at least to question the nice pictures drawn by chemists. The main contribution of quantum mechanics was to provide a quantitative theory that "explains" why the electron cannot fall into the nucleus. In brief, when the electron is close to the nucleus its kinetic energy—which could be zero classically—is forced to increase in such a way that the total energy (1.11) goes to $+\infty$ as the average distance $|x|$ goes to zero. This property is known as the **uncertainty principle**.

PART II. QUANTUM MECHANICS IN A NUTSHELL

Schrödinger's answer to the problem of classical mechanics was the following. While an electron is truly a point particle, its state at any given time cannot be described by a point $x \in \mathbf{R}^3$ and a momentum $p \in \mathbf{R}^3$ (or velocity $v = \frac{1}{m}p$) as in the classical view. Instead *the state of an electron is a (complex valued) function* ψ *in* $L^2(\mathbf{R}^3)$. Any ψ will do provided it satisfies the normalization condition

$$\|\psi\|_2^2 = \int_{\mathbf{R}^3} |\psi(x)|^2 \, dx = 1. \tag{2.1}$$

(Actually, this statement is not accurate; an electron has a property called **spin**, and the mathematical expression of this fact is that ψ is really in $L^2(\mathbf{R}^3; \mathbf{C}^2)$, i.e., $\psi(x)$ is a two-component spinor, $\psi(x) = (\psi_1(x), \psi_2(x))$ with each $\psi_i \in L^2(\mathbf{R}^3)$. This complication—which does not affect the present discussion very much—will frequently be ignored here.) Thus, the state of an electron is a point ψ in an infinite dimensional Hilbert space instead of a point (p, x) in $\mathbf{R}^6$.

The interpretation of ψ, due to M. Born [BM] (Jammer's book [JM] can be consulted for historical details) is that

$$\rho_\psi(x) \equiv |\psi(x)|^2 \tag{2.2}$$

is *the probability density for finding the electron at* x. The expected value of the potential energy $W(x)$ in the state ψ of one electron is then

$$W_\psi = \int_{\mathbf{R}^3} \rho_\psi(x) W(x)\, dx. \tag{2.3}$$

To localize an electron at a point x_0 would require $|\psi|^2 = \delta_{x_0}$, where δ_{x_0} is Dirac's delta measure, but this is obviously not the square of an $L^2(\mathbf{R}^3)$ function. Thus, it no longer makes sense to speak of an electron unambiguously located at a single point x_0.

What about the kinetic energy $T = p^2/2m$? To define T we first consider the Fourier transform of ψ scaled by the number h and defined by

$$\widehat{\psi}(p) = h^{-3/2} \int_{\mathbf{R}^3} \psi(x) \exp\left[-\frac{2\pi i}{h}(p, x)\right] dx. \tag{2.4}$$

Usually h is taken to be 1 or 2π in textbooks but any h can be used. Although $\widehat{\psi}$ depends on the choice of h, Plancherel's formula is always the following:

$$\int_{\mathbf{R}^3} |\widehat{\psi}(p)|^2 dp = 1. \tag{2.5}$$

The constant h relevant for quantum mechanics is Planck's constant; it is not arbitrary and has to be determined experimentally in order to satisfy the following physical interpretation of $|\widehat{\psi}(p)|^2$. In view of (2.5), $|\widehat{\psi}(p)|^2$ can be regarded as a probability density – it is *the probability density of the electron having a momentum* p. With this interpretation, the kinetic energy of the electron in the state ψ is

$$T_\psi = \frac{1}{2m} \int_{\mathbf{R}^3} |\widehat{\psi}(p)|^2 p^2 dp = \frac{1}{2m} \hbar^2 \int_{\mathbf{R}^3} |\nabla \psi(x)|^2\, dx, \tag{2.6}$$

where, in a customary notation, $\hbar = h/2\pi$. The role of h in the definition (2.4) is brought into focus by the right side of (2.6). Numerically $\hbar = 1.05 \times 10^{-27}$ erg seconds.

Another way to formulate (2.3) to (2.6) is to think of $L^2(\mathbf{R}^3)$ as a Hilbert space with inner product $(\psi, \phi) = \int \overline{\psi}(x)\phi(x)\,dx$ so that

$$W_\psi = (\psi, W\psi). \tag{2.7}$$

The momentum p is replaced by the self adjoint operator

$$p = -i\hbar\nabla \quad \text{and} \quad p^2 = -\hbar^2\Delta$$

with Δ being the Laplacian $\partial^2/\partial x^2 + \partial^2/\partial y^2 + \partial^2/\partial z^2$. Then the kinetic energy is (after integrating by parts)

$$2mT_\psi = (p\psi, p\psi) = \hbar^2\|\nabla\psi\|_2^2 = (\psi, p^2\psi) \tag{2.8}$$

These concepts can be generalized to the **many-body case** of N electrons by replacing $\mathbf{R}^3$ by $\mathbf{R}^{3N}$. Then $\psi(x)$ is replaced by $\psi(\underline{X}) = \psi(x_1, \dots, x_N)$ in $\bigotimes_1^N L^2(\mathbf{R}^3) \approx L^2(\mathbf{R}^{3N})$. (If spin is included then we use $\bigotimes_1^N L^2(\mathbf{R}^3, \mathbf{C}^2) \approx L^2(\mathbf{R}^{3N}; \mathbf{C}^{2^N})$.) A different generalization that might leap to the reader's mind would be to replace $\psi(x)$ by an N-tuple of functions on $\mathbf{R}^3$; that would lead to a nonlinear theory and would be wrong. The normalization condition is

$$\int_{\mathbf{R}^{3N}} |\psi(\underline{X})|\,dx_1 \cdots dx_N = 1 \tag{2.9}$$

and $|\psi(\underline{X})|^2$ is the joint probability density for finding the electrons at $x_1, \dots, x_N$. The potential and kinetic energies are given by

$$W_\psi = \int_{\mathbf{R}^{3N}} W(\underline{X})|\psi(\underline{X})|^2\,dx_1 \cdots dx_N \tag{2.10}$$

$$T_\psi = \frac{1}{2m}\hbar^2 \sum_{i=1}^N \int_{\mathbf{R}^{3N}} |(\nabla_{x_i}\psi)(\underline{X})|^2\,dx_1 \cdots dx_N. \tag{2.11}$$

The **energy** of ψ is then

$$E_\psi = T_\psi + W_\psi. \tag{2.12}$$

The problem that will concern us here is to find a lower bound to the **ground state energy**

$$E = E(N, k, \underline{Z}, \underline{R}) = \inf_\psi E_\psi, \tag{2.13}$$

Bull. Amer. Math. Soc. *22*, 1-49 (1990)

because the boundedness of E precludes the kind of collapse that Jeans was worried about. (Recall that E_ψ depends on $N, k, \underline{Z}$ and $\underline{R}$ since $W(\underline{X})$ does.) In (2.13) ψ is required to satisfy the normalization condition (2.9). It will also be required to satisfy another condition—the **Pauli exclusion principle**—which will be explained later.

The **absolute ground state energy** is

$$E(N, k, \underline{Z}) = \inf_{\underline{R}} E(N, k, \underline{Z}, \underline{R}) \tag{2.14}$$

and it is the ground state energy when the nuclei are placed in the most favorable locations.

There are two notions of stability

(A) **STABILITY OF THE FIRST KIND.** $E(N, k, \underline{Z}, \underline{R})$ is finite for every $N, k, \underline{Z}$ and $\underline{R}$.

(B) **STABILITY OF THE SECOND KIND.** There is a nonnegative function $z \mapsto A(z)$ such that for all N and k

$$E(N, k, \underline{Z}) \geq -A(z)(N+k) \tag{2.15}$$

provided each $z_j \leq z$. Recall that $z_j \leq 92$ in nature. The significance of the linear law (2.15) will be discussed later in Part V.

The infimum in (2.13) may or may not be attained by some ψ. If it is, a minimizing ψ, which may not be unique, is called a **ground state**. It satisfies the linear Euler-Lagrange equation (also known as the **time independent Schrödinger equation**)

$$H\psi = E\psi \tag{2.16}$$

where H is the **Schrödinger Hamiltonian**, namely the selfadjoint second order elliptic differential operator

$$H = -\frac{1}{2m}\hbar^2 \sum_{i=1}^{N} \Delta_{x_i} + W(\underline{X}). \tag{2.17}$$

In the previous Hilbert space notation, $E_\psi = (\psi, H\psi)$.

Quantum mechanics is the study of the eigenvalues of H, i.e., solutions of $H\psi = \lambda\psi$ with ψ in L^2, and also the associated (time dependent) Schrödinger equation of the time evolution

$$H\psi = -i\hbar\frac{\partial\psi}{\partial t} \tag{2.18}$$

which is important in several contexts, especially in the subject of scattering theory. The eigenvalues λ are called **energy levels** and differences between any two, $\lambda_i - \lambda_j$, is the energy carried away or absorbed by a photon when the system makes a "quantum jump" from eigenstate ψ_1 to eigenstate ψ_2. The fact that the system "jumps" from one eigenstate to another is a mysterious axiom of quantum mechanics (more precisely, quantum measurement theory) that need not concern us here, but it has the following important physical implication. According to Einstein, the frequency ν of a photon of energy e is always $\nu = e/h$. Thus, by energy conservation the photon emitted or absorbed has frequency $\nu = (\lambda_i - \lambda_j)/h$. Since the λ_i's are discrete, the frequencies (or spectral lines) are also discrete. With classical mechanics we would expect to find *all* frequencies, and the experimental existence of discrete spectral lines is regarded in many textbooks as being the crucial problem that classical physics could not explain. However, Jeans's question about the existence of stable atoms capable of emitting these discrete spectral lines can evidently be regarded as a question of higher priority.

In this lecture we shall be interested only in the lowest eigenvalue of H, which is E. Equation (2.18) is therefore irrelevant here. Also (2.16) will not be invoked because we shall deal directly with the minimization problem (2.13) which, incidentally, avoids the technically difficult question of the selfadjointness of H.

PART III. THE HYDROGENIC ATOM AND SOBOLEV'S INEQUALITY

In Parts III-VI we shall use units in which $\hbar^2/2m = h^2/8\pi^2 m = 1$ and $e = 1$. This can always be done by changing the length scale $x \to (\hbar^2/2me^2)x'$. The original energy E is related to the new energy E' (which will henceforth be denoted simply as E) by $E = (2me^2/\hbar^2)E'$.

The problem for hydrogen is to minimize

$$(3.1)\quad E_\psi = T_\psi + W_\psi = \int_{\mathbf{R}^3} |\nabla\psi(x)|^2 dx - z\int_{\mathbf{R}^3} |\psi(x)|^2 |x|^{-1} dx$$

with $\int |\psi|^2 = 1$. We can take ψ to be real, as is true in all the problems considered here, because the real and imaginary parts of ψ appear independently in (3.1) and (2.1). Of course $z = 1$ for hydrogen but it will be useful later to consider (3.1) with arbitrary $z > 0$.

The fact that T_ψ controls W_ψ, thereby making E finite, is most clearly seen with the Sobolev inequality

$$(3.2)\quad T_\psi = \|\nabla\psi\|_2^2 \ge S\|\psi\|_6^2 = S\left\{\int_{\mathbf{R}^3} \rho_\psi(x)^3 dx\right\}^{1/3} = S\|\rho_\psi\|_3$$

for any $\psi \in L^2(\mathbf{R}^3)$ (not necessarily normalized). Here $S = 3(\pi/2)^{4/3} \approx 5.5$ and we recall (2.2). The important point about (3.2) is that it bounds T_ψ in terms of ρ_ψ; the derivatives have been eliminated. The ground state energy can be bounded from below using (3.2), and only the fact that $\int \rho_\psi = 1$, by

$$E \geq \inf_\rho \left\{ S\|\rho\|_3 - z \int_{\mathbf{R}^3} |x|^{-1} \rho(x) dx \right\} \tag{3.3}$$

where the infimum is over all nonnegative ρ with $\int \rho = 1$. This minimization problem is an easy exercise and the answer is

$$E \geq -\frac{1}{3} z^2. \tag{3.4}$$

This is a remarkably good bound since the correct answer, which is obtained by solving (2.16), is $E = -z^2/4$.

Before moving on, it is important to understand heuristically why (3.1) leads to $E \approx -z^2$. If ψ is some nice function with $\|\psi\|_2 = 1$ and whose width is about L (for example, the Gaussian $\psi(x) = (\text{const.}) \exp[-x^2/L^2])$, then T_ψ is roughly L^{-2}. On the other hand, W_ψ is roughly $-zL^{-1}$. Therefore, the problem of minimizing E_ψ is roughly the same as

$$E \approx \min_L \left\{ \frac{1}{L^2} - \frac{z}{L} \right\} \tag{3.5}$$

The minimizing L is $L = 2/z$ and the minimum in (3.5) is $-z^2/4$, which is the right answer. Indeed, the minimizing ψ for (3.1) is

$$\phi(x) = (\text{const.}) e^{-z|x|/2} \tag{3.6}$$

which shows that the width of ψ is, in fact, about z^{-1}. From this calculation, and recalling that $|\psi(x)|^2$ is the probability density for the electron, we learn that *the size of a one-electron atom with nuclear charge z decreases with z like z^{-1}*.

There is another version of (3.2) that is technically weaker, but more transparent and eventually more useful. Since $\int \rho_\psi = 1$, a simple application of Hölder's inequality to the right side of (3.2) yields

$$T_\psi \geq K^1 \int_{\mathbf{R}^3} \rho_\psi(x)^{5/3} dx. \tag{3.7}$$

with $K^1 = S$. The constant K^1 can be improved from S to 9.57, but that is conceptually unimportant. Instead of (3.3) we then have

$$E \geq \inf_{\rho} \left\{ \int_{\mathbf{R}^3} [K^1 \rho(x)^{5/3} - |x|^{-1} \rho(x)] dx \right\} \tag{3.8}$$

and this yields a result only slightly worse than (3.4), namely $-0.35z^2$ with $K^1 = 9.57$. The important point is that (3.7) (with K^1 or with S) is just as good for our purposes as (3.3) because it yields the z^2 law, but the important advantage of (3.7) over (3.3) is that the integral comes with the exponent 1 in (3.7) instead of the exponent 1/3 in (3.2). This gives (3.7) the form of the integral of an energy density $\rho_\psi(x)^{5/3}$. Independent "bumps" in ρ_ψ give additive contributions to the right side of (3.7); in (3.2), by contrast, we have to take the cube root—which spoils the additivity.

The inequality (3.7) can be stated poetically as follows. *An electron is like a rubber ball, or a fluid, with an energy density proportional to* $\rho_\psi^{5/3}$. *It costs energy to squeeze it and this accounts for the stability of atoms.*

PART IV. LARGE ATOMS AND THE PAULI EXCLUSION PRINCIPLE

Suppose that $N = z$ and z is large, say $z = 50$. Since $N > 1$, W_ψ has two terms which come from $-A$ and B in (1.4). The constant U is zero since there is only one nucleus. The B term is positive but it turns out to be small compared to the A term and, for a preliminary orientation, can be neglected. We would then try to minimize

$$\widetilde{E}_\psi = \sum_{i=1}^{N} \int_{\mathbf{R}^{3N}} \{|\nabla_{x_i} \psi(\underline{X})|^2 - z|x_i|^{-1}|(\psi(\underline{X})|^2\} dx_1 \cdots dx_N \tag{4.1}$$

The minimum with $\|\psi\|_2 = 1$ is then achieved by

$$\psi(\underline{X}) = \prod_{i=1}^{N} \phi(x_i), \tag{4.2}$$

where $\phi(x)$ is the minimizer for (3.1) and is given in (3.6). This leads to

$$\widetilde{E} = \inf_\psi \widetilde{E}_\psi = -Nz^2/4 = -z^3/4. \tag{4.3}$$

From this calculation we would be led to two conclusions: (1) The energy of an atom is finite (stability of the first kind) but it grows like z^3. (ii) Because the scale of ϕ is z^{-1}, the size of a large atom decreases with z like z^{-1}. Conclusion (ii) is especially troublesome: *large atoms are smaller than small atoms.*

Both conclusions are false because of an additional axiom introduced by Pauli [P] in 1925 (and interpreted in the context of Schrödinger's wave mechanics by Dirac and Heisenberg in (1926)), namely the **exclusion principle**. This states that the allowed ψ's in the minimization problem (2.13) are the **antisymmetric functions** in $L^2(\mathbf{R}^{3N}) \approx \bigotimes^N L^2(\mathbf{R}^3)$. (The reason for the word exclusion will be explained shortly.) In other words $\psi \in \bigwedge^N L^2(\mathbf{R}^3)$, the antisymmetric tensor product of $L^2(\mathbf{R}^3)$. More explicitly

$$\psi(x_1, \dots, x_i, \dots, x_j, \dots, x_N) = - \psi(x_1, \dots, x_j, \dots, x_i, \dots, x_N)$$

for every $i \neq j$. If spin is taken into account then the condition is $\psi \in \bigwedge^N L^2(\mathbf{R}^3; \mathbf{C}^2)$. (Equivalently, ψ is an antisymmetric function in $L^2(\mathbf{R}^3, \mathbf{C}^{2^N})$.) The physical significance of this peculiar antisymmetry restriction will not be discussed here, but it is related to the fact that electrons are indistinguishable.

A function in $\bigwedge^N L^2(\mathbf{R}^3; \mathbf{C}^2)$ may be hard to think about—in which case the reader should ignore the complication of spin and think instead about $\bigwedge^N L^2(\mathbf{R}^3)$ without any essential loss. The following remarks may be useful, however. Elementary manipulations with the symmetric (or permutation) group S_N show that our minimization problem (2.13) (or even the Schrödinger equation (2.16) or (2.18)) in $\bigwedge^N L^2(\mathbf{R}^3; \mathbf{C}^2)$ is equivalent to solving the problem in a subset of the more familiar space $L^2(\mathbf{R}^{3N})$. This is the subset consisting all function $\psi = \psi(x_1, \dots, x_N)$ with the property that there is some integer $1 \leq J \leq N$ such that ψ is antisymmetric in the variables $x_1, \dots, x_J$ and also antisymmetric in the variables $x_{J+1}, \dots, x_N$ if $J < N$. No assertion is made about permuting variables x_i and x_j when $i \leq J < j$. Thus, it is as though there are two species of *spinless* particles (J of one kind and $N - J$ of the other kind), each of which satisfies the Pauli exclusion principle for spinless particles. The terminology that is employed to describe these "two kinds" of electrons is "electrons with spin up" and "electrons with spin down".

Particles such as electrons, protons and neutrons (and some, but not all nuclei) that satisfy the Pauli exclusion principle are called **fermions**. They always have spin. There is another kind of particle in nature—**bosons**. These have the restriction that ψ is symmetric (there is only $L^2(\mathbf{R}^3)$ this time, not $L^2(\mathbf{R}^3; \mathbf{C}^2)$). *Every* elementary particle is *either* a boson or a fermion. Fortunately, there are no negatively charged bosons which do not decay rapidly, for matter composed of negative bosons and positive bosons is not stable of the second kind—as will be seen later in Part VI.

It is a very general fact that the infimum in (2.13) over the full tensor product $\bigotimes^N L^2(\mathbf{R}^3)$ is the infimum over symmetric ψ's, i.e., bosons. To prove this fact, first replace ψ (which is real and normalized) by $|\psi|$, which does not increase E_ψ because $|\psi|^2$ and $|\nabla\psi|^2$ remain the same. Next, for *any* function $\Phi(\underline{X})$ we can construct a symmetric function $\Phi_\sigma(\underline{X})$ by summing over permutations, i.e.,

$$\Phi_\sigma(\underline{X}) = (N!)^{-1}\sum_{\pi\in S_N}\Phi(\pi\underline{X}).$$

One easily checks that Φ_σ is automatically orthogonal in $\bigotimes^N L^2(\mathbf{R}^3)$ to the function $\Phi_\alpha \equiv \Phi - \Phi_\sigma$. For the same reason (symmetrizing the integrand in (2.10), (2.11)) E_Φ splits as $E_\Phi = E_{\Phi_\alpha} + E_{\Phi_\sigma}$. Applying this construction to our ψ we easily conclude that

$$\begin{aligned}|||\psi_\sigma|||_2^2 &= (N!)^{-2}\sum_{\mu,\pi\in S_N}(|\psi|(\mu\cdot), |\psi|(\pi\cdot))\\ &\geq (N!)^{-2}\sum_{\pi\in S_N}\|\psi\|_2^2 = (N!)^{-1}\|\psi\|_2^2 = (N!)^{-1}.\end{aligned}$$

Let E be the unrestricted infimum and let E^σ denote the infimum restricted to symmetric functions. Then $E^\sigma - E \geq 0$ and

(4.4)
$$\begin{aligned}E &= \inf_\psi E_\psi = \inf_\psi E_{|\psi|} = \inf_\psi\{E_{|\psi|_\alpha} + E_{|\psi|_\sigma}\}\\ &\geq \inf_\psi\{|||\psi|_\alpha\|_2^2 E + |||\psi|_\sigma\|_2^2 E^\sigma\}\\ &= \inf_\psi\{(1 - |||\psi|_\sigma\|_2^2)E + |||\psi|_\sigma\|_2^2 E^\sigma\} \geq E + (N!)^{-1}[E^\sigma - E].\end{aligned}$$

Hence $E = E^\sigma$ as claimed.

In brief, *the imposition of the Pauli exclusion principle raises* E. The miracle is that it raises E enough so that stability of the second kind holds. While it is easy to state that ψ must be antisymmetric (here we return to $\bigwedge^N L^2(\mathbf{R}^3)$ for simplicity) it is not easy to quantify the effect of antisymmetry. Even the experts have difficulty, for it is not easy to think of an antisymmetric function of a large number of variables.

The most dramatic effect of antisymmetry concerns T_ψ. In Part III we used the Sobolev inequality to bound T_ψ in terms of

Bull. Amer. Math. Soc. *22*, 1-49 (1990)

$\rho_\psi(x) = |\psi(x)|^2$ and we would like now to find a similar bound in the N-particle case. It is *not* useful to define $\rho_\psi(\underline{X}) = |\psi(\underline{X})|^2$ because this is a function of N variables and is quite unmanageable. A more useful definition is the **one-particle density** $\rho_\psi : \mathbf{R}^3 \to \mathbf{R}^+$ given by

$$\rho_\psi(x) = N \int_{\mathbf{R}^{3(N-1)}} |\psi(x, x_2, \dots, x_N)|^2 dx_2 \dots dx_N, \tag{4.5}$$

from which it follows that $\int_{\mathbf{R}^3} \rho_\psi = N$ and $\rho_\psi(x)$ is the **density of electrons** at $x \in \mathbf{R}^3$. (Note that, since ψ is antisymmetric, $|\psi|^2$ is symmetric and therefore it is immaterial which variable is set equal to x in (4.5).)

It is an easy exercise using Minkowski's inequality to deduce from (3.2) that

$$T_\psi \geq S\|\rho_\psi\|_3 \tag{4.6}$$

in the N-particle case, but this inequality is not very useful; it does not distinguish the fact that ψ is antisymmetric instead of symmetric. The right side of (4.6) has the following property. Suppose ρ is a smooth nonnegative function on $\mathbf{R}^3$ with $\int \rho = 1$ and suppose also that ρ_ψ for an N particle ψ is $N\rho$. Then the right side of (4.6) is proportional to N. Without the imposition of antisymmetry nothing more can be asserted, but with it T_ψ grows like $N^{5/3}$, as Theorem 1 below shows.

Let us return to (3.7), the weakened form of Sobolev's inequality for one variable. Again, Minkowski's inequality can be used to translate (3.7) into an inequality for N variables:

$$T_\psi \geq N^{-2/3} K^1 \int_{\mathbf{R}^3} \rho_\psi(x)^{5/3} dx. \tag{4.7}$$

The right side of (4.7) also shows a linear dependence on N when $\rho_\psi = N\rho$. But antisymmetry comes to the rescue in the form of the following inequality of Lieb and Thirring [LT1, LT2, L1, L2].

Theorem 1 (The 5/3 law for the kinetic energy of fermions). *There is a universal constant* K *such that for all* N *and all antisymmetric (complex)* $\psi \in L^2(\mathbf{R}^{3N})$ *with* $\|\psi\|_2 = 1$

$$T_\psi \geq 2^{2/3} K \int_{\mathbf{R}^3} \rho_\psi(x)^{5/3} dx. \tag{4.8}$$

If ψ *is an antisymmetric function in* $L^2(\mathbf{R}^{3N}; \mathbf{C}^{2^N})$ *then*

$$T_\psi \geq K \int_{\mathbf{R}^3} \rho_\psi(x)^{5/3} dx. \tag{4.9}$$

$K = (2.7709)2^{-2/3} = 1.7455$ *will do.*

Contrast (4.9) with (4.7); the $N^{-2/3}$ is gone. The poetic remark at the end of Part III is still correct for many fermions, but ρ_ψ is now the total density which, very heuristically, is N times the density of one particle.

Remarks. (1) The sharp K is not known; it is conjectured [LT2] to be $K^c = 3(3\pi^2)^{2/3}/5 = 5.7425$. This number is the "classical" value of K and it arises in the following way. Take a cube $\Gamma_L \subset \mathbf{R}^3$ of length L and then compute $\tau(n, L)$, the minimum of T_ψ over all antisymmtric n-particle ψ's with support in $(\Gamma_L)^n$. The best ψ has the form (4.10) below in which each ϕ_i satisfies $-\Delta\phi_i = \lambda_i\phi_i$ in Γ_L and $\phi_i = 0$ on the boundary of Γ_L. One finds, for large n, that $\tau(n, L) \approx 2^{2/3}K^c$ (resp. K^c) times $n^{5/3}/L^2$ and $\rho_\psi(x) \approx n/L^3$ for $x \in \Gamma_L$ and $\rho_\psi(x) = 0$ for $x \notin \Gamma_L$. These values of $T_\psi = \tau(n, L)$, ρ_ψ and K^c then give equality in (4.8) and (4.9).

(2) Theorem 1 is stated above for the case in which n (the dimension of each variable) is 3. A similar theorem holds for all $n \geq 1$ (unlike Sobolev's inequality which holds only when $n \geq 3$) if $5/3$ is replaced by $1 + 2/n$ and if K is replaced by a constant K_n depending on n (but not on N). See [LT2 and **L2**].

To explore the significance of Theorem 1, and also the significance of the word "exclusion", let us examine the simplest kind of antisymmetric ψ, namely a **determinantal function**. Let $\phi_1, \dots, \phi_N$ be any set of N orthonormal functions in $L^2(\mathbf{R}^3)$ and let

$$\psi(\underline{X}) = (N!)^{-1/2} \det\{\phi_i(x_j)\}_{i,j=1}^N. \tag{4.10}$$

This ψ is antisymmetric and has $\|\psi\|_2 = 1$. The word "exclusion" comes from the fact that the ϕ_i's have to be distinct—in fact orthogonal to each other. The right side of (4.10) is the closest we can come to thinking in terms of "one $L^2(\mathbf{R}^3)$ function for each electron". When a physicist or chemist uses a determinantal ψ in calculations the ϕ_i's are called *orbitals*, and one says that "only one electron can be in each orbital". (In the case of $L^2(\mathbf{R}^3; \mathbf{C}^2)$ one says that it is possible to have "at most two electrons in each orbital".) In fact it was on the basis of this simple heuristic applied to Bohr's "old quantum mechanics" that Pauli [P] was able to explain the periodic table of the elements in 1925—before Schrödinger invented his equation and hence before the relation between antisymmetry and "exclusion" was understood.

For the ψ given by (4.10), the ρ_ψ in (4.5) and T_ψ are easily computed to be

$$\begin{aligned}\rho_\psi(x) &= \sum_{i=1}^{N} |\phi_i(x)|^2 = \sum_{i=1}^{N} \rho_{\phi_i}(x)\\ T_\psi &= \sum_{i=1}^{N} \int_{\mathbf{R}^3} |\nabla\phi_i(x)|^2 dx = \sum_{i=1}^{N} T_{\phi_i}.\end{aligned} \tag{4.11}$$

For each ϕ_i, (3.7) says that

$$\int_{\mathbf{R}^3} |\nabla\phi_i(x)|^2 dx \geq K^1 \int_{\mathbf{R}^3} |\phi_i(x)|^{10/3} dx. \tag{4.12}$$

As an illustration of Theorem 1, consider the case that $|\phi_i(x)|^2 = \rho(x)$ for each i, where ρ is a given nonnegative function with $\int \rho = 1$. We could take $\phi_1(x) = \rho(x)^{1/2}$, but then the remaining ϕ_j's would have to be $\phi_j(x) = \rho(x)^{1/2}\exp[i\theta_j(x)]$, with the θ_j's real and chosen to insure orthogonality. As N increases, the θ_j's have to "wiggle" more and more to insure this orthogonality, so that T_{ϕ_j} increases with j. How fast? According to Theorem 1, with $\rho_\psi = N\rho$ in this case,

$$T_\psi \geq K N^{5/3} \int_{\mathbf{R}^3} \rho(x)^{5/3} dx \tag{4.13}$$

and this shows that T_{ϕ_j} increases with j like $j^{2/3}$, on the average.

Armed with Theorem 1, let us return to the question of finding a lower bound to E; this time the B term will not be ignored. Using (4.9) we have for any admissible ψ,

$$E_\psi \geq K \int_{\mathbf{R}^3} \rho_\psi(x)^{5/3} dx + \int W(\underline{X})|\psi(\underline{X})|^2 dx_1 \cdots dx_N \tag{4.14}$$

$$\begin{aligned}&= K \int_{\mathbf{R}^3} \rho_\psi(x)^{5/3} dx - z \int_{\mathbf{R}^3} |x|^{-1} \rho_\psi(x) dx\\ &\quad + \int B(\underline{X})|\psi(\underline{X})|^2 dx_1 \cdots dx_N.\end{aligned} \tag{4.15}$$

Note that the A term of (1.4) is simply expressible in terms of ρ_ψ, as in (4.15). The B term is more complicated. One would guess by analogy with a fluid of density ρ_ψ that roughly

$$\int B(\underline{X})|\psi(\underline{X})|^2 dx_1 \cdots dx_N = \frac{1}{2} \int_{\mathbf{R}^3}\int_{\mathbf{R}^3} \rho_\psi(x)\rho_\psi(y)|x-y|^{-1} dx dy. \tag{4.16}$$

This is not exactly correct, of course, but it can be proved [LO] that for every normalized ψ (with or without the Pauli exclusion principle), the left side minus the right side of (4.16) is bounded below by— $(1.68)\int \rho_\psi^{4/3}(x)dx$. This error term is small; in any case it can be bounded using the Schwarz inequality by $N^{1/2}(\int \rho_\psi^{5/3})^{1/2}$ and this can easily be controlled by the kinetic energy term $\int \rho_\psi^{5/3}$. Therefore, up to controllable errors

$$E_\psi \geq \mathscr{E}^{TF}(\rho_\psi),$$

where

$$\begin{aligned}\mathscr{E}^{TF}(\rho) = K\int_{\mathbf{R}^3}\rho(x)^{5/3}dx - z\int_{\mathbf{R}^3}|x|^{-1}\rho(x)dx \\ + \frac{1}{2}\int_{\mathbf{R}^3}\int_{\mathbf{R}^3}\rho(x)\rho(y)|x-y|^{-1}dxdy.\end{aligned} \tag{4.17}$$

This functional, $\mathscr{E}^{TF}$, which is defined for all nonnegative ρ in $L^{5/3}(\mathbf{R}^3)\cap L^1(\mathbf{R}^3)$, is called the **Thomas-Fermi energy functional.** It was introduced independently by Thomas [T] and Fermi [FE] shortly after Schrödinger's discovery, but with the constant K replaced by $K^c = 3(3\pi^2)^{2/3}/5$ which, as remarked above, is conjectured to be the sharp constant in Theorem 1. They were, of course, unaware of Theorem 1 which was proved only in 1975, but they proposed that the **Thomas-Fermi ground state energy**

$$E^{TF} = E^{TF}(N, z) \equiv \inf\left\{\mathscr{E}^{TF}(\rho) : \int \rho = N\right\} \tag{4.18}$$

should be a good approximation to E. We have seen that, apart from minor errors, $E^{TF} \leq E$ when K is used instead of K^c.

The minimization problem (4.18) is an interesting problem in itself and has been analyzed in great detail [LS, L3]. Note that it is defined for all $N \geq 0$, not just integral N. A simple scaling $\rho(x) \to z^2\tilde{\rho}(z^{1/3}x)$ replaces z by 1, N by N/z and $E^{TF}(N, z) = z^{7/3}E^{TF}(N/z, 1)$. It turns out that the absolute minimum of $\mathscr{E}^{TF}(\rho)$ without regard to the value of N always occurs when $N = z$, i.e., for the neutral atom. There is a unique minimizing ρ^{TF} for (4.18) if $N \leq z$ and there is no minimizer if $N > z$. In either case

$$E^{TF}(N, z) \geq E^{TF}(z, z) = -(\text{const.})z^{7/3} \tag{4.19}$$

with (const.) $= -E^{TF}(1, 1) > 0$.

The remarks in the preceding paragraph apply only to the TF energy and not to the Schrödinger energy but, since $E > E^{TF}$

(with K) to within small errors, we have that the energy of a large atom is bounded below by $-(\text{const.})z^{7/3}$. That this is indeed the correct power law was shown by Lieb and Simon [LS]. See also [L3], where a simpler proof is given.

Theorem 2 (TF Theory is asymptotically exact). *Fix any* $0 < \lambda \le 1$ *and let* $z_N = N/\lambda$ *for* $N = 1, 2, 3, \ldots$. *Let* $E(N, z_N)$ *be the Schrödinger ground state energy* (2.13) *for an atom with nuclear charge* z_N *and* N *electrons and let* $\rho_N(x)$ *in* (4.5) *be the density for any ground state* ψ_N. *Let* $E^{TF}(N, z_N)$ *and* ρ_N^{TF} *be the corresponding Thomas-Fermi energy and density with* $K = K^c$ *(not the smaller* K *of Theorem* 1). *Then*

$$\begin{aligned}\lim_{N\to\infty} & E(N, z_N)/E^{TF}(N, z_N) \\ & = \lim_{N\to\infty} E(N, z_N)/[z^{7/3}E^{TF}(\lambda, 1)] = 1.\end{aligned} \tag{4.20}$$

In the sense of weak $L^1(\mathbf{R}^3)$ *convergence on compact sets,*

$$\lim_{N\to\infty} z^{-2}\rho_N(z^{-1/3}x) = \lim_{N\to\infty} z^{-2}\rho_N^{TF}(z^{-1/3}x) = \tilde{\rho}(x) \tag{4.21}$$

where $\tilde{\rho}$ *is the TF minimizer for* $z = 1$ *and* $N = \lambda$.

If $\lambda > 1$ *is fixed and* $z_N = N/\lambda$ *then* (4.20) *and* (4.21) *hold with the following replacements*:

$$E^{TF}(N, z_N) \to E^{TF}(z_N, z_N), \qquad E^{TF}(\lambda, 1) \to E^{TF}(1, 1), \quad \rho_N^{TF} \to \rho_{z_N}^{TF}$$

and $\tilde{\rho}$ *is the minimizer for* $z = 1$ *and* $\lambda = 1$. *In other words, the Schrödinger quantities for an atom with net negative charge converge to the corresponding TF quantitites for a neutral atom.*

From these considerations we can conclude

(1) *The energy of a large, neutral atom grows with* z *like* $z^{7/3}$. This is significantly different from the earlier result (4.3) without the Pauli exclusion principle because it means that the average energy of each electron is $-z^{4/3}$ instead of $-z^2$.

(2) *For large* z *almost all of the electrons are located at a distance* $z^{-1/3}$ *from the nucleus.* The average electron density in this ball of radius $z^{-1/3}$ is z^2. Without the Pauli exclusion principle we found the distance to be z^{-1} and the electron density to be z^4.

Conclusion (2) is that even with the Pauli exlcusion principle a large z atom is smaller than a small z atom. This seems to contradict the experimental fact that "The large range of atomic masses is not accompanied by a correspondingly large or systematic variation in size. Atomic radii all lie between 0.5×10^{-8} cm and 2.5×10^{-8} cm, with no marked increase from the lightest to the heaviest [FR]." While it is not possible to give an unambiguous definition of the radius of something as fuzzy as an atom, it appears [AI] that the largest atom is cesium $(z = 55)$ and not uranium $(z = 92)$.

The paradox can be resolved by a closer examination of the meaning of "atomic size". If we define the radius to be the radius of the ball in which most of the electrons reside then $z^{-1/3}$ is, indeed, the correct answer. But that is not the radius that the chemist sees or the distance between atoms in a solid. This "chemical radius" *can be* defined, for example, as the radius R such that

$$\int_{|x|>R} \rho_\psi(x)dx = \frac{1}{4} \tag{4.22}$$

for a ground state ψ. In Thomas-Fermi theory this radius is independent of z for large z (because for large $|x|$, $\rho^{TF}(x) = C|x|^{-6}$ when $z = N$, with C independent of z). Theorem 2 (4.21) tells us what ρ_N looks like on distance scales of order $z^{-1/3}$ where ρ_N is of the order z^2, but it says nothing about distance scales of order one where ρ_N is *presumably* of order one. Nothing is known rigorously about this latter distance scale (and even numerical calculations are uncertain here), but it is surely the case that the true density ρ_N in an atom with $z = N$ and with z large looks like that shown *very* schematically in Figure 1. Also shown in Figure 1 is the innermost region of radius z^{-1}. Thomas-Fermi theory states that $\rho^{TF}(x) = (\text{const.})\, z^3$ when $|x| = z^{-1}$ and $\rho^{TF}(0) = \infty$. It is unproved, but undoubtedly true, that when z is large $\rho(x) = Cz^3[1 - O(z|x|)]$ for $0 \le |x| \le z^{-1}$, and there is even a conjecture about the precise value of C. The investigation of the inner and outer parts of large atoms is now a subject of active research.

Using poetic license again, we can say:

A large atom is like a galaxy. It has a small, high density, energetic core in which most of the electrons are to be found and whose size decreases with z like $z^{-1/3}$. The electron-electron electrostatic repulsion always manages to push a few electrons out to a distance of order one which is roughly independent of z; this is the radius one observes if one tries to "touch" an atom.

Bull. Amer. Math. Soc. *22*, 1-49 (1990)

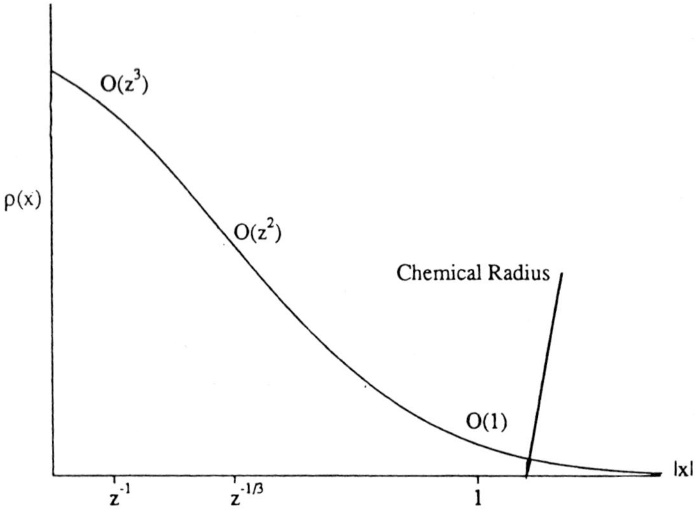

FIGURE 1.

The electron density $\rho(x)$ for a large z atom is crudely plotted as a function of $|x|$, the distance from the nucleus. The graph is intended to show three regimes. (i) At distances of order $z^{-1/3}$, $\rho(x)$ is of order z^2 and most of the electrons are to be found here. (ii) At the very short distance z^{-1}, $\rho(x)$ is of order z^3. (iii) At distances of order 1, $\rho(x)$ is of order 1 and chemistry takes place. These last two assertions have not yet been established rigorously.

PART V. BULK MATTER: MANY ELECTRONS AND MANY NUCLEI

The analysis given in Part IV can be applied to the general case of N electrons, k nuclei with charges $\underline{Z}$ and locations $\underline{R}$, and with the full potential energy $W(\underline{X})$ in (1.4). The goal is to prove stability of the second kind: $E(N, k, \underline{Z}) \geq -A(z)(N+k)$ for a suitable z-dependent constant $A(z)$, cf. (2.14). This is the theorem first proved by Dyson and Lenard [DL, LE], but with $A(z)$ unrealistically large—of the order $10^{14}z^2$.

Before giving the proof let us discuss the crucial importance of this linear law. For simplicity, assume that all the z_j's have the same value z and that $N = kz$, i.e., that the system is neutral. First of all, by a suitable choice of a comparison function ψ in

(2.13) on the one hand, and using the linear lower bound plus Newton's theorem on the other hand, Lebowitz and Lieb proved the following [LL].

Theorem 3 (The existence of the thermodynamic limit).
There is a a number $A'(z)$ such that

$$\lim_{N\to\infty} N^{-1}E(N, k = N/z, Z) = -A'(z).$$

The linear law is not just a lower bound it is, indeed, the correct *asymptotic law.* This fact is crucial for the following argument. (There are historical and physical reasons that the words thermodynamic limit instead of asymptotic law are used to describe Theorem 3.)

Suppose now that we have two large collections of N electrons and $k = N/z$ nuclei which are far apart from each other. This means that the nuclear coordinates $\underline{R} = (R_1, \dots, R_{2N})$ satisfy $|R_i - R_{N+j}| > d$ for all $i, j \in \{1, \dots, N\}$ and where d is some very large number. Think of two well separated stones. The ground state energy would then be nearly $2E(N)$ where $E(N) \equiv E(N, k, Z)$. If the two stones are allowed to come together, the ground state energy becomes $E(2N)$ and the energy released is $\delta E = 2E(N) - E(2N)$. With the linear law in the form of Theorem 3, $\delta E = o(N)$; in fact it turns out that $\delta E \sim N^{2/3}$, which is the attractive surface energy that makes the two stones eventually adhere to each other. If, on the other hand, $E(N)$ is proportional to $-N^p$ with $p > 1$ then $\delta E = -(2^p - 2)E(N)$. Thus, the released energy would be of the order of the energy in each stone which, using the actual physical value of $E(N)$, is more than 1000 times larger than the energy released in an explosion of TNT of the same mass. As we shall see in Part VI, hypothetical matter composed solely of bosons would have $E(N) \approx -N^{7/5}$. Using this value of $E(N)$ instead of the energy in a physical stone enhances the energy by a factor $N^{2/5}$; the energy in a smallish stone, $N = 10^{23}$, would then be that of a hydrogen bomb [D]. Such "matter" would be very unpleasant stuff to have lying around the house.

A somewhat less dramatic, but more profound way to state the importance of the thermodynamic limit, Theorem 3, is that it is part of the foundation of thermodynamics [LL]. If this $N \to \infty$ limit did not exist the argument in the preceding paragraph shows that matter would not behave the way we expect it to behave—*even if stability of the second kind is assumed to hold.*

Returning to the problem of proving stability of the second kind, we can use Theorem 1 together with the replacement (4.16) (which yields a controllable error) to deduce that $E(N, k, \underline{Z}, \underline{R}) \geq E^{TF}(N, k, \underline{Z}, \underline{R})$ when E^{TF} is given by (4.18) but now

$$\mathscr{E}^{TF}(\rho) = K\int_{\mathbf{R}^3} \rho(x)^{5/3}dx - \int_{\mathbf{R}^3} V(x)\rho(x)dx + \frac{1}{2}\int_{\mathbf{R}^2}\int_{\mathbf{R}^3} \rho(x)\rho(y)|x-y|^{-1}dxdy + U. \tag{5.1}$$

The Thomas-Fermi functional in (5.1) is like that in (4.17) except that $z|x|^{-1}$ is replaced by $V(x)$ and U is added (cf. (1.4)–(1.8)). These depend on $\underline{Z}$ and $\underline{R}$.

A lower bound for E^{TF} can be obtained by omitting the condition that $\int \rho = N$. The absolute minimum occurs [LS, L3] for the neutral case

$$N = \sum_{j=1}^{k} z_j, \tag{5.2}$$

as it did for the atom in (4.17). One can struggle with minimizing (5.1) or, preferably, one can use a theorem of Teller (see [LS, L3]) which states that *atoms do not bind in Thomas-Fermi theory.* For arbitrary N, this means that there are k positive numbers $N_1, \ldots, N_k$ with $\sum_{j=1}^{k} N_j = N$ such that

$$E^{TF}(N, k, \underline{Z}, \underline{R}) \geq \sum_{j=1}^{k} E_{atom}^{TF}(N_j, z_j), \tag{5.3}$$

where E_{atom}^{TF} is the solution to the atomic TF minimization problem (4.18). In particular, under the assumption of neutrality (5.2), the optimum choice is $N_j = z_j$ and thus

$$E^{TF}\left(\sum_{j=1}^{k} z_j, k, \underline{Z}, \underline{R}, \right) \geq \sum_{j=1}^{k} E_{atom}^{TF}(z_j, z_j). \tag{5.4}$$

The right side of (5.4) was already computed in (4.19) and therefore, assuming (4.16),

$$E(N, k, \underline{Z}) \geq -(\text{const.})\sum_{j=1}^{k} z_j^{7/3}. \tag{5.5}$$

This is our desired goal of stability of the second kind with $A(z) = (\text{const.})z^{7/3}$. However, errors were made when (4.16) was used. If these are included and bounded in a straightforward way, it turns out that

$$E(N,k,\underline{Z}) \geq -(\text{const.})\left\{N + \sum_{j=1}^{k} z_j^{7/3}\right\}, \tag{5.6}$$

which also implies stability of the second kind. The constant in (5.6) turns out to be about 5 instead of 10^{14} as in [DL].

Incidentally, it should not be inferred that instability of the second kind is necessarily associated with singularities and/or slow fall-off of the potential at infinity – as is the case with the electrostatic potential $|x|^{-1}$. In other words, seemingly "nice" potentials can fail to produce stability of the second kind. Indeed, Thirring [TW, p. 258] shows that if the function $|x|^{-1}$ is replaced everywhere in (1.4)–(1.8) by the seemingly harmless function $(1+2|x|)\exp[-|x|]$ then stability of the second kind does not hold for the modified system.

Having proved stability of the second kind with electrostatic forces, we are left with another question which was stated by Ehrenfest in a 1931 address to Pauli (quoted in [D]):

> "We take a piece of metal. Or a stone. When we think about it, we are astonished that this quantity of matter should occupy so large a volume. Admittedly, the molecules are packed tightly together, and likewise the atoms within each molecule. But why are the atoms themselves so big?
>
> Consider for example the Bohr model of an atom of lead. Why do so few of the 82 electrons run in the orbits close to the nucleus? The attraction of the 82 positive charges in the nucleus is so strong. Many more of the 82 electrons could be concentrated into the inner orbits, before their mutual repulsion would become too large. What prevents the atom from collapsing in this way? Answer: only the Pauli principle, 'No two electrons in the same state.' That is why atoms are so unnecessarily big, and why metal and stone are so bulky.
>
> You must admit, Pauli, that if you would only partially repeal your prohibitions, you could relieve many of our practical worries, for example the traffic problem on our streets."

One simple measure of the "size" of our many-body system in a state ψ is the average radius $R_p(\psi)$ defined by

$$\begin{aligned}[R_p(\psi)]^p &= \frac{1}{N}\int_{\mathbf{R}^3} |x|^p \rho_\psi(x)dx \\ &= \frac{1}{N}\int \sum_{i=1}^{N} |x_i|^p |\psi(\underline{X})|^2 dx_1\cdots dx_N.\end{aligned} \tag{5.7}$$

Here, p is any convenient fixed, positive number. According to our experience with ordinary matter, we should expect $R_p(\psi)$ to grow with N like $N^{1/3}$, i.e., the volume grows like N, when ψ is taken to be a ground state for N electrons. Indeed this surmise is correct—in fact it is correct, as will now be proved, even for some ψ's which are quite far from being ground states [L1] (see also [LT2]).

Assume that the locations of the nuclei and a normalized ψ satisfy only the condition

$$E_\psi \leq 0, \tag{5.8}$$

which certainly includes the absolute ground states because $E(N, k, \underline{Z}) < 0$ always. Then

$$E_\psi = \frac{1}{2} T_\psi + E'_\psi \tag{5.9}$$

where E'_ψ is the energy in (2.12) but with T_ψ replaced by $\frac{1}{2}T_\psi$ (which amounts to doubling the electron mass). It is easy to see, by a simple scaling $\psi(\underline{X}) \to 2^{3N/2}\psi(2\underline{X})$ and $R_j \to \frac{1}{2}R_j$ that $E'(N, k, \underline{Z}, 2\underline{R}) = 2E(N, k, \underline{Z}, \underline{R})$ and therefore $E'(N, k, \underline{Z}) = 2E(N, k, \underline{Z})$. By (5.8) and (5.9) and stability of the second kind

$$T_\psi \leq 4|E(N, k, \underline{Z})| \leq 4A(z)(N + k). \tag{5.10}$$

An easy general inequality is the existence of a positive constant C_p such that

(5.11)

$$\left\{\int_{\mathbf{R}^3} \rho(x)^{5/3} dx\right\} \left\{\int_{\mathbf{R}^3} |x|^p \rho(x) dx\right\}^{2/p} \geq C_p \left\{\int_{\mathbf{R}^3} \rho(x) dx\right\}^{5/3+2/p}$$

for all nonnegative ρ. Using the basic kinetic energy inequality (4.9) together with (5.10) and the fact that $\int \rho_\psi = N$, we conclude that

$$4A(z)(N + k) \geq T_\psi \geq KC_p N^{5/3} R_p(\psi)^{-2}. \tag{5.12}$$

We do not want to impose neutrality (5.2) but let us assume $N \leq 2\sum_j z_j$; one can show [L4] that this is the only case that needs to be treated. Then (5.12) implies that for each $p > 0$

$$R_p(\psi) \geq C'_p N^{1/3} \tag{5.13}$$

as required. Here C'_p depends on p and on z, the maximum of the nuclear charges.

In summary, *bulk matter is stable, and has a volume proportional to the number of particles, because of the Pauli exclusion principle for fermions (i.e., the electrons). Effectively the electrons behave like a fluid with energy density* $\rho_\psi^{5/3}$, *and this limits the compression caused by the attractive electrostatic forces.*

PART VI. BOSONS IN BULK

The previous paragraph states *inter alia* that the Pauli exclusion principle is essential for stability of the second kind. The verification of this assertion and the dependence of the energy on N without the Pauli principle is itself an interesting mathematical problem. As explained in Part IV, omitting the Pauli principle is equivalent to studying the minimization problem (2.13) for *symmetric* functions, i.e., bosons.

THE $N^{5/3}$ LAW FOR BOSONS. First, let us see what happens if we try to use the Thomas-Fermi energy as a lower bound—as was done in Part V. Theorem 1 is not available for bosons and (4.7) must be used in place of (4.9). The difference between K^1 and K is not important, but the factor $N^{-2/3}$ is crucial. For fixed N, the substitution $K \to N^{-2/3}K^1$ in (5.1) is mathematically equivalent to replacing the electron mass (which is 1 in our units) by $m' = N^{2/3}K/K^1$. By the scaling $\underline{R} \to \underline{R}/m'$ and $\psi(\underline{X}) \to (m')^{3N/2}\psi(m'\underline{X})$ for the Schrödinger theory or $\rho(x) \to m'\rho(m'x)$ in TF theory one easily finds, in both theories, that $E(N, k, \underline{Z}) \to m'E(N, k\underline{Z})$. Thus, using (5.6),

$$E(N, k, Z) \geq -(\text{const.})N^{2/3}\left(N + \sum_{j=1}^{k} z_j^{7/3}\right) \tag{6.1}$$

for bosons.

This certainly violates stability of the second kind. If we take $z_j = z$ and $Nk = z$ as before then $E(N) \geq -(\text{const.})N^{5/3}$. Is this result correct or is it simply a bad lower bound? The answer is that *the $N^{5/3}$ law for bosons with fixed nuclei* is correct. It can easily be proved by the following simple choice [L5] of a comparison function ψ for (2.13) and nuclear coordinates $R_1, \dots, R_k$.

Let $f: \mathbf{R}^1 \to \mathbf{R}$ be given by $f(x) = \sqrt{3/2}[1 - |x|]$ if $|x| \leq 1$ and $f(x) = 0$ otherwise. For each $\lambda > 0$ define $\phi_\lambda: \mathbf{R}^3 \to \mathbf{R}$ by

$$\phi_\lambda(x^1, x^2, x^3) = \lambda^{3/2} f(\lambda x^1) f(\lambda x^2) f(\lambda x^3) \tag{6.2}$$

with $x = (x^1, x^2, x^3)$, so that for all λ

$$\|\phi_\lambda\|_2 = 1 \quad \text{and} \quad T_{\phi_\lambda} = 9\lambda^2. \tag{6.3}$$

The support of ϕ_λ is the cube $\Gamma_\lambda = \lambda^{-1}\cdot[-1, 1]^3$ of volume $8\lambda^{-3}$. Our comparison function will be

$$\psi_\lambda(\underline{X}) = \prod_{i=1}^{N} \phi_\lambda(x_i), \tag{6.4}$$

so that

$$\rho_{\psi_\lambda}(x) = N\phi_\lambda(x)^2 \quad \text{and} \quad T_{\psi_\lambda} = 9N\lambda^2. \tag{6.5}$$

For this ψ_λ, (4.16) turns out to be nearly exact.

$$\begin{aligned} &\int B(\underline{X})\psi_\lambda(\underline{X})^2 dx_1 \cdots dx_N \\ &\quad = \frac{1}{2}\frac{N-1}{N}\int_{\mathbf{R}^3}\int_{\mathbf{R}^3} \rho_{\psi_\lambda}(x)\rho_{\psi_\lambda}(y)|x-y|^{-1} dx dy \end{aligned} \tag{6.6}$$

and the last integral is proportional to N^2/λ. Now assume for simplicity that N/z is an integer and take $k = N/z$ nuclei of charge z. It is possible [L5] to place these nuclei in Γ (roughly in a periodic arrangement) so that the *total* potential energy, $W = -A + B + U$, is equal to $-Cz^{2/3}N^{4/3}\lambda$, where C is some positive constant. (To understand this, note that in a volume which is roughly $d^3 = 1/k\lambda^3$ there is one nucleus and z units of negative charge. The total potential energy is then roughly $-kz^2/d$.) With these favorable locations of the nuclei,

$$E_\psi = 9N\lambda^2 - Cz^{2/3}N^{4/3}\lambda. \tag{6.7}$$

When this is minimized with respect to λ, one finds $\lambda = \frac{C}{18}z^{2/3}N^{1/3}$ and therefore the upper bound

$$E(N, k, \underline{Z}) \le -\frac{C^2}{36}z^{4/3}N^{5/3}. \tag{6.8}$$

Thus bosonic matter is not only unstable of the second kind but also its radius, λ^{-1}, decreases with N like $N^{-1/3}$. We see here an example of the fact that lack of stability is intimately connected with collapse in the geometric sense.

THE $N^{7/5}$ LAW FOR BOSONS. The $N^{5/3}$ law, which was derived above, is not the end of the story. Observe that it was crucially important that the nuclei were fixed points that could be located at will in space. Suppose we dispense with that approximation and treat the nuclei also as dynamical quantum mechanical particles. Of course they will also be assumed to be bosons, for if they are fermions then we would merely have a charge reversal of the situation described in Part V, i.e., stability of the second kind would hold.

For simplicity we take $z = 1$ and assume that these positive bosons also have unit mass. The Hamiltonian is then, with x_i denoting the coordinates of the negative bosons and R_i denoting the coordinate of the positive bosons,

$$H_{N,k} = -\sum_{i=1}^{N} \Delta_{x_i} - \sum_{j=1}^{k} \Delta_{R_i} + W(\underline{X}, \underline{R}). \tag{6.9}$$

The potential function $W(\underline{X}, \underline{R})$ is that given in (1.4)–(1.8) with $e = 1$; there it was denoted simply by $W(\underline{X})$. The operator $H_{N,k}$ acts on functions $\psi(\underline{X}, \underline{R}) = \psi(x_1, \dots, x_N, R_1, \dots, R_k)$ in $\bigotimes^N \bigotimes^k L^2(\mathbf{R}^3)$. The ground state energy is given by (2.13), as usual, with $E_\psi \equiv (\psi, H_{N,k}\psi)$. As explained in Part IV, the optimum choice for ψ, neglecting symmetry considerations, is always a function that is symmetric in the x_i's and symmetric in the R_i's. But this is precisely the statement that both kinds of particles are bosons.

It is *much more difficult* to estimate the ground state energy for $H_{N,k}$ in (6.9) than for our previous problem which uses the H in (2.17). The first person to do this rigorously was Dyson [D] who proved the following in 1967.

Theorem 4 (Upper bound for the energy of bosons).
Without loss of generality assume that $N \le k$. Then the ground state energy, $E(N, k)$, for bosons with the Hamiltonian $H_{N,k}$ satisfies

$$E(N, k) \le -1.32 \times 10^{-6} N^{7/5} \tag{6.10}$$

for large N. (It is asserted [D] *that this constant can be improved substantially.)*

An extended comment is in order here. While the $N^{7/5}$ law (6.10) may seem superficially to be only a small improvement over the $N^{5/3}$ law, the conceptual difference is enormous. The proof in Part IV of stability of the second kind for fermions (electrons) and fixed positive particles (nuclei) could be understood solely in terms of what is called semi-classical considerations, i.e., Thomas-Fermi theory. In (5.1) we see that the last three terms on the right

Bull. Amer. Math. Soc. *22*, 1-49 (1990)

side are simply the classical electrostatic energy of a charged fluid of density $\rho(x)$ interacting with the fixed nuclei. Once we accept the input from quantum mechanics that the kinetic energy is like that of a fluid with energy density $\rho(x)^{5/3}$, the whole energy has a simple classical interpretation—in other words the one-particle density $\rho(x)$ determines the energy with reasonable accuracy.

On the contrary, Theorem 4 cannot be understood this way. In order to achieve an energy as low as $-N^{7/5}$, intricate correlations between the positive and negative particles have to be built into a comparison $\psi(\underline{X}, \underline{R})$, and then the potential energy cannot be expressed solely in terms of ρ_ψ. A delicate balance between potential and kinetic energies is needed and, in the end, it is impossible to think of (6.10) in a simple way—although Dyson [D] does try to give a heuristic explanation of (6.10). Dyson's comparison ψ that leads to (6.10) is very complicated and five pages are needed just to compute E_ψ. This ψ was suggested by work of Bogolubov in 1947 on the superfluidity of liquid helium and it is similar in many ways to the ψ used by Bardeen, Cooper and Schrieffer in their 1957 Nobel prize winning work on superconductivity. Therefore, if the $N^{7/5}$ law is really correct, and not just an upper bound, this kind of highly correlated ψ will be validated in some weak sense as a good approximation to the true ground state.

Two decades later the requisite lower bound was proved by Conlon, Lieb and H-T. Yau [CLY]. The proof is too involved to explain here, even heuristically, but the main result is the following.

Theorem 5 (Lower bound for the energy of bosons).
Without loss of generality assume that $N \le k$. Then the ground state energy $E(N, k)$ for bosons with the Hamiltonian $H_{N,k}$ satisfies

$$E(N, k) \ge -AN^{7/5} \tag{6.11}$$

for some universal constant A. If $N = k$ (neutrality) and N is large then A can be taken to be 0.79 *in* (6.11).

PART VII. RELATIVISTIC MATTER

According to Einstein's 1905 special theory of relativity, the **relativistic kinetic energy** as a function of the momentum $p \in \mathbf{R}^3$ should be

$$T(p) = \sqrt{p^2c^2 + m^2c^4} - mc^2 \tag{7.1}$$

instead of $p^2/2m$ as in (1.11). Here c is the speed of light. The quantity mc^2 is called the **rest energy** and it has been substracted off in (7.1). If $p^2 \ll mc^2$ then $T(p) \approx p^2/2m$ as before. Using the first part of (1.13) the relation between p and the velocity v is

$$v = pc^2(p^2c^2 + m^2c^4)^{-1/2}. \tag{7.2}$$

When $p^2 \ll mc^2$ this gives $v \approx p/m$ as before, but if $p^2 \to \infty$ then $v^2 \to c^2$. Thus no particle can move faster than the speed of light, and a measure of the importance of relativistic effects is the ratio $|v|/c$.

To gain an understanding of the possible effect of relativistic mechanics on atoms it is convenient to introduce the number

$$\alpha = e^2/\hbar c. \tag{7.3}$$

This is a *dimensionless* constant whose numerical value is 1/137 and it is called **the fine structure constant**. (The reason for this appelation is that α also governs certain small effects in atoms that are connected with the electron spin and magnetism, and that have not been discussed here; these effects cause a small splitting of each spectral line into several nearby lines—that experimentalists call the fine structure of the line.)

What is the value of $|v|/c$ for an electron in a hydrogenic atom? We can take v^2 in a state ψ to be $(2/m)$ times the kinetic energy T_ψ (which is $p^2/2m$ classically). Using the ground state ψ given by (3.6), and restoring all units (recall that we took $\hbar^2/2m = 1$ and $e = 1$), one easily finds from the definition (2.6) that

$$|v|/c = \alpha z. \tag{7.4}$$

For hydrogen $(z = 1)$, $|v|/c = 1/137$ so relativistic effects are unimportant. But when z is large (7.4) correctly gives $|v|/c$ for the "innermost" electrons, and we see that it is not small.

The problem of combining relativistic mechanics with quantum mechanics is an old and extremely difficult one. It is not even easy at the classical mechanics level. At that level one *could* do the following. Return to the classical Hamiltonian function in (1.11) and simply replace $p_i^2/2m$ by $T(p_i)$ and then use the equations of motion (1.13). While such a dynamical theory makes sense mathematically, it is *not* a relativistic theory. The reason is that it is not invariant under Lorentz transformations of space-time unless $W \equiv 0$. To remedy this defect it is necessary to give up the idea of particles interacting instantaneously by a force (1.9) which depends only on their locations in $\mathbf{R}^3$. Instead, it is necessary to invoke the intervention of the full electromagnetic field, to introduce a piece of the Hamiltonian for this field (whose equations of motion in the spirit of (1.13) turn out to be Maxwell's equations), and to introduce another piece of the Hamiltonian which gives the interaction of the particles with the field. All this is very complicated and the final equations of motion do not even make strict mathematical sense for point particles.

The situation is even worse when quantum mechanics is introduced because then the electromagnetic field also has to be "quantized." This is the enormous subject of *quantum electrodynamics* (Q.E.D) which, in turn, is part of an even larger subject—*quantum gauge field theory*. (Then there is superstring theory which is still more complicated.)

When $\alpha = 0$ (equivalently $e = 0$) Q.E.D. is trivial because then the electromagnetic field is decoupled from the particles and there are no interactions. This suggests that one can solve problems in Q.E.D. by making a power series expansion in α since α is small. Indeed, such a "perturbation theory" has been investigated in great detail and many of its predictions are confirmed incredibly well by experiments. At present, however, nobody knows in what sense this power series converges, if at all, or how to find any except the first few terms in the series, or what a "nonperturbative" theory would predict. In particular, what happens when there is a very large number of electrons and nuclei, in which case a perturbative treatment is inadequate? Does Q.E.D. predict the stability of matter?

It should be stated that the particle aspect of Q.E.D. is not built on the Schrödinger $p^2 = -\hbar^2\Delta$, which is a second order elliptic operator, but instead on the *Dirac operator* which is a quartet of first order operators which acts on four-component spinors. The Dirac operator is relativistic (i.e., it is invariant under Lorentz transformations), but if we simply add the potential energy term $W(\underline{X})$ to it the theory will still not be relativistic for the same reason as before (i.e., instantaneous interactions are not Lorentz invariant). Nevertheless, this kind of "mixed" theory is often used instead of the Schrödinger Hamiltonian because it presumably takes account of most of the relativistic corrections. However, from our point of view the Dirac Hamiltonian has a much more serious defect: the ground state energy is *not* given by a minimization problem as in (2.13). In fact the spectrum of the Dirac Hamiltonian is not bounded below and Dirac had to introduce an extra axiom in order to decide which of the many eigenvalues should be considered as the ground state energy. (This axiom is known as "filling the negative energy sea;" unfortunately, it is ambiguous in the many-body case. The axiom also led to the prediction of certain kinds of elementary particles called *antimatter*; the first one to be discovered was the positron (antielectron) by Anderson in 1932.)

In order to investigate the stability of matter with relativistic kinetic energy, $T(p)$, in a mathematically rigorous way, we are led to study the following theory which uses a "relativistic" modification of the Schrödinger energy. This theory is a caricature of the proper theory, but it will have the advantage of being a well-posed minimization problem as before, and one which is tractable—unlike Q.E.D. At present it is the best that can be done to analyze the stability of matter question rigorously.

With ψ as before, replace the kinetic energy of (2.11) by

$$T_\psi = \sum_{i=1}^{N} (\psi, T_{op}(p_i)\psi) \tag{7.5}$$

where $T_{op}(p)$ is given by (7.1), but with $p_i = -i\hbar\nabla_{x_i}$ as before. In other words, T_{op} is the operator

$$T_{op} = \sqrt{-\hbar^2 c^2 \Delta + m^2 c^4} - mc^2 \tag{7.6}$$

which is perfectly respectable although, unlike the operator $-\hbar^2\Delta/2m$, it is not local, i.e., $(T_{op}\psi)(x)$ is *not* determined by ψ in an infinitesimally small neighborhood of x. In terms of the Fourier transform $\widehat{\psi}$ given by (2.4) for one particle,

$$T_\psi = \int_{\mathbf{R}^3} \left\{ \sqrt{p^2c^2 + m^2c^4} - mc^2 \right\} |\widehat{\psi}(p)|^2 dp \tag{7.7}$$

replaces (2.6). The definitions (2.12)–(2.14) for the energy are unaltered.

For the purpose of investigating stability, a simplification can be made, and will be made in this part but not in the next. Since

$$c|p| - mc^2 \le T(p) \le c|p| \tag{7.8}$$

the difference of the operators T_{op} and $c\hbar\sqrt{-\Delta}$ is a bounded operator. Therefore, for *both* stability of the first and second kind it suffices to replace T_{op} by the operator $c\hbar\sqrt{-\Delta}$, i.e., we can set $m = 0$. Then T_ψ becomes, for one particle,

$$T_\psi = c\int_{\mathbf{R}^3} |p||\widehat{\psi}(p)|^2 dp. \tag{7.8}$$

The advantage of this replacement is a scaling invariance that will become apparent in (7.14) and (7.15). It also brings into focus the essential feature of any theory of relativistic quantum mechanics (Q.E.D. or the Dirac equation or the Klein-Gordon equation or whatever) which can be stated as follows. The central fact about quantum mechanics is that p^2 becomes the operator $-\hbar^2\Delta$ and this can be thought of heuristically (as in (3.5)) as the reciprocal of a length squared—the length being essentially the width of a function ψ. In nonrelativistic quantum mechanics this operator is also the kinetic energy and it handily controls the potential energy which is proportional to the reciprocal of the same length. In relativistic quantum mechanics, on the other hand, the kinetic energy is essentially $c\sqrt{-\hbar^2\Delta}$ and this is only the reciprocal of the length—not the length squared. Thus, both the potential and kinetic energies are on the same footing in a relativistic quantum theory, and we therefore have what would usually be termed "the critical case".

A word about constants is required here. In the nonrelativistic case we used scaling, as stated at the beginning of Part III, to eliminate all constants except for the nuclear charge numbers z_j. This cannot be done now because T_ψ and W_ψ scale in the same way. We can, however, use units in which $\hbar c = 1$. Then the energy becomes

$$E_\psi = T_\psi + \alpha W_\psi \tag{7.9}$$

where W_ψ is given by (2.10) and (1.4) with e^2 set equal to one there. T_ψ is given by

(7.10)

$$T_\psi = \sum_{i=1}^{N} \left(\psi, \sqrt{-\Delta_{x_i}}\,\psi\right) = \int_{\mathbf{R}^{3N}} \sum_{i=1}^{N} |p_i||\widehat{\psi}(p_1, \ldots, p_N)|^2 dp_1 \cdots dp_N$$

with $\widehat{\psi}$ being the $\mathbf{R}^{3N}$ Fourier transform. The Hamiltonian that replaces (2.17) is

$$H = \sum_{i=1}^{N} \sqrt{-\Delta_{x_i}} + \alpha W(\underline{X}). \tag{7.11}$$

Thus there are now two constants in the problem: the fine structure constant α and z, the maximum nuclear charge.

Let us begin by analyzing the **hydrogenic atom**. In this case only the combination $z\alpha$ enters because $U = 0$. Heuristic considerations, like those in Part III, would lead us to investigate (cf. (3.5))

$$\min_L \left\{ \frac{1}{L} - \frac{z\alpha}{L} \right\} \tag{7.12}$$

as an approximation to E. This quantity is zero if $z\alpha \leq 1$ and $-\infty$ if $z\alpha > 1$. Indeed this conclusion is qualitatively correct becuase of an inequality of Kato [K] and Herbst [H] (see also [LY1])

$$\int_{\mathbf{R}^3} |p||\widehat{\psi}(p)|^2 dp \geq \frac{2}{\pi} \int_{\mathbf{R}^3} |x|^{-1} |\psi(x)|^2 dx \tag{7.13}$$

in which $2/\pi$ is the sharp constant. Consequently, the hydrogenic atom has the following ground state energy

$$\begin{aligned} E &= 0, \quad \text{if} \quad z\alpha \leq \tfrac{2}{\pi} \\ &= -\infty \quad \text{if} \quad z\alpha > \tfrac{2}{\pi} \end{aligned} \tag{7.14}$$

and *stability of the first kind holds if and only if* $z\alpha \leq 2/\pi$, i.e., $z \leq 87$ with $\alpha = 1/137$.

Relativistic quantum mechanics stabilizes an atom only if $z\alpha$ is small enough.

It is to be noted that a similar catastrophe occurs with the Dirac Hamiltonian for an atom [KS]. Again, only $z\alpha$ enters and there is a critical value $z\alpha = 1$, instead of $2/\pi$. For $z\alpha \leq \sqrt{3}/2$ this operator is essentially selfadjoint on C_c^∞, the infinitely differentiable functions of compact support. When $z\alpha > \sqrt{3}/2$ the deficiency indices are $(+1, +1)$ and so there is a selfadjoint extension, but only for $\sqrt{3}/2 < z\alpha < 1$ is there a *distinguished*, physical selfadjoint extension. It is distinguished either by analytic continuation from the $z\alpha \leq \sqrt{3}/2$ case or by finiteness of the potential energy; in other words, when $z\alpha \geq 1$ the kinetic and potential energies cannot be defined separately. When $z\alpha = 1$ the hydrogenic ground state energy "falls into the negative energy sea" and the Dirac operator then ceases to make good physical sense. The underlying reason for the catastrophe is the same in both cases: the kinetic energy is $1/L$ instead of $1/L^2$.

The next step is to investigate **large atoms**, as in Part VI. Clearly, *stability of the first kind holds here if and only if* $z\alpha \leq 2/\pi$ because, as a simple comparison function $\psi(\underline{X})$ demonstrates, the positive electron-electron repulsion $B(\underline{X})$ of (1.7) cannot overcome the $-\infty$ energy to be gained by letting one electron "fall into the nucleus" when $z\alpha > 2/\pi$.

In fact conclusion (7.14) holds in the large atom case as well. This is where the nice scaling property of $\sqrt{-\Delta}$ and $|x|^{-1}$ come in. To jump ahead for the moment, we can always say, when $m = 0$, that in the *general case* of many electrons and many nuclei there are precisely two alternatives:

$$\text{(7.15)} \qquad \textit{either} \quad E(N, k, \underline{Z}) = 0 \quad \textit{or} \quad E(N, k, \underline{Z}) = -\infty,$$

which tells us that *stability of the first kind is equivalent to stability of the second kind for relativistic matter.* The proof of (7.15) is a simple one: If $E_\psi < 0$ for some ψ then, by setting $\psi_\lambda(\underline{X}) = \lambda^{-3N/2}\psi(\lambda\underline{X})$, we have that $E_{\psi_\lambda} = \lambda E_\psi$, and this can be driven to $-\infty$ by letting $\lambda \to \infty$. If, on the other hand, $E_\psi > 0$ then we can drive E_{ψ_λ} to 0 by letting $\lambda \to 0$.

Apart from the stability question we can also ask about the *structure* of large atoms—as in Part IV. Is there an appropriate Thomas-Fermi theory in this case? The answer is *No* and it is instructive to understand why.

We can start by asking for an analogue of Theorem 1, which is the basic kinetic energy estimate for fermions. There is one—as discovered by Daubechies [DA]: *Theorem 1 holds for the* $\sqrt{-\Delta}$ *kinetic energy if* $5/3$ *is replaced everywhere by* $4/3$, *if* $2^{2/3}$ *is replaced by* $2^{1/3}$ *in (4.8) and if* K *is changed.* ($K = 1.26$ *will do.*)

So far, so good. Next we form the Thomas-Fermi functional as in (4.17):

$$\text{(7.16)} \qquad \begin{aligned} \mathscr{E}^{TF}(\rho) = K \int_{\mathbf{R}^3} \rho(x)^{4/3} dx - z\alpha \int_{\mathbf{R}^3} |x|^{-1} \rho(x) dx \\ + \frac{1}{2}\alpha \int_{\mathbf{R}^3} \int_{\mathbf{R}^3} \rho(x)\rho(y)|x-y|^{-1} dx dy. \end{aligned}$$

This, unfortunately, is unbounded below for any choice of $N = \int \rho$. (The reader can easily verify that $\int \rho^{4/3}$ cannot control the $|x|^{-1}$ singularity, and the last term in (7.16) does not save the situation.) The conclusion is that Thomas-Fermi theory is useless for large atoms (and hence also for the case of many electrons and many nuclei), but in Part VIII we shall see that it *is* useful for describing the relativistic matter interacting by gravitational forces in a star.

Finally, we turn to **bulk matter** as in Part V. Now, however, even stability of the first kind is problematic for one electron and many nuclei; it already is problematic even for one electron ($N = 1$) and one nucleus ($k = 1$). The following additional concern now presents itself. Suppose there are two nuclei with $z\alpha = 3/2\pi$, located at R_1 and R_2 in $\mathbf{R}^3$, and suppose there is but one electron. If the two points are coincident, say $R_1 = R_2 = 0$, we have a "bomb." While each nucleus is subcritical the combined nucleus is supercritical, $\tilde{z}\alpha = 3/\pi$, and the electron can fall into it, releasing an infinite energy (7.15). What prevents this from happening? Answer: the nucleus-nucleus repulsive energy

$$U = \alpha z^2 |R_1 - R_2|^{-1} = \frac{(z\alpha)^2}{\alpha} |R_1 - R_2|^{-1}, \tag{7.17}$$

which goes to $+\infty$ as $R_1 \to R_2$ and which is one part of $E(1, 2, \underline{Z}, \underline{R})$. There are now genuinely *two* parameters in the problem, which can conveniently be taken to be $z\alpha$ and α. By length scaling, all the energies are proportional to $|R_1 - R_2|^{-1}$, so $E(1, 2, \underline{Z}, \underline{R}) = C|R_1 - R_2|^{-1}$. If $z\alpha$ is fixed at $3/2\pi$, is the constant $1/\alpha$ in (7.16) large enough so that the repulsion U wins, i.e., so that $C \geq 0$?

This exercise with $N = 1$ and $k = 2$ tells us that stability will require not only an upper bound on $z_j\alpha$ of $2/\pi$ for each nucleus but *also* an upper bound, α_c, for α. This is clarified in Theorem 6 below. Furthermore, we might fear that this α_c goes to zero as $z\alpha \to 2/\pi$. Here, $z = \max_j\{z_j\}$ as before. It might also happen that α_c depends on N and/or k. Fortunately, neither of these two fears materializes.

The first result on this problem was by Daubechies and Lieb [DAL] who proved stability for one electron and arbitrarily many nuclei, namely,

$$E(1, k, \underline{Z}) = 0 \tag{7.18}$$

provided $z_j\alpha \leq 2/\pi$ for each j and provided $\alpha \leq 1/3\pi$.

The big breakthrough was by Conlon [CO] who proved for all N and k that

$$E(N, k, \underline{Z}) = 0 \tag{7.19}$$

provided $z_j = 1$ for all j and provided $\alpha \leq 10^{-200}$ [*sic*] and provided that spin is omitted, i.e., $\psi \in \bigwedge^N L^2(\mathbf{R}^3)$ instead of $\psi \in \bigwedge^N L^2(\mathbf{R}^3; \mathbf{C}^2)$. This was vastly improved by Fefferman and de la Llave [FD] to $\alpha \leq 1/2.06\pi$ with the other conditions remaining the same.

The results of Conlon, Fefferman and de la Llave fall short of the critical case $z\alpha = 2/\pi$. They also do not include all ψ's because the inclusion of spin is not as easy as it was in the nonrelativistic case. The matter was finally settled by Lieb and H-T. Yau [LY1] who treated a slightly more general problem. The method of proof in [LY1] is very different from that in [C] and [FD].

Theorem 6 (Stability of relativistic matter up to $z\alpha = 2/\pi$). *Let N and k be arbitrary and let $z_j\alpha \le 2/\pi$ for each $j = 1, \dots, k$. Let $E_q(N, k, \underline{Z})$ be the infimum of the energy* (7.9) *as in* (2.11) *and* (2.14), *but with $\psi \in \bigwedge_1^N L^2(\mathbf{R}^3; \mathbf{C}^q)$. ($q = 1$ is the simple antisymmetric case, $q = 2$ is the physical case and $q > 2$ is for fun.) Then $E_q(N, k, \underline{Z}) = 0$ if $q\alpha \le 1/47$. In particular, stability holds in the physical case: $\alpha = 1/137$.*

Is this limitation on α (especially the $1/q$ dependence) merely an artifact of the proof in [LY1]? What about the stability of relativistic bosonic matter? These questions are answered in the following two theorems [LY1]. Roughly, the proof of Theorem 7 uses the same elementary ideas as in the proof of (6.7) and (6.8).

Theorem 7 (Instability of bulk matter for large α). *With the definition of E_q as in Theorem 6, assume there are N electrons and k nuclei, each with the same charge $z > 0$.*

(1) q **and** z **independent bound.** *If $\alpha > 128/15\pi$, if $N \ge 1$ and if $k \ge 1 + 1/z$ then there is collapse for all q, i.e., $E_q(N, k, \underline{Z}) = -\infty$.*

(2) q **and** z **dependent bound.** *If $\alpha > 36q^{-1/3}z^{2/3}$, if $N \ge q$ and if $k \ge q/z$ then $E_q(N, k, \underline{Z}) = -\infty$.*

Thus, if α is large one electron can cause collapse no matter how small z is, provided enough nuclei are used in order to make the system approximately neutral. When q is large the critical α decreases with q at least as fast as $q^{-1/3}$. This contrasts with the q^{-1} dependence of Theorem 6; the exact dependence of the critical α on q is not known.

One reason for mentioning the q dependence is that the choice $q = N$ is equivalent to omitting the Pauli exclusion principle altogether. The interested reader can easily deduce this from the discussion of symmetry in Part IV. This means that the case $q = N$ includes the boson case, and as a corollary of Theorem 7 we have

Theorem 8 (Instability of relativistic bosons with fixed nuclei). *Let $\alpha > 0$ and assume that all nuclei have a common charge $z > 0$. If the dynamic, negative particles are bosons instead of fermions, and if the nuclei are fixed as before, then collapse always occurs for sufficiently large N and k, i.e., $E(N, k, \underline{Z}) = -\infty$. The choice $N \ge (36)^3 z^2 \alpha^{-3}$ and $k \ge (36)^3 z\alpha^{-3}$ suffices.*

PART VIII. THE STABILITY AND INSTABILITY OF COLD STARS

Stars shine because they are a vast, continuous nuclear explosion. Nevertheless, the effect of attractive gravitational forces in such a huge body is not negligible—a fact that can easily be appreciated by noting that the sun maintains a spherical shape despite the nuclear violence. The byproducts of the nuclear reactions are light, heat and various kinds of particles, all of which produce an internal pressure that keeps the star expanded before these byproducts can leak out through the surface.

After the nuclear reactions are finished, in what can be called the post-twinkle phase, the star eventually cools, and it might be supposed that it would then collapse because of the gravitational attraction. Sometimes this does happen with great rapidity—for that is what a supernova is; the enormous gravitational energy is then converted into the production of a vast amount of light and numerous particles, mostly neutrinos. But sometimes the cold star, or the remnant of a supernova explosion, which is also a cold star, merely attains a new, much smaller radius and is quite stable—although lightless. (Another conceivable final state of a supernova is a black hole, in which case the collapse might be silent because black holes do not permit light to escape; I thank F. Dyson for this observation.)

The determining factor for total collapse or stability is the stellar mass; the dividing line is several solar masses. What mechanism is available, in the stable case, to provide the internal pressure that prevents the gravitational attraction from crushing the cold star? Answer: *The quantum mechanical kinetic energy of fermions.*

Actually there are two kinds of cold stars: **neutron stars** and **white dwarfs**. The latter kind will be discussed at the end. The former kind, which are remnants of supernovae, are composed mostly of chargeless neutrons. (In Part I it was stated that a free neutron decays into a proton and an electron in about 13 minutes, but in the very dense interior of such a star the decay ceases for several reasons.) With $\underline{X} = (x_1, \dots, x_N)$ denoting the coordinates of N neutrons, the classical potential energy is given by Newton's gravitational formula (cf. (1.2) and (1.9))

$$W(\underline{X}) = -\kappa \sum_{1 \leq i < j \leq N} |x_i - x_j|^{-1}. \tag{8.1}$$

Here, $\kappa = GM_n^2$, with G being Newton's gravitational constant and M_n being the neutron mass (which is just a little bigger than the proton mass, M_p). There is no electrostatic potential energy since the neutron has no electric charge.

Neutron stars were mentioned as a possibility by Landau in 1932, after the discovery of the neutron by Chadwick in the same year [B, ST]. In 1934 Baade and Zwicky proposed their connection with supernovae. Their actual existence was "established" by the 1968 discovery of pulsars (which mostly emit accurately timed pulses of radio waves) and Gold's identification of them as rotating neutron stars. The supposed facts about these stars are amazing [B, ST]. Their mass is a few solar masses (about 10^{57} neutrons) but their radius is only about 10 km. Thus the gravitational force at their surface is about 10^{12} times that on Earth. The internal structure of such an object is hard to guess, because at these densities—especially the central density—neutrons will cease to look like individual particles and might become some sort of quark soup. Moreover, this large mass in such a small volume will cause space to be "curved" in accordance with the principles of Einstein's general theory of relativity. These complications will be ignored here and we shall suppose that a neutron star is just a collection of N particles with the gravitational potential energy function of (8.1).

The model to be discussed here is standard and it is built along the lines of Part VII. With an abuse of history we shall call it the Chandrasekhar model, for a reason that will become clear at the end. Since the neutron star is cold we can suppose that it is in its quantum mechanical ground state. The energy of a wave function ψ is given, as usual, by $E_\psi = T_\psi + W_\psi$ with T_ψ defined in (7.5) with $m = M_n$ and W_ψ defined in (2.10) with $W(\underline{X})$ given by (8.1). Adopting henceforth units in which $\hbar = c = 1$, the Hamiltonian is

$$(8.2) \qquad H_N = \sum_{i=1}^{N} \left\{ \sqrt{-\Delta_{x_i} + m^2} - m \right\} - \kappa \sum_{1 \le i < j \le N} |x_i - x_j|^{-1}.$$

with $\kappa = GM_n^2$ and $m = M_n$. The problem is to compute

$$(8.3) \qquad E(N) = \inf_\psi E_\psi = \inf_\psi (\psi, H_N \psi)$$

and to delineate the properties of the density ρ_ψ for a minimizing ψ.

Two things are to be noticed. One is that the neutrons are being treated (special) relativistically with $T(p) = \sqrt{p^2c^2 + m^2c^4} - mc^2$. This is obviously important to do since $|v|/c$ will turn out to be quite large at these densities. As in Part VII, the question of collapse or no collapse can be decided by the simpler choice $T(p) = c|p|$, but if we want to compute the density ρ_ψ in the stable phase it is essential to keep the full expression (8.2).

The second point is that $W(\underline{X})$ has the same form as the electron-electron repulsion term $B(\underline{X})$ in (1.7) except for one thing—the *minus sign.* For the repulsive case we were able to approximate this energy in terms of ρ_ψ as in (4.16), because, as discussed there, the left side of (4.16) minus the right side is bounded *below* [LO] by $-(1.68)\int \rho_\psi^{4/3}$. But because of the minus sign in $W(\underline{X})$ we now require an *upper* bound for the difference in (4.16) in order to achieve a lower bound to E_ψ. Unfortunately the only upper bound is $+\infty$. Therefore, controlling W_ψ by T_ψ is not at all a simple matter in this case.

Nevertheless, there is a simple, well-defined Thomas-Fermi approximation to this problem, as in Part IV, that is obtained using (4.16). Whether it is correct or not remains to be seen. Its construction requires approximating the relativistic kinetic energy (7.5) in terms of ρ_ψ in analogy with Theorem 1 or Theorem 2. Imitating the discussion in Remark 1 after (4.9), we place n particles in a cube of size L and compute the minimum of T_ψ. With $\rho = n/L^3$ and $\eta \equiv (6\pi^2\rho/q)^{1/3}$ it turns out to be (for large n) equal to the volume L^3 times the quantity

$$j(\rho) = \frac{q}{2\pi^2}\int_0^\eta \{(p^2+m^2)^{1/2} - m\}p^2\,dp. \tag{8.4}$$

(Recall that $q = 1$ for $\psi \in \bigwedge^N L^2(\mathbf{R}^3)$ and $q = 2$ for the physical case $\psi \in \bigwedge_1^N L^2(\mathbf{R}^3; \mathbf{C}^2)$.) This $j(\rho)$ reduces to $(2/q)^{2/3}K^c\rho^{5/3}$ when ρ is small (corresponding to the nonrelativistic kinetic energy) and to $(3(6\pi^2)^{1/3}/4q^{1/3})\rho^{4/3}$ when ρ is large (corresponding to the relativistic case). It should be remarked that Daubechies [DA], who extended Theorem 1 to $\sqrt{-\Delta}$, as noted before (7.16), also extended it to $\sqrt{-\Delta + m^2} - m$.

We can now form the Thomas-Fermi functional for a neutron star.

$$\mathscr{E}^{TF}(\rho) = \int_{\mathbf{R}^3} j(\rho(x))dx - \frac{\kappa}{2}\int_{\mathbf{R}^3}\int_{\mathbf{R}^3} \rho(x)\rho(y)|x-y|^{-1}dx\,dy \tag{8.5}$$

and the Thomas-Fermi energy is

$$E^{TF}(N) = \inf\left\{\mathscr{E}(\rho) : \int \rho = N\right\}. \tag{8.6}$$

This problem can also be conveniently reformulated by setting

$$\tilde{\rho}(x) = \rho(N^{1/3}x) \Rightarrow \int_{\mathbf{R}^3} \tilde{\rho}(x)dx = 1 \tag{8.7}$$

so that, with

$$\tilde{\mathscr{E}}^{TF}(\rho) = \int_{\mathbf{R}^3} j(\rho(x)dx - \frac{1}{2}\kappa N^{2/3} \int_{\mathbf{R}^3}\int_{\mathbf{R}^3} \rho(x)\rho(y)|x-y|^{-1}dxdy, \tag{8.8}$$

the energy becomes

$$\frac{1}{N}E^{TF}(N) = \inf\left\{\tilde{\mathscr{E}}^{TF}(\rho) : \int \rho = 1\right\} \equiv e^{TF}(\kappa N^{2/3}). \tag{8.9}$$

The advantage of (8.7)–(8.9) is that it makes clear that the relevant parameter in the problem is $\tau = \kappa N^{2/3}$. Numerically, $\kappa = Gm_N^2$ is about 10^{-38} in our units and $N \approx 10^{57}$, so that $\tau \approx 1$. This suggests that it is an excellent approximation, indeed, to consider the limit $\kappa \to 0$, $N \to \infty$ but with $\kappa N^{2/3} = \tau$ held fixed.

The first attempt [LT3] relate $E(N)$ to $E^{TF}(N)$ succeeded to within a factor of 4. The following theorem of Lieb and H-T. Yau [LY2] finally settled the question.

Theorem 9 (Thomas-Fermi theory is asymptotically exact for cold stars). *Fix* $\tau = \kappa N^{2/3}$ *and let* $\kappa \to 0$ *and* $N \to \infty$. *Then*

$$\lim_{\kappa\to 0,\, N\to\infty} E(N)/N = e^{TF}(\tau). \tag{8.10}$$

(*The error, for finite* N, *can be bounded.*) *Furthermore, there is a critical number,* τ_c, *such that the Thomas-Fermi minimization problem* (8.8), (8.9) *satisfies*

(1) *If* $\tau < \tau_c$ *then* $e^{TF}(\tau)$ *is finite and there is a unique minimizing* ρ_τ^{TF}.

(2) *If* $0 < \tau < \tau_c$ *and if* ψ_N *is a minimizer for the quantum mechanical problem with density* ρ_{ψ_N} *then, weakly in* $L^{4/3}(\mathbf{R}^3) \cap L^1(\mathbf{R}^3)$,

$$\lim_{\kappa\to 0,\, N\to\infty} \rho_{\psi_N}(N^{1/3}x) = \rho_\tau^{TF}(x). \tag{8.11}$$

(3) *The function* $N \mapsto E^{TF}(N)$, *for fixed* κ, *is concave and decreasing.*

(4) *If* $\tau > \tau_c$ *then* $e^{TF}(\tau) = -\infty$, *while* $e^{TF}(\tau_c)$ *is finite.*

Thus, *a neutron star is stable if* $\kappa N^{2/3} < \tau_c$ *and it collapses if* $\kappa N^{2/3} > \tau_c$. It is not hard to see from (8.5) that such a collapse occurs and, indeed, to infer the value of τ_c. For large ρ (which is relevant here) $j(\rho) = C\rho^{4/3}$ with $C = 3(6\pi^2)^{1/3}/4q^{1/3}$. The *sharp* inequality

$$A\left\{\int_{\mathbf{R}^3}\rho(x)dx\right\}^{2/3}\int_{\mathbf{R}^3}\rho(x)^{4/3}dx \geq \frac{1}{2}\int_{\mathbf{R}^3}\int_{\mathbf{R}^3}\rho(x)\rho(y)|x-y|^{-1}dxdy \tag{8.12}$$

is well known, with $A = 1.092$. Therefore

$$\tau_c = C/A. \tag{8.13}$$

The Euler-Lagrange equation for the problem (8.8), (8.9) is

$$j'(\rho(x)) = [\eta(x)^2 + m^2]^{1/2} - m = \{\tau|x|^{-1} * \rho - \mu\}_+, \tag{8.14}$$

where $\eta(x) = (6\pi^2\rho(x)/q)^{1/3}$, $*$ denotes convolution, $\mu > 0$ is a Lagrange multiplier (which has to be adjusted so that $\int\rho = 1$) and $\{A\}_+ = \max(A, 0)$. The optimum solution ρ is spherically symmetric, decreasing and has *compact support* in a ball B_R of radius R. The Euler-Lagrange equation can be converted into a differential equation by applying the Laplacian, Δ, to (8.14).

$$-\Delta\Theta = 4\pi[\Theta^2 + 2\Theta]^{3/2} \quad \text{in } B_R \tag{8.15}$$

with $\Theta = \kappa|x|^{-1} * \rho$ and $\Theta(x) = 0$ when $|x| = R$.

The existence of a solution to (8.14), (8.15) was proved by Auchmuty and Beals [AB]. The analysis of this equation is, in itself, an interesting mathematical problem. The following has been proved [LY2].

Theorem 10 (Properties of the Thomas-Fermi density). *There is exactly one nonnegative radial solution to* (8.14) *for each* $\mu > 0$ *and* $0 < \tau < \tau_c$. *It has compact support. For* $\tau \geq \tau_c$ *there is no solution* (*even though* $e^{TF}(\tau_c)$ *is finite*). *If* $\tau < \tau_c$ *there is exactly one* μ *so that* $\int\rho = 1$. *Both* ρ *and* μ *are continuous functions of* τ *and* $\rho(0)$ *is an increasing function of* τ, *while the radius* R *is a decreasing function with* $R \to 0$ *as* $\tau \to \tau_c$ *and* $R \to \infty$ *as* $\tau \to 0$. *If* $\tau_1 < \tau_2 < \tau_c$ *then the solutions* ρ_1 *and* ρ_2 *satisfy* $\rho_1(x) = \rho_2(x)$ *for exactly one value of* $|x|$.

These properties show that the star has a unique shape that evolves continuously with τ, i.e., there are no jumps.

What would happen if the neutrons, which are fermions, are replaced by **bosons**? Elementary particles that are bosons, and that are stable have never been seen, as stated earlier, but they are posited to exist in certain theories in order to account for the "missing mass" in the universe. They are called axions and are not expected to condense into stars because they cannot get rid of their gravitational energy easily, but it is amusing to consider the possibility anyway. See [RB].

For bosons we might imagine that a simple product function $\psi(\underline{X}) = \Pi\phi(x_i)$ with $\phi \in L^2(\mathbf{R}^3)$, as in (4.2), would adequately describe the ground state. Then $\rho_\psi(x) = N|\phi(x)|^2$ and with $\phi(x) \geq 0$ (which is the optimum choice),

$$(8.16) \quad T_\psi = N(\phi, T_{op}\phi) = \left(\sqrt{\rho_\psi}, T_{op}\sqrt{\rho_\psi}\right)$$

$$(8.17) \quad W_\psi = \frac{\kappa}{2}\frac{N-1}{N}\int_{\mathbf{R}^3}\int_{\mathbf{R}^3}\rho_\psi(x)\rho_\psi(y)|x-y|^{-1}dxdy.$$

Adding these (and supposing N to be large) we obtain a **Hartree type functional**

(8.18)

$$\mathscr{E}^H(\rho) = \left(\sqrt{\rho}, T_{op}\sqrt{\rho}\right) - \frac{\kappa}{2}\int_{\mathbf{R}^3}\int_{\mathbf{R}^3}\rho(x)\rho(y)|x-y|^{-1}dxdy$$

for nonnegative ρ with $\int\rho = N$. This differs from the Thomas-Fermi functional in the replacement of $\int j(\rho)$ by (8.16). As usual, we define the **Hartree energy** to be

$$(8.19) \qquad E^H(N) = \inf\left\{\mathscr{E}^H(\rho): \int\rho = N\right\}.$$

The scaling $\rho(x) \to N\rho(x)$ reduced the problem to one in which $\int\rho = 1$, $\kappa \to \kappa N \equiv \omega$ and $E \to NE$. From this we learn that the parameter $\omega = \kappa N$ is the crucial one for bosons, not $\tau = \kappa N^{2/3}$.

The minmization of (8.18) leads now to a genuine differential equation for $f = \sqrt{\rho}$

$$(8.20) \qquad \{T_{op} - \omega|x|^{-1} * f^2\}f = -\mu f$$

on all of $\mathbf{R}^3$ and with μ chosen so that $\int f^2 = N$. It is proved in [LY2] that again there is a critical value ω_c such that when $\omega < \omega_c$ a minimizer exists and it is spherically symmetric and decreasing (although not with compact support). If $\omega > \omega_c$ then $E^H(N) = -\infty$. If we fix ω and let $\kappa \to \infty$ and $N \to \infty$ with $\omega = \kappa N$ then, as in the previous fermionic case, the solution to the quantum problem converges to the solution to the Hartree problem, i.e., $E(N)/E^H(N) \to 1$ and $\frac{1}{N}\rho_\psi - \frac{1}{N}\rho^H \to 0$.

Thus, *bosonic stars can collapse as well*, but whereas in the fermionic case the critical number N_c is proportional to $\kappa^{-3/2} \approx 10^{57}$, for bosons N_c is proportional to $\kappa^{-1} \approx 10^{38}$. The former defines an object which is the size of a *star*, while the latter defines only an object as massive as a *mountain* (assuming that the same value is used for the mass of the constituent particles.)

It is possible to understand this κ^{-1} behavior in a simple way—which also explains the central difficulty in proving Theorem 9. In (7.14) we saw that the kinetic energy of a particle ceases to control a $-C/|x|$ singularity whenever $C > 2/\pi$. In our bosonic star case we have many particles with the tiny constant $C \equiv \kappa$ between each pair. Suppose now that $(N-1)\kappa$ exceeds $2/\pi$ and that $N-1$ particles come together at a common point. Then the Nth particle feels an over-critical attraction and "falls in the hole." But then every particle is trapped and none can escape. Something like this artificial scenario is what happens in the bosonic case and leads to collapse if $\omega = N\kappa$ is big enough. (It is not claimed that $\omega_c = 2/\pi$; in fact it is known only that $4/\pi < \omega_c < 2.7$.) The interesting fact about fermions is that the Pauli exclusion principle prevents this scenario from happening. Since $N \approx \kappa^{3/2}$ in this case, it would require the coalescence of only a tiny fraction of the particles (namely $N^{2/3}/N$) in order to form a "trap" for the remaining particles—but such a "fluctuation" does not occur with any significant probability.

Now let us return to the second kind of cold star—the **white dwarf.** In 1914 Adams discovered that while the companion star of Sirius has a mass about equal to that of the sun (which was known from its perturbation of Sirius) its radius had to be very small [ST]. He inferred this from the fact that the star is hot (in fact its light is whiter than that of most stars—hence the name) but it has a very small total luminosity. Many such stars are now known and they are thought to be burned out stars consisting of ordinary electrons and nuclei such as we discussed in Part VII. Their radius is bigger than that of a neutron star—about 5000 km—for a reason that will soon be apparent, and this means that general relativistic effects are not important here. The ones we can see optically shine because they have not yet rid themselves of all the energy of gravitational collapse.

For some time it was a real puzzle to explain why this dead star did not collapse entirely. We have seen that the quantum mechanical kinetic energy provides the requisite "internal pressure", and the person who modelled this correctly was Chandrasekhar in

Bull. Amer. Math. Soc. *22*, 1-49 (1990)

1931 [C]. There is, however, a slight twist from the neutron star model, which requires some explanation.

The main force among electrons and nuclei is electrostatic—not gravitational. But with a grain of faith we can suppose that the electric potential is cancelled locally, i.e., local neutrality is strongly enforced. This leaves only the gravitational potential, which is additive because there are no $\pm$ signs. Ideally we would like to incorporate both forces, but we shall retain only the gravitational force. The nuclei, as before, can be considered to be almost motionless but the electrons move with high speeds (because of the high density) and therefore have to be treated relativistically.

With this information we can, with another grain of faith, suppose that the Hamiltonian H_N in (8.2) is relevant, but with the following identifications. The number of electrons is N and the mass m that appears in the kinetic energy, T_{op}, is the *electron* mass. On the other hand, the gravitational force comes mostly from the nuclei (because of their large mass), and therefore $\kappa = GM^2$ with M now being the mass in the star per electron, i.e., $M = M_{nucleus}/z_{nucleus}$, which is much bigger than the electron's mass and which is typically about twice M_n. It is the presence of the very small m in the kinetic energy which make the radius bigger for the solution of (8.5), (8.6); by scaling, the radius is proportional to $1/m$ for a fixed N. The critical τ does not change (except insofar as $M \neq M_n$).

The proper model to analyze, of course, would be one with two kinds of particles (electrons and nuclei) and with two kinds of forces (electrostatic and gravitational), but this remains an open problem. There are some remarks about this problem in [LT3]. Another thing one would like to analyze is the effect of positive temperature. For *nonrelativistic* particles, both kinds of modifications have been achieved, in a certain limit, by Hertel, Narnhofer and Thirring [HT, HNT, TW]. See also Messer's book [M]. The relativistic case with both kinds of forces is still an open problem.

Chandrasekhar, of course, did not analyze the Hamiltonian H_N as we did here. He started instead with the Euler-Lagrange equation of the Thomas-Fermi approximation, (8.14) or (8.15). In fact, it was the following physical interpretation of (8.14) that motivated Chandrasekhar: If we take the gradient of both sides and then multiply by $-\rho(x)$, the right side becomes the gravitational force at a point x in the star. The left side can be interpreted as the gradient of the pressure, P, of a "quantum mechanical fluid." This fluid has an "equation of state" (i.e., the pressure, P as a function of the density, ρ) given by the following formula that is valid for all fluids.

$$P(\rho) = \rho j'(\rho) - j(\rho) = \rho^2 \frac{\partial}{\partial \rho} \frac{j(\rho)}{\rho}. \tag{8.21}$$

This balancing of forces, i.e., the "gravitational-hydrostatic equilibrium," was Chandrasekhar's starting point.

From this point of view there is an interesting contrast between the bosonic case (well approximated by the Hartree theory (8.18)) and the fermionic case (well approximated by the Thomas-Fermi theory (8.5)). The latter energy can be thought of as that of a simple fluid which has an "equation of state." This is how physicists think of the matter, even though it is quantum mechanics that produces the pressure. The bosonic energy (8.18) has no such interpretation. Quantum mechanics shows itself to the bitter end because the Hartree energy still has gradients in it. The concept of an "equation of state" cannot be used for high density bosonic matter in its ground state.

This brings us to the end of the stability of matter story. If one more hyperbolic remark be permitted, it can be said that

Quantum mechanics is a bizarre theory, invented to explain atoms. As far as we know today it is capable of explaining everything about ordinary matter (chemistry, biology, superconductivity), sometimes with stunning numerical accuracy. But it also says something about the occurrence of the most spectacular event in the cosmos—the supernova. The range is 57 orders of magnitude!

References

[AI] *American Institute of Physics Handbook*, McGraw-Hill, New York, 1972 third ed., p. 7–6.

[AM] P. Armbruster and G. Münzenberg, *Creating superheavy elements*, Scientific American **260** (1989), 66–72.

[AB] J. Auchmuty and R. Beals, *Variational solution of some nonlinear free boundary problems*, Arch. Rat. Mech. Anal. **43** (1971), 255–271. See also *Models of rotating stars*, Astrophys. J. **165** (1971), L79-L82.

[B] G. Baym, *Neutron stars*, in Enclyclopedia of Physics, (R. G. Lerner and G. L. Trigg eds.) Addison-Wesley, London, 1981, pp. 659–660.

[BM] M. Born, *Quantenmechanik der Stossvorgänge*, Z. Phys. **38** (1926), 803–827.

[CH] S. Chandrasekhar, *The maximum mass of ideal white dwarfs*, Astrophys. J. **74** (1931), 81–82. See also *On stars, their evolution and stability*, Rev. Mod. Phys. **56** (1984), 137–147.

[CO] J. Conlon, *The ground state energy of a classical gas*, Comm. Math. Phys. **94** (1984), 439–458.

[CLY] J. G. Conlon, E. H. Lieb and H-T. Yau, *The $N^{7/5}$ law for charged bosons*, Comm. Math. Phys. **116** (1988), 417–448.

[DA] I. Daubechies, *An uncertainity principle for fermions with generalized kinetic energy*, Comm. Math. Phys. **90** (1983), 511–520.

[DAL] I. Daubechies and E. H. Lieb, *One electron relativistic molecules with Coulomb interactions*, Comm. Math. Phys. **90** (1983), 497–510.

[D] F. J. Dyson, *Ground state energy of a finite system of charged particles*, J. Math. Phys. **8** (1967), 1538–1545.

[DL] F. J. Dyson and A. Lenard, *Stability of matter*. I and II, J. Math. Phys. **8** (1967), 423–434; ibid **9** (1968), 698–711.

[FD] C. Fefferman and R. de la Llave, *Relativistic stability of matter*. I., Rev. Math. Iberoamericana **2** (1986), 119–215.

[FE] E. Fermi, *Un metodo statistico per la determinazione di alcune priorieta dell'atomo*, Atti Acad. Naz. Lincei, Rend. **6** (1927), 602–607.

[FR] A. P. French, *Atoms*, in Encyclopedia of Physics, (R. G. Lerner and G. L. Trigg eds.), Addison-Wesley, London (1981), p. 64.

[H] I. Herbst, *Spectral theory of the operator* $(p^2+m^2)^{1/2}-ze^2/r$, Comm. Math. Phys. **53** (1977), 285–294. Errata, ibid. **55** (1977), 316.

[HNT] P. Hertel, H. Narnhofer and W. Thirring, *Thermodynamic functions for fermions with gravostatic and electrostatic interactions*, Comm. Math. Phys. **28** (1972), 159–176.

[HT] P. Hertel and W. Thirring, *Free energy of gravitating fermions*, Comm. Math. Phys. **24** (1971), 22–36.

[J] J. H. Jeans, *The mathematical theory of electricity and magnetism*, Cambridge Univ. Press, Cambridge, third edition, 1915, p. 168.

[JM] M. Jammer, *The conceptual development of quantum mechanics*, McGraw-Hill, New York, 1966.

[K] T. Kato, *Perturbation theory for linear operators*, Springer-Verlag, Heidelberg, 1966. See Remark 5.12 on p. 307.

[KS] H. Kalf, U.-W. Schminke, J. Walter and R. Wüst, *On the spectral theory of Schrödinger and Dirac operators with strongly singular potentials*, Lecture Notes in Math., vol. **448** Springer-Verlag, Berlin and New York, 1974, pp. 182–226.

[LE] A. Lenard, *Lectures on the Coulomb stability problem*, Lecture Notes in Physics **20** (1973), 114–135.

[L1] E. H. Lieb, *Stability of matter*, Rev. Mod. Phys. **48** (1976), 553–569.

[L2] ——, *On characteristic exponents in turbulence*, Comm. Math. Phys. **92** (1984), 473–480.

[L3] ——, *Thomas-Fermi and related theories of atoms and molecules*, Rev. Mod. Phys. **53** (1981), 603-641; errata ibid **54** (1982), 311.

[L4] ——, *Bound on the maximum negative ionization of atoms and molecules*, Phys. Rev. **29A** (1984), 3018–3028. A summary is in Phys. Rev. Lett. **52** (1984), 315–317.

[L5] ——, *The* $N^{5/3}$ *law for bosons*, Phys. Lett. A **70** (1979), 71–73.

[LL] E. H. Lieb and J. L. Lebowitz, *The constitution of matter: existence of thermodynamics for systems composed of electrons and nuclei*, Adv. in Math. **9** (1972), 316–398.

[LO] E. H. Lieb and S. Oxford, *An improved lower bound on the indirect Coulomb energy*, Int. J. Quant. Chem. **19** (1981), 427–439.

[LS] E. H. Lieb and B. Simon, *The Thomas-Fermi theory of atoms, molecules and solids*, Adv. in Math. **23** (1977), 22–116.

[LT1] E. H. Lieb and W. E. Thirring, *Bound for the kinetic energy of fermions which proves the stability of matter*, Phys. Rev. Lett. **35** (1975), 687–689. Errata ibid. **35** (1975), 1116.

[LT2] ——, *Inequalities for the moments of the eigenvalues of the Schrödinger Hamiltonian and their relation to Sobolev inequalities*, in *Studies in Mathematical Physics*, (E. Lieb, B. Simon and A. Wightman, eds.), Princeton Univ. Press, Princeton, New Jersey, 1976, pp. 269–330.

[LT3] ——, *Gravitational collapse in quantum mechanics with relativistic kinetic energy*, Ann. of Phys. (NY) **155** (1984), 494–512.

[LY1] E. H. Lieb and H-T. Yau, *The stability and instability of relativistic matter*, Comm. Math. Phys. **118** (1988), 177–213. A summary is in *Many-body stability implies a bound on the fine structure constant*, Phys. Rev. Lett. **61** (1988), 1695–1697.

[LY2] ——, *The Chandrasekhar theory of stellar collapse as the limit of quantum mechanics*, Comm. Math. Phys. **112** (1987), 147–174. A summary is in *A rigorous examination of the Chandrasekhar theory of stellar collapse*, Astrophys. J. **323** (1987), 140–144.

[M] J. Messer, *Temperature dependent Thomas-Fermi theory*, Lectures Notes in Physics no. 147, Springer-Verlag, Berlin and New York, 1981.

[P] W. Pauli, *Über den Zusammenhang des Abschlusses der Elektronengruppen im Atom mit der Komplexstruktur der Spektren*, Z. Phys. **31** (1925), 765–785.

[RB] R. Ruffini and S. Bonazzola, *Systems of self-gravitating particles in general relativity and the concept of equation of state*, Phys. Rev. **187** (1969), 1767–1783.

[SE] E. Schrödinger, *Quantisierung als Eigenwertproblem*, Ann. Phys. **79** (1926), 361–376. See also ibid. **79** (1926), 489–527; **80** (1926), 437–490; **81** (1926), 109–139.

[ST] S. L. Shapiro and S. A. Teukolsky, *Black holes, white dwarfs and neutron stars*, Wiley, New York, 1983.

[T] L. H. Thomas, *The calculation of atomic fields*, Proc. Cambridge Philos. Soc. **23** (1927), 542–548.

[TW] W. Thirring, *A course in mathematical physics*, vol. 4, Springer-Verlag, Berlin and New York, 1983.

Departments of Mathematics and Physics, Princeton University, Princeton, New Jersey 08544

Part II

Exact Results on Atoms

Lower bound to the energy of complex atoms

Peter Hertel

Institut für Theoretische Physik der Universität Wien, A-1090 Wien, Boltzmanngasse 5, Austria

Elliot H. Lieb*

Departments of Mathematics and Physics, Princeton University, Princeton, New Jersey 08540

Walter Thirring

Institut für Theoretische Physik der Universität Wien, A-1090 Wien, Boltzmanngasse 5, Austria

(Received 31 January 1975)

There are methods available for calculating rather precise lower bounds for the energy of simple atoms or molecules.[1] If complex atoms or molecules are to be investigated, these methods become inapplicable, or impracticable. Here we propose a new method which, for the above cases, promises to become as accurate as the Hartree–Fock procedure for the upper bound. Although applicable to a wider class of problems, we shall demonstrate it for an atom with N electrons.

Our aim is to find a lower bound to the Coulomb repulsion energy

$$V=\frac{1}{2}\sum_{\substack{a,b=1\\a\neq b}}^{N}\frac{1}{|\mathbf{x}_a-\mathbf{x}_b|} \tag{1}$$

of the form

$$V_{LB}=\sum_{a=1}^{N}\Phi(\mathbf{x}_a)-\Lambda\,, \tag{2}$$

where Λ is a constant. To achieve this we replace the Coulomb potential $1/|\mathbf{x}|$ by a subsidiary potential v with the following properties:

$$\frac{1}{|\mathbf{x}|}\geq v(\mathbf{x})=\int\frac{d^3k}{(2\pi)^3}e^{i\mathbf{k}\mathbf{x}}\tilde{v}(\mathbf{k})\quad\text{with}\quad\tilde{v}(\mathbf{k})\geq 0\,. \tag{3}$$

$\tilde{v}$, the Fourier transform of v, is assumed to be integrable so that v is continuous everywhere. The repulsion potential Φ should be the Fourier transform of a real integrable function:

$$\Phi(\mathbf{x})=\int\frac{d^3k}{(2\pi)^3}e^{i\mathbf{k}\mathbf{x}}\tilde{\Phi}(\mathbf{k})\quad\text{with}\quad\Phi(\mathbf{k})=\Phi(-\mathbf{k})=\tilde{\Phi}^*(\mathbf{k})\,. \tag{4}$$

From (3) we deduce

$$V\geq\frac{1}{2}\sum_{a,b=1}^{N}v(\mathbf{x}_a-\mathbf{x}_b)-\tfrac{1}{2}Nv(0)\,. \tag{5}$$

With Jensen's or Schwarz's inequality

$$\int d\mu(x)|f(x)|^2\geq\frac{|\int d\mu(x)f(x)|^2}{\int d\mu(x)} \tag{6}$$

(μ a measure) we deduce with (3)

$$\sum_{a,b=1}^{N}v(\mathbf{x}_a-\mathbf{x}_b)=\int\frac{d^3k}{(2\pi)^3}\tilde{v}(\mathbf{k})\left|\sum_{a=1}^{N}e^{i\mathbf{k}\mathbf{x}_a}\right|^2$$
$$=\int\frac{d^3k}{(2\pi)^3}\frac{\tilde{\Phi}(\mathbf{k})^2}{\tilde{v}(\mathbf{k})}\left|\sum_{a=1}^{N}e^{i\mathbf{k}\mathbf{x}_a}\frac{\tilde{v}(\mathbf{k})}{\tilde{\Phi}(\mathbf{k})}\right|^2\geq\frac{1}{c}U^2\,, \tag{7}$$

where

$$U=\sum_{a=1}^{N}\int\frac{d^3k}{(2\pi)^3}e^{i\mathbf{k}\mathbf{x}_a}\tilde{\Phi}(\mathbf{k})=\sum_{a=1}^{N}\Phi(\mathbf{x}_a)\,, \tag{8}$$

$$c=\int\frac{d^3k}{(2\pi)^3}\frac{\tilde{\Phi}(\mathbf{k})^2}{\tilde{v}(\mathbf{k})} \tag{9}$$

Note that U is Hermitian and c real.

By subtracting the nonnegative operator $(1/c)(U-c)^2$ from the rhs of (7) we finally arrive at the desired bound:

$$V\geq\sum_{a=1}^{N}\Phi(\mathbf{x}_a)-\frac{1}{2}\int\frac{d^3k}{(2\pi)^3}\left\{\frac{\tilde{\Phi}(\mathbf{k})^2}{\tilde{v}(\mathbf{k})}+Nv(\mathbf{k})\right\}\,. \tag{10}$$

There are two functions to play with: a repulsion potential Φ (essentially arbitrary) and a subsidiary potential v [restricted by (3)] which enters the constant only.

A special choice for the subsidiary potential is

$$v(\mathbf{x})=\frac{1-e^{-\mu|\mathbf{x}|}}{|\mathbf{x}|}\,. \tag{11}$$

Defining n by

$$\Phi(\mathbf{x})=\int d^3x'\frac{n(\mathbf{x}')}{|\mathbf{x}-\mathbf{x}'|} \tag{12}$$

we can write

$$\Lambda=\frac{1}{2}\left\{\int d^3x\Phi(\mathbf{x})n(\mathbf{x})+4\pi\mu^{-2}\int d^3xn(\mathbf{x})^2+N\mu\right\} \tag{13}$$

for the constant in the rhs of (10). For fixed n (or Φ) the optimal subsidiary potential of type (11) is obtained for

$$\mu^3=\frac{8\pi}{N}\int d^3xn(\mathbf{x})^2\,. \tag{14}$$

To check whether (10) can give good lower bounds we have tried the Tietz approximation[2] to the Thomas–Fermi potential for neutral atoms:

$$\Phi_T(\mathbf{x})=(N-1)\frac{r+2(9/2N)^{1/3}}{[r+(9/2N)^{1/3}]^2}\qquad(r=|\mathbf{x}|)\,. \tag{15}$$

The factor $N-1$ (instead of N) in (15) has been chosen in order to make the potential of the lower bound Hamiltonian to go as $-1/r$ for large r. Equation (15) with (14) gives

$$\Lambda=\frac{2}{5}\left(\frac{2}{9}\right)^{1/3}\left(1-\frac{1}{N}\right)^2N^{7/3}+\frac{3}{2}\left(\frac{2}{7}\right)^{1/3}\left(1-\frac{1}{N}\right)^{2/3}N^{5/3}\,. \tag{16}$$

It remains to solve for the energy levels of the one-particle

Schrödinger equation with spherically symmetric potential $-N/|\mathbf{x}| + \Phi_T(\mathbf{x})$ and to fill the N electrons into the lowest levels in accordance with the exclusion principle. The lower bound E_{LB} for the nonrelativistic ground state energy of neutral atoms as calculated from (10) with (11), (14), and (15) is compared with an upper bound[3] E_{UB} for a few neutral atoms in the following table:

Atom	N	$-N^{-7/3}E_{LB}$	$-N^{-7/3}E_{UB}$
Ne	10	0.755	0.594
Ca	20	0.726	0.620
Zr	40	0.714	0.646
Nd	60	0.711	0.648
Hg	80	0.711	0.656

Energies are in units of $me^4/\hbar^2 = 27.21$ eV. The uncorrected Thomas–Fermi value (which becomes exact[4] in the limit $N\to\infty$) is $N^{-7/3}E_{TF} = -0.769$. The trivial lower bound (by neglecting V altogether) is $-(3/2)^{1/3}N^{7/3}$ for large N. Note that the lower and upper bounds are still far away from the Thomas–Fermi value.

The increase with N of the lower bounds is due to the fact that the potential (15) is tailored for large atoms. By adjusting some parameters the lower bounds for light atoms can be improved so that they decrease with N.

*On leave from the Departments of Mathematics and Physics, M.I.T., Cambridge, MA 02139. Work partially supported by National Science Foundation Grant GP-31674X.

[1]For a summary see A. Weinstein and W. Stenger, *Methods of Intermediate Problems for Eigenvalues* (Academic, New York, 1972).

[2]T. Tietz, J. Chem. Phys. **25**, 787 (1956). See S. Flügge, *Practical Quantum Mechanics II* (Springer, Berlin, 1971), p. 128, for the choice of the coefficient $(9/2)^{1/3}$.

[3]From R. O. Mueller, A. R. P. Rau, and L. Spruch, Phys. Rev. A 8, 1186 (1973), Table II.

[4]E. H. Lieb and B. Simon, Phys. Rev. Lett. **31**, 681 (1973).

With S. Oxford in Int. J. Quant. Chem. *19*, 427-439 (1981)

Improved Lower Bound on the Indirect Coulomb Energy*

ELLIOTT H. LIEB AND STEPHEN OXFORD
Departments of Mathematics and Physics, Princeton University, Princeton, New Jersey 08544, U.S.A.

Abstract

For a Coulomb system of particles of charge e, it has previously been shown that the indirect part of the repulsive Coulomb energy (exchange plus correlation energy) has a lower bound of the form $-Ce^{2/3}\int \rho(x)^{4/3}\,dx$, where ρ is the single particle charge density. Here we lower the constant C from the 8.52 previously given to 1.68. We also show that the best possible C is greater than 1.23.

1. Introduction

In the study of quantum Coulomb systems of charged particles (atoms, molecules, and solids), it is frequently desirable to estimate various energies in terms of the (diagonal) single particle charge density $\rho_\psi(x)$ belonging to a given state ψ of N particles. We will be concerned with the repulsive Coulomb energy

$$I_\psi = \langle\psi|\sum_{i<j}^{N} e_i e_j |x_i - x_j|^{-1}|\psi\rangle, \tag{1}$$

where $e_i > 0$ are the particle charges (by a trivial change we could assume all $e_i < 0$). ψ is *any* normalized state, not necessarily an eigenstate of any given Hamiltonian; it could also be a density matrix. Our goal is to find a lower bound to I_ψ. [See note added in proof.]

We *define*

$$D_\psi = D(\rho_\psi, \rho_\psi) \tag{2}$$

to be the *direct part* of I_ψ, where, in general, we have

$$D(f, g) = \tfrac{1}{2}\iint f(x)g(y)|x-y|^{-1}\,dx\,dy. \tag{3}$$

The density ρ_ψ is the sum $\sum_{i=1}^{N}$ of the N individual single particle charge densities defined by

$$\rho_\psi^i(x) = e_i \sum_{\alpha_1,\ldots,\alpha_N} \int |\psi(x_1, \ldots, x_{i-1}, x, x_{i+1}, \ldots, x_N, \alpha_1, \ldots, \alpha_N)|^2\,d\hat{x}_i. \tag{4}$$

* Work partially supported by U.S. National Science Foundation Grant No. PHY-7825390 A01.

Here the α_j are any quantum numbers (e.g., spin) and $d\hat{x}_i$ means integration over all variables except x_i. The *indirect part* E_ψ of I_ψ is defined by

$$E_\psi + D_\psi = I_\psi. \tag{5}$$

E_ψ includes all "exchange" and "correlation" effects.

It was shown by one of us [1], that for all ψ and $C = 8.52$, we have

$$E_\psi \geq -Ce^{2/3} \int \rho_\psi(x)^{4/3}\, dx. \tag{6}$$

A more complicated bound was given when the e_i are not identical.

The main result of this paper is given in Section 2. We show that

$$E_\psi \geq -C\left[\int \left(\sum_{i=1}^{N} e_i \rho_\psi^i(x)\right)^{4/3} dx\right]^{1/2} \left[\int \rho_\psi(x)^{4/3}\, dx\right]^{1/2}, \tag{7}$$

even in the case that the e_i may be distinct, with $C = 1.68$. With this improved value of C, the bound for E_ψ may have computational value as well as theoretical value.

Inequality (7) reduces to the form of Eq. (6) when all $e_i = e$. Our proof differs from that in Ref. 1 chiefly in avoiding the use of the Hardy–Littlewood maximal function. We also show in Section 3 that the best (i.e., smallest) possible C, in Eq. (6), must be greater than 1.23 by explicitly calculating E for a two-particle wave function ψ.

It should be noted that Eq. (6) is a special case of the lower bound for N-particle states ψ,

$$E_\psi \geq -C_p e^{(5p-6)/(3p-3)} \left(\int \rho_\psi^p\right)^{1/(3p-3)} N^{(3p-4)/(3p-3)}, \tag{8}$$

which is a consequence of Thomas–Fermi theory for $p > \frac{3}{2}$ (see Refs. 1–3). Hölder's inequality applied to Eq. (6) gives Eq. (8) for all $p \geq \frac{4}{3}$ with the same C as in Eq. (6). However, the best C_p in Eq. (8) will in general be smaller than this C and it would be interesting to have an estimate for it. [In Ref. 1 it was incorrectly stated that Thomas–Fermi theory gives Eq. (8) for $p > \frac{4}{3}$. This is wrong because Thomas–Fermi theory is bounded below only for $p > \frac{3}{2}$. Equation (8), for $p = \frac{5}{3}$, was first given in Ref. 3, where it was used to prove the stability of matter.]

The bound, equations (6) or (7), can be examined in a number of ways. One might try to find a constant C in Eq. (6) only for ψ symmetric or antisymmetric, or only for ψ with a particular spin value. However any such restriction on ψ cannot improve the constant in Eq. (6). Let all $e_i = e$. Suppose we take an arbitrary $\psi(x_1, \ldots, x_N; \alpha_1, \ldots, \alpha_N)$, where $\alpha_1, \ldots, \alpha_N$ are arbitrary quantum numbers. Let

$$f_\psi(x_1, \ldots, x_N) = \sum_{\alpha_1 \cdots \alpha_N} |\psi(x_1, \ldots, x_N, \alpha_1, \ldots, \alpha_N)|^2,$$

and let the symmetrized version of f_ψ be given as

$$F(x_1,\ldots,x_N)=\frac{1}{N!}\sum_{P\in S_N} f_\psi(x_{P1},\ldots,x_{PN}).$$

We can define a symmetric ψ^S such that $E_{\psi^S}=E_\psi$, $I_{\psi^S}=I_\psi$, and $\rho_{\psi^S}(x)=\rho_\psi(x)$. We merely take $\psi^S(x_1,\ldots,x_N)=F(x_1,\ldots,x_N)^{1/2}$. This shows that C cannot be improved by excluding bosonic ψ. In a similar way, we define an antisymmetric ψ^a,

$$\psi^a=F(x_1,\ldots,x_N)^{1/2}\theta(x_1,\ldots,x_N),$$

where θ is any antisymmetric function which takes on the values ± 1 except on a set of measure zero. We see that we cannot improve the constant by excluding ψ with Fermi statistics. A similar construction shows that restrictions on the spin quantum number cannot improve C.

It will be noted that the right-hand side of Eq. (6) is of the form given by Dirac [4] to approximate the exchange energy. There is, however, a difference between the two if spin is taken into account. The Dirac approximation is

$$E_\psi \simeq -Cq^{-1/3}\int \rho_\psi(x)^{4/3}\,dx, \tag{9}$$

with $C=0.93$ and q is the number of spin states ($q=2$ for electrons). Dirac computed Eq. (9) using a plane wave determinant for ψ. This determinant depends upon q. In view of Eq. (9) one might infer that a good lower bound to E_ψ should have a $q^{-1/3}$ factor. This, as we have just noted above, is not the case. Another way to say this is the following: a "diagonal" operator, such as the Coulomb repulsion cannot distinguish the spin of a particle. To "see" the spin it is necessary to examine an off-diagonal operator such as the kinetic energy T_ψ. A lower bound for T_ψ does exist [3] and it does have a factor $q^{-2/3}$. In order to have a lower bound for E_ψ that measures q it would be necessary to use an expectation value such as T_ψ that is "off-diagonal." A useful bound of this kind might exist, but a bound such as ours that involves only the "diagonal" quantity ρ_ψ has no q dependence.

It *is* true however, that the constant in Eq. (6) or (7) can be improved if we specify the number of particles. Consider the case of equal charges. Define C_N to be the best constant for Eq. (6) when we consider only N-particle states ψ, i.e.,

$$C_N=\sup_{N\text{ particle }\psi}\left(-\int \rho_\psi(x)^{4/3}\,dx\right)^{-1} E_\psi.$$

In Appendix A we compute $C_1=1.092$, following a treatment by Gadre, Bartolotti, and Handy [5]. One notes that C_1 is less than the constant 1.23 computed for the variational two-particle wave function of Section 3, which is itself less than C_2. In Appendix B, we show that this is a general phenomenon, i.e., $C_N\le C_{N+1}$. Presumably, one might take advantage of this if one were only dealing with a fixed low number of particles. However, we have no method available for computing an upper bound for C_N, $N\ge 2$, except 1.68, which holds for all N.

Inequality (7) might be improved upon in another way in the case of unequal charges. One might try to prove that a universal constant C existed such that the following is true (for fixed $0<\alpha\leq 1$):

$$E_\psi \geq -C\left[\int\left(\sum_{i=1}^{N} e_i^{1/2\alpha}\rho_\psi^i(x)\right)^{4/3} dx\right]^{\alpha}\left[\int \rho(x)^{4/3}\,dx\right]^{1-\alpha}. \tag{10}$$

It is an open question whether a bound exists when $\alpha=1$ in Eq. (10), called the symmetric form. Hölder's inequality implies that for Eq. (10), (right-hand side, $\alpha=1)\geq$(right-hand side, $\alpha\leq 1$); thus the symmetric form gives the best lower bound. In Ref. 1 a bound with constant 8.52 was proved for $\alpha=\frac{3}{4}$ and in this paper we sacrifice even more "symmetry" in the bound to show that Eq. (10) holds for $\alpha=\frac{1}{2}$ with the improved constant 1.68.

In Ref. 5, the comment was made that no one has yet produced any upper bound for E_ψ. The following simple remark is relevant. There certainly cannot be any upper bound of form $C\int\rho(x)^{4/3}\,dx$. The reason is that $E_\psi/\int\rho(x)^{4/3}\,dx$ can be made arbitrarily positive simply by taking a ψ for two particles of the form

$$\psi = A[k\cdot(x_1-x_2)]\exp\left(-|x_1|^2-|x_2|^2-\lambda|x_1-x_2|^2\right.,$$

where k is some fixed vector. As λ tends to infinity, E_ψ will tend to $+\infty$ while $\int\rho_\psi^{4/3}\,dx$ will remain bounded. This ψ is antisymmetric.

2. A Lower Bound for E_ψ

We will use an argument similar to that given in Ref. 1 to derive Eq. (7) for $C=1.68$.

We first fix charges $e_1\geq 0,\ldots,e_N\geq 0$. Let $f_\psi(x_1,\ldots,x_N)$ be the particle density associated with an N-particle wave function ψ, as given in Section 1. Let ρ_ψ be the associated single-particle charge density, equation (4), and let $x_1,\ldots,x_N$ be distinct but otherwise arbitrary points in $\mathbf{R}^3$. We take $\mu(y)$ to be some function satisfying the following: (i) μ is non-negative; (ii) μ is spherically symmetric about the origin and $\mu(x)=0$ if $|x|>1$; and (iii) $\int\mu(y)\,dy=1$. Let λ be a positive constant to be determined later.

We now define a function

$$\mu_x(\cdot):\qquad \mu_x(y)\equiv\lambda^3\rho_\psi(x)\mu(\lambda\rho_\psi(x)^{1/3}(y-x)), \tag{11}$$

if $\rho_\psi(x)>0$, and $\mu_x(y)\equiv 0$ if $\rho_\psi(x)=0$. We see that μ_x is a non-negative function which satisfies (i) its integral (with respect to y) is 1 if $\rho_\psi(x)>0$; (ii) it is spherically symmetric about x; (iii) $\mu_x(y)=0$ if $|y-x|>\lambda^{-1}\rho_\psi(x)^{-1/3}$.

We observe that Lemma 1 of Ref. 1 may be applied to this choice of μ. Namely, we prove

Lemma 1:

$$\sum_{1\leq i<j\leq N} e_ie_j|x_i-x_j|^{-1}\geq -D(\rho_\psi,\rho_\psi)+2\sum_{i=1}^{N} D(\rho_\psi, e_i\mu_{x_i})-\sum_{i=1}^{N} D(e_i\mu_{x_i}, e_i\mu_{x_i}). \tag{12}$$

Proof. It is a well-known fact [1] that the potential ϕ generated by a non-negative spherically symmetric charge distribution of total charge 1 satisfies $\phi(x) \leq 1/|x|$. In particular, taking x_i to be the origin, $\int \mu_{x_i}(y)|x-y|^{-1}\,dy \leq |x-x_i|^{-1}$. Hence

$$\begin{aligned} D(e_i\mu_{x_i}, e_j\mu_{x_j}) &= \tfrac{1}{2}e_ie_j \int \mu_{x_i}(y)|x-y|^{-1}\mu_{x_j}(x)\,dx\,dy \\ &\leq \tfrac{1}{2}e_ie_j \int |x-x_i|\mu_{x_j}(x)\,dx \leq \tfrac{1}{2}e_ie_j|x_i-x_j|^{-1}. \end{aligned} \tag{13}$$

We now observe that

$$D\Big(\rho_\psi - \sum_{i=1}^N e_i\mu_{x_i}, \rho_\psi - \sum_{i=1}^N e_i\mu_{x_i}\Big) \geq 0 \tag{14}$$

by the positive definiteness of the Coulomb kernel. Expanding the left-hand side of Eq. (14) and rearranging, one has that

$$2\sum_{i<j} D(e_i\mu_{x_i}, e_j\mu_{x_j}) \geq -D(\rho_\psi, \rho_\psi) + 2\sum_{i=1}^N D(\rho_\psi, e_i\mu_{x_i}) - \sum_{i=1}^N D(e_i\mu_{x_i}, e_i\mu_{x_i}). \tag{15}$$

The lemma follows by applying Eq. (13) to the left-hand side of Eq. (15). ■

Now let δ_{x_i} be a point charge distribution of charge one centered at x_i. Adding and subtracting terms on the right-hand side of inequality (12), we get

$$\begin{aligned} \sum_{i<j} e_ie_j|x_i-x_j|^{-1} \geq{}& -D(\rho_\psi, \rho_\psi) + 2\sum_{i=1}^N D(\rho_\psi, e_i\delta_{x_i}) \\ & -\Big(2\sum_{i=1}^N D(\rho_\psi, e_i\delta_{x_i} - e_i\mu_{x_i}) + \sum_{i=1}^N D(e_i\mu_{x_i}, e_i\mu_{x_i})\Big). \end{aligned} \tag{16}$$

We now integrate Eq. (16) against $f_\psi(x_1, \ldots, x_N)$, whence

$$\begin{aligned} I_\psi \geq D(\rho_\psi, \rho_\psi) - \Big(\sum_{i=1}^N &\int 2D(\rho_\psi, \delta_{x_i} - \mu_{x_i})\rho_\psi^i(x_i)\,dx_i \\ &+ \sum_{i=1}^N \int D(\mu_{x_i}, \mu_{x_i})e_i\rho_\psi^i(x_i)\,dx_i\Big). \end{aligned} \tag{17}$$

We wish to find an upper bound of the appropriate kind to the expression in large parentheses in Eq. (17). Let us denote the first sum in large parentheses as (*) and the second as (**). We rewrite (*) as follows:

$$(*) = \int \rho_\psi(x)2D(\rho_\psi, \delta_x - \mu_x)\,dx = \iint \rho_\psi(y)F_\lambda(\rho_\psi(x), |x-y|)\,dx\,dy, \tag{18}$$

where we define

$$F_\lambda(a, r) = [ar^{-1} - \lambda a^{4/3}\phi(\lambda a^{1/3}r)], \tag{19}$$

and ϕ is the potential generated by our fixed μ. Hereafter we require that μ be bounded. In this case, $F_\lambda(a, r)$, considered as a function of a on $(0, \infty)$, satisfies the following for $r>0$; (*i*) it is continuously differentiable; (ii) $F_\lambda(0,r)=0$ and $F_\lambda(a,r)=0$ if $\lambda a^{1/3}r>1$. These properties follow from the continuity and differentiability of ϕ and the relation $\phi(x)=1/|x|$ for $|x|>1$. For $a>0$, let $\chi_a(x)=\theta[\rho_\psi(x)-a]$, where $\theta(t)=0$ if $t\leq 0$, and $\theta(t)=1$ if $t>0$. χ_a is the characteristic function of the set $\rho_\psi(x)>a$. By Fubini's theorem and the fundamental theorem of calculus, one has that

$$\int_0^\infty da \int \chi_a(x)\frac{\partial}{\partial a}F_\lambda(a,r)\,dx=\int dx\int_0^{\rho(x)}\frac{\partial}{\partial a}F_\lambda(a,r)\,da=\int dx\,F_\lambda(\rho(x),r), \tag{20}$$

and thus

$$(*)=\int_0^\infty\int_0^\infty da\,db\int dx\,dy\,\chi_a(x)\chi_b(y)\frac{\partial}{\partial a}F_\lambda(a,|x-y|)$$

where we have used the representation $\rho(y)=\int_0^\infty db\,\chi_b(y)$.

We bound $(*)$ as follows: Let $(y)_+=y$ if $y\geq 0$, and $(y)_+=0$ if $y\leq 0$. Then

$$\begin{aligned}(*)&\leq\int_0^\infty\int_0^\infty da\,db\int dx\,dy\,\chi_a(x)\chi_b(y)\left(\frac{\partial}{\partial a}F_\lambda(a,|x-y|)\right)_+\\ &\leq\int_{a>b}da\,db\int dx\,dy\,\chi_a(x)\left(\frac{\partial}{\partial a}F_\lambda(a,|x-y|)\right)_+ \qquad (21)\\ &\quad+\int_{b<a}da\,db\int dx\,dy\,\chi_b(y)\left(\frac{\partial}{\partial a}F_\lambda(a,|x-y|)\right)_+.\end{aligned}$$

By scaling properties of $(\partial/\partial a)F_\lambda(a,|x-y|)$, one has that

$$\int\left(\frac{\partial}{\partial a}F_\lambda(a,|x-y|)\right)_+dx=\int\left(\frac{\partial}{\partial a}F_\lambda(a,|x-y|)\right)_+dy=\lambda^{-2}Ka^{-2/3}, \tag{22}$$

where $K=\int[(\partial/\partial a)F_1(1,|z|)]_+\,dz$ and K only depends on the original choice of μ.

We, therefore, have that

$$\begin{aligned}(*)&\leq\lambda^{-2}K\int dx\left(\int_0^\infty da\,\chi_a(x)a^{-2/3}\int_0^a db+\int_0^\infty\chi_b(x)\,db\int_0^b a^{-2/3}\,da\right)\\ &=4\lambda^{-2}K\int dx\int_0^\infty\chi_a(x)a^{1/3}\,da=3\lambda^{-2}K\int\rho_\psi^{4/3}(x)\,dx,\end{aligned} \tag{23}$$

where we have used the representation $\rho_\psi^{4/3}(x)=(4/3)\int_0^\infty a^{1/3}\chi_a(x)\,da$.

The second sum $(**)$ in the large parentheses of (17) can be written

$$\begin{aligned}(**)&=\sum_{i=1}^N\int D(\mu_x,\mu_x)e_i\rho_\psi^i(x)\,dx\\ &=\lambda D(\mu,\mu)\sum_{i=1}^N\int e_i\rho_\psi^i(x)\rho_\psi(x)^{1/3}\,dx \qquad (24)\\ &\leq\lambda D(\mu,\mu)\left[\int\left(\sum_{i=1}^N e_i\rho_\psi^i(x)\right)^{4/3}dx\right]^{3/4}\left[\int\rho_\psi(x)^{4/3}\,dx\right]^{1/4}. \qquad (25)\end{aligned}$$

Equation (24) follows from simple scaling and Eq. (25) is the Hölder inequality.

Optimizing Eqs. (24) and (25) with respect to λ yields

$$(*)+(**)\leq \tfrac{3}{2}[6KD(\mu,\mu)^2]^{1/3}\left[\int\left(\sum_{i=1}^{N} e_i\rho^i(x)\right)^{4/3}dx\right]^{1/2}\left[\int \rho(x)^{4/3}\,dx\right]^{1/2}. \tag{26}$$

A variational argument shows that the optimum choice of μ would be the uniform ball *if* $[(\partial/\partial a)F_\lambda(a,r)]_+$ *were replaced by* $(\partial/\partial a)F_\lambda(a,r)$ [in which case the constant in Eq. (26) would be 1.45]. However, trial and error indicates this choice is also approximately best with the cutoff. We find that $[\partial F_1(1,r)/\partial a]_+=\partial F_1(1,r)/\partial a$ if and only if $r\leq R$ with $R=(5^{1/2}-1)/2$. Then $K=0.6489$ and $D(\mu,\mu)=\frac{3}{5}$. The constant in Eq. (26) is then 1.68.

Thus we reach the conclusion that

$$E_\psi\geq -1.68\left[\int\left(\sum_{i=1}^{N} e_i\rho_\psi^i(x)\right)^{4/3}dx\right]^{1/2}\left[\int \rho_\psi(x)^{4/3}\,dx\right]^{1/2}. \tag{27}$$

3. A Lower Bound for C_2

We now exhibit a lower bound to C_2 [and thus to the best possible C in Eq. (6)] which is greater than C_1.

We choose a singular $\psi(x,y)$, and take $e_1=e_2=1$. Let $t\doteq|x|$ and $s\doteq|y|$ for $x,y\in\mathbf{R}^3$, and let h and e be unit vectors $e=x/|x|$ and $h=y/|y|$. We define

$$|\psi|^2[(t,e),(s,h)]\equiv f\equiv(15/4\pi^2)\delta(1-t-s)\delta(e\cdot h+1)\theta(1-t)\theta(1-s).$$

We check the following:

$$\begin{aligned}\int f[(t,e),(s,h)]s^2\,ds\,dh&=\frac{15}{4\pi^2}\int_0^1\delta(1-t-s)s^2\,dx\\&\quad\times\int_0^{2\pi}\int_0^{\pi}\delta(\cos\psi+1)\sin\psi\,d\psi\,d\phi\,\theta(1-t)\\&=\frac{15}{2\pi}(1-t)^2\theta(1-t)\equiv\rho^1(t).\end{aligned}$$

We have used spherical coordinates to evaluate the above integral with the north pole in the (fixed) "e" direction. Similarly, the s marginal is $\rho^2(s)=\rho^1(s)$. One checks that $\int\rho^1=1$ and hence f is properly normalized. We have that $\rho(t)=2\rho^1(t)$. Trivially, $I_\psi=1$ since the particles are always one unit apart. We have that

$$\int\rho(x)^{4/3}\,dx=4\pi\left(\frac{15}{\pi}\right)^{4/3}\int_0^1(1-t)^{8/3}t^2\,dt=2.084.$$

By Newton's theorem

$$D(\rho,\rho)=(4\pi)^2\int_0^\infty\rho(t)t\int_0^t\rho(s)s^2=3.572.$$

With S. Oxford in Int. J. Quant. Chem. *19*, 427-439 (1981)

The result is that

$$C \geq C_2 \geq -E_\psi \Big/ \int \rho_\psi(x)^{4/3}\, dx = 1.234.$$

Appendix A: Evaluation of C_1

For one particle $I_\psi = 0$. Therefore C_1 is the maximum of $H(\rho)$ given in Eq. (28). We prove the following theorem and then compute C_1. This was done in Ref. 5. The purpose of this appendix is only to give a rigorous justification of the calculation done in Ref. 5.

Theorem. There exists a symmetric decreasing function ρ which maximizes the functional

$$H(\rho) = \left(\int \rho(x)^{4/3}\, dx \right)^{-1} D(\rho, \rho) \tag{28}$$

over the set $A = \{\rho(x) | \rho(x) \geq 0, \rho \in L^{4/3}(\mathbf{R}^3), \int \rho(x)\, dx = 1\}$.

Proof. For any $\rho \geq 0$ and $\rho \in L^{4/3}(\mathbf{R}^3) \cap L^1(\mathbf{R}^3)$, let $(\rho)_\lambda$ be a scaled version of ρ, i.e.,$(\rho)_\lambda(x) = \lambda^3 \rho(\lambda x)$. It is simple to check that

$$\int (\rho)_\lambda(x)\, dx = \int \rho(x)\, dx,$$

$$\int (\rho)_\lambda(x)^{4/3}\, dx = \lambda \int \rho(x)^{4/3}\, dx,$$

and

$$H[(\rho)_\lambda] = H(\rho). \tag{29}$$

Let $\rho_j \in A, j = 1, 2, \ldots$ be such that $\lim_j H(\rho_j) = \sup_{\rho \in A} H(\rho)$. This supremum may *a priori* be infinite, but we will see that it is finite. By scaling ρ_j and using Eq. (29), we may assume henceforth that $\int \rho_j(x)^{4/3}\, dx = 1$. By the Riesz inequality for symmetric decreasing rearrangements [6],† we have $H(\rho^*) \geq H(\rho)$, where ρ^* is the symmetric decreasing rearrangement of ρ. One also has that $\int \rho^*(x)\, dx = \int \rho(x)\, dx$, $\int \rho^*(x)^{4/3}\, dx = \int \rho(x)^{4/3}\, dx$; therefore, by replacing ρ_j by ρ_j^* if necessary, we may also assume that the ρ_j are symmetrically decreasing.

We now use an idea used in Ref. 7. By the symmetric decreasing property of ρ_j, we have [writing $\rho_j(x) = \rho_j(|x|)$]

$$\frac{4\pi}{3} R^3 \rho_j(R) \leq \int_{|x| \leq R} \rho_j(x)\, dx \leq \int \rho_j(x)\, ds = 1.$$

† The 3-dimensional proof can be found in Brascamp, Lieb, and Luttinger, J. Funct. Anal. **17**, 227 (1974).

Hence

$$\rho_j(R)\le k_1/R^3, \text{ all } j. \tag{30}$$

Similarly,

$$\frac{4\pi}{3}R^3\rho_j^{4/3}(R)\le\int\rho_j(x)^{4/3}\,dx=1,$$

$$\rho_j(R)\le k_2/R^{9/4}, \text{ all } j. \tag{31}$$

We define $f(R)\equiv\min(k_1/R^3, k_2/R^{9/4})$. Since the ρ_j are symmetric decreasing and uniformly bounded by f (which is finite except at 0), by a variant of Helley's theorem [8], some subsequence of the ρ_j (which we continue to denote by ρ_j) converges pointwise almost everywhere to some symmetric decreasing $\rho(x)$ and $\rho(x)\le f(x)$. We will see that $\rho(x)\ne 0$.

We now show that the ρ we have found satisfies the conditions of the theorem. By calculation $D(f,f)<\infty$. We therefore apply the dominated convergence theorem to conclude that

$$\lim_j D(\rho_j,\rho_j)=D(\rho,\rho)<\infty. \tag{32}$$

In particular,

$$0<\sup_A H(\rho)=\lim D(\rho_j,\rho_j)<\infty \text{ and thus } \rho\ne 0. \tag{33}$$

Furthermore, by Fatou's lemma we have that

$$\int\rho(x)^{4/3}\,dx\le\lim_j\int\rho_j(x)^{4/3}\,dx=1, \tag{34}$$

$$\int\rho(x)\,dx\le\lim_j\int\rho_j(x)\,dx=1. \tag{35}$$

Therefore by Eqs. (32)–(34), $H(\rho)\ge\sup_{\rho\in A}H(\rho)$. By Eq. (35) we can multiply ρ by a scalar $\lambda\ge 1$ so that $\int\lambda\rho(x)\,dx=1$. By definition of A, $\lambda\rho\in A$ and

$$H(\lambda\rho)=\lambda^{2/3}H(\rho)\ge H(\rho)\ge\sup_{\rho\in A}H(\rho). \tag{36}$$

It must be that the inequalities in Eq. (36) are actually equalities and thus that $\lambda=1$. We therefore have that ρ belongs to A and maximizes $H(\cdot)$ on that set. ■

We shall now show that the constant C_1 can be calculated. By usual variational arguments [7], one knows that the optimizing ρ satisfies the following variational equations:

$$\phi(x)-\tfrac{4}{3}C_1\rho(x)^{1/3}+\lambda\begin{cases}=0 & \text{if } \rho(x)>0, & (37)\\ \le 0 & \text{if } \rho(x)=0, & (38)\end{cases}$$

where $\phi(x)$ is the potential generated by ρ,

$$\phi(x)=\int \rho(y)|x-y|^{-1}\,dy, \tag{39}$$

and where λ is a real-valued Lagrange multiplier.

We first note that Eqs. (37) and (38) imply that ρ is compactly supported. If not, then one must have that $\rho(x)>0$ for all x by the symmetric decreasing property of ρ. By letting $|x|\to\infty$ one has that $\lambda=0$, since ρ and ϕ tend to zero in Eq. (37). We then would have by Eq. (37), $\rho(x)=(\text{constant})\phi(x)^3$, where the constant is positive. For sufficiently large $|x|$, we see that $\rho(x)\geq\frac{1}{2}(\text{constant})|x|^{-3}$. This implies that ρ is not integrable, contradicting the fact that $\int\rho(x)\,dx=1$. Let r_0 be the distance at which ρ first vanishes.

We now apply the Laplacian to Eq. (39) and use Eq. (37),

$$-(1/4\pi)\Delta\phi(x)=\rho(x)=\begin{cases}\{\frac{3}{4}C_1^{-1}[\phi(x)+\lambda]\}^3, & |x|\leq r_0,\\ 0, & |x|\geq r_0.\end{cases} \tag{40}$$

Let $f(r)=(3\pi/C_1)^{3/2}[\phi(x)+\lambda]/4\pi$. We rewrite Eq. (40) in spherical coordinates

$$r^{-1}\frac{d^2}{dr^2}rf(r)=-f(r)^3,\quad r<r_0. \tag{41}$$

Equations (40) and (41) hold in the distributional sense.

We now argue that $f(r)$ is continuously differentiable and that $f'(0)=0$. This will also imply that Eq. (41) is supplemented by $f(r_0)=0$. In spherical coordinates, one can write

$$\phi(r)=\begin{cases}4\pi r^{-1}\int_0^r\rho(s)s^2\,ds+4\pi\int_r^{r_0}\rho(s)s\,ds, & r\leq r_0,\\ r^{-1}, & r\geq r_0.\end{cases} \tag{42}$$

We apply Hölder's inequality to the first integral and use the fact that $\rho\in L^{4/3}$,

$$\int_0^r\rho(s)s^2\,ds\leq\left(\int_0^r\rho(s)^{4/3}s^2\,ds\right)^{3/4}\left(\int_0^r s^2\,ds\right)^{1/4}\leq(\text{constant})r^{3/4}.$$

This and a similar inequality satisfied by the second integral imply that $\phi(r)=O(r^{-1/4})$ near the origin. By Eq. (37) one has that $\rho(r)=O(r^{-3/4})$ near $r=0$. This in turn implies that ϕ is bounded at the origin, and therefore by Eq. (37), ρ is also bounded at the origin. Since ϕ is the potential of a bounded, compactly supported charge distribution, it is also continuous, hence ρ is continuous by Eqs. (37) and (38).

One now can see that ϕ (hence f) is C^1 by examining Eq. (42). Since ρ is continuous and bounded, the first term of Eq. (42) is of the form $r^{-1}g(r)$, where $g(r)$ is continuously differentiable for $r>0$, $g(r)=O(r^3)$ and $g'(r)=O(r^2)$ near the origin. Hence the first term is continuously differentiable for $r\geq0$, and has vanishing derivative at $r=0$. The preceding statement is true of the second term in

Eq. (42) by inspection. Thus $\phi(r)$ is continuously differentiable for $r \geq 0$ and $\phi'(0) = 0$. Equation (41) holds in the strong sense because its right-hand side is C^1.

As first noted by Gadre, Bartolotti, and Handy [5], Eq. (41) is the Emden equation of order 3. One may rescale $\rho(x) \to \alpha^3 \rho(\alpha x)$ to ensure that $f(0) = 1$. The two conditions $f(0) = 1$ and $f'(0) = 0$ uniquely determine the solution of the ordinary differential equation (41).

If r_0 is the first zero of the solution, we have that $\rho(r) = 0$ if $r \geq r_0$ and $\rho(r) = (3\pi)^{-3/2} C_1^{3/2} f(r)^3$ if $r \leq r_0$. In Ref. 5 it was noted that this equation determines the constant C_1. Namely, we have that

$$\begin{aligned} 1 = 4\pi \int_0^{r_0} \rho(r) r^2 \, dr &= 4 \cdot 3^{-3/2} \pi^{-1/2} C_1^{3/2} \int_0^{r_0} r^2 f(r)^3 \, dr \\ &= -4 \cdot 3^{-3/2} \pi^{-1/2} C_1^{3/2} \int_0^{r_0} r [r f(r)]'' \, dr \\ &= -4 \cdot 3^{-3/2} \pi^{-1/2} C_1^{3/2} r_0^2 f'(r_0). \end{aligned} \tag{43}$$

Emden functions are tabulated [9]. We find that $r_0 = 6.89684$, $f'(r_0) = -0.04243$. Equation (43) then gives $C_1 = 1.092$.

Appendix B: Monotonicity of C_N

We show that $C_N \leq C_{N+1}$, where C_N is defined in Section 1 as the best constant in Eq. (6) for an N-particle state. We consider the case $e_i = e$.

Let $\varepsilon > 0$ be arbitrary but fixed. We let $f_N(x_1, \ldots, x_N)$ be an N-particle density which vanishes for $|x_i| > L$ for $1 \leq i \leq N$, where L is some finite number, and furthermore, let f_N have the property that

$$-\left(e^{2/3} \int \rho_{f_N}(x)^{4/3} \, dx\right)^{-1} E_{f_N} \geq C_N - \varepsilon. \tag{44}$$

A simple approximation argument using dominated convergence shows that L and f_N can be found satisfying Eq. (44).

Let $x_0 \in \mathbf{R}^3$ be chosen such that $|x_0| \geq L + 2R$, where R will be determined later. We define a one-particle density $f_1(x) = (\frac{4}{3}\pi R^3)^{-1} \theta(R - |x - x_0|)$ and we also define the $(N+1)$-particle density $f_{N+1}(x_1, \ldots, x_{N+1}) = f_N(x_1, \ldots, x_N) f_1(x_{N+1})$. One sees that $\rho_{f_{N+1}}(x) = \rho_{f_N}(x) + e f_1(x)$. Since ρ_N and f_1 are never simultaneously nonzero, we have

$$\begin{aligned} e^{2/3} \int \rho_{f_{N+1}}(x)^{4/3} \, dx &= e^{2/3} \left(\int \rho_{f_N}(x)^{4/3} \, dx + e^{4/3} \int f_1(x)^{4/3} \, dx \right) \\ &= e^{2/3} \int \rho_{f_N}(x)^{4/3} \, dx + \frac{e^2 (3/4\pi)^{1/3}}{R}. \end{aligned} \tag{45}$$

We also have that

$$I_{f_{N+1}}=I_{f_N}+e^2\sum_{i=1}^{N}\int f_N(x_1,\ldots,x_N)f_1(x_{N+1})|x_i-x_{N+1}|^{-1}\,dx_1,\ldots,dx_{N+1}$$
$$\le I_{f_N}+e^2N/R, \tag{46}$$

by the definition of f_1. The evident inequality $D_{f_{N+1}}\ge D_{f_N}$ together with Eq. (46) implies that $E_{f_{N+1}}\le E_{f_N}+e^2N/R$.

This and Eq. (45) imply that

$$C_{N+1}\ge -\left(e^{2/3}\int \rho_{f_{N+1}}(x)^{4/3}\,dx\right)^{-1}E_{f_{N+1}}$$
$$\ge -\left(e^{2/3}\int \rho_{f_N}(x)^{4/3}\,dx+\frac{e^2(3/4\pi)^{1/3}}{R}\right)^{-1}\left(E_{f_N}+\frac{e^2N}{R}\right). \tag{47}$$

We now choose R so large that the right-most term in Eq. (47) is greater or equal to

$$-\left(e^{2/3}\int \rho_{f_N}(x)^{4/3}\,dx\right)^{-1}E_{f_N}-\varepsilon.$$

Recalling Eq. (44), we have the result that $C_{N+1}\ge C_N-2\varepsilon$. Since ε was arbitrary, $C_{N+1}\ge C_N$.

In the case of distinct e_i's, one may define (for some fixed α, $0<\alpha\le 1$)

$$C_N(e_1,\ldots,e_N)=\sup_{f_N(x_1,\ldots,x_N)} -\left[\int\left(\sum_{i=1}^{N} e_i^{1/2\alpha}\rho^i_{f_N}(x)\right)^{4/3}dx\right]^{-\alpha}$$
$$\times\left(\int \rho_{f_N}(x)^{4/3}\,dx\right)^{\alpha-1}E_{f_N}.$$

A similar argument shows that these constants also increase, i.e., $C_N(e_1,\ldots,e_N)\le C_{N+1}(e_1,\ldots,e_N,e_{N+1})$, where $e_{N+1}\ge 0$ is arbitrary. Of course Section 2 shows that $C_N(e_1,\ldots,e_N)\le 1.68$ for all N and e_i when $\alpha=\frac{1}{2}$.

Note added in proof:

In the text we proved the inequalities, Eqs. (6) and (7), when ψ is a wave function (pure state), and remarked that the inequalities also hold for a density matrix. To prove this, note that any density matrix, μ, can be written as $\mu=\sum_\beta \psi^\beta><\psi^\beta$. In the definition, Eq. (4), simply regard β as just one more quantum number to sum over—on the same footing as the α's. The rest of the proof is then the same as in the pure state case.

Bibliography

[1] E. H. Lieb, Phys. Lett. **70A**, 444 (1979).
[2] E. H. Lieb, Rev. Mod. Phys. **48**, 553 (1976).
[3] E. H. Lieb and W. E. Thirring, Phys. Rev. Lett. **35**, 687 (1975); **35**, 1116 (1975) (errata).
[4] P. A. M. Dirac, Proc. Cambridge Philos. Soc. **26**, 376 (1930).

[5] S. R. Gadre, L. J. Bartolotti, and N. C. Handy, J. Chem. Phys. **72**, 1034 (1980).
[6] F. Riesz, J. London Math. Soc. **5**, 162 (1930).
[7] E. H. Lieb, Stud. Appl. Math. **57**, 93 (1977).
[8] W. Feller, *An Introduction to Probability Theory and its Applications*, (*Wiley, New York*, 1966), Vol. 2, p. 261.
[9] *British Association for the Advancement of Science Mathematical Tables*, (Office of the British Assoc., Burlington House, London, 1932) Vol. 2.

Received June 10, 1980.
Accepted for publication September 18, 1980

LETTER TO THE EDITOR

Monotonicity of the molecular electronic energy in the nuclear coordinates

Elliott H Lieb†

Departments of Mathematics and Physics, Princeton University, POB 708, Princeton, NJ 08544, USA

Received 3 December 1981

Abstract. Let $e(R_1, \ldots, R_k)$ be the electronic contribution to the ground-state energy of a molecule consisting of one electron and k nuclei located at $R_1, \ldots, R_k$. It is shown that $e(R_1', \ldots, R_k') \geqslant e(R_1, \ldots, R_k)$ if $|R_i' - R_j'| \geqslant |R_i - R_j|$ for all pairs i, j.

Consider the non-relativistic Schrödinger ground-state energy, E, of a molecule consisting of N electrons and k fixed nuclei with positive charges $z_1, \ldots, z_k$ located at distinct points $R_1, \ldots, R_k \in \mathbb{R}^3$. (The approximation that the nuclei are fixed is sometimes called the Born–Oppenheimer approximation.) Since we shall be interested in the dependence of E on the R_i, with the z_i fixed, we use the notation $\underset{\sim}{R} = (R_1, \ldots, R_k)$ to indicate the relevant variables. Further, we assume $\hbar^2/2m = 1$ and $e = -1$ where m and e are the electron mass and charge.

$$E(\underset{\sim}{R}) = e(\underset{\sim}{R}) + U(\underset{\sim}{R}) \tag{1}$$

$$U(\underset{\sim}{R}) = \sum_{1 \leqslant i < j \leqslant k} z_i z_j |R_i - R_j|^{-1}. \tag{2}$$

$U(\underset{\sim}{R})$ is the nuclear–nuclear potential energy while $e(\underset{\sim}{R})$ is the electronic contribution to the energy and is the ground-state energy (inf spectrum) of the Hamiltonian

$$H(\underset{\sim}{R}) = -\sum_{i=1}^{N} (\Delta_i + V(x_i, \underset{\sim}{R})) + \sum_{1 \leqslant i < j \leqslant N} |x_i - x_j|^{-1} \tag{3}$$

with

$$V(x, \underset{\sim}{R}) = \sum_{j=1}^{k} z_j |x - R_j|^{-1} \tag{4}$$

being the negative of the electron–nuclear interaction.

An old conjecture is that when the R_i are moved apart, in some sense that has to be made precise, and if N is not too large, $e(\underset{\sim}{R})$ *increases*. ($U(\underset{\sim}{R})$ obviously *decreases*.) In other words $e(\underset{\sim}{R})$ is attractive. Thus, binding (i.e. a minimum for $E(\underset{\sim}{R})$) could be viewed as a competition between the electronic attraction and the nuclear–nuclear repulsion.

† Work partially supported by US National Science Foundation Grant PHY-7825390 A02.

Three years ago, Lieb and Simon (1978) proved† that $e(\underline{R})$ is indeed attractive when $N=1$ (and for any choice of the positive charges z_i) in the following sense:

Theorem 1. Let $\underline{R}$ and $\underline{R}'$ be two sets of nuclear coordinates such that for some $\lambda>1$ and all i, $R'_i=\lambda R_i$ (i.e. uniform dilation). Then $e(\underline{R})\leqslant e(\underline{R}')$.

The reader is referred to Lieb and Simon (1978) for a discussion of the problems that arise in trying to extend this theorem to $N>1$, and for a proof of the fact that for *all* N, $e(\underline{R})$ achieves its minimum value when $R_1=R_2=\ldots=R_k$.

The main point of the present letter is theorem 2. For $N=1$, $e(\underline{R})$ is attractive for a more general class of deformations of $\underline{R}$ than simple dilation. This larger class is quite natural, for it is more nearly the class of deformations for which $U(\underline{R})$ is repulsive (i.e. $U(\underline{R})$ decreases).

Definition 1. Let R and R' be two sets of k coordinates in $\mathbb{R}^3$. $\underline{R}'$ is said to be *bigger than* $\underline{R}$, ($\underline{R}'>\underline{R}$) if $|R'_i-R'_j|\geqslant|R_i-R_j|$ for every pair (i,j) of coordinates. (Note: It is *not* necessary that there be a continuous deformation of $\underline{R}$ into $\underline{R}'$ such that the pairwise distances $|R_i-R_j|$ increase during the deformation. The stated inequality is the only condition.)

Clearly,

$$\underline{R}'>\underline{R} \;\Rightarrow\; U(\underline{R}')\leqslant U(\underline{R}). \tag{5}$$

Theorem 2. Let $N=1$ and let the $z_i>0$ be arbitrary but fixed. If $\underline{R}'$ is bigger than $\underline{R}$ then $e(\underline{R}')\geqslant e(\underline{R})$.

Remark. Obviously the electron–electron repulsion $|x_i-x_j|^{-1}$ plays no role in these theorems since $N=1$. The only relevant function is the elementary electron–nuclear potential, $w(x)$, which in our case is $w(x)=|x|^{-1}$. Theorem 1 holds if $|x|^{-1}$ in equation (4) is replaced by any w satisfying the following condition:

(C1). For every real number s, the set $B_s=\{x|w(x)>s\}$ is convex and balanced (i.e. $x\in B_s\Rightarrow -x\in B_s$). In particular, theorem 1 holds under the following condition.

(C2). The function $w(x)$ is spherically symmetric and is a decreasing function of $r=|x|$.

While it is plausible that theorem 2 holds under condition (C2) we are unable to prove it by the methods given here. The sort of problems that arises in following the 'log concavity' proof of theorem 1 in Lieb and Simon (1978) is similar to the problem of proving the Klee (1979) conjecture about the intersection of spheres.

The proof of theorem 2 given here does show that theorem 2 holds if $|x|^{-1}$ is replaced by any w satisfying both of the following more stringent conditions:

(C3). The function w is spherically symmetric.

(C4). If $u(r)\equiv w(r^{1/2})$ then $u(r)$ is completely monotone (see definition 2). For the Coulomb case, $u(r)=r^{-1/2}$ and this is completely montone. More generally, each of the k potentials in equation (4) can be replaced by a different w satisfying (C3) and (C4).

† A different proof for the case of two nuclei was later given in Hoffmann-Ostenhof T 1980 *J. Phys. A: Math. Gen.* **13** 417–24.

Definition 2. A function $u(r)$, defined for $0<r<\infty$, is *completely monotone* if u is infinitely differentiable and $(-1)^j u^{(j)}(r)\geqslant 0$ for all r and all $j=0,1,2,\ldots$ (with $u^{(j)}$ denoting the jth derivative).

The following lemma due to Bernstein (1928) (see also Donoghue 1974) is required for the proof of theorem 2.

Lemma 3. The function $u(r)$ is completely monotone if and only if $u(r)$ is the (one-sided) Laplace transform of a positive measure, i.e.

$$u(r)=\int \exp(-tr)\,\mathrm{d}\mu(t) \tag{6}$$

where μ is a (positive) measure supported on $[0,\infty)$.

Clearly, if $u(r)$ is completely monotone then so is $\exp(bu(r))$ for every $b\geqslant 0$. Lemma 3 thus yields the following representation for all $b\geqslant 0$

$$\exp(b|x-R|^{-1})=\int \exp(-t|x-R|^2)\,\mathrm{d}\mu(t;b). \tag{7}$$

A similar representation holds if $|x-R|^{-1}$ is replaced by $w(x-R)$ on the left side of equation (7). The reader may wish to have an explicit formula for μ in equation (7). This is provided by the formulae (12.2.1) on p 72 and (29) on p 171 in Roberts and Kaufman (1966)

$$\mathrm{d}\mu(t;b)=\left(t^{-3/2}(b/\pi)^{1/2}\int_0^\infty u^2 I_1(2b^{1/2}u)\exp(-u^4/4t)\,\mathrm{d}u+\delta(t)\right)\mathrm{d}t$$

where $\delta(t)\,\mathrm{d}t$ is the Dirac delta measure.

The second tool that will be needed is the following simple lemma.

Lemma 4. Let $A=\{A_{ij}\}$ be a positive-semidefinite real symmetric N-square matrix satisfying (i) $\Sigma_{i=1}^N A_{ij}=0$ and (ii) $A_{ij}\leqslant 0$ when $i\neq j$. Let $M>0$ be an integer and let $R_{ij}(i=1,\ldots,N$ and $j=1,\ldots,M)$ be NM vectors in $\mathbb{R}^3$ (the choice of three dimensions is made for simplicity). Denote these collectively by $\underline{R}$. Let $\lambda_{ij}\geqslant 0$ be a collection of NM real numbers satisfying $\Lambda_i\equiv\Sigma_{j=1}^M\lambda_{ij}>0$. Let $x_1,\ldots,x_N\in\mathbb{R}^3$ be N three-dimensional variables. Let

$$Z(\underline{R})\equiv\int \mathrm{d}x_1\ldots\mathrm{d}x_N\exp\left(-\sum_{i,j=1}^N x_i\cdot x_jA_{ij}-\sum_{n=1}^N\sum_{j=1}^M\lambda_{ij}|x_i-R_{ij}|^2\right). \tag{8}$$

If $\underline{R}'>\underline{R}$ (i.e. $|R'_{ij}-R'_{kl}|\geqslant|R_{ij}-R_{kl}|$ for all i,j,k,l) then $Z(\underline{R}')\leqslant Z(\underline{R})$.

Proof. The integral for Z is Gaussian and easily evaluated. Let Λ denote the diagonal matrix with the values Λ_i. By (i), $\Sigma_{j=1}^N(A+\Lambda)_{ij}^{-1}\Lambda_j=1$ (since $\Sigma_j(A+\Lambda)_{ij}=\Lambda_i$). Using this fact, a little algebra shows that

$$Z(\underline{R})=\pi^{3N/2}\exp\left(-\tfrac{1}{2}\sum_{i,j=1}^N\sum_{k,l=1}^M(A+\Lambda)_{ij}^{-1}\lambda_{ik}\lambda_{jl}|R_{ik}-R_{jl}|^2-\tfrac{3}{2}\ln\mathrm{Det}(A+\Lambda)\right).$$

The lemma will be proved if $(A+\Lambda)_{ij}^{-1}\geqslant 0$ for all i,j. But (ii) states that $A+\Lambda$ has negative off-diagonal matrix elements. The inverse of a positive definite matrix P with this property has $P_{ij}^{-1}\geqslant 0$ (Ostrowski 1937). QED.

J. Phys. B: At. Mol. Phys. *15*, L63–L66 (1982)

With this preparation we can now prove theorem 2. In fact, something stronger will be proved.

Theorem 5. For $T>0$ let $G(x, y, T, \underline{R}) = \exp(-TH(\underline{R}))(x, y)$ be the kernel of the operator $\exp(-TH(\underline{R})$. Let $(\underline{R}, x, y)$ denote the $k+2$ vectors $R_1, \ldots, R_k, x, y$. Suppose that (R', x', y') is bigger than $(\underline{R}, x, y)$. Then $G(x' y', T, \underline{R}') \leqslant G(x, y, T, \underline{R})$.

Proof. First, regularise $w(x) = |x|^{-1}$ by $w_\varepsilon(x) = w((x^2+\varepsilon)^{1/2}) = (x^2+\varepsilon)^{-1/2}$ for $\varepsilon > 0$. This amounts to replacing $d\mu(t; b)$ by $\exp(-\varepsilon t)\, d\mu(t; b)$ in equation (7). Second, use the Trotter product formula (see Lieb and Simon 1978)

$$G(x, y, T, \underline{R}) = \lim_{n\to\infty} [\exp(T\Delta/n)\exp(TV(\underline{R})/n)]^n(x, y)$$

where $V(\underline{R})$ is given in equation (4) (with w_ε).

$$\exp(T\Delta/n)(x, y) = (4\pi T/n)^{-3/2} \exp[-n(x-y)^2/4T]$$

is a Gaussian with negative off-diagonal matrix elements (i.e. $n/4T$). The total Gaussian has the exponent $\alpha + \beta$ with

$$\alpha = -(n/4T) \sum_{i=1}^{n-2} (x_{i+1} - x_i)^2$$

$$\beta = -(n/4T)[(x_1 - x)^2 + (x_{n-1} - y)^2]$$

where the x_i are the intermediate integration variables. The first term, α, should be regarded as the 'A' term in equation (8). The second term, β, should be regarded as part of the second term in equation (8). Use the representation, equation (7), nk times. The theorem (for w_ε) then follows from lemma 4. Finally, we let $\varepsilon \to 0$. (Remark: $G(x, y, T, \underline{R})$ is monotone decreasing in ε since w_ε is decreasing.) QED.

Corollary 6. For the simple hydrogen atom (with $V(x) = |x|^{-1}$), $G(x', y', T) \leqslant G(x, y, T)$ if $|x'| \geqslant |x|$, $|y'| \geqslant |y|$ and $|x' - y'| \geqslant |x - y|$.

Proof of theorem 2. Take $y = x$, $y' = x'$ and choose these such that $|x' - R_i'| \geqslant |x - R_i|$, $i = 1, \ldots, k$. Since $e(\underline{R}) = -\lim_{T\to\infty} T^{-1} \ln G(x, x, T, \underline{R})$ (for any x), theorem 5 implies theorem 2. QED.

References

Bernstein S 1928 *Acta Math.* **52** 1–66
Donoghue W F Jr 1974 *Monotone Matrix Functions and Analytic Continuation* (New York: Springer) p 13
Klee V 1979 *Math. Mag.* **52** 131–145 (see problem 7 on p 142)
Lieb E H and Simon B 1978 *J. Phys. B: At. Mol. Phys.* **11** L537–42
Ostrowski A 1937 *Comm. Math. Helv.* **10** 69–96
Roberts G E and Kaufman H 1966 *Table of Laplace Transforms* (Philadelphia: Saunders)

VOLUME 50, NUMBER 22 PHYSICAL REVIEW LETTERS 30 MAY 1983

Proof of the Stability of Highly Negative Ions in the Absence of the Pauli Principle

Rafael Benguria
Departamento de Fisica, Universidad de Chile, Casilla 5487, Santiago, Chile

and

Elliott H. Lieb
Departments of Mathematics and Physics, Princeton University, New Jersey 08544
(Received 31 January 1983)

It is well known that ionized atoms cannot be both very negative and stable. The maximum negative ionization is only one or two electrons, even for the largest atoms. The reason for this phenomenon is examined critically and it is shown that electrostatic considerations and the uncertainty principle cannot account for it. The exclusion principle plays a crucial role. This is shown by proving that when Fermi statistics is ignored, then the degree of negative ionization is at least of order z, the nuclear charge, when z is large.

PACS numbers: 31.10.+z, 03.65.Ge

One of the interesting and important facts about atoms is that they cannot be very negatively ionized (in a stable state, as distinguished from metastable state). For a nucleus of charge z, let $N_c(z)$ denote the maximum number of electrons that can be bound to this nucleus (*in vacuo*, not in water or other matter). Experiments indicate that $N_c(z)-z$ is one, or possibly two, as z varies over the periodic table. It is often said that this striking fact, which begs for an explanation, is a consequence of electrostatics; namely, if an atom has a net negative charge then an additional electron will not bind because the electron can lower its energy by escaping to infinity. The purpose of this note is to examine this simple, but important physical problem in a critical way and to show that the correct explanation does not lie with electrostatics alone—*the Pauli exclusion principle plays a central role in the correct explanation.* We are not able to offer an explanation of the phenomenon, but we thought it worthwhile at least to expose the fallacy in the "simple electrostatic" explanation and thereby show that the phenomenon is really a deep consequence of quantum mechanics.

To prove that the Pauli principle is essential we shall consider an atom in which the electrons are spinless bosons. (Since the ground state of a many-body system is nodeless, it is automatically symmetric; therefore "bosons" and "ignoring statistics" are synonymous.) We shall prove that in this model, $N_c(z)-z \geq \gamma z$ when z is large, and where $\gamma > 0$ is some fixed constant. We do not know the numerical value of γ, except that $0<\gamma < 1$. It can be found by solving an equation [namely (11) with $\mu=0$] on a computer, if there is sufficient interest in doing so. The exact numerical value of γ is not as important as the fact that "bosonic" atoms would not obey the $N_c(z)-z\approx 1$ rule for sufficiently large z. Just how large z has to be in order to violate the rule substantially, we do not know. Equation (11) has to be solved to answer the question.

One can adopt different points of view about this. It is possible that the rule is not really a rule at all for fermions, and that $N_c(z)-z$ grows at least as fast as z for large z. In this case the fact that $N_c(z)-z\approx 1$ within the periodic table is fortuitous, and in reality bosons and fermions are qualitatively similar as far as the phenomenon goes. Another possibility is that $N_c(z)/z - 1 \to 0$ as $z\to\infty$ [thereby allowing the possibility that $N_c(z)-z\approx z^{1/2}$, for example], in which case the Pauli principle is crucial. We, of course, do not know which point of view will ultimately prevail. It is to be hoped that if it is the second one then someone will find a simple, but rigorous explanation. In any case, the phenomenon should not be left merely as a numerical statement about the periodic table but should be understood on a deeper level.

While we use the nonrelativistic Schroedinger equation and we regard the nucleus as fixed, it will be clear, at least intuitively, that our conclusions are not limited by these approximations. The lack of the Pauli principle is, however, crucial.

Before turning to the mathematical proof, let us consider the problem from a heuristic viewpoint. Suppose N electrons are bound to the nucleus. With neglect of many-body effects, the effective potential that an $(N+1)$th electron feels is approximately

$$\varphi_\rho(\vec{x}) = V(\vec{x}) + \int |\vec{x} - \vec{y}|^{-1}\rho(\vec{y})d^3y\,, \tag{1}$$

where $V(\vec{x}) = -z/|\vec{x}|$, and units in which the electron charge is unity are used. $\rho(\vec{x})$ is the density of the N electrons, $\int \rho = N$. The effective Hamiltonian for the $(N+1)$th electron is approximately

$$h_{\rm eff} = -m^{-1}\Delta + \varphi_\rho(\vec{x}), \tag{2}$$

where m is the electron mass and $\hbar^2 = 2$.

As N increases from zero, φ_ρ increases in some average sense. When $N > z$, $\varphi_\rho(\vec{x})$ is positive for large $|\vec{x}|$ [by Newton's theorem, $\varphi_\rho(\vec{x}) \approx (N-z)/|\vec{x}|$ for large $|\vec{x}|$]. However, $\varphi_\rho(\vec{x})$ is still very large and negative for x near zero, namely, $-z/|\vec{x}|$. The uncertainty principle is crucial here; it prevents $\rho(\vec{x})$ from being a delta function and thereby screening the nucleus for small $|\vec{x}|$. Thus, even if $N > z$, $h_{\rm eff}$ might have a genuine bound state and the $(N+1)$th electron might be bound. Eventually, of course, the region of negative φ_ρ will be too small and binding will cease. Implicit in this discussion is the fact that the $(N+1)$th electron is allowed to go into any available bound state of $h_{\rm eff}$. In other words, the argument works if statistics is ignored, which is the same thing as saying that we are dealing with bosons. If, on the other hand, the electrons are fermions then, for binding, $h_{\rm eff}$ must have something like $N+1$ bound states in order that the $(N+1)$th electron can go into an orbital that is orthogonal to the previously occupied orbitals. It is a remarkable fact about Fermi statistics that the $(N+1)$th bound state of $h_{\rm eff}$ disappears when $N \approx z$, if the $N_c(z) - z \approx 1$ rule is obeyed.

The above argument is not completely convincing, even on the heuristic level, because it is not obvious that $h_{\rm eff}$ indeed has a bound state when $N = (1+\gamma)z$.

A proper proof, starting from the correct Schroedinger equation, and without any approximations, will now be given. Let

$$H(N) = \sum_{i=1}^{N}[-m^{-1}\Delta_i + V(\vec{x}_i)] + \sum_{1\le i<j\le N}|\vec{x}_i - \vec{x}_j|^{-1} \tag{3}$$

be the Hamiltonian for N bosonic electrons [$V(\vec{x}) = -z/|\vec{x}|$]. Let $E(N)$ be the ground-state energy of $H(N)$. We shall find two bounding functions, $E_\pm(N)$, with $E_-(N) \le E(N) \le E_+(N)$ and

$$E_+(N) = -mz^3 e(N/z)\,, \tag{4a}$$

$$E_-(N) = -mz^3[e(N/z)^{1/2} + bz^{-3/2}N^{5/6}]^2. \tag{4b}$$

Here, $b = 0.36$ and $e(t)$ is a monotone nondecreasing, concave function ($\dot{e} \ge 0$, $\ddot{e} \le 0$), defined for all real $t \ge 0$, with $e(0) = 0$. $E_+(N)$ is essentially the Hartree energy. The *crucial point* about $e(t)$ is that it is *strictly increasing* in t up to some $t_c = 1+\gamma$ and $0 < \gamma < 1$. We do not know the numerical value of γ, but that is unimportant. [It can be found by solving Eq. (11), as explained below.] From Eq. (4) we see that if $N < \overline{N}$, where

$$E_-(\overline{N}) = E_+(zt_c) \tag{5}$$

(which means that $\overline{N}$ is not necessarily an integer), then the energy can be lowered by adding some number of electrons because $E(N) \ge E_-(N) > E_-(\overline{N}) = E_+(zt_c) = E_+([zt_c]) \ge E([zt_c])$. Here, $[x]$ denotes the smallest integer $\ge x$. [Note that Eq. (5) has a unique solution; $E_-(N)$ is monotone in N since $e(N/z)$ is monotone in N.] In other words

$$N_c(z) \ge [\overline{N}]. \tag{6}$$

As $z \to \infty$, $\overline{N} \to z(1+\gamma)$ because (5) reads (with $\overline{N}/z \equiv t$)

$$e(t)^{1/2} + bz^{-2/3}t^{5/6} = e(t_c)^{1/2}. \tag{7}$$

Since the solution to Eq. (7) satisfies $t < t_c < 2$ and since $e(t)$ is continuous, we have that $t \to t_c$ as $z \to \infty$. Thus, our main point is proved, namely, asymptotically

$$N_c(z) - z \ge \gamma z\,, \text{ large } z\,. \tag{8}$$

An *upper* bound for $N_c(z)$ was first given by Ruskai in the form $N_c(z) \le (\text{const})z^2$ for bosons.[1] Sigal[2] proved for fermions that $N_c(z) \le cz$ for z sufficiently large, with $c > 2$ being some constant. For fermions Ruskai[3] proved that $N_c(z) \le (\text{const}) \times z^{6/5}$. Sigal[4] improved his result for fermions to $N_c(z) \le \alpha(z)z$, with $\alpha(z) \le 12$ and $\alpha(z) \to 2$ as $z \to \infty$.

Now we turn to the proof of Eq. (4).

Upper bound for E(N).—We take a product trial function $\psi = \prod_{i=1}^{N} f(\vec{x}_i)$, with $\int f^2 = 1$ and with $f(\vec{x})$ real and nonnegative, and use the variational principle:

$$E(N) \le \langle\psi|H(N)|\psi\rangle \le \int\{m^{-1}|\nabla\rho(\vec{x})^{1/2}|^2 + V(\vec{x})\rho(\vec{x})\}d^3x + \tfrac{1}{2}\int\int\rho(\vec{x})\rho(\vec{y})|\vec{x}-\vec{y}|^{-1}d^3x\,d^3y \equiv L(\rho)\,, \tag{9}$$

VOLUME 50, NUMBER 22 PHYSICAL REVIEW LETTERS 30 MAY 1983

where $\rho(\vec{x}) \equiv Nf(\vec{x})^2$. [We could insert a factor $(N-1)/N$ before the last integral, but choose not to do so because we want $L(\rho)$ to be independent of N.] Next we define

$$E_+(N) = \inf\{L(\rho) \mid \int\rho = N, \rho(\vec{x}) \geq 0\}. \quad (10)$$

Equation (10) means that we try to minimize $L(\rho)$ under the stated conditions. The minimum may not be achieved by any ρ (it is achieved if and only if $N \leq 1+\gamma$ as stated below), but in any case E_+ cannot exceed the "greatest lower bound" or "infimum" of $L(\rho)$. The problem posed by Eq. (10) is a special case of the generalized Thomas-Fermi-von Weizsaecker problem analyzed earlier.[5,6] In our case, by the simple scaling $\rho(\vec{x}) \to m^3z^4\rho(mz\vec{x})$ the problem can be reduced to the case in which $m=1$, $z=1$, and $\int\rho = N/z$. Equation (4a) is thus seen to hold. Moreover, the m dependence of e is $e(N/z, m) = me(N/z, m=1)$ as in Eq. (4a). (Incidentally, this scaling shows that the radius of bosonic atoms *shrinks* as z^{-1}, whereas it *shrinks* as $z^{-1/3}$ for fermions.) There is a minimizing ρ for $L(\rho)$ (and it is unique) if and only if $N/z \leq t_c$ for some definite number $t_c > 1$. (Note that t_c is independent of m.) This ρ satisfies

$$[-m^{-1}\Delta + \varphi_\rho(\vec{x})]\rho(\vec{x})^{1/2} = -\mu\rho(\vec{x})^{1/2} \quad (11)$$

with $\mu \geq 0$, and $\mu = 0$ when $N = t_c z$. The proof that $t_c > 1$ is given in Ref. 5, lemma 13, and Ref. 6, theorem 7.16 (note: in these proofs take $p = \frac{5}{3}$ and $\gamma = 0$). The proof that $t_c < 2$ is given in Ref. 6, theorem 7.23. [The reason that $t_c > 1$ is that when $N = z$ then $\varphi_\rho(\vec{x}) < 0$ and one can prove that this potential has a bound state; thus μ cannot be zero. To prove that $t_c < 2$, set $\mu = 0$, multiply Eq. (11) by $|\vec{x}|\rho(\vec{x})^{1/2}$ and integrate. One can show, by partial integration, that $\int|\vec{x}|\rho(\vec{x})^{1/2}\Delta\rho(\vec{x})^{1/2} \leq 0$. Obviously, $\int V(\vec{x})|\vec{x}|\rho(\vec{x}) = -zN$. Finally, $I \equiv \iint|\vec{x}|\rho(\vec{x})\rho(\vec{y})|\vec{x}-\vec{y}|^{-1} = \frac{1}{2}\iint\rho(\vec{x})\rho(\vec{y})|\vec{x}-\vec{y}|^{-1}(|\vec{x}| + |\vec{y}|)$. But $(|\vec{x}|+|\vec{y}|)|\vec{x}-\vec{y}|^{-1} \geq 1$, so that $I \geq N^2/2$. Since $I \leq zN$, $N \leq 2z$.] That $e(t)$ is concave (and strictly concave when $t \leq t_c$) is a consequence of the fact that when $\int\rho$ is increased the additional density can be placed, if need be, at infinity where its energy contribution is zero (see Refs. 5 and 6).

Lower bound for $E(N)$.—Let $\psi(\vec{x}_1, \ldots, \vec{x}_N)$ be *any* normalized function. We want to show that

$$F(\psi) \equiv \langle\psi|H(N)|\psi\rangle \geq \text{right side of (4b)}. \quad (12)$$

Let

$$\rho_\psi(\vec{x}) = \sum_{i=1}^{N} \int|\psi(\vec{x}_1, \ldots, \vec{x}_{i-1}, \vec{x}, \vec{x}_{i+1}, \ldots, \vec{x}_N)|^2 d^3x_1 \cdots d^3x_{i-1}d^3x_{i+1}\cdots d^3x_N \quad (13)$$

be the single-particle density associated with ψ; $\int\rho_\psi = N$. We shall use several known inequalities. The first is the kinetic energy inequality of Hoffmann-Ostenhof[7] (see also Ref. 8 for a further discussion of kinetic energy inequalities):

$$\langle\psi| - \sum_{i=1}^{N}\Delta_i|\psi\rangle \geq \int|\nabla\rho_\psi(\vec{x})^{1/2}|^2 d^3x. \quad (14)$$

This follows by taking the gradient in (13) and then using the Schwarz inequality. The second is the "exchange and correlation" inequality[9,10]:

$$\langle\psi| \sum_{1\leq i<j\leq N} |\vec{x}_i - \vec{x}_j|^{-1}|\psi\rangle \geq \tfrac{1}{2}\iint\rho_\psi(\vec{x})\rho_\psi(\vec{y})|\vec{x}-\vec{y}|^{-1}d^3x\,d^3y - (1.68)\int\rho_\psi(\vec{x})^{4/3}d^3x. \quad (15)$$

Inserting Eqs. (14) and (15) in (12) we have, for any ψ,

$$F(\psi) \geq L(\rho_\psi) - (1.68)\int\rho_\psi(\vec{x})^{4/3}\,d^3x. \quad (16)$$

To bound the right side of (16) from below, let us first explicitly denote the m dependence of $L(\rho)$ by $L(\rho, m)$. Choose $\epsilon > 0$ and let $m_\epsilon = (1+\epsilon)m$. Then Eq. (16) reads

$$F(\psi) \geq L(\rho_\psi, m_\epsilon) + P(\rho_\psi, m_\epsilon), \quad (17)$$

$$P(\rho_\psi, m_\epsilon) = \epsilon m_\epsilon^{-1}\int|\nabla\rho_\psi(\vec{x})^{1/2}|^2 d^3x - (1.68)\int\rho_\psi(\vec{x})^{4/3}\,d^3x. \quad (18)$$

We have already seen that [see Eq. (10) and the following remark about scaling]

$$L(\rho_\psi, m_\epsilon) \geq (m_\epsilon/m)E_+(N) = (1+\epsilon)E_+(N). \quad (19)$$

VOLUME 50, NUMBER 22 PHYSICAL REVIEW LETTERS 30 MAY 1983

To bound P, we use the Sobolev inequality[6,11]

$$\int |\nabla g(\mathbf{x})|^2\, d^3x \geq 3(\pi/2)^{4/3}\{\int |g(\mathbf{x})|^6 d^3x\}^{1/3} \tag{20}$$

for any g. Thus

$$P(\rho_\psi, m_\epsilon) \geq 3(\pi/2)^{4/3}\epsilon m_\epsilon^{-1}\{\int \rho_\psi(\mathbf{x})^3 d^3x\}^{1/3} - (1.68)\int \rho_\psi(\mathbf{x})^{4/3} d^3x. \tag{21}$$

By Hoelder's inequality, $\int\rho^{4/3} \leq X\{\int\rho\}^{5/6}$ with $X = \{\int\rho^3\}^{1/6}$. Inserting this in (21), and then minimizing the right side with respect to the unknown X, we have

$$P(\rho_\psi, m_\epsilon) \geq -b^2 m_\epsilon N^{5/3}/\epsilon, \tag{22}$$

$$b = (1.68)2^{-1/3}3^{-1/2}\pi^{-2/3} = 0.36. \tag{23}$$

Inserting Eqs. (19) and (22) in (17) we have, for any $\epsilon > 0$, and any normalized ψ,

$$\langle\psi|H(N)|\psi\rangle \geq (1+\epsilon)E_+(N) - (1+1/\epsilon)mb^2N^{5/3}. \tag{24}$$

Maximizing this with respect to ϵ yields Eq. (4b).

One of the authors (E.L.) acknowledges gratefully the hospitality and support of the Departamento de Física, Universidad de Chile, where this work was carried out. He also acknowledges the support of the U. S. National Science Foundation under Grant No. PHY-8116101 A01. We thank Barry Simon for valuable comments.

[1]M. B. Ruskai, Commun. Math. Phys. 82, 457 (1982).

[2]I. M. Sigal, Commun. Math. Phys. 85, 309 (1982). See also *Mathematical Problems in Theoretical Physics*, Lecture Notes in Theoretical Physics Vol. 153 (Springer-Verlag, Berlin, 1982), pp. 149–156.

[3]M. B. Ruskai, Commun. Math. Phys. 85, 325 (1982).

[4]I. M. Sigal, Institute Mittag-Leffler Report No. 12, 1982 (unpublished).

[5]R. Benguria, H. Brezis, and E. H. Lieb, Commun. Math. Phys. 79, 167 (1981).

[6]E. H. Lieb, Rev. Mod. Phys. 53, 603 (1981).

[7]M. Hoffmann-Ostenhof and T. Hoffmann-Ostenhof, Phys. Rev. A 16, 1782 (1977).

[8]E. H. Lieb, *Density Functionals for Coulomb Systems*, in *Physics as Natural Philosophy: Essays in Honor of Laszlo Tisza on His 75th Birthday*, edited by A. Shimony and H. Feshbach (MIT Press, Cambridge, Mass., 1982), pp. 111–149.

[9]E. H. Lieb, Phys. Lett. 70A, 444 (1979).

[10]S. Oxford and E. H. Lieb, Int. J. Quantum Chem. 19, 427 (1981).

[11]E. H. Lieb, Rev. Mod. Phys. 48, 553 (1976).

PHYSICAL REVIEW LETTERS

VOLUME 52 30 JANUARY 1984 NUMBER 5

Atomic and Molecular Negative Ions

Elliott H. Lieb
Departments of Mathematics and Physics, Princeton University, Princeton, New Jersey 08544
(Received 16 November 1983)

An upper bound is given for the maximum number, N_c, of negative particles (fermions or bosons or a mixture of both) of charge $-e$ that can be bound to an atomic nucleus of charge $+ze$. If z is integral then $N_c \leq 2z$. In particular, this is the first proof that H^{--} is not stable. For a molecule, $N_c \leq 2Z+K-1$, where K is the number of atoms in the molecule and Z is the total nuclear charge.

PACS numbers: 03.65.Ge, 31.10.+z

One of the striking, nonperiodic facts about the periodic table is that the maximum number of electrons, N_c, that can be bound to a nucleus of charge z is never more than $z+1$. Recently, several authors[1-8] have attempted to find bounds on N_c; one of the strongest results so far[8] (for fermions) is that $\lim_{z\to\infty} N_c/z = 1$. For *bosons*,[5] however, $N_c > 1.2z$ for large z. Thus, the value of N_c is very dependent on the statistics of the bound particles.

The purpose of this note is to announce a theorem about N_c, the full details of which will appear elsewhere.[9] The theorem applies to *any mixture* of bound particles, with possibly different statistics, masses, and charges (as long as they are all negative), and even with possibly different magnetic fields acting on the various particles. (Naturally, symmetry requires that particles of the same type have the same mass, etc.) The theorem also applies to a molecule. The usual approximation that the nuclei be fixed (or infinitely massive) is important, but if they are not fixed a weaker theorem holds. The same theorems hold in the Hartree-Fock (restricted or unrestricted) and Hartree approximations to the ground-state energy.[9]

Suppose that we have a molecule with K nuclei of charges $z_1, \ldots, z_K > 0$ (units are used in which the electron charge is unity) located at fixed, distinct positions $\vec{R}_1, \ldots, \vec{R}_K$. The electric potential of these nuclei is

$$V(\vec{x}) = \sum_{j=1}^{K} z_j |\vec{x} - \vec{R}_j|^{-1}. \tag{1}$$

Let there be N negative particles with masses $m_1, \ldots, m_N$ and charges $-q_1, \ldots, -q_N < 0$ (in the usual case each $q_i = 1$) and let each be subject to (possibly different) magnetic fields $\vec{A}_1(\vec{x}), \ldots, \vec{A}_N(\vec{x})$. (The generality of allowing nonintegral nuclear and negative particle charges may have some physical relevance because, as pointed out to me by W. Thirring, particles in solids such as semiconductors may have nonintegral effective charges due to dielectric effects.) The Hamiltonian is

$$H_N = \sum_{j=1}^{N} \{T_j - q_j V(\vec{x}_j)\} + \sum_{1\leq i<j\leq N} q_i q_j |\vec{x}_i - \vec{x}_j|^{-1}. \tag{2}$$

Here, T_j is the kinetic energy operator for the jth particle and it is one of the following (possibly different for different j) two types (nonrelativistic or relativistic):

$$T_j = [\vec{p}_j - q_j \vec{A}_j(\vec{x})/c]^2/2m_j, \tag{3}$$

$$T_j = \{[\vec{p}_j c - q_j \vec{A}_j(\vec{x})]^2 + m_j^2 c^4\}^{1/2} - m_j c^2. \tag{4}$$

Let q denote the maximum of the q_j, let $Q=\sum_{j=1}^N q_j$ be the total negative charge, and let $Z=\sum_{j=1}^K z_j$ be the total nuclear charge. Let E_N denote the ground-state energy of H_N [$\equiv$ inf spec(H_N)].

Theorem 1.—If the above system is bound (meaning that E_N is an eigenvalue of H_N) then, necessarily,

$$Q < 2Z + qK. \tag{5}$$

In the atomic case ($K=1$) this can be strengthened to

$$Q < 2Z + \sum_{j=1}^{N} q_j^2/Q. \tag{6}$$

The strict inequality in Eqs. (5) and (6) is important; in the atomic case with $q=1$ and z integral, Eq. (5) implies

$$N_c \leq 2z. \tag{7}$$

For a hydrogen atom ($z=1$), Eq. (7) implies that $N_c=2$ (since it is known that two electrons can, in fact, be bound). H^{--} is *not* stable. This result had not been proved before, although there exist partial results in this direction.[6]

Although Eq. (7) is far from optimal when z is large [in view of the $N \approx z + O(1)$ conjecture], it is the strongest explicit estimate obtained so far and it is of the right order of magnitude for bosons (recalling the $N_c > 1.2z$ result[5]).

The theorem holds even if the nuclei are not points, but are spherical charge distributions, i.e., $|\vec{x}-\vec{R}|^{-1}$ is replaced by $\int d\mu(\vec{y})|\vec{y}-\vec{x}-\vec{R}|^{-1}$ in Eq. (1), where μ is any positive, spherical measure of unit total charge.

If the nuclear coordinates, $\vec{R}_j$, are dynamical instead of fixed, a weaker theorem holds. Let $\tilde{H}_N = H_N + T_{\text{nucl}} + U(\vec{R})$, where T_{nucl} is the nuclear kinetic energy [consisting of terms of the form in Eqs. (3) and (4)] and $U(\vec{R})$ is a potential depending on the nuclear coordinates. Let $\tilde{E}_N$ be the ground-state energy of $\tilde{H}_N$ and let $\tilde{E}_{N,j}^{\infty}$ be the ground-state energy when the nuclear masses are infinite (i.e., T_{nucl} is omitted) and the negative particle j is removed. If $\tilde{E}_{N,j}$ is the ground-state energy when particle j is removed, but T_{nucl} is retained, it is easy to see that $\tilde{E}_{N,j} \geq \tilde{E}_N$ and $\tilde{E}_{N,j} \geq \tilde{E}_{N,j}^{\infty}$. The theorem for dynamical nuclei, which assumes an additional inequality, is the following.

Theorem 2.—If the N-particle system is bound and if, in addition, $\tilde{E}_N \leq \tilde{E}_{N,j}^{\infty}$ for all $j=1,\ldots,N$, then Eq. (5) [respectively Eq. (6) for $K=1$] holds.

The proof of the theorems is simple enough to be given in an elementary quantum mechanics course—at least in the atomic case with fixed nucleus and with $T_j = p_j^2/2m_j$ (all j). The proof (ignoring some technical fine points) in this atomic case is the following: Take $\hbar^2/2m=1$ and $\vec{R}=0$. Pick some j and write $H_N = H_{N,j} + h_j$, where $H_{N,j}$ is the Hamiltonian for the remaining $N-1$ particles and

$$h_j = T_j - q_j z|\vec{x}_j|^{-1} + \sum_{k\neq j} |\vec{x}_j - \vec{x}_k|^{-1} q_j q_k. \tag{8}$$

Assume that the system is bound and let ψ be the ground state (which is real). Multiply the Schrödinger equation, $H_N\psi = E_N\psi$, by $|\vec{x}_j|\psi$ and integrate over all N variables. Let X_j denote all the $N-1$ variables other than $\vec{x}_j$. For the $H_{N,j}$ term, do the $d^{3(N-1)}X_j$ integration first; by the variational principle, the X_j integral is, for each fixed $\vec{x}_j$, not less than $E_{N,j}$ ($\equiv$ the ground-state energy of $H_{N,j}$) times the same integral without $H_{N,j}$. This inequality is preserved after the $\vec{x}_j$ integration since $|\vec{x}_j|$ is a positive weight. Thus

$$\langle |\vec{x}_j|\psi|H_{N,j}|\psi\rangle \geq E_{N,j}\langle |\vec{x}_j|\psi|\psi\rangle. \tag{9}$$

Recalling the easily proved fact that $E_N \leq E_{N,j}$, we have

$$\langle |\vec{x}_j|\psi|h_j|\psi\rangle \leq 0. \tag{10}$$

The claim is that Eq. (10) cannot hold for all j if condition (6) is violated.

First, the term $t_j = \langle |\vec{x}_j|\psi|p_j^2|\psi\rangle$ is positive. To see this, do the $\vec{x}_j$ integration first and note that it then suffices to prove the following for any function, f, of one variable:

$$t = -\int |\vec{x}| f(\vec{x}) \nabla^2 f(\vec{x}) d^3x > 0. \tag{11}$$

Write $g(\vec{x}) = |\vec{x}| f(\vec{x})$ and integrate by parts:

$$t = \int \nabla g(\vec{x}) \cdot \{|\vec{x}|^{-1}\nabla g(\vec{x}) + g(\vec{x})\nabla|\vec{x}|^{-1}\} d^3x$$

$$= \int \{|\nabla g(\vec{x})|^2 |\vec{x}|^{-1} - \tfrac{1}{2} g(\vec{x})^2 \nabla^2 |\vec{x}|^{-1}\} d^3x > 0$$

since $\nabla^2|\vec{x}|^{-1} \leq 0$. (The fact that $g\nabla g = \nabla g^2/2$, together with another integration by parts, was used for the second term.)

The second term in h_j is easy:

$$A_j \equiv q_j \langle |\vec{x}_j|\psi|V(\vec{x}_j)|\psi\rangle = q_j z\langle\psi|\psi\rangle = q_j z,$$

with the assumption that ψ is normalized.

The third term is

$$R_j \equiv \int \psi(X)^2 |\vec{x}_j| \sum_{k\neq j} q_j q_k |\vec{x}_j - \vec{x}_k|^{-1} d^{3N}X,$$

where X denotes all the N variables.

If there is binding then Eq. (10) holds for all j and hence, summing over j and using $t_j > 0$, we have that $A \equiv \sum_j A_j > \sum_j R_j \equiv R$. On the one hand,

VOLUME 52, NUMBER 5 PHYSICAL REVIEW LETTERS 30 JANUARY 1984

$A = z\sum_j q_j = zQ$. On the other hand,

$$R = \tfrac{1}{2}\int \psi(X)^2 \sum_{k\neq j}\sum q_j q_k |\vec{x}_j - \vec{x}_k|^{-1}(|\vec{x}_j| + |\vec{x}_k|)\, d^{3N}X.$$

But $|\vec{x}_j| + |\vec{x}_k| \geq |\vec{x}_j - \vec{x}_k|$ (triangle inequality), so that

$$R \geq \tfrac{1}{2}\sum_{k\neq j}\sum q_j q_k = \tfrac{1}{2}Q^2 - \tfrac{1}{2}\sum_j q_j^2.$$

Hence, binding implies that

$$Q^2 - \sum_j q_j^2 < 2zQ,$$

which is precisely Eq. (6). Q.E.D.

This work was partially supported by U. S. National Science Foundation Grant No. PHY-8116101-A01.

[1] M. B. Ruskai, Commun. Math. Phys. 82, 457 (1982).

[2] I. M. Sigal, Commun. Math. Phys. 85, 309 (1982). See also *Mathematical Problems in Theoretical Physics*, Lecture Notes in Theoretical Physics Vol. 153, edited by G. Dell-Antonio, S. Doplicher, and R. Jona-Lasinio (Springer-Verlag, Berlin, 1982), pp. 149–156.

[3] M. B. Ruskai, Commun. Math. Phys. 85, 325 (1982).

[4] I. M. Sigal, Institute Mittag Leffler Report No. 12, 1982, revised 1983 (to be published).

[5] R. Benguria and E. H. Lieb, Phys. Rev. Lett. 50, 1771 (1983).

[6] R. N. Hill, in *Mathematical Problems in Theoretical Physics*, edited by K. Osterwalder, Lecture Notes in Physics Vol. 116 (Springer-Verlag, New York, 1979), pp. 52–56.

[7] G. Zhislin, Tr. Mosk. Mat. Obshch. 9, 81 (1960).

[8] E. H. Lieb, I. M. Sigal, B. Simon, and W. E. Thirring, "Asymptotic Bulk Neutrality of Large-Z Ions" (to be published), and to be published.

[9] E. H. Lieb, "Bound on the Maximum Negative Ionization of Atoms and Molecules," Phys. Rev. A (to be published).

Phys. Rev. A29, 3018-3028 (1984)

Bound on the maximum negative ionization of atoms and molecules

Elliott H. Lieb
Departments of Mathematics and Physics, Princeton University, P.O. Box 708, Princeton, New Jersey 08544
(Received 14 November 1983)

It is proved that N_c, the number of negative particles that can be bound to an atom of nuclear charge z, satisfies $N_c < 2z+1$. For a molecule of K atoms, $N_c < 2Z+K$ where Z is the total nuclear charge. As an example, for hydrogen $N_c=2$, and thus H^{--} is not stable, which is a result not proved before. The bound particles can be a mixture of different species, e.g., electrons and π mesons; statistics plays no role. The theorem is proved in the static-nucleus approximation, but if the nuclei are dynamical, a related, weaker result is obtained. The kinetic energy operator for the particles can be either $[p-eA(x)/c]^2/2m$ (nonrelativistic with magnetic field) or $\{[pc-eA(x)]^2+m^2c^4\}^{1/2}-mc^2$ (relativistic with magnetic field). This result is not only stronger than that obtained before, but the proof (at least in the atomic case) is simple enough to be given in an elementary quantum-mechanics course.

I. INTRODUCTION

One of the nonperiodic facts about the Periodic Table is that the number of electrons that can be bound to an atomic nucleus of charge ze is at most $z+1$, at least as far as present confirmed experimental data go. The theoretical proof of this fact, starting with the Schrödinger equation, is a challenge that has drawn the attention of several authors in recent years.[1-13]

The problem can obviously be extended in two ways: (i) The electrons, which are fermions, can be replaced by bosons or, more generally, by a mixture of particles of different species. (Because of spin, the two-electron problem is well known to be the same as the two-boson problem.) (ii) Instead of a single atom, a molecule can be considered. That these problems are difficult is shown by the fact that it was only recently proved by Ruskai[1] (for bosons) and later Sigal[2] and Ruskai[4] (for fermions) that the number of bound particles is not infinite.

The following is a summary of rigorous results to date. Our notation for the maximum particle number is N_c.

(1) Ruskai[1] proved that $N_c \le (\text{const})z^2$ for bosons. Recently, Sigal[3] proved for bosons that for every $\epsilon>0$ there is a constant C_ϵ such that $N_c \le C_\epsilon z^{1+\epsilon}$.

(2) For fermions, Sigal[2] proved that $N_c \le cz$ with c being some constant. Ruskai[4] proved that $N_c \le (\text{const})z^{6/5}$. Sigal[3] improved his result to $N_c \le \alpha(z)z$ with $\alpha(z) \le 12$ and $\alpha(z) \to 2$ as $z\to\infty$.

(3) For bosons, Benguria and Lieb[5] proved that $N_c \ge \beta(z)z$ with $\beta(z)\to 1+\gamma$ as $z\to\infty$. Here, γ is some number satisfying $0<\gamma<1$ and is obtained by solving a Hartree equation. This equation was subsequently solved numerically by Baumgartner[6,11] with the result $\gamma=0.21$. Thus, bosons strongly violate the $z+1$ rule; the Pauli exclusion principle plays a key role in the electron problem.

(4) In a related development, Benguria and Lieb[7] studied the Thomas–Fermi–von Weizsäcker (TFW) equation—a well-known density-functional equation which is supposed to imitate the Schrödinger equation for fermions. They proved that $N_c \le z+0.73$ in the TFW model of an atom. (In TFW theory, the electronic charge is not quantized.) Thus, on the one hand, TFW theory really imitates the Schrödinger equation and, on the other hand, the TFW result supports the conjecture that $N_c \le z+\text{const}$ for the Schrödinger equation for arbitrarily large z. Earlier,[14] it had been shown that $Z<N_c<2Z$ in TFW theory, even for a molecule. Here Ze is the total nuclear charge,

$$Z=\sum_{j=1}^{K} z_j \ , \tag{1.1}$$

for a molecule of K atoms with nuclear charges ez_j. Note that in Thomas–Fermi (TF) theory one always has $N_c=Z$ for any molecule.[14]

(5) Lieb, Sigal, Simon, and Thirring,[8] using Sigal's method,[3] have proved that $N_c/z\to 1$ as $z\to\infty$ in the fermion case.

(6) For $z=1$ (hydrogen), Hill[9] proved that three electrons cannot be bound in a quartet ($S=\frac{3}{2}$) state. There does not seem to be any proof of nonbinding for the doublet ($S=\frac{1}{2}$) state with $N=3$.

All of the above results are for a single atom. The Ruskai and Sigal methods can be extended to the molecular case; this was explicitly done for bosons.[1] The Benguria-Lieb result[7] in (4) extends to a molecule: $N_c < Z+0.73$ K. Furthermore, all the results apply to the fixed-nucleus (sometimes called Born-Oppenheimer) approximation.

(7) Zhislin[10] proved that $N_c \ge z$ for an arbitrary mixture of particles (with any statistics) and including nuclear motion. This result extends to a molecule, $N_c \ge Z$.

In this paper it will be proved (Theorem 1) that

$$N_c < 2z+1 \tag{1.2}$$

for a fixed-nucleus atom and with any mixture of bound particles (with possibly different masses and statistics). Equation (1.2) holds if all the charges are $-e$, but a similar result holds with nonconstant but negative charges [see Eq. (2.11)]. If z is an integer, as in the physical case, (1.2) implies

$$N_c \le 2z \ . \tag{1.3}$$

This completes the story for hydrogen, i.e., $N_c = 2$, since it is well known that two electrons can indeed be bound [see (6) above]. H^{--} is *not* stable. (Incidentally, I am not aware of any proof of the obvious assertion that if N electrons cannot be bound then $M > N$ electrons cannot be bound. Thus, even if the $S = \frac{1}{2}$ case in (6) above were settled, it would not immediately follow from this that $N_c = 2$.) Equation (1.3) also states that two π^- mesons and a muon cannot be bound with $z = 1$. Equation (1.2) implies that the critical z to bind two particles is at least 0.5. The exact value[12] is 0.9112.

For large z, Eq. (1.3) is hardly optimal in view of the conjecture that $N_c - z$ is of order unity. For bosons, however, Eq. (1.3) gives the right order of magnitude since $N_c \geq 1.2z$ for large z [see (3) above].

In the case of a molecule of K atoms it will be proved that

$$N_c < 2Z + K \tag{1.4}$$

for fixed nuclei. Thus, for example, the hydrogen molecule cannot bind more than five particles. Again, this holds for arbitrary negative particles with common charge $-e$. For nonconstant charges see Eq. (2.10). A summary of the results of this paper is in Ref. 13.

As stated above, Theorem 1 does not require that the nuclear or the negative particle charges be integral. This generality may be relevant physically because, as pointed out to me by W. Thirring (private communication), particles in solids such as semiconductors may have nonintegral effective charges because of dielectric effects.

A remark should be made about the meaning of "fixed nuclei" in the molecular case. There are two possible interpretations.

Case A: The nuclei, of charges $z_1, \ldots, z_K > 0$, have coordinates $R_1, \ldots, R_K$ which are arbitrary but which are fixed once and for all, independent of the particle number N.

Case B: For each particle number, the nuclear coordinates R_j are adjusted to minimize the total energy, namely,

$$\hat{E}_N \equiv \inf_{\underline{R}} E_N(\underline{R}) + U(\underline{R}) , \tag{1.5}$$

where $\underline{R}$ denotes $\{R_1, \ldots, R_K\}$ and $E_N(\underline{R})$ is the electronic ground-state energy depending on $\underline{R}$, and $U(\underline{R})$ is the internuclear interaction. Usually $U(\underline{R})$ is the Coulomb energy

$$e^2 \sum_{1 \leq i < j \leq K} z_i z_j |R_i - R_j|^{-1} ,$$

but for our purposes $U(\underline{R})$ can be anything.

The result in Eq. (1.4), and more generally in Theorem 1, holds for both Case A and Case B. Case B is, of course, more physical. However, Theorem 1 in Case A implies Theorem 1 in Case B; this is obvious since if binding does not occur for every choice of $\underline{R}$ it certainly does not occur for the minimizing $\underline{R}$. It is Case A (fixed nuclei) that we shall actually consider henceforth.

The case of truly dynamical nuclei cannot be treated as definitively by the method presented here. Nevertheless, some information about this system is contained in Sec. VI. Section VI also contains the extension of the previously cited results to the case of smeared, but spherical, nuclear charge densities. These results are also shown to hold in the Hartree-Fock theory.

Finally, it will also be proved that Eqs. (1.2) and (1.4) hold if the particle kinetic energy operator, which is

$$p^2/2m = -(\hbar^2/2m)\Delta \tag{1.6}$$

in the usual Schrödinger equation, is replaced by the relativistic expression

$$(p^2c^2 + m^2c^4)^{1/2} - mc^2 = (-c^2\hbar^2\Delta + m^2c^4)^{1/2} - mc^2 . \tag{1.7}$$

Another variation for which (1.2) and (1.4) hold is the inclusion of a magnetic field in either Eq. (1.6) or (1.7):

$$p^2 \rightarrow [\vec{p} - e\vec{A}(x)/c]^2 , \tag{1.8}$$

where $\vec{A}$ is a bounded vector potential. One can even have different $\vec{A}$'s for different particles. Other generalizations of the kinetic energy are also possible, as well as generalizations to potentials other than the Coulomb potential (see the remark in Appendix A).

Not only are the results proved here stronger than that obtained previously [except for (5) above], but the proof itself is much simpler. For the atomic case with one species of fermions or bosons the proof is so short that it can be given in an elementary quantum-mechanics course. In order to display the basic idea as clearly as possible, this atomic case will be treated first in Sec. IV. The method of proof borrows heavily from the proof in Ref. 14, Theorem 7.23 of $N_c < 2Z$ for the TFW theory. (In Ref. 14, N_c was called λ_c.) As mentioned there, the basic idea for the TFW proof in the atomic case is due to Benguria.

II. PRELIMINARIES AND NOTATION

The Hamiltonian for N particles is

$$H_N = \sum_{i=1}^{N} [T_i - q_i V(x_i)] + \sum_{1 \leq i < j \leq N} q_i q_j |x_i - x_j|^{-1} , \tag{2.1}$$

where

$$V(x) = \sum_{j=1}^{K} z_j |x - R_j|^{-1} \tag{2.2}$$

is the electric potential produced by K fixed nuclei of charges

$$\underline{z} = \{z_1, \ldots, z_K\}$$

located at $\underline{R} = \{R_1, \ldots, R_K\}$ with $R_i \in \mathbb{R}^3$. Units are used in which the electron charge e is unity, and $\hbar = 1$. We assume $z_i > 0$, all $i = 1, \ldots, K$. The number $-q_i$ is the charge of the ith particle, and we assume $q_i \geq 0$ for $i = 1, \ldots, N$.

The operator T_i is the kinetic energy operator for the ith particle and it is assumed to have one of the following three forms:

$$T_i^{(1)} = -\Delta_i / 2m_i , \tag{2.3}$$

$$T_i^{(2)}=[i\vec{\nabla}_i+\vec{A}_i(x)]^2/2m_i\ , \tag{2.4}$$

$$T_i^{(3)}=c\{[i\vec{\nabla}_i+\vec{A}_i(x)]^2+m_i^2c^2\}^{1/2}-m_ic^2\ . \tag{2.5}$$

Here, $m_i>0$ is the mass of the ith particle, $c\vec{A}_i(x)/q_i$ is the vector potential applied to the ith particle, and c is the speed of light. $\vec{A}_i(x)$ is assumed to be bounded and to go to zero as $|x|\to\infty$. Equation (2.3) is obviously a special case of (2.4) but, for simplicity, it is treated separately. It must be noted, however, that if the form $T_i^{(3)}$ is used even for just one particle (for example, i), then every nuclear charge z_j must satisfy $\alpha q_i z_j \le 2/\pi$, where $\alpha=e^2/\hbar c$ is the fine-structure constant. The reason is that the single-particle operator $T^{(3)}-e^2qV(x)$ is bounded below (as a quadratic form) if and only if $e^2qz_j\le 2/\pi$ for every j. The situation is not changed by the addition of the third term in Eq. (2.1). (See Ref. 15.)

The ith particle can have any one of the three forms of T_i, independently. The m_i, q_i, and $\vec{A}_i$ need not be related for different i. There is one proviso, however. If several particles are of the same type (bosons or fermions) then that group must, of course, have the same m_i, q_i, and T_i. Spin can be included in the usual way. H_N is spin independent. The easiest way to treat spin is to think of the spin coordinate as merely labeling a particle type. Thus, for spin-$\frac{1}{2}$ electrons, there are two kinds of fermions: those with spin up and those with spin down.

The ground-state energy, is defined by

$$E_N=\inf \mathrm{Spec}(H_N) \tag{2.6}$$

$$=\inf_\psi \frac{\langle\psi|H_N|\psi\rangle}{\langle\psi|\psi\rangle}\ , \tag{2.7}$$

the latter being the variational principle. The admissible ψ in Eq. (2.7) must, of course, satisfy the required statistics for the various groups of particles. E_N need not be an eigenvalue, and it will not be if the system is not bound. When a particle is added to the system it can always be placed at infinity (i.e., arbitrarily far away and with arbitrarily small kinetic and potential energy), and we thus have the relation

$$E_{N,j}\ge E_N,\quad j=1,\dots,N\ , \tag{2.8}$$

where N,j denotes the $(N-1)$-particle system with the jth particle removed.

Definition: The N-particle system is said to be *bound* if and only if E_N is an eigenvalue, i.e.,

$$H_N\psi=E_N\psi \tag{2.9}$$

for some $\psi\in L^2(\mathbb{R}^{3N})$.

In this definition it is not required that $E_{N,j}$ be an eigenvalue for any j or that $E_N<E_{N,j}$ for any j. If $E_N<E_{N,j}$ for some j then E_N is automatically an eigenvalue. The word "bound" is *not applicable* to a system that merely has an eigenvalue in the continuum (i.e., E_N is not an eigenvalue but H_N has an eigenvalue greater than $E_{N,j}$ for some j); such a system would only be metastable. Our main result is the following.

Theorem 1: Let $Z=\sum_{j=1}^K z_j$ as before. Let $Q=\sum_{i=1}^N q_i$ be the negative of the total particle charge and let q be the maximum of the q_i. Then, if binding occurs, the following must be satisfied:

$$Q<2Z+qK\ . \tag{2.10}$$

In the atomic case ($K=1$), Eq. (2.10) can be replaced by the slightly stronger requirement

$$Q<2Z+\sum_{i=1}^N q_i^2/Q\ . \tag{2.11}$$

III. STRATEGY OF THE PROOF

Given Eq. (2.9) select one variable, for example, j, and a positive function of one variable, denoted by $1/\phi(x)$, to be determined appropriately. Multiply Eq. (2.9) by

$$\psi^*(x_1,\dots,x_N)/\phi(x_j)$$

and integrate over all the variables, and then take the real part. On the left-hand side there will be four terms:

$$h_j=\langle[\psi/\phi(x_j)]\,|\,H_{N,j}\,|\,\phi\rangle\ , \tag{3.1}$$

$$t_j=\mathrm{Re}\langle[\psi/\phi(x_j)]\,|\,T_j\,|\,\psi\rangle\ , \tag{3.2}$$

$$-a_j=-q_j\langle[\psi/\phi(x_j)]\,|\,V(x_j)\,|\,\psi\rangle\ , \tag{3.3}$$

$$r_j=\left\langle[\psi/\phi(x_j)]\,\middle|\,\sum_{\substack{i\\ i\neq j}} q_iq_j\,|x_i-x_j|^{-1}\,\middle|\,\psi\right\rangle . \tag{3.4}$$

[The right-hand sides of Eqs. (3.1), (3.3), and (3.4) are automatically real.] On the right-hand side there will be E_NI_j with

$$I_j=\langle[\psi/\phi(x_j)]\,|\,\psi\rangle\ . \tag{3.5}$$

Let us assume, provisionally, that all these terms are finite.

Because of Eq. (2.8)

$$E_NI_j\le h_j\ . \tag{3.6}$$

To see this, let X_j denote the variables $x_1,\dots,x_N$ with x_j excluded. Consider

$$P_j(X_j,X_j')=\int \psi(x_j,X_j)[\psi^*(x_j,X_j')/\phi(x_j)]d^3x_j\ , \tag{3.7}$$

so that

$$\Gamma_j(X_j,X_j')\equiv P_j(X_j,X_j')/I_j$$

is a properly normalized density matrix. (The positivity of ϕ is crucial here.) Moreover, Γ_j satisfies the correct statistics in the X_j variables (since ψ does) and thus, by the variational principle,

$$h_{N,j}/I_j=\mathrm{Tr}(H_{N,j}\Gamma_j)\ge E_{N,j}\ . \tag{3.8}$$

This, together with Eq. (2.8), proves (3.6). Thus, binding cannot occur if

$$t_j-a_j+r_j>0\ . \tag{3.9}$$

Let $\phi(x)$ be any function of the form

$$\phi(x)=\int |x-y|^{-1}d\mu(y)+C\ , \tag{3.10}$$

where $d\mu$ is a (positive) measure with $0<\int d\mu<\infty$ and $C\ge 0$ is a constant. In our application we shall choose ϕ

Phys. Rev. A29, 3018-3028 (1984)

to be of the same form as V but with different coefficients $\mu_j>0$:

$$\phi(x)=\sum_{j=1}^{k}\mu_j|x-R_j|^{-1}\ . \tag{3.11}$$

It will be proved that for every j

$$t_j>0 \tag{3.12}$$

for any ψ and ϕ and choice of T_j [see Eqs. (2.3)–(2.5)]. Consequently, binding cannot occur if, for any $j=1,\ldots,N$,

$$r_j\geq a_j\ . \tag{3.13}$$

Actually, we shall not prove Eq. (3.13) for any particular j, but shall prove instead that when Eq. (2.10) or (2.11) is violated then

$$R\equiv\sum_{j=1}^{N}r_j\geq A\equiv\sum_{j=1}^{N}a_j\ . \tag{3.14}$$

[In fact, we shall actually prove that $R>A$ and, therefore, that strict inequality is not really needed in Eq. (3.12). The reason that $R>A$ is given in the second part of Appendix A.] Clearly, Eq. (3.14) implies that Eq. (3.13) holds for some j, and thus Theorem 1 will be proved.

To prove (3.12) we note that the X_j integration in (3.2) can be done after the x_j integration. Therefore, (3.12) will be true if it holds for each X_j, i.e., if it holds for any function of the one variable x_j. Thus, if $f(x)\in L^2(R^3)$ we want to prove that

$$t\equiv\mathrm{Re}\int f^*(x)[(Tf)(x)/\phi(x)]d^3x>0\ . \tag{3.15}$$

When $T=p^2/2m$, (3.15) was first proved for f spherically symmetric and real by Benguria, and then for f real by Lieb. This was given in Ref. 14, Lemma 7.21. Baumgartner[11] found a more direct proof and also extended (3.15) to complex f. Baumgartner's proof easily extends to $T^{(2)}$, but the proof for $T^{(3)}$ is very different and is given in Appendix A.

In Appendix A a proof of (3.15) under carefully stated conditions on f is given. The following technical point, which is also discussed in Appendix A, has to be considered: We assumed that all the quantities in Eqs. (3.1)–(3.5) are finite. By the condition stated after Eq. (3.10), r_j and a_j are automatically finite. Conceivably, h_j, t_j, and I_j could be infinite. If so, this can be remedied by replacing $\phi(x)$ by $\phi(x)+C$ and then letting $C\to 0$ at the end. This procedure is also discussed in Appendix A.

Since one of our stated goals is to present a proof of Theorem 1 in the atomic case that is simple enough to be given in an elementary quantum-mechanics course, let us temporarily suspend any reservations about technicalities and give the following proof of (3.15), following Baumgartner's method,[11] when $T=p^2/2m$.

The key fact is that $\Delta\phi\leq 0$. Given f and ϕ, define $g(x)=f(x)/\phi(x)$. We then require that

$$2mt=-\mathrm{Re}\int g^*(x)\Delta[g(x)\phi(x)]d^3x>0\ . \tag{3.16}$$

By partial integration,

$$\begin{aligned}2mt&=\mathrm{Re}\int\vec\nabla g^*(x)\cdot\vec\nabla[\phi(x)g(x)]d^3x\\&=\mathrm{Re}\int\vec\nabla g^*(x)\cdot[\phi(x)\vec\nabla g(x)+g(x)\vec\nabla\phi(x)]d^3x\\&=\int\phi(x)|\vec\nabla g(x)|^2d^3x+\tfrac12\int\vec\nabla\phi(x)\cdot[g(x)\vec\nabla g^*(x)+g^*(x)\vec\nabla g(x)]d^3x\\&=\int\phi(x)|\vec\nabla g(x)|^2d^3x+\tfrac12\int\vec\nabla\phi(x)\cdot\vec\nabla[|g(x)|^2]d^3x\\&=\int\phi(x)|\vec\nabla g(x)|^2d^3x-\tfrac12\int|g(x)|^2\Delta\phi(x)d^3x>0\ .\end{aligned} \tag{3.17}$$

[Note that we have greater than 0 instead of greater than or equal to 0 because $\int\phi|\vec\nabla g|^2>0$ since $\phi(x)>0$ for all x; if $\vec\nabla g(x)\equiv 0$ then $f=(\mathrm{const})\phi$, but $\phi\notin L^2$ and $f\in L^2$.]

In the rest of this paper (except Appendix A) we shall assume that (3.15) holds and shall concentrate on proving that Eq. (3.14) holds. As mentioned before, all r_j and a_j are necessarily real and finite.

IV. ATOMS WITH IDENTICAL PARTICLES

Take the nuclear coordinate $R_1=0$, $z_1\equiv z$ and let $\phi(x)=1/|x|$. Assume that the particles are identical (bosons or fermions) so that $q_j=q$, $r_j=r$, and $a_j=a$ are independent of j. (Note: If the particles are fermions they are allowed to have spin, in which case $\int dx$ in the following should be understood as $\sum_\sigma\int dx$, where σ is the spin variable.)

We denote $\{x_1,\ldots,x_N\}$ by X and assume that ψ is normalized. Taking $j=1$ we have

$$a=zq\int[|\psi(X)|^2|x_1|/|x_1|]d^{3N}X=zq\ , \tag{4.1}$$

$$r=q^2\sum_{j=2}^{N}\int|\psi(X)|^2|x_1||x_1-x_j|^{-1}d^{3N}X \tag{4.2}$$

$$=q^2(N-1)\int|\psi(X)|^2|x_1||x_1-x_2|^{-1}d^{3N}X\ . \tag{4.3}$$

In going from Eq. (4.2) to (4.3) the fact that $|\psi|^2$ is symmetric was used. This symmetry also implies that the integral in Eq. (4.3) is not changed if $|x_1|$ is replaced by $|x_2|$. Thus,

$$r=\tfrac12q^2(N-1)\int|\psi(X)|^2[|x_1|+|x_2|]\times|x_1-x_2|^{-1}d^{3N}X\geq\tfrac12q^2(N-1) \tag{4.4}$$

since

$$|x_1|+|x_2|\geq|x_1-x_2|$$

by the triangle inequality.

Therefore,

$$r-a\geq\tfrac{1}{2}q^2(N-1)-zq \tag{4.5}$$

and, by Eq. (3.13), binding will not occur if $Q\geq 2z+q$ with $Q=Nq$. This proves Theorem 1 in this special case.

V. GENERAL CASE

Our goal here is to prove (3.14) if (2.11) (atomic case) or (2.10) (molecular case) is violated. In the atomic case without particle symmetry, we cannot use symmetry to go from Eq. (4.2) to (4.3). In the molecular case, even if particle symmetry exists, the obvious choice $\phi(x)=V(x)$ will not work and we shall have to use ϕ as given in Eq. (3.11).

In the atomic case

$$\phi(x)=1/|x|=V(x)/z$$

does work. Assume ψ to be normalized. For each j, Eqs. (4.1) and (4.2) become

$$a_j=zq_j\,, \tag{5.1}$$

$$r_j=\sum_{i\,(\neq j)} q_iq_j\int|\psi(X)|^2|x_j|\,|x_i-x_j|^{-1}d^{3N}X\,. \tag{5.2}$$

Now sum these over j to obtain [recalling Eq. (3.14)]

$$R-A=\tfrac{1}{2}\sum_{\substack{i,j\\ i\neq j}}q_iq_j\int|\psi(X)|^2(|x_i|+|x_j|)\times|x_i-x_j|^{-1}d^{3N}X-Qz \tag{5.3}$$

$$\geq\tfrac{1}{2}\sum_{\substack{i,j\\ i\neq j}}q_iq_j-Qz=\tfrac{1}{2}Q^2-\tfrac{1}{2}\sum_i q_i^2-Qz\,. \tag{5.4}$$

Again, the triangle inequality $|x|+|y|\geq|x-y|$ has been used. Clearly, if Eq. (2.11) is violated then $R\geq A$ and this proves Theorem 1 in the atomic case.

In the molecular case the following is needed. It does not depend on the dimension d being 3.

Lemma 1: Let $\psi(X)$ be any normalized function in $L^2(R^{Nd})$ (without any particular symmetry). For $j=1,\ldots,N$ let

$$\rho_j(x)=\int|\psi(x,X_j)|^2d^{Nd-d}X_j \tag{5.5}$$

be the one-particle density for particle j. Let $g_1(x),\ldots,g_N(x)$ be any given functions of one variable such that $g_j(x_j)\psi(X)$ is in $L^2(\mathbb{R}^{Nd})$ and define

$$\langle g_j\rangle=\int\rho_j(x)g_j(x)d^dx=\int|\psi(X)|^2g_j(x_j)d^{Nd}X\,. \tag{5.6}$$

Then

$$\mathrm{Re}\int|\psi(X)|^2\sum_{1\leq i<j\leq N}g_i^*(x_i)g_j(x_j)d^{Nd}X\geq\tfrac{1}{2}\left|\sum_{j=1}^N\langle g_j\rangle\right|^2-\tfrac{1}{2}\sum_{j=1}^N\langle|g_j|^2\rangle\,. \tag{5.7}$$

This is proved simply by noting that

$$\mathrm{Re}\sum_{\substack{i,j\\ i<j}}g_i^*g_j=\tfrac{1}{2}\left|\sum_i g_i\right|^2-\tfrac{1}{2}\sum_i|g_i|^2$$

and then using the Schwarz inequality on the first term.

To apply Lemma 1 to the general case, let

$$\tilde\rho(x)\equiv\sum_{j=1}^N q_j\rho_j(x) \tag{5.8}$$

be the negative of the single particle *charge* density for ψ [see Eq. (5.5)]. Let $\phi(x)$ be the potential in Eq. (3.11). We then have

$$A=\sum_{j=1}^N a_j=\sum_{s=1}^K z_s\gamma_s\,, \tag{5.9}$$

with

$$\gamma_s=\int\tilde\rho(x)|x-R_s|^{-1}\phi(x)^{-1}d^3x\,. \tag{5.10}$$

$$R=\sum_{j=1}^N r_j=\sum_{1\leq i<j\leq N}q_iq_j\int|\psi(X)|^2|x_i-x_j|^{-1}\times[\phi(x_i)^{-1}+\phi(x_j)^{-1}]d^{3N}X\,. \tag{5.11}$$

Let us write [following an idea in Ref. 11)]

$$\phi(x)^{-1}+\phi(y)^{-1}=[\phi(x)\phi(y)]^{-1}[\phi(x)+\phi(y)]=\sum_{j=1}^K\mu_j[\phi(x)|x-R_j|]^{-1}[\phi(y)|y-R_j|]^{-1}(|x-R_j|+|y-R_j|)\,. \tag{5.12}$$

Again, noting that

$$|x-R_j|+|y-R_j|\geq|x-y|$$

we have, upon inserting Eq. (5.12) in (5.11),

$$R\geq\sum_{s=1}^K\mu_s\int|\psi(X)|^2\sum_{1\leq i<j\leq N}g_i^s(x_i)g_j^s(x_j)d^{3N}X\,, \tag{5.13}$$

with

$$g_i^s(x)=q_i[\phi(x)|x-R_s|]^{-1}\,. \tag{5.14}$$

Using Lemma 1 for each s we have

$$R\geq\tfrac{1}{2}\sum_{s=1}^K\mu_s\left[\gamma_s^2-\sum_{i=1}^N q_i^2\int\rho_i(x)\phi(x)^{-2}\times|x-R_s|^{-2}d^3x\right]. \tag{5.15}$$

However,

$$\mu_s/[\,|x-R_s|\,\phi(x)]\le 1\,,$$

and $q_i \le q$ (by definition). Hence,

$$R \ge \tfrac{1}{2}\sum_{s=1}^{N}(\mu_s\gamma_s^2 - q\gamma_s)\,. \tag{5.16}$$

We shall have $R \ge A$ if

$$\sum_{s=1}^{K}\gamma_s\{\mu_s\gamma_s - q - 2z_s\} \ge 0\,, \tag{5.17}$$

and our aim is to choose the $\{\mu_s\}$ so that Eq. (5.17) is satisfied when condition (2.10) is violated. (Note that γ_s depends on $\{\mu_s\}$.) To this end, let

$$\delta_s \equiv \mu_s\gamma_s/Q\,, \tag{5.18}$$

$$\beta_s \equiv (2z_s+q)/(2Z+qK)\,.$$

Note that [see Eqs. (5.8) and (5.10)]

$$\sum_{s=1}^{K}\delta_s = \sum_{s=1}^{K}\beta_s = 1\,. \tag{5.19}$$

Suppose we can choose $\{\mu_s\}$ such that

$$\delta_s = \beta_s,\quad s=1,\dots,K\,. \tag{5.20}$$

Then the left-hand side of Eq. (5.17) becomes

$$\sum_{s=1}^{K}\gamma_s\delta_s[Q-(2Z+qK)]\,,$$

and this is non-negative if condition (2.10) is violated.

Thus, showing that the K equations (5.20) in K unknowns have a solution proves Theorem 1. This is done in Appendix B.

VI. THREE GENERALIZATIONS OF THEOREM 1

A. Smeared nuclei

Suppose that the nuclear charge densities, instead of being points, are smeared into *spherically symmetric* distributions about R_j, namely, $z_j|x-R_j|^{-1}$ in Eq. (2.2) is replaced by

$$V_j(x) = \int |x-y-R_j|^{-1}d\mu_j(y)\,, \tag{6.1}$$

where $d\mu_j$ is a spherically symmetric (positive) measure with $\int d\mu_j(y)=z_j$. Then *Theorem 1 continues to hold without modification.* The proof is as before—with the same $\phi(x)$. One merely has to note that

$$V_j(x) \le z_j|x-R_j|^{-1} \tag{6.2}$$

for all x and, hence, a_j, defined by Eq. (3.3), is not greater than it would be for the point nucleus.

B. Dynamical nuclei

Suppose that the K nuclear coordinates $R_1,\dots,R_K$ are dynamical variables and the Hamiltonian is

$$\widetilde{H}_N = H_N + T_{\text{nuc}} + U(\underline{R})\,, \tag{6.3}$$

and

$$T_{\text{nuc}} = \sum_{i=1}^{K} T_i\,, \tag{6.4}$$

where H_N is given by Eq. (2.1) and each T_i is one of the operators given in Eqs. (2.3)–(2.5). $U(\underline{R})$ is a potential that depends on the nuclear coordinates, as explained after Eq. (1.5). If $U(\underline{R})$ is translation invariant, the eigenvalue equation

$$\widetilde{H}_N\psi = \widetilde{E}_N\psi\,, \tag{6.5}$$

with $\widetilde{E}_N = \inf\operatorname{Spec}(\widetilde{H}_N)$ and $\psi=\psi(X,\underline{R})$, would not generally be expected to have a solution in $L^2(\mathbb{R}^{3N+3K})$; if there are no magnetic fields present then it certainly would not. With magnetic fields present one cannot simply remove the center of mass motion, and the situation is complicated. To avoid technical complications it will be assumed that $U(\underline{R})$ also contains one-body terms which serve to contain the nuclei, and therefore that Eq. (6.5) indeed has an L^2 solution. Physically, this is no real restriction because the confining potential could, for example, be an infinitely high walled box of arbitrarily large size. Because the *negative* particles are *not confined* it is still true that $\widetilde{E}_N \le \widetilde{E}_{N,j}$.

In this case a weakened form of Theorem 1 holds. Let us define $E_{N,j}^{\infty}$ to be the inf Spec $(\widetilde{H}_{N,j}^{\infty})$, with negative particle j removed as before, but with all the nuclear masses set equal to infinity. Alternatively,

$$E_{N,j}^{\infty} = \inf_{\underline{R}} E_{N,j}(\underline{R}) + U(\underline{R})\,, \tag{6.6}$$

where $E_{N,j}(\underline{R})$ is the ground-state energy of $H_{N,j}$ as defined in Eq. (2.7). Clearly,

$$E_{N,j}^{\infty} < \widetilde{E}_{N,j} = \inf\operatorname{Spec}(\widetilde{H}_{N,j})\,. \tag{6.7}$$

Theorem 2: Suppose that the system is bound [i.e., Eq. (6.5) holds] and the binding energy satisfies

$$\widetilde{E}_N - \widetilde{E}_{N,j} \le E_{N,j}^{\infty} - \widetilde{E}_{N,j} \tag{6.8}$$

for all $j=1,\dots,N$. Then Eq. (2.10) holds in the molecular case $K>1$ and Eq. (2.11) holds in the atomic case $K=1$.

In the physical situation, the right-hand side of Eq. (6.9) is numerically small, but it is a challenge to eliminate this condition.

To prove Theorem 2 we multiply Eq. (6.5) by

$$\psi^*(X,\underline{R})/\phi(x_j,\underline{R})$$

and integrate over all the variables. Here $\phi(x,\underline{R})$ is as in the proof of Theorem 1, namely, Eq. (3.11) with *fixed constants* μ_j. [It is important that the μ_j do not depend on $\underline{R}$, otherwise the dependence of ϕ on each R_j would not have the form of Eq. (3.10), and thus the positivity of the integrated T_{nuc} term might be lost.]

Let us write (for any fixed j, $1\le j\le N$)

$$\widetilde{H}_N = \widetilde{H}_{N,j}^{\infty} + T_{\text{nuc}} + T_j - q_jV(x_j,\underline{R}) + \sum_{i\,(\ne j)}|x_i-x_j|^{-1}\,. \tag{6.9}$$

The first term, $\widetilde{H}_{N,j}^{\infty} = \widetilde{H}_{N,j}^{\infty}(X_j,\underline{R})$, satisfies (as an operator) $\widetilde{H}_{N,j}^{\infty} \ge E_{N,j}^{\infty}$. Therefore, moving this term to the right-hand side of Eq. (6.5) we obtain less than or equal to 0 as before [using Eq. (6.8)].

The term involving T_j is positive as before. The T_{nuc} term is positive since $\phi(x_j,\underline{R})$, with all variables except R_i fixed, has the correct form [as given by Eq. (3.10)] as a function of R_i. The third term is (after summing on j) the same as in Eq. (5.9) but with

$$\gamma_s = \int\int |\psi(X,\underline{R})|^2 \sum_{j=1}^{N} [q_j |x_j - R_s|^{-1}/\phi(x_j,\underline{R})] \times d^{3N}X\, d^{3K}R \ . \qquad (6.10)$$

In the atomic case, $\phi(x,R) = |x-R|^{-1}$ and the proof proceeds exactly as before. In the molecular case, Eq. (5.16) holds with γ_s given by Eq. (6.10). [An obvious modification of Lemma 1 and Eqs. (5.11)—(5.15) is needed to include the $\underline{R}$ dependence; the basic observation is that the Schwarz inequality used in Lemma 1 is still applicable.]

To complete the proof in the molecular case, we require that Eq. (5.20) have a solution with the new definition of γ_s, Eq. (6.10). The proof in Appendix B is easily modified in terms of the appropriate matrix M [which has strictly positive elements and satisfies Eq. (B8)]:

$$M_{sj} = Q^{-1}\int\int |\psi(X,\underline{R})|^2 \times \sum_{k=1}^{N} q_k |x_k - R_s|^{-1} |x_k - R_j|^{-1} \times \phi(x_k,\underline{R})^{-2} d^{3N}X\, d^{3K}R \ . \qquad (6.11)$$

C. Hartree and Hartree-Fock theories

Theorem 1 applies to the Schrödinger equation, but the conclusion remains true in both the Hartree and Hartree-Fock (HF) (either restricted or unrestricted) approximations. Here, the proof for unrestricted HF theory will be given; the proof for the Hartree theory is very similar. It will also be assumed that all the N charges, masses, and kinetic energy operators are identical. We take $q_i = 1$.

In the HF theory, the ground-state energy is defined as in Eq. (2.7), but with ψ restricted to the class of determinantal functions:

$$\psi = (N!)^{-1/2} \det u_i(x_j,\sigma_j) \qquad (6.12)$$

and the u_i are orthonormal. As in the earlier definition, the N-particle system is said to be bound if and only if there is a ψ that actually minimizes the energy expression (2.7). As before, $E_N \le E_{N,j}$. The reader is referred to Ref. 16 for details; in particular, for the proof that binding occurs if $N < Z+1$ (when $T = T^{(1)}$).

If there is a minimum, the u_i (after possibly an $N \times N$ unitary transformation) satisfy the N coupled HF equations:

$$hu_i = \epsilon_i u_i \ , \qquad (6.13)$$

where h is the single-particle operator

$$h = T - V(x) + U - K \ . \qquad (6.14)$$

Here, T is one of the operators in Eqs. (2.3)—(2.5), U is the direct part of the Coulomb repulsion, and K is the exchange part.

Since ψ minimizes the energy, each $\epsilon_i \le 0$ in Eq. (6.13). The reason is that the dependence of the numerator in Eq. (2.7) on any one u_i is constant plus quadratic, and the latter term is just ϵ_i. If $\epsilon_1 > 0$, for example, then it is easy to see that the $(N-1)$-particle ψ composed of $u_2,\dots,u_N$ would have an energy strictly below E_N, and this contradicts $E_{N,1} \ge E_N$.

Let $\phi(x)$ be as in Eq. (3.11), multiply Eq. (6.13) by $u_i^*(x)/\phi(x)$, integrate over x, and sum over spins. Then sum over i. The right-hand side is nonpositive. On the left-hand side, the terms involving T are positive, as before. The $V(x)$ term is

$$A = N\int |\psi(X)|^2 V(x_1)/\phi(x_1) d^{3N}X \ .$$

The repulsion term, $U-K$, is

$$R = \tfrac{1}{2}N(N-1)\int |\psi(X)|^2 |x_1 - x_2|^{-1} \times [\phi(x_1)^{-1} + \phi(x_2)^{-1}] d^{3N}X \ .$$

(A summation on spins is understood in these two integrals.) The rest of the analysis proceeds as before.

ACKNOWLEDGMENTS

This work was partially supported by U.S. National Science Foundation Grant No. PHY-81-16101-A01.

APPENDIX A: KINETIC-ENERGY INEQUALITY

1. Proof of Eq. (3.15)

Our first goal is to prove Eq. (3.15) when T is one of the three forms in Eqs. (2.3)—(2.5). The potential ϕ is defined by

$$\phi(x) = \int |x-y|^{-1} d\mu(y) + C \ , \qquad (A1)$$

with $d\mu$ a positive measure, $0 < \int d\mu < \infty$, and $C \ge 0$. The function f satisfies

$$f \text{ and } Tf \in L^2(R^3) \ . \qquad (A2)$$

Note that if $C > 0$ then $1/\phi(x)$ is a bounded function and $f(x)/\phi(x)$ is automatically in L^2.

The following Lemma 2 validates the assertions used in Sec. III provided

$$\psi(X)/\phi(x_j) \in L^2(R^{3N}) \ .$$

Subsection II below shows how to deal with the case $\psi/\phi \notin L^2$ by taking $C > 0$ and then letting $C \to 0$. [Technical remark: In the nonrelativistic case a solution to Eq. (2.9) automatically satisfies $T_j\psi \in L^2(R^{3N})$ and hence $Tf \in L^2(R^3)$. The analogous statement is known to hold in the relativistic case if $e^2 q_j z_i < \frac{1}{2}$ (for all i), which is less than the critical value $2/\pi$. Thus, there is possibly a minor technical gap in the relativistic case.] Lemma 2 in the relativistic case was originally proved only for $\vec{A}(x) \equiv 0$. The extension to $\vec{A}(x) \ne 0$ follows from Lemma 3, which is a joint work with Michael Loss. Lemma 3 shows that Lemma 2 (relativistic) follows automatically from Lemma 2 (nonrelativistic). Nevertheless, the original proof of Lemma 2 for $T^{(3)}$ with $\vec{A}(x) \equiv 0$ is not without interest and is therefore given below.

Lemma 2: Assume Eqs. (A1), (A2), and

$$f(x)/\phi(x) \in L^2(\mathbf{R}^3) .$$

Let T be any one of the operators in Eqs. (2.3)–(2.5), with $\vec{A}(x)$ bounded in Eqs. (2.4) and (2.5). Then

$$h(x) \equiv f^*(x)(Tf)(x)/\phi(x) \tag{A3}$$

is in L^1 and

$$t = \mathrm{Re} \int h(x) d^3x > 0 . \tag{A4}$$

Proof: At first assume $C > 0$ so that $1/\phi$ is bounded. By a simple density argument, we can suppose $f \in C_0^\infty$ (infinitely differentiable functions of compact support). We can also assume [by replacing $d\mu$ by $\exp(-\epsilon x^2) * d\mu$, for example, and then using dominated convergence] that ϕ and $1/\phi \in C^\infty \cap L^\infty$. Unfortunately, the limiting argument just cited will only allow us to conclude that $t \geq 0$. The proof that $t > 0$ is given at the very end.

$T^{(1)} = -\Delta$: The proof given in Eqs. (3.16) and (3.17) is completely rigorous for such functions. Another proof is given below in connection with $T^{(3)}[\vec{A}(x) \equiv 0]$.

$T^{(2)} = [i\vec{\nabla} + \vec{A}(x)]^2$: We set $g(x) = f(x)/\phi(x)$ as before and follow the same manipulation as in Eq. (3.17), but replacing $\vec{\nabla}$ by $\vec{\nabla} - i\vec{A}(x)$. The only essential difference is that we get an extra term

$$i \int \vec{\nabla}\phi(x) \cdot \vec{A}(x) \, |g(x)|^2 d^3x ,$$

but this vanishes when the real part is taken. Strict positivity follows from the fact that $(i\vec{\nabla} - \vec{A})g$ cannot vanish identically.

$T^{(3)} = (-\Delta + 1)^{1/2} - 1$: Here we have to work in momentum space. $T^{(3)}f$ is defined to be the function whose Fourier transform $(T^{(3)}f)\hat{}$ is

$$[(k^2+1)^{1/2} - 1]\hat{f}(k) ,$$

with

$$\hat{f}(k) = \int \exp(ik \cdot x) f(x) d^3x .$$

Again, let $g = f/\phi$. Since $(ab)\hat{} = \hat{a} * \hat{b}$, and since $(|x|^{-1})\hat{} = 4\pi/k^2$, we have

$$\begin{aligned} 2\pi^2 t = \mathrm{Re} \int d\mu(y) \int \hat{g}^*(k) \, |k - q|^{-2} e^{iy \cdot (k-q)} \\ \times m(q) \hat{g}(q) d^3k \, d^3q \\ + (C/4\pi) \int |g(k)|^2 m(k) d^3k , \end{aligned} \tag{A5}$$

with $m(q) = (q^2 + 1)^{1/2} - 1$. Clearly, the second term on the right-hand side of Eq. (A5) is positive. As for the first term, it is sufficient to prove strict positivity for each y but, since

$$\exp[iy \cdot (k - q)]$$

is a product function, it suffices to prove that the kernel

$$K(k,q) = |k - q|^{-2}[m(k) + m(q)] \tag{A6}$$

is positive definite.

Let us temporarily return to $T^{(1)} = -\Delta$. In this case $m(k)$ in Eq. (A6) is replaced by k^2. However,

$$k^2 + q^2 = |k - q|^2 + 2k \cdot q \tag{A7}$$

and thus

$$K(k,q) = 1 + 2k \cdot q \, |k - q|^{-2} . \tag{A8}$$

The kernel 1 in Eq. (A8) is clearly positive semidefinite. For the second term, note that $|k - q|^{-2}$ is positive definite and $k \cdot q$ is a product function; thus, K is positive definite and we have a second proof for the $T^{(1)}$ case.

Returning to $T^{(3)}$, let us write

$$\begin{aligned} K(k,q) = [n(k) + n(q)]^{-1} |k - q|^{-2} [n(k) + n(q)] \\ \times [m(k) + m(q)] , \end{aligned} \tag{A9}$$

with $n(k) = (k^2 + 1)^{1/2}$. The last two factors in Eq. (A9) are

$$B(k,q) = k^2 + q^2 + 2[n(k) - 1][n(q) - 1] . \tag{A10}$$

Now the first factor can be written as an integral over product functions, namely,

$$\int_0^\infty ds \exp\{-s[n(k) + n(q)]\} ,$$

and this does not affect the positive definiteness. As for the rest we have two terms, namely,

$$|k - q|^{-2}(k^2 + q^2) ,$$

which is positive definite as we just proved for $T^{(1)}$, and

$$|k - q|^{-2}[n(k) - 1][n(q) - 1] .$$

But this last term is a positive definite kernel times a product function, so it too is positive definite. This concludes the proof that $t > 0$ if $C > 0$ and $\vec{A}(x) \equiv 0$ for $T^{(3)}$.

Thus far we have proved that $t > 0$ when (i) $f \in C_0^\infty$, $\phi \in C^\infty$ and (ii) $C > 0$. In all cases we found that $t = Q(g,g)$ with Q being a positive definite quadratic form and $g = f/\phi$. Let us first remove condition (i). There exist sequences f^n and $d\mu^n$ such that $\phi^n \to \phi$ pointwise almost everywhere and $f^n \to f$, $Tf^n \to Tf$, and

$$g^n = f^n/\phi^n \to g = f/\phi$$

in L^2. Then

$$t = \lim_{n\to\infty} t^n = \lim_{n\to\infty} Q(g^n, g^n) \geq Q(g,g) > 0 .$$

Finally, we want to let $C \to 0$ (if that is the case at hand). With f and ϕ fixed, let $\phi_C = \phi + C$, $g_C = f/(\phi + C)$. With $Y_C = \phi/(\phi + C)$, we have that $0 < Y_C \leq 1$, and $Y_C \to 1$ pointwise almost everywhere. Then $t_C \to t$ by dominated convergence. Also, $g_C \to g = f/\phi$ and $\hat{g}_C \to \hat{g}$ in L^2. Again,

$$t = \lim_{C\to 0} t_C = \lim_{C\to 0} Q(g_C, g_C) \geq Q(g,g) > 0 .$$

This completes the proof for $T^{(1)}$, $T^{(2)}$, and $T^{(3)}$ $[\vec{A}(x) \equiv 0]$. The general case $T^{(3)}$ $[\vec{A}(x) \not\equiv 0]$ follows from the $T^{(2)}$ case and Lemma 3, in which C is the multiplication operator $1/\phi(x)$ and B is the operator $T^{(2)}$, and the integral representation

$$(B+1)^{1/2}-1=\pi^{-1}B\int_1^\infty (x-1)^{1/2}x^{-1}(x+B)^{-1}dx\ , \tag{A11}$$

Q.E.D.

Lemma 3 (in collaboration with M. Loss): Let H be a Hilbert space with an inner product $(\cdot,\cdot)$ and let B and C be non-negative, self-adjoint linear operators with domain $D(B)$,$D(C)$. Suppose that

(i) $(B+x)^{-1}$: $D(C)\to D(C)$, all $x>0$

(ii) $\mathrm{Re}(B\phi,C\phi)>0$ (respectively, ≥ 0) ,

all $0\neq\phi\in D(B)\cap D(C)$

(iii) $g(\lambda)=\int d\mu(y)(s\lambda+t)(\lambda+y)^{-1}$ (A12)

with $s,t\geq 0$, $s+t>0$, and $\mu\not\equiv 0$ a non-negative Borel measure on R with $\mu\{(-\infty,0]\}=0$ and

$$\int d\mu(y)(1+y)^{-1}<\infty\ .$$

Then

$$\mathrm{Re}(g(B)\phi,\ C\phi)>0 \text{ (respectively, } \geq 0) ,$$

$$\text{all } \phi\in D(g(B))\cap D(C)\ . \tag{A13}$$

Proof: First, consider the special case

$$g(\lambda)=g_x(\lambda)=(s\lambda+t)(\lambda+x)^{-1}$$

for some fixed $x>0$. Then $G_x=g_x(B)$ is bounded and we want to prove that $I=(G_x\phi,C\phi)$ satisfies $\mathrm{Re}I>0$ (respectively, ≥ 0) for all $0\neq\phi\in D(C)$. Let $\psi=(B+x)^{-1}\phi$, whence

$$0\neq\psi\in D(B)\cap D(C)\ .$$

Since $(B+x)\psi\in D(C)$ then ψ and $B\psi\in D(C)$. Thus,

$$I=((sB+t)\psi,\ C(B+x)\psi)=s(B\psi,\ CB\psi)+sx(B\psi,\ C\psi)$$
$$+t(\psi,\ CB\psi)+tx(\psi,\ C\psi)\ .$$

Since $(\psi,\ CB\psi)=(C\psi,\ B\psi)=(B\psi,\ C\psi)^*$, we have

$$\mathrm{Re}I\geq(sx+t)\mathrm{Re}(B\psi,\ C\psi)>0 \text{ (respectively, } \geq 0)\ .$$

Now let $I=(g(B)\phi,\ C\phi)$ with

$$\phi\in D(g(B))\cap D(C)$$

and, for $0<\epsilon<1$,

$$I_\epsilon=(g^\epsilon(B)\phi,\ C\phi)$$

with

$$g^\epsilon(\lambda)=\int d\mu^\epsilon(y)(s\lambda+t)/(\lambda+y)$$

and where μ^ϵ is μ restricted to the interval $(\epsilon,1/\epsilon)$. Clearly, $g^\epsilon(B)$ is bounded and

$$I_\epsilon=\int d\nu(\phi,C\phi;\lambda)\int d\mu^\epsilon(y)(s\lambda+t)/(\lambda+y)\ ,$$

where $\nu(\phi,\phi';\cdot)$ is the spectral measure of B associated with ϕ,ϕ'. We want to show that

$$I_\epsilon=\int d\mu^\epsilon(y)\int d\nu(\phi,C\phi;\lambda)(s\lambda+t)/(\lambda+y)\ .$$

Fubini's Theorem cannot be used since ν is not positive. However, by the polarization identity,

$$d\nu(\phi,C\phi;\lambda)=\tfrac{1}{4}d\nu(a,a;\lambda)-\tfrac{1}{4}d\nu(b,b;\lambda)$$
$$+\tfrac{1}{4}id\nu(c,c;\lambda)-\tfrac{1}{4}id\nu(d,d;\lambda)\ ,$$

where $a=\phi+C\phi$, $b=\phi-C\phi$, $c=\phi-iC\phi$, and $d=\phi+iC\phi$. Each of these four measures is non-negative and, since $g^\epsilon(B)$ is bounded, each integral is finite. Thus, we can exchange the order of integration and

$$I_\epsilon=\int d\mu^\epsilon(x)M(x)$$

with

$$M(x)=(G_x\phi,C\phi)$$

and $\mathrm{Re}M(x)>0$ (respectively, ≥ 0). Therefore, $\lim_{\epsilon\to 0}\mathrm{Re}I_\epsilon>0$ (respectively, ≥ 0). On the other hand,

$$I-I_\epsilon=([g(B)-g^\epsilon(B)]\phi,\ C\phi)\ .$$

Since $\phi\in D(g(B))$, it is easy to see that $[g(B)-g_\epsilon(B)]\phi\to 0$. Thus,

$$I=\lim_{\epsilon\to 0}I_\epsilon\ .$$

Q.E.D.

Remark: Suppose that $\tilde{g}(\lambda)$ is another function with the same kind of representation as in Eq. (A12). Then, starting with Eq. (A13) and with the pair $g(B),C$ instead of B,C, one can apply Lemma 3 to $\tilde{g}(C)$ and deduce that

$$\mathrm{Re}(g(B)\phi,\ \tilde{g}(C)\phi)>0 \text{ (respectively, } \geq 0) ,$$

$$\text{all } \phi\in D(g(B))\cap D(\tilde{g}(C))\ . \tag{A14}$$

It is merely necessary to verify that for all $x>0$,

$$(C+x)^{-1}:\ D(g(B))\to D(g(B))\ .$$

This implies the following generalization of the results of this paper:

(i) The relativistic kinetic energy (with magnetic field) can be generalized to any function g of $[\vec{p}-\vec{A}(x)]^2$ that has the form of Eq. (A12).

(ii) The Coulomb potential $1/|x|$ can be replaced (everywhere) in Eq. (2.1) by $v(x)=1/w(|x|)$ for any function w with the representation

$$w(|x|)=\int d\mu(y)(s|x|)(|x|+y)^{-1}\ . \tag{A15}$$

With $s>0$ and $\mu\geq 0$. For example, $1/|x|\to|x|^{-p}$, $0<p<1$ is allowed. It is easy to check that $C=w(|x|)$ satisfies

$$(C+\lambda)^{-1}:\ D(T^{(i)})\to D(T^{(i)}),\ i=1,2,3$$

for $\lambda>0$. It is also necessary to check that the "triangle inequality"

$$w(|x|)+w(|z|)\geq w(|x-z|)$$

holds, and this is easily seen to be the case from Eq. (A15). [It is the requirement of the triangle inequality that dictates $s|x|$, instead of $s|x|+t$, in Eq. (A15).]

Phys. Rev. A29, 3018-3028 (1984)

2. Eliminating infinity

After Eq. (3.5) we made the assumption that all quantities in Eqs. (3.1)–(3.5) were finite. Conceivably, this need not be true with ϕ given by Eq. (3.11). To remedy this defect replace ϕ in Eq. (3.11) by

$$\phi_C(x)=\phi(x)+C,\quad C>0\ .$$

Then all quantities are finite. Denote R (respectively, A) with ϕ_C by R_C (respectively, A_C). Binding cannot occur if $R_C>A_C$ for any C (here, the fact that $t>0$ is ignored).

As $C\rightarrow 0$, R_C and A_C have finite limits R and A which, by dominated convergence, are the R and A given in Eq. (3.14). Thus, it suffices to show that $R>A$ when condition (2.10) or (2.11) is violated. In the earlier proof in Secs. IV and V it was shown that $R\geq A$ by using the triangle inequality

$$|x_i-x_j|\leq|x_i|+|x_j|\ .$$

This inequality will now be investigated more closely to show that, in fact, $R>A$.

If we look at Eq. (5.11), for example, we have, after integrating over the variables other than $x_i=x$ and $x_j=y$, an expression of the form

$$L\equiv\int f(x,y)(|x|+|y|)|x-y|^{-1}d^3x\,d^3y\ , \tag{A16}$$

and we wish to show that

$$L>M\equiv\int f(x,y)d^3x\,d^3y>0\ . \tag{A17}$$

Note that $f(x,y)$ is a non-negative *function* in L^1, and not a distribution. The function

$$g(x,y)=(|x|+|y|)|x-y|^{-1}-1$$

satisfies $g\geq 0$ and $g=0$ if and only if $y=-bx$ with $b\geq 0$. The set on which this occurs has six-dimensional Lebesgue measure zero. Thus, $g>0$ almost everywhere. Since $f>0$ on a set of positive measure, $\int fg>0$ and hence Eq. (A17) holds.

APPENDIX B: SOLUTION OF EQ. (5.20)

Let μ denote $(\mu_1,\ldots,\mu_K)$ and consider the function

$$F(\mu)=\sum_{s=1}^{K}[\delta_s(\mu)-\beta_s]^2 \tag{B1}$$

defined on the positive orthant D: $\mu_i\geq 0$, but excluding the origin $\mu=0$. The β_s are fixed, strictly *positive* constants satisfying $\sum_s\beta_s=1$. δ_s is of the form

$$\delta_s=\int\rho(x)[\mu_s|x-R_s|^{-1}/\phi(x)]d^3x\ , \tag{B2}$$

with $\int\rho=1$, and

$$\phi(x)=\sum_s\mu_s|x-R_s|^{-1}\ .$$

Equation (B2) implies $\sum\delta_s=1$.

Now $\delta_s(\mu)$ is continuous on D (in particular, $\delta_s(\mu)=0$ if $\mu_s=0$) and homogeneous of degree zero, i.e., $\delta_s(\lambda\mu)=\delta_s(\mu)$. Therefore, $F(\mu)$ has a minimum on D. We want to show that this minimum is zero, whence $\delta_s(\mu)=\beta_s$ for all s. Let μ be a minimum point and suppose (without loss of generality) that $\mu_1,\ldots,\mu_t>0$, $\mu_{t+1},\ldots,\mu_K=0$. (Not all the μ_s can vanish since 0 is not in D.) For $1\leq s\leq t$, $\delta_s(\mu)$ is differentiable in $\mu_1,\ldots,\mu_t$ and for $t+1\leq s\leq K$, $\delta_s(\mu)\equiv 0$ in a (t-dimensional) neighborhood of this point. Therefore, at the minimum,

$$0=\frac{\partial F}{\partial\mu_j}=2\sum_{s=1}^{t}\mu_sM_{sj}(\delta_j-\beta_j-\delta_s+\beta_s) \tag{B3}$$

for $1\leq j\leq t$, where M is the t-square matrix

$$M_{sj}=\int\rho(x)|x-R_s|^{-1}|x-R_j|^{-1}\phi(x)^{-2}d^3x\ . \tag{B4}$$

Clearly, M is symmetric and positive semidefinite and, most importantly, M has positive matrix elements.

Eq. (B3) can be rewritten in the following way (recalling that $\mu_s>0$ for $1\leq s\leq t$):

$$Nv=v\ , \tag{B5}$$

where N is a matrix and v is a vector given by

$$v_s=(\delta_s-\beta_s)\delta_s^{1/2}\ , \tag{B6}$$

$$N_{sj}=M_{sj}\mu_s\mu_j(\delta_s\delta_j)^{-1/2}\ . \tag{B7}$$

The fact that

$$\sum_s\mu_s\mu_jM_{sj}=\delta_j \tag{B8}$$

has been used which, in terms of N, reads

$$Nw=w\ , \tag{B9}$$

with

$$w_s=\delta_s^{1/2}\ . \tag{B10}$$

Now N is symmetric and has strictly positive matrix elements. By the Perron-Frobenius theorem, N has a *unique* eigenvalue of largest modulus λ. Moreover, this eigenvalue is positive and has only one eigenvector u, which (up to a phase) has *strictly positive* components. Equation (B9) implies that $\lambda=1$ and $u=w$ for, otherwise, taking the inner product of Eq. (B9) with u we would obtain $(\lambda-1)(u,w)=0$, which is impossible since $(u,w)>0$. Thus, the solution to Eq. (B5) is

$$v=cw\ , \tag{B11}$$

where c is a constant. This means that

$$\delta_s-\beta_s=c$$

for

$$1\leq s\leq t\ . \tag{B12}$$

Summing this on s we obtain (since $\delta_s=0$ for $s>t$)

$$1-\sum_{s=1}^{t}\beta_s=ct\ . \tag{B13}$$

If $t=K$, the left-hand side of Eq. (B13) is zero and we are finished. If $t<K$, then $c>0$. In the latter case, replace $\mu_s=0$ by $\mu_s=\epsilon$ for $t<s\leq K$ and with $\epsilon>0$. It is easy to see that δ_s decreases for $1\leq s\leq t$ and δ_s becomes strictly positive for $t<s\leq K$. If ϵ is small enough, $(\delta_s-\beta_s)^2$ will decrease for all $1\leq s\leq K$. Thus, $F(\mu)$ will decrease; as this is a contradiction, $t=K$ and the proof is complete.

[1]M. B. Ruskai, Commun. Math. Phys. 82, 457 (1982).

[2]I. M. Sigal, Commun. Math. Phys. 85, 309 (1982). See also *Mathematical Problems in Theoretical Physics,* Vol. 153 of *Lecture Notes in Theoretical Physics* (Springer, Berlin, 1982), pp. 149–156.

[3]I. M. Sigal, Institute Mittag Leffler Report No. 12 (1982, revised in 1983), Ann. Phys. (to be published).

[4]M. B. Ruskai, Commun. Math. Phys. 85, 325 (1982).

[5]R. Benguria and E. H. Lieb, Phys. Rev. Lett. 50, 1771 (1983).

[6]B. Baumgartner, J. Phys. A (to be published).

[7]R. Benguria and E. H. Lieb, (unpublished).

[8]E. H. Lieb, I. M. Sigal, B. Simon, and W. E. Thirring (unpublished). See also Phys. Rev. Lett. 52, 994 (1984).

[9]R. N. Hill, *Mathematical Problems in Theoretical Physics. Proceedings of the International Conference on Mathematical Physics, Lausanne, 1979,* Vol. 116 of *Lecture Notes in Physics* edited by K. Osterwalder (Springer, New York, 1979), pp. 52–56.

[10]G. Zhislin, Trudy, Mosk. Mat. Obšč. 9, 81 (1960).

[11]B. Baumgartner, Lett. Math. Phys. 7, 439 (1983).

[12]F. H. Stillinger and D. K. Stillinger, Phys. Rev. A 10, 1109 (1974); F. H. Stillinger, J. Chem. Phys. 45, 3623 (1966).

[13]E. H. Lieb, Phys. Rev. Lett. 52, 315 (1984).

[14]E. H. Lieb, Rev. Mod. Phys. 53, 603 (1981). Erratum: 54, 311(E) (1982).

[15]I. Daubechies and E. H. Lieb, Commun. Math. Phys. 90, 511 (1983).

[16]E. H. Lieb and B. Simon, J. Chem. Phys. 61, 735 (1974); Commun. Math. Phys. 53, 185 (1977). Theorem 2.4 states, inter alia, that $\epsilon_1, \ldots, \epsilon_N$ are the N lowest points of the spectrum of h. This is correct, but the proof is wrong. A correct proof is given in the earlier summary: E. H. Lieb, in *Proceedings of the International Congress of Mathematicians, Vancouver, 1974,* Vol. 2, pp. 383–386.

Approximate Neutrality of Large-Z Ions*

Elliott H. Lieb[1], Israel M. Sigal[2], Barry Simon[3] and Walter Thirring[4]

1 Departments of Mathematics and Physics, Princeton University, Princeton, NJ 08544, USA
2 Department of Mathematics, University of Toronto, Toronto, Canada M5S 1A1
3 Division of Physics, Mathematics and Astronomy, California Institute of Technology, Pasadena, CA 91125, USA
4 Institute for Theoretical Physics, University of Vienna, Vienna, Austria

Abstract. Let $N(Z)$ denote the number of electrons which a nucleus of charge Z can bind in non-relativistic quantum mechanics (assuming that electrons are fermions). We prove that $N(Z)/Z \to 1$ as $Z \to \infty$.

1. Introduction

This paper is a contribution to the exact study of Coulombic binding energies in quantum mechanics. Let $H(N,Z)$ denote the Hamiltonian

$$H(N,Z) = \sum_{i=1}^{N} (-\Delta_i - Z|x_i|^{-1}) + \sum_{i<j} |x_i - x_j|^{-1},$$

and let $E(N,Z)$ denote its minimum over all fermion states (we suppose there are two spin states allowed, although any fixed number could be accommodated). For comparison purpose, we let $E_b(N,Z)$ denote the same minimum, but over all states (taken on a totally symmetric wave function, hence b for boson).

It is a fundamental result of Ruskai [9] for bosons, and Sigal [11] for fermions (see also Ruskai [10]) that there exists $N(Z), N_b(Z)$ so that, for all $j = 0, 1, \ldots,$

$$E(N(Z),Z) = E(N(Z)+j, Z); \quad E_b(N_b(Z),Z) = E_b(N_b(Z)+j, Z).$$

We let $N(Z)$ (respectively $N_b(Z)$) denote the smallest number for which the first (respectively second) equality holds for all j. Sigal [12] showed that

$$\overline{\lim}\,[N(Z)/Z] \leqq 2, \quad \lim\,[\ln N_b(Z)/\ln Z] \leqq 1, \tag{1.1}$$

and then Lieb [6, 7] proved the bounds

$$N(Z) < 2Z+1, \quad N_b(Z) < 2Z+1 \tag{1.2}$$

which implies, in particular, that a doubly ionized hydrogen atom is unstable.

* Research partially supported by the NSERC under Grant NA7901 and by the USNSF under Grants DMS-8416049 and PHY 85-15288-A01

Zhislin [15] proved that $N(Z) \geqq Z$ and $N_b(Z) \geqq Z$. A more detailed review of the history and status of this and related problems is given in [7].

Our main goal in this paper is to show that

Theorem 1.1. $\lim_{Z \to \infty} [N(Z)/Z] = 1$.

That is, asymptotically, the excess charge in negative ions is a small fraction of the total charge. While this is physically reasonable, and partially captures the observed fact that in nature there are no highly negative ions, it is not as "obvious" as it might appear at first. For Benguria and Lieb [1] have shown that

$$\lim N_b(Z)/Z > 1.$$

(They actually prove that lim is at least the critical charge for the Hartree equation which is rigorously known to lie between 1 and 2, and numerically [16] is about 1.2.) Thus, the Pauli principle will enter into our proof of Theorem 1.1.

Part of our argument closely follows that in Sigal [12] (see also Cycon et al. [13]). We differ from Sigal in one critical aspect. He gets a factor of 2 in (1.1) by using the obvious fact that if one has $2Z+1$ electrons surrounding a nucleus of charge Z, one can always gain classical energy by taking the electron farthest from the nucleus to infinity. We will exploit the fact that if you have $Z(1+\varepsilon)$ electrons and Z is large, classical energy can be gained by taking *some* electron to infinity. We will prove precisely this fact in Sect. 3. Actually, for technical reasons, we will need a slightly stronger result, also proven there. Unfortunately, our proof of this key classical fact is by contradiction, using a compactness result. Hence our proof is non-constructive, which means that we have no estimates on how large Z has to be for $N(Z)/Z$ to be bounded by $1+\varepsilon$ for any given ε.

The theorem we prove in Sect. 3 says that, with $N = Z(1+\varepsilon)$ point electrons, one gains classical energy by taking some electron to infinity. We prove this by appealing to an analogous result for a "fluid" of negative charge: if $\int d\rho(x) \geqq Z(1+\varepsilon)$, then for some x_0 in supp ρ one has $Z|x_0|^{-1} - \int |x_0 - y|^{-1} d\rho(y) \geqq 0$. This fact is proven in Sect. 2. It is interesting that our results in quantum potential theory require various results in classical potential theory.

In Sect. 4, we construct a partition of unity in $\mathbb{R}^{3N}$-the result of Sect. 3 is needed only to assure that certain sets over $\mathbb{R}^{3N}$. Given this partition, the actual proof of Theorem 1.1. in Sect. 5 follows Sigal [12]. Section 6 provides some additional remarks.

2. Classical Continuum Theorem

Theorem 2.1. *Let ρ be a nonzero finite (positive) measure on $\mathbb{R}^3$ which is not a point mass at 0, and let ϕ_ρ be its potential,* i.e.,

$$\phi_\rho(x) = \int |x-y|^{-1} d\rho(y). \tag{2.1}$$

Then, for any $\varepsilon > 0$, the set of points $x \neq 0$ such that

$$\phi_\rho(x) \geqq (1-\varepsilon)|x|^{-1} \rho(\mathbb{R}^3) \tag{2.2}$$

has positive ρ measure.

Proof. Let T_ε denote the set of points in $\mathbb{R}^3\backslash\{0\}$ such that (2.2) holds. Our goal is to show that $\rho(T_\varepsilon) > 0$. First, let us eliminate any possible small point mass at 0 by defining $\tilde{\rho} = \rho - c\delta(x)$. For some $0 \leqq c < 1, \tilde{\rho}(\{0\}) = 0$. Additionally, defining $\tilde{\varepsilon} = \varepsilon\rho(\mathbb{R}^3)/\tilde{\rho}(\mathbb{R}^3)$, one sees that proving the theorem for ε and ρ is equivalent to proving it for $\tilde{\varepsilon}$ and $\tilde{\rho}$ with $(1-\tilde{\varepsilon})\rho(\mathbb{R}^3) > c$. It suffices to assume, therefore, that $\tilde{\varepsilon} = \varepsilon, \tilde{\rho} = \rho$ and $\rho(\{0\}) = 0$, and we shall do so henceforth.

Let B denote the set of $x \neq 0$ for which $\phi_\rho(x) = \infty$. If $\rho(B) > 0$ we are trivially done, so assume $\rho(B) = 0$. Since $\rho(\{0\}) = 0$, this means that ϕ_ρ is finite ρ-a.e. and we can apply Baxter's Theorem 2 [17]. If we define the measure $\mu = (1-\varepsilon)\{\int d\rho\}\delta(x)$, this theorem asserts the existence of a (positive) measure γ such that

(a) $\gamma \leqq \rho$ and $\gamma(\mathbb{R}^3) \geqq \rho(\mathbb{R}^3) - \mu(\mathbb{R}^3) = \varepsilon \int d\rho > 0$,
(b) $\phi_\rho(x) = \phi_\gamma(x) + \phi_\mu(x)$ γ.a.e..

(In Baxter's notation, $\rho = \nu, \rho - \gamma = \lambda$ and $\mu = \mu$.) Thus, $\rho(T_\varepsilon) \geqq \gamma(T_\varepsilon) = \gamma(\mathbb{R}^3) > 0$. □

Remark. One can also prove this theorem by appealing to Choquet's theorem [2, 5].

3. Classical-Discrete Theorem

Theorem 3.1. *For any ε, there exists N_0 so that, for all sets $\{\vec{x}_a\}_{a=1}^N$ of $N \geqq N_0$ points, we have*

$$\max_b \left\{ \sum_{a \neq b} \frac{1}{|\vec{x}_a \rightarrow \vec{x}_b|} - \frac{(1-\varepsilon)N}{|\vec{x}_b|} \right\} \geqq 0.$$

Remarks. This is clearly a classical analog of the quantum theorem that we are seeking. It says that if the electron excess over the nuclear charge above Z is more than $\varepsilon(1-\varepsilon)^{-1}Z$, then one gains energy by moving at least one of the electrons off to infinity.

2. Unfortunately, our proof is by contradiction, and therefore non-constructive. The fact that we cannot make our estimates explicit, even in principle, comes from this fact.

Proof. Suppose not. Then, there is $\varepsilon_0 > 0$ and sequences points $\{x_a^{(n)}\}_{a=1}^{N_n}$ with $N_n \rightarrow \infty$ and

$$\sum_{a \neq b} \frac{1}{|x_b^{(n)} - x_a^{(n)}|} < \frac{(1-\varepsilon_0)N_n}{|x_b^{(n)}|} \tag{3.1}$$

for all n and all $1 \leqq b \leqq N_n$.

Equation (3.1) is invariant under rotations and scaling of the x's as well as relabelling. Thus, without loss we can suppose that

$$x_1^{(n)} = (1, 0, 0) = x_0,$$
$$|x_a^{(n)}| \leqq 1 \quad \text{all} \quad 1 \leqq a \leqq N_n.$$

The measures

$$\rho^{(n)} = N_n^{-1} \sum_{a=1}^{N_n} \delta(x - x_a^{(n)})$$

are probability measures on the unit ball. Thus, by passing to a subsequence if necessary, we can suppose that ρ_n converges in the $C(\mathbb{R}^3)$-weak topology to a probability measure $d\rho$. We will show that $d\rho$ violates Theorem 2.1. If y is a limit point of $x_{i_n}^{(n)}$ and $g(z)=(|z|^2+M^2)^{-1/2}$, then since g is C^1 with bounded derivatives

$$\lim \int g(x-x_i^{(n)})d\rho_n(x)=\int g(x-y)d\rho(x).$$

Thus, by (3.1):

$$\int g(x-y)d\rho(x) \leqq (1-\varepsilon_0)|y|^{-1}.$$

By the monotone convergence theorem, we can take M to zero to obtain

$$\int |x-y|^{-1}d\rho(x) \leqq (1-\varepsilon)|y|^{-1}. \tag{3.2}$$

We have just proven (3.2) for any y in the limit set of the $\{x_a^{(n)}\}$. Any $y \in \operatorname{supp} \rho$ is such a limit point so (3.2) holds for all $y \in \operatorname{supp} \rho$. Thus we will have a contradiction with Theorem 2.1 if we show that $\rho \neq \delta_0$. But since $x_1^{(n)}=x_0$, we have (3.2) for $y=x_0$, i.e.,

$$\int |x-x_0|^{-1}d\rho(x) \leqq (1-\varepsilon). \tag{3.3}$$

Since, for $d\rho=\delta_0$, the left side is 1, we can conclude that $d\rho \neq \delta_0$. □

We will actually need an extension of Theorem 3.1 to potentials cut off at short distances, but in a way that may seem unnatural at first. Define

$$G_\alpha(x,y)=\begin{cases} |x-y|^{-1} & \text{if } |x-y| \geqq \alpha|x| \\ \alpha^{-1}|x|^{-1} & |x-y| \leqq \alpha|x|. \end{cases} \tag{3.4}$$

For a set of points $\{x_a\}_{a=1}^N$ in $\mathbb{R}^3$, define $|x|_\infty \equiv \sup_a |x_a|$.

Theorem 3.2. *Let $\alpha_N \to 0$ as $N \to \infty$. Then, for any ε, there exists N_0 and $\delta>0$ so that, for any $N \geqq N_0$ and any set of points $\{x_a\}_{a=1}^N$, there is a point x_a with $|x_a| \geqq \delta|x|_\infty$ and*

$$\sum_{j \neq a} G_{\alpha_N}(x_a,x_j) \geqq \frac{(1-\varepsilon)N}{|x_a|}. \tag{3.5}$$

Proof. G is defined to be invariant under scaling (which is why we took the cutoff to be $\alpha|x|$, not just α) and rotations. Also, the condition $|x_a| \geqq \delta|x|_\infty$ has the same invariance. Thus, if the result is false, we can find $\varepsilon_0>0$, $\delta_N \to 0$ and a sequence with $|x|_\infty=1$; $x_1^{(n)}=(1,0,0)$ so that (3.5) fails. Taking the limit, we get the same contradiction as in the proof of Theorem 3.1. □

4. A Partition of Unity

As noted in Sect. 1, the key element in the proof of Sigal, which we will mimic, is the construction of a partition of unity. Here we will construct such a partition which we will use in the next section. The preliminaries in the last section will be relevant precisely in order to be sure that certain sets cover $\mathbb{R}^{3N}$.

Theorem 4.1. *For all $\varepsilon>0$, there exists N_0 and $\delta>0$, and for each $N \geqq N_0$ and each $R>0$, a family $\{J_a\}_{a=0}^N$ of C^∞ functions on $\mathbb{R}^{3N}$ so that:*

(1) J_0 *is totally symmetric,* $\{J_a\}_{a \neq 0}$ *is symmetric in* $\{x_b\}_{b \neq a}$.

(2) $\sum_a J_a^2 = 1$.

(3) $\operatorname{supp} J_0 \subset \{\{x_a\} \mid |x|_\infty < R\}$.

(4) $\operatorname{supp} J_a \subset \{x \mid |x|_\infty \geqq (1-\varepsilon)R,\ |x_a| \geqq (1-2\varepsilon)\delta |x|_\infty$ *and* $\sum_{b \neq a} |x_b - x_a|^{-1} \geqq (1-2\varepsilon)N|x_a|^{-1}\}$.

(5) *For a constant C, depending only on* ε, $\sum_a |\nabla J_a|^2 \leqq CN^{1/2}(\ln N)^2 |x|_\infty^{-1} R^{-1}$.

$$(4.1)$$

Proof. Without loss, we can take $R = 1$ since the result for $R = 1$ implies the result for all R by scaling. Moreover, we can prove (4.1) with $|x|_\infty^{-1}$ replaced by $|x|_\infty^{-2}$. For the left-hand side of (4.1) (when $R = 1$) is supported in the region where $|x|_\infty \geqq (1-\varepsilon)$, and in that region $|x|_\infty^{-2} \leqq (1-\varepsilon)^{-1}|x|_\infty^{-1}$.

Next, we note that instead of finding J_a's obeying (1)–(5), it suffices to find F_a's obeying (1′), (2′), (3), (4) and (5′):

(2′) $\sum_a F_a^2 \geqq \frac{1}{2}$,

(5′) $\left(\sum_a (\nabla F_a)^2\right) \Big/ \left(\sum_a F_a^2\right) \leqq CN^{1/2}(\ln N)^2 |x|_\infty^{-2}$,

and for any permutation π:

(1′) $F_{\pi(a)}(x_{\pi(1)}, \ldots, x_{\pi(N)}) = F_a(x_1, \ldots, x_N)$.

For if $J_a = F_a \Big/ \left(\sum_a F_a^2\right)^{1/2}$, then J_a has the same symmetry and support properties as F has, and

$$\sum_a (\nabla J_a)^2 \leqq \sum_a (\nabla F_a)2 / \sum F_a^2. \tag{4.2}$$

To understand (4.2), think of $F = (F_0, \ldots, F_N)$ as a function from $\mathbb{R}^{3N}$ to $\mathbb{R}^{N+1}$, in which case J is the "angular" part of F, and (4.2) is a standard inequality on the gradient of the angular part.

Now we concentrate on constructing the F_a's. Let ψ be a C^∞ function on $[0, \infty)$ with

$$\begin{aligned} \psi(y) &= 0 \qquad && y \leqq 1 - 2\varepsilon \\ &= 1 && y \geqq 1 - \varepsilon \\ &\in [0,1] && \text{all } y \end{aligned}$$

and define $\varphi = \psi^2$. Let $\alpha_N \equiv (\ln N)^{-1}$ and choose N_0, δ as given by Theorem 3.2. As a preliminary, take F_0, F_a as follows. These functions are not C^∞, but are continuous and are C^1 off the set $\{x \mid |x_a| = |x_b|$ for some $a \neq b\} \cup \{x \mid |x_a - x_b| = \alpha_N |x_a|$ some $a, b\}$ with discontinuities of the gradients allowed on that set:

$$F_0(x) = 1 - \psi(|x|_\infty),$$

$$F_a(x) = \varphi(|x|_\infty)\varphi(|x_a|/\delta|x_\infty|)\varphi\left(N^{-1}|x_a| \sum_{b \neq a} G_{\alpha_N}(x_a, x_b)\right),$$

where G_α is given by (3.3). The symmetry condition (1) is obvious, and (3) holds since $\varphi(y)=1$ if $y \geqq 1$.

$F_a \neq 0$ implies that $|x|_\infty \geqq 1-2\varepsilon$, $|x_a| \geqq (1-\varepsilon)\delta|x|_\infty$, and

$$\sum_{b \neq a} G_{\alpha_N}(x_a, x_b) \geqq (1-2\varepsilon)N|x_a|^{-1},$$

since $\varphi(y) \neq 0$ implies $|y| \geqq 1-2\varepsilon$. Since $|x-y|^{-1} \geqq G_\alpha(x,y)$, we have proven (4). That leaves the key conditions (5′) and (2′).

For (2′), we use Theorem 3.2. This guarantees us that there is an a where the last two factors in F_a are 1, and so there is an a with $F_a(x)=\varphi(|x|_\infty)$. Since

$$\theta^2 + (1-\theta)^2 \geqq \tfrac{1}{2}$$

for all θ, for this a, $F_0(x)^2 + F_a(x)^2 \geqq \frac{1}{2}$, proving (2′).

As a preliminary to (5′), we want to note that, for some constant C and all γ:

$$|\varphi'|^2 \leqq \gamma + C\gamma^{-1}|\varphi|^2. \tag{4.3}$$

For

$$|\varphi'|^2 = 4|\psi|^2|\psi'|^2 \leqq \gamma + 4\gamma^{-1}|\psi|^4|\psi'|^4,$$

proving (4.3) with $C = 4\|\psi'\|_\infty^4$.

Away from points where $|x_a| = |x_b|$ for some $a \neq b$,

$$\nabla\varphi(|x|_\infty) = \varphi'(|x|_\infty)\nabla|x|_\infty.$$

Since $|x|_\infty$ is some $|x_a|$ and

$$\nabla_a|x_b| = \delta_{ab}x_a/|x_a|,$$

we see that

$$|\nabla\varphi(|x|_\infty)| = \varphi'(|x|_\infty). \tag{4.4}$$

Similarly,

$$\nabla\varphi(|x_a|/\delta|x|_\infty) = \varphi'(|x_a|/\delta|x|_\infty)\left[-\frac{|x_a|}{\delta|x|_\infty^2}\nabla|x|_\infty + \delta^{-1}|x|_\infty^{-1}\nabla|x_a|\right]. \tag{4.5}$$

Finally, if

$$\eta_a = \varphi(N^{-1}|x_a|\sum_{b \neq a} G_{\alpha_N}(x_a, x_b)),$$

then, for $b \neq a$

$$|\nabla_b\eta_a| \leqq \varphi'(N^{-1}|x_a|\sum_{b \neq a} G_{\alpha_N}(x_a, x_b))N^{-1}|x_a||G_{\alpha_N}(x_a, x_b)|^2,$$

since

$$\nabla_b G_\alpha = [G_\alpha^2](x_b - x_a)/|x_b - x_a| \quad \text{or} \quad 0.$$

Recall that $G_\alpha(x,y) \leqq \alpha^{-1}|x|^{-1}$, so since $\alpha_N^{-1} = \ln N$,

$$\sum_{b \neq a}|\nabla_b\eta_a|^2 \leqq \left[\varphi'\left(N^{-1}|x_a|\sum_{b \neq a} G_{\alpha_N}(x_a, x_b)\right)\right]^2 N^{-2}|x_a|^2(\ln N)^3|x_a|^{-3}\sum_{b \neq a} G_{\alpha_N}(x_a, x_b).$$

However, $\operatorname{supp}\varphi' \subset [1-2\varepsilon, 1-\varepsilon]$, so on $\operatorname{supp}\varphi'\left(N^{-1}|x_a|\sum_{b \neq a} G_{\alpha_N}(x_a, x_b)\right)$, we have that

$$\sum_{b \neq a} G_{\alpha_N}(x_a, x_b) \leqq |x_a|^{-1} N.$$

Thus

$$\sum_{b \neq a} |\nabla_b \eta_a|^2 \leqq N^{-1} (\ln N)^3 |x_a|^{-2} \left[\varphi'\left(N^{-1} |x_a| \sum_{b \neq a} G_{\alpha_N}(x_a, x_b) \right) \right]^2. \tag{4.6}$$

As a final gradient estimate,

$$|\nabla_a \eta_a| \leqq \varphi'\left(N^{-1} |x_a| \sum_{b \neq a} G_{\alpha_N}(x_a, x_b) \right) \left[N^{-1} \sum_{b \neq a} G_{\alpha_N}(x_a, x_b) + N^{-1} \sum_{b \neq a} |x_a| G_{\alpha_N}(x_a, x_b)^2 \right],$$

since $|\nabla_a |x_a|| = 1$ and

$$|\nabla_a G_\alpha(x_a, x_b)| = G_\alpha(x_a, x_b)^2 \quad \text{or} \quad \alpha G_\alpha(x_a, x_b)^2$$

and $\alpha \leqq 1$. Thus, since $|x_a| \sum_{b \neq a} G_{\alpha_N}(x_a, x_b) \leqq N$ when $\varphi'(\cdot) \neq 0$,

$$|\nabla_a \eta_a| \leqq \varphi'\left(N^{-1} |x_a| \sum_{b \neq a} G_{\alpha_N}(x_a, x_b) \right) (1 + (\ln N)) |x_a|^{-1}. \tag{4.7}$$

Since $|x_a| \geqq (1 - 2\varepsilon)\, \delta |x|_\infty$ on supp F_a, we see that, for $a \neq 0$.

$$\sum_b |\nabla_b F_a|^2 \leqq c_1 (\gamma + c_2 \gamma^{-1} F_a^2)(\ln N)^2 |x|_\infty^{-2},$$

by using (4.3)–(4.7). Thus

$$\sum_{a,b} |\nabla_b F_a|^2 \leqq c_1 \left(\gamma N + c_2 \gamma^{-1} \sum_a F_a^2 \right) (\ln N)^2 |x|_\infty^{-2} + c_3 |x|_\infty^{-2}.$$

Since $\sum F_a^2 \geqq \frac{1}{2}$, we can take $\gamma = N^{-1/2}$ and obtain:

$$\sum_{a,b} |\nabla_b F_a|^2 \leqq c_4 N^{1/2} (\ln N)^2 |x|_\infty^{-2} \sum_a F_a^2,$$

as required.

The J's constructed in this way are continuous but are only piecewise C^1. By convoluting with a smooth, totally symmetric function of very small support, we can arrange for C^∞ J's which still obey the required properties. □

Remark. By using $\varphi = \psi^m$ in the above construction for m suitable, we can reduce $N^{1/2}$ to any desired positive power of N.

5. The Main Theorem

Here we will prove Theorem 1.1. Given the construction in the last section, this follows Sigal [12] fairly closely. Pick $\varepsilon > 0$. We shall prove $\overline{\lim}\, N(Z)/Z \leqq (1 - 3\varepsilon)^{-1}$.

Let $\{J_a\}$ be as in Theorem 4.1, and let

$$L = \sum_{a=0}^{N} |\nabla J_a|^2.$$

By the IMS localization formula (see Chap. 3 of [3]),

$$H = \sum_{a=0}^{N} J_a H J_a - L = \sum_{a=0}^{N} J_a (H - L) J_a. \tag{5.1}$$

By condition (5) of Theorem 4.1,

$$L \leqq C N^{1/2} (\ln N)^2 |x|_\infty^{-1} R^{-1}.$$

For $a > 0$, $\operatorname{supp} J_a \subset \{x \,|\, |x_a| \geqq (1-2\varepsilon)\delta |x|_\infty\}$, and thus:

$$J_a L J_a \leqq c_1 N^{1/2} (\ln N)^2 |x_a|^{-1} R^{-1} \quad (C_1 = c\delta^{-1}(1-2\varepsilon)^{-1}). \tag{5.2}$$

Since $\cup \operatorname{supp}(\nabla J_a) \subset \{x \,|\, |x|_\infty \geqq (1-2\varepsilon) R\}$:

$$J_0 L J_0 \leqq c_2 N^{1/2} (\ln N)^2 R^{-2} \quad (c_2 = (1-2\varepsilon)^{-1} C) \tag{5.3}$$

Let $H_a(N-1, Z)$ be the $(N-1)$ electron Hamiltonian obtained by removing from $H(N,Z)$ all terms involving x_a, so:

$$H(N,Z) = H_a(N-1,Z) - \Delta_a - |x_a|^{-1} Z + \sum_{b \neq a} |x_b - x_a|^{-1}.$$

Since $H_a(N-1,Z) \geqq E(N-1,Z)$ and $-\Delta_a \geqq 0$, taking into account (5.2) and the support property of J_a, we have that

$$J_a (H(N,Z) - L) J_a \geqq J_a [E(N-1,Z) + |x_a|^{-1} d(Z,N,R)] J_a, \tag{5.4a}$$

where

$$d(Z,N,R) = -Z - c_1 N^{1/2} (\ln N)^2 R^{-1} + (1-2\varepsilon) N. \tag{5.4b}$$

R_N has not been introduced up to now.

By solving for a Bohr atom (and this is where the Pauli principle enters):

$$\sum_{i=1}^{N} (-\Delta_i - Z|x_i|^{-1}) \geqq -c_3 Z^2 N^{1/3},$$

so since $|x_a - x_b| \leqq 2R$ on $\operatorname{supp} J_0$:

$$J_0 (H(N,Z) - L) J_0 \geqq J_0 [-c_3 Z^2 N^{1/3} - c_2 N^{1/2} (\ln N)^2 R^{-2} + \tfrac{1}{4} R^{-1} N(N-1)] J_0. \tag{5.5}$$

Choose $R = N^{-2/5}$. Then, for $N \geqq (1-3\varepsilon)^{-1} Z$ and large Z, $d(Z,N,R) > 0$ since $\frac{1}{2} + \frac{2}{5} < 1$. Moreover, $J_0(H(N,Z) - L) J_0 \geqq 0 \geqq J_0 E(N-1,Z) J_0$ since $N^{12/5}$ dominats $N^{7/3}$ and $N^{13/10} (\ln N)^2$ and $E(N-1,Z) \leqq 0$.

Thus,

$$H(H,Z) \geqq E(N-1,Z)$$

if $N \geqq (1-3\varepsilon)^{-1} Z$ and Z is large, i.e., for Z large

$$N(Z) \leqq (1-3\varepsilon)^{-1} Z.$$

Since ε is arbitrary:

$$\overline{\lim}\, N(Z)/Z \leqq 1.$$

It is well known (see [15, 13]) that $H(Z,Z)$ has bound states, i.e., that $N(Z) \geqq Z$. □

Remark. Without the Pauli principle, $Z^2 N^{1/3}$ becomes $Z^2 N$, so one must take $R_N = cN^{-1}$, in which case the localization term $N^{1/2}(\ln N)^2 R_N^{-1}$ in (5.4b) becomes uncontrollable. Our proof must, of course, fail without the Pauli principle because of the result in [1].

6. Extensions

Our result extends easily to accommodate arbitrary magnetic fields (the same for all electrons) and/or a finite nuclear mass.

The exact form of the electron kinetic energy entered only in two places: in the IMS localization formula and in the positivity of $-\Delta$, both of which hold in an arbitrary magnetic field. We also used the Bohr atom binding energy, but that only decreases in a magnetic field (i.e., $-c_3 N^2 Z^{1/3}$ is a lower bound for all fields). Thus, we obtain a magnetic field independent bound $\tilde{N}(Z)$ with

$$\tilde{N}(Z)/Z \to 1 \quad \text{as} \quad Z \to \infty.$$

As for finite nuclear mass, let x_0 be the nuclear coordinate, and use $J_a(x_b - x_0)$ in place of $J_a(x_b)$. With this change, the nuclear coordinates pass through all proofs with essentially no change at all.

Acknowledgement. This work was begun while I.S. was at the Weizmann Institute, and B.S. would like to thank H. Dym and I. Sigal for the hospitality of that Institute. W.T. would like to thank E. Lieb for the hospitality of Princeton University, and M. Goldberger and R. Vogt for the hospitality of Caltech. An announcement appeared in [8].

References

1. Benguria, R., Lieb, E.: Proof of the stability of highly negative ions in the absence of the Pauli principle. Phys. Rev. Lett. **50**, 1771 (1983)
2. Choquet, G.: Sur la fondements de la théorie finie du potential. C.R. Acad. Sci. Paris **244**, 1606 (1957)
3. Cycon, H., Froese, R., Kirsch, W., Simon, B.: Schrödinger operators with application to quantum mechanics and global geometry. Berlin, Heidelberg, New York: Springer 1987
4. Evans, G.: On potentials of positive mass, I. Trans. AMS **37**, 226 (1935)
5. Helms, L.: Introduction to potential theory. New York: Wiley 1966
6. Lieb, E.: Atomic and molecular ionization. Phys. Rev. Lett. **52**, 315 (1984)
7. Lieb, E.: Bound on the maximum negative ionization of atoms and molecules. Phys. Rev. **A29**, 3018–3028 (1984)
8. Lieb, E., Sigal, I. M., Simon, B., Thirring, W.: Asymptotic neutrality of large-*Z* ions. Phys. Rev. Lett. **52**, 994 (1984)
9. Ruskai, M.: Absence of discrete spectrum in highly negative ions. Commun. Math. Phys. **82**, 457–469 (1982)
10. Ruskai, M.: Absence of discrete spectrum in highly negative ions, II. Commun. Math. Phys. **85**, 325–327 (1982)

11. Sigal, I. M.: Geometric methods in the quantum many-body problem. Nonexistence of very negative ions. Commun. Math. Phys. **85**, 309–324 (1982)
12. Sigal, I. M.: How many electrons can a nucleus bind? Ann. Phys. **157**, 307–320 (1984)
13. Simon, B.: On the infinitude or finiteness of the number of bound states of an N-body quantum system, I. Helv. Phys. Acta **43**, 607–630 (1970)
14. Vasilescu, F.: Sur la contribution du potential á traverse des masses et la démonstration d'une lemme de Kellogg. C.R. Acad. Sci. Paris **200**, 1173 (1935)
15. Zhislin, G.: Discussion of the spectrum of Schrödinger operator for systems of many particles. Tr. Mosk. Mat. Obs. **9**, 81–128 (1960)
16. Baumgartner, B.: On Thomas–Fermi–von Weizsäcker and Hartree energies as functions of the degree of ionisation. J. Phys. **A17**, 1593–1602 (1984)
17. Baxter, J.: Inequalities for potentials of particle systems, Ill. J. Math. **24**, 645–652 (1980)

Communicated by A. Jaffe

Received December 7, 1987

Universal nature of van der Waals forces for Coulomb systems

Elliott H. Lieb
Department of Mathematics and Department of Physics, Princeton University, Jadwin Hall, P.O. Box 708, Princeton, New Jersey 08544

Walter E. Thirring
Institut für Theoretische Physik, Universität Wien, Boltzmanngasse 5, A-1090 Vienna, Austria
(Received 2 December 1985)

The nonrelativistic Schrödinger equation is supposed to yield a pairwise R^{-6} attractive interaction among atoms or molecules for large separation, R. Up to now this attraction has been investigated only in perturbation theory or else by invoking various assumptions and approximations. We show rigorously that the attraction is at least as strong as R^{-6} for any shapes of the molecules, independent of other features such as statistics or sign of charge of the particles. More precisely, we prove that two neutral molecules can always be oriented such that the ground-state energy of the combined system is less than the sum of the ground-state energies of the isolated molecules by a term $-cR^{-6}$ provided R is larger than the sum of the diameters of the molecules. When several molecules are present, a pairwise bound of this kind is derived. In short, we prove that in the quantum mechanics of Coulomb systems everything binds to everything else if the nuclear motion is neglected.

I. INTRODUCTION

Our purpose is to explore and to answer some elementary but fundamental questions about the binding of neutral atoms and molecules. To simplify matters, we shall use the infinite nuclear-mass approximation in which the nuclei are held fixed, but many of our mathematical constructions can, with additional work and appropriate changes, be carried over to the more realistic case of dynamic nuclei. In our fixed-nuclei approximation we do *not* assume that the nuclei are necessarily in the configuration that minimizes the energy of the molecule.

Consider two neutral molecules (or atoms) labeled α and β, with respective diameters $2r^\alpha$ and $2r^\beta$, and whose centers are separated by a distance $R^{\alpha\beta}>r^\alpha+r^\beta$. (The precise definition of r^α, r^β, and $R^{\alpha\beta}$ will be given in Sec. III.) Let the ground-state energies of the isolated molecules and of the combined system be e^α, e^β, and $e(R^{\alpha\beta})$, respectively. The question we shall address is this: Is it possible to orient the nuclear coordinates at the two neutral molecules with respect to each other (with $R^{\alpha\beta}$ fixed) so that after an appropriate readjustment of the electronic wave function

$$e(R^{\alpha\beta})<e^\alpha+e^\beta\,? \tag{1.1}$$

In particular, is there an upper bound of the van der Waals form

$$e(R^{\alpha\beta})\leq e^\alpha+e^\beta-C(R^{\alpha\beta})^{-6} \tag{1.2}$$

for a constant $C>0$ which depends on the intrinsic properties of the two molecules, but not on $R^{\alpha\beta}$?

We shall prove, using a variational argument, that Eq. (1.2) is true; in other words, we prove that "everything binds to everything else" when the nuclear kinetic energy is neglected. First, several remarks are in order.

(1) Equation (1.2), or even Eq. (1.1), implies binding in the fixed-nuclei approximation. When the nuclear kinetic energy is added, the uncertainty principle may destroy the binding, as is probably the case for He_2. Thus we can only say that sufficiently heavy isotopes will always bind.

(2) Density-functional theories (at least the ones known to us) fail to predict Eq. (1.2). Although a density functional that predicts Eq. (1.2) exists in principle,[1,2] no one has actually constructed one. In Thomas-Fermi theory even Eq. (1.1) fails because Teller's theorem[3] states that in Thomas-Fermi theory $e(R^{\alpha\beta})>e^\alpha+e^\beta$ always. When gradient corrections are added, as in Thomas–Fermi–von Weizsäcker theory, Eq. (1.1) holds[3] when $R^{\alpha\beta}\approx r^\alpha+r^\beta$ but Eq. (1.2) fails when $R^{\alpha\beta}\gg r^\alpha+r^\beta$. The reason for this failure of (local) density-functional theory (as explained in Refs. 2 and 3) is the following.

The R^{-6} attraction comes from a dipole-dipole interaction but (in the combined system) there is almost no static dipole moment in each molecule (if both molecules were free of static dipole moments in their ground states). The interaction energy of dipole moments $\mathbf{d}^\alpha$ and $\mathbf{d}^\beta$ on the respective molecules is proportional to $-d^\alpha d^\beta(R^{\alpha\beta})^{-3}$. Density-functional theory, since it deals only with single-particle densities, can produce these only as static moments and at an energy cost of $\frac{1}{2}c^\alpha(d^\alpha)^2+\frac{1}{2}c^\beta(d^\beta)^2$. Thus, when $R^{\alpha\beta}>(c^\alpha c^\beta)^{1/6}$ the optimum choice is $d^\alpha=d^\beta=0$ and there is no attraction.

The true source of the R^{-6} term in Eq. (1.2) is a correlation effect between the electrons in molecule α and those in molecule β. It is essential to think of electrons as particles and not as a simple fluid. In the language of quantum mechanics the molecules make a virtual transition (simultaneously and not separately) to an excited state. The energy to create d^α and d^β is then (from second-order perturbation theory) proportional to $(d^\alpha d^\beta)^2$. The minimum with respect to d^α and d^β for fixed $R^{\alpha\beta}$ of

$(d^\alpha d^\beta)^2 - d^\alpha d^\beta (R^{\alpha\beta})^{-3}$ is the required $-(R^{\alpha\beta})^{-6}$. Apart from the fact that quantum mechanics is needed in order to give meaning to the concept of the ground state, the effect is classical insofar as no interference effect is involved. There is simply a coherence in the motion of the electrons in each molecule so that the (time) average dipole-dipole interaction is not zero even though the average dipole moment of each molecule is essentially zero.

(3) When one studies the question of the additivity of the van der Waals forces between pairs of molecules the correlation effect appears even more strikingly. The Coulomb interaction correlates the dipole moments $\mathbf{d}^\alpha$ and $\mathbf{d}^\beta$ with the displacement $\mathbf{R}^{\alpha\beta}$ between them in such a way as to give the directional factor $-3(\mathbf{d}^\alpha\cdot\mathbf{R}^{\alpha\beta})(\mathbf{d}^\beta\cdot\mathbf{R}^{\alpha\beta})+(\mathbf{d}^\alpha\cdot\mathbf{d}^\beta)\,|\mathbf{R}^{\alpha\beta}|^2$ as function of direction its minimal value, $-2\,|\mathbf{d}^\alpha|\,|\mathbf{d}^\beta|\,|\mathbf{R}^{\alpha\beta}|^2$, *simultaneously* for all pairs. Thus the nonstatic dipole moments $\mathbf{d}^\alpha$ do not depend on the *single* molecules, but only on the *pairs* of molecules which interact with each other. As a consequence, our bounds on the effective interaction potentials will add together like scalar potentials and not like dipole potentials. This does not, of course, imply that the true effective interaction potential has this property.

(4) The analysis in remark (2) was based on the Coulomb potential, but presumably this is not essential. If the electron-electron interaction were r^{-p} instead of r^{-1}, we should expect the appropriate modification of Eq. (1.2) would have R^{-2p-4} in place of R^{-6}. We shall not pursue this aspect of the problem, however, and will confine our attention to the Coulomb potential.

(5) It is not at all essential that the dynamic particles are electrons. They could be any mixture of bosons and fermions. Also, for example, matter and antimatter will bind in the infinite nuclear-mass approximation—which carries the physical implication that there is no quantum-mechanical Coulomb barrier to the annihilation process. There are several variational calculations of the dependence of the energy on the nuclear separation, but with different conclusions.[4–6]

(6) No assumption is made about the spherical symmetry of the two molecules and they could have permanent electric dipole or higher-pole moments in their ground states (but not monopole moments). Parity conservation does not preclude this since the nuclear coordinates are fixed. A feature of Eq. (1.2) is that it is independent of any assumption about the permanent moments; if any exist then the binding could be stronger, but not weaker than R^{-6}.

(7) There is, of course, an enormous amount of literature about van der Waals forces (see, e.g., Refs. 7–9). In a certain sense our results are thus not new, but from another point of view they are new. The drawback to the usual theories is that they are always based on perturbation theory in two ways: (i) One assumes that R is sufficiently large so that the $1/r$ Coulomb interaction can be expanded up to the dipole-dipole order, and all higher terms ignored. Although it can be shown that this expansion is asymptotically exact,[10] we are usually not told how large this R has to be. (ii) One uses second-order quantum perturbation theory—and this is usually calculated with some unverified assumptions about the excited-state molecular or atomic wave functions. Our point is that none of this is necessary. While we make no pretense to getting the correct constant C in Eq. (1.2), we do get a lower bound to the binding energy of the correct form (when $R > r^\alpha + r^\beta$) by a fairly simple and direct variational argument.

(8) Another point about the standard theory that needs to be addressed is the well-known effect of retardation discovered by Casimir and Polder[11] and elaborated by Lifshitz.[12] The R^{-6} term in Eq. (1.2) is replaced by R^{-7} when R is "large." However, large means (Bohr radius)/(fine-structure constant) and this is huge compared to $r^\alpha + r^\beta$ (for molecules that are not too large). Thus, for distances of major interest for binding, it is physically correct to use the Schrödinger equation without retardation, and hence Eq. (1.2) is meaningful. For small R, nuclear recoil effects may play a role (see Ref. 13).

The calculation and notation in this paper will seem to the reader to be complex. Actually, the complexity is more apparent than real, and a few words about the strategy of our proof may be helpful. The implementation of the strategy will be given in detail later.

We start with several small units (molecules) which we call clusters. There are $\mathscr{C}$ clusters and $\alpha = 1, 2, \ldots, \mathscr{C}$ designates the cluster. The ground-state energy and wave function of each cluster is e^α and ϕ_0^α. A conveniently chosen point in each cluster, called the "center," is denoted by $\mathbf{R}^\alpha$. The problem is to construct a variational trial function ψ (of all the variables in the system) whose energy is lower than $\sum e^\alpha$ by an amount const$\times \sum (R^{\alpha\beta})^{-6}$.

Step 1: Apply a cutoff to each ϕ_0^α in a large ball centered at $\mathbf{R}^\alpha$, in such a way that the balls are disjoint. By the variational principle the energy must *increase*, but since ϕ_0^α decays exponentially, this increase of the energy will only be by an amount proportional to $\sum \exp(-R_\alpha)$, where R_α is the radius of the cutoff. The cutoff function is denoted by ϕ^α.

Step 2: Let $\psi_{\mathbf{m}}^\alpha = \mathbf{m}\cdot\sum_{i=1}^{n^\alpha} \nabla_i \phi^\alpha$ where n^α is the number of electrons in cluster α and $|\mathbf{m}|=1$. The trial function is given by

$$\psi = \prod_{\alpha=1}^{\mathscr{C}} \phi^\alpha + \sum_{\substack{\alpha,\beta \\ (1\le\alpha<\beta\le\mathscr{C})}} \lambda_{\alpha\beta}\psi_{\mathbf{m}_{\alpha\beta}}^\alpha \psi_{\mathbf{n}_{\alpha\beta}}^\beta \prod_{\gamma\,(\neq\alpha,\beta)}'' \phi^\gamma . \tag{1.3}$$

The $\lambda_{\alpha\beta}$ are adjustable constants and the $\mathbf{m}_{\alpha\beta}$ and $\mathbf{n}_{\alpha\beta}$ are adjustable vectors. Since $\langle \psi_{\mathbf{m}_{\alpha\beta}}^\alpha | \phi^\alpha \rangle = 0$, the normalization $\langle \psi | \psi \rangle$ will be of the form

$$1 + \sum_{\substack{\alpha,\beta \\ (\alpha<\beta)}} \lambda_{\alpha\beta}^2 \times \text{const} .$$

However, $\langle \psi | H\psi \rangle$ will have a cross term proportional to the $\lambda_{\alpha\beta}$ coming from the *inter*cluster Coulomb potential. We should *like* $\langle \psi | H\psi \rangle$ to be of the form

$$\left[\sum_\alpha e^\alpha\right]\langle \psi | \psi \rangle + \sum \lambda_{\alpha\beta}^2 \times \text{const} + \sum \lambda_{\alpha\beta}(R^{\alpha\beta})^{-3} \times \text{const} .$$

Minimizing the ratio $\langle \psi | H\psi \rangle / \langle \psi | \psi \rangle$ with respect to each $\lambda_{\alpha\beta}$ would then lead (more or less) to $\lambda_{\alpha\beta}\sim(R^{\alpha\beta})^{-3}$ and hence to

$$E<\sum_{\alpha} e^{\alpha}-\text{const}\times \sum_{\substack{\alpha,\beta \\ (\alpha<\beta)}} (R^{\alpha\beta})^{-6} .$$

Step 3: Unfortunately, the intercluster Coulomb interaction is not simply $\lambda_{\alpha\beta}(R^{\alpha\beta})^{-3}$. Its calculation is complicated by the fact that each cluster is not spherically symmetric. To avoid this difficulty and to get a rigorous bound to the $\lambda_{\alpha\beta}$ term, we consider all spatial orientations of the clusters and orientations of the $\mathbf{m}_{\alpha\beta}$ and $\mathbf{n}_{\alpha\beta}$ (which enters in the definition of ψ). By averaging over the orientations of the clusters and also averaging over certain selected orientations of the $\mathbf{m}_{\alpha\beta}$ and $\mathbf{n}_{\alpha\beta}$, a bound of the desired form is obtained. This implies that there exists *some* orientation of the clusters and the $\mathbf{m}$'s and $\mathbf{n}$'s so that $\langle\psi\,|\,H\psi\rangle\leq$(average over orientations)$\leq(\sum e^{\alpha}+$ van der Waals term$)\langle\psi\,|\,\psi\rangle$.

In any particular case, the wave function ψ which we construct is not the best possible one within the framework of functions of the type (1.3). We use it because we are striving for simplicity and generality—not for good constants in the energy bound.

II. THE GENERAL PROCEDURE

First, some more notation is needed. There are $\mathscr{C}$ clusters indexed by α, $\alpha=1,2,\ldots,\mathscr{C}$ and $\mathbf{R}^{\alpha}\in\mathbb{R}^{3}$ denotes the center of the cluster α. $\mathbf{X}_i^{\alpha}$, $i=1,2,\ldots,M^{\alpha}$ are the coordinates of the nuclei in α relative to $\mathbf{R}^{\alpha}$. This is also denoted collectively by $\underline{X}^{\alpha}$. Similarly, $Z_i^{\alpha}\in\mathbb{N}$ and $\underline{Z}^{\alpha}$ denote the charges of these nuclei. We suppose that all the nuclear coordinates $\underline{X}^{\alpha}$ are contained in a ball, B^{α}, of radius $R_{\alpha}>0$, centered at $\mathbf{R}^{\alpha}$ and that the $\mathscr{C}$ balls are disjoint. In fact, we can define R_{α} by

$$R_{\alpha}=\tfrac{1}{2}\min_{\substack{\alpha,\beta \\ \beta\neq\alpha}}\{\,|\mathbf{R}^{\alpha\beta}|\,\}\,, \tag{2.1}$$

with $\mathbf{R}^{\alpha\beta}\equiv\mathbf{R}^{\alpha}-\mathbf{R}^{\beta}$. Later on we shall specify how large R_{α} must be in order that our bounds have a simple form.

If only cluster α were present, it would have $n^{\alpha}=\sum_i Z_i^{\alpha}$ electrons to be neutral. The coordinates of these n^{α} electrons (relative to $\mathbf{R}^{\alpha}$) are denoted by $\mathbf{x}_i^{\alpha}$ and collectively by $\underline{x}^{\alpha}$. The wave functions we shall use will be sums of functions of the form $\Phi=\phi^{1}(\underline{x}^{1})\phi^{2}(\underline{x}^{2})\cdots\phi^{\mathscr{C}}(\underline{x}^{\mathscr{C}})$ with $\phi^{\alpha}(\underline{x}^{\alpha})=0$ if any x_i^{α} is outside of B^{α}. Each ϕ^{α} satisfies the Pauli principle but Φ does not. (The electron spins should be included but, since they play no essential role, we omit explicit reference to the spins for simplicity of notation.) However, if Φ is antisymmetrized no cross terms appear in $\langle\Phi\,|\,H\Phi\rangle$ or in $\langle\Phi\,|\,\Phi\rangle$ (because the B^{α} are disjoint) which means that we can simply ignore the antisymmetrization of Φ. Therefore it makes sense in the full problem to continue to speak of n^{α} electrons being associated with cluster α.

Because of translation invariance, $\mathbf{R}^{\alpha}$ will not appear in the Hamiltonian H^{α} of cluster α but only in the combination $\mathbf{R}^{\alpha\beta}=\mathbf{R}^{\alpha}-\mathbf{R}^{\beta}$ in the interaction between cluster α and cluster β. In atomic units, $\hbar=e=2m_e=1$, and, in the above notation, H^{α} is written as

$$\begin{aligned}H^{\alpha}=&\sum_{i=1}^{n^{\alpha}}-\Delta_{i,\alpha}+\sum_{\substack{i,k \\ (1\leq i<k\leq n^{\alpha})}}|\mathbf{x}_i^{\alpha}-\mathbf{x}_k^{\alpha}|^{-1}\\ &-\sum_{i=1}^{n^{\alpha}}\sum_{j=1}^{M^{\alpha}}Z_j^{\alpha}|\mathbf{x}_i^{\alpha}-\mathbf{X}_j^{\alpha}|^{-1}\\ &+\sum_{\substack{i,k \\ (1\leq i<k\leq M^{\alpha})}}Z_i^{\alpha}Z_k^{\alpha}|\mathbf{X}_i^{\alpha}-\mathbf{X}_k^{\alpha}|^{-1}\,.\end{aligned} \tag{2.2}$$

The interaction between two clusters again involves an electron-electron, an electron-nucleus, and a nucleus-nucleus interaction,

$$\begin{aligned}V^{\alpha\beta}=&\sum_{i,j}|\mathbf{x}_i^{\alpha}-\mathbf{x}_j^{\beta}+\mathbf{R}^{\alpha\beta}|^{-1}-\sum_{i,j}|\mathbf{x}_i^{\alpha}-\mathbf{X}_j^{\beta}+\mathbf{R}^{\alpha\beta}|^{-1}Z_i^{\beta}\\ &-\sum_{i,j}|\mathbf{x}_j^{\beta}-\mathbf{X}_i^{\alpha}+\mathbf{R}^{\alpha\beta}|^{-1}Z_i^{\alpha}\\ &+\sum_{i,j}|\mathbf{X}_i^{\alpha}-\mathbf{X}_j^{\beta}+\mathbf{R}^{\alpha\beta}|^{-1}Z_i^{\alpha}Z_j^{\beta}\,.\end{aligned} \tag{2.3}$$

The total Hamiltonian consists of H^0, the sum of the cluster Hamiltonians, and the interaction V between them

$$H(\underline{X}^{\alpha},\underline{R}^{\alpha\beta})=H^0+V=\sum_{\alpha=1}^{\mathscr{C}}H^{\alpha}+\sum_{\substack{\alpha,\beta \\ (1\leq\alpha<\beta\leq\mathscr{C})}}V^{\alpha\beta}\,. \tag{2.4}$$

To formalize the averaging procedure denote by $\mathscr{R}^{\alpha}\mathbf{X}_k^{\alpha}$ a rotation of the coordinate $\mathbf{X}_k^{\alpha}$ around the center $\mathbf{R}^{\alpha}$ and by $d\mathscr{R}^{\alpha}$ the normalized volume element (Haar measure) of the rotations. We denote $\prod_{\alpha}d\mathscr{R}^{\alpha}$ by $d\mathscr{R}$. What we shall show is that for some trial functions $\psi(\underline{X}^{\alpha},\underline{m})$, which in addition to the nuclear coordinates also depend on some collection of polarization vectors, $\underline{m}$, we have for sufficiently large $|\mathbf{R}^{\alpha\beta}|$,

$$\int d\mathscr{R}\sum_{\underline{m}}\left[\langle\psi(\mathscr{R}^{\alpha}\underline{X}^{\alpha},\underline{m})\,|\,H(\mathscr{R}^{\alpha}\underline{X}^{\alpha},\underline{R}^{\alpha\beta})\,|\,\psi(\mathscr{R}^{\alpha}\underline{X}^{\alpha},\underline{m})\rangle-\|\psi(\mathscr{R}^{\alpha}\underline{X}^{\alpha},\underline{m})\|^2\left[\sum_{\alpha=1}^{\mathscr{C}}e^{\alpha}-\sum_{\substack{\alpha,\beta \\ \alpha>\beta}}v_{\alpha\beta}(|\underline{R}^{\alpha\beta}|)\right]\right]\leq 0\,, \tag{2.5}$$

where $v_{\alpha\beta}(R)=C_1/(R^6+C_2)$. We shall use the symbol $\sum_{\underline{m}}$ for an average over the polarization directions, whose precise nature will be given in Sec. VI. Since the average in (2.5) is negative, we reach the conclusion that there are some orientations such that the clusters attract each other at least as much as the van der Waals energy.

III. THE TRIAL FUNCTION

For our results we require the clusters to be clearly separated. By this we mean that the wave functions should not overlap. This, of course, is not true for the ground-state wave functions which decay only exponentially, but may be achieved by a slight adjustment of the wave function which will not cost too much energy. If ϕ_0^α is the (real-valued) ground-state wave function of H^α,

$$H^\alpha \phi_0^\alpha = e^\alpha \phi_0^\alpha \,, \tag{3.1}$$

we shall use a function

$$\phi^\alpha(\underline{x}^\alpha, \underline{X}^\alpha) = \phi_0^\alpha(\underline{x}^\alpha, \underline{X}^\alpha) f(\underline{x}^\alpha) \,, \tag{3.2}$$

where

$$f(\underline{x}^\alpha) = \prod_{k=1}^{n_\alpha} \chi(|\mathbf{x}_k^\alpha|) \,, \tag{3.3}$$

where $\chi(s)$ is a smooth function which is 1 for $s < R_\alpha - 1$ and 0 for $s > R_\alpha$, and such that $|\chi'(s)|, |\chi''(s)| \le 4$ for $R_\alpha - 1 < s < R_\alpha$ and zero otherwise. For ϕ^α we have

$$H^\alpha \phi^\alpha = e^\alpha \phi^\alpha - \sum_{k=1}^{n^\alpha} (2 \nabla_k^\alpha \phi_0^\alpha \cdot \nabla_k^\alpha f + \phi_0^\alpha \Delta_k^\alpha f) \,, \tag{3.4}$$

and, by some partial integrations,

$$\langle \phi^\alpha | H^\alpha \phi^\alpha \rangle = e^\alpha + \sum_{k=1}^{n^\alpha} \int d\underline{x}^\alpha |\phi_0^\alpha|^2 |\nabla_k^\alpha f|^2 \,, \tag{3.5}$$

where we assumed $\langle \phi^\alpha | \phi^\alpha \rangle = 1$.

The radius R_α was defined in (2.1) and it increases with increasing separation of the clusters. It will turn out that the error in the energy caused by replacing ϕ_0^α by ϕ^α depends only on the single-particle electron density $\rho^\alpha(\mathbf{x})$, defined in the usual way [see (5.2) but omit the spherical average $\int d\mathscr{R}$]. It is known that $\rho^\alpha(\mathbf{x})$ decays exponentially, which means that for some r^α and $c > 0$,

$$\rho^\alpha(\mathbf{x}) < \exp[c(r^\alpha - |\mathbf{x}|)] \quad \text{for } |\mathbf{x}| > r^\alpha \,.$$

We take r^α as the *definition* of the *cluster radius*. The additional terms in (3.4) and (3.5), which involve only the region $R_\alpha - 1 < |x| < R_\alpha$, will thus be exponentially small (as a function of R_α) when $R_\alpha > r^\alpha + 1$.

Next, we need a wave function that describes a polarized cluster and there are many choices possible. We found the most convenient form to be

$$\psi_{\mathbf{m}}^\alpha = \mathbf{m} \cdot \sum_{k=1}^{n^\alpha} \nabla_k^\alpha \phi^\alpha, \quad \mathbf{m} \in \mathbb{R}^3, \ |\mathbf{m}| = 1 \,. \tag{3.6}$$

It is easily seen to be orthogonal to ϕ^α,

$$\langle \psi_{\mathbf{m}}^\alpha | \phi^\alpha \rangle = 0 \,, \tag{3.7}$$

and its norm is the expectation value of the square of the total momentum of all electrons in cluster α in the direction of $\mathbf{m}$,

$$\langle \psi_{\mathbf{m}}^\alpha | \psi_{\mathbf{m}}^\alpha \rangle = \int d\underline{x}^\alpha \left| \mathbf{m} \cdot \sum_{k=1}^{n^\alpha} \nabla_k^\alpha \phi^\alpha \right|^2 . \tag{3.8}$$

This quantity still depends on the direction $\mathbf{m}$. However, in our results only the scalar product averaged over rotations of the nuclear coordinates will appear and

$$\int d\mathscr{R} \langle \psi_{\mathbf{m}}^\alpha | \psi_{\mathbf{m}}^\alpha \rangle \equiv \tau^\alpha$$

is independent of $\mathbf{m}$. At the end of the paper we shall collect simple estimates for the various constants such as τ^α which will enter our result. For now we go on to exhibit the trial function ψ for the total system (the products are always in the sense of tensor products)

$$\psi = \prod_{\alpha=1}^{\mathscr{C}} \phi^\alpha + \sum_{\substack{\alpha,\beta \\ (1 \le \alpha < \beta \le \mathscr{C})}} \lambda_{\alpha\beta} \phi^1 \cdots \phi^{\alpha-1} \psi_{\mathbf{m}_{\alpha\beta}}^\alpha \phi^{\alpha+1} \cdots \phi^{\beta-1} \times \psi_{\mathbf{n}_{\alpha\beta}}^\beta \phi^{\beta+1} \cdots \phi^{\mathscr{C}}$$

$$\equiv \Phi + \sum_{\substack{\alpha,\beta \\ (\alpha<\beta)}} \lambda_{\alpha\beta} \psi^{\alpha\beta} \,. \tag{3.9}$$

Here $\lambda_{\alpha\beta} \in \mathbf{R}$ and $\mathbf{m}_{\alpha\beta}, \mathbf{n}_{\alpha\beta} \in S^2$ are variational parameters to be chosen later. Notice that the polarization vectors $\mathbf{m}$ and $\mathbf{n}$ of the clusters depend on the *pair* of clusters α and β. Because of the orthogonality (3.7) there are no terms linear in the λ's in the norm of ψ,

$$\langle \psi | \psi \rangle = 1 + \sum_{\substack{\alpha,\beta \\ (\alpha<\beta)}} \lambda_{\alpha\beta}^2 \|\psi_{\mathbf{m}_{\alpha\beta}}^\alpha\|^2 \|\psi_{\mathbf{n}_{\alpha\beta}}^\beta\|^2 \,. \tag{3.10}$$

IV. THE EXPECTATION VALUE OF H^0

In calculating $\langle \psi | H^\alpha \psi \rangle$ we first note that the term independent of λ is almost e^α except for the exponentially decreasing contribution in (3.5),

$$\langle \Phi | H^\alpha \Phi \rangle = e^\alpha + \sum_{k=1}^{n^\alpha} \int d\underline{x}^\alpha |\phi_0^\alpha|^2 |\nabla_k^\alpha f|^2 \equiv e^\alpha + b_1^\alpha \,, \tag{4.1}$$

where the boundary term b_1^α will be less than $c \exp(-R_\alpha)$. The orthogonality (3.7) makes the contributions which are linear in λ vanish,

$$\langle \Phi | H^\alpha \psi^{\alpha\beta} \rangle = \langle \phi^\alpha \cdot \phi^\beta | H^\alpha \psi_{\mathbf{m}_{\alpha\beta}}^\alpha \psi_{\mathbf{n}_{\alpha\beta}}^\beta \rangle = \langle \phi^\alpha | H^\alpha \psi_{\mathbf{m}_{\alpha\beta}}^\alpha \rangle \langle \phi^\beta | \psi_{\mathbf{n}_{\alpha\beta}}^\beta \rangle = 0 \,. \tag{4.2}$$

The terms quadratic in λ require some rearrangement because we want to compare $\langle \psi | H | \psi \rangle$ with $\|\psi\|^2 \sum_\alpha e^\alpha$ and not just with $\sum_\alpha e^\alpha$, which we get from (4.1). To achieve this we use the following identity, where we momentarily abbreviate $\sum_{k=1}^{n^\alpha} \mathbf{m} \cdot \nabla_k^\alpha$ by $P^\alpha = -(P^\alpha)^*$,

$$\begin{aligned} &\langle \psi^{\alpha\beta} | H^\alpha | \psi^{\alpha\beta} \rangle \\ &= -\langle \phi^\alpha | P^\alpha H^\alpha P^\alpha | \phi^\alpha \rangle \|\psi_{\mathbf{n}_{\alpha\beta}}^\beta\|^2 \\ &= -\tfrac{1}{2} \langle \phi^\alpha | [P^\alpha, [H^\alpha, P^\alpha]] + H^\alpha (P^\alpha)^2 + (P^\alpha)^2 H^\alpha | \phi^\alpha \rangle \\ &\quad \times \|\psi_{\mathbf{n}_{\alpha\beta}}^\beta\|^2 \,. \end{aligned} \tag{4.3}$$

Using (3.4), $-H^\alpha (P^\alpha)^2 - (P^\alpha)^2 H^\alpha$ indeed yields $e^\alpha \|\psi^{\alpha\beta}\|^2$

and some boundary terms, which we call b_2^α, and which are also exponentially decreasing. As regards the double commutator, we first see from (2.2) that the only term in H^α that is not invariant under a common displacement of all electrons is the electron-nucleus attraction

$$[P^\alpha,H^\alpha]=-\sum_{k,j}\mathbf{m}_{\alpha\beta}\cdot\nabla_k^\alpha Z_j^\alpha\,|\,\mathbf{x}_k^\alpha-\mathbf{X}_j^\alpha\,|^{-1}\,, \tag{4.4}$$

$$[P^\alpha,[P^\alpha,H^\alpha]]=-\sum_{k,j}(\mathbf{m}_{\alpha\beta}\cdot\nabla_k^\alpha)^2 Z_j^\alpha\,|\,\mathbf{x}_k^\alpha-\mathbf{X}_j^\alpha\,|^{-1}\,. \tag{4.5}$$

A further simplification of the expectation value of this expression results if we carry out the integrations $\int d\mathscr{R}^\alpha \int d\mathscr{R}^\beta$. The integration $\int d\mathscr{R}^\beta$ makes $\|\psi^\beta_{\mathbf{n}_{\alpha\beta}}\|^2$ independent of $\mathbf{n}_{\alpha\beta}$ and $\int d\mathscr{R}^\alpha$ makes the expectation value of (4.5) independent of $\mathbf{m}_{\alpha\beta}$. The reason is that after rotating the nuclei there is no distinguished direction left. Since the result is independent of $\mathbf{m}_{\alpha\beta}$, we may freely average over three orthogonal directions of $\mathbf{m}_{\alpha\beta}$ and thus replace $-(\mathbf{m}\cdot\nabla_k^\alpha)^2$ by $-\frac{1}{3}\Delta_k^\alpha$ which just gives $(4\pi/3)\delta(\mathbf{x}_k^\alpha-\mathbf{X}_j^\alpha)$ in (4.5). Upon collecting the contributions we end up with

$$\int d\mathscr{R}^\alpha \int d\mathscr{R}^\beta \langle \psi^{\alpha\beta}\,|\,H^\alpha\,|\,\psi^{\alpha\beta}\rangle = e^\alpha\|\psi^{\alpha\beta}\|^2 + \left[b_2^\alpha+\frac{2\pi}{3}\sum_{k,j}\langle\phi^\alpha\,|\,Z_j^\alpha\delta(\mathbf{x}_k^\alpha-\mathbf{X}_j^\alpha)\,|\,\phi^\alpha\rangle\right]\tau^\beta$$

$$\equiv e^\alpha\|\psi^{\alpha\beta}\|^2+(b_2^\alpha+Q^\alpha)\tau^\beta \tag{4.6}$$

where Q^α is $2\pi/3$ times the sum of the electron densities at the nuclei in cluster α.

V. THE EXPECTATION VALUE OF V

For the evaluation of $\langle\psi\,|\,V\,|\,\psi\rangle$ we shall heavily use Newton's theorem according to which the potential of a spherically symmetric charge distribution with net charge zero vanishes outside the support of this charge distribution. This theorem makes the $\langle\Phi\,|\,V\,|\,\Phi\rangle$ term vanish upon rotation of the nuclei because it becomes the electrostatic interaction between two nonoverlapping spherical charge distributions of net charge zero. To demonstrate this formally we note that

$$\int d\mathscr{R}\langle\Phi(\underline{x},\mathscr{R}\underline{X})\,|\,V(\underline{x},\mathscr{R}\underline{X})\,|\,\Phi(\underline{x},\mathscr{R}\underline{X})\rangle$$

is a sum of terms

$$\int d\mathscr{R}^\alpha\int d\mathscr{R}^\beta\langle\phi^\alpha(\underline{x}^\alpha,\mathscr{R}^\alpha\underline{X}^\alpha)\phi^\beta(\underline{x}^\beta,\mathscr{R}^\beta\underline{X}^\beta)\,|\,V^{\alpha\beta}(\underline{x}^\alpha,\underline{x}^\beta,\mathscr{R}^\alpha\underline{X}^\alpha,\mathscr{R}^\beta\underline{X}^\beta,\mathbf{R}^{\alpha\beta})\,|\,\phi^\alpha(\underline{x}^\alpha,\mathscr{R}^\alpha\underline{X}^\alpha)\phi^\beta(\underline{x}^\beta,\mathscr{R}^\beta\underline{X}^\beta)\rangle\,. \tag{5.1}$$

They involve only the one-electron density which upon integrating over $\mathscr{R}$ becomes spherically symmetric (around the center R^α) since $\phi^\alpha(\underline{x}^\alpha,\mathscr{R}^\alpha\underline{X}^\alpha)=\phi^\alpha((\mathscr{R}^\alpha)^{-1}\underline{x}^\alpha,\underline{X}^\alpha)$. The one-electron density is

$$\rho^\alpha(\mathbf{x}_1^\alpha)\equiv n^\alpha\int d\mathscr{R}^\alpha\int dx_2^\alpha\cdots\int dx_{n^\alpha}^\alpha\,|\,\phi^\alpha(\mathbf{x}_1^\alpha,\mathbf{x}_2^\alpha,\ldots,\mathbf{x}_{n^\alpha}^\alpha,\mathscr{R}^\alpha\underline{X}^\alpha)\,|^2\,. \tag{5.2}$$

For $|\,\mathbf{x}\,|<R^\alpha-1$, we have $\rho^\alpha(\mathbf{x})=[1+C\exp(-R_\alpha)]\rho_0^\alpha(\mathbf{x})$ for some $C>0$, where $\rho_0^\alpha(\mathbf{x})$ is the "true" electron density calculated with ϕ_0^α.

With the averaged nuclear charge

$$n^\alpha(\,|\,\mathbf{x}\,|\,)=\int d\mathscr{R}^\alpha\sum_{j=1}^{M^\alpha}\delta(\mathbf{x}-\mathscr{R}^\alpha\mathbf{X}_j^\alpha)\,, \tag{5.3}$$

and similar definitions for cluster β, (5.1) becomes

$$\int d\mathbf{x}\int d\mathbf{y}\,[\rho^\alpha(\mathbf{x})-n^\alpha(\mathbf{x})][\rho^\beta(\mathbf{y})-n^\beta(\mathbf{y})]\,|\,\mathbf{x}-\mathbf{y}+\mathbf{R}^{\alpha\beta}\,|^{-1}=0\,. \tag{5.4}$$

In the terms proportional to λ we have the factor $\phi^\alpha P^\alpha\phi^\alpha=\frac{1}{2}P^\alpha(\phi^\alpha)^2$. After partial integration, the gradient acts on the potential, and thus we obtain the dipole-dipole interaction directly:

$$\int d\mathscr{R}^\alpha\int d\mathscr{R}^\beta\langle\phi^\alpha\cdot\phi^\beta\,|\,V^{\alpha\beta}\,|\,\psi^\alpha\cdot\psi^\beta\rangle$$

$$=\int d\mathscr{R}^\alpha\int d\mathscr{R}^\beta\int d\underline{x}^\alpha\int d\underline{x}^\beta\frac{1}{4}\sum_i\mathbf{m}_{\alpha\beta}\cdot\nabla_i^\alpha\,|\,\phi^\alpha(\underline{x}^\alpha,\mathscr{R}^\alpha\underline{X}^\alpha)\,|^2\sum_j\mathbf{n}_{\alpha\beta}\cdot\nabla_j^\beta\,|\,\phi^\beta(\underline{x}^\beta,\mathscr{R}^\beta\underline{X}^\beta)\,|^2V(\underline{x}^\alpha,\underline{x}^\beta,\mathscr{R}^\alpha\underline{X}^\alpha,\mathscr{R}^\beta\underline{X}^\beta,\mathbf{R}^{\alpha\beta})$$

$$=\frac{1}{4}\int d\mathbf{x}\int d\mathbf{y}\,\rho^\alpha(\mathbf{x})\rho^\beta(\mathbf{y})(\mathbf{m}_{\alpha\beta}\cdot\nabla_{\mathbf{x}})(n_{\alpha\beta}\cdot\nabla_{\mathbf{y}})\,|\,\mathbf{x}-\mathbf{y}+\mathbf{R}^{\alpha\beta}\,|^{-1}$$

$$=\frac{1}{4}n^\alpha n^\beta[3(\mathbf{m}_{\alpha\beta}\cdot\mathbf{R}^{\alpha\beta})(\mathbf{n}_{\alpha\beta}\cdot\mathbf{R}^{\alpha\beta})-(\mathbf{m}_{\alpha\beta}\cdot\mathbf{n}_{\alpha\beta})\,|\,\mathbf{R}^{\alpha\beta}\,|^2]\,|\,\mathbf{R}^{\alpha\beta}\,|^{-5}\,. \tag{5.5}$$

Newton's theorem has been used again in order to perform the $d\mathbf{x}$ and dy integrations. The contribution $\langle \psi^{\alpha\beta} | V^{\alpha\beta} | \psi^{\alpha\beta} \rangle$ is analogous to the term $\langle \phi | V^{\alpha\beta} | \phi \rangle$ except that the one-electron density is now calculated with $\psi^\alpha = \mathbf{m} \cdot \sum_{j=1}^{n^\alpha} \nabla_j^\alpha \phi^\alpha$ instead of ϕ^α. If we call this density $\rho_{\mathbf{m}}^\alpha(\mathbf{x})$ it will not be spherically symmetric, but $\rho_{\mathscr{R}\mathbf{m}}^\alpha(\mathscr{R}\mathbf{x}) = \rho_{\mathbf{m}}^\alpha(\mathbf{x})$. Since it is quadratic in $\mathbf{m}$ it must be of the form $|\mathbf{m}|^2 f_1(|\mathbf{x}|) + (\mathbf{m}\cdot\mathbf{x})^2 f_2(|\mathbf{x}|)$. Adding in the averaged charge of the nuclei will give a charge distribution $Q(x)$ with no net charge, and no dipole moment since $Q(x) = Q(-x)$. However, it may have a quadrupole moment and we may pick up a term

$$\langle \psi^{\alpha\beta} | V^{\alpha\beta} | \psi^{\alpha\beta} \rangle = c^{\alpha\beta} |\mathbf{R}^{\alpha\beta}|^{-5} \tag{5.6}$$

of unknown sign.

VI. CHOICE OF THE POLARIZATION DIRECTIONS

Our goal is to make the term proportional to λ, which will be negative for a suitable choice of λ, as big as possible while the term proportional to λ^2 should be kept small. Only (5.5) contributes linearly in λ and its maximal value $2n^\alpha n^\beta |\mathbf{R}^{\alpha\beta}|^{-3}$ is reached if we choose $\mathbf{m}_{\alpha\beta}$ and $\mathbf{n}_{\alpha\beta}$ in the direction of $\mathbf{R}^{\alpha\beta}$. However, this leaves us with the quadrupole term (5.6). We can get rid of the latter if we average over $\mathbf{m}_{\alpha\beta}$ and $\mathbf{n}_{\alpha\beta}$ in the following way. Let

$$\begin{pmatrix}1\\0\\0\end{pmatrix}$$

be the unit vector in the direction of $\mathbf{R}^{\alpha\beta}$ and

$$\begin{pmatrix}0\\1\\0\end{pmatrix}, \quad \begin{pmatrix}0\\0\\1\end{pmatrix}$$

two directions completing the orthogonal basis. Then average the polarizations over the following nine pairs of vectors:

$$\begin{pmatrix}1\\0\\0\end{pmatrix}, \begin{pmatrix}1\\0\\0\end{pmatrix}; \quad \begin{pmatrix}1\\0\\0\end{pmatrix}, \begin{pmatrix}0\\-1\\0\end{pmatrix}; \quad \begin{pmatrix}1\\0\\0\end{pmatrix}, \begin{pmatrix}0\\0\\-1\end{pmatrix};$$

$$\begin{pmatrix}0\\1\\0\end{pmatrix}, \begin{pmatrix}1\\0\\0\end{pmatrix}; \quad \begin{pmatrix}0\\1\\0\end{pmatrix}, \begin{pmatrix}0\\-1\\0\end{pmatrix}; \quad \begin{pmatrix}0\\1\\0\end{pmatrix}, \begin{pmatrix}0\\0\\-1\end{pmatrix};$$

$$\begin{pmatrix}0\\0\\1\end{pmatrix}, \begin{pmatrix}1\\0\\0\end{pmatrix}; \quad \begin{pmatrix}0\\0\\1\end{pmatrix}, \begin{pmatrix}0\\-1\\0\end{pmatrix}; \quad \begin{pmatrix}0\\0\\1\end{pmatrix}, \begin{pmatrix}0\\0\\-1\end{pmatrix} .$$

In this way the quadrupole term (5.6) vanishes since $Q(x)$ becomes spherically symmetric. On the other hand (5.5) averages to $\frac{1}{9} n^\alpha n^\beta |\mathbf{R}^{\alpha\beta}|^{-3}$. Since λ will be proportional to R^{-3} the quadrupole term (5.6) will contribute to the order R^{-11} and the first choice (i.e., no averaging) will give a better estimate for large R. On the other hand, for smaller distances averaging may be better, as (5.6) is not easy to estimate accurately.

VII. CHOICE OF THE $\lambda_{\alpha\beta}$

So far we have evaluated expressions such as $\langle \psi^{\alpha\beta} | V^{\alpha'\beta'} | \psi^{\alpha''\beta''} \rangle$ where all three index pairs $(\alpha\beta)$, $(\alpha'\beta')$, and $(\alpha''\beta'')$ are equal. The orthogonality (3.7) insures that terms in which they are not equal vanish. Similarly $\langle \psi^{\alpha\beta} | H^\gamma | \psi^{\alpha'\beta'} \rangle$ is not zero only if $(\alpha\beta) = (\alpha'\beta')$ and γ is either α or β. Thus collecting our results we arrive at

$$\sum_{\mathbf{m}} \int d\mathscr{R} \langle \psi | H | \psi \rangle = \sum_{\mathbf{m}} \int d\mathscr{R} \Bigg[\sum_{\alpha=1}^{\mathscr{C}} \left[e^\alpha ||\psi||^2 + \sum b_1^\alpha \right] + \frac{1}{9} \sum_{\substack{\alpha,\beta \\ (1 \le \alpha < \beta \le \mathscr{C})}} \lambda_{\alpha\beta} n^\alpha n^\beta |\mathbf{R}^{\alpha\beta}|^{-3}$$
$$+ \sum_{\substack{\alpha,\beta \\ (1 \le \alpha < \beta \le \mathscr{C})}} \lambda_{\alpha\beta}^2 [(b_2^\alpha + Q^\alpha)\tau^\beta + (b_2^\beta + Q^\beta)\tau^\alpha] \Bigg] . \tag{7.1}$$

To get an upper bound for the energy we have to extract a factor $||\psi||^2 = 1 + \sum_{\alpha<\beta} \lambda_{\alpha\beta}^2 \tau_\alpha \tau_\beta$ from the term in large square brackets. The minimization with respect to the $\lambda_{\alpha\beta}$'s then leads to a coupled system of cubic equations which cannot be solved analytically. For our purpose, however, it is sufficient to minimize the last two terms by putting

$$\lambda_{\alpha\beta} = -\tfrac{1}{18} n^\alpha n^\beta |\mathbf{R}^{\alpha\beta}|^{-3} [(Q^\alpha + b_2^\alpha)\tau^\beta + (Q^\beta + b_2^\beta)\tau^\alpha]^{-1} . \tag{7.2}$$

By extracting

$$||\psi||^2 = 1 + \frac{1}{(18)^2} \sum (n^\alpha n^\beta)^2 \tau^\alpha \tau^\beta |\mathbf{R}_{\alpha\beta}|^{-6} [(Q^\alpha + b_2^\alpha)\tau^\beta + (Q^\beta + b_2^\beta)\tau^\alpha]^{-2} ,$$

we obtain

$$\sum_m \int d\mathscr{R}\langle\psi|H|\psi\rangle=\sum_m \int d\mathscr{R}\|\psi\|^2\left[\sum_{\alpha=1}^{\mathscr{C}}(e^\alpha+b_1^\alpha/\|\psi\|^2) - \sum_{\substack{\alpha,\beta\\(1\le\alpha<\beta\le\mathscr{C})}} \frac{(n^\alpha n^\beta)^2}{(18)^2[(Q^\alpha+b_1^\alpha)\tau^\beta+(Q^\beta+b_2^\beta)\tau^\alpha]} \times \left[|\mathbf{R}^{\alpha\beta}|^6+\sum_{\substack{\sigma,\rho\\(1\le\sigma<\rho\le\mathscr{C})}}\frac{(n^\sigma n^\rho)^2\tau^\sigma\tau^\rho}{(18)^2[(Q^\sigma+b_2^\sigma)\tau^\rho+(Q^\rho+b_2^\rho)\tau^\sigma]}\right]^{-1}\right] . \quad (7.3)$$

Taking $\mathbf{m}_{\alpha\beta}$ and $\mathbf{n}_{\alpha\beta}$ in the direction of $\mathbf{R}^{\alpha\beta}$ would change the factor $(18)^2$ into 4^2 and add $c^{\alpha\beta}|\mathbf{R}^{\alpha\beta}|^{-5}$ to $(Q^\alpha+b_2^\alpha)\tau^\beta+(Q^\beta+b_2^\beta)\tau^\alpha$.

VIII. SOME ESTIMATES

Our result (7.3) contains two parameters, the squared center-of-mass momentum τ, and Q, the density of the electrons at the nuclei. For the former we have the trivial estimate (where T^α is the kinetic energy in the ground state)

$$\tau^\alpha \le n^\alpha\left\langle\phi^\alpha\middle|-\sum_{k=1}^{n^\alpha}\Delta_k^\alpha\middle|\phi^\alpha\right\rangle \equiv n^\alpha T^\alpha . \quad (8.1)$$

From the stability of matter one knows that for states of negative energy the kinetic energy is bounded by n^α,[14] so that $\tau^\alpha \le c(n^\alpha)^2$. We *conjecture* that actually $\tau^\alpha \le T^\alpha$. Similarly, for the Q^α the bound from Ref. 15 for the atomic case can be extended to the molecular case to give

$$Q^\alpha \le \tfrac{1}{3}\sum_{k,j}\langle\phi^\alpha|Z_j^\alpha|x_k^\alpha-X_j^\alpha|^{-2}|\phi^\alpha\rangle . \quad (8.2)$$

Using the fact that $|\mathbf{x}|^{-2}\le -4\Delta$, the right side of (8.2) can be bounded by the kinetic energy. Again, keeping $\max_{\alpha,j}\{Z_j^\alpha\}$ fixed, we get a bound proportional to $(n^\alpha)^2$, whereas we *conjecture* it should be proportional to n^α. From the *proven* bounds the coefficient in front of the large parentheses in (7.3) becomes independent of the n^α. For the *conjectured* bounds it would be proportional to $n^\alpha n^\beta$, thereby indicating a linearity of the van der Waals forces with respect to the electron number in each cluster.

If one calculates our constants τ and Q for a hydrogen atom one gets a constant in front of R^{-6} which is about an order of magnitude below the known constant for the van der Waals force between two hydrogen atoms. The reason is that our way of generating polarization by a rigid shift is a rather brutal act. In particular, shifting the electrons near the nucleus is energetically costly; it is much better to adjust only the outer parts of the electron cloud. The best way to do this will depend on the exact shape of the cluster, and a simple, general bound which is numerically good in all cases seems to be beyond reach. Our result shows that irrespective of these details one can always get some R^{-6} attraction by a rigid shift.

It will be noted that the factor in large parentheses in (7.3) contains a constant term, $\sum$, in addition to $|\mathbf{R}^{\alpha\beta}|^6$. While this term is asymptotically negligible for large $|\mathbf{R}^{\alpha\beta}|$, it unfortunately grows with $\mathscr{C}$. Our result is therefore useless if $\mathscr{C}$ is very large, e.g., in a crystal where $\mathscr{C}\sim 10^{23}$. To circumvent this difficulty a better trial function Ψ is needed. We believe that the following choice is adequate, but we have not actually pursued the matter: Define the "operator" $D^{\alpha\beta}$ by $D^{\alpha\beta}\phi^\alpha\phi^\beta=\psi^\alpha\psi^\beta, D^{\alpha\beta}\phi^\alpha\psi^\beta=D^{\alpha\beta}\psi^\alpha\phi^\beta=D^{\alpha\beta}\psi^\alpha\psi^\beta=0$. Then a natural generalization of (3.9) is

$$\Psi=\prod_{\substack{\alpha,\beta\\(\alpha<\beta)}}(1+\lambda_{\alpha\beta}D^{\alpha\beta})\prod_{\alpha=1}^{\mathscr{C}}\phi^\alpha(\underline{x}^\alpha,\underline{X}^\alpha) . \quad (8.3)$$

ACKNOWLEDGMENTS

This work was partially supported by U.S. National Science Foundation Grant No. PHY-81-16101-A03.

[1] P. Hohenberg and W. Kohn, Phys. Rev. **136**, B864 (1964).
[2] E. H. Lieb, Int. J. Quantum Chem. **24**, 243 (1983).
[3] E. H. Lieb, Rev. Mod. Phys. **53**, 603 (1981); **54**, 311(E) (1982).
[4] B. R. Junker and J. N. Bardsley, Phys. Rev. Lett. **28**, 1227 (1972).
[5] D. L. Morgan, Jr. and V. W. Hughes, Phys. Rev. A **7**, 1811 (1973); Phys. Rev. D **2**, 1389 (1970).
[6] W. Kołos, D. L. Morgan, D. M. Schrader, and L. Wolniewicz, Phys. Rev. A **6**, 1792 (1975).
[7] G. Feinberg and J. Sucher, Phys. Rev. A **2**, 2395 (1970).
[8] H. Margenau and N. R. Kestner, *The Theory of Intermolecular Forces* (Pergamon, New York, 1971).
[9] Yu. S. Barash and V. L. Ginzburg, Usp. Fiz. Nauk **143**, 345 (1984) [Sov. Phys.—Usp. **27**, 7 (1984)].
[10] B. Simon and J. Morgan, Int. J. Quantum Chem. **17**, 1143 (1980).
[11] H. G. Casimir and D. Polder, Phys. Rev. **73**, 360 (1948).
[12] E. M. Lifshitz, Zh. Eksp. Teor. Fiz. **29**, 94 (1955).
[13] J. R. Manson and R. H. Ritchie, Phys. Rev. Lett. **54**, 785 (1985).
[14] E. H. Lieb, Rev. Mod. Phys. **48**, 553 (1976).
[15] M. Hoffmann-Ostenhof, T. Hoffmann-Ostenhof, and W. Thirring, J. Phys. B **11**, L571 (1978).

PHYSICAL REVIEW A VOLUME 52, NUMBER 5 NOVEMBER 1995

Electron density near the nucleus of a large atom

Ole J. Heilmann
Chemistry Laboratory 3, H.C. Orsted Institute, University of Copenhagen, Universitetsparken 5, DK2100 Copenhagen, Denmark

Elliott H. Lieb
Departments of Physics and Mathematics, Princeton University, P.O. Box 708, Princeton, New Jersey 08544-0708
(Received 1 May 1995)

The density of electrons on a distance scale $1/Z$ near the nucleus of a large atom with nuclear charge Ze is given (asymptotically as $Z\to\infty$) by the sum of the squares of all the hydrogenic bound-state functions (with nuclear charge Ze). This density function, which is an important limiting function in quantum chemistry, is investigated here in detail. Several analytic results are found: In particular, the asymptotic expansion for large r is derived and it is shown that the function falls off as $r^{-3/2}$ for large r; this behavior coincides with the Thomas-Fermi density for *small* r. "Shell structure" is visible, but barely so.

PACS number(s): 31.10.+z, 31.15.An, 31.15.Ew

I. INTRODUCTION AND BASIC DEFINITIONS

In attempting to understand the ground state of an atom (for the nonrelativistic Schrödinger equation with an infinitely massive fixed nucleus) it is useful to explore the properties of this ground state as the nuclear charge Ze tends to infinity. In this study a function of fundamental importance arises, and it is our purpose here to explore its properties. That function is the sum of the squares of all the bound-state eigenfunctions of the hydrogen atom.

The *Schrödinger Hamiltonian* for an atom with N electrons is

$$H_{N,Z}=-\frac{\hbar^2}{2m}\sum_{i=1}^{N}\Delta_i-Ze^2\sum_{i=1}^{N}|x_i|^{-1}+e^2\sum_{1\le i<j\le N}|x_i-x_j|^{-1}. \tag{1.1}$$

Here $-e$ and m are, respectively, the charge and mass of an electron; the x_i's are the electron coordinates and Δ is the three-dimensional Laplacian. The first operator in (1.1) is the kinetic energy, the second is the attraction of the electrons to the nucleus (located at the origin in $\mathbb{R}^3$), and the third term is the electron-electron repulsion.

The ground-state energy $E_{N,Z}$ is defined to be

$$E_{N,Z}=\inf_{\psi}\langle\psi|H_{N,Z}|\psi\rangle, \tag{1.2}$$

where the infimum is taken over all functions $\psi(x_1,\ldots,x_n;\sigma_1,\ldots,\sigma_N)$ of N space and spin variables (x_i,σ_i) that are normalized, i.e.,

$$\sum_{\sigma_1=1}^{2}\cdots\sum_{\sigma_N=1}^{2}\int|\psi(x_1,\ldots,x_N;\sigma_1,\ldots,\sigma_N)|^2\times d^3x_1\cdots d^3x_n=1, \tag{1.3}$$

and that are antisymmetric [i.e., ψ changes sign if (x_i,σ_i) is interchanged with (x_j,σ_j) for any $i\neq j$]. For any normalized ψ (not necessarily a ground state ψ), there is the *single particle density* ρ_ψ defined by

$$\rho_\psi(x)=N\sum_{\sigma_I=1}^{2}\cdots\sum_{\sigma_N=1}^{2}\int|\psi(x,x_2,\ldots,x_N;\sigma_1,\ldots,\sigma_N)|^2\times d^3x_2\cdots d^3x_N, \tag{1.4}$$

which satisfies

$$\int\rho_\psi(x)d^3x=N. \tag{1.5}$$

Note that because of the antisymmetry of ψ it does not matter which x_i is set equal to x in (1.4).

If $N\le Z$, then [1] there is at least one square integrable ψ that actually minimizes the right-hand side of (1.2). Such a ψ is a *ground state* and satisfies Schrödinger's equation

$$H_{N,Z}\psi=E_{N,Z}\psi. \tag{1.6}$$

Of course a ground state might exist when $N>Z$ (i.e., for a negative ion), but it is not known how large N can be (see [2] for a review). In particular it is known [2] that N must satisfy $N<2Z+1$ and [3] that as $Z\to\infty$ the maximum N satisfies $N/Z\to1$. In fact [4], $|N/Z-1|<Z^{-1/7}$.

Our goal here is to study $\rho_\psi(x)$ for a ground state ψ when Z is large. To be precise, fix some number λ with $0<\lambda\le1$, let N take the values $N=1,2,3,\ldots$, and for each N let Z_N be defined by $\lambda Z_N=N$. For each integer N let $\psi_{N,Z}$ be a ground state and let $\rho_{N,Z}$ be given by (1.4) with ψ replaced by $\psi_{N,Z}$. Of course $\psi_{N,Z}$ (and hence $\rho_{N,Z}$) may not be unique, but nevertheless the following is true [5,6], for any choice of the $\psi_{N,Z}$'s: As N and Z tend to infinity, with $N=\lambda Z_N$ and with λ fixed

$$Z_N^{-2}\rho_{N,Z}(Z_N^{-1/3}x)\to\rho_\lambda^{\mathrm{TF}}(x). \tag{1.7}$$

1050-2947/95/52(5)/3628(16)/$06.00 52 3628

Here $\rho_\lambda^{\rm TF}$ is the Thomas-Fermi (TF) density. It is the unique non-negative function that minimizes the TF energy functional for a unit nuclear charge,

$$\mathcal{E}^{\rm TF}(\rho)=\frac{3}{5}\frac{\hbar^2}{2m}(3\pi^2)^{2/3}\int\rho(x)^{5/3}d^3x-e^2\int\rho(x)\frac{1}{|x|}d^3x+\frac{1}{2}e^2\int\int\rho(x)\rho(y)\frac{1}{|x-y|}d^3x\,d^3y, \quad (1.8)$$

under the condition that

$$\int\rho(x)d^3x=\lambda. \quad (1.9)$$

The unique minimizing $\rho_\lambda^{\rm TF}(x)$ is the unique solution to the TF equation

$$\frac{\hbar^2}{2m}(3\pi^2)^{2/3}[\rho_\lambda^{\rm TF}(x)]^{2/3}=e^2\left[\frac{1}{|x|}-\int\frac{1}{|x-y|}\rho_\lambda^{\rm TF}(y)d^3y-\mu\right]_+, \quad (1.10)$$

where $[t]_+$ equals t if $t>0$ and 0 if $t\leq 0$. In (1.10), μ is the chemical potential [given by $-dE^{\rm TF}(\lambda)/d\lambda$] and $E^{\rm TF}(\lambda)$ is the TF energy given by

$$E^{\rm TF}(\lambda)=\mathcal{E}^{\rm TF}(\rho_\lambda). \quad (1.11)$$

The convergence as $N\to\infty$ in (1.7) occurs in several senses. First, there is the energy convergence

$$Z^{-7/3}E_{N,Z}\to E^{\rm TF}(\lambda)\quad \text{as } N\to\infty, \quad (1.12)$$

where $N/\lambda=Z$ as before. This means that TF theory exactly gives the leading term in the ground-state energy. Second, $Z_N^{-2}\rho_{N,Z}(Z^{-1/3}x)$ converges to $\rho_\lambda^{\rm TF}(x)$ in the following sense as $N\to\infty$:

$$\int_A[Z_N^{-2}\rho_{N,Z}(Z^{-1/3}x)-\rho_\lambda^{\rm TF}(x)]d^3x\to 0 \quad (1.13)$$

for any measurable set A in $\mathbb{R}^3$. Note that for each N

$$\int Z_N^{-2}\rho_{N,Z}(Z_N^{-1/3}x)d^3x=\frac{N}{Z_N}=\lambda=\int\rho_\lambda^{\rm TF}(x)d^3x, \quad (1.14)$$

which means that the TF density correctly accounts for *all* of the electrons. Equation (1.13) implies that "most of the electrons" in a large atom are at a distance of order $Z^{-1/3}$ (or $N^{-1/3}$) from the nucleus. Thus, in one sense at least, *a large atom has a smaller radius than a small atom.* This radius $Z^{-1/3}$ answers the question, where are most of the electrons, but it is not the *chemical radius*, which can be defined as the radius beyond which there is one electron. This chemical radius is presumably of the order unity (i.e., independent of N), but no one has succeeded in proving this so far.

There is, however, another important radius, namely, Z^{-1}. This is the radius of the innermost, or K-shell, electrons. Quantum effects at this inner radius give rise to the next correction to the ground-state energy, i.e.,

$$E_{N,Z}=Z^{7/3}E^{\rm TF}(\lambda)+Z^2/4+(\text{lower order terms}). \quad (1.15)$$

The second term $Z^2/4$ is the correction predicted by Scott [7], which has been proved to be correct by Siedentop and Weikard [8,9] and Hughes [10]. It is important to note that this second term is *independent of* λ and is therefore consistent with the view that as long as N is of the order Z (i.e., $\lambda>0$) the innermost shells are always totally filled and the contributions of these shells (of radius of order Z^{-1}) to the energy and density are not affected by the bulk of the electrons, which are much farther out at a radius of order $Z^{-1/3}$.

How many electrons are in the inner shells? Thomas-Fermi theory would predict [from (1.7) and (1.10)] that when $|x|\ll Z^{-1/3}$ and $N=\lambda Z$

$$\rho_{N,Z}(x)\approx(2m/\hbar^2)^{3/2}(3\pi^2)^{-1}(Z/|x|)^{3/2}. \quad (1.16)$$

Note that this form, (1.16), depends only on Z and not on N and therefore is consistent with the remark in the preceding paragraph. When $|x|\approx Z^{-1}$, then (1.16) predicts that $\rho_{N,Z}(x)\approx Z^3$, which is reasonable, but as $|x|\to 0$ the right-hand side of (1.16) tends to infinity and this cannot be a correct conclusion. In other words, the TF prediction (1.16) has to be modified for $|x|$ of the order Z^{-1} or smaller and it is this modification that ultimately leads to the Scott correction to the energy $Z^2/4$.

There was a conjecture [6] about the appropriate corrections to (1.16) when $|x|$ is small. This is the *strengthened form of the Scott correction* and recently it has been proved [11]. The purpose of this paper is to try to elucidate the consequences of the strengthened form of the Scott correction, which is the following. For distances that are small compared to the Thomas-Fermi scale $Z^{-1/3}$, the density $\rho_N(x)$ is given for large N and Z by

$$\rho_{N,Z}(x)\approx\rho_Z^B(x), \quad (1.17)$$

where ρ_Z^B is the density in an infinite "Bohr atom" of nuclear charge Z, which depends on Z, but not on N. By this we mean the density in an atom with infinitely many electrons in which the electron-electron repulsion [i.e., the third term in (1.1) is omitted]. The precise statement of (1.17) is given in (1.21) below. This density ρ_Z^B can be calculated explicitly and we do so in the following. While the formal definition of ρ_Z^B is easy to obtain, the actual elucidation of the properties of this function, which is the purpose of this paper, turns out to be surprisingly difficult.

A. Definition of the hydrogenic density

Let $\psi_{nlm}(x)$, with x in $\mathbb{R}^3$, denote the normalized bound-state Bohr orbitals of a hydrogen atom (with unit nuclear charge) and in atomic units in which $\hbar^2/m=1$ and $e=1$, i.e.,

$$\left[-\frac{1}{2}\Delta-\frac{1}{|x|}\right]\psi_{nlm}=e_n\psi_{nlm}. \tag{1.18}$$

Here n is the principal quantum number, l is the angular momentum, and m is the azimuthal quantum number. The energy is $e_n=-1/(2n^2)$. Then

$$\rho^{\mathrm{H}}(x)=\sum_{n=1}^{\infty}\sum_{l=0}^{n-1}\sum_{m=-l}^{l}|\psi_{nlm}(x)|^2 \tag{1.19}$$

is the sum of the square of all these bound-state eigenfunctions. This sum, as will be seen later, is finite for all x. Note also that, since we sum on m in (1.19), $\rho^{\mathrm{H}}(x)$ is a function only of the radius $|x|=r$. We shall abuse notation by writing $\rho^{\mathrm{H}}(r)$ instead of $\rho^{\mathrm{H}}(x)$.

The relation between ρ_Z^B and ρ^{H} is

$$\rho_Z^B(x)=2\left[\frac{Ze^2m}{\hbar^2}\right]^3\rho^{\mathrm{H}}\left(\frac{Ze^2m}{\hbar^2}x\right), \tag{1.20}$$

in which the factor 2 arises from the two spin states of the electron. The precise statement of (1.17) is that as N and Z tend to infinity with $N/Z=\lambda$,

$$\left[\frac{\hbar^2}{Z_Ne^2m}\right]^3\rho_{N,Z}\left(\frac{\hbar^2}{Z_Ne^2m}x\right)-2\rho^{\mathrm{H}}(x)\to 0 \tag{1.21}$$

in two senses. One is the three-dimensional integral sense as in (1.13). The second and stronger sense is pointwise (with respect to radius) on spheres: If S_R is a sphere of radius $R>0$ centered at the origin and if A is any (two-dimensional) measurable subset of S_R, then the (two-dimensional) integral over A of the left-hand side of (1.21) converges to zero. This is proved in [11].

B. Definition of ψ_{nlm}

In polar coordinates $x=r\,(\sin\theta\cos\phi,\ \sin\theta\sin\phi,\ \cos\theta)$,

$$\psi_{nlm}(x)=R_{nl}(r)Y_{lm}(\theta,\phi). \tag{1.22}$$

Here Y_{lm} is a normalized spherical harmonic

$$\int_0^{2\pi}\int_0^{\pi}|Y_{lm}(\theta,\phi)|^2\sin\theta\,d\theta\,d\phi=1. \tag{1.23}$$

The radial function is

$$R_{nl}(r)=N_{nl}(2r/n)^l e^{-r/n}L_{n-l-1}^{2l+1}(2r/n). \tag{1.24}$$

The normalization constant N_{nl} is given by

$$(N_{nl})^2=4n^{-4}\frac{(n-l-1)!}{(n+l)!} \tag{1.25}$$

and is chosen so that

$$\int_0^{\infty}R_{nl}(r)^2r^2dr=1. \tag{1.26}$$

The function L_n^m is the Laguerre polynomial (see [12], Sec. 10.12)

$$L_n^m(z)=\sum_{j=0}^{n}\binom{n+m}{n-j}\frac{1}{j!}(-z)^j. \tag{1.27}$$

The summation on m in (1.19) can be done easily:

$$\sum_{m=-l}^{l}|Y_{lm}(\theta,\phi)|^2=\frac{1}{4\pi}(2l+1)\ ; \tag{1.28}$$

therefore

$$\rho^{\mathrm{H}}(r)=\frac{1}{4\pi}\sum_{n=1}^{\infty}\sum_{l=0}^{n-1}(2l+1)[R_{nl}(r)^2]. \tag{1.29}$$

The simplest case of R_{nl} is $l=n-1$, where

$$R_{n,n-1}(r)=2n^{-2}[(2n-1)!]^{-1/2}(2r/n)^{n-1}e^{-r/n}, \tag{1.30}$$

which is easily seen to satisfy the normalization condition (1.26) and the Schrödinger equation (1.18) with $e_n=1/(2n^2)$.

II. ANALYTIC PROPERTIES OF THE HYDROGENIC DENSITY ρ^{H}

A. Large distance asymptotics

The function $\rho^{\mathrm{H}}(x)$ is a fixed function, but the right-hand side of (1.20) defines a function that has a scale length Z^{-1} and a typical amplitude Z^3. The Thomas-Fermi prediction (1.7), namely, $Z^2\rho_\lambda^{\mathrm{TF}}(Z^{1/3}x)$, is a function with scale length $Z^{-1/3}$ and a typical amplitude Z^2. Since the strong form of the Scott correction is correct these two functions must come together smoothly. This means that the *large distance* behavior of $\rho^{\mathrm{H}}(x)$ must coincide with the *short distance* behavior of $\rho^{\mathrm{TE}}(x)$, which is given by (1.16) with $\hbar/m=Z=1$. This is indeed true.

Theorem 1 (large distance behavior of ρ^{H}). *As $r\to\infty$, $\rho^H(r)$ has the asymptotic expansion*

$$\rho^{\mathrm{H}}(r)=\frac{1}{\pi^2\sqrt{2}}r^{-3/2}\left[\sum_{j=0}^{\infty}a_j(8r)^{-j}-\sin(\sqrt{32r})\sum_{j=1}^{\infty}b_j(8r)^{-j}+\cos(\sqrt{32r})\sum_{j=1}^{\infty}c_j(8r)^{-j-1/2}\right]. \tag{2.1}$$

The first few coefficients are

$$a_0=2/3,\quad a_1=-1/12,\quad a_2=79/960,$$

$$b_1=3/2,\quad b_2=-140\ 589/11\ 200,$$

$$c_1=141/40,\quad c_2=-2\ 028\ 627/44\ 800.$$

In particular,

$$\rho^{\mathrm{H}}(r)=2^{1/2}(3\pi^2)^{-1}r^{-3/2}+O(r^{-5/2}), \tag{2.2}$$

which is equivalent to

$$\lim_{r\to\infty}r^{3/2}\rho^B(r)=\lim_{r\to 0}r^{3/2}\rho_\lambda^{\mathrm{TF}}(r) \tag{2.3}$$

for each $0<\lambda\leq 1$.

Equation (2.3) is thus consistent with (1.16), (1.17), (1.20), and (1.21). The truth of Eq. (2.3) is motivated by the physical considerations at the beginning of this section. In Appendix F we derive the asymptotic expansion, the coefficients of which can, in principle, be calculated to arbitrarily high order.

Two important facts should be noted about this theorem. The first is that the oscillations of $\rho^{\rm H}(r)$ (which is often called the "shell structure") are quite small. They hardly leap to the eye in the graph of $\rho^{\rm H}(r)$ (Fig. 1). The second fact is that the period of the oscillations (as a function of $\sqrt{r}$, not r) is $2\pi/\sqrt{32}=1.11$. This is what one would expect from the fact that the average potential energy $-1/r$ for principle quantum number n is $2e_n=n^{-2}$. The fact that $\sqrt{r}$ does not peak at n but rather at slightly larger value $1.11n$ is consistent with Hölder's inequality applied to expectation values $\langle\ \rangle_n$ at the nth level; this inequality is, in general. $\langle r^{1/2}\rangle\ \langle r^{-1}\rangle^{1/2}\geqslant 1$. What is a bit unexpected is that the increase (from 1 to 1.11) is *independent* of n (see Fig. 3). The following is needed for the proof of Theorem 1.

Lemma 1 (first integral representation).

$$\rho^{\rm H}(r)=\pi^{-2}(2r)^{-3/2}\int_0^\infty xe^{-x}\phi(x)\int_0^\pi w(x,\theta)\times J_3(2\phi(x)[2r\,w(x,\theta)]^{1/2})d\theta\,dx, \tag{2.4}$$

where J_3 is a Bessel function and

$$\phi(x)=[x/(1-e^{-x})]^{1/2} \tag{2.5}$$

and

$$w(x,\theta)=1+e^{-x}-2e^{-x/2}\cos\theta. \tag{2.6}$$

Part of the proof of Lemma 1, which is given in Appendix B, rests on the fact that we are able to do the sum on l in (1.29) in closed form as follows.

Lemma 2 (the sum on a row). *For all complex numbers z*

$$S_n(z):=\sum_{l=0}^{n-1}(2l+1)\frac{(n-l-1)!}{(n+l)!}z^{2l}[L_{n-l-1}^{2l+1}(z)]^2 \tag{2.7}$$

satisfies

$$S_n(z)=\sum_{m=0}^{n-1}(m!)^{-2}z^{2m}L_{n-m-1}^{2m+1}(2z)=n[L_{n-1}^1(z)]^2+n[L_{n-2}^1(z)]^2+(z-2n)L_{n-1}^1(z)L_{n-2}^1(z). \tag{2.8}$$

Moreover,

$$\frac{d}{dz}e^{-z}S_n(z)=-e^{-z}[L_{n-1}^1(z)]^2, \tag{2.9}$$

which implies that $e^{-z}S_n(z)$ is monotonically decreasing in z for real z.

Using the definitions (1.24) and (1.29) we have

$$\rho^{\rm H}(r)=\frac{1}{\pi}\sum_{n=1}^\infty n^{-4}e^{-2r/n}S_n(2r/n), \tag{2.10}$$

$$\rho^{\rm H}(r)=\frac{1}{\pi}\sum_{n=1}^\infty n^{-4}e^{-2r/n}[nL_{n-1}^1(2r/n)^2+nL_{n-2}^1(2r/n)^2+2(r/n-n)L_{n-1}^1(2r/n)L_{n-2}^1(2r/n)] \tag{2.11}$$

and

$$\frac{d}{dr}\rho^{\rm H}(r)=-\frac{2}{\pi}\sum_{n=1}^\infty n^{-5}e^{-2r/n}[L_{n-1}^1(2r/n)]^2, \tag{2.12}$$

which shows that $\rho^{\rm H}(r)$ is monotonically decreasing in r and hence that $\rho^{\rm H}(r)$ is finite for all r and achieves its maximum at $r=0$:

$$\rho^{\rm H}(r)\leqslant\rho^{\rm H}(0)=\frac{1}{\pi}\sum_{n=1}^\infty n^{-3}\simeq 0.383. \tag{2.13}$$

Lemma 2 will be proved in Appendix A. It will be useful for us to recast Lemma 1 in the following form, which will be proved in Appendix C.

Lemma 3 (second integral representation). *The density ρ^H can be written as*

$$\rho^{\rm H}(r)=\frac{2}{\pi^2(2r)^{3/2}}\int_0^\infty J_3(y\sqrt{8r})y^{-2}\int_{\psi_3(y)}^{\psi_1(y)}[\phi_3(x)]^2\times[y^2-\phi_1(x)^2]^{-1/2}[\phi_2(x)-y^2]^{-1/2}\,x\,dx\,dy, \tag{2.14}$$

where

$$\phi_1(x)=[x\tanh(x/4)]^{1/2}, \tag{2.15}$$

$$\phi_2(x)=[x\coth(x/4)]^{1/2}, \tag{2.16}$$

$$\phi_3(x)=\frac{x}{2\sinh(x/2)}, \tag{2.17}$$

$$\psi_3(y)=\begin{cases}0 & \text{if } 0\leqslant y\leqslant 2\\ \psi_2(y) & \text{if } 2\leqslant y,\end{cases}$$

and where ψ_1 and ψ_2 are the inverse functions of ϕ_1 and ϕ_2, e.g., $\psi_i(y)=x$ when $y=\phi_i(x)$, $i=1,2$.

The second integral representation brings us to the following problem, which can be stated generally as follows. Find an asymptotic expansion, for large x, of integrals of Bessel functions of the form

$$\int_a^b J_\nu(xt)f(t)dt. \tag{2.18}$$

We formulate our results as two lemmas, which are proved in Appendix D and E, respectively, and we emphasize that our proofs are an adaptation of Olver's general method [13].

Lemma 4 [asymptotics of (2.18) with a=0]. *Let $f(t)$ have the expansion*

$$f(t)\approx t^{\mu-\nu-1}\sum_{j=0}^{\infty} a_j t^j, \quad \mu>0 \tag{2.19}$$

in the limit $t\to 0+$, valid in the sense that (2.19) represents both $f(t)$ and all its derivatives $f^{(n)}(t)$ asymptotically correctly in the limit $t\to 0+$. Assume further that $f(t)$ is continuously differentiable infinitely many times in the open interval $0<t<b$ and that all the limits

$$\lim_{t\uparrow b} f^{(n)}(t) =: f^{(n)}(b), \quad n=0,1,2,\ldots \tag{2.20}$$

exist and are finite. Then we have the asymptotic expansion for $x\to +\infty$

$$\int_0^b J_\nu(xt)f(t)dt=\sum_{j=0}^{\infty} a_j x^{\nu-\mu-j}2^{\mu+j-\nu-1} \times\Gamma((\mu+j)/2)/\Gamma(\nu+1-(\mu+j)/2) +G_+(1,b^-,x), \tag{2.21}$$

where

$$G_\pm(\lambda,b,x)=\left(\frac{2}{\pi}\right)^{1/2}\sum_{s=0}^{\infty} x^{-s-\lambda-1/2}b^{-s-1/2} \times\cos\left[xb-\left(\nu+s\pm\lambda+\frac{1}{2}\right)\frac{\pi}{2}\right] \times\sum_{j=0}^{s}\frac{f^j(b)}{j!}b^j\sum_{l=0}^{s-j}(-1)^{j+l}\frac{\Gamma(s-l+\lambda)}{l!(s-j-l)!} \times\frac{\Gamma(s-j+\frac{1}{2})\Gamma(\nu+l+\frac{1}{2})}{\Gamma(l+\frac{1}{2})!\Gamma(\nu-l+\frac{1}{2})}2^{-l}. \tag{2.22}$$

Lemma 5 [asymptotics of (2.18) with $a>0$]. *Let $f(t)$ have the expansion*

$$f(t)\approx\sum_{j=0}^{\infty} a_j(t-a)^{j+\lambda-1}, \quad \lambda>0 \tag{2.23}$$

in the limit $t\to a+$, valid in the sense that (2.23) represents both $f(t)$ and all its derivatives asymptotically correctly in the limit $t\to a+$. Assume further that $f(t)$ is continuously differentiable infinitely many times in the open interval $a<t<b$ and that all the limits

$$\lim_{t\uparrow b} f^{(n)}(t) =: f^{(n)}(b), \quad n=0,1,2,\ldots \tag{2.24}$$

exists and are finite. Then we have the asymptotic expansion for $x\to +\infty$

$$\int_a^b J_\nu(xt)f(t)dt=G_-(\lambda,a+,x)+G_+(1,b-,x), \tag{2.25}$$

where the functions $G_\pm$ are given by Eq. (2.22).

If a and b are interchanged (i.e., $b<a$), then the condition in the limit $t\to a-$ becomes

$$f(t)\approx\sum_{j=0}^{\infty}(-1)^j a_j(a-t)^{j+\lambda-1}, \quad \lambda>0 \tag{2.26}$$

and the result becomes

$$\int_b^a J_\nu(xt)f(t)dt=G_-(1,b+,x)+G_+(\lambda,a-,x) \tag{2.27}$$

If we take the limit $b\to\infty$ in (2.25), we get the asymptotic expansion

$$\int_a^\infty J_\nu(xt)f(t)dt=G_-(\lambda,a+,x) \tag{2.28}$$

for $x\to +\infty$, provided $f(t)$ is continuously differentiable infinitely many times in the open interval $a<t<\infty$ and Eq. (2.23) still holds in the limit $t\to a+$, while the requirements in the limit $t\to +\infty$ are absolute integrability of $f(t)$ and

$$\lim_{t\to\infty} t^{-1/2}f^{(n)}(t)=0, \quad n=1,2,\ldots. \tag{2.29}$$

We conclude with a heuristic derivation of the origin of the oscillatory terms in Theorem 2.1.

B. Large n asymptotics of $S_n(r)$

According to (2.9) the function of interest is

$$W_n(z)=e^{-z/2}L_{n-1}^1(z) \tag{2.30}$$

(but recall that we are interested in this for $z=2r/n$). According to Erdelyi *et al.* [12], there is an asymptotic formula due to Tricomi that is useful for numerical computations:

$$W_n(z)=n^{1/2}z^{-1/2}\sum_{m=0}^{\infty} A_m(z/4n)^{m/2}J_{m+1}(\sqrt{4nz}), \tag{2.31}$$

in which J_{m+1} is a Bessel function and

$$A_0=1,\quad A_1=0,\quad A_2=1,$$

$$A_m=A_{m-2}-(2n/m)A_{m-3},\quad m>2. \tag{2.32}$$

This series is valid for $z<$ const $\times n^{1/3}$.

The Bessel functions $J_{m+1}(x)$ are oscillatory for large x with period 2π and it is this oscillation that gives rise to the oscillation shown in Theorem 1. For large n and fixed $z>0$, we have from (2.31) that [Fejér's formula, see [12], Eq. (10.15.1)]

$$W_n(z)=\pi^{-1/2}z^{-3/4}n^{1/4}\cos[2(nz)^{1/2}-3\pi/4]+O(n^{-1/4}). \tag{2.33}$$

By substituting $z=2r/n$ in (2.31) and then squaring W_n (which halves the period) we would conclude that the correction term in $\rho^{\mathrm{H}}(r)$ has period $\pi/\sqrt{8}$ (in $\sqrt{r}$), as given in Theorem 1. This argument requires additional justification, however, for the following reason. We require $W_n(2r/n)$ for n of the order $\sqrt{r}$, which is much smaller than r when r is large. This means that $z=2r/n$ is not always small and therefore the asymptotic formula (2.31) is not necessarily valid in the region of interest. Nevertheless, the conclusion just stated is correct. The oscillation period in $\sqrt{r}$ is indeed $\pi/\sqrt{8}$ as

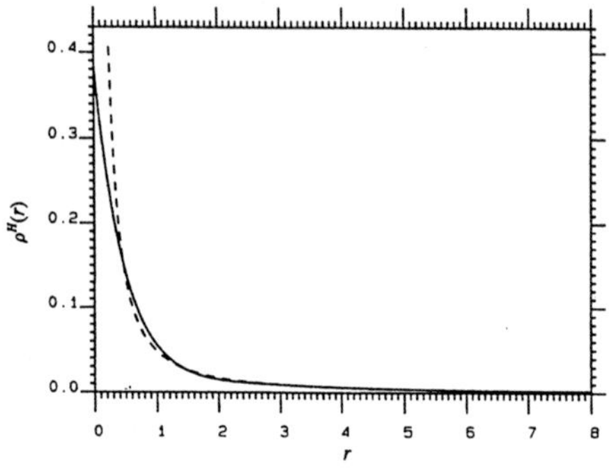

FIG. 1. $\rho^{\mathrm{H}}(r)$ as a function of r. Also shown (dashed curve) is the Thomas-Fermi asymptotic expression $2^{1/2}(3\pi^2)^{-1}r^{-3/2}$.

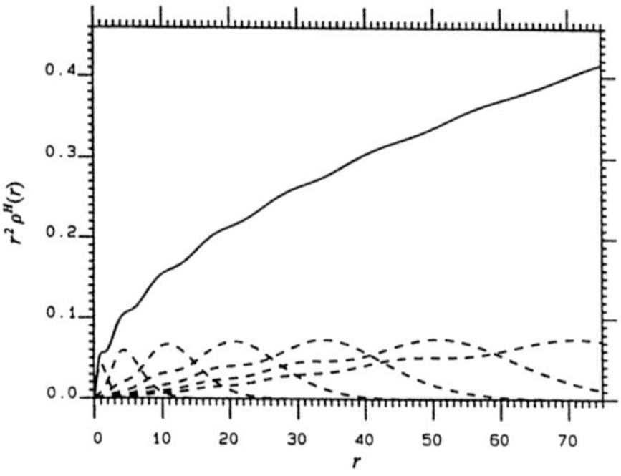

FIG. 3. $r^2\rho^{\mathrm{H}}(r)$ as a function of r. Also shown (dashed curves) are $r^2 S_n(2r/n)n^{-4}e^{-2r/n}/\pi$ for $n = 1-7$.

(2.31) would seem to imply, but we do *not* assert that the amplitudes A_m are the correct ones. The correct formula is in Theorem 1.

III. SUMMARY OF RESULTS

We have analyzed the function $\rho^{\mathrm{H}}(r)$, which is the sum of the squares of the hydrogenic bound-state functions. This function is the density that would be seen in a very large Z atom on a distance scale Z^{-1} from the nucleus. It is very nicely behaved, being finite everywhere and monotonically decreasing. We have derived integral representations for this function that are useful for analytical and numerical analysis. We have analyzed the large r behavior of this function in detail and discovered that, contrary to our expectation at least, there is very little oscillatory structure. In fact, it is necessary to take two derivatives with respect to r in order to make the oscillatory terms as large as the nonoscillatory ones. In short, shell structure is not a prominent property of this universal atomic function.

In order to clarify the nature of $\rho^{\mathrm{H}}(r)$ we give a graph of this function together with its asymptotic Thomas-Fermi form $2^{1/2}(3\pi^2)^{-1}r^{-3/2}$ in Fig. 1. The difference between the two functions, which is the part of the density that is not semiclassical, is also plotted in Fig. 2. The conventional function $r^2\rho^{\mathrm{H}}(r)$ is plotted in Fig. 3.

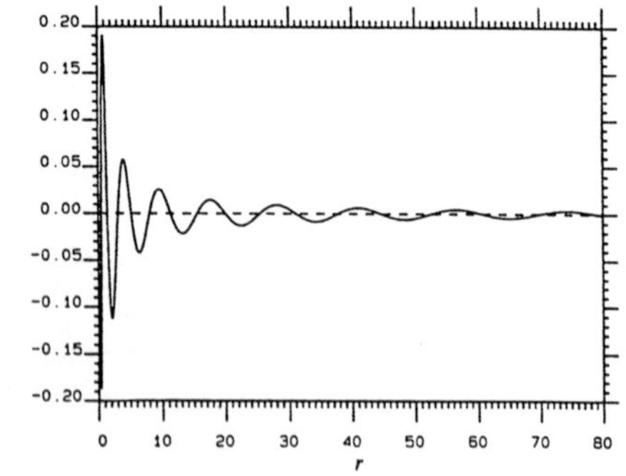

FIG. 2. Relative difference $\rho^{\mathrm{H}}(r)/[2^{1/2}(3\pi^2)^{-1}r^{-3/2}]-1$ as a function of r.

Note: Appendices A-F, which contain the proofs of Lemmas 1-5 and Theorem 1, have been omitted in this reproduction. They can be found in the original publication.

[1] G. Zhislin, Tr. Mosk. Mat. Obshch. **9**, 81 (1960).
[2] E. H. Lieb, Phys. Rev. A **29**, 3018 (1984).
[3] E. H. Lieb, I. M. Sigal, B. Simon, and W. Thirring, Phys. Rev. Lett. **52**, 994 (1980); Commun. Math. Phys. **116**, 635 (1988).
[4] C. L. Fefferman and L. A. Seco, Commun. Math. Phys. **128**, 109 (1990).
[5] E. H. Lieb and B. Simon, Adv. Math. **23**, 22 (1977).
[6] E. H. Lieb, Rev. Mod. Phys. **53**, 603 (1981); **54**, 311(E) (1982).
[7] J. M. C. Scott, Philos. Mag. **43**, 859 (1952).
[8] H. Siedentop and R. Weikard, Invent. Math. **97**, 213 (1989).
[9] H. Siedentop and R. Weikard, Commun. Math. Phys. **112**, 471 (1987).
[10] W. Hughes, Adv. Math. **79**, 213 (1990).
[11] A. Iantchenko, E. H. Lieb, and H. Siedentop, J. Reine Angew. Math. (to be published).
[12] A. Erdelyi, W. Magnus, F. Oberhettinger, and F. G. Tricomi, *Higher Transcendental Functions* (McGraw-Hill, New York, 1953).
[13] F. W. J. Olver, SIAM J. Math. **5**, 19 (1974).
[14] G. N. Watson, *A Treatise on the Theory of Bessel Functions* (Cambridge University Press, Cambridge, 1962).

J. reine angew. Math. **472** (1996), 177—195

Journal für die reine und angewandte Mathematik

Proof of a conjecture about atomic and molecular cores related to Scott's correction

By *Alexei Iantchenko* at Rennes, *Elliott H. Lieb*[1]) at Princeton and *Heinz Siedentop*[2]) at Oslo

1. Introduction

A great deal is known about the ground states of large atoms in the framework of the non-relativistic Schrödinger equation, with fixed (i.e., infinitely massive) nuclei. The leading term, in powers of the nuclear charge Z, is given exactly by Thomas-Fermi theory, as was proved by Lieb and Simon [12]; see [11] for a review. This leading term in the energy is proportional to $Z^{7/3}$, with the proportionality constant depending on the ratio of N/Z, which is assumed to be held fixed as $Z \to \infty$. Here, N is the electron number. Neutrality, i.e., $N = Z$, is not required, even though it is the case of primary physical interest. The characteristic length scale for the electron density (in the sense that all the electrons can be found on this scale in the limit $Z \to \infty$) is $Z^{-1/3}$. The fact that the true quantum-mechanical electron density, ϱ_d, converges (after suitable scaling) to the Thomas-Fermi density, ϱ^{TF}, as $Z \to \infty$ with N/Z fixed was proved in [12]. The chemical radius, which is another length altogether, is believed, but not proved, to be order Z^0 as $Z \to \infty$.

The first correction to the $Z^{7/3}$ law is not, as was formerly supposed, the $Z^{5/3}$ corrections arising from exchange and correlation effects and kinetic energy corrections on the $Z^{-1/3}$ scale. Instead it is $Z^2 = Z^{6/3}$ and arises from extreme quantum mechanical effects on the innermost electrons, which are at a distance Z^{-1} from the nucleus. Among these the most important are the K shell electrons. It was Scott [19] who pointed this out and he also gave a formula for the correction term to the energy,

$$E^{\text{Scott}} = \frac{q}{8} Z^2, \tag{1}$$

where q is the number of spin states per electron (of course $q = 2$ in nature). It is noteworthy that E^{Scott} does not depend on the fixed ratio N/Z, provided $N/Z \neq 0$. This fact agrees with the idea that E^{Scott} arises from the innermost electrons, whose energies, in leading

[1]) Work partially supported by US National Science Foundation grant PHY90 19433 A04.
[2]) Work partially supported by Norges forskningsråd, grant 412.92/008.

order, are independent of the presence of the electrons that are further from the nucleus. The truth of (1) (i.e., the statement that the energy is $E^{\mathrm{TF}} + E^{\mathrm{Scott}} + o(Z^2)$ for fixed $N/Z \neq 0$) was proved in [22], [23] (upper and lower bound) and by Hughes [9] (lower bound) in the atomic case and by Bach [1] in the ionic case. (For different proofs and extensions of this result see [24], Fefferman and Seco [6], [2], [3], [4], [5], [7], and Ivrii and Sigal [10].)

The conjecture about the density, which we prove here, was made later by Lieb in [11], Equation (5.34). It concerns the electron density ϱ_d at distances of order Z^{-1} from the nucleus and states that in limit $Z \to \infty$ a suitably scaled ϱ_d converges to the sum of the squares of *all* the hydrogenic bound states ϱ^H. Because of it's relation to the Scott correction we will use henceforth the term "strong Scott correction".

The function ϱ^H is defined in Section 2 below and is extensively analyzed in [8]. (Previously, an upper bound for ϱ_d at the origin of the correct form, namely $O(Z^3)$, was derived in [20], [21].) We prove this convergence in several senses, one of which is a "pointwise" convergence on spheres. In fact we go further and show that the individual angular momentum densities converge to the hydrogenic values, thereby giving a somewhat more refined picture of the ground state.

Our main proof strategy is the usual one. We add ε times a one-body test potential to our Hamiltonian and then differentiate the ground state energy with respect to ε at $\varepsilon = 0$ in order to find ϱ_d. To obtain pointwise convergence the test potential is a radial delta-function. To control the energy we rely, in part, on the results and methods in [22], [23].

In the following we shall state and prove our theorems for the neutral case $N = Z$. We do so to avoid the notational complexity and additional discussion required for $N/Z \neq 1$. It is straightforward, however, to generalize our results to $N/Z \neq 1$.

In the next section precise definitions, as well as our main theorems are given. Section 3 contains two lemmata about the difference in energies with and without the test potential. Since there are infinitely many hydrogenic bound states, we need these estimates in order to be able to show that the sum of the derivatives (with respect to ε) equals the derivative of the sum. The strong Scott conjecture for atoms is proved in Section 4. Section 5 contains the obvious extension to molecules. The Appendix contains a few needed facts about ground state energies.

2. Definitions and main results

The Hamiltonian of an atom of N electrons with q spin states each and a fixed nucleus of charge Z located at the origin is given by

$$H_{N,Z} = \sum_{\nu=1}^{N} \left(-\Delta_\nu - \frac{Z}{|\mathfrak{r}_\nu|}\right) + \sum_{1 \leq \mu < \nu \leq N} \frac{1}{|\mathfrak{r}_\mu - \mathfrak{r}_\nu|}, \tag{2}$$

in units in which $\hbar^2/2m = 1$ and $|e| = 1$. It is self-adjoint in the Hilbert space $\mathfrak{H}_N := \bigwedge_{\nu=1}^{N} (L^2(\mathbb{R}^3) \otimes \mathbb{C}^q)$, i.e., antisymmetric functions of space and spin. A general ground state density matrix, denoted by d, can be written as

(3) $$d = \sum_{\nu=1}^{M} w_\nu |\psi_\nu\rangle\langle\psi_\nu|,$$

where the ψ_ν constitute an orthonormal basis for the ground state eigenspace and where the w_ν are nonnegative weights such that $\sum_{\nu=1}^{M} w_\nu = 1$. It is well known that the ground state can be degenerate, e.g., it is for the carbon atom. The corresponding one-electron density is the diagonal part of the one-electron density matrix and is, by definition,

(4) $$\varrho_d(\mathfrak{r}) = N \sum_{\nu=1}^{M} w_\nu \sum_{\sigma_1,\ldots,\sigma_N=1}^{q} \int_{\mathbb{R}^{3(N-1)}} |\psi_\nu(\mathfrak{r}, \mathfrak{r}_2, \ldots, \mathfrak{r}_N; \sigma_1, \ldots, \sigma_N)|^2 d\mathfrak{r}_2 \ldots d\mathfrak{r}_N.$$

The density $\varrho_{l,d}$ of angular momentum l electrons at radius r from the nucleus is given in terms of the normalized spherical harmonics $Y_{l,m}(\omega)$.

(5) $$\varrho_{l,d}(r) = N \sum_{m=-l}^{l} \sum_{\nu=1}^{M} w_\nu \sum_{\sigma_1,\ldots,\sigma_N=1}^{q} \int_{\mathbb{R}^{3(N-1)}} \left| \int_{\mathbb{S}^2} \overline{Y_{lm}(\omega)} \psi_\nu(r\omega, \mathfrak{r}_2, \ldots, \mathfrak{r}_N; \sigma_1, \ldots, \sigma_N) d\omega \right|^2 d\mathfrak{r}_2 \ldots d\mathfrak{r}_N$$

where we write $\mathfrak{r} = r\omega$ and $d\omega$ denotes the usual unnormalized surface measure on the two dimensional sphere $\mathbb{S}^2$ with $(4\pi)^{-1} \int_{\mathbb{S}^2} d\omega = \int_{\mathbb{S}^2} |Y_{l,m}|^2 d\omega = 1$.

Throughout the paper we will write $\varphi_Z^{\mathrm{TF}}(r)$ for the Thomas-Fermi potential of electron number $N = Z$ and nuclear charge Z, i.e.,

$$\varphi_Z^{\mathrm{TF}}(r) = Z/r - |\mathfrak{r}|^{-1} * \varrho_Z^{\mathrm{TF}},$$

where ϱ_Z^{TF} is the nonnegative minimizer of the Thomas-Fermi functional

$$\int_{\mathbb{R}^3} \left(\frac{3}{5} (6\pi^2/q)^{2/3} \varrho^{5/3}(\mathfrak{r}) - \frac{Z}{|\mathfrak{r}|} \varrho(\mathfrak{r}) \right) d\mathfrak{r} + D(\varrho, \varrho)$$

under the condition $\int \varrho = N = Z$ and with

$$D(\varrho, \varrho) := \frac{1}{2} \int_{\mathbb{R}^6} \frac{\overline{\varrho(\mathfrak{r})} \varrho(\mathfrak{s})}{|\mathfrak{r} - \mathfrak{s}|} d\mathfrak{r}\, d\mathfrak{s}.$$

Both φ_Z and ϱ_Z^{TF} are spherically symmetric, i.e., they depend only on $r = |\mathfrak{r}|$. There is a scaling relation $\varphi_Z^{\mathrm{TF}}(r) = Z^{4/3} \varphi_1^{\mathrm{TF}}(Z^{1/3} r)$, where φ_1^{TF} is the Thomas-Fermi potential for $Z = 1$ and "electron number" equal to 1. Similarly, $\varrho_Z^{\mathrm{TF}}(r) = Z^2 \varrho_1^{\mathrm{TF}}(Z^{1/3} r)$. This scaling shows that the "natural" length in an atom is $Z^{-1/3}$. Note that the Thomas-Fermi functional has a unique minimizer [12].

The Scott conjecture, on the other hand, concerns the length scale Z^{-1}, where we expect the density to be of order Z^3 instead of Z^2. In terms of the "true" density defined in (4), we now define

(6) $$\varrho_Z(\mathfrak{r}) := Z^{-3}\varrho_d(\mathfrak{r}/Z).$$

Likewise, we define the angular momentum density

(7) $$\varrho_{l,Z}(r) := Z^{-3}\varrho_{l,d}(r/Z).$$

To formulate the strong Scott conjecture we consider the angular momentum l states of a hydrogen atom ($Z = 1$) with radial Hamiltonian

(8) $$h_l^H := -\frac{d^2}{dr^2} + \frac{l(l+1)}{r^2} - \frac{1}{r}$$

with normalized eigenfunctions $\psi_{l,n}^H$ (that vanish at 0 and ∞) corresponding to *negative* eigenvalues $e_{l,n}^H$. (The superscript H denotes "Hydrogen" and distinguishes h_l^H from other radial Hamiltonians to be considered later.) The normalization is $\int_0^\infty |\psi_{n,l}(r)|^2 dr = 1$. We define the corresponding density in the channel l to be

(9) $$\varrho_l^H(r) := q(2l+1) \sum_{n=0}^{\infty} |\psi_{l,n}^H(r)|^2/(4\pi r^2);$$

the total density is then

(10) $$\varrho^H(r) = \sum_{l=0}^{\infty} \varrho_l^H(r).$$

Although we shall not be interested in detailed properties of ϱ^H, we note the following proved in [8]: The sum over l and n defining $\varrho^H(r)$ is pointwise convergent for all r. It is monotone decreasing and it decays asymptotically for large r as $1/(6\pi^2 r^{3/2})$. This large r asymptotics meshes nicely with the *small* r behavior of $\frac{1}{q}\varrho_1^{\mathrm{TF}}(r)$.

Note: In [8] the operator h^H is defined using atomic units $\hbar^2/m = 1$, i.e., with $\frac{1}{2}(-d^2/dr^2 + l(l+1)/r^2)$ instead of $-d^2/dr^2 + l(l+1)/r^2$. Note also that we have included the factor q in the definition of ϱ which was not done in [8]. Thus some care is needed in comparing formulae there with formulae here.

Various notions for the convergence of the rescaled density are possible. Our precise statements are Theorems 1 and 2 below and Theorem 3 in Section 5.

Theorem 1 (Convergence in angular momentum channels). *Fix the angular momentum l_0.*

1. *For positive r*

(11) $$\lim_{Z\to\infty} \varrho_{l_0,Z}(r) = \varrho_{l_0}^H(r)$$

(*pointwise convergence*).

2. *Let V be an integrable function on the positive real line. Then we have the weak convergence*

$$\lim_{Z\to\infty} \int_0^\infty rV(r)\varrho_{l_0,Z}(r)\,dr = \int_0^\infty rV(r)\varrho^H_{l_0}(r)\,dr. \tag{12}$$

Theorem 2 (Convergence of the total density). 1. *Let W be a bounded (not necessarily constant) function on the unit sphere and r positive. Then, as $Z \to \infty$,*

$$\int_{\mathbb{S}^2} W(\omega)\varrho_Z(r\omega)\,d\omega \to \varrho^H(r)\int_{\mathbb{S}^2} W(\omega)\,d\omega \tag{13}$$

(pointwise convergence of spherical averages).

2. *Let V be a locally bounded, integrable function on $\mathbb{R}^3$. Then, as $Z \to \infty$,*

$$\int_{\mathbb{S}^3} |\mathfrak{r}|V(\mathfrak{r})\varrho_Z(\mathfrak{r})\,d\mathfrak{r} \to \int_{\mathbb{R}^3} |\mathfrak{r}|V(\mathfrak{r})\varrho^H(|\mathfrak{r}|)\,d\mathfrak{r}. \tag{14}$$

Remarks. 1. It is not really necessary to take a sequence of ground state density matrices. We could take just a sequence of states, $d_{N,Z}$, that is an *approximate ground state* in the sense that

$$\frac{\operatorname{tr}(H_{N,Z}d_{N,Z}) - E_{N,Z}}{Z^2} \to 0$$

as $Z \to \infty$. Here $E_{N,Z}$ is the bottom of the spectrum of $H_{N,Z}$. It might not be an eigenvalue, and it certainly will not be one if N/Z is larger than 2.

2. It is important to note that W and V in (13) and (14) need not be spherically symmetric. It might appear that only the spherical averages of W and V are relevant, but this would miss the point. Theorem 2 says, that in the limit $Z \to \infty$ there is no way to construct a ground state or approximate ground state that is not spherically symmetric on a length scale Z^{-1}. For example, in the case of carbon there are ground states that are not spherically symmetric and for which replacing W by its spherical average changes the left side of (13).

3. A word about pointwise convergence. The one-body density *matrix* $\gamma(\mathfrak{r},\mathfrak{r}')$ which is defined as in (4) but with $|\psi_\nu(\cdots)|^2$ replaced by

$$\psi_\nu(\mathfrak{r},\mathfrak{r}_2,\ldots,\mathfrak{r}_N;\sigma_1,\ldots,\sigma_N)\overline{\psi(\mathfrak{r}',\mathfrak{r}_2,\ldots,\mathfrak{r}_N;\sigma_1,\ldots,\sigma_N)},$$

is easily seen to be in the Sobolev space $H^1(\mathbb{R}^3)$ when γ is considered as a function of each variable separately. The trace theorem in Sobolev spaces then implies that the function of ω on the sphere $\mathbb{S}^2$, $\gamma(r\omega, r'\omega')$ is in $L^q(\mathbb{S}^2)$ for all $q \leq 4$. Thus, the integrals in (4) are well defined and $\varrho_d(r\omega) = \gamma(r\omega, r\omega)$ is in $L^2(\mathbb{S}^2)$. It is also easy to see that $\sqrt{\varrho_{l,Z}(r)}$ is in $H^1(0,\infty)$ and hence it is a continuous function of r. Since $\varrho_d(\mathfrak{r})$ is in $L^2(\mathbb{S}^2)$ the integrals in (11) and (13) are well defined when $W \in L^2(\mathbb{S}^2)$.

If ϱ and γ belong to a ground state of $H_{N,Z}$ with $N \leq Z$ then they are even continuous functions in all variables. This follows from the regularity theorem of Kato and Simon

(Reed and Simon [18], Theorem XIII.39) and the uniform exponential decay of ground state eigenfunctions. This decay is implied as follows. By Zhislin's theorem the atomic Hamiltonian has infinitely many eigenvalues below the essential spectrum and the ground state eigenspace has finite dimension. This implies that the ground state energy is always a discrete eigenvalue which, in turn, implies exponential decay of the ground state eigenfunctions according to Theorem XIII.42 of [18].

3. Eigenvalue differences of Schrödinger operators perturbed on the scale $1/Z$

It is well known, and will be seen more explicitly in Section 4, that the eigenvalues of $H_{N,Z}$ can be controlled to within an accuracy of $o(Z^2)$ by considering a one-body Schrödinger operator with the spherically symmetric potential given by Thomas-Fermi theory. In the angular momentum l channel, this is

$$h_{l,Z}^{\mathrm{TF}} = -\frac{d^2}{dr^2} + \frac{l(l+1)}{r^2} - \varphi_Z^{\mathrm{TF}}(r). \tag{15}$$

(We suppress the dependence on N in $h_{l,Z}^{\mathrm{TF}}$ since $N = Z$.) Closely related to $h_{l,Z}^{\mathrm{TF}}$ is the unscreened hydrogenic Hamiltonian

$$h_{l,Z}^{H} = -\frac{d^2}{dr^2} + \frac{l(l+1)}{r^2} - \frac{Z}{r}. \tag{16}$$

In this section we want to study how the spectra of these operators are shifted by the addition of a perturbing potential of the form

$$\varepsilon U_Z(r) = \varepsilon Z^2 U(Zr)$$

where ε is a small parameter and where U is some fixed function. In particular, U will be a radial delta function, $U(r) = \delta(r-a)$ for some $a > 0$.

Both cases, h^{TF} and h^{H}, will be considered together and we write

$$h_{l,\varepsilon,Z} = -\frac{d^2}{dr^2} + V_{l,Z}(r) - \varepsilon U_Z(r), \tag{17}$$

in which $V_{l,Z} = -Z/r + l(l+1)/r^2$ or $V_{l,Z} = -\varphi_Z^{\mathrm{TF}}(r) + l(l+1)/r^2$.

Our first lemma estimates the difference in the spectra of $h_{l,0,Z}$ and $h_{l,\varepsilon,Z}$ by the difference in the trace (tr) of the negative parts $(h_{l,\varepsilon,Z})_-$ and $(h_{l,0,Z})_-$ (i.e., the sums of the negative eigenvalues). This lemma will later on allow us to interchange the limits $Z \to \infty$ and $\varepsilon \to 0$ with the l summation.

Lemma 1. *Set $U(r) = \delta(r-a)$, $U_Z(r) = Z^2 U(Zr)$ and assume $|\varepsilon| \leqq \pi/(16a)$. Then*

$$|\operatorname{tr}(h_{l,0,Z})_- - \operatorname{tr}(h_{l,\varepsilon,Z})_-| \leqq |\varepsilon| \frac{9aZ^2}{(l+1)^2(2l+1)}.$$

Proof. By the minimax principle we have for $\varepsilon > 0$

$$0 \leqq s_{\varepsilon,l,Z} := \operatorname{tr}(h_{l,0,Z})_- - \operatorname{tr}(h_{l,\varepsilon,Z})_- \leqq \varepsilon \operatorname{tr}(U_Z d_{l,\varepsilon,Z}). \tag{18}$$

Inserting the identity twice in the right side of (10) we have

$$s_{\varepsilon,l,Z} \leqq \varepsilon \operatorname{tr}(A \circ B \circ C \circ B^* \circ A^*) \leqq \varepsilon \|A\|_\infty^2 \|B\|_\infty^2 \operatorname{tr} C \tag{19}$$

with

$$A := d_{l,\varepsilon,Z}(h_{l,\varepsilon,Z} + c_{l,Z})^{1/2} \geqq 0,$$

$$B := (h_{l,\varepsilon,Z} + c_{l,Z})^{-1/2}(H_{0,l} + c_{l,Z})^{1/2},$$

$$C := (H_{0,l} + c_{l,Z})^{-1/2} U_Z (H_{0,l} + c_{l,Z})^{-1/2} \geqq 0,$$

where $c_{l,Z}$ is any positive number bigger than $|\inf \sigma(h_{l,\varepsilon,Z})|$, where $\sigma(h)$ denotes the spectrum of h. We also define $H_{0,l} := -d^2/dr^2 + l(l+1)/r^2$ to be the free operator in the angular momentum channel l. Since $\varphi_Z^{\mathrm{TF}}(r) \leqq Z/r$ and since $\inf \sigma(H_{0,l} - Z/r) = -Z^2/[4(l+1)^2]$ we can take $c_{l,Z} := 2Z^2/(l+1)^2$ provided ε is not too large.

We now estimate these norms individually:

Because $c_{l,Z}$ is bigger than the modulus of the lowest spectral point of $h_{l,\varepsilon,Z}$ and $d_{l,\varepsilon,Z}$ is the projection onto the negative spectral subspace of $h_{l,\varepsilon,Z}$ we have

$$\|A\|_\infty \leqq \sqrt{c_{l,Z}}. \tag{20}$$

For B we get

$$\begin{aligned} \|B\phi\|^2 &= \|(h_{l,\varepsilon,Z} + c_{l,Z})^{-1/2}(H_{0,l} + c_{l,Z})^{1/2}\phi\|^2 \\ &= (\phi, (H_{0,l} + c_{l,Z})^{1/2}(h_{l,\varepsilon,Z} + c_{l,Z})^{-1}(H_{0,l} + c_{l,Z})^{1/2}\phi) \\ &= \left(\phi, \frac{1}{1 - W_{l,\varepsilon,Z}}\phi\right) \end{aligned} \tag{21}$$

with

$$W_{l,\varepsilon,Z} := (H_{0,l} + c_{l,Z})^{-1/2}(\varphi_Z + \varepsilon U_Z)(H_{0,l} + c_{l,Z})^{-1/2}.$$

We will then have

$$\|B\| \leqq \sqrt{2} \tag{22}$$

if we can show that $W_{l,\varepsilon,Z}$ is bounded above by $\frac{1}{2}$. To this end we note that $H_{0,l} + c_{l,Z}$ is invertible, so that we can write any normalized $\phi \in L^2(\mathbb{R}^+)$ as

$$\phi := (H_{0,l} + c_{l,Z})^{1/2}\psi / \|(H_{0,l} + c_{l,Z})^{1/2}\psi\|$$

with ψ in the domain of $H_{0,l}$. Thus, we have to show that

$$(\phi, W_{l,\varepsilon,Z}\,\phi) = (\psi, (\varphi_Z + \varepsilon U_Z)\psi)/(\psi, (H_{0,l} + c_{l,Z})\psi) \leqq \frac{1}{2},$$

which is equivalent to

$$\frac{1}{2}(H_{0,l} + c_{l,Z}) - \varphi_Z - \varepsilon U_Z \geqq 0. \tag{23}$$

Since $\varphi_Z^{\mathrm{TF}}(r) \leqq Z/r$ and since

$$(\psi, U_Z\psi) = Z|\psi(a/Z)|^2 \leqq Z\int_0^{a/Z} |\psi|^{2'}(r)\,dr \leqq 2Z\Re\int_0^{a/Z} \psi(r)\psi'(r)\,dr \tag{24}$$
$$\leqq 2Z\|\psi'\|_2\left\{\int_0^{a/Z}\psi(r)^2 dr\right\}^{1/2}$$

we have that $(\psi, U_Z\psi) \leqq (4a/\pi)\|\psi'\|_2^2$. (Here we use the inequality, $\int_0^L \psi'^2 \geqq (\pi/2L)^2\int_0^L \psi^2$, when $\psi(0) = 0$.) Thus, (23) is implied by

$$-4\frac{Z^2}{4(l+1)^2} + \frac{c_{l,Z}}{2} + \inf\sigma\left(\left(\frac{1}{4} - \frac{4\varepsilon a}{\pi}\right)H_{0,l}\right) \geqq 0. \tag{25}$$

The sum of the first two terms in (25) vanishes because of our choice of $c_{l,Z}$ and last term in (25) is zero when $\varepsilon \leqq \pi/(16a)$. This is true by hypothesis, and the bound on the norm of B is proved.

Finally the trace of C is computed easily, since it is of rank one. Since the kernel $(H_{0,l} + c_{l,Z})^{-1}(r, r')$, is a positive, continuous function in both variables,

$$\operatorname{tr} C = Z(H_0 + c_{l,Z})^{-1}(a/Z, a/Z).$$

A well known calculation yields

$$(H_{0,l} + c_{l,Z})^{-1}(r, r') = \sqrt{r}\,K_{l+\frac{1}{2}}(\sqrt{c_{l,Z}}\,r_>)\,I_{l+\frac{1}{2}}(\sqrt{c_{l,Z}}\,r_<)\sqrt{r'},$$

where $r_> = \max\{r, r'\}$ and $r_< = \min\{r, r'\}$. Thus

$$\operatorname{tr} C = a K_{l+\frac{1}{2}}\left(\sqrt{c_{l,Z}}\,\frac{a}{Z}\right) I_{l+\frac{1}{2}}\left(\sqrt{c_{l,Z}}\,\frac{a}{Z}\right).$$

The modified Bessel functions $I_{l+\frac{1}{2}}$ and $K_{l+\frac{1}{2}}$ are both positive and the following uniform asymptotic expansions hold. (See Olver [14] for a proof of the estimates of the remainder terms, [15], p. 6 for the remainder in the form used here, [17], Chapter 10, Paragraph 7 for a review; see also Olver [16], section 9.7.)

$$K_n(nx) = \sqrt{\frac{\pi t}{2n}}\,e^{-n\xi}[1 + \varepsilon_{0,2}(n, t)], \tag{26}$$

$$I_n(nx) = \sqrt{\frac{t}{2\pi n}} \frac{e^{n\xi}}{1-\varepsilon_{0,1}(n,0)} [1+\varepsilon_{0,1}(n,t)], \tag{27}$$

where

$$\xi := \frac{1}{t} - \frac{1}{2} \log\frac{1+t}{1-t},$$

$$t := (1+x^2)^{-\frac{1}{2}}$$

and

$$|\varepsilon_{0,1}(n,t)| \leqq \frac{n_0}{n-n_0},$$

$$|\varepsilon_{0,2}(n,t)| \leqq \frac{n_0}{n-n_0}$$

with $n_0 := \frac{1}{6\sqrt{5}} + \frac{1}{12} \leqq \frac{1}{6}$. Thus

$$K_n(nx) I_n(nx) \leqq \frac{1}{2n(1+x^2)^{\frac{1}{2}}} \frac{n^2}{(n-2n_0)(n-n_0)} \leqq \frac{9}{4n},$$

where the last inequality holds for $n \geqq \frac{1}{2}$. Thus

$$\operatorname{tr} C \leqq \frac{9}{2} \frac{a}{(2l+1)}. \tag{28}$$

Putting (19), (20), (22), and (28) together yields

$$s_{\varepsilon,l,Z} \leqq \varepsilon \frac{9c_{l,Z}a}{(2l+1)} \leqq \varepsilon \frac{9aZ^2}{(l+1)^2(2l+1)}$$

which is more than the desired result for $\varepsilon > 0$.

If ε is negative we have, again by the minimax principle,

$$0 \geqq s_{\varepsilon,l,Z} := \operatorname{tr}(h_{l,0,Z})_- - \operatorname{tr}(h_{l,\varepsilon,Z})_- \geqq \varepsilon \operatorname{tr}(U_Z d_{l,0,Z}). \tag{29}$$

Similar to the previous analysis, we have

$$s_{\varepsilon,l,Z} \geqq \varepsilon \operatorname{tr}(D \circ E \circ C \circ E^* \circ D^*) \geqq \varepsilon \|D\|_\infty^2 \|E\|_\infty^2 \operatorname{tr} C \tag{30}$$

with $D := d_{l,0,Z}(h_{l,0,Z} + c_{l,Z})^{\frac{1}{2}}$ and $E = (h_{l,0,Z} + c_{l,Z})^{-\frac{1}{2}}(H_{l,0} + c_{l,Z})^{\frac{1}{2}}$. As for A above, we have $\|D\|_\infty \leqq \sqrt{c_{l,Z}}$ and $\|E\| \leqq \sqrt{2}$. Putting this together with (28) gives

$$s_{\varepsilon,l,Z} \geqq \varepsilon \frac{9c_{1,Z}a}{(2l+1)} \geqq \varepsilon \frac{9aZ^2}{(l+1)^2(2l+1)}$$

which is the desired result for negative ε. □

The next result will later on allow us to interchange the limits $Z \to \infty$ and $\varepsilon \to 0$ with the n summation for fixed l.

Lemma 2. *Set* $U(r) = \delta(r-a)$ *and assume* $|\varepsilon| \leqq \pi/(4a)$, $a > 0$. *Let*

$$h_{l,\varepsilon} := -\frac{d^2}{dr^2} + \frac{l(l+1)}{r^2} - \frac{1}{r} - \varepsilon U(r) \tag{31}$$

with form domain $H_0^1(0,\infty)$. *Let* $e_{n,l,\varepsilon}$ *denote the n-th eigenvalue of* $h_{l,\varepsilon}$. *Then*

$$|e_{n,l,0} - e_{n,l,\varepsilon}| \leqq \frac{1}{(n+l)^2} \frac{|\varepsilon| a}{\pi - 4\varepsilon a}. \tag{32}$$

Proof. For any ψ in $H_0^1(0,\infty)$ we have

$$|\psi(a)|^2 \leqq 2\frac{2}{\pi} a \|\psi'\|_2^2,$$

as proved in (24) of Lemma 1. Thus, for $\varepsilon > 0$,

$$h_{l,\varepsilon} \geqq \left(1 - \frac{4\varepsilon a}{\pi}\right)\left[-\frac{d^2}{dr^2} + \frac{l(l+1)}{r^2} - \frac{1}{\left(1-\frac{4\varepsilon a}{\pi}\right)r}\right]. \tag{33}$$

This implies

$$e_{n,l,\varepsilon} \geqq \left(1 - \frac{4\varepsilon a}{\pi}\right)\tilde{e}_{n,l,0}$$

where $\tilde{e}_{n,l,0}$ is the n-th eigenvalue of [ ] in (33), i.e., where the potential r^{-1} is replaced by $(1-4\varepsilon a/\pi)^{-1} r^{-1}$. Thus,

$$0 \leqq e_{n,l,0} - e_{n,l,\varepsilon} \leqq \frac{1}{4(n+l)^2}\left(-1 + \left(1 - \frac{4\varepsilon a}{\pi}\right)^{-1}\right) = \frac{1}{(n+l)^2}\frac{\varepsilon a}{\pi - 4\varepsilon a},$$

which proves the claim when $0 < \varepsilon < \pi/(4a)$.

If ε is negative we have

$$h_{l,\varepsilon} \leqq \left(1 - \frac{4\varepsilon a}{\pi}\right)\left[-\frac{d^2}{dr^2} + \frac{l(l+1)}{r^2} - \frac{1}{\left(1-\frac{4\varepsilon a}{\pi}\right)r}\right]$$

which again proves the claim (by the same argument) when $0 > \varepsilon > -\pi/(4a)$. □

4. Proof of the strong Scott conjecture

We are now able to give the proofs of our theorems. We begin with the proof of Theorem 1 and begin with the first statement:

1. *The proof of the convergence of the spherical averages.* Set $U(r) := \delta(r-a)$ for $a > 0$ and $U_Z(r) := Z^2 U(Zr) = Z\delta\left(r - \frac{a}{Z}\right)$. Fix l_0 and let

$$H^{\varepsilon}_{N,Z} := H_{N,Z} - \varepsilon \sum_{\nu=1}^{N} U_Z(r_\nu)\Pi(l_0) \tag{34}$$

where $\Pi(l_0)$ denotes the projection onto angular momentum l_0. We define $\lambda(Z)$ – which does not depend on ε – by

$$\lambda(Z) := a^2 \varrho_{l_0}(a) = \frac{\operatorname{tr}(H_{N,Z}d) - \operatorname{tr}(H^{\varepsilon}_{N,Z}d)}{\varepsilon Z^2}. \tag{35}$$

Let us define $e^H_{n,l,\gamma,Z}$ and $e_{n,l,\gamma,Z}$, $n = 1, 2, \ldots$, $\varepsilon \in \mathbb{R}$, to be the negative eigenvalues of the operators

$$H^H_{l,\varepsilon,Z} := -\frac{d^2}{dr^2} + \frac{l(l+1)}{r^2} - \frac{Z}{r} - \varepsilon U_Z \delta_{l,l_0}, \tag{36}$$

$$H_{l,\varepsilon,Z} := -\frac{d^2}{dr^2} + \frac{l(l+1)}{r^2} - \varphi^{\mathrm{TF}}_Z - \varepsilon U_Z \delta_{l,l_0} \tag{37}$$

with zero Dirichlet boundary on $(0, \infty)$. To obtain an upper bound for $\lambda(Z)$ we pick ε positive and estimate as follows: by (63) we have the upper bound

$$\operatorname{tr}(H_{N,Z}d) \tag{38}$$
$$\leqq \sum_{l=0}^{L-1} q(2l+1)\sum_n e^H_{n,l,0,Z} + \sum_{l=L}^{\infty} q(2l+1)\sum_n e_{n,l,0,Z} - D(\varrho_{\mathrm{TF}}, \varrho_{\mathrm{TF}}) + \mathrm{const}\, Z^{\frac{47}{24}}$$

where $L = [Z^{1/9}]$.

To obtain a lower bound on $\operatorname{tr}(H^{\varepsilon}_{N,Z}d)$ we first use the lower bound ([13], [11]) on the correlations, namely $-\mathrm{const}\,[N \int \varrho_d^{5/3}]^{1/2}$, to reduce it to a radial problem. Using the fact that $Z/r \geqq \varphi^{\mathrm{TF}}_Z(r)$ for $r > 0$ it follows from this that

$$\operatorname{tr}(H^{\varepsilon}_{N,Z}d) \tag{39}$$
$$\geqq \sum_{l=0}^{L-1} q(2l+1)\sum_n e^H_{n,l,\varepsilon,Z} + \sum_{l=L}^{\infty} q(2l+1)\sum_n e_{n,l,\varepsilon,Z} - D(\varrho_{\mathrm{TF}}, \varrho_{\mathrm{TF}}) - \mathrm{const}\, Z^{\frac{5}{3}}.$$

Note that (39) arises from a relatively simple lower bound calculation. Part of the proof of the Scott conjecture amounts to proving that the right hand of (39) is accurate to $o(Z^2)$. This proof was carried out in [23] (see also [9], [24]). We are *not* rederiving the Scott correction for the energy, and it is not necessary for us to do so here.

Define

$$\theta(n) := \begin{cases} 1, & n > 0, \\ 0, & n \leqq 0. \end{cases}$$

Since the eigenvalues of the perturbed problem ($\varepsilon \neq 0$) are equal to the unperturbed one ($\varepsilon = 0$) except for $l = l_0$, we get the inequality

$$\text{(40)} \qquad \limsup_{Z\to\infty} \lambda(Z)$$

$$\leqq \liminf_{\varepsilon\searrow 0} \limsup_{Z\to\infty} \left[q(2l_0+1) \frac{\operatorname{tr}(H^H_{l_0,0,Z})_- - \operatorname{tr}(H^H_{l_0,\varepsilon,Z})_-}{\varepsilon Z^2} \theta(L-l_0) \right.$$

$$\left. + \frac{\operatorname{tr}(H_{l_0,0,Z})_- - \operatorname{tr}(H_{l_0,\varepsilon,Z})_-}{\varepsilon Z^2} \theta(l_0 - L) + \operatorname{const} Z^{-\frac{1}{24}} \varepsilon^{-1} \right].$$

Because L eventually becomes larger than the fixed l_0,

$$\text{(41)} \quad \limsup_{Z\to\infty} \lambda(Z) \leqq q(2l_0+1) \liminf_{\varepsilon\searrow 0} \limsup_{Z\to\infty} \frac{\operatorname{tr}(H^H_{l_0,0,Z})_- - \operatorname{tr}(H^H_{l_0,\varepsilon,Z})_-}{\varepsilon Z^2}$$

$$= q(2l_0+1) \liminf_{\varepsilon\searrow 0} \frac{\operatorname{tr}(H^H_{l_0,0,1})_- - \operatorname{tr}(H^H_{l_0,\varepsilon,1})_-}{\varepsilon},$$

where the last equation holds since because of the scaling of $h^H_{l_0,0,Z}$ and $h_{l_0,\varepsilon,Z}$. Therefore

$$\text{(42)} \qquad \limsup_{Z\to\infty} \lambda(Z) \leqq q(2l_0+1) \liminf_{\varepsilon\searrow 0} \sum_n \frac{e^H_{n,l_0,0,1} - e^H_{n,l_0,\varepsilon,1}}{\varepsilon}$$

$$\text{(43)} \qquad = q(2l_0+1) \sum_n \liminf_{\varepsilon\searrow 0} \frac{e^H_{n,l_0,0,1} - e^H_{n,l_0,\varepsilon,1}}{\varepsilon}$$

$$\text{(44)} \qquad = a^2 \varrho^H_{l_0}(a).$$

To exchange the limit $\varepsilon \searrow 0$ with the summation in (42) we use Lemma 2, which provides a summable majorant for the series that is uniform in ε and thus allows us to fulfil Weierstraß' criterion for uniform convergence. Finally, to deduce (44) from (43) we use the fact that the one-dimensional delta potential is a relatively form bounded perturbation, i.e., defines an analytic family in the sense of Kato.

To obtain a lower bound for $\lambda(Z)$ we pick ε negative instead of positive and take the limit $\limsup_{\varepsilon\nearrow 0}$ and $\liminf_{Z\to\infty}$ instead of $\liminf_{\varepsilon\searrow 0}$ and $\limsup_{Z\to\infty}$. Repeating the same steps gives the same result except for reversing the inequalities, thereby yielding the same bound (44) from below. This establishes the first claim of the theorem.

2. *Proof of the weak convergence.* Because of the linear dependence of the right and left hand side of (12), it suffices to prove the claim for the positive and negative parts of V separately. Thus we may – and shall – assume that V is positive. We can now roll the

proof back to the previous case as follows: First we pick Z large enough so that $l_0 < L$. It is convenient now, to replace ε by ε/a in order that the right side of (32) in Lemma 2 is uniformly bounded in a and ε for all $a \in (0, \infty)$ and for $|\varepsilon| \leqq \pi/8$. Then we integrate the inequality

$$(45)\quad a^2 \varrho_{l_0}(a) \leqq \frac{\mathrm{tr}(H^H_{l_0,0,Z})_- - \mathrm{tr}(H^H_{l_0,\varepsilon/a,Z})_-}{(\varepsilon/a)Z^2}\,\theta(L-l) + \mathrm{const}/((\varepsilon/a)Z^{\frac{1}{24}}),$$

i.e.,

$$(46)\quad a\varrho_{l_0}(a) \leqq \frac{\mathrm{tr}(H^H_{l_0,0,Z})_- - \mathrm{tr}(H^H_{l_0,\varepsilon/a,Z})_-}{\varepsilon Z^2}\,\theta(L-l) + \mathrm{const}/(\varepsilon Z^{\frac{1}{24}})$$

against $V(a)$ from 0 to ∞. Thanks to Lemma 1 the right side of (46) is bounded by const a and hence the integral is finite. Next we write out the traces appearing in (44) in terms of the eigenvalues and then use Lemma 2 to provide a bound that is summable (over the eigenvalues) and integrable (from 0 to ∞), if $|\varepsilon| < \pi/8$. This bound is uniformly bounded in ε, and so, by dominated convergence, we can take the limit $\varepsilon \searrow 0$ term by term. Using the result (11), which we established above, Equation (12) is now verified. □

Proof of Theorem 2. As was the case in the proof of Theorem 2 we shall assume that W and V are nonnegative. For Part 1 we proceed as for Theorem 2 and define $H^{\varepsilon}_{N,Z}$ as in (34), but with $\Pi(l_0)$ replaced by $W(\omega)$. First we treat the case $W(\omega) = 1$. We follow the proof of Theorem 1 up to equations (36) and (37) (with δ_{l,l_0} replaced by W). Then we obtain analogously

$$(47)\quad \limsup_{Z\to\infty} \lambda(Z)$$

$$(48)\quad \leqq \liminf_{\varepsilon\searrow 0} \limsup_{Z\to\infty} \Bigg[\sum_{l=0}^{\infty} q(2l+1)\,\frac{\mathrm{tr}(H^H_{l,0,Z})_- - \mathrm{tr}(H^H_{l,\varepsilon,Z})_-}{\varepsilon Z^2}\,\theta(L-l)$$

$$(49)\quad + \sum_{l=0}^{\infty} q(2l+1)\,\frac{\mathrm{tr}(H_{l,0,Z})_- - \mathrm{tr}(H_{l,\varepsilon,Z})_-}{\varepsilon Z^2}\,\theta(l-L) + \mathrm{const}\, Z^{-\frac{1}{24}}\varepsilon^{-1}$$

$$(50)\quad = \sum_{l=0}^{\infty} q(2l+1) \liminf_{\varepsilon\searrow 0} \limsup_{Z\to\infty} \frac{\mathrm{tr}(H^H_{l,0,Z})_- - \mathrm{tr}(H^H_{l,\varepsilon,Z})_-}{\varepsilon Z^2}$$

$$(51)\quad = \sum_{l=0}^{\infty} q(2l+1) \liminf_{\varepsilon\searrow 0} \frac{\mathrm{tr}(H^H_{l,0,1})_- - \mathrm{tr}(H^H_{l,\varepsilon,1})_-}{\varepsilon}$$

$$(52)\quad = a^2 \sum_{l=0}^{\infty} \liminf_{\varepsilon\searrow 0} \varrho^H_l(a)$$

$$(53)\quad = a^2 \varrho^H(a).$$

To obtain (49) we use inequalities (38) and (39). To obtain (50) we use the fact that Lemma 1 provides a majorant uniform in ε and Z that is absolutely summable with respect to $\sum_{l=0}^{\infty} q(2l+1)$, i.e., fulfills the Weierstraß criterion for uniform convergence (or the hypothesis of Lebesgue's dominated convergence theorem), and therefore allows the interchange of the limit and the l summation, and that the second sum tends term by term to zero. To obtain (51) we use the fact that the eigenvalues of the bare problem scale like Z^2. Finally, the convergence result of Theorem 2 was used to obtain (52).

To obtain a lower bound we pick ε negative instead of positive and take the limit $\limsup_{\varepsilon \nearrow 0}$ and $\liminf_{Z \to \infty}$ instead of $\liminf_{\varepsilon \searrow 0}$ and $\limsup_{Z \to \infty}$. Repeating the same steps gives the same result except for reversing the inequalities thereby yielding the bound from below.

Let W now be a general bounded, measurable function on the unit sphere which we may – according to the remarks in the beginning – assume to be positive. We take $\|W\|_\infty = 1$.

Let us try to imitate the steps (47) to (53). As before we are faced with estimating the eigenvalues of the one-body operators

$$(54) \quad H^H_{\varepsilon,Z} := -\Delta - Z/|\cdot| - \varepsilon Z W \delta_{\frac{a}{Z}} \quad \text{and} \quad H^{\mathrm{TF}}_{\varepsilon,Z} = -\Delta - \varphi^{\mathrm{TF}}_Z - \varepsilon Z W \delta_{\frac{a}{Z}}$$

but unlike the previous case they cannot be simply indexed by the angular momentum l when $\varepsilon \neq 0$; indeed the one-body operators cannot be reduced to a direct sum of radial Schrödinger operators as in (36) and (37). However, the eigenvalues are real analytic functions of ε and we can label the eigenvalues by the l-value they have when ε tends to zero. In short, the only change needed in (47) to (53) is to replace $(2l+1)e^H_{n,l,\varepsilon,Z}$ by the sum of the eigenvalues in the multiplet of $H^H_{\varepsilon,Z}$ that converge to $e^H_{n,l,0,Z}$ as ε tends to zero. Since W is bounded by 1, all our previous bounds for eigenvalue differences (Lemmata 1 and 2) continue to hold and we are finally led to the $\liminf_{\varepsilon \searrow 0}$ in (51).

The crucial point is this: Even if W is not spherical symmetric, the sum of the eigenvalues in any multiplet is rotationally invariant to first order in ε in the following sense. The only property of W that matters – to first order – is the average $W_{\mathrm{average}} := (4\pi)^{-1} \int W(\omega)\,d\omega$.

Reversing the sign of ε again gives the lower bound.

2. *Proof of the weak convergence.* The proof can be rolled back to the previous case as follows: First we assume that V is spherically symmetric and integrate the inequality

$$a^2 \varrho_Z(a\omega) \leqq \Bigg[\sum_{l=0}^{\infty} q(2l+1) \frac{\operatorname{tr}(H^H_{l,0,Z})_- - \operatorname{tr}(H^H_{l,\varepsilon,Z})_-}{\varepsilon Z^2} \theta(L-l)$$

$$+ \sum_{l=0}^{\infty} q(2l+1) \frac{\operatorname{tr}(H_{l,0,Z})_- - \operatorname{tr}(H_{l,\varepsilon,Z})_-}{\varepsilon Z^2} \theta(l-L) \Bigg] + \mathrm{const}/(Z^{\frac{1}{24}})$$

against $aV(a)$ from 0 to ∞. Observe that because of Lemma 1 the summand of the sum on the right side of this integrated inequality is uniformly bounded by

$$\frac{9}{(l+1)^2(2l+1)} \int_0^\infty V(a) a^2 \, da$$

which, when multiplied by $(2l+1)$, is summable. Again, the same argument holds when expressing the traces as sum over eigenvalues. Thus we are allowed to take the limits term by term for the differences of the eigenvalues giving the desired result as above.

The extension to the non-spherical case is as in Part 1. □

5. Extensions to molecules

The ground state energy of a neutral molecule with nuclear charges $Z_1 = \lambda z_1, \dots, Z_K = \lambda z_K$ and positions of the nuclei at $\mathfrak{R}_1, \dots, \mathfrak{R}_K$ is given as

$$E(N, \vec{Z}) = \inf\{\inf\sigma(H_{N,\vec{Z},\vec{R}}) \mid \vec{R} \in \mathbb{R}^{3K}\} \tag{55}$$

where

$$H_{N,\vec{Z},\vec{R}} = \sum_{\nu=1}^{N}\left(-\Delta_\nu - \sum_{\kappa=1}^{K}\frac{Z_\kappa}{|\mathfrak{r}_\nu - \mathfrak{R}_\kappa|}\right) + \sum_{\substack{\mu,\nu=1\\ \mu<\nu}}^{N}\frac{1}{|\mathfrak{r}_\mu - \mathfrak{r}_\nu|} + \sum_{\substack{\kappa,\kappa'=1\\ \kappa<\kappa'}}\frac{Z_\kappa Z_{\kappa'}}{|\mathfrak{R}_\kappa - \mathfrak{R}_{\kappa'}|} \tag{56}$$

self-adjointly realized in $\mathfrak{H}_N$. Here $\vec{Z}$ denotes the K-tuple $(Z_1, \dots, Z_K)$ and $\vec{R}$ the $3K$-tuple $(\mathfrak{R}_1, \dots, \mathfrak{R}_K)$. We also set $\vec{z} := (z_1, \dots, z_K)$. Solovej [25] showed recently that for arbitrary but fixed $\vec{z}$ and $N = Z_1 + \cdots + Z_K$

$$E(N, \vec{Z}) = \sum_{\kappa=1}^{K} E(Z_\kappa, Z_\kappa) + o(\lambda^{\frac{5}{3}}) \tag{57}$$

holds as λ tends to infinity and that the minimizing inter-nuclear distances are of order $\lambda^{-5/21}$ or bigger. These results imply among other things not only that the atomic Scott correction and Schwinger correction implies the molecular one but allows us to generalize Theorem 2 as well: The molecular density in the vicinity of each nucleus converges in the sense of Theorem 1 to the hydrogen density at each of the centers. Our precise result is:

Theorem 3. *Assume that $E(N, \vec{Z})$ as defined in* (55) *is equal to*

$$\inf\{\inf\sigma(H_{N,\vec{Z},\vec{R}}) \mid \vec{R} \in \mathbb{R}^{3K}, \forall_{1 \leq \kappa < \kappa' \leq K} |\mathfrak{R}_\kappa - \mathfrak{R}_{\kappa'}| \geq R := \mathrm{const}\,\lambda^\gamma\} \tag{58}$$

with $\gamma > -1/4$. Assume $N = Z_1 + \cdots + Z_k$, $Z_1 = \lambda z_1, \dots, Z_K = \lambda z_K$ with given fixed $z_1, \dots, z_K$. Furthermore fix $\kappa_0 \in 1, \dots, K$ and pick a sequence of ground state density matrices d_c of $H_{N,\vec{Z},\vec{R}}$ with densities ϱ_{d_c}. Define $\varrho_{\lambda,\kappa_0}(\mathfrak{r}) := \varrho_\lambda((\mathfrak{r} - \mathfrak{R}_{\kappa_0})/\lambda)/\lambda^3$. Finally assume $W \in L^2(\mathbb{S}^2)$. Then for $r > 0$

$$\int_{\mathbb{S}^2} W(\omega)\varrho_{\lambda,\kappa_0}(r\omega) \to q\varrho^H(r)\int_{\mathbb{S}^2} W \tag{59}$$

as $\lambda \to \infty$.

Proof. First note that by suitable relabeling we can always assume that $\kappa_0 = 1$. Set $H^\varepsilon_{N,\vec{Z},\vec{R}} := H_{N,\vec{Z},\vec{R}} - \sum_{\nu=1}^{N} \varepsilon U_\lambda(\mathfrak{r}_\nu)$ where U_λ is defined as in the atomic case, i.e.,

$$U_\lambda(\mathfrak{r}) := \lambda^2 U(\lambda(\mathfrak{r} - \mathfrak{R}_1)), \quad U(\mathfrak{r}) := W(\omega)\delta(r - a),$$

and W is a square integrable function on the unit sphere. Because of (57) it suffices that

$$\operatorname{tr}(dH^\varepsilon_{N,\vec{Z},\vec{R}}) \geq \operatorname{tr}(H^{\mathrm{TF}}_{\varepsilon,Z_1})_- - D(\varrho^{\mathrm{TF}}_{Z_1}, \varrho^{\mathrm{TF}}_{Z_1}) + \sum_{\kappa=2}^{N} \left(\operatorname{tr}(H^{\mathrm{TF}}_{0,Z_\kappa})_- - D(\varrho^{\mathrm{TF}}_{Z_\kappa}, \varrho^{\mathrm{TF}}_{Z_\kappa})\right) - \operatorname{const} Z^{2-\delta}$$

for some positive δ and an approximate ground state d where $H^{\mathrm{TF}}_{\varepsilon,Z}$ is defined as in (54).

To this end we introduce the localizing functions

$$v_\kappa(\mathfrak{r}) := \cos\left(\psi(|\mathfrak{r} - \mathfrak{R}_\kappa|/R)\right)$$

where $\psi(t)$ is some continuous, piecewise differentiable, monotone decreasing function which vanishes, if $t < 1/4$, and which is $\pi/2$, if $t > 1/2$. Note that the supports of these functions have at most finitely many points in common because R is the minimal nuclear distance. We also define

$$v_0 := \sqrt{1 - \sum_{\kappa=1}^{K} v_\kappa^2}.$$

Now pick the density $\varrho(\mathfrak{r}) := \sum_{\kappa=1}^{K} \varrho^{\mathrm{TF}}_{Z_\kappa}(|\mathfrak{r} - \mathfrak{R}_\kappa|)$ and denote the one-particle density matrix belonging to d by d_1. Note that $\operatorname{tr} d_1 = N$. By the correlation inequality ([13], [11]) and the localization formula using the above decomposition of unity we have

$$\text{(60)} \qquad \operatorname{tr}(H^\varepsilon_{N,\vec{Z},\vec{R}}\, d)$$
$$\geqq \operatorname{tr}\left\{\left[-\Delta - \sum_{\kappa=1}^{K} \varphi^{\mathrm{TF}}_{Z_\kappa}(\cdot - \mathfrak{R}_\kappa) - \varepsilon U_\lambda\right] d_1\right\}$$
$$- D(\varrho, \varrho) + \sum_{\substack{\kappa,\kappa'=1 \\ \kappa<\kappa'}}^{K} \frac{Z_\kappa Z_{\kappa'}}{|\mathfrak{R}_\kappa - \mathfrak{R}_{\kappa'}|} - \operatorname{const} \lambda^{\frac{5}{3}}$$
$$\geqq \operatorname{tr}\left\{\sum_{\kappa=0}^{K} v_\kappa \left[-\Delta - \sum_{\kappa'=1}^{K} \varphi^{\mathrm{TF}}_{Z_\kappa}(|\cdot - \mathfrak{R}_{\kappa'}|) - \varepsilon U_\lambda\right] v_\kappa d_1\right\}$$
$$- \left\| \sum_{\kappa=0}^{K} |\operatorname{grad} v_\kappa|^2 \right\|_\infty N - \sum_{\kappa=1}^{K} D(\varrho^{\mathrm{TF}}_{Z_\kappa}, \varrho^{\mathrm{TF}}_{Z_\kappa}) - \operatorname{const} \lambda^{\frac{5}{3}}.$$

To obtain (60) we used the spherical symmetry of $\varphi_1, \ldots, \varphi_K$ to show that

$$D(\varrho^{\mathrm{TF}}_{Z_\kappa}, \varrho^{\mathrm{TF}}_{Z_{\kappa'}}) \leqq Z_\kappa Z_{\kappa'} |\mathfrak{R}_\kappa - \mathfrak{R}_{\kappa'}|^{-1}.$$

Now pick any arbitrary pair of different indices $\kappa, \kappa' \in \{1, \ldots, K\}$. On the support of v_κ we have

$$\varphi^{\mathrm{TF}}_{Z_{\kappa'}}(|\mathfrak{r} - \mathfrak{R}_{\kappa'}|) \leqq \frac{2^2 3^4 \pi^2}{q^2 (R/2)^4}$$

where we use the fact that the Sommerfeld solution of the Thomas-Fermi equation is a pointwise upper bound of the Thomas-Fermi potential ([12], Section V.2). Thus on the support of v_κ we have

$$\sum_{\kappa'=1,\kappa'\neq\kappa}^{K} \varphi_{Z_{\kappa'}}^{\mathrm{TF}}(|\mathfrak{r}-\mathfrak{R}_{\kappa'}|) \leqq \frac{2^6 3^4 (K-1)}{q^2 R^4}.$$

For the derivative of the v_κ, which governs the localization error, we have the following uniform estimate

$$\sum_{\kappa=0}^{K} |\mathrm{grad}\, v_\kappa|^2 = |\psi'(|\mathfrak{r}-\mathfrak{R}_\kappa|)|^2/R^2 = 4\pi^2/R^2$$

where, for definiteness, we picked ψ to be the linear function interpolating between 0 and $\pi/2$ on the interval $[1/4, 1/2]$. Note that outside the annuli of thickness $R/2$ centered at the nuclei the derivatives vanish, in fact, whereas in these annuli the bound is actually an equality.

Next we show that there is no relevant contribution to the energy stemming from regions which are not close to at least one of the nuclei, i.e., from the support of v_0. To this end we first remark that the supports of U_λ and v are disjoint for λ large enough. We set $\Upsilon := \mathrm{supp}\, v_0 \cap (\mathrm{supp}\, v_1 \cup \cdots \cup v_K)$. Then, by the Lieb-Thirring inequality,

$$\mathrm{tr}\left\{\left[v_0\left(-\Delta - \sum_{\kappa=1}^{K} \varphi_{Z_\kappa}^{\mathrm{TF}} - \chi_\Upsilon \frac{4\pi^2}{R^2}\right) v_0\right] d_1\right\}$$

$$\geqq -\mathrm{const}\Big(\int_{|\mathfrak{r}|>\frac{R}{4}} |\mathfrak{r}|^{-10} d\mathfrak{r} + \int_{\frac{R}{2}>|\mathfrak{r}|>\frac{R}{4}} d\mathfrak{r}\Big) \geqq -\mathrm{const}(R^{-7}+R^3) \geqq -\mathrm{const}\,\lambda^{7/4}.$$

This yields

$$\mathrm{tr}(H_{N,\underline{Z},\underline{R}}^{\varepsilon} d)$$

$$\geqq \sum_{\kappa=1}^{K} \{\mathrm{tr}[-\Delta - \varphi_{Z_\kappa}^{\mathrm{TF}} - \frac{4\pi^2}{R^2} + \frac{2^6 3^4 (K-1)}{q^2 R^4} \chi_{B_{\frac{R}{2}}(0)} - \varepsilon\delta_{1,\kappa}\lambda^2 U(\lambda\cdot) d_1] - D(\varrho_{Z_\kappa}^{\mathrm{TF}}, \varrho_{Z_\kappa}^{\mathrm{TF}})\} - \mathrm{const}\,\lambda^{\frac{7}{4}}$$

$$\geqq \sum_{\kappa=1}^{K} \{\mathrm{tr}(-\Delta - \varphi_{Z_\kappa}^{\mathrm{TF}} - \varepsilon\delta_{1,\kappa} U_\lambda)_- - D(\varrho_{Z_\kappa}^{\mathrm{TF}}, \varrho_{Z_\kappa}^{\mathrm{TF}})\} - N\left(\frac{4\pi^2}{R^2} + \frac{2^6 3^4 (K-1)}{q^2 R^4}\right) - \mathrm{const}\,\lambda^{\frac{7}{4}}$$

$$\geqq \inf\sigma(H_{\varepsilon,Z_1}^{\mathrm{TF}}) - D(\varrho_{Z_1}^{\mathrm{TF}}, \varrho_{Z_1}^{\mathrm{TF}}) + \sum_{\kappa=2}^{K} [\inf\sigma(H_{0,Z_\kappa}^{\mathrm{TF}}) - D(\varrho_{Z_\kappa}^{\mathrm{TF}}, \varrho_{Z_\kappa}^{\mathrm{TF}})] - \mathrm{const}\left(\frac{\lambda}{R^4} + \lambda^{\frac{7}{4}}\right)$$

$$\geqq \inf\sigma(H_1^{\varepsilon}) - D(\varrho_{Z_\kappa}^{\mathrm{TF}}, \varrho_{Z_\kappa}^{\mathrm{TF}}) + \sum_{\kappa=2}^{K} [\inf\sigma(H_\kappa) - D(\varrho_{Z_\kappa}^{\mathrm{TF}}, \varrho_{Z_\kappa}^{\mathrm{TF}})] - \mathrm{const}\,\lambda^{2-\delta}$$

for some sufficiently small but positive δ. Combining this with Solovej's upper bound reduces the convergence question to that of the one-center case. □

A. Appendix: Facts about the atomic ground state energy

According to [22] we have

$$E_{Z,Z} \leqq E_{\mathrm{TF}}(Z,Z) + \frac{q}{8} Z^2 + \mathrm{const}\, Z^{\frac{47}{24}}, \tag{61}$$

and according to [23] (see also [24] and Hughes [9])

$$\begin{aligned} E_{Z,Z} &\geqq \sum_{l=0}^{L-1} q(2l+1)\,\mathrm{tr}(H^H_{l,0,Z})_- \\ &\quad + \sum_{l=L}^{\infty} q(2l+1)\,\mathrm{tr}(H_{l,0,Z})_- - D(\varrho_{\mathrm{TF}}, \varrho_{\mathrm{TF}}) - \mathrm{const}\, Z^{\frac{5}{3}} \\ &\geqq E_{\mathrm{TF}}(Z,Z) + \frac{q}{8} Z^2 - \mathrm{const}\, Z^{\frac{17}{9}} \log Z \end{aligned} \tag{62}$$

with $L = [Z^{\frac{1}{9}}]$. Combining (61) and (62) gives

$$\begin{aligned} E_{Z,Z} &= \sum_{l=0}^{L-1} q(2l+1)\,\mathrm{tr}(H^H_{l,0,Z})_- \\ &\quad + \sum_{l=L}^{\infty} q(2l+1)\,\mathrm{tr}(H_{l,0,Z})_- - D(\varrho_{\mathrm{TF}}, \varrho_{\mathrm{TF}}) + O(Z^{\frac{47}{24}}). \end{aligned} \tag{63}$$

References

[1] *V. Bach*, A proof of Scott's conjecture for ions, Rep. Math. Phys. **28** (2) (1989), 213–248.

[2] *C. Fefferman* and *L. Seco*, Eigenfunctions and eigenvalues of ordinary differential operators, Adv. Math. **95**(2) (1992), 145–305.

[3] *C. Fefferman* and *L. Seco*, The density of a one-dimensional potential, Adv. Math. **107**(2) (1994), 187–364.

[4] *C. Fefferman* and *L. Seco*, The eigenvalue sum of a one-dimensional potential, Adv. Math. **108**(2) (1994), 263–335.

[5] *C. Fefferman* and *L. Seco*, On the Dirac and Schwinger corrections to the ground-state energy of an atom, Adv. Math. **107**(1) (1994), 1–188.

[6] *C. Fefferman* and *L. Seco*, The density in a three-dimensional radial potential, Adv. Math. **111** (1) (1995), 88–161.

[7] *C.L. Fefferman* and *L.A. Seco*, Aperiodicity of the Hamiltonian flow in the Thomas-Fermi potential, Rev. Math. Iberoamer. **9**(3) (1993), 409–551.

[8] *O.J. Heilmann* and *E.H. Lieb*, Electron density near the nucleus of a large atom, Phys. Rev. A **52** (5) (1995), 3628–3643.

[9] *W. Hughes*, An atomic lower bound that agrees with Scott's correction, Adv. Math. **79** (1990), 213–270.

[10] *V.J. Ivrii* and *I.M. Sigal*, Asymptotics of the ground state energies of large Coulomb systems, Ann. Math. **138**(2) (1993), 243–335.

[11] *E.H. Lieb*, Thomas-Fermi and related theories of atoms and molecules, Rev. Mod. Phys. **53**(4) (1981), 603–641.

[12] *E.H. Lieb* and *B. Simon*, The Thomas-Fermi theory of atoms, molecules and solids, Adv. Math. **23** (1977), 22–116.

[13] *E.H. Lieb* and *W.E. Thirring*, Bound for the kinetic energy of Fermions which proves the stability of matter, Phys. Rev. Lett. **35**(11) (1975), 687–689.

[14] *F.W.J. Olver*, Error bounds for the Lioville-Green (or WKB) approximation, Proc. Camb. Phil. Soc. **57** (1961), 790–810.
[15] *F.W.J. Olver*, Tables for Bessel Functions of Moderate or Large Orders, volume 6 of Mathematical Tables, Her Majesty's Stationary Office, London, 1 edition, 1962.
[16] *F.W.J. Olver*, Bessel functions of integer order, in: Milton Abramowitz and Irene A. Stegun, editors, Handbook of Mathematical Functions with Formulas, Graphs, and Mathematical Tables, chapter 9, pages 355–433, Dover Publications, New York, 5 edition, 1968.
[17] *F.W.J. Olver*, Asymptotics and Special Functions, Academic Press, New York, 1 edition, 1974.
[18] *M. Reed* and *B. Simon*, Methods of Modern Mathematical Physics, volume 4: Analysis of Operators, Academic Press, New York, 1 edition, 1978.
[19] *J.M.C. Scott*, The binding energy of the Thomas-Fermi atom, Phil. Mag. **43** (1952), 859–867.
[20] *H. Siedentop*, An upper bound for the atomic ground state density at the nucleus, Lett. Math. Phys. **32**(3) (1994), 221–229.
[21] *H. Siedentop*, Bound for the atomic ground state density at the nucleus, CRM Proc. Lect. Notes **8** (1995), 271–275.
[22] *H. Siedentop* and *R. Weikard*, On the leading energy correction for the statistical model of the atom: Interacting case, Comm. Math. Phys. **112** (1987), 471–490.
[23] *H. Siedentop* and *R. Weikard*, On the leading correction of the Thomas-Fermi model: Lower bound – with an appendix by A.M.K. Müller, Invent. Math. **97** (1989), 159–193.
[24] *H. Siedentop* and *R. Weikard*, A new phase space localization technique with application to the sum of negative eigenvalues of Schrödinger operators, Ann. Sc. Éc. Norm. Sup. **24**(2) (1991), 215–225.
[25] *Jan Philip Solovej*, In preparation.

Institut de Recherche Mathématique de Rennes, U.R.A. 305 – CNRS –Université de Rennes I, F-35042 Rennes Cedex, France

Departments of Mathematics and Physics, Princeton University, Princeton, NJ 08544-0708, USA

Matematisk institutt, Universitetet i Oslo, Postboks 1053, N-0316 Oslo, Norway

Eingegangen 7. März 1995, in revidierter Fassung 21. August 1995

Asymptotics of Natural and Artificial Atoms in Strong Magnetic Fields*

Elliott H. Lieb
Princeton University

Jan Philip Solovej
Aarhus University

Jakob Yngvason
University of Iceland

1 Introduction

Magnetic fields in terrestrial experiments have only tiny effects on the ground--state properties of conventional atoms. The reason is that the natural atomic unit for magnetic field strength, $B_0 = m^2e^3c/\hbar^3 = 2.35 \times 10^5$ Tesla, is enormous compared with laboratory fields, which are seldom larger than 10 T. Here m denotes the electron mass, e the elementary charge, and $\hbar$ and c have their usual meaning. The unit B_0 is the field strength B at which the magnetic length $\ell_B = (\hbar c/(eB))^{1/2}$ ($\sim$ cyclotron radius for an electron in the lowest Landau level) is equal to the Bohr radius $a_0 = \hbar^2/(me^2)$. Equivalently, at $B = B_0$ the Landau energy $\hbar\omega_B$, with $\omega_B = eB/(mc)$ the cyclotron frequency, becomes equal to the Rydberg energy e^2/a_0. For $B \ll B_0$ distortions of ground-state wave functions and energy level shifts due to the magnetic field will therefore be small, and their standard treatment by means of perturbation theory is completely adequate.

Magnetic fields comparable to and even much larger than B_0, however, exist around cosmic bodies like white dwarfs and neutron stars [1]. In fields of such strength the magnetic forces are no longer a small perturbation of the Coulomb forces, and may drastically alter the structure of atoms and matter in bulk. The discovery of pulsars in 1967 spurred the interest of astrophysicists in the properties of atoms in high magnetic fields, and a considerable number of papers devoted to this subject have appeared since the early 1970's (see [2,3] for a history and a list of references.)

---

*This is the text of a lecture given by J. Yngvason at the XIth International Congress of Mathematical Physics, Paris 1994. It appeared originally in the proceedings of the congress, ed. by D. Iagolnitzer, pp. 185–205, International Press 1995, but has been updated and slightly expanded for the present publication. It summarizes the joint work [2], [3] and [5] of the three authors and is presented here in lieu of [2] and [3] which are too long to be included in this volume.

In recent years remarkably small two-dimensional structures, called *quantum dots*, have been fabricated by semiconductor technology. In quantum dots electrons, ranging in number from zero to several thousand, are confined within regions of diameter $\sim$ 10–1000 nm. Quantum dots have many properties in common with natural atoms and can justly be regarded as two-dimensional *artificial atoms* (see [4] for a review). Because the electrons interact with the semiconductor crystal their effective parameters can differ markedly from the free values. Thus the effective mass m_* in GaAs is approximately $0.07\,m$ and the effective charge $e_* = e/\sqrt{\epsilon} \approx 0.3\,e$, where $\epsilon \approx 13$ is the dielectric constant. Corresponding to these values there is an effective Bohr radius, $a_* = \hbar^2/(m_* e_*^2) \approx 185\,a_0$. The magnetic length ℓ_B becomes equal to a_* for $B = B_* = (a_0/a_*)^2 B_0 \approx 7$ T, and hence effects that for natural atoms require magnetic fields of astronomical strength, can for artificial atoms be studied in any well-equipped laboratory. [1]

In the papers [2,3,5] (see also [6–8]) we have studied the quantum mechanical ground states of natural and artificial atoms in homogeneous magnetic fields of arbitrary strength. Our main results are limit theorems for the ground-state energy and electronic density as the number of electrons, N, and the strength of the attractive potential, measured by the nuclear charge Z (natural atoms) or a coupling constant K (quantum dots), tend to infinity with N/Z or N/K fixed. The ground states can in this limit be evaluated *exactly* by five nonlinear functionals for natural atoms and three for quantum dots, corresponding to different physics at different scales of the magnetic field B as measured by powers of N. The asymptotic theories are amenable to computer studies and results of numerical computations carried out by K. Johnsen and Ö. Rögnvaldsson will be presented below.

Owing to their higher dimensionality, natural atoms have a richer structure than quantum dots, and the larger part of this review is devoted to the former. Parts of the analysis are similar for both cases and after having discussed natural atoms one can be more brief about the dots. There are some marked differences between the two cases, however. In particular, the repulsive interaction of the electrons in a quantum dot is the three-dimensional Coulomb potential, but the motion of the electrons is two-dimensional. This gives rise to somewhat peculiar electrostatics, since Newton's law for the field produced by a spherically symmetric charge distribution no longer applies. Also, the estimate of the kinetic energy in terms of a functional of the density is harder in 2D than in 3D, and for strong fields the treatment of the indirect part of the Coulomb interaction requires different techniques in the two cases.

[1] Natural atoms in highly *excited* states ("Rydberg atoms") can also be strongly affected by magnetic fields of just a few T, giving rise to various forms of "quantum chaos". This subject will not be discussed here since we deal exclusively with ground-state properties.

2 Heavy Atoms in High Magnetic Fields

2.1 The Hamiltonian

The starting point for the investigation of natural, three-dimensional atoms is the Hamiltonian for N electrons in the Coulomb field of a nucleus with charge Ze and a homogeneous magnetic field $\mathbf{B} = (0, 0, B)$:

$$H_{N,B,Z} = \sum_{i=1}^{N} \left\{ [(\mathbf{p}^{(i)} + \mathbf{A}(x^{(i)})) \cdot \sigma^{(i)}]^2 - Z|x^{(i)}|^{-1} \right\} + \sum_{1 \le i < j \le N} |x^{(i)} - x^{(j)}|^{-1} \tag{1}$$

Here $\mathbf{A}(x) = (1/2)(-x_2 B, x_1 B, 0)$ with $x = (x_1, x_2, x_3) \in \mathbf{R}^3$ is the vector potential, $\sigma = (\sigma_1, \sigma_2, \sigma_3)$ is the vector of Pauli matrices, and the units are chosen such that $\hbar = e = 2m_e = 1$, $c = 1/\alpha \approx 137$. The Hamiltonian $H_{N,B,Z}$ operates on antisymmetric wave functions $\Psi \in \bigwedge_1^N L^2(\mathbf{R}^3; \mathbf{C}^2)$ of space and spin variables. The ground-state energy is

$$E^{\mathrm{Q}}(N, B, Z) = \inf_{(\Psi,\Psi)=1} (\Psi, H_{N,B,Z}\Psi), \tag{2}$$

and the ground-state electron density

$$\rho^{\mathrm{Q}}_{N,B,Z}(x) = N \sum_{s^{(i)}=\pm\frac{1}{2}} \int \left| \Psi_0(x, x^{(2)}, \ldots, x^{(N)}; s^{(1)}, \ldots, s^{(N)}) \right|^2 dx^{(2)} \cdots dx^{(N)}, \tag{3}$$

where Ψ_0 is a ground-state wave function. We study the asymptotic behavior of the energy and density as $N, Z \to \infty$ with N/Z fixed. The magnetic field B is allowed to vary with N as well. If $B = 0$ the leading contribution is well known ([9,10]): It is given by Thomas–Fermi (TF) theory, which for $Z > 20$ is accurate within a few percent as far as the bulk of the electrons is concerned. The theories described here are the generalizations to the case that $B \to \infty$ as $N, Z \to \infty$. This case is important for the study of neutron star surfaces, which are believed to consist mostly of iron atoms ($Z = 26$) in magnetic fields as large as $10^7 - 10^9$ T.

Before discussing the asymptotics of (2) and (3) let us recall the basic facts about the one-particle "kinetic energy" operator (including the interaction of the electron magnetic moment with the field)

$$H_{\mathrm{A}} = [(\mathbf{p} + \mathbf{A}(x)) \cdot \sigma]^2 = (\mathbf{p}_\perp + \mathbf{A}(x))^2 + B\sigma_3 + p_3^2. \tag{4}$$

Here $\mathbf{p}_\perp$ and p_3 denote respectively the momentum perpendicular and parallel to the magnetic field. The spectrum of H_{A} decomposes into Landau bands $\nu = 0, 1, \ldots$ with energy range $[2\nu B, \infty[$, and we have

$$\Pi_\nu H_{\mathrm{A}} \Pi_\nu = -\partial_3^2 + 2\nu B \tag{5}$$

where Π_ν is the projector on the eigenfunctions in the νth band, and $\partial_3 := \partial/\partial x_3$. The degeneracy of the Landau level with energy $2\nu B$ is, per unit area perpendicular to the field, $d_0(B) = (\pi\ell_B^2)^{-1} = B/(2\pi)$ for $\nu = 0$ and $d_\nu(B) = 2(\pi\ell_B^2)^{-1} = B/\pi$ for $\nu \geq 1$.

2.2 The Five Asymptotic Regions

In the following the symbols $\ll$, $\sim$ and $\gg$ are meant to indicate the asymptotic behavior of N-dependent quantities as $N \to \infty$. Thus $B \ll Z^p$, $B \sim Z^p$, and $B \gg Z^p$ mean, respectively, that $B/Z^p \to 0$, $B/Z^p \to (\text{const.}) \neq 0$ and $B/Z^p \to \infty$ as $N \to \infty$ with $Z \sim N$. As already noted in [11–14] the exponents $p = 4/3$ and $p = 3$ are of special importance for the asymptotic properties of the ground states of (1). The analysis in [2] and [3] distinguishes the following parameter regions and establishes their precise status:

Region 1: $B \ll Z^{4/3}$ *Region 2:* $B \sim Z^{4/3}$ *Region 3:* $Z^{4/3} \ll B \ll Z^3$
Region 4: $B \sim Z^3$ *Region 5:* $B \gg Z^3$

A first orientation about the different physics in the different regions may be obtained by the following heuristic reasoning.

By the Pauli principle each electron can be thought of as occupying a "private room". At $B = 0$ its spatial extension a is independent of direction and the kinetic energy is $\varepsilon_{\text{kin}} \sim 1/a^2$. To minimize the Coulomb energy due to the attraction of the nucleus the electrons arrange themselves in a sphere around the nucleus. If R denotes the approximate radius of the sphere the potential energy of an electron is $\varepsilon_{\text{pot}} \sim -Z/R \sim -Z/(N^{1/3}a)$ since $Na^3 \sim R^3$. Optimizing $\varepsilon = \varepsilon_{\text{kin}} + \varepsilon_{\text{pot}}$ with respect to a gives

$$a \sim Z^{-1}N^{1/3} \sim Z^{-2/3} \quad \text{and} \quad R \sim Z^{-1}N^{2/3} \sim Z^{-1/3}. \tag{6}$$

The energy per electron, ε, and the ground-state energy, $E = N\varepsilon$, are

$$\varepsilon \sim -Z^2N^{-2/3} \sim -Z^{4/3} \quad \text{and} \quad E \sim -Z^2N^{1/3} \sim -Z^{7/3}. \tag{7}$$

The repulsive energy, $\varepsilon_{\text{rep}} \sim N/R$, has been ignored since it does not affect the rough estimates of energies and sizes presented here, although it is of course crucial for other questions, e.g., the maximal number of electrons that a nucleus can bind.

Since the effects of a magnetic field on the energy are of the order B, one expects only a small perturbation of the $B = 0$ picture as long as $B \ll |\varepsilon|$, i.e., $B \ll Z^{4/3}$. The wave functions are little affected by the magnetic field, provided $a \ll \ell_B$. Since $a \sim Z^{-2/3}$ and $\ell_B \sim B^{-1/2}$, this leads to the same condition as before. On the other hand, for $B \sim Z^{4/3}$ the distance between electrons is comparable with the magnetic length and the field will start to have effects to leading order. These contributions will involve all Landau levels.

At $B \gg Z^{4/3}$ energy differences between Landau levels are much larger than the Coulomb energy scale, and the electrons will essentially be confined to the lowest Landau band. The private room of an electron is now a cylinder with radius $\sim \ell_B \sim B^{-1/2}$ and length $L \gg \ell_B$. The kinetic energy is $\varepsilon_{\text{kin}} \sim 1/L^2$. A spherically symmetric arrangement of the electrons around the nucleus is still optimal and possible, provided the condition $L \ll R$, with R the radius of the sphere, is compatible with $NL\ell_B^2 \sim R^3$ and $1/L^2 - Z/R =$ minimum. The last two conditions give

$$R \sim Z^{-1/5} N^{2/5} B^{-2/5} \sim Z^{1/5} B^{-2/5} = Z^{-1/3}(B/Z^{4/3})^{-2/5} \tag{8}$$

and hence

$$L \sim Z^{-3/5} N^{1/5} B^{-1/5} \sim ((BZ^2/N^5)^{1/5})R \sim (B/Z^3)^{1/5} R. \tag{9}$$

The condition for sphericity, $L \ll R$, thus becomes $B \ll Z^3$. The energy is

$$E \sim -Z^{6/5} N^{3/5} B^{2/5} \sim -Z^{9/5} B^{2/5} = -Z^{7/3}(B/Z^{4/3})^{2/5}. \tag{10}$$

At $B \sim Z^3$ the atom begins to deviate from spherical shape because the condition $L \ll R$ no longer holds for the optimal L.

If $B \gg Z^3$ the extension of an atom along the field is the same as for one electron, i.e., the atom is approximately a cylinder of length L and radius

$$R \sim (N\ell_B^2)^{1/2} \sim (N/B)^{1/2} \ll L. \tag{11}$$

The Coulomb energy of an electron is now $\varepsilon_{\text{pot}} \sim -(Z/L)\ln(L/R)$ and the optimum of $\varepsilon = \varepsilon_{\text{kin}} + \varepsilon_{\text{pot}}$ with $\varepsilon_{\text{kin}} \sim 1/L^2$ is obtained for

$$L \sim Z^{-1}[\ln(B/(Z^2N))]^{-1} \sim Z^{-1}[\ln(B/(Z^3))]^{-1}. \tag{12}$$

We see that $L/R \sim (B/(Z^2N))^{1/2}\ln(B/(Z^2N))$, so we indeed have $L \gg R$ for $B \gg Z^3$. The energy $E = N\varepsilon$ is

$$E \sim -Z^2 N \ln[(B/(Z^2N)] \sim -Z^3 \ln(B/Z^3). \tag{13}$$

To summarize, the heuristic considerations have disclosed the asymptotic regions $B \ll Z^{4/3}$, $Z^{4/3} \ll B \ll Z^3$ and $B \gg Z^3$, where the simple expressions (7), (10) and (13) for the ground-state energy are plausible, and the transition regions $B \sim Z^{4/3}$ and $B \sim Z^3$, where a more complex behavior is expected.

It is a long way from the rough estimates given above to rigorous theorems about the ground-state asymptotics of (1). Clearly the cases $B \sim Z^{4/3}$ and $B \sim Z^3$ are the most challenging, and in fact the other three cases can be regarded as limits of these two. If $B \sim Z^{4/3}$ the asymptotics turns out to be given by a B-dependent density functional theory of the TF type, that was first introduced by Tomishima and Yonei [15]. Its limit for $B/Z^{4/3} \to 0$ is the usual $B = 0$ TF theory, whereas for $B/Z^{4/3} \to \infty$, but $B/Z^3 \to 0$, it passes into another TF-type theory first considered by Kadomtsev [11]. For $B \sim Z^3$ a new type of functional, depending on density *matrices* rather than the density alone gives the correct limit. If $B/Z^3 \to \infty$ it simplifies to a one-dimensional theory that is solvable *in closed form.*

2.3 The Semiclassical Theories

The density functional that correctly describes the ground state for $B \sim Z^{4/3}$ is the *magnetic Thomas–Fermi functional* (MTF)

$$\mathcal{E}^{\mathrm{MTF}}[\rho; B, Z] = \int \tau_B(\rho(x))dx - Z\int |x|^{-1}\rho(x)dx + D(\rho,\rho). \tag{14}$$

Here $\rho \geq 0$ is the electron density,

$$D(\rho,\rho) = \tfrac{1}{2}\int\int \rho(x)\rho(y)|x-y|^{-1}dxdy \tag{15}$$

is the Coulomb repulsion and τ_B is the kinetic energy density of a free-electron gas in magnetic field B. This function is the Legendre transform of the pressure P_B and is most conveniently characterized by the fact that $\tau_B(0) = 0$ and the derivative τ_B' is the inverse of the derivative of the pressure, i.e.,

$$P_B'(\tau_B'(\rho)) = \rho \tag{16}$$

with

$$P_B'(w) = (2\pi^2)^{-1}B[w^{1/2} + 2\sum_{\nu\geq 1}[w - 2B\nu]_+^{1/2}]. \tag{17}$$

The function $\tau_B(\rho)$ is for large ρ bounded above by (const.) $\rho^{5/3}$, and for $B \to 0$ it tends to the kinetic energy density at $B = 0$, $\tau_0(\rho) = (3/5)(3\pi^2)^{2/3}\rho^{5/3}$. Standard TF theory, where τ_B is replaced by τ_0, is thus a limiting case of (14). In sufficiently strong fields only the contribution from the lowest Landau level in (17) is relevant, and τ_B is replaced by

$$\tau_B^{\mathrm{STF}}(\rho) = (4\pi^4/3)\rho^3/B^2. \tag{18}$$

The corresponding theory will be referred to as the STF theory; it was first studied in [11,14,16].

For each N there is a minimal energy $E^{\mathrm{MTF}}(N,B,Z)$ and a unique minimizing density $\rho^{\mathrm{MTF}}_{N,B,Z}$ for the functional (14):

$$E^{\mathrm{MTF}}(N,B,Z) = \inf\{\mathcal{E}^{\mathrm{MTF}}[\rho] \mid \textstyle\int\rho = N\} = \mathcal{E}^{\mathrm{MTF}}[\rho^{\mathrm{MTF}}_{N,B,Z}]. \tag{19}$$

This is proved by the methods of [9] and [10]. Corresponding quantities for the TF and STF theories are denoted $E^{\mathrm{TF}}(N,Z)$, $\rho^{\mathrm{TF}}_{N,Z}$, $E^{\mathrm{STF}}(N,B,Z)$ and $\rho^{\mathrm{STF}}_{N,B,Z}$. The following scaling laws for the energies hold with $\lambda = N/Z$, $\beta = B/Z^{4/3}$:

$$E^{\mathrm{TF}}(N,Z) = Z^{7/3}E^{\mathrm{TF}}(\lambda,1) \tag{20}$$

$$E^{\mathrm{MTF}}(N,B,Z) = Z^{7/3}E^{\mathrm{MTF}}(\lambda,\beta,1) \tag{21}$$

$$E^{\mathrm{STF}}(N,B,Z) = Z^{7/3}\beta^{2/5}E^{\mathrm{STF}}(\lambda,1,1) \tag{22}$$

The corresponding relations for the densities are

$$\rho^{\mathrm{TF}}_{N,Z}(x) = Z^2\rho^{\mathrm{TF}}_{\lambda,1}(Z^{1/3}x) \tag{23}$$

$$\rho^{\mathrm{MTF}}_{N,B,Z}(x) = Z^2\rho^{\mathrm{MTF}}_{\lambda,\beta,1}(Z^{1/3}x) \tag{24}$$

$$\rho^{\mathrm{STF}}_{N,B,Z}(x) = Z^2\beta^{6/5}\rho^{\mathrm{STF}}_{\lambda,1,1}(Z^{1/3}\beta^{2/5}x). \tag{25}$$

Thus the TF theory has one nontrivial parameter, λ, MTF has two parameters, λ and β, and STF again only one, λ.

The basic quantum mechanical limit theorems for the three regions $B \ll Z^{4/3}$, $B \sim Z^{4/3}$ and $Z^{4/3} \ll B \ll Z^3$ are

Theorem 2.1 (Energy asymptotics in Regions 1–3) *Let N, $Z \to \infty$ with N/Z fixed. If $B/Z^3 \to 0$, then*

$$E^{\mathrm{Q}}(N,B,Z)/E^{\mathrm{MTF}}(N,B,Z) \to 1. \tag{26}$$

If $B/Z^{4/3} \to 0$, then

$$E^{\mathrm{Q}}(N,B,Z)/E^{\mathrm{TF}}(N,Z) \to 1. \tag{27}$$

Finally, if $B/Z^{4/3} \to \infty$, but $B/Z^3 \to 0$, then

$$E^{\mathrm{Q}}(N,B,Z)/E^{\mathrm{STF}}(N,B,Z) \to 1. \tag{28}$$

Theorem 2.2 (Density asymptotics in Regions 1–3) *Let N, $Z \to \infty$ with $\lambda = N/Z$ and $\beta = B/Z^{4/3}$ fixed. Then*

$$Z^{-2}\rho^{\mathrm{Q}}(Z^{-1/3}x) \to \rho^{\mathrm{MTF}}_{\lambda,\beta,1}(x). \tag{29}$$

If $B/Z^{4/3} \to 0$,

$$Z^{-2}\rho^{\mathrm{Q}}(Z^{-1/3}x) \to \rho^{\mathrm{TF}}_{\lambda,1}(x). \tag{30}$$

If $B/Z^{4/3} \to \infty$, but $B/Z^3 \to 0$, then

$$Z^{-2}\beta^{-6/5}\rho^{\mathrm{Q}}(Z^{-1/3}\beta^{-2/5}x) \to \rho^{\mathrm{STF}}_{\lambda,1,1}(x). \tag{31}$$

The limits are in the sense of weak convergence in $L^{5/3}_{\mathrm{loc}}(\mathbf{R}^3)$.

Theorems 2.1–2 are proved in [3]. A brief discussion of the ingredients for the proofs will be given in Sect. 2.6. See also [17] for a special case and [18], [19] for related results and partial generalizations.

The theorems allow us to draw various conclusions about basic properties of heavy atoms in Regions 1–3, i.e., for $B \ll Z^3$. First, the theorems confirm the heuristic arguments given in Sect. 2.2 that in these parameter regions *atoms are spherical* to leading order with radius $\sim Z^{-1/3}(1+(B/Z^{4/3}))^{-2/5}$. Second, it is a general property of TF-type theories that the maximum number of electrons that can be bound is $N = Z$. In other words, *negative ions do not exist* in

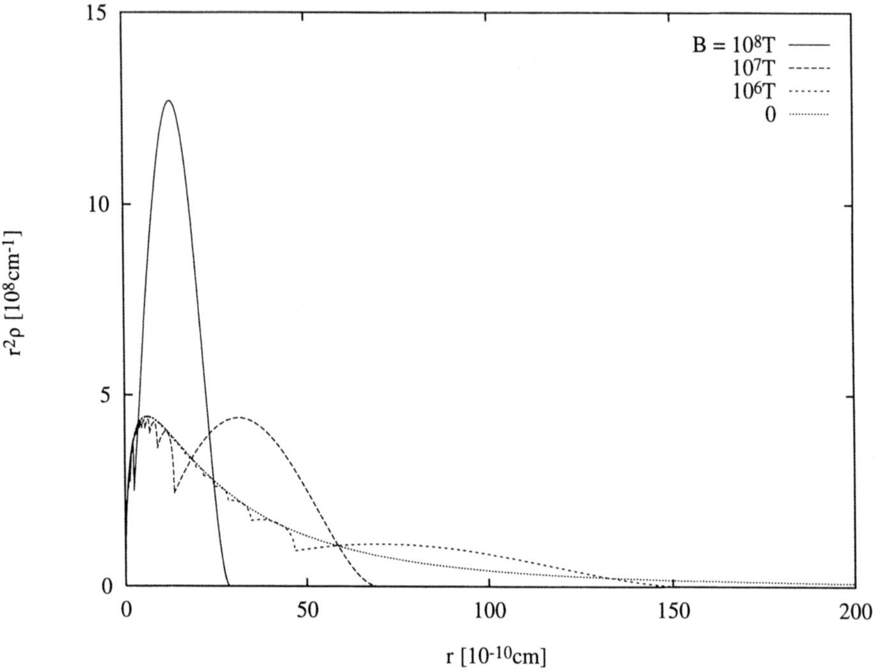

Figure 1: Plots of $|x|^2\rho(x)$ in MTF theory for iron atoms in magnetic fields $B = 0,\ 10^6,\ 10^7,\ 10^8$ T.

the limit considered. Third, *molecules do not bind* in TF-type theories [9,10] which in the present context means that in Regions 1–3 the quantum mechanical binding energies of molecules are of lower order than ground state energies of atoms, which are of the order $Z^{7/3}(1 + B/Z^{4/3})^{2/5}$.

For further discussions of MTF theory and its astrophysical applications see [20,21] and references quoted therein. Figure 1, computed by Ö. Rögnvaldsson, shows the density ρ^{MTF} for various values of B.

2.4 The Density Matrix Theory

In Region 4, i.e., for $B \sim Z^3$, atoms cease to be spherical and a new type of functional takes over from the semiclassical theories of the last subsection. The variable of this functional is not a density but a mapping $x_\perp \to \Gamma_{x_\perp}$ from $x_\perp = (x_1, x_2) \in \mathbf{R}^2$ into density matrices, i.e., nonnegative trace class operators on $L^2(\mathbf{R}, dx_3)$, satisfying the condition

$$0 \leq \Gamma_{x_\perp} \leq (B/(2\pi))I \tag{32}$$

for all $x_\perp$. Let $\Gamma_{x_\perp}(x_3, y_3)$ denote the integral kernel of $\Gamma_{x_\perp}$ and put $\rho_\Gamma(x) = \Gamma_{x_\perp}(x_3, x_3)$. The *Density matrix functional* (DM) is defined as follows:

$$\mathcal{E}^{\mathrm{DM}}[\Gamma; B, Z] = \int \left[-\partial_3^2 \Gamma_{x_\perp}(x_3, y_3) \right]_{x_3=y_3} dx_3 dx_\perp$$
$$-Z \int |x|^{-1} \rho_\Gamma(x) dx + D(\rho_\Gamma, \rho_\Gamma). \tag{33}$$

The corresponding energy $E^{\mathrm{DM}}(N, B, Z)$ and minimizer $\Gamma^{\mathrm{DM}}_{N,B,Z} \equiv \Gamma^{\mathrm{DM}}$ (which can be shown to exist and be unique) are defined by

$$E^{\mathrm{DM}}(N, B, Z) = \inf\{\mathcal{E}^{\mathrm{DM}}[\Gamma] : \textstyle\int \rho_\Gamma \leq N\} = \mathcal{E}^{\mathrm{DM}}[\Gamma^{\mathrm{DM}}]. \tag{34}$$

To understand the motivations for (33) and the condition (32) recall first that for $B \gg Z^{4/3}$ the electrons in an approximate ground state Ψ of (1) are expected to be confined to the lowest Landau band. On may then, with small error, replace the kinetic energy operator (4) by $\Pi_0 H_{\mathrm{A}} \Pi_0 = -\partial_3^2$; cf. (5). Also, one expects that for large N and Z the exchange-correlation part of the Coulomb repulsion is small compared with the direct part $D(\rho_\Psi, \rho_\Psi)$, where $\rho_\Psi(x)$ denotes the density corresponding to Ψ. Defining

$$\Gamma^\Psi_{x_\perp}(x_3, y_3) = N \int \overline{\Psi(x_\perp, x_3; x^{(2)}, \ldots, x^{(N)})}$$
$$\times \Psi(x_\perp, y_3; x^{(2)}, \ldots, x^{(N)}) dx^{(2)} \cdots dx^{(N)} \tag{35}$$

(summation over spins is understood) we have $\rho_\Psi(x) = \Gamma^\Psi_{x_\perp}(x_3, x_3)$. Thus,

$$(\Psi, H_{N,B,Z}\Psi) \approx \mathcal{E}^{\mathrm{DM}}[\Gamma^\Psi; B, Z]. \tag{36}$$

The condition (32) can be traced to the fact that the density of states per unit area in the lowest Landau level is $d_0(B) = B/(2\pi)$. If Ψ is composed of wave functions in the lowest Landau band it is easy to deduce from this that Γ^Ψ must satisfy (32).

The DM functional (33) treats the electrostatic interactions classically but the kinetic energy for the motion along the magnetic field quantum mechanically by the $-\partial_3^2$ term. In directions perpendicular to the field the motion is restricted by the "hard core" condition (32). Note that without (32) the functional would not be bounded below.

The DM theory has two parameters, $\lambda = N/Z$ and $\eta = B/Z^3$. This is manifested in the scaling relations

$$E^{\mathrm{DM}}(N, B, Z) = Z^3 E^{\mathrm{DM}}(\lambda, \eta, 1) \tag{37}$$

and

$$\rho^{\mathrm{DM}}_{N,B,Z}(x) = Z^4 \rho^{\mathrm{DM}}_{\lambda,\eta,1}(Zx), \tag{38}$$

where ρ^{DM} is the density corresponding to the minimizer Γ^{DM}. The $\eta \to \infty$ limit will be considered in the next section.

The main quantum mechanical limit theorems for strong fields are

Theorem 2.3 (Energy asymptotics in Regions 3–5) *Let N, $Z \to \infty$ with N/Z fixed. If $B/Z^{4/3} \to \infty$, then*

$$E^{\mathrm{Q}}(N,B,Z)/E^{\mathrm{DM}}(N,B,Z) \to 1. \tag{39}$$

Theorem 2.4 (Density asymptotics in Region 4) *Let N, Z and $B \to \infty$ with $N/Z = \lambda$ and $B/Z^3 = \eta$ fixed. Then*

$$Z^{-4}\rho^{\mathrm{Q}}_{N,B,Z}(Z^{-1}x) \to \rho^{\mathrm{DM}}_{\lambda,\eta,1}(x) \tag{40}$$

weakly in L^1_{loc}.

Note that Theorem 2.3 overlaps with Theorem 2.1 since Region 3 is covered by both. A limit theorem for the density in Region 5 will be formulated in the next subsection.

Because of Theorems 2.3–4, DM theory is relevant for heavy atoms such as iron ($Z = 26$) in fields of neutron star strength. (At $B = 10^9$ T one has $\beta \approx 14$, $\eta \approx 0.06$ if $Z = 26$.) Numerical studies of the DM theory have been carried out by K. Johnsen [22], and Fig. 2 shows contour plots of the density ρ^{DM} for $N = Z = 26$ and three values of B: 10^8, 10^9 and 10^{10} T. As apparent from these figures, the atomic core is still approximately spherical at the weakest field but the atom shrinks and becomes increasingly elongated as the field goes up. This is in accord with the order of magnitude estimates (11) and (12). Another noteworthy point is that for the strongest field the shape of the atom is simpler than for the weaker field, where the atom is composed of several cylindrical shells. The reason for this is that minimizers for the DM functional are obtained by seeking at each $x_\perp$ the lowest eigenfunctions for a one-dimensional Schrödinger Hamiltonian $-\partial_3^2 + V^{\mathrm{DM}}_{x_\perp}(x_3)$ where $V^{\mathrm{DM}}_{x_\perp}$ is the self-consistent potential generated by the nucleus and ρ^{DM}. The number of the cylindrical shells reflects the number of eigenfunctions that contribute to Γ^{DM}. At $B = 10^{10}$ T this number has dropped to one, in which case $\Gamma^{\mathrm{DM}}_{x_\perp}(x_3, y_3) = \sqrt{\rho^{\mathrm{DM}}(x)}\sqrt{\rho^{\mathrm{DM}}(y)}$ and the DM theory simplifies to a density functional theory. The corresponding functional, denoted by $\mathcal{E}^{SS}$ (SS stands for *super-strong*), is

$$\mathcal{E}^{\mathrm{SS}}[\rho; B, Z] = \int \left[\partial_3\sqrt{\rho}\right]^2 dx - Z\int |x|^{-1}\rho(x)dx + D(\rho,\rho) \tag{41}$$

and condition (32) becomes

$$\int \rho(x_\perp, x_3)\, dx_3 \leq B/(2\pi) \tag{42}$$

for all $x_\perp \in \mathbf{R}^2$. We denote the ground-state energy of (41) by $E^{\mathrm{SS}}(N,B,Z)$. The transition from DM to SS occurs at a certain critical value $\eta_c(\lambda)$ of the parameter $\eta = B/Z^3$, i.e., $E^{\mathrm{DM}}(N,B,Z) = E^{\mathrm{SS}}(N,B,Z)$ for $\eta \geq \eta_c$. Numerical studies [22] give the value $\eta_c = 0.148$ at $\lambda = 1$. This corresponds to $B = 2.44 \times 10^9$ T for iron atoms.

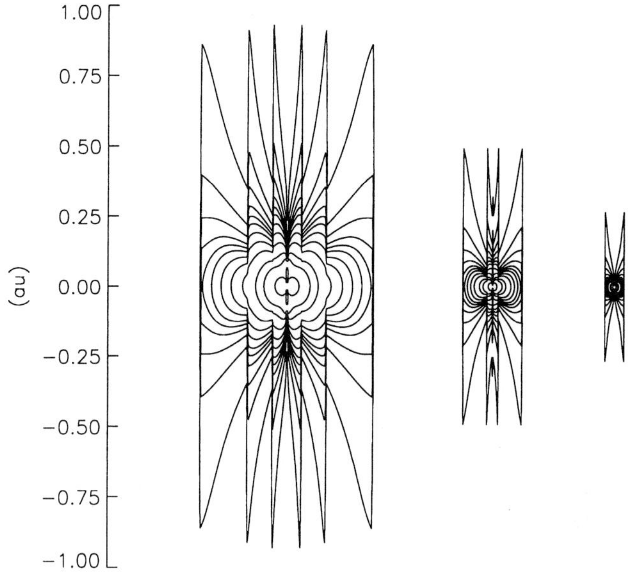

Figure 2: Contour plots of iron atoms in DM theory in magnetic fields $B = 10^8$, 10^9 and 10^{10} T.

In contrast to the semiclassical theories of Sect. 2.3 *negative ions and molecular binding are possible* in DM theory. In fact, in the extreme limit $\eta \to \infty$ considered in the next section the maximum value of λ is $\lambda_c = 2$, and the binding energy of a diatomic molecule is six times as large as the ground-state energy of individual atoms! At $B = 10^8$ T and $Z = 26$ the computations in [22] give $\lambda_c \approx 1.04$ and a ratio of about 0.18 between the binding energy of a diatomic molecule and the ground-state energy of an atom.

2.5 The Limit of Hyper-Strong Fields

According to (11) and (12) the length of the atom is expected to scale as $[Z \ln(B/Z^3)]^{-1}$ and the diameter as $(Z/B)^{-1/2}$ when $\eta = B/Z^3 \to \infty$. This suggests to change variables in the functional $\mathcal{E}^{\mathrm{SS}}$ in accordance with these length scales and look for the $\eta \to \infty$ limit of the resulting functional. Closer examination of $\mathcal{E}^{\mathrm{SS}}$ [22] shows that convergence is more rapid if $\ln \eta$ is replaced by another function with the same asymptotic behavior, namely the solution

$L(\eta)$ of the equation

$$(\eta/2)^{1/2} = L(\eta)\sinh(L(\eta)/2). \tag{43}$$

The rescaled Coulomb potential becomes

$$V_\eta(x) = L(\eta)^{-1}(\eta^{-1}L(\eta)^2 x_\perp^2 + x_3^2)^{-1/2} \tag{44}$$

and the rescaled SS functional is defined as

$$\mathcal{E}_\eta^{\rm SS}(\rho) = \int \left[\partial_3\sqrt{\rho}\right]^2 dx - \int V_\eta(x)\rho(x)dx + \int V_\eta(x-y)\rho(x)\rho(y)\,dxdy. \tag{45}$$

The energy

$$E_\eta^{\rm SS}(\lambda) = \inf\{\mathcal{E}_\eta^{\rm SS}(\rho) | \textstyle\int \rho dx \le \lambda, \int \rho dx_3 \le 1\} \tag{46}$$

is related to $E^{\rm SS}$ by

$$E^{\rm SS}(N,B,Z) = Z^3 L(\eta)^2 E_\eta^{\rm SS}(\lambda). \tag{47}$$

Now $V_\eta(x) \to \delta(x_3)$ in distributional sense as $\eta \to \infty$. Defining $\bar\rho(x_3) = \int \rho(x_\perp, x_3)dx_\perp$ and the *Hyper-strong density functional* by

$$\mathcal{E}^{\rm HS}[\bar\rho] = \int \left[\partial_3\sqrt{\bar\rho}\right]^2 dx_3 - \bar\rho(0) + \int \bar\rho(x_3)^2 dx_3 \tag{48}$$

it follows that $\mathcal{E}_\eta^{\rm SS}[\rho] \to \mathcal{E}^{\rm HS}[\bar\rho]$ for all ρ. With a little more effort one shows that $E_\eta^{\rm SS}(\lambda) = E^{\rm HS}(\lambda) + O(L(\eta)^{-1})$, with $E^{\rm HS}(\lambda)$ the ground-state energy of $\mathcal{E}^{\rm HS}$, and thus by (47)

$$E^{\rm SS}(N,B,Z) = Z^3 L(\eta)^2 E^{\rm HS}(\lambda) + Z^3 O(L(\eta)). \tag{49}$$

Since $L(\eta)$ increases only logarithmically with η the convergence of SS to HS is fairly slow.

The hyper-strong functional is a rough approximation of the SS functional at all but extremely large η, but it has the interesting property that it can be minimized in closed form. One finds

$$E^{\rm HS}(\lambda) = -\tfrac{1}{4}\lambda + \tfrac{1}{8}\lambda^2 - \tfrac{1}{48}\lambda^3 \tag{50}$$

for $\lambda \le 2$ and $E^{\rm HS}(\lambda) = E^{\rm HS}(2) = -1/6$ for $\lambda \ge 2$. Writing the minimizer $\bar\rho^{\rm HS}(x_3)$ as $\psi^{\rm HS}(x_3)^2$ one has

$$\psi^{\rm HS}(x) = \frac{\sqrt{2}(2-\lambda)}{4\sinh[\frac{1}{4}(2-\lambda)|x| + c]} \quad \text{for } \lambda < 2 \tag{51}$$

$$\psi^{\rm HS}(x) = \sqrt{2}(2+|x|)^{-1} \quad \text{for } \lambda \ge 2 \tag{52}$$

where $\tanh c = (2-\lambda)/2$.

Combining Theorem 2.3 with (49) one obtains

Theorem 2.5 (Energy asymptotics in Region 5) *Let N, $Z \to \infty$ with $N/Z = \lambda$ fixed. If $\eta = B/Z^3 \to \infty$, then*

$$E^{\mathrm{Q}}(N,B,Z)/(Z^3 L(\eta)^2 E^{\mathrm{HS}}(\lambda)) \to 1 \tag{53}$$

where $L(\eta)$ and $E^{\mathrm{HS}}(\lambda)$ are given by (43) and (50).

There is also a limit theorem for the density:

Theorem 2.6 (Density asymptotics in Region 5) *Let N, $Z \to \infty$ with $N/Z = \lambda$ fixed. If $\eta = B/Z^3 \to \infty$ then*

$$\left[Z^2 L(\eta)\right]^{-1} \int \rho(x_\perp, [ZL(\eta)]^{-1} x_3) dx_\perp \to \bar{\rho}^{\mathrm{HS}}(x_3) \tag{54}$$

weakly in $L^1_{\mathrm{loc}}(\mathbf{R}, dx_3)$.

From these limit theorems and the explicit solution (50)-(52) of the HS theory one also obtains results about enhanched negative ionization and molecular binding of atoms in hyperstrong magnetic fields.

Theorem 2.7 (Negative ionization) *The maximal number, N_c^{Q}, of electrons that can be bound to an atom of nuclear charge Z, defined by*

$$N_c^{\mathrm{Q}} = \max\{N : E^{\mathrm{Q}}(N,B,Z) < E^{\mathrm{Q}}(N-1,B,Z)\}, \tag{55}$$

satisfies

$$\liminf N_c^{\mathrm{Q}}/Z \geq 2 \tag{56}$$

as $Z \to \infty$ and $B/Z^3 \to \infty$.

The result about molecular binding is based on the fact that the nuclei of molecules coalesce in the HS limit, despite the nuclear repulsion. The striking fact is that *the binding energy is of the order of the atomic energy itself*—contrary to the 'ordinary' situation in which it is only a tiny perturbation. To state this precisely, let $E^{\mathrm{Q}}(N,B,Z_1,\ldots,Z_K)$ denote the quantum mechanical ground-state energy of a molecule of K nuclei with charges $Z_1,\ldots,Z_K$ and N electrons. This energy includes the nuclear repulsion energy $\sum_{k<\ell} Z_k Z_\ell |R_k - R_\ell|^{-1}$, where the R_k denote the positions of the nuclei (that are assumed to be infinitely heavy), and it is understood that the infimum over all nuclear positions has been taken. We have

Theorem 2.8 (Bound atoms are isocentric for large B) *Put $Z_{\mathrm{tot}} = Z_1 + \ldots + Z_K$. If $B/Z_{\mathrm{tot}}^3 \to \infty$ as $N, Z_{\mathrm{tot}} \to \infty$ with N/Z_{tot} fixed, then*

$$E^{\mathrm{Q}}(N,B,Z_1,\ldots,Z_K)/E^{\mathrm{Q}}(N,B,Z_{\mathrm{tot}}) \to 1. \tag{57}$$

The molecular binding energy is defined by

$$E_{\rm b}^{\rm Q}(N,B,Z_1,\ldots,Z_K) = \min\{E^{\rm Q}(N^{\{a\}},B,Z^{\{a\}}) + E^{\rm Q}(N^{\{b\}},B,Z^{\{b\}}) - E^{\rm Q}(N,B,Z_1,\ldots,Z_K)\}. \tag{58}$$

The minimum is over decompositions of $\{1,\ldots,K\}$ into two clusters, $\{a\}$ and $\{b\}$, and $Z^{\{a\}}$ and $Z^{\{b\}}$ stand for n-tuples of nuclear charges corresponding to each cluster, while $N^{\{a\}} + N^{\{b\}} = N$. For neutral molecules of identical atoms we obtain as an immediate corollary of Theorems 2.5 and 2.8:

Theorem 2.9 (Strong molecular binding) *The binding energy of a neutral molecule of K nuclei with charge Z satisfies*

$$\frac{E_{\rm b}^{\rm Q}(KZ,B,Z,\ldots,Z)}{K|E^{\rm Q}(Z,B,Z)|} \to 3[K/2](K-[K/2]) \tag{59}$$

as $Z\to\infty$ and $B/Z^3\to\infty$. Here $[K/2]$ denotes the integer part of $K/2$.

2.6 Ingredients for the Proofs

The chain of arguments leading to the proofs of the quantum mechanical limit theorems in Sects. 2.3–5 is rather long and it is not possible here to mention but the main links. A first remark is that it suffices to prove the theorems about the *energy* asymptotics, but with slightly more general potentials than $-Z/|x|$, for the limit theorems for the density can then be obtained in a standard way by variation with respect to the potential [9]. To prove Theorems 2.1 and 2.3 one must derive upper and lower bounds on the quantum mechanical energy in terms of ground-state energies of the relevant functionals with controlled errors. The upper bounds are easier than the lower bounds and are obtained using the variational principle of Lieb [23]. This principle implies that for any kernel $\mathcal{K}(x,s;x',s')$ in the space and spin variables, satisfying $0\le\mathcal{K}\le I$ as an operator on $L^2(\mathbf{R}^3;\mathbf{C}^2)$ as well as $\mathrm{Tr}\,\mathcal{K}\le N$, one has

$$E^{\rm Q}(N,B,Z) \le \mathrm{Tr}[(H_{\rm A} - Z/|x|^{-1})\mathcal{K}] + D(\rho_{\mathcal{K}},\rho_{\mathcal{K}}) \tag{60}$$

where $\rho_{\mathcal{K}}(x) = \sum_s \mathcal{K}(x,s;x,s)$. Suitable kernels $\mathcal{K}$ are in the semiclassical Regions 1–3 constructed by means of *coherent states*, which may be regarded as a rigorous version of the "private rooms" used in the heuristic arguments in Sect. 2.2. Coherent states are usually understood as maps from the classical phase space of a system into wave functions, or rather into rank one projectors, having optimal localization around the classical values. The objects used here are slightly more general and are reminiscent of the coherent operators considered in [24]. They are defined by the integral kernels

$$\Pi_{\nu,u,p}(x,s;x',s') = g_r(x-u)\Pi_\nu^\perp(x_\perp,x'_\perp)e^{ip(x_3-x'_3)}g_r(x'-u)\delta_{ss'}. \tag{61}$$

Here ν is a Landau level index, $u \in \mathbf{R}^3$, $p \in \mathbf{R}$, $\Pi_\nu^\perp(x_\perp, x'_\perp)$ is the integral kernel of the projector on the νth Landau level in $L^2(\mathbf{R}^2, dx_\perp)$, and $g_r(x) = r^{-3/2}g(x/r)$ with $\int g^2 = 1$ is a localization function. These operators were first used in [17] to prove Theorem 2.1 for the special case of Regions 1 and 2. They are positive, although not projectors, and satisfy the coherent operator identities

$$(2\pi)^{-1}\sum_\nu \int\int \Pi_{\nu,u,p}\, dudp = I \quad \text{and} \quad \mathrm{Tr}\,(\Pi_{\nu,u,p}) = d_\nu(B). \tag{62}$$

Moreover, they are approximate eigenoperators for both H_{A} and local potentials V in the sense that

$$\mathrm{Tr}\,(H_{\mathrm{A}}\Pi_{\nu,u,p}) = d_\nu(B)\varepsilon_{\nu,p} + d_\nu(B)\, r^{-2}\int(\nabla g)^2, \tag{63}$$

where $\varepsilon_{\nu,p} = 2\nu B + p^2$ is a (generalized) eigenvalue of H_{A}, and

$$\mathrm{Tr}\,(V\Pi_{\nu,u,p}) = d_\nu(B)\, V * g_r^2. \tag{64}$$

The test kernel $\mathcal{K}$ is defined by

$$\mathcal{K}(x,s;x',s') = \delta_{ss'}\sum_\nu \int\int \Pi_{\nu,u,p}(x,s;x',s')\Theta\left(\tau'_B\left(\rho^{\mathrm{MTF}}(u)\right) - \varepsilon_{\nu,p}\right) dudp, \tag{65}$$

where Θ is the Heaviside function, $\Theta(t) = 1$ for $t \geq 0$ and 0 for $t < 0$. Using (58), (59) and the properties of ρ^{MTF} it is easy to show that if the localization length r is chosen $\sim (Z^{-1/3}(1+\beta)^{-2/5})^\gamma$ with $0 < \gamma < 1$ the errors in the upper bound (55) are small compared to the main contribution $E^{\mathrm{MTF}}(N, B, Z)$.

For the upper bound in Regions 4 and 5 one cannot use the coherent states (56) because the localization error due to the function g becomes too large. Instead one makes use of the fact that the the wave functions in the lowest Landau level are themselves localized on a length scale $\sim \ell_B$, which, by the heuristic arguments of Sect. 2.2, is small compared with the atomic dimensions perpendicular to the field. The kernel $\mathcal{K}$ is here defined as

$$\mathcal{K}(x,s;x',s') = \delta_{ss'}\, d_0(B)^{-1}\int \Pi_0^\perp(x_\perp, y_\perp)\Gamma_{y_\perp}^{\mathrm{DM}}(x_3, x'_3)\Pi_0^\perp(y_\perp, x'_\perp)dy_\perp. \tag{66}$$

To prove that the error is of lower order than the ground-state energy one uses an independent simple upper bound for E^{Q} in terms of the energy without electronic repulsion but with a reduced nuclear charge.

The lower bounds on the quantum mechanical energy require more tools. In Regions 1–3 one uses coherent states in a similar way as in the upper bound and, as an essential new ingredient, a generalization of the Lieb–Thirring estimate on the sum of negative eigenvalues of Schrödinger Hamiltonians [25].

Theorem 2.10 (Magnetic LT-inequality) *Let* $|V|_- \in L^{5/2}(\mathbf{R}^3) \cap L^{3/2}(\mathbf{R}^3)$, *where* $|V(x)|_- = |V(x)|$ *for* $V \leq 0$ *and* 0 *otherwise, and let* $e_j(B,V)$, $j = 1, 2, \ldots$ *denote the negative eigenvalues of the operator* $H = H_{\mathrm{A}} + V(x)$. *Then*

$$\sum_j |e_j(B,V)| \leq L_1 B \int |V(x)|_-^{3/2} dx + L_2 \int |V(x)|_-^{5/2} dx, \tag{67}$$

for certain constants L_1 *and* L_2.

For generalizations to inhomogeneous field and a discussion of the constants in (62) see [26]. The Lieb–Thirring inequality (62) implies via Legendre transformation an estimate of the kinetic energy $T_\Psi = (\Psi, \sum_i H_{\mathrm{A}}^{(i)} \Psi)$ from below by (const.)$\int \tau_B(\rho_\Psi)$. These estimates are needed both to control the localization error, when the Coulomb singularity is smeared as in (59) by g_r^2, and the indirect (i.e., exchange-correlation) part of the Coulomb repulsion, which, by the Lieb–Oxford inequality [27] is bounded below by $-$(const.)$\int \rho_\Psi^{4/3}$.

In Regions 4 and 5 there are two main ingredients for the lower bound. The first is a theorem on the confinement of the electrons in the lowest Landau band as $B/Z^{4/3} \to \infty$.

Theorem 2.11 (Confinement to lowest Landau band) *Define*

$$E_{\mathrm{conf}}^{\mathrm{Q}}(N,Z,B) = \inf_{(\Psi,\Psi)=1} (\Psi, \Pi_0^N H_{N,B,Z} \Pi_0^N \Psi) \tag{68}$$

where Π_0^N *is the projector on the lowest Landau band for the* N*-electron system. Then*

$$E^{\mathrm{Q}}/E_{\mathrm{conf}}^{\mathrm{Q}} \to 1 \tag{69}$$

as $B/Z^{4/3} \to \infty$ *with* N/Z *fixed.*

The second ingredient is an estimate of the indirect part of the Coulomb repulsion for strong magnetic fields. The Lieb–Oxford inequality, although universally true, is not strong enough for $B \sim Z^5$ and larger. In fact, if ρ is essentially concentrated in a cylinder of radius $\sim (Z/B)^{1/2}$ and length $\sim [Z \ln \eta]^{-1}$, as ρ_Ψ will eventually be according to the heuristics in Sect. 2.2, then $\int \rho^{4/3}$ is of the order $Z^{4/3} B^{1/3} [\ln \eta]^{1/3}$. Only for $B \ll Z^5$ will this be smaller than the expected ground-state energy $\sim Z^3 [\ln \eta]^2$. The indirect part of the Coulomb energy is essentially N times the self-energy of a unit charge smeared over the private room of an electron and should accordingly be of the order $Z^2 \ln(B/Z^3) \ln(B/Z^2)$. The estimate derived in [3] does not quite reach this ideal. The bound is obtained by introducing two ultraviolet cut-offs in the Coulomb kernel, one for the directions perpendicular to the field and another in the direction of the field, and it involves the kinetic energy. The result is

Theorem 2.12 (Bound on the indirect Coulomb energy)

$$\begin{aligned} H_{N,B,Z} \; &\geq \; \sum_{i=1}^{N} \Big((1 - Z^{-1/3}) H_{\mathrm{A}}^{(i)} + V^{\mathrm{DM}}(x^{(i)}) \Big) - D(\rho^{\mathrm{DM}}, \rho^{\mathrm{DM}}) \\ &- \; C_\lambda (1 + \lambda^5)(1 + Z^{8/3})(1 + [\ln \eta]^2) \end{aligned}$$

where V^{DM} is the potential generated by the nucleus and ρ^{DM}, and C_λ is a constant depending only on $\lambda = N/Z$.

In addition to Theorems 2.7–9 good control over the asymptotic MTF and DM theories is important for the proofs of the limit theorems. In the MTF case the methods of [9] and [10] can be used, but the DM theory requires new ideas. In particular, establishing the existence and uniqueness of the minimizer Γ^{DM} is not entirely straightforward.

3 Quantum Dots

3.1 The Hamiltonian

A quantum dot is modeled as a two-dimensional system of N electrons with effective mass m_* in a continuous potential V with $V(x) \to \infty$ for $|x| \to \infty$. (For instance, $V(x) = K|x|^2$.) The two-dimensional medium has dielectric constant ϵ and the effective charge for the electronic Coulomb interactions is $e_* = e/\sqrt{\epsilon}$. A magnetic field of strength B points in the direction perpendicular to the two-dimensional plane. Units are chosen such that $\hbar = 2m_* = e_* = 1$, and B is measured in units of $4B_*$ with $B_* = e_*^3 m_*^2 c/(\epsilon^{1/2}\hbar^2)$. To simplify notation the electron spin will here be ignored. The Hamiltonian can then be written

$$H_{N,B,V} = \sum_{i=1}^{N} \left[\left(\mathbf{p}^{(i)} + \mathbf{A}(x^{(i)}) \right)^2 - B + V(x^{(i)}) \right] + \sum_{i<j} |x^{(i)} - x^{(j)}|^{-1} \qquad (70)$$

where $\mathbf{p} = -i(\partial_1, \partial_2)$, $x = (x_1, x_2) \in \mathbf{R}^2$ and $\mathbf{A}(x) = \frac{1}{2}(-x_2 B, x_1 B)$ as before. The spectrum of the kinetic energy operator $H_{\mathrm{kin}} = (\mathbf{p} + \mathbf{A})^2 - B$ consists of the Landau levels $2\nu B$, $\nu = 0, 1, \ldots$, where B has been subtracted for convenience in order that the spectrum of H_{kin} starts at 0. We write the potential as $V(x) = Kv(x)$ with a coupling constant K, and regard v as fixed, while K varies proportionally to N. The ground-state energy and electronic density for the Hamiltonian (65) are denoted respectively by $E^{\mathrm{Q}}(N, B, K)$ and $\rho^{\mathrm{Q}}_{N,B,K}(x)$.

3.2 The Asymptotic Theories

The main features of the asymptotics of $E^{\mathrm{Q}}(N, B, K)$ and $\rho^{\mathrm{Q}}_{N,B,K}(x)$ can be made plausible by simple heuristic arguments in the spirit of Sect. 2.2. Here, however, it is necessary to include the Coulomb repulsion between the electrons to get the right picture of the size of the dot for strong B. The reason is twofold: First, the kinetic energy vanishes in the lowest Landau level since there is no degree of freedom along the field as in the three dimensional case. Second, the potential V is nonsingular by assumption and the repulsion is therefore not compensated by an attractive singularity as for natural atoms.

Starting at $B = 0$, the kinetic energy of an electron is $\varepsilon_{\text{kin}} \sim N/R^2$, with R the radius of the dot, the potential energy from the confining V is $\varepsilon_{\text{pot}} \sim Kv(R)$ and the repulsive energy $\varepsilon_{\text{rep}} \sim N/R$. If $K \to \infty$ with $K/N = k$ fixed as $N \to \infty$, all these terms are proportional to N and the optimal radius is thus to leading order *independent of* N. (It depends on k though, and tends to infinity if $k \to 0$.) The energy of an electron is $\sim N$ and the total ground-state energy is thus $E \sim N^2$. The division line between weak and strong fields is clearly $B \sim N$: for $B \ll N$ the energy differences between Landau levels are much smaller than the other energies, whereas for $B \gg N$ the electrons will essentially sit in the lowest Landau level and the kinetic energy will be close to zero. An interesting point is that the electronic repulsion and the absence of attractive singularities of V prevents the dot from shrinking indefinitely as $B/N \to \infty$, contrary to what happens if $B/Z^3 \to \infty$ for natural, 3D atoms. In fact, dropping the kinetic term in (65) altogether leads to a model of N classical point particles interacting by Coulomb repulsion and with V. The radius of this system is approximately the value of R that minimizes $kv(R)+1/R$, and while it is slightly smaller than the $B = 0$ value, where the kinetic term $1/R^2$ pushes the radius up, it is independent of B. Note also that the for $B \gg N$ the electrons are localized on the scale $\ell_B \sim B^{-1/2}$ which is much smaller than the mean electronic distance $\sim N^{-1/2}$, while in the 3D case, with the potential $-Z/|x|$, the hard core bound $R \geq N^{1/2}\ell_B$ is saturated in the asymptotic ground state.

The density functionals, that enable to turn the heuristic reasoning above into precise theorems about the $N, K \to \infty$ limit, are defined as follows. For weak fields, $B \ll N$, a standard two-dimensional Thomas–Fermi energy functional applies,

$$\mathcal{E}^{\text{TF}}[\rho; K] = 2\pi \int \rho(x)^2 dx + \int V(x)\rho(x)dx + D(\rho,\rho). \tag{71}$$

At moderate fields, $B \sim N$, the correct asymptotics is given by a two-dimensional magnetic TF functional

$$\mathcal{E}^{\text{MTF}}[\rho; B, K] = \int j_B(\rho(x))dx + \int V(x)\rho(x)dx + D(\rho,\rho) \tag{72}$$

where j_B is a piece-wise linear function, defined by $j_B'(\rho) = [2\pi\rho/B]$ where $[t]$ denotes the integral part of t. This functional was first applied to quantum dots in [28] and is capable of modeling interesting details in the conductivity of large quantum dots in magnetic fields [29]. The last asymptotic density functional, the "classical continuum model", applies to strong fields, $B \gg N$. It contains only classical interactions of the charge density and no kinetic energy term:

$$\mathcal{E}^{\text{C}}[\rho; K] = \int V(x)\rho(x)dx + D(\rho,\rho). \tag{73}$$

In addition to the three density functionals the model of N point charges with energy

$$\mathcal{E}^{\text{P}}(x_1,\ldots,x_N) = \sum_i V(x_i) + \sum_{i<j} |x_i - x_j|^{-1} \tag{74}$$

is also relevant. The ground-state energies of (66)–(68), i.e. the infima (in fact minima) of the functionals for densities with given $\int \rho = N$, are denoted $E^{\rm TF}(N,K)$, $E^{\rm MTF}(N,B,K)$ and $E^{\rm C}(N,K)$, and an analogous notation is used for the minimizing densities. The infimum of (69) for given N is denoted $E^{\rm P}(N,K)$. The energies and densities scale as expected from the heuristic arguments, e.g., $E^{\rm TF}(N,K) = N^2 E^{\rm TF}(1,K/N)$, $E^{\rm C}(N,K) = N^2 E^{\rm C}(1,K/N)$, $E^{\rm MTF}(N,B,K) = N^2 E^{\rm MTF}(1,B/N,K/N)$, $\rho^{\rm MTF}_{N,B,K}(x) = N\rho^{\rm MTF}_{1,B/N,K/N}(x)$ etc.

The difference between $E^{\rm C}(N,K)$ and $E^{\rm P}(N,K)$ can be estimated, using an electrostatic lemma of Lieb and Yau [30], as

$$E^{\rm C}(N,K) - aN^{3/2} \geq E^{\rm P}(N,K) \geq E^{\rm C}(N,K) - bN^{3/2}, \tag{75}$$

where $a, b > 0$ depend only on $k = K/N$. Note that the order $N^{3/2} = N \times (N^{-1/2})^{-1}$ is precisely what what is expected from the indirect part of the Coulomb interaction (which is absent in $E^{\rm C}$).

In order to guarantee that the minimizer of (68) is an L^1 function and not just a measure, V is taken to be in the class $C^{1,\alpha}$ of once differentiable functions whose derivatives are Lipschitz continuous of some order $\alpha > 0$. For the important special case $V(x) = K|x|^2$ the minimizer $\rho^{\rm C}_{N,K}$ can be computed explicitly; it has the shape of a half-ellipsoid:

$$\rho^{\rm C}_{N,V}(x) = \frac{3}{2\pi} N R^{-2} \sqrt{1 - R^{-2}|x|^2} \tag{76}$$

if $|x| \leq R = (3\pi N/8K)^{1/3}$ and 0 otherwise.

The shape of the electronic density in the three asymptotic theories is shown in Fig. 3 for a quadratic potential $V(x) = K|x|^2$ and different values of B. The pictures were prepared by K. Johnsen. At $B = 0$ we have the usual Thomas–Fermi model, which may be regarded as a limit of the MTF model with infinitely many Landau levels occupied. As the field increases the number of occupied Landau levels decreases and the full levels show up as plateaus at integer values of $2\pi\rho/B$. At $B = 2\,\rm T$ three Landau levels are full and electrons in the central dome occupy the fourth level. At $B = 4\,\rm T$ most electrons are in the lowest level, but some are in the next highest level. At $B = 8\,\rm T$, the density is a half ellipsoid, given by the minimizer (71) of the classical functional (68). Note that the diameter of the dot at $B = 8\,\rm T$ is slightly smaller than at $B = 0$, but as soon as the region of the model (68) has been reached the density does not change further although B is increased.

3.3 The Limit Theorem

The analogue for quantum dots of the limit theorems in Sects. 2.3–5 is

Theorem 3.1 (Limit theorem for the energy and density) *Let $V = Kv$ with v a fixed function in $C^{1,\alpha}$, $0 < \alpha$. Then, if $N, K \to \infty$ with $K/N = k$ fixed,*

$$E^{\rm Q}(N,B,K)/E^{\rm MTF}(N,B,K) \to 1 \tag{77}$$

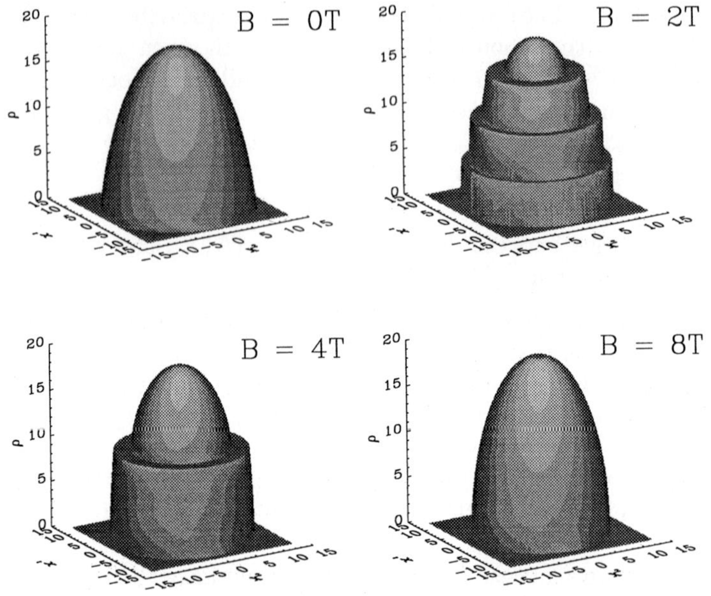

Figure 3: Quantum dots at various magnetic field strengths. The potential is $V(x) = (1/2)m_*\omega^2|x|^2$, with $m_* = 0.67m$, $\hbar\omega = 3.37$ meV, and $N = 50$. The coordinate axes are displayed in units of 10^{-8} m and the density ρ in the units $10^{-14}\,\mathrm{m}^{-2}$.

uniformly in B. *Moreover,*

$$E^{\mathrm{Q}}(N,B,K)/E^{\mathrm{TF}}(N,K) \to 1 \qquad \text{if } B/N \to 0 \tag{78}$$

and

$$E^{\mathrm{Q}}(N,B,K)/E^{\mathrm{C}}(N,K) \to 1 \qquad \text{if } B/N \to \infty. \tag{79}$$

The densities converge also:

$$\frac{1}{N}\rho^{\mathrm{Q}}_{N,B,K} \to \rho^{\mathrm{MTF}}_{1,B/N,k} \tag{80}$$

uniformly in B, *and*

$$\frac{1}{N}\rho^{\mathrm{Q}}_{N,B,K} \to \rho^{\mathrm{TF}}_{1,k} \qquad \text{if } B/N \to 0 \tag{81}$$

$$\frac{1}{N}\rho^{\mathrm{Q}}_{N,B,K} \to \rho^{\mathrm{C}}_{1,k} \qquad \text{if } B/N \to \infty. \tag{82}$$

The convergence is in the weak L^1 *sense.*

The proof of this theorem proceeds partly along similar lines as in the 3D case. Thus the upper bound on the energy is obtained using coherent states as in (56) (without the $\exp ip(x_3 - x_3')$ factor, of course). For the lower bound one treats the cases $B \ll N$ and $B \sim N$ differently from $B \gg N$. For strong fields the Hamiltonian is estimated from below by the point charge functional (69), which in turn is bounded by the classical functional with a small error according to (70). For weak and moderate fields one uses coherent states, and the indirect Coulomb energy is estimated by $-(\text{const.}) \int \rho^{3/2}$. The following inequality is needed to control this term.

Theorem 3.2 (2D magnetic Lieb–Thirring inequality) *Let W be locally integrable, and denote by $e_1(W), e_2(W), \ldots$ the negative eigenvalues (if any) of the operator $H = (\mathbf{p} + \mathbf{A})^2 - B + W$. Then for all $0 < \lambda < 1$ we have the estimate*

$$\sum_j |e_j(W)| \leq \lambda^{-1} \frac{B}{4\pi} \int_{\mathbf{R}^2} |W(x)|_- dx + \tfrac{3}{8}(1-\lambda)^{-2} \int_{\mathbf{R}^2} |W(x)|_-^2 dx. \quad (83)$$

The proof is slightly more complicated than in the 3D case because the quantum mechanical kinetic energy vanishes in the lowest Landau level.

Acknowledgments. Thanks are due to K. Johnsen and Ö. Rögnvaldsson for the pictures presented here, and to I. Fushiki, E. H. Gudmundsson, V. Gudmundsson, J. Kinaret, M. Loss and C. Pethick for valuable discussions. This work was supported by US NSF grants Nos. PHY90-19433 (EHL) and DMS 92-03829 (JPS), and by the Icelandic Science Foundation (JY).

References

[1] G. Chanmugam, *Magnetic Fields of Degenerate Stars*, Ann. Rev. Astron. Astrophys. **30**, 143–184 (1992)

[2] E.H. Lieb, J.P. Solovej and J. Yngvason, *Asymptotics of Heavy Atoms in High Magnetic Fields: I. Lowest Landau Band Regions*, Commun. Pure Appl. Math. **47**, 513–591 (1994)

[3] E.H. Lieb, J.P. Solovej and J. Yngvason, *Asymptotics of Heavy Atoms in High Magnetic Fields: II. Semiclassical Regions*, Commun. Math. Phys **161**, 77–124 (1994)

[4] M.A. Kastner, *Artificial atoms*, Phys. Today **46**, 24–31 (1993)

[5] E.H. Lieb, J.P. Solovej and J. Yngvason, *The Ground States of Large Quantum Dots in Magnetic Fields*, Phys. Rev. B **51**, 10646–10665 (1995)

[6] E.H. Lieb, J.P. Solovej and J. Yngvason, *Heavy Atoms in the Strong Magnetic Field of a Neutron Star*, Phys. Rev. Lett. **69**, 749–752 (1992)

[7] E.H. Lieb and J.P. Solovej, *Atoms in the Magnetic Field of a Neutron Star*, in: Differential Equations with Applications to Mathematical Physics, W.F. Ames, J.V. Herod and E.M. Harrell II, eds., pp. 221–237, Academic Press 1993

[8] E.H. Lieb, J.P. Solovej and J. Yngvason, *Quantum Dots*, in: Proceedings of the Conference on Partial Differential Equations and Mathematical Physics, University of Alabama, Birmingham, 1994, I. Knowles, ed., pp. 157–172, International Press 1995

[9] E.H. Lieb and B. Simon, *The Thomas–Fermi Theory of Atoms, Molecules and Solids*, Adv. in Math. **23**, 22–116 (1977)

[10] E.H. Lieb, *Thomas–Fermi and related theories of atoms and molecules*, Rev. Mod. Phys. **53**, 603–641 (1981); *Erratum*, Rev. Mod. Phys. **54**, 311 (1982)

[11] B.B. Kadomtsev, *Heavy Atoms in an Ultrastrong Magnetic Field*, Soviet Phys. JETP **31**, 945–947 (1970)

[12] B.B. Kadomtsev and V.S. Kudryavtsev, *Atoms in a superstrong magnetic field*, JETP Lett. **13**, 42–44 (1971)

[13] M. Ruderman, *Matter in Superstrong Magnetic Fields: The Surface of a Neutron Star*, Phys. Rev. Lett. **27**, 1306–1308 (1971).

[14] R.O. Mueller, A.R.P. Rau and L. Spruch, *Statistical Model of Atoms in Intense Magnetic Fields*, Phys. Rev. Lett. **26**, 1136–1139 (1971)

[15] Y. Tomishima and K. Yonei, *Thomas–Fermi Theory for Atoms in a Strong Magnetic Field*, Progr. Theor. Phys. **59**, 683–696 (1978)

[16] B. Banerjee, D.H. Constantinescu, and P. Rehák, *Thomas–Fermi and Thomas–Fermi–Dirac calculations for atoms in a very strong magnetic field*, Phys. Rev. D **10**, 2384–2395 (1974)

[17] J. Yngvason, *Thomas–Fermi Theory for Matter in a Magnetic Field as a Limit of Quantum Mechanics*, Lett. Math. Phys. **22**, 107–117 (1991)

[18] V. Ivrii, *Semiclassical Microlocal Analysis and Spectral Asymptotics*, Springer (to be published)

[19] A. Sobolev, *The quasi-classical asymptotics of local Riesz means for the Schrödinger operator in a strong homogeneous magnetic field*, Duke Math. J. **74**, 319–429 (1994)

[20] I. Fushiki, E.H. Gudmundsson, C.J. Pethick, and J. Yngvason, *Matter in a Magnetic Field in the Thomas–Fermi and Related Theories*, Ann. Phys. **216**, 29–72 (1992)

[21] Ö.E. Rögnvaldsson, I. Fushiki, C.J. Pethick, E.H. Gudmundsson and J. Yngvason, *Thomas–Fermi Calculations of Atoms and Matter in Magnetic Neutron Stars: Effects of Higher Landau Bands*, Astrophys. J. **216**, 276–290 (1993)

[22] K. Johnsen, MS thesis, Univ. of Iceland, 1994. See also: K. Johnsen and J. Yngvason, *Density Matrix Functional Calculations for Matter in Strong Magnetic Fields: Ground States of Heavy Atoms*, Phys. Rev. A, in press (1996)

[23] E.H. Lieb, *A Variational Principle for Many-Fermion Systems*, Phys. Rev. Lett. **46**, 457–459; Erratum **47**, 69 (1981)

[24] E.H. Lieb, J.P. Solovej, *Quantum coherent operators: A generalization of coherent states*, Lett. Math. Phys. **22**, 145–154 (1991)

[25] E.H. Lieb and W.E. Thirring, *Bound for the Kinetic Energy of Fermions Which Proves the Stability of Matter*, Phys. Rev. Lett. **35**, 687–689 (1975)

[26] L. Erdős, *Magnetic Lieb–Thirring inequalities*, Comm. Math. Phys. **170**, 629–669 (1995)

[27] E.H. Lieb and S. Oxford, *An Improved Lower Bound on the Indirect Coulomb Energy*, Int. J. Quant. Chem. **19**, 427–439 (1981)

[28] P.L. McEuen, E.B. Foxman, J. Kinaret, U. Meirav, M.A. Kastner, N.S. Wingreen and S.J. Wind, *Self consistent addition spectrum of a Coulomb island in the quantum Hall regime*, Phys. Rev. B **45**, 11419–11422 (1992)

[29] N.C. van der Vaart, M.P. de Ruyter van Steveninck, L.P. Kouwenhoven, A.T. Johnson, Y.V. Nazarov, and C.J.P.M. Harmans, *Time-Resolved Tunneling of Single Electrons between Landau Levels in a Quantum Dot*, Phys. Rev. Lett. **73**, 320–323 (1994)

[30] E.H. Lieb and H.-T. Yau, *The stability and instability of relativistic matter*, Commun. Math. Phys. **118**, 177–213 (1988)

PHYSICAL REVIEW B VOLUME 51, NUMBER 16 15 APRIL 1995-II

Ground states of large quantum dots in magnetic fields

Elliott H. Lieb and Jan Philip Solovej
Department of Mathematics, Fine Hall, Princeton University, Princeton, New Jersey 08544

Jakob Yngvason
Science Institute, University of Iceland, Dunhaga 3, IS-107 Reykjavik, Iceland

(Received 1 December 1994)

The quantum-mechanical ground state of a two-dimensional (2D) N-electron system in a confining potential $V(x)=Kv(x)$ (K is a coupling constant) and a homogeneous magnetic field B is studied in the high-density limit $N\to\infty$, $K\to\infty$ with K/N fixed. It is proved that the ground-state energy and electronic density can be computed *exactly* in this limit by minimizing simple functionals of the density. There are three such functionals depending on the way B/N varies as $N\to\infty$: A 2D Thomas-Fermi (TF) theory applies in the case $B/N\to 0$; if $B/N\to\text{const}\neq 0$ the correct limit theory is a modified B-dependent TF model, and the case $B/N\to\infty$ is described by a classical continuum electrostatic theory. For homogeneous potentials this last model describes also the weak-coupling limit $K/N\to 0$ for arbitrary B. Important steps in the proof are the derivation of a Lieb-Thirring inequality for the sum of eigenvalues of single-particle Hamiltonians in 2D with magnetic fields, and an estimation of the exchange-correlation energy. For this last estimate we study a model of classical point charges with electrostatic interactions that provides a lower bound for the true quantum-mechanical energy.

I. INTRODUCTION

In the past few years considerable experimental and theoretical work has been devoted to the study of quantum dots, which are atomiclike two-dimensional systems, confined within semiconductor heterostructures. The number of articles on this subject is by now quite large. See, e.g., Refs. 1 and 2 for reviews, Refs. 3–7 for recent measurements of conductivity and capacity of quantum dots, and Refs. 8–18 for various theoretical aspects and further references. The parameters of such artificial atoms may differ appreciably from their natural counterparts because of the interactions of the electrons with the crystal where they reside. In a quantum dot the natural atomic unit of length is $a_*=\epsilon\hbar^2/(m_*e^2)$, where ϵ is the dielectric constant and m_* is the effective electron mass. Compared with the usual Bohr radius $a_0=\hbar^2/(me^2)$, the length a_* is typically large, e.g., $a_*\approx 185a_0$ in GaAs. The corresponding natural unit B_* with which we measure the magnetic field B is the field at which the magnetic length $l_B=\hbar e/(B^{1/2}c)$ equals a_*, i.e., $B_*=(a_0/a_*)^2B_0$, where $B_0=e^3m^2c/\hbar^3=2.35\times 10^5$ T is the value corresponding to free electrons. If a_0/a_* is small, B_* can be much smaller than B_0. Thus $B_*\approx 7$ T in GaAs. This makes it possible to study in the laboratory effects which, for natural atoms, require the magnetic fields of white dwarfs or even neutron stars.

The ground-state properties of natural atoms in high magnetic fields have recently been analyzed rigorously in the asymptotic limit where the number of electrons and the nuclear charge are large.[19–21] For artificial atoms one may expect asymptotic analysis to be even more useful because the accuracy increases with the number of electrons, and a quantum dot can easily accommodate several hundred or even a thousand electrons. In the present paper we carry out such an analysis of the ground state of a quantum dot in a magnetic field. One of our conclusions is that the self-consistent model introduced by McEuen *et al.*[3,13] is a rigorous limit of quantum mechanics. This model has recently been applied to explain interesting features of the addition spectra of large quantum dots in strong magnetic fields.[15,6]

Before discussing our results for dots we summarize, for comparison, the main findings about atoms in Refs. 19–21. The quantum-mechanical ground-state energy and electronic density of a natural atom or ion with electron number N and nuclear charge Z in a homogeneous magnetic field B can, in the limit $N\to\infty$, $Z\to\infty$ with Z/N fixed, be described exactly by functionals of the density, or, in one case, of density matrices. There are five different functionals, depending on the way B varies with N as $N\to\infty$. In each of the cases $B\ll N^{4/3}$ (with B measured in the natural unit B_0), $B\sim N^{4/3}$, and $N^{4/3}\ll B\ll N^3$ the correct asymptotics is given by an appropriate functional of the semiclassical Thomas-Fermi type. For $B\sim N^3$ a modified functional, depending on density matrices, is required, whereas the case $B\gg N^3$ is described by a density functional that can be minimized in closed form.

A review of our results about quantum dots was given in Ref. 22. Due to the reduced dimensionality of the electronic motion, there are only three different asymptotic theories for quantum dots instead of five for natural atoms. These three theories are given by simple functionals of the density and correspond, respectively, to the cases $B\ll N$, $B\sim N$, and $B\gg N$ (B measured in units of B_*) as $N\to\infty$ with V/N fixed, where V is the attractive exterior potential that restricts the two-dimensional motion of the electrons. This potential, which plays the same role as the nuclear attraction in a natural atom, is generated in a quantum dot by exterior gates, and thus is adjustable to a certain extent. In the course of proving the asymptotic limits we shall also consider, in addition to the density functionals, a model of classical point charges in two dimensions that gives a lower bound to the quantum-mechanical energy.

Some of the methods and results of the present paper contrast markedly with those of our earlier work.[19–21] From a mathematical point of view the most interesting feature of quantum dots compared to natural atoms is the somewhat peculiar electrostatics that appear because the interaction between the electrons is given by the three-dimensional Coulomb potential although the motion is two dimensional. Also, the fact that the kinetic energy vanishes in the lowest Landau level requires additional mathematical effort in order to bound the kinetic energy from below by a functional of the density. We now describe in more detail the limit theorems to be proved in the sequel. A quantum dot with N electrons in a confining potential V and a homogeneous magnetic field B is modeled by the following Hamiltonian:

$$H_N=\sum_{j=1}^{N} H_1^{(j)}+\frac{e^2}{\epsilon}\sum_{1\le i<j\le N}|x_i-x_j|^{-1}\,, \tag{1.1}$$

with $x_i\in\mathbf{R}^2$ and where H_1 is the one-body Hamiltonian

$$H_1=\frac{\hbar^2}{2m_*}\left[i\nabla-\frac{e}{\hbar c}\mathbf{A}\right]^2+g_*\left[\frac{\hbar e}{2mc}\right]\mathbf{S}\cdot\mathbf{B}-\left[\frac{\hbar e}{2mc}\right]\left[\frac{m}{m_*}-\frac{|g_*|}{2}\right]B+V(x)\,. \tag{1.2}$$

As before, e and m denote the charge and mass of a (free) electron, ϵ is the dielectric constant, m_* is the effective mass, and g_* is the effective g factor. The magnetic vector potential is $\mathbf{A}(x)=\frac{1}{2}(-Bx^2,Bx^1)$ [with $x=(x^1,x^2)\in\mathbf{R}^2$], $\mathbf{B}=(0,0,B)$, and $\mathbf{S}$ is the vector of electron spin operators. The potential $V(x)$ is supposed to be continuous and confining, which is to say that $V(x)\to\infty$ as $|x|\to\infty$. It is not assumed to be circularly symmetric. The constant term in (1.2), $-[\hbar e/(2mc)][(m/m_*)-|g_*|/2]B$, is included in order that the kinetic-energy operator $H_{\rm kin}=H_1-V(x)$ has a spectrum starting at zero. The Hilbert space is that appropriate for fermions with spin, the antisymmetric tensor product $\wedge_1^N L^2(\mathbf{R}^2;\mathbf{C}^2)$.

We define an effective charge by $e_*=e/\sqrt{\epsilon}$ and choose units such that $\hbar=m_*=e_*=1$. The unit of length is then the effective Bohr radius $a_*=\hbar^2/(m_*e_*^2)$ and the unit of energy is $E_*=e_*^2/a_*=e_*^4m_*/\hbar^2$. Moreover, the unit B_* for the magnetic field is determined by $\hbar eB_*/(m_*c)=E_*$, so $B_*=e_*^3m_*^2c/(\epsilon^{1/2}\hbar^3)$. The values for GaAs are $a_*=9.8$ nm, $E_*=12$ meV, and $B_*=6.7$ T.

The true quantum-mechanical ground-state energy of H_N is denoted by $E^Q(N,B,V)$ and the true ground-state electron density by $\rho^Q_{N,B,V}(x)$. The density functionals that describe the asymptotics of E^Q and ρ^Q are of three types. The first is a standard two-dimensional Thomas-Fermi energy functional

$$\mathcal{E}^{\rm TF}[\rho;V]=(\pi/2)\int\rho(x)^2dx+\int V(x)\rho(x)dx+D(\rho,\rho) \tag{1.3}$$

with

$$D(\rho,\rho)=\tfrac{1}{2}\int\int\frac{\rho(x)\rho(y)}{|x-y|}dx\,dy\,. \tag{1.4}$$

Here ρ is a non-negative density on $\mathbf{R}^2$ and all integrals are over $\mathbf{R}^2$ unless otherwise stated. The second functional is a two-dimensional magnetic Thomas-Fermi functional

$$\mathcal{E}^{\rm MTF}[\rho;B,V]=\int j_B[\rho(x)]dx+\int V(x)\rho(x)dx+D(\rho,\rho)\,, \tag{1.5}$$

where j_B is a piecewise linear function that will be defined precisely in the next section. This functional is the two-dimensional analog of the three-dimensional magnetic Thomas-Fermi functional that was introduced in Ref. 23 and further studied in Refs. 24, 25, and 21. The present two-dimensional version was first stated in Ref. 3; these authors call it the self-consistent (SC) model. The repulsion term considered in Ref. 3 is slightly different from $D(\rho,\rho)$, since it has cutoffs at long and short distances. It is still positive definite as a kernel and our methods can easily be adapted to prove Theorems 1.1 and 1.2 with such cutoff Coulomb kernels.

The last asymptotic functional will be called the classical functional, since the kinetic-energy term is absent and only classical interactions remain:

$$\mathcal{E}^{\rm C}[\rho;V]=\int V(x)\rho(x)dx+D(\rho,\rho)\,. \tag{1.6}$$

The functionals (1.3) and (1.6) are in fact limiting cases of (1.5) for $B\to0$ and $B\to\infty$, respectively. As discussed in detail later, for each functional there is a unique density that minimizes it under the constraint $\int\rho=N$. We denote these densities, respectively, by $\rho^{\rm TF}_{N,V}(x)$, $\rho^{\rm MTF}_{N,B,V}(x)$, $\rho^{\rm C}_{N,V}(x)$, and the corresponding minimal energies by $E^{\rm TF}(N,V)$, $E^{\rm MTF}(N,B,V)$, and $E^{\rm C}(N,V)$.

In order to relate E^Q to these other energies we take a high-density limit. This is achieved by letting N tend to infinity (which is a reasonable thing to do physically, since N can be several hundred) and we let V tend to infinity. The latter statement means that we fix a potential v and set $V=Nv$. With this understanding of $N,V\to\infty$ our main results are summarized in the following two theorems. [In order to prove these theorems we need to assume that V is sufficiently regular. The technical requirement is that V belongs to the class $C^{1,\alpha}_{\rm loc}$ (see Theorem 3.2 for the definition of $C^{1,\alpha}_{\rm loc}$)].

1.1 THEOREM (limit theorem for the energy). Let $V=Nv$ with v a fixed function in $C^{1,\alpha}_{\rm loc}$. Then

$$\lim_{N\to\infty}E^Q(N,B,V)/E^{\rm MTF}(N,B,V)=1 \tag{1.7}$$

uniformly in B. Moreover,

$$\lim_{N\to\infty}E^Q(N,B,V)/E^{\rm TF}(N,V)=1\ \text{ if } B/N\to0 \tag{1.8}$$

and

$$\lim_{N\to\infty} E^Q(N,B,V)/E^C(N,V)=1 \quad \text{if } B/N\to\infty \ . \quad (1.9)$$

1.2 THEOREM (limit theorem for the density). Let $V=Nv$ with v *a fixed function in* $C^{1,\alpha}_{\text{loc}}$. *Then*

$$\frac{1}{N}\rho^Q_{N,B,V}\to\rho^{\text{MTF}}_{1,B/N,v} \quad (1.10)$$

uniformly in B, and

$$\frac{1}{N}\rho^Q_{N,B,V}\to\rho^{\text{TF}}_{1,v} \quad \text{if } B/N\to 0 \ , \quad (1.11)$$

$$\frac{1}{N}\rho^Q_{N,B,V}\to\rho^{C}_{1,v} \quad \text{if } B/N\to\infty \ . \quad (1.12)$$

The convergence is in the weak L^1 *sense.* [By definition, a sequence of functions f_n converges to a function f in weak L^1 sense if $\int f_n g\to\int fg$ for all bounded (measurable) functions g.]

Let us add a few comments on these results. As discussed in the Sec. II, the energy E^{MTF} has the scaling property

$$E^{\text{MTF}}(N,B,V)=N^2E^{\text{MTF}}(1,B/N,V/N) \ . \quad (1.13)$$

Thus (1.7) is equivalent to

$$E^Q(N,V,B)=N^2E^{\text{MTF}}(1,B/N,V/N)+o(N^2) \ , \quad (1.14)$$

where the error term is uniformly bounded in B for V/N fixed. One expects the error to be $O(N^{3/2})$, which is the order of the exchange contribution to the Coulomb interaction, but our methods do not quite allow us to prove this. We do, however, show that for B/N larger than a critical value (depending on V/N) one has $E^{\text{MTF}}(N,B,V)=E^C(N,V)$ and

$$E^Q(N,V,B)\geq N^2E^C(1,V/N)-bN^{3/2} \ , \quad (1.15)$$

where the coefficient b depends only on V/N.

The condition that V/N is fixed as $N\to\infty$ guarantees that the diameter of the electronic density distribution stays bounded as $N\to\infty$; thus the limit we are considering is really a high-density limit rather than simply a large-N limit. On the other hand, for a homogeneous potential V (e.g., quadratic, as is often assumed) one obtains also a nontrivial $N\to\infty$ limit for V *fixed,* if the lengths are suitably scaled. In fact, this limit is given by the classical functional (1.6). Intuitively this is easy to understand, for if an increase in N is not compensated by an increase in V the charge density spreads out and the kinetic-energy terms in (1.5) and (1.3) become negligible compared with the other terms. (The result again requires V to be in $C^{1,\alpha}_{\text{loc}}$.)

1.3 THEOREM (energy limit with a homogeneous potential). Assume that v *is homogeneous of degree* $s\geq 1$, *i.e.,*

$$v(\lambda x)=\lambda^s v(x) \ .$$

Then

$$\lim_{N\to\infty} E^Q(N,B,Kv)/E^{\text{MTF}}(N,B,Kv)=1 \quad (1.16)$$

uniformly in B and in K as long as K/N is bounded above. Moreover, if $K/N\to 0$ *as* $N\to\infty$, *then*

$$\lim_{N\to\infty} E^Q(N,B,Kv)/E^C(N,B)=1 \quad (1.17)$$

uniformly in B.

One can also prove a limit theorem for the density in the case of homogeneous potentials. Since the formulation of such a theorem becomes somewhat complicated we refrain from doing this, but refer to Eqs. (2.14)–(2.16) below for the scaling of the MTF functional with $k=K/N$ and to (3.24) for the weak-coupling limit of the MTF density.

The proof of the limit theorems involves the following steps. In Secs. II and III we discuss the basic properties of the functionals (1.3)–(1.6). In Sec. IV we consider the energy of a system of classical point charged particles in $\mathbf{R}^2$ in the exterior potential V as a function of the positions of the charges. This energy has a minimum, denoted by $E^P(N,V)$ (with P denoting "particle"). A significant remark is that the charge configuration, for which the minimum is obtained, is confined within a radius independent of the total charge N for fixed V/N. This finite-radius lemma, which also holds for the charge densities minimizing the functionals (1.3)–(1.6), is proved in the Appendix. Using this and an electrostatics lemma of Lieb and Yau[26] we derive the bounds

$$E^C(N,V)-aN^{3/2}\geq E^P(N,V)\geq E^C(N,V)-bN^{3/2} \ , \quad (1.18)$$

where a and b depend only on V/N. These bounds are of independent interest apart from their role in the proof of the limit theorems where, in fact, only the latter inequality is needed. Upper and lower bounds to the quantum-mechanical energy $E^Q(N,B,V)$ in terms of $E^{\text{MTF}}(N,B,V)$ with controlled errors are derived in Sec. V. The upper bound is a straightforward variational calculation using magnetic coherent states in the same way as in Ref. 21. For the lower bound one treats the cases of large B and small B separately. The estimate for large B is obtained by first noting that obviously $E^Q(N,B,V)\geq E^P(N,V)$, because the kinetic energy is non-negative, and then using (1.18). For small B two auxiliary results are required: a generalization of the magnetic Lieb-Thirring equality considered in Ref. 21, and an estimate of the correlation energy. Once these have been established the proof of Theorem 1.1 is completed by a coherent state analysis. The limit theorem for the density follows easily from the limit theorem for the energy by perturbing V with bounded functions.

II. THE MTF THEORY: ITS DEFINITION AND PROPERTIES

By employing the natural units defined in the Introduction, the kinetic-energy operator can be written

$$H_{\text{kin}}=\tfrac{1}{2}(i\nabla-\mathbf{A})^2+\gamma\mathbf{S}\cdot\mathbf{B}-\tfrac{1}{2}(1-|\gamma|)B \quad (2.1)$$

with $\gamma=g_*m_*/(2m)$. The spectrum of H_{kin} is

$$\varepsilon_{n,\sigma}=(n+\gamma\sigma+\tfrac{1}{2}|\gamma|)B \quad (2.2)$$

with $n=0,1,\ldots,\sigma=\pm\frac{1}{2}$. We write the energy levels (2.2) in *strictly* increasing order as $\varepsilon_\nu(B)$, $\nu=0,1,\ldots$. The degeneracy of each level per unit area is

$d_\nu(B)=B/(2\pi)$, except if, by coincidence, γ happens to be an integer; in that case $d_\nu(B)=B/(2\pi)$ for $\nu=0,\ldots,|\gamma|-1$, while $d_\nu(B)$ is twice as large for the higher levels. It is worth recalling that if $V(x)=K|x|^2$, the spectrum of the one-body Hamiltonian H_1 in (1.2) is solvable. The spectrum of H_1 was determined by Fock[27] in 1928, two years before Landau's paper on the spectrum of $(i\nabla-\mathbf{A})^2$. For the Hamiltonian without spin, namely, $\frac{1}{2}(i\nabla-\mathbf{A})^2+K|x|^2$, the spectrum is given by

$$E=\tfrac{1}{2}(n_1-n_2)B+\tfrac{1}{2}(n_1+n_2+1)[4K+B^2]^{1/2}$$

with $n_1,n_2=0,1,2,\ldots$. It is remarkable that this simple spectrum gives a qualitatively good fit to some of the data.[8]

For a gas of noninteracting fermions with the energy spectrum (2.2) the energy density j_B as a function of the particle density ρ is given by

$$j_B(0)=0\ ,$$

$$j_B'(\rho)=\varepsilon_\nu(B)\ \text{ if } D_\nu(B)<\rho<D_{\nu+1}(B),\ \ \nu=0,1,\ldots,$$

where $j_B'=dj_B/d\rho$ and

$$D_\nu(B)=\sum_{\nu'=0}^{\nu} d_{\nu'}(B)\ .$$

More explicitly,

$$j_B(\rho)=\sum_{\nu=0}^{\nu_{\max}}\varepsilon_\nu(B)d_\nu(B)+[\rho-D_{\nu_{\max}}(B)]\varepsilon_{\nu_{\max}+1}(B)\ , \tag{2.3}$$

where $\nu_{\max}=\nu_{\max}(\rho,B)$ is defined by

$$D_{\nu_{\max}}(B)\le\rho<D_{\nu_{\max}+1}(B)\ .$$

Thus j_B is a convex, piecewise linear function with $j_B(\rho)=0$ for $0\le\rho\le d_1(B)$. It has the scaling property

$$j_B(\rho)=B^2j_1(\rho/B)\ . \tag{2.4}$$

As $B\to0$, j_B becomes a quadratic function of the density:

$$\lim_{B\to0}j_B(\rho)=j_0(\rho)=\frac{\pi}{2}\rho^2\ . \tag{2.5}$$

Moreover,

$$j_B(\rho)\le j_0(\rho) \tag{2.6}$$

for all ρ and B (see Fig. 1).

Given an exterior potential V the MTF functional is defined by (1.5). We assume that V is continuous (V measurable and locally bounded would suffice) and tends to ∞ as $|x|\to\infty$. In particular, V is bounded below and, by adding a constant if necessary, we may assume that $V(x)\ge0$ everywhere. Because of (2.6) the functional (1.5) is defined for all non-negative functions ρ such that $\int\rho V<\infty$, $\int\rho^2<\infty$, and $D(\rho,\rho)<\infty$. Since $V\ge0$ the functional is non-negative. If N is some positive number we denote

$$\mathcal{C}_N=\left\{\rho:\rho\ge0,\int\rho V<\infty,\int\rho^2<\infty,D(\rho,\rho)<\infty,\int\rho=N\right\}$$

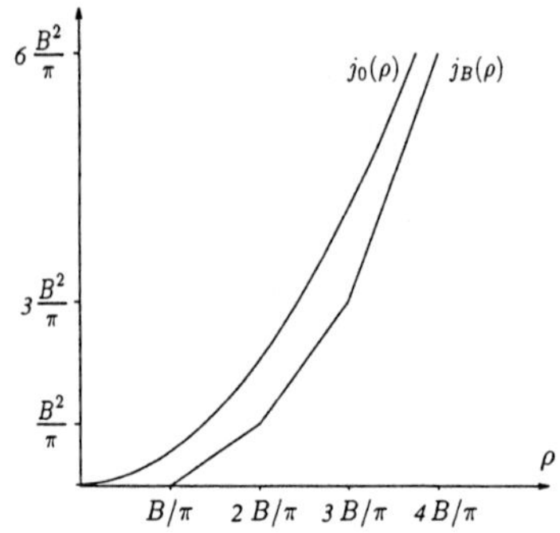

FIG. 1. The kinetic-energy densities $j_B(\rho)$ and $j_0(\rho)$ in the special case where $\gamma=0$.

and define the MTF energy by

$$E^{\rm MTF}(N,B,V)=\inf_{\rho\in\mathcal{C}_N}\mathcal{E}[\rho;B,V]\ . \tag{2.7}$$

Because of (2.4) the energy satisfies the scaling relation

$$E^{\rm MTF}(N,B,V)=N^2E^{\rm MTF}(1,B/N,V/N)\ . \tag{2.8}$$

In the limit $B\to0$ the kinetic-energy density (2.3) converges to $j_0(\rho)=(\pi/2)\rho^2$ and (1.5) converges to the energy functional (1.3) of two-dimensional TF theory at $B=0$. It is easy to see that also $\lim_{B\to0}E^{\rm MTF}(N,B,V)=E^{\rm TF}(N,V)$, where $E^{\rm TF}$ is defined in the same way as $E^{\rm MTF}$ with (1.3) replacing (1.5). We can thus consider the TF theory as a special case of MTF theory. In the opposite limit, $B\to\infty$, the kinetic-energy term vanishes altogether and one obtains a classical electrostatic model (1.6) that we shall study in Sec. III. Note also that since $j_B\le j_0$ for all B it follows that $E^{\rm MTF}(N,B,V)\le E^{\rm TF}(N,V)$ for all B. In particular $E^{\rm MTF}(1,\beta,v)$ is uniformly bounded in the parameter $\beta=B/N$ for fixed $v=V/N$.

For fixed B and V, $E^{\rm MTF}(N,B,V)$ is a convex, continuously differentiable function of N and, since $V\ge0$, it is monotonically increasing. By the methods of Refs. 28, 29, and 21 (see also Ref. 30) it is straightforward to prove the existence and uniqueness of a minimizer for the variational problem (2.7).

2.1 THEOREM (minimizer). There is a unique density $\rho^{\rm MTF}_{N,B,V}\in\mathcal{C}_n$ *such that*

$$E^{\rm MTF}(N,B,V)=\mathcal{E}^{\rm MTF}(\rho^{\rm MTF}_{N,B,V})\ .$$

Note that the existence of a minimizing density with $\int\rho=N$ is guaranteed for all N because $V(x)\to\infty$ as $|x|\to\infty$. This condition on V also implies that $\rho_{N,B,V}$ vanishes outside a ball of finite radius, cf. Lemma A1 in the Appendix. The scaling relation for the minimizing density is

$$\rho^{\text{MTF}}_{N,B,V}(x)=N\rho^{\text{MTF}}_{1,B/N,V/N}(x)\ . \tag{2.9}$$

Theorem 2.1 includes the TF theory as a special case. In the same way as in Prop. 4.14 in Ref. 21 one shows that $\rho^{\text{MTF}}_{N,B,V}\to\rho^{\text{TF}}_{N,V}$ weakly in L^1 as $B\to 0$.

The shape of the electronic density (computed by Kristinn Johnsen) in the case of a quadratic potential $V(x)=K|x|^2$ and $\gamma=0$ is shown in Fig. 2 for different values of B. At the highest value of B (8 T), the density is everywhere below $d_0(B)$ and given by the minimizer (3.15) of the classical functional (1.6). At $B=7$ T, all the electrons are still in the lowest Landau level, but that level is full around the middle of the dot where the density is anchored at $d_0(B)$. As the field gets weaker it becomes energetically favorable for electrons at the boundary of the dot, where the potential is high, to move into the next Landau level close to the minimum of the potential. A dome-shaped region then arises above the plateau at $\rho=d_0(B)=D_0(B)$, but eventually the density hits the next plateau at $\rho=D_1(B)$. This gradual filling of levels continues as the field strength goes down. At $B=2$ T three Landau levels are full and electrons in the central dome are beginning to occupy the fourth level. Finally, at $B=0$, we have the usual Thomas-Fermi model, which may be regarded as a limiting case with infinitely many Landau levels occupied.

In order to state the variational equation for the minimization problem it is convenient to define the derivative $j_B'=dj_B/d\rho$ of the kinetic-energy density everywhere, including points of discontinuity, as a *set valued* function (cf. Ref. 30), namely,

$$j_B'(\rho)=\begin{cases}\{\varepsilon_\nu(B)\} & \text{for } D_\nu(B)<\rho<D_{\nu+1}(B),\ \nu=0,1,\dots\\ [\varepsilon_\nu(B),\varepsilon_{\nu+1}(B)] & \text{for } \rho=D_{\nu+1}(B),\ \nu=0,1,\dots\ .\end{cases} \tag{2.10}$$

With this notation the Thomas-Fermi equation for the functional (1.5) may be written as follows.

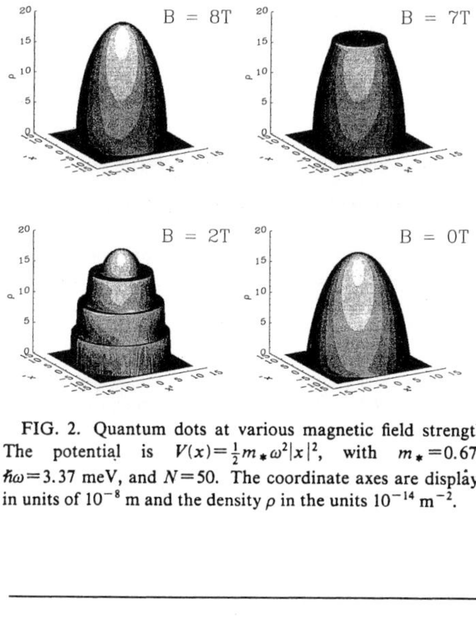

FIG. 2. Quantum dots at various magnetic field strengths. The potential is $V(x)=\frac{1}{2}m_*\omega^2|x|^2$, with $m_*=0.67m$, $\hbar\omega=3.37$ meV, and $N=50$. The coordinate axes are displayed in units of 10^{-8} m and the density ρ in the units 10^{-14} m^{-2}.

2.2 THEOREM (Thomas-Fermi equation). There is a non-negative number $\mu=\mu(N,B,V)$ such that the minimizer $\rho=\rho^{\text{MTF}}_{N,B,V}$ satisfies

$$\mu-V(x)-\rho*|x|^{-1}\begin{cases}\in j_B'[\rho(x)] & \text{if } \rho(x)>0\\ \le 0 & \text{if } \rho(x)=0\ .\end{cases} \tag{2.11}$$

The quantity μ appearing in the TF equation is the physical chemical potential, i.e.,

$$\mu=\partial E(N,B,V)/\partial N\ . \tag{2.12}$$

Since E is convex as a function of N, μ is monotonically increasing with N for fixed B and V. It satisfies

$$\mu(N,B,V)=N\mu(1,B/N,V/N)\ . \tag{2.13}$$

The derivation of the TF equation is analogous to that in Ref. 29. It is also true that if (ρ,μ) is *any* solution pair for (2.11), then ρ is the minimizer of $\mathcal{E}^{\text{MTF}}$ for some N and $\mu=\mu(N,B,V)$. The proof of this is a bit trickier than in the standard case,[29] because j_B is not continuously differentiable. It has been carried out by Lieb and Loss.[30]

Finally we discuss the relationship between the MTF theory and the classical theory defined by the functional (1.6). We of course have that

$$\mathcal{E}^{\text{MTF}}[\rho;B,V]=\int j_B[\rho(x)]dx+\mathcal{E}^{C}[\rho;V]\ .$$

From the definition of j_B one expects that the kinetic-energy term above can be neglected for large B and hence that $\lim_{B\to\infty}E^{\text{MTF}}=E^{C}$. The rigorous proof of this fact relies on a careful study of the classical problem. This analysis is far from trivial and is postponed to the next section.

There is another case where the MTF energy can be related to the classical energy. Namely, for a homogeneous exterior potential, i.e.,

$$V(\lambda x)=\lambda^s V(x)$$

for all $\lambda>0$ with some $s>0$. We consider the potentials $kV(x)$ with $k>0$ and are interested in the dependence of the MTF energy and density on the coupling constant k.

Writing

$$\rho(x)=k^{2/(s+1)}\hat{\rho}(k^{1/(s+1)}x) \tag{2.14}$$

we have the scaling

$$\begin{aligned}\mathcal{E}^{\text{MTF}}[\rho;B,kV]&=k^{2/(s+1)}\int j_b(\hat{\rho})+\mathcal{E}^{C}[\rho;kV]\\ &=k^{1/(s+1)}\left[k^{1/(s+1)}\int j_b(\hat{\rho})+\mathcal{E}^{C}[\hat{\rho};V]\right],\end{aligned} \tag{2.15}$$

where

$$b=Bk^{-2/(s+1)}\ . \tag{2.16}$$

Changing k is thus equivalent to changing the kinetic energy by a multiplicative factor and rescaling the magnetic field, keeping the potential fixed. We shall show in the next section that for k small E^{C} is a good approximation to E^{MTF}.

III. THE CLASSICAL CONTINUOUS MODEL: A LIMIT OF MTF THEORY

For densities ρ small enough $[\rho(x) \le d_1(B)$ for all $x]$ the kinetic energy $j_B(\rho)$ vanishes. It is therefore natural to consider the resulting classical energy functional defined by (1.6), i.e.,

$$\mathcal{E}^C[\rho;V] = \int \rho(x)V(x)dx + \tfrac{1}{2}\int\int \rho(x)|x-y|^{-1}\rho(y)dx\,dy\ . \tag{3.1}$$

The corresponding classical energy is

$$E^C(N,V) = \inf\left\{\mathcal{E}^C[\rho;V] : \rho \ge 0,\ \int\rho = N\right\}. \tag{3.2}$$

In this section we analyze this functional and prove that it is, indeed, the large-B limit of MTF theory.

As before we assume that the confining potential V is positive and that $V(x)\to\infty$ as $|x|\to\infty$. Moreover, we shall here assume that V is continuous (in fact, we shall make an even more stringent regularity assumption in Theorem 3.2 below).

We begin by showing the existence of a minimizer for (3.2). For general continuous V (without further assumptions) we must take into account the possibility that the minimizing ρ may be a measure. In (3.2) we therefore minimize over all positive measures ρ with $\int\rho = N$. It follows from the finite radius lemma given in the Appendix that

$$E^C(N,V) = \inf\left\{\mathcal{E}^C[\rho,V] : \text{support } \rho \subset \{x:|x|\le R_v\},\ \rho\ge 0,\ \int\rho = N\right\}. \tag{3.3}$$

Here R_v depends only on $v=V/N$. Later on we shall show that the minimizer is, indeed, a function, and hence that (3.2) does give us the large-B limit of MTF theory for suitable V.

3.1 PROPOSITION (existence and uniqueness of a minimizing measure). Let V be continuous. Then there is a unique positive measure $\rho^C_{N,V}$ with $\int\rho^C_{N,V}=N$ such that $E^C(N,V)=\mathcal{E}^C[\rho^C_{N,V};V]$.

Proof. [Note: we write measures as $\rho(x)dx$, even if they are not absolutely continuous with respect to Lebesgue measure.] By (3.3) we can choose a sequence of positive measures, $\rho_1,\rho_2,\ldots$, supported in $\{x:|x|\le R_v\}$ with $\int\rho_n = N$, such that $\lim_{n\to\infty}\mathcal{E}^C[\rho_n;V]=E^C(N,V)$. The bounded measures are the dual of the continuous functions, and so, by the Banach-Alaoglu Theorem, we may assume (by possibly passing to a subsequence) that ρ_n converges weakly to a positive measure ρ still supported in $\{x:|x|\le R_v\}$. In particular it follows that $\int\rho=N$ and $\int\rho_n V\to\int\rho V$. Moreover, the product measures $\rho_n\times\rho_n\to\rho\times\rho$ weakly. Hence

$$\begin{aligned}\int\int\rho(x)\rho(y)|x-y|^{-1}dx\,dy &= \lim_{\delta\to 0}\int\int\rho(x)\rho(y)(|x-y|+\delta)^{-1}dx\,dy\\ &= \lim_{\delta\to 0}\lim_{n\to\infty}\int\int\rho_n(x)\rho_n(y)(|x-y|^{-1}+\delta)^{-1}dx\,dy\\ &\le \liminf_{n\to\infty}\int\int\rho_n(x)\rho_n(y)|x-y|^{-1}dx\,dy\ .\end{aligned} \tag{3.4}$$

The first equality follows by the Lebesgue's monotone convergence theorem. The last inequality is an immediate consequence of the pointwise bound $(|x-y|+\delta)^{-1} \le |x-y|^{-1}$.

We conclude from (3.4) that

$$E^C(N,V)\le\mathcal{E}^C[\rho;V]\le\liminf_{n\to\infty}\mathcal{E}^C[\rho_n;V]=E^C(N,V)\ , \tag{3.5}$$

and hence that ρ is a minimizer.

The uniqueness of ρ follows from strict convexity of $D(\rho,\rho)$. Q.E.D.

The next theorem gives conditions which are perfectly adequate for the physical applications under which the minimizer is a function and not just a measure. Moreover, that function has certain nice integrability properties.

3.2 THEOREM (the minimizer is a function). Assume that the potential V is in the class $C^{1,\alpha}_{\text{loc}}$ for some $0<\alpha\le 1$ (i.e., V is once continuously differentiable and for each $R>0$ its derivative satisfies

$$|\nabla V(x)-\nabla V(y)|\le c_R|x-y|^\alpha\ , \tag{3.6}$$

inside the ball of radius R centered at the origin for some constant $c_R>0$). Then the minimizing measure $\rho^C_{N,V}$ of Proposition 3.1 is a function. It has the properties (with $\hat{\rho}$ being the Fourier transform of ρ)

$$\int|\hat{\rho}^C_{N,V}(p)|^2|p|^r dp < C_1,\quad -1\le r<\alpha \tag{3.7}$$

$$\rho^C_{N,V} * |x|^{-1} \text{ is continuous ,} \tag{3.8}$$

$$\int \rho^C_{N,V}(x)^q dx < C_2, \quad 1 \le q < \frac{4}{2-\alpha} , \tag{3.9}$$

where C_1 and C_2 are constants (implicitly computed below) that depend only on the constants c_R, q, r, α and on N.

Proof. We write $\rho^C_{N,V}=\rho$. We know that $\int \rho = N$ and that ρ has compact support. From the former fact we conclude that $\hat\rho$ is well defined, continuous, and bounded.

Let g_a be the function with Fourier transform

$$\hat g_a(p) = \int g_a(x) e^{ipx} dx = \begin{cases} 1, & |p| \le a \\ 0, & |p| \ge a . \end{cases} \tag{3.10}$$

Then g_a is continuous, $\int g_a = 1$, and $\int y g_a(y) dy = 0$.

Let ρ_a be the convolution $\rho * g_a$, so that $\int \rho_a = N$.

Since ρ is a minimizer, $\mathcal{E}^C[\rho;V] \le \mathcal{E}^C[\rho_a;V]$. Explicitly this inequality is

$$\int (\rho_a - \rho) V + D(\rho_a,\rho_a) - D(\rho,\rho) \ge 0 . \tag{3.11}$$

Since $\int V\rho_a = \int (V * g_a)\rho$ we can write the first term as

$$\int\int [V(x-y) - V(x)] \rho(x) g_a(y) dx\, dy .$$

By integrating ∇V along the line from $x-y$ to x, and using (3.6), we have

$$|V(x-y) - V(x) + y\cdot\nabla V(x)| \le c|y|^{\alpha+1} .$$

(Note that by the finite radius Lemma A1 all integrals are restricted to a finite ball.) Using the fact that $\int y g_a(y) = 0$ we can estimate the first term in (3.11) as follows:

$$\int (\rho_a - \rho) V \le C \int |y|^{\alpha+1} |g_a(y)| dy = \text{const } a^{-\alpha-1} .$$

The last two terms in (3.11) are

$$\text{const} \int |\hat\rho(p)|^2 |p|^{-1} [|\hat g_a(p)|^2 - 1] dp = \text{const} \int_{|p|\ge a} |\hat\rho(p)|^2 |p|^{-1} dp .$$

(Recall that the Fourier transform of $|x|^{-1}$ is equal to const $|p|^{-1}$ in two dimensions.) The inequality (3.11) thus implies

$$\int_{|p|\ge a} |\hat\rho(p)|^2 |p|^{-1} dp \le \text{const } a^{-\alpha-1} . \tag{3.12}$$

Using (3.12) and $\hat\rho(p) \le \int \rho = N$ we can now prove (3.7) as follows:

$$\begin{aligned}\int |\hat\rho(p)|^2 |p|^r dp &= \int_{|p|\le 1} |\hat\rho(p)|^2 |p|^r dp + \sum_{n=0}^{\infty} \int_{2^n \le |p| \le 2^{n+1}} |\hat\rho(p)|^2 |p|^r dp \\ &\le N^2 \int_{|p|\le 1} |p|^r dp + \text{const} \sum_{n=0}^{\infty} 2^{(r+1)(n+1)} \int_{2^n \le |p| \le 2^{n+1}} |\hat\rho(p)|^2 |p|^{-1} dp \\ &\le (\text{const}) N^2 + \text{const} \sum_{n=0}^{\infty} 2^{(r+1)(n+1) - n(\alpha+1)} < \infty\end{aligned}$$

if $r < \alpha$.

To prove that $\rho * |x|^{-1}$ is continuous is now easy. We simply prove that its Fourier transform is integrable. The Fourier transform of $\rho * |x|^{-1}$ is (const) $\hat\rho(p)|p|^{-1}$ and

$$\int |\hat\rho(p)||p|^{-1} dp \le \int_{|p|\le 1} |\hat\rho(p)||p|^{-1} dp + \left[\int_{|p|\ge 1} |\hat\rho(p)|^2 |p|^r dp\right]^{1/2} \left[\int_{|p|\ge 1} |p|^{-r-2} dp\right]^{1/2} < \infty .$$

Finally, we prove (3.9). For $1 \le q \le 2$ there is no problem because we know that $\int \rho^2 = \int \hat\rho^2$. Hence ρ is a square integrable function and, since $\int \rho = N$, we conclude by Hölder's inequality that (3.9) holds for $1 \le q \le 2$. For $q > 2$ we will prove that

$$\int |\hat\rho|^t < \infty \quad \text{with} \quad \frac{4}{\alpha+2} < t = \frac{q}{q-1} < 2 .$$

This will prove (3.9) by the Hausdorff-Young inequality, which states that $(\int |\hat\rho|^t)^{1/t} \ge (\int \rho^q)^{1/q}$ when $1 \le t \le 2$.

We write $|\hat\rho(p)|^t = \{|\hat\rho(p)|^t (1+|p|)^m\}\{(1+|p|)^{-m}\}$ and then use Hölder's inequality with $a^{-1} + b^{-1} = 1$ to conclude that

$$\int |\hat\rho|^t \le \left[\int |\hat\rho(p)|^{ta} (1+|p|)^{ma} dp\right]^{1/a} \times \left[\int (1+|p|)^{-mb} dp\right]^{1/b} .$$

Thus $\int |\hat\rho|^t < \infty$ if we can satisfy $ta = 2$, $ma < \alpha$, and $mb > 2$, in addition to $a^{-1} + b^{-1} = 1$. This requires $\alpha/a > m > 2/b$, or $1 \le a < 1 + \frac{1}{2}\alpha$. Thus, we require $t = 2/a > 4/(\alpha+2)$ which, since $q = t/(t-1)$, means $q < 4/(2-\alpha)$. Q.E.D.

Corresponding to the minimization (3.2) there is a variational equation satisfied by the minimizer ρ. In the general case in which ρ might be a measure, the variational equation exists but is slightly complicated to state.

In physically interesting cases V is certainly in $C^{1,\alpha}$, in which case Theorem 3.2 tells us that ρ is a function and that $\rho * |x|^{-1}$ is continuous. Hence the total potential

$$V^C = V + \rho * |x|^{-1} \tag{3.13}$$

is continuous. It is then easy to derive by standard arguments, as in Sec. I, that ρ is the unique non-negative solution to the variational equation

$$V(x)+\rho * |x|^{-1}=\mu \quad \text{if } \rho(x)\neq 0 ,$$
$$V(x)+\rho * |x|^{-1}\geq \mu \quad \text{if } \rho(x)=0 , \tag{3.14}$$

for a unique $\mu>0$. As usual the chemical potential μ is a monotone function of the particle number $N=\int\rho$. In the special case of a parabolic confining potential the solution to (3.14) can be given in closed form.

3.3 PROPOSITION (minimizer for the parabolic exterior potential). If $V(x)=K|x|^2$ then the minimizer of $\mathcal{E}^C_{N,V}$ is

$$\rho^C_{N,V}(x)=\begin{cases}\dfrac{3}{2\pi}N\lambda\sqrt{1-\lambda|x|^2} & \text{if } |x|\leq\lambda^{-1}\\ 0 & \text{if } |x|>\lambda^{-1} ,\end{cases} \tag{3.15}$$

where $\lambda=(8K/3\pi N)^{2/3}$. In fact, $\rho^C_{N,V}$ is the solution to (3.14) with $\mu=(3\pi/4)N\lambda^{1/2}$.

Proof. This solution (3.15) was certainly known before; see, e.g., Ref. 10. We give the proof here for the convenience of the reader. We only have to show that $\rho=\rho^C_{N,V}$ is the solution to (3.14). It is enough to consider the case $\lambda=1$ and $N=1$. Then $V(x)=(3\pi/8)|x|^2$ and $\mu=3\pi/4$. We may compute $\rho*|x|^{-1} =3/2\pi\int\sqrt{1-|x-y|^2}|y|^{-1}dy$ by writing y in polar coordinates $(|y|,\theta)$ and performing the $|y|$ integration first:

$$\begin{aligned}\rho*|x|^{-1}&=\frac{3}{2\pi}\int\int[(1-|y|-|x|\cos\theta)^2\\&\qquad-|x|^2\sin^2\theta]^{1/2}d\theta\, d|y|\\&=\frac{3}{2\pi}\int(1-|x|^2\sin^2\theta)^{1/2}\\&\qquad\times\int\left[1-\frac{(|y|-|x|\cos\theta)^2}{1-|x|^2\sin^2\theta}\right]^{1/2}d|y|\,d\theta ,\end{aligned}$$

the integrations are over the intervals in θ and $|y|$ for which the integrands are real. Introducing the variable $t=(|y|-|x|\cos\theta)(1-|x|^2\sin^2\theta)^{-1/2}$ we obtain

$$\begin{aligned}\rho*|x|^{-1}&=\frac{3}{2\pi}\int_{-\theta_m(x)}^{\theta_m(x)}(1-|x|^2\sin^2\theta)d\theta\int_{-1}^{1}(1-t^2)^{1/2}dt\\&=\frac{3}{4}\int_{-\theta_m(x)}^{\theta_m(x)}(1-|x|^2\sin^2\theta)d\theta ,\end{aligned} \tag{3.16}$$

where

$$\theta_m(x)=\begin{cases}\pi/2 & \text{if } |x|\leq 1\\ \sin^{-1}\dfrac{1}{|x|} & \text{if } |x|>1 .\end{cases}$$

Thus

$$\rho*|x|^{-1}=\frac{3\pi}{4}-\frac{3\pi}{8}|x|^2 \quad \text{if } |x|\leq 1$$

and

$$\rho*|x|^{-1}\geq\frac{3\pi}{4}-\frac{3\pi}{8}|x|^2 \quad \text{if } |x|>1 .$$

[The last inequality comes from the fact that the integral from θ_m to $\pi/2$ in (3.16) is negative.] Q.E.D.

The energy function $E^C(N,V)$ has the simple scaling:

$$E^C(N,V)=N^2E^C\left[1,\frac{V}{N}\right] . \tag{3.17}$$

The minimizing density $\rho^C_{N,V}$ for (3.2) scales as

$$\rho^C_{N,V}(x)=N\rho^C_{1,v}(x)$$

where $v=V/N$.

We shall now make precise in what sense the classical energy E^C is a limit of the MTF energy. In fact, in two different limits (the large-B limit and the low coupling limit) the MTF energy will converge to the classical energy. We first treat the large-B limit.

3.4 THEOREM (large-B limit of MTF*). If the exterior potential V is in the class $C^{1,\alpha}_{\mathrm{loc}}$ we have, as $B\to\infty$,*

$$E^{\mathrm{MTF}}(N,B,V)\to E^C(N,V) \tag{3.18}$$

and

$$\rho^{\mathrm{MTF}}_{N,B,V}(x)\to\rho^C_{N,V} , \tag{3.19}$$

in the weak L^1 sense.

Proof. If we use $\rho^{\mathrm{MTF}}_{N,B,V}$ as a trial density in $\mathcal{E}^C$ and recall that $j_B\geq 0$ we immediately obtain $E^C(N,V)\leq E^{\mathrm{MTF}}(N,B,V)$.

For the bound in the opposite direction we use $\rho^C_{N,V}$ as a trial density for $\mathcal{E}^{\mathrm{MTF}}$. In order to do this it is, however, important that we know (from Theorem 3.2) that $\rho^C_{N,V}$ is a function. Hence $j_B(\rho^C_{N,V})$ is well defined. Moreover, from the definition of j_B, $j_B(\rho^C_{N,V})\to 0$ almost everywhere as $B\to\infty$ and $j_B(\rho^C_{N,V})\leq j_0(\rho^C_{N,V})$. Since $j_0(\rho)=(\pi/2)(\rho)^2$ we know from (3.9) that $j_0(\rho^C_{N,V})$ is integrable. The limit in (3.18) is therefore an immediate consequence of Lebesgue's dominated convergence theorem.

The convergence of the densities in (3.19) follows in a standard way by replacing V by $V+\varepsilon f$ with f a bounded (measurable) function and differentiating with respect to ε, see e.g., Ref. 29. Q.E.D.

We point out that if $\rho^C_{N,V}$ is a bounded function, as it is, e.g., for $V=K|x|^2$, then

$$E^{\mathrm{MTF}}(N,B,V)=E^C(N,V) , \tag{3.20}$$

for B large enough, because in that case $j_B(\rho^C_{N,V})$ vanishes for B large.

Finally, we now discuss the weak-coupling limit in the case of homogeneous exterior potentials. Suppose V is a homogeneous function of x, $V(\lambda x)=\lambda^sV(x)$, $s>0$. If we consider the exterior potentials $kV(x)$ with $k>0$ the classical energy and density obey the scalings

$$E^C(N,kV)=k^{1/(s+1)}E^C(N,V) \tag{3.21}$$

and

$$\rho^C_{N,kV}(x)=k^{2/(s+1)}\rho^C_{N,V}(k^{1/(s+1)}x) . \tag{3.22}$$

If k is small we see from (3.22) that the minimizing density for the MTF functional will spread out and its kinetic energy will be negligible compared with the classical terms. We prove this rigorously now.

3.5 THEOREM (weak-coupling limit of MTF with homogeneous potentials). Let V be $C^{1,\alpha}_{\rm loc}$ and homogeneous of degree s. If $k \to 0$ then

$$\frac{E^{\rm MTF}(N,B,kV)}{E^C(N,kV)} \to 1 \tag{3.23}$$

and

$$k^{-2/(s+1)}\rho^{\rm MTF}_{N,B,kV}(k^{-1/(s+1)}x) \to \rho^C_{N,V}(x) \;, \tag{3.24}$$

in weak L^1 sense. Both limits are uniform in B.

Proof. As above we may use $\rho^{\rm MTF}_{N,B,V}$ as a trial density in $\mathcal{E}^C$ to conclude that $E^C(N,V) \le E^{\rm MTF}(N,B,V)$.

To prove the bound in the opposite direction we again use $\rho^C_{N,V}$ as a trial density for $\mathcal{E}^{\rm MTF}$. We then obtain from (2.15) and the scaling (3.22) that

$$E^{\rm MTF}(N,B,kV) \le k^{2/(s+1)} \int j_0(\rho^C_{N,V}) + E^C(N,kV) \;, \tag{3.25}$$

where we used that $j_b \le j_0$. If we compare this with the scaling in (3.21) we see that $E^{\rm MTF}/E^C \to 1$ as $k \to 0$ since $j_0(\rho^C_{N,V}) = (\pi/2)(\rho^C_{N,V})^2$ is integrable.

The convergence of the densities follows again by replacing V by $V+\varepsilon f$ and differentiating with respect to ε. Q.E.D.

In the same way as for the large-B limit (3.23) becomes an identity for small k if $\rho^C_{N,V}$ is a bounded function.

We may of course also introduce the scaling $V = Nv$ when v is homogeneous of degree s. Then $kV = Kv$, where $K = kN$, and the limit in (3.23) is uniform in N. The limit in (3.24) is uniform if we formulate it as

$$N^{-1}k^{-2/(s+1)}\rho^{\rm MTF}_{N,B,kV}(k^{-1/(s+1)}x) \to \rho^C_{1,v}(x) \;. \tag{3.26}$$

We remark that if a potential W is asymptotically homogeneous in the sense that there is a homogeneous potential V with $\lim_{|x|\to\infty} W(x)/V(x) = 1$, then

$$\lim_{k\to 0} E(N,B,kW)/k^{1/(1+s)} = E^C(N,V) \tag{3.27}$$

uniformly in B, where s is the degree of homogeneity of V.

IV. THE CLASSICAL POINT CHARGE MODEL: A LIMIT OF QUANTUM MECHANICS

Another model that sheds some light on the physics of our problem—and that will also be important for bounding the difference between the TF theory and the original quantum theory in Sec. V—is the classical particle model. In this model the kinetic energy is simply omitted altogether, but the pointlike nature of the electrons is retained.

4.1 DEFINITION (classical particle energy). With $V(x)$ being the confining potential the *classical particle energy* for N points in $\mathbf{R}^2$ is defined by

$$\mathcal{E}^P(x_1,\ldots,x_N;V) = \sum_{i=1}^N V(x_i) + \sum_{1\le i<j\le N} |x_i - x_j|^{-1} \;. \tag{4.1}$$

The *minimum classical particle energy* for N point particles in $\mathbf{R}^2$ is

$$E^P(N,V) = \inf\{\mathcal{E}^P(x_1,\ldots,x_N;V) : x_i \in \mathbf{R}^2\} \;. \tag{4.2}$$

We shall estimate the particle energy $E^P(N,V)$ in terms of the classical continuum energy $E^C(N,V)$. We first show that $E^C(N,V)$ gives an exact upper bound on $E^P(N,V)$.

4.2 LEMMA (upper bound for E^P). For all N we have

$$E^P(N,V) \le E^C(N,V) - N^{3/2}/(8R_v) \;, \tag{4.3}$$

where R_v is the maximal radius given in Lemma A1.

Proof. First, let us give a very simple argument that yields an error term proportional to N instead of $N^{3/2}$. The energy $E^P(N,V)$ is bounded above by

$$\int \mathcal{E}^P(x_1,\ldots,x_N;V)\Phi(x_1,\ldots,x_N)dx_1\cdots dx_N \;,$$

for any non-negative function Φ with $\int\Phi = 1$. We take $\Phi(x_1,\ldots,x_N) = \prod_{i=1}^N \rho_{[1]}(x_i)$, where for simplicity we have introduced the notation $\rho_{[1]}$ for the minimizer $\rho^C_{1,V/N}$ for $\mathcal{E}^C[\rho;V/N]$ with $\int \rho^C_{1,V/N} = 1$. Note that $\rho_{[1]}$ depends only on $v = V/N$. We obtain

$$\begin{aligned} E^P(N,V) &\le \int \mathcal{E}^P(x_1,\ldots,x_N;V)\prod_{i=1}^N \rho_{[1]}(x_i)dx_1\cdots dx_N \\ &= N\int V(x)\rho_{[1]}(x)dx \\ &\quad + \frac{N(N-1)}{2}\int\int \rho_{[1]}(x)|x-y|^{-1} \\ &\quad \times \rho_{[1]}(y)dx\,dy \;. \end{aligned}$$

Recalling that the minimizer of $\mathcal{E}^C$ is $\rho^C_{N,V}(x) = N\rho_{[1]}(x)$, we get an error term $-aN$, with $a = \frac{1}{2}\int\int \rho_{[1]}(x)\rho_{[1]}(y)|x-y|^{-1}dx\,dy$.

Now we turn to a proof of (4.3) which, obviously, has to be more complicated than the previous discussion. By Lemma A1 there is a fixed square Q centered at the origin, whose width W equals $2R_v$, such that the minimizer $\rho = \rho^C_{N,V}$ for $\mathcal{E}^C$ is supported in Q. For simplicity we suppose that $\sqrt{N}$ is an integer; if this is not so the following proof can be modified in an obvious way.

First, cut Q into $\sqrt{N}$ vertical, disjoint strips, $S_1, S_2, \ldots, S_{\sqrt{N}}$ such that $\int_{S_j}\rho = \sqrt{N}$ for all j. Let t_j denote the width of S_j, so that $\sum_{j=1}^{\sqrt{N}} t_j = W$. Next, make $\sqrt{N}-1$ horizontal cuts in each S_j so that the resulting rectangles R_{jk} for $k = 1,\ldots,\sqrt{N}$ satisfy $\int_{R_{jk}}\rho = 1$. Denote the height of these rectangles by h_{jk}, so that $\sum_{k=1}^{\sqrt{N}} h_{jk} = W$ for each j. Having done this we note, by convexity, that for each j

$$\begin{aligned} N^{-1/2}\sum_{k=1}^{\sqrt{N}}(t_j+h_{jk})^{-1} &\ge \left[N^{-1/2}\sum_{k=1}^{\sqrt{N}}(t_j+h_{jk})\right]^{-1} \\ &= [t_j + N^{-1/2}W]^{-1} \;. \end{aligned}$$

Again, using the same convexity argument for the j summation, we have that

$$\sum_{j=1}^{\sqrt{N}}\sum_{k=1}^{\sqrt{N}}(t_j+h_{jk})^{-1} \ge \frac{N^{3/2}}{2W} \;. \tag{4.4}$$

Let ρ_{jk} be the minimizing density ρ restricted to the rectangle R_{jk}, i.e., $\rho_{jk}(x)=1$ if $x\in R_{jk}$ and $=0$ otherwise. Thus, $\int\rho_{jk}=1$. We denote these N functions by ρ^i, $i=1,\dots,N$. Define $\Phi(x_1,\dots,x_N):=\prod_{i=1}^N\rho^i(x_i)$ and, as in the previous proof, a simple computation yields

$$E^P(N,V)\le\int\mathcal{E}^P\Phi=\mathcal{E}^C[\rho;V]-\sum_{i=1}^N D(\rho^i,\rho^i)\ ,$$

with $D(f,g)=\frac{1}{2}\int\int f(x)g(y)|x-y|^{-1}dx\,dy$.

To complete our proof we note that as long as x and y are in R_{jk} we have that $|x-y|^{-1}\ge(t_j+h_{jk})^{-1}$. Thus, $\sum_{j,k}D(\rho_{jk},\rho_{jk})\ge N^{3/2}/4W$ by (4.4) and the fact that $\int\rho_{jk}=1$. Q.E.D.

4.3 LEMMA (lower bound for E^P). Assume that V is a potential in $C^{1,\alpha}_{\rm loc}$. Then for all N we have

$$E^P(N,V)\ge E^C(N,V)-bN^{3/2}\ . \tag{4.5}$$

with

$$b=\tfrac{4}{3}\sqrt{\tfrac{2}{3}}\int[\rho^C_{1,v}(x)]^{3/2}dx+\left[\frac{2\pi}{2-p}\right]^{1/p}(2R_v)^{-1+(2/p)}\left[\int(\rho^C_{1,v})^q\right]^{1/q} \tag{4.6}$$

and where q is any number satisfying $2<q<4/(2-\alpha)$, R_v is the maximal radius given in Lemma A1 and $p=q/(q-1)<2$. As we explained in Theorem 3.2, our hypothesis that $V\in C^{1,\alpha}_{\rm loc}$ implies that for $q<4/(2-\alpha)$, $[\int(\rho^C_{1,v})^q]^{1/q}$ is less than some constant that depends only on q, α, and $v=V/N$. In particular, b depends only on v.

We see from (4.3) and (4.5) that when $V\in C^{1,\alpha}_{\rm loc}$ the power $\frac{3}{2}$ in the error term is optimal.

In order to prove (4.5) we need the following electrostatics lemma of Lieb and Yau.[26] The original version was for $\mathbf{R}^3$; we state it here for $\mathbf{R}^2$ solely for the convenience of our present application.

4.4 LEMMA (the interaction of points and densities). Given points $x_1,\dots,x_N$ in $\mathbf{R}^2$, we define Voronoi cells $\Gamma_1,\dots,\Gamma_N\subset\mathbf{R}^2$ by

$$\Gamma_j=\{y\in\mathbf{R}^2:|y-x_j|\le|y-x_k|\ \text{ for all } k\ne j\}\ .$$

These Γ_j have disjoint interiors and their union covers $\mathbf{R}^2$. We also define R_j to be the distance from x_j to the boundary of Γ_j, i.e., R_j is half the distance of x_j to its nearest neighbor. Let ρ be any (not necessarily positive) function on $\mathbf{R}^2$. (In general, ρ can be replaced by a measure, but it is not necessary for us to do so.) Then [with $D(f,g)=\frac{1}{2}\int\int f(x)g(y)|x-y|^{-1}dx\,dy$]

$$\sum_{1\le i<j\le N}|x_i-x_j|^{-1}\ge-D(\rho,\rho)+\sum_{j=1}^N\int_{\mathbf{R}^2}\rho(y)|y-x_j|^{-1}dy+\tfrac{1}{8}\sum_{j=1}^N R_j^{-1}-\sum_{j=1}^N\int_{\Gamma_j}\rho(y)|y-x_j|^{-1}dy\ . \tag{4.7}$$

Proof of (4.5). We choose ρ in (4.7) to be the minimizer $\rho^C_{N,V}$ for the functional $\mathcal{E}^C$ and we choose the x_i's to be any (not necessarily minimizing) configuration for $\mathcal{E}^P$. It is important for us that a minimizer exists for $\mathcal{E}^C$, for then ρ satisfies the Euler-Lagrange equation (3.14). Since $\int\rho=N$, we conclude from (3.14) that

$$\sum_{j=1}^N\left[\int\rho(y)|y-x_j|^{-1}dy+V(x_j)\right]=\sum_{j=1}^N V^C(x_j)\ge N\mu=\int V^C\rho=2D(\rho,\rho)+\int V\rho\ .$$

Thus, if we add $\sum_j V(x_j)$ to both sides of (4.7), we have that

$$E^P(N,V)\ge E^C(N,V)+\sum_{j=1}^N\tfrac{1}{8}R_j^{-1}-\int_{\Gamma_j}\rho(y)|y-x_j|^{-1}dy\ . \tag{4.8}$$

Our goal will be to control the rightmost term in (4.8) by the R_j^{-1} term.

We split each region Γ_j into two disjoint subregions, $\Gamma_j=A_j\cup B_j$, where

$$A_j:=\{x:|x-x_j|<R_j\},\quad B_j:=\{x\in\Gamma_j:|x-x_j|\ge R_j\}\ .$$

Then, by Hölder's inequality

$$\sum_{j=1}^N\int_{B_j}|y-x_j|^{-1}\rho(y)dy\le\left[\sum_j\int_{\Gamma_j}\rho^{3/2}\right]^{2/3}\left[\sum_j\int_{|y-x_j|\ge R_j}|y-x_j|^{-3}dy\right]^{1/3}$$
$$=\left[\int\rho^{3/2}\right]^{2/3}\left[2\pi\sum_j R_j^{-1}\right]^{1/3}. \tag{4.9}$$

If we define $X:=\sum_j R_j^{-1}$ we can rewrite (4.9) plus the $X/8$ term in (4.8) as $\frac{1}{8}X-(\int\rho^{3/2})^{2/3}X^{1/3}$. The minimum of this quantity, over all values X, is $\frac{4}{3}\sqrt{2/3}\int\rho^{3/2}$, and thus we have accounted for the first error term in (4.6).

To estimate the term $I:=\sum_{j=1}^N\int_{A_j}\rho(y)|y-x_j|^{-1}dy$ some control is needed over the possible singularities of ρ. Let $p=q/(q-1)$ be the dual of q. Then

$$I\le\left[\int\rho^q\right]^{1/q}\left[\sum_j\int_{A_j}|y-x_j|^{-p}dy\right]^{1/p}$$

$$=\left[\int\rho^q\right]^{1/q}\left[\frac{2\pi}{2-p}\sum_j R_j^{2-p}\right]^{1/p}. \tag{4.10}$$

We note that, since $1\le p<2$,

$$\left[\sum_{j=1}^N R_j^{2-p}\right]^{1/p}\le\left[\sum_{j=1}^N R_j^2\right]^{1/p-1/2}N^{1/2}$$

by Hölder's inequality. Now πR_j^2 is the area of the disc A_j and thus $\pi\sum_j R_j^2$ is the total area of all these disjoint discs. How large can this area be? To answer this we recall Lemma A1 in the Appendix, which states that for the purpose of finding a set of points that minimizes the classical particle energy $\mathcal{E}^P$ we can restrict attention to a disc of radius R_v, centered at the origin. We may therefore assume that our x_j's satisfy $|x_j|\le R_v$. This we can do whether or not an energy minimizing configuration exists. Having done so and assuming that $N\ge 2$ we have that $R_j<R_v$ for all j, and hence all our discs are contained in a disc of radius $2R_v$ centered at the origin. Thus $\sum_j R_j^2\le(2R_v)^2$, and our second error term, (4.10), is bounded above by

$$\left[\int\rho^q\right]^{1/q}\left[\frac{2\pi}{2-p}\right]^{1/p}(2R_v)^{-1+2/p}N^{1/2}.$$

This yields (4.6). Q.E.D.

V. MTF THEORY IS THE HIGH-DENSITY LIMIT OF QUANTUM MECHANICS

In this section we prove that the quantum energy and the quantum density are given by the corresponding MTF quantities to leading order for large N. These are the statements of Theorems 1.1 and 1.2. We shall not prove Theorem 1.2 since it follows from Theorem 1.1 in a standard way by replacing V by $V+\varepsilon f$ with f a function in $C^{1,\alpha}_{\rm loc}$ and differentiating w.r.t. ε, see, e.g., Ref. 29.

We shall prove Theorem 1.1 by giving sharp upper and lower bounds to the quantum ground-state energy. The upper bound is obtained by a variational calculation using the magnetic coherent states introduced in Refs. 25 and 21. The lower bound is more difficult. Besides the results of the previous sections several ingredients are needed. The first ingredient is a kinetic energy inequality of the Lieb-Thirring type.[31,32,21,33] Such an inequality estimates the kinetic energy of a many-body wave function from below in terms of a functional of the density. The proof of this inequality in the two-dimensional case considered here is harder than the three-dimensional case treated in Ref. 21 and involves some new mathematical ideas. The second ingredient is a lower bound on the exchange-correlation energy. The proof of this inequality is similar to that given in Refs. 34 and 35 for the three-dimensional case.

Once the kinetic energy and exchange-correlation inequalities have been established the proof of the lower bound is completed by a coherent states analysis.

We start by discussing the magnetic coherent states used in the proofs of both the upper and lower bounds. They are constructed from the kernels

$$\Pi_{\alpha\sigma}(x\sigma',y\sigma'')=\frac{B}{2\pi}\exp\{i(x\times y)\cdot B-|x-y|^2B/4\}$$
$$\times L_\alpha(|x-y|^2B/2)\delta_{\sigma\sigma'}\delta_{\sigma\sigma''} \tag{5.1}$$

of the projection operators onto the Landau levels $\alpha=0,1,2,\ldots$ with z-component of spin $\sigma=\pm\frac{1}{2}$. Here L_α are Laguerre polynomials normalized by $L_\alpha(0)=1$. In fact, all that matters are the projectors Π_ν on the states with energy $\varepsilon_\nu(B)$; these are given by a sum of at most two of the projections $\Pi_{\alpha\sigma}$. More precisely,

$$\Pi_\nu=\sum_{\substack{\alpha,\sigma\\ \alpha+\frac{1}{2}+\gamma\sigma=\varepsilon_\nu(B)/B}}\Pi_{\alpha\sigma}. \tag{5.2}$$

We shall not need the explicit form (5.1). The three important properties of Π_ν that we use are the following:

$$\sum_\nu\Pi_\nu(x\sigma',y\sigma'')=\delta(x-y)\delta_{\sigma'\sigma''}, \tag{5.3}$$

$$\sum_{\sigma'}\Pi_\nu(x\sigma',x\sigma')=d_\nu(B), \tag{5.4}$$

$$H_{\rm kin}\Pi_\nu=\varepsilon_\nu(B)\Pi_\nu, \tag{5.5}$$

where $H_{\rm kin}$ is given by (2.1).

Let g be a real continuous function on $\mathbf{R}^2$, with $g(x)=0$ for $|x|>1$, $\int g^2=1$, and $\int(\nabla g)^2<\infty$. [The optimal choice that minimizes $\int(\nabla g)^2$ is the Bessel function J_0, suitably scaled and normalized.] Define $g_r(x)=r^{-1}g(x/r)$, with $0<r<1$ to be specified later. For each $u\in\mathbf{R}^2$, $\nu=0,1,2$, we define the operator $\Pi_{\nu u}$—the coherent "operator"—with kernel

$$\Pi_{\nu u}(x\sigma',y\sigma'')=g_r(x-u)\Pi_\nu(x\sigma',y\sigma'')g_r(y-u). \tag{5.6}$$

It easily follows from (5.3) and (5.4) and the properties of g that these kernels satisfy the coherent operator identities[36]

$$\sum_\nu\int\Pi_{\nu u}(x\sigma',y\sigma'')du=\delta(x-y)\delta_{\sigma'\sigma''}, \tag{5.7}$$

$${\rm Tr}\Pi_{\nu u}=\sum_\sigma\int\Pi_{\nu\mu}(x\sigma',x\sigma')dx\doteq d_\nu(B). \tag{5.8}$$

Moreover, a simple computation gives, using (5.5),

$${\rm Tr}[H_{\rm kin}\Pi_{\nu u}]=d_\nu(B)\left[\varepsilon_\nu(B)+\int(\nabla g_r)^2\right], \tag{5.9}$$

$${\rm Tr}[V\Pi_{\nu u}]=d_\nu(B)V*g_r^2(u), \tag{5.10}$$

where V is a (continuous) potential and $*$ denotes convolution. Likewise, for all f with $\langle f|f\rangle=1$

$$\langle f|H_{\text{kin}}|f\rangle=\sum_{\nu}\int \varepsilon_\nu(B)\langle f|\Pi_{\nu u}|f\rangle du-\int(\nabla g_r)^2\ , \tag{5.11}$$

$$\langle f|V*g_r^2|f\rangle=\sum_{\nu}\int V(u)\langle f|\Pi_{\nu u}|f\rangle du\ . \tag{5.12}$$

Equations (5.9) and (5.10) will be used in proving the upper bound, while (5.11) and (5.12) are needed for the lower bound.

A. The upper bound

We use the variational principle of Ref. 37. According to this principle

$$E^Q(N,B,V)\le \text{Tr}[(H_{\text{kin}}+V)K] +\tfrac{1}{2}\sum_{\sigma',\sigma'}\int_{\mathbf{R}^2}\int_{\mathbf{R}^2}\frac{K(x\sigma,x\sigma)K(y\sigma',y\sigma')}{|x-y|}\times dx\,dy \tag{5.13}$$

for all operators K, with kernel $K(x\sigma,y\sigma')$, satisfying

$$0\le\langle f|K|f\rangle\le\langle f|f\rangle \tag{5.14}$$

for all f, and

$$\text{Tr}[K]=\sum_{\sigma}\int_{\mathbf{R}^2}K(x\sigma,x\sigma)dx=N\ . \tag{5.15}$$

We shall choose K as follows. Let ρ^{MTF} be the MTF density, i.e., the minimizer of the functional (1.5) with $\int\rho^{\text{MTF}}=N$. Denote by $\nu_{\max}(x)$ the highest filled level. Then

$$0\le\rho^{\text{MTF}}(x)-\sum_{\nu\le\nu_{\max}(x)}d_\nu(B)<d_{\nu_{\max}(x)+1}(B)\ . \tag{5.16}$$

We introduce the filling factors

$$f_\nu(x)=\begin{cases}1, & \nu\le\nu_{\max}(x)\\ [\rho^{\text{MTF}}(x)-\sum_{\nu\le\nu_{\max}(x)}d_\nu(B)]/d_{\nu_{\max}(x)+1}(B), & \nu=\nu_{\max}(x)+1\\ 0, & \nu>\nu_{\max}(x)+1\end{cases} \tag{5.17}$$

and define

$$K(x\sigma,y\sigma')=\sum_{\nu}\int f_\nu(u)\Pi_{\nu u}(x\sigma,y\sigma')du\ , \tag{5.18}$$

with $\Pi_{\nu u}$ as in (5.6). It follows from (5.7) that K satisfies (5.14) and from (5.8), (5.16), and (5.17) that $\text{Tr}[K]=\int\rho^{\text{MTF}}(u)du=N$.

Note that (5.4) and (5.6) imply

$$\sum_{\sigma}K(x\sigma,x\sigma)=\rho^{\text{MTF}}*g_r^2(x)\ . \tag{5.19}$$

Hence, the last term in (5.13) is $D(\rho^{\text{MTF}}*g_r^2,\ \rho^{\text{MTF}}*g_r^2)$, where the functional D was defined in (1.4). By convexity of D we find that

$$D(\rho^{\text{MTF}}*g_r^2,\rho^{\text{MTF}}*g_r^2)\le D(\rho^{\text{MTF}},\rho^{\text{MTF}})\ .$$

From (5.9), (5.10), and (5.13) we obtain

$$\begin{aligned}E^Q(N,B,V)&\le\mathcal{E}^{\text{MTF}}(\rho^{\text{MTF}})+N\int(\nabla g_r)^2dx\\ &\quad+\int[V*g_r^2(*)-V(x)]\rho^{\text{MTF}}(x)dx\\ &\le E^{\text{MTF}}(N,V,B)+Nr^{-2}\int[\nabla g(x)]^2dx\\ &\quad+N^2\sup_{|x|<R}[v*g_r^2(x)-v(x)]\ ,\end{aligned} \tag{5.20}$$

where $R=R_v$ is the finite radius and we have written $V=Nv$. Since v is in $C^{1,\alpha}$

$$\sup_{|x|<R}|v*g_r^2(x)-v(x)|\le(\text{const})r\ .$$

We can choose $r=r_N$ such that $r_N\to0$ and $r_N^{-2}/N\to0$ as $N\to\infty$. This means that r_N should go to zero but still be large compared with the average spacing $N^{-1/2}$ between electrons. The optimal choice is of the order $r=(\text{const})N^{-1/3}$. Thus the error

$$[E^Q(N,B,V)-E^{\text{MTF}}(N,B,V)]N^{-2}$$

is bounded above by a function $\varepsilon_N^+(v)=c^+(v)N^{-1/3}$ (independent of B). This finishes the proof of the upper bound.

5.1 THEOREM (Lieb-Thirring inequality in two dimensions). Let $H_{\mathbf{A}}=\frac{1}{2}(i\nabla-\mathbf{A})^2+\mathbf{S}\cdot\mathbf{B}$. [*This is the operator* H_{kin} *from (2.1) with* $\gamma=1$.] *Let* W *be a locally integrable function and denote by* $e_1(W)$, $e_2(W)$,... *the negative eigenvalues (if any) of the operator* $H=H_{\mathbf{A}}-W$ *defined on* $L^2(\mathbf{R}^2;\mathbf{C}^2)$, *the space of wave functions of a single spin-*$\frac{1}{2}$ *particle. Define* $|W|_+(x)=\frac{1}{2}[|W(x)|+W(x)]$. *For all* $0<\lambda<1$ *we then have the estimate*

$$\sum_j|e_j(W)|\le\lambda^{-1}\frac{B}{2\pi}\int_{\mathbf{R}^2}|W|_+(x)dx+\tfrac{3}{4}(1-\lambda)^{-2}\int_{\mathbf{R}^2}|W|_+^2(x)dx\ .$$

Proof. For any self-adjoint operator A we denote by $N_\alpha(A)$ the number of eigenvalues of A greater than or equal to α.

Since replacing W by its positive part $|W|_+$ will only enhance the sum of the negative eigenvalues we shall henceforth assume that W is positive, i.e., $W=|W|_+$.

We consider the Birman-Schwinger kernel

$$K_E = W^{1/2}(H_{\mathbf{A}}+E)^{-1}W^{1/2} .$$

According to the Birman-Schwinger principle (see, e.g., Ref. 38, p. 89) the number $N_E(-H)$ of eigenvalues of H below $-E$ is equal to the number $N_1(K_E)$ of eigenvalues of K_E greater than or equal to 1. We find

$$\sum_j |e_j(W)| = \int_0^\infty N_E(-H)dE = \int_0^\infty N_1(K_E)dE .$$

In order to estimate $N_1(K_E)$ we decompose the Birman-Schwinger kernel into a part K_E^0 coming from the lowest Landau level and a part $K_E^>$ coming from the higher levels. If Π_0 is the projection onto the lowest Landau band these two parts are defined by

$$K_E^0 = W^{1/2}\Pi_0(H_{\mathbf{A}}+E)^{-1}\Pi_0 W^{1/2} = E^{-1}W^{1/2}\Pi_0 W^{1/2} \tag{5.21}$$

and

$$K_E^> = W^{1/2}(I-\Pi_0)(H_{\mathbf{A}}+E)^{-1}(I-\Pi_0)W^{1/2} .$$

Since Π_0 commutes with $H_{\mathbf{A}}$ we have $K_E = K_E^0 + K_E^>$.

Now we use Fan's theorem,[39] which states that if $\mu_1(X) \ge \mu_2(X) \ge \cdots$ denote the eigenvectors of a self-adjoint compact operator X then $\mu_{n+m+1}(X+Y) \le \mu_{n+1}(X) + \mu_{m+1}(Y)$ for $n,m \ge 0$. From this we have $N_1(X+Y) \le N_\lambda(X) + N_{1-\lambda}(Y)$ which, in our case, reads as follows:

$$N_1(K_E) \le N_\lambda(K_E^0) + N_{1-\lambda}(K_E^>) \quad \text{for } 0 \le \lambda \le 1 .$$

This inequality permits us to consider the two parts of K_E separately. We first consider the contribution from the lowest level: $N_\lambda(K_E^0) = N_{\lambda E}(W^{1/2}\Pi_0 W^{1/2})$. We get

$$\begin{aligned}\int_0^\infty N_\lambda(K_E^0)dE &= \int_0^\infty N_{\lambda E}(W^{1/2}\Pi_0 W^{1/2})dE \\ &= \lambda^{-1}\int_0^\infty N_E(W^{1/2}\Pi_0 W^{1/2})dE \\ &= \lambda^{-1}\mathrm{Tr}(W^{1/2}\Pi_0 W^{1/2}) \\ &= \lambda^{-1}\int \Pi_0(x,x)W(x)dx \\ &= \lambda^{-1}\frac{B}{2\pi}\int W(x)dx .\end{aligned}$$

The second part is straightforward. We first notice that $H_{\mathbf{A}}(I-\Pi_0) \ge B(I-\Pi_0)$. Hence $H_{\mathbf{A}}(I-\Pi_0) \ge \frac{2}{3}(H_{\mathbf{A}} + \frac{1}{2}B)(I-\Pi_0) \ge \frac{1}{3}(i\nabla - \mathbf{A})^2(I-\Pi_0)$. [Note that $(i\nabla-\mathbf{A})^2$ commutes with Π_0.] Since the operator inequality $0 < X \le Y$ implies $X^{-1} \le Y^{-1}$ we have that

$$\begin{aligned}K_E^> &\le W^{1/2}(I-\Pi_0)[\tfrac{1}{3}(i\nabla-\mathbf{A})^2+E]^{-1}(I-\Pi_0)W^{1/2} \\ &\le W^{1/2}[\tfrac{1}{3}(i\nabla-\mathbf{A})^2+E]^{-1}W^{1/2} .\end{aligned}$$

We conclude that $N_{1-\lambda}(K_E^>) \le N_1(\tilde{K}_E)$, where

$$\tilde{K}_E = [(1-\lambda)^{-1}W]^{1/2}[\tfrac{1}{3}(i\nabla-\mathbf{A})^2+E]^{-1} \times [(1-\lambda)^{-1}W]^{1/2}$$

is the Birman-Schwinger kernel for the operators $\tilde{H} = \frac{1}{3}[(i\nabla-\mathbf{A})^2 - 3(1-\lambda)^{-1}W]$. The Birman-Schwinger principle implies that $\int_0^\infty N_1(\tilde{K}_E)dE$ is the sum of the negative eigenvalues of $\tilde{H}$. An estimate on this quantity follows from the standard Lieb-Thirring inequality, i.e.,

$$\begin{aligned}\int_0^\infty N_{1-\lambda}(K_E^>)dE &\le (0.24)\tfrac{1}{3}\int \left|\frac{3}{(1-\lambda)}W(x)\right|^2 dx \\ &\le \tfrac{3}{4}(1-\lambda)^{-2}\int W(x)^2 dx .\end{aligned}$$

The constant 0.24 can be found as $L_{1,2}$ in Ref. 40, Eq. (51). It was improved slightly by Blanchard and Stubbe,[41] see also Ref. 42. In these references only the case $\mathbf{A}=0$ was considered. It is, however, a simple consequence of the diamagnetic inequality (see Ref. 43) that the constant is independent of $\mathbf{A}$. Q.E.D.

The Lieb-Thirring inequality in Theorem 5.1 implies an estimate on the kinetic energy

$$T_\psi = \left\langle \psi \left| \sum_{j=1}^N H_{\mathrm{kin}}^{(j)} \right| \psi \right\rangle$$

in terms of the one-particle density

$$\rho_\psi(x) = N \sum_{\sigma_1=\pm 1/2} \cdots \sum_{\sigma_N=\pm 1/2} \int_{\mathbf{R}^{2(N-1)}} |\psi(x,x_2,\ldots,x_N;\sigma_1,\ldots,\sigma_N)|^2 dx_2 \cdots dx_N .$$

Here ψ is a normalized N-particle fermionic wave function.

5.2 COROLLARY (kinetic-energy inequality in two dimensions). Let T_ψ and ρ_ψ be defined as above. Then for all $0<\lambda<1$ we have

$$T_\psi \ge \begin{cases} 0 & \text{if } \rho_\psi \le \lambda^{-1}\dfrac{B}{\pi} \\ \tfrac{1}{3}(1-\lambda)^2\displaystyle\int \left[\rho_\psi(x) - \lambda^{-1}\frac{B}{\pi}\right]^2 dx & \text{if } \rho_\psi \ge \lambda^{-1}\dfrac{B}{\pi} . \end{cases} \tag{5.22}$$

Proof. The inequality in Theorem 5.1 holds for the operator $H_{\mathrm{kin}} - W$ if $|\gamma| \ge 1$. If $|\gamma| < 1$, however, one should choose Π_0 in the proof of Theorem 5.1 as the projection onto the levels $\nu=0$ and $\nu=1$ (not only onto $\nu=0$). Equation (5.21) is then no longer an identity but a bound. In this way one concludes that the negative eigenvalues $e_1(W)$, $e_2(W),\ldots$ for $H_{\text{kin}}-W$ satisfy

$$\sum_j |e_j(W)| \le \alpha \int_{\mathbf{R}^2} |W|_+(x)dx + \beta \int_{\mathbf{R}^2} |W|_+^2(x)dx ,$$

with $\alpha=\lambda^{-1}B/\pi$ and $\beta=\frac{3}{4}(1-\lambda)^{-2}$. This bound is clearly valid for all γ.

The proof of (5.22) now follows by a standard Legendre transformation. In fact, if $W\ge 0$ we have

$$T_\psi = \left\langle \psi \left| \sum_j H_{\text{kin}}^{(j)} - W(x_j) \right| \psi \right\rangle + \int W\rho_\psi \ge \int_{\mathbf{R}^2} [W\rho_\psi - \alpha W - \beta W^2] . \tag{5.23}$$

Since the Legendre transformation of the function $W \mapsto \alpha W + \beta W^2$ is the function

$$\rho \mapsto \sup_{W\ge 0} [\rho W - \alpha W - \beta W^2] = \begin{cases} 0 & \text{if } \rho \le \alpha \\ (4\beta)^{-1}(\rho-\alpha)^2 & \text{if } \rho > \alpha , \end{cases}$$

we see that (5.22) follows by making the optimal choice for W in (5.23). Q.E.D.

5.3 LEMMA (exchange inequality in two dimensions). Let $\psi \in \otimes^N L^2(\mathbf{R}^2;\mathbf{C}^2)$ be any normalized N-particle wave function (not necessarily fermionic) and let

$$\rho_\psi(x) = \sum_{i=1}^N \sum_{\sigma_1=\pm 1/2} \cdots \sum_{\sigma_N=\pm 1/2} \int_{\mathbf{R}^{2N}} |\psi(x_1,\ldots,x_{i-1},x,x_{i+1}^2,\ldots,x_N;\sigma_1,\ldots,\sigma_N)|^2 dx_1 \cdots dx_{i-1} dx_{i+1} \cdots dx_N$$

be the corresponding one-particle density. Then

$$\sum_{\sigma_1} \cdots \sum_{\sigma_N} \int_{\mathbf{R}^{2N}} |\psi|^2 \sum_{1\le i<j\le N} |x_i-x_j|^{-1} dx_1 \cdots dx_N \ge \tfrac{1}{2} \int_{\mathbf{R}^2} \int_{\mathbf{R}^2} \rho_\psi(x)\rho_\psi(y)|x-y|^{-1} dx\, dy - 192(2\pi)^{1/2} \int_{\mathbf{R}^2} \rho_\psi(x)^{3/2} dx . \tag{5.24}$$

Proof. The proof is essentially the same as in Ref. 34, where the three-dimensional equivalent of (1.1) was proved. Our presentation is inspired by Ref. 44.

We use the representation (in three dimensions a similar representation was originally used by Fefferman and de La Llave[45])

$$|x-y|^{-1} = \pi^{-1} \int_{\mathbf{R}^2} \int_{\mathbf{R}} \chi_R(x-z)\chi_R(y-z) R^{-4} dR\, dz , \tag{5.25}$$

where χ_R is the characteristic function of the ball of radius R centered at the origin. If we use (5.25) to represent $\sum_{i<j} |x_i-x_j|^{-1}$ we can estimate the integrand as follows:

$$\begin{aligned}
\sum_{1\le i<j\le N} \chi_R(x_i-z)\chi_R(x_j-z) &= \tfrac{1}{2} \left[\sum_i \chi_R(x_i-z) \right]^2 - \tfrac{1}{2} \sum_i \chi_R(x_i-z) \\
&= \tfrac{1}{2} \left[\sum_i \chi_R(x_i-z) - \int \rho_\psi(y)\chi_R(y-z)dy \right]^2 \\
&\quad + \sum_i \chi_R(x_i-z) \int \rho_\psi(z)\chi_R(y-z)dy - \tfrac{1}{2} \left[\int \rho_\psi(y)\chi_R(y-z)dy \right]^2 - \tfrac{1}{2} \sum_i \chi_R(x_i-z) \\
&\ge \sum_i \chi_R(x_i-z) \int \rho_\psi(y)\chi_R(y-z)dy - \tfrac{1}{2} \left[\int \rho_\psi(y)\chi_R(y-z)dy \right]^2 - \tfrac{1}{2} \sum_i \chi_R(x_i-z) .
\end{aligned}$$

If we integrate this inequality over the measure $R^{-4}dR\, dz$, the last term $\frac{1}{2}\sum_i \chi_R(x_i-z)$ will give a divergent integral. For the purpose of a lower bound, however, we can restrict the integration in (5.25) to $R > r(z)$, where $r(z)>0$ is some specific function we shall choose below. Using the fact that

$$\sum_{\sigma_1} \cdots \sum_{\sigma_N} \int_{\mathbf{R}^{2N}} |\psi|^2 \sum_i \chi_R(x_i-z) dx_1 \cdots dx_N = \int_{\mathbf{R}^2} \rho_\psi(y)\chi_R(y-z)dy ,$$

we obtain

$$\sum_{\sigma_1}\cdots\sum_{\sigma_N}\int_{\mathbf{R}^{2N}}|\psi|^2\sum_{i<j}|x_i-x_j|^{-1}dx_1\cdots dx_N \geq \tfrac{1}{2}\pi^{-1}\int_{\mathbf{R}^2}\int_{R>r(z)}\left[\int \rho_\psi(y)\chi_R(y-z)dy\right]^2 R^{-4}dR\,dz$$
$$-\tfrac{1}{2}\pi^{-1}\int_{\mathbf{R}^2}\int_{R>r(z)}\int \rho_\psi(y)\chi_R(y-z)dy\,R^{-4}dR\,dz$$
$$\geq \tfrac{1}{2}\pi^{-1}\int\int \rho_\psi(x)\rho_\psi(y)|x-y|^{-1}dx\,dy$$
$$-\tfrac{1}{2}\pi^{-1}\int_{\mathbf{R}^2}\int_{R<r(z)}\pi^2\rho_\psi^*(z)^2dR\,dz$$
$$-\tfrac{1}{2}\pi^{-1}\int_{\mathbf{R}^2}\int_{R>r(z)}R^{-2}\pi\rho_\psi^*(z)dR\,dz\,. \tag{5.26}$$

Here we have introduced the Hardy-Littlewood maximal function

$$\rho_\psi^*(z)=\sup_R(\pi R^2)^{-1}\int \rho_\psi(y)\chi_R(y-z)dy\;,$$

which, viewed as a map from $L^p(\mathbf{R}^2)$ to $L^p(\mathbf{R}^2)$, is a bounded map for all $p>1$ (see Ref. 46, pp. 54–58). The error terms in (5.26) can be computed as

$$\tfrac{1}{2}\pi^{-1}\int_{\mathbf{R}^2}\left[\int_{R<r(z)}\pi^2\rho_\psi^*(z)^2dR+\int_{R>r(z)}R^{-2}\pi\rho_\psi^*(z)dR\right]dz=\tfrac{1}{2}\pi^{-1}\int_{\mathbf{R}^2}[r(z)\pi^2\rho_\psi^*(z)^2+r(z)^{-1}\pi\rho_\psi^*(z)]dz\;.$$

The optimal choice for $r(z)$ is $r(z)=[\pi\rho^*(z)]^{-1/2}$. This means that the error is $\pi^{1/2}\int_{\mathbf{R}^2}\rho_\psi^*(z)^{3/2}dz$, but this can be estimated by the maximal inequality to be less than $192(2\pi)^{1/2}\int_{\mathbf{R}^2}\rho(z)^{3/2}dz$. Q.E.D.

B. The lower bound

Our goal here is to give a lower bound to $E^Q(N,B,V)$ in terms of $E^{\mathrm{MTF}}(N,B,V)$ with errors of lower order than N^2 as N tends to infinity. It is important here that $V=Nv$, where v is fixed. To be more precise we shall prove that

$$N^{-2}[E^Q(N,B,V)-E^{\mathrm{MTF}}(N,B,V)]\geq -\varepsilon_N^-(v)\;, \tag{5.27}$$

where $\varepsilon_N^-(v)$ is a non-negative function which tends to 0 as $N\to\infty$ for fixed v. Note, however, that $\varepsilon_N^-(v)$ does not depend on B.

We shall treat the cases of large B and small B separately. In the large-B regime we prove (5.27) by a comparison with the classical models discussed in Secs. III and IV. In the small-B regime we use magnetic coherent states, Theorem 5.1 and Lemma 5.3.

The dividing line between large and small B is determined as follows. If the minimizer $\rho_{1,v}^C$ of $\mathcal{E}^C$, with $\int\rho_{1,v}^C=1$ and confining potential v, is bounded [e.g., for $v(x)=|x|^2$], then we define small B to mean

$$B/N\leq\beta_c:=2\pi\sup_x\rho_{1,v}^C(x)\;. \tag{5.28}$$

As explained in (3.20) we have for $\beta\geq\beta_c$ that $E^{\mathrm{MTF}}(1,\beta,v)=E^C(1,v)$.

For the general class of v where we do not know the minimizer $\rho_{1,v}^C$ is bounded we simply define

$$\beta_c=N^{1/3}\;.$$

By Theorem 3.4 we then have that the function

$$\delta(N,v)=\sup_{\beta\geq\beta_c}|E^{\mathrm{MTF}}(1,\beta,v)-E^C(1,v)| \tag{5.29}$$

tends to zero as N tends to infinity.

Case 1, $B/N\geq\beta_c$*:* By simply ignoring the kinetic-energy operator, which we had normalized to be positive, we have the obvious inequality $E^Q(N,B,V)\geq E^P(N,V)$ where E^P is the energy of the classical point problem.

From Lemma 4.3 we can therefore conclude that

$$E^Q(N,B,V)\geq E^P(N,V)\geq E^C(N,V)-b(v)N^{3/2}\;. \tag{5.30}$$

Since $E^C(N,V)=N^2E^C(1,v)$ and $E^{\mathrm{MTF}}(N,B,V)=N^2E^{\mathrm{MTF}}(1,B/N,v)$ we have from (5.29) that

$$E^Q(N,B,V)\geq E^{\mathrm{MTF}}(N,B,V)-\delta(N,v)N^2-b(v)N^{3/2}\;. \tag{5.31}$$

Thus (5.27) holds with $\varepsilon_N^-(v)=\delta(N,v)+b(v)N^{-1/2}$.

We emphasize again that if $\rho_{1,v}^C$ is bounded (e.g., for $v=k|x|^2$) then $\delta(N,v)$ is not needed.

Case 2, $B/N\leq\beta_c$*:* In this case we use inequality (5.24) to reduce the many-body problem to a one-body problem.

Let ψ be the many-body ground state for H_N. (Since the exterior potential V tends to infinity at infinity H_N will have a ground state.) The correlation estimate (5.24) gives

$$E^Q(N,B,V)=\langle\psi|H_N|\psi\rangle$$
$$\geq\sum_{j=1}^N\langle\psi|H_{\mathrm{kin}}^{(j)}+V(x_j)|\psi\rangle+D(\rho_\psi,\rho_\psi)$$
$$-C\int_{\mathbf{R}^2}\rho_\psi^{3/2}(x)dx\;. \tag{5.32}$$

We first estimate the last term in (5.32) in terms of the kinetic energy

$$T_\psi=\left\langle\psi\left|\sum_{j=1}^N H_{\mathrm{kin}}^{(j)}\right|\psi\right\rangle$$

of ψ. According to (5.22) we have

$$\int \rho_\psi^{3/2} \le (\text{const})\, B^{1/2} \int_{\rho_\psi < (\text{const})B} \rho_\psi + (\text{const}) \left[\int \rho_\psi \right]^{1/2} \left[\int_{\rho_\psi \ge (\text{const})B} \rho_\psi^2 \right]^{1/2} \le (\text{const})(\sqrt{\beta_c} N^{3/2} + \sqrt{T_\psi} N^{1/2}) . \tag{5.33}$$

Hence, for all $0<\varepsilon<1$ (we shall later choose $\varepsilon \sim N^{-1/2}$) we have

$$E^Q(N,B,V) \ge \sum_{j=1}^N \langle \psi | (1-\varepsilon) H_{\text{kin}}^{(j)} + V(x_j) | \psi \rangle + D(\rho_\psi, \rho_\psi) - (\text{const})\, (\sqrt{\beta_c} N^{3/2} + \varepsilon^{-1} N) , \tag{5.34}$$

where we have used the fact that

$$\varepsilon T_\psi - \text{const} \sqrt{T_\psi} N^{1/2} \ge -(\text{const}) \varepsilon^{-1} N . \tag{5.35}$$

To relate (5.34) to the MTF problem we use the inequality

$$0 \le D(\rho_\psi - \rho^{\text{MTF}}, \rho_\psi - \rho^{\text{MTF}}) = -\left\langle \psi \left| \sum_{j=1}^N \rho^{\text{MTF}} * |x_j|^{-1} \right| \psi \right\rangle + D(\rho_\psi, \rho_\psi) + D(\rho^{\text{MTF}}, \rho^{\text{MTF}}) , \tag{5.36}$$

which is a consequence of the positive definiteness of the kernel $|x-y|^{-1}$. Inserting this in (5.34) gives

$$E^Q(N,B,V) \ge \sum_{j=1}^N \langle \psi | (1-\varepsilon) H_{\text{kin}}^{(j)} + V(x_j) + \rho^{\text{MTF}} * |x_j|^{-1} | \psi \rangle - D(\rho^{\text{MTF}}, \rho^{\text{MTF}}) - (\text{const})(\sqrt{\beta_c} N^{3/2} + \varepsilon^{-1} N) . \tag{5.37}$$

Since we have normalized the potential to be positive we have that $(1-\varepsilon)^{-1} V(x) \ge V(x)$ and also $(1-\varepsilon)^{-1} \rho^{\text{MTF}} * |x|^{-1} \ge \rho^{\text{MTF}} * |x|^{-1}$.

Hence

$$E^Q(N,B,V) \ge (1-\varepsilon) \left\langle \psi \left| \sum_{j=1}^N [H_{\text{kin}}^{(j)} + V(x_j) + \rho^{\text{MTF}} * |x_j|^{-1}] \right| \psi \right\rangle - D(\rho^{\text{MTF}}, \rho^{\text{MTF}}) - (\text{const})(\sqrt{\beta_c} N^{3/2} + \varepsilon^{-1} N) . \tag{5.38}$$

Obviously

$$\left\langle \psi \left| \sum_{j=1}^N [H_{\text{kin}}^{(j)} + V(x_j)] + \rho^{\text{MTF}} * |x_j|^{-1} \right| \psi \right\rangle \ge \sum_{j=1}^N e_j , \tag{5.39}$$

where $e_1, e_2, \ldots, e_N$ are the N lowest eigenvalues of the one-particle Hamiltonian

$$H_1^{\text{MTF}} = H_{\text{kin}} + V(x) + \rho^{\text{MTF}} * |x|^{-1} = H_{\text{kin}} + V^{\text{MTF}}(x) . \tag{5.40}$$

We shall estimate $\sum_{j=1}^N e_j$ by a straightforward coherent states analysis.

Let $f_1, \ldots, f_N$ be the N lowest normalized eigenfunctions of H_1^{MTF}. For technical reasons we introduce a modified operator $\tilde{H}_1^{\text{MTF}}$ which is obtained from H_1^{MTF} by replacing V^{MTF} by the truncated potential

$$\tilde{V}^{\text{MTF}}(x) = \begin{cases} V^{\text{MTF}}(x) , & |x| \le R_v \\ CN , & |x| \ge R_v , \end{cases}$$

where R_v is the finite radius given in the Appendix and $C = \inf_{|x|>R_v} V^{\text{MTF}}(x)/N$ is independent of N by the scaling (2.9) of MTF theory. Note that $V^{\text{MTF}} \ge \tilde{V}^{\text{MTF}}$. Then from (5.11) and (5.12) we have

$$\sum_{j=1}^N e_j = \sum_{j=1}^N \langle f_j | H_1^{\text{MTF}} | f_j \rangle \ge \sum_{j=1}^N \langle f_j | \tilde{H}_1^{\text{MTF}} | f_j \rangle = \sum_v \int [\varepsilon_v(B) + \tilde{V}^{\text{MTF}}(u)] \sum_{j=1}^N \langle f_j | \Pi_{vu} | f_j \rangle du - r^{-2} N \int (\nabla g)^2 + \sum_{j=1}^N \langle f_j | \tilde{V}^{\text{MTF}} - \tilde{V}^{\text{MTF}} * g_r^2 | f_j \rangle . \tag{5.41}$$

We first consider the last term. Writing $\sum_{j=1}^N |f_j(x)|^2 = \tilde{\rho}(x)$ we have

$$\sum_{j=1}^N \langle f_j | \tilde{V}^{\text{MTF}} - \tilde{V}^{\text{MTF}} * g_r^2(x) | f_j \rangle = \int_{|x|<R_v+r} [\tilde{V}^{\text{MTF}}(x) - \tilde{V}^{\text{MTF}} * g_r^2(x)] \tilde{\rho}(x) dx \ge \int_{|x| \le R_v - r} [V^{\text{MTF}}(x) - V^{\text{MTF}} * g_r^2(x)] \tilde{\rho}(x) dx - \int_{R_v - r \le |x| \le R_v + r} \tilde{V}^{\text{MTF}} * g_r^2(x) \tilde{\rho}(x) dx .$$

Since $V^{\text{MTF}} = N v^{\text{MTF}}$ and $\int \tilde{\rho} = N$ we have

$$\sum_{j=1}^N \langle f_j | \tilde{V}^{\text{MTF}} - \tilde{V}^{\text{MTF}} * g_r^2 | f_j \rangle \ge -N^2 \sup_{|x|<R_v} |v^{\text{MTF}}(x) - v^{\text{MTF}} * g_r^2(x)| - \left[\sup_{|x|<R_v+r} v^{\text{MTF}}(x) \right] R_v^2 r N^2 . \tag{5.42}$$

We can then write (5.41) as

$$\sum_{j=1}^{N} e_j \geq \sum_{\nu} \int [\varepsilon_\nu(B)+\tilde{V}^{\rm MTF}(u)] \times \sum_{j=1}^{N} \langle f_j|\Pi_{\nu u}|f_j\rangle du - N^2\tilde{\varepsilon}_N(v) , \quad (5.43)$$

where

$$\tilde{\varepsilon}_N(v)= \sup_{|x|<R_v} |v^{\rm MTF}(x)-v^{\rm MTF}*g_r^2(x)| + \left[\sup_{|x|<R_v+r} v^{\rm MTF}(x)\right]R_v^2 r - N^{-1}r^2\int(\nabla g)^2$$
$$\leq C(v)N^{-1/3} . \quad (5.44)$$

For the last step we made the choice $r\sim N^{-1/3}$.

We focus next on the first term in (5.43). It has the form

$$\int [\varepsilon_\nu(B)+\tilde{V}^{\rm MTF}(u)]\rho_\nu(u)du , \quad (5.45)$$

where we have denoted $\sum_{j=1}^{N}\langle f_j|\Pi_{\nu u}|f_j\rangle$ by $\rho_\nu(u)$. These functions satisfy

$$0\leq \rho_\nu(u)\leq {\rm Tr}\Pi_{\nu u}=d_\nu(B) \quad (5.46)$$

and

$$\sum_\nu \int \rho_\nu(u)du = N . \quad (5.47)$$

We obtain a lower bound to (5.45) by minimizing over all functions ρ_ν satisfying (5.46) and (5.47).

Minimizers ρ_ν can be constructed as follows. There is a $\mu>0$ such that

$$\rho_\nu(u)= \begin{cases} d_\nu(B) & \text{if } \varepsilon_\nu(B)+\tilde{V}^{\rm MTF}(u)<\mu \\ 0 & \text{if } \varepsilon_\nu(B)+\tilde{V}^{\rm MTF}(u)>\mu \\ \leq d_\nu(B) & \text{if } \varepsilon_\nu(B)+\tilde{V}^{\rm MTF}(u)=\mu . \end{cases} \quad (5.48)$$

All families ρ_ν satisfying (5.48) and the constraint (5.47) are minimizers. Note that it is possible that $\varepsilon_\nu(B)+\tilde{V}^{\rm MTF}(u)=\mu$ on an open set of u values. The minimizers are therefore not necessarily unique. The chemical potential μ is uniquely determined by (5.48) and the condition (5.47).

We shall now prove that $\mu=\mu^{\rm MTF}$. All we have to show is that we can find functions ρ_ν satisfying (5.47) and (5.48) with $\mu=\mu^{\rm MTF}$.

We know from the MTF equation, Theorem 2.2, that if $\rho^{\rm MTF}(u)=0$ then $V^{\rm MTF}(u)\geq\mu^{\rm MTF}$. Since $\tilde{V}^{\rm MTF}$ differs from $V^{\rm MTF}$ only on the set $\rho^{\rm MTF}=0$ we may in (5.48) when $\mu=\mu^{\rm MTF}$ replace $\tilde{V}^{\rm MTF}$ by $V^{\rm MTF}$. We then know from the MTF equation that there are *unique* functions ρ_ν satisfying (5.48) with $\mu=\mu^{\rm MTF}$ and $\sum_\nu\rho_\nu(u)=\rho^{\rm MTF}(u)$. In fact, in terms of the filling factors (5.17) we have $\rho_\nu(u)=f_\nu(u)d_\nu(B)$. Since $\int\rho^{\rm MTF}(u)du=N$ we have produced the functions ρ_ν allowing us to conclude that μ is indeed equal to $\mu^{\rm MTF}$. If we insert these functions in (5.45) we obtain

$$\sum_\nu \int \varepsilon_\nu(B)\rho_\nu(u)du + \int V^{\rm MTF}(u)\rho^{\rm MTF}(u)du = \int \{j_B[\rho^{\rm MTF}(u)]+V^{\rm MTF}(u)\}du , \quad (5.49)$$

where the identity follows from (2.3).

We can now combine (5.38), (5.39), (5.43), (5.44), and (5.49) to arrive at

$$E^Q(N,B,V)\geq(1-\varepsilon)E^{\rm MTF}(N,B,V)-\varepsilon D(\rho^{\rm MTF},\rho^{\rm MTF}) - ({\rm const})(\sqrt{\beta_c}N^{3/2}+\varepsilon^{-1}N)-C(v)N^{5/3} . \quad (5.50)$$

Hence, since $D(\rho^{\rm MTF},\rho^{\rm MTF})\leq E^{\rm MTF}(N,B,V)$ we have

$$N^{-2}[E^Q(N,B,V)-E^{\rm MTF}(N,B,V)] \geq -2\varepsilon E^{\rm MTF}(1,B/N,v) - ({\rm const})[\sqrt{\beta_c}N^{-1/2}+\varepsilon^{-1}N^{-1}+c(v)N^{-1/3}] .$$

Note that $E^{\rm MTF}(1,B/N,v)$ is bounded by a constant depending only on v. If we choose $\varepsilon\sim N^{-1/2}$ we find

$$N^{-2}[E^Q(N,B,V)-E^{\rm MTF}(N,B,V)] \geq -c^-(v)[\sqrt{\beta_c}N^{-1/2}+N^{-1/3}] .$$

This is equivalent to (5.27) with $\varepsilon_N^-(v) =c^-(v)[\sqrt{\beta_c}N^{-1/2}+N^{-1/3}]$. In the case when ρ^C is bounded, β_c is a constant, otherwise we chose it to be $\beta_c=N^{1/3}$. In both cases $\varepsilon_N^-(v)$ will tend to zero as N tends to infinity. This finishes the proof of the lower bound.

We have proved (1.7). The limits in (1.8) and (1.9) follow immediately from the corresponding results for $E^{\rm MTF}$ proved in Secs. II and III.

C. Homogeneous exterior potentials

Finally, we shall show how to prove the stronger result Theorem 1.3 for homogeneous exterior potentials.

In this case we do not have that the minimizing MTF density is supported within a fixed ball. In fact, the density will spread out as the coupling constant becomes small.

We shall prove that given $\varepsilon>0$ and k_0 there is an N_ε independent of B such that for $N\geq N_\varepsilon$ and $K/N\leq k_0$

$$|E^Q(N,B,Kv)/E^{\rm MTF}(N,B,Kv)-1|<\varepsilon . \quad (5.51)$$

We consider large and small K in very much the same way as we did for B in the lower bound above. We shall see below that we can find a k_c (depending only on v and ε) such that (5.51) holds for $K/N\leq k_c$.

In the case of large K, i.e., $K/N\geq k_c$ (but $K/N\leq k_0$) the proof of (5.51) is then identical to the proof of Theorem 1.1 given above.

For small K we again consider the upper and lower bounds separately. We begin with the upper bound. We proceed as in Sec. V A. We define the trial operator as in (5.17) and (5.18) except that we replace $\rho^{\rm MTF}$ by $\rho^C_{N,Kv}$. The estimate (5.20) now becomes

$$E^Q(N,B,Kv)\le \mathcal{E}^{\rm MTF}[\rho^C_{N,Kv};B,Kv] +Nr^{-2}\int[\nabla g(x)]^2dx +NK\sup_{|x|\le R_K}[v*g_r^2(x)-v(x)]\ .$$

Here R_k is the radius of the ball containing the support of $\rho^C_{N,KV}$ for $K=kN$. According to (3.22) $R_k=k^{-1/(s+1)}R_1$, where R_1 is the radius for $k=1$, which depends only on v.

Using the homogeneity of v we have

$$v*g_r^2(x)-v(x)=|x|^s\int\{v[(x-y)/|x|] -v(x/|x|)\}g_r^2(y)dy \le c_3(v)r|x|^{s-1}\ .$$

We obtain since $s\ge 1$

$$\begin{aligned}E^Q(N,B,Kv)-\mathcal{E}^{\rm MTF}[\rho^C;B,Kv]&\le c_4(v)N[r^{-2}+KrR^{s-1}]\\ &\le c_5(v)N[r^{-2}+Kr(K/N)^{-(s-1)/(s+1)}]\\ &\le c_5(v)N^{2/3}(K/N)^{4/[3(s+1)]}\ ,\end{aligned}$$

with the choice $r=(K/N)^{-2/[3(s+1)]}$. We also know from (3.25) that

$$\mathcal{E}^{\rm MTF}[\rho^C;B,Kv]\le (K/N)^{2/(s+1)}N^2\int j_0(\rho^C_{1,v})+E^C(N,Kv)\ .$$

Hence, from (3.21) and (3.17) we obtain

$$\begin{aligned}E^Q(N,B,Kv)&\le\left\{1+\frac{c_6(v)N^2}{E^C(N,Kv)}[N^{-1/3}(K/N)^{4/[3(s+1)]}+(K/N)^{2/(s+1)}]\right\}E^C(N,Kv)\\ &\le(1+c_7(v)[N^{-1/3}(K/N)^{1/[3(s+1)]}+(K/N)^{1/(s+1)}])E^C(N,Kv)\end{aligned}$$

It therefore follows from Theorem 3.5 that we can find k_c depending only on ε (but not on B) such that $E^Q(N,B,Kv)/E^{\rm MTF}(N,B,Kv)\le 1+\varepsilon$ for $K/N\le k_c$.

We turn next to the lower bound. As in Sec. V B (in the case $B/N\ge\beta_c$) we may ignore the kinetic-energy operator, which we had normalized to be positive. We then have the obvious inequality $E^Q(N,B,V)\ge E^P(N,V)$ where E^P is the energy of the classical point problem.

We shall use Lemma 4.3 to compare $E^P(N,Kv)$ to $E^C(N,Kv)$. We must, however, first discuss the scaling of $E^P(N,Kv)$. It is clear that if v is homogeneous of degree s then

$$\mathcal{E}^P(x_1,\dots,x_N;Kv) =k^{1/(s+1)}\mathcal{E}^P(k^{1/(s+1)}x_1,\dots,k^{1/(s+1)}x_N;v)\ .$$

Therefore, $E^P(N,kV)=k^{1/(s+1)}E^P(N,V)$, i.e., $E^P(N,kV)$ has the same scaling as $E^C(N,Kv)$ [see (3.21)].

From Lemma 4.3 we thus find that

$$\begin{aligned}E^P(N,Kv)&\ge E^C(N,Kv)-b(v)(K/N)^{1/(s+1)}N^{3/2}\\ &\ge E^C(N,Kv)[1-c_8(v)N^{-1/2}]\ .\end{aligned}\tag{5.52}$$

According to Theorem 3.5 we may thus assume that k_c is such that

$$E^Q(N,B,Kv)/E^{\rm MTF}(N,B,Kv) \ge(1-\varepsilon/2)[1-c_8(v)N^{-1/2}]\tag{5.53}$$

for $K/N\le k_c$. We can therefore clearly find N_ε such that the right-hand side of (5.53) is greater than $1-\varepsilon$ for $N\ge N_\varepsilon$.

VI. CONCLUSION

We have analyzed the ground state of a two-dimensional gas of N electrons interacting with each other via the (three-dimensional) Coulomb potential and subject to a confining exterior potential $V(x)=Kv(x)$ where K is an adjustable coupling constant. The electrons are also subject to a uniform magnetic field B perpendicular to the two-dimensional plane.

We have found the exact energy and electron density function $\rho(x)$ to leading order in $1/N$, i.e., in the high-density limit. This limit is achieved by letting K be proportional to N as $N\to\infty$, thus effectively confining the electrons to a fixed region of space, independent of N.

It turns out that the answer to the problem depends critically on the behavior of B as $N\to\infty$. There are three regimes.

(i) If $B/N\to 0$, i.e., $N\gg B$ in appropriate units, then normal (two-dimensional) Thomas-Fermi theory gives the exact description. Correlations can be ignored to leading order in this high-density situation.

(ii) If $B/N=$const, a modified TF theory in which the "kinetic-energy density" is changed from (const)ρ^2 to a certain B-dependent function of ρ [called $j_B(\rho)$] is exact.

(iii) If $B/N\to\infty$ then the kinetic-energy term can be omitted entirely and a classical continuum electrostatics theory emerges as the exact theory. This electrostatics problem is mathematically interesting in its own right and can be solved in closed form for the customary choice $v(x)=|x|^2$.

Related to the continuum problem is an electrostatics problem for point charged particles. Apart from its mathematical interest, it provides a crucial lower bound to the energy in case (iii). Another technical point of some interest is the extension of the Lieb-Thirring inequality to two-dimensional particles in a magnetic field which involves dealing with a continuum of zero energy modes (i.e., the lowest Landau level).

ACKNOWLEDGMENTS

We thank Vidar Gudmundsson and Jari Kinaret for valuable discussions. We thank Kristinn Johnsen for allowing us to use his graphs of the electron density (Fig. 2). E.H.L. was partially supported by U.S. National Science Foundation Grant No. PHY90-19433 A03. J.P.S. was partially supported by U.S. National Science Foundation Grant No. DMS 92-03829. J.Y. was partially supported by the Icelandic Science Foundation and the Research Fund of the University of Iceland.

APPENDIX

Here we prove that the minimizers for our three semiclassical problems can be sought among densities that vanish outside some finite radius—for which we give an upper bound. This lemma is in an appendix because it pertains to several sections of the paper.

A.1 LEMMA (finite radius of minimizers). Consider the three cases: (a) The classical energy; (b) the classical particle energy; (c) the MTF energy. Let $V(x)$ be the confining potential. We assume that $V(x)\to+\infty$ as $|x|\to\infty$ in the sense that the number $W(R) := \inf\{V(x) : |x|\geq R\}$ tends to ∞ as $R\to\infty$.

Then there is a radius R_v, depending only on $v=V/N$ such that

$$E^P(N,V)=\inf\{\mathcal{E}^P(x_1,\ldots,x_N) : |x_i|\leq R_v \text{ for all } i\}$$

$$E^C(N,V)=\inf\left\{\mathcal{E}^C[\mu;V] : \text{support } \mu\subset\{x:|x|\leq R_v\},\ \int d\mu=N\right\}$$

$$E^{\mathrm{MTF}}(N,B,V)=\inf\left\{\mathcal{E}^{\mathrm{MTF}}[\rho;B,V] : \rho(x)=0 \text{ for } |x|>R_v,\ \int\rho=N\right\}. \tag{A1}$$

Furthermore, any minimizing particle distribution measure or density satisfies the conditions given in braces in (A1).

A choice for R_v, which is far from optimal, is any R satisfying the inequality

$$\frac{1}{N}W(R)\geq(2+\pi^{-1})+\frac{1}{N}\langle V\rangle_1\,, \tag{A2}$$

with $\langle V\rangle_1$ being the average of V in the unit disk:

$$\langle V\rangle_1=\frac{1}{\pi}\int_{|x|<1}V(x)dx\ .$$

Proof: Particle case. Suppose that $|x_1|>R_v$. Then we move particle 1 inside D, the unit disc centered at the origin. The point y to which we move particle 1 is not known, so we average the energy over all choices of $y\in D$. If we show that this average energy is less than the original energy then we know that there is some point $y\in D$ such that the energy is lowered. Thus, we have to show that

$$V(x_1)+\sum_{j=2}^{N}|x_1-x_j|^{-1}>\langle V\rangle_1+\frac{1}{\pi}\sum_{j=2}^{N}\int_D|y-x_j|^{-1}dy\ .$$

Noting that $\int_D|y-x|^{-1}dy\leq\int_D|y|^{-1}dy=2\pi$, by a simple rearrangement inequality, we see that it suffices to have $W(R_v)>\langle V\rangle_1+2N$, which agrees with (A2).

The classical case. If μ is any measure with $\int d\mu=N$, we define μ^+ to be μ restricted to the complement of the closed disc of radius R_v centered at the origin. Thus $\mu^+(A)=\mu(A\cap\{x:|x|>R_v\})$. Similarly, μ^- is μ restricted to the disc, so that $\mu=\mu^++\mu^-$. Assuming that $\mu^+\neq0$, we replace μ by $\mu_\varepsilon:=(1-\varepsilon)\mu^++\mu^-+\delta\nu$, where ν is Lebesgue measure restricted to the unit disc D, and where $\pi\delta=\varepsilon\int d\mu^+$. Thus $\int d\mu_\varepsilon=N$. The change in energy, to $O(\varepsilon)$ as $\varepsilon\downarrow0$, is easily seen to be

$$\delta\int_D V(x)dx-\varepsilon\int V(x)\mu^+(dx)+\delta\int_D\int_{\mathbb{R}^2}|x-y|^{-1}dx\,\mu(dy)$$
$$-\varepsilon\int\int|x-y|^{-1}\mu^+(dx)\mu(dy)<\pi\delta\langle V\rangle_1-\varepsilon W(R_v)\int d\mu^++2\pi\delta N\ ,$$

which is negative by (A1).

The MTF case. This is similar to the classical case, but with two differences: (i) The measure μ is replaced by a function ρ with $\int \rho(x)dx = N$ and (ii) a "kinetic-energy" term $\int j_B[\rho(x)]dx$ is added to the energy. Point (i) only simplifies matters. For point (ii) we note the simple fact that $j_B(\rho)$ is bounded above by $\pi\rho^2/2$ and its derivative, $j_B'(\rho)$, is bounded above by $\pi\rho$; this is true for all B. Let us assume that $d\mu^+ := \rho^+(x)dx$ is not zero, with $\rho^+(x) = \rho(x)$ for $|x| > R_v$ and $\rho^+(x) = 0$ otherwise. The argument is as before, but now we must take into account the change in kinetic energy which, to leading order in ε, is

$$\delta \int_D j_B'[\rho(x)]dx - \varepsilon \int \rho^+(x) j_B'[\rho^+(x)]dx < \delta\pi \int_D \rho(x)dx \le \delta N = \varepsilon \int \rho^+(x)dx \; .$$

The total energy change is then negative by (A1). Q.E.D.

[1]H. van Houten, C. W. J. Beenakker, and A. A. M. Staring, in *Single Charge Tunneling,* edited by H. Grabert, J. M. Martinis, and M. H. Devoret (Plenum, New York, 1991).
[2]M. A. Kastner, Rev. Mod. Phys. **64**, 849 (1992).
[3]P. L. McEuen, E. B. Foxman, J. Kinaret, U. Meirav, M. A. Kastner, N. S. Wingreen, and S. J. Wind, Phys. Rev. B **45**, 11 419 (1992).
[4]R. C. Ashoori, H. L. Stormer, J. S. Weiner, L. N. Pfeiffer, S. J. Pearton, K. W. Baldwin, and K. W. West, Phys. Rev. Lett. **68**, 3088 (1992).
[5]R. C. Ashoori, H. L. Stormer, J. S. Weiner, L. N. Pfeiffer, K. W. Baldwin, and K. W. West, Phys. Rev. Lett. **71**, 613 (1993).
[6]N. C. van der Vaart, M. P. de Ruyter van Steveninck, L. P. Kouwenhoven, A. T. Johnson, Y. V. Nazarov, and C. J. P. M. Harmans, Phys. Rev. Lett. **73**, 320 (1994).
[7]O. Klein, C. Chamon, D. Tang, D. M. Abush-Magder, X.-G. Wen, M. A. Kastner, and S. J. Wind (unpublished).
[8]A. Kumar, S. E. Laux, and F. Stern, Phys. Rev. B **42**, 5166 (1990).
[9]C. W. J. Beenakker, Phys. Rev. B **44**, 1646 (1991).
[10]V. Shikin, S. Nazin, D. Heitmann, and T. Demel, Phys. Rev. B **43**, 11 903 (1991).
[11]V. Gudmundsson and R. R. Gerhardts, Phys. Rev. B **43**, 12 098 (1991).
[12]A. H. MacDonald, S. R. Eric Yang, and M. D. Johnson, Aust. J. Phys. **46**, 345 (1993).
[13]P. L. McEuen, N. S. Wingreen, E. B. Foxman, J. Kinaret, U. Meirav, M. A. Kastner, and S. J. Wind, Physica B **189**, 70 (1993).
[14]S.-R. Eric Yang, A. H. MacDonald, and M. D. Johnson, Phys. Rev. Lett. **71**, 3194 (1993).
[15]J. M. Kinaret and N. S. Windgreen, Phys. Rev. B **48**, 11 113 (1993).
[16]D. Pfannkuche, V. Gudmundsson, and P. A. Maksym, Phys. Rev. B **47**, 2244 (1993).
[17]D. Pfannkuche, V. Gudmundsson, P. Hawrylak, and R. R. Gerhardts, Solid State Electron. **37**, 1221 (1994).
[18]M. Ferconi and G. Vignale (unpublished).
[19]E. H. Lieb, J. P. Solovej, and J. Yngvason, Phys. Rev. Lett. **69**, 749 (1992).
[20]E. H. Lieb, J. P. Solovej, and J. Yngvason, Commun. Pure Appl. Math. **47**, 513 (1994).
[21]E. H. Lieb, J. P. Solovej, and J. Yngvason, Commun. Math. Phys. **161**, 77 (1994).
[22]E. H. Lieb, J. P. Solovej, and J. Yngvason, in *Proceedings of the Conference on Partial Differential Equations and Mathematical Physics, Birmingham, AL, 1994,* edited by I. Knowles (International Press, Cambridge, MA, in press).
[23]Y. Tomishima and K. Yonei, Progr. Theor. Phys. **59**, 683 (1978).
[24]I. Fushiki, E. H. Gudmundsson, C. J. Pethick, and J. Yngvason, Ann. Phys. (N.Y.) **216**, 29 (1992).
[25]J. Yngvason, Lett. Math. Phys. **22**, 107 (1991).
[26]E. H. Lieb and H.-T. Yau, Commun. Math. Phys. **118**, 177 (1988).
[27]V. Fock, Z. Phys. **47**, 446 (1928).
[28]E. H. Lieb, Rev. Mod. Phys. **53**, 603 (1981); **54**, 311(E) (1982).
[29]E. H. Lieb and B. Simon, Adv. Math. **23**, 22 (1977).
[30]E. H. Lieb and M. Loss (unpublished).
[31]E. H. Lieb and W. E. Thirring, Phys. Rev. Lett. **35**, 687 (1975).
[32]E. H. Lieb and W. E. Thirring, in *Studies in Mathematical Physics: Essays in Honor of Valentine Bargmann,* edited by E. H. Lieb, B. Simon, and A. Wightman (Princeton University Press, Princeton, NJ, 1976), pp. 269–303.
[33]L. Erdös, Commun. Math. Phys. (to be published).
[34]E. H. Lieb, Phys. Lett. **70A**, 444 (1979).
[35]E. H. Lieb and S. Oxford, Int. J. Quantum Chem. **19**, 427 (1981).
[36]E. H. Lieb and J. P. Solovej, Lett. Math. Phys. **22**, 145 (1991).
[37]E. H. Lieb, Phys. Rev. Lett. **46**, 457 (1981); **47**, 69(E) (1981).
[38]B. Simon, *Functional Integration and Quantum Physics* (Academic, New York, 1979).
[39]K. Fan, Proc. Natl. Acad. Sci. U.S.A. **37**, 760 (1951).
[40]E. H. Lieb, Commun. Math. Phys. **92**, 57 (1984).
[41]Ph. Blanchard and J. Stubbe (unpublished).
[42]A. Martin, Commun. Math. Phys. **129**, 161 (1990).
[43]E. H. Lieb, in *Schrödinger Operators,* Proceedings of a Conference in Sønderborg, Denmark, 1988, edited by H. Holden and A. Jensen, Springer Lecture Notes in Physics Vol. 345 (Springer, Berlin, 1989), pp. 371–382.
[44]V. Bach, Commun. Math. Phys. **147**, 527 (1992).
[45]C. Fefferman and R. de la Llave, Rev. Iberoamericana **2**, 119 (1986).
[46]E. M. Stein and G. Weiss, *Fourier Analysis on Euclidean Spaces* (Princeton University Press, Princeton, NJ, 1971).

Part III
General Results with Applications to Atoms

Schrödinger Operators, H. Holden and A. Jensen eds., Springer Lect. Notes Phys. *345*, 371-382 (1989)

KINETIC ENERGY BOUNDS AND THEIR APPLICATION TO THE STABILITY OF MATTER

Elliott H. Lieb
Departments of Mathematics and Physics, Princeton University
P.O. Box 708, Princeton, NJ 08544

The Sobolev inequality on $\mathbf{R}^n, n \geq 3$ is very important because it gives a lower bound for the kinetic energy $\int |\nabla f|^2$ in terms of an L^p norm of f. It is the following.

$$\int_{\mathbf{R}^n} |\nabla f|^2 \geq S_n \left\{ \int_{\mathbf{R}^n} |f|^{2n/(n-2)} \right\}^{(n-2)/n} = S_n \|f\|^2_{2n/(n-2)}. \tag{1}$$

Applying Hölder's inequality to the right side we obtain the following modification of (1).

$$\int_{\mathbf{R}^n} |\nabla f|^2 \geq K^1_n \left\{ \int_{\mathbf{R}^n} \rho^{(n+2)/n} \right\} / \left\{ \int_{\mathbf{R}^n} \rho \right\}^{2/n} = K^1_n \|\rho\|^{(n+2)/n}_{(n+2)/n} \|\rho\|_1^{-2/n} \tag{2}$$

with $\rho(x) \equiv |f(x)|^2$. The superscript 1 on K^1_n indicates that in (2) we are considering only one function, f. Hölder's inequality implies that $K^1_n \geq S_n$ but, in fact, the *sharp* value of K^1_n (which can be obtained by solving a nonlinear PDE) is larger than S_n. In particular, $K^1_n > 0$ for *all* $n \geq 1$, even though $S_n = 0$ for $n < 3$.

Inequality (2), unlike (1) has the following important property: The non-linear term $\int \rho^{(n+2)/n}$ enters with the power 1 (and not $(n-2)/n$) and is therefore "extensive." The price we have to pay for this is the factor $\|f\|_2^{4/n} = \|\rho\|_1^{2/n}$ in the denominator, but since we shall apply (2) to cases in which $\|f\|_2 = 1$ (L^2 normalization condition) this is not serious.

Inequality (2) is equivalent to the following: Consider the Schrödinger operator on $\mathbf{R}^n$

$$H = -\Delta - V(x) \tag{3}$$

and let $e_1 = \inf \operatorname{spec}(H)$. (We assume H is self-adjoint.) Let $V_+(x) \equiv \max\{V(x), 0\}$. Then

$$e_1 \geq -L^1_{1,n} \int V_+(x)^{(n+2)/n} dx = -L^1_{1,n} \|V_+\|^{(n+2)/2}_{(n+2)/2} \tag{4}$$

with

$$L^1_{1,n} = \left(\frac{n}{2K^1_n} \right)^{n/2} \left(1 + \frac{n}{2} \right)^{-(n+2)/n}. \tag{5}$$

The reason for the subscript 1 in $L^1_{1,n}$ will be clarified in eq. (8).

Here is the proof of the equivalence. We have

$$e_1 \geq \inf_f \left\{ \int |\nabla f|^2 - \int \rho V_+ \mid \|f\|_2 = 1 \text{ and } \rho = |f|^2 \right\}.$$

Use (2) and Hölder to obtain (with $X = \|\rho\|_{(n+2)/n}$)

$$e_1 \geq \inf_X \left\{ K_n^1 X^{(n+2)/n} - \|V_+\|_{(n+2)/2} X \right\} \tag{6}$$

Minimizing (6) with respect to X yields (4). To go from (4) to (2), take $V = V_+ = \alpha |f|^{4/n} = \alpha \rho^{2/n}$ in (3). Then

$$-L_{1,n}^1 \alpha^{(n+2)/2} \int \rho^{(n+2)/n} \leq e_1 \leq (f, Hf) = \int |\nabla f|^2 - \alpha \int \rho^{(n+2)/n}.$$

Optimizing this with respect to α yields (2).

So far this is trivial, but now we turn to a more interesting question. Let $e_1 \leq e_2 \leq \ldots \leq 0$ be the negative spectrum of H (which may be empty). Is there a bound of the form

$$\sum e_i \geq -L_{1,n} \int V_+(x)^{(n+2)/2} dx \tag{7}$$

for some universal, V independent, constant $L_{1,n} > 0$ (which, of course, is $\geq L_{1,n}^1$)? The point is that the right side of (7) has the same form as the right side of (4). More generally, given $\gamma \geq 0$, does

$$\sum |e_i|^\gamma \leq L_{\gamma,n} \int V_+(x)^{\gamma + \frac{n}{2}} \tag{8}$$

hold for suitable $L_{\gamma,n}$? When $\gamma = 0, \sum |e_i|^0$ is interpreted as the number of $e_i \leq 0$.

The answer to these questions is yes in the following cases:

<u>$n = 1$</u>: All $\gamma > \frac{1}{2}$. The case $\gamma = 1/2$ is unsettled. For $\gamma < \frac{1}{2}$, examples show there can be no bound of the form (8).

<u>$n = 2$</u>: All $\gamma > 0$. There can be no bound when $\gamma = 0$.

<u>$n \geq 3$</u>: All $\gamma \geq 0$.

The cases $\gamma > 0$ were first done in [10], [11]. The $\gamma = 0$ case for $n \geq 3$ was done in [3], [6], [14], with [6] giving the best estimate for $L_{0,n}$. For a review of what is currently known about these constants and conjectures about the sharp values of $L_{\gamma,n}$, see [8]. The proof of (8) is involved (especially when $\gamma = 0$) and will not be given here. It uses the Birman-Schwinger kernel, $V_+^{1/2}(-\Delta + \lambda)^{-1} V_+^{1/2}$.

There is a natural "guess" for $L_{\gamma,n}$ in terms of a semiclassical approximation (and which is not unrelated to the theory of pseudodifferential operators):

$$\sum |e_i|^\gamma \approx (2\pi)^{-n} \int_{\mathbf{R}^n \times \mathbf{R}^n, p^2 \le V(x)} [V(x) - p^2]^\gamma dp dx \tag{9}$$

$$= L^c_{\gamma,n} \int V_+(x)^{\gamma+n/2} dx. \tag{10}$$

From (9),

$$L^c_{\gamma,n} = (4\pi)^{-n/2} \Gamma(\gamma+1)/\Gamma(1+\gamma+n/2). \tag{11}$$

It is easy to prove that

$$L_{\gamma,n} \ge L^c_{\gamma,n}. \tag{12}$$

The evaluation of the sharp $L_{\gamma,n}$ is an interesting open problem – especially $L_{1,n}$. In particular, for which γ, n is $L_{\gamma,n} = L^c_{\gamma,n}$? It is known [1] that for each fixed n, $L_{\gamma,n}/L^c_{\gamma,n}$ is nonincreasing in γ. Thus, if $L_{\gamma_0,n} = L^c_{\gamma_0,n}$ for some γ_0, then $L_{\gamma,n} = L^c_{\gamma,n}$ for all $\gamma > \gamma_0$. In particular, $L_{3/2,1} = L^c_{3/2,1}$ [11], so $L_{\gamma,1} = L^c_{\gamma,1}$ for $\gamma \ge 3/2$. No other sharp values of $L_{\gamma,n}$ are known. It is also known [11] that $L_{\gamma,1} > L^c_{\gamma,1}$ for $\gamma < 3/2$ and $L_{\gamma,n} > L^c_{\gamma,n}$ for $n = 2, 3$ and small γ.

Just as (4) is related to (2), inequality (7) is related to a generalization of (2). (The proof is basically the same.) Let $\phi_1, \ldots, \phi_N$ be any set of L^2 orthonormal functions on $\mathbf{R}^n (n \ge 1)$ and define

$$\rho(x) \equiv \sum_{i=1}^N |\phi_i(x)|^2. \tag{13}$$

$$T \equiv \sum_{i=1}^N \int |\nabla \phi_i|^2 \tag{14}$$

Then we have **The Main Inequality**

$$T \ge K_n \int \rho(x)^{1+2/n} dx \tag{15}$$

with K_n related to $L_{1,n}$ as in (5), i.e.

$$L_{1,n} = \left(\frac{n}{2K_n}\right)^{n/2} \left(1 + \frac{n}{2}\right)^{-(n+2)/2} \tag{16}$$

The best current value of K_n, for $n = 1, 2, 3$ is in [8]; in particular $K_3 \ge 2.7709$. We might call (15) *a Sobolev type inequality for orthonormal functions.* The point is that if the ϕ_i are merely normalized, but not orthogonal, then the best one could say is

$$T \ge N^{-2/n} K^1_n \int \rho(x)^{1+2/n} dx. \tag{17}$$

The orthogonality eliminates the factor $N^{-2/n}$, but replaces K_n^1 by the slightly smaller value K_n.

One should notice, especially, the N dependence in (15). The right side, loosely speaking, is proportional to $N^{(n+2)/n}$, whereas the right side of (17) appears, falsely, to be proportional to N^1, which is the best one could hope for without orthogonality. The difference is crucial for applications. In fact, if one is willing to settle for N^1 one can proceed directly from (1) (for $n \geq 3$). One then has (with $p = n/(n-2)$)

$$T \geq S_n \left\{ \int \rho(x)^p dx \right\}^{1/p}, \quad (n \geq 3). \tag{17a}$$

This follows from $\sum \|\phi_i^2\|_p \geq \left\| \sum |\phi_i|^2 \right\|_p$.

Eq. (11) gives a "classical guess" for $L_{1,n}$. Using that, together with (16), we have a "classical guess" for K_n, namely

$$\begin{aligned} K_n^c &= 4\pi n \Gamma \left(\frac{n+2}{2} \right)^{2/n} / (2+n) \\ &= \frac{3}{5} (6\pi^2)^{2/3} = 9.1156 \text{ for } n = 3. \end{aligned} \tag{18}$$

Since $L_{1,n} \geq L_{1,n}^c$, we have $K_n \leq K_n^c$. A conjecture in [11] is that $K_3 = K_3^c$, and it would be important to settle this.

Inequality (15) can be easily extended to the following: Let $\psi(x_1, \ldots, x_N) \in L^2((\mathbf{R}^n)^N), x_i \in \mathbf{R}^n$. Suppose $\|\psi\|_2 = 1$ and ψ is *antisymmetric* in the N variables, i.e., $\psi(x_1, \ldots, x_i, \ldots, x_j, \ldots, x_N) = -\psi(x_1, \ldots x_j, \ldots, x_i, \ldots, x_N)$. Define

$$\rho_i(x) = \int \left| \psi(x_1, \ldots, x_{i-1}, x, x_{i+1}, \ldots, x_N) \right|^2 dx_1 \ldots \widehat{dx_i} \ldots dx_N \tag{19}$$

$$T_i(x) = \int |\nabla_i \psi|^2 dx_1 \ldots dx_N \tag{20}$$

$$\rho(x) = \sum_{i=1}^N \rho_i(x) \qquad T = \sum_{i=1}^N T_i. \tag{21}$$

(Note that $\rho(x) = N\rho_1(x)$ and $T = NT_1$ since ψ is antisymmetric, but the general form (19)-(21) will be used in the next paragraph.) **Then (15) holds with ρ and T given by (19)-(21)** (with the same K_n as in (15)). This is a generalization of (13)-(15) since we can take

$$\psi(x_1, \ldots, x_N) = (N!)^{-1/2} \det \left\{ \phi_i(x_j) \right\}_{i,j=1}^N,$$

which leads to (13) and (14).

A variant of (15) is given in (52) below. It is a consequence of the fact that (17) and (17a) *also hold* with the definitions (19)-(21). *Antisymmetry of ψ is not required.* The proof of (17a) just uses (1) as before plus Minkowski's inequality, namely for $p \geq 1$

$$\int \left\{ \int |F(x,y)|^p dy \right\}^{1/p} dx \geq \left\{ \int \left\{ \int |F(x,y)| dx \right\}^p dy \right\}^{1/p}.$$

We turn now to some applications of these inequalities.

Application 1. Inequality (15) can be used to bound L^p **norms of Riesz and Bessel potentials of orthonormal functions** [7]. Again, $\phi_1, \ldots, \phi_N$ are L^2 orthonormal and let

$$u_i = (-\Delta + m^2)^{-1/2} \phi_i \tag{22}$$

$$\rho(x) \equiv \sum_{i=1}^{N} |u_i(x)|^2. \tag{23}$$

Then there are constants L, B_p, A_n (independent of m) such that

$$n = 1: \quad \|\rho\|_\infty \leq L/m, \qquad m > 0 \tag{24}$$

$$n = 2: \quad \|\rho\|_p \leq B_p m^{-2/p} N^{1/p}, \qquad 1 \leq p < \infty, m > 0 \tag{25}$$

$$n \geq 3: \quad \|\rho\|_p \leq A_n N^{1/p}, \qquad p = n/(n-2), m \geq 0. \tag{26}$$

If the orthogonality condition is dropped then the right sides of (24)-(26) have to be multiplied by $N, N^{1-1/p}, N^{1-1/p}$ respectively. Possibly the absence of N in (24) is the most striking. Similar results can be derived [7] for $(-\Delta + m^2)^{-\alpha/2}$ in place of $(-\Delta + m^2)^{-1/2}$, with $\alpha < n$ when $m = 0$.

Inequality (15) also has applications in mathematical physics.

Application 2. (Navier-Stokes equation.) Suppose $\Omega \subset \mathbf{R}^n$ is an open set with finite volume $|\Omega|$ and consider

$$H = -\Delta - V(x)$$

on Ω with Dirichlet boundary conditions. Let $\lambda_1 \leq \lambda_2 \leq \ldots$ be the eigenvalues of H. Let $\bar{N}$ be the smallest integer, N, such that

$$E_N \equiv \sum_{i=1}^{N} \lambda_i \geq 0. \tag{27}$$

We want to find an upper bound for $\bar{N}$.

If $\phi_1, \phi_2, \ldots$ are the normalized eigenfunctions then, from (13)-(15) with $\phi_1, \ldots, \phi_N$,

$$E_N = T - \int \rho V \geq K_n \int \rho^{1+n/2} - \int V_+ \rho \geq G(\rho), \tag{28}$$

where (with $p = 1 + n/2$ and $q = 1 + 2/n$)

$$G(\rho) \equiv K_n \|\rho\|_p^p - \|V_+\|_q \|\rho\|_p. \tag{29}$$

Thus, for all N,

$$E_N \geq \inf\{G(\rho) |\ \|\rho\|_1 = N,\ \rho(x) \geq 0\}. \tag{30}$$

But $\|\rho\|_p |\Omega|^{1/q} \geq \|\rho\|_1 = N$ so, with $X \equiv \|\rho\|_p$,

$$E_N \geq \inf\{J(X) \mid X \geq N|\Omega|^{-1/q}\} \tag{31}$$

where

$$J(X) \equiv K_n X^p - \|V_+\|_q X. \tag{32}$$

Now $J(X) \geq 0$ for $X \geq X_0 = \{\|V_+\|_q / K_n\}^{1/(p-1)}$, whence we have the following implication:

$$N \geq |\Omega|^{1/q} \{\|V_+\|_q / K_n\}^{1/(p-1)} \Longrightarrow E_N \geq 0. \tag{33}$$

Therefore

$$\bar{N} \leq |\Omega|^{1/q} \{\|V_+\|_q / K_n\}^{1/(p-1)}. \tag{34}$$

The bound (34) can be applied [8] (following an idea of Ruelle) to the Navier-stokes equation. There, $\bar{N}$ is interpreted as the Hausdorff dimension of an attracting set for the N-S equation, while $V(x) \equiv \nu^{-3/2} \varepsilon(x)$, where $\varepsilon(x) = \nu |\nabla v(x)|^2$ is the average energy dissipation per unit mass in a flow v. ν is the viscosity.

Application 3. (Stability of matter.) This is the original application [10,11]. In the quantum mechanics of Coulomb systems (electrons and nuclei) one wants a lower bound for the Hamiltonian operator:

$$H = -\sum_{i=1}^{N} \Delta_i - \sum_{i=1}^{N} \sum_{j=1}^{K} z_j |x_i - R_j|^{-1} + \sum_{1 \leq i < j \leq N} |x_i - x_j|^{-1} + \sum_{1 \leq i < j \leq K} z_i z_j |R_i - R_j|^{-1} \tag{35}$$

on the L^2 space of *antisymmetric* functions $\psi(x_1, \ldots, x_N), x_i \in \mathbf{R}^3$. Here, N is the number of electrons (with coordinates x_i) and $R_1, \ldots, R_K \in \mathbf{R}^3$ are fixed vectors representing the locations of fixed nuclei of charges $z_1, \ldots, z_K > 0$. The desired bound is linear:

$$H \geq -A(N+K) \tag{36}$$

for some A independent of $N, K, R_1, \ldots, R_K$ (assuming all $z_i <$ some $\bar{z}$).

The main point is that antisymmetry of ψ is crucial for (36) and this is reflected in the fact that (15) holds with antisymmetry, but only (17) holds without it. Without the antisymmetry condition, H would grow as $-(N+K)^{5/3}$. This is discussed in Application 6 below. By using (15) one can eliminate the differential operators Δ_i. The functional $\psi \to (\psi, H\psi)$, with $(\psi, \psi) = 1$ can be bounded below using (15) by a functional (called the Thomas-Fermi functional) involving only $\rho(x)$ defined in (21). The minimization of this latter functional with respect to ρ is tractable and leads to (36).

Application 4. (Stellar structure.) Going from atoms to stars, we now consider N neutrons which attract each other gravitationally with a coupling constant $\kappa = Gm^2$, where G is the gravitational constant and m is the neutron mass. There are no Coulomb forces. Moreover, a "relativistic" form is assumed for the kinetic energy, which means that $-\Delta$ is replaced by $(-\Delta)^{1/2}$. Thus (35) is replaced by

$$H_N = \sum_{i=1}^{N} (-\Delta_i)^{1/2} - \kappa \sum_{1 \leq i < j \leq N} |x_i - x_j|^{-1} \tag{37}$$

(again on antisymmetric functions). One finds asymptotically for large N, that

$$\begin{aligned} \inf \operatorname{spec}(H_N) &= 0 \quad \text{if} \quad \kappa \leq CN^{-2/3} \\ &= -\infty \quad \text{if} \quad \kappa > CN^{-2/3} \end{aligned} \tag{38}$$

for some constant, C. Without antisymmetry, $N^{-2/3}$ must be replaced by N^{-1}. Equation (38) is proved in [12]. An important role is played by Daubechies's generalization [4] of (15) to the operator $(-\Delta)^{1/2}$ on $L^2(\mathbf{R}^N)$, namely (for antisymmetric ψ with $\|\psi\|_2 = 1$)

$$\left(\psi, \sum_{i=1}^{N} (-\Delta_i)^{1/2} \psi\right) \geq B_n \int \rho(x)^{1+1/n} \tag{39}$$

with ρ given by (19), (21). In general, one has

$$\left(\psi, \sum_{i=1}^{N} (-\Delta)^p \psi\right) \geq C_{p,n} \int \rho(x)^{1+2p/n} dx. \tag{40}$$

Recently [13] there has been considerable progress in this problem beyond that in [12]. Among other results there is an evaluation of the sharp asymptotic C in (38), i.e. if we first define $\kappa^c(N)$ to be the precise value of κ at which $\inf \operatorname{spec}(H_N) = -\infty$, we then define

$$C = \lim_{N\to\infty} N^{2/3}\kappa^c(N). \tag{41}$$

Let B_n^c be the "classical guess' in (39). This can be calculated from the analogue of (9) (using $|p|$ instead of p^2, and which leads to $\sum |e_i| \approx \widetilde{L} \int V_+^{1+n}$), and then from the analogue of (16), namely $\widetilde{L} = C_n B_n^{-n}$. One finds $B_3^c = (3/4)(6\pi^2)^{1/3}$ (cf. (18)). Using B_3^c, we introduce the functional

$$\mathcal{E}(\rho) = B_3^c \int \rho^{4/3} - \frac{1}{2}\kappa \int\int \rho(x)\rho(y)|x-y|^{-1}dxdy \tag{42}$$

for $\rho \in L^1(\mathbf{R}^3) \cap L^{4/3}(\mathbf{R}^3)$ and define the energy

$$E^c(N) = \inf\{\mathcal{E}(\rho) | \int \rho = N\}. \tag{43}$$

One finds there is a finite $\alpha^c > 0$ such that $E^c(N) = 0$ if $\kappa N^{2/3} \leq \alpha^c$ and $E^c(N) > -\infty$ if $\kappa N^{2/3} > \alpha^c$. (This α^c is found by solving a Lane-Emden equation.)

Now (42) and (43) constitute the semiclassical approximation to H_N in the following sense. We expect that if we set $\kappa = \alpha N^{-2/3}$ in (37), with α fixed, then if $\alpha < \alpha^c$

$$\lim_{N\to\infty} \inf \operatorname{spec}(H_N) = 0 \tag{44}$$

while if $\alpha > \alpha^c$ there is an N_0 such that

$$\inf \operatorname{spec}(H_N) = -\infty \quad \text{if} \quad N > N_0. \tag{45}$$

Indeed, (44) and (45) are true [13], and thus α^c is the sharp asymptotic value of C in (38).

An interesting point to note is that Daubechies's B_3 in (39) is about half of B_3^c. The sharp value of B_3 is unknown. Nevertheless, with some additional tricks one can get from (37) to (42) with B_3^c and not B_3. Inequality (39) plays a role in [13], but it is not sufficient.

Application 5. (Stability of atoms in magnetic fields.) This is given in [9]. Here $\psi(x_1, \ldots, x_N)$ becomes a spinor-valued function, i.e. ψ is an antisymmetric function in $\bigwedge_1^N L^2(\mathbf{R}^3; \mathbf{C}^2)$. The operator H of interest is as in (35) but with the replacement

$$-\Delta \to \{\sigma \cdot (i\nabla - A(x))\}^2 \tag{46}$$

where $\sigma_1, \sigma_2, \sigma_3$ are the 2×2 Pauli matrices (i.e. generators of $SU(2)$) and $A(x)$ is a given vector field (called the magnetic vector potential). Let

$$E_0(A) = \inf \operatorname{spec}(H) \tag{47}$$

after the replacement of (46) in (35). As $A \to \infty$ (in a suitable sense), $E_0(A)$ can go to $-\infty$. The problem is this: Is

$$\widetilde{E}(A) \equiv E_0(A) + \frac{1}{8\pi} \int (\text{curl } A)^2 \tag{48}$$

bounded below for all A? In [9] the problem is resolved for $K = 1$, all N and $N = 1$, all K. It turns out that $\widetilde{E}(A)$ is bounded below in these cases if and only if all the z_i satisfy $z_i < z^c$ where z^c is some fixed constant independent of N and K. The problem is still open for all N and all K.

One of the main problems in bounding $\widetilde{E}(A)$ is to find a lower bound for the kinetic energy (the first term in (35) after the replacement given in (46)) for an antisymmetric ψ. First, there is the identity

$$\left(\psi, \sum_{i=1}^{N} \{\sigma \cdot (i\nabla - A(x_i)\}^2 \psi\right) = T(\psi, A) - \left(\psi, \sum_{i=1}^{N} \sigma \cdot B(x_i) \psi\right) \tag{49}$$

with $B = $ curl A being the magnetic field and

$$T(\psi, A) = \left(\psi, \sum_{i=1}^{N} |i\nabla - A(x)|^2 \psi\right). \tag{50}$$

The last term on the right side of (49) can be controlled, so it will be ignored here. The important term is $T(\psi, A)$. Since Pauli matrices do not appear in (50) we can now let ψ be an ordinary complex valued (instead of spinor valued) function.

It turns out that (8), and hence (15), hold with some $\widetilde{L}_{\gamma,n}$ which is *independent* of A. The T in (15) is replaced, of course, by the $T(\psi, A)$ of (50). To be more precise, the *sharp* constants $L_{\gamma,n}$ and $\widetilde{L}_{\gamma,n}$ are unknown (except for $\gamma \geq 3/2, n = 1$ in the case of $L_{\gamma,n}$) and conceivably $\widetilde{L}_{\gamma,n} > L_{\gamma,n}$. However, all the current bounds for $L_{\gamma,n}$ (see [8]) also hold for $\widetilde{L}_{\gamma,n}$. Thus, for $n = 3$ we have

$$T(\psi, A) \geq K_3 \int \rho^{5/3} \tag{51}$$

with K_3 being the value given in [8], namely 2.7709.

However, in [9] another inequality is needed

$$T(\psi, A) \geq C\left\{\int \rho^2\right\}^{2/3}. \tag{52}$$

It seems surprising that we can go from an $L^{5/3}$ estimate to an L^2 estimate, but the surprise is diminished if (17a) with its L^3 estimate is recalled. First note that (1) holds (with the same S_n) if $|\nabla f|^2$ is replaced by $|[i\nabla - A(x)]f|^2$. (By writing $f = |f|e^{i\theta}$ one finds that $|\nabla|f||^2 \leq |[i\nabla - A(x)]f|^2$.) Then (17a) holds since only convexity was used. Thus, using the mean of (15) and (17a),

$$T(\psi, A) \geq (S_n K_n)^{1/2} \|\rho\|_3^{1/2} \|\rho\|_{5/3}^{5/6}. \tag{53}$$

An application of Hölders inequality yields (52) with $C^2 = S_n K_n$.

Application 6. (Instability of bosonic matter.) As remarked in Application 3, dropping the antisymmetry requirement on ψ (the particles are now bosons) makes $\inf\operatorname{spec}(H)$ diverge as $-(N+K)^{5/3}$. The extra power 2/3, relative to (36) can be traced directly to the factor $N^{-2/3}$ in (17).

An interesting problem is to allow the positive particles also to be movable and to have charge $z_i = 1$. This should raise $\inf\operatorname{spec} H$, but by how much? For $2N$ particles the new H is

$$\widetilde{H} = -\sum_{i-1}^{2N} \Delta_i + \sum_{1\leq i<j\leq 2N} e_i e_j |x_i - x_j|^{-1} \tag{54}$$

with $e_i = +1$ for $1 \leq i \leq N$ and $e_i = -1$ for $N+1 \leq i \leq 2N$. $\widetilde{H}$ acts on $L^2(\mathbf{R}^{3N})$ without any symmetry requirements.

Twenty years ago Dyson [5] proved, by a variational calculation, that $\inf\operatorname{spec}(\widetilde{H}) \leq -AN^{7/5}$ for some $A > 0$. Thus, stability (i.e. a linear law (36)) is not restored, but the question of whether the correct exponent is 7/5 or 5/3, or something in between, remained open. It has now been proved [2] that $N^{7/5}$ is correct, $\inf\operatorname{spec}(\widetilde{H}) \geq -BN^{7/5}$. The proof is much harder than for (36) because no simple semiclassical theory (like Thomas-Fermi theory) is a good approximation to $\widetilde{H}$. Correlations are crucial.

Application 7. (Stability of relativistic matter.) Let us return to Coulomb systems (electrons and nuclei) as in application 3, but with (35) replaced by

$$H = \sum_{i=1}^{N} \{(-\Delta_i + m^2)^{1/2} - m\} + \alpha V_c(x_1, \ldots, x_N; R_1, \ldots, R_K) \tag{55}$$

with $\alpha = e^2$ = electron charge squared (and $\hbar = c = 1$) and where

$$V_c = -\sum_{i=1}^{N}\sum_{j=1}^{K} z_j |x_i - R_j|^{-1} + \sum_{1\leq i<j\leq N} |x_i - x_j|^{-1} + \sum_{1\leq i<j\leq K} z_i z_j |R_i - R_j|^{-1} \tag{56}$$

is the Coulomb potential. The electron charge, $\alpha^{1/2}$, is explicitly displayed in (55) for a reason to be discussed presently. Also (55) differs from (35) in that the kinetic energy operator $-\Delta$ is replaced by the relativistic form $(-\Delta + m^2)^{1/2} - m$, where m is the electron mass. Since $-\Delta - m \leq (-\Delta + m^2)^{1/2} - m \leq -\Delta$, the difference of these two operators is a bounded operator and therefore, as far as the stability question is concerned, we may as well use the simplest operator $(-\Delta)^{1/2}$ in (55), which will be done henceforth. This, in fact, was already done in (37). We define

$$E_{N,K}(R_1, \ldots, R_K) = \inf \mathrm{spec} H \tag{57}$$

$$E_{N,K} = \inf_{R_1,\ldots,R_K} E_{NK}(R_1, \ldots, R_K) \tag{58}$$

$$E = \inf_{N,K} E_{NK} \tag{59}$$

Under scaling (dilation of coordinates in $\mathbf{R}^{3N+3K}$) the operators $(-\Delta)^{1/2}$ and $|x|^{-1}$ behave the same (proportional to length)$^{-1}$ and hence we conclude that

$$E_{N,K} = 0 \qquad \text{or} \qquad -\infty. \tag{60}$$

The system is said to be **stable** if $E = 0$. For simplicity of exposition let us take all z_j to be some common value, z.

For the hydrogenic atom $N = K = 1$ the only constant that appears is the combination $z\alpha$. It is known that $E_{1,1} = 0$ if and only if $z\alpha \leq 2/\pi$. In the many-body case there are two constants (which can be taken to be $z\alpha$ and α) and the question is whether the system is stable *all the way up to* $z\alpha = 2/\pi$ for α less than some small, but fixed $\alpha_c > 0$. The answer will depend on q, the number of spin states allowed for the fermionic electrons. (Note: in application 3 we implicitly took $q = 1$. In fact $q = 2$ in nature. To say that there are q spin states means that under permutations $\psi(x_1, \ldots, x_N)$ belongs to a Young's tableaux of q or fewer columns.)

This problem is resolved in [15] where it is shown that stability occurs if

$$q\alpha \leq 1/47. \tag{61}$$

The kinetic energy bound (39) plays a crucial role in the proof (but, of course, many other inequalities are also needed). It is also shown in [15] that stability definitely fails to occur if

$$\alpha > 36q^{-1/3}z^{2/3} \tag{62}$$

or if

$$\alpha > 128/15\pi. \tag{63}$$

If (63) holds then instability occurs for *every* $z > 0$, no matter how small.

REFERENCES

[1] M. Aizenman and E.H. Lieb, On semiclassical bounds for eigenvalues of Schrödinger operators, Phys. Lett. *66*A, 427-429 (1978).

[2] J. Conlon, E.H. Lieb and H.-T. Yau, The $N^{7/5}$ law for bosons, Commun. Math. Phys. (submitted).

[3] M. Cwikel, Weak type estimates for singular values and the number of bound states of Schrödinger operators, Ann. Math. *106*, 93-100 (1977).

[4] I. Daubechies, Commun. Math. Phys. *90*, 511-520 (1983).

[5] F.J. Dyson, Ground state energy of a finite systems of charged particles, J. Math. Phys. *8*, 1538-1545 (1967).

[6] E.H. Lieb, The number of bound states of one-body Schrödinger operators and the Weyl problem, A.M.S. Proc. Symp. in Pure Math. *36*, 241-251 (1980). The results were announced in Bull. Ann. Math. Soc. *82*, 751-753 (1976).

[7] E.H. Lieb, An L^p bound for the Riesz and Bessel potentials of orthonormal functions, J. Funct. Anal. *51*, 159-165 (1983).

[8] E.H. Lieb, On characteristic exponents in turbulence, Commun. Math. Phys. *92*, 473-480 (1984).

[9] E.H. Lieb and M. Loss, Stability of Coulomb systems with magnetic fields: II. The many-electron atom and the one-electron molecule, Commun. Math. Phys. *104*, 271-282 (1986).

[10] E.H. Lieb and W.E. Thirring, Bounds for the kinetic energy of fermions which proves the stability of matter, Phys. Rev. Lett. *35*, 687-689 (1975). Errata *35*, 1116 (1975).

[11] E.H. Lieb and W.E. Thirring, "Inequalities for the moments of the eigenvalues of the Schrödinger Hamiltonian and their relation to Sobolev inequalities" in *Studies in Mathematical Physics* (E. Lieb, B. Simon, A. Wightman eds.) Princeton University Press, 1976, pp. 269-304.

[12] E.H. Lieb and W.E. Thirring, Gravitational collapse in quantum mechanics with relativistic kinetic energy, Ann. of Phys. (NY) *155*, 494-512 (1984).

[13] E.H. Lieb and H.-T. Yau, The Chandrasekhar theory of stellar collapse as the limit of quantum mechanics, Commun. Math. Phys. *112*, 147-174 (1987).

[14] G.V. Rosenbljum, Distribution of the discrete spectrum of singular differential operators. Dokl. Akad. Nauk SSSR *202*, 1012-1015 (1972). (MR 45 #4216). The details are given in Izv. Vyss. Ucebn. Zaved. Matem. *164*, 75-86 (1976). (English trans. Sov. Math. (Iz VUZ) *20*, 63-71 (1976).)

[15] E.H. Lieb and H.T. Yau, The stability and instability of relativistic matter, Commun. Math. Phys. *118*, 177-213 (1988). For a short summary see: Many body stability implies a bound on the fine structure constant, Phys. Rev. Lett. *61*, 1695-1697 (1988).

With W. Thirring in Studies in Mathematical Physics, Princeton University Press, 269-303 (1976)

INEQUALITIES FOR THE MOMENTS OF THE EIGENVALUES OF THE SCHRÖDINGER HAMILTONIAN AND THEIR RELATION TO SOBOLEV INEQUALITIES

Elliott H. Lieb*
Walter E. Thirring

1. *Introduction*

Estimates for the number of bound states and their energies, $e_j \leq 0$, are of obvious importance for the investigation of quantum mechanical Hamiltonians. If the latter are of the single particle form $H = -\Delta + V(x)$ in $\mathbf{R}^n$, we shall use available methods to derive the bounds

$$\sum_j |e_j|^\gamma \leq L_{\gamma,n} \int d^n x\, |V(x)|_-^{\gamma+n/2}, \qquad \gamma > \max(0, 1-n/2). \quad (1.1)$$

Here, $|V(x)|_- = -V(x)$ if $V(x) \leq 0$ and is zero otherwise.

Of course, in many-body theory, one is more interested in Hamiltonians of the form $-\sum_i \Delta_i + \sum_{i>j} v(x_i - x_j)$. It turns out, however, that the energy bounds for the single particle Hamiltonian yield a lower bound for the kinetic energy, T, of N fermions in terms of integrals over the single particle density defined by

$$\rho(x) \equiv N \int |\psi(x, x_2, \cdots, x_N)|^2\, d^n x_2 \cdots d^n x_N \,, \quad (1.2)$$

where ψ is an antisymmetric, normalized function of the N variables $x_i \in \mathbf{R}^n$. Our main results, in addition to (1.1), will be of the form

*Work supported by U. S. National Science Foundation Grant MPS 71-03375-A03.

$$T \equiv \sum_{i=1}^{N} \int |\vec{\nabla}_i \psi(x_1 \cdots x_N)|^2 \, d^n x_1 \cdots d^n x_N$$

$$\geq K_{p,n} \left[\int d^n x \, \rho(x)^{p/(p-1)} \right]^{2(p-1)/n} \qquad (1.3)$$

when $\max\{n/2, 1\} \leq p \leq 1 + n/2$.

For $N = 1$, $p = n/2$, (1.3) reduces to the well-known Sobolev inequalities. (1.3) is therefore a partial generalization of these inequalities, and we shall expand on this in Section 3.

Our constants $K_{p,n}$ are not always the best possible ones, but nevertheless, they may be useful for many purposes. In particular, in ref. [1], a special case of (1.3) was used to give a simple proof of the stability of matter, with a constant of the right order of magnitude. The result for q species of fermions $(2m = e = \hbar = 1)$ moving in the field of M nuclei with positive charges Z_j is

$$H \geq -1.31\, q^{2/3} N \left[1 + \left(\sum_{j=1}^{M} Z_j^{7/3}/N \right)^{1/2} \right]^2 . \qquad (1.4)$$

In particular, if $q = 2$ (spin 1/2 electrons), we have a bound $\sim N$, and if we set $q = N$, we get a bound $\sim N^{5/3}$ if no symmetry requirement is imposed on the wave function; *a fortiori* this is a bound for *bosons*. Our bound implies stability of matter in its intuitive meaning such that the volume occupied by N particles will be $\sim N$ (Bohr radius)^3. To give a formal demonstration of this fact, one might use a method which gives lower bounds for the radii of complex atoms (compare Equation (3.6, 38) of ref. [20]). As a first observation, one calculates the ground state energy of N electrons (with spin) in a harmonic potential. Filling the oscillator levels, one finds

$$\frac{1}{2} \sum_{i=1}^{N} (-\Delta_i + \omega^2 \vec{x}_i^2) \geq \omega\, N^{4/3} \, \frac{3^{4/3}}{4} (1 + O(N^{-1/3})) . \qquad (1.5)$$

Next, take the expectation value of this operator inequality with the ground state of H, set

$$\omega = \frac{4}{3^{4/3}} < -\sum_i \Delta_i > \frac{1}{N^{4/3}} \tag{1.6}$$

and use the virial theorem

$$< -\sum_i \Delta_i > = -E_0 \leq 2.08\,N \left[1 + \left(\sum_{j=1}^{M} Z_j^{7/3}/N \right)^{1/2} \right]^2 . \tag{1.7}$$

Altogether we find

$$< \sum_{i=1}^{N} \vec{x}_i^2 > \geq \frac{(3N)^{8/3}}{16 < -\sum_i \Delta_i >} \geq \frac{3^{8/3} N^{5/3}}{16 \cdot 2.08 \left[1 + \left(\sum_j Z_j^{7/3}/N \right)^{1/2} \right]^2} . \tag{1.8}$$

Thus we have proved that

$$< \vec{x}_i^2 >^{1/2} \geq c\,N^{1/3} , \qquad c = \frac{.75}{\left[1 + \left(\sum_j Z_j^{7/3}/N \right)^{1/2} \right]} . \tag{1.9}$$

Therefore, if the system is not compressed by other forces, so that the virial theorem is valid, it will not collapse, but will adjust its volume to a size proportional to the number of particles. Regarding the Z-dependence, we see that with $Z = Z_j = N/M$ we have (for large Z)

$$< \vec{x}_i^2 >^{1/2} \sim M^{1/3} Z^{-1/3} .$$

That is, the mean atomic radius is predicted to be $\geq Z^{-1/3}$. A better result can hardly be expected since for $M = 1$, this is the correct Z-dependence for large Z.

Although we have no results on the best possible constants, $K_{p,n}$, except in a few special cases, experience drawn from computer calculations suggests that there is a critical value $\gamma_{c.n}$ above which the classical value gives a bound:

$$\left(\sum |e_j|^\gamma\right)_{\text{classical}} = (2\pi)^{-n} \int d^np\, d^nx\, |p^2 + V(x)|_-^\gamma$$

$$\equiv L^C_{\gamma,n} \int |V(x)|_-^{\gamma+n/2}\, d^nx \ ,$$

$$\gamma \geq \gamma_{c,n} \ , \qquad (1.10)$$

and where $L^C_{\gamma,n}$, given by the above integral, is

$$L^C_{\gamma,n} = 2^{-n}\pi^{-n/2}\,\Gamma(\gamma+1)/\Gamma(\gamma+1+n/2)\ . \qquad (1.11)$$

We conjecture $\gamma_{c,1} = 3/2$, $\gamma_{c,3} \cong .863$ and $\gamma_{c,n} = 0$, all $n \geq 8$. If this conjecture were to be true, the constants in (1.3, 1.4) could be further improved.

In the next section we shall deduce bounds for $\sum_j |e_j|^\gamma$ and use them in Section 3 to derive (1.3). In Section 4 we shall discuss our conjectures and support them for $n = 1$ with results from the Korteweg-de Vries equation. Section 5 contains new results added in proof. In Appendix A, generously contributed by J. F. Barnes, further evidence from computer studies is presented. We are extremely grateful to Dr. Barnes for taking an interest in this problem, for without his results we would have been hesitant to put forth our conjectures.

2. *Bounds for Moments of the Eigenvalues*

In this section we shall deduce bounds of the form (1.1), and we shall compare our $L_{\gamma,n}$ with the classical values which one gets by replacing

$$\sum_j |e_j|^\gamma \quad \text{by} \quad (2\pi)^{-n} \int d^nx\, d^np\, |p^2 + V(x)|_-^\gamma\ .$$

For $n \sim 3$ and $\gamma \sim 1$, the latter are smaller by about an order of magnitude.

Our inequalities are based on the Birman-Schwinger [2, 3] method for estimating N_E, the number of bound states of $H = -\Delta + V(x)$ having an energy $\leq E$. Since

$$\frac{\partial}{\partial E} N_E = \sum_j \delta(E - e_j)$$

we have

$$\sum_j |e_j|^\gamma = \gamma \int_0^\infty d\alpha\, \alpha^{\gamma-1} N_{-\alpha} \,. \tag{2.1}$$

Now, according to Birman-Schwinger [2, 3], for all $\alpha \geq 0$, $m \geq 1$ and $t \in [0, 1]$,

$$N_{-\alpha} \leq \operatorname{Tr}(|V + (1-t)\alpha|_-^{1/2}(-\Delta + t\alpha)^{-1}|V + (1-t)\alpha|_-^{1/2})^m \,. \tag{2.2}$$

REMARKS ABOUT (2.2):

1. We are only interested in potentials such that $V_- \in L^{\gamma+n/2}(R^n)$ for $\gamma \geq \min(0, 1 - n/2)$. For such potentials (2.2) is justified, and a complete discussion is given in Simon [4, 5]. Moreover, it is sufficient to consider $V \in C_0^\infty(R^n)$ in (2.2), and in the rest of this paper, and then to use a limiting argument. Such potentials have the advantage that they have only a finite number of bound states [5].

2. Since we are interested in maximizing $\Sigma |e_j|^\gamma / \int |V|_-^{\gamma+n/2}$, we may as well assume that $V(x) \leq 0$, i.e. $V = -|V|_-$. This follows from the max-min principle [4] which asserts that $e_j(V) \geq e_j(-|V|_-)$, all j, including multiplicity.

To evaluate the trace in (2.2), we use the inequality

$$\operatorname{Tr}(B^{1/2} A B^{1/2})^m \leq \operatorname{Tr} B^{m/2} A^m B^{m/2} \tag{2.3}$$

when A, B are positive operators and $m \geq 1$. When m is integral and A, B is of our special form, (2.3) is a consequence of Hölder's inequality. For completeness, we shall give a more general derivation of (2.3) in Appendix B.

To calculate

$$\mathrm{Tr}\,|V+(1-t)\alpha|_-^m(-\Delta+t\alpha)^{-m}\,, \tag{2.4}$$

we shall use an x-representation where $(-\Delta+t\alpha)^{-m}$ is the kernel

$$G_{t\alpha}^{(m)}(x-y) = (2\pi)^{-n}\int d^np\,(p^2+t\alpha)^{-m}\,e^{ip(x-y)} \tag{2.5}$$

if $m > n/2$. Using

$$d^np = \frac{2\pi^{n/2}}{\Gamma(n/2)}\int_0^\infty dp\;p^{n-1}\,, \tag{2.6}$$

we easily compute

$$G_{t\alpha}^{(m)}(0) = (2\pi)^{-n}\,\frac{2\pi^{n/2}}{\Gamma(n/2)}\,(t\alpha)^{-m+n/2}\int_0^\infty p^{n-1}(p^2+1)^{-m}\,dp$$

$$= (4\pi)^{-n/2}\,\frac{\Gamma(m-n/2)}{\Gamma(m)}\,(t\alpha)^{-m+n/2} \tag{2.7}$$

if $m > n/2$. Thus,

$$N_{-\alpha} \le (4\pi)^{-n/2}\,\frac{\Gamma(m-n/2)}{\Gamma(m)}\,(t\alpha)^{-m+n/2}\int d^nx\,|V(x)+(1-t)\alpha|_-^m. \tag{2.8}$$

Next, we substitute (2.8) into (2.1). If we impose the condition that $t < 1$, it is easy to prove that one can interchange the α and the x integration. Changing variables $\alpha \to (1-t)^{-1}\,|V(x)|\_\beta$, leads to

$$\sum |e_j|^\gamma \le \gamma(4\pi)^{-n/2}\,t^{-m+n/2}(1-t)^{m-\gamma-n/2}\,\frac{\Gamma(\gamma-m+n/2)\Gamma(m-n/2)}{\Gamma(\gamma+1+n/2)}\,m\int d^nx$$
$$\times\,|V(x)|_-^{\gamma+n/2} \tag{2.9}$$

provided $n/2 < m < n/2+\gamma$, $m \ge 1$ and $0 < t < 1$. The optimal t is $t = (m-n/2)/\gamma$.

If we put our results together, we obtain the following (see note added in proof, Section 5).

THEOREM 1. *Let* $V_- \in L^{\gamma+n/2}(R^n)$, $\gamma \geq \max(0, 1-n/2)$. *Let* $H = -\Delta + V(x)$, *and let* $e_j \leq 0$ *be the negative energy bound states of* H. *Then*

$$\sum |e_j|^\gamma \leq L_{\gamma,n} \int |V(x)|_-^{\gamma+n/2} \tag{2.10}$$

where

$$L_{\gamma,n} \leq \tilde{L}_{\gamma,n} \equiv \min_m (4\pi)^{-n/2}\gamma^{\gamma+1} \frac{m}{\Gamma\left(\gamma+\frac{n}{2}+1\right)} F\left(m-\frac{n}{2}\right) F\left(\gamma+\frac{n}{2}-m\right), \tag{2.11}$$

and where $F(x) = \Gamma(x)x^{-x}$, $\max\{1, n/2\} \leq m < n/2 + \gamma$.

REMARKS:

1. When $\gamma = 0$, $\Sigma|e_j|^0$ means the number of bound states, including zero energy states. For $n \geq 2$, our $\tilde{L}_{0,n} = \infty$. In Section 4, we shall discuss the $\gamma = 0$ case further. See also Section 5.

2. In (2.11), $\tilde{L}_{\gamma,n}$ is the bound we have obtained using the Birman-Schwinger principle. We shall henceforth reserve the symbol $L_{\gamma,n}$ for the quantity

$$L_{\gamma,n} \equiv \sup_V \sum |e_j|^\gamma \Big/ \int |V|_-^{\gamma+n/2} . \tag{2.12}$$

Optimization with respect to m in (2.11) can be done either numerically or analytically in the region where Stirling's formula

$$F(x) \sim e^{-x}\sqrt{2\pi/x} \tag{2.13}$$

can be applied. In [1], for $n = 3$, $\gamma = 1$, we used the value 2 for m. A marginal improvement can be obtained with $m = 1.9$.

If (2.13) were exact, the best m would be

$$\bar{m} = n(\gamma+n/2)/(n+\gamma) . \tag{2.14}$$

Note that as $\gamma \to \infty$, $\bar{m}$ is bounded by n. Using $\bar{m}$, together with (2.13), which is valid when $\gamma n(\gamma+n)^{-1}$ is large,

$$\tilde{L}_{\gamma,n} \sim (4\pi)^{1-n/2} \frac{\gamma^\gamma e^{-\gamma}}{\Gamma(\gamma+n/2)} \left[\frac{n/2}{\gamma+n/2}\right]^{1/2} . \tag{2.15}$$

Finally, we want to compare our bounds with their classical values, $L^C_{\gamma,n}$. From the results of Martin [6] and Tamura [7], one has the following

THEOREM 2. *If* $V(x) \leq 0$ *and* $V \in C_0^\infty(R^n)$, *then*

$$\lim_{\lambda\to\infty} \sum_j |e_j(\lambda V)|^\gamma \Big/ \int |\lambda V|^{\gamma+n/2} = L^C_{\gamma,n} . \tag{2.16}$$

COROLLARY.

$$L_{\gamma,n} \geq L^C_{\gamma,n} . \tag{2.17}$$

Our $\tilde{L}_{\gamma,n}$ satisfies (2.17), in particular in the asymptotic region (2.15), we find

$$\tilde{L}_{\gamma,n}/L^C_{\gamma,n} \approx [4\pi n(\gamma+n/2)]^{1/2} \gamma^{-1/2} . \tag{2.18}$$

We conjecture in Section 4 that for γ sufficiently large, the best possible $L_{\gamma,n}$ should be $L^C_{\gamma,n}$, a result which does not follow from the Birman-Schwinger method employed here. For small γ, we know that $L^C_{\gamma,n}$ is not a bound.

We conclude this section with a theorem about $L_{\gamma,n}$ which will be useful in the discussion of the one-dimensional case in Section 4.

THEOREM 3. *Let* $\gamma \geq 1 + \max(0, 1-n/2)$. *Then*

$$L_{\gamma,n} \leq L_{\gamma-1,n}[\gamma/(\gamma+n/2)] . \tag{2.19}$$

PROOF. Choose $\varepsilon > 0$. We can find a $V \in C_0^\infty(R^n)$, with $V \leq 0$, such that

$$L_{\gamma,n}(V) = \sum_j |e_j(V)|^\gamma \Big/ \int |V|^{\gamma+n/2} \geq L_{\gamma,n} - \varepsilon .$$

Let $g \in C_0^\infty(R^n)$ be such that $0 \leq g(x) \leq 1$, $\forall x$, and $V(x) \neq 0$ implies $g(x) = 1$. Let $V_\lambda(x) = V(x) - \lambda g(x)$, $\lambda \leq 0$. The functions $|e_j(V_\lambda)|$ are continuous and monotone increasing in λ. Furthermore, there are a finite number of values $-\infty < \lambda_1 \leq \lambda_2 \leq \cdots \leq \lambda_k \leq 0$ with λ_j being the value of λ at which $e_j(V_\lambda)$ first appears. λ_1 is finite because V_λ is non-negative for λ sufficiently negative. $e_j(V_\lambda)$ is continuously differentiable on $\Lambda = \{\lambda | 0 \geq \lambda > \lambda_1, \lambda \neq \lambda_i,\ i = 1, \cdots, k\}$ and

$$de_j(V_\lambda)/d\lambda = -\int |\psi_j(x; V_\lambda)|^2 g(x)\, d^n x$$

by the Feynman-Hellman theorem. It is easy to prove that if $f, g \in L^p(\mathbf{R}^n)$, $p > 1$, then

$$h(\lambda) \equiv \int |f(x) - \lambda g(x)|_-^p \, d^n x$$

is differentiable, $\forall \lambda$ and

$$dh/d\lambda|_{\lambda=0} = p \int |f(x)|_-^{p-1} g(x)\, d^n x \, .$$

Thus $L_{\gamma,n}(V_\lambda)$ is piecewise C^1 on Λ and its derivative, $\dot{L}_{\gamma,n}$, is given by

$$\dot{L}_{\gamma,n} = \left[\int |V_\lambda|_-^{\gamma+n/2}\right]^{-1} \left\{ \gamma \sum_j |e_j(V_\lambda)|^{\gamma-1} \int g(x) \psi_j(x; V_\lambda)|^2 d^n x - (\gamma+n/2) L_{\gamma,n}(V_\lambda) \right.$$

$$\left. \cdot \int |V_\lambda(x)|_-^{\gamma+n/2-1} g(x)\, d^n x \right\} .$$

By the stated properties of $L_{\gamma,n}$, there exists a $\lambda \in (\lambda_1, 0]$ such that

(i) $\dot{L}_{\gamma,n}(V_\lambda) \geq 0$;

(ii) $L_{\gamma,n}(V_\lambda) \geq L_{\gamma,n} - 2\varepsilon$.

Thus, using the properties of g,

$$0 \leq \gamma \sum_j |e_j(V_\lambda)|^{\gamma-1} - L_{\gamma,n}(V_\lambda)(\gamma + n/2) \int |V_\lambda|_-^{\gamma+n/2-1} \, . \tag{2.20}$$

Since ε was arbitrary, (2.20) implies the theorem.

If we use (2.17) together with the fact that $L^C_{\gamma,n} = L^C_{\gamma-1,n}[\gamma/(\gamma+n/2)]$, we have

COROLLARY. *If for some* $\gamma \geq \max(0, 1-n/2)$, $L_{\gamma,n} = L^C_{\gamma,n}$, *then*

$$L_{\gamma+j,n} = L^C_{\gamma+j,n}, \qquad j = 0, 1, 2, 3, \cdots .$$

REMARK. By the same proof

$$L^1_{\gamma,n} \leq L^1_{\gamma-1,n}[\gamma/(\gamma+n/2)] \tag{2.21}$$

(see (3.1) for the definition of $L^1_{\gamma,n}$).

3. *Bounds for the Kinetic Energy*

In this section, we shall use Theorem 1 to derive inequalities of the type (1.3). We recall the definition (2.14) and we further define

$$L^1_{\gamma,n} \equiv \sup_V |e_1|^\gamma \Big/ \int |V|_-^{\gamma+n/2} . \tag{3.1}$$

Clearly,

$$L^1_{\gamma,n} \leq L_{\gamma,n} . \tag{3.2}$$

If $\psi \in \mathcal{H}_{N,n,q}$ = the N-fold antisymmetric tensor product of $L^2(R^n; C^q)$, we can write ψ pointwise as $\psi(x_1, \cdots, x_N; \sigma_1, \cdots, \sigma_N)$ with $x_j \in R^n$, $\sigma_j \in \{1, 2, \cdots, q\}$ and $\psi \to -\psi$ if (x_i, σ_i) is permuted with (x_j, σ_j). $q = 2$ for spin 1/2 fermions. We can extend the definition (1.2) to

$$\rho_\sigma(x) \equiv N \sum_{\sigma_2=1}^{q} \cdots \sum_{\sigma_N=1}^{q} \int |\psi(x, x_2, \cdots, x_N; \sigma, \sigma_2, \cdots, \sigma_N)|^2 \, d^n x_2 \cdots d^n x_N . \tag{3.3}$$

We also define

$$T_\psi \equiv \sum_{j=1}^{N} \sum_{\sigma_1=1}^{q} \cdots \sum_{\sigma_N=1}^{q} \int |\nabla_j \psi(\underline{x};\underline{\sigma})|^2 \, d^{nN}\underline{x} \,, \tag{3.4}$$

$$\|\psi\|_2^2 \equiv \sum_{\sigma_1=1}^{q} \cdots \sum_{\sigma_N=1}^{q} \int |\psi(\underline{x};\underline{\sigma})|^2 \, d^{nN}\underline{x} \,. \tag{3.5}$$

Our result is

THEOREM 4. *Let* p *satisfy* $\max\{n/2, 1\} \leq p \leq 1 + n/2$ *and suppose that* $L_{p-n/2,n} < \infty$. *If* $\|\psi\|_2 = 1$, *then, except for the case* $n = 2$, $p = 1$, *there exists a positive constant* $K_{p,n}$ *such that*

$$T_\psi \geq K_{p,n} \sum_{\sigma=1}^{q} \left[\int \rho_\sigma(x)^{p/(p-1)} \, d^n x \right]^{2(p-1)/n} \tag{3.6}$$

and

$$K_{p,n} \geq \frac{1}{2} np^{-2p/n} (p-n/2)^{-1+2p/n} (L^1_{p-n/2,n}/L_{p-n/2,n})^{-1+2p/n} L^1_{p-n/2,n}{}^{-2/n} . \tag{3.7}$$

Before giving the proof of Theorem 4, we discuss its relation to the well-known Sobolev inequalities [9, 10]:

THEOREM 5 (Sobolev-Talenti-Aubin). *Let* $\nabla\psi \in L^r(R^n)$ *with* $1 < r < n$. *Let* $t = nr/(n-r)$. *Then*

$$\int |\nabla\psi|^r \geq C_{r,n} \left(\int |\psi|^t \right)^{r/t} \tag{3.8}$$

for some $C_{r,n} > 0$.

Talenti [11] and Aubin [21] have given the best possible $C_{r,n}$ (for $n = 3$, $r = 2$, $t = 6$, $C_{2,n}$ is also given in [8] and [12]):

$$C_{r,n} = n\pi^{r/2}\left(\frac{r-1}{n-r}\right)^{1-r}\left\{\frac{\Gamma(1+n-n/r)\Gamma(n/r)}{\Gamma(n)\Gamma(1+n/2)}\right\}^{r/n} . \tag{3.9}$$

Our inequality (3.6) relates only to the $r = 2$ case in (3.8), in which case $t = 2n/(n-2)$. Consider (3.8) with $r = 2$ and $\|\psi\|_2 = 1$. Using Hölder's inequality on the right side of (3.8), one gets

$$\int |\nabla\psi|^2 \geq C_{2,n}\left[\int |\psi|^{2p/(p-1)}\right]^{2(p-1)/n}\left[\int |\psi|^2\right]^{-2(p-n/2)/n} \tag{3.10}$$

whenever $n > 2$ and $p \geq n/2$. However, $C_{2,n}$ is not necessarily the best constant in (3.10) when $p \neq n/2$ ($p = n/2$ corresponds to $r = 2$ in (3.8)). Indeed, Theorem 4 says something about this question.

In the case that $N = 1$ and $q = 1$, Theorem 4 is of the same form as (3.10) (since $\rho = |\psi|^2$ and $\|\psi\|_2 = 1$). We note two things:

1. For $n > 2$ and $p = n/2$, (3.6) agrees with (3.8) except, possibly, for a different constant. We have, therefore, an alternative proof of the usual Sobolev inequality (for the $r = 2$ case). As we shall also show $K_{n/2,n} = C_{2,n}$, so we also have the best possible constant for this case.

2. If $\max\{n/2, 1\} < p \leq 1 + n/2$, Theorem 4 gives an improved version of (3.10), *even if* $n = 1$ or 2 (in which cases $C_{2,n} = 0$, but $K_{p,n} > 0$). For $p > 1 + n/2$, one can always use Hölder's inequality on the $p = 1+n/2$ result to get a nontrivial bound of the form (3.10). However, in Theorem 4, the restriction $p \leq 1 + n/2$ is really necessary. This has to do with the dependence of T_ψ on N rather than on n, as we shall explain shortly.

Next we turn to the case $N > 1$. To illustrate the nature of (3.6), we may as well suppose $q = 1$. To fix ideas, we take a special, but important form for ψ, namely

$$\psi(x_1,\cdots,x_N) = (N!)^{-1/2}\,\mathrm{Det}\,\{\phi^i(x_j)\}_{ij=1}^{N} \tag{3.11}$$

and where the ϕ^i are orthonormal functions in $L^2(R^n)$. Then, suppressing the subscript σ because $q = 1$,

$$\rho(x) = \sum_{i=1}^{N} \rho^i(x) ,$$

$$\rho^i(x) = |\phi^i(x)|^2 ,$$

$$T_\psi = \sum_{i=1}^{N} t^i ,$$

$$t^i = \int |\nabla\phi^i|^2 . \tag{3.12}$$

Theorem 4 says that

$$\sum_i t^i \geq K_{p,n}\left\{\int \left[\sum_i \rho^i(x)\right]^{p/(p-1)} d^n x\right\}^{2(p-1)/n} . \tag{3.13}$$

If we did not use the orthogonality of the ϕ^i, all we would be able to conclude, using (3.6) with $N = 1$, N times, would be

$$\sum_i t^i \geq K_{p,n} \sum_i \left[\int \rho^i(x)^{p/(p-1)} d^n x\right]^{2(p-1)/n} . \tag{3.14}$$

If $p = n/2$, then (3.14) is better than (3.13), by convexity. In the opposite case, $p = 1 + n/2$, (3.13) is superior. For in between cases, (3.13) is decidedly better if N is large and if the ρ^i are close to each other (in the $L^{p/(p-1)}(\mathbf{R}^n)$ sense). Suppose $\rho^i(x) = \rho(x)/N$, $i = 1,\cdots,N$. Then the right side of (3.13) is proportional to $N^{2p/n}$ while the right side of (3.14) grows only as N. This difference is caused by the orthogonality of the ϕ^i, or the Pauli principle.

In fact, the last remark shows why $p \leq 1 + n/2$ is important in Theorem 4. If $\rho^i = \rho/N$, all i, then the best bound, insofar as the N dependence is concerned, occurs when p is as large as possible. It is easy to see by example, however, that the largest growth for T_ψ due to the orthogonality condition can only be $N^{(n+2)/n}$.

PROOF OF THEOREM 4. Let $V(x) \leq 0$ be a potential in $\mathbf{R}^n$ with at least one bound state. If $e_1 = \min\{e_i\}$, then, for $\gamma \in [0,1]$,

$$\sum_j |e_j|^\gamma \geq |e_1|^{\gamma-1} \sum_j |e_j| \; .$$

Using the definition (2.14) and (3.2), we have that

$$\sum_j |e_j| \leq A_{\gamma,n} \left\{ \int |V|^{\gamma+n/2} \right\}^{1/\gamma} \tag{3.15}$$

$$A_{\gamma,n} = L_{\gamma,n}(L^1_{\gamma,n})^{-1+1/\gamma} \tag{3.16}$$

when $1 \geq \gamma \geq \max(0, 1-n/2)$. (3.15) holds even if V has no bound state. Let π_σ, $\sigma = 1,\cdots,q$, be the projection onto the state σ, i.e. for $\psi \in L^2(\mathbf{R}^n;\mathbf{C}^q)$, $(\pi_\nu\psi)(x,\sigma) = \psi(x,\nu)$ if $\sigma = \nu$ and zero otherwise. Choose $\gamma = p - n/2$. Let $\{\rho_\sigma\}_{\sigma=1}^q$ be given by (3.3) and, for $a_\sigma \geq 0$, $\sigma = 1,\cdots,q$, define

$$h = -\Delta - \sum_{\sigma=1}^{q} a_\sigma \rho_\sigma(x)^{1/(\gamma+n/2-1)} \pi_\sigma \tag{3.17}$$

to be an operator on $L^2(\mathbf{R}^n;\mathbf{C}^q)$ in the usual way. Define

$$H_N = \sum_{i=1}^{N} h_i \tag{3.18}$$

where h_i means h acting on the i-th component of $\mathcal{H}_{N,n,q}$. Finally, let $E = \inf \operatorname{spec} H_N$.

Now, by the Rayleigh-Ritz variational principle

$$E \leq (\psi, H_N \psi) = T_\psi - \sum_{\sigma=1}^{q} \alpha_\sigma \int \rho_\sigma^{p/(p-1)} . \qquad (3.19)$$

On the other hand, $E \geq$ the sum of all the negative eigenvalues of h

$$\geq -A_{\gamma,n} \sum_{\sigma=1}^{q} \alpha_\sigma^{p/\gamma} \left\{ \int \rho_\sigma^{p/(p-1)} \right\}^{1/\gamma} \qquad (3.20)$$

by (3.15). Combining (3.19) and (3.20) with

$$\alpha_\sigma = \left\{ \int \rho_\sigma^{p/(p-1)} \right\}^{2(\gamma-1)/n} \left\{ \frac{\gamma}{\gamma + n/2} \frac{1}{A_{\gamma,n}} \right\}^{2\gamma/n} ,$$

the theorem is proved.

Note that when $p = 1 + n/2$ (corresponding to $\gamma = 1$ in the proof), $L^1_{1,n}$ does not appear in (3.7). In this case, the right side of (3.7) is the best possible value of $K_{1+n/2,n}$, as we now show.

LEMMA 6. *From* (3.7), *define*

$$L^*_{1,n} \equiv [n/(2K_{1+n/2,n})]^{n/2} (1 + n/2)^{-1-n/2} .$$

Then $L^*_{1,n} = L_{1,n}$.

PROOF. By (3.7), we only have to prove that $L^*_{1,n} \geq L_{1,n}$. Let $V \leq 0$, $V \in C_0^\infty(R^n)$ and let $H = -\Delta + V$. Let $\{\phi_i, e_i\}_{i=1}^N$ be the bound state eigenfunctions and eigenvalues of H. Let ψ and ρ^i be as defined in (3.11), (3.12). Then

$$\sum_i |e_i| = -\int V\rho - T_\psi \leq \|V\|_p \|\rho\|_{p/(p-1)} - T_\psi$$

with $p = 1 + n/2$. Using Theorem 4 for T_ψ, one has that

$$\sum_i |e_i| \leq \max_{y>0} \{\|V\|_p y - K_{1+n/2,n} y^{2p/n}\} = L^*_{1,n} \|V\|_p^p .$$

We conclude with an evaluation of $K_{n/2,n}$ for $n > 2$ as promised. By a simple limiting argument

$$K_{n/2,n} \geq \lim_{p \downarrow n/2} \text{(right side of (3.7))} . \tag{3.21}$$

Our bound (2.11) on $L_{p-n/2,n}$ shows that

$$\lim_{p \downarrow n/2} (L_{p-n/2,n})^{-1+2p/n} = 1 . \tag{3.22}$$

Hence

$$K_{n/2,n} \geq (L^1_{0,n})^{-2/n} . \tag{3.23}$$

On the other hand, by the method of Lemma 6 applied to the $N = 1$ case, $K_{n/2,n} \leq (L^1_{0,n})^{-2/n}$. The value of $L^1_{0,n}$ is given in (4.24). To be honest, its evaluation requires the solution of the same variational problem as given in [8, 11, 12]. Substitution of (4.24) into (3.23) yields the required result

$$K_{n/2,n} = C_{2,n} = \pi n(n-2)[\Gamma(n/2)/\Gamma(n)]^{2/n} . \tag{3.24}$$

If we examine (3.23) when $n = 2$, one gets $K_{1,2} \geq 0$ since $L^1_{0,2} = \infty$. This reflects the known fact [5] that an arbitrarily small $V < 0$ always has a bound state in two dimensions. This observation can be used to show that

$$K_{1,2} = 0 . \tag{3.25}$$

When $n = 1$, the smallest allowed p is $p = 1$. In this case, (3.6) reads

$$T_\psi \geq K_{1,1} \sum_{\sigma=1}^{q} \|\rho_\sigma\|_\infty^2 . \tag{3.26}$$

Using (3.7) and (4.20),

$$K_{1,1} \geq [2L_{1/2,1}]^{-1} . \tag{3.27}$$

If one accepts the conjecture of Section 4 that $L_{1/2,1} = L^1_{1/2,1} = 1/2$, then

$$K_{1,1} = 1 . \tag{3.28}$$

The reason for the equality in (3.28) is that $K_{1,1} = 1$ is well known to be the best possible constant in (3.26) when $q = 1$ and $N = 1$.

4. *Conjecture About* $L_{\gamma,n}$

We have shown that for the bound state energies $\{e_j\}$ of a potential V in n dimensions and with

$$L_{\gamma,n}(V) \equiv \sum_j |e_j|^\gamma \Big/ \int |V|^{\gamma+n/2} , \tag{4.1}$$

then

$$L_{\gamma,n} \equiv \sup_{V \in L^{\gamma+n/2}} L_{\gamma,n}(V) \tag{4.2}$$

is finite whenever $\gamma + n/2 > 1$ and $\gamma > 0$. The "boundary points" are

$$\begin{aligned} \gamma &= 1/2 & n &= 1 \\ \gamma &= 0 & n &\geq 2 . \end{aligned} \tag{4.3}$$

We showed that for $n = 1$, $L_{1/2,1} < \infty$. For $\gamma < 1/2$, $n = 1$, there cannot be a bound of this kind, for consider $V_L(x) \equiv -1/L$ for $|x| < L$ and zero otherwise. For $L \to 0$, this converges towards $-2\delta(x)$ and thus has a bound state of finite energy (which is -1 for $-2\delta(x)$). On the other hand,

$$\lim_{L \to 0} \int dx\, |V_L|^{1/2+\gamma} = 0 \quad \text{for} \quad \gamma < 1/2 .$$

For $n = 2$, $\gamma = 0$ is a "double boundary point" and $L_{0,2} = \infty$, i.e. there is no upper bound on the number of bound states in two dimensions. (Cf. [5].)

For $n \geq 3$, $L_{0,n}$ is *conjectured* to be finite (see note added in proof, Section 5); for $n = 3$, this is the well-known $\int |V|_-^{3/2}$ conjecture on the number, $N_0(V)$, of bound states (cf. [5]). The best that is known at present is that

$$N_0(V) \leq c \left[\int |V|_-^{3/2} \right]^{4/3}, \tag{4.4}$$

but for spherically symmetric V, a stronger result is known [8]:

$$N_0(V) \leq I\left(1 + \frac{1}{4} \ln I\right),$$

$$I = 4(3\pi^2\, 3^{1/2})^{-1} \int |V|_-^{3/2} . \tag{4.5}$$

In (1.4) and (3.1), we introduced L^C and L^1 and showed that

$$L_{\gamma,n} \geq \max(L^1_{\gamma,n}, L^C_{\gamma,n}) . \tag{4.6}$$

A parallel result is Simon's [22] for $n \geq 3$:

$$N_0(V) \leq D_{n,\varepsilon}(\|V_-\|_{\varepsilon+n/2} + \|V_-\|_{-\varepsilon+n/2})^{n/2}$$

with $D_{n,\varepsilon} \to \infty$ as $\varepsilon \to 0$.

In our previous paper [4], we conjectured that $L_{1,3} = L^C_{1,3}$, and we also pointed out that $L^1_{1,1} > L^C_{1,1}$. A remark of Peter Lax (private communication), which will be explained presently, led us to the following:

CONJECTURE. *For each* n, *there is a critical value of* $\gamma, \gamma_{c,n}$, *such that*

$$L_{\gamma,n} = L^C_{\gamma,n} \qquad \gamma \geq \gamma_c$$

$$L_{\gamma,n} = L^1_{\gamma,n} \qquad \gamma \leq \gamma_c$$

γ_c is defined to be that γ for which $L^C_{\gamma,n} = L^1_{\gamma,n}$; the uniqueness of this γ_c is part of the conjecture. Furthermore, $\gamma_{c,1} = 3/2$, $\gamma_{c,2} \sim 1.2$, $\gamma_{c,3} \sim .86$ and the smallest n such that $\gamma_{c,n} = 0$ is $n = 8$.

(A) Remarks on $L^1_{\gamma,n}$

We want to maximize

$$\left|\int [|\psi|^2 V + |\nabla\psi|^2]\, d^n x\right|^{\gamma} \Big/ \int |V|^{\gamma+n/2} \tag{4.7}$$

with respect to V, and where $\int |\psi|^2 = 1$ and $(-\Delta + V)\psi = e_1\psi$. By the variational principle, we can first maximize (4.7) with respect to V, holding ψ fixed. Hölder's inequality immediately yields

$$V(x) = -\sigma |\psi(x)|^{2/(\gamma+n/2-1)}$$

with $\sigma > 0$. The kinetic energy, $\int |\nabla\psi|^2$, is not increased if $\psi(x)$ is replaced by $|\psi(x)|$ and, by the rearrangement inequality [13], this is not increased if $|\psi|$ is replaced by its symmetric decreasing rearrangement. Thus, we may assume that $|V|$ and $|\psi|$ are spherically symmetric, non-increasing functions.

By the methods of [8] or [11], (4.7) can be shown to have a maximum when $\gamma + n/2 > 1$. The variational equation is

$$-\Delta\psi(x) - \sigma\psi(x)^{(\gamma+n/2+1)/(\gamma+n/2-1)} = e_1\psi(x) \tag{4.8}$$

with

$$\sigma = \left\{\frac{\gamma |e_1|^{\gamma-1}}{(\gamma+n/2)\, L^1_{\gamma,n}}\right\}^{1/(\gamma+n/2-1)} . \tag{4.9}$$

Equation (4.8) determines ψ up to a constant and up to a change of scale in x. The former can be used to make $\int \psi^2 = 1$ and the latter leaves (4.7) invariant.

Equation (4.8) can be solved analytically in two cases, to which we shall return later:

(i) $n = 1$, all $\gamma > 1/2$

(ii) $n \geq 3$, $\gamma = 0$.

(B) The One-Dimensional Case

Lax's remark was about a result of Gardner, Greene, Kruskal and Miura [14] to the effect that

$$L_{3/2,1} = L^C_{3/2,1} = 3/16 \, . \tag{4.10}$$

To see this, we may assume $V \in C_0^\infty(R)$, and use the theory of the Korteweg-de Vries (KdV) equation [14]:

$$W_t = 6WW_x - W_{xxx} \, . \tag{4.11}$$

There are two remarkable properties of (4.11):

(i) As W evolves in time, t, the eigenvalues of $-d^2/dx^2 + W$ remain invariant.

(ii) $\int W^2 \, dx$ is constant in time.

Let $W(x,t)$ be given by (4.11) with the initial data

$$W(x,0) = V(x) \, .$$

Then $L_{3/2,1}(W(\cdot\,,t))$ is independent of t, and may therefore be evaluated by studying its behavior as $t \to \infty$.

There exist traveling wave solutions to (4.11), called solitons, of the form

$$W(x,t) = f(x-ct) \, .$$

Equation (4.11) becomes

$$-cf_x = -f_{xxx} + 6ff_x \, . \tag{4.12}$$

The solutions to (4.12) which vanish at ∞ are

$$f_a(x) = -2a^2 \cosh^{-2}(ax)$$

$$c = 4a^2 \, . \tag{4.13}$$

Any solution (4.13), regarded as a potential in the Schrödinger equation has, as we shall see shortly, exactly one negative energy bound state with energy and wave function

$$e = -a^2$$

$$\psi(x) = \cosh^{-1}(ax) \ . \tag{4.14}$$

Now the theory of the KdV equation says that as $t \to \infty$, W evolves into a sum of solitons (4.13) plus a part that goes to zero in $L^\infty(\mathbf{R})$ norm (but not necessarily in $L^2(\mathbf{R})$ norm). The solitons are well separated since they have different velocities. Because the number of bound states is finite, the non-soliton part of W can be ignored as $t \to \infty$. Hence, for the initial V,

$$\sum |e_j|^{3/2} = \sum_{\text{solitons}} a^3 \tag{4.15}$$

while

$$\int V(x)^2\, dx \geq \sum_{\text{solitons}} \int f_a(x)^2\, dx \ . \tag{4.16}$$

Since $4\int_{-\infty}^{\infty} \cosh^{-4}(x)\, dx = 16/3$, we conclude that

$$L_{3/2,1} = L^C_{3/2,1} = 3/16 \tag{4.17}$$

with equality if and only if $W(x,t)$ is composed purely of solitons as $t \to \infty$. For the same reason,

$$L^1_{3/2,1} = L^C_{3/2,1} \tag{4.18}$$

(cf. (4.21)).

Not only do we have an evaluation of $L_{3/2,1}$, (4.17), but we learn something more. When $\gamma = 3/2$, there is an infinite family of potentials for which $L_{3/2,1}(V) = L_{3/2,1}$, and these may have any number of bound states = number of solitons.

What we believe to be the case is that when $\gamma < 3/2$, the optimizing potential for $L_{\gamma,n}$ has only one bound state, and satisfies (4.8). When

$\gamma > 3/2$, the optimizing potential is, loosely speaking, infinitely deep and has infinitely many bound states; thus $L_{\gamma,n} = L^C_{\gamma,n}$.

An additional indication that the conjecture is correct is furnished by the solution to (4.8). When $\gamma = 3/2$, this agrees with (4.14). In general, one finds that, apart from scaling, the nodeless solution to (4.8) is

$$\psi_\gamma(x) = \Gamma(\gamma)^{1/2}\pi^{-1/4}\Gamma(\gamma-1/2)^{-1/2}\cosh^{-\gamma+1/2}(x)$$

$$V_\gamma(x) = -(\gamma^2-1/4)\cosh^{-2}(x)$$

$$e_1 = -(\gamma-1/2)^2 \ . \qquad (4.19)$$

Thus,

$$L^1_{\gamma,1} = \pi^{-1/2}\frac{1}{\gamma-1/2}\frac{\Gamma(\gamma+1)}{\Gamma(\gamma+1/2)}\left(\frac{\gamma-1/2}{\gamma+1/2}\right)^{\gamma+1/2} \ . \qquad (4.20)$$

When $L^1_{\gamma,1}$ is compared with $L^C_{\gamma,1}$, one finds that

$$L^1_{\gamma,1} \geq L^C_{\gamma,1} \qquad \gamma \leq 3/2$$

$$L^1_{\gamma,1} \leq L^C_{\gamma,1} \qquad \gamma \geq 3/2 \ . \qquad (4.21)$$

This confirms at least part of the conjecture.

However, more is true. For $\gamma = 3/2$, V_γ has a zero energy single node bound state

$$\phi(x) = \tanh(x) \ .$$

Since V_γ is monotone in γ, it follows that V_γ has only one bound state for $\gamma < 3/2$ and at least two bound states for $\gamma > 3/2$. The (unnormalized) second bound state can be computed to be

$$\phi(x) = \sinh(x)\cosh^{-\gamma+1/2}(x)$$

$$e_2 = -(\gamma-3/2)^2 \ . \qquad (4.22)$$

In like manner, one can find more bound states as γ increases even further.

Thus we see that the potential that optimizes the ratio $|e_1|^{\gamma}/\int|V|^{\gamma+1/2}$ automatically has a second bound state when $\gamma > \gamma_c$.

Finally, we remark that Theorem 3, together with (4.10), shows that

THEOREM 7.

$$L_{\gamma,1} = L^{C}_{\gamma,1} \quad \textit{for} \quad \gamma = 3/2,\ 5/2,\ 7/2,\ \textit{etc.}$$

An application of Theorem 7 to scattering theory will be made in Section 4(D).

(C) Higher Dimensions

We have exhibited the solution to the variational equation (4.8) for $L^1_{\gamma,1}$.

When $n > 2$ and $\gamma = 0$, we clearly want to take $e_1 = 0$ in order to maximize $L^1_{0,n}(V)$. (4.8) has the zero energy solution

$$\phi(x) = (1+|x|^2)^{1-n/2}$$

$$V(x) = a\phi(x)^{2/(\gamma+n/2-1)} = n(n-2)(1+|x|^2)^{(2-n)/(n/2-1)} \qquad (4.23)$$

(note: $\phi \in L^2(R^n)$ if and only if $n > 4$, but $V \in L^{n/2}(R^n)$ always). This leads to

$$L^1_{0,n} = [\pi n(n-2)]^{-n/2}\,\Gamma(n)/\Gamma(n/2)\ . \qquad (4.24)$$

The smallest dimension for which $L^1_{0,n} \leq L^C_{0,n}$ is $n = 8$.

If we suppose that the ratio $L^1_{\gamma,n}/L^C_{\gamma,n}$ is monotone decreasing in γ (as it is when $n = 1$ and as it is when $n = 3$ on the basis of the numerical solution of (4.10) by J. F. Barnes, given in Appendix A), and if our conjecture is correct, then $L_{\gamma,n} = L^C_{\gamma,n}$ for $n \geq 8$. The value of γ_c obtained numerically is

$$\gamma_c = 1.165 \qquad n = 2$$

$$\gamma_c = .863 \qquad n = 3\ . \qquad (4.26)$$

The other bit of evidence, apart from the monotonicity of $L^1_{\gamma,n}/L^C_{\gamma,n}$, for the correctness of our conjecture is a numerical study of the energy levels of the potential

$$V_\lambda(x) = \lambda e^{-|x|}\,, \qquad \lambda > 0\,,$$

in three dimensions. This is given in Appendix A. The energy levels of the square well potential are given in [15, 16]. In both cases, one finds that

$$\lim_{\lambda\to\infty} L_{1,3}(V_\lambda) = L^C_{1,3}$$

and the limit is approached from below. Unfortunately, it is not true, as one might have hoped, that $L_{1,3}(V_\lambda)$ is monotone increasing in λ.

(D) Bounds on One-Dimensional Scattering Cross-Sections

In their study of the KdV equation, (4.11), Zakharov and Fadeev [17] showed how to relate the solution $W(x,t)$ to the scattering reflection coefficient $R(k)$ and the bound state eigenvalues $\{e_j\}$ of the initial potential $V(x)$. There are infinitely many invariants of (4.11) besides $\int W^2$ and these have simple expressions in terms of $R(k)$, $\{e_j\}$.

Thus, for any potential V,

$$\int V^2 = (16/3)\sum |e_j|^{3/2} - 4\int_{-\infty}^{\infty} k^2\, T(k)\,dk \tag{4.27}$$

$$\int V^3 + \frac{1}{2} V_x^2 = -(32/5)\sum |e_j|^{5/2} - 8\int_{-\infty}^{\infty} k^4\, T(k)\,dk \tag{4.28}$$

$$\int V^4 + 2VV_x^2 + \frac{1}{5} V_{xx}^2 = (256/35)\sum |e_j|^{7/2} - (64/5)\int_{-\infty}^{\infty} k^6\, T(k)\,dk \tag{4.29}$$

where

$$T(k) = \pi^{-1}\ln(1-|R(k)|^2) \le 0\,. \tag{4.30}$$

These are only the first three invariants; a recursion relation for the others can be found in [17].

Notice that 3/16, 5/32, 35/256 are, respectively $L^C_{3/2,1}$, $L^C_{5/2,1}$, $L^C_{7/2,1}$. Since $\int V^2 \geq \int |V_-|^2$, (4.27) establishes that $L_{3/2,1} = L^C_{3/2,1}$, as mentioned earlier. For the higher invariants, the signs in (4.28) and (4.29) are not as fortunately disposed and we cannot use these equations to prove Theorem 7. But, given that Theorem 7 has already been proved, we can conclude that

THEOREM 8. *For any nonpositive potential* V(x),

$$\int V_x^2 \geq -16 \int_{-\infty}^{\infty} k^4\, T(k)\, dk\,. \tag{4.31}$$

For any potential V(x),

$$2\int VV_x^2 + (1/5)\int V_{xx}^2 \leq -(64/5)\int_{-\infty}^{\infty} k^6\, T(k)\, dk\,. \tag{4.32}$$

The first inequality, (4.31), is especially transparent: If V(x) is very smooth, it cannot scatter very much.

5. *Note Added in Proof*

After this paper was written, M. Cwikel and Lieb, simultaneously and by completely different methods, showed that the number of bound states, $N_0(V)$ for a potential, V, can be bounded (when $n \geq 3$) by

$$N_0(V) \leq A_n \int |V(x)|_-^{n/2}\, d^n x\,. \tag{5.1}$$

Cwikel exploits the weak trace ideal method of Simon [22]; his method is more general than Lieb's, but for the particular problem at hand, (5.1), his

A_n does not seem to be as good. Lieb's method uses Wiener integrals and the general result is the following:

$$N_{-\alpha}(V) \leq \int d^n x \int_0^\infty dt\, t^{-1} e^{-\alpha t} (4\pi t)^{-n/2} f(t|V(x)|_-) \tag{5.2}$$

for any non-negative, convex function $f:[0,\infty) \to [0,\infty)$ satisfying

$$1 = \int_0^\infty t^{-1} f(t) e^{-t} dt \,. \tag{5.3}$$

For $\alpha = 0$, one can choose $f(t) = c(t-b)$, $t \geq b$, $f(t) = 0$, $t \leq b$. This leads to (5.1), and optimizing with respect to b, one finds that

$$A_3 = 0.116\,, \quad A_4 = 0.0191 \tag{5.4}$$

and, as $n \to \infty$,

$$A_n/L^C_{0,n} = (n\pi)^{1/2} + O(n^{-1/2}) \,. \tag{5.5}$$

Note that $A_3/L^1_{0,3} = 1.49$; i.e. A_3 exceeds $L_{0,3}$ by at most 49%.

Since $N_{-\alpha}(V) \leq N_0(-|V+\alpha|_-)$, one can use (5.1) and (2.1) to deduce that for $\gamma \geq 0$ and $n \geq 3$,

$$L_{\gamma,n} \leq L^C_{\gamma,n}(A_n/L^C_{0,n}) \,. \tag{5.6}$$

This is better than (2.11), (2.18). In particular, for $n = 3$, $\gamma = 1$, the improvement of (5.6) over (2.11) with $m = 2$ is a factor of 1.83. The factor 1.31 in Equation (1.4) can therefore be replaced by $1.31\ (1.83)^{-2/3} = 0.87$.

APPENDIX A. NUMERICAL STUDIES

John F. Barnes
Theoretical Division
Los Alamos Scientific Laboratory
Los Alamos, New Mexico 87545

I. *Evaluation of* $L^1_{\gamma,n}$, $n = 1, 2, 3$

The figure shows the numerical evaluation of $L^1_{\gamma,n}$ as well as $L^C_{\gamma,n}$. The latter is given in (1.11)

$$L^C_{\gamma,n} = 2^{-n}\pi^{-n/2}\,\Gamma(\gamma+1)/\Gamma(\gamma+1+n/2)\ .$$

The former is obtained by solving the differential equation (4.8) in polar coordinates and choosing σ such that $\psi(x) \to 0$ as $|x| \to \infty$. Note that by scaling, one can take $e_1 = -1$, whence

$$(L^1_{\gamma,n})^{-1} = \sigma^{(\gamma+n/2)} \int |\psi(x)|^{(2\gamma+n)/(\gamma-1+n/2)}\, d^n x\ .$$

In one dimension, $L^1_{\gamma,1}$ is known analytically and is given in (4.20). Another exact result, (4.24), is

$$L^1_{0,3} = 4\pi^{-2}\,3^{-3/2} = 0.077997\ .$$

The critical values of γ, at which $L^1_{\gamma,n} = L^C_{\gamma,n}$ are:

$$\gamma_{c,1} = 3/2$$

$$\gamma_{c,2} = 1.165$$

$$\gamma_{c,3} = 0.8627\ .$$

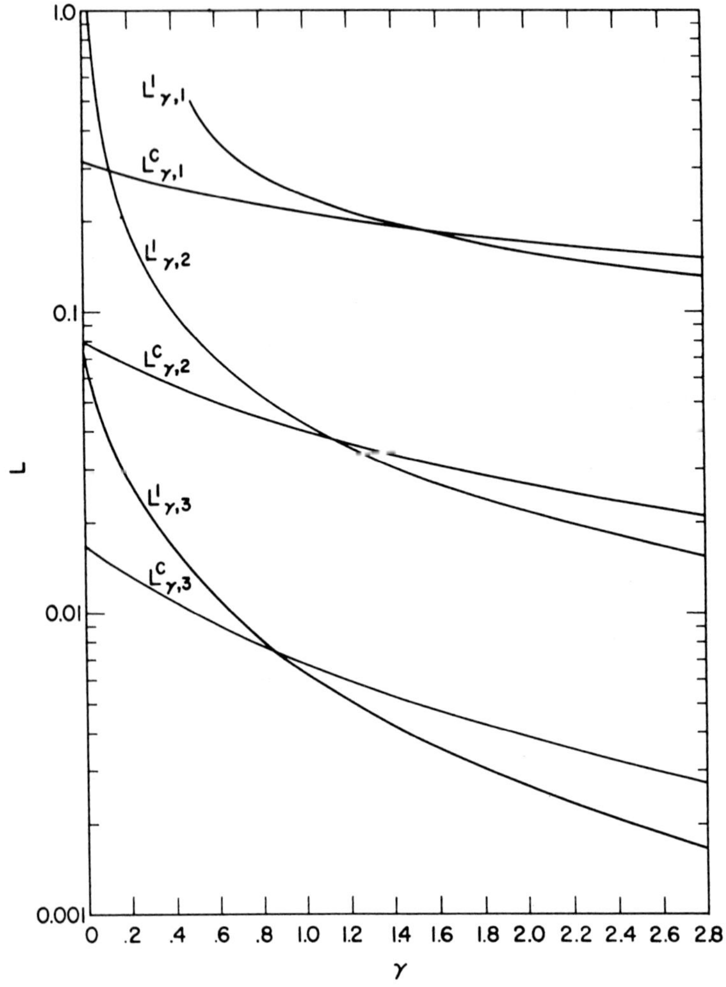
1.0
0.1
0.01
0.001
L
$L^{I}_{\gamma,1}$
$L^{C}_{\gamma,1}$
$L^{I}_{\gamma,2}$
$L^{C}_{\gamma,2}$
$L^{I}_{\gamma,3}$
$L^{C}_{\gamma,3}$
0 .2 .4 .6 .8 1.0 1.2 1.4 1.6 1.8 2.0 2.2 2.4 2.6 2.8
γ

II. *The Exponential Potential*

To test the conjecture that $L_{1,3} = L^C_{1,3}$, the eigenvalues of the potential $V_\lambda = -\lambda \exp(-|x|)$ in three dimensions were evaluated for $\lambda = 5, 10, 20, 30, 40, 50,$ and 100. These are listed in the table according to angular momentum and radial nodes. These numbers have been corroborated by H. Grosse, and they can be used to calculate $L_{\gamma,3}(V_\lambda)$ for any γ. The final column gives $L_{1,3}(V_\lambda)$, since $\int |V_\lambda|^{5/2} = \lambda^{5/2}(64\pi)/125$. It is to be noted that the classical value $L^C_{1,3} = 0.006755$, is approached from below, in agreement with the conjecture, but *not monotonically*.

$$V_\lambda = -\lambda e^{-r}$$

	ℓ	$\lvert e\rvert$	nodes	states	$\sum \lvert e\rvert$	$\dfrac{\sum \lvert e\rvert}{\lambda^{5/2}\,\frac{64\pi}{125}}$
$\lambda = 5$	0	0.55032	0	$\frac{1}{1}$	$\frac{0.55032}{0.55032}$	0.006120
$\lambda = 10$	0	0.06963	1			
		2.18241	0	2	2.2520	
	1	0.33405	0	$\frac{3}{5}$	$\frac{1.0022}{3.2542}$	0.006398
$\lambda = 20$	0	0.00869	2			
		1.42562	1			
		6.62410	0	3	8.0584	
	1	0.16327	1			
		2.71482	0	6	8.6342	
	2	0.43136	0	$\frac{5}{14}$	$\frac{2.1568}{18.8494}$	0.006551

$V_\lambda = -\lambda e^{-r}$ (continued)

	ℓ	$\|e\|$	nodes	states	$\sum \|e\|$	$\frac{\sum \|e\|}{\lambda^{5/2} \frac{64\pi}{125}}$
$\lambda = 30$	0	0.58894	2			
		3.83072	1			
		11.84999	0	3	16.270	
	1	1.39458	1			
		6.12302	0	6	22.553	
	2	0.00593	1			
		2.36912	0	10	11.875	
	3	0.07595	0	7	0.532	0.006461
				26	51.230	
$\lambda = 40$	0	0.07676	3			
		1.86961	2			
		6.88198	1			
		17.53345	0	4	26.362	
	1	0.41991	2			
		3.35027	1			
		10.13596	0	9	41.718	
	2	0.93459	1			
		5.03378	0	10	29.842	
	3	1.54738	0	7	10.832	0.006682
				30	108.754	

$V_\lambda = -\lambda e^{-r}$ (continued)

	ℓ	$\lvert e\rvert$	nodes	states	$\sum \lvert e\rvert$	$\frac{\sum \lvert e\rvert}{\lambda^{5/2}\frac{64\pi}{125}}$
$\lambda = 50$	0	0.60190	3			
		3.66447	2			
		10.39110	1			
		23.53215	0	4	38.190	
	1	1.43321	2			
		5.81695	1			
		14.56904	0	9	65.458	
	2	0.07675	2			
		2.45887	1			
		8.19840	0	15	53.670	
	3	0.26483	1			
		3.61626	0	14	27.168	
	4	0.49009	0	9	4.411	
				51	188.897	0.006643
$\lambda = 100$	0	0.39275	5			
		2.91408	4			
		8.29231	3			
		17.44909	2			
		32.07168	1			

$V_\lambda = -\lambda e^{-r}$ (continued)

ℓ	$\|e\|$	nodes	states	$\sum \|e\|$	$\frac{\sum \|e\|}{\lambda^{5/2} \frac{64\pi}{125}}$
	56.28824	0	6	117.41	
1	1.10170	4			
	4.76748	3			
	11.62740	2			
	22.79910	1			
	40.45495	0	15	242.25	
2	0.02748	4			
	2.04022	3			
	6.85633	2			
	15.22147	1			
	28.46495	0	25	263.05	
3	0.22692	3			
	3.14743	2			
	9.13429	1			
	19.04073	0	28	220.85	
4	0.52962	2			
	4.37856	1			
	11.56470	0	27	148.26	

$V_\lambda = -\lambda e^{-r}$ (continued)

ℓ	$\lvert e\rvert$	nodes	states	$\sum \lvert e\rvert$	$\dfrac{\sum \lvert e\rvert}{\lambda^{5/2}\,\frac{64\pi}{125}}$
5	0.88997	1			
	5.69707	0	22	72.46	
6	1.26789	0	13 / 136	16.48 / 1080.76	0.006719

APPENDIX B: PROOF OF (2.3)

THEOREM 9. *Let $\mathcal{H}$ be a separable Hilbert space and let* A, B *be positive operators on $\mathcal{H}$. Then, for* $m \geq 1$,

$$\operatorname{Tr}(B^{1/2} A B^{1/2})^m \leq \operatorname{Tr} B^{m/2} A^m B^{m/2} . \qquad \text{(B.1)}$$

REMARK. When $\mathcal{H} = L^2(R^n)$ and A is a kernel $a(x-y)$ and B is a multiplication operator $b(x)$ (as in our usage (2.2)), Seiler and Simon [19] have given a proof of (B.1) using interpolation techniques. Simon (private communication) has extended this method to the general case. Our proof is different and shows a little more than just (B.1).

PROOF. For simplicity, we shall only give the proof when A and B are matrices; for the general case, one can appeal to a limiting argument. For $m = 1$, the theorem is trivial, so assume $m > 1$. Let $C = A^m$ and $f(C) \equiv g(C) - h(C)$, where $g(C) \equiv \operatorname{Tr}(B^{1/2} C^{1/m} B^{1/2})^m$ and $h(C) = \operatorname{Tr} B^{m/2} C B^{m/2}$. Let M^+ be the positive matrices. Clearly $M^+ \ni C \to h(C)$ is linear. Epstein [18] has shown that $M^+ \ni C \to g(C)$ is concave (actually, he showed this for m integral, but his proof is valid generally for $m \geq 1$). Write $C = C^D + C^0$ where C^D is the diagonal part of C in a basis in which B is diagonal. $C_\lambda \equiv C^D + \lambda C^0 = \lambda C + (1-\lambda) C^D$ is in M^+ for $\lambda \in [0,1]$, because $C^D \in M^+$. Then $\lambda \to f(C_\lambda) \equiv R(\lambda)$ is concave on $[0,1]$. Our goal is to show that $R(1) \leq 0$. Since $[C^D, B] = 0$, $R(0) = 0$ and, by concavity, it is sufficient to show that $R(\lambda) \leq 0$ for $\lambda > 0$ and λ small. $h(C_\lambda) = h(C^D)$ for $\lambda \in [0,1]$. Since $f(C)$ is continuous in C, we can assume that C^D is nondegenerate and strictly positive, and that C_λ is positive when $\lambda \geq -\varepsilon$ for some $\varepsilon > 0$. Then $R(\lambda)$ is defined and concave on $[-\varepsilon, 1]$. $\lambda \to C_\lambda^{1/m}$ is differentiable at $\lambda = 0$ and its derivative at $\lambda = 0$ has zero diagonal elements. (To see this, use the representation $C^{1/m} = K\int_0^\infty dx\, x^{-1+1/m} \cdot C(C+xI)^{-1}$.) Likewise, the derivative of $(B^{1/2}(D + \lambda O)B^{1/2})^m$ at $\lambda = 0$ has zero diagonal elements when O has and when D is diagonal. Thus

$$dR(\lambda)/d\lambda|_{\lambda=0} = 0 \,.$$

Acknowledgment

One of the authors (Walter Thirring) would like to thank the Department of Physics of the University of Princeton for its hospitality.

ELLIOTT H. LIEB
DEPARTMENTS OF MATHEMATICS AND PHYSICS
PRINCETON UNIVERSITY
PRINCETON, NEW JERSEY

WALTER E. THIRRING
INSTITUT FÜR THEORETISCHE PHYSIK
DER UNIVERSITÄT WIEN, AUSTRIA

REFERENCES

[1] E. H. Lieb and W. E. Thirring, Phys. Rev. Lett. *35*, 687 (1975). See Phys. Rev. Lett. *35*, 1116 (1975) for errata.

[2] M. S. Birman, Mat. Sb. *55 (97)*, 125 (1961); Amer. Math. Soc. Translations Ser. 2, *53*, 23 (1966).

[3] J. Schwinger, Proc. Nat. Acad. Sci. *47*, 122 (1961).

[4] B. Simon, "Quantum Mechanics for Hamiltonians Defined as Quadratic Forms," Princeton University Press, 1971.

[5] B. Simon, "On the Number of Bound States of the Two Body Schrödinger Equation – A Review," in this volume.

[6] A. Martin, Helv. Phys. Acta *45*, 140 (1972).

[7] H. Tamura, Proc. Japan Acad. *50*, 19 (1974).

[8] V. Glaser, A. Martin, H. Grosse and W. Thirring, "A Family of Optimal Conditions for the Absence of Bound States in a Potential," in this volume.

[9] S. L. Sobolev, Mat. Sb. *46*, 471 (1938), in Russian.

[10] ________, Applications of Functional Analysis in Mathematical Physics, Leningrad (1950), Amer. Math. Soc. Transl. of Monographs, 7 (1963).

[11] G. Talenti, Best Constant in Sobolev's Inequality, Istituto Matematico, Universitá Degli Studi Di Firenze, preprint (1975).

[12] G. Rosen, SIAM Jour. Appl. Math. *21*, 30 (1971).

[13] H. J. Brascamp, E. H. Lieb and J. M. Luttinger, Jour. Funct. Anal. *17*, 227 (1974).

[14] C. S. Gardner, J. M. Greene, M. D. Kruskal and R. M. Miura, Commun. Pure and Appl. Math. *27*, 97 (1974).

[15] S. A. Moszkowski, Phys. Rev. *89*, 474 (1953).

[16] A. E. Green and K. Lee, Phys. Rev. *99*, 772 (1955).

[17] V. E. Zakharov and L. D. Fadeev, Funkts. Anal. i Ego Pril. *5*, 18 (1971). English translation: Funct. Anal. and its Appl. *5*, 280 (1971).

[18] H. Epstein, Commun. Math. Phys. *31*, 317 (1973).

[19] E. Seiler and B. Simon, "Bounds in the Yukawa Quantum Field Theory," Princeton preprint (1975).

[20] W. Thirring, T7 Quantenmechanik, Lecture Notes, Institut für Theoretische Physik, University of Vienna.

[21] T. Aubin, C. R. Acad. Sc. Paris *280*, 279 (1975). The results are stated here without proof; there appears to be a misprint in the expression for $C_{r,n}$.

[22] B. Simon, "Weak Trace Ideals and the Bound States of Schrödinger Operators," Princeton preprint (1975).

ON SEMI-CLASSICAL BOUNDS FOR EIGENVALUES OF SCHRÖDINGER OPERATORS☆

Michael AIZENMAN
Department of Physics, Princeton University, Princeton, NJ 08540, USA

and

Elliott H. LIEB
Departments of Mathematics and Physics, Princeton University Princeton, NJ 08540, USA

Received 27 April 1978

Our principal result is that if the semiclassical estimate is a bound for some moment of the negative eigenvalues (as is known in some cases in one-dimension), then the semiclassical estimates are also bounds for all higher moments.

Bounds on the moments of energy levels of Schrödinger operators have been the object of several studies [1, 2, 5–8]. In [1] such bounds were used to obtain a lower bound for the kinetic energy of fermions in terms of their one particle density and thereby prove the stability of matter.

In the notation of [2],

$$\sum_j |e_j(V)|_-^\gamma \leqslant L_{\gamma,n} \int \mathrm{d}^n x\, |V(x)|_-^{\gamma+n/2} \tag{1}$$

where $e_j(V)$ are the eigenvalues of $-\Delta + V(x)$ defined in $L^2(\mathbf{R}^n)$ and $|y|_- \equiv \max(-y, 0)$. $L_{\gamma,n}$ denotes the smallest number for which (1) holds independently of V. The case $\gamma = 1$ is the one needed for the kinetic energy bound. It was shown in [2] that $L_{\gamma,n} < \infty$ for $\gamma > \max(0, 1 - n/2)$. Eq. (1) also holds for $\gamma = 0$, $n \geqslant 3$ but the proof is quite different (see refs. [5, 6, 8]).

For $\gamma \geqslant 0$ we use the notation

$$H(x,p) = p^2 + V(x)$$

$$R_n(\gamma, V) = \sum_j |e_j(V)|_-^\gamma \Big/ \int \mathrm{d}^n z\, |H(x,p)|_-^\gamma \tag{2}$$

$$R_n(\gamma) = \sup_V \{R_n(\gamma, V)\}, \tag{3}$$

and $\mathrm{d}^n z = \mathrm{d}^n x\, \mathrm{d}^n p\, (2\pi)^{-n}$. $R_n(\gamma, V)$ is the ratio of the moments of the binding energies of a quantum mechanical hamiltonian to the moments of its classical analog. The integral in (1) comes from doing the $\mathrm{d}^n p$ integration in (2). In the notation of [2]

$$R_n(\gamma) = L_{\gamma,n}/L^c_{\gamma,n}\,. \tag{4}$$

For $V \leqslant 0$, $V \in C_0^\infty(\mathbf{R}^n)$ it is known [3,4,9] that

$$R_n(\gamma, \lambda V) \to 1 \tag{5}$$

as $\lambda \to \infty$, which is the semiclassical limit. Thus

$$R_n(\gamma) \geqslant 1\,. \tag{6}$$

In [2] it was conjectured that $R_n(\gamma) = 1$ for certain γ and n, in particular for $\gamma = 1$, $n = 3$ which is the case of primary physical interest. $R_3(1) = 1$ would imply that the Thomas–Fermi theory of atoms and molecules (together with a modified treatment of the electron-electron repulsion) gives a lower bound to the true Schrödinger ground state energy (see [1]). The only cases where the value of $R_n(\gamma)$ is known are $n = 1$, $\gamma = 3/2, 5/2, 7/2, \ldots$, where $R_1(\gamma) = 1$. Part (a) of the following theorem, together with (6), settles the question for $n = 1, \gamma \geqslant 3/2$.

Theorem:

(a) For any n, $R_n(\gamma)$ is a monotone nonincreasing function of γ.

☆ Work partly supported by U.S. National Science Foundation grant MCS 75-21684 A02.

(b) If, for some $\gamma > \max(0, 1 - n/2)$, the supremum in (3) is attained, i.e. $R_n(\gamma) = R_n(\gamma, V)$ for some V with $|V|_- \in L^{\gamma+n/2}$, then $R_n(\cdot)$ is strictly decreasing from the left at γ. In fact

$$\liminf_{\delta \to 0^+} \, [R_n(\gamma - \delta) - R_n(\gamma)]/\delta > 0 .$$

Proof:
(a) Fix V. For $\gamma \geqslant 0$, $\delta > 0$, let

$$I(\gamma, \delta) = \int_0^1 \mathrm{d}\lambda\, \lambda^{-1+\delta}\, |1 - \lambda|_-^\gamma < \infty . \tag{7}$$

By scaling, for any $e \in R$

$$|e|_-^{\gamma+\delta} = I(\gamma, \delta)^{-1} \int_0^\infty \mathrm{d}\lambda\, \lambda^{-1+\delta}\, |e + \lambda|_-^\gamma . \tag{8}$$

Thus, for any Schrödinger potential V,

$$\sum_j |e_j(V)|_-^{\gamma+\delta} = I(\gamma, \delta)^{-1} \int_0^\infty \mathrm{d}\lambda\, \lambda^{-1+\delta} \sum_j |e_j(V) + \lambda|_-^\gamma . \tag{9}$$

However, $e_j(V) + \lambda$ are the eigenvalues of the potential $V(x) + \lambda$. Therefore, by definition (3),

$$\sum_j |e_j(V) + \lambda|_-^\gamma \leqslant R_n(\gamma) \int \mathrm{d}^n z\, |H(x,p) + \lambda|_-^\lambda \tag{10}$$

and, by substitution in (9) and using (8),

$$\sum_j |e_j(V)|_-^{\gamma+\delta} \leqslant I(\gamma, \delta)^{-1} \int_0^\infty \mathrm{d}\lambda\, \lambda^{-1+\delta} R_n(\gamma) \times \int \mathrm{d}^n z\, |H(x,p) + \lambda|_-^\gamma = R_n(\gamma) \int \mathrm{d}^n z\, |H(x,p)|_-^{\gamma+\delta} . \tag{11}$$

Hence

$$R_n(\gamma + \delta, V) \leqslant R_n(\gamma)$$

and therefore

$$R_n(\gamma + \delta) \leqslant R_n(\gamma) . \tag{12}$$

(b) Let $\gamma > \max(0, 1 - n/2)$ and assume that for some V with $|V|_- \in L^{\gamma+n/2}(\mathbf{R}^n)$

$$R_n(\gamma, V) = R_n(\gamma) . \tag{13}$$

In particular (13) implies that

$$0 > e_0 \equiv \inf \operatorname{spec}(-\Delta + V) > \operatorname{ess\,inf} \{V(x)\} . \tag{14}$$

We shall prove that $R_n(\cdot)$ is strictly decreasing from the left at γ by showing that

$$\liminf_{\delta \to 0^+} \, [R_n(\gamma - \delta) - R_n(\gamma)]/\delta \geqslant R_n(\gamma) \, \frac{\int \mathrm{d}^n z\, |H(x,p)|_-^\gamma\, \mu_\gamma(|e_0/H|)}{\int \mathrm{d}^n z\, |H(x,p)|_-^\gamma} > 0 , \tag{15}$$

where

$$\mu_\gamma(t) = \int_t^1 \mathrm{d}\lambda (1 - \lambda)^\gamma/\lambda > (1 - t)^{1+\gamma}/(1 + \gamma) . \tag{16}$$

The key fact which will be used is that the integral in (9) can be cut off from above at $|e_0|$.
For any $\delta > 0$

$$\begin{aligned}
\sum_j |e_j(V)|_-^\gamma &= I(\gamma - \delta, \delta)^{-1} \times \int_0^{|e_0|} \mathrm{d}\lambda\, \lambda^{-1+\delta} \sum_j |e_j(V) + \lambda|_-^{\gamma-\delta} \\
&\leqslant R_n(\gamma - \delta)\, I(\gamma - \delta, \delta)^{-1} \times \int \mathrm{d}^n z \int_0^{|e_0|} \mathrm{d}\lambda\, \lambda^{-1+\delta} |H(x,p) + \lambda|_-^{\gamma-\delta} \\
&= R_n(\gamma - \delta) \int \mathrm{d}^n z\, |H(x,p)|_-^\gamma \\
&\quad - R_n(\gamma - \delta)\, I(\gamma - \delta, \delta)^{-1} \times \int \mathrm{d}^n z \int_{|e_0|}^\infty \mathrm{d}\lambda\, \lambda^{-1+\delta}\, |H(x,p) + \lambda|_-^{\gamma-\delta} .
\end{aligned} \tag{17}$$

Therefore, using (2), (13) and $R_n(\gamma - \delta) \geqslant R_n(\gamma)$,

$$[R_n(\gamma-\delta)-R_n(\gamma)]/\delta \geqslant [R_n(\gamma)/\delta I(\gamma-\delta,\delta)] \times \int_{H<e_0} \mathrm{d}^n z\, |H(x,p)|_-^{\gamma} \times \int_{|e_0/H|}^{1} \mathrm{d}\lambda\, \lambda^{-1+\delta}(1-\lambda)^{\gamma-\delta} \Big/ \int \mathrm{d}^n z\, |H(x,p)|_-^{\gamma} . \tag{18}$$

Using $\lim_{\delta\to 0^+} \delta I(\gamma-\delta,\delta)=1$ and Fatou's lemma, we obtain (15). ∎

In view of (16) the theorem implies

Corollary 1: If for some γ, $R_n(\gamma)=1$ then $R_n(\bar{\gamma}) = 1$ for all $\bar{\gamma} > \gamma$. Moreover, for $\bar{\gamma} > \gamma$ the supremum in (3) is not attained by any potential.

This proves part of a conjecture made in [2] (another part of the conjecture was disproved, for $n \geqslant 7$, in [7]).

In one dimension we can say even more since it is known, [2], that $R_1(3/2)=1$ ($R_1(\gamma)>1$ for $\gamma<3/2$).

Corollary 2: For $\gamma \geqslant 3/2$, $R_1(\gamma)=1$.

One may also study bounds like (3) for some restricted classes of potentials, V, as was done in [7] for the spherically symmetric ones (the constants thus obtained are no larger than $R_n(\gamma)$ but it is not known whether any of them are strictly smaller). The theorem and its proof extend to such bounds as long as the class of potentials is closed under the addition of constants.

References

[1] E.H. Lieb and W.E. Thirring, Phys. Rev. Lett. 35 (1975) 687. See Phys. Rev. Lett. 35 (1975) 1116 for errata. Also E.H. Lieb, Rev. Mod. Phys. 48 (1976) 553.
[2] E.H. Lieb and W.E. Thirring, in: Studies in mathematical physics, Essays in honor of V. Bargmann (Princeton Univ. Press, Princeton, N.J., 1976).
[3] A. Martin, Helv. Phys. Acta 45 (1972) 140.
[4] H. Tamura, Proc. Japan Acad. 50 (1974) 19.
[5] M. Cwikel, Ann. of Math. 106 (1977) 93.
[6] E.H. Lieb, Bull. Amer. Math. Soc. 82 (1976) 751.
[7] V. Glaser, H. Grosse and A. Martin, Bounds on the Number of Eigenvalues of the Schrödinger Operator, CERN preprint TH2432 (1977).
[8] G.V. Rosenblum, The distribution of the discrete spectrum for singular differential operators, Isvestia Math. 164 No. 1 (1976) 75.
[9] M.S. Birman and V.V. Borzov, On the asymptotics of the discrete spectrum of some singular differential operators, Topics in Math. Phys. 5 (1972) 19.

Proceedings of the Math. Soc. Symposia in Pure Math. *36*, 241-252 (1980)

THE NUMBER OF BOUND STATES OF ONE-BODY SCHROEDINGER OPERATORS AND THE WEYL PROBLEM

Elliott H. Lieb[1]

ABSTRACT. If $\tilde{N}(\Omega,\lambda)$ is the number of eigenvalues of $-\Delta$ in a domain Ω in a suitable Riemannian manifold of dimension n, we derive bounds of the form $\tilde{N}(\Omega,\lambda) \leq D_n \lambda^{n/2}|\Omega|$ for *all* Ω, λ, n, Likewise, if $N_\alpha(V)$ is the number of nonpositive eigenvalues of $-\Delta + V(x)$ which are $\leq \alpha \leq 0$, then $N_\alpha(V) \leq L_n \int_M [V - \alpha]_-^{n/2}$ for *all* α and V and $n \geq 3$.

I. INTRODUCTION AND BACKGROUND.

Two closely related problems will concern us here : One is to bound the nonpositive eigenvalues of the one-body Schroedinger operator

$$H = -\Delta + V(x). \tag{1.1}$$

The other is to find an upper bound for

$$\tilde{N}(\Omega,\lambda) = \text{number of eigenvalues of } -\Delta \leq \lambda \tag{1.2}$$

in a domain Ω with Dirichlet boundary conditions.

In both cases the setting is a Riemannian manifold, M, and Δ is the Laplace-Beltrami operator. The only way in which the properties of the manifold will appear in our results will be through the fundamental solution of the heat equation

$$G(x, y; t) = [\exp(t\Delta)](x, y) \tag{1.3}$$

1980 Mathematics Subject Classification 35P15.
[1] Work supported by U.S.National Foundation grants PHYS-7825390 and INT 78-01160.

Proceedings of the Math. Soc. Symposia in Pure Math. *36*, 241-252 (1980)

evaluated on the diagonal x = y, or else through the Green function

$$(-\Delta + e)^{-1}(x, y) = \int_0^\infty dt\, e^{-et}\, G(x, y; t) \qquad (1.4)$$

for $e \geq 0$. (Note: G is defined for the *whole* manifold, not the subdomain Ω in the case of (1.2). One could of course, use G defined for the domain Ω in all the formulas, but then the dependence of the result on Ω will be complicated. It is precisely to avoid this complication that we use the G for the whole of M).

Let us begin with the problem defined by (1.2), which we may term the *Weyl problem*. Weyl [1] proved the asymptotic formula (for suitable domains) :

$$\tilde{N}(\Omega,\lambda) \approx C_n\, \lambda^{n/2} |\Omega| \qquad (1.5)$$

where $|\Omega|$ is the Riemannian volume of Ω, n is the dimension of M and

$$C_n = (4\pi)^{-n/2}\, \Gamma(1 + n/2)^{-1} \qquad (1.6)$$
$$= (2\pi)^{-n}\, \tau_n ,$$

where τ_n is the volume of the unit ball in $\mathbb{R}^n$. The constant C_n is called the *classical constant* for reasons which will become clear later. This result (1.5) is discussed in [2] and [3], §XIII.15, vol.4. The proof uses Dirichlet-Neumann bracketing.

Polya's conjecture is that (1.5) holds in $\mathbb{R}^n$ for all λ and Ω, not just asymptotically. Here we will prove

THEOREM 1. *For all* λ *and* Ω *there exist constants* D_n *and* E_n *(depending on the manifold M) such that*

$$\tilde{N}(\Omega,\lambda) \leq D_n\, \lambda^{n/2} |\Omega| \qquad (1.7)$$

if

$$G(x, x; t) \leq A_n\, t^{-n/2}, \ \forall x \in M,\ \forall t > 0 \qquad (1.8)$$

for some $A_n < \infty$, *while*

$$\tilde{N}(\Omega,\lambda) \leq (D_n\, \lambda^{n/2} + E_n) |\Omega| \qquad (1.9)$$

if

$$G(x, x; t) \leq A_n t^{-n/2} + B_n,\ \forall x \in M,\ \forall t > 0, \qquad (1.10)$$

D_n *and* E_n *are proportional to* A_n *and* B_n *respectively.*

In particular (1.8) and (1.7) hold for $\mathbb{R}^n$ (with $A_n = (4\pi)^{-n/2}$) and for many noncompact M, e.g. homogeneous spaces with curvature ≤ 0. (1.10) and (1.9) hold for compact M.

Next we turn to the Schroedinger problem (1.1). Let $E_1(V) \leq E_2(V) \leq \ldots \leq 0$ be the nonpositive eigenvalues of H on $L^2(M)$. If we write

$$V = V_+ - V_- \qquad (1.11)$$

with

$$V_\pm(x) = |V(x)| , \quad \text{when } V(x) \gtrless 0$$
$$= 0, \text{ otherwise} \qquad (1.12)$$

then the negative spectrum of H is discrete if $V_- \epsilon L^{n/2}$, for example.

DEFINITION. $N_\alpha(V)$ *is the number of eigenvalues of H which are* $\leq \alpha \leq 0$.

Our main result is

THEOREM 2. *Suppose.* (1.8) *holds and suppose* $n \geq 3$. *Then*

$$N_0(V) \leq L_n \int_M V_-(x)^{n/2}\, dx \tag{1.13}$$

for some constant L_n *depending on* M.

There are many remarks to be made about Theorem 2 and its connection with Theorem 1. First, the history of (1.13). Rosenbljum [4] first announced Theorem 2 in 1972. Unaware of this, Simon [5] proved an inequality of the form

$$N_0(V) \leq S_{n,\varepsilon}[\, \| V_- \|_{n/2+\varepsilon} + \| V_+ \|_{n/2+\varepsilon}]^{n/2}$$

with $S_{n,\varepsilon} \to \infty$ as $\varepsilon \downarrow 0$. Also unaware of [4], Cwikel and myself [6,7] simultaneously found a proof of Theorem 2. Reed and Simon [3] call Theorem 2 the Cwikel - Lieb - Rosenbljum bound. Cwikel's method exploits some ideas in [5]. All three methods are different, Cwikel's and Rosenbljum's are applicable to a wider class of operators than the Schroedinger operator, but my method [7], based on Wiener integrals, which is the one presented here, gives the best constant by far. This result was announced in [7] and the proof was written up in [3], Theorem XIII.12 and in [8]. Because all the technical details can easily be found in [8], the presentation given here will ignore technicalities . Not only am I indebted to B.Simon for his help with the technical details, as just mentioned, but I also wish to acknowledge his role in stimulating my interest in the problem, and for his constant encouragement while the ideas were taking shape.

The connection between the two theorems is

PROPOSITION 3. *Let* $\alpha \leq 0$. *Then for all* M

$$\tilde{N}(\Omega,\lambda) \leq N_\alpha(\,(\alpha-\lambda)\,\chi_\Omega) \tag{1.14}$$

where χ_Ω *is the characteristic function of* Ω.

PROOF. Let ψ_j be the j th eigenfunction of $-\Delta$ in Ω and let $\tilde{\psi}_j$, defined on all of M, be ψ_j in Ω and zero outside. (1.14) is obtained by using the $\tilde{\psi}_j$ as variational functions for $-\Delta + (\alpha - \lambda)\,\chi_\Omega$ in the Rayleigh - Ritz variational principle. QED.

Similarly, we have

PROPOSITION 4. *if* $0 \leq \beta \leq -\alpha$., *then for all* M

$$N_\alpha(V) \leq N_{\alpha+\beta}(-[V+\beta]_-). \tag{1.15}$$

PROOF. Same as for proposition 3. Alternatively, one can remark that $V(x) \geq -[V+\beta]_-(x) - \beta$, and adding a positive operator can never decrease any eigenvalue. QED.

Proceedings of the Math. Soc. Symposia in Pure Math. *36*, 241-252 (1980)

In particular, $N_\alpha(V) \leq N_0(-[V-\alpha]_-)$ and $N_0(V) \leq N_0(-V_-)$, whence we have

COROLLARY 5. *If* $n \geq 3$ *and* (1.13) *holds, then*

$$N_\alpha(V) \leq L_n \int_M [V-\alpha]_-(x)^{n/2}\, dx. \tag{1.16}$$

Moreover to prove (1.13) *it is sufficient to consider* V *satisfying* $V(X) \leq 0$, $\forall x$.

Proposition 3 will be used to derive Theorem 1 from Theorem 2 (actually from a generalization of Theorem 2, namely Theorem 8 and (4.3), which holds for all n). However, at this point we can, under the assumption that (1.8) holds, deduce (1.7) of Theorem 2 from proposition 3 and Theorem 2. Choosing $\alpha = 0$, we have

$$\begin{aligned}\tilde{N}(\Omega,\lambda) &\leq L_n \lambda^{n/2} \int_M \chi_\Omega^{n/2}\, dx \\ &= L_n \lambda^{n/2}|\Omega|\end{aligned} \tag{1.17}$$

for $n \geq 3$.

It is to be emphasized that Theorem 2 is more delicate than Theorem 1. For one thing Theorem 1 holds for all n, whereas Theorem 2 holds only for $n \geq 3$. The analogue of Theorem 2 is definitely false for $n = 1$ and 2. In $\mathbb{R}^n$, at least, an arbitrarily small, nonpositive potential always has a negative eigenvalue when $n < 3$; cf. [3], Theorem XIII.11. For another thing, the best constant in (1.7) is , according to the Polya conjecture, the classical constant C_n (in $\mathbb{R}^n$ at least). However, the best constant L_n is bigger than C_n in general. It is easy to prove [9] (by considering a very "large" V and using Dirichlet-Neumann bracketing) that $L_n \geq C_n$. In [9] it was shown that $L_n > C_n$ for $3 \leq n \leq 7$ and recently [16] it was shown that $L_n > C_n$ for $n \geq 7$; in both cases explicit examples were constructed.

It is somewhat ironic that although Theorem 2 starts to hold only for $n = 3$, Theorem 1 is easiest to prove for $\mathbb{R}^n$ with $n = 1$. In that case the only domain that need be considered is a finite interval, and there the eigenvalues of $-\Delta$ can be computed explicitly. The Polya conjecture is easily seen to be true.

The intuition behind Theorem 2, and thereby the reason for calling C_n the classical constant is important. In the semiclassical picture of quantum mechanics in $\mathbb{R}^n$, which is similar to WKBJ theory, one has the mystical postulate that "each nice set in phase space $\Gamma = \{(p,x) \mid p \in \mathbb{R}^n,\ x \in \mathbb{R}^n\}$ of volume $(2\pi)^n$ can accomodate one eigenstate of H". This postulate can be made more precise by mean of the Dirichlet-Neumann bracketing method mentioned before. In any event, since the "eigenvalues" of $-\Delta$ are p^2 and the "eigenvalues" of V are V(x), the postulate implies that

$$N_\alpha(V) \approx (2\pi)^{-n} \int_{\mathbb{R}^n\times\mathbb{R}^n} dpdx\ \Theta(\alpha - (p^2+V(x))) \tag{1.18}$$

with $\Theta(a) = 1$, for $a \geq 0$ and $= 0$ for $a < 0$. The p integration in (1.18) for fixed x, is easy to do, namely

$$\int_{p^2\leq \alpha-V(x)} dp = \tau_n\ [V - \alpha]_-(x)^{n/2} \quad . \tag{1.19}$$

Thus, (1.18) yields

$$N_\alpha(V) \approx C_n \int [V - \alpha]_-(x)^{n/2}\ dx \quad . \tag{1.20}$$

While the chief purpose of this paper is to prove Theorems 1 and 2, quantities of no less interest are the moments of the nonpositive eigenvalues of H.

DEFINITION. *For* $\gamma > 0$

$$\begin{aligned} I_\gamma(V) &= \sum_i |E_i(V)|^\gamma \\ &= \gamma \int_{-\infty}^0 |\alpha|^{\gamma-1}\ N_\alpha d\alpha \end{aligned} \tag{1.21}$$

$I_0(V)$ *is defined to be* $N_0(V)$.

For $n \geq 3$ we can use Corollary 5 and Fubini's theorem to obtain

$$\begin{aligned} I_\gamma(V) &\leq \gamma\ L_n \int_M dx \int_{-\infty}^0 |\alpha|^{\gamma-1}[V - \alpha\ ]_-(x)^{n/2}\ d\alpha \\ &= \gamma\ L_n \int_M dx \int_{-V_-(x)}^0 d\alpha\ |\alpha|^{\gamma-1}(V_-(x) + \alpha)^{n/2} \\ &= L_{\gamma,n} \int_M V_-(x)^{\gamma+n/2}\ dx \end{aligned} \tag{1.22}$$

with

$$L_{\gamma,n} = L_n\ \Gamma(\gamma+1)\ \Gamma(1+n/2)\ \Gamma(1+\gamma+n/2)^{-1} \tag{1.23}$$

There are several things to be said about (1.22). Although it was derived from Corollary 5 under the assumptions (1.13) and $n \geq 3$, it holds much more generally. For example it holds in $\mathbb{R}^n$ for $n = 2$, $\gamma > 0$ and for $n = 1$, $\gamma > 1/2$ provided $\alpha > 0$. This was first given in [9]. In [9] it was stated that it holds for $n = 1$ and $\gamma = 1/2$. That was an error ; it is not known if (1.22) holds for $n = 1$, $\gamma = 1/2$ but it is known [9] that (1.22) does not hold for $n = 1$, $\gamma < 1/2$. In section II we shall briefly mention how to deduce (1.22).

The best constant $L_{\gamma,n}$ in (1.22) is not given by (1.23), as the foregoing remark already indicates. If we use C_n in place of $L_{\gamma,n}$ in (1.23) we have the classical value of $L_{\gamma,n}$ namely,

$$L^C_{\gamma,n} = (4\pi)^{-n/2}\ \Gamma(\gamma+1)\ \Gamma(1+\gamma+n/2)^{-1} \tag{1.24}$$

As in the case of L_n, it is easy to prove that $L_{\gamma,n} \geq L^C_{\gamma,n}$. The classical constant $L^C_{\gamma,n}$ can also be "derived" from the semiclassical assumption as in (1.18), (1.19), namely

$$\sum_i |E_i|^\gamma \approx (2\pi)^{-n} \int_{\mathbb{R}^n \times \mathbb{R}^n} dp dx \, |p^2 + V(x)|^\gamma \, \Theta(-p^2 - V(x)). \tag{1.25}$$

If the p integration is done in (1.25), the result is (1.22) with $L^C_{\gamma,n}$.

An important question is :

When is $L_{\gamma,n} = L^C_{\gamma,n}$?

It seems to be true that $L_{\gamma,n} = L^C_{\gamma,n}$ for γ large enough, depending on n. This is known to be true [9, 10] for $n = 1$ and $\gamma \geq 3/2$. In fact [10], if $L_{\gamma_0,n} = L^C_{\gamma_0,n}$ for some γ_0, then equality holds for all $\gamma \geq \gamma_0$.

The case of primary *physical interest* is $\gamma = 1$, $n = 3$, where it is conjectured [9] that equality holds. If this were so, it would have important consequences for physics and it is hoped that someone will be motivated to solve the problem.

We now turn to the proof of Theorems 1 and 2 in the next three sections.

II. THE BIRMAN-SCHWINGER KERNEL

As stated in Corollary 5 we can assume $V(x) \leq 0$. Therefore write $V(x) = -U(x)$, $U \geq 0$.

A useful device for studying the nonpositive eigenvalues of $-\Delta - U$ was discovered by Birman [11] and Schwinger [12]. If $(-\Delta - U)\psi = E\psi$, $E \leq 0$, then

$$\psi = (-\Delta + |E|)^{-1} U\psi \cdot \tag{2.1}$$

Defining $U^{1/2}\psi = \phi$, and multiplying (2.1) by $U^{1/2}$, we have

$$\phi = K_{|E|}(U)\phi \tag{2.2}$$

where $K_e(U)$, for $e \geq 0$, is the positive *Birman-Schwinger Kernel* given explicitly by

$$K_e(x, y; U) = U(x)^{1/2}(-\Delta + e)^{-1}(x, y)U^{1/2}(y) . \tag{2.3}$$

What (2.2) says is that for every nonpositive eigenvalue, E, of $-\Delta - U$, $K_{|E|}(U)$ has an eigenvalue 1. The converse is also easily seen to hold (see [3,8] for more details). $K_e(U)$ is to be thought of as an operator on L^2 ; we will see that it is compact, when $e > 0$ at least, and U is in a suitable L^p space.

In addition to the advantage that the study of the E's reduces to the study of a compact operator, there is the following important fact: Since $(-\Delta + e)^{-1}$ is operator monotone decreasing as a function of e, so is K_e. Hence (with $V = -U$),

$$N_\alpha(V) = k_{|\alpha|}(U) \equiv \textit{number of eigenvalues of } K_{|\alpha|}(U) \geq 1. \tag{2.4}$$

(2.4) will be exploited in the following way.

PROPOSITION 6. *Let* $F : \mathbb{R}^+ \to \mathbb{R}^+$ *be any function such that* $F(x) \geq 1$ *for* $x \geq 1$.

Then

$$\operatorname{Tr} F(K_e(U)) = \sum_i F(l_e^i(U))$$
$$\geq k_e(U) = N_{-e}(V) \tag{2.5}$$

where Tr *means trace and the* $l_e^i(U)$ *are the eigenvalues of* $K_e(U)$.

For example, consider $\mathbb{R}^n$, $n \leq 3$, and $F(x) = x^2$. Then

$$\operatorname{Tr} K_e(U)^2 = \iint U(x)\, U(y)\, [(-\Delta + e)^{-1}(x - y)]^2\, dxdy$$
$$\leq \|U\|_2^2\, \|(-\Delta + e)^{-1}(x)\|_2^2 \tag{2.6}$$

by Young's inequality. The last factor is of the form

$$\|(-\Delta + e)^{-1}(x)\|_2^2 = h_n e^{-2+n/2} \tag{2.7}$$

This shows that $K_e(U)$ is Hilbert-Schmidt when $U \in L^2$ and $e > 0$. When $n > 3$, $h_n = \infty$ but one can show that $K_e(U)$ is compact by considering the trace of a higher power of $K_e(U)$, cf. [9].

At this point we can derive the aforementioned bound, (1.22), on $I_\gamma(V)$. If we use (2.5) and (2.6) and insert the latter in (1.21), the α integration will diverge. The trick [9] is to use Proposition 4 with $2\beta = e = -\alpha$. Thus

$$N_{-e}(V) \leq \operatorname{Tr} K_{e/2}([V + e/2]_-)$$
$$\leq \|[U - e/2]_+\|_2^2\, h_n (e/2)^{-2+n/2} \tag{2.8}$$

Inserting (2.8) into (1.21), and doing the α integration first, we obtain

$$I_\gamma(V) \leq 2^{2-n/2}\, \gamma h_n \int dx \int_{-2U(x)}^0 d\alpha\, |\alpha|^{\gamma-3+n/2} [U(x) + \alpha/2]^2$$
$$= L_{\gamma,n} \int U(x)^{\gamma+n/2} dx \tag{2.9}$$

if $\gamma > 2 - n/2$ and $n \leq 3$.

(2.9) can be extended to other values of n and γ (but with $\gamma > 1/2$ for $n = 1$), cf. [9].

There is no way, however, to make this method work with $F(x) = x^a$ when $\gamma = 0$ and $n \geq 3$. Quiet reflection shows that if theorem 2 is to be provable by this method then we need $x^{-n/2} F(x) \to 0$ as $x \to 0$ but $x^{-1}F(x)$ remains bounded as $x \to \infty$. The tool we will use to bound $\operatorname{Tr}F(K_e(U))$ for such F's is the Wiener integral. That is the subject of the next section, which is really the main point of this paper.

III. THE WIENER INTEGRAL REPRESENTATION OF $F(K_e(U))$

Let $d\mu_{x,y;t}$ be *conditional* Wiener measure on paths $\omega(t)$ with $\omega(0) = x$ and $\omega(t) = y$. This measure gives a representation for G (cf. (3)) by

$$\int d\mu_{x,y;t}(\omega) = G(x,y;t) = e^{t\Delta}(x, y). \tag{3.1}$$

G itself has the *semigroup property*

Proceedings of the Math. Soc. Symposia in Pure Math. *36*, 241-252 (1980)

$$\int_M G(x,\ y;\ t)\ G(y,\ z;\ s)\ dy = G(x,\ z;\ t+s). \tag{3.2}$$

The well known Feynman-Kac formula [13] is

$$\int d\mu_{x,y;t}(\omega)\ \exp\ [-\lambda\int_0^t U(\omega(s))ds] = e^{-t(-\Delta\ +\ \lambda U)}(x,\ y). \tag{3.3}$$

(Note the signs in (3.3).). Take $e \geq 0$, $\lambda \geq 0$, and $U \geq 0$, multiply (3.3) by $U(x)^{1/2}U(y)^{1/2}$ exp $(-et)$ and integrate with respect to t. The result is

$$\begin{aligned} A &= U(x)^{1/2}U(y)^{1/2}\int_0^\infty dt\ e^{-et}\int d\mu_{x,y;t}(\omega)\ \exp\ [-\lambda\int_0^t U(\omega(s))ds] \\ &= \{U^{1/2}(-\Delta + \lambda U + e)^{-1}\ U^{1/2}\}\ (x,\ y). \end{aligned} \tag{3.4}$$

Now

$$(-\Delta + e)^{-1} = (-\Delta + \lambda U + e)^{-1} + \lambda(-\Delta + \lambda U + e)^{-1}\ U(-\Delta + e)^{-1} \tag{3.5}$$

Multiplying (3.5) on both sides by $U^{1/2}$ we obtain

$$A = \{K_e(U)\ [1 + \lambda K_e(U)]^{-1}\}(x,\ y). \tag{3.6}$$

(3.6) can be cast in a more general form. If $g(x) = \exp(-\lambda x)$ and $F(x) = x(1 + \lambda x)^{-1}$ then

$$F(x) = x\int_0^\infty dy\ e^{-y}\ g(xy) \tag{3.7}$$

and

$$\begin{aligned} U(x)^{1/2}U(y)^{1/2}\int_0^\infty dt\ e^{-et}\int d\mu_{x,y;t}\ (\omega)\ g(\int_0^t U(\omega(s))ds) \\ = F(K_e(U))\ (x,\ y). \end{aligned} \tag{3.8}$$

Next we want to take the trace of both sides of (3.8). This is obtained by setting $y = x$ and integrating. (To be precise, one must first take $U \in C^\infty$ and then use a limiting argument.) The point to notice is that the x dependence occurs through

$$\int_M dx\ U(x)\ d\mu_{x,x;t}\ (\omega) = \int_M dx\ d\mu_{x,x;t}\ (\omega)\ U(\omega(0)). \tag{3.9}$$

By the semigroup property of $\exp[-t(-\Delta + \lambda U)]$, however, the same result is obtained (after the $d\mu$ integration) if $U(\omega(0))$ in (3.9) is replaced by $U(\omega(s))$ for any $0 \leq s \leq t$. Thus,

$$\int_0^\infty dt\ t^{-1}\ e^{-et}\int_M dx\int d\mu_{x,x;t}\ (\omega)\ f(\int_0^t U(\omega(s))ds) = \mathrm{Tr}\ F(K_e(U)). \tag{3.10}$$

with

$$f(x) = xg(x), \tag{3.11}$$

i.e.

$$F(x) = \int_0^\infty dy\ y^{-1}\ e^{-y}\ f(xy) \tag{3.12}$$

Now the relations between F, f and g are linear, as are the relations (3.8) and (3.10). The latter therefore extend to a large class of f's. Since we are here not particularly interested in the operator version, (3.8), we will concentrate on the trace version (3.10). The g's of the form $\exp(-\lambda x)$ are norm dense in the continuous functions which vanish at infinity. Furthermore, using the semigroup property of $\exp[-t(-\Delta + U)]$, one can see explicitly that (3.8) and (3.10) hold for f's of the form $x^k \exp(-\lambda x)$, with k a positive integer. By a monotone convergence arguement one arrives at

THEOREM 7. *Let* f *be a nonnegative lower semicontinuous function on* $[0, \infty)$ *satisfying:*

(i) $f(0) = 0$

(ii) $x^p f(x) \to 0$ *as* $x \to \infty$ *for some* $p < \infty$.

Let $U \geq 0$ *and* $U \in L^p + L^q$ *with* $p = n/2$ $(n \geq 3)$, $p > 1$ $(n = 2)$, $p = 1$ $(n = 1)$ *and* $p < q < \infty$ *Then* (3.10) *holds, with* F *given by* (3.12), *in the sense that both sides may be* $+\infty$

The reader is referred to [8] for details. Obviously the class of f's in Theorem 7 is not the largest possible, but it is more than adequate for our intended application.

The remark that allows us to bound the left side of (3.10) is the following.

THEOREM 8. *Suppose that* f *satisfies the conditions of Theorem* 7 *and that* f *is also convex. Then*

$$\operatorname{Tr} F(K_e(U)) \leq \int_0^\infty dt\, t^{-1} e^{-et} \int_M dx\, G(x, x; t)\, f(tU(x)). \tag{3.13}$$

PROOF. By Jensen's inequality, for any fixed path $t \to \omega(t)$,
$f(\int_0^t U(\omega(s))\, ds) = f(\int_0^t (t^{-1}ds)\, t\, U(\omega(s))) \leq \int_0^t (t^{-1}ds)\, f(tU(\omega(s)))$.
By the same remark as that preceding (3.10), $\int_M dx \int d\mu_{x,x;t}(\omega)\, h(\omega(s))$ is independent of s for any fixed function h. Inserting this in (3.10) gives (3.11). QED.

IV. APPLICATIONS OF THEOREM 8

PROOF OF THEOREM 1: We use Proposition 3 and Proposition 6. f is chosen to be of the following form for some $0 < a < b$

$$\begin{aligned} f(x) &= 0, & 0 \leq x \leq a \\ &= b(x - a), & a \leq x. \end{aligned} \tag{4.1}$$

F given by (3.12) is monotone, and the condition that $F(x) = 1$ is that a and b are related by

$$1 = h(a, b) = b\int_a^\infty dy(1 - a/y)e^{-y} = be^{-a} - abE_1(a) \tag{4.2}$$

where E_1 is the exponential integral. Assuming (1.10), (1.9) will be proved; (1.7) follows from the special assumption $B_n = 0$ in what follows.

Proceedings of the Math. Soc. Symposia in Pure Math. *36*, 241-252 (1980)

In Proposition 3 write $\alpha = -e$ so that $\tilde{N}(\Omega, \lambda) <$ (3.13) with $U = (e + \lambda)\chi_\Omega$. The x integration can be done last in (3.13) in which case $f(tU(x)) = 0$ if $x \notin \Omega$. For $x \in \Omega$ we change variables to $t \to t/(e + \lambda)$. Thus,

$$b|\Omega| \int_a^\infty dt(1 - a/t)(A_n(e + \lambda)^{n/2}t^{-n/2} + B_n)\exp[-et/(e + \lambda)] \geq \tilde{N}(\Omega, \lambda) \ . \tag{4.3}$$

The simplest choice for e, which is arbitrary, is $e = 0$. This will work only when $n \geq 3$ and $B_n = 0$. But for all cases we can choose $e = c\lambda$, $c > 0$ and thereby prove Theorem 1. QED.

PROOF OF THEOREM 2: Proposition 6 is used again and the proof parallels that of Theorem 1. (4.1) and (4.2) are assumed. We change variables in (3.13) to $t \to tU(x)^{-1}$ if $U(x) \neq 0$. If $e = 0$ the result is (1.13) with

$$L_n = A_n \int_0^\infty dt\, t^{-1-n/2} f(t) = 4A_n b a^{1-n/2}[n(n-2)]^{-1} \ . \qquad \text{QED.} \tag{4.4}$$

REMARKS: (i) If $\alpha = -e \neq 0$ then the only estimate we have for $N_\alpha(V)$ is contained in Corollary 5, which is valid only for $n \geq 3$. Alternatively, one could try to estimate (3.13) directly with $e \neq 0$, but this is messy. As stated earlier, no inequality of the form (1.16) holds for all α, V when $n = 1$ or 2. But recently Itō [14] has bounded (3.13) when $e \neq 0$ and $n = 2$. He uses the fact that $f(x) \leq bx$ for $x \geq a$ and obtains a complicated upper bound for $N_\alpha(V)$ in terms of $\|V_-\|_2$ and $\|V_-|\ln V_-|^{1/2}\|_2$.

(ii) If $n \geq 3$ and $B_n = 0$ we can choose $e = 0$ in (4.3). This estimate for D_n is, of course, the same as L_n given by (4.4).

As an illustration of how good our bound is let us consider the case of $\mathbb{R}^3$, where $A_3 = (4\pi)^{-3/2}$. We choose $a = 0.25$ in (4.1) and find that $E_1(.25) = 1.0443$ and $b = 1.9315$ according to (4.2). Using (4.4),

$$D_3 = L_3 = 0.1156 \ . \tag{4.5}$$

This value of D_3 can be used in (1.7). When compared with $C_3 = 0.0169$, which is supposed to be the sharp constant, it is not very good. The estimate for D_3 can be improved by using (4.3) with $e = c\lambda$, $c > 0$.

If, however, the same number, L_3, is used in (1.13) the result is quite good. As already stated, the best $L_3 > C_3$. In fact, by an explicit example,

$$L_3 \geq (3\pi)^{-3/2}\Gamma(3)/\Gamma(3/2) = 0.0780, \tag{4.6}$$

cf. [9], eqn. (4.24). It is conjectured that the right side of (4.6) is, in fact, the sharp constant in (1.13) for $\mathbb{R}^3$. In any case our result, (4.5), is off by at most 49%.

As stated in Section 1, a quantity of physical interest is I_1, the sum of the absolute values of the eigenvalues, in $\mathbb{R}^3$. Using the bound (1.22), (1.23), together with (4.5), we have

$$L_{1,3} \le (2/5)L_3 = .04624 \tag{4.7}$$

This result was announced in [15].

BIBLIOGRAPHY

1. H. Weyl, "Das asymptotische Verteilungsgesetz der Eigenwerte Linearer partieller Differentialgleichungen", Math. Ann. 71 (1911), 441-469.

2. M. Kac, "Can one hear the shape of a drum?", Slaught Memorial Papers, no. 11, Amer. Math. Monthly 73 (1966), no. 4, part II, 1-23.

3. M. Reed and B. Simon, Methods of Modern Mathematical Physics, Acad. Press, N. Y., 1978.

4. G. V. Rosenbljum, "Distribution of the discrete spectrum of singular differential operators", Dokl. Aka. Nauk SSSR, 202 (1972), 1012-1015 (MR 45 #4216). The details are given in "Distribution of the discrete spectrum of singular differential operators", Izv. Vyss. Ucebn. Zaved. Matematika 164 (1976), 75-86. [English trans. Sov. Math. (Iz. VUZ) 20 (1976), 63-71.]

5. B. Simon, "Weak trace ideals and the number of bound states of Schroedinger operators", Trans. Amer. Math. Soc. 224 (1976), 367-380.

6. M. Cwikel, "Weak type estimates for singular values and the number of bound states of Schroedinger operators", Ann. Math. 106 (1977), 93-100.

7. E. Lieb, "Bounds on the eigenvalues of the Laplace and Schroedinger operators", Bull. Amer. Math. Soc. 82 (1976), 751-753.

8. B. Simon, Functional Integration and Quantum Physics, Academic Press, N. Y., to appear 1979.

9. E. Lieb and W. Thirring, "Inequalities for the moments of the eigenvalues of the Schroedinger equation and their relation to Sobolev inequalities", in Studies in Mathematical Physics: Essays in Honor of Valentine Bargmann (E. Lieb, B. Simon and A. Wightman eds.), Princeton Univ. Press, Princeton, N. J., 1976. These ideas were first announced in "Bound for the kinetic energy of fermions which proves the stability of matter", Phys. Rev. Lett. 35 (1975), 687-689, Errata 35 (1975), 1116.

10. M. Aizenman and E. Lieb, "On semi-classical bounds for eigenvalues of Schroedinger operators", Phys. Lett. 66A (1978), 427-429.

11. M. Birman, "The spectrum of singular boundary problems", Math. Sb. 55 (1961), 124-174. (Amer. Math. Soc. Trans. 53 (1966), 23-80).

12. J. Schwinger, "On the bound states of a given potential", Proc. Nat. Acad. Sci. U.S.A. 47 (1961), 122-129.

Proceedings of the Math. Soc. Symposia in Pure Math. *36*, 241-252 (1980)

13. M. Kac, "On some connections between probability theory and differential and integral equations", Proceedings of the Second Berkeley Symposium on Mathematical Statistics and Probability, Univ. of Calif. Press, Berkeley, 1951, 189-215.

14. K. R. Ito, "Estimation of the functional determinants in quantum field theories", Res. Inst. for Math. Sci., Kyoto Univ. (1979), preprint.

15. E. Lieb, "The stability of matter", Rev. Mod. Phys. 48 (1976), 553-569.

16. V. Glaser, H. Grosse and A. Martin, "Bounds on the number of eigenvalues of the Schroedinger operator", Commun. Math. Phys. 59 (1978), 197-212.

DEPARTMENTS OF MATHEMATICS AND PHYSICS
PRINCETON UNIVERSITY
JADWIN HALL
P.O.BOX 708
PRINCETON, N. J. 08544

PHYSICAL REVIEW LETTERS

VOLUME 46 16 FEBRUARY 1981 NUMBER 7

Variational Principle for Many-Fermion Systems

Elliott H. Lieb
Departments of Mathematics and Physics, Princeton University, Princeton, New Jersey 08544
(Received 10 December 1980)

If ψ is a determinantal variational trial function for the N-fermion Hamiltonian, H, with one- and two-body terms, then $e_0 \leq \langle\psi, H\psi\rangle = E(K)$, where e_0 is the ground-state energy, K is the one-body reduced density matrix of ψ, and $E(K)$ is the well-known expression in terms of direct and exchange energies. If an *arbitrary* one-body K is given, which does not come from a determinantal ψ, then $E(K) \geq e_0$ does not necessarily hold. It is shown, however, that if the two-body part of H is positive, then in fact $e_0 \leq e_{\mathrm{HF}} \leq E(K)$, where e_{HF} is the Hartree-Fock ground-state energy.

PACS numbers: 05.30.Fk, 21.60.Jz, 31.15.+q

The variational principle is useful for obtaining accurate upper bounds to the ground-state energy e_0 of an N-particle fermion Hamiltonian, H_N. A normalized wave function ψ_N (or density matrix ρ_N) which *satisfies the Pauli principle* is required; then $e_0 \leq e(\rho_N) \equiv \mathrm{Tr}\rho_N H_N$ (with $\rho_N = |\psi_N\rangle \times\langle\psi_N|$ being a pure state in the wave-function case). In practice, however, it is often possible to make a good guess for ρ_N^1, the reduced single-particle density matrix, but the evaluation of $e(\rho_N)$ is complicated by the reconstruction problem for ρ_N: To evaluate $e(\rho_N)$ we first have to know ρ_N. In the simplest case ρ_N^1 is an N-dimensional projection and ρ_N is a pure state, with ψ_N being a determinantal, or Hartree-Fock (HF) function. Otherwise, ρ_N is a very complicated (and, in general, a nonunique) function of ρ_N^1, and the calculation of $e(\rho_N)$ can be extremely difficult because of the "orthogonality problem." For this reason most variational calculations do not depart very far from a HF calculation.

It is the purpose of this note to show that under a positivity condition on the two-body part of H_N (which, fortunately, holds for one case of major interest—the Coulomb potential) it is possible to obtain an upper bound to e_0 which involves only ρ_N^1; *the reconstruction problem is eliminated.* In the HF case, this bound agrees with $e(\rho_N)$. Moreover, our bound, E, satisfies $E \geq e_{\mathrm{HF}}$, where e_{HF} is the *lowest* HF energy. While our bound is thus not superior to the best HF bound, it may be superior in practice because the exact HF orbitals are unknown in general. A possible application might occur in the theory of itinerant ferromagnetism.

Let us make some definitions. Let $z = (x, \sigma)$ denote a single-particle space-spin variable and $\int dz \equiv \sum_\sigma \int dx$. A *single*-particle operator $K(z;z')$ is called *admissible* if it is positive semidefinite and

$$\mathrm{Tr}K = N,\ K \leq I,\ \text{i.e.,}$$

$$\langle\psi, K\psi\rangle \leq \langle\psi,\psi\rangle \text{ for all } \psi. \tag{1}$$

Given ρ_N satisfying the Pauli prinicple,

$$\rho_N^1(z;z') \equiv N\int\rho_N(z, z_2, \ldots, z_N; z', z_2, \ldots, z_N) \times dz_2 \cdots dz_N. \tag{2}$$

Any such ρ_N^1 is admissible. Conversely, given an admissible K there is always at least one ρ_N with $\rho_N^1 = K$. In the HF case $\rho_N = |\psi_N\rangle\langle\psi_N|$ and

$$\psi_N = (N!)^{-1/2}\det[f_i(z_j)],$$
$$\rho_N^1(z;z') = \sum_{j=1}^N f_j(z) f_j^*(z'), \tag{3}$$

with $f_1, \ldots, f_N$ being any N orthonormal functions.

Phys. Rev. Lett. *46*, 457-459 (1981)

Consider now Hamiltonians of the form

$$H_N = \sum_{j=1}^{N} h_j + \sum_{1 \le i < j \le N} v_{ij}, \tag{4}$$

where h and v are self-adjoint operators. v is the two-body part and h is the one-body part [usually $-(\hbar^2/2m)\Delta + U(z)$]. Our method has obvious extensions to higher than two-body interactions, but for brevity only (4) will be considered. Normally v is diagonal [a local potential, such that $v_{ij} = v(z_i, z_j)$], but this is not necessary. *If* ρ_N is of the HF type, then ρ_N^2, defined analogously to (2), satisfies $\rho_N^2 = K_2$, where

$$K_2(z,w;z',w') \equiv K(z;z')K(w;w') - K(z;w')K(w;z'), \tag{5}$$

with $K = \rho_N^1$. In this HF case $e(\rho_N) = E(K)$, where

$$E(K) \equiv \mathrm{Tr}(Kh) + \tfrac{1}{2}\mathrm{Tr}(K_2 v) \tag{6}$$

and

$$\mathrm{Tr}(K_2 v) = \iint v(z,w) K_2(z,w;z,w)\, dz\, dw \tag{7}$$

in the diagonal case. This formula is well known. For *any* admissible K, (5) and (6) *define* $E(K)$.

The problem addressed here is the following: Given an arbitrary, admissible K, does a ρ_N exist such that $\rho_N^1 = K$ and $e(\rho_N) \le E(K)$? If $v \ge 0$, the answer is yes! Note, however, that $\mathrm{Tr}\rho_N^2 = N(N-1)$ but that $\mathrm{Tr}K_2 = (\mathrm{Tr}K)^2 - \mathrm{Tr}K^2$, and this is $N(N-1)$ if and only if K is an N-dimensional projection as in (3). Otherwise, $\mathrm{Tr}K_2 > N(N-1)$. Therefore the ρ_N which I wish to construct cannot simply satisfy $\rho_N^2 = K_2$. The solution must be more complicated than that.

Our main result is stated as follows:

Theorem.—Let v be positive semidefinite [i.e., in the diagonal case, $v(z,w) \ge 0$ for all z,w], and let K be any admissible single-particle operator. Then (i) there exists a density matrix ρ_N satisfying the Pauli principle such that $\rho_N^1 = K$ and

$$e_0 \le e(\rho_N) \le E(K); \tag{8}$$

(ii) there exists a normalized determinantal function ψ_N such that

$$e_0 \le \langle \psi_N, H_N \psi_N \rangle \le e(\rho_N) \le E(K). \tag{9}$$

Note that in (ii) it is not claimed that if $\tilde{\rho}_N \equiv |\psi_N\rangle\langle\psi_N|$ then $\tilde{\rho}_N^1 = K$. However, (9) does say that, among all admissible K, an HF-type K [N-dimensional projection (3)] gives the lowest value of $E(K)$. The proof requires the following:

Lemma.—Let $c_1 \ge c_2 \ge \cdots$ be an infinite sequence with $0 \le c_i \le 1$ and $\sum_1^\infty c_i = N$, where N is an integer. Then there exist N orthonormal vectors $V^1, \ldots, V^N$ in l^2, **the Hilbert space of square-summable sequences indexed by the positive integers (i.e.,** $V \in l^2$ **means** $\sum_{j=1}^\infty |V_j|^2 < \infty$**), such that** $\sum_{i=1}^N |V_j^i|^2 = c_j$.

Proof: Induction on N is used. For $N=1$, choose $V_j^1 = c_j^{1/2}$. Assume that the lemma holds for $N = n-1$; the lemma will first be proved for n under the assumption that $c_j = 0$ for $j \ge 2n$. Define $d_j \equiv 1 - c_j$ $(1 \le j < 2n)$, and $d_j = 0$ $(j \ge 2n)$. The d_j satisfy the hypothesis for $n-1$, so that there exist orthonormal $W^1, \ldots, W^n$ with $\sum_{i=1}^{n-1} |W_j^i|^2 = d_j$. Let $v^1, \ldots, v^n$ be n orthonormal, $(2n-1)$-dimensional vectors which are orthogonal to $W^1, \ldots, W^{n-1}$ [thought of as $(2n-1)$-dimensional vectors]. Then the n vectors $V_j^i = v_j^i$ $(1 \le j < 2n)$ [$V_j^i = 0$ $(j \ge 2n)$] satisfy the lemma. Next, suppose $c_j = 0$ for $j \ge J$. I use induction on J starting with $J = 2n$. Note that $c_{l-1} + c_l \le 1$ when $l \ge 2n$. For $J+1$, apply the lemma (with n and J) to the sequence $b_j = c_j$ $(1 \le j < J-1)$, $b_{J-1} = c_{J-1} + c_J$, and $b_j = 0$ $(j \ge J)$. This sequence may not be decreasing, but that is irrelevant. Let $W^1, \ldots, W^n$ be the orthonormal vectors. The required vectors, V^i, for $J+1$, are given by $V_j^i = W_j^i$ $(1 \le j < J-1)$, $V_{J-1}^i = W_{J-1}^i (c_{J-1}/b_{J-1})^{1/2}$, $V_J^i = W_{J-1}^i (c_J/b_{J-1})^{1/2}$, $V_j^i = 0$ $(j > J)$. Finally, if $c_j > 0$ for all j, choose L so that $b_L \equiv \sum_{j=L}^\infty c_j \le 1$. Then apply the lemma to the finite sequence of length L: $b_j = c_j$ $(1 \le j < L)$, b_L. If $W^1, \ldots, W^n$ are the orthonormal vectors, let $V_j^i = W_j^i$ $(1 \le j < L)$, $V_j^i = W_L^i (c_j/b_L)^{1/2}$ for $j \ge L$. Q.E.D.

Proof of Theorem: Write $K(z;z') = \sum_{j=1}^\infty c_j f_j(z) \times f_j(z')^*$, where the f_j are the orthonormal eigenfunctions of K ("natural orbitals") and the eigenvalues of K, c_j, satisfy the hypothesis of the lemma. Let $V^1, \ldots, V^N$ be the vectors of the lemma and let $\theta = \{\theta_1, \theta_2, \ldots\}$ be any infinite sequence of reals. The N functions $F_k^\theta(z) = \sum_{j=1}^\infty e^{i\theta_j} V_j^k f_j(z)$ are orthonormal for any θ. Let $\rho_N^\theta = |\psi_N^\theta\rangle\langle\psi_N^\theta|$, where $\psi_N^\theta(z_1, \ldots, z_N) = (N!)^{-1/2} \det[F_k^\theta(z_j)]$. Let $\langle \cdots \rangle_\theta$ denote the average {over $[0, 2\pi)^{Z_+}$} with respect to all the θ_j. (Formally, this requires infinitely many integrations $\int_0^{2\pi} d\theta_j / 2\pi$, but one can easily make sense of this by taking suitable limits.) It is easy to check, with use of the property of the V^i, that $\rho_N = \langle \rho_N^\theta \rangle_\theta$ satisfies $\rho_N^1 = \langle \rho_N^{\theta,1} \rangle_\theta = K$. Now $\rho_N^2 \ne K_2$, as stated before, but $\rho_N^2 = \langle \rho_N^{\theta,2} \rangle_\theta = K_2 - L_2$, with

$$2L_2(z,w;z',w') = \sum_{a,b=1}^{\infty} W_{ab} [f_a(z) f_b(w) - f_b(z) f_a(w)][f_a^*(z') f_b^*(w') - f_b^*(z') f_a^*(w')]$$

and $W_{ab}=|\sum_{i=1}^{N}V_a{}^iV_b{}^{i*}|^2\geq 0$. Thus, $e(\rho_N)=E(K)-D$, with $2D=\mathrm{Tr}(L_2v)$. But clearly L_2 is positive semidefinite, so that $D\geq 0$. This proves (i). To prove (ii), note that $E(K)\geq e(\rho_N)=\langle G^\theta\rangle_\theta$, where $G^\theta=\langle\psi_N{}^\theta,H_N\psi_N{}^\theta\rangle$ is real for each θ. Hence, for some θ, $G^\theta\leq e(\rho_N)$. Q.E.D.

A very useful discussion with Professor J. K. Percus is gratefully acknowledged. This work was partially supported by the National Science Foundation under Grant No. PHY-78-25390-A01.

Note added.—After reading this manuscript, Professor M. B. Ruskai kindly pointed out that the lemma is essentially a consequence of Horn's theorem[1]: Let $y_1\geq y_2\geq\cdots\geq y_M$ and $x_1\geq x_2\geq\cdots\geq x_M$ be two sets of reals. Then there exists an $M\times M$ hermitean matrix B with eigenvalues $\{x_i\}$ and diagonal elements $B_{ii}=y_i$ if and only if $\sum_{i=1}^{t}(x_i-y_i)\geq 0$ for all $1\leq t\leq M$, and with equality for $t=M$. The existence of B is equivalent to $y_j=\sum_{i=1}^{M}|U_{ij}|^2x_i$ for some unitary U. To apply this to the lemma, suppose that $c_j=0$ for $j>M\geq N$ and take $y_j=c_j$ (for $j\leq M$) and $x_1=x_2=\cdots=x_N=1$, and $x_j=0$ for $j>N$. The required orthonormal vectors V^i are then $V_j{}^i=U_{ij}$ for $j\leq M$ and $V_j{}^i=0$ for $j>M$. Finally, if $c_j>0$ for all j, then an argument such as that given at the end of the proof of the lemma, or something similar, must be used.

[1]A. Horn, Am. J. Math. 76, 620 (1954).

This paper is most properly cited as

Elliott H. Lieb, Phys. Rev. Lett. 46, 457 (1981), and 47, 69(E) (1981).

All corrections in the Erratum have been incorporated in this version of the reprints.

VOLUME 47, NUMBER 1 PHYSICAL REVIEW LETTERS 6 JULY 1981

ERRATUM

VARIATIONAL PRINCIPLE FOR MANY-FERMION SYSTEMS. Elliott H. Lieb [Phys. Rev. Lett. 46, 457 (1981)].

The name of the author was misspelled in the printed version. The correct spelling is as given above.

Since the meaning of the abstract was adversely affected by editorial processing, the entire abstract is reproduced below for clarification:

If ψ is a determinantal variational trial function for the N-fermion Hamiltonian, H, with one- and two-body terms, then $e_0 \leqslant \langle\psi, H\psi\rangle = E(K)$, where e_0 is the ground-state energy, K is the one-body reduced density matrix of ψ, and $E(K)$ is the well known expression in terms of direct and exchange energies. If an *arbitrary* one-body K is given, which does not come from a determinantal ψ, then $E(K) \geqslant e_0$ does not necessarily hold. It is shown, however, that if the two-body part of H is positive, then in fact $e_0 \leqslant e_{\mathrm{HF}} \leqslant E(K)$, where e_{HF} is the Hartree-Fock ground-state energy.

Considerable distortions of the correct forms in the original manuscript have necessitated the following changes in the printed version:

On page 457, first column, line 5 should read "... which *satisfies the Pauli principle* is required."

On page 458, second column, lines 5 and 6 should read "... N orthonormal vectors $V^1, \ldots, V^N$ in l^2, the Hilbert space of square-summable sequences indexed by the positive integers (i.e., $V \in l^2$ means $\sum_{j=1}^{\infty} |V_j|^2 < \infty$), such that $\sum_{i=1}^{N} |V_j^i|^2 = c_j$.

On page 458, second column, lines 5 and 6 in the proof of the theorem should read "Let $V^1, \ldots, V^N$ be the vectors of the lemma... ."

On page 458, second column, line 11 in the proof of the theorem should read "... over $[0, 2\pi)^{Z_+}$... ."

In addition, the following correction should be noted:

On page 457, the c_j on the right-hand side of Eq. (3) should be deleted.

Part IV

Thomas-Fermi and Related Theories

Rev. Mod. Phys. *53*, 603-641 (1981)

Thomas-fermi and related theories of atoms and molecules*

Elliott H. Lieb

Departments of Mathematics and Physics, Princeton University, POB 708, Princeton, New Jersey 08544

This article is a summary of what is know rigorously about Thomas-Fermi (TF) theory with and without the Dirac and von Weizsäcker corrections. It is also shown that TF theory agrees asymptotically, in a certain sense, with nonrelativistic quantum theory as the nuclear charge z tends to infinity. The von Weizsäcker correction is shown to correct certain undesirable features of TF theory and to yield a theory in much better agreement with what is believed (but as yet unproved) to be the structure of real atoms. Many open problems in the theory are presented.

CONTENTS

*This article appears in the Proceedings of the NATO Advanced Study Institute on Rigorous Atomic and Molecular Physics held at Erice in June, 1980, edited by G. Velo and A. S. Wightman, and published by Plenum Corporation. The present Rev. Mod. Phys. version contains corrections of some errors in the Plenum version.

I. INTRODUCTION

In recent years some of the properties of the Thomas-Fermi (TF) and related theories for the ground states of nonrelativistic atoms and molecules with fixed nuclei have been established in a mathematically rigorous way. The aim of these notes is to summarize that work to date—at least as far as the author's knowledge of the subject goes. In addition, some open problems in the subject will be stated.

TF theory was invented independently by Thomas (1927) and Fermi (1927). The exchange correction was introduced by Dirac (1930), and the gradient correction to the kinetic energy by von Weizsäcker (1935).

No attempt will be made to summarize the voluminous subject of TF theory. Such a summary would have to include many varied applications, many formulations of related theories (e.g., relativistic corrections to TF theory, nonzero temperature TF theory) and reams of data and computations. Some reviews exist (March, 1957; Gombás, 1949; Torrens, 1972), but they are either not complete or not up to date.

We shall concentrate on nonrelativistic TF and related theories for the ground state with the following goals in mind:

(1) The definition of TF and related theories (i.e., the von Weizsäcker and Dirac corrections). The main question here is whether the theories are well defined mathematically and whether the equations to which they give rise have (unique) solutions.

(2) Properties of TF and related theories. It turns out that, unlike the correct Schrödinger, quantum (Q) theory, the TF and related theories have many interesting physical properties that can be deduced without computation. Some of these properties are physically realistic and some are not, e.g., Teller's no-binding theorem. As will be seen, however, the no-binding result is natural and correct if TF theory is placed in its correct physical context as a large-Z (=nuclear-charge) theory.

(3) The relation of TF theory to Q theory. The main result will be that TF theory is exact in the large-Z (nuclear-charge) limit. For this reason, TF theory should be taken seriously as one of the cornerstones of atomic physics. The only other regime in which it is possible to make simple, exact statements is the one-electron hydrogenic atom. The natural open question is to find the leading correction, in Z, beyond TF theory. This will lead to a discussion of the Scott correction

(Scott, 1952) which, while it is very plausible, has not yet been proved. It turns out that Thomas–Fermi–von Weizsäcker (TFW) theory has precisely the properties that Scott predicts for Q theory. Moreover, TFW theory remedies some defects of TF theory: It displays atomic binding, it gives exponential falloff of the density at large distances, it yields a finite density at the nucleus, and negative ions are stable (i.e., bound).

The work reported here originated in articles by Lieb and Simon, 1973 and 1977 (hereafter LS). Subsequently, the ideas were developed by, and in collaboration with, Benguria and Brezis. I am deeply indebted to these coworkers.

Since many unsolved problems remain, these notes are more in the nature of a progress report than a textbook. The proofs of many theorems are sketchy, or even absent, but it is hoped that the interested reader can fill in the details with the help of the references. Unless clearly stated otherwise, however, everything presented here is meant to be rigorous.

II. THOMAS-FERMI THEORY

The theories will be stated in this section purely as mathematical problems. Their physical motivation from Q theory will be explained in Sec. V. In order to present the basic ideas as clearly as possible, only TF theory will be treated in this section; the variants will be treated in Secs. VI, VII and VIII. However, the basic definitions of all the theories will be given in Sec. II.A, and there will be some mention of Thomas-Fermi-Dirac (TFD) theory in Sec. II.B and Sec. III.

A. The definitions of Thomas-Fermi and related theories

All the theories we shall be concerned with start with some *energy functional* $\mathcal{E}(\rho)$, where ρ is a non-negative function on three-space, $\mathbb{R}^3$. ρ is called a *density* and physically is supposed to be the electron density in an atom or molecule.

The functionals will involve the following function V and constant U:

$$V(x)=\sum_{j=1}^{k} z_j|x-R_j|^{-1}, \tag{2.1}$$

$$U=\sum_{1\le i<j\le k} z_i z_j|R_i-R_j|^{-1}. \tag{2.2}$$

$V(x)$ is the electrostatic potential of k nuclei of charges (in units in which the electron charge $e=-1$) $z_1,\ldots z_k>0$, and located at $R_1,\ldots,R_k\in\mathbb{R}^3$. The R_i are distinct. The positivity of the z_i is important for many of the theorems; while TF theory makes mathematical sense when some $z_i<0$, it has not been investigated very much in that case. U is the repulsive electrostatic energy of the nuclei.

TF-type theories can, of course, be defined for potentials that are not Coulombic, but many of the interesting properties presented here rely on potential theory and hence will not hold for non-Coulombic potentials. This is discussed in Sec. III. There is, however, one generalization of Eqs. (2.1) and (2.2) that can be made without spoiling the theory, namely, that the nuclei can be "smeared out," i.e., the following replacements can be made:

$$z_j|x-R_j|^{-1}\to\int dm_j(y)|x-R_j+y|^{-1}, \tag{2.3}$$

$$z_i z_j|R_i-R_j|^{-1}\to\int dm_i(y)dm_j(w)|y-w-R_i+R_j|^{-1}, \tag{2.4}$$

where m_i is a positive measure (not necessarily spherically symmetric) of mass z_i.

The functional for TF theory is

$$\mathcal{E}(\rho)=\tfrac{3}{5}\gamma\int\rho(x)^{5/3}dx-\int\rho(x)V(x)dx+D(\rho,\rho)+U, \tag{2.5}$$

where

$$D(g,f)=\tfrac{1}{2}\int\int g(x)f(y)|x-y|^{-1}dx\,dy\,.$$

All integrals are three dimensional.

γ is an arbitrary positive constant, but to establish contact with Q theory we must choose

$$\gamma_p=(6\pi^2)^{2/3}\hbar^2(2mq^{2/3})^{-1}, \tag{2.6}$$

where $\hbar=h/2\pi$, $h=$ Planck's constant, and m is the electron mass. q is the number of spin states (=2 for electrons).

U appears in $\mathcal{E}$ as a constant, ρ-independent term. It is unimportant for the problem of minimizing $\mathcal{E}$ with respect to ρ. Nevertheless U will be very important when we consider how the minimum depends on the R_i, e.g., in the no-binding theorem (Sec. III.C).

For the Thomas-Fermi-Dirac (TFD) theory

$$\mathcal{E}^{\mathrm{TFD}}(\rho)=\mathcal{E}(\rho)-\tfrac{3}{4}C_e\int\rho(x)^{4/3}dx\,, \tag{2.7}$$

with C_e a positive constant. In the original theory (Dirac, 1930), the value $C_e=(6/\pi q)^{1/3}$ was used for reasons which will be explained in Sec. VI. This value is not sacrosanct, however, and it is best to leave C_e as an adjustable constant.

The Thomas–Fermi–von Weizsäcker theory (TFW) is given by (von Weizsäcker, 1935)

$$\mathcal{E}^{\mathrm{TFW}}(\rho)=\mathcal{E}(\rho)+\delta\int[(\nabla\rho^{1/2})(x)]^2dx\,, \tag{2.8}$$

with $\delta=A\hbar^2/2m$, and A an adjustable constant. Originally, A was taken to be unity, but in Sec. VII.D it will be seen that $A=0.186$ is optimum from one point of view.

The most complicated, and least analyzed, case is the combination of all three (Sec. VIII):

$$\mathcal{E}^{\mathrm{TFDW}}(\rho)=\mathcal{E}(\rho)-\tfrac{3}{4}C_e\int\rho(x)^{4/3}dx+\delta\int[(\nabla\rho^{1/2})(x)]^2dx\,. \tag{2.9}$$

The first question to face is the following.

B. Domain of definition of the energy functional

Since ρ is supposed to be the electron density we require $\rho(x)\ge 0$ and

$$\int\rho(x)dx=\lambda=\text{electron number} \tag{2.10}$$

is finite. In addition we require $\rho \in L^{5/3}$ in order that the first term in $\mathcal{E}(\rho)$ (called the kinetic energy term) be finite. λ is not necessarily an integer.

Definition. A function f is said to be in L^p if $[\int |f(x)|^p dx]^{1/p} \equiv \|f\|_p$ is finite, $1 \leq p < \infty$. $\|f\|_\infty \equiv \text{ess sup}|f(x)|$ (see Theorem 3.12).

If $f \in L^p \cap L^q$ with $p < q$ then $f \in L^t$ for all $p < t < q$. $\|f\|_t \leq \|f\|_p^\lambda \|f\|_q^{1-\lambda}$, where $\lambda p^{-1} + (1-\lambda)q^{-1} = t^{-1}$.

Proposition 2.1. *If $\rho \in L^{5/3} \cap L^1$ then all the terms in $\mathcal{E}$ and $\mathcal{E}^{\rm TFD}$ are finite. If $\int\rho \leq \lambda$ then $\mathcal{E}(\rho)$ and $\mathcal{E}^{\rm TFD}(\rho)$ are bounded below by some constant $C(\lambda)$. Furthermore, for all λ, $\mathcal{E}(\rho) > C > -\infty$ for some fixed C.*

Proof. The first part is an easy application of Young's and Hölder's inequalities. The second part requires a slightly more refined estimate of the Coulomb energies (cf. LS). ■

Remark. Although $\mathcal{E}^{\rm TFW}$ will be seen to be also bounded below by a constant independent of λ, neither $\mathcal{E}^{\rm TFD}$ nor $\mathcal{E}^{\rm TFDW}$ is so bounded. This fact leads to an amusing unphysical consequence of the D theories which will be mentioned later.

A very important fact (which, incidentally, is not true for Hartree-Fock theory) is the following.

Proposition 2.2. *$\rho \to \mathcal{E}(\rho)$ is strictly convex, i.e., $\mathcal{E}(\lambda\rho_1 + (1-\lambda)\rho_2) < \lambda\mathcal{E}(\rho_1) + (1-\lambda)\mathcal{E}(\rho_2)$ for $0 < \lambda < 1$ and $\rho_1 \neq \rho_2$.*

Proof. ρ^p is strictly convex for $p > 1$. $\int V\rho$ is linear in ρ and hence convex. $D(\rho,\rho)$ is strictly convex since the Coulomb kernel $|x-y|^{-1}$ is positive definite. ■

Remark. $\mathcal{E}^{\rm TFW}$ is also strictly convex, but the functionals $\mathcal{E}^{\rm TFD}$ and $\mathcal{E}^{\rm TFDW}$ are not convex because of the $-\int\rho^{4/3}$ term. However, $\mathcal{E}^{\rm TFD}$ can be "convexified" in a manner to be described in Sec. VI.

C. Minimization of the energy functional

The central problem is to compute

$$E(\lambda) = \inf\left\{\mathcal{E}(\rho) \,|\, \rho \in L^{5/3} \cap L^1, \int \rho = \lambda\right\} \tag{2.11}$$

and

$$e(\lambda) = E(\lambda) - U. \tag{2.12}$$

$E(\lambda)$ is the TF energy for a given electron number, λ, and $e(\lambda)$ is the *electronic contribution to the energy.* The "inf" in Eq. (2.11) is important because, as we shall see, the minimum is not always achieved, although the inf always exists by Prop. 2.1.

Theorem 2.3. *$e(\lambda)$ is convex, negative if $\lambda > 0$, nonincreasing and bounded below. Furthermore,*

$$E(\lambda) = \inf\left\{\mathcal{E}(\rho) \,|\, \rho \in L^{5/3} \cap L^1, \int \rho \leq \lambda\right\}. \tag{2.13}$$

Proof. The first part follows from Prop. 2.2 together with the observation that $V(x) \to 0$ as $|x| \to \infty$. This means that if λ increases we can add some $\delta\rho$ arbitrarily far from the origin so that $\mathcal{E}(\rho + \delta\rho) - \mathcal{E}(\rho) < \varepsilon$ for any $\varepsilon > 0$. Equation (2.13) is a simple consequence of the monotonicity of $e(\lambda)$ and $E(\lambda)$. (cf. LS). ■

Equation (2.13) has an important advantage over (2.11), as Theorem 2.4 shows.

Theorem 2.4. *There exists a unique ρ that minimizes $\mathcal{E}(\rho)$ on the set $\int\rho \leq \lambda$.*

Note. Uniqueness means, of course, that ρ is determined only almost everywhere (a.e.).

Proof. (See LS.) Since $\mathcal{E}(\rho)$ is *strictly* convex, a minimum, if there is one, must be unique. Let $\rho^{(n)}$ be a minimizing sequence for $\mathcal{E}$, namely $\mathcal{E}(\rho^{(n)}) \to E(\lambda)$ and $\int\rho^{(n)} \leq \lambda$. It is easy to see that $\int(\rho^{(n)})^{5/3} \leq c$, where c is some constant; this in fact comes out of the simple estimates used in the proof of Prop. 2.1. We should like to extract a convergent subsequence from the given $\rho^{(n)}$. This cannot be done *a priori* in the strong topology, but the Banach-Alaoglu theorem tells us that a $L^{5/3}$ weakly convergent subsequence can be found; this will be denoted by $\rho^{(n)}$. We should like to prove

$$\liminf \mathcal{E}(\rho^{(n)}) \geq \mathcal{E}(\rho). \tag{2.14}$$

Since $\rho^{(n)} \to \rho$ weakly in $L^{5/3}$ we have (by the Hahn-Banach theorem, for example) that

$$\liminf \int [\rho^{(n)}]^{5/3} \geq \int \rho^{5/3}, \tag{2.15}$$

$$\liminf D(\rho^{(n)}, \rho^{(n)}) \geq D(\rho,\rho). \tag{2.16}$$

The term $-\int V\rho$ requires slightly more delicate treatment. Write $|x-R|^{-1} = f(x) + g(x)$, where $f(x) = |x-R|^{-1}$ for $|x-R| \leq 1$ and $f(x) = 0$ otherwise. $f \in L^{5/2}$ and $\int f\rho^{(n)} \to \int f\rho$ by weak convergence. On the other hand, $g \in L^{3+\varepsilon}$ for all $\varepsilon > 0$. $\rho^{(n)}$ is bounded in $L^{5/3}$ and in L^1 (by λ), so it is bounded in all L^q with $1 \leq q \leq \frac{5}{3}$ and therefore $\rho^{(n)} \to \rho$ weakly in L^q as well as in $L^{5/3}$. Fix $\infty > \varepsilon > 0$ and let q be dual to $3+\varepsilon$. Then $\int g\rho^{(n)} \to \int g\rho$. This proves Eq. (2.14) which, since $E(\lambda) = \liminf \mathcal{E}(\rho^{(n)})$ and $E(\lambda) \leq \mathcal{E}(\rho)$, implies that ρ is minimizing provided we can show $\int\rho \leq \lambda$. This follows from the fact that if $\int\rho > \lambda$ then there is a bounded set A such that $\int_A \rho > \lambda$. If α is the characteristic function of A then $\alpha \in L^{5/2}$ and $\lambda \geq \int \alpha\rho^{(n)} \to \int\alpha\rho$ by weak $L^{5/3}$ convergence. ■

Remark. The proof of Theorem 2.4 can be considerably shortened by using Mazur's (1933) theorem. $\rho \to \mathcal{E}(\rho)$ is obviously norm continuous and hence norm lower semicontinuous. Mazur's theorem says that the convexity of $\mathcal{E}(\rho)$ then automatically implies weak lower semicontinuity since norm closed convex sets are automatically weakly closed. The proof given above has the virtue of an explicit demonstration of the weak lower semicontinuity.

Remark. The analogous proof in TFD theory will be harder, since $\rho^{5/3} - \rho^{4/3}$ is not convex, monotone, or positive; hence we cannot say that

$$\liminf \int [\rho^{(n)}]^{5/3} - [\rho^{(n)}]^{4/3} \geq \int \rho^{5/3} - \rho^{4/3}.$$

However, in TFW theory a different strategy, using Fatou's lemma, will be employed to deal with these terms. The strategy also works for TFDW theory. Thus the introduction of the W term (2.8) makes part of the proof easier. It would be desirable to know how to

Rev. Mod. Phys. *53*, 603-641 (1981)

use Fatou's lemma (which does not require convexity) in the TF and TFD proofs.

Since $E(\lambda)$ is nonincreasing, bounded, and convex (and hence continuous) we can make the following definition in TF theory.

Definition. λ_c, *the critical* λ, is the largest λ with the property that for all $\lambda' < \lambda$, $E(\lambda') > E(\lambda)$. Equivalently, if $E(\infty) = \lim_{\lambda\to\infty} E(\lambda)$ then $\lambda_c = \inf\{\lambda \mid E(\lambda) = E(\infty)\}$. In principle λ_c could be $+\infty$, but this will not be the case. In TFD and TFDW theories $E(\lambda)$ is not bounded and the above definition has to be generalized. λ_c is the largest λ with the property that $2E(\lambda) < E(\lambda - \varepsilon) + E(\lambda + \varepsilon)$ for all $0 < \varepsilon < \lambda$. In other words, $E(\lambda) = E(\lambda_c) + (\text{const})(\lambda - \lambda_c)$ for $\lambda \geq \lambda_c$. The j model in TFD theory is bounded, so the first definition is applicable to that model. λ_c will be shown to be $Z = \sum z_j$ in TF and TFD theory, Theorem 3.18. In TFW theory, $\lambda_c > Z$ (Theorem 7.19).

Theorems 2.3, 2.4, and Proposition 2.2 yield the following picture of the minimization problem in TF theory.

Theorem 2.5. *For* $\lambda \leq \lambda_c$ *there exists a unique minimizing* ρ *with* $\int \rho = \lambda$. *On the set* $[0, \lambda_c]$, $E(\lambda)$ *is strictly convex and monotone decreasing. For* $\lambda > \lambda_c$ *there is no minimizing* ρ *with* $\int \rho = \lambda$, *and* $E(\lambda) = E(\lambda_c)$; *the minimizing* ρ *in Theorem 2.4 is the* ρ *for* λ_c.

Proof. For $\lambda \leq \lambda_c$ use the ρ given by Theorem 2.4 and note that if $\lambda' = \int \rho < \lambda$ then $E(\lambda') = \mathcal{E}(\rho) = E(\lambda)$. The strict convexity is trivial: if $\lambda = a\lambda_1 + (1-a)\lambda_2$ use $a\rho_1 + (1-a)\rho_2$ as a trial function for λ. On the other hand, for $\lambda > \lambda_c$ the ρ given by Theorem 2.4 will have $\int \rho = \lambda_c$ because if a minimum existed with $\int \rho = \lambda' > \lambda_c$ then $\tilde{\rho} \equiv (\rho + \rho_c)/2$ (with ρ_c being the ρ for λ_c) would satisfy $\lambda_c < \int \tilde{\rho} = \frac{1}{2}(\lambda' + \lambda_c) < \lambda'$ but, by strict convexity,

$$\mathcal{E}(\tilde{\rho}) < [\mathcal{E}(\rho) + \mathcal{E}(\rho_c)]/2 = E(\lambda_c),$$

which is a contradiction. ■

The general situation is shown in Fig. 1. There Z is shown as less than λ_c; while that is the case for TFW theory, in TF and TFD theory $\lambda_c = Z$. The straight portion to the right of λ_c is horizontal for TF and TFW, but has a negative slope for TFD and TFDW. The slope at the origin is infinite for TF and TFD but finite for TFW and TFDW.

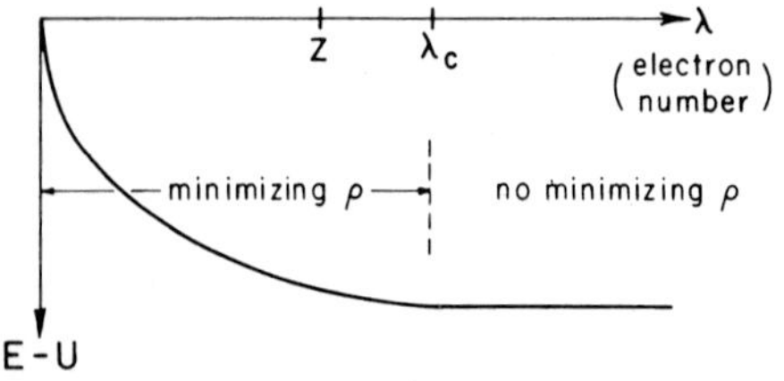

FIG. 1. "The electronic part" of the TF energy, $E-U$, is shown schematically as a function of the "electron number", $\lambda = \int \rho$. For $\lambda \leq \lambda_c$ there is a unique ρ that minimizes the TF energy $\mathcal{E}(\rho)$. For $\lambda > \lambda_c$ there is no such ρ. In TF theory $\lambda_c = Z = \sum_{j=1}^{k} z_j$ = total nuclear charge. $E-U$ is constant for $\lambda \geq \lambda_c$. These features are different for TF, TFD, and TFDW theories (see text).

D. The Thomas-Fermi equation and properties of the density

The variational derivative of $\mathcal{E}(\rho)$ is $\delta\mathcal{E}/\delta\rho = \gamma\rho^{2/3}(x) - \phi_\rho(x)$ where

$$\phi_\rho(x) = V(x) - \int \rho(y)|x-y|^{-1}dy. \tag{2.17}$$

A Lagrange multiplier μ should be added to $\delta\mathcal{E}/\delta\rho$ to insure that $\int \rho = \lambda$. It is then expected that $\delta\mathcal{E}/\delta\rho + \mu = 0$ if $\rho(x) > 0$, but $\delta\mathcal{E}/\delta\rho + \mu \geq 0$ if $\rho(x) = 0$ because negative variations of $\rho(x)$ are not allowed. The two situations can be written as

$$\gamma\rho^{2/3}(x) = \max[\phi_\rho(x) - \mu, 0] \equiv [\phi_\rho(x) - \mu]_+. \tag{2.18}$$

This is *the TF equation.* (Note that the $[\;]_+$ is very important.) This formal manipulation is, indeed, correct.

Theorem 2.6. *If* ρ *minimizes* $\mathcal{E}(\rho)$ *with* $\int \rho = \lambda \leq \lambda_c$ *then* ρ *satisfies Eq. (2.18) for some (unique)* $\mu(\lambda)$. *Conversely if* ρ, μ *satisfy Eq. (2.18) and* $\rho \in L^1 \cap L^{5/3}$ *then* ρ *minimizes* $\mathcal{E}(\rho)$ *for* $\lambda = \int \rho$. *Hence (2.18) can have at most one solution* ρ, μ *with* $\int \rho = \lambda$. *If* $\lambda = \lambda_c$ *then* $\mu = 0$.

Proof. The first part is standard in the calculus of variations. Now let $\rho_i, \mu_i, i = 1, 2$, satisfy Eq. (2.18) with the same λ. Let $F_i(h) = (3\gamma/5)\int h^{5/3} - \int \phi_i h$. It is easy to check that $F_i(h)$ has a unique minimum, F_i, on the set $\int h = \lambda$, $h \geq 0$; the minimizing h_i is ρ_i. However, $F_1(\rho_2) + F_2(\rho_1) = F_1 + F_2 - D(\rho_1 - \rho_2, \rho_1 - \rho_2)$. This is a contradiction unless $\rho_1 = \rho_2$ (and hence $\mu_1 = \mu_2$). The last part (i.e., $\mu = 0$) follows by considering the absolute minimum of $\mathcal{E}(\rho)$, in which case no μ is necessary. But this is equivalent to setting $\mu = 0$. This minimum occurs for $\lambda \geq \lambda_c$ but as we have shown, only at λ_c is there a minimizing ρ (cf. LS). ■

Remarks. In Sec. III a proof of the uniqueness part of Theorem 2.6 which uses only potential theory will be given. It should be noted that we arrived at the existence of a solution to Eq. (2.18) by first considering the minimization problem. A direct attack on (2.18) is rather difficult. Such a direct approach was carried out by Hille (1969) in the atomic case, but even in that case he did not prove that the spherically symmetric solution is the only one; our uniqueness result guarantees that.

Theorem 2.7. $E(\lambda)$ *is continuously differentiable and* $dE/d\lambda = -\mu(\lambda)$ *if* $\lambda \leq \lambda_c$. $dE/d\lambda = 0$ *if* $\lambda \geq \lambda_c$. *Thus* $-\mu(\lambda)$ *is the* chemical potential.

Proof. The convexity and boundedness of $\mathcal{E}(\rho)$ is used. (See LS, Theorem II.10 and Lemma II.27.) ■

It will be noted that we have not used the fact that V is Coulombic, only that it vanishes at ∞. Likewise, the only property of the kernel $|x-y|^{-1}$ that was used was its positive definiteness. In Sec. III we shall exploit the fact that $|x-y|^{-1}$ is Coulombic and, to a lesser extent, the fact that V is superharmonic. Also, it will be shown that $\lambda_c = Z = \sum_1 z_j$.

Definition. A function $f(x)$ defined on an open set $\Omega \subset \mathbb{R}^3$ is *superharmonic on* Ω if, for almost all $x \in \Omega$ and for almost all spheres centered at x, but contained in Ω, $f(x) \geq$ (the average of f on the sphere), i.e., $f(x) \geq (4\pi)^{-1} \times \int_{|y|=R} f(x+y)dy$. This is the same as $\Delta f \leq 0$ (in the sense of distributions) in Ω. f is *subharmonic* if $-f$ is superharmonic. f is *harmonic* if it is both subharmonic and superharmonic.

In Sec. III potential theory will shed considerable light on the solution to Eq. (2.18). Here we shall concentrate on some other aspects of (2.18).

Let us assume that $V(x)=\sum z_j|x-R_j|^{-1}$. ϕ denotes ϕ_ρ for the solution to Eq. (2.18). In Sec. III we show $\phi(x)>0$. As a distribution,

$$\begin{aligned}-\Delta\phi(x)/4\pi&=\sum z_j\delta(x-R_j)-\rho(x)\\&=\sum z_j\delta(x-R_j)-\gamma^{-3/2}(\phi(x)-\mu)_+^{3/2}.\end{aligned}\tag{2.19}$$

This is the TF *differential equation* and is *equivalent* to Eq. (2.18). It involves ϕ alone. Since $\rho\in L^{5/3}\cap L^1$, ϕ is continuous away from the R_j (Lemma 3.1) and goes to zero as $|x|\to\infty$. The fact that ϕ goes to zero at infinity is understood as a boundary condition in Eq. (2.19).

Theorem 2.8 (LS Theorem IV.5).

(a) Near each R_j

$$\rho(x)=(z_j/\gamma)^{3/2}|x-R_j|^{-3/2}+O(|x-R_j|^{-1/2})$$

(b) $\rho(x)\to0$ as $|x|\to\infty$.

(c) ρ and ϕ are real analytic on $A=\{x\,|\,x\neq R_j$ all j, $\rho(x)>0\}$.

(d) In the neutral case ($\mu=0$) $\rho(x)>0$, all x.

(e) In the ionic case ($\lambda<Z$, $\mu>0$) ρ has compact support and ρ and ϕ are C^1 away from the R_j.

Proof. (a) and (b) follow directly from Eq. (2.18). ϕ continuous $\Rightarrow\rho$ continuous away from the $R_j\Rightarrow\phi$ is C^1 everywhere. Then ρ is C^1 away from the R_j. [Note $(\phi-\mu)_+^{3/2}$ is C^1 if ϕ is C^1]. By a bootstrap argument ρ is C^∞ on A. By Theorem 5.8.6 in Morrey (1966), ϕ is real analytic away from the R_j and where $\phi>\mu$, namely A. Finally, since $\phi(x)\to0$ as $|x|\to\infty$, ρ has compact support when $\mu>0$. The positivity of ϕ is established in Sec. III, so $\rho>0$ in the neutral case. ■

In the ionic case ($\lambda<Z$) the set $\Omega=\{x\,|\,\rho(x)>0\}$ is bounded. What can be said about its boundary, $\partial\Omega$? In the atomic case $\partial\Omega$ is, of course, a sphere. In the general case, the TF equation (2.19) is a "free boundary problem" about which Caffarelli and Friedman (1979) have proved the following result among others.

Theorem 2.9. *Consider the generalized TF problem with $\frac{5}{3}$ replaced by p and $\frac{3}{2}<p<2$. There are at most a finite number of open C^1 curves $\Gamma_1,\dots,\Gamma_l$ such that $\partial\Omega\backslash\{\cup_{i=1}^l\bar\Gamma_i\}$ is a $C^{3+\alpha}$ manifold with $\alpha=(2-p)/(p-1)$.*

The next question to consider is the asymptotics of ρ, in the neutral case ($\mu=0$), as $|x|\to\infty$. This involves finding universal bounds on ρ. The function $\psi(x)=\gamma^3(3/\pi)^2|x|^{-4}$ satisfies Eq. (2.19) for $|x|>0$ and $x\neq R_j$. It is the only "power law" that does so. This was noted by Sommerfeld, who concluded that $\psi(x)$ is the asymptotic form of ϕ. Hille (1969), who was possibly the first to make a serious mathematical study of the TF equation, proved this asymptotic law in the atomic case. It is remarkable that ψ, the asymptotic form of ϕ, is *independent* of z, and it is just as remarkable that the same form holds even in the molecular case.

Theorem 2.10 (LS Sec. V.2). *Suppose $\mu=0$ and $|R_j|<R$, for all j and some R. For $r=|x|>R$ let $\phi_+(r)$ (resp. $\phi_-(r)$) be the max (resp. min) of $\phi(x)$ on $|x|=r$ and $C_\pm(r)=\phi_\pm(r)/\psi(r)$ with $\psi(r)=\gamma^3(3/\pi r^2)^2$. Then $C_\pm(r)\to1$ as $r\to\infty$. Furthermore, if $R<r$ then*

(i) $C_+(R)\geq1\Rightarrow C_+(r)\leq C_+(R)$,

(ii) $C_+(R)\leq1\Rightarrow C_+(r)\leq1$,

(iii) $C_-(R)\leq1\Rightarrow C_-(r)\geq C_-(R)$,

(iv) $C_-(R)\geq1\Rightarrow C_-(r)\geq1$.

Proof. If f,g are continuous, positive functions on $|x|\geq R$ which go to zero as $|x|\to\infty$, and if $\gamma^{3/2}\Delta f\leq4\pi f^{3/2}$, $\gamma^{3/2}\Delta g\geq4\pi g^{3/2}$ for $|x|>R$, and if $f(x)\geq g(x)$ for $|x|=R$, then $f(x)\geq g(x)$ for all $|x|>R$. This is easily proved by a "maximum argument" as in Sec. III. ϕ is of this type with $\gamma^{3/2}\Delta\phi=4\pi\phi^{3/2}$. If $C_+(R)\geq1$, compare $\phi(x)$ with $C_+(R)\psi(x)$. Then $C_+(r)\leq C_+(R)$, all $r>R$. This proves (i) and similarly (iii). To prove (ii) and (iv) compare ϕ with ψ. It remains to show that $C_\pm(r)\to1$. C_+ is continuous. We shall show that $\limsup C_+(r)\leq1$; by a similar argument $\liminf C_-(r)\geq1$. This will complete the proof. If $C_+(R)>1$, let $R_0=\sup\{r\,|\,C_+(r)\geq1\}$ whence $C_+(r)\geq1$ for $R\leq r\leq R_0$ and $C_+(r)\leq1$ for $r\geq R_0$. It is then only necessary to consider $R_0=\infty$. Then, since $C_+(r)$ is decreasing, $C(\infty)=\lim C_+(r)$ exists. Assume $C(\infty)>1$. Pick $\varepsilon>0$ and choose R_1 so that $C_+(R_1)<C(\infty)+\varepsilon$. Consider

$$f(x)=\gamma^3(3/\pi)^2(1+2b/3)^2(|x|-bR_1)^{-4}$$

for $|x|\geq R_1$ and $b<1$. $\gamma^{3/2}\Delta f\leq4\pi f^{3/2}$. Choose $b<1$ such that $(1+2b/3)^2=C_+(R_1)(1-b)^4$. Then $f\geq\phi$ for $|x|>R_1$ since $f\geq\phi$ when $|x|=R_1$. But this means $C(\infty)\leq(1+2b/3)^2=[C(\infty)+\varepsilon](1-b)^4$. Since $b>B>0$ satisfying $(1+2B/3)^2=C(\infty)(1-B)^4$, and ε is arbitrary, $C(\infty)\leq1$. For the C_- problem use $g(x)=\gamma^3(3/\pi)^2(1-2b/3)^2(|x|+bR_1)^{-4}$. ■

There are some interesting facts about the possible singularities of TF-type differential equations in a ball. These are related to and complement Theorem 2.10.

Theorem 2.11. *Let $B=\{x\,|\,0<|x|<R\}$ and suppose ϕ satisfies $\Delta\phi(x)=G(\phi(x))$ in the sense of distributions on B, and G is C^1. Then*

(a) If $\phi\in L^\infty_{\rm loc}(B)$ and G satisfies $G(t)>\varepsilon t^3$ as $t\to\infty$, $G(t)<\varepsilon t^3$ as $t\to-\infty$, $\varepsilon>0$, there exists a C^2 function on $0\leq|x|<R$ which agrees with ϕ a.e. in B. Any singularity is thus removable. In particular $-\Delta\phi+\phi^3=\delta(x)$ has no solution in $B'=\{x\,|\,0\leq x<R\}$.

(b) If $\phi\in C^2(B)$, $\phi>0$, and $G(t)=t^q$, $1<q<3$, one of the following is true:

(i) ϕ has a C^2 extension on B'.

(ii) $\phi(x)\sim C|x|^{-1}$ as $|x|\to0$, $C>0$ arbitrary

(iii) $\phi(x)\sim l|x|^{-a}$ as $|x|\to0$ with $a=2/(q-1)$, $l^{q-1}=a(a-1)$. This is called the "strong singularity."

(c) Let $q>1$ and $B'=\{x\,|\,0\leq|x|<R\}$. Let $\phi\in L^\infty_{\rm loc}(B')$ satisfy $\Delta\phi=|\phi|^{q-1}\phi$ in B' in the sense of distributions. There is a universal constant $C_q<\infty$ such that $|\phi(0)|\leq C_qR^{-2/(q-1)}$. This implies that if $\phi\in L^\infty_{\rm loc}(B)$ satisfies this equation in B, then $|\phi(x)|\leq C_q|x|^{-2/(q-1)}$ for $2|x|<R$. A stronger bound than this is given by Veron (1979) and Brezis and Veron (1980) for $1<q<3$.

Proof. (a) is given in Brezis and Veron, 1980, and (b) and (c) are given in Veron, 1979. (c) was given earlier

Rev. Mod. Phys. *53*, 603-641 (1981)

for $q=\frac{3}{2}$ in Brezis and Lieb, 1979. ■

There are other theorems of this type in Veron, 1979 and Brezis and Lieb, 1979.

See Sec. IV.C for an application of the strong singularity.

There is another property of ρ which can be derived directly from the variational principle, namely,

Theorem 2.12. *In the atomic case $\rho(x)$ is a symmetric, decreasing function.*

Proof. Assume the nucleus is at the origin and let ρ^* be the symmetric, decreasing rearrangement of ρ (for a definition see Lieb, 1977). We claim that if $\rho \neq \rho^*$ then $\mathcal{E}(\rho^*) < \mathcal{E}(\rho)$, thereby proving the theorem. $\int(\rho^*)^p = \int \rho^p$, all p. For the Coulomb terms note that when $\int\rho \leq z$ then $f(x)=z|x|^{-1} - |x|^{-1} * (\rho^*)$ is a symmetric, decreasing function; hence $\int f\rho \leq \int f\rho^*$. Thus $P(\rho) \equiv D(\rho,\rho) - \int V\rho = D(\rho-\rho^*, \rho-\rho^*) - \int \rho f - D(\rho^*,\rho^*)$ and thus $P(\rho) > P(\rho^*)$ if $\rho \neq \rho^*$. ■

Notation. $f*g$ denotes convolution, namely $(f*g)(x) \equiv \int f(x-y)g(y)dy$.

Remarks. (i) The same theorem (and proof) holds for the TFD, TFW, and TFDW theories *provided* $\lambda = \int\rho \leq z$. The only additional fact needed for the W theories is that $\int(\nabla\psi)^2 \geq \int(\nabla\psi^*)^2$ (see Lieb, 1977, appendix). In fact, Theorem 2.12 holds for all λ in TFW theory (Theorem 7.26).

(ii) The spherically symmetric (but not the decreasing) property of ρ also follows from the uniqueness of ρ which, in turn, follows from the strict convexity of $\mathcal{E}$. The decreasing property also follows from Eq. (2.18) since ϕ is decreasing by Newton's theorem.

E. The virial and related theorems

Let us generalize the TF functional $\mathcal{E}$ by multiplying the term $D(\rho,\rho)$ in Eq. (2.5) by a parameter $\beta > 0$. $e(\lambda) = E(\lambda) - U$ in Eq. (2.12) is then a function of γ, $\{z_j\}$, and β. Define

$$K=\tfrac{3}{5}\gamma\int\rho^{5/3}, \quad R=\beta D(\rho,\rho), \quad A=\int V\rho\,, \tag{2.20}$$

with ρ being the minimizing ρ for $\int\rho=\lambda$ with $\lambda \leq \lambda_c$. [By scaling, $\lambda_c(\beta)=\lambda_c(\beta=1)/\beta$.]

Theorem 2.13. *$e(\lambda,\gamma,\{z_j\},\beta)$ is a C^1 function of its $k+3$ arguments (assuming all are >0, except for β which is ≥ 0, and $\lambda \leq \lambda_c$). e is convex in λ and jointly concave in $(\gamma,\{z_j\},\beta)$. Moreover, $\partial e/\partial\gamma = K/\gamma$, $\partial e/\partial\beta = R/\beta$, $\partial e/\partial\lambda = -\mu$ $\partial e/\partial z_j = -\int\rho(x)|x-R_j|^{-1}dx$. This implies*

$$\partial E/\partial z_j = \lim_{x\to R_j}\{\phi(x) - z_j|x-R_j|^{-1}\}\,. \tag{2.21}$$

Proof. See LS. The proof uses the convexity of $\rho \to \mathcal{E}(\rho)$. The concavity in the parameters is a trivial consequence of the variational principle and the linearity of $\mathcal{E}$ in the parameters. ■

Now we return to $\beta=1$.

Theorem 2.14. *(a)* $5K/3 = A - 2R - \mu\lambda$,
(b) for an atom $(k=1)$, $2K = A - R$.

Proof. (a) Simply multiply the TF equation (2.18) by ρ and integrate. Alternatively, note that ρ minimizes $G(\rho) = \mathcal{E}(\rho) + \mu\int\rho$ on all of $L^{5/3}\cap L^1$. Therefore $f(t) \equiv G(\rho_t)$, with $\rho_t(x) = t\rho(x)$, has its minimum at $t=1$. But $df/dt = 0$ gives (a).

(b) Here, scaling is essential. Consider $\rho_t(x) = t^3\rho(tx)$, so that $\int\rho_t = \lambda$. Then $f(t) = \mathcal{E}(\rho_t)$ has its minimum at $t=1$ and $df/dt = 0$ gives (b). ■

Remark. (b) is called the *Virial theorem. A priori* there is an analog of (b) for a molecule. Suppose that, with λ fixed, e is stationary with respect to all R_j, i.e., $\nabla_{R_j} e = 0$. Then, by the same scaling argument together with $R_j \to tR_j$, one would conclude that $2K = A - R - U$, equivalently $K + E = 0$. See Fock, 1932 and Jensen, 1933. The difficulty with this is that *there are no stationary points* for $k \geq 2$. The no-binding Theorem 3.23 shows that there are no global minima, and the positivity of the pressure proved in Sec. IV.B shows that there are no local minima (at least for neutral molecules). There it will be shown that for $k \geq 2$, the pressure P satisfies

$$3P = K + E > 0 \quad \text{for neutral molecules}\,. \tag{2.22}$$

For non-neutral molecules, a sharpening of Theorem 4.7 into a strict inequality for the derivative would suffice to show the absence of local minima.

For a neutral atom, (a) and (b) combine to give the following simple ratios:

$$R:K:-e:A = 1:3:3:7\,. \tag{2.23}$$

The energy of a neutral atom is

$$e = E = -3.678\,74 z^{7/3}/\gamma.$$

I thank D. Liberman for this numerical value.

Scaling. Suppose the nuclear coordinates R_i are replaced by lR_i with $l>0$. If $\underline{z}$, $\underline{R}$ denote the nuclear charges and coordinates, and if $E(\underline{z},\lambda,l\underline{R})$, $-\mu(\underline{z},\lambda,l\underline{R})$, $\rho(\underline{z},\lambda,l\underline{R};x)$, and $\phi(\underline{z},\lambda,l\underline{R};x)$ denote the TF energy, chemical potential, density, and potential with $\int\rho=\lambda$, then

$$\begin{aligned} E(\underline{z},\lambda,l\underline{R}) &= l^{-7}E(l^3\underline{z},l^3\lambda,\underline{R})\,,\\ \mu(\underline{z},\lambda,l\underline{R}) &= l^{-4}\mu(l^3\underline{z},l^3\lambda,\underline{R})\,,\\ \rho(\underline{z},\lambda,l\underline{R};x) &= l^{-6}\rho(l^3\underline{z},l^3\lambda,\underline{R};l^{-1}x)\,,\\ \phi(\underline{z},\lambda,l\underline{R};x) &= l^{-4}\phi(l^3\underline{z},l^3\lambda,\underline{R};l^{-1}x)\,. \end{aligned} \tag{2.24}$$

This is a trivial consequence of the scaling properties of $\mathcal{E}(\rho)$.

F. The Thomas-Fermi theory of solids

A solid is viewed as a large molecule with the nuclei arranged periodically. For simplicity, but not necessity, let us suppose that there is one nucleus of charge z per unit cell located on the points of $\mathbb{Z}^3 \subset \mathbb{R}^3$. ($\mathbb{Z}^3$ consists of the points with integer coordinates.) If Λ is a finite subset of $\mathbb{Z}^3$ we want to know if, as $\Lambda\to\infty$ in a suitable sense, the energy/unit volume $|\Lambda|^{-1}E_\Lambda$ has a limit E, and ρ_Λ has a limit ρ, which is a periodic function. Here, $|\Lambda|$ is the volume of Λ. If so, the equation for ρ and an expression for E in terms of ρ is required. Naturally, it is necessary to consider only *neutral* systems, for otherwise $|\Lambda|^{-1}E_\Lambda\to\infty$. Everything works out as expected except for one mildly surprising thing; a quantity ψ_0 appears in the equation for ρ which, while it looks like a chemical potential, and is often assumed to be one, is not a chemical potential. ψ_0 is the average electric potential in the solid. All of this is proved in LS, Sec. VI.

Definition. A sequence of domains $\{\Lambda_i\}$ in Z^3 is said to *tend to infinity* (denoted by $\Lambda\to\infty$) if

(i) $\cup_{i=1}^{\infty}\Lambda_i = Z^3$, (ii) $\Lambda_{i+1}\supset\Lambda_i$, (iii) $\Lambda_i^h\subset Z^3$ is the set of points not in Λ_i, but whose distance to Λ_i is less than h. Then $|\Lambda_i^h|/|\Lambda_i|\to 0$ for each $h>0$. $\Gamma=\{x\in R^3 \,\|\, x^i| < \frac{1}{2}\}$ is the elementary cube centered at the origin.

Theorem 2.15. *As $\Lambda\to\infty$ the following limits exist and are independent of the sequence Λ_i:*

(i) $\phi(x)=\lim_{\Lambda\to\infty}\phi_\Lambda(x)$.

ϕ is periodic, $\gamma\rho(x)^{2/3}\equiv\phi(x)$, and the convergence is uniform on compacts in R^3.

(ii) $\phi(x)=\lim_{\Lambda\to\infty}|\Lambda|^{-1}\sum_{y\in\Lambda}\phi_\Lambda(x+y)$,

(iii) $\lim_{x\to 0}\phi(x)-z|x|^{-1}=\lim_{\Lambda\to\infty}|\Lambda|^{-1}\sum_{y\in\Lambda}\lim_{x\to y}\phi_\Lambda(x)-z\times|x-y|^{-1}$,

(iv) $\int_\Gamma\rho=\lim_{\Lambda\to\infty}\int_\Gamma\rho_\Lambda=z$,

(v) $\int_\Gamma\rho^{5/3}=\lim_{\Lambda\to\infty}|\Lambda|^{-1}\int_{R^3}\rho_\Lambda^{5/3}$,

(vi) $E=\lim_{\Lambda\to\infty}|\Lambda|^{-1}E_\Lambda$.

Definition. $G(x)$ is the *periodic Coulomb potential.* It is defined up to an unimportant additive constant in Γ by $-\Delta G/4\pi=\delta(x)-1$. A specific choice is

$$G(x)=\pi^{-1}\sum_{\substack{k\in Z^3\\ k\neq 0}}|k|^{-2}\exp[2\pi i k\cdot x]\,.$$

Theorem 2.16. *ϕ, ρ and E satisfy*

$$\text{(i) } E=(\gamma/10)\int_\Gamma\rho^{5/3}+(z/2)\lim_{x\to 0}\{\phi(x)-z|x|^{-1}\}\,, \tag{2.25}$$

$$\text{(ii) } \phi(x)=zG(x)-\int_\Gamma G(x-y)\rho(y)+\psi_0 \tag{2.26a}$$

for some ψ_0. Alternatively,

$$-\Delta\phi(x)/4\pi=\sum_{y\in Z^3}z\delta(x-y)-\rho(x)\,, \tag{2.26b}$$

(iii) *ϕ and ρ are real analytic on $R^3\backslash Z^3$.*

(iv) *There is a unique pair ρ,ψ_0 that satisfies Eq. (2.26) with $\gamma\rho^{2/3}=\phi$ and $\int\rho=z$ (cf. Theorem 2.6).*

Formula (2.25) may appear strange but it is obtained simply from the TF equation; an analogous formula also holds for a finite molecule.

Equation (2.26), together with $\gamma\rho^{2/3}=\phi$, is the *periodic TF equation.* ψ_0 is *not* a chemical potential. The chemical potential, $-\mu$, is zero because μ_Λ is zero for every finite system. If (2.26) is integrated over Γ we find, since $\int\rho=z$, that $\psi_0=\int_\Gamma\phi=$ average electric potential. It might be thought that ψ_0 could be calculated in the same way that the Madelung potential is calculated: In each cubic cell there is (in the limit) a charge density $z\delta(x)-\rho(x)$. Therefore if

$$\bar\phi(x)\equiv\sum_{y\in Z^3}g(x-y)\,,$$

with

$$g(x)=z|x|^{-1}-\int_\Gamma\rho(y)|x-y|^{-1}dy\,,$$

it might be expected that $\bar\phi=\phi$. The correct statement is that $\phi(x)=\bar\phi(x)+d$ and $d\neq 0$ in general. One can show that $\int_\Gamma\bar\phi=2\pi\int_\Gamma x^2\rho(x)dx$ (see LS). The fact that $d\neq 0$, precludes having a simple expression for ψ_0. Why is $d\neq 0$, i.e., why is $\phi\neq\bar\phi$? The reason is that the charge density in the cell centered at $y\in Z^3$ is $z\delta(x-y)-\rho(x-y)$ only in the limit $\Lambda\to\infty$. For any finite Λ there are cells near the surface of Λ that do not yet have this charge distribution. Thus $d\neq 0$ essentially because of a neutral double layer of charge on the surface.

In LS asymptotic formulas as $z\to 0$ and ∞ are given for the various quantities.

Theorems 2.15 and 2.16 will not be proved here. Teller's lemma, which implies that $\phi_\Lambda(x)$ is monotone increasing in Λ, is used repeatedly. Apart from this, the analysis is reasonably straightforward.

G. The Thomas-Fermi theory of screening

Another interesting solid-state problem is to calculate the potential generated by one impurity nucleus, the other nuclei being smeared out into a uniform positive background (jellium model). If Λ is any bounded, measurable set in R^3, and if $\rho_B=(\text{const})>0$ is the charge density of the positive background in Λ, and if the impurity nucleus has charge $z>0$ and is located at 0, then the potential is

$$V_\Lambda(x)=z|x|^{-1}+\rho_B\int_\Lambda|x-y|^{-1}dy\,. \tag{2.27}$$

The TF energy functional, without the nuclear repulsion, and with $\gamma=1$, is

$$\mathcal{E}_\Lambda(\rho)=\tfrac{3}{5}\int\rho^{5/3}-\int V_\Lambda\rho+D(\rho,\rho)\,. \tag{2.28}$$

The integrals are over R^3, not Λ. Let $\rho_\Lambda(x)$ be the neutral minimizing ρ (so that $\int\rho_\Lambda=z+\rho_B|\Lambda|$).

Definition. A sequence of domains Λ in R^3 is said to *tend to infinity weakly* if every bounded subset of R^3 is eventually contained in Λ.

Remark. This is an extremely weak notion of $\Lambda\to\infty$.

It is intuitively clear that if $\Lambda\to\infty$ weakly and $z=0$ then $\rho_\Lambda(x)\to\rho_B$. For $z\neq 0$, $\rho_\Lambda(x)-\rho_B$ is expected to approach some function which looks like a Yukawa potential for large $|x|$. This is stated in many textbooks and is correct except for one thing: The coefficient of the Yukawa potential is *not* z but is some smaller number. In TF theory there is *over-screening* because of the nonlinearities.

Theorem 2.17. *Let $\Lambda\to\infty$ weakly and $z=0$. Then $\phi_\Lambda(x)\to\rho_B^{2/3}$ uniformly on compacts in R^3.*

The theorem is another example of the effects of "surface charge." Since $\rho_\Lambda\to\rho_B$ and $\phi_\Lambda=\rho_\Lambda^{2/3}$, the result is

Rev. Mod. Phys. *53*, 603-641 (1981)

natural. But it means that the average potential is not zero. If, on the other hand, the integrals in Eq. (2.28) are restricted to Λ then $\rho_\Lambda(x)=\rho_B$ for all Λ and $x\in\Lambda$, and $\phi_\Lambda(x)=0$.

Theorem 2.18. *Let $\Lambda\to\infty$ weakly and $z>0$. Let*

$$f(x)=\lim_{\Lambda\to\infty}\phi_\Lambda(x)-\rho_B^{2/3}$$

and

$$g(x)=\lim_{\Lambda\to\infty}\rho_\Lambda(x)-\rho_B\,.$$

(i) *these limits exist uniformly on compacts,*
(ii) $g\in L^1\cap L^{5/3}$,
(iii) $0\le f(x)\le\phi^{\text{atom}}(x)$,
(iv) *f and g are strictly positive and real analytic away from $x=0$,*
(v) *$f(x)$ is monotone increasing in z,*
(vi) *These limits satisfy the TF equation*

$$f(x)=z|x|^{-1}-\int|x-y|^{-1}g(y)dy\,, \qquad (2.29)$$

$$[\rho_B^{2/3}+f(x)]^{3/2}-\rho_B=g(x)\,, \qquad (2.30)$$

$$\int g=z\,, \qquad (2.31)$$

(vii) *Assuming only that $g\in L^1\cap L^{5/3}$ and $f(x)\ge-\rho_B^{2/3}$ there is only one solution to Eqs. (2.29) and (2.30) [without assuming (2.31)].*

There is a scaling relation:

$$f(x;z)=\rho_B^{2/3}F(\rho_B^{1/6}|x|;\rho_B^{-1/2}z)$$

$$g(x;z)=\rho_BG(\rho_B^{1/6}|x|;\rho_B^{-1/2}z)\,.$$

Let us write $F(r;z)=q(r;z)Y(r)$ where $Y(r)=(1/r)\times\exp\{-(6\pi)^{1/2}r\}$ is the Yukawa potential.

Theorem 2.19. (i) *$q(r;z)$ is monotone decreasing in r and increasing in z;*
(ii) $q(0;z)=z$;
(iii) *$Q(z)=\lim_{r\to\infty}q(r;z)$ exists. $0<Q(z)<z$ and Q is monotone increasing. $\limsup_{z\to\infty}Q(z)(bz)^{-2/3}<1$ with $b=1.039$.*

LS contains graphical plots of $Q(z)$ and $q(r;53.7)$. An asymptotic formula for $Q(z)$ has *not* been given. In the linearized approximations found in textbooks, $Q(z)=z$, but we see that this is false.

H. The Firsov variational principle

The problem of minimizing $\mathcal{E}(\rho)$ is a convex minimization problem. It has a dual which we now explore. The advantage of the dual problem is that it gives a *lower bound* to E. The principle was first given and applied in (Firsov, 1957) in the neutral case ($\mu=0$) and was first rigorously justified in that case by Benguria (1979). Here we shall also state and prove the principle for non-neutral systems; furthermore, in the neutral case our (and Benguria's) principle will contain a slight improvement over Firsov's.

The dual functional to be considered is

$$\mathfrak{F}_\mu(f)=-(8\pi)^{-1}\int|\nabla f(x)|^2dx$$
$$-\tfrac{2}{5}\gamma^{-3/2}\int[V(x)-f(x)-\mu]_+^{5/2}dx+U\,, \qquad (2.32)$$

where μ is a real parameter. The domain of $\mathfrak{F}_\mu$ is $B=\{f\,|\,\nabla f\in L^2, |f(x)|<c|x|^{-1}$ for some $c<\infty$ and for $|x|>R$ for some $R\}$. V is assumed to go to zero at ∞ and is such that the TF problem has a minimum for that V, and the minimizing ρ (with $\int\rho\le\lambda$) satisfies the TF equation (2.18) (for all λ). We define

$$E^F(\mu)=\sup\{\mathfrak{F}_\mu(f)\,|\,f\in B\} \qquad (2.33)$$

Remark. When $\mu=0$, Firsov imposed the additional constraint $V\ge f$. This, as we shall see, is unnecessary provided $[\ ]_+^{5/2}$ is used as in Eq. (2.32).

Theorem 2.20. *If $\mu<0$ then $E^F(\mu)=-\infty$. If $\mu\ge0$ then there is a unique maximizing f for $\mathfrak{F}_\mu$. This f is $f_\mu\equiv|x|^{-1}*\rho_\mu$ where ρ_μ is the unique solution to Eq. (2.18). If $\lambda=\int\rho_\mu$ then (see remark below)*

$$E^F(\mu)=E(\lambda)+\mu\lambda\,. \qquad (2.34)$$

Proof. Suppose $\mu<0$. Since V and any $f\in B\to0$ as $|x|\to\infty$, the second term in (2.32) is $-\infty$. Suppose $\mu\ge0$. Let $\tilde E_\mu=$ right side of (2.34). Clearly $\mathfrak{F}_\mu(f_\mu)=\tilde E_\mu$ by the TF equation (2.18). $f\to\mathfrak{F}_\mu(f)$ is strictly concave because $\int(\nabla f)^2$ is strictly convex. Thus there can be at most one maximizing f, and we therefore must show that if $f\ne f_\mu$ then $\mathfrak{F}_\mu(f)\le\tilde E_\mu$. By Minkowski's inequality ($|ab|\le2|a|^{5/2}/5+3|b|^{5/3}/5$) we have

$$(\tfrac{2}{5})(V-f-\mu)^{5/2}\ge-(\tfrac{3}{5})(V-f_\mu-\mu)^{5/2}$$
$$+(V-f-\mu)\,(V-f_\mu-\mu)^{3/2}.$$

But $(V-f-\mu)_+\ge V-f-\mu$, so $\mathfrak{F}_\mu(f)\le\tilde E_\mu+h(f)$ where

$$h(f)=-(8\pi)^{-1}\int(\nabla f)^2+\int f\rho_\mu-D(\rho_\mu,\rho_\mu)\,.$$

By standard methods (e.g., Fourier transforms), $h(f)\le0$. Furthermore, $h(f)=0$ only for $f=f_\mu$, which shows once again that the maximizing f is uniquely f_μ. ■

It should be noted that $E^F(\mu)$ is the Legendre transform of $E(\lambda)$. Namely $\lambda\to E(\lambda)$ is convex and

$$E^F(\mu)=\inf_{\lambda\ge0}[E(\lambda)+\lambda\mu]\,,\quad \text{all }\mu\in\mathbb{R}\,. \qquad (2.35)$$

This shows that $E^F(\mu)$ is *concave* in μ. On the other hand, Theorem 2.20 displays $E^F(\mu)$ as the supremum (not infimum) of a family of concave functions. Furthermore, since $E(\lambda)$ is convex and bounded it is its own double Legendre transform, viz.

$$E(\lambda)=\sup_\mu[E^F(\mu)-\mu\lambda]\,. \qquad (2.36)$$

Theorem 2.21. *Fix $\lambda\ge0$. Then [by Eq. (2.36)]*

$$\sup\{\mathfrak{F}_\mu(f)-\mu\lambda\,|\,f\in B,\mu\in\mathbb{R}\}=E(\lambda)\,. \qquad (2.37)$$

Remark. In Theorem 2.20 we refer to the unique ρ_μ satisfying Eq. (2.18) for $\mu\ge0$. This requires some explanation. If $V(x)$ is unbounded (e.g., point nuclei), then as μ goes from ∞ to 0, λ goes from 0 to λ_c and $\rho_\mu(\lambda)$ minimizes $\mathcal{E}$ on $\int\rho=\lambda$. If $\operatorname{ess\,sup}V(x)=v<\infty$, then $\rho_\mu\equiv0$ [and $E^F(\mu)=0$] for $\infty>\mu\ge v$. In this range $\lambda(\mu)=0$. Then, as μ goes from v to 0, λ goes from 0 to λ_c, and $\rho_{\mu(\lambda)}$ minimizes $\mathcal{E}$ on $\int\rho=\lambda$. (ess sup is defined in Theorem 3.12).

III. THE "NO-BINDING" AND RELATED POTENTIAL-THEORETIC THEOREMS

The no-binding theorem was discovered by Teller (1962) and is one of the most important facts about the TF and TFD theories of atoms and molecules. It "explained" the absence of binding found numerically by Sheldon (1955). That this crucial theorem was not proved until 1962—after 35 years of intensive study of TF theory—is remarkable. It can be considered to be a prime example of the fact that pure analysis can sometimes be superior to numerical studies.

While Teller's ideas were correct, his proof was questioned on grounds of rigor. Balàzs (1967) found a different proof for the special case of the symmetric diatomic molecule. A rigorous transcription of Teller's ideas was given in LS. In any case, all proofs of the theorem rely heavily on the fact that the potential is Coulombic.

There are really two kinds of theorems. An example of the first kind is "Teller's lemma," which states that the potential increases when nuclear charge is added. The second, "Teller's theorem" is the no-binding Theorem 3.23. The second, but not the first, requires the nuclear repulsion U. If U is dropped then the theorem goes the other way. The proof of Teller's theorem given in LS is complicated in the non-neutral case, but recently Baxter (1980) found a much nicer proof—one which actually produces a variational ρ that lowers the energy for separated molecules. Baxter's proposition (proposition 3.24) will appear again in Lemma 7.22.

In this section we shall consider general V and assume that

$$\mathcal{E}(\rho)=\int j(\rho(x))dx-\int V(x)\rho(x)dx+D(\rho,\rho)\,, \qquad (3.1)$$

where j is a C^1 convex function with $j(0)=j'(0)=0$. Note that in this section (only) $\mathcal{E}(\rho)$ does not contain U. This is done partly for convenience, but mainly for the reason that since V is not necessarily Coulombic the definition of U would have no clear meaning.

The Euler-Lagrange equation for (3.1) and $\rho(x)\geq 0$ is (with $\phi_\rho=V-|x|^{-1}*\rho$):

$$\begin{aligned}\phi_\rho(x)-\mu&=j'(\rho(x)) \text{ a.e. when } \rho(x)>0\,,\\ &\leq 0 \text{ a.e. when } \rho(x)=0\,.\end{aligned} \qquad (3.2)$$

Any solution to (3.2) is determined only almost everywhere (a.e.).

We could, in fact, allow more general j's of the form $j(\rho,x)$ [and $\int j(\rho(x),x)dx$ in $\mathcal{E}$] with $j(\cdot,x)$ having the above properties for all x, but we shall not do so. An annoying case we must consider, however, is $j'(\rho)=0$ for $0<\rho<\rho_0$ and $j'(\rho)>0$ for $\rho>\rho_0$. This is discussed in some detail in Sec. III.C and is needed for TFD theory (Sec. VI). If $j'(\rho)>0$, all $\rho>0$, as it is in TF theory with $j'(\rho)=\gamma\rho^{2/3}$, then Eq. (3.2) can be written as

$$(\phi_\rho(x)-\mu)_+\equiv\max[\phi_\rho(x)-\mu,0]=j'(\rho(x))\,, \qquad (3.2')$$

but otherwise (3.2) is stronger than (3.2′).

One aim of this section is to study solutions of Eq. (3.2) without considering whether or not (3.2) truly comes from minimizing (3.1) or assuming uniqueness.

Definition. $\mathcal{C}=\{\rho\,|\,\rho(x)\geq 0,\ \rho\in L^1$, and $\int\rho(y)|x-y|^{-1}dy$ is a bounded, continuous function which goes to zero as $x\to\infty\}$.

We shall be concerned only with solutions to (3.2) in $\mathcal{C}$.

The following lemma (LS, II.25) is useful, in the cases of interest, to guarantee that $\rho\in\mathcal{C}$.

Lemma 3.1. *If $f\in L^p, g\in L^{p'}$, $1/p+1/p'=1$, $p,p'>1$ then $f*g$ is a bounded, continuous function which goes to zero as x goes to infinity. In particular, if $\rho\in L^{3/2+\varepsilon}\cap L^1$ then $\rho\in L^{3/2+\varepsilon}\cap L^{3/2-\varepsilon}$. Since $|x|^{-1}\in L^{3+\varepsilon}+L^{3-\varepsilon}$, $\rho\in\mathcal{C}$.*

It will always be assumed that $V(x)\to 0$ as $|x|\to\infty$ (this always means uniformly with respect to direction). Hence μ cannot be negative in Eq. (3.2), for otherwise $\rho\notin L^1$.

A. Some variational principles and Teller's lemma

At first it will *not* be assumed that V is Coulombic.

Theorem 3.2. *Fix $\lambda>0$ and suppose that ρ_λ,μ_λ satisfy Eq. (3.2) with $\int\rho_\lambda=\lambda$. Let $\phi_\lambda=\phi_{\rho_\lambda}$ and assume that $\rho_\lambda\in\mathcal{C}$. Then, for all x,*

(a) $\phi_\lambda(x)-\mu_\lambda=\sup_\rho\Big\{\phi_\rho(x)-\mu\,\Big|\,\phi_\rho(y)-\mu\leq j'(\rho(y))$ a.e. y, $\int\rho\leq\lambda$, $\rho\in\mathcal{C}\Big\}$,

(b) $\phi_\lambda(x)-\mu_\lambda=\inf_\rho\Big\{\phi_\rho(x)-\mu\,\Big|\,\phi_\rho(y)-\mu\geq j'(\rho(y))$ a.e. y when $\rho(y)>0$, $\int\rho\geq\lambda$, $\rho\in\mathcal{C}\Big\}$

(c) $\phi_\lambda(x)=\sup_\rho\{\phi_\rho(x)\,|\,\phi_\rho(y)-\mu_\lambda\leq j'(\rho(y))$, a.e. y, $\rho\in\mathcal{C}\}$

(d) $\phi_\lambda(x)=\inf_\rho\{\phi_\rho(x)\,|\,\phi_\rho(y)-\mu_\lambda\geq j'(\rho(y))$. a.e. y when $\rho(y)>0$, $\rho\in\mathcal{C}\}$.

Furthermore, in (a) [*resp.* (b)] *there is no ρ satisfying the conditions on the right when $\mu<\mu_\lambda$ [resp. $\mu>\mu_\lambda$]. Note that in* (a) *and* (b) *μ is arbitrary (including $\mu<0$) and ρ is constrained, while in* (c) *and* (d) *the opposite is true [except, of course, $\rho(x)\geq 0$].*

In the following, a statement such as $\Delta\phi/4\pi=\rho$ is always meant in the distributional sense. We shall need Lemma II.26 from LS.

Lemma 3.3. *Let $\rho_1,\rho_2\in L^1$ with $\rho_i(x)\geq 0$ and $\psi_i=|x|^{-1}*\rho_i$. If $\psi_1(x)\geq\psi_2(x)$, all x, then $\int\rho_1\geq\int\rho_2$.*

Proof. Suppose $\int\rho_2-\rho_1=4\varepsilon>0$. There exists a ball, B, of radius R, such that $\int\rho_1(1-\Theta)<\varepsilon$ and $\int\rho_2(1-\Theta)<\varepsilon$, where $\Theta(x)=1$ for $x\leq R$ and zero otherwise. Compute the spherical average of $\psi_2-\psi_1$ on the sphere of radius R; it cannot be positive. The contribution from inside B is, by Newton's theorem, $R^{-1}\int(\rho_2-\rho_1)\Theta>2\varepsilon/R$. The contribution from outside B is at least $-R^{-1}\int\rho_1(1-\Theta)>-\varepsilon/R$. Adding these gives a contradiction. ■

Remark. Even if $\psi_1(x)>\psi_2(x)$ for all x, we cannot conclude that $\int\rho_1>\int\rho_2$.

Proof of Theorem 3.2. (a) will be proved here; (b), (c), and (d) follow similarly. Since ρ_λ gives equality, $\phi_\lambda-\mu_\lambda\leq\sup\{\ \}$. We have to show that if $(\phi_\rho-\mu)_+\leq j'(\rho)$

Rev. Mod. Phys. 53, 603-641 (1981)

a.e. and if $\int\rho \leqslant \lambda$ then

(i) $\phi_\lambda(x) - \mu_\lambda \geqslant \phi_\rho(x) - \mu$,
(ii) $\mu \geqslant \mu_\lambda$.

First suppose $\mu \geqslant \mu_\lambda$ and let $\psi(x) = \phi_\rho(x) - \phi_\lambda(x) + \mu_\lambda - \mu$. Let $B = \{x \mid \psi(x) > 0\}$. B is open since ψ is continuous. As a distribution $-(4\pi)^{-1}\Delta\psi(x) = \rho_\lambda(x) - \rho(x) \leqslant 0$ a.e. on B since $j'(\rho) \geqslant \phi_\rho - \mu$, $j'(\rho_\lambda) = \phi_\lambda - \mu_\lambda$ when $\rho_\lambda > 0$ and j' is nondecreasing. Hence ψ is subharmonic on B and takes its maximum on ∂B, the boundary of B, or at ∞. $\psi = 0$ on ∂B. At ∞, $\psi = \mu_\lambda - \mu \leqslant 0$. Hence B is empty and (i) is proved. Suppose now that $\mu_\lambda - \mu = \delta > 0$. Then $j'(\rho) \geqslant \phi_\rho - \mu > \phi_\rho - \mu_\lambda$ and, by the previous proof (applied to $\mu = \mu_\lambda$), $\phi_\lambda(x) \geqslant \phi_\rho(x)$. By Lemma 3.3, $\int\rho \geqslant \int\rho_\lambda = \lambda$. Hence $\int\rho = \lambda$. At this point there are two possible strategies.

(i) If we assume that $\mathcal{E}(\rho)$ has a minimum that satisfies Eq. (3.2) for all $\mu \geqslant 0$, then we can use the fact [which follows from the strict convexity of $\mathcal{E}(\rho)$] that μ_λ is a continuous decreasing function of λ. Then there exists $\gamma > \lambda$ with $\mu_\lambda > \mu_\gamma > \mu$. Since $j'(\rho) \geqslant \phi_\rho - \mu_\gamma$, $\int\rho = \gamma$ by what we just proved. But this is a contradiction.

(ii) There is a purely potential theoretic argument without invoking Eq. (3.1). There is a (not necessarily unique) f which satisfies $j'(f(x)) = [\phi_\rho(x) - \mu/2 - \mu_\lambda/2]_-$ and $f(x) = 0$ when $[\ ] = 0$. Hence $f(x) < \rho(x)$ a.e. when $\rho(x) > 0$, and $f(x) = 0$ when $\rho(x) = 0$. Thus $f \in \mathcal{C}$. Since $\int\rho = \lambda > 0$, $\int f < \lambda$. Let $g = (1-\varepsilon)\rho + \varepsilon f$, $0 < \varepsilon < 1$. Since $|x|^{-1} * \rho$ (and hence $|x|^{-1} * f$) are bounded, $\phi_g(x) \leqslant \phi_\rho(x) + \varepsilon C$ for some constant C. Choose $\varepsilon > 0$ so that $\varepsilon C < \delta/2$. Then $j'(g) \geqslant j'(f) \geqslant \phi_g - \mu_\lambda$. Since $\int g < \lambda$, g satisfies the condition in (a) with $\mu = \mu_\lambda$ but, as we have seen, this implies $\int g = \lambda$. ■

Teller's lemma (Theorem 3.4) is closely related to Theorem 3.2.

Definition. We say $V \in \mathfrak{D}$ if $V \neq 0$ and V is superharmonic, vanishing at ∞ (and hence $V > 0$). Moreover, the set $\{x \mid V(x) = \infty\} = S_V$ (called the *singularities* of V) is closed, V is continuous on the complement of S_V, and $V(x) \to \infty$ as $x \to S_V$.

Theorem 3.4. *Suppose V is replaced by $V' = V + W$ with $W \in \mathfrak{D}$. (In the case of interest $W = z|x - R|^{-1}$, which means that we add, or increase, a nuclear charge.) Suppose that for some common μ there are solutions to Eq. (3.2) $0 \leqslant \rho$, $\rho' \in \mathcal{C}$ with V and with V'. Then $\phi'(x) \geqslant \phi(x)$ all x and, if j' is strictly monotone or if $\phi - \phi' \in H^2$ (i.e., $\phi - \phi'$ and its first two derivatives are in L^2) away from S_W, then $\rho'(x) \geqslant \rho(x)$, a.e.*

Proof. Let $\psi = \phi' - \phi$ and $B = \{x \mid \psi(x) < 0\}$. Clearly $B \cap S_W = \varnothing$ so B is open. As a distribution, $\Delta\psi/4\pi \leqslant \rho' - \rho \leqslant 0$, so ψ is superharmonic on B. Thus B is empty and $\phi' \geqslant \phi$. The proof that $\rho' \geqslant \rho$ is trickier. If j' is strictly monotone it is obvious. Otherwise it can be shown (see Benguria, 1979) that for suitable V, W, and j', $\psi \in H^2$ away from S_W. Assuming $\psi \in H^2$, if $\rho'(x) < \rho(x)$ then $x \in C = \{x \mid \psi(x) = 0\}$. On C, $\Delta\psi = 0$, a.e. (see Benguria, 1979, Theorems 2.19 and 3.3). Let $D = C \cap \{x \mid \rho'(x) < \rho(x)\}$. On D, $0 = \Delta\psi/4\pi \leqslant \rho' - \rho < 0$ a.e., so D has zero measure. ■

Remark. If $j'(s)$ is strictly monotone and $W \neq 0$ then $\phi'(x) > \phi(x)$ for all $x \notin S_W$.

A similar proof yields

Theorem 3.5. *If $V \in \mathfrak{D}$ then $\phi_\lambda(x) \geqslant 0$. Consequently if $V(x) = \int dM(y)|x - y|^{-1}$, $dM \geqslant 0$, and $\int dM = Z$, then there is no solution if $\lambda > Z$ because then $\phi_\lambda(x) < 0$ for some large x. Cf. Theorem 6.7.*

There are many easy, but important corollaries of Theorem 3.2. We stress that V need not be Coulombic; the important ingredient is that the electron-electron repulsion is Coulombic.

Definition. j' is said to be *subadditive* if $j'(\rho_1 + \rho_2) \leqslant j'(\rho_1) + j'(\rho_2)$. j' is subadditive in the TF case.

Corollary 3.6. *Suppose $V = V_1 + V_2$ and μ is fixed. Let ϕ, ϕ_1, ϕ_2 be solutions to Eq. (3.2) for this μ with V, V_1, V_2, respectively. Suppose $\phi_i \geqslant 0$ (e.g., $V_i \in \mathfrak{D}$) and suppose j' is subadditive. Then $\phi(x) \leqslant \phi_1(x) + \phi_2(x)$, all x.*

Proof. Use Theorem 3.2 (d) with $\rho_1 + \rho_2$ on the right side. ■

Corollary 3.7. *Let $\lambda > 0$. There can be at most one pair ρ, μ satisfying Eq. (3.2) with $\rho \in \mathcal{C}$ (in particular for $\rho \in L^{5/3} \cap L^1$) and $\int\rho = \lambda$.*

Proof. If ρ_1, ρ_2 are two solutions, use Theorem 3.2(a) twice with ρ_1 and ρ_2 to deduce $\phi_1 - \mu_1 = \phi_2 - \mu_2$. This implies $\mu_1 = \mu_2$ and hence $\phi_1 = \phi_2$. But then $0 = \Delta(\phi_1 - \phi_2) = 4\pi(\rho_2 - \rho_1)$. ■

This uniqueness result was proved earlier, Theorem 2.6, using the strict convexity of $\mathcal{E}(\rho)$.

Corollary 3.8. *If $0 < \lambda' < \lambda$ then*

(i) $\phi_{\lambda'} \geqslant \phi_\lambda$
(ii) $\mu_{\lambda'} \geqslant \mu_\lambda$
(iii) $\phi_{\lambda'} - \mu_{\lambda'} \leqslant \phi_\lambda - \mu_\lambda$.

Proof. For (iii) use Theorem 3.2(b) with $\rho_\lambda, \mu_\lambda$ as trial function for the λ' problem. (iii) $\Rightarrow$ (ii). For (i) use (c) with ρ_λ as variational function for the λ' problem. ■

Corollary 3.9. *Suppose ρ_1, μ and ρ_2, μ (same μ) are two solutions to Eq. (3.2) with $\int\rho_1, \int\rho_2 > 0$. Then $\phi_1 = \phi_2$ and $\rho_1 = \rho_2$ a.e. Therefore, by Corollary 3.8, whenever $\lambda_2 > \lambda_1$ then $\mu_2 < \mu_1$ (i.e., $\mu_2 = \mu_1$ cannot occur).*

Proof. Using Theorem 3.2(d), $\phi_1 = \phi_2$. Then

$$0 = \Delta(\phi_1 - \phi_2)/4\pi = \rho_1 - \rho_2 \text{ a.e. } ■$$

Corollary 3.10. *Suppose $j_1'(\rho) \leqslant j_2'(\rho)$, all ρ. Let $\rho_\lambda^1, \mu_\lambda^1$ and $\rho_\lambda^2, \mu_\lambda^2$ be corresponding solutions to Eq. (3.2) with fixed λ, and ρ^1, ρ^2 solutions with fixed μ. $\phi_{(\lambda)}^i(x)$ are the corresponding potentials. Then*

(i) $\phi_\lambda^1 - \mu_\lambda^1 \leqslant \phi_\lambda^2 - \mu_\lambda^2$,
(ii) $\mu_\lambda^1 \geqslant \mu_\lambda^2$,
(iii) $\phi^1 \leqslant \phi^2$.

Proof. For (i) use Theorem 3.2(a) with $\rho_\lambda^1, \mu_\lambda^1$ as trial function for the 2 problem. (i) $\Rightarrow$ (ii). For (iii) use (d) with ρ_2 as trial function for the 1 problem. ■

Lemma 3.11. *When $\mu > 0$, ρ has compact support.*

Remark. As will be seen in Sec. VI, ρ has compact support in TFD theory even when $\mu = 0$. See Theorem 6.6.

Among the most important consequences of Theorem

3.2 are the *variational principles for the chemical potential* [LS].

Theorem 3.12. *Define the functionals*

$$T(\rho)=\operatorname{ess\,sup}_x\{\phi_\rho(x)-j'(\rho(x))\},$$
$$S(\rho)=\operatorname*{ess\,inf}_{x:\rho(x)>0}\{\phi_\rho(x)-j'(\rho(x))\}. \tag{3.3}$$

(ess sup *means supremum modulo sets of measure zero*). *Then, whenever there is a solution to (3.2) with* $\int\rho=\lambda>0$,

$$\mu_\lambda=\inf\left\{T(\rho)\,|\,\rho\in\mathcal{C},\int\rho\le\lambda\right\} \tag{3.4}$$

$$\mu_\lambda=\sup\left\{S(\rho)\,|\,\rho\in\mathcal{C},\int\rho\ge\lambda\right\}. \tag{3.5}$$

Corollary 3.13. *If* $j'(\rho)$ *is concave (as in TF theory with* $j'=\rho^{2/3}$*) then* μ_λ *and* $\mu_\lambda-\phi_\lambda(x)$*, for each x, are jointly convex functions of V and* λ.

Corollary 3.14. *If* λ *is fixed and* $V_1(x)\ge V_2(x)$*, all x, then* $\mu_\lambda(1)\ge\mu_\lambda(2)$.

By Corollaries 3.9 and 3.14 we know that increasing V increases μ while increasing λ decreases μ. What happens if V and λ are both increased, in particular if we scale up the size of a molecule by $V\to\alpha V$, $\lambda\to\alpha\lambda$? A partial answer is given by the following two corollaries.

Corollary 3.15. *Let* $V_1, V_2\in\mathcal{D}$ *and* $V=V_1+V_2$. *Assume* j' *is subadditive and suppose Eq. (3.2) has solutions to the three problems* (V_1,λ_1), (V_2,λ_2), (V,λ) *with* $\lambda=\lambda_1+\lambda_2$. *Then* $\mu\ge\min(\mu_1,\mu_2)$.

Proof. In general, if $W\in\mathcal{D}$ and ρ is a solution to (3.2) with W then $\phi_\rho(x)-j'(\rho(x))=\mu$ a.e. if $\rho(x)>0$ and ≥ 0 a.e. if $\rho(x)=0$ (Theorem 3.5). From this remark it follows that $S_V(\rho_1+\rho_2)\ge\min(\mu_1,\mu_2)$. [Here, $S_V(\rho)$ refers to Eq. (3.3) with V.] ■

Corollary 3.16. *Let* $\alpha>1$ *and suppose Eq. (3.2) has solutions with* (V,λ,μ) *and* $(\alpha V,\alpha\lambda,\mu(\alpha))$. *Assume* j' *satisfies* $j'(\alpha t)\le\alpha j'(t)$*, all t (this holds in TF theory). Then* $\mu(\alpha)\ge\alpha\mu$.

Proof. If ρ is the solution to (V,λ,μ) then $S_V(\rho)=\mu$. But $S_{\alpha V}(\alpha\rho)\ge\alpha S_V(\rho)$. ■

Corollary 3.17. *Suppose there is a solution to Eq. (3.2) for all* $\lambda\in(a,b)$ *with* $a<b$. *Then* μ_λ *is continuous on this interval.*

Proof. Let $\lambda_2=\lambda_1+\varepsilon$. By Corollary 3.9, $\mu_1>\mu_2$. Let $\rho=\rho_1+\varepsilon\chi$ with $\int\chi=1$, $0\le\chi(x)\le b$ for some b, $\chi(x)=0$ if $\rho_1(x)=0$, and $\chi(x)=0$ if $\rho_1(x)>a$ for some a. Then $\chi\in\mathcal{C}$. Since j' is continuous, $S(\rho)\ge\mu_1-Q(\varepsilon)$ where $Q(\varepsilon)\downarrow 0$ as $\varepsilon\downarrow 0$. ■

Theorem 3.18. *Let* $V\in\mathcal{D}$, $V(x)=\int dM(y)\,|y-x|^{-1}$, $dM\ge 0$, $\int dM=Z>0$. *Suppose that for large t,* $j'(t)>ct^{(1/2)+\varepsilon}$*, with* $c,\varepsilon>0$. *By a simple modification of the method of Theorems 2.4, 2.5, and 2.6,* $\mathcal{E}(\rho)$ *has a unique minimum on the set* $\mathcal{C}$ *with* $\int\rho\le\lambda$. *This* ρ *satisfies Eq. (3.2) and* $\int\rho=\lambda$ *if* $\lambda\le\lambda_c$*, whereas* $\int\rho=\lambda_c$ if $\lambda>\lambda_c$. *Now assume, in addition, that* $j'(t)<dt^{\varepsilon+1/3}$, $\varepsilon>0$, *for* small t *(this is true in all cases of interest). Then* $\lambda_c=Z$.

Proof. If $\lambda>Z$ there is no solution by Theorem 3.5. Now suppose $\mu=0$; we claim $\lambda\ge Z$, and hence that $\lambda=Z$. If so, we are done because $\mathcal{E}(\rho)$ has an absolute minimum. This minimum corresponds to $\mu=0$ and has $\lambda=\lambda_c$; but $\mu=0$ implies $\lambda=Z$. Now, to prove that $\lambda\ge Z$, let ϕ be the solution. If $\lambda=Z-3\delta$, let χ be the characteristic function of a ball centered at the origin such that $\int\chi dM>Z-\delta$. Then

$$\phi(x)>\psi(x)=\int[\chi(y)dM(y)-\rho(y)dy]\,|x-y|^{-1}.$$

For $|x|>$ some R, $\psi(x)<2Z|x|^{-1}$. Also $[\psi(x)]=$ (spherical average of ψ) $>2\delta|x|^{-1}$ for $x>R$. For a given $|x|=r>R$ let $\Omega_+(r)$ be the proportion of the sphere of radius r such that $2Z>r\psi(x)>\delta$, and let $\Omega_-(r)$ be the complement. Then $2\delta<r[\psi(x)]<2Z\Omega_++\delta\Omega_-=\delta+(2Z-\delta)\Omega_+$. Thus $\Omega_+(r)>\delta/(2Z-\delta)$ for all r. On Ω_+, $\rho^{(1/3)+\varepsilon}>\delta|x|^{-1}$ for large $|x|$, and therefore $\rho\notin L^1$ if $\delta>0$. ■

Brezis and Benilan (Brezis, 1978, 1980) have generalized this. Even if $j(\rho)\sim\rho^{(4/3)-\varepsilon}$ for large ρ there is a solution to Eq. (3.2) if $\lambda\le Z$, and no solution otherwise. This is noteworthy, since if $j(\rho)\sim\rho^a$ for large ρ with $a\le\frac{3}{2}$ then $\mathcal{E}(\rho)$ has no lower bound for point nuclei. There are similar results for other potentials, V, in LS, Theorem II.18.

There is also an "energetic," as distinct from potential theoretic, reason that there is no solution if $\lambda>Z$. A solution to Eq. (3.2) implies a minimum for the functional $\mathcal{E}(\rho)$, by strict convexity. If $\lambda=\int\rho>Z$ then ϕ_ρ is negative in some set A of positive measure. Then it is easy to see that if ρ is decreased slightly in A to $\tilde\rho$, then $\mathcal{E}(\tilde\rho)<\mathcal{E}(\rho)$. But $\int\tilde\rho<\lambda$ and $E(\lambda)$ is nonincreasing.

In the variational principle, Theorem 3.12, ρ_λ gives equality, i.e., $T(\rho_\lambda)=S(\rho_\lambda)=\mu_\lambda$. Is this the only ρ with this property? If $\lambda>Z$ there are many ρ's with $T(\rho)=0$ and no ρ with $S(\rho)=0$ (cf. LS). In Brezis, 1980, Sec. 4, it is shown that if j' is concave (as in TF theory) and V has suitable properties (satisfied for $V\in\mathcal{D}$) then when $\lambda<Z$ only ρ_λ satisfies either $T(\rho)=\mu_\lambda$ or $S(\rho)=\mu_\lambda$. If $\lambda=Z$ this uniqueness is lost in general!

Asymptotics of the chemical potential.

Theorem 3.12 can be used to obtain bounds on μ_λ. In the TF case with point nuclei, the asymptotic formula

$$\mu_\lambda\sim\gamma^{-1}(\pi^2\textstyle\sum z_j^3/4\lambda)^{2/3} \tag{3.6}$$

holds for λ small (LS, Theorem II.31). For λ near Z LS (Theorems IV.11, 12) find upper and lower bounds for μ_λ of the form $\alpha_\pm(Z-\lambda)^{4/3}$ with $Z=\sum z_j$. Brezis and Benilan (unpublished) have shown that

$$\alpha=\lim_{\lambda\uparrow Z}\mu_\lambda(Z-\lambda)^{-4/3}\text{ exists} \tag{3.7}$$

and is given by solving some differential equation. α is independent of the number of nuclei and their individual coordinates and charges!

Equation (3.7) implies that there is a well defined *ionization potential I* in TF theory (although it probably has nothing to do with the true Schrödinger ionization energy). First observe that if we start with $\sum z_j=1$ and then replace z_j by Zz_j, R_j by $Z^{-1/3}R_j$, and λ by $Z\lambda$, then by scaling Eq. (2.24),

$$\mu_{Z\lambda}=Z^{4/3}\mu_\lambda\,. \tag{3.8}$$

Rev. Mod. Phys. 53, 603-641 (1981)

Therefore, by Eq. (3.7), if we let $\lambda = Z - \varepsilon$ with $\varepsilon > 0$ fixed, and let $Z \to \infty$, then

$$\lim_{Z\to\infty} \mu_{Z-\varepsilon} = \alpha\varepsilon^{4/3}. \tag{3.9}$$

The ionization potential is defined to be

$$I = E(\lambda = Z-1) - E(\lambda = Z). \tag{3.10}$$

By integrating (3.9), and appealing to dominated convergence,

$$I \to 3\alpha/7 \quad \text{as } Z \to \infty. \tag{3.11}$$

Another implication of Eq. (3.7) is that an ionized *atom* has a well defined radius as $Z \to \infty$. This question was raised by Dyson. Suppose $V(x) = Z|x|^{-1}$ and $\lambda = Z - \varepsilon$. The density ρ will have support in a ball of radius $R(Z, \varepsilon)$. At $|x| = R$, $\phi(x) = \mu$. But since ρ is spherically symmetric, $R\phi(x) = Z - \lambda = \varepsilon$ by Newton's theorem. Thus the atomic radius satisfies

$$R = \varepsilon/\mu \quad \text{for all atoms} \tag{3.12}$$

and, by Eq. (3.9),

$$\lim_{Z\to\infty} R(Z,\varepsilon) = (\alpha\varepsilon^{1/3})^{-1}. \tag{3.13}$$

There are other ways in which TF theory yields a well defined atomic radius. See Sec. V.C (6).

B. The case of flat j' (TFD)

In TFD theory, as will be seen in Sec. VI, we have to consider

$$\begin{aligned} j'(\rho) &= 0, \quad 0 \le \rho \le \rho_0 = (5C_e/8\gamma)^3 \\ &= \gamma\rho^{2/3} - C_e\rho^{1/3} + 15C_e^2/4^3\gamma, \quad \rho_0 \le \rho. \end{aligned} \tag{3.14}$$

j' satisfies all necessary conditions. It is neither concave nor subadditive, however. Let us consider V of the form

$$V(x) = \int dm(y)|x-y|^{-1}, \tag{3.15}$$

with m being a measure that is not necessarily positive. In the primary case of interest, $dm(x) = \sum z_j \delta(x - R_j)$.

The question we address here (and which will be important in Sec. VI) is this: Does $\rho(x)$ [the solution to Eq. (3.2)] take values in $(0, \rho_0)$? It may or may not, depending on m and λ.

Example. Suppose $dm(x) = g(x)dx$ with $g(x) \in (0, \rho_0)$ and $\int g = Z < \infty$. Then $\rho(x) = g(x)$ satisfies Eq. (3.2) with $\lambda = Z$, and thus $\rho(x) \in (0, \rho_0)$. This ρ also clearly minimizes $\mathcal{E}(\rho)$ in Eq. (3.1).

Nevertheless, in some circumstances $\rho \notin (0, \rho_0)$.

Theorem 3.19. *Suppose $j'(\rho) = \alpha = constant$ for $\rho \in F \equiv (\rho_1, \rho_0]$ with $0 \le \rho_1 \le \rho_0 < \infty$, and $j'(\rho) > c\rho^{(1/2)+\varepsilon}$ for large ρ. Let V be given by Eq. (3.15) and let A be a bounded open set such that as distributions on A either $\rho_0 dx < dm < (\rho_0 + const)dx$ or $dm < \rho_1 dx$. Let $\rho \in \mathfrak{C}$ satisfy Eq. (3.2). Then $\rho(x) \notin F$ a.e. (with respect to Lebesgue measure) on A.*

Proof. Cf Benguria, 1979, Lemmas 2.19, 3.2. First, it can be shown that $\phi_\rho \in H^2(A)$ (Sobolev space). Let $B = \{x | \rho(x) \in F\} \cap A$. On B, $\phi_\rho - \mu = \alpha$ and since $\phi_\rho \in H^2(A)$, $\Delta\phi_\rho = 0$ a.e. on B. But $\Delta\phi_\rho/4\pi = \rho - dm/dx$. ■

Remark. Since a solution to Eq. (3.2) is determined only a.e., $\rho(x)$ can be chosen $\notin F$ for all $x \in A$.

Corollary 3.20. *Consider the TFD problem (3.14) with $V(x) = \sum z_j |x - R_j|^{-1}$. Then any solution to (3.2) can be modified on a set of measure zero so that $\rho(x) \notin (0, \rho_0]$ for all x.*

C. No-binding theorems

Henceforth it will be assumed, as in Theorem 3.18, that j is such that Eq. (3.1) has a minimum for $\lambda \le \lambda_c$ which satisfies Eq. (3.2). We shall be interested in comparing three (nonzero) potentials, V_1, V_2, and $V_{12} = V_1 + V_2$ with $V_i \in \mathfrak{D}$. At first we shall consider what happens when the repulsion U is absent. As usual we define $e_a(\lambda) \equiv \inf \mathcal{E}_a(\rho)$ with $\lambda = \int\rho$ and $\mathcal{E}_a$ having V_a. There is no U term in $\mathcal{E}_a$, Eq. (3.1). Define

$$\Delta e(\lambda) = e_{12}(\lambda) - \min_{\lambda_1 + \lambda_2 = \lambda} e_1(\lambda_1) + e_2(\lambda_2). \tag{3.16}$$

Definition. If $\Delta e < 0$ (resp. ≥ 0) we say that *in the absence of the repulsion U there is binding* (resp. *no binding*).

Theorem 3.21. *Suppose j satisfies*

$$j(a+b) \le j(a) + j(b) + aj'(b) + bj'(a), \quad a, b \ge 0. \tag{3.17}$$

[If j' is subadditive then Eq. (3.17) is satisfied. $j(t) = t^{5/3}$ satisfies (3.17).] Then $\Delta e < 0$.

Proof. For $i = 1, 2$ let λ_i minimize in Eq (3.16) and let ρ_i be the minimizing ρ for $\mathcal{E}_i$ with $\int\rho_i \le \lambda_i$. Recall $e_a(\lambda)$ is monotone nonincreasing. Let $\rho \equiv \rho_1 + \rho_2$ be a trial function for e_{12} in $\mathcal{E}_{12}$ and use the variational equations (3.2) for ρ_i and the fact that $\phi_i(x) \ge 0$. ■

Remark. The condition (3.17) is satisfied in TF theory but not in TFD theory.

Theorem 3.21 says we can [and do, if j satisfies Eq. (3.17)] have binding if the repulsion U is absent. The no-binding theorem, which we turn to now, relies on the addition of U which, by itself without $\mathcal{E}$, obviously has the no-binding property.

Proposition 3.22. *If j is convex and $j(0) = 0$, then j has the superadditivity property: $j(a+b) \ge j(a) + j(b)$. If j' is strictly monotone, then the foregoing inequality is strict when $a, b \ne 0$.*

Note. We assumed that j is convex in all cases. Therefore Theorem 3.23 holds in all cases.

Definition. Let

$$V_i = \frac{1}{|x|} * m_i \quad (m_i \text{ a measure})$$

be in $\mathfrak{D}$. Then

$$D(m_1, m_2) \equiv \tfrac{1}{2}\int dm_1(x)dm_2(y)|x-y|^{-1}.$$

Theorem 3.23 (no binding). *Let m_i, $i = 1, 2$ be nonnegative measures of finite mass $z_i > 0$ and $V_i \in \mathfrak{D}$. Then*

$$\Delta E(\lambda) \equiv \Delta e(\lambda) + 2D(m_1, m_2) \ge 0. \tag{3.18}$$

If j is strictly superadditive then > 0 holds.

Remarks. Obviously $\Delta E(\lambda)$ is the energy difference when the repulsion U is included. *Binding never occurs.* In particular, if

$$V_1=\sum_{j=1}^{n} z_j|x-R_j|, \qquad V_2=\sum_{j=n+1}^{k} z_j|x-R_j|^{-1}$$

then

$$m_1=\sum_{j=1}^{n} z_j\delta(x-R_j), \qquad m_2=\sum_{j=n+1}^{k} z_j\delta(x-R_j).$$

In TFD theory j is not strictly superadditive. As we shall see in Sec. IV.C, it is possible to have a neutral diatomic molecule for which equality holds in Eq. (3.18).

Proof. We give two proofs. The LS proof in the neutral case $\lambda=z_1+z_2$ is the following: Clearly $\lambda_1=z_1$, $\lambda_2=z_2$, $\mu_1=\mu_2=\mu_{12}=0$. Consider $m_1\to\alpha m_1$, $\lambda_1\to\alpha z_1$, $0\le\alpha\le 1$. By Theorem 2.13 we have

$$\partial e_1/\partial\alpha=-\int V_1\rho_1$$

and

$$\partial e_{12}/\partial\alpha=-\int V_1\rho_{12}.$$

Thus

$$\partial(e_{12}-e_1+2D(\alpha m_1,m_2))/\partial\alpha=\int dm_1(x)[\phi_{12}(x)-\phi_1(x)]\,.$$

But $\phi_{12}(x)\ge\phi_1(x)$, all x (Theorem 3.4 and following remark). When $\alpha=0$, $\Delta E=0$, so this proves the theorem. In the non-neutral case the $\mu_a\ne 0$ and it is necessary to take into account the change of μ_a with α. This is complicated (see LS).

The second proof is due to Baxter (1980). For any ρ_{12} with $\int\rho_{12}=\lambda$ we can, by Prop. 3.24, find g, $0\le g(x)\le\rho_{12}(x)$, and $h(x)\equiv\rho_{12}(x)-g(x)$ such that $\psi_g(x)=\psi_{m_1}(x)=V_1(x)$ a.e. when $h(x)>0$, and $\psi_g(x)\le V_1(x)$ a.e. when $h(x)=0$.

Let $a=\int g$, $b=\int h$. Then

$$\begin{aligned}\min\{e_1(\lambda_1)+e_2(\lambda_2)\,|\,\lambda_1+\lambda_2=\lambda\}&\le e_1(a)+e_2(b)\le\mathcal{E}_1(g)+\mathcal{E}_2(h)\\&\le\mathcal{E}_{12}(\rho_{12})+2D(m_1,m_2)+\int h(V_1-\psi_g)dx-\int(V_1-\psi_g)dm_2\\&\le e_{12}(\lambda)+2D(m_1,m_2)\,.\end{aligned}\tag{3.19}$$

The third inequality uses the superadditivity of j. If j' is strictly monotone this superadditivity is strict (and so is the final inequality) provided $g\ne\rho_{12}$. If $g=\rho_{12}$ a.e. then $\psi_{\rho_{12}}\le V_1$ and hence $\lambda\le z_1$ must hold. Choose $\lambda_1=\lambda$, $\lambda_2=0$ and note that $e_1(\lambda)<\mathcal{E}_1(\rho_{12})$ because ρ_{12} does not satisfy Eq. (3.2) since $V_2\ne 0$. Equation (3.19) then gives strict inequality. ■

Proposition 3.24 (Baxter, 1980). *Let $V\in\mathfrak{D}$ and let $\rho(x)\ge 0$ be a given function with $|x|^{-1}*\rho\equiv\psi_\rho\in\mathfrak{D}$. Assume $\rho\in L^p$ for some $p>\frac{3}{2}$ and $D(\rho,\rho)<\infty$. Then there exists g with $0\le g(x)\le\rho(x)$ such that $\psi_g=|x|^{-1}*g$ satisfies $\psi_g(x)=V(x)$ a.e. when $\rho(x)-g(x)>0$ and $\psi_g(x)\le V(x)$ a.e.*

Proof. Baxter proves this when ρ and g are measures. We give a simpler proof for functions. Consider $\mathcal{E}(g)\equiv D(g,g)-\int Vg$ and $E=\inf\{\mathcal{E}(g)\,|\,0\le g(x)\le\rho(x)\}$. Let g^n be a minimizing sequence. There exists a subsequence that converges weakly in L^p to some g and, by Mazur's theorem (1933), there exists a sequence h^n of convex combinations of the g^n that converges strongly to g in L^p. Then a subsequence of the h^n converges a.e. to g. Clearly, $0\le h^n(x)\le\rho(x)$. Since $\mathcal{E}(\cdot)$ is convex (this is crucial) $\mathcal{E}(h^n)\to E$ but, by dominated convergence, $\mathcal{E}(h)\to\mathcal{E}(g)$. So g minimizes and satisfies (a.e.): $V(x)=\psi_g(x)$ when $0<g(x)<\rho(x)$; $\psi_g(x)\le V(x)$ when $g(x)=\rho(x)$ and $\rho(x)>0$; $\psi_g(x)\ge V(x)$ when $g(x)=0$ and $\rho(x)>0$. We have to eliminate the possibility $\psi_g(x)-V(x)\equiv f(x)>0$ when $g(x)=0$. We claim $\psi_g\in\mathfrak{D}$ and hence f is continuous and goes to zero at ∞. Since $g\le\rho$, $\psi_g\le\psi_\rho$ so $\psi_g\to 0$ at infinity. To examine the continuity at $x=0$, write $\psi_g=h+(\psi_g-h)$ with $h=|x|^{-1}*(\chi g)$ and χ is the characteristic function of the ball $|x|<1$. Clearly ψ_g-h is continuous at $x=0$. Moreover, $\chi g\in L^p\cap L^1$ so $h\in\mathfrak{D}$ by Lemma 3.1. (It is here that $p>\frac{3}{2}$ is used.) Now, since $f\in\mathfrak{D}$, $B=\{x\,|\,f(x)>0\}$ is open and, since $x\in B\Rightarrow g(x)=0$, f is subharmonic on B. But f vanishes on the boundary of B and at infinity, so B is empty. ■

IV. DEPENDENCE OF THE THOMAS-FERMI ENERGY ON THE NUCLEAR COORDINATES

In the previous sections TF theory was analyzed when the nuclear coordinates $\{R_j\}$ are held fixed. The one exception was Teller's theorem (Theorem 3.23) which states that the TF energy is greater than the TF energy for isolated atoms (which is the same as the energy when the R_j are infinitely far apart). Here, more detailed information about the dependence of E on the R_j is reviewed.

Note that in this section (and henceforth) E refers to the total energy, [Eq. (2.11)], including the repulsion U. This is crucial.

Although several unsolved problems remain, a fairly complete picture will emerge. The principal open problem is to prove the positivity of the pressure (Sec. IV.B) for subneutral molecules, and to prove it for deformations more general than uniform dilation. The results of this section have been proved only for TF theory, and it is not known which ones extend to the variants (see the discussion of TFD theory in Sec. IV.C).

A. The many-body potentials

The results here are from Benguria and Lieb, 1978a. As usual, the two-body atomic energy is *defined* to be the difference between the energy of a diatomic molecule (with nuclear separation R) and the energy of isolated atoms. Teller's theorem states that this is always positive. We shall now investigate the k-body energy which can be defined similarly. The three-body energy will be shown to be negative, the four-body positive, etc. In all cases, only neutral systems will be considered; in this case there is a unique way to apportion the electron charge among the isolated atoms, namely, make them all neutral. An interesting problem is to treat the k-body energy for subneutral systems.

Definitions. When $c=\{c_1,c_2,\ldots,c_k\}$ is a finite subset of the positive integers with $|c|=k$ elements, $E(c)$ denotes the TF energy for a neutral molecule consisting of nuclear charges $z_{c_i}>0$ located R_{c_i}. $\phi(c,x)$ denotes the TF potential for this molecule. The z's can all be different.

$$\varepsilon(c)=\sum_{b\subseteq c}(-1)^{|b|+|c|}E(b) \tag{4.1}$$

is the $|c|$ body energy for this molecule. Thus, if $c=\{1,2\}$, $|c|=2$ and the two-body energy is $\varepsilon(1,2)=E(1,2)-E(1)-E(2)$ as explained above. If $c=\{1,2,3\}$, $|c|=3$ and the three-body energy is

$$\varepsilon(1,2,3)=E(1,2,3)-[E(1,2)+E(1,3)+E(2,3)]+E(1)+E(2)+E(3).$$

$E(1),E(2),E(3)$ are atomic energies, of course. From Eq. (4.1)

$$E(c)=\sum_{b\subseteq c}\varepsilon(b). \tag{4.2}$$

It is worth remarking that the many-body energies (4.1) are defined in terms of the *total* energy E. It is equally possible to use $e=E-U$ on the right side of Eq. (4.1). e is the *electronic contribution* to E, so the corresponding ε's would be the *electronic contribution to the many-body potential*. However, note that U contains only two-body pieces, $z_iz_j|R_i-R_j|^{-1}$. Therefore the two sets of ε's agree whenever $|c|\geq 3$, i.e., the three- and higher-body ε's are entirely electronic. As far as the two-body energy is concerned, $\varepsilon(1,2)_{\text{tot}}>0$ (Teller) but

$$\varepsilon(1,2)_{\text{elec}}=\varepsilon(1,2)_{\text{tot}}-U(1,2)<0$$

(Theorem 3.21).

In the following $b\subset c$ means b is a subset of c and $b\neq c$.

Theorem 4.1 (Sign of the many-body potential). *If c is not empty*

$$(-1)^{|c|}\varepsilon(c)>0.$$

More generally, if $b\subset c$ and either $|c\backslash b|\geq 2$ or else $|b|=0$ and $|c|>0$

$$\bar{E}(b,c)\equiv\sum_{b\subseteq a\subseteq c}(-1)^{|b|+|a|}E(a)>0.$$

Theorem 4.2 (Remainder Theorem). *If $2\leq\beta\leq|c|$ then the sign of*

$$E(c)-\sum_{\substack{b\subset c\\ |b|<\beta}}\varepsilon(b)$$

is $(-1)^\beta$. In other words, if, in Eq. (4.2), we sum only over the terms smaller than β-body, the sign of the error is the sign of the first omitted terms.

Theorem 4.3 (Monotonicity of the many-body potential). *Suppose that $b\subset c$ and $|b|\geq 2$. Then*

$$(-1)^{|b|}\varepsilon(b)>(-1)^{|c|}\varepsilon(c).$$

Theorems 4.1 and 4.3 imply, for example,

$$0>\varepsilon(1,2,3)>-\min[\varepsilon(1,2),\varepsilon(1,3),\varepsilon(2,3)].$$

Theorem 4.4. *If $b\subset c$ and c is not empty*

$$\bar{\phi}(b,c,x)\equiv\sum_{b\subseteq a\subseteq c}(-1)^{|a|+|b|}\phi(a,x)<0.$$

Partial Proof. Basically Theorems 4.1, 4.2, and 4.3 are corollaries of Theorem 4.4 through the relation, for $j\in c$,

$$\frac{\partial E(c)}{\partial z_j}=\lim_{x\to R_j}\{\phi(c,x)-z_j|x-R_j|^{-1}\},\quad\text{(Theorem 2.13)}.$$

As an illustration we shall prove here that $\varepsilon(1,2,3)<0$; surprisingly, the proof is much more complicated when $|c|>3$. The proof for $|c|=3$ only uses that the function $(j')^{-1}$ is convex [cf. Eq. (3.1)]. The proof for $|c|>3$ requires that $j(\rho)=\rho^k$ with $\frac{3}{2}\leq k\leq 2$.

First note that $\varepsilon(1,2,3)=0$ when $z_3=0$. Thus it suffices to prove that $\partial\varepsilon(1,2,3)/\partial z_3=F(R_3)<0$, where

$$F(x)\equiv\phi(1,2,3,x)-\phi(1,3,x)-\phi(2,3,x)+\phi(3,x).$$

Now

$$\Delta F=4\pi[\rho(1,2,3,x)-\rho(1,3,x)-\rho(2,3,x)+\rho(3,x)]$$

and $\rho=(\phi/\gamma)^{3/2}$. Let $B=\{x\,|F(x)>0\}$. F is continuous, so B is open. We claim F is subharmonic on B, which implies B is empty. What is needed is the fact that $a-b-c+d\geq 0\Rightarrow a^{3/2}-b^{3/2}-c^{3/2}+d^{3/2}\geq 0$ under the conditions that $a\geq b\geq d\geq 0$ and $a\geq c\geq d\geq 0$ (Theorem 3.4). But this is an elementary exercise in convex analysis. Finally, as in the strong form of Theorem 3.4, one can prove that F is strictly negative. ■

It is noteworthy that *all* the many-body potentials fall off at the same rate, R^{-7}. This will be shown in Sec. IV.C.

B. The positivity of the pressure

Teller's theorem (Theorem 3.23) suggests that the nuclear repulsion dominates the electronic attraction and therefore a molecule in TF theory should be unstable under local as well as global dilations.

Let us fix the nuclear charges $\underline{z}=\{z_1,\ldots,z_k\}$ and move the R_i keeping λ fixed. Under which deformations does E decrease? We can also ask when $e=E-U$, *the electronic contribution to the energy*, decreases. A natural *conjecture* is the following: Suppose $R_i\to R_i'$ with $|R_i'-R_j'|\geq|R_i-R_j|$ for every pair i,j. Then

(i) E decreases and e increases.

(ii) Furthermore, if $\lambda_1<\lambda_2$ then the decrease (increase) in $E(e)$ is smaller (larger) for λ_2 than for λ_1. There is one case in which this conjecture can be proved; it is given in Theorem 4.7 due to Benguria (1981).

One interesting case is that of uniform dilation in

which each $R_i \to lR_i$. For this case we define the *pressure* and *reciprocal compressibility* to be

$$P(l) = -(3l^2)^{-1}dE(l)/dl \tag{4.3}$$

$$\kappa^{-1} = -(l/3)dP(l)/dl, \tag{4.4}$$

where $E(l)$ is the energy. This definition comes from thinking of the "volume" as proportional to l^3. If $K(l)$ is the kinetic energy [Eq. (2.20)] then

$$3l^3P(l) = E(l) + K(l).$$

To see this, define $E(\gamma, l)$ to be the energy with the parameter γ thought of as a variable (but with λ fixed). Then, by setting $\rho(x, l) = l^{-3}\tilde{\rho}(x/l, l)$, one easily sees that $E(\gamma, l) = l^{-1}E(\gamma/l, 1)$ and $K(\gamma, l) = l^{-1}K(\gamma/l, 1)$. Equation (4.4) follows from this and Theorem 2.13.

Note that Eq. (4.4) is true (for the same reason) in Q theory and also in TFD, TFW, and TFDW theories provided K is interpreted as Eq. (2.20) in TFD and as (2.20) + $\delta\int[\nabla\rho^{1/2}]^2$ in TFDW and TFW.

That $e = E - U$ increases under dilation has also been conjectured to hold in Q theory when $\lambda \leq Z$. It is known to hold for one electron, but an arbitrary number of nuclei (Lieb and Simon, 1978). There is one simple statement that can be made (in all theories): The (unique) minimum of e occurs when $l = 0$ (for any $\lambda > 0$), i.e., all the nuclei are at one point. To prove this, assume $R_1, \ldots, R_k$ are not all identical and let ρ be the minimizing solution. Let $\psi = |x|^{-1} * \rho$. ψ has a maximum at some point R_0. Now place all the nuclei at R_0 and use the same ρ as a variational ρ for this problem. Then, trivially, $e(R_0, \ldots, R_0) < e(R_1, \ldots, R_k)$, with the strict inequality being implied by the fact that this ρ does not satisfy the variational equation for $R_0, \ldots, R_0$.

It is useful to have a formula for the variation of e with R_i. A natural extension of Theorem 2.13 (a "Feynman-Hellman"-type theorem) would be the following: Suppose $V_1, \ldots, V_k \in \mathfrak{D}$ with

$$V_i(x) = \int dm_i(y)|y - x|^{-1} \tag{4.5}$$

and with m_i a positive measure of mass z_i. Take

$$V(x) = \sum_{i=1}^{k} V_i(x - R_i).$$

Then e is a C^1 function of the R_i and

$$\nabla_{R_i} e = \int \nabla V_i(x - R_i)\rho(x)dx = -\int dm_i(y)\nabla\psi(y + R_i), \tag{4.6}$$

with $\psi = |x|^{-1} * \rho$. Equation (4.6) is clearly true, and easy to prove if the m_i are suitably bounded. Benguria (unpublished) proved (4.6) when $V_i(x) = z_i|x|^{-1}$ for $|x| \geq a$ and $V_i(x) = z_i a^{-1}$ for $|x| \leq a$, with $a > 0$, i.e., $dm_i(y) = z_i(\text{const})\delta(|y| - a)$. In this case, the last equality in Eq. (4.6) follows from LS, Lemma IV.4.

For point nuclei, on the other hand, (4.6) has not been proved; indeed, the quantities in (4.6) are not even well defined. We conjecture that the following is true when $V_i(x) = z_i|x|^{-1}$: e is a C^1 function of the R_i on the set where $R_i \neq R_j$, for all $i \neq j$, and

$$\nabla_{R_i} e = -z_i \lim_{a \downarrow 0} \int_{|x - R_i| > a} (x - R_i)|x - R_i|^{-3}\rho(x)\,dx \tag{4.7a}$$

$$= -\lim_{x \to R_i} \nabla_x\{\psi(x) + (z_i/\gamma)^{3/2}(16\pi/3)|x - R_i|^{1/2}\}. \tag{4.7b}$$

Equation (4.7a) makes sense because, by Theorem 2.8,

$$\rho(x) = (z_i/\gamma)^{3/2}|x - R_i|^{-3/2} + O(|x - R_i|^{-1/2})$$

near R_i; the angular integration over the first term vanishes. This leading term in ρ implies that near R_i, $\psi(x) \approx (\text{const}) - (z_i/\gamma)^{3/2}(16\pi/3)|x - R_i|^{1/2}$. The nondifferentiable, but spherically symmetric term in ψ is subtracted in Eq. (4.7b).

The following theorems have been proved so far. (Theorems 4.5 and 4.6 are in Benguria and Lieb, 1978b; Theorem 4.7 is in Benguria, 1981.)

Theorem 4.5 (Uniform dilation). *Replace each R_i by lR_i and call the energy $E(\lambda, l)$. If $\lambda = Z$ then $E(\lambda, l)$ is strictly monotone decreasing and convex in l. In particular, the pressure and compressibility are positive.*

Remarks. (i) If $\lambda = 0$ the conclusion is obviously also true. In Benguria and Lieb (1978b) it is conjectured that this theorem holds for all λ. That $e = E - U$ is *monotone increasing* is also conjectured there.

(ii) In Benguria and Lieb (1978b) several interesting subadditivity and convexity properties of the energy and potential are also proved.

Theorem 4.6 (Molecule with planar symmetry). *Suppose the molecule is symmetric with respect to the plane $P = \{(x^1, x^2, x^3)\,|\,x^1 = 0\}$ and suppose no nucleus lies in the plane. Neutrality is not assumed. Let R_i^1 denote the 1 coordinate of nucleus i and, for all i, replace R_i^1 by $R_i^1 \pm l$, with $\pm$ if $R_i^1 \gtrless 0$, and $l \geq 0$. Then for all fixed $\lambda \leq Z$, E is decreasing in l.*

Remark. For a homopolar diatomic molecule the dilations in Theorems 4.5 and 4.6 are the same. Balàzs (1967) first proved Theorem 4.6 in this case. For a general diatomic molecule, Benguria's Theorem 4.7 is the strongest theorem.

Theorem 4.7. *Suppose there exists a plane P containing $R_1, \ldots, R_m$ and such that all the other R_j (with $j = m+1, \ldots, k$) are on one (open) side of P (call this side P^+). Assume the nuclei at $R_1, \ldots, R_m$ are point nuclei, but the nuclei at $R_{m+1}, \ldots, R_k$ are anything in $\mathfrak{D}$ and given by Eq. (4.5) with the supports of $m_i \in P^+$ (this includes point nuclei). Let $\mathbf{n}$ be the normal to P pointing away from P^+. Let $l_1, \ldots, l_m \geq 0$ be given and let $R_i \to R_i + l_i\mathbf{n}$ for $i = 1, \ldots, m$. Let $E(\lambda, l)$ denote the energy for fixed $\lambda \leq Z$ and let $\Delta E(\lambda, l) = E(\lambda, l) - E(\lambda, 0)$ denote the change in energy. Likewise define $\Delta e(\lambda, l) = \Delta E(\lambda, l) - \Delta U$. Then*

(i) $\Delta e(\lambda, l) \geq 0$,

(ii) $\Delta E(\lambda, l) \leq 0$,

(iii) $\Delta E(\lambda_1, l) \leq \Delta E(\lambda_2, l)$ *if* $\lambda_1 \leq \lambda_2$,

(iv) $\Delta e(\lambda_1, l) \leq \Delta e(\lambda_2, l)$ if $\lambda_1 \leq \lambda_2$.

To prove Theorem 4.7 the following Lemma 4.8, which is of independent interest, is needed.

Lemma 4.8. *Assume the plane P, with $R_1, \ldots, R_m$ in P and $R_{m+1}, \ldots, R_k$ in P^+ as in Theorem 4.7. However, point nuclei are not assumed. Instead, assume each V_i*

$\in D$ and given by Eq. (4.5), with m_i required to be spherically symmetric for $i=1,\ldots,m$. This includes point nuclei. Assume also that the support of $m_i \subset P^+$ for $i=m+1,\ldots,k$. If $x\in P^+$ then x^* is defined to be the reflection of x through P. Let ϕ be the potential. For $x\in P^+$, let $\phi_-(x)=\phi(x^*)$ and $f(x)=\phi(x)-\phi_-(x)$. Then

(i) $f(x)>0$ for $x\in P^+$.

(ii) For each $x\in P^+$, $f(x)$ strictly decreases when λ increases.

(iii) $\rho(x)-\rho(x^*)\geq 0$ for $x\in P^+$.

Question. Is it true that $\rho(x)-\rho(x^*)$ is a monotone increasing function of λ?

Proof. (i) Clearly $f(x)=0$ on $\partial P^+=P$ and at ∞. Let $B=\{x\in P^+ | f(x)<0\}$. Since each $V_i(x)$ is symmetric decreasing the singularities of V are not in B. Thus B is open. On B, $-\Delta f(x)/4\pi \geq -\rho(x)+\rho(x^*)>0$. Thus f is superharmonic on B so B is empty. By the strong maximum principle $f(x)>0$, in fact, for $x\in P^+$.

(ii) Let $\lambda'<\lambda$ with corresponding f' and f. We want to prove $B=\{x\in P^+|f(x)-f'(x)>0\}$ is empty. B is open and $f-f'=0$ on P and at ∞. $\Delta(f-f')/4\pi=a_+^{3/2}-b_+^{3/2}-c_+^{3/2}+d_+^{3/2}\equiv h$, where $a=\phi-\mu, b=\phi'-\mu', c=\phi_--\mu$, $d=\phi'_--\mu'$. By (i) and Corollary 3.8, $a\geq b>d$ and $a>c\geq d$ for all $x\in P^+$. In B, $a+d>b+c$. Thus $h\geq 0$ in B, whence $f-f'$ is subharmonic on B and hence B is empty. Again, one can prove the stronger result that $f-f'<0$ for $x\in P^+$. Trivially, (i)$\Rightarrow$ (iii) through the TF equation. ■

Proof of Theorem 4.7. We may assume all the l_i are equal to some common l, for otherwise if $l_1\leq l_2\leq\cdots\leq l_m$ we could first move all the m nuclei by l_1, then move $R_2,\ldots,R_m$ by l_2-l_1, etc. Next, replace all the point nuclei at $R_1,\ldots,R_m$ by smeared potentials given by Eq. (4.5) with $dm_i(x)=z_i g^{(n)}(x)dx$ where $g^{(n)}(x)\in C_0^\infty$ and $g^{(n)}$ is symmetric decreasing and with sufficiently small support such that the supports of dm_i $(i=1,\ldots,m)$ are pairwise disjoint and also disjoint from the supports of dm_i $(i=m+1,\ldots,k)$. Under these conditions, e is C^1 in $R_1,\ldots,R_m$ in some neighborhood of the original $R_1,\ldots,R_m$ with derivatives given by Eq. (4.6). We shall prove

(i)' $\mathbf{n}\cdot\nabla_{R_i}e\geq 0$,

(ii)' $\mathbf{n}\cdot\nabla_{R_i}E\leq 0$,

and that *(iii)* and *(iv)* hold for these derivatives. Then the theorem is proved because the original point potentials $z_i|x|^{-1}$ can be approximated in $L^{5/2}$ norm by these smeared potentials $z_i|x|^{-1}*g^{(n)}$, and the energies $e^{(n)}$ and $E^{(n)}$ converge to e and E by LS, Theorem II.15. If *(i)'* holds for $e^{(n)}$, then $(d/dl)e^{(n)}(\lambda,l)\geq 0$ with $R_i\to R_i+l\mathbf{n}$, $i=1,\ldots,m$, and, by integration, *(i)* holds for $e^{(n)}$. Then, when $n\to\infty$, *(i)* holds for e. The same applies to *(ii)*–*(iv)*. Henceforth the superscript (n) will be suppressed.

Assume $\mathbf{n}=(1,0,0)$, $P=\{x|x^1=0\}$, and thus $(R_i)^1=0$ for $i=1,\ldots,m$. Since g is symmetric decreasing,

$$(\partial g/\partial x^1)(x^1,x^2,x^3)=-x^1h(x^1,x^2,x^3)$$

with $h(x)\geq 0$ and

$$h(x^1,x^2,x^3)=h(-x^1,x^2,x^3).$$

Likewise,

$$(\partial V_i/\partial x^1)(x^1,x^2,x^3)=-z_ix^1p(x^1,x^2,x^3),$$

and p has the same properties as h. To prove *(i)'* use Eq. (4.6) whence

$$\mathbf{n}\cdot\nabla_{R_i}e/z_i=-\int x^1p(x-R_i)\rho(x)dx$$

$$=-\int_{x^1\leq 0} p(x-R_i)[\rho(x)-\rho(x^*)]x^1dx\geq 0$$

by Lemma 4.8. To prove *(ii)'* use the second integral in Eq. (4.6), whence

$$B_i\equiv\mathbf{n}\cdot\nabla_{e_i}E=\int dm_i(y)\mathbf{n}\cdot\nabla\phi(y+R_i),$$

where ϕ is the potential. [Note: $V_i(x-R_i)$ is symmetric in x about R_i, so the term $\nabla V_i(x-R_i)$ does not contribute to this integral.] Since V_i is C^∞ it is easy to see that ϕ is also C^∞ near R_i. Now integrate by parts:

$$B_i=-\int\mathbf{n}\cdot\nabla g(y)\phi(y+R_i)dy$$

$$=\int y^1h(y)\phi(y+R_i)dy$$

$$=\int_{y^1\leq 0} y^1h(y)[\phi(y+R_i)-\phi_-(y+R_i)]\leq 0$$

by Lemma 4.8. To prove *(iii)* note that the last quantity [ ] decreases when λ increases by Lemma 4.8. Clearly *(iii)* is equivalent to *(iv)*. ■

Proof of Theorem 4.6. Let $\rho(x)$ be the density when $l=0$. For $l>0$ use the variational $\tilde\rho$ given by $\tilde\rho(x^1,x^2,x^3)=\rho(x^1\mp l,x^2,x^3)$ if $x^1\gtrless l$ and $\tilde\rho(x)=0$ otherwise. Then all terms in the energy $\mathcal{E}(\tilde\rho)$ remain the same except for the Coulomb interaction of the two charge distributions on either side of the plane P. This term is of the form

$$W(l)=\int_{x^1,y^1\geq 0} d^3x\,d^3y\,f(x)f(y)\times[(x^1+y^1+2l)^2+(x^2-y^2)^2+(x^3-y^3)^2]^{-1/2},$$

where $f(x)=-\rho(x)+\sum' z_j\delta(x-R_j)$ and the $\sum'$ is over those R_i with $R_i^1>0$. Since the Coulomb potential is reflection positive (Benguria and Lieb, 1978, Lemma B.2), $W(l)$ is a decreasing, log convex function of l. ■

Proof of Theorem 4.5. Let $\underline{z}=(z_1,\ldots,z_k)$ and write $E(\underline{z})$, $K(\underline{z})$, $A(\underline{z})$, and $R(\underline{z})$ for the energy and its components (cf. Sec. II.E) of a neutral molecule. These functions are defined on $\mathbb{R}_+^k$. For an atom $3P=E+K=0$ (Theorem 2.14). By Theorem 3.23, $E\geq\sum_1^k E^{\rm atom}(z_j)$ and, by Theorem 4.10, $K\geq\sum_1^k K^{\rm atom}(z_j)$. This shows $P\geq 0$. Likewise, by Theorem 4.12, $\kappa^{-1}\geq 0$ and $E(\underline{z},l)$ is convex in l (equivalently l^2P is decreasing in l). ■

Definition. Let f be a real valued function on $\mathbb{R}_+^k$ and $\underline{z}_1,\underline{z}_2,\underline{z}_3\in\mathbb{R}_+^k$. Then f is

(i) *weakly superadditive* (WSA)$\Leftrightarrow f(\underline{z}_1+\underline{z}_2)\geq f(\underline{z}_1)+f(\underline{z}_2)$ whenever $(z_1)_i(z_2)_i=0$, all i,

(ii) *superadditive* (SA)$\Leftrightarrow f(\underline{z}_1+\underline{z}_2)\geq f(\underline{z}_1)+f(\underline{z}_2)$,

(iii) strongly superadditive (SSA)$\Leftrightarrow f(\underline{z}_1+\underline{z}_2+\underline{z}_3)+f(\underline{z}_1)\geq f(\underline{z}_1+\underline{z}_2)+f(\underline{z}_1+\underline{z}_3)$.

Theorems 4.9–4.12 are for neutral molecules.

Theorem 4.9. *As a function of $\underline{z}\in\mathbb{R}_+^k$, for each fixed*

$x \in \mathbb{R}^3$,

(i) $-\phi(\underline{z}, x)$ *is SSA, convex, and decreasing (the latter is Teller's lemma),*

(ii) $\phi(\underline{z}, x) \in C^1(\mathbb{R}_+^k)$ *and* $\in C^2(\mathbb{R}_+^k \backslash 0)$,

(iii) $\phi_i(\underline{z}, x)$ *is decreasing in* $\underline{z}$ *and* > 0,

(A subscript i denotes $\partial/\partial z_i$.)

(iv) $\phi_{ij}(\underline{z}, x) \leq 0$ *(all i,j) and is negative semidefinite as a k×k matrix.*

Remark. It is easy to prove that when $f \in C^2(\mathbb{R}_+^k)$ then SSA is equivalent to $f_{ij} \geq 0$ for all i, j. See Benguria and Lieb, 1978b, for this and similar equivalences.

Theorem 4.10. $K(\underline{z}) \in C^1(\mathbb{R}_+^k)$ *and* $\in C^2(\mathbb{R}_+^k \backslash 0)$

(i) $K_i(\underline{z}) = 3 \lim_{x \to R_i} \left\{ \phi(\underline{z}, x) - \sum_{j=1}^k z_j \phi_j(\underline{z}, x) \right\}$,

(ii) $K_{ij}(\underline{z}) = -3 \sum_{p=1}^k z_p \phi_{ij}(\underline{z}, R_p)$,

(iii) $K(\underline{z})$, $R(\underline{z})$, *and* $A(\underline{z})$ *are SSA and SA and convex,*

(iv) $E(\underline{z})$ *is WSA (Teller's theorem).*

Definition. $X(\underline{z}) \equiv 3K(\underline{z}) - \sum_{i=1}^k z_i K_i(\underline{z})$.

Theorem 4.11. $X(\underline{z})$ *is SA and SSA and ray convex. I.e.,* $X(\lambda \underline{z}_1 + (1-\lambda)\underline{z}_2) \leq \lambda X(\underline{z}_1) + (1-\lambda) X(\underline{z}_2)$, $0 \leq \lambda \leq 1$, *when* $\underline{z}_1, \underline{z}_2 \in \mathbb{R}_+^k$ *and either* $\underline{z}_1 - \underline{z}_2$ *or* $\underline{z}_2 - \underline{z}_1 \in \mathbb{R}_+^k$.

Theorem 4.12. *(i)* $3l^3 P = E + K$,

(ii) $9l^3 \kappa^{-1} = 6l^3 P + 2E + 3X$,

(iii) P *and* κ^{-1} *are WSA and non-negative,*

(iv) $l^2 P$ *is decreasing in l. Equivalently, E is convex in l. Equivalently,* $2E + 3X \geq 0$

[*note:* $\partial(l^2 P)/\partial l = 2lP - 3l\kappa^{-1} = -(\frac{1}{3})(2E + 3X)$].

Proof of (iv). $2E + 3X = 0$ for an atom. By Theorem 4.10, $2E + 3X \geq 0$. ■

The proofs of Theorems 4.9–4.12 are complicated. However, if all necessary derivatives are assumed to exist, then an easy heuristic proof can be given (see Benguria and Lieb, 1978b). We illustrate this for K being SSA, which is equivalent to $K_{ij} \geq 0$, all i, j. This will then prove $P \geq 0$, since $K(0) = 0$. First we show $\phi_{ij} \leq 0$ and then Theorem 4.10 (ii).

Differentiate the TF differential equation [$\Delta\phi/4\pi = -\sum z_j \delta(x - R_j) + (\phi/\gamma)^{3/2}$, which holds for any *neutral* system] with respect to z_i and then z_j:

$$\mathcal{L}\phi_i = \delta(x - R_i), \tag{4.8}$$

$$\mathcal{L}\phi_{ij} = -(3/4\gamma^{3/2})\phi^{-1/2}\phi_i\phi_j, \tag{4.9}$$

with $\mathcal{L} = -\Delta/4\pi + (3\gamma^{-3/2}/2)\phi(x)^{1/2}$. The kernel for $\mathcal{L}^{-1}$ is a positive function, so $\phi_i \geq 0$. Likewise $\phi_{ij} \leq 0$ and ϕ_{ij} is a negative semidefinite matrix.

Next, $K = (3\gamma^{-3/2}/5)\int \phi^{5/2}$, so

$$K_{ij} = (3\gamma^{3/2}/2)\left\{ \int \phi^{3/2}\phi_{ij} + \frac{3}{2}\int \phi^{1/2}\phi_i\phi_j \right\}.$$

Using Eq. (4.9) and integrating by parts,

$$K_{ij} = 3\int \phi_{ij}[\Delta\phi/4\pi - (\phi/\gamma)^{3/2}]$$

$$= -3\sum_{p=1}^k z_p \phi_{ij}(R_p) \geq 0.$$

C. The long-range interaction of atoms

In Sec. IV.B it was shown that the energy of a molecule decreases monotonically under dilation (at least for neutral molecules). If the $R_i \to lR_i$ then, for small l, E is dominated by U, so $E \approx l^{-1}$. To complete the picture it is necessary to know what happens for large l. We define

$$\Delta E = E^{\text{mol}} - \sum_{j=1}^k E^{\text{atom}}. \tag{4.10}$$

For large l it is reasonable to consider only neutral molecules, for otherwise $\Delta E \approx l^{-1}$ because of the unscreened Coulomb interaction. In the neutral case $\Delta E \approx l^{-7}$, as proved by Brezis and Lieb (1979). This result (l^{-7}) is not easy to ascertain numerically (Lee, Longmire, and Rosenbluth, 1974), so once again the importance of pure analysis in the field is demonstrated. Some heuristic remarks about the result are given at the end of this section.

A surprising result is that *all* the many-body potentials are $\approx l^{-7}$. Thus in TF theory it is *not* true that the interaction of atoms may be approximated purely by pair potentials at large distances.

An interesting open problem is to find the long-range interaction of polyatomic molecules of fixed shape. Presumably this is also $\approx l^{-7}$.

Theorem 4.13. *For a neutral molecule, let the nuclear coordinates be* lR_i *with* $\{R_i, z_i\} = (\underline{R}, \underline{z})$ *fixed and* $z_i > 0$. *Then*

$$\Delta E(l, \underline{z}, \underline{R}) \equiv l^{-7} C(l, \underline{z}, \underline{R}),$$

where C is increasing in l and has a finite limit, $\Gamma(\underline{R}) > 0$ *as* $l \to \infty$. Γ *is independent of* $\underline{z}$. *Furthermore, if A denotes a subset of the nuclei (with coordinates* $\underline{R}_A$*), and* $\varepsilon(A)$ *is the many-body potential of Eq. (4.1), then, by (4.1), for* $|A| \geq 2$

$$l^7 \varepsilon(A) \to \sum_{B \subseteq A} (-1)^{|A|-|B|}\Gamma(\underline{R}_B) \tag{4.11}$$

and the right side of Eq. (4.11) is strictly positive (negative) if $|A|$ *is even (odd).*

Proof of first part. By scaling, Eq. (2.24), we find that

$$\Delta E(l, \underline{z}, \underline{R}) = l^{-7}\{E(l^3\underline{z}, \underline{R}) - \sum_j E^{\text{atom}}(l^3 z_j)\}.$$

Therefore, C increasing is equivalent to $f = E^{\text{mol}} - \sum E^{\text{atom}}$ increasing in $\underline{z}$. But $\partial f/\partial z_j = \lim_{x \to R_j}\phi^{\text{mol}}(x) - \phi^{\text{atom}}(x)$, and this is positive by Teller's lemma. All that has to be checked is that C is bounded above. This is done by means of a variational ρ for E^{mol}. Let B_i be a ball of radius lr_i centered at lR_i; the r_i are chosen so that the B_i are disjoint. Let $\rho_i(x) = \rho^{\text{atom}}(x - lR_i)$ be the TF atomic densities for z_i, and let $\rho(x) = \rho_i(x)$ in B_i and $\rho(x) = 0$ otherwise. Of course $\int\rho < \sum z_j$ but this is immaterial for a variational calculation, since the minimum molecular energy occurs when $\int\rho = \sum z_j$. It is easy to check that $f < (\text{const})l^{-7}$. Finally, since f is monotone in each z_i, $\lim_{l\to\infty} f$ must be independent of the z_j. ■

Remarks. (i) The variational calculation shows clearly why Γ is independent of the z_j. The long-range interaction comes, in some sense, from the tails of the atomic ρ's, but these tails are independent of z, namely $\rho(x) \approx (3\gamma/\pi)^3|x|^{-6}$. (See Theorem 2.10.)

(ii) At first sight it might appear counterintuitive that the interaction is $+l^{-7}$ and not $-l^{-6}$, as would be obtained from a dipole-dipole interaction. The following heuristic remark might be useful in this respect. Consider two neutral atoms separated by a large distance R. In the quantum theory, as in all the theories discussed in this paper, there is almost no *static* polarization of the atoms; i.e., there is no polarization of the single-particle density ρ. TF theory is therefore correct as far as the density is concerned. The reason there is no polarization is that the formation of a dipole moment d increases the atomic energy by $+\alpha d^2$ with $\alpha>0$. The dipole-dipole energy gain is $-(\text{const})d^2R^{-3}$. Hence, if R is large enough, the formation of dipoles does not decrease the energy. In quantum theory there is, in fact, a $-R^{-6}$ dipolar energy, but this effect is a correlation, and not a static effect. There are two ways to view it. In second-order perturbation theory there are *virtual* transitions to excited, polarized states. Alternatively, the electrons in each atom are correlated so that they go around their respective atoms in phase, but spherically symmetrically. This correlated motion increases the internal atomic energy only by αd^4, not d^2. In short, the $-R^{-6}$ interaction arises from the fact that the density ρ is not that of a structureless "fluid" but is the average density of many separate particles which can be correlated. This fact poses a serious problem for any "density functional approach." It is necessary to predict a $-R^{-6}$ dipolar interaction, yet predict essentially zero static polarization.

An explicit formula for $\Gamma(\underline{R})$ does not seem to be easy to obtain. Two not very explicit formulas are given in Brezis and Lieb, 1979. One is simply to integrate the formula for $\partial f/\partial l = 3l^2\sum z_j\partial f/\partial z_j$ given in the above proof. Another is obtained by noting that Γ is related to ϕ in the limit $z\to\infty$. This limiting ϕ can be defined, and satisfies the TF differential equation, but with a *strong singularity* at R_i instead of the usual $z|x-R_i|^{-1}$ singularity. As we saw in Theorem 2.11, the only other singularity allowed for the TF equation is $\phi(x)\approx\gamma^3(3/\pi)^2|x-R_i|^{-4}$. Therefore that peculiar solution to the TF equation does have physical interest; it is related to the asymptotic behavior of the interatomic interaction.

TFD theory. Here the interaction for large l is *precisely zero* and not l^{-7}. To be precise, $\Delta E=0$ when the spacing between each pair $|R_i-R_j|$ exceeds a critical length, $L(z_i)+L(z_j)$. The same is *a fortiori* true for the many-body potentials ε.

The reason is the following. In TFD theory an atomic ρ has compact support, namely a ball of radius $L(z)$. See Theorem 6.6. When $|R_i-R_j|>L(z_i)+L(z_j)$, then $\rho(x)=\sum_j\rho(x-R_j;z_j)$ where $\rho(\cdot\,;\cdot)$ is the TFD atomic ρ. Since each atom is neutral, there is then no residual interaction, by Newton's theorem. One may question whether the ρ just defined is correct. It is trivial to check that it satisfies the TFD equation and, since the solution is unique, this must be the correct ρ.

V. THOMAS-FERMI THEORY AS THE $Z\to\infty$ LIMIT OF QUANTUM THEORY

Our goal in this section is to show that TF theory is the $Z\to\infty$ limit of Q theory and that it correctly describes the cores of heavy atoms. This is the perspective from which to view TF theory, and in this light it is seen to be a cornerstone of many-body theory, just as the theory of the hydrogen atom is an opposite cornerstone useful for thinking about light atoms. We shall not review the stability of matter question here (see Lieb, 1976).

In units in which $\hbar^2/2m=1$ and $|e|=1$ the Hamiltonian for N electrons is

$$H_N=\sum_{i=1}^{N}\{-\Delta_i+V(x_i)\}+\sum_{1\le i<j\le N}|x_i-x_j|^{-1}+U. \quad (5.1)$$

E_N, $\rho_N(x)$, and μ will denote the TF energy, ρ and μ corresponding to this problem with $\lambda=N$ electrons if

$$N\le Z=\sum_{j=1}^{k} z_j.$$

Of course, γ is taken to be γ_p [see Eq. (2.6)]. If $N>Z$ then these quantities are *defined* to be the corresponding TF quantities for $N=Z$. E_N^Q denotes the ground-state energy of H_N (*defined* to be inf specH_N) on the physical Hilbert space $\mathcal{H}_N=\wedge_1^N L^2(\mathbb{R}^3;\mathbb{C}^q)$ (antisymmetric tensor product). q is the number of spin states (= 2 for electrons), but it is convenient to have it arbitrary, but fixed. The TF quantities also depend on q through γ_p.

A. The $Z\to\infty$ limit for the energy and density

Let us first concentrate on the energy; later on we shall investigate the meaning of $\rho(x)$. For simplicity the number of nuclei is fixed to be k; it is possible to derive theorems similar to the following if $k\to\infty$ in a suitable way (e.g., a solid with periodically arranged nuclei), but we shall not do so here. In TF theory the relevant scale length is $Z^{-1/3}$ and therefore we shall consider the following limit.

Fix $\{\underline{z}^0,\underline{R}^0\}=\{z_j^0,R_j^0\}_{j=1}^{k}$ and $\lambda>0$. For each $N=1,2,\ldots,$ define a_N by $\lambda a_N=N$, and in H_N, replace z_j by $a_N z_j^0$ and R_j by $a_N^{-1/3}R_j^0$. Thus $\lambda=Z^0N/Z$, and a_N is the scale parameter. The TF quantities scale as [Eq. (2.24)]:

$$\begin{aligned}&E_{\lambda a}(a\underline{z}^0,a^{-1/3}\underline{R}^0)=a^{7/3}E_\lambda(\underline{z}^0,\underline{R}^0),\\ &\rho_{\lambda a}(a^{-1/3}x,a\underline{z}^0,a^{-1/3}\underline{R}^0)=a^2\rho_\lambda(x,\underline{z}^0,\underline{R}^0).\end{aligned} \quad (5.2)$$

In this limit the nuclear spacing decreases as $a_N^{-1/3}\sim N^{-1/3}\sim Z^{-1/3}$. This should be viewed as a refinement rather than as a necessity. If instead the R_j are fixed $=R_j^0$, then in the limit one has isolated atoms. All that really matters are the limits $N^{1/3}|R_i-R_j|$.

Theorem 5.1 (LS Sec. III). *With $N=\lambda a_N$ as above*

$$\lim_{N\to\infty} a_N^{-7/3}E_N^Q(a_N\underline{z}^0,a_N^{-1/3}\underline{R}^0)=E_\lambda(\underline{z}^0,\underline{R}^0).$$

The proof is via upper and lower bounds for E_N^Q. The upper bound is greater than the Hartree-Fock energy, which therefore proves that Hartree-Fock theory is correct to the order we are considering, namely $N^{7/3}$.

1. Upper bound for E_N^Q

The original LS proof used a variational calculation with a determinantal wave function; this is cumbersome. Baumgartner (1976) gave a simpler proof (both upper and lower bounds) which intrinsically relied on the same Dirichlet-Neumann bracketing ideas as in LS. Here, we give a new upper bound (Lieb, 1981a) that uses coherent states; these will also be very useful for obtaining a lower bound.

Let $y=(x,\sigma)$ denote a single space-spin pair and $\int dy \equiv \sum_{\sigma=1}^{q}\int dx$. Let $K(y,y')$ be any *admissible single-particle density matrix* for N fermions, namely $0 \le K \le I$ [as an operator on $L^2(\mathbb{R}^3;\mathbb{C}^q)$] and $\mathrm{Tr}K=N$. Let h be the single-particle operator $-\Delta+V(x)$. Then (Lieb, 1981a)

$$E_N^Q \le E_N^{\mathrm{HF}} \le \tilde{E}(K), \tag{5.3}$$

with

$$\tilde{E}(K)=\mathrm{Tr}Kh+\tfrac{1}{2}\int\int dy\,dy'|x-x'|^{-1} \times\{K(y,y)K(y',y')-|K(y,y')|^2\}. \tag{5.4}$$

In Eq. (5.3), E_N^{HF} is the Hartree-Fock energy. Since $|x-x'|^{-1}$ is positive we can drop the "exchange term," $-|K|^2$, in Eq. (5.4) for the purposes of an upper bound.

First, suppose $N\le Z$. To construct K, let $g(x)$ by any function on $\mathbb{R}^3$ such that $\int|g|^2=1$ and let $M(p,r)$ be any function on $\mathbb{R}^3\times\mathbb{R}^3$ such that $0\le M(p,r)\le 1$ and $(2\pi)^{-3}\times\int M\,dp\,dr=N/q$. Then the coherent states in $L^2(\mathbb{R}^3)$ which we shall use are

$$f_{pr}(x)=g(x-r)\exp[ip\cdot x] \tag{5.5}$$

and

$$K(y,y')=I_\sigma(2\pi)^{-3}\int dp\,dr\,g(x-r)g(x'-r)^*M(p,r) \times\exp[ip\cdot(x-x')]. \tag{5.6}$$

I_σ is the identity operator in spin space. It is easy to check that $\mathrm{Tr}K=N$ and that for any normalized ϕ in L^2, $(\phi,K\phi)\le 1$ by using Parseval's theorem and the properties of g and M. Thus K is admissible.

We choose [with $\rho=\rho_{\min(N,Z)}$ in Eqs. (5.7)–(5.26)]

$$M(p,r)=\theta(\gamma_p\rho(r)^{2/3}-p^2), \tag{5.7}$$

where $\theta(t)=1$ if $t\ge 0$ and $\theta(t)=0$ otherwise. γ_p is given in Eq. (2.6). One easily computes

$$K(y,y)=q^{-1}I_\sigma\rho_g(x), \tag{5.8}$$

$$\mathrm{Tr}(-\Delta)K=(3\gamma_p/5)\int\rho(x)^{5/3}dx+N\int|\nabla g(x)|^2dx, \tag{5.9}$$

$$\mathrm{Tr}\,VK=\int V_g(x)\rho(x)dx, \tag{5.10}$$

where $\rho_g=|g|^2*\rho$ and $V_g=V*|g|^2$.

For $g(x)$ we choose

$$g(x)=(2\pi R)^{-1/2}|x|^{-1}\sin(\pi|x|/R) \tag{5.11}$$

for $|x|\le R$, and $g=0$ otherwise, and with $R=N^{2/5}Z^{-1}$. Then

$$\int|\nabla g|^2=\pi^2/R^2=\pi^2Z^2N^{-4/5}.$$

The electron-electron interaction term in Eq. (5.4) is less than $D(\rho,\rho)$ because, as an operator (and function),

$$[|g|^2*|x|^{-1}*|g|^2](x-x')<|x-x'|^{-1}.$$

To see this, use Fourier transforms. Thus

$$E_N^Q\le\tilde{E}(K)\le E_N+\pi^2N^{1/5}Z^2 +\int[V(x)-V_g(x)]\rho(x)dx. \tag{5.12}$$

To bound the last term in Eq. (5.12) note that, by Newton's theorem, $|x|^{-1}-|g|^2*|x|^{-1}=0$ for $x\ge R$. Furthermore, with the scaling we have employed, $|R_i-R_j|>2R$ for all $i\ne j$ and N large enough. Since $\gamma_p\rho^{2/3}(x)<V(x)$, then for sufficiently large N and for $|x-R_i|\le R$ we have $\gamma_p\rho(x)^{2/3}<2z_i|x-R_i|^{-1}$. Thus the last integral in Eq. (5.12) is bounded above for large N by

$$3\gamma_p^{-3/2}\sum_{j=1}^{k}z_j^{5/2}A$$

with

$$A=\int_{|x|\le R}|x|^{-5/2}dx=8\pi R^{1/2}=8\pi N^{1/5}Z^{-1/2}.$$

If $N\le Z$, we have established an adequate upper bound, namely,

$$E_N^Q-E_N\le(\mathrm{const})N^{1/5}Z^2. \tag{5.13}$$

Since $Z\approx N$, this error is $\approx N^{11/5}$, and this is small compared to E, which is $\approx N^{7/3}$.

If $N>Z$ we use $K=K^1+K^\infty$ where K^1 is given above (with $N=Z$) and K^∞ is a density matrix (really, a sequence of density matrices) whose trace is $N-Z$ and whose support is a distance d arbitrarily far away from the origin. K^∞ does not contribute to $\tilde{E}(K)$ in the limit $d\to\infty$. ■

2. Lower bound for E_N^Q

In LS a lower bound was constructed by decomposing $\mathbb{R}^3$ into boxes and using Neumann boundary conditions on these boxes. However, control of the singularities of V caused unpleasant problems. Here we use coherent states again (cf. Thirring, 1981).

Let $\psi(x_1,\ldots,x_N;\sigma_1,\ldots,\sigma_N)$ be any normalized function in $\mathcal{H}_N$ and let

$$\rho_\psi(x)=N\sum_{\sigma=1}^{q}\int|\psi(x,x_2,\ldots,x_N;\sigma_1,\ldots,\sigma_N)|^2dx_2\cdots dx_N \tag{5.14}$$

$$E_\psi=(\psi,H_N\psi) \tag{5.15}$$

$$T_\psi=\left(\psi,-\sum\Delta_i\psi\right). \tag{5.16}$$

It is known that (Lieb, 1979; Lieb and Oxford, 1981)

$$I_\psi=\left(\psi,\sum_{i<j}|x_i-x_j|^{-1}\psi\right) \ge D(\rho_\psi,\rho_\psi)-(1.68)\int\rho_\psi(x)^{4/3}dx. \tag{5.17}$$

Choose any $\tilde{\rho}(x)\ge 0$ and $\tilde{\phi}=V-|x|^{-1}*\tilde{\rho}$. Since $D(\rho_\psi-\tilde{\rho},\rho_\psi-\tilde{\rho})\ge 0$, we have for any $0\le\varepsilon<1$

$$E_\psi \geq \left(\psi, \sum_{i=1}^N h_i\psi\right) + U - D(\tilde\rho,\tilde\rho) - (1.68)\int \rho_\psi^{4/3} + \varepsilon T_\psi, \tag{5.18}$$

with

$$h = -(1-\varepsilon)\Delta - \tilde\phi(x) \tag{5.19}$$

being a single-particle operator.

We shall choose $\tilde\rho$ to be the TF density for the problem with γ_p replaced by $(1-\varepsilon)\gamma_p$, and with the same $\lambda = \min(N,Z)$. $-\tilde\mu$ and $\tilde E$ are the corresponding chemical potential and energy.

Let f_{pr} be the coherent states in $L^2(R^3)$ given by Eq. (5.5) and π_{pr} = (projection onto f_{pr}) $\otimes I_\sigma$. For any function $m(y) = m(x,\sigma)$ in $L^2(R^3; C^q)$ we easily compute:

$$(m,m) = (2\pi)^{-3}\int dp\,dr\,(m,\pi_{pr}m),$$

$$\int |\nabla m|^2 dz = (2\pi)^{-3}\int dp\,dr\,p^2(m,\pi_{pr}m) - (m,m)\int |\nabla g(x)|^2 dx,$$

$$\int |m|^2\tilde\phi_g(x)dz = (2\pi)^{-3}\int dp\,dr\,\tilde\phi(r)(m,\pi_{pr}m), \tag{5.20}$$

with $\tilde\phi_g = |g|^2 * \tilde\phi$.

Write $\tilde\phi = \tilde\phi_g + (\tilde\phi - \tilde\phi_g)$ and $h^g = -(1-\varepsilon)\Delta - \tilde\phi_g(x)$. Let us first concentrate on $e_1 = \inf e_1(\psi)$, where

$$e_1(\psi) = \left(\psi, \sum_{i=1}^N h_i^g\psi\right).$$

Since $\sum h^g$ is a sum of single-particle operators we need only consider ψ's which are determinants of N orthonormal single-particle functions. If $m_1,\ldots,m_N$ are such, then

$$M(p,r) = \sum_{i=1}^N (m_i, \pi_{pr}m_i)$$

has the property that $0 \leq M(p,r) \leq q$ and

$$(2\pi)^{-3}\int dp\,dr\,M(p,r) = N.$$

Therefore

$$e_1(\psi) = (2\pi)^{-3}\int\int dp\,dr\{(1-\varepsilon)p^2 - \tilde\phi(r)\}M(p,r) - N\int |\nabla g|^2. \tag{5.21}$$

The minimum of the right side of Eq. (5.21) over all M with the stated properties is given as follows:

$$M(p,r) = q\theta(\tilde\phi(r) - (1-\varepsilon)p^2 - \mu)$$

for some $\mu \geq 0$. μ is the smallest μ such that $(2\pi)^{-3} \times \int M(p,r) \leq N$. Since $\tilde\phi$ is the TF potential [for $(1-\varepsilon)\gamma_p$] we see that $\mu = \tilde\mu$ and

$$e_1 - D(\tilde\rho,\tilde\rho) + U \geq \tilde E - N\int |\nabla g|^2. \tag{5.22}$$

Next, let us consider the missing piece $e_2 = -\int(\tilde\phi - \tilde\phi_g)\rho_\psi$. The second piece of $\tilde\phi$, namely $-\tilde\psi = -|x|^{-1} * \rho$, has the property that $\tilde\psi - |g|^2 * \tilde\psi \geq 0$ since ψ is superharmonic and $|g|^2$ is spherically symmetric. Therefore we can ignore this piece in e_2. $V - V_g$ is bounded above, as before, by

$$\sum z_j|x-R_j|^{-1}\theta(R - |x-R_j|).$$

For large N, $|R_i - R_j| > 2R$ and, using Hölder's inequality,

$$e_2 \geq -\|\rho_\psi\|_{5/3}\left[8\pi R^{1/2}\sum z_j^{5/2}\right]^{2/5}$$
$$\geq -\|\rho_\psi\|_{5/3}\{(8\pi)^{2/5}R^{1/5}Z\}. \tag{5.23}$$

The negative term e_2 is controlled by the εT_ψ term through an inequality of Lieb and Thirring (1975 and 1976; see also Lieb, 1976):

$$T_\psi \geq L\int \rho_\psi(x)^{5/3}dx, \tag{5.24}$$

with $L = \frac{3}{5}(3\pi/2q)^{2/3}$. Furthermore, by the Schwarz inequality, $\int\rho_\psi^{4/3} \leq \{N\int\rho_\psi^{5/3}\}^{1/2}$. If we write $\int\rho_\psi^{5/3} = X$ then $e_2 \geq -X^{3/5}D$, with $D = \{\,\}$ in Eq. (5.23), and

$$e_2 + \varepsilon T_\psi - (1.68)\int \rho_\psi^{4/3}$$
$$\geq \min_{X>0} -DX^{3/5} + \varepsilon LX - (1.68)N^{1/2}X^{1/2} \equiv Y. \tag{5.25}$$

Equation (5.22) contains $\tilde E$ instead of E; we must bound the difference. Using $\tilde\rho$ as a trial function for E, $E \leq \tilde E + [\varepsilon/(1-\varepsilon)]\tilde K$, where

$$\tilde K = [3(1-\varepsilon)/5]\gamma_p\int \tilde\rho^{5/3}.$$

Choose $\varepsilon = Z^{-1/30}$ (this is not optimum). For large Z, $\varepsilon < \frac{1}{2}$ and it is easy to see that $\tilde K < (\text{const})Z^{7/3}$ for all N, Z. Thus

$$0 > \tilde E - E > -(\text{const})Z^{7/3}Z^{-1/30}. \tag{5.26}$$

Choose $R = Z^{-1/2}$, which is a different choice from the upper bound calculation. Then $D \approx Z^{9/10}$ and

$$Y \geq -(\text{const})Z^{7/3}Z^{-1/30}. \tag{5.27}$$

[It is easy to see that the term $-(1.68)N^{1/2}X^{1/2}$ is negligible as long as N/Z is fixed.]

Finally $-N\int|\nabla g|^2 \approx -NR^{-2} \approx Z^2$. Combining all these bounds, we find

$$E_N^Q - E > -(\text{const})Z^{7/3-1/30}$$

which is the desired result. ■

Clearly there is room for a great deal of improvement, for it is believed that $E^Q - E > 0$ as explained in Sec. V.B. But first let us turn to the correlation functions.

3. Correlation functions

In analogy with Eq. (5.14) we define

$$\rho_\psi^j(x_1,\ldots,x_j) = j!\binom{N}{j}\sum_\sigma\int |\psi(x_1,\ldots,x_N;\sigma_1,\ldots,\sigma_N)|^2 \times dx_{j+1}\cdots dx_N. \tag{5.28}$$

We wish to obtain a limit theorem for ρ_ψ^j when ψ is a ground state of H_N. But there may be no ground state (inf specH_N may not be an eigenvalue) or there may be several. In any case, it is intuitively clear that the limit of ρ_ψ^j should not depend upon ψ being exactly a ground state, but only upon ψ being "nearly" a ground state.

Definition. Let $\psi_1, \psi_2, \ldots$ be a sequence of normalized functions with $\psi_N \in \mathcal{H}_N$ for N particles. This *sequence* is called an *approximate ground state* if $|\psi_N, H_N\psi_N) - E_N^Q| a_N^{-7/3} \to 0$ as $N \to \infty$. H_N always has k nuclei.

Theorem 5.2. *Let* $\{\psi_N\}$ *be an approximate ground state with the scaling given before Eq. (5.2), and let* $\rho_N^j(x)$ *be given by Eq. (5.28) with* ψ_N, *and*

$$\hat\rho_N^j(x_1,\ldots,x_j) \equiv a_N^{-2j}\rho_N^j(a_N^{-1/3}x_1,\ldots,a_j^{-1/3}x_j).$$

Let $\rho^j(x_1,\ldots,x_j) = \rho(x_1)\cdots\rho(x_j)$ *with* ρ *being the solution to the TF problem for* λ *and* $\{z_j^0, R_j^0\}$. *(Note that* $\lambda = N/Z$ *is now fixed.) Then*

$$\Big(\psi_N, -\sum \Delta_i\psi_N\Big)a_N^{-7/3} \to \tfrac{3}{5}\gamma_p\int \rho(x)^{5/3}dx\,,$$

$$\Big(\psi_N, \sum V(x_i)\psi_N\Big)a_N^{-7/3} \to \int \rho V\,,$$

$$\Big(\psi_N, \sum |x_i - x_j|^{-1}\psi_N\Big)a_N^{-7/3} \to D(\rho,\rho)\,.$$

Moreover, $\hat\rho_N^j(x) \to \rho^j(x)$ *in the sense that if* Ω *is any bounded set in* $\mathbb{R}^{3j}$ *then*

$$\int_\Omega \hat\rho_N^j(x)d^{3j}x \to \int_\Omega \rho^j(x)d^{3j}x\,.$$

If $\lambda \le Z = \sum z_j$, *the restriction that* Ω *be bounded can be dropped and* $\hat\rho_N^j \to \rho^j$ *in the weak* L^1 *sense.*

Proof. The reader is referred to LS, Theorem III.5 for details. The basic idea is to consider a function $U(x_1,\ldots,x_j) \in C_0^\infty(\mathbb{R}^{3j})$ and add $\alpha\int\rho^j U d^{3j}x$ to the TF functional, $\mathcal{E}(\rho)$. On the other hand, the potential

$$\alpha a_N^{4j/3}\sum_{\substack{i_1\cdots i_j\\ \text{unequal}}} U(a_N^{1/3}x_{i_1},\ldots,a_N^{1/3}x_{i_j})$$

is added to H_N. By the aforementioned methods the energies are shown to converge on the scale of $a_N^{7/3}$. But $\partial E/\partial\alpha|_{\alpha=0} = \int\rho^j U$. By concavity of $E(\alpha)$ the derivatives and the limits $a_N \to \infty$ can be interchanged. Thus, for all such U, $\int\hat\rho_N^j U \to \int\rho^j U$. ∎

One of the assertions of Theorem 5.2 is that, as $N\to\infty$, correlations among any finite number of electrons disappear. *A posteriori* this is the justification for replacing the electron-electron repulsion $\sum|x_i - x_j|^{-1}$ by $D(\rho,\rho)$ in TF theory.

B. The Scott conjecture for the leading correction

We have seen that $E^{\mathrm{TF}} = -CZ^{7/3}$ under the assumption that the nuclear coordinates R_j and charges z_j scale as $Z^{-1/3}R_j^0$ and Zz_j^0, $\sum z_j^0 = 1$, and $\lambda = N/Z > 0$ is fixed. C depends on $\lambda, \underline{z}^0, \underline{R}^0$. What is the next correction to the energy? While this question takes us to some extent outside TF theory, we should like to mention briefly the interesting conjecture of Scott (1952) and a generalization of that conjecture. None of these conjectures have been proved.

The basic idea of Scott is that in the *Bohr atom* (no electron repulsion) the electrons close to the nuclei *each* have an energy $\sim -Z^2$. This should also be true in some sense even with electron repulsion. Since TF theory cannot yield exactly the right energy near the singularities of V, the leading correction should be $O(Z^2)$.

The leading correction should have three properties.

(i) It is the *same* with or without electron repulsion because the repulsive part of $\phi(x)$, namely $|x|^{-1}*\rho$, is $O(Z^{4/3})$ for all x.

(ii) It is independent of N/Z, provided $N/Z > 0$ and fixed. This is so because the correction comes from the core electrons whose distance from the nucleus is $O(Z^{-1})$. The number of electrons thus involved is small compared to Z.

(iii) It should be additive over a molecule. If the correction is Dz^2 for an atom then the total leading correction should be

$$\Delta E = D\sum_{j=1}^{k} z_j^2 \tag{5.29}$$

and

$$E^Q = E^{\mathrm{TF}} + \Delta E + o(Z^2)\,. \tag{5.30}$$

Of course E^{TF} depends on whether electron repulsion is present or not, but ΔE supposedly does not change. To calculate D let us first calculate E^{TF} for an atom without repulsion. The general theory goes through as before, but now the TF equation is $\gamma\rho^{2/3} = (V-\mu)_+$, $V(x) = z/|x|$, $\int\rho = N$, and $\mu > 0$, even when $N = z$. It is found (Lieb, 1976, p. 560) that $\mu = z/R$, $R = 3\gamma(4N/\pi^2)^{2/3}/5z$, and $E_{\mathrm{Bohr}}^{\mathrm{TF}} = -3z^2N^{1/3}(\pi^2/4)^{2/3}/\gamma$. Using γ_p,

$$E_{\mathrm{Bohr}}^{\mathrm{TF}} = -z^{7/3}(3N/z)^{1/3}(2mq^{2/3}/\hbar^2)/4\,.$$

The quantum energy is computed by adding up the Bohr levels. For each principal quantum number n, the energy is $e_n = m/2\hbar^2 n^2$ and it is qn^2-fold degenerate. The result (Lieb, 1976) is

$$E_{\mathrm{Bohr}}^{Q} = E_{\mathrm{Bohr}}^{\mathrm{TF}} + qz^2/8 + O(z^{5/3})\,, \tag{5.31}$$

thus

$$D = qz^2/8 \tag{5.32}$$

in the Scott conjecture. Scott's (1952) derivation was slightly different from the above, but his basic idea was the same.

The Scott conjecture about the energy can be supplemented by the following about the density. Let $f_{nlm}(z,x)$ be the normalized *bound-state* eigenfunctions for the hydrogenic atom with nuclear charge z, and define

$$\rho^H(z,x) = q\sum_{nlm}|f_{nlm}(z,x)|^2\,. \tag{5.33}$$

This sum converges and represents the quantum density for a Bohr atom with *infinitely many electrons*. It is being tabulated and studied by Heilmann and Lieb. It is monotone decreasing and a graphical plot of ρ^H shows that it has almost no discernible shell structure. Clearly $\rho^H(z,x) = z^3\rho^H(1,zx)$ and is spherically symmetric. By our previous analysis of the $z\to\infty$ limit (which strictly speaking is not applicable when $N=\infty$, but which can be suitably modified)

$$z^{-2}\rho^H(z, z^{-1/3}x) \to z^{-2}\rho_{\mathrm{Bohr}}^{\mathrm{TF}}(z, z^{-1/3}x) \tag{5.34}$$

Rev. Mod. Phys. *53*, 603-641 (1981)

as $z \to \infty$. But

$$\rho^{\mathrm{TF}}_{\mathrm{Bohr}}(z,x) = (z/\gamma_p |x|)^{3/2}$$

when $\mu = 0$, as we have just seen. Thus

$$\rho^H(1,y) \to (\gamma_p |y|)^{-3/2} \tag{5.35}$$

as $y \to \infty$. Equation (5.35) is not obvious, but it can be directly proved from (5.33).

Thus $\rho^H(z,x)$, whose scale length is z^{-1}, agrees nicely with $\rho^{\mathrm{TF}}(z,x)$, whose scale length is $z^{-1/3}$, in the overlap region $z^{-1} \ll |x| \ll z^{-1/3}$. This is true even when electron repulsion is included in ρ^{TF} because of Theorem 2.8(a). The common value is $\rho(z,x) = (z/\gamma_p |x|)^{3/2}$. Because of this we are led to the following.

Conjecture. Suppose the sequence $\{\psi_N\} \in \mathcal{H}_N$ is an *approximate ground state* for a molecule (with repulsion) in the *strong sense* that

$$|(\psi_N, H_N\psi_N) - E^Q_N| a_N^{-2} \to 0 \quad \text{as } N \to \infty$$

Let $\rho^Q_N(x)$ be given by Eq. (5.14). Recall that $R_j = a_N^{-1/3} R^0_j$. Fix $\lambda = N/Z > 0$ and $x \neq R^0_j$, all j. Then, as $N \to \infty$,

$$a_N^{-2} \rho^Q_N(a_N^{-1/3} x) \to \rho^{\mathrm{TF}}(x), \tag{5.36}$$

where ρ^{TF} is the TF density for λ, z^0_j, R^0_j. On the other hand, for all fixed y,

$$a_N^{-3} \rho^Q_N(a_N^{-1/3} R^0_j + a_N^{-1} y) \to (z^0_j)^3 \rho^H(1, z^0_j y). \tag{5.37}$$

Equation (5.36) has already been proved in Sec. V.A.

TFW Theory. It is a remarkable fact that the TFW correaction, which has no strong *a priori* justification, has, as its chief effect, precisely the kind of correction (i), (ii), (iii) above predicted by Scott. If δ is chosen correctly in Eq. (2.8), even the constant D in Eq. (5.32) can be duplicated. This will be elucidated in Sec. VII. TFW theory also (accidentally?) improves TF theory in two other ways: negative ions can be supported and binding occurs.

C. A picture of a heavy atom

With the real and imagined information at our disposal we can view the energy and density profile of a heavy, neutral, nonrelativistic atom as being composed of seven regions.

(1) The inner core. Distances are $O(z^{-1})$ and ρ is $O(z^3)$. For large σ, the number of electrons out to $R = \sigma/z$ is $\sim \sigma^{3/2}$, while the energy $\sim z^2 \sigma^{1/2}$. If $1 \ll zr \ll z^{2/3}$, $\rho(r)$ is well approximated by $(z/\gamma_p r)^{3/2}$. $\rho(r)$ is infinity on a scale of z^2 which is the appropriate scale for the next, or TF region. The leading corrections, beyond TF theory, come from this region. None of this has been proved.

(2) The core. Distances are $O(z^{-1/3})$ and ρ is $O(z^2)$. TF theory is exact to leading order. The energy is $E^{\mathrm{TF}} \sim -z^{7/3}$ and almost all the electrons are in this region. This is proved.

(3) The core mantle. Distances are of order $\sigma z^{-1/3}$ with $\sigma \gg 1$. $\rho(r) = (3\gamma_p/\pi)^3 r^{-6}$, the Sommerfeld asymptotic formula. ρ is still $O(z^2)$. This is proved.

(4) A transition region to the outer shell. This region may or may not exist.

(5) The outer shell. In the Bohr theory, $z^{1/3}$ shells are filled. The outer shell, if it can be defined, would presumably contain $O(z^{2/3})$ electrons and each electron in the shell would "see" an effective nuclear charge of order $z^{2/3}$. This picture would give a radius unity for the last shell and an *average* density $\sim z^{2/3}$ in the shell. On the same basis the *average* electron energy would be $O(z^{2/3})$ and thus the energy in the shell would be $O(z^{4/3})$. All this is conjectural, for reliable estimates are difficult to obtain.

(6) The surface. Here the potential is presumably $O(1)$, and so is the energy of each electron. Chemistry takes place here.

TF theory, which is unreliable in this region, nevertheless predicts a surface radius of $O(1)$. We thank J. Morgan for this remark. His idea is that if the surface radius R_s is defined to be such that outside R_s there is one unit of electron charge, then $R_s = O(1)$ because the TF density is $\rho(r) = (3\gamma_p/\pi)^3 r^{-6}$, independent of z, for large r. Likewise, if R_0 is defined such that between R_0 and R_s there are $z^{2/3}$ electrons, then the average TF density in this "outer shell" is $z^{2/3}$ in conformity with *(5)*. Finally, the energy needed to remove one electron is $O(1)$ as Eq. (3.11) shows. The radius of this ionized atom is also $O(1)$ as Eq. (3.13) shows.

In no sense is it being claimed that TF theory is reliable at the surface, or even that the existence of the surface, as described, is proved. We are only citing an amusing coincidence. It is quite likely that the surface radius of a large atom has a weak dependence on z.

(7) The region of exponential falloff. $\rho(r) \sim K \times \exp[-2(2me/\hbar^2)^{1/2}(r-R)]$, where e is the ionization potential, K is the density at the surface, and R is the surface radius. An upper bound for ρ of this kind has been proved by many people, of whom the first was O'Connor (1973). See also Deift, Hunziker, Simon, and Vock, 1978, and M. Hoffmann-Ostenhof, T. Hoffmann-Ostenhof, R. Ahlrichs, and J. Morgan, 1980, for recent developments and bibliographies of earlier work.

The density profile of a heavy atom, as described above, is shown schematically in Fig. 2.

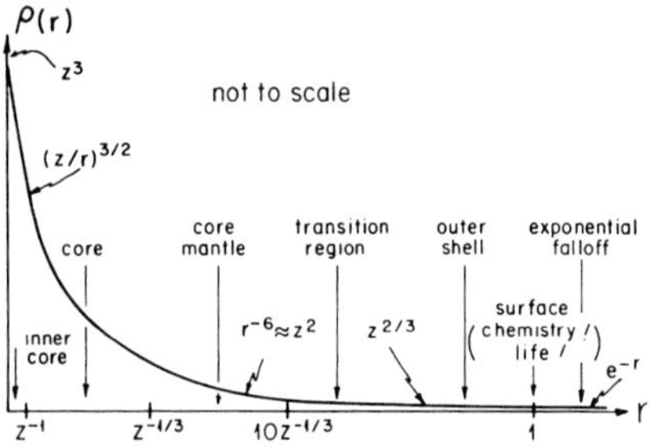

FIG. 2. Schematic plot of the electron density $\rho(r)$ in a neutral heavy atom of charge z. The inner core extends to distances of order z^{-1}; the core to order $z^{-1/3}$; the mantle to $z^{-1/3}$ times a large number. The core and its mantle are correctly described by TF theory. The outer shell extends to distances of order z^p where p is near zero. Finally, there is the surface, and then the region of exponential falloff. The surface thickness is not shown.

VI. THOMAS-FERMI-DIRAC THEORY

The previous sections contain most of the mathematical tools for the analysis of this model; the main new mathematical idea to be introduced here will be the j model and its relation to TFD theory.

The TFD functional is

$$\mathcal{E}^{\mathrm{TFD}}(\rho)=\int J(\rho(x))dx-\int V\rho+D(\rho,\rho)+U, \tag{6.1}$$

and

$$J(\rho)=\tfrac{3}{5}\gamma\rho^{5/3}-\tfrac{3}{4}C_e\rho^{4/3}. \tag{6.2}$$

The term $-D=-(3C_e/4)\int\rho^{4/3}$ was suggested by Dirac (1930) to account for the "exchange" energy. The true electron repulsion I in (5.17) is expected to be less than $D(\rho,\rho)$ because the electrons are correlated. For an ideal Fermi gas at constant density, I is computed to be $D(\rho,\rho)-D$ with $C_e=(6/\pi q)^{1/3}$. There is, however, no fundamental justification for the Dirac approximation; it can even lead to unphysical results, as will be seen shortly. In particular, I is always positive but $D(\rho,\rho)-D$ can be arbitrarily negative. As remarked in (5.17), there is a *lower* bound of this form $D(\rho,\rho)-D$ (Lieb, 1979; Lieb and Oxford, 1981) with $3C_e/4=1.68$ (independent of q). In any event, it should be remembered that *D is part of the Coulomb energy* even though it is mathematically convenient to combine it with the kinetic energy as in Eq. (6.2).

For simplicity we assume

$$V(x)=\sum_{i=1}^{k}V_i(x-R_i), \tag{6.3}$$

with $V_i\in\mathfrak{D}$: $V_i=|x|^{-1}*m_i$ (with m_i a non-negative measure) and $|m_i|=z_i$.

Henceforth the superscript TFD will be omitted. All quantities in this section refer to TFD, and not TF, theory, unless otherwise stated.

A. The TFD minimization problem

The function space is the same as for TF theory, namely

$$\mathcal{J}=\{\rho\mid\rho\in L^1\cap L^{5/3},\rho(x)\geq 0\}. \tag{6.4}$$

The energy is

$$E(\lambda)=\inf\left\{\mathcal{E}(\rho)\,\Big|\int\rho=\lambda,\rho\in\mathcal{J}\right\}. \tag{6.5}$$

Theorem 6.1. *$E(\lambda)$ is finite, nonincreasing in λ, and*

$$E(\lambda)=\inf\left\{\mathcal{E}(\rho)\,\Big|\int\rho\leq\lambda,\rho\in\mathcal{J}\right\}. \tag{6.6}$$

Moreover, $e(\lambda)\equiv E(\lambda)-U<0$ when $\lambda>0$.

Proof. Same as Proposition 2.1 and Theorem 2.3. The crucial fact to note is that $J(0)=J'(0)=0$, which permits us to place "surplus charge density" at infinity. ■

It is not immediately obvious that $E(\lambda)$ is convex because J is not convex. The proof of convexity is complicated and will be given later (Theorem 6.9).

A second difficulty is that $E(\lambda)$ is not bounded below for all λ. This is so because J is not positive. This latter difficulty can be dealt with in the following way. Introduce

$$\mathcal{E}_\alpha(\rho)=\mathcal{E}(\rho)+\alpha\int\rho \tag{6.7}$$

with

$$\alpha=15C_e^2(64\gamma)^{-1}. \tag{6.8}$$

This amounts to replacing J by

$$J_\alpha(\rho)=J(\rho)+\alpha\rho. \tag{6.9}$$

Note that $J_\alpha(\rho)\geq 0$ and

$$J_\alpha(\rho_0)=0=J'_\alpha(\rho_0) \tag{6.10}$$

for $\rho_0=(5C_e/8\gamma)^3$. $-\alpha$ and ρ_0 are the minimum value and the minimum point of the function $J(\rho)/\rho$. Correspondingly, introduce

$$E_\alpha(\lambda)=\inf\left\{\mathcal{E}_\alpha(\rho)\,\Big|\int\rho=\lambda,\rho\in\mathcal{J}\right\}. \tag{6.11}$$

Theorem 6.2. *$E_\alpha(\lambda)$ is nonincreasing in λ and has a lower bound, independent of λ. Moreover,*

$$E_\alpha(\lambda)=E(\lambda)+\alpha\lambda \tag{6.12}$$

and

$$E_\alpha(\lambda)=\inf\left\{\mathcal{E}_\alpha(\rho)\,\Big|\int\rho\leq\lambda,\rho\in\mathcal{J}\right\}. \tag{6.13}$$

$e_\alpha(\lambda)\equiv E_\alpha(\lambda)-U=e(\lambda)+\alpha\lambda<0$ when $\lambda>0$, and $e_\alpha(\lambda)\to\inf_\lambda e_\alpha(\lambda)\equiv e_\alpha(\infty)$ as $\lambda\to\infty$.

Proof. Again the proof is the same as for Proposition 2.1 and Theorem 2.3. Here, however, $J'_\alpha(0)>0$; the fact that $J_\alpha(\rho_0)=J'_\alpha(\rho_0)=0$ is used instead. The fact that $J_\alpha\geq 0$ is responsible for the lower bound. ■

Remark. One consequence of Theorem 6.2 is that $dE(\lambda)/d\lambda\leq-\alpha$ (if the derivative exists). Another is that when λ is large enough so that $e_\alpha(\lambda)=e_\alpha(\infty)$ then $e(\lambda)=e_\alpha(\infty)-\alpha\lambda$. As will be seen, this happens when $\lambda\geq\lambda_c=Z=\sum z_j$. Thus the graph of $e_\alpha(\lambda)$ is similar to that for $e^{\mathrm{TF}}(\lambda)$ in Fig. 1. $e(\lambda)$ then has a negative slope, $-\alpha$, at λ_c and afterwards $e(\lambda)$ has the same constant negative slope. This is a highly unphysical feature of TFD theory which arises from the fact that one can have spatially small "clumps" of density in which $\rho=\rho_0$, arbitrarily far apart. These "clumps" have an energy approximately $-\alpha\rho_0\cdot$(volume) and are physically nonsensical because the $-\rho^{4/3}$ term, which causes this effect, is a gross underestimate of the positive electron repulsion which it is meant to represent. There is no minimizing ρ for these "clumps" because for no ρ is the energy exactly $-\alpha\rho_0\cdot$(volume). The "inf" in Eq. (6.5) is crucial.

B. The j model

Now we must deal with the fact that J_α is not convex. To this end we follow Benguria (1979), who introduced the "convexified" j model. With its aid, Benguria was the first to place the TFD theory on a rigorous basis for a certain class of amenable potentials in Eq. (6.3), which is defined in Sec. VI.C. This class includes the point nuclei. It will turn out that the j model also permits us to analyze TFD theory for all potentials, not just the amenable class. However, for nonamenable potentials, the analysis is complicated and the final result has an unexpected feature, namely, that a minimizing ρ for E may not exist, even if $\lambda<\lambda_c$. The j model is explored in

Rev. Mod. Phys. *53*, 603-641 (1981)

detail here because, as will be seen in Sec. VI.C, *its energy is the same as* $E_\alpha(\lambda)$ for the TFD model. Moreover, for amenable potentials the density ρ of the two models is also the same.

Definition.

$$j(\rho)=J_\alpha(\rho),\quad \rho\geq\rho_0=(5C_e/8\gamma)^3$$
$$=0,\quad 0\leq\rho\leq\rho_0\,. \tag{6.14}$$

The derivative of this convex function is given in Eq. (3.14). $\mathcal{E}_j(\rho)$ is given by Eq. (6.1) with J replaced by j. $E_j(\lambda)$ is defined by Eq. (6.5) with $\mathcal{E}$ replaced by $\mathcal{E}_j$.

By the methods of Secs. II and III the j model has many of the same properties as TF theory.

Theorem 6.3. *If V is given by Eq. (6.3) and if $\mathcal{E}$ is replaced by $\mathcal{E}_j$, E by E_j, and e by $e_j=E_j-U$, then the following results of TF theory hold for the j model (they also hold for TF theory, of course, with this V) (Ignore any mention of* TFD *and* TFDW *theory in the cited theorems.):*

Propositions 2.1 and 2.2; Theorems 2.3 and 2.4; the definition of λ_c; Theorem 2.5; Theorem 2.6 [with Eq. (2.18) replaced by (3.2)]; Theorem 2.7; Theorem 2.12 (for a point nucleus); Theorem 2.13 without the γ dependence (for point nuclei; the last two equations in this theorem have an obvious generalization for non-point nuclei.); Theorem 2.14 (for point nuclei) is changed to (a) $5K/3=A-2R-\mu\lambda-\alpha\lambda+4D/3$, (b) $2K=A-R+D$ for an atom, with $D=(3C_e/4)\int\rho^{4/3}$ (note that Theorem 3.19 must be used in the proof); Equation (2.22); Theorem 3.2; Theorems 3.4, 3.5 [Benguria (1979) *has shown that if W is the potential of point nuclei then $\phi'-\phi\in H^2$ away from S_W]; Corollaries 3.7, 3.8, 3.9, and 3.10 (note, in particular, that $\phi^{\rm TF}>\phi'^{\,\rm model}$ for fixed μ); Lemma 3.11; Theorem 3.12, Corollaries 3.14 and 3.17; Theorem 3.18 (i.e., $\lambda_c=Z$); Equation (3.6); Sec. III.B; Theorem 3.23 (but note that equality can occur. See remark at the end of Sec. IV.C).*

Remarks. (i) Theorem 2.8(a) holds in the sense that $\rho(x)\approx(z_j/\gamma)^{3/2}|x-R_j|^{-3/2}$ near R_j.

(ii) There is no simple scaling for the j model, as in Eq. (2.24) for TF theory.

(iii) We emphasize that a minimizing ρ exists if and only if $\lambda\leq Z$. This ρ is unique and satisfies the *Thomas-Fermi-Dirac equation* (3.2).

(iv) *Question.* Under what conditions do the conclusions of Corollaries 3.13, 3.15, and 3.16 and Theorem 3.21 hold for the j model? *Question.* To what extent do the results of Sec. IV carry over to the j model?

(v) To prove the analogue of Eq. (2.15), Mazur's theorem can be used, as in the proof of Proposition 3.24.

There are some useful additional facts about the j model not mentioned in Theorem 6.3.

Theorem 6.4. *If C_e increases then (i) $\phi(x)-\mu(\lambda)$ decreases and $\mu(\lambda)$ increases, for fixed λ; (ii) $\phi(x)$ decreases for fixed μ.*

Proof. By Corollary 3.10, since $j'(\rho)$ decreases with C_e for fixed ρ. ■

Theorem 6.5. *For all λ, $E^{\rm TF}(\lambda)>E_j(\lambda)>E(\lambda)$ since $J(\rho)<j(\rho)<3\gamma\rho^{5/3}/5$. On the other hand, suppose V is the potential of k point nuclei as in Eq. (2.1). Then for $\lambda\leq\lambda_c=Z$,*

$$E^{\rm TF}(\lambda)\leq E_j(\lambda)-\alpha\lambda+(3C_e/4)\lambda^{1/2}\{(5\varepsilon_1/2\gamma)Z^{7/3}\}^{1/2}$$
$$+27C_e^2\lambda/(10\gamma)\,, \tag{6.15}$$

where $-\varepsilon_1$ is the TF energy for a neutral atom with $z=1$ [see Eq. (7.15)].

Remarks. (i) When $\lambda>Z$ then $(E^{\rm TF}-E_j)(\lambda)=(E^{\rm TF}-E_j)(Z)$.

(ii) Clearly Eq. (6.15) can be improved. But it does show that the effect of the Dirac term is to decrease the energy by $O(Z^{5/3})$ for large Z. (Note: by Theorem 6.8, $E_j-\alpha\lambda=E$.)

Proof. Let ρ be the minimizing density for E_j, and $\rho^{\rm TF}$ that for $E^{\rm TF}$. Use ρ as a trial function for $E^{\rm TF}$. Noting that $\rho(x)\notin(0,\rho_0]$ a.e. (Theorem 3.19), we have $E^{\rm TF}\leq\mathcal{E}^{\rm TF}(\rho)=E_j-\alpha\lambda+(3C_e/4)\int\rho^{4/3}$. By Theorem 3.19, $\gamma\rho^{2/3}-C_e\rho^{1/3}+\alpha=\phi-\mu$ when $\rho>0$. But by Corollary 3.10, $\phi-\mu\leq\phi^{\rm TF}-\mu^{\rm TF}\leq\gamma(\rho^{\rm TF})^{2/3}$. Thus $\rho^{2/3}\leq(\rho^{\rm TF})^{2/3}+C_e/\gamma)\rho^{1/3}$. Squaring this and using $\int(\rho^{\rm TF})^{2/3}\rho^{1/3}\leq\lambda$ (Hölder), and $\int\rho^{2/3}\leq\lambda(\rho_0)^{-1/3}$, and $\int(\rho^{\rm TF})^{4/3}\leq[\lambda\int(\rho^{\rm TF})^{5/3}]^{1/2}$, we obtain Eq. (6.15), but with $5K^{\rm TF}/3\gamma$ in place of $\{\,\}$. By Theorem 2.14(a), $2K^{\rm TF}/3<-e^{\rm TF}$, and by the remark preceding Eq. (4.5), $e^{\rm TF}>e^{\rm TF}$ (all nuclei at one point). ■

The next theorem states that ρ always has compact support, even when $\lambda=\lambda_c$. When $\lambda<\lambda_c$ this is also true in TF theory (Lemma 3.11). The proof we give seems unnecessarily complicated; a simpler one must be possible.

Theorem 6.6. *Suppose $V=|x|^{-1}*m\in\mathfrak{D}$, with m a nonnegative measure of compact support and $\int dm=Z$. Let ρ be the minimizing j model density for $\lambda\leq\lambda_c=Z$. Then ρ has compact support. Moreover, suppose* supp(m) $\subset B_R$*, the ball of radius R centered at 0. Then* supp(ρ) $\subset B_r$ *for some r depending on R and Z, but independent of λ. $r\leq 2R+tZ(\rho_0R^2)^{-1}$ for some universal constant t, independent of all parameters.*

Proof. The strategy is to construct a function f such that supp$(\rho)\subset$ supp(f). Let $S_R=\partial B_R$ be the sphere of radius R. There exists a function (surface charge distribution) σ on S_{2R} such that V_σ, the potential of σ outside B_{2R}, is V, i.e.,

$$V(x)=V_\sigma(x)\equiv 4\pi(2R)^2\int d\Omega\sigma(\Omega)\,|x-2R\Omega|^{-1}$$

for $|x|>2R$, where Ω denotes a point on S_1 and $\int d\Omega=1$. It is easy to see that σ is a bounded, continuous function since supp$(m)\subset B_R$, and $|\sigma(\Omega)|\leq sZR^{-2}$ for some universal constant s. Let $\Sigma(\Omega)=-\sigma(\Omega)+sZR^{-2}\geq 0$, and let

$$dM(x)=dm(x)+\Sigma(x/2R)\delta(|x|-2R)dx\,.$$

If $V_M=|x|^{-1}*M$, we see that $V_M(x)\geq V(x)$, all x, and $V_M(x)=Q|x|^{-1}$ for $|x|>2R$, with $0\leq Q\leq(1+16\pi s)Z$. V_M-V is a bounded function in $\mathfrak{D}$. Now let $f(x)=\rho_0$ for $2R<|x|\leq r$ and $f(x)=0$ otherwise, with $\int f=Q$. Let $\psi=|x|^{-1}*\rho$, $g=|x|^{-1}*f$, $\phi=V-\psi$, $h=V_M-g$ where ρ satisfies Eq. (3.2). We claim $u(x)\equiv h(x)-\phi(x)+\mu\geq 0$. If not, let $B=\{x\,|\,u(x)<0\}$. B is open since $h-\phi$ is continuous. $-\Delta u/4\pi \geq -f+\rho$. But $f \leq \rho$ on B since, for $|x| \leq 2R$, $f(x)=0$; for $|x| > 2R$, $\rho(x)$ is either 0 or $> \rho_0$ a.e. by Theorem 3.19. If $\rho(x)=0$ then $\phi - \mu \leq 0$ by (3.2) and $h(x)$ is clearly ≥ 0, so $x \notin B$; if $\rho(x) \geq \rho_0$, $\rho - f \geq 0$. Thus u is superharmonic on B and, since $u=0$ on ∂B and $u=\mu \geq 0$ at infinity, B is empty. Now consider $A = \{x \mid r < |x|\}$. In A, $-\Delta u = 4\pi\rho \geq 0$ and $u \geq 0$ on ∂A and at infinity. Therefore either (i) $\rho=0$ a.e. in A or (ii) $u > 0$ everywhere in A. In case (ii), $\phi < \mu$ in A because $h=0$ in A. But then, by Eq. (3.2), $\rho=0$ a.e. in A. The bound for r is obtained by $4\pi(r-2R)(2R)^2 < (4\pi/3)[r^3 - (2R)^3] = Q/\rho_0$. ■

Remark. For an atom with nucleus located at the origin, R can be chosen to be any positive number. If the inequality for r is minimized then we find $\rho(x)=0$ for $|x| > 3(tZ/\rho_0)^{1/3}$.

Theorem 6.7. *Suppose $V \in \mathfrak{D}$ and $\rho \in \mathfrak{C}$ are such that the second line of Eq. (3.2) holds with $\mu = 0$, in the sense that $\phi_\rho(x) \leq 0$ a.e. when $\rho(x)=0$ and $\rho(x)=0$ a.e. when $\phi_\rho(x) < 0$. Let A be the complement of the support of ρ. Then $\phi \equiv 0$ on $\bar{A}$, the closure of A.*

Remark. Theorem 6.7 does not mention j. However, the theorem is meaningful only if supp(ρ) is not all of R^3. This does not happen in TF theory when $\mu=0$, but it does happen for the j model if the hypothesis of Theorem 6.6 holds. The significance of Theorem 6.7 is that there is *total shielding* in TFD theory. This is in contrast to TF theory, where there is *under-screening* in the neutral case in the sense that the potential falls off only with a power law. One consequence of Theorem 6.7 is that two or more molecules, each of fixed shape, do not interact when their supports are disjoint. See the remark at the end of Sec. IV.

Proof. Let $B = \{x \mid \phi(x) < 0\}$. Clearly the singularities of V are not in B, so B is open. On B, $\Delta\phi \leq 0$ since $\rho=0$ a.e. in B. But $\phi=0$ on ∂B and at infinity so B is empty. Therefore $\phi \geq 0$ everywhere. Let $D = \{x \mid \rho(x) = 0\}$. $\phi \leq 0$ a.e. on D. Since $A \subset D$ is open, and ϕ is continuous on $\{x \mid \phi(x) < 1\}$, $\phi \equiv 0$ on A, and hence on $\bar{A}$. ■

C. The relation of the j model to TFD theory

We shall show that the energy of the j model is exactly $E_\alpha(\lambda) = E(\lambda) + \alpha\lambda$ for the TFD problem. Thus all the facts about the energy in Theorems 6.3 and 6.5 hold for TFD theory. However, the densities may be different!

Let us start with the simplest case studied by Benguria (1979).

Definition. A (non-negative) measure m is said to be *amenable* if

$$dm(x) = \sum_{i=1}^{k} z_i \delta(x - R_i)dx + g(x)dx$$

with $z_i > 0$ and g satisfies: (i) $g \geq 0$, (ii) $g \in L^\infty_{\rm loc}$, (iii) If $A = \{x \mid g(x) = 0\}$ and $\sim A$ is its complement then $R^3 \setminus [\text{Interior}(A) \cup \text{Interior}(\sim A)]$ has zero Lebesgue measure, (iv) $g(x) \geq \rho_0$ for $x \notin A$. (v) $V = |x|^{-1} * m \in \mathfrak{D}$. m is *strongly amenable* if $g(x) > \rho_0$ for $x \notin A$.

Remark. This amenable class is more restrictive than necessary for Theorem 6.8. Technicalities aside, (iv) is the crucial point. δ measures (corresponding to point nuclei) are strongly amenable.

Theorem 6.8. *Suppose that in Eq. (6.1) $V = |x|^{-1} * m$ and m is amenable. Then $E_\alpha(\lambda) = E(\lambda) + \alpha\lambda$ for the TFD problem equals $E_j(\lambda)$ for the j model. Moreover, there is a minimizing ρ for the TFD problem if and only if $\lambda \leq \lambda_c = Z = \int dm$. This ρ is unique and is the same as the ρ for the j model. It satisfies Eq. (3.2).*

Proof. Clearly $E_\alpha \geq E_j$ since $J_\alpha(\rho) \geq j(\rho)$. First suppose $\lambda \leq \lambda_c = Z$ and let ρ be the unique minimum for the j problem. By Theorem 3.19, $\rho(x) \notin (0, \rho_0)$ so $E_\alpha(\lambda) \leq \mathcal{E}_\alpha(\rho) = \mathcal{E}_j(\rho) = E_j(\lambda)$. Thus $E_\alpha(\lambda) = E_j(\lambda)$. Let ρ satisfy $\int\rho = \lambda$ and $\mathcal{E}_\alpha(\rho) = E_\alpha(\lambda)$. Then since $\mathcal{E}_\alpha(\rho) \geq \mathcal{E}_j(\rho) \geq E_j(\lambda)$ we conclude that ρ minimizes $\mathcal{E}_j(\rho)$. But there is only one such ρ. Next, suppose $\lambda > \lambda_c$. Then $E_j(\lambda) = E_j(\lambda_c) = E_\alpha(\lambda_c)$. But $E_\alpha(\lambda) \leq E_\alpha(\lambda_c)$ by Theorem 6.2, and $E_\alpha(\lambda) \geq E_j(\lambda)$. Hence $E_\alpha(\lambda) = E_j(\lambda)$. By the above argument, any minimizing ρ for $\mathcal{E}_\alpha$ would have to minimize $\mathcal{E}_j$, but no such ρ exists. ■

Remark. By Theorem 3.19, $\rho(x) \notin (0, \rho_0)$ a.e. if m is amenable, and $\rho(x) \neq \rho_0$ a.e. if m is strongly amenable. If ρ is merely amenable, $\rho(x)$ can be ρ_0 with positive measure. An example is $dm(x) = \rho_0 B_R(x)dx$, with B_R being the characteristic function of a ball of radius R centered at 0. Then $\rho_\lambda(x) = \rho_0 B_r(x)$ with $4\pi\rho_0 r^3/3 = \lambda$ for $\lambda \leq \lambda_c = 4\pi\rho_0 R^3/3$. This ρ_λ is easily seen to satisfy Eq. (3.2).

If m is not amenable the situation is much more complicated, but more amusing mathematically. First let us consider the energy.

Theorem 6.9. *If $V = |x|^{-1} * m \in \mathfrak{D}$, then $E_\alpha(\lambda) = E_j(\lambda)$ for all λ. In particular, $\lambda_c = Z = \int dm$ and E_α is convex in λ. If there is a minimizing ρ for $E(\lambda)$, it is unique and it is the ρ for the j model.*

A number of lemmas are needed for the proof.

Lemma 6.10. *Let $A \subset R^3$ be a measurable set and let ρ be a function in L^1 with $0 \leq \rho(x) \leq 1$ for $x \in A$, and $\rho(x) = 0$ for $x \notin A$. (This implies $\rho \in L^p$, all p.) Then there exists a sequence of functions $f^n \in L^1$ such that (i) $f^n \to \rho$ weakly in every L^p with $1 < p < \infty$; (ii) f^n is the characteristic function of some measurable set $F^n \subset A$; (iii) $\int f^n = \int \rho$.*

Proof: For $\delta > 0$ and $y \in Z^3$ let $B(\delta, y) = \{x \in R^3 \mid -\delta/2 < x^i - \delta y^i \leq \delta/2\}$ be the elementary cubes of side δ. Let $C(\delta, y) = A \cap B(\delta, y)$. Partition $C(\delta, y)$ into two disjoint measurable sets, C^+ and C^-, such that $|C^+(\delta, y)| = \int B(\delta, y)\rho$. Let $F^\delta = \cup_{y \in Z^3} C^+(\delta, y)$, and let f^δ be the characteristic function of F^δ. Clearly $f^\delta \in L^p$ with norm $(\int\rho)^{1/p}$, and f^δ satisfies (ii) and (iii). Let $1/p + 1/q = 1$. Since C_0, the continuous functions of compact support, are dense in L^q, and $\|f^\delta\|_p^p = $ constant, it suffices to prove that $I(\delta, g) = \int g(f^\delta - \rho) \to 0$ as $\delta \to 0$ for every $g \in C_0$. But g is uniformly continuous, so for any $\varepsilon > 0$, $|g(x) - g(\delta y)| < \varepsilon$ (uniformly) for $x \in B(\delta, y)$ when δ is small enough. Since

$$\int_{B(\delta, y)} (f^\delta - \rho) = 0, \quad |I(\delta, g)| < 2\varepsilon \int \rho. \; ■$$

Lemma 6.11. *Suppose $\rho \in \mathcal{G}$ and*

$$A(\rho)=\{x\,|\,0<\rho(x)<\rho_0\} \tag{6.16}$$

has positive measure. Then there exists $\tilde{\rho}\in\mathcal{G}$ *satisfying (i)* $\mathcal{E}(\tilde{\rho})<\mathcal{E}(\rho)$; *(ii)* $A(\tilde{\rho})$ *is empty; (iii)* $\tilde{\rho}(x)=\rho(x)$ *if* $x\notin A(\rho)$; *(iv)* $\int\tilde{\rho}=\int\rho$.

Proof. Apply Lemma 6.10 to the function $h(x)=\rho(x)/\rho_0$ if $x\in A(\rho)$, $h(x)=0$ otherwise. Let $\tilde{\rho}^n(x)=\rho_0 f^n(x)$ if $x\in A(\rho)$, $\tilde{\rho}^n(x)=\rho(x)$ otherwise. Then $\int\tilde{\rho}^n=\int\rho$ and $\int J_\alpha(\tilde{\rho}^n)-\int J_\alpha(\rho)=-K$ with $K>0$ and independent of n [since $J_\alpha(t)>0$ for $0<t<\rho_0$, and $0=J_\alpha(0)=J_\alpha(\rho_0)$]. Now, as in the proof of Theorem 2.4, $\int V\tilde{\rho}^n\to\int V\rho$. $\lim D(\tilde{\rho}^n,\tilde{\rho}^n)=D(\rho,\rho)$ (easy proof). Hence for any $\varepsilon>0$ there is some n such that $\int V\tilde{\rho}^n>\int V\rho-\varepsilon$ and $D(\tilde{\rho}^n,\tilde{\rho}^n)<D(\rho,\rho)+\varepsilon$. ■

The following is a corollary of Lemma 6.11.

Theorem 6.12. *(a) If* $\rho\in\mathcal{G}$ *minimizes* $\mathcal{E}(\rho)$ *on* $\int\rho=\lambda$, *then* measure $\{A(\rho)\}=0$. *(b) Even if there is no minimizing* ρ, *a minimizing sequence for* $E(\lambda)$ *with* $\int\rho=\lambda$ *can be chosen such that* $A(\rho^n)$ *is empty for all n.*

Theorem 6.13. *Let* ρ *minimize* $\mathcal{E}_j$ *for* $\int\rho=\lambda$. $A(\rho)$ *may not be empty, but for any* $\varepsilon>0$ *there exists a* $\tilde{\rho}$ *such that* $\int\tilde{\rho}=\lambda$, $A(\tilde{\rho})$ *is empty, and* $\mathcal{E}_j(\tilde{\rho})<E_j(\lambda)+\varepsilon$.

Proof. Again, use Lemma 6.10 and mimic the proof of Theorem 6.11. ■

Proof of Theorem 6.9. That $E_\alpha(\lambda)=E_j(\lambda)$ follows from Theorem 6.13 and an imitation of the argument in Theorem 6.8. By Theorem 6.12 any minimizing ρ has $A(\rho)$ empty and thus minimizes $\mathcal{E}_j(\rho)$. There can be only one such ρ, since the minimizing ρ for $\mathcal{E}_j$ is unique. ■

In summary: $E(\lambda)=E_j(\lambda)-\alpha\lambda$ always, but a minimizing ρ may or may not exist for the TFD problem. It exists if and only if the minimizing ρ for the j model (which always exists when $\lambda\leq\lambda_c$) satisfies $\rho(x)\notin(0,\rho_0)$ a.e. A sufficient condition on V for this to occur is that V be amenable; a necessary condition seems to be difficult to find. The example of Sec. III.B illustrates the nonexistence phenomenon: If $dm(x)=g(x)dx$ with $g(x)\in(0,\rho_0)$ and $g\in L^1$, then $\rho_j(x)=g(x)$ in the neutral case, $\lambda=\lambda_c=\int g$. But $g(x)$ does not minimize $\mathcal{E}(\rho)$. A *sequence* of minimizing ρ's for $\mathcal{E}(\rho)$ are functions which on the average locally imitate g but which oscillate rapidly between the two values 0 and ρ_0.

VII. THOMAS-FERMI-VON WEIZSÄCKER THEORY

TFW theory was originally suggested by von Weizsäcker (1935) with $\delta=\delta_1\equiv\hbar^2/2m$ in Eq. (2.8). There is no fundamental justification for TFW theory in the sense that there is for TF theory, i.e., there is no theorem that the correction to the energy or density caused by the Weizsäcker term with $\delta=\delta_1$ agrees with the real quantum problem. We shall see, however, that if $\delta=(0.186)\,\delta_1$ then the energy correction proposed by Scott (Sec. V.B), but not the density correction conjectured in Eq. (5.37), is realized in TFW theory. If Scott is correct then *a posteriori* TFW theory has some fundamental meaning for atoms and molecules.

We were able to make a great deal of progress with TF and TFD theories essentially because of the pointwise relation between $\phi(x)$ and $\rho(x)$. In TFW theory this relation is lost, and therefore TFW theory is much more difficult mathematically. However, the physical consequences of TFW theory are much richer and *qualitatively* more nearly parallel the physics of real atoms and molecules. In addition to the above mentioned Z^2 energy correction, TFW theory remedies three defects of TF (and TFD) theory:

(i) ρ will be finite at the nuclei.

(ii) binding of atoms occurs and negative ions are stable (i.e., $\lambda_c>Z$). These two facts are closely related.

(iii) ρ has exponential falloff if $\lambda<\lambda_c$, e.g., for neutral atoms and molecules.

The theory presented here was begun by Benguria (1979) and then further developed by Benguria, Brezis, and Lieb (1981) (BBL), to which we shall refer for technical details. Some newer results will also be given, especially that $\lambda_c>Z$ for molecules, the Z^2 correction to the energy (Sec. VII.D), and the binding of equal atoms. Many interesting problems are still open, however.

The TFW energy functional (see Note (iv) below) is

$$\mathcal{E}(\rho)=A\int[\nabla\rho^{1/2}(x)]^2dx+(\gamma/p)\int\rho(x)^p dx$$
$$-\int V(x)\rho(x)dx+D(\rho,\rho)+U. \tag{7.1}$$

This agrees with Eq. (2.8) in units in which $\hbar^2/2m=1$. A closely related functional, obtained by writing $\psi^2=\rho$, is

$$\mathcal{E}'(\psi)=A\int(\nabla\psi)^2+(\gamma/p)\int\psi^{2p}$$
$$-\int V\psi^2+D(\psi^2,\psi^2)+U. \tag{7.2}$$

Note. (i) In this section all quantities refer to TFW theory unless otherwise stated.

(ii) Equation (7.1) is defined for $\rho(x)\geq 0$ while in Eq. (7.2) $\psi(x)$ only has to be real.

(iii) As in Sec. VI, we shall assume for simplicity that V is given by Eq. (6.3) *et seq.* Later on a slightly stronger hypothesis (7.12) will be used.

(iv) $p>1$ is a parameter; it will not be indicated explicitly unless necessary. Recall that E^{TF} was finite for point nuclei only if $p>\frac{3}{2}$. E is finite in TFW theory for *all* $p>0$. We need $p>1$ for Theorems 7.1 and 7.2, among other reasons. Even though we are interested in $p=\frac{5}{3}$, we allow p to be arbitrary because the dependence on p is interesting. It will turn out that $p=\frac{5}{3}$, the case of physical interest, is special—at least it is so for the proof that $\lambda_c>Z$.

A. The TFW minimization problem

The function space is

$$G_p'=\{\psi\,|\,\nabla\psi\in L^2,\ \psi\in L^6\cap L^{2p},\ D(\psi^2,\psi^2)<\infty\}. \tag{7.3}$$

We say $\rho\in G_p$ if $\rho(x)\geq 0$ and $\rho^{1/2}\in G_p'$. G_p' contains

$$F_p'=G_p'\cap L^2=\{\psi\,|\,\nabla\psi\in L^2,\ \psi\in L^6\cap L^{2p}\cap L^2\}. \tag{7.4}$$

Even though we are interested in $\rho\in L^1$ (or $\psi\in L^2$), the larger space G_p' is used for technical reasons in order to prove that $\lambda_c < \infty$; in other words, we shall eventually find that all ρ's of interest are in F_p (defined analogously to G_p).

Remark. By Sobolev's inequality, $\|\nabla\psi\|_2 \geq L\|\psi\|_6$ with $L = 3^{1/2}(\pi/2)^{2/3}$ (cf. Lieb, 1976) when $\psi \in L^q$, $1 \leq q < \infty$. Thus when $\psi \in F_p'$ the restrictions that $\psi \in L^6$ and $\psi \in L^{2p}$ (if $p \leq 3$) are unnecessary. In short, $F_p' = H^1 \cap L^{2p}$, where $H^1 = \{\psi | \nabla\psi$ and $\psi \in L^2\}$.

As usual,

$$E(\lambda) = \inf\left\{\mathcal{E}(\rho) \,|\, \rho \in F_p, \int \rho = \lambda\right\},$$
$$E'(\lambda) = \inf\left\{\mathcal{E}'(\psi) \,|\, \psi \in F_p', \int \psi^2 = \lambda\right\}. \tag{7.5}$$

Theorem 7.1. *$\mathcal{E}(\rho)$ is strictly convex in ρ.*

Proof. The only term that has to be checked is $\int(\nabla\rho^{1/2})^2$. If $\psi_i = \rho_i^{1/2}$, $i = 1, 2$, and $\psi = (\sum \alpha_i^2\rho_i)^{1/2}$, $\sum\alpha_i^2 = 1$, then

$$\psi\nabla\psi = \sum (\alpha_i\psi_i)(\alpha_i\nabla\psi_i)$$

and

$$(\psi\nabla\psi)^2 \leq \left(\sum \alpha_i^2\psi_i^2\right)\left(\sum \alpha_i^2(\nabla\psi_i)^2\right).$$

Assuming $\psi(x) > 0$ everywhere, we are done. Otherwise, the result follows by approximation. ∎

Remark. $\mathcal{E}'(\psi)$ is *not convex* in ψ because of the $-\int V\psi^2$ term.

Theorem 7.2. *For all finite λ*

(*i*) $E(\lambda) = E'(\lambda)$;

(*ii*) $E(\lambda) = \inf\left\{\mathcal{E}(\rho) \,|\, \rho \in F_p, \int \rho \leq \lambda\right\}$

and similarly for $E'(\lambda)$;

(*iii*) *$E(\lambda)$ is convex and monotone nonincreasing in λ.*

Proof. (i) Given ρ, we can always construct $\psi = \rho^{1/2}$, so $E'(\lambda) \leq E(\lambda)$. Conversely, given ψ let $f = |\psi|$. But $\nabla f = (\nabla\psi)(\text{sgn}\psi)$ so $\int(\nabla f)^2 = \int(\nabla\psi)^2$. Thus $\mathcal{E}'(f) = \mathcal{E}'(\psi)$. Choosing $\rho = f^2$, $E(\lambda) \leq E'(\lambda)$. (ii) As before, "excess charge" can be put at "infinity." Here $p > 1$ is essential. (iii) $\mathcal{E}(\rho)$ is convex so $E(\lambda)$ is convex. Monotonicity is implied by (ii). ∎

Remark. (i) relates the two problems defined by Eqs. (7.1) and (7.2). To obtain the convexity (iii), $\mathcal{E}$ and Theorem 7.1 were used. We shall use $\mathcal{E}'$ to obtain the existence of a minimum, and then the TFW equation for this minimum.

Lemma 7.3. *Let $V = |x|^{-1} * m$, with m a measure and $|m| = Z < \infty$. Let $\rho(x) \geq 0$. Then there exists a constant C independent of m and ρ such that for every $\varepsilon > 0$*

$$\int V\rho \leq \varepsilon Z\|\rho\|_3 + Z\varepsilon^{-1/2}CD(\rho,\rho)^{1/2}.$$

Proof. By regarding R^3 as the union of balls of unit radius centered on the points of $(\frac{1}{2})\mathrm{Z}^3$ it suffices to assume $\text{supp}(m) \subset B_1$, where $B_R = \{x \,|\, |x| \leq R\}$ and χ_R is the characteristic function of B_R. In the following, irrelevant constants will be suppressed. Write $V = V_- + V_+$ where $V_- = V\chi_2$ and similarly for ρ. First consider $I_- = \int V_-\rho_-$. By Young's inequality (and writing $|x-y|^{-1} = |x-y|^{-1}\chi_4(x-y)$ for $x, y \in B_2$) $\|V_-\|_2 \leq Z$. Thus

$$I_- \leq Z\|\rho_-\|_2 \leq Z\|\rho\|_3^{3/4}\|\rho_-\|_1^{1/4}$$
$$\leq Z\{\|\rho\|_3 + \|\rho_-\|_1\}.$$

But

$$D(\rho,\rho) \geq D(\rho_-,\rho_-) \geq \|\rho_-\|_1^2$$

since $|x-y|^{-1} > \frac{1}{8}$ in B_2. Now, outside B_2, $V(x) < 2Z/|x| \equiv W(x)$. Let $Q = q\chi_2$ be the constant charge distribution such that $|x|^{-1} * Q = W(x)$ outside B_2. Then $I_+ \leq 2D(Q,\rho_+) \leq 2D(Q,Q)^{1/2}D(\rho_+,\rho_+)^{1/2}$. But $D(Q,Q) = Z^2$. Therefore (on the whole of R^3) $I \leq Z\|\rho\|_3 + ZCD(\rho,\rho)^{1/2}$. Now replace $\rho(x)$ by $\varepsilon^3\rho(\varepsilon x)$ and $dm(x)$ by $\varepsilon^3 dm(\varepsilon x)$. Then $\varepsilon I = I_\varepsilon \leq Z\varepsilon^2\|\rho\|_3 + ZC\varepsilon^{1/2}D(\rho,\rho)$. ∎

From the Sobolev inequality, the following is obtained:

Corollary 7.4. *There are constants a and $b > 0$ such that for every $\psi \in G_p'$*

$$\mathcal{E}'(\psi) \geq a[\|\nabla\psi\|_2^2 + \|\rho\|_p^p + \|\rho\|_3 + D(\rho,\rho)] + U - b, \tag{7.6}$$

with $\rho = \psi^2$. In particular, $\mathcal{E}'$ is bounded below on G_p' and $E(\lambda)$ is bounded below.

It is obvious that $E(\lambda) \geq E^{\text{TF}}(\lambda)$, with the same p. If $p > \frac{3}{2}$, $E^{\text{TF}}(\lambda)$ is finite for point nuclei. The following illustrates the sort of lower bound for $E(\lambda)$ that can be obtained with the Sobolev inequality.

Theorem 7.5. *Let $p = \frac{5}{3}$. Let $E^{\text{TFW}}(A,\gamma,\lambda)$ denote the TFW energy and $E^{\text{TF}}(\gamma,\lambda)$ denote the TF energy. Let $\bar{L} = 9.578$. Then*

$$E^{\text{TFW}}(A,\gamma,\lambda) \geq E^{\text{TF}}(\gamma + A\bar{L}\lambda^{-2/3},\lambda). \tag{7.7}$$

In particular, for an atom with a point nucleus, $E^{\text{TF}}(\gamma,\lambda) \sim \gamma^{-1}$ whence, for an atom,

$$E^{\text{TFW}}(A,\gamma,\lambda) \geq \gamma(\gamma + A\bar{L}\lambda^{-2/3})^{-1}E^{\text{TF}}(\gamma,\lambda). \tag{7.8}$$

Proof.

$$\int |\nabla\psi|^2 \geq \bar{L}\left(\int |\psi|^{10/3}\right)\left(\int |\psi|^2\right)^{-2/3}.$$

See Lieb, 1976. ∎

Remark. The right side of Eq. (7.8) has two properties: (i) Its slope at $\lambda = 0$ is finite. (ii) It is strictly monotone decreasing for all λ. To some extent, E^{TFW} will be seen to mimic this: E^{TFW} has a finite slope at $\lambda = 0$ and is strictly decreasing up to $\lambda_c > Z$.

Theorem 7.6. *(i) $\mathcal{E}'(\psi)$ has a minimum on the set $\psi \in F_p'$ and $\int\psi^2 \leq \lambda$.*

(ii) $\mathcal{E}'(\psi)$ has a minimum on G_p'.

(iii) The same is true for $\mathcal{E}(\rho)$ on $F_p(\int\rho \leq \lambda)$ and G_p. Furthermore ρ and ψ are related by $\rho(x) = \psi(x)^2$. The minimizing ρ is unique.

Proof. The proof we give is different from the proof of Theorem 2.4 because Fatou's lemma will be used, as stated in the remark after Theorem 2.4. Let ψ^n be a minimizing sequence. By Corollary 7.4 all quantities in Eq. (7.6) are bounded. By passing to a subsequence we can demand, by the Banach-Alaoglu theorem, that $\nabla\psi^n \to f$ weakly in L^2 and $\rho^n \to \rho$ weakly in L^3 and in L^p [where

$\rho^n=(\psi^n)^2]$. Furthermore, for any bounded ball B, $\psi \in L^2(B)$ since $\psi\in L^6(B)$. Moreover, $H^1(B)$ is relatively (norm) compact in $L^2(B)$. Thus, by passing to a further subsequence, we may assume $\psi^n\to\psi$ strongly in $L^2(B)$ for every B and pointwise a.e. Then it is clear that $\rho=\psi^2$ and $f=\nabla\psi$. As before, $\liminf\|\nabla\psi^n\|_2\geq\|\nabla\psi\|_2$. From the pointwise convergence and Fatou's lemma, $\liminf D(\rho^n,\rho^n)\geq D(\rho,\rho)$ and $\liminf\|\rho^n\|_p\geq\|\rho\|_p$. For the V term we write $m=m_1+m_2$ with $m_1=m\chi_R$ and choose R large enough so that $|m_2|<\delta$ (since $|m|=Z<\infty$). If $V_2=m_2*|x|^{-1}$ then $\int V_2|\rho-\rho^n|<\delta(\text{const})$ by Lemma 7.3 (with $\varepsilon=1$). Next, write $V_1=V_-+V_+$ with $V_-=V_1\chi_r$. If $r>2R$, $V_+(x)<2Z/|x|$. Let Q_r be the uniform charge distribution inside B_r so that $Q_r*|x|^{-1}=2Z/|x|$ outside B_r. Then $\int V_+|\rho-\rho^n|\leq D(Q_r,Q_r)^{1/2}D(|\rho-\rho^n|,|\rho-\rho^n|)^{1/2}$. Choose r large enough so that $D(Q_r,Q_r)<\delta^2$. $V_-\in L^{3/2}$, so $\int V_-(\rho-\rho^n)\to 0$. Since δ was arbitrary, $\int V(\rho-\rho^n)\to 0$. Combining all this, $\liminf\mathcal{E}'(\psi^n)\geq\mathcal{E}'(\psi)$. Finally, if $\int\rho^n\leq\lambda$ then $\int\rho\leq\lambda$ as in the proof of Theorem 2.4 (but using L^3). As remarked in the proof of Theorem 7.2, we can choose $\psi^n(x)\geq 0$ everywhere; hence $\psi(x)\geq 0$ and $\rho(x)=\psi(x)^2$ minimizes $\mathcal{E}(\rho)$. The uniqueness of ρ follows from the strict convexity of $\mathcal{E}(\rho)$. ■

Definition. λ_c can be defined as in Sec. II, namely, $\lambda_c=\sup\{\lambda|E(\lambda)=\lim_{\lambda\to\infty}E(\lambda)\}$. A simple variational calculation, which exploits the fact that $V(x)\approx -Z/|x|$ for large $|x|$, shows that $\lambda_c>0$.

Theorem 7.7. *There is a minimizing ρ on F_p' with $\int\rho=\lambda$ if and only if $\lambda\leq\lambda_c$. The minimizing ρ in Theorem 7.6 when $\lambda>\lambda_c$ is the ρ for λ_c. $E(\lambda)$ is strictly convex on $[0,\lambda]$.*

Proof. Same as for Theorem 2.5. ■

Theorem 7.8. *(i) Any minimizing $\psi\in F_p'$ for $\mathcal{E}'(\psi)$ on the set $\int\psi^2\leq\lambda$ satisfies the* TFW equation *(in the sense of distributions)*:

$$[-A\Delta+W_\psi(x)]\psi(x)=-\mu\psi(x)\,, \tag{7.9}$$

where

$$W_\psi(x)=\gamma\psi(x)^{2p-2}-\phi_\rho(x) \tag{7.10}$$

and

$$\phi_\rho=V-|x|^{-1}*\rho\quad with\ \rho=\psi^2.$$

(ii) If ψ minimizes $\mathcal{E}'(\psi)$ on G_p', then ψ satisfies Eq. (7.9) with $\mu=0$.

(iii) $E(\lambda)$ is continuously differentiable and $-\mu=dE/d\lambda$ for $\lambda\leq\lambda_c$, while $0=dE/d\lambda$ for $\lambda\geq\lambda_c$. In particular, $\mu\geq 0$.

(iv) If $\rho\in G_p$ satisfies Eq. (7.9) and $\int\rho=\lambda$ (possibly ∞), then ρ minimizes $\mathcal{E}(\cdot)$ on the set $\int\rho\leq\lambda$.

(v) Fix λ. There can be at most one pair ρ,μ [with $\rho(x)\geq 0$] that satisfies Eq. (7.9) with $\int\rho=\lambda$.

Proof. (i) and (ii) are standard. Just consider $\psi+\varepsilon f$ with $f\in C_0^\infty$ and $(f,\psi)=0$ and set $d\mathcal{E}/d\varepsilon=0$. For the absolute minimum we do not require $(f,\psi)=0$. The proof of (iii) is as in Theorem 2.7 (cf. LS Theorem II.10 and Lemma II.27). The proof of (iv) and (v) imitates that of Theorem 2.6. ■

We shall eventually prove that the minimizing ψ is unique (we already know that ρ, and hence ψ^2 is unique), but Theorem 7.9 is needed first. We also want to prove that $\lambda_c<\infty$, i.e., the ψ that satisfies Eq. (7.9) with $\mu=0$ satisfies $\int\psi^2<\infty$. This will be done in Sec. VII.B.

Theorem 7.9. *If $\psi\in G_p'$ satisfies Eq. (7.9) (as a distrubution) and $\psi(x)\geq 0$ for all x, then (i) ψ is continuous. More precisely, $\psi\in C^{0,\alpha}$ for every $\alpha<1$ [i.e., for every bounded ball B, $|\psi(x)-\psi(y)|<M|x-y|^\alpha$ for some M and all $x,y\in B$]. (ii) If V is C^∞ on some open set Ω, then ψ is C^∞ on Ω. For point nuclei, $\Omega=\mathbb{R}^3\backslash\{R_i\}$. (iii) Either $\psi\equiv 0$ or $\psi(x)>0$ everywhere. (iv) $W_\psi\in L^{3-\varepsilon}_{\rm loc}$ for every $\varepsilon>0$.*

Proof. Clearly $V\in L^{3-\varepsilon}_{\rm loc}$ (all $\varepsilon>0$) and, since $\psi\in L^6$, $-A\Delta\psi\leq f$ with $f=V\psi\in L^{2-\varepsilon}_{\rm loc}$ (all $\varepsilon>0$). Choosing $\varepsilon<\frac{1}{2}$, we can apply a result of Stampacchia (1965, Theorem 5.2) to conclude $\psi\in L^\infty_{\rm loc}$ and hence $\psi^{2p-1}\in L^{3-\varepsilon}_{\rm loc}$ (all $\varepsilon>0$). Now, $g=|x|^{-1}*\rho\in L^6$ [since $\Delta g=-4\pi\rho\Rightarrow K\|g\|_6^2\leq\int(\nabla g)^2=8\pi D(\rho,\rho)$]. Therefore $-\Delta\psi\in L^{3-\varepsilon}_{\rm loc}$ (all $\varepsilon>0$). Then (Adams, 1975, p. 98) $\psi\in C^{0,\alpha}$. (ii) follows by a bootstrap argument as in Theorem 2.8. For (iii) we note that $-\Delta\psi=b\psi$ and $b\in L^q_{\rm loc}$, $q>\frac{3}{2}$. The conclusion follows from Harnack's inequality (Gilbarg and Trudinger, 1977). ■

We know that $\rho^{1/2}\geq 0$ satisfies Eq. (7.9), so $\rho^{1/2}$ enjoys the above properties. Since ρ is unique we shall henceforth denote Eq. (7.10) simply by W. We shall also use the notation

$$H=-A\Delta+W\,. \tag{7.11}$$

Theorem 7.10. *The minimizing ψ is unique up to a sign which is fixed by $\psi(x)=\rho(x)^{1/2}>0$ everywhere. ψ is also the unique ground-state eigenfunction of $H=-A\Delta+W(x)$ and μ is its ground-state eigenvalue.*

Proof. If ψ is minimizing then $\psi^2=\rho$ and H are uniquely determined. $f=\rho^{1/2}$ satisfies $Hf=-\mu f$. Since f is nonnegative, it is the ground state of H, and the ground state of H is unique up to sign (cf. Reed and Simon, 1978, Sec. XIII.12). ■

Remarks. (i) It is *not* claimed that the TFW equation (7.9) and (7.10) has no solution other than the positive one. Infinitely many other solutions probably exist. They have been found for certain nonlinear equations which have some resemblance to the TFW equation (Berestycki and Lions, 1980), but the TFW equation itself has not been analyzed in this regard. These other solutions correspond, in some vague sense, to "excited states."

(ii) The interplay between $\mathcal{E}'(\psi)$ and $\mathcal{E}(\rho)$ should be noted. Apart from the somewhat pedantic question of the uniqueness of ψ, $\mathcal{E}$ was used to get the uniqueness of $\rho=\psi^2$ and the convexity of $E(\lambda)$. $\mathcal{E}'$ was used to get the TFW equation in which it is not necessary to distinguish between $\rho(x)>0$ and $\rho(x)=0$ as in the TF equation (2.18). The ψ of interest automatically turns out to be positive. For purposes of comparison, the TF equation is $(W+\mu)\psi=0$ if $\psi>0$, and $(W+\mu)\geq 0$ if $\psi=0$. The TFW equation is $(W+\mu)\psi=A\Delta\psi$ everywhere.

(iii) Note that there is a solution even for $\mu=0$. For this ρ, $H=-A\Delta+W$ has zero as its ground-state eigenvalue with an L^2 eigenfunction, ψ (Theorem 7.12). This is unusual. Zero is also the bottom of the essential spectrum of H.

To complete the picture of $E(\lambda)$ we have to know how $E(\lambda)$ behaves for small λ. Since μ is a decreasing function of λ (by convexity of E), μ has its maximum at $\lambda=0$.

Theorem 7.11. *$\mu(\lambda=0)=-e_0$ where $e_0<0$ is the ground-state energy of the Hamiltonian $H_0=-A\Delta-V(x)$. In particular, for a point nucleus $\mu(\lambda=0)=z^2/(4A)$.*

Proof. $\mu(0)=\lim_{\lambda\to 0}E(0)-E(\lambda)$. $E(0)=U$. Let f be the normalized ground state of H_0: $H_0 f=e_0 f$. Let $\bar\rho=\lambda f^2$. Clearly, $\mathcal{E}(\bar\rho)=\lambda e_0+U+o(\lambda)$, since $p>1$. On the other hand, for any ρ with $\int\rho=\lambda$, $\mathcal{E}(\rho)\geq\lambda e_0+U$. ■

In Sec. VII.B we shall see that $Z<\lambda_c<\infty$. Therefore the behavior of $E(\lambda)$ can be summarized as follows: $e(\lambda)=E(\lambda)-U$ in TFW theory looks like Fig. 1 with two important changes: (i) $\lambda_c>Z$ (at least for $p\geq\frac{5}{3}$). $e(\lambda)$ is strictly convex for $0\leq\lambda\leq\lambda_c$. (ii) The slope at $\lambda=0$ is finite. [In TF theory $e(\lambda)\approx\lambda^{1/3}$.]

B. Properties of the density and λ_c

Our main concern here will be to estimate λ_c. For energetic reasons, it is intuitively clear that $\lambda_c\geq Z$ for large enough p because otherwise the energy could always be lowered by adding some additional charge far out. Benguria (1979) proved this for $p\geq\frac{4}{3}$. We shall also see that $\lambda_c>Z$ for $p\geq\frac{5}{3}$.

What is far from obvious, however, is that λ_c is finite. There is no energetic reason why $E(\lambda)$ could not steadily decrease (and be bounded, of course). It is easy to construct a $\rho(x)$ with $\int\rho=\infty$ so that all the terms in the energy and also $\phi(x)$, except at the nuclei, are finite. $\rho(x)=(1+x^2)^{-3/2}$ is an example. That $\lambda_c<\infty$ is a subtle fact. The same question arises in quantum theory, and it has only recently been proved there that λ_c is finite. Ruskai (1981) proved this when the "electrons" are bosons. I. M. Sigal later found a proof (by a different method) for fermions (paper in preparation).

In the following, ψ always means $+\rho^{1/2}$. For simplicity we shall henceforth assume the following condition in addition to $V\in\mathfrak{D}$:

$$V(x)\leq C/|x| \tag{7.12}$$

for some $C<\infty$ and for all $|x|>$ some R. The fact that $V=|x|^{-1}*m$ and $|m|=Z$ does not guarantee Eq. (7.12). If, however, m has compact support, then (7.12) holds.

Theorem 7.12. *$\lambda_c<\infty$ for all $p>1$.*

Proof. Let ρ give the absolute minimum of $\mathcal{E}(\rho)$ on G_p. ψ satisfies Eq. (7.9) with $\mu=0$. We shall prove that this ρ has $\lambda=\int\rho<\infty$, thereby proving that $E(\lambda)$ has an absolute minimum at λ, and hence that $\lambda_c=\lambda$. Assume $\lambda=\infty$. Then for $|x|\geq$ some R [which is bigger than the R in Eq. (7.12)], $|x|^{-1}*\rho>2C/|x|$. Thus, for $|x|>R$, $-A\Delta\psi\leq-C\psi/|x|$. Now we use a comparison argument. Let

$$f(x)=M\exp\{-2[C|x|/A]^{1/2}\}$$

with $M>0$. f satisfies $-A\Delta f\geq-Cf/|x|$, for $|x|\neq 0$, so $-A\Delta(\psi-f)\leq-C(\psi-f)/|x|$. Fix M by $f(x)\geq\psi(x)$ for $|x|=R$. If we knew that $\psi(x)\to 0$ as $|x|\to\infty$ we could conclude, from the maximum principle, that $\psi<f$ for $|x|\geq R$. This implies that $\psi\in L^2$. Unfortunately, we only know that $\psi(x)\to 0$ in a weak sense (namely, L^6). This, it turns out, is good enough. See BBL for details. ■

Now that we know $\int\rho<\infty$, even for the absolute minimum ($\mu=0$), we can prove

Theorem 7.13. *ψ is bounded on $\mathbb{R}^3$ and $\psi(x)\to 0$ as $|x|\to\infty$. Also, $\psi\in H^2$ (i.e., $\psi,\nabla\psi$ and $\Delta\psi\in L^2$).*

Proof. $-A\Delta\psi<V\psi$ so $(-A\Delta+1)\psi\leq(V+1)\psi$. Since $(V+1)\psi\in L^2$, $\psi\leq(-A\Delta+1)^{-1}[(V+1)\psi]$ and this is bounded and goes to zero as $|x|\to\infty$ (Lemma 3.1). Finally, $\psi^{2p-1}\leq d\psi$ for some d, and $g=|x|^{-1}*\rho\in L^6$ together with $\psi\in L^6$ imply $g\psi\in L^2$. Hence $\Delta\psi\in L^2$. ■

Theorem 7.14. *If $p\geq\frac{3}{2}$ then, for all x,*

$$\gamma\rho(x)^{p-1}\leq V(x). \tag{7.13}$$

In particular, if $p=\frac{5}{3}$, $\rho(x)<[V(x)/\gamma]^{3/2}$.

Proof. The essential point is that since V is superharmonic, so is V^t for $t\leq 1$. Let $f=\psi-(V/\gamma)^t$ with $t=1/(2p-2)$. Let $B=\{x\,|\,f(x)>0\}$. Since ψ and V are continuous on B, B is open. On B, $W>0$ so $-\Delta f<0$. $f=0$ at ∞ and on ∂B, so B is empty. ■

Remark. The bound in Eq. (7.13) also holds trivially in TF theory from Eq. (2.18).

Theorem 7.15. *If $p\geq\frac{4}{3}$ then $\lambda_c\geq Z$.*

Proof. Suppose $\lambda_c=Z-\varepsilon$. Since $H=-A\Delta+W$ has zero as its ground-state energy, $(f,Hf)\geq 0$ for any $f\in C_0^\infty$. Let $f_1(x)\not\equiv 0$ be spherically symmetric with support in $1\leq|x|\leq 2$, $f_1(x)\leq 1$, and $f_n(x)=f(x/n)$. Then $\int f_n^2\phi=\int f_n^2[\phi]$, where $[\phi]$, is the spherical average of ϕ. It is easy to see that for $|x|\geq$ some R, $[\phi]\geq\varepsilon/2|x|$ since $\int\rho=Z-\varepsilon$. Therefore $\int f_n^2\phi\geq(\text{const})n^2$ for large n. $\int(\nabla f_n)^2=(\text{const})n$. The crucial quantity is $D_n=\int f_n^2\rho^{p-1}$. If $p\geq 2$, $D_n\leq(\text{const})\int\rho$. If $p<2$ use Hölder's inequality: $D_n\leq X_n^{p-1}Y_n^{2-p}$, where $X_n=\int f_n^2\rho$ and $Y_n=\int f_n^2$. Clearly $X_n\to 0$ as $n\to\infty$ since $\rho\in L^1$. $Y_n=(\text{const})n^3$. Now let $n\to\infty$, whence $(f_n,Hf_n)\to-\infty$. ■

Remarks. (i) The basic reason that $p\geq\frac{4}{3}$ is needed in the proof of Theorem 7.15 is that we want to be able to ignore the ρ^{p-1} term in W and thereby obtain a negative-energy bound state for H when $\lambda<Z$. However, if $\rho\in L^1$ then (essentially) $\rho(x)\sim|x|^{-3}f(x)$, where $f(x)$ can be slowly decreasing. Hence we can be certain that ρ^{p-1} is small compared to $|x|^{-1}$ only if $3(p-1)\geq 1$.

(ii) In Theorems 7.16 and 7.19 we prove that $\lambda_c>Z$. The underlying idea is that to have a zero-energy L^2 bound state, $W(x)$ has to be *positive* for large $|x|$. Essentially, $W(x)$ has to be as big as $|x|^{-2}$; this requirement is clear if we assume $\psi(x)\sim|x|^{-a}$ for large $|x|$. If $\lambda_c=Z$, then ϕ is (essentially) positive for large $|x|$, so the repulsion has to come from ρ^{p-1}. But if $p-1\geq\frac{2}{3}$ then ρ^{p-1} cannot be sufficiently big since $\rho\in L^1$. The theorem that $\lambda_c>Z$ when $p\geq\frac{5}{3}$ was proved for an atom in BBL. We give that proof first in Theorem 7.16 in order to clarify the ideas. Then, after Lemma 7.18, we give a proof (which is not in BBL) of the general case in Theorem 7.19. Some condition on p really is needed to have $\lambda_c>Z$. In BBL it is proved that if $p=\frac{3}{2}$, $\gamma=1$, V is given by Eq. (2.1), and $A\leq 1/16\pi$, then $\lambda_c=Z$.

Theorem 7.16. *Suppose $p\geq\frac{5}{3}$ and suppose $V=|x|^{-1}*m$*

where m is a non-negative measure that satisfies the following conditions: (i) m is spherically symmetric; (ii) the support of m is contained in some ball, $B_R = \{x \mid |x| < R\}$. *Then* $\lambda_c > Z$.

Proof. Assume that $\lambda_c \leq Z$. By Newton's theorem, $\phi(x) \geq 0$ for $|x| > R$. Then when $\lambda = \lambda_c$, $-\Delta\psi \geq -\gamma\psi^{2p-1}$ for $|x| > R$. ψ is spherically symmetric and $\psi(R) > 0$. Let $f(x) = C|x|^{-3/2}$, which satisfies $-A\Delta\psi \geq -\gamma\psi^{2p-1}$ for $|x| \geq R$ provided $0 < C \leq D$ with $D^{2p-2} = (3A/4\gamma)R^{3p-5}$. Let $C = \min[D, \psi(R)]$. Then $\psi(x) \geq f(x)$ for all $|x| > R$, because $-A\Delta(\psi - f) \geq -\gamma(\psi^{2p-1} - f^{2p-1})$, which would imply that $\psi - f$ is superharmonic on the set where $\psi - f < 0$. Since ψ and f go to zero at infinity, and $\psi - f \geq 0$ at $|x| = R$, this is impossible. Hence $\psi \notin L^2$, which contradicts $\|\psi\|_2 \leq Z$. ■

In the foregoing we used a comparison argument which, in turn, relied on the fact that the positive part of W, namely ρ^{p-1}, was simply related to ψ. In the proof of Theorem 7.19 we shall not have that luxury, and so the more powerful Lemma 7.18 is needed.

Lemma 7.17. *Let* S_R *denote the sphere* $\{x \mid |x| = R\}$ *and let* $d\Omega$ *be the normalized, invariant, spherical measure on* S_1. *For any function h, let* $[h](r) = \int h(r,\Omega)d\Omega$ *be the spherical average of h. Now suppose* $\psi(x) > 0$ *is* C^2 *in a neighborhood of* S_R. *Let* $f(r) = \exp\{[\ln\psi](r)\}$. *Then, for all r in some neighborhood of R,*

$$[\Delta\psi/\psi](r) \geq (\Delta f/f)(r)$$

$$= \{d^2f/dr^2 + (2/r)df/dr\}/f(r).$$

Proof. Let $g(x) = \ln\psi(x)$. Then $\Delta\psi/\psi = \Delta g + (\nabla g)^2$. Clearly $[\Delta g] = \Delta[g]$. Moreover, $(\nabla g)^2 \geq \{\partial g(r,\Omega)/\partial r\}^2$, and $[(\partial g/\partial r)^2] \geq (d[g]/dr)^2$ by the Schwarz inequality. Thus $[\Delta\psi/\psi] \geq \Delta[g] + (\nabla[g])^2 = \Delta f/f$. ■

Lemma 7.18. *Suppose* $\psi(x) > 0$ *is a* C^2 *function in a neighborhood of the domain* $D = \{x \,\|\, x| > R\}$ *and* ψ *satisfies* $\{-A\Delta + W(x)\}\psi(x) \geq 0$ *on D. Let* $[W]$ *be the spherical average of W and write* $[W] = [W]_+ - [W]_-$ *with* $[W]_+(x) = \max[[W](x), 0]$. *Suppose* $[W]_+ \in L^{3/2}(D)$. *Then* $\psi \notin L^2(D)$. *(Note: no hypothesis is made about* $[W]_-$.*) See note added in proof below.*

Remarks. Simon (1981, Appendix 3) proves a similar theorem for $D = R^3$, except that $[W] = [W]_+ - [W]_-$ is replaced by $W = W_+ - W_-$ with $W_+ = \max(W, 0)$. Simon does not require the technical restrictions that $\psi(x) > 0$ and ψ is C^2. Simon's theorem will be used in our proof. Lemma 7.18 improves Simon's result in two ways: (i) It is sufficient to consider D and not all of R^3. (ii) It is only necessary that $[W]_+$, and not W_+, be in $L^{3/2}$; the latter distinction is important. As an example, suppose that for large $|x|$ the potential W is that of a dipole, i.e., $W(x_1,x_2,x_3) = x_1|x|^{-3}$. $W_+ \notin L^{3/2}$ but, since $[W]_+ = 0$, Lemma 7.18 says that this W cannot have a zero-energy L^2 bound state.

Proof. Let $f = \exp\{[\ln\psi]\}$ as in Lemma 7.17. Then $-A\Delta f/f + [W] \geq [-A\Delta\psi/\psi + W] \geq 0$. By Jensen's inequality $\int f^2 \leq \int \psi^2$, so if $f \notin L^2$ then $\psi \notin L^2$. Therefore it suffices to consider $\{-A\Delta + [W](x)\}f \geq 0$ and to prove $f \notin L^2$ under the stated condition on $[W]$. First, suppose $D = R^3$. Then this is just Simon's (1981) theorem. (However, since we are now dealing with spherically symmetric $[W]$ and f, it is likely that a direct, ordinary differential equation proof can be found to replace Simon's proof.) Next, suppose $R > 0$. Let $g(x) > 0$ be any C^2 function defined in R^3 such that $g(x) = f(x)$ for $|x| \geq R$. Then $\{-A\Delta + U(x)\}g \geq 0$ on R^3 where $U = [W]$ for $|x| \geq R$ and U is bounded for $|x| \leq R$. Clearly $[W]_+ \in L^{3/2}(D)$ if and only if $U_+ \in L^{3/2}(R^3)$. Apply Simon's theorem to U. ■

Note added in proof. H. Brezis (private communication) has found a direct ordinary differential equation proof. Moreover, under the hypotheses of Lemma 7.18, $\psi \notin L^{3-\varepsilon}$ for all $\varepsilon > 0$.

Theorem 7.19. *Let the hypothesis be the same as in Theorem 7.16 except that (i) is omitted. (In other words, a molecule is now being considered.) Then* $\lambda_c > Z$.

Proof. For $|x| > R$, $V(x)$ is C^∞ so $\psi(x) > 0$ and $\psi \in C^2$ by Theorem 7.9. Assume $\lambda_c \leq Z$. The hypotheses of Lemma 7.18 are satisfied with $[-A\Delta + W(x)]\psi = 0$. To obtain a contradiction we have to show $[W]_+ \in L^{3/2}$. Consider ϕ. Even if ϕ is negative somewhere, $[\phi](r) > 0$ in D by Newton's theorem. Therefore it suffices to show $[\rho^{p-1}] \in L^{3/2}$. If $p \geq \frac{5}{3}$ then $p - 1 \geq \frac{2}{3}$ and $[\rho^{p-1}](r) < C[\rho^{2/3}](r)$, since ρ is bounded. But $[\rho^{2/3}](r) \leq \{[\rho](r)\}^{2/3}$, by Hölder, and $\int [\rho]^{(2/3)(3/2)} = \int \rho < \infty$. ■

We know that $\lambda_c > Z$ in the physically interesting case $p = \frac{5}{3}$. *How large is* $\lambda_c - Z$? In other words, *how negative can ions be*? This seems to be a very difficult question, even for an atom. To obtain qualitative agreement with quantum theory, it would be desirable if $\lambda_c - Z \sim 1$, at least for Z up to 100, say. The only available bound, at present, is Theorem 7.23. First Lemmas 7.20, 7.21, and 7.22 are needed. The lemmas were inspired by the work of R. Benguria (private communication), who proved the lemmas and Theorem 7.23 in the spherically symmetric case (which corresponds to the atom in TFW theory).

Lemma 7.20. *Let* ψ *and f be two real valued functions on* R^3 *which satisfy* $-\Delta\psi = f$ *in the sense of distributions. Let r denote the function* $|x|$. *Suppose* $\psi \in L^2$, $f \in L^2$, *and* $r\psi f \in L^1$. *Then, for any constant* $d \geq 0$,

$$\int (|x|^2 + d)^{1/2}\psi(x)f(x)dx \geq 0.$$

Proof. Using dominated convergence, it is sufficient to consider only $d > 0$. Let $R = (r^2 + d)^{1/2} \in C^\infty$. We have $\Delta R = 2R^{-1} + dR^{-3}$ and $|(R/r)\nabla R|^2 = 1$. Suppose $\phi \in C_0^\infty$ (infinitely differentiable functions of compact support). We claim $I = -\int R\phi\Delta\phi \geq 0$. To see this, integrate by parts: $I = A + B$ with $A = \int (\nabla\phi)^2 R$ and

$$B = \int \phi\nabla\phi\cdot\nabla R = \int (\nabla\phi\cdot\nabla R)(R/r)(r/R)\phi.$$

By Schwarz, and $\{(\nabla\phi\cdot\nabla R)(R/r)\}^2 \leq (\nabla\phi)^2$, we have $B^2 \leq AC$ with

$$2C = 2\int \phi^2 r^2 R^{-3} \leq \int \phi^2 \Delta R.$$

However,

$$2B = \int \nabla\phi^2\cdot\nabla R = -\int \phi^2\Delta R,$$

and hence $|B| \leqslant A$, which proves the lemma. Now, suppose ψ and $f \notin C_0^\infty$ have compact support. Given $\varepsilon > 0$ there exists $g \in C_0^\infty$ such that $\|\psi - g\|_2 < \varepsilon$, $\|\nabla\psi - \nabla g\|_2 < \varepsilon$, and $\|\Delta\psi - \Delta g\|_2 < \varepsilon$. (Note: since ψ and $\Delta\psi \in L^2$, so is $\nabla\psi$.) Then $gR \in C_0^\infty$ and

$$\int gRf = -\int gR\Delta\psi \equiv -\int \psi\Delta(Rg)$$
$$= -\int g\Delta(Rg) - M = -\int Rg\Delta g - M,$$

with $M = \int (\psi - g)\Delta(Rg)$. It suffices to show that $M \to 0$ as $\varepsilon \to 0$ because $\int gRf \to \int \psi Rf$. But

$$M = \int (\psi - g)\{g\Delta R + 2\nabla g \cdot \nabla R + R\Delta g\}.$$

We can assume supp(g) is in some fixed ball, independent of ε. Since g, ∇g, and Δg are uniformly L^2 bounded, $M \to 0$. For the general case, let $h \in C_0^\infty$ satisfy $1 \geqslant h \geqslant 0$, $(\nabla h)(0) = 0$, $h(0) = 1$, $h(x) = 0$ for $|x| > 1$. Let $h_n(x) = h(x/n)$ and $\psi_n = h_n\psi$. Then, as a distribution,

$$-\Delta\psi_n = h_n f - 2\nabla h_n \cdot \nabla\psi - \psi\Delta h_n \equiv K_n.$$

By the previous result, $T_n \equiv \int R\psi_n K_n \geqslant 0$. But $\int h_n^2 \psi f R \to \int \psi f R$ by dominated convergence. $Rh_n \Delta h_n = n^{-1}P_n(x/n)$ with

$$P_n(x) = (|x|^2 + dn^{-2})^{1/2}(h\Delta h)(x) < a,$$

so $\int R\psi_n \psi\Delta h_n \to 0$, since $\psi \in L^2$. Similarly, $R\nabla h_n^2 \equiv L_n$ is uniformly bounded and converges pointwise to zero. Since ψ and $\nabla\psi \in L^2$, $\int \psi\nabla\psi L_n \to 0$ by dominated convergence. ■

Remarks. (i) Lemma 7.20 is useful for L^2 solutions to the Schrödinger equation $[-\Delta + W(x)]\psi = -\mu\psi$. Then $\int \psi^2(W + \mu)r \leqslant 0$ under some mild conditions on W [e.g., $W(x) < c/r$ for large r, $W \in L^2_{\text{loc}}$ and $r\psi^2 \in L^1$ if $\mu \neq 0$].

(ii) The essential properties of R that were used were $R > 0$ and $R\Delta R \geqslant 2(\nabla R)^2$. Therefore Lemma 7.20 will hold for functions other than $(r^2 + d)^{1/2}$ having these properties. Formally, this means that if $R(x) = 1/V(x)$ then we require $V > 0$ and $\Delta V < 0$. We state this as Lemma 7.21, whose proof imitates the proof of Lemma 7.20.

Lemma 7.21. *Suppose $V = |x|^{-1} * m$, with m a nonnegative measure, $|m| = Z < \infty$, and V satisfies Eq. (7.12). Then if ψ and f satisfy the hypotheses of Lemma 7.20,*

$$\int dx\, \psi(x) f(x)/V(x) \geqslant 0.$$

Lemma 7.22. *Let $\rho(x) \geqslant 0$, $\int \rho = \lambda < \infty$, and $V = |x|^{-1} * m$ where m is any non-negative measure with $|m| = Z < \infty$. Then*

$$I \equiv \int \rho(x)\rho(y)|x - y|^{-1}V(x)^{-1}dx\,dy \geqslant \lambda^2/2Z.$$

Proof. Take $Z = 1$ and let $0 < \varepsilon < 1$. By Proposition 3.24, $\rho = \rho_1 + \rho_2$ with $\rho_1, \rho_2 \geqslant 0$, and $H_i = |x|^{-1} * \rho_i$ satisfies $H_1 \leqslant \varepsilon\lambda V$ and $H_1 = \varepsilon\lambda V$ when $\rho_2 > 0$. Clearly, $\int \rho_1 \leqslant \varepsilon\lambda$ by Lemma 3.3. Then $\int \rho_2 \geqslant (1 - \varepsilon)\lambda$ and

$$I \geqslant \int \rho_2(H_1 + H_2)/V \geqslant \varepsilon(1 - \varepsilon)\lambda^2 + \int \rho_2 H_2/V.$$

Repeat the argument with ρ_2 [using $\int \rho_2 \geqslant (1 - \varepsilon)\lambda$], and so on *ad infinitum*. Then

$$I \geqslant \lambda^2\varepsilon(1 - \varepsilon)\sum_{j=0}^{\infty}(1 - \varepsilon)^{2j} = \lambda^2(1 - \varepsilon)/(2 - \varepsilon).$$

Now let $\varepsilon \to 0$. ■

Remark. Benguria proved Lemma 7.22 when $V = 1/r$. In this case one can simply use the fact that $(|x| + |y|)|x - y|^{-1} \geqslant 1$.

Theorem 7.23. *Assume V satisfies Eq. 7.12. Then $\lambda_c < 2Z$, for all $p > 1$.*

Proof. We know $\lambda_c < \infty$. Let ψ be the minimizing solution of Eq. (7.9) with $\mu = 0$. Then, by Theorem 7.13, ψ and $f = (\gamma\rho^{p-1} - \phi)\psi$ satisfy the hypotheses of Lemma 7.21. Thus $0 < \int \rho\phi/V = \lambda_c - I$ with $I = \int H\rho/V$ and $H = |x|^{-1} * \rho$. But $I \geqslant \lambda_c^2/2Z$. ■

Remark. This bound does not involve the value of A in Eq. (7.1). It also does not utilize the $\gamma\rho^{p-1}$ term in W. There is considerable room for improvement.

The next two theorems are about the asymptotics of ψ.

Theorem 7.24. *Let ψ be the positive solution to Eq. (7.9), for any p.*

(i) Let $\mu > 0$. Then for every $t < \mu$ there exists a constant M such that

$$\psi(x) \leqslant M\exp[-(t/A)^{1/2}|x|].$$

(ii) Let $\mu = 0$ (i.e., $\lambda = \lambda_c$), and assume $\lambda_c > Z$, as is certainly the case when $p \geqslant \frac{4}{3}$. Assume also that m has compact support. Then for every $t < \lambda_c - Z$ there is a constant M such that

$$\psi(x) \leqslant M\exp[-2(t|x|/A)^{1/2}].$$

Proof. (i) is standard. Since ψ and $V \to 0$ as $|x| \to \infty$, we have $\psi = -(-A\Delta + t)^{-1}(W + \mu - t)\psi$. For $|x| >$ some R, $W + \mu - t > 0$. Therefore, since $\psi > 0$,

$$\psi(x) \leqslant \int_{|y| \leqslant R} Y(x - y)[W(y) + \mu - t]\psi(y)dy,$$

where

$$Y(x) = (4\pi A|x|)^{-1}\exp[-(t/A)^{1/2}|x|].$$

The proof of (ii) is the same as the proof of Theorem 7.12. It is only necessary to note that, since m has compact support (in B_R, say), $V(x) \leqslant Z/(|x| - R)$ for $|x| > R$, and this is $\leqslant (Z + \varepsilon)/|x|$ for $|x|$ large enough. ■

The next theorem is the well known cusp condition (Kato, 1957).

Theorem 7.25. *Let $V(x) = \sum z_j|x - R_j|^{-1}$ be the potential of point nuclei. Then at each R_j*

$$z_j\psi(R_j) = -2A\lim_{r \downarrow 0}\int_{|x - R_j| = r} r^{-1}(x - R_j) \cdot \nabla\psi(x)d\Omega,$$

where $d\Omega$ is the normalized uniform measure on the sphere. This holds for any λ. In particular, for an atom with nuclear charge z located at the origin, ψ is spherically symmetric and

$$z\psi(0) = -2A\lim_{r \downarrow 0}(d\psi/dr)(r).$$

Proof. Recall that, by Theorem 7.9, ψ is C^∞ away from the R_j and ψ is Hölder continuous everywhere. The theorem is proved by integrating Eq. (7.9) in a small ball B_r and then integrating by parts. The spherical symmetry in the atomic case is implied by uniqueness. ■

Theorem 7.26. *Let $V(x)=z/|x|$ be the potential of an atom with a point nucleus. Then, for any λ, $\psi(r)$ is a strictly decreasing function of r.*

Proof. In Theorem 2.12 and the remark following it we proved this for $\lambda \leq z$ by using rearrangement inequalities. Here we give a different proof for $\lambda \leq z$ which extends to $\lambda > z$. Recall that ψ is continuous and positive and that ψ is C^∞ for $r>0$. Also, $\mu \geq 0$. Let $Q(r)=\int \chi_r \rho$ be the electronic charge inside the ball B_r. By Newton's theorem, the potential ϕ satisfies:

(i) $\phi(r) \leq [z-Q(r)]/r$.

(ii) $\dot{\phi}=[Q(r)-z]/r^2$ (dots denote d/dr).

(iii) If $\lambda \leq z$, $\phi(r) \geq 0$ and $\phi(r)$ is decreasing.

(iv) If $\lambda > z$ there is a unique $R>0$ such that $\phi(r) \geq 0$ and decreasing for $r \leq R$, and $\phi(r)<0$ for $r>R$. $Q(R) < z$.

$\lambda \leq z$: By Theorem 7.24 $\dot{\psi}(r)<0$ near $r=0$. If ψ is not monotone, then since $\psi(r)\to 0$ as $r\to\infty$, there are two points $0<r_1<r_2$ such that $\psi(r_1)\leq\psi(r_2)$, $\ddot{\psi}(r_1)\geq 0$, $\ddot{\psi}(r_2) \leq 0$, and $\dot{\psi}(r_1)=\dot{\psi}(r_2)=0$. Since ψ does not have compact support, $Q(r)<z$, all r. Hence $W(r_2)>W(r_1)$. Since

$$\ddot{\psi}(r_1)=[W(r_1)+\mu]\psi(r_1)\geq 0\,,$$

we have $W(r_1)+\mu\geq 0$. But then

$$0\geq\ddot{\psi}(r_2)=[W(r_2)+\mu]\psi(r_2)$$

is impossible.

$\lambda > z$: There is an $\varepsilon>0$ such that $W(r)>0$ for $r>R-\varepsilon$. Let $D_r=\{x\in\mathbb{R}^3 \,\|\, x|>r\}$. Take $r>R-\varepsilon$. On D_r, $-\Delta\psi<0$. Since $\psi>0$ is subharmonic on D_r, ψ has its unique maximum on ∂D_r, namely, $|x|=r$. This proves the theorem on the domain $\{r|r>R-\varepsilon\}$. To prove the theorem on the domain $\{r|0\leq r<R\}$ the argument in the $\lambda\leq z$ case can be used, since $Q(r)<z$ in this domain. ■

Conjecture. In the point nucleus, atomic case ψ is convex, possibly even log convex.

C. Binding in TFW theory

In TF theory binding never occurs when the repulsion U is included. In TFW theory binding is a common phenomenon. We *conjecture* that every neutral system (molecule or atom) binds to every other neutral system. In the following, the occurrence of binding will be proved in enough cases to render the conjecture plausible. It will also be seen that binding in TFW theory is intimately connected with the existence of negative ions, i.e., $\lambda_c > Z$. We shall assume here that $\lambda_c > Z$ for all the systems under consideration. $p\geq\frac{5}{3}$ guarantees this, but no requirement on p other than $\lambda_c>Z$ will be made. $V=|x|^{-1}*m$, with m a non-negative measure of compact support [so that Eq. (7.12) is satisfied], $|m|=Z$, and m spherically symmetric in the atomic case.

First, let us define what binding means. Suppose we have two systems (not necessarily atoms) with potentials V_1 and V_2 and a combined system with $V(x)=V_1(x)+V_2(x-R)$ for some vector R. The combined system is neutral, i.e., $\lambda=Z\equiv Z_1+Z_2$. Let $E(R)$ denote the energy of the combined system and $E_i(\lambda)$ denote the energies of the subsystems with arbitrary electron charge λ and $E_i=E_i(\lambda=Z_i)$ (note the difference in notation). Then

$$E(\infty)\equiv\min_{0\leq\lambda\leq Z} E_1(\lambda)+E_2(Z-\lambda)\,. \tag{7.14}$$

Let μ_i be the chemical potentials of the subsystems when they are neutral, i.e., $\lambda_i=Z_i$. We know that $\mu_i>0$. If $\mu_1=\mu_2$ then $E(\infty)=E_1+E_2$. Otherwise, $E(\infty)<E_1+E_2$. In general, λ in Eq. (7.14) is determined by $\mu_1(\lambda)=\mu_2(Z-\lambda)$ if this equation has a solution for $0\leq\lambda\leq Z$; otherwise, $\lambda=0$ if $\mu_1(0)\leq\mu_2(Z)$ and $\lambda=Z$ if $\mu_2(0)\leq\mu_1(Z)$. (Recall that $\mu_i(\lambda)$ is monotone.)

If $\mu_1\neq\mu_2$ then the subsystems *spontaneously ionize* when they are infinitely far apart. This is *not* considered to be binding. For real atoms the phenomenon of spontaneous ionization apparently never occurs, because it seems to be the case that the lowest ionization potential among *all* atoms is less than the largest electron affinity. (I thank J. Morgan III for pointing this out to me.) In real atoms, λ and $Z-\lambda$ in Eq. (7.14) are restricted to be integral, but no such restriction occurs in TFW theory. In TF theory the phenomenon never occurs because μ_i is always zero.

In TFW theory it is possible for λ to be zero in Eq. (7.14), i.e., one subsystem is completely stripped of electrons. Let $V_i=z_i/r$ with $z_2\gg z_1$. If $\mu_1(\lambda=0)\leq\mu_2(\lambda=z_1+z_2)$, then $\lambda=0$ in Eq. (7.14). By Theorem 7.11, $\mu_1(\lambda)\leq\mu_1(0)=z_1^2/4A$. Since $\lambda_c(2)>z_2$ and $\mu_2>0$, the above inequality will hold for any fixed z_2 if z_1 is chosen small enough. This case was cited in BBL as an example where binding occurs (see Theorem 7.27).

Definition. Binding is said to occur if $E(R)<E(\infty)$ for some R.

Theorem 7.27. *Suppose the chemical potentials of the neutral subsystems are unequal, i.e., $\mu_1\neq\mu_2$. Then binding occurs. (This holds for all $p>1$, even if $\lambda_c=Z$ for one or more of the three systems.)*

Proof. Suppose $\lambda<Z_1$ in Eq. (7.14). Then when $R=\infty$ subsystem 1 is positively charged (with charge $Q=Z_1-\lambda$) and subsystem 2 has charge $-Q$. Let ρ_i be the TFW densities (with λ and $Z-\lambda$, respectively). By Theorem 7.24, ρ_i has exponential falloff. For the combined system (at R) consider the variational ρ defined by $\rho(x)=\rho_1(x)+\rho_2(x-R)$. The first term in Eq. (7.1) is subadditive (by convexity). For large R the total Coulomb energy decreases essentially by $-Q^2/R$ because of the exponential falloff of each ρ_i. The $\int\rho^p$ term is superadditive, but it increases only by a term of order $\exp[-(\text{const})R]$ for large R. We omit the easy proof of these last two assertions. Thus for large enough, but finite, R, $E(R)<E(\infty)$. ■

The difficult case is $\mu_1=\mu_2$. Henceforth we confine our attention to atoms.

Conjecture. If $z_1<z_2$ for two atoms with point nuclei, then $\mu_1(\lambda=bz_1)<\mu_2(\lambda=bz_2)$ for all $b\leq 1$. In particular $\mu_1<\mu_2$. Moreover, $\lambda_c(1)-z_1<\lambda_c(2)-z_2$.

If this conjecture is correct then only the homopolar case has to be considered for point nuclei. In Theorem 7.28 we prove binding for the homopolar molecule, even for "smeared" nuclei. However, we have already shown that binding occurs if $z_1 \ll z_2$, so it is likely that binding always occurs, even if the conjecture is wrong.

Theorem 7.28. *Binding occurs for two equal atoms for any nuclear charge* z *and for any* $p>1$ *provided* $\lambda_c > z$ *for the atom.*

Proof. We shall construct a variational $\tilde{\rho}$ for the combined system, with $\int\tilde{\rho}=2z$, such that $\mathcal{E}(\tilde{\rho})$ for the combined system at some R is less than $E(\infty)=2E_1$. First, consider the atom with the nucleus at the origin and with $\lambda=z+\varepsilon$, where $z<\lambda<\lambda_c$. Let ρ be the TFW density. Denote E_1 by E and μ_1 by μ. Center the nucleus at the point $(-R,0,0)$, where $R>0$, depending on ε, is such that $\int\chi_-\rho=z$, with χ_- being the characteristic functions of the half space, $H=\{(x_1,x_2,x_3)\,|\,x_1\le 0\}$. Assume, for the moment, that the nuclear m has support in $B_{R/2}$, i.e., the displaced m has support in $\{x_1\le -R/2\}$. Center the second atom at $(R,0,0)$. Its corresponding density is ρ^*, where the asterisk means reflection through the plane $x_1=0$. Choose the variational $\tilde{\rho}=\rho_-+\rho_-^*$ with $\rho_-=\chi_-\rho$. Clearly $\tilde{\rho}$ is continuous across the plane $x_1=0$, and $\int\tilde{\rho}=2z$, so it is a valid variational function. In the following bookkeeping of $\mathcal{E}(\tilde{\rho})$ we use the terminology "energy gain" (resp. "loss") to mean that the contribution to $\mathcal{E}(\tilde{\rho})$ is negative (resp. positive) relative to $2E$. Before the χ_- cutoff, we start with $2E_1(\lambda)\le 2E-(2\varepsilon)(\mu/2)$ if ε is small enough, so we have gained $\varepsilon\mu$. This linear term in ε is the crucial point; it exists because $\lambda_c>z$. After the cutoff we gain the kinetic energy [first two terms in Eq. (7.1)] contributions from the missing pieces of ρ and ρ^*. Next, we lose on the $-\int V\rho$ term (for each atom separately) because of the missing pieces. Each missing charge is ε and its distance to its atomic origin is R. Since the atomic $V(r)\le z/r$, the energy loss is at most $2(\varepsilon z/R)$. Clearly we gain on the missing atomic repulsion, $D(\rho,\rho)$ term. Finally, if $dM(x)=dm(x+R)-\rho_-(x)dx$ is the total charge density in H, we lose the atom-atom interaction $\Delta=2D(M,M^*)$. By reflection positivity, $\Delta\ge 0$. (See Benguria and Lieb (1978b), Lemma B.2.) On balance, the net energy gain is at least $\varepsilon(\mu-2z/R)-\Delta$.

Now we claim two things: (i) As $\varepsilon\to 0$, $R\to\infty$. (ii) $\Delta<C\varepsilon z/R$ for some constant C. [Actually, it is possible to prove $\Delta<o(\varepsilon)z/R$.] Using (i) and (ii) we are done, because for sufficiently small ε the gain is positive and the assumption on supp(m) is justified.

Proof of (i). Let ρ_n be the atomic density for $\lambda=z+\varepsilon_n$, with $\varepsilon_n\to 0$. As in the proof of Theorem 7.6, we can find a weakly convergent subsequence, $\rho_n\to\bar{\rho}$, so that $E=\lim\mathcal{E}(\rho_n)\ge\mathcal{E}(\bar{\rho})$. But $\int\bar{\rho}\le z$, so $\bar{\rho}$ must be ρ_z, the atomic density with $\lambda=z$. If R_n does not tend to ∞ then, for large enough n, $\int\chi_-\rho_n<z$ by the weak convergence, which is a contradiction.

Proof of (ii). This is messy. Let B_r be the ball of radius r centered at $(-R,0,0)$ and let χ_r be its characteristic function. Write $\rho=\rho^a+\rho^b$, where $\rho^a=\chi_{3R/2}\rho$. By elementary geometry, $d\int\rho^b\ge\int\rho_-^b$ with $d<1$. Since

$$\int\rho^b(1-\chi_-)<\varepsilon,\quad\int\rho^b<\varepsilon/(1-d).$$

Let $t=d/(1-d)$. The contribution of ρ^b to Δ is

$$-4D(\rho_-^b,M^*)-2D(\rho_-^b,\rho_-^{b*})\le -4D(\rho_-^b,M^*)\le 4D(\rho_-^b,\rho_-^*)$$
$$\le 4D(\rho^b,\rho_-^*)\le 2z\left(\int\rho^b\right)\Big/R,$$

since the potential of ρ^b is everywhere less than $(3R/2)^{-1}\int\rho^b$. Henceforth we can assume $\rho_-=\rho_-^a$ and $z>\int\rho_-^a>z-t\varepsilon$. This assumption changes M to M^a. Let $d\bar{M}(x)=dm(x+R)-\rho^a(x)dx$. [Note: supp($\bar{M}$) extends outside H, but is inside $\{x_1\le R/2\}$.] $\phi\equiv|x|^{-1}*M^{a*}$ is subharmonic on supp($\bar{M}$) and harmonic on supp(m) so

$$D(\bar{M},M^{a*})\le\left(\int d\bar{M}\right)D(\delta,M^{a*}),$$

where δ is a delta function at $(-R,0,0)$. This is

$$\le\left|\int d\bar{M}\right|z/R\le t\varepsilon z/R,$$

since the distance of supp(M^{a*}) to $(-R,0,0)$ is R. Finally,

$$D(M^a-\bar{M},M^{a*})=D(\rho^a-\rho_-^a,M^{a*})\le D(\rho^a-\rho_-^a,m^*)$$
$$=D(\rho^a-\rho_-^a,z\delta^*)\le 2\varepsilon z/R$$

since $\int\rho^a-\rho_-^a\le\varepsilon$ and the distance of supp(ρ^a) to $(R,0,0)$ is $R/2$. ■

I thank J. Morgan III for valuable discussions about Theorem 7.28. Balázs (1967) gave a heuristic argument for the binding of two equal atoms with point nuclear charges.

D. The Z^2 correction and the behavior near the nuclei

Here we consider point nuclei with potential given by Eq. (2.1). The question we address is what is the principal correction to the TF energy and density caused by the first term in Eq. (7.1)? This term, $A\int(\nabla\rho^{1/2})^2$, will henceforth be denoted by T. For simplicity we confine our attention to $p=\frac{5}{3}$, the physical value of p.

$E^{\mathrm{TF}}\sim Z^{7/3}$. In particular, for a neutral atom,

$$E^{\mathrm{TF}}=-3.678\,74z^{7/3}/\gamma \tag{7.15}$$

(I thank D. Liberman for this numerical value). At first sight, it might be thought that the leading energy correction is $O(z^{5/3})$. If $\rho^{\mathrm{TF}}(z,r)=z^2\rho^{\mathrm{TF}}(1,z^{1/3}r)$ is inserted into T, then, by scaling, $T(z)=z^{5/3}T(z=1)$. But $T(z=1)=\infty$ since $\rho^{\mathrm{TF}}\sim r^{-3/2}$ for small r. Thus, for point nuclei, T cannot be regarded as a small perturbation.

The actual correction is $+O(z^2)$ and bounds of this form can easily be found. The following bounds are for an atom, and can obviously be generalized for molecules.

Upper bound: Use a variational $\tilde{\rho}_{\mathrm{TF}}$ for TFW of the form $\tilde{\rho}(r)=\rho^{\mathrm{TF}}(r)$ for $r>1/z$ and $\tilde{\rho}(r)=\rho^{\mathrm{TF}}(1/z)$ for $r<1/z$.

Lower bound: Let $b>0$ and write $V(r)=\tilde{V}(r)+H(r)$, where $H(r)=z/r-z^2/b$ for $zr<b$ and $H(r)=0$ otherwise. For small enough b, $-A\Delta+H>0$, since $\|H\|_{3/2}\sim b$. Now $\tilde{V}=|x|^{-1}*m$, with $m\ge 0$ and $|m|=z$. Let $\tilde{\rho}$ minimize $\mathcal{E}^{\mathrm{TF}}(\tilde{V},\rho)$ with energy $E^{\mathrm{TF}}(\tilde{V})$. Then $E^{\mathrm{TFW}}\ge E^{\mathrm{TF}}(\tilde{V})$. But $E^{\mathrm{TF}}(V)\le\mathcal{E}^{\mathrm{TF}}(V,\tilde{\rho})=E^{\mathrm{TF}}(\tilde{V})-\int\tilde{\rho}H$. It is not hard to prove, from the TF equation with $\tilde{V}$, that this last integral is $O(z^2)$.

The foregoing calculations show that the main correction in TFW theory comes from distances of order z^{-1} near the nuclei. The calculations, if carried out for arbitrary λ, also show that the correction is essentially independent of λ. We now show how this correction can be exactly computed to leading order in z, namely, $O(z^2)$.

Let us begin by considering the atom without electron-electron repulsion. The TF theory of such an atom was presented in Sec. V.B following Eq. (5.30). The analogous TFW equation (with $\delta = A\hbar^2/2m$ and $\hbar^2/2m = 1$) is

$$[-A\Delta + W(x)]\psi = -\mu\psi, \tag{7.16}$$

with $W(x) = \gamma\rho(x)^{2/3} - z/|x|$, and $\rho = \psi^2$. The absolute minimum, which corresponds to $\lambda = \infty$, has $\mu = 0$, namely,

$$(-A\Delta + \gamma|\psi|^{4/3} - z|x|^{-1})\psi = 0. \tag{7.17}$$

The first task is to analyze Eq. (7.17). By simple scaling, any solution scales with A, γ, and z as

$$\psi(z,\gamma,A;x) = (z^2/A\gamma)^{3/4}\psi(1,1,1;zx/A). \tag{7.18}$$

Up to Eq. (7.28) we take $z=\gamma=A=1$. Consider the functional

$$\mathfrak{F}'(\psi) = T(\psi) + P(\psi), \tag{7.19}$$

$$T(\psi) = \int (\nabla\psi)^2, \quad P(\psi) = \int k(\psi(x), x)dx, \tag{7.20}$$

$$k(\psi, x) = 3|\psi|^{10/3}/5 + 2|x|^{-5/2}/5 - |\psi|^2|x|^{-1}. \tag{7.21}$$

Note that $k \geq 0$ and, for each x, k has a minimum at $\psi = |x|^{-3/4}$. The function space for $\mathfrak{F}'$ is

$$G' = \{\psi | \nabla\psi \in L^2, P(\psi) < \infty\}. \tag{7.22}$$

G' is not convex since $0 \notin G'$. Clearly, Eq. (7.17) is the variational equation for $\mathfrak{F}'$. We can also define $G = \{\rho | \rho \geq 0, \rho^{1/2} \in G'\}$ and $\mathfrak{F}(\rho) = \mathfrak{F}'(\rho^{1/2})$. G is convex and $\rho \to \mathfrak{F}(\rho)$ is convex.

Theorem 7.29. *$\mathfrak{F}'(\psi)$ has a minimum on G'. This minimizing ψ is unique, except for sign, and satisfies: (i) $\psi > 0$. (ii) ψ is spherically symmetric. (iii) ψ satisfies Eq. (7.17). (iv) ψ is the only non-negative solution to Eq. (7.17) in G'. (v) ψ is C^∞ for $|x| > 0$. (vi) ψ satisfies the cusp condition $2(d\psi/dr)(0) = -\psi(0)$. (vii) for large $r = |x|$, ψ has the asymptotic expansion [which can be formally deduced from Eq. (7.17)],*

$$\psi(r) = r^{-3/4} - \tfrac{9}{64}r^{-7/4} - \tfrac{3}{2}\left(\tfrac{21}{64}\right)^2 r^{-11/4} + O(r^{-15/4}), \tag{7.23}$$

$$\rho(r) = r^{-3/2} - \tfrac{9}{32}r^{-5/2} - (621/2^{11})r^{-7/2} + O(r^{-9/2}). \tag{7.24}$$

(viii) Any solution f to Eq. (7.17) in G' satisfies $|f(x)| < |x|^{-3/4}$. (ix) By (viii), ψ is superharmonic, and thus $\psi(r)$ is decreasing.

The proof of Theorem 7.29 follows the methods of Secs. VII.A and VII.B, and is given in Lieb, 1981b. The following numerical values, together with a tabulation of ψ, are in Liberman and Lieb, 1981. $\rho = \psi^2$.

$$\begin{aligned} &\psi(0) = 0.970\,133\,0, \\ &I_1 = \int (\nabla\psi)^2 = 8.583\,819\,7, \\ &I_2 = \int \{r^{-5/2} - \rho^{5/3}\} = 42.92, \\ &I_3 = \int [r^{-3/2} - \rho]/r = 34.34. \end{aligned} \tag{7.25}$$

From Eq. (7.17) one has $I_1 + I_3 = I_2$. By dilating $\psi(r) \to t^{3/4}\psi(tr)$ in Eq. (7.19), a "virial theorem" is obtained: $5I_1 + 3I_2 = 5I_3$. Thus

$$I_1 : I_2 : I_3 = 1:5:4, \tag{7.26}$$

$$\Delta E \equiv \mathfrak{F}(\psi) = I_1 - 3I_2/5 + I_3 = 2I_1. \tag{7.27}$$

If the parameters are reintroduced

$$I_1(z,\gamma,A) = A\int (\nabla\psi)^2 = z^2 A^{1/2}\gamma^{-3/2} I_1. \tag{7.28}$$

Let us denote the ρ we have just obtained in Theorem 7.29 [with the parameters reintroduced according to Eq. (7.18)] by ρ_∞. The scale length of ρ_∞ is z^{-1} and, for *large r*, $\rho_\infty(r)$ agrees to leading order with $\rho^{\mathrm{TF}}(r)$ for *small r* (on a scale of $z^{-1/3}$), namely, $(z/\gamma r)^{3/2}$. We claim that ρ_∞ can be spliced together with ρ^{TF} in the overlap region, $r = O(z^{-2/3})$, and the result is ρ^{TFW} to leading order in z. The splicing is independent of λ provided $\lambda/z >$ (const) >0. The change in energy for an atom is then, to leading order, ΔE of Eq. (7.27), and is independent of λ. An analogous situation holds for a molecule; near each nucleus ρ^{TF} is spliced together with ρ_∞ for the appropriate z_j. This is formalized in the following theorem.

Theorem 7.30. *Let $V(x) = \sum z_j |x - R_j|^{-1}$. Consider the $Z \to \infty$ limit with the scaling given before Eq. (5.2), except that the electron charge N is not restricted to be integral. $\lambda = N/Z > 0$ is fixed. $z_j = az_j^0$, $R_j = a^{-1/3}R_j^0$, with $a\lambda = N$. Then, as $N \to \infty$,*

$$(i)\ E^{\mathrm{TFW}}(N) = E^{\mathrm{TF}}(N) + D\sum_{j=1}^{k} z_j^2 + O(a^2), \tag{7.29}$$

with $D = 2A^{1/2}\gamma^{-3/2}I_1$.

$$(ii)\ a^{-4/3}\mu^{\mathrm{TFW}}(N) \to \mu^{\mathrm{TF}}(\lambda, \underline{z}^0, \underline{R}^0). \tag{7.30}$$

(iii) Fix x. Then

$$a^{-2}\rho^{\mathrm{TFW}}(N, \underline{z}, \underline{R}; a^{-1/3}x) \to \rho^{\mathrm{TF}}(\lambda, \underline{z}^0, \underline{R}^0; x), \tag{7.31}$$

with convergence in the sense of weakly in L^1 if $\lambda \leq Z$ and weakly in L^1_{loc} if $\lambda > Z$.

(iv) Fix y. For each j

$$z_j^{-3}\rho^{\mathrm{TFW}}(N, \underline{z}, \underline{R}; R_j + z_j^{-1}y) \to (A\gamma)^{-3/2}\psi^2(y/A), \tag{7.32}$$

where ψ is the solution to Eq. (7.17) with $A = z = \gamma = 1$ given by Theorem 7.29. The convergence is pointwise and in L^1_{loc}. A refinement of Eq. (7.32) is given in Theorems 7.32–7.35.

Before proving Theorem 7.30 let us comment on its significance.

(i) Equation (7.29) states that the energy correction in TFW theory is exactly of the form of the quantum correction conjectured by Scott [Eq. (5.29)]. In particular, since $\gamma^{3/2} \sim q^{-1}$, the q dependence is the same. In order to obtain the conjectured coefficient $\frac{1}{8}$ of Eq. (5.32), with $\gamma = \gamma_p$, we must choose

$$A = q^2\gamma_p^3[16I_1]^{-2} = 0.185\,909\,19. \tag{7.33}$$

This number was mentioned after Eq. (2.8).

Yonei and Tomishima (1965) also realized that $A = 1/5$ is a good choice. They analyzed the TFW atom without electron repulsion, namely Eq. (7.16), and compared the

TFW energy with the quantum Bohr energy, Eq. (5.31), for neutral atoms with z up to 100. They did not seem to notice that this choice for A is valid even if $\lambda = N/z < 1$. Yonei (1971) analyzed TFDW theory with electron repulsion and again advocated $A = 1/5$. This is not surprising since Theorem 7.30 says that the electron repulsion does not affect ΔE to $O(z^2)$ and Theorem 6.5 (suitably modified) says that the Dirac correction changes the energy to $O(z^{5/3})$. Yonei (1971) claims that the dissociation energy and the equilibrium internuclear distance for the nitrogen molecule, calculated with this TFDW theory, are in good agreement with experiment.

(ii) The density, on a length scale $Z^{-1/3}$ agrees with quantum (and TF) theory, Theorem 5.2.

(iii) On a length scale z^{-1} near each nucleus, Eq. (7.32) states that ρ^{TFW} converges to a universal function. This phenomenon is the same as we conjectured in Eq. (5.37) for quantum theory. The universal functions are not exactly the same, but they are very close. For large values of the argument they agree, namely, $(\gamma_p y)^{-3/2}$, independent of A. Since the convergence in Eq. (7.32) is *pointwise*, it makes sense to ask what happens at $y=0$. Using γ_p and A given by Eq. (7.33), the right side of (7.32) is obtained from (7.18) and (7.25) as

$$q^{-1}z_j^{-3}\rho^{\mathrm{TFW}}(x=R_j) \to 0.198\,271\,49\,. \tag{7.34}$$

On the other hand, ρ^H in Eq. (5.33) can be evaluated at $x=0$, since only S waves contribute. At $x=0$, $f_{n00}(0)^2 = (8\pi n^3)^{-1}$. Thus Eq. (5.37), if correct, would state that

$$q^{-1}z_j^{-3}\rho^{Q}(x=R_j) \to \zeta(3)/8\pi = 0.047\,828\,325\,. \tag{7.35}$$

To prove Theorem 7.30, Theorems 7.32–7.35, which are independently interesting, are needed. To prove them we need the following comparison theorem which was proved by Morgan (1978) in the spherically symmetric case and by T. Hoffmann-Ostenhof (1980) in the general case.

Lemma 7.31. *Let $B \subset \mathbb{R}^3$ be open, and let f and g be continuous functions on the closure of B that satisfy Δf and $\Delta g \in L^1(B)$ and $f(x)$ and $g(x) \to 0$ as $|x| \to \infty$ if B is unbounded. Assume $\Delta f \leq Ff$ and $\Delta g \geq Gg$ as distributions on B, where F, G are functions satisfying $F(x) < G(x)$ a.e. in B. Assume $f(x) > 0$ in B and $f(x) \geq g(x)$ for all $x \in \partial B$. Then $f(x) \geq g(x)$ for all $x \in B$.*

Theorem 7.32. *Let $V = z/|x|$, and let ψ_∞ be the positive solution to Eq. (7.17) given in Theorem (7.29). Let ψ be the positive solution to the TFW equation, (7.9), for some $\mu \geq 0$ and $p = \frac{5}{3}$. Then, for all x, $\psi(x) \leq \psi_\infty(x)$.*

Proof. Let $B = \{x \mid \psi(x) - \psi_\infty(x) > 0\}$. Take $f = \psi_\infty$ and $g = \psi$ in Lemma 7.31. Since f and g are continuous and, by Theorem 7.24, $g(x) < f(x)$ for large $|x|$, B is open and bounded. Hence Δf and $\Delta g \in L^1(B)$. On B, $A\Delta f = Ff$, $A\Delta g > Gg$ with $F = -V + f^{4/3}\gamma$ and $G = -V + g^{4/3}\gamma$. Since $F < G$ in B and $f = g$ on ∂B, $f \geq g$ in B. Therefore, B is empty. ■

For a molecule, an upper bound to ψ, which is not as nice as Theorem 7.32 but which is sufficient for Theorem 7.30, can also be obtained. We always assume $p = \frac{5}{3}$.

Theorem 7.33. *Let V be as in Theorem 7.30 with the scaling given there. Let ψ be the positive solution to the TFW equation for $\mu \geq 0$ and let B be the ball $\{x \mid Z^{-2/3} > |x - R_1|\}$. Then, for sufficiently large a,*

$$\psi(x) \leq \psi_\infty(x - R_1) \quad \textit{for } x \in B \tag{7.36}$$

where ψ_∞ is the positive solution to Eq. (7.17) with

$$z = z_1 + dZ^{2/3}$$

and

$$d = 1 + 2(Z^0)^{-1/3}/\min\{|R_j^0 - R_1^0| \mid j = 2, \ldots, k\}\,.$$

Proof. By Theorem 7.14 and Eq. (7.24), we can choose a large enough so that $\psi_\infty(x - R_1) > \psi(x)$ when $x \in \partial B$ and so that $R_2, \ldots, R_k \notin B$. The proof is then the same as for Theorem 7.32, with $f = \psi_\infty$ and $g = \psi$, provided we can verify that $M(x) \equiv z|x - R_1|^{-1} - V(x) > 0$ when $x \in B$. But M, being superharmonic in B, has its minimum on ∂B. This minimum is positive for large enough a. ■

To obtain a lower bound to ψ, the following is needed.

Theorem 7.34. *Assume the hypothesis of Theorem 7.30 with $\lambda > 0$ and let ψ be the positive solution to the TFW equation. Then there is a constant d, independent of λ, such that*

(i) $h(x) \equiv |x|^{-1} * \rho < da^{4/3}$.
(ii) $\mu < da^{4/3}\lambda^{-2/3}$.

Proof. For (i) we use Theorem 7.14 together with the fact that $\int \rho = a\lambda$. For any x, let B be the ball of radius $a^{-1/3}$ centered at x. The contribution to h from $\chi_B \rho$ is bounded by $(\mathrm{const})(Z^0)^{3/2}a^{4/3}$. The contribution from $(1 - \chi_B)\rho$ is bounded by $a^{1/3}\int \rho$. For (ii), since μ is decreasing in λ, $\mu \leq -e(N)/N$. However, $e(N) > e^{\mathrm{TF}}(N)$. But $-e^{\mathrm{TF}}(N)$ scales as $a^{7/3}f(\lambda)$ and $f(\lambda) \leq (\mathrm{const})\lambda^{1/3}$ by Eq. (3.6). ■

Theorem 7.35. *Assume the same hypothesis as in Theorem 7.33. Then, for sufficiently large a,*

$$\psi(x) \geq \psi_\infty(x - R_1)\sigma(x - R_1) \quad \textit{for all } x\,, \tag{7.37}$$

where ψ_∞ is the positive solution to Eq. (7.17) with $z = z_1 - 4ta^{2/3}A$ and

$$\sigma(x) = [1 - a^{2/3}t|x|]\exp(-a^{2/3}t|x|)\,.$$

Here, $At^2 = d(1 + \lambda^{-2/3})$ with d given in Theorem 7.34.

Proof. Let $f = \psi$ and $g =$ right side of Eq. (7.37). We have to verify (7.37) only in $B = \{x \mid a^{2/3}t|x - R_1| < 1\}$ because $g \leq 0$ otherwise. Since, by Theorem 7.29, both ψ_∞ and σ are symmetric decreasing, $\Delta g > \sigma\Delta\psi_\infty + \psi_\infty\Delta\sigma$. But

$$(\Delta\sigma)(x) = (a^{4/3}t^2 - 4a^{2/3}t/|x|)\sigma\,,$$

and $\psi_\infty^{4/3} \geq g^{4/3}$ since $\sigma \leq 1$. Therefore, to imitate the proof of Theorem 7.32, it is only necessary to verify that $a^{4/3}At^2 > h(x) + \mu$, but this is clearly true. ■

Proof of Theorem 7.30. (iv) is a trivial consequence of Theorems 7.33 and 7.35. (iii) is proved in the same way as Theorem 5.2 if we note that the energy can be controlled to $O(z^{7/3})$ by the variational upper bound given in the paragraph after Eq. (7.15). (ii) is proved by noting that, by the proof of (iii) just given,

$$a^{-7/3}E^{\mathrm{TFW}}(a, N=a\lambda) \to E^{\mathrm{TF}}(a=1, \lambda).$$

The limit of the derivative of a sequence of convex (in λ) functions is the derivative of the limit function.

The proof of (i) is complicated. Upper and lower bounds to E of the desired accuracy, $O(Z^2)$, are needed. First, let us make a remark. Consider E as a function of A. By standard arguments used earlier, $E(A)$ is monotone increasing, concave, and hence differentiable almost everywhere for $A>0$. $dE/dA = T/A$ a.e., and $E^{\mathrm{TFW}} - E^{\mathrm{TF}} = \int_0^A (T/A)dA$. If we can find a lower and upper bound to T/A of the form

$$T/A = A^{-1/2}\gamma^{-3/2}I_1 \sum z_j^2 + \text{(lower order)}$$

then Eq. (7.29) will be proved. We can, indeed, find a lower bound of this form, and hence a lower bound to E. We cannot find an upper bound of this form and therefore must resort to a direct variational calculation to obtain an upper bound to E.

Upper bound. By the monotonicity of E in N, it is only necessary that $\int\rho \le N$. There are several ways to construct a variational ρ, which we call f. The details of the calculation of $\mathcal{E}(f)$ are left to the reader. One construction is to define $B=\{x\,|\,\rho^{\mathrm{TF}}(x) > Z^{5/2}\}$. For large a, B is the union of k connected components which are approximately spheres centered at R_j. Call these B_j. Let $\psi_{\infty j}$ be the solution to Eq. (7.17) centered at R_j and with $z=z_j - ta^{2/3}$. Let $C_j=\{x\,|\,\psi_{\infty j} > Z^{5/4}\}$. For large enough, but fixed t, $C_j \subset B_j$ for large a. The variational f is defined by $f(x)=\rho^{\mathrm{TF}}(x)$ for $x\notin B$, $f(x)=Z^{5/2}$ for $x \in B_j\backslash C_j$, and $f(x)=\psi_{\infty j}(x)^2$ for $x\in C_j$.

Lower bound. We construct a lower bound to T/A. Suppose $P_1,\ldots,P_k$ are orthogonal, vector valued functions. Then $T/A \ge \sum_j L_j^2/\int P_j^2$, where $L_j=\int \nabla\psi\cdot P_j$. We take $P_j(x)=\nabla\psi_{\infty j}(x)\chi_j(x)$, where χ_j is the characteristic function of $D_j=\{x\,|\,|x-R_j| < tz_j^{-2/3}\}$, and t is some fixed constant. For large a, the D_j are disjoint so the P_j are orthogonal. Clearly, $\int P_j^2 = \int \nabla\psi_{\infty j}^2 + o(Z^2)$. Now multiply Eq. (7.17) for $\psi_{\infty j}$ by ψ and integrate over D_j. Then

$$L_j = -A^{-1}\int W_{\infty j}\psi_{\infty j}\psi\chi_j + \int \psi\nabla\psi_{\infty j}\cdot \mathbf{n}ds.$$

By the bound (7.37), the first integral is $(T_{\infty j}/A)+o(Z^2)$. It is not difficult to show that the second integral is $o(Z^2)$. This can be done by using Eq. (7.24), whence, for some $t\in[\frac{1}{2},1]$, $d\psi_{\infty j}/dr > -10z_j^{3/4}\gamma^{-7/4}$ at $r=tz_j^{-2/3}$. ∎

VIII. THOMAS-FERMI-DIRAC-VON WEIZSÄCKER THEORY

This theory has not been as extensively studied as the other theories. The results presented here are from unpublished work by Benguria, Brezis, and Lieb done in connection with their 1981 paper.

The energy functional is

$$\mathcal{E}'(\psi) = A\int(\nabla\psi)^2 + \int J(\psi^2) - \int V\psi^2 + D(\psi^2,\psi^2) + U \tag{8.1}$$

in units in which $\hbar^2/2m = 1$.

$$J(\rho) = (\gamma/p)\rho^p - (3C_e/4)\rho^{4/3}. \tag{8.2}$$

For convenience we assume $p>\frac{4}{3}$ (not $p>1$). $\mathcal{E}(\rho) \equiv \mathcal{E}'(\rho^{1/2})$. The function space for ψ is the same as for TFW theory, namely, G_p' of Eq. (7.3). Note that $\mathcal{E}(\rho)$ is not convex because of the $-\int\rho^{4/3}$ term.

As in TFD theory Eqs. (6.7)–(6.10), we introduce

$$J_\alpha(\rho) = J(\rho) + \alpha\rho, \tag{8.3}$$

and α is chosen so that $J_\alpha(\rho)\ge 0$ and $J_\alpha(\rho_0) = 0 = J_\alpha'(\rho_0)$ for some ρ_0, namely,

$$\rho_0^{p-4/3} = C_e p[4\gamma(p-1)]^{-1}$$
$$\alpha = (3p-4)[4(p-1)]^{-1}\rho_0^{1/3}C_e. \tag{8.4}$$

The necessity of $p>\frac{4}{3}$ for this construction is obvious. $\mathcal{E}_\alpha'$ and $\mathcal{E}_\alpha$ are defined by using J_α in Eq. (8.1).

The energy for $\lambda \ge 0$ is

$$E(\lambda) = \inf\left\{\mathcal{E}(\rho)\,|\,\rho\in G_p, \int\rho=\lambda\right\}, \tag{8.5}$$

and similarly for $E_\alpha(\lambda)$ and $E'(\lambda)$, $E_\alpha'(\lambda)$ using $\mathcal{E}'$. If the condition $\int\rho=\lambda$ is omitted in (8.5) we obtain E, E_α, E', E_α'.

Theorem 8.1. *(i) The four functions $E(\lambda)$, $E_\alpha(\lambda)$, $E'(\lambda)$, and $E_\alpha'(\lambda)$ are finite, continuous, and satisfy*

$$E(\lambda) = E'(\lambda) = E_\alpha(\lambda) - \alpha\lambda = E_\alpha'(\lambda) - \alpha\lambda. \tag{8.6}$$

(ii) E_α is finite.

(iii) ρ minimizes $\mathcal{E}(\rho)$ on $\int\rho=\lambda$ if and only if $\psi=\rho^{1/2}$ minimizes $\mathcal{E}'(\psi)$ on $\int\psi^2=\lambda$. This ρ and ψ also obviously minimize $\mathcal{E}_\alpha$ and $\mathcal{E}_\alpha'$.

Proof. The same as for Theorems 2.1, 6.2, and 7.2. Note that $[\int\rho^{4/3}]^{3p-3} \le [\int\rho]^{3p-4}\int\rho^p$ (by Hölder). ∎

Theorem 8.2. *Let ψ minimize $\mathcal{E}'(\psi)$ on the set $\int\psi^2=\lambda$. Then ψ satisfies the* TFDW equation:

$$[-A\Delta + W(x)]\psi = -\mu\psi, \tag{8.7}$$

in the sense of distributions, with

$$W = \gamma\rho^{p-1} - C_e\rho^{1/3} - \phi + \alpha, \tag{8.8}$$

$\phi = V - |x|^{-1} * \rho$, *and* $\rho=\psi^2$. *Apart from a sign, $\psi(x)>0$ for all x, and ψ satisfies the conclusions of Theorem 7.9. ψ is the unique ground state of $H=-A\Delta+W(x)$ and μ is its ground-state eigenvalue. E is differentiable at λ and $\mu = -dE_\alpha/d\lambda = -dE/d\lambda - \alpha \ge -\alpha$. $\mu=0$ if $E_\alpha(\lambda)$ has an absolute minimum at this λ.*

Proof. The proof is basically the same as for Theorems 7.8–7.10. Although it is not known that $\rho=\psi^2$ is unique, this is not really necessary. By considering the variation of $\mathcal{E}'(\psi)$, ψ satisfies Eqs. (8.7) and (8.8). If ψ is minimizing, then so is $|\psi|$ (cf. Theorem 7.2). Hence $|\psi|$ satisfies Eq. (8.7) with the same W. But, as in Theorem 7.10, the ground state of $H=-A\Delta+W$ is unique and non-negative and therefore ψ may be taken to be ≥ 0 for all x. The rest follows by the methods of Theorem 7.9. (Note: $\rho^{1/3}\psi\in L^3$ since $\psi\in L^6\cap L^2$.) ∎

Remark. As in Sec. VII, the role of $\mathcal{E}'$, as distinct from $\mathcal{E}$, is solely to prove Eq. (8.7), in which no explicit reference to $\rho\ge 0$ is made.

Remark. Theorem 8.2 does not assert the existence of

a minimizing ψ with $\int \psi^2 = \lambda$.

Now we turn to a difficult and serious problem. We do not know that $E_\alpha(\lambda)$ is monotone nonincreasing. Therefore, if we *define*

$$\hat{E}_\alpha(\lambda) = \inf\left\{\mathcal{E}_\alpha(\rho)\,|\,\rho \in G_p, \int \rho \leq \lambda\right\}, \tag{8.9}$$

we do not know that $E_\alpha(\lambda) = \hat{E}_\alpha(\lambda)$. By definition, $\hat{E}_\alpha(\lambda)$ is monotone nonincreasing. The source of the difficulty is this: Although $J_\alpha(\rho_0) = J'_\alpha(\rho_0) = 0$ (as in TFD theory), we cannot simply add small clumps of charge, of amplitude ρ_0, at ∞. This is so because such a clump would then have $\int (\nabla\psi)^2 = \infty$. Nevertheless, we can add clumps with $\mathcal{E}_\alpha$ energy strictly less than $\alpha \int \rho$, as the following theorem shows.

Theorem 8.3. *Set* $V = 0$ *in* $\mathcal{E}'$. *There are* C^∞ *functions of compact support such that* $\mathcal{E}'(\psi) < 0$.

Proof. Let f be any function in C_0^∞ and let $\psi(x) = b^2 f(bx)$. For some sufficiently small, but positive b, $\mathcal{E}'(\psi) < 0$. To see this, note that $\int (\nabla\psi)^2$ scales as b^3, $\int \rho^{5/3}$ as $b^{11/3}$, $D(\rho,\rho)$ as b^3, while $\int \rho^{4/3}$ scales as $b^{7/3}$. ■

As a corollary we have the following.

Theorem 8.4. $E(\lambda)$ *is strictly monotone decreasing in* λ. *Hence*

$$E(\lambda) = \inf\left\{\mathcal{E}(\rho)\,|\,\rho \in G_p, \int \rho \leq \lambda\right\}. \tag{8.10}$$

$$E = \inf_\lambda E(\lambda) = -\infty\,.$$

We conjecture that $E(\lambda)$ is convex. Unfortunately, the "convexification" trick of Sec. VI, in which J_α is replaced by j, is not helpful. Because of the gradient term, any minimizing ψ will be continuous, and therefore ψ cannot omit the values $(0, \rho_0)$, even for point nuclei. While the energy for j is, indeed, convex, it is strictly smaller than $E_\alpha(\lambda)$ for all λ.

Theorem 8.5 states that E_α and E_α have absolute minima at some common, finite λ. For all we know, there may be several such λ, but all these λ are bounded. Furthermore, for every λ there is a minimizing ρ for $\hat{E}_\alpha(\lambda)$. Unfortunately, for no λ are we able to infer that $\int \rho = \lambda$.

Theorem 8.5. (*i*) *There exists a minimizing* ρ *for* $\mathcal{E}_\alpha(\rho)$ *on* G_p, *and* $\psi = \rho^{1/2}$ *minimizes* $\mathcal{E}'_\alpha(\psi)$. *Every such* $\rho \in L^1$, *and* $\int \rho \leq$ *some constant which is independent of* ρ.

(*ii*) *There exists a minimizing* ρ *for* $\mathcal{E}_\alpha(\rho)$ *on the set* $\int \rho \leq \lambda$.

Remark. It is not claimed (but it is conjectured) that the minimizing ρ is unique.

Proof. The proofs of (i) and (ii) are the same, so we concentrate on (i). The proof merely imitates the proof of Theorem 7.6. The only new point is that $\rho \in L^1$. Each term in $\mathcal{E}(\rho)$ is finite and, in particular, $I = \int J_\alpha(\rho) < \infty$. But $J_\alpha(\rho) \geq k\rho$ when $0 \leq \rho \leq \beta$ for some $k, \beta > 0$. If χ is the characteristic function of $\{x\,|\,\rho(x) \leq \beta\}$, $k\int \chi\rho < \infty$. On the other hand, $\beta^2(1-\chi)\rho \leq \rho^3$, so $\beta^2 \int (1-\chi)\rho \leq \int \rho^3 < \infty$ since $\rho \in L^3$. It is easy to see from Eq. (7.6) that the bound on $\int \rho$ is independent of ρ. ■

Remark. It is surprising that the fact that $\lambda < \infty$ for any absolute minimum is obtained so easily. Recall that in TFW theory the proof of this fact (Theorem 7.12) required analysis of the TFW equation.

An important question is whether λ, for an absolute minimum, always satisfies $\lambda \geq Z$. A few things can be said about the properties of any minimizing ρ on $\int \rho = \lambda$.

Theorem 8.6. *In the atomic case,* $V(r) = z/r$, *any minimizing* ψ *is symmetric decreasing when* $\lambda \leq z$. (*Conjecture: this also holds for all* λ.)

Proof. The rearrangement inequality proof of Theorem 2.12 is applicable. ■

Theorem 8.7. *The conclusions of Theorem* 7.13 *hold for any minimizing* ψ. *Moreover, for every* $t < \mu + \alpha$ *there exists a constant* M *such that*

$$\psi(x) \leq M \exp[-(t/A)^{1/2}|x|]\,.$$

Proof. Same as for Theorems 7.13 and 7.24. ■

Theorem 8.8. *Every minimizing* ψ *satisfies Theorem 7.25.*

Plainly, TFDW theory is not in a satisfactory state from the mathematical point of view. In TFD theory we were able to deal with the lack of convexity by means of the J_α trick. In TFW theory, the presence of the gradient term does not spoil the general theory because $\mathcal{E}$ is convex. When taken together, however, the two difficulties present an unsolved mathematical problem.

ACKNOWLEDGMENTS

I am grateful to the U. S. National Science Foundation (Grant No. PHY-7825390-A02) for supporting this work. Thanks go to Freeman Dyson and Barry Simon for a critical reading of the manuscript.

REFERENCES

Adams, R. A., 1975, *Sobolev Spaces* (Academic, New York).

Balàzs, N., 1967, "Formation of stable molecules within the statistical theory of atoms," Phys. Rev. **156**, 42–47.

Baumgartner, B., 1976, "The Thomas–Fermi theory as result of a strong-coupling limit," Commun. Math. Phys. **47**, 215–219.

Baxter, J. R., 1980, "Inequalities for potentials of particle systems," Ill. J. Math. **24**, 645–652.

Benguria, R., 1979, "The von Weizsäcker and exchange corrections in Thomas–Fermi theory," Ph.D. thesis, Princeton University (unpublished).

Benguria, R., 1981, "Dependence of the Thomas–Fermi Energy on the Nuclear Coordinates," Commun. Math. Phys., to appear.

Benguria, R., H. Brezis, and E. H. Lieb, 1981, "The Thomas–Fermi–von Weizsäcker theory of atoms and molecules," Commun. Math. Phys. **79**, 167–180.

Benguria, R., and E. H. Lieb, 1978a, "Many-body potentials in Thomas–Fermi theory," Ann. of Phys. (N.Y.) **110**, 34–45.

Benguria, R., and E. H. Lieb, 1978b, "The positivity of the pressure in Thomas–Fermi theory," Commun. Math. Phys. **63**, 193–218, Errata **71**, 94 (1980).

Berestycki, H., and P. L. Lions, 1980, "Existence of stationary states in nonlinear scalar field equations," in *Bifurcation Phenomena in Mathematical Physics and Related Topics*, edited by C. Bardos and D. Bessis (Reidel, Dordrecht), 269–292. See also "Nonlinear Scalar field equations, Parts I and II," Arch. Rat. Mech. Anal., 1981, to appear.

Brezis, H., 1978, "Nonlinear problems related to the Thomas–Fermi equation," in *Contemporary Developments in Continuum Mechanics and Partial Differential Equations*, edited by

G. M. de la Penha, and L. A. Medeiros (North-Holland, Amsterdam), 81–89.

Brezis, H., 1980, "Some variational problems of the Thomas–Fermi type," in *Variational Inequalities and Complementarity Problems: Theory and Applications*, edited by R. W. Cottle, F. Giannessi, and J-L. Lions (Wiley, New York), 53–73.

Brezis, H., and E. H. Lieb, 1979, "Long range atomic potentials in Thomas–Fermi theory," Commun. Math. Phys. **65**, 231–246.

Brezis, H., and L. Veron, 1980, "Removable singularities of nonlinear elliptic equations," Arch. Rat. Mech. Anal. **75**, 1–6.

Caffarelli, L. A., and A. Friedman, 1979, "The free boundary in the Thomas–Fermi atomic model," J. Diff. Equ. **32**, 335–356.

Deift, P., W. Hunziker, B. Simon, and E. Vock, 1978, "Pointwise bounds on eigenfunctions and wave packets in N-body quantum systems IV," Commun. Math. Phys. **64**, 1–34.

Dirac, P. A. M., 1930, "Note on exchange phenomena in the Thomas–Fermi atom," Proc. Cambridge Philos. Soc. **26**, 376–385.

Fermi, E., 1927. "Un metodo statistico per la determinazione di alcune priorieta dell'atome," Rend. Accad. Naz. Lincei **6**, 602–607.

Firsov, O. B., 1957, "Calculation of the interaction potential of atoms for small nuclear separations," Zh. Eksper. i Teor. Fiz. **32**, 1464. [English transl. Sov. Phys.—JETP **5**, 1192–1196 (1957)]. See also Zh. Eksp. Teor. Fiz. **33**, 696 (1957); **34**, 447 (1958) [Sov. Phys.—JETP **6**, 534–537 (1958); **7**, 308–311 (1958)].

Fock, V., 1932, "Uber die Gültigkeit des Virialsatzes in der Fermi–Thomas'schen Theorie," Phys. Z. Sowjetunion **1**, 747–755.

Gilbarg, D., and N. Trudinger, 1977, *Elliptic Partial Differential Equations of Second Order* (Springer Verlag, Heidelberg).

Gombas, P., 1949, *Die statistischen Theorie des Atomes und ihre Anwendungen* (Springer Verlag, Berlin).

Hille, E., 1969, "On the Thomas–Fermi equation," Proc. Nat. Acad. Sci. (USA) **62**, 7–10.

Hoffmann-Ostenhof, M., T. Hoffmann-Ostenhof, R. Ahlrichs, and J. Morgan, 1980, "On the exponential falloff of wave functions and electron densities," *Mathematical Problems in Theoretical Physics, Proceedings of the International Conference on Mathematical Physics held in Lausanne, Switzerland, August 20–25, 1979*, Springer Lectures Notes in Physics, edited by K. Osterwalder (Springer-Verlag, Berlin, Heidelberg, New York, 1980), Vol. 116, 62–67.

Hoffmann-Ostenhof, T., 1980, "A comparison theorem for differential inequalities with applications in quantum mechanics," J. Phys. A **13**, 417–424.

Jensen, H., 1933, "Über die Gültigkeit des Virialsatzes in der Thomas–Fermischen Theorie," Z. Phys. **81**, 611–624.

Kato, T., 1957, "On the eigenfunctions of many-particle systems in quantum mechanics," Commun. Pure Appl. Math. **10**, 151–171.

Lee, C. E., C. L. Longmire, and M. N. Rosenbluth, 1974, "Thomas–Fermi calculation of potential between atoms," Los Alamos Scientific Laboratory Report No. LA-5694-MS.

Liberman, D. A., and E. H. Lieb, 1981, "Numerical calculation of the Thomas–Fermi–von Weizsäcker function for an infinite atom without electron repulsion," Los Alamos National Laboratory Report in preparation.

Lieb, E. H., 1974, "Thomas–Fermi and Hartree–Fock theory," in *Proceedings of the International Congress of Mathematicians, Vancouver*, Vol. 2, 383–386.

Lieb, E. H., 1976, "The stability of matter," Rev. Mod. Phys. **48**, 553–569.

Lieb, E. H., 1977, "Existence and uniqueness of the minimizing solution of Choquard's nonlinear equation," Stud. in Appl. Math. **57**, 93–105.

Lieb, E. H., 1979, "A lower bound for Coulomb energies," Phys. Lett. A **70**, 444–446.

Lieb, E. H., 1981a, "A variational principle for many-fermion systems," Phys. Rev. Lett. **46**, 457–459; Erratum **47**, 69 (1981).

Lieb, E. H., 1981b, "Analysis of the Thomas–Fermi–von Weizsäcker equation for an atom without electron repulsion," in preparation.

Lieb, E. H., and S. Oxford, 1981, "An improved lower bound on the indirect Coulomb energy," Int. J. Quantum Chem. **19**, 427–439.

Lieb, E. H., and B. Simon, 1977, "The Thomas–Fermi theory of atoms, molecules and solids," Adv. in Math. **23**, 22–116. These results were first announced in "Thomas–Fermi theory revisited," Phys. Rev. Lett. **31**, 681–683 (1973). An outline of the proofs was given in Lieb, 1974.

Lieb, E. H., and B. Simon, 1978, "Monotonicity of the electronic contribution to the Born–Oppenheimer energy," J. Phys. B **11**, L537–542.

Lieb, E. H., and W. Thirring, 1975, "Bound for the kinetic energy of fermions which proves the stability of matter," Phys. Rev. Lett. **35**, 687–689; Errata **35**, 1116 (1975).

Lieb, E. H., and W. Thirring, 1976, "A bound for the moments of the eigenvalues of the Schroedinger Hamiltonian and their relation to Sobolev inequalities," in *Studies in Mathematical Physics: Essays in Honor of Valentine Bargmann*, edited by E. H. Lieb, B. Simon, and A. S. Wightman (Princeton University Press, Princeton), 269–303.

March, N. H., 1957, "The Thomas–Fermi approximation in quantum mechanics," Adv. in Phys. **6**, 1–98.

Mazur, S., 1933, "Über konvexe Mengen in linearen normierten Räumen," Studia Math. **4**, 70–84. See p. 81.

Morgan, J., III., 1978, "The asymptotic behavior of bound eigenfunctions of Hamiltonians of single variable systems," J. Math. Phys. **19**, 1658–1661.

Morrey, C. B., Jr., 1966, *Multiple integrals in the calculus of variations* (Springer, New York).

O'Connor, A. J., 1973, "Exponential decay of bound state wave functions," Commun. Math. Phys. **32**, 319–340.

Reed, M., and B. Simon, 1978, *Methods of Modern Mathematical Physics* (Academic, New York), Vol. 4.

Ruskai, M. B., 1981, "Absence of discrete spectrum in highly negative ions," Commun. Math. Phys. (to appear).

Scott, J. M. C., 1952, "The binding energy of the Thomas–Fermi atom," Philos. Mag. **43**, 859–867.

Sheldon, J. W., 1955, "Use of the statistical field assumption in molecular physics," Phys. Rev. **99**, 1291–1301.

Simon, B., 1981, "Large time behavior of the L^p norm of Schroedinger semigroups," J. Func. Anal. **40**, 66–83.

Stampacchia, G., 1965, *Equations elliptiques du second ordre a coefficients discontinus* (Presses de l'Universite, Montreal).

Teller, E., 1962, "On the stability of molecules in the Thomas–Fermi theory," Rev. Mod. Phys. **34**, 627–631.

Thirring, W., 1981, "A lower bound with the best possible constant for Coulomb Hamiltonians," Commun. Math. Phys. **79**, 1–7 (1981).

Thomas, L. H., 1927, "The calculation of atomic fields," Proc. Camb. Philos. Soc. **23**, 542–548.

Torrens, I. M., 1972, *Interatomic Potentials* (Academic, New York).

Veron, L., 1979, "Solutions singulières d'equations elliptiques semilinéaire," C. R. Acad. Sci. Paris **288**, 867–869. This is an announcement; details will appear in "Singular solutions of nonlinear elliptic equations," J. Non-Lin. Anal., in press.

von Weizsäcker, C. F., 1935, "Zur Theorie der Kernmassen," Z. Phys. **96**, 431–458.

Yonei, K., and Y. Tomishima, 1965, "On the Weizsäcker correction to the Thomas–Fermi theory of the atom," Jour. Phys. Soc. Japan **20**, 1051–1057.

Yonei, K., 1971, "An extended Thomas–Fermi–Dirac theory for diatomic molecules," Jour. Phys. Soc. Japan **31**, 882–894.

INDEX

L/R refer to left/right column

Rev. Mod. Phys. *54*, 311 (1982)

Erratum: Thomas-Fermi and related theories of atoms and molecules [Rev. Mod. Phys. 53, 603–641 (1981)]

Elliott H. Lieb

Department of Mathematics and Physics, Princeton University, POB 708, Princeton, New Jersey 08544

Please note the following corrections:
Page 604, line after Eq. (2.8): A an adjustable *positive* constant.
Page 606: The last equality in Eq. (2.18) should read $[\phi_\rho(x)-\mu]_+$. In the line after Eq. (2.18), [ ] should read $[\ ]_+$. *)
Page 620: In Eq. (5.1) change $+V(x)$ to $-V(x)$.
Page 621, line before Eq. (5.3): Change $+V(x)$ to $-V(x)$.
Page 623, line 7: $|\psi_N,H_N\psi_N)$ should read $|(\psi_N,H_N\psi_N)$.
Page 623: Eq. (5.32) should read $D=q/8$.

*) To avoid further confusion this error has been corrected in this re-edition of the article on page 194 of this volume.

With B. Simon in Commun. Math. Phys. *53*, 185-194 (1977)

The Hartree-Fock Theory for Coulomb Systems

Elliott H. Lieb*

Departments of Mathematics and Physics, Princeton University, Princeton, New Jersey 08540, USA

Barry Simon**

Department of Physics, Yeshiva University, New York, New York 10033, USA

Abstract. For neutral atoms and molecules and positive ions and radicals, we prove the existence of solutions of the Hartree-Fock equations which minimize the Hartree-Fock energy. We establish some properties of the solutions including exponential falloff.

§ 1. Introduction

In this paper we discuss the Hartree (H) and Hartree-Fock (HF) theories associated with the purely Coulombic Hamiltonian of electrons interacting with static nucleii. Our purpose will be to prove that these theories exist (in the sense that the equations have solutions which minimize the H or HF energy) whenever the system has an excess positive charge after the removal of one electron. An announcement of these results was given in [22] and an outline of the proof was given in [19].

The precise quantum system is described by the Hamiltonian

$$H = -\sum_{i=1}^{N} \Delta_i + \sum_{i=1}^{N} V(x_i) + \sum_{i<j} |x_i - x_j|^{-1}\,, \tag{1}$$

where

$$V(x) = -\sum_{j=1}^{k} z_j |x - R_j|^{-1} \tag{2}$$

acting on the Hilbert space $\mathscr{H} = L_a^2(\mathbb{R}^{3N}; \mathbb{C}^{2N})$. We assume $z_j > 0$, all j. The subscript a on L^2 indicates that we are to consider functions in L^2 as $\Psi(x_1, \sigma_1; \ldots; x_N, \sigma_N)$ with $x_i \in \mathbb{R}^3$, $\sigma_i \in \pm 1/2$ and only allow those Ψ antisymmetric under interchanges of i and j. The particles have two spin states, but we could allow q spin states in our analysis below with only notational changes. The physically correct Fermi statistics

* Research partially supported by U.S. National Science Foundation Grant MCS-75-21684

** Research partially supported by U.S. National Science Foundation under Grants MPS-75-11864 and MPS-75-20638. On leave from Departments of Mathematics and Physics, Princeton University, Princeton, NJ 08540, USA

(antisymmetric functions) which we impose turns out to be the most difficult; our method would apply equally well to any other kind of statistics.

In (2), the z_j are the charges of the nucleii at positions R_j. By a famous theorem of Kato [16], H is essentially self-adjoint on $C_0^\infty(\mathbb{R}^{3N};\mathbb{C}^{2N})_a=\mathscr{D}_{\text{phys}}$, the C^∞ functions of compact support.

We set:

$$E_N^Q(z_i,R_i)\equiv\inf\{(\Psi,H\Psi)|\Psi\in\mathscr{D}_{\text{phys}};\|\Psi\|=1\} \tag{3}$$

which is defined to be the quantum ground state energy.

In 1928, Hartree [14] introduced an approximate method for finding E_N^Q. He apriori ignored the spin variables and the Pauli principle and considered product wave functions.

$$\Psi(x_1,\dots,x_N)=\prod_{i=1}^N u_i(x_i)\,. \tag{4a}$$

Minimization of the functional

$$\mathscr{E}_{\text{H}}(u_1,\dots,u_N)=(\Psi,H\Psi) \tag{4b}$$

with the constraint $\|u_i\|=1$ then leads to the Euler-Lagrange equation

$$\tilde{h}_iu_i=\varepsilon_iu_i\,, \tag{5a}$$

where the ε_i are Lagrange multipliers and

$$(\tilde{h}_iw)(x)=(-\Delta w)(x)+V(x)w(x)+R_\Psi^i(x)w(x)\,, \tag{5b}$$

$$R_\Psi^i(x)=\sum_{j\neq i}\int|x-y|^{-1}|u_j(y)|^2d^3y\,. \tag{5c}$$

Note that in the H equations, (5), the h_i depend non-trivally on i. This is to be contrasted with the HF equations (7) where h is independent of i. Of course, the equations (5) formally only correspond to stationary points of $\mathscr{E}_{\text{H}}$ so there should be solutions corresponding to u's that do not minimize $\mathscr{E}_{\text{H}}$. Hartree attempted to take the Pauli principle into account by seeking solutions with $u_1=u_2$ and u_3 "approximately orthogonal to u_1", $u_3=u_4$ etc. [We should also mention that Hartree's derivation of (5) did not go through a minimization in the variational principle–this is a refinement due to Slater [30] which led him to the HF equations.] A more systematic and satisfactory way to take the Pauli principle into account was discovered in 1930 independently by Fock [10] and Slater [30] yielding equations now called Hartree-Fock (HF) equations. One considers trial functions $u_i(x_i,\sigma_i)$; $i=1,\dots,N$ with $(u_i,u_j)=\delta_{ij}$ and the Slater determinant

$$\Psi(x_1,\sigma_1,\dots,x_N,\sigma_N)=(N!)^{-1/2}\det(u_i(x_j,\sigma_j)) \tag{6a}$$

and minimizes

$$\mathscr{E}_{\text{HF}}(u_1,\dots,u_N)=(\Psi,H\Psi) \tag{6b}$$

with the constraint $(u_i, u_j)=\delta_{ij}$. The corresponding Euler-Lagrange equations are:

$$hu_i = \varepsilon_i u_i \,, \tag{7a}$$

$$(hw)(x) = (-\Delta w)(x) + V(x)w(x) + U_\Psi(x)w(x) - (K_\Psi w)(x) \,, \tag{7b}$$

$$U_\Psi(x) = \sum_{j=1}^{N} \int |x-y|^{-1} |u_j(y)|^2 d^3y \,, \tag{7c}$$

$$(K_\Psi w)(x) = \sum_{j=1}^{N} u_j(x) \int |x-y|^{-1} \overline{u_j(y)} w(y) d^3y \,. \tag{7d}$$

U_Ψ is the "direct" interaction and K_Ψ is the "exchange" interaction. We will show that minimizing solutions of (7a) exist whenever $N<Z+1$ where Z is the nuclear charge

$$Z = \sum_{j=1}^{k} z_j \,.$$

We make the convention that when u's depending on spin are involved, as in (7c) and (7d) the symbol $\int \text{—} d^3y$ indicates also a sum over the spin variable attached to y. [We note that the naive Euler-Lagrange equations are more complicated than (7) but after a unitary change, $u_i^{\text{New}} = \sum a_{ij} u_j^{\text{Old}}$ with a_{ij} a unitary $N \times N$ matrix, (7) results. The Slater determinant (6a) is unaffected by the change so that (6b) is unaffected. This is proved in Lemma 2.3 and is further discussed in many texts, e.g. Bethe-Jackiw [6]; it plays an important role in § 2 below.]

Irrespective of the physical content of the H and HF equations, (5), (7), it is far from evident that there exist any solutions of them, let alone minimizing solutions, for they are clearly complicated non-linear equations. Because the full N-body Schrödinger equation is, at present, virtually inaccessible to computer calculation while the HF equation, especially in the spherical approximation [6], is ideal for computer iterative solution, the HF equations are extensively used in quantum chemistry [27].

Before our work, the only existing theorems were for the Hartree equation (5) as follows: Reeken [26] considered the *restricted* Hartree equations for Helium, i.e. he considered (5) with $k=1$, $z_1=2$ and the additional restriction $u_1=u_2$. He found a solution for this case with $u_i \geqq 0$ pointwise; his method works for any $z>1$. Independently Gustafson and Sather [12] found solutions for the restricted two electron problems for sufficiently large z (they state their results for $z=2$ but with $\|u_i\|$ sufficiently small rather than 1. Since we insist on the normalization condition $\|u_i\|=1$, we scale coordinates to translate their result into a large z, $\|u_i\|=1$ result). These authors all use a bifurcation analysis further discussed in Stuart [32], and depend on the fact that they seek spherically symmetric solutions so that methods of ordinary differential equations are available. Properties of their solutions are further discussed in [3, 4]. Relations between the restricted Hartree two electron problem and the unrestricted problem appear to present some interesting mathematical phenomena and we hope to return to them in a future publication.

Using a Schauder-Tychonoff theorem, Wolkowisky [34] found ground state and excited solutions of the Hartree equation in the spherical approximation (see e.g. Bethe-Jackiw [6] for a discussion of the approximation).

All these authors attack the equations directly as fixed point equations in some sense. The reason we are able to go further is that we exploit the form of the equations as gradient maps, i.e. as Euler-Lagrange equations and directly attempt to find solutions by finding minimizing u's for $\mathscr{E}_{\mathrm{H}}$ and $\mathscr{E}_{\mathrm{HF}}$. (This method has already been used successfully in [23] to find solutions of another of the non-linear equations of atomic physics: the Thomas-Fermi [9, 33] model.)

These results for the H and HF equations, which we give in § 2 were announced in [22] and sketched in [19]; seemingly unaware of our work, Bader [2] has recently presented a similar method to obtain similar results for the H (but not HF) equations. We note that prior to our work, solutions of the HF equations for a class of potentials *excluding* Coulomb potentials were found by Fonte et al. [11]. Recently, several authors [7, 8] have proved existence of the time-dependent HF equations.

In § 3, we establish various "regularity" properties of any u's (not necessarily minimizing ones), which solve the H and HF equation. Among these is the exponential falloff of the u's announced in [22]; after our announcement, similar results *for the* H *equations* were obtained by [5].

In § 4, we repeat the remark already made in [23] that our proof that HF theory is "exact" in the $Z\to\infty$ limit implies the same result for HF theory.

§ 2. Solutions of the H and HF Equations

While one could present the existence theory for the H and HF equations as two cases of one general result, we present the two theories in sequence to illustrate the extra difficulties in the HF case. The basic strategy is (cf. [23]) to introduce a weak topology on the trial functions in which the trial functions are precompact and then to prove that the functional one wishes to minimize is lower semicontinuous. This establishes that the functional is minimized at some point in the closure of the trial functions. In many cases, additional arguments are then available to prove that the minimizing point belongs to the original trial functions rather than merely to the closure.

Theorem 2.1 (H Theory). *Fix* N, k; $z_1, \ldots, z_k$, $R_1, \ldots, R_k$. *There exist functions* $u_1, \ldots, u_N \in L^2(\mathbb{R}^3; \mathbb{C}^2)$ *with* $u_i \in Q(-\Delta)$, *the quadratic form domain of* $-\Delta$, *such that the* u_i *minimize*

$$\tilde{\mathscr{E}}_{\mathrm{H}}(u_1, \ldots, u_N) = \sum_{i=1}^{N} (u_i, (-\Delta + V)u_i) + \sum_{i<j} \int |u_i(x)|^2 |u_j(y)|^2 |x-y|^{-1} d^3x d^3y \tag{8}$$

with the subsidiary conditions, $u_i \in Q(-\Delta)$ *and*

$\|u_i\| \leqq 1$.

The u_i*'s satisfy* (5a) *with the additional condition, that for each i, either* $\varepsilon_i \leqq 0$ *or* $u_i = 0$. *In either event* $\varepsilon_i = \inf \operatorname{spec}(\tilde{h}_i)$ *and if* $\varepsilon_i < 0$, $\|u_i\| = 1$. *If, moreover,* $N < Z+1$, *then all* $\varepsilon_i < 0$ *and each* $\|u_i\| = 1$. *In all cases*

$$E_{\mathrm{H}} \equiv \min \{\tilde{\mathscr{E}}_{\mathrm{H}}(u_1, \ldots, u_N) \mid \|u_i\| \leqq 1, u_i \in Q(-\Delta)\}$$

is finite.

Remark. We have introduced the function $\tilde{\mathscr{E}}_{\mathrm{H}}$ which agrees with $\mathscr{E}_{\mathrm{H}}$ *only* when all $\|u_i\| = 1$. Theorem 2.1 says that $\tilde{\mathscr{E}}_{\mathrm{H}}$ always has a minimum if we only impose $\|u_i\| \leqq 1$. When the minimum of $\tilde{\mathscr{E}}_{\mathrm{H}}$ occurs for $\|u_i\| = 1$, all i, as we assert it does if $Z+1>N$, then, of course, these u_i also minimize $\mathscr{E}_{\mathrm{H}}$ subject to $\|u_i\| = 1$.

Proof. By a well-known result of Kato [18], for any $\varepsilon > 0$,

$$(u, Vu) \leqq \varepsilon(u, -\Delta u) + C_\varepsilon(u, u)$$

from which it follows that

$$E_{\mathrm{H}} = \inf\{\tilde{\mathscr{E}}_{\mathrm{H}}(u_1, \ldots, u_N) \mid \|u_i\| \leqq 1, u_i \in Q(-\Delta)\}$$

is finite and that for some K:

$$\tilde{\mathscr{E}}_{\mathrm{H}}(u_1, \ldots, u_N) \leqq E_H + 1\,;\ \|u_j\| \leqq 1 \Rightarrow \|\nabla u_i\| \leqq K\,. \tag{9}$$

Now pick sets $u_i^{(n)}$, $1 \leqq i \leqq N$, $n = 1, \ldots$ so that $\tilde{\mathscr{E}}_{\mathrm{H}}(u_i^{(n)}) \leqq E_{\mathrm{H}} + 1/n$. By (9), the $u_i^{(n)}$'s lie in a fixed ball in the Sobolev space [1], $H^1 = \{u \mid |||u||| \equiv (\|u\|^2 + \|\nabla u\|^2)^{1/2} < \infty\}$. Thus, by the Banach-Alaoglu theorem, there exists a subsequence such that $u_i^{(n)} \to u_i^{(\infty)}$ in the weak-H^1 topology. Clearly $\|u_i^{(\infty)}\| \leqq 1$. We claim that $\tilde{\mathscr{E}}_{\mathrm{H}}(u_i^{(\infty)}) \leqq \underline{\lim}\, \tilde{\mathscr{E}}_{\mathrm{H}}(u_i^{(n)}) = E_{\mathrm{H}}$, whence it follows that the $u_i^{(\infty)}$ minimize $\tilde{\mathscr{E}}_{\mathrm{H}}$. Positive definite quadratic forms are always non-increasing under weak-limits (see e.g. [23]) so that

$$(u_i^{(\infty)}, -\Delta u_i^{(\infty)}) \leqq \underline{\lim}\,(u_i^{(n)}, -\Delta u_i^{(n)})$$

$$(u_i^{(\infty)} u_j^{(\infty)}, |x_i - x_j|^{-1} u_i^{(\infty)} u_j^{(\infty)}) \leqq \underline{\lim}\,(u_i^{(n)} u_j^{(n)}, |x_i - x_j|^{-1} u_i^{(n)} u_j^{(n)})$$

since $u_i^{(n)} u_j^{(n)} \xrightarrow{w} u_i^{(\infty)} u_j^{(\infty)}$ in $L^2(R^6)$. Finally, because [18, 25] V is relatively $-\Delta$ form *compact* [i.e. $(\Delta+1)^{-1/2} V(-\Delta+1)^{-1/2}$ is compact] $(u_i^{(n)}, V u_i^{(n)}) \to (u_i^{(\infty)}, V u_i^{(\infty)})$. It follows that $\underline{\lim}\, \tilde{\mathscr{E}}_{\mathrm{H}}(u_i^{(n)}) \geqq \tilde{\mathscr{E}}_{\mathrm{H}}(u_i^{(\infty)})$. Henceforth, u_i is used to denote this $u_i^{(\infty)}$.

To see that the u_i's satisfy (5a), fix $u_1, \ldots, u_{i-1}, u_{i+1}, \ldots, u_N$ and let

$$\begin{aligned} f(u) &= \tilde{\mathscr{E}}_{\mathrm{H}}(u_1, \ldots, u_{i-1}, u, u_{i+1}, \ldots, u_N) \\ &= \text{const} + (u, \tilde{h}_i u)\,. \end{aligned}$$

Since $f(u)$ is minimized by $u = u_i$ subject to $\|u\| \leqq 1$, we conclude that either $\tilde{h}_i \geqq 0$, $u_i = 0$, or $\tilde{h}_i u_i = \varepsilon_i u_i$ with $\varepsilon_i = \inf \operatorname{spec} \tilde{h}_i \leqq 0$.

Now suppose that $N < Z+1$. Let v be a spherically symmetric function on $\mathbb{R}^3$. Then $(v, \tilde{h}_i v) = (v, -\Delta v) + (v, Kv)$, where

$$K(r) = -\sum_{j=1}^{k} z_j(\max(r, |R_j|))^{-1} + \sum_{j \neq i} \int |u_j(k)|^2 (\max(x, r))^{-1} dx\,.$$

Since $\|u_j\| \leqq 1$, we have that

$$K(r) \leqq -[Z-(N-1)]|r|^{-1} \quad \text{when} \quad r > \max(|R_j|)\,.$$

It is easy to see (use explicit hydrogenic wave functions, or a scaling argument [28]), that $(v, h_i v) < 0$ for suitable v's. It follows that $\varepsilon_i < 0$ so that $\|u_i\| = 1$. □

Remark. 1) In particular, in the neutral case $\sum z_j = N$, a solution of the H equation exists.

2) Notice that no assertion is made about uniqueness.

3) In the above proof, we used $|x_i - x_j|^{-1} \geqq 0$ pointwise. In distinction, at the

analogous point in the TF theory we used the fact that $|x|^{-1}$ is positive definite.

4) The above method fails for the Hartree-like Choquard functional $\mathscr{E}(u,v)=\|\nabla u\|^2+\|\nabla v\|^2-\int |u(x)|^2|v(y)|^2|x-y|^{-1}dxdy$, because the last term is negative instead of positive. Nevertheless, alternate methods involving rearrangement inequalities can be used to prove that minimizing u and v exist, see Lieb [21].

5) Since the u_i's are ground states of h_i, they are pointwise positive [25].

To prove the existence of solutions of the HF equation, we must extend $\mathscr{E}_{\text{HF}}$ in a manner analogous to (8); we define:

$$\tilde{\mathscr{E}}_{\text{HF}}(u_1,\ldots,u_N)=\sum_{i=1}^{N}(u_i,(-\Delta+V)u_i)+\sum_{1\leqq i<j\leqq N}((ij)_2,|x_i-x_j|(ij)_2)\,, \tag{10}$$

where

$$(ij)_2(x_i,x_j)=2^{-1/2}[u_i(x_i)u_j(x_j)-u_j(x_i)u_i(x_j)]\,.$$

For future reference we note that

$$\begin{aligned}
&((ij)_2,|x_i-x_j|^{-1}(ij)_2)=D_{ij}-E_{ij}\\
&D_{ij}=\int |x-y|^{-2}|u_i(x)|^2|u_j(y)|^2dxdy\\
&E_{ij}=\int |x-y|^{-1}\overline{\varrho(x)}\varrho(y)dxdy\\
&\varrho(x)=u_i(x)\overline{u_j(x)}\,.
\end{aligned} \tag{11}$$

The critical element in the extension will be to locate the weak closure of $\{(u_1,\ldots,u_N)|(u_i,u_j)=\delta_{ij}\}$:

Lemma 2.2. *Let $u_i^{(n)}\to u_i(i=1,\ldots,N)$ weakly with $(u_i^{(n)},u_j^{(n)})=\delta_{ij}$. Then $(u_i,u_j)=M_{ij}$ is an $N\times N$ matrix with $0\leqq M\leqq 1$. More generally the conclusion remains true if the weaker hypothesis $(u_i^{(n)},u_j^{(n)})=M_{ij}^{(n)}$ with $0\leqq M^{(n)}\leqq 1$ is imposed.*

Remark. The point is that it is easy to see that every $(u_1,\ldots,u_N)$ with M_{ij} obeying $0\leqq M\leqq 1$ arises as a weak limit of orthonormal N-tuples. Since we do not need this below, we do not give the easy proof of this converse which is based on diagonalizing M.

Proof. Let $z\in\mathbb{C}^N$. Then $(z,Mz)\equiv\sum\bar{z}_iM_{ij}z_j=(u(z),u(z))=(w\text{-}\lim u^{(n)}(z),\,w\text{-}\lim u^{(n)}(z))\leqq\sum|z_i|^2$ where $u^{(n)}(z)=\sum z_iu_i^{(n)}$. The last inequality follows from the fact that balls are weakly closed and the calculation $(u^{(n)}(z),u^{(n)}(z))=\sum|z_i|^2$. Thus $M\leqq 1$. $M\geqq 0$ is trivial. □

We will also need the elementary observation:

Lemma 2.3. *Let $\tilde{u}_i=\sum_j a_{ij}u_j$ where $A=\{a_{ij}\}_{1\leqq i,j\leqq N}$ is a unitary $N\times N$ matrix then*

$$\tilde{\mathscr{E}}_{\text{HF}}(\tilde{u}_i)=\tilde{\mathscr{E}}_{\text{HF}}(u_i)\,.$$

Proof. Let $K_{ij}=(u_i,(-\Delta+V)u_j)$, $\tilde{K}_{ij}=(\tilde{u}_i,(-\Delta+V)\tilde{u}_j)$, $R_{i_1i_2j_1j_2}=((i_1j_2)_2,|x-y|^{-1}(j_1j_2)_2)$, etc. Then $\tilde{K}=A^*KA$ so $\sum\tilde{K}_{ii}=\text{Tr}(\tilde{K})=\text{Tr}(K)=\sum K_{ii}$. Similarly, by taking traces on the antisymmetric tensor product of $\mathbb{C}^N$ with itself

$$\sum_{i<j}\tilde{R}_{ij,ij}=\sum_{i<j}R_{ij,ij}\,.\quad\square$$

Theorem 2.4 (HF Theory). *Fix $N, k; z_1, \ldots, z_k, R_1, \ldots, R_k$. There exist functions $u_1, \ldots, u_N \in L^2(\mathbb{R}^3; \mathbb{C}^2)$ such that $u_i \in Q(-\Delta)$, the quadratic form domain of $-\Delta$, and such that the u_i minimize $\tilde{\mathscr{E}}_{\mathrm{HF}}$ (given by (10)) with the subsidiary condition.*

$$M_{ij} \equiv (u_i, u_j) \quad \textit{obeys} \quad 0 \leqq M \leqq 1 .$$

Moreover, the u's obey $(u_i, u_j) = \lambda_i \delta_{ij}$ and satisfy the HF equations (7) with the additional condition that either $\varepsilon_i \leqq 0$ or $u_i = 0$ for each i. $\varepsilon_i, \ldots, \varepsilon_N$ are the N lowest points of the spectrum of h and if $\varepsilon_i < 0$, $\lambda_i = 1$. If moreover, $N < Z + 1$, then all $\varepsilon_i < 0$ and $(u_i, u_j) = \delta_{ij}$.

Proof. By mimicking the proof of Theorem 2.1, we find $u_i^{(\infty)}$ obeying $0 \leqq M^{(\infty)} \leqq 1$ which minimize $\tilde{\mathscr{E}}_{\mathrm{HF}}$. In this proof, we use Lemma 2.2 to be sure that $0 \leqq M^{(\infty)} \leqq 1$ and the fact that $(\mathrm{ij})^{(n)} \to (\mathrm{ij})^{(\infty)}$ if $u_i^{(n)} \to u_i^{(\infty)}$.

Choose a unitary $N \times N$ matrix A so that $A^* M A$ is diagonal and let $u_i = \sum_j a_{ij} u_j^{(\infty)}$. By Lemma 2.3, $\{u_i\}$ minimizes $\tilde{\mathscr{E}}_{\mathrm{HF}}$ also, and clearly $(u_i, u_j) = \lambda_i \delta_{ij}$. Now $F(u) \equiv \tilde{\mathscr{E}}_{\mathrm{HF}}(u_1, \ldots, u_{i-1}, u, u_{i+1}, \ldots, u_{\mathrm{N}}) = \mathrm{const} + (u, hu)$ so since $u = u_i$ minimizes $F(u)$ subject to $(u, u_j) = 0 (j \neq i)$, $(u, u) \leqq 1$, u_i must be a linear combination of the N smallest eigenvectors of h with only eigenvalues $\leqq 0$ allowed. Since each u_i has this property, by further unitary change, the u_i's can be made to obey $hu_i = \varepsilon_i u_i$.

To complete the proof we need only show that if $N - 1 < Z$, then h has N points of its spectrum below zero, i.e. $\dim P_{(-\infty, 0)}(h) \geqq N$. Now write $h = h_0 - K_\Psi$ with K_Ψ given by (7d). By the positive definitness of $|x|^{-1}$, K_Ψ is a positive operator, so we need only show that h_0 has N negative eigenvalues. This follows, as in the proof of Theorem 2.1 by considering spherically symmetric trial functions. □

As already remarked in [22], the above method also yields solutions of *modified* HF equations in which we restrict the u's to lie in certain sets with suitable properties. Because of the spin independence of the assumed Hamiltonian, H, we can obtain several solutions of the actual HF equation by taking each u_i to be a product of a space and spin function, for example, and demanding a particular symmetry of the spinor functions. We emphasize that the "true" minimum of Theorem 2.4 is *not known* to come from a set of u's which are product functions. An additional restriction which is often made and for which our method applies, is to demand the spatial functions be real.

§ 3. Properties of Solutions

Theorem 3.1. *The solutions of the* H *equation* (5) *constructed in Theorem* 2.1 *obey:*

a) *The u_i are globally Lipschitz and lie in $D(\tilde{h}_i) = D(-\Delta)$.*

b) *Away from the points $r = R_j$, the u_i are infinitely differentiable.*

c) *Exponential falloff: For any $\alpha < |\varepsilon_i|^{1/2}$ there exists a C_α so that*

$$|u_i(x)| \leqq C_\alpha \exp(-\alpha |x|) .$$

Proof. $\tilde{h}_i = -\Delta + \tilde{V}_i$, with $\tilde{V}_i = V + R_\Psi^i$. Since $u_j \in L^2 \cap Q(-\Delta)$, by a Sobolov estimate [31] (for a simple discussion of this point, see [20]) it follows that $u_j^2 \in L^1 \cap L^{3/2 - \varepsilon}$, so by Young's inequality, $R_\Psi^i \in L^\infty$. Thus $D(\tilde{h}_i) = D(-\Delta)$. Since $u_i \in D(\tilde{h}_i)$, we have u_i

continuous so that R^i_Ψ is continuous. a) Now follows by a result of Kato [17] and c) by a result of Simon [29] (both specialized to the two-body case; see the original papers for other references including earlier results for the two body case). b) Follows by a bootstrap argument reminiscent of our argument in [23]: We exploit the following facts (see e.g. [24], Section IX.6):

a) Let Ω be a bounded open set. If $u\in L^2(\Omega)$; $-\Delta u=Wu$ and $W\in C^k(\Omega)$, then $D^\alpha u\in L^2$ for all multiindices α with $|\alpha|\leqq k+2$.

b) If $D^\alpha u\in L^2(\Omega)$ for all α, then u is C^∞ on Ω.

The additional fact which we need is that if $u\in L^2(\mathbb{R}^3)$ and if $D^\alpha u\in L^2(\Omega)$, $|\alpha|\leqq m$, then $g(x)=\int |u(y)|^2|x-y|^{-1}dy$ is C^m on Ω. This follows by writing $|x|^{-1}=\varphi_1+\varphi_2$ with $\varphi_2\in C^\infty$ with support outside a ball of radius $\varepsilon/2$ and φ_1 supported in a ball of radius ε. Then $g(x)=\int |u(y)|^2\varphi_1(x-y)dy+\int |u(y)|^2\varphi_2(x-y)dy\equiv g_1+g_2$. The g_2 term is C^∞ on all of $\mathbb{R}^3$. The g_1 term is easily seen to be C^m on those x such that $\{y\,|\,|x-y|<\varepsilon\}\subset\Omega$.

With a), b), and the above fact, u_j is C^∞ away from the R_j by an obvious inductive argument. □

At first sight, the methods of Theorem 3.1 appear to be inapplicable to the HF case because of the non-local term. However, an elementary trick allows one to write the HF equations in local form; namely we consider the operator A on $\bigoplus_{j=1}^{N} L^2(R^3;\mathbb{C}^2)$ given by:

$$A_{ij}=\delta_{ij}(-\Delta+V(x)+R(x)-\varepsilon_i)+Q_{ij}(x)$$

with

$$Q_{ij}(x)=-\int |x-y|^{-1}\overline{u_j(y)}u_i(y)dy\,.$$

Then $A\Psi=0$ where Ψ is the vector with $\Psi_i(x)=u_i(x)$. With this remark, the following can be proven by following the proof of Theorem 3.1:

Theorem 3.2. *The solutions of the* HF *equation* (7) *constructed in Theorem* 2.4 *obey:*

a) *The u_i are globally Lipschitz and lie in $\mathscr{D}(h)=D(-\Delta)$.*

b) *Away from the points $r=R_j$, the u_i are C^∞.*

c) *Let $k_0=\min_i |\varepsilon_i|^{1/2}$. Then for any $\alpha<k_0$:*

$$|u_i(x)|\leqq C_\alpha\exp(-\alpha|x|),\qquad \textit{all } i\,.$$

Remarks. 1) The common exponential rate of falloff which was obtained comes about because we have written the HF equation as a single multicomponent equation. However, it is evident that barring some miraculous cancellation, the u_i's should have the same rate of falloff because in the HF equations $[(-\Delta+V)u_i](x)$ is a sum of terms containing all the $u_j(x)$'s in them. After making this remark in [22], we learned that it had already been made in the chemical physics literature [13].

2) By following the method of Kato [17] (or an alternative of Jensen [15]) one can give the precise singularity in the first derivatives of the u_i at the points R_j.

3) We believe that the u_i's are real analytic away from the R_j's.

4) In both Theorems 3.1 and 3.2, only the form of the equations and $u_i\in Q(-\Delta)\cap L^2$ is used. *Any* solutions satisfying this $Q(-\Delta)$ condition will obey the conclusions of the theorems.

§ 4. Connection with the Quantum Theory

We make explicit a remark of ours in [23]:

Theorem 4.1. *Let* $E_N^Q(z_i, R_i)$ *be given by* (3) *and define*

$$E_N^{\mathrm{HF}}(z_i, R_i) \equiv \inf\{(\Psi, H\Psi) | \Psi \in \mathscr{D}_{\mathrm{phys}}; \|\Psi\| = 1; \Psi \text{ a Slater determinant}\} .$$

Let R_i, z_i *be* N *dependent in the following manner:*

a) $z_i/N \to \lambda_i$.

b) $R_1 = 0$; $R_j N^{-1/3} \to r_j$ *or to* ∞.

Then

$$\lim_{N\to\infty} E_N^{\mathrm{HF}}(z_i, R_i)/E_N^Q(z_i, R_i) = 1 .$$

Proof. $E_N^Q \leqq E_N^{\mathrm{HF}} < 0$ by the variational principle so clearly the $\overline{\lim}$ is $\leqq 1$. In our proof that TF theory is asymptotically correct (§ III of [23]) we constructed an explicit Slater determinant so that as $N \to \infty$, $(\Psi, H\Psi)/E_N^{TF} \to 1$ (where E^{TF} is the Thomas-Fermi energy). Since $E_N^Q/E_N^{TF} \to 1$ by [23], and $E_N^{\mathrm{HF}} \leqq (\Psi, H\Psi)$, the $\underline{\lim} \geqq 1$. □

Remarks. 1) As explained in [23], we expect $E_N^{\mathrm{HF}} - E_N^Q = o(N^{5/3})$ and $E_N^Q = aN^{7/3} + bN^2 + cN^{5/3} + o(N^{5/3})$. The proof of these facts seems to us to be an important problem in understanding the bulk properties of large Z atoms and molecules. All we were able to obtain rigorously is the leading term $aN^{7/3}$. In [23], a conjecture, due to Scott, is made about the next term bN^2 (see [20] for more details).

2) We emphasize that Theorem 4.1 is only a limit theorem about total binding energy. It is physically more important to prove that HF theory gives asymptotically correct ionization energies.

References

1. Adams, R.: Sobolev spaces. New York: Academic Press 1976
2. Bader, P.: Méthode variationelle pour l'équation de Hartree. E.P.F. Lausanne Thesis
3. Bazley, N., Seydel, R.: Existence and bounds for critical energies of the Hartree operator. Chem. Phys. Letters **24**, 128—132 (1974)
4. Behling, R., Bongers, A., Kuper, T.: Upper and lower bounds to critical values of the Hartree operator. University of Köln (preprint)
5. Benci, V., Fortunato, D., Zirilli, F.: Exponential decay and regularity properties of the Hartree approximation to the bound state wavefunctions of the helium atom. J. Math. Phys. **17**, 1154—1155 (1976)
6. Bethe, H., Jackiw, R.: Intermediate quantum mechanics. New York: Benjamin 1969
7. Bove, A., DaPrato, G., Fano, G.: An existence proof for the Hartree-Fock time dependent problem with bounded two-body interaction. Commun. math. Phys. **37**, 183—192 (1974)
8. Chadam, J.M., Glassey, R.T.: Global existence of solutions to the Cauchy problem for time-dependent Hartree equations. J. Math. Phys. **16**, 1122—1130 (1975)
9. Fermi, E.: Un metodo statistico per la determinazione di alcune priorietà dell atome. Rend. Acad. Nat. Lincei **6**, 602—607 (1927)
10. Fock, V.: Näherungsmethode zur Lösing des quantenmechanischen Mehrkörperproblems. Z. Phys. **61**, 126—148 (1930)
11. Fonte, G., Mignani, R., Schiffrer, G.: Solution of the Hartree-Fock equations. Commun. math. Phys. **33**, 293—304 (1973)

12. Gustafson, K., Sather, D.: Branching analysis of the Hartree equations. Rend. di Mat. **4**, 723—734 (1971)
13. Handy, N.C., Marron, M.T., Silverstom, H.J.: Long range behavior of Hartree-Fock orbitals. Phys. Rev. **180**, 45—47 (1969)
14. Hartree, D.: The wave mechanics of an atom with a non-coulomb central field. Part I. Theory and methods. Proc. Comb. Phil. Soc. **24**, 89—132 (1928)
15. Jensen, R.: Princeton University Senior Thesis, 1976
16. Kato, T.: Fundamental properties of Hamiltonian operator of Schrödinger type. Trans. Am. Math. Soc. **70**, 195—211 (1951)
17. Kato, T.: On the eigenfunctions of many particle systems in quantum mechanics. Comm. Pure Appl. Math. **10**, 151—177 (1957)
18. Kato, T.: Perturbation theory for linear operators. Berlin-Heidelberg-New York: Springer 1966
19. Lieb, E.H.: Thomas-Fermi and Hartree-Fock theory, Proc. 1974 International Congress Mathematicians, Vol. II, pp. 383—386
20. Lieb, E.H.: The stability of matter. Rev. Mod. Phys. **48**, 553 – 569 (1976)
21. Lieb, E.H.: Existence and uniqueness of minimizing solutions of Choquard's non-linear equation. Stud. Appl. Math. (in press)
22. Lieb, E.H., Simon, B.: On solutions to the Hartree Fock problem for atoms and molecules. J. Chem. Phys. **61**, 735—736 (1974)
23. Lieb, E.H., Simon, B.: The Thomas-Fermi theory of atoms, molecules, and solids. Adv. Math. **23**, 22 – 116 (1977)
24. Reed, M., Simon, B.: Methods of modern mathematical physics. II. Fourier analysis, self-adjointness. New York: Academic Press 1975
25. Reed, M., Simon, B.: Methods of modern mathematical physics. IV. Analysis of operators. New York: Academic Press 1977
26. Reeken, M.: General theorem on bifurcation and its application to the Hartree equation of the helium atom. J. Math. Phys. **11**, 2505—2512 (1970)
27. Schaefer III, H.F.: The electronic structure of atoms and molecules. Reading: Addison Wesley 1972
28. Simon, B.: On the infinitude vs. finiteness of the number of bound states of an N-body quantum system. Helv. Phys. Acta **43**, 607—630 (1970)
29. Simon, B.: Pointwise bounds on eigenfunctions and wave packets in N-body quantum systems. I. Proc. Am. Math. Soc. **42**, 395—401 (1974)
30. Slater, J.C.: A note on Hartree's method. Phys. Rev. **35**, 210—211 (1930)
31. Stein, E.: Singular integrals and differentiability properties of functions. Princeton: University Press 1970
32. Stuart, C.: Existence theory for the Hartree equation. Arch. Rat. Mech. Anal. **51**, 60—69 (1973)
33. Thomas, L.H.: The calculation of atomic fields. Proc. Comb. Phil. Soc. **23**, 542—548 (1927)
34. Wolkowisky, J.: Existence of solutions of the Hartree equations for N electrons. An application of the Schauder-Tychonoff theorem. Ind. Univ. Math. Journ. **22**, 551—558 (1972)

Communicated by J. Glimm

Received December 17, 1976

VOLUME 72, NUMBER 19 PHYSICAL REVIEW LETTERS 9 MAY 1994

There Are No Unfilled Shells in Unrestricted Hartree-Fock Theory

Volker Bach,[1,*] Elliott H. Lieb,[1,2] Michael Loss,[3] and Jan Philip Solovej[2]

[1]*Department of Physics, Jadwin Hall, Princeton University, P.O. Box 708, Princeton, New Jersey 08544*
[2]*Department of Mathematics, Fine Hall, Princeton University, Princeton, New Jersey 08544*
[3]*School of Mathematics, Georgia Institute of Technology, Atlanta, Georgia 30332*

(Received 29 July 1993)

We prove that in an exact, unrestricted Hartree-Fock calculation each energy level of the Hartree-Fock equation is either completely filled or completely empty. The only assumption needed is that the two-body interaction is—like the Coulomb interaction—repulsive; it could, however, be more complicated than a simple potential; e.g., it could have tensor forces and velocity dependence. In particular, the Hartree-Fock energy levels of atoms and molecules, often called shells, are never partially filled.

PACS numbers: 05.30.Fk, 31.10.+z, 31.20.Lr

The Hartree-Fock (HF) variational calculation provides an approximate determination of ground states and ground state energies of quantum mechanical systems such as atoms, molecules, solids, superconductors, nuclei, etc., and is widely used in physics and chemistry. Physicists often turn to it for qualitative, if not quantitative, guidance because particle correlations can sometimes be mimicked by a suitable one-body mean-field operator. Sometimes HF theory is used as the beginning of a more accurate calculational scheme, but our interest here is on qualitative features rather than numerics. The picture of quantum systems that HF theory yields is one of independent particle levels which are the eigenvalues of the HF operator; these levels are often called shells and, *a priori*, they may or may not be completely filled.

At the outset we should clarify our definition of HF theory. What we mean is the totally unrestricted theory in which one searches for the energetically very best determinantal wave function with no *a priori* assumption whatsoever on the orbitals (which are allowed to be general functions of space and spin). The orbitals are neither required to have any symmetry properties (such as having a well-defined angular momentum) nor are they required to be functions of space time functions of spin. In *restricted* HF theory, in which, for example, rotational symmetry is preserved by requiring every orbital to have a definite angular momentum, it is evident that there can be unfilled shells. It is true that the unrestricted HF ground state may break the symmetry of the Hamiltonian. There are, however, several examples where this reflects the correct physics. One such example is the Hubbard model at half filling. Indeed, in this case, the HF ground state has Neél antiferromagnetic order which would be lost if one insisted on preserving translational invariance [1].

In addition, the unrestricted HF theory yields a better estimate for the ground state energy than the restricted theory, of course, and, if conservation of symmetry is considered desirable, there is the assurance that at least one of the projections of the HF wave function onto its symmetry irreducible components (which will not be HF functions in general) will have an equal or better energy than the unrestricted energy itself. This follows from the fact that if $\psi=\sum_j\psi_j$, with ψ_j being irreducible components, then

$$0=\langle\psi|H|\psi\rangle-E_{\rm HF}\langle\psi|\psi\rangle=\sum_j\left[\langle\psi_j|H|\psi_j\rangle-E_{\rm HF}\langle\psi_j|\psi_j\rangle\right].$$

A very simple proof, given here, shows that the orbitals of the unrestricted HF ground state *fully* occupy every level of the HF operator up to the highest filled level; i.e., the degeneracy, if any, of the highest filled level of the HF operator is always exactly what is needed to accommodate the assumed number of particles. In brief, *shells are always filled in HF theory.* The idea of the proof is to assume that the level is not filled and then to use one of the remaining eigenfunctions of the HF operator to construct a new Slater determinant which has a strictly lower energy than the HF ground state. We also obtain a crude lower bound on the size of the gap, and there is no indication that this is a small number in the atomic case when it is compared to atomic energy scales.

The theorem and its proof given below obviously generalize to any system in which the two-body interaction V is repulsive, i.e., positive definite as an operator on the two-particle Hilbert space. In particular, V is allowed to be spin dependent, to contain projection operators, and to be velocity dependent. The electronic Coulomb repulsion, for example, satisfies this positivity condition. The one-body part of the Hamiltonian can be arbitrary. For convenience and because of its familiarity, we use a molecular Hamiltonian with fixed nuclei as an illustration—but only as an illustration. No symmetry is assumed to be present.

Thus, we consider a Hamiltonian

$$H=\sum_{i=1}^N\left(-\frac{\hbar^2}{2m}\Delta_i+U(\mathbf{r}_i)\right)+\tfrac{1}{2}\sum_{i\neq j}V(\mathbf{r}_i,\mathbf{r}_j)$$

acting on N-electron wave functions, i.e., wave functions $\Psi(\mathbf{r}_1,\sigma_1;\ldots;\mathbf{r}_N,\sigma_N)$ that are antisymmetric with respect to interchanging $(\mathbf{r}_i,\sigma_i)$ with $(\mathbf{r}_j,\sigma_j)$. In the example of a molecule with K nuclear charges Z_je located at posi-

0031-9007/94/72(19)/2981(3)$06.00

tions $\mathbf{R}_j$, U and V would be given by $U(\mathbf{r}) = -\sum_{j=1}^K Z_j \times e^2|\mathbf{r}-\mathbf{R}_j|^{-1}$ and $V(\mathbf{r},\mathbf{r}') = e^2|\mathbf{r}-\mathbf{r}'|^{-1}$. In the general setting, V may also depend on spin coordinates as well, i.e., $V = V(\mathbf{r},\sigma;\mathbf{r}',\sigma')$. Indeed, the operator V could even be allowed to be more general than a simple multiplication operator, but to keep our notation simple we shall not consider this case explicitly.

To obtain an approximate value for the ground state energy, $E_Q = \min\langle\Psi|H|\Psi\rangle/\langle\Psi|\Psi\rangle$, the HF calculation restricts attention to Slater determinants, i.e., wave functions of the form

$$\Phi(\mathbf{r}_1,\sigma_1;\dots;\mathbf{r}_N,\sigma_N) = (N!)^{-1/2}\operatorname{Det}\{f_i(\mathbf{r}_j,\sigma_j)\} \tag{1}$$

in which $f_1,\dots,f_N$ are orthonormal functions of space and spin: $\langle f_i|f_j\rangle = \delta_{ij}$. The approximate ground state energy is then given by the HF energy, which is defined to be

$$E_{\mathrm{HF}} = \min\langle\Phi|H|\Phi\rangle\ [\Phi \text{ has the form (1)}]. \tag{2}$$

Any minimizer, Φ_{HF}, i.e., a determinantal function satisfying $E_{\mathrm{HF}} = \langle\Phi_{\mathrm{HF}}|H|\Phi_{\mathrm{HF}}\rangle$, is a HF ground state. It may not be unique. We remark that mathematical precision actually requires an "infimum" rather than "minimum" in (2) because a HF ground state may not exist. This will be the case, e.g., for a molecule with $N > 2Z+K$, i.e., a very negative ion [2]. For neutral or positively ionized atoms and molecules, however, it was proved in [3] that a HF ground state does exist and, at least in this case, the word "minimum" in (2) is justified.

If a HF ground state does exist, it necessarily obeys the HF (or self-consistent field) eigenfunction equations

$$h_\Phi\varphi_k = \varepsilon_k\varphi_k \tag{3}$$

for all $1 \le k \le N$, where h_Φ is the one-body operator defined by its action on an arbitrary function of one space-spin variable by

$$\begin{aligned}(h_\Phi f)(\mathbf{r},\sigma) = &\left\{-\frac{\hbar^2}{2m}\Delta + U(\mathbf{r}) + \int\sum_{\tau=\pm1}\sum_{j=1}^N|\varphi_j(\mathbf{r}',\tau)|^2V(\mathbf{r},\sigma;\mathbf{r}',\tau)d^3\mathbf{r}'\right\}f(\mathbf{r},\sigma)\\ &-\sum_{\tau=\pm1}\sum_{j=1}^N\varphi_j(\mathbf{r},\sigma)\int\varphi_j(\mathbf{r}',\tau)^*f(\mathbf{r}',\tau)V(\mathbf{r},\sigma;\mathbf{r}',\tau)d^3\mathbf{r}',\end{aligned} \tag{4}$$

and where $\varphi_1,\dots,\varphi_N$ denote the special N orthonormal functions comprising the energy minimizing Slater determinant Φ_{HF}. The eigenvalues, ε_k, of h_Φ give us some insight into the possible energy levels for binding an extra electron, but that is not our concern here.

Theorem.—Assume that V is positive definite, i.e., for every nonzero function ψ of two space-spin variables,

$$\sum_{\sigma,\sigma'=\pm1}\int|\psi(\mathbf{r},\sigma;\mathbf{r}',\sigma')|^2V(\mathbf{r},\sigma;\mathbf{r}',\sigma')d^3r\,d^3r' > 0.$$

Let φ be an eigenfunction of h_Φ with eigenvalue ε (i.e., $h_\Phi\varphi = \varepsilon\varphi$) that is orthogonal to the minimizing set $\varphi_1,\dots,\varphi_N$, i.e., $\langle\varphi|\varphi_k\rangle = 0$ for all $1\le k\le N$. Then $\varepsilon > \varepsilon_k$ for all $1 \le k \le N$.

Before proving this theorem let us point out its main corollaries. First, it implies that the functions $\varphi_1,\dots,\varphi_N$ comprising Φ_{HF} occupy the N *lowest* energy levels of h_Φ, which is a fact that was also noted in [3] and [4]; the reader may or may not find this surprising, but we point out that there is no proof of this assertion without the assumption that $V \ge 0$. *Our main point, however, is the second implication, which is more surprising: Φ_{HF} does not leave any degenerate level unfilled. There is a gap because $\varepsilon > \varepsilon_k$ for all $k = 1,2,\dots,N$.*

Proof of the theorem—For notational convenience assume $\varepsilon_1 \le \varepsilon_2 \le \cdots \le \varepsilon_N$. We shall derive a contradiction to the assumption that $\varepsilon \le \varepsilon_N$. First, we introduce some more notation. Denote ε by ε_{N+1} and φ by φ_{N+1}. Further for all $1 \le k \le N+1$ and $1 \le l \le N+1$, we define

$$h_k = \langle\varphi_k| - (\hbar^2/2m)\Delta - U(\mathbf{r})|\varphi_k\rangle$$

and

$$V_{k,l} = \sum_{\sigma,\sigma'=\pm1}\int\tfrac12|\varphi_k(\mathbf{r},\sigma)\varphi_l(\mathbf{r}',\sigma') - \varphi_l(\mathbf{r},\sigma)\varphi_k(\mathbf{r}',\sigma')|^2 \times V(\mathbf{r},\sigma;\mathbf{r}',\sigma')d^3r\,d^3r'.$$

Notice that $V_{k,k} = 0$ and $V_{l,k} = V_{k,l} > 0$ if $k \ne l$ since V is positive definite.

Now let $\bar\Phi$ be the Slater determinant built from $\varphi_1,\dots,\varphi_{N-1},\varphi_{N+1}$, as in (1). One easily checks that

$$\langle\Phi_{\mathrm{HF}}|H|\Phi_{\mathrm{HF}}\rangle = \sum_{k=1}^N h_k + \tfrac12\sum_{k,l=1}^N V_{k,l},$$

$$\langle\bar\Phi|H|\bar\Phi\rangle = \sum_{k=1}^{N-1} h_k + \tfrac12\sum_{k,l=1}^{N-1}V_{k,l} + h_{N+1} + \sum_{l=1}^{N-1}V_{l,N+1},$$

and, for $1 \le k \le N+1$,

$$h_k + \sum_{l=1}^N V_{k,l} = \langle\varphi_k|h_\Phi|\varphi_k\rangle = \varepsilon_k. \tag{5}$$

Notice that the term $l = k$ in the sum in (5) does not contribute since $V_{k,k} = 0$.

Now

$$\begin{aligned}\langle\bar\Phi|H|\bar\Phi\rangle &= \langle\Phi_{\mathrm{HF}}|H|\Phi_{\mathrm{HF}}\rangle + h_{N+1} - h_N + \sum_{l=1}^{N-1}(V_{l,N+1} - V_{l,N})\\ &= \langle\Phi_{\mathrm{HF}}|H|\Phi_{\mathrm{HF}}\rangle + \varepsilon_{N+1} - \varepsilon_N - V_{N,N+1}\\ &\le \langle\Phi_{\mathrm{HF}}|H|\Phi_{\mathrm{HF}}\rangle - V_{N,N+1}.\end{aligned}$$

The last inequality uses the assumption $\varepsilon_{N+1} \le \varepsilon_N$, but we then have a contradiction since $\tilde{\Phi}$ has an energy which is below the Hartree-Fock energy by the amount $V_{N,N+1} > 0$. QED.

The proof does not give a rigorous numerical estimate of the gap $\varepsilon_{N+1} - \varepsilon_N$, but it does show that the gap is at least $V_{N,N+1}$, which is usually not a tiny quantity for small systems. For such systems, even an "approximate" degeneracy is unlikely.

This work was supported by the U.S. National Science Foundation through the following grants: PHY90-19433 A03 (V.B. and E.H.L.), DMS92-07703 (M.L.), and DMS92-03829 (J.P.S.).

Current address: FB Mathematik MA 7-2, Technische Universität Berlin, Strasse des 17 Juni 136, D-W-1000 Berlin, Germany.

[1] V. Bach, E. H. Lieb, and J. P. Solovej, "Generalized Hartree-Fock Theory and the Hubbard Model," J. Stat. Phys. (to be published).

[2] E. H. Lieb, Phys. Rev. A **29**, 3018 (1984).

[3] E. H. Lieb and B. Simon, Commun. Math. Phys. **53**, 185 (1977); see also E. H. Lieb, in *Proceedings of the 1974 International Congress of Mathematicians* (Canadian Mathematical Congress, 1975), Vol. 2, p. 383.

[4] O. A. Pankratov and P. P. Poparov, Phys. Lett. A **134**, 339 (1989).

With R. Benguria in Ann. Phys. (N.Y.) *110*, 34–45 (1978)

Many-Body Atomic Potentials in Thomas–Fermi Theory

RAFAEL BENGURIA*,†

Department of Physics, Princeton University, Princeton, New Jersey 08540

AND

ELLIOTT H. LIEB†

Departments of Mathematics and Physics, Princeton University, Princeton, New Jersey 08540

Received May 6, 1977

Many-body atomic potentials, ϵ, are functions of the nuclear coordinates, and are defined by differences of ground state energies, E, e.g., $\epsilon(1, 2) \equiv E(1, 2) - E(1) - E(2)$. We prove that in Thomas–Fermi theory the n-body potential always has the sign $(-1)^n$ for all coordinates. We also prove that the remainder in the expansion of the total energy E in terms of the ϵ's, when truncated at the n-body terms, has the sign $(-1)^{n+1}$.

1. INTRODUCTION

Many-body potentials are a useful concept to describe the interaction of atoms. These quantities, which we denote by ϵ, are defined in terms of ground state energies, denoted by E, as follows:

To each positive integer j we associate a fixed nucleus of charge $z_j > 0$ located at R_j. The z_j's can all be different. $E(j_1, ..., j_n)$ denotes the ground state energy of an isolated molecule composed of n nuclei $j_1, ..., j_n$, the electron–electron and nuclear–nuclear Coulomb repulsion being included. The molecule is assumed to be neutral so that the number of electrons is the sum of the z_j. Then the one-, two-, and three-body potentials are defined to be

$$\begin{aligned} &\text{one body:} & \epsilon(j) &= E(j), \\ &\text{two body:} & \epsilon(j, k) &= E(j, k) - [E(j) + E(k)], \\ &\text{three body:} & \epsilon(j, k, l) &= E(j, k, l) - [E(j, k) + E(j, l) + (k, l)] \\ &&&\quad + [E(j) + E(k) + E(l)], \end{aligned} \tag{1.1a}$$

* On leave from the Department of Physics, Universidad de Chile, Santiago, Chile.

† Work partially supported by U.S. National Science Foundation grant MCS 75-21684 A01.

and so forth. The inverse of (1.1a) is

$$\begin{aligned} E(j) &= \epsilon(j), \\ E(j,k) &= [\epsilon(j)+\epsilon(k)]+\epsilon(j,k), \\ E(j,k,l) &= [\epsilon(j)+\epsilon(k)+\epsilon(l)]+[\epsilon(j,k)+\epsilon(j,l)+\epsilon(k,l)]+\epsilon(j,k,l). \end{aligned} \tag{1.1b}$$

The generalization of (1.1) is given in (1.8) and (1.9). Note that the n-body potential $\epsilon(j_1,..., j_n)$ depends *only* on the charges and coordinates of nuclei $j_1,..., j_n$. Equation (1.1) should not be confused with a virial expansion. It will be our goal to show that in Thomas–Fermi (TF) theory, the sign of the n-body ϵ is always $(-1)^n$; moreover, if the series (1.1b) is truncated at the n-body terms (usually only the two-body terms are retained in practice) then the remainder has the sign of the first omitted terms, namely $(-1)^{n+1}$.

It is, of course, very difficult to calculate the E's in the Schrödinger theory, especially if non-Coulombic effects are taken into account. With only pure Coulomb forces, even the two-body energy $E(1, 2)$ is difficult to estimate, except for small z. In fact the sign of the ϵ's is unknown. One of the few things that can be reliably stated is that the two-body $\epsilon(1, 2)$ goes to $+\infty$ like $z_1z_2 \mid R_1 - R_2\mid^{-1}$ as $R_1 \to R_2$; the other ϵ's are bounded functions of the R_j.

It is an interesting, and perhaps useful, fact that if the Schrödinger energies are replaced by Thomas–Fermi energies, then the sign of the ϵ's and the sign of the remainder in the expansion of E in terms of ϵ can be determined. These facts are Theorems 1 and 2 of this paper. The TF theory is an approximation to the Schrödinger theory and it is asymptotically exact [1] in the limit $z \to \infty$; therefore these theorems may not be entirely irrelevant. TF theory also plays a role in a recent proof of the stability of matter [2]. See also note added in proof.

Before proceeding we note that knowledge of the ε's (or E's) is, strictly speaking, insufficient to determine all quantities of physical interest. First, real nuclei are not static (i.e., infinitely massive) and the nuclear kinetic energy cannot be strictly separated from the electronic energy. Second, the electrons will not be in their ground state except at zero temperature. A refinement would be to use the free energy, F, in place of the ground state energy, E (either in Schrödinger or TF theory). The difficulty with this is that $F = -\infty$ unless the electrons are confined to a box (for the ground state no such confinement is necessary because the ground state neutral atoms are always bound states). Thus $\varepsilon(1, 2)$, for example, would depend upon the size of the box (i.e., the density) and not merely upon nuclei 1 and 2. Third, one may wish to consider atoms that are not neutral, but then $\varepsilon(1, 2. 3)$, for example, would not be defined by (1, 1) until a prescription is given for apportioning the electrons among the subsystems (1.2), (1, 3), and (2, 3). These remarks notwithstanding, the many-body potentials, as we have defined them, are useful for a wide variety of applications. The above-mentioned difficulties will not be considered further in this paper.

The TF energy for nuclei of charges $z_i > 0$ (which need not be integral) located at

R_i, $i = 1,\ldots, n$ is defined as follows. In units in which $h^2(8m)^{-1}(3/\pi)^{2/3} = 1$ (m = electron mass) and $|e| = 1$, one introduces the energy functional

$$\mathcal{E}(\rho) = \tfrac{3}{5}\int \rho(x)^{5/3}\, d^3x - \int V(x)\,\rho(x)\, d^3x + \tfrac{1}{2}\iint \rho(x)\,\rho(y)\,|x-y|^{-1}\, d^3x\, d^3y + \sum_{1\leqslant i<j\leqslant n} z_i z_j\,|R_i - R_j|^{-1}. \quad (1.2)$$

Here $\rho(x) \geqslant 0$ is the electron density, and

$$V(x) = \sum_{j=1}^{n} z_j\,|x - R_j|^{-1}. \quad (1.3)$$

The minimum of $\mathcal{E}(\rho)$ over *all* nonnegative functions ρ occurs for a unique ρ [1]. This ρ has the property

$$\int \rho(x)\, d^3x = \sum_{j=1}^{n} z_j \equiv Z \quad (1.4)$$

and satisfies

$$\rho(x) = \phi(x)^{3/2}, \quad (1.5a)$$

$$0 \leqslant \phi(x) = V(x) - \int \rho(y)\,|x-y|^{-1}\, d^3y. \quad (1.5b)$$

Equation (1.4) states that a minimum energy TF molecule is always neutral. Equation (1.5) is the TF equation; its solution is unique given that $\int\rho$ and $\int\rho^{5/3}$ are finite. $\mathcal{E}(\rho)$ does have a unique minimizing ρ under the additional restriction that $\int\rho = \lambda$ and $\lambda < Z$, but this ρ does not satisfy (1.5). There is no minimizing ρ under the condition $\lambda > Z$, i.e., negative molecules do not exist in TF theory. Finally, the TF energy is given by

$$E(1,\ldots, n) = \min_{\rho} \mathcal{E}(\rho) = \mathcal{E}(\bar\rho) \quad (1.6)$$

where $\bar\rho$ is the unique solution to (1.5).

The energy of an isolated TF atom is

$$E(1) = -Kz_1^{7/3} \quad (1.7)$$

where $K = 3.678$ by numerical computation.

Our theorems do not depend on the fact that the nuclei are point charges. They can be smeared out in any way provided the nuclear charge *densities* are nonnegative. This generalization requires an obvious redefinition of (1.2) and (1.3), but (1.4), (1.5), and Theorems 1, 2, 3, 4 still hold.

To state the theorems we must introduce some set theoretic notation. If $b = (b_1, b_2,\ldots, b_k)$ is a finite subset of the positive integers, $E(b)$ denotes the TF energy (given by (1.2) and (1.6)) for the k nuclei of charges $z_{b_1},\ldots, z_{b_k}$ located at $R_{b_1},\ldots, R_{b_k}$. $|b| \equiv k$ is the cardinality of b. Φ denotes the empty set, and we define $E(\Phi) = 0$. $b \subset c$ means that b is a proper subset of c, while $b \subseteq c$ means that b is a subset of c and can be c itself. $c \backslash b$ is the set of points which are in c and not in b. The generalization of (1.1) is

$$\epsilon(c) = \sum_{b \subseteq c} (-1)^{|b|+|c|} E(b) \tag{1.8}$$

from which it follows that

$$E(c) = \sum_{b \subseteq c} \epsilon(b) \tag{1.9}$$

since, if $a \subseteq c$,

$$\sum_{a \subseteq b \subseteq c} (-1)^{|b|} = (-1)^{|c|} \delta(a, c) \tag{1.10}$$

with δ being the Kronecker delta.

THEOREM 1 (Sign of the many-body potentials). *For all choices of the nuclear z's and R's and all sets, c,*

$$(-1)^{|c|} \epsilon(c) \geqslant 0. \tag{1.11a}$$

More generally, if $b \subset c$ then

$$\tilde{E}(b, c) \equiv \sum_{b \subseteq a \subseteq c} (-1)^{|b|+|a|} E(a) \geqslant 0 \tag{1.11b}$$

whenever $|c \backslash b| \geqslant 2$ or $b = \Phi$ (this latter case is (1.11a)).

Remarks. (i) This can be extended to $(-1)^{|c|} \epsilon(c) > 0$, as in Ref. [1, Section V.2], when c is nonempty.

(ii) The special case $\epsilon(1, 2) > 0$ is Teller's theorem (cf. Ref. [1]).

THEOREM 2 (Remainder Theorem). *For all nuclear z's and R's, all sets c, and all $2 \leqslant \gamma \leqslant |c|$, the sign of*

$$E(c) - \sum_{\substack{b \subset c \\ |b| < \gamma}} \epsilon(b)$$

is $(-1)^\gamma$. In other words, if, in (1.9), we sum over all smaller than γ-body terms, then the remainder has the same sign as the first omitted terms.

THEOREM 3 (Monotonicity of the many-body potentials). *Suppose that $b \subset c$ and $|b| \geqslant 2$. Then for all nuclear z's and R's*

$$(-1)^{|b|} \epsilon(b) \geqslant (-1)^{|c|} \epsilon(c).$$

Remark. Again, the inequality can be shown to be strict. As an application of Theorems 1 and 3, the three-body potential satisfies

$$0 \geqslant \epsilon(1, 2, 3) \geqslant -\min[\epsilon(1, 2), \epsilon(1, 3), \epsilon(2, 3)].$$

A more general theorem which, as we shall see, implies the others, concerns the TF potential itself.

DEFINITION. Let c be a nonempty subset of the positive integers. $\phi(c, x)$ denotes the TF potential, given by (1.5b), at the point x for the corresponding collection of nuclei. If $c = \Phi$, $\phi(\Phi, x) \equiv 0$.

THEOREM 4. *Suppose $b \subset c$. Then for all z's and R's and all x*

$$\tilde{\phi}(b, c, x) \equiv \sum_{b \subseteq a \subseteq c} (-1)^{|a|+|b|} \phi(a, x) \leqslant 0. \tag{1.12}$$

Remark. As in Ref [1, Section V.2], strict inequality can be proved here, when c is nonempty.

Theorem 4 will be proved in the next section. The proof we give is patterned after the proof of Teller's lemma [1], but an additional combinatorial lemma is needed. All the necessary combinatorial facts are given in Section 3, and the required lemma is Lemma 13. It is important that the exponent p in (1.5a), namely $\frac{3}{2}$, satisfies $1 \leqslant p \leqslant 2$. Lemma 13 holds for $1 \leqslant p \leqslant 2$ [which would correspond to replacing $\int \rho^{5/3}$ by $\int \rho^k$, $\frac{3}{2} \leqslant k \leqslant 2$ in $\mathscr{E}(\rho)$], but Lemma 13 is false for $p < 1$ or $p > 2$. It is amusing that $p = 2$ corresponds to $k = \frac{3}{2}$, and TF theory does not exist for $k \leqslant \frac{3}{2}$ (because $\mathscr{E}(\rho)$ is then not bounded below).

We conclude this section with the mention of a basic fact [1] that relates the TF potential to the TF energy. Theorem 4 will follow from Lemmas 5 and 13.

LEMMA 5. *Let c be nonempty and $j \in c$. Then the derivative of $E(c)$ with respect to the nuclear charge z_j is given by*

$$\partial E(c)/\partial z_j = \lim_{x \to R_j} \{\phi(c, x) - z_j \mid x - R_j \mid^{-1}\}.$$

This lemma is proved [1] by differentiating (1.2) with respect to z_j and using the fact that $\mathscr{E}(\rho)$ is stationary with respect to the minimizing ρ.

2. PROOFS OF THEOREMS 1, 2, 3, AND 4

In this section we give the proof of the theorems stated in the Introduction. We begin with the general Theorem 4 about the TF potential. The other theorems are a consequence of this one. The proof of Theorem 4 is based on the proof of Teller's lemma (see Theorem V.5, Ref. [1]) and the combinatorial Lemma 13 proved in Section 3. Logically, Section 3 should be inserted at this point, but the only things needed are (i) the definition of the transform (3.1); (ii) the definition of an anticanonical function, (iii) Lemma 13.

Proof of Theorem 4. Insert (1.5b) into the right side of (1.12). If $j \in c$, it is easy to check, using (1.10), that all terms of the type $z_j \mid x - R_j \mid^{-1}$ either cancel (if $c \neq$

$b \cup \{j\}$) or have a coefficient -1 (if $c = b \cup \{j\}$). The (distributional) Laplacian of $\tilde{\phi}$ therefore satisfies

$$(4\pi)^{-1} \Delta\tilde{\phi}(b, c, x) \geqslant \sum_{b \subseteq a \subseteq c} (-1)^{|a|+|b|} \phi^{3/2}(a, x)$$

$$= \widetilde{\phi^{3/2}}(b, c, x). \tag{2.1}$$

Let

$$D(b, c) = \{x \mid \tilde{\phi}(b, c, x) > 0\}.$$

Suppose we can show that

$$x \in D(b, c) \quad \text{implies} \quad \widetilde{\phi^{3/2}}(b, c, x) \geqslant 0. \tag{2.2}$$

Then $\tilde{\phi}(b, c, \cdot)$ is subharmonic on $D(b, c)$ and therefore takes its maximum on the boundary of $D(b, c)$ or at infinity. If $|c \backslash b| > 1$, $\tilde{\phi}$ is continuous on all of $\mathbb{R}^3$; if $c = b \cup \{j\}$, $\tilde{\phi}$ is continuous on $\mathbb{R}^3 \backslash R_j$ and $\tilde{\phi}(b, c, x) \to -\infty$ as $x \to R_j$ [1]. Hence $\tilde{\phi}$ is zero on the boundary of D. Since $\tilde{\phi}$ goes to zero as $x \to \infty$ [1], we conclude that D is empty. This potential theoretic argument is the main idea in the proof of Teller's lemma [1]. Thus, the problem is reduced to proving assertion (2.2). To do so we use induction on the cardinality of $c \backslash b$. Let $P(n)$, for $n \geqslant 1$, be the proposition: Theorem 4 is true when $1 \leqslant |c \backslash b| \leqslant n$, i.e., $\tilde{\phi}(b, c, x) \leqslant 0$ for all x. If $n = 1$ then $\tilde{\phi}(b, c, x) = -\phi(c, x) + \phi(b, x)$, and (2.2) is trivially verified. Hence $P(1)$ is true. (Note: This case is just Teller's original lemma.) $P(n+1)$ will be proved by showing that $P(n)$ implies (2.2). $D(b, c)$ is empty when $b \neq \Phi$. For $x \in D(\Phi, c)$ fixed, consider $d \to \phi(d, x)$ as a positive function on the power set 2^c. In the terminology of Section 3, $\phi(\cdot, x)$ is anticanonical. Then Lemma 13 is precisely (2.2). ∎

Proof of Theorem 1 (sign *of the many-body potentials*). Let $j \in c \backslash b$, and consider the dependence of $\tilde{E}(b, c)$ on z_j. When $z_j = 0$, $E(a) = E(a \backslash \{j\})$ if $j \in a$, and it is then easy to check that

$$\tilde{E}(b, c; z_j = 0) = 0. \tag{2.3}$$

By Lemma 5

$$\partial\tilde{E}(b, c)/\partial z_j = \lim_{x \to R_j} \sum_{b \cup \{j\} \subseteq a \subseteq c} (-1)^{|b|+|a|} \{\phi(a, x) - z_j |x - R_j|^{-1}\}. \tag{2.4}$$

Using (1.10) one sees that the terms on the right side of (2.4) proportional to $z_j |x - R_j|^{-1}$ either cancel (if $|c \backslash b| \geqslant 2$) or else have a coefficient $+1$ (if $c = b \cup \{j\}$). The terms involving $\phi(a, x)$ are, by definition, $-\tilde{\phi}(b \cup \{j\}, c, x)$. If $|c \backslash b| \geqslant 2$, the right side of (2.4) is then $-\lim_{x \to R_j} \tilde{\phi}(b \cup \{j\}, c, x)$ and this is nonnegative by Theorem 4. If $b = \Phi$ and $c = \{j\}$, the right side of (2.4) is $\lim_{x \to R_j} \{z_j |x - R_j|^{-1} - \phi(\{j\}, x)\}$ and this is positive by (1.5b). In either case. therefore,

$$\partial\tilde{E}(b, c)/\partial z_j \geqslant 0. \tag{2.5}$$

Equations (2.3) and (2.5) prove the theorem. ∎

Proof of Theorem 3 (*monotonicity of the many-body potentials*). This is really a simple corollary of Theorem 1. Obviously it is sufficient to consider the case $c = b \cup \{j\}$ and $j \notin b$. Let $a = \{j\}$, whence $|c \backslash a| = |b| \geqslant 2$. By Theorem 1, $\tilde{E}(a, c) \geqslant 0$. Inserting (1.9) into the definition of $\tilde{E}(a, c)$, and interchanging the summation order, we have that

$$\begin{aligned} 0 \leqslant \tilde{E}(a, c) &= \sum_{h \subseteq c} \Bigg(\sum_{a \cup h \subseteq d \subseteq c} (-1)^{|d|} \Bigg) (-1)^{|a|} \epsilon(h) \\ &= \sum_{h \subseteq c} \delta(a \cup h, c)(-1)^{|c|+|a|} \epsilon(h) \\ &= \sum_{c \backslash a \subseteq h \subseteq c} (-1)^{|c|+|a|} \epsilon(h) \\ &= (-1)^{|b|} \epsilon(b) - (-1)^{|c|} \epsilon(c). \quad \blacksquare \end{aligned} \tag{2.6}$$

Proof of Theorem 2 (*remainder theorem*). For $b \subseteq c$ and $\gamma \geqslant 0$ introduce

$$l_\gamma(b, c) = (-1)^{\gamma+|b|} \sum_{\substack{b \subseteq f \subseteq c \\ |f \backslash b| \geqslant \gamma}} \epsilon(f). \tag{2.7}$$

We want to show that $l_\gamma(\Phi, c) \geqslant 0$ for $\gamma \geqslant 2$. First consider l_0. In general,

$$l_0(b, c) = \tilde{E}(c \backslash b, c) \tag{2.8}$$

by the calculation in (2.6) (which holds for *all* $a \subseteq c$). Since $c \backslash (c \backslash b) = b$, we conclude (from Theorem 1) that $l_0(b, c) \geqslant 0$ if either $c = b$ or $|b| \geqslant 2$. Now it is easy to check that for $\gamma \geqslant 1$ and $j \in c \backslash b$:

$$l_\gamma(b, c) = l_\gamma(b, c \backslash \{j\}) + l_{\gamma-1}(b \cup \{j\}, c). \tag{2.9}$$

When $\gamma \geqslant 1$, $l_\gamma(c, c) = 0$ (all c) by (2.7). If $\gamma = 1$ and $b \neq \Phi$, a simple induction on $|c \backslash b|$ using (2.9) and the fact that $|b \cup \{j\}| \geqslant 2$, shows that $l_1(b, c) \geqslant 0$. Now suppose $\gamma \geqslant 2$ (all b, c). Using (2.8) and induction on γ followed by induction on $|c \backslash b|$ together with the fact that $b \cup \{j\} \neq \Phi$ in (2.9), one has that $l_\gamma(b, c) \geqslant 0$ for all b, c when $\gamma \geqslant 2$. ∎

3. Functions Defined On The Power Set Of A Given Set

In this section we present some general properties of positive real functions defined on the power set of a given set.

Definition. Consider a fixed, nonempty, finite set S and its power set 2^S, that is the set of all subsets of S (including the empty set Φ). To every function f from 2^S to the complex numbers $\mathbb{C}$ we associate the function ("transform of f") $\tilde{f}$: $2^s \times 2^s \to \mathbb{C}$, defined by

$$\tilde{f}(a, b) = \sum_{a \subseteq d \subseteq b} (-1)^{|a|+|d|} f(d) \tag{3.1}$$

when $a \subseteq b \subseteq S$. If $a \not\subseteq b$ $\tilde{f}$ is defined to be zero. Note that $\tilde{f}(a, a) = f(a)$.

The following lemma provides a convolution formula for the "transforms"

LEMMA 6. *Let f, g be functions on 2^s. Then*

$$\widetilde{f \cdot g}(a, b) = \sum_{a \subseteq d \subseteq b} \tilde{f}(a, d)\, \tilde{g}(d, b). \tag{3.2}$$

Remark. Although the left side of (3.2) is invariant under the interchange of f and g, the right side is not manifestly invariant.

Proof. Inserting definition (3.1) into the RHS of (3.2) and interchanging the sum orders we get

$$\begin{aligned} \text{RHS (3.2)} &= \sum_{a \subseteq d \subseteq b} \Big(\sum_{a \subseteq e \subseteq d} (-1)^{|a|+|e|} f(e) \Big)\Big(\sum_{d \subseteq h \subseteq b} (-1)^{|d|+|h|} g(h) \Big) \\ &= \sum_{a \subseteq e \subseteq b} \sum_{e \subseteq h \subseteq b} \Big(\sum_{e \subseteq d \subseteq h} (-1)^{|d|} \Big) (-1)^{|a|+|e|+|h|} f(e)\, g(h) \\ &= \sum_{a \subseteq e \subseteq b} \sum_{e \subseteq h \subseteq b} (-1)^{|a|+|e|}\, \delta(e, h)\, f(e)\, g(h) \\ &= \sum_{a \subseteq e \subseteq b} (-1)^{|a|+|e|} f(e)\, g(e) = \widetilde{f \cdot g}(a, b). \end{aligned}$$

In the third equality we have used Eq. (1.10).

DEFINITION. $P(S)$ denotes the nonnegative functions f: $2^s \to \mathbb{R}^+$. If $f \in P(S)$, note that $\tilde{f}(a, a) \geqslant 0$. We call $f \in P(S)$ *normal* if $\tilde{f}(a, b) \leqslant 0$ for every $a \subset b \subseteq S$ such that $b \backslash a \neq S$. f is *canonic* (resp. *anticanonic*) if f is normal and $\tilde{f}(\Phi, S) \leqslant 0$ (resp. if $\tilde{f}(\Phi, S) \geqslant 0$).

We will now prove some properties of these special functions.

LEMMA 7. *Suppose that $f \in P(S)$ and either* (i) *f is canonic and $|S| \geqslant 1$ or* (ii) *f is normal and $|S| \geqslant 2$. Then $a \subset b \subseteq S$ implies* that $f(a) \leqslant f(b)$.

Proof. It is sufficient to consider $a \subset S$ and $b = a \cup \{x\}$ with $x \in S$, $x \notin a$. Then, in both cases,

$$0 \geqslant \tilde{f}(a, b) = f(a) - f(b). \quad \blacksquare$$

LEMMA 8. *If $f \in P(S)$ is normal* (*resp. canonic*) *then $f^{1/2}$ is normal* (*resp. canonic*).

Proof. By hypothesis, $\tilde{f}(a, b) \leqslant 0$ whenever $a \subset b \subseteq S$ and $|b \backslash a| < |S|$. We want to show that under the same conditions $\widetilde{f^{1/2}}(a, b) \leqslant 0$. The proof is by induction on the cardinality of $|b \backslash a|$. Consider the proposition $P(n)$ for $n \leqslant |S|$:

$$\widetilde{f^{1/2}}(a, b) \leqslant 0 \qquad \text{for} \quad a \subset b \subseteq S, \quad |b \setminus a| \leqslant n.$$

$P(1)$ is true by inspection. We will show $P(n)$ implies $P(n+1)$ when $(n+1) < |S|$. The convolution formula (3.2) can be written as

$$\tilde{f}(a, b) = \sum_{a \subset k \subset b} \widetilde{f^{1/2}}(a, k)\, \widetilde{f^{1/2}}(k, b) + \widetilde{f^{1/2}}(a, b)[\widetilde{f^{1/2}}(a, a) + \widetilde{f^{1/2}}(b, b)]. \quad (3.3)$$

Assume $|b \backslash a| = n+1$. The sum appearing in Eq. (3.3) is nonnegative by $P(n)$. Moreover $\tilde{f}(a, b) \leqslant 0$ (if $n+1 < |S|$). Therefore $\widetilde{f^{1/2}}(a, b) \leqslant 0$ if $f(a) + f(b) > 0$; otherwise let $f(c) \to f(c) + x$ and use continuity in x. Finally, (3.3) implies $P(|S|)$ if f is canonic. ▮

The following is an immediate consequence of Lemma 8.

COROLLARY 9. *If f is normal (resp. canonic), then f^p with $p = 2^{-k}$, k a positive integer, is normal (resp. canonic).*

Proof. By Lemma 8 and induction on k. ▮

Remarks. (i) Note that if f is anticanonic the only thing we can say about $f^{1/2}$ is that it is normal.

(ii) Our goal is to extend Lemma 8 to f^p, $p \in [0, 1]$ and, indeed to any positive Pick function. This is Lemma 12. Lemma 8 is not needed for the proof of Lemma 12, but we presented it for two reasons: (a) the case $p = \frac{1}{2}$ is what is needed for TF theory: (b) the proof just given for $p = \frac{1}{2}$ (and hence $p = 2^{-k}$) is simpler than the proof of Lemma 12.

LEMMA 10. *If f is normal and $f(a) > 0$, all $a \subseteq S$ then $\widetilde{f^{-1}}(a, b) > 0$ for $a \subseteq b \subseteq S$, and $b \backslash a \neq S$. If f is also canonic, then $\widetilde{f^{-1}}(a, b) \geqslant 0$ for $a \subseteq b \subseteq S$.*

Proof. Again we use induction on the cardinality of $b \backslash a$ and the convolution Eq. (3.2).

$$\delta(a, b) = \tilde{1}(a, b) = \sum_{a \subset d \subset b} \tilde{f}(a, d)\, \widetilde{f^{-1}}(d, b) + \tilde{f}(a, a)\, \widetilde{f^{-1}}(a, b) + \tilde{f}(a, b)\, \widetilde{f^{-1}}(b, b).$$

The proof is now a straightforward imitation of the proof of Lemma 8. ▮

Consider $f \in P(S)$ and its transform $\tilde{f}$. If $x \in P(S)$ is a constant mapping (i.e., $x(a) = x \geqslant 0$, all $a \subseteq S$) then

$$\widetilde{(f + x)}(a, b) = \tilde{f}(a, b), \qquad a \subset b \subseteq S \quad (3.4)$$

as can be easily checked using (1.10) and the definition of the transform (3.1). Therefore if f is respectively canonic, anticanonic, or normal, so is $f + x$.

LEMMA 11. *Let $f \in P(S)$ be normal (resp. canonic). Let $x > 0$. Then the function g given by $g(a) = f(a)[f(a) + x]^{-1}$ is normal (resp. canonic).*

Proof. We have to show that $\tilde{g}(a, b) \leqslant 0$ for $a \subset b \subseteq S$ (with the restriction $b \backslash a \neq S$ if f is only normal).

$$\tilde{g}(a, b) = \tilde{1}(a, b) - \widetilde{x(f + x)^{-1}}\,(a, b).$$

The lemma is proved by using (3.4), Lemma 10, and $\tilde{1}(a, b) = \delta(a, b)$. ▮

Finally we state the generalization of Lemma 8.

LEMMA 12. *If f is canonic (resp. normal), then f^p is canonic (resp. normal) for $0 \leqslant p \leqslant 1$.*

Proof. If $p = 0$ or 1 the proof is trivial. For $0 < p < 1$, use the representation

$$f^p = K_p \int_0^\infty dx\, x^{p-1} f(f + x)^{-1} \tag{3.5}$$

(with $K_p = \pi^{-1} \sin(p\pi)$, $0 < p < 1$). Taking the transform on both sides of (3.6) the lemma follows from Lemma 11. ▮

Remark. Lemma 12 obviously remains true if f^p is replaced by $h(f)$, where h: $\mathbb{R}^+ \to \mathbb{R}^+$ has the representation

$$h(x) = a + bx + \int_0^\infty x(x + y)^{-1}\, d\mu(y)$$

with $a, b \geqslant 0$ and μ a positive Borel measure on $[0, \infty]$. Such functions are Pick (or Herglotz) functions [3] on $\mathbb{R}^+$.

With the help of the previous lemmas we can now prove the following theorem which is needed in the previous section.

LEMMA 13. *If f is anticanonic and $0 \leqslant p \leqslant 1$ then $\widetilde{f^{1+p}}(\Phi, S) \geqslant 0$.*

Proof. If $|S| = 1$ the proof is by inspection. If $|S| \geqslant 2$ let us define g by

$$f(a) = g(a) + f(\Phi).$$

Then $g \in P(S)$, by Lemma 7 and, moreover, $g(\Phi) = 0$. We call $f(\Phi) = x_0$ $(x_0 \geqslant 0)$. Define the function $g_x = g + x$ on 2^s. By (3.4), g_x is anticanonic for all $x \geqslant 0$. Consider h: $\mathbb{R}^+ \to \mathbb{R}$ defined by

$$h(x) = \widetilde{g_x^{1+p}}(\Phi, S). \tag{3.6}$$

Since $f = g_x$ for $x = x_0$, we want to show that $h(x) \geqslant 0$ for every nonnegative x. From (3.6)

$$h'(x) = (1 + p)\, \widetilde{g_x^p}(\Phi, S). \tag{3.7}$$

Again using the convolution Eq. (3.2) we have that

$$h(x) = \sum_{\Phi \subset a \subset S} \tilde{g}_x(\Phi, a)\, \widetilde{g_x^p}'(a, S) + (1 + p)^{-1}\, h'(x)\, \tilde{g}_x(\Phi, \Phi) + \tilde{g}_x(\Phi, S)\, \widetilde{g_x^p}(S,S). \tag{3.8}$$

Note that $\tilde{g}_x(\Phi, \Phi)$ and $\tilde{g}_x(S, S)$ are nonnegative. Furthermore, the sum in (3.8) is nonnegative by Lemma 12. If $p = 1$ then $(1 + p)^{-1}\, h'(x) = \tilde{g}_x(\Phi, S)$ and the two last terms in (3.8) are $\tilde{g}_x(\Phi, S)[g_x(\Phi, \Phi) + \tilde{g}_x(S, S)] \geqslant 0$ for all x. This proves the lemma when $p = 1$, and henceforth we assume $p < 1$. Now

$$h(0) = \sum_{\Phi \subset a \subset S} \tilde{g}(\Phi, a)\, \widetilde{g^p}(a, S) + \tilde{g}(\Phi, S)\, \widetilde{g^p}(S, S)$$

which is positive because g is anticanonic (and therefore normal). As $x \to \infty$, $g_x^{1+p} \sim x^{1+p} + (1 + p)\, x^p g + o(1)$. Hence, for large x,

$$h(x) \sim (1 + p)\, x^p \tilde{g}(\Phi, S) \geqslant 0$$

because g is anticanonic. $h(x)$ is a continuously differentiable function of x on $(0, \infty)$, $h(0) \geqslant 0$ and, as $x \to \infty$, either $h(x) \to 0$ or $h(x) \to +\infty$. Therefore, either $h(x) \geqslant 0$ for every positive x or there must exist $y > 0$ such that $h(y) < 0$ and $h'(y) = 0$. But $h'(y) = 0$ implies $h(y) > 0$ (by Eq. (3.8)). $\blacksquare$

Remarks.   (i) If p is outside the interval $[0, 1]$ and $|S| \geqslant 3$ the statement given in Lemma 13 is definitely false. Consider the example

$$S = \{1, 2, 3\}$$

$$f(1, 2, 3) = 1, \qquad\qquad f(\Phi) = 0,$$

$$f(1, 2) = f(1, 3) = \tfrac{3}{4}, \qquad f(2, 3) = \tfrac{1}{2},$$

$$f(1) = \tfrac{1}{2}, \qquad\qquad f(2) = f(3) = \tfrac{1}{4}.$$

Here f is anticanonic but $\widetilde{f^{2+\epsilon}}(\Phi, S) < 0$ for every $\epsilon > 0$.

(ii) If $|S| = 2$, a simple convexity argument shows that the statement made in Lemma 13 is valid for any $p > 0$. In the case $S = \{1, 2\}$ f anticanonic is equivalent to $f(1, 2) \geqslant f(1) \geqslant f(\Phi)$; $f(1, 2) \geqslant f(2) \geqslant f(\Phi)$; $f(1, 2) + f(\Phi) \geqslant f(1) + f(2)$. We want to show that $f(1, 2)^q + f(\Phi)^q \geqslant f(1)^q + f(2)^q$, when $q = 1 + p \geqslant 1$. Now $x \to x^q$ is convex and monotone increasing for $x \geqslant 0$. For any such function, g, $w \leqslant x \leqslant y \leqslant z$, and $w + z \geqslant x + y$ imply $g(w) + g(z) \geqslant g(x) + g(y)$.

Note added in proof. The dependence of E on the R_i, for fixed z_i, is also an interesting question in TF theory. We have been able to prove [4] that the pressure and compressibility (defined by the change of E under uniform dilation) are positive, and that the kinetic energy is superadditive. This is true even for finite molecules. Thus, problems 6, 7, and 8 of Ref. [1, p. 33] and the problems posed in Ref. [1, pp. 104–105] have been solved affirmatively.

References

1. E. H. Lieb and B. Simon, *Advances in Math.* **23** (1977), 22–116. See also, E. H. Lieb and B. Simon, *Phys. Rev. Lett.* **31** (1973), 681–683; E. H. Lieb, "Proc. Int. Congress of Math.," Vancouver (1974); *Rev. Modern Phys.* **48** (1976), 553–569.
2. E. H. Lieb and W. E. Thirring, *Phys. Rev. Lett.* **35** (1975), 687–689. See also, E. H. Lieb, *Rev. Modern. Phys.* **48** (1976), 553–569.
3. W. F. Donoghue, Jr., "Monotone Matrix Functions and Analytic Continuation," Springer, New York, 1974.
4. R. Benguria and E. H. Lieb, The positivity of the pressure in Thomas–Fermi theory, *Commun. Math. Phys.*, to be submitted.

With R. Benguria in Commun. Math. Phys. *63*, 193–218 (1978)

The Positivity of the Pressure in Thomas Fermi Theory*

R. Benguria[1]** and E. H. Lieb[2]

[1] Department of Physics and
[2] Departments of Mathematics and Physics, Princeton University, Princeton, New Jersey 08 540, USA

Abstract. We prove the positivity of the pressure and compressibility for neutral systems in the Thomas-Fermi theory of molecules. Our results include some new properties of the Thomas-Fermi potential and a proof that the kinetic energy is superadditive.

I. Introduction

The Thomas-Fermi (TF) theory of atoms, molecules and solids has been given a firm mathematical foundation and many of the qualitative properties of the theory are understood and have been proven [1] (see also [2]; properties of the many-body TF potential are proved in [3]). There were, however, some open questions in [1], one of which we solve in this paper: the positivity of the pressure and compressibly for neutral systems.

The TF theory is defined by the energy functional (in units in which $h^2(8m)^{-1}(3/\pi)^{2/3}=1$ and $|e|=1$, where e and m are the electron charge and mass)

$$\xi(\varrho)=K(\varrho)-A(\varrho)+R(\varrho)+U \tag{1.1}$$

$$K(\varrho)=\tfrac{3}{5}\int \varrho(x)^{5/3}dx$$

$$A(\varrho)=\int V(x)\varrho(x)dx$$

$$V(x)=\sum_{j=1}^{k} z_j|x-R_j|^{-1}$$

$$R(\varrho)=\tfrac{1}{2}\iint \varrho(x)\varrho(y)|x-y|^{-1}dxdy$$

$$U=\sum_{1\leq i<j\leq k} z_iz_j|R_i-R_j|^{-1}. \tag{1.2}$$

Here $z_1,\ldots,z_k\geq 0$ are the charges of k fixed nuclei located at $R_1,\ldots,R_k$. $\int dx$ is always a three-dimensional integral. $\xi(\varrho)$ is defined for electron densities $\varrho(x)\geq 0$ such that $\int\varrho$ and $\int\varrho^{5/3}$ are finite.

---

* Work partially supported by U.S. National Science Foundation grant MCS 75-21684 A02

** On leave from Department of Physics, Universidad de Chile, Santiago, Chile

The TF energy for λ (not necessarily integral) electrons is defined by

$$e(\lambda)=\inf\{\xi(\varrho)|\int\varrho=\lambda\} \tag{1.3}$$

It is known [1] that for $\lambda\leqq Z\equiv\sum_{j=1}^{k}z_j$ there is a unique minimizing ϱ for (1.3). It is the unique solution to the TF equation

$$\varrho(x)^{2/3}=\max[\phi(x)-\Phi_0,0] \tag{1.4a}$$

for some $\Phi_0\geqq 0$, and with

$$\phi(x)\equiv V(x)-\int|x-y|^{-1}\varrho(y)dy. \tag{1.4b}$$

$-\Phi_0$ is the chemical potential [1], i.e.

$$\frac{de(\lambda)}{d\lambda}=-\Phi_0. \tag{1.5}$$

For $\lambda\leqq Z$, $\phi(x)>0$, all x. $\Phi_0=0$ if and only if $\lambda=Z$ and hence, for the *neutral case* the TF equation is

$$\varrho^{2/3}(x)=\phi(x). \tag{1.4c}$$

If $\lambda>Z$, there is no minimizing ϱ for (1.3), and $e(\lambda)=e(Z)$ in this case.

There are various possible definitions of the pressure. The one we shall use is the "*change in energy under uniform dilation*" defined as follows: Replace each R_i by lR_i, l being a scale factor, and let $e(\lambda,l)$ be the TF energy for a given λ and l. Then $P=-\partial e/\partial V$ which we interpret as

$$P=-(3l^2)^{-1}\partial e(\lambda,l)/\partial l. \tag{1.6}$$

The reciprocal compressibility, κ^{-1}, should be $-V\dfrac{\partial P}{\partial V}$ which we interpret as

$$\kappa^{-1}=-(l/3)\partial P/\partial l. \tag{1.7}$$

We shall prove that in the neutral case P and κ^{-1} are nonnegative (in the atomic case they are, of course, zero). In the process of doing so, we shall prove several interesting facts about the dependence of $\phi(x)$, K, A and R on the z_i. (Note: here and in the sequel, $\phi(x)$, K, A, R, etc. mean the respective quantities evaluated at the unique, minimizing TF density, ϱ.) We are not able to prove that P and κ are non-negative in the ionic (i.e. subneutral) case but *conjecture* that they are. The only thing we shall have to say about the ionic case except for appendix B is to give a formula (1.14) for P in terms of e and K. We are led to make the further *conjecture* that P is a decreasing function of λ and thus that the neutral case is the worst case. When $\lambda=0$, $P>0$ and $\kappa>0$ because $e=l^{-1}\sum_{i<j}z_iz_j|R_i-R_j|^{-1}$. In other words, the pressure is positive because the nuclear repulsion dominates the attractive forces; this repulsion presumably grows stronger as electrons are removed from the system.

The above definitions (1.6, 1.7) of P and κ carry over, in the thermodynamic limit, to the ordinary definitions for a solid (see [1], Sect. VI). There are, however, two other useful definitions which we will not touch upon in the main text (see Appendix B, however) except to *conjecture* that P and κ are non-negative for these definitions as well.

(i) *Dilation in one direction:* Let $R_i \to (lR_i^1, R_i^2, R_i^3)$, instead of $R_i \to lR_i$. Since this is a one dimensional expansion it seems appropriate to define $P = -\partial e(\lambda, l)/\partial l$ and $\kappa^{-1} = -l\partial P/\partial l$. As far as nonnegativity is concerned, this new definition changes κ but not P.

(ii) *Separation relative to a plane:* choose any plane which does not contain nuclei. For convenience it may be assumed to be the $x-y$ plane $\{(x^1, x^2, x^3) | x^3 = 0\}$. If $R_i = (R_i^1, R_i^2, R_i^3)$, replace R_i^3 by $R_i^3 + l$ if $R_i^3 > 0$ and by $R_i^3 - l$ if $R_i^3 < 0$. Note that in this case we shift by l instead of dilate by l. Again, $P = -\partial e(\lambda, l)/\partial l$ and $\kappa^{-1} = -l\partial P/\partial l$. In appendix B we will prove that $P > 0$ if the plane is a symmetry plane. This latter case was also proved by Balàsz [4] but our proof is somewhat different; it uses reflection positivity. Balàsz assumed there were only two nuclei, but his method works for any symmetric situation. One reason for being interested in this special case is that our (and Balàsz') proofs are valid for the *ionic* case as well.

The definitions we shall work with (1.6, 1.7) have one virtue, namely the dependence of e on l can be converted into a dependence of e on the z_i. This is a consequence of the following *scaling properties:*

Henceforth, $R_1, R_2, \ldots, R_k$ are fixed (with $R_i \neq R_j$ if $i \neq j$). We denote the k-tuple $z_1, \ldots, z_k$ simply by $\underline{z}$. Let $e(\underline{z}, \lambda, l)$ be the energy with the uniform dilation l. Then, from (1.1),

$$e(\underline{z}, \lambda, l) = l^{-7} e(l^3 \underline{z}, l^3 \lambda, 1) \tag{1.8}$$

and the minimizing TF density satisfies

$$\varrho(\underline{z}, \lambda, l; x) = l^{-6} \varrho(l^3 \underline{z}, l^3 \lambda, 1; x/l). \tag{1.9}$$

Substituting (1.8) in (1.6, 1.7) yields (assuming that all derivatives exist)

$$3l^{10} P = 7e - 3l^3 \sum_{i=1}^{k} z_i e_i - 3l^3 \lambda e_2 \tag{1.10}$$

$$\begin{aligned} 9l^{10} \kappa^{-1} &= 70e - 42l^3 \sum_{i=1}^{k} z_i e_i - 42l^3 \lambda e_2 \\ &\quad + 9l^6 \sum_{i,j=1}^{k} z_i z_j e_{ij} + 9l^6 \lambda^2 e_{22} + 18l^6 \lambda \sum_{i=1}^{k} z_i e_{2i}. \end{aligned} \tag{1.11}$$

In (1.10, 1.11) the notation is the following:

$$e_i \equiv \partial e(\underline{x}, y, 1)/\partial x_i, \quad e_2 \equiv \partial e(\underline{x}, y, 1)/\partial y, \quad \text{etc.}$$

These quantities are evaluated at $\underline{x} = l^3 \underline{z}$ and $y = l^3 \lambda$. A numerical error in the expression for κ^{-1} was made in Ref. [1], Eq. (145).

A more convenient form for P is obtained by noting that $e = K - A + R + U$. Furthermore, if (1.4) is multiplied by $\varrho(x)$ and integrated over the set on which $\varrho(x) \geqq 0$, one obtains

$$(5/3)K = A - 2R - l^3 \lambda \Phi_0 \tag{1.12}$$

moreover, $e_2 = -\Phi_0$ (cf. (1.5)). Finally,

$$l^3 \sum_{i=1}^{k} z_i e_i = 2U - A = 2e - (1/3)K + l^3 \lambda \Phi_0 \tag{1.13}$$

([1], Theorem II.16 or Lemma V.7). Combining these facts and then using (1.8), yields, for all λ,

$$3l^3 P(\underline{z}, \lambda, l) = e(\underline{z}, \lambda, l) + K(\underline{z}, \lambda, l). \tag{1.14}$$

For an atom, $2K = A - R$ (Virial Theorem, [1], Theorem II.22), $U = 0$ and $e = K - A + R$. Thus (1.14) gives $P = 0$ for all λ, as it should in this case.

The conjecture stated above, that the neutral case is the worst can be given a more transparent form:

Conjecture 1. $e + K$ is a decreasing function of λ for fixed $\underline{z}$ and R_i.

In this paper we will prove the positivity of P and κ for the neutral case. In the next Section the list of theorems to be proved is given. These theorems have an easy heuristic proof and these are given in Sect. III. We do so because these heuristic proofs are a guide to the proper proofs given in Sect. IV, and because they may be a useful guide to future work.

II. Theorems to Be Proved

We will be concerned only with the neutral case and use the notation $\phi(\underline{z}, x)$, $\varrho(\underline{z}, x)$, $e(\underline{z})$, $K(\underline{z})$, $A(\underline{z})$, $R(\underline{z})$, $U(\underline{z})$ to denote the TF potential and density at the point $x \in \mathbb{R}^3$, the total TF energy, the kinetic energy, the attractive energy, the electron repulsion and the nuclear repulsion, respectively, (cf. (1.1)), for the unique TF ϱ that satisfies the TF equation (1.4b, 1.4c). $\underline{z} \in \mathbb{R}^k_+ \equiv \{(z_1, \ldots, z_k) | z_i \geqq 0\}$. The R_i are fixed and distinct.

Definitions. If f is a real valued function on $\mathbb{R}^k_+$ then:

(i) f is *weakly superadditive* (WSA)$\Leftrightarrow f(\underline{z}_1 + \underline{z}_2) \geqq f(\underline{z}_1) + f(\underline{z}_2)$, $\forall \underline{z}_1, \underline{z}_2 \in \mathbb{R}^k_+$, such that $(\underline{z}_1, \underline{z}_2) = 0$, i.e. $(z_1)_i (z_2)_i = 0$, $\forall i$.

(ii) f is *superadditive* (SA)$\Leftrightarrow f(\underline{z}_1 + \underline{z}_2) \geqq f(\underline{z}_1) + f(\underline{z}_2)$, $\forall \underline{z}_1, \underline{z}_2 \in \mathbb{R}^k_+$.

(iii) f is *strongly superadditive* (SSA)$\Leftrightarrow f(\underline{z}_1 + \underline{z}_2 + \underline{z}_3) - f(\underline{z}_1 + \underline{z}_2) - f(\underline{z}_1 + \underline{z}_3) + f(\underline{z}_1) \geqq 0$, $\forall \underline{z}_1, \underline{z}_2, \underline{z}_3 \in \mathbb{R}^k_+$.

(iv) f is *ray convex*$\Leftrightarrow f(\lambda \underline{z}_1 + (1-\lambda)\underline{z}_2) \leqq \lambda f(\underline{z}_1) + (1-\lambda) f(\underline{z}_2)$, $\forall \lambda \in [0, 1]$, $\underline{z}_1, \underline{z}_2 \in \mathbb{R}^k_+$ and either $\underline{z}_1 - \underline{z}_2 \in \mathbb{R}^k_+$ or $\underline{z}_2 - \underline{z}_1 \in \mathbb{R}^k_+$.

(v) f is *ray concave*$\Leftrightarrow -f$ is ray convex.

(vi) f is *increasing*$\Leftrightarrow f(\underline{z}_1 + \underline{z}_2) \geqq f(\underline{z}_1)$, $\forall \underline{z}_1, \underline{z}_2 \in \mathbb{R}^k_+$.

Obviously,

$$f \text{ is SSA and } f(0) \leqq 0 \Rightarrow f \text{ is SA} \Rightarrow f \text{ is WSA}. \tag{2.1}$$

Further relations among these definitions are proved in Appendix A. These are the following ($C^p(\mathbb{R}^k_+)$ denotes the p-fold continuously differentiable functions and subscripts denote partial derivatives):

Lemma 2.1. (i) *If* $f \in C^2(\mathbb{R}^k_+)$ *then* f *is SSA*$\Leftrightarrow f_{ij} \geqq 0$, $\forall i, j$.

(ii) *If* $f \in C^1(\mathbb{R}^k_+)$ *then* f *is* SSA$\Leftrightarrow f_i$ *is increasing.*

Lemma 2.2. (i) *If* $f \in C^2(\mathbb{R}_+^k)$, $f(0)=0$, *and* $f_{ij} \geqq 0\ \forall i \neq j$, *then* f *is* WSA.

(ii) *If* $f \in C^1(\mathbb{R}_+^k)$, $f(0)=0$ *and* f_i *is an increasing function of* z_j *for* $j \neq i$, *then* f *is* WSA.

Remark. The converse implication is false as the WSA function $f(\underline{z}) = z_1 \sin^2(z_2)$ on $\mathbb{R}_+^2$ shows.

Lemma 2.3. f *is* SSA *implies* f *is weakly ray-convex, i.e.* f *satisfies definition* (iv) *with* $\lambda = 1/2$.

Remark. The converse implication is false, even if SSA is replaced by WSA. On $\mathbb{R}_+^2$, $f(\underline{z}) = |z_1 - z_2| + z_1 + z_2$ is convex (not merely ray-convex), increasing and $f(0)=0$, but $f(1,1) < f(1,0) + f(0,1)$.

Lemma 2.4. *If* f *is ray-convex and* $f \in C^1(\mathbb{R}_+^k)$ *then* f_i *is increasing.*

Corollary 2.5. *If* f *is ray-convex and* $f \in C^1(\mathbb{R}_+^k)$ *then*

$$\sum_{i=1}^{k} z_i' f_i(\underline{z}) \leqq f(\underline{z} + \underline{z}') - f(\underline{z}) \leqq \sum_{i=1}^{k} z_i' f_i(\underline{z} + \underline{z}').$$

The theorems to be proved can now be stated.

Properties of $\phi(z, x)$ *(neutral case):*

Theorem 2.6. *For each fixed* $x \in \mathbb{R}^3$, *different from* $R_1, \ldots, R_k$, $\underline{z} \mapsto \phi(\underline{z}, x)$ *is in* $C^1(\mathbb{R}_+^k)$ *and* $C^2(\mathbb{R}_+^k \setminus \underline{0})$. $\underline{z} \mapsto \phi_i(\underline{z}, x)$ *and* $\underline{z} \mapsto \phi_{ij}(\underline{z}, x)$ *are equicontinuous in* x. *Furthermore,*

(i) $\phi_{ij}(\underline{z}, x) \leqq 0$, $\forall i, j$, *and is negative semidefinite as a matrix, i.e.* $\sum_{i,j=1}^{k} \bar{c}_i c_j \phi_{ij}(\underline{z}, x) \leqq 0$ *for all* $\underline{c} \in C^k$.

(ii) $\phi_{ij}(\underline{z}, R_p) = \lim_{x \to R_p} \phi_{ij}(\underline{z}, x)$ *exists* ($\underline{z} \neq \underline{0}$).

(iii) $\underline{z} \mapsto \phi_i(\underline{z}, x) \geqq 0$ *and is ray-convex,* $\forall i$.

(iv) $\lim_{x \to R_i} \{\phi_i(\underline{z}, x) - |x - R_i|^{-1}\} \leqq 0$ *exists.* $\phi_i(\underline{z}, x) < |x - R_i|^{-1}$.

(v) $\phi_i(\underline{z}, R_j) = \lim_{x \to R_j} \phi_i(\underline{z}, x)$ *exists for* $i \neq j$. *Moreover,* $\phi_i(\underline{z}, R_j) = \phi_j(\underline{z}, R_i)$.

(vi) *For every* $\alpha < (1 + \sqrt{73})/2$, *there exist an* $R(\alpha) < \infty$ *and finite numbers* $M(\alpha)$ *and* $B(\alpha)$ *such that* $\phi_i(\underline{z}, x) \leqq M(\alpha)|x|^{-\alpha}$, $-\phi_{ij}(\underline{z}, x) \leqq B(\alpha)|x|^{-\alpha}$, ($\underline{z} \neq 0$), *hold when* $|x| > R(\alpha)$.

Using Lemma 2.1, we have

Corollary 2.7. *For each fixed* $x \in \mathbb{R}^3$, *different from* $R_1, R_2, \ldots, R_k$,

(i) $-\phi(\underline{z}, x)$ *is* SSA.

(ii) $\phi(\underline{z}, x)$ *is concave (not merely ray-concave).*

(iii) $-\phi_i(\underline{z}, x)$ *and* $(\phi(\underline{z}, x)$ *are strictly increasing.*

Remark. That $\phi(\underline{z}, x)$ is increasing is Teller's Lemma [5], [Theorem V.5, [1]].

Properties of K, A, R and e (neutral case):

Theorem 2.8. *$K(\underline{z}) \in C^1(\mathbb{R}^k_+)$ and $C^2(\mathbb{R}^k_+ \backslash \underline{0})$ and:*

$$\text{(i)}\ K_i(\underline{z}) = 3 \lim_{x \to R_i} \left[\phi(\underline{z}, x) - \sum_{j=1}^{k} z_j \phi_j(\underline{z}, x) \right] \tag{2.2}$$

$$\text{(ii)}\ K_{ij}(\underline{z}) = -3 \sum_{p=1}^{k} z_p \phi_{ij}(\underline{z}, R_p). \tag{2.3}$$

Remark. The limit in (2.2) exists by Theorem 2.6, and by $e_i = \lim_{x \to R_i} \{\phi(\underline{z}, x) - z_i |x - R_i|^{-1}\}$ ([1], Theorem II.16, Lemma V.7).

Using Theorem 2.6 we have

Corollary 2.9. (i) *$K_{ij}(\underline{z}) \geqq 0$ and is positive semidefinite as a matrix,*
(ii) *$K(\underline{z})$ is convex (not merely ray-convex) and* SSA *on $\mathbb{R}^k_+$,*
(iii) *$K(\underline{0}) = 0$, which implies $K(\underline{z})$ is* SA.

Theorem 2.10. (i) *$R(\underline{z})$ and $A(\underline{z})$ are convex (not merely ray convex) and* SSA *on $\mathbb{R}^k_+$.*
(ii) *$e(\underline{z})$ is* WSA *on $\mathbb{R}^k_+$.*

Remark. (ii) is just Teller's Theorem [5], [1, Theorem V.1], $e(\underline{z})$ is not SA. For $k=1$, $e(\underline{z}) = -(\text{const.}) z^{7/3}$, [1], and this is not SA. However we make the following.

Conjecture 2. Let $\tilde{e}(\underline{z}) = \sum_{j=1}^{k} e^{at}(z_j)$, where $e^{at}(z)$ is the TF energy of an isolated atom of charge z. Then $e(\underline{z}) - \tilde{e}(\underline{z})$ is SA.

Remark. $e - \tilde{e}$ is not SSA because

$$(\partial^2/\partial z_i^2)(e - \tilde{e}) = \lim_{x \to R_i} (\phi_i(\underline{z}, x) - (\partial \phi^{at}/\partial z)(z_i, x)),$$

and this is negative if some $z_j \neq 0$ $(j \neq i)$ by Corollary 2.7 (iii). It is obvious that $e - \tilde{e}$ is WSA since e and $-\tilde{e}$ are both WSA.

Definition. $$X(\underline{z}) \equiv 3K(\underline{z}) - \sum_{i=1}^{k} z_i K_i(\underline{z}). \tag{2.4}$$

Theorem 2.11. *$X(\underline{z})$ is* SSA *and $X(\underline{0}) = 0$. Hence X is* SA. *Moreover $X(\underline{z})$ is ray convex (as follows from Lemma 2.3 and Theorem 2.8).*

These theorems can be combined to yield the desired results about the pressure and compressibility.

Theorem 2.12. *For the neutral molecule, the pressure and compressibility as given by* (1.6), (1.7) *exist and satisfy:*

$$\text{(i)}\ 3l^3 P(\underline{z}) = e(\underline{z}) + K(\underline{z}), \tag{2.5}$$

$$\text{(ii)}\ 9l^3 \kappa^{-1}(\underline{z}) = 6l^3 P(\underline{z}) + 2e(\underline{z}) + 3X(\underline{z}) \ (cf.\ (2.4)), \tag{2.6}$$

(iii) *P and κ^{-1} are* WSA *and non-negative,*
(iv) *$l^2 P(\underline{z}, l)$ is a decreasing function of l. Equivalently, $e(\underline{z}, l)$ is a convex function of l.*

Proof. We can write (1.8) in the form $e(\underline{z}, l) = l^{-7} e(l^3 \underline{z})$, where $e(l^3\underline{z}) \equiv e(l^3\underline{z}, 1)$. Hence $3l^{10}P = 7e - 3l^3 \sum_{i=1}^{k} z_i e_i$ since $z_i e_i$ exists [1]. (1.12), (1.13) are true [1], and so (2.5) is proved [cf. (1.14)]. Since e and K are WSA, so is $e+K$ and $e+K \geqq$ (sum of the $e+K$ for isolated atoms) $=0$. Using scaling again on the right side of (2.5) ($K(\underline{z}, l) = l^{-7}K(l^3\underline{z}, 1)$ also), and Theorem 2.8(i), we can differentiate (2.5). Again using (1.13) and rescaling, (2.6) is obtained. By Theorem 2.11, (iii) is true. To prove (iv) note that for an atom, $e = -(\text{const.}) z^{7/3}$ and $K = -e$; hence $2e + 3X = 0$ in this case. Since e and X are WSA, $2e + 3X \geqq 0$. Thus $-l\dfrac{\partial P}{\partial l} \geqq 2P$. If one writes $P(\underline{z}, l) = l^{-7}\pi(\underline{z}, l)$, then $\partial\pi/\partial l \leqq 0$. □

The following conjecture, if true, would show that $l^4 P$ is decreasing, for the right side of (2.6) is $12l^3 P(\underline{z}) + \tilde{X}(\underline{z})$. It would also show that $K(\underline{z}, l)$ is decreasing in l.

Conjecture 3. $\tilde{X}(\underline{z}) \equiv 3X(\underline{z}) - 2K(\underline{z})$ is WSA.

Remark. $\tilde{X}(\underline{z}) = 0$ for an atom.

Let us define $E(\underline{z}) = e(\underline{z}) - U(\underline{z})$. It has been proved ([1], Theorem V.3) that $-E(\underline{z})$ is WSA. We conjecture that something stronger holds, namely

Conjecture 4. $E(\underline{z}, l)$ is monotone increasing in l, for fixed $\underline{z}$.

Remark. It is easy to check that Conjecture 4 is implied by Conjecture 1. Conjecture 4 means that the pressure of a molecule in which the *nuclear-nuclear repulsion is neglected* is negative instead of positive. Some results in this direction for the Schrödinger theory are given in [11].

III. Heuristic Proofs

In this section we give simple, but non-rigorous proofs that $K(\underline{z})$ and $X(\underline{z})$, (2.4), are SSA and $K(\underline{z})$ is convex. From this, Theorem 2.12 on the positivity of P and K follows, as mentioned in Sect. II. We think it is important to provide these "proofs" because the main line of the argument may be obscure in the proper proofs given in the next section. These "proofs" assume that all necessary derivatives exist. Let us begin with some facts about the TF potential $\phi(\underline{z}, x)$. Hereafter we refer only to the neutral case. By (1.4) $\phi(\underline{z}, x)$ satisfies the TF equation

$$-(4\pi)^{-1}\Delta\phi(\underline{z}, x) + \phi(\underline{z}, x)^{3/2} = \sum_{i=1}^{k} z_i \delta(x - R_i). \tag{3.1}$$

The kernel for $[-(4\pi)^{-1}\Delta + \phi^{1/2}]^{-1}$ is positive, and $z_i\delta(x - R_i)$ are positive "functions". Therefore $\phi(\underline{z}, x) \geqq 0$ all x. Differentiating (3.1) twice with respect to the z's we formally get

$$[-(4\pi)^{-1}\Delta + (3/2)\phi(\underline{z}, x)^{1/2}]\,\phi_i(\underline{z}, x) = \delta(x - R_i) \tag{3.2a}$$

and

$$[-(4\pi)^{-1}\Delta + (3/2)\phi(\underline{z}, x)^{1/2}]\,\phi_{ij}(\underline{z}, x) = -(3/4)\phi(\underline{z}, x)^{-1/2}\phi_i(\underline{z}, x)\phi_j(\underline{z}, x). \tag{3.2b}$$

From Eq. (3.2a) we have $\phi_i \geqq 0$, since the kernel for $[-(4\pi)^{-1}\Delta + (3/2)\phi^{1/2}]^{-1}$ is positive. For the same reason, $\phi_{ij} \leqq 0$, all i,j, and therefore $-\phi$ is SSA. Multiplying (3.2b) by $\bar{c}_i, c_j$, with $c_j \in C$, and summing over i,j we get,

$$[-(4\pi)^{-1}\Delta + (3/2)\phi(\underline{z},x)^{1/2}] \sum_{i,j=1}^{k} \bar{c}_i \phi_{ij} c_j = -(3/4)\phi^{-1/2} \left| \sum_{i=1}^{k} c_i \phi_i \right|^2 .$$

Therefore the quadratic form $\sum_{i,j=1}^{k} \bar{c}_i \phi_{ij}(\underline{z},x) c_j$ is non-positive for all $\underline{c} \in C^k$. Hence ϕ is concave in $\mathbb{R}_+^k$. Finally differentiating (3.2b) with respect to z_l we have,

$$\begin{aligned} &[-(4\pi)^{-1}\Delta + (3/2)\phi^{1/2}]\phi_{ijl} \\ &\quad = (3/8)\phi^{-3/2}\phi_i\phi_j\phi_l - (3/4)\phi^{-1/2}[\phi_{il}\phi_j + \phi_{ij}\phi_l + \phi_{jl}\phi_i], \end{aligned} \tag{3.2c}$$

which in turn implies $\phi_{ijl} \geqq 0$ all i,j,l. Indeed the following is formally true: $(-1)^{n+1}\phi_{i_1 i_2 \ldots i_n} \geqq 0$ for all i_j and all $n \geqq 1$.

Remark. If one assumes that the derivative $\phi_{i_1 \ldots i_n}$ exists, then Theorem 4 of Ref. [3] shows that the sign is indeed $(-1)^{n+1}$. To use Theorem 4 for this purpose it is necessary to choose $R_i = R_j$ for some $i \neq j$, but this is allowed, as explained in [3] in the paragraph after (1.6). Theorem 4 of [3] directly gives the SSA of $-\phi$ without going through Lemma 2.1. Indeed, Theorem 4 is a generalization of SSA; for example $\phi(\underline{z}_1 + \underline{z}_2 + \underline{z}_3 + \underline{z}_4, x) - \phi(\underline{z}_1 + \underline{z}_2 + \underline{z}_3, x) - \phi(\underline{z}_1 + \underline{z}_3 + z_4, x) - \phi(\underline{z}_1 + \underline{z}_2 + \underline{z}_4, x) + \phi(\underline{z}_1 + \underline{z}_2, x) + \phi(\underline{z}_1 + \underline{z}_3, x) + \phi(\underline{z}_1 + \underline{z}_4, x) - \phi(\underline{z}_1, x) \geqq 0$. However, we are obliged to prove the existence of the first two derivatives of $\phi(\underline{z},x)$ because we need them in our proof that $K(\underline{z})$ and $X(\underline{z})$ are SSA.

From $\phi_{ijl} \geqq 0$ follows the ray-convexity of ϕ_i because the quadratic form $\sum_{j,l=1}^{k} (\phi_i)_{jl} z_j z_l$ is non-negative for all $\underline{z} \in \mathbb{R}_+^k$.

Now, let us formally show that K is SSA. We have to prove that $K_{ij} \geqq 0$ all i,j (see Lemma 2.1). For the neutral molecule the kinetic energy is given by

$$K(\underline{z}) = (3/5) \int \phi(\underline{z},x)^{5/2} dx . \tag{3.3}$$

Differentiating (3.3) twice with respect to the z's we get,

$$\begin{aligned} K_{ij}(\underline{z}) = (3/2) [&\int \phi(\underline{z},x)^{3/2} \phi_{ij}(\underline{z},x) dx \\ &+ (3/2) \int \phi(\underline{z},x)^{1/2} \phi_i(\underline{z},x) \phi_j(\underline{z},x) dx] . \end{aligned} \tag{3.4}$$

Introducing (3.2b) in (3.4), partial integration yields

$$K_{ij} = 3 \int \phi_{ij}((4\pi)^{-1}\Delta\phi - \phi^{3/2}) dx = -3 \sum_{l=1}^{k} z_l \phi_{ij}(R_l),$$

where the last equality is a consequence of (3.1). But $\phi_{ij} \leqq 0$, all i,j and therefore $K_{ij} \geqq 0$ and K is SSA. Furthermore $[\phi_{ij}]$ is negative semi-definite [recall the discussion after (3.2b)]; hence $[K_{ij}]$ is positive semi-definite and K is convex.

It remains to be shown that $X_{ij} \geqq 0$, all i,j, that is, X is SSA (Lemma 2.1). Differentiating X twice with respect to z_i, z_j we have

$$\begin{aligned} X_{ij} &= K_{ij} - \sum_{l=1}^{k} z_l K_{ijl} \\ &= 3 \sum_{l,m=1}^{k} z_l z_m \phi_{ijm}(R_l), \end{aligned}$$

where the last equality follows from (3.5). Therefore $X_{ij} \geqq 0$, all i,j because $\phi_{ijm}(x) \geqq 0 \forall x, i, j, m$.

IV. Proof of Theorems 2.6–2.11

Here we give the rigorous proofs of the theorems enunciated in Sect. II. Only neutral systems are considered. Let us begin by recalling some of the known facts about the TF potential $\phi(\underline{z}, x)$ that we are going to need in our proofs:

(P-1) $\phi(\underline{z}, x)$ satisfies,

$$\phi(\underline{z}, x) = \sum_{i=1}^{k} z_i |x - R_i|^{-1} - \int |x-y|^{-1} \phi(\underline{z}, y)^{3/2} dy. \tag{4.1}$$

(P-2) $\phi(\underline{z}, x)$ is bounded and continuous on any open subset of $\mathbb{R}^3$ which is at non-zero distance from all the R_i ([1], Theorem IV.1). In fact, the TF potential is real analytic away from all the R_j, on all of $\mathbb{R}^3$ ([1], Theorem IV.6). $\left(\phi(\underline{z}, x) - \sum_{i=1}^{k} z_i |x - R_i|^{-1} \text{ is continuous for all } x.\right)$

(P-3) $\phi(\underline{z}, x)$ is strictly positive for $\underline{z} \neq \underline{0}$ ([1], Theorem IV.3).

(P-4) $|x|^4 \phi(\underline{z}, x) \to 9\pi^{-2}$ as $|x| \to \infty$, uniformly with respect to direction. (This is Sommerfeld's formula, [1], Theorem IV.10.) Moreover, for every $c < 3\pi^{-1} \exists R(c) < \infty$ such that $\phi(\underline{z}, x) \geqq c^2 |x|^{-4}$ when $|x| \geqq R(c)$ ([1], Theorems IV.8, IV.10).

(P-5) Properties (P-1) and (P-2) imply that $\phi(\underline{z}, x) = z_j |x - R_j|^{-1} + g(x)$ near R_j, where g is a continuous function.

(P-6) By the foregoing $\phi(\underline{z}, x) \in L^p$ for every $p \in [1, 3)$, and $\phi(\underline{z}, x)^{1/4} \in L^p$ for every $p \in (3, 12)$.

(P-7) $\phi(\underline{z}, x)$ is increasing in $\underline{z}$ for every $x \in \mathbb{R}^3$. (This is Teller's lemma, [1], Theorem V.5.)

(P-8) $\phi(\underline{z}, x)$ is strongly subadditive in $\underline{z}$ for every $x \in \mathbb{R}^3$ ([3], Theorem 4). In particular $\phi(\underline{z}, x)$ is subadditive.

(P-9) $\phi(\underline{z}, x)$ is concave in $\underline{z}$.

Proof of (P-9). Let $\psi(x) \equiv (\phi(\underline{z}, x) - \alpha\phi(\underline{z}_1, x) - (1-\alpha)\phi(\underline{z}_2, x)$, with $\underline{z} \equiv \alpha \underline{z}_1 + (1-\alpha)\underline{z}_2$, $0 \leqq \alpha \leqq 1$. By (P-2) ψ is continuous for all x and by (P-4) ψ goes to zero at infinity, hence $S = \{x | \psi(x) < 0\}$ is open and $\psi = 0$ on $\partial S \cup \{\infty\}$. On S, $-(4\pi)^{-1} \Delta\psi = -\phi(\underline{z}, x)^{3/2} + \alpha\phi(\underline{z}_1, x)^{3/2} + (1-\alpha)\phi(\underline{z}_2, x)^{3/2} \geqq -\phi(\underline{z}, x)^{3/2} + \phi(\underline{z}, x)^{3/2} = 0$ because $t \mapsto t^{3/2}$ is convex. "Hence ψ is superharmonic on S and thus ψ takes its minimum on $\partial S \cup \{\infty\}$ where it is zero. Then S is empty." □

Remark. Since the argument between the apostrophes in the last paragraph repeats several times throughout this paper we will denote it by MMP (maximum modulus principle) to abbreviate.

If we call $\phi_i(\underline{z}, x)$ the derivative of $\phi(\underline{z}, x)$ with respect to z_i, we have formally

$$\phi_i(\underline{z}, x) = |x - R_i|^{-1} - \tfrac{3}{2} \int |x-y|^{-1} \phi(\underline{z}, y)^{1/2} \phi_i(\underline{z}, y) dy. \tag{4.2}$$

Our first task will be to investigate the general properties of equations like (4.2).

IV.1. General Properties of an Integral Equation [Eq. (4.4)]

We deal here with L^p spaces ($\mathbb{R}^3$ always being understood) and with the weak L^p_w spaces:

Definition. $f \in L^p_w$ $(p>0)$ if and only if there is a constant $c<\infty$ such that $D_f(a) \equiv \mu\{x \,|\, |f(x)|>a\} \leqq c^p a^{-p}$, all $a>0$, where μ is Lebesgue measure. The infimum of all such c is denoted by $\|f\|_{p,w}$.

Remarks. (1) $L^p \subset L^p_w$ (for $p \geqq 0$). If $f \in L^p$, $\|f\|_{p,w} \leqq \|f\|_p$.

(2) $\|f\|_{p,w}$ is not a norm since it fails to satisfy the triangle inequality.

(3) It can be easily checked from the definition that $|x|^{-1} \in L^3_w$.

We will need the following later.

Lemma 4.1. *If $f, g \in L^p_w$ and L^q_w respectively then $f \cdot g \in L^r_w$ and $\|f \cdot g\|_{r,w} \leqq 2^{1/r} \|f\|_{p,w} \|g\|_{q,w}$, with $r^{-1} = p^{-1} + q^{-1}$ and $0<p,q,r<\infty$.*

Proof. Without loss we can assume $\|f\|_{p,w} = \|g\|_{q,w} = 1$.

$$\{x \,|\, |f(x)g(x)|>a\} \subset \{x \,|\, |f(x)|>a^{r/p}\} \cup \{x \,|\, |g(x)|>a^{r/q}\}.$$

Therefore $D_{fg}(a) \leqq D_f(a^{r/p}) + D_g(a^{r/q})$. But $D_f(a^{r/p}) \leqq a^{-r} \|f\|^p_{p,w}$ and $D_g(a^{r/q}) \leqq a^{-r} \|g\|^q_{q,w}$, whence $D_{fg}(a) \leqq 2a^{-r}$. □

Notes. (i) The constant $2^{1/r}$ is not the best possible. It is easy to find a better one, namely $p^{1/p} q^{1/q} / r^{1/r}$.

(ii) For more details about L^p_w spaces the reader can consult [6].

The main tool to show existence and uniqueness of solutions to equations like (4.2) in some function spaces is given by:

Theorem 4.2. *Let $w \in L^3_w(\mathbb{R}^3)$, w real, and let $\frac{3}{2} < p < 3$. Then the map*

$$T_w : g \mapsto w(x) \int |x-y|^{-1} w(y) g(y) dy, \tag{4.3}$$

is a bounded map from $L^p(\mathbb{R}^3) \to L^p(\mathbb{R}^3)$.

Note. Theorems of this kind have been proved by Faris [9] and Strichartz [10].

Proof. By the previous lemma $A_w : g \mapsto wg$ is a bounded map from $L^p_w \to L^r_w$ with $r^{-1} = p^{-1} + 1/3$. Also A_w restricted to L^p is a bounded map by Remark (1). Now, $B : h \mapsto |x|^{-1} * h$ is a bounded map from $L^r_w \to L^t_w$ with $1 + t^{-1} = r^{-1} + (1/3)$ (since $|x|^{-1} \in L^3_w$, and the weak form of Young's inequality, [6]), when $t>1$, $1<r<3/2$. Therefore $T_w = A_w B A_w$ is a bounded map from $L^p_w \to L^p_w$ for all $p \in (3/2, 3)$. Finally by the Marcinkiewicz-Zygmund interpolation theorem T_w extends to a bounded map from $L^p \to L^p$, $3/2<p<3$. □

If we restrict the domain of T_w to L^2, T_w is a bounded operator from the Hilbert space L^2 into itself. Moreover T_w is self-adjoint and positive since the kernel $|x-y|^{-1}$ is positive definite. Hence we have,

Corollary 4.3. *The equation* $(T_w+1)g=u$ *with* T_w *defined by* (4.3) *and* $w\in L^3_w$, $u\in L^2$ *has a unique* L^2 *solution,* g.

We now obtain the main result of this section:

Theorem 4.4. *Let* $w\in L^3_w$ *and* $wv\in L^2$. *Then there is a unique* f *(defined a.e.) which satisfies the equation*

$$f(x)=v(x)-\int|x-y|^{-1}w(y)^2f(y)dy \tag{4.4}$$

a.e. and such that $wf\in L^2$.

Proof. (i) Existence: Define

$$f(x)\equiv v(x)-\int|x-y|^{-1}w(y)g(y)dy, \tag{4.5}$$

where g is the unique L^2 solution (by Corollary 4.3) to

$$(T_w+1)g=vw. \tag{4.6}$$

From (4.5) and (4.6) we have $wf=g$ and therefore the f defined in this way satisfies (4.4) and also $wf\in L^2$. (ii) Uniqueness: Assume there are two solutions f_1,f_2 to Eq. (4.4) such that wf_1 and $wf_2\in L^2$. Both wf_1 and wf_2 satisfy Eq. (4.6) which has a unique solution. Therefore $wf_1=wf_2$ a.e., and hence $f_1=f_2$ a.e. [using (4.4)]. □

Having shown the existence of a unique solution to (4.4) for some class of v and w, we next specialize to the particular v of interest.

Theorem 4.5. *Let* $u\in\mathbb{R}^3$ *and let* f_u *be the solution to the equation*

$$f_u(x)=|x-u|^{-1}-\int|x-y|^{-1}w(y)^2f_u(y)dy, \tag{4.7}$$

with $w\in L^3_w$ *and* $h_u(x)\equiv w(x)|x-u|^{-1}\in L^2(\mathbb{R}^3)$ *for all* $u\in\mathbb{R}^3$. *Then the integral in* (4.7) *is finite for all* $x\in\mathbb{R}^3$ *and thus* $f_u(x)$ *is defined by the right side of* (4.7) *for all* $x\neq u$. *Furthermore, if* $u\neq t\in\mathbb{R}^3$, $f_u(t)=f_t(u)$.

Proof. Since $w\in L^3_w$ and $h_u(x)\in L^2$. Theorem 4.4 implies that there is a unique (a.e.) function solving (4.7) and satisfying $wf_u\equiv g_u\in L^2$. Since $wf_u\in L^2$ and $h_x(y)\in L^2$ for all x, the integrand in (4.7) is absolutely integrable for all x. Now, g_u satisfies $(1+T_w)g_u=h_u$ and $(g_t,h_u)=(g_t,(1+T_w)g_u)=((1+T_w)g_t,g_u)=(h_t,g_u)$ since T_w [defined in (4.3)] is self-adjoint. Explicitly, this says

$$\int w(x)^2[f_t(x)|x-u|^{-1}-f_u(x)|x-t|^{-1}]dx=0.$$

Using (4.7) this implies $0=f_u(t)-|t-u|^{-1}-f_t(u)+|u-t|^{-1}$. □

Up to now the only assumption on w was $w\in L^3_w$. We will now make a stronger assumption about w in order to obtain continuity of the solution to (4.7). First, a preliminary remark:

Lemma 4.6. *If $f \in L^p$ and $g \in L^q$ with p, q dual indices different from* 1 *and* ∞ *then $f * g$ is a bounded continuous function going to zero at infinity.*

Proof. This result is standard. See [1], Lemma II.25. □

Lemma 4.7. *Let $w \in L^3_w$ and such that $w \in L^{6-\varepsilon} \cap L^{6+\varepsilon}$ (for some $\varepsilon > 0$). Let v be such that $vw \in L^2$ and let f denote the solution to* (4.4). *Then the integral in* (4.4) *(namely $f - v$) is a bounded continuous function going to zero at infinity.*

Proof. By Theorem 4.4 the solution f exists and satisfies $g \equiv wf \in L^2$. Using Hölder's inequality, $wg \in L^{p_-} \cap L^{p_+}$ with $p_\pm = 2(6 \pm \varepsilon)(8 \pm \varepsilon)^{-1}$. We can always decompose $|x|^{-1} = |x|^{-1}_< + |x|^{-1}_>$ with $|x|^{-1}_< \in L^{3-\eta_+}$, $|x|^{-1}_> \in L^{3+\eta_-}$ ($\eta_\pm$, positive). Choose $\eta_\pm = \varepsilon(4 \pm \varepsilon)^{-1}$; then $(3 \mp \eta_\pm)$ is the dual of $p_\pm$. But $f - v = |x|^{-1} * gw = |x|^{-1}_< * gw + |x|^{-1}_> * gw$, and hence this lemma follows from Lemma 4.6. □

Remark. The w which will eventually be used is simply $\phi^{1/4}(\underline{z}, x)$. This satisfies the conditions of Lemma 4.7 by (P-6).

We now study the dependence of the solution f on v and w.

Lemma 4.8. *Let $w \in L^3_w$ and such that $w \in L^{6-\varepsilon} \cap L^{6+\varepsilon}$ (for some $\varepsilon > 0$). Let $u \in \mathbb{R}^3$ be a parameter, and let*

$$v_u(x) = |x - u|^{-1} + V(x), \tag{4.8}$$

where $V(x)$ is a continuous superharmonic function, bounded and going to zero at infinity such that $wV \in L^2$, then:

(i) *The solution f_u to* (4.4) *is non-negative for all x.*

(ii) *If v_u is fixed and if $w_1(x)^2 \geqq w_2(x)^2$ all x, the corresponding solutions f_1, (resp. f_2) to* (4.4) *with $w = w_1$ (resp. $w = w_2$) are such that $f_1(x) \leqq f_2(x)$ all x.*

(iii) *Now keep w fixed. Let v_{1u}, v_{2u} be of the form* (4.8) *with $v_{1u} - v_{2u}$ superharmonic, then the corresponding solutions f_{1u}, f_{2u} are such that $f_{1u}(x) \geqq f_{2u}(x)$ all x.*

Proof. Since $w \in L^{6-\varepsilon} \cap L^{6+\varepsilon}$ and $|x - u|^{-1} \in L^{3-\eta_+} + L^{3+\eta_-}$ with $\eta_\pm = \varepsilon(4 \pm \varepsilon)^{-1}$, using Hölder's inequality we have $w(x)|x - u|^{-1} \in L^2$. Therefore $v_u w \in L^2$ and, by Theorem 4.4, there is a unique solution f_u to Eq. (4.4), with this v_u, satisfying $wf_u \in L^2$. Moreover by Lemma 4.6 and the properties of v_u, f_u [defined by the right side of (4.4)] is continuous away from u and goes to zero at infinity. (i) Let $S = \{x | f(x) < 0\}$. Since $f_u \to \infty$ as $x \to u$, S is disjoint from u and open (since f_u is continuous away from u). On S, the distributional laplacian of f_u is given by

$$-(4\pi)^{-1} \Delta f_u = -w^2 f_u - (4\pi)^{-1} \Delta V \geqq -w^2 f_u \geqq 0.$$

Then (i) follows from MMP. (ii) Call $\psi = f_2 - f_1$. ψ is continuous everywhere and goes to zero at infinity. Let $S = \{x | \psi(x) < 0\}$. S is open and $\psi = 0$ on $\partial S \cup \{\infty\}$. On S, $-(4\pi)^{-1} \Delta \psi = w_1^2 f_1 - w_2^2 f_2 \geqq -w_2^2 \psi > 0$ and (ii) follows using MMP. (iii) is a consequence of (i) and the linearity of f_u in v_u. □

Theorem 4.9. *(Asymptotic Behavior of $f(x)$). Consider $v(x) = |x|^{-1}$ and w as in Lemma 4.8 and, moreover, $w(x)^2 \geqq c|x|^{-2}$ for $|x| > R$ and some $c > 0$. Then $f(x) \leqq M(c)|x|^{-\alpha(c)}$ for $|x| > R$ where $\alpha(c) = (1 + \sqrt{1 + 16\pi c})/2$ and $M(c) = \alpha(c)^{-1} R^{\alpha(c)-1}$.*

Proof. Take w_1 defined by $w_1(x)^2 = c|x|^{-2}$ for $|x| > R$ and $w_1^2 = 0$ for $|x| < R$, the solution to (4.4) corresponding to this w is given by $f_1(x) = M(c)|x|^{-\alpha(c)}$ for $|x| > R$ and $f_1(x) = |x|^{-1}(1 + a|x|)$ for $|x| < R$ with $M(c) = \alpha(c)^{-1} R^{\alpha(c)-1}$ and $a = (1-\alpha(c))/(\alpha(c)R)$. We have $w(x)^2 \geqq w_1(x)^2$ all x and hence, by Lemma 4.8(ii) $f(x) \leqq f_1(x)$. In particular $f(x) \leqq M(c)|x|^{-\alpha(c)}$ for $|x| > R$. □

To close this section we prove the following,

Lemma 4.10. *If* $v(x) = |x|^{-1}$, *and* w *as in Lemma* 4.8, *then* $\int w(x)^2 f(x) dx \leqq 1$, *If, moreover,* $w(x)^2 \geqq c|x|^{-2}$ *for* $|x| > R$ *(for some* c, R*), then* $\int w(x)^2 f(x) dx = 1$.

Proof. Assume $\int w^2 f > 1$. Define the spherical average $[f](r) \equiv (4\pi)^{-1} \int_{S_2} f(r\Omega) d\Omega$. From (4.4) we have $[f](r) = r^{-1} - \int w(y)^2 f(y) \left[(4\pi)^{-1} \int_{S_2} d\Omega |r\Omega - y|^{-1}\right] dy$. Using the well known formula $(4\pi)^{-1} \int_{S_2} d\Omega |r\Omega - y|^{-1} = \{\max(r, |y|)\}^{-1}$ we get $[f](r) \leqq r^{-1}(1 - \int_{|y| \leqq r} w(y)^2 f(y) dy)$. Therefore for r large enough $[f](r) < 0$ which contradicts Lemma 4.8(i). Hence $\int w^2 f \leqq 1$. Let us now consider w such that $w(x)^2 \geqq c|x|^{-2}$ for $|x| > R$. $[f](r) \geqq r^{-1}(1 - \int w^2 f)$ by the same arguments as above. If $\int w^2 f < 1$, then $[f](r) \geqq d r^{-1}$ for some positive d which contradicts Theorem 4.9. □

IV.2. Proof of Theorem 2.6: Properties of the TF *Potential*

The strategy to prove that $\underline{z} \to \phi(\underline{z}, x) \in C^1$ is the following: we first show that a unique solution to Eq. (4.2) exists (Lemma 4.11) and is continuous in $\underline{z}$ uniformly with respect to x (Lemma 4.13). We then show that $\phi_i^\varepsilon(\underline{z}, x) \equiv \varepsilon^{-1}[\phi(\underline{z} + \varepsilon \underline{e}_i, x) - \phi(\underline{z}, x)]$, with $\underline{e}_i = (\delta_i^j)$ a unit vector in $\mathbb{R}_+^k$ along z_i, converges to $\phi_i(\underline{z}, x)$ as $\varepsilon \to 0$ uniformly in x. (Lemma 4.14). We then imitate the same argument to show that $\phi \in C^2$.

In what follows we study the equation

$$\phi_u(\underline{z}, x) = |x - u|^{-1} - (3/2) \int dy |x - y|^{-1} \phi(\underline{z}, y)^{1/2} \phi_u(\underline{z}, y). \tag{4.9}$$

Note that $w \equiv (3/2)^{1/2} \phi^{1/4} \in L_w^3$ (since w goes as $|x|^{-1}$ at infinity) and $w \in L^p$ for all $p \in (3, 12)$ [(P-4), (P-6)]. In particular $w \in L^{6-\varepsilon} \cap L^{6+\varepsilon}$, for some $\epsilon > 0$, therefore $|x - u|^{-1} w \in L^2$ as discussed in the proof of Lemma 4.8.

Lemma 4.11. *(Existence of* $\phi_u(\underline{z}, x)$*). There is a unique* $\phi_u(\underline{z}, x)$ *satisfying* Eq. (4.9) *with* $\phi_u \phi^{1/4} \in L^2$, *and it has the following properties:*

(i) $\phi_u(\underline{z}, x) - |x - u|^{-1}$ *is a bounded continuous function going to zero at infinity.*

(ii) $\underline{z} \to \phi_u(\underline{z}, x)$ *is non-negative and decreasing*

(iii) $\underline{z} \to \phi_u(\underline{z}, x)$ *is ray-convex.*

(iv) *For every* $\alpha < (1 + \sqrt{73})/2 \simeq 4.77$, *there exists an* $R(\alpha) < \infty$ *and a finite number* $M(\alpha)$ *such that* $\phi_u(\underline{z}, x) \leqq M(\alpha)|x|^{-\alpha}$ *for* $|x| \geqq R(\alpha)$.

Proof. Since $w \in L_w^3$ and $|x - u|^{-1} w \in L^2$, Theorem 4.4 implies the existence of a unique $\phi_u(\underline{z}, x)$ satisfying (4.9) with $w\phi_u \in L^2$. (i) follows from Lemma 4.7, since

$w \in L^{6-\varepsilon} \cap L^{6+\varepsilon}$. As for (ii), Lemma 4.8(i) implies that $\underline{z} \mapsto \phi_u(\underline{z}, x) \geqq 0$ all x; (P-7) together with Lemma 4.8(ii) imply that $\underline{z} \mapsto \phi_u(\underline{z}, x)$ is decreasing. To prove (iii), let $\underline{z}_1, \underline{z}_2 \in \mathbb{R}_+^k$ with $\underline{z}_1 - \underline{z}_2 \in \mathbb{R}_+^k$ and define $\underline{z} = \lambda \underline{z}_1 + (1-\lambda)\underline{z}_2$ with $0 \leqq \lambda \leqq 1$. Define $\psi(x) \equiv \lambda\phi_u(\underline{z}_1, x) + (1-\lambda)\phi_u(\underline{z}_2, x) - \phi_u(\underline{z}, x)$. Because of (i) $\psi(x)$ is continuous everywhere and goes to zero at infinity. Then $S = \{x | \psi(x) < 0\}$ is open and $\psi = 0$ on $\partial S \cup \{\infty\}$. From (4.9)

$$-(4\pi)^{-1}\Delta\psi = \tfrac{3}{2}\{-\lambda\phi(\underline{z}_1, x)^{1/2}\phi_u(\underline{z}_1, x) - (1-\lambda)\phi(\underline{z}_2, x)^{1/2}\phi_u(\underline{z}_2, x) + \phi(\underline{z}, x)^{1/2}\phi_u(\underline{z}, x)\}.$$

Because of (P-9), (P-7) and part (ii) (since $\underline{z}_1 - \underline{z}_2 \in \mathbb{R}_+^k$) we have $-(4\pi)^{-1}\Delta\psi \geqq 0$. Hence (iii) follows using MMP. (iv) given $\alpha < (1+\sqrt{73})/2$ (i.e. $c < (9/2\pi)$) there exists $R(c) < \infty$ such that $w(x)^2 \equiv (3/2)\phi(\underline{z}, x)^{1/2} \geqq c|x|^{-4}$ (P-4). Hence, by Theorem 4.9, $\phi_u(\underline{z}, x) \leqq M(\alpha)|x|^{-\alpha}$ for $|x| \geqq R(c)$. □

Remark. In the atomic case, Hille [7] used methods of ordinary differential equations to prove that the asymptotic formula with $\alpha = (1+\sqrt{73})/2$ was exact [[7], Eq. (4.5)].

We now prove a general theorem that we will later need:

Theorem 4.12. *Let f be a real (or complex) function on $\mathbb{R}_+^k$. Suppose f satisfies the following condition:*

$$|f(\underline{z}_1) - f(\underline{z}_2)| < K\|\underline{z}_1 - \underline{z}_2\|_2^\alpha, \tag{4.10}$$

for all $\underline{z}_1, \underline{z}_2 \in \mathbb{R}_+^k$ such that $\underline{z}_1 - \underline{z}_2 \in \mathbb{R}_+^k$, for some $\alpha > 0$ and some $K > 0$, then $\underline{z} \mapsto f(\underline{z})$ is continuous in the whole of $\mathbb{R}_+^k$.

Proof. Assume first that $\underline{z} \in \mathrm{Int}(\mathbb{R}_+^k)$. Let $\underline{n} = (1, 1, \ldots, 1)$, and $\underline{z}_0 \equiv \underline{z} - \delta\underline{n}$, with $\delta \leqq \min_{1 \leqq i \leqq k}(z_i)$ (i.e. $\underline{z}_0 \in \mathbb{R}_+^k$). Let $\underline{z}' \in B(\underline{z}, \delta)$, the ball of radius δ centered at $\underline{z}$. Applying (4.10) twice we get $|f(\underline{z}') - f(\underline{z})| \leqq (\|\underline{z}' - \underline{z}_0\|_2^\alpha + \|\underline{z} - \underline{z}_0\|_2^\alpha)K$, because $\underline{z}' - \underline{z}_0 \in \mathbb{R}_+^k$, $\underline{z} - \underline{z}_0 \in \mathbb{R}_+^k$. But, as $\delta \to 0$, $\|\underline{z} - \underline{z}_0\|_2 \to 0$ and $\|\underline{z}' - \underline{z}_0\| \to 0$ uniformly in $B(\underline{z}, \delta)$, so f is continuous at $\underline{z}$. Now, if $\underline{z}$ is in one face, F, of $\mathbb{R}_+^k$ (of dimension $0 \leqq l < k$) the same argument can be repeated using $\tilde{\underline{n}} \equiv$ projection of $\underline{n}$ on F. □

Lemma 4.13. *(Continuity of $\phi_u(\underline{z}, x)$ in $\underline{z}$). $\phi_u(\underline{z}, x)$ defined as the solution to (4.9) (Satisfying $\phi^{1/4}\phi_u \in L^2$) is continuous for all $\underline{z} \in \mathbb{R}_+^k$ uniformly with respect to x.*

Proof. We divide the proof into two steps. First we prove continuity at $\underline{z} \neq \underline{0}$, and then at $\underline{z} = \underline{0}$. (i) $\underline{z} \neq \underline{0}$. There is a $\underline{z}^* \in \mathbb{R}_+^k$, $\underline{z}^* \neq \underline{0}$ such that $\underline{z} - \underline{z}^* \in \mathbb{R}_+^k$. Let $\underline{z}_1, \underline{z}_2 \in (\underline{z}^* + \mathbb{R}_+^k)$ with $\underline{z}_1 - \underline{z}_2 \in \mathbb{R}_+^k$. From Eq. (4.9) we get,

$$\phi_u(\underline{z}_2, x) - \phi_u(\underline{z}_1, x) = (3/2)\int |x-y|^{-1}\phi_u(\underline{z}_1, y)[\phi(\underline{z}_1, y)^{1/2} - \phi(\underline{z}_2, y)^{1/2}]dy + (3/2)\int |x-y|^{-1}\phi(\underline{z}_2, y)^{1/2}[\phi_u(\underline{z}_1, y) - \phi_u(\underline{z}_2, y)]. \tag{4.11}$$

Since $\underline{z}_1 - \underline{z}_2 \in \mathbb{R}_+^k$, $\phi_u(\underline{z}_2, x) - \phi_u(\underline{z}_1, x) \geqq 0$ because of Lemma 4.11(ii). Hence, (4.11) implies

$$|\phi_u(\underline{z}_2, x) - \phi_u(\underline{z}_1, x)| < (3/2)\int |x-y|^{-1}\phi_u(\underline{z}^*, y)(\phi(\underline{z}_1, y)^{1/2} - \phi(\underline{z}_2, y)^{1/2}).$$

Because of (P-8) we have,

$$\phi(\underline{z}_1, y)^{1/2} - \phi(\underline{z}_2, y)^{1/2} \leqq \phi(\underline{z}_1 - \underline{z}_2, y)^{1/2} \leqq \|\underline{z}_1 - \underline{z}_2\|_2^{1/2} \sum_{i=1}^{k} \frac{1}{|y - R_i|^{1/2}}, \tag{4.12}$$

where the last inequality follows from Eq. (4.1). Hence

$$|\phi_u(\underline{z}_2, x) - \phi_u(\underline{z}_1, x)| < \|\underline{z}_1 - \underline{z}_2\|_2^{1/2} g(x), \tag{4.13a}$$

where

$$g(x) = (3/2)\int |x-y|^{-1} \sum_{i=1}^{k} \left(\frac{1}{|y - R_i|^{1/2}}\right) \phi_u(\underline{z}^*, y). \tag{4.13b}$$

By Young's inequality $g(x) \in L^\infty$ because $|x|^{-1} \in L^4 + L^{5/2}$ and $|y - R_i|^{-1/2}\phi_u(\underline{z}^*, y) \in L^p$ for any $1 \leqq p < 2$, in particular for $p = 4/3$ and $p = 5/3$. [Lemma 4.11(i), (iv)]. Theorem 4.12 and Eq. (4.13) then imply that $\phi_u(\underline{z}, x)$ is continuous in $\underline{z}$, uniformly with respect to x, for all $\underline{z} \in (\underline{z}^* + \mathbb{R}_+^k)$. But $\bigcup_{z^*} (\underline{z}^* + \mathbb{R}_+^k) = \mathbb{R}_+^k \setminus \{\underline{0}\}$. (ii) $\underline{z} = \underline{0}$, Equation (4.9) and Lemma 4.11(ii) imply

$$|\phi_u(\underline{z}, x) - \phi_u(\underline{0}, x)| \leqq h(x) \equiv \int |x-y|^{-1} \phi(\underline{z}, y)^{1/2} |y-u|^{-1} dy. \tag{4.14}$$

Using Young's inequality we get,

$$\begin{aligned} \|h(x)\|_\infty \leqq \|\phi(\underline{z}, \cdot)^{1/4}\|_6 (\| |x|_<^{-1} \|_{42/17} \|\phi^{1/4}|y-u|^{-1}\|_{7/3} \\ + \| |x|_>^{-1}\|_4 \|\phi^{1/4}|y-u|^{-1}\|_{12/7}) \end{aligned}$$

and thus $\|h(x)\|_\infty \leqq c\|\phi^{1/4}\|_6$, with $c < \infty$ because $\phi^{1/4}|y-u|^{-1} \in L^p$ for any $(3/2) < p < (12/5)$ and $|x|_<^{-1} \in L^p (p < 3)$, $|x|_>^{-1} \in L^p (p > 3)$. From (4.14) we finally get $\|\phi_u(\underline{z}, x) - \phi_u(\underline{0}, x)\|_\infty \leqq c\|\phi^{1/4}\|_6 = c(\int \varrho)^{1/6} = cz^{1/6}$, where $z = \sum_{i=1}^{k} z_i$. □

Let us define $\phi_i(\underline{z}, x)$ to be $\phi_u(\underline{z}, x)$ with $u = R_i$. Then the last step to prove that $\phi(\underline{z}, x) \in C^1(\mathbb{R}_+^k)$, uniformly with respect to x, is the following:

Lemma 4.14. *(Convergence of $\phi_i^\varepsilon(\underline{z}, x)$ to $\phi_i(\underline{z}, x)$). Let $\phi_i^\varepsilon(\underline{z}, x) \equiv \varepsilon^{-1}[\phi(\underline{z} + \varepsilon \underline{e}_i, x) - \phi(\underline{z}, x)]$ with $\underline{e}_i = (\delta_i^j)$ being a unit vector in $\mathbb{R}_+^k$, along z_i and $\varepsilon \geqq -z_i$. Then $\phi_i^\varepsilon(\underline{z}, x) \to \phi_i(\underline{z}, x)$ as $\varepsilon \to 0$, uniformly with respect to x.*

Proof. (i) Consider first $\varepsilon > 0$. We will prove the following,

$$\phi_i^\varepsilon(\underline{z}, x) - \phi_i(\underline{z} + \varepsilon \underline{e}_i, x) \geqq 0, \tag{4.15a}$$

$$\phi_i^\varepsilon(\underline{z}, x) - \phi_i(\underline{z}, x) \leqq 0. \tag{4.15b}$$

Consider $\psi(x) \equiv \phi_i^\varepsilon(\underline{z}, x) - \phi_i(\underline{z} + \varepsilon \underline{e}_i, x)$. By Lemma 4.11 and (P-2), ψ is continuous for all x and goes to zero at infinity. Then $S \equiv \{x | \psi < 0\}$ is open and $\psi = 0$ on $\partial S \cup \{\infty\}$. On S,

$$\begin{aligned} -(4\pi)^{-1}\Delta\psi &= \varepsilon^{-1}[\phi(\underline{z}, x)^{3/2} - \phi(\underline{z} + \varepsilon \underline{e}_i, x)^{3/2}] \\ &\quad + (3/2)\phi_i(\underline{z} + \varepsilon \underline{e}_i, x) \cdot \phi(\underline{z} + \varepsilon \underline{e}_i, x)^{1/2} \\ &\geqq (2\varepsilon)^{-1} \phi(\underline{z} + \varepsilon \underline{e}_i, x)^{1/2} \phi(\underline{z}, x)[\mu^2 - 3 + 2\mu^{-1}], \end{aligned}$$

where $\mu^2 \equiv \phi(\underline{z}+\varepsilon\underline{e}_i, x)/\phi(\underline{z}, x) \geqq 1$ [by (P-7)]. Hence $\mu^2 - 3 + 2\mu^{-1} \geqq 0$. MMP then implies (4.15a). The proof of (4.15b) is analogous. From (4.15) and Lemma 4.13, $\|\phi_i^\varepsilon(\underline{z}, x) - \phi_i(\underline{z}, x)\|_\infty \to 0$ as $\varepsilon \downarrow 0$. (ii) If $-z_i < \varepsilon < 0$, (4.15a, b) imply

$$\phi_i(\underline{z}, x) \leqq \phi_i^\varepsilon(\underline{z}, x) \leqq \phi_i(\underline{z}+\varepsilon\underline{e}_i, x) \tag{4.16}$$

which in turn implies $\|\phi_i^\varepsilon(\underline{z}, x) - \phi_i(\underline{z}, x)\|_\infty \to 0$ as $\varepsilon \uparrow 0$. □

If we denote by $\phi_{ij}(\underline{z}, x)$ the derivative of $\phi_i(\underline{z}, x)$ with respect to z_j, we formally get (from (4.2)):

$$\begin{aligned} -\phi_{ij}(\underline{z}, x) &= (3/4)\int |x-y|^{-1}\phi(\underline{z}, y)^{-1/2}\phi_i(\underline{z}, y)\phi_j(\underline{z}, y)dy \\ &\quad -(3/2)\int |x-y)^{-1}\phi(\underline{z}, y)^{1/2}(-\phi_{ij}(\underline{z}, y))dy\,. \end{aligned} \tag{4.17}$$

As we have already mentioned, the strategy to prove that $\phi(\underline{z}, x)$ is in $C^2(\mathbb{R}^k_+ \setminus \underline{0})$, uniformly with respect to x, will be the same as before. Now there will be an additional difficulty, namely the control of $\phi(\underline{z}, y)^{-1/2}$. Let us start proving that a solution to (4.17) indeed exists.

Lemma 4.15. *(Existence of $\phi_{ij}(\underline{z}, x)$). For $\underline{z} \neq \underline{0}$, there is a unique $\phi_{ij}(\underline{z}, x)$ satisfying Eq. (4.17) and such that $\phi_{ij}\phi^{1/4} \in L^2$. Moreover:*

(i) *$\phi_{ij}(\underline{z}, x)$ is continuous for all x. It is bounded and goes to zero at infinity.*

(ii) *$-\phi_{ij}(\underline{z}, x)$ is non-negative and so is $\sum_{1 \leqq i,j \leqq k} \bar{c}_i(-\phi_{ij}(\underline{z}, x))c_j$, and $\underline{c} \in C^k$.*

(iii) *$-\phi_{ij}(\underline{z}, x)$ is a decreasing function of $\underline{z}$.*

Proof. Note first that, for $\underline{z} \neq \underline{0}$, $\phi(\underline{z}, \cdot)^{-1/4}\phi_i(\underline{z}, \cdot) \in L^q$ for any $1 \leqq q < 4$. In fact, for $\underline{z} \neq \underline{0}$, ϕ is strictly positive (P-3) and $\phi(\underline{z}, x) \geqq \tilde{c}|x|^{-4}$ for $|x| > R \equiv 2\max_i |R_i|$ and some positive constant $\tilde{c}$ (P-4). Because of Lemma 4.11(iv), $\phi_i\phi^{-1/4} \leqq c_1|x|^{1-\alpha}$ for $|x| > R(\alpha)$ with $4 < \alpha < 4.77$. Then if $B(0, R(\alpha)) = \{x \,|\, |x| \leqq R(\alpha)\}$, $\phi_i\phi^{-1/4} \in L^p(\mathbb{R}^3 \setminus B(0, R(\alpha)))$, $\forall p \geqq 1$. Inside $B(0, R(\alpha))$ and away from R_i $\phi^{-1/4}\phi_i$ is bounded since B is compact, (P-4) and Lemma 4.11(i). In a neighborhood of R_i $\phi_i\phi^{-1/4}$ behaves like $|x-R_i|^{-3/4}$, hence $\phi_i\phi^{-1/4} \in L^q$ for any $1 \leqq q < 4$. Therefore, $\phi_i\phi^{-1/4}\phi_j\phi^{-1/4} \in L^s$ for $1 \leqq s < 2$ and, since $|x|^{-1} \in L^3_w$, $v_{ij} \equiv (3/4)|x|^{-1} * (\phi_i\phi^{-1/4}\phi_j\phi^{-1/4}) \in L^t$ for any $3 < t < \infty$ (by the generalized Young's inequality). Moreover, $v_{ij}\phi^{1/4} \in L^2$ because $\phi^{1/4} \in L^p$, $3 < p < 12$. Finally, since $\phi^{1/4} \in L^3_w$, Theorem 4.4 implies that there is a unique solution $\phi_{ij}(\underline{z}, x)\,(\underline{z} \neq \underline{0})$ to Eq. (4.17) satisfying $\phi^{1/4}\phi_{ij} \in L^2$. (i) By Lemma 4.7, $\phi_{ij} + v_{ij}$ is a bounded continuous function going to zero at infinity. Lemma 4.6 shows that v_{ij} is continuous, bounded and goes to zero at infinity because $(\phi^{-1/4}\phi_i)\,(\phi^{-1/4}\phi_j) \in L^s$ for any $1 \leqq s < 2$ and $|x|^{-1} \in L^4 + L^{5/2}$. (ii) Follows from Lemma 4.8 because v_{ij} is superharmonic and so is $\sum \bar{c}_i v_{ij} c_j$. Lemma 4.11(ii) and (P-7) imply that $v_{ij}(\underline{z}_1, x) - v_{ij}(\underline{z}_2, x)$ is superharmonic if $\underline{z}_1 - \underline{z}_2 \in \mathbb{R}^k_+$. Then (iii) follows from Lemma 4.8(ii) and (iii). □

In order to prove the asymptotic behavior of $\phi_{ij}(\underline{z}, x)$ we will need the following comparison Lemma. See also [[1], Theorem 4.7].

Lemma 4.16. *Assume that f_1, f_2 are continuous positive functions on $\{x \,|\, |x| \geqq R\}$ with the following properties:*

(i) *$f_1(x)$, $f_2(x) \to 0$ as $|x| \to \infty$.*

(ii) $-(4\pi)^{-1}\Delta f_1=\varrho_1-w_1 f_1$, $-(4\pi)^{-1}\Delta f_1=\varrho_2-w_2 f_2$, *where the derivatives and equalities are in distributional sense, and* $\varrho_1(x)\geqq\varrho_2(x)$, $0\leqq w_1(x)\leqq w_2(x)$.

(iii) $f_1(x)\geqq f_2(x)$ *for all x such that* $|x|=R$. *Then* $f_1(x)\geqq f_2(x)$ *for all x such that* $|x|\geqq R$.

Proof. Define $\psi=f_1-f_2$. Let $S=\{x|\psi(x)<0\}$, which is open. On S $-(4\pi)^{-1}\Delta\psi=(\varrho_1-\varrho_2)-w_1\psi+f_2(w_2-w_1)\geqq 0$. The Lemma follows from MMP since, by (i) and (iii), $\psi=0$ at ∞ and $\psi>0$ on ∂S. □

Lemma 4.17. *(Asymptotic Behavior of* $\phi_{ij}(\underline{z},x)$*). Let* $\alpha<(1+\sqrt{73})/2\simeq 4.77$ *and let* $\phi_{ij}(\underline{z},x)$ *be the solution to* (4.17) *satisfying* $\phi^{1/4}\phi_{ij}\in L^2$. *Then there exists an* $R(\alpha)<\infty$ *such that* $-\phi_{ij}(\underline{z},x)\leqq B(\alpha)|x|^{-\alpha}$ *for some* $B(\alpha)>0$, *when* $|x|>R(\alpha)$.

Remark. The remark below Lemma 4.11 also applies to $-\phi_{ij}$.

Proof. Consider the equation,

$$-(4\pi)^{-1}\Delta f=br^{2(1-\alpha)}-dr^{-2}f, \quad \text{for} \quad r=|x|>R \tag{4.18}$$

where $\alpha(\alpha-1)=4\pi d$, together with the boundary condition $f(x)=N$ for $|x|=R$. The solution $f(x)$ to (4.18), going to zero at infinity is,

$$f(x)=N(R/r)^{\alpha}+4\pi b(3\alpha-5)^{-1}(\alpha-4)^{-1}r^{-\alpha}R^{4-\alpha}(1-(R/r)^{\alpha-4}). \tag{4.19}$$

Given any $\alpha(c)<(1+\sqrt{73})/2$ there exists $R(c)<\infty$ such that $w(x)^2\equiv(3/2)\phi(\underline{z},x)^{1/2}\geqq c|x|^{-4}$ (P-4) and $\phi_i(\underline{z},x)\leqq M(\alpha(c))|x|^{-\alpha(c)}$ (Lemma 4.11(iv)) for $|x|\geqq R(c)$. Hence $(3/4)\phi^{-1/2}\phi_i\phi_j\leqq(3/4)M(c)^2c^{-1}|x|^{2-2\alpha(c)}$ for $|x|\geqq R(c)$. Using (4.17), (4.18) and the comparison Lemma we get:

$$-\phi_{ij}(\underline{z},x)\leqq[3\pi M^2c^{-1}(3\alpha-5)^{-1}(\alpha-4)^{-1}]r^{-\alpha}R^{4-\alpha}(1-(R/r)^{\alpha-4})+N(R/r)^{\alpha}$$

for $|x|\geqq R$, with $N=\max_{|x|=R}(-\phi_{ij}(\underline{z},x))$ which is finite because ϕ_{ij} is bounded (Lemma 4.15(i)). □

Lemma 4.18. *(Continuity of* $\phi_{ij}(\underline{z},x)$*).* $\phi_{ij}(\underline{z},x)$ *is continuous in* $\underline{z}$ *for all* $\underline{z}\in\mathbb{R}^k_+\backslash\{\underline{0}\}$, *uniformly in x.*

Proof. Let $\underline{z}^*\in\mathbb{R}^k_+\backslash\{\underline{0}\}$ such that $\underline{z}-\underline{z}^*\in\mathbb{R}^k_+$. Let $\underline{z}_1$, $\underline{z}_2\in(\underline{z}^*+\mathbb{R}^k_+)$ with $(\underline{z}_1-\underline{z}_2)\in\mathbb{R}^k_+$. Lemma 4.15(iii) and Eq. (4.17) imply

$$0\leqq\phi_{ij}(\underline{z}_1,x)-\phi_{ij}(\underline{z}_2,x)\leqq I(x)+J(x), \tag{4.20a}$$

with

$$\begin{aligned} I(x)=(3/4)\int|x-y|^{-1}\{&\phi(\underline{z}_2,y)^{-1/2}\phi_i(\underline{z}_2,y)\phi_j(\underline{z}_2,y)\\ &-\phi(\underline{z}_1,y)^{-1/2}\phi_i(\underline{z}_1,y)\phi_j(\underline{z}_1,y)\}dy \end{aligned} \tag{4.20b}$$

and

$$J(x)=(3/2)\int|x-y|^{-1}(-\phi_{ij}(\underline{z}^*,y))[\phi(\underline{z}_1,y)^{1/2}-\phi(\underline{z}_2,y)^{1/2}]dy. \tag{4.20c}$$

To estimate $J(x)$ we use (4.12) to get.

$$J(x)\leqq\|\underline{z}_1-\underline{z}_2\|_2^{1/2}g_1(x), \tag{4.21}$$

where $g_1(x)=(3/2)\int|x-y|^{-1}(-\phi_{ij}(z^*, y))\sum_{n=1}^{k}|y-R_n|^{-1/2}dy$. Since for $\underline{z}\neq\underline{0}$ ϕ_{ij} is bounded everywhere [Lemma 4.15(i)] and since it goes to zero at least as fast as $r^{-4.5}$ at infinity (Lemma 4.17), $|y|^{-1/2}\phi_{ij}\in L^p$ for any $1\leqq p<6$. In particular $|y-R_n|^{-1/2}\phi_{ij}\in L^{4/3}\cap L^2$ and therefore, by Young's inequality $g_1\in L^\infty$ because $|x|^{-1}\in L^4+L^2$. $I(x)$ is decomposed as,

$$I(x)=I_1(x)+I_2(x)+I_3(x) \tag{4.22}$$

where

$$I_1(x)=(3/4)\int|x-y|^{-1}(\phi(\underline{z}_2,y)^{-1/2}-\phi(\underline{z}_1,y)^{-1/2})\phi_i(\underline{z}_2,y)\phi_j(\underline{z}_2,y)dy\,, \tag{4.23a}$$

$$\begin{aligned}I_2(x)=(3/4)\int|x-y|^{-1}(\phi_i(\underline{z}_2,y)\\ -\phi_i(\underline{z}_1,y))\phi(\underline{z}_1,y)^{-1/2}\phi_j(\underline{z}_2,y)dy\,,\end{aligned} \tag{4.23b}$$

$$\begin{aligned}I_3(x)=(3/4)\int|(x-y|^{-1}(\phi_j(\underline{z}_2,y)\\ -\phi_j(\underline{z}_1,y))\phi(\underline{z}_1,y)^{-1/2}\phi_i(\underline{z}_1,y)dy\,.\end{aligned} \tag{4.23c}$$

To find a bound for $I_1(x)$ we use Lemma 4.11(ii) and the following estimate:

$$\begin{aligned}&\phi(\underline{z}_2,y)^{-1/2}-\phi(\underline{z}_1,y)^{-1/2}\\ &\quad=[\phi(\underline{z}_1,y)-\phi(\underline{z}_2,y)]\phi(\underline{z}_1,y)^{-1}\phi(\underline{z}_2,y)^{-1}[\phi(\underline{z}_1,y)^{-1/2}+\phi(\underline{z}_2,y)^{-1/2}]^{-1}\\ &\quad\leqq\|\underline{z}_1-\underline{z}_2\|_2\left[\sum_{i=1}^{k}|x-R_i|^{-1}\right](1/2)\phi(\underline{z}^*,y)^{-3/2}\,.\end{aligned}$$

which follows from Eq. (4.1) and (P-7). Hence we have

$$I_1(x)\leqq\|\underline{z}_1-\underline{z}_2\|_2 g_2(x)\,, \tag{4.24}$$

with $g_2(x)=(3/8)\int|x-y|^{-1}\phi(\underline{z}^*,y)^{-3/2}\phi_i(\underline{z}^*,y)\phi_j(z^*,y)\sum_{n=1}^{k}|x-R_n|^{-1}dy$. Since $\phi_i\phi^{-1}\in L^p$ for any $p\geqq 6$, $(\phi_i\phi^{-1})(\phi_j\phi^{-1})\in L^p$, $p\geqq 3$. Also, $\phi^{1/2}\sum_{n=1}^{k}|x-R_n|^{-1}\in L^q$, $1<q<2$ and therefore $\phi^{-3/2}\phi_i\phi_j\sum_{n=1}^{k}|x-R_n|^{-1}\in L^s$, $1\leqq s<2$. Since $|x|^{-1}\in L^4+L^{5/2}$ $g_2\in L^\infty$ by Young's inequality. Equations (4.12) and (4.23) and Lemma (4.13) imply

$$I_2(x)\leqq(3/4)\int|x-y|^{-1}g(y)\,\|\underline{z}_1-\underline{z}_2\|_2^{1/2}\phi(z^*,y)^{-1/2}\phi_j(z^*,y)dy$$

with

$$g(x)=(3/2)\int|x-y|^{-1}\sum_{i=1}^{k}|y-R_i|^{-1/2}\phi_i(z^*,y)dy\in L^\infty$$

because of Lemma 4.13. Then

$$I_2(x)\leqq\|\underline{z}_1-\underline{z}_2\|_2^{1/2}g_3(x) \tag{4.25}$$

with

$$g_3(x)=(3/4)\,\|g\|_\infty\int|x-y|^{-1}\phi(\underline{z}^*,y)^{-1/2}\phi_j(\underline{z}^*,y)dy\,.$$

Since $\phi^{-1/2}\phi_j \in L^p$ for any $(3/2.5) \leqq p < 6$ [Lemma 4.11(iv) and (P-4)] and $|x|^{-1} \in L^4 + L^2$ we have $g_3 \in L^\infty$. Then the lemma follows from Eqs. (4.20a), (4.21), (4.24), (4.25) and Theorem 4.12. □

We conclude with the proof that $\phi \in C^2(\mathbb{R}_+^k \backslash \underline{0})$ uniformly in x, with the following

Lemma 4.19. *(Convergence of $\phi_{ij}^\varepsilon(\underline{z}, x)$ to $\phi_{ij}(\underline{z}, x)$): Let*

$$\phi_{ij}^\varepsilon(\underline{z}, x) \equiv \varepsilon^{-1}[\phi_i(\underline{z} + \varepsilon \underline{e}_j, x) - \phi_i(\underline{z}, x)]$$

with $\underline{e}_j = [\delta_j^i]$ unit vector in $\mathbb{R}_+^k$ along z_j and $\varepsilon \geqq -z_j$. Then $\phi_{ij}^\varepsilon(\underline{z}, x) \to \phi_{ij}(\underline{z}, x)$ as $\varepsilon \to 0$, uniformly in x.

Proof. (i) Consider first $\varepsilon > 0$. As in Lemma 4.14 we prove first:

$$\phi_{ij}(\underline{z}, x) \leqq \phi_{ij}^\varepsilon(\underline{z}, x) \leqq \phi_{ij}(\underline{z} + \varepsilon \underline{e}_j, x)\,. \tag{4.26}$$

Let $\psi(x) = \phi_{ij}^\varepsilon(\underline{z}, x) - \phi_{ij}(\underline{z}, x)$. By Lemmas 4.11(i) and 4.15(i), ψ is continuous everywhere and goes to zero at infinity. Then $S = \{x \mid \psi < 0\}$ is open and $\psi = 0$ on $\partial S \cup \{\infty\}$. Since $\varepsilon > 0$, using Lemma 4.11(ii) we get,

$$\begin{aligned} -(4\pi)^{-1}\Delta\psi &\geqq -(3/2)\phi(\underline{z}, x)^{1/2}\psi \\ &+ (3/2\varepsilon)\phi_i(\underline{z} + \varepsilon\underline{e}_j, x)[\phi(\underline{z}, x)^{1/2} - \phi(\underline{z} + \varepsilon\underline{e}_j, x)^{1/2} + (\varepsilon/2)\phi(\underline{z}, x)^{-1/2}\phi_j(\underline{z}, x)]\,. \end{aligned}$$

Moreover, $\phi(\underline{z}, x)^{1/2} - \phi(\underline{z} + \varepsilon\underline{e}_j, x)^{1/2} + (1/2)\phi(\underline{z}, x)^{-1/2}\phi_j(\underline{z}, x) \geqq 0$ because $\phi^{1/2}(\underline{z}, x)$ is concave (P-9), and ϕ is $C^2(\mathbb{R}_+^k)$ for each x. Therefore, on S $-(4\pi)^{-1}\Delta\psi \geqq 0$ and by the MMP the first inequality in (4.26) follows. The other one is proved in the same way. Lemma 4.18 and (4.26) then imply $\|\phi_{ij}^\varepsilon(\underline{z}, x) - \phi_{ij}(\underline{z}, x)\|_\infty \to 0$ as $\varepsilon \downarrow 0$ for $\underline{z} \neq \underline{0}$.

(ii) If $-z_j < \varepsilon < 0$, (4.26) is replaced by $\phi_{ij}(\underline{z} + \varepsilon\underline{e}_j, x) \leqq \phi_{ij}^\varepsilon(\underline{z}, x) \leqq \phi_{ij}(\underline{z}, x)$ and the lemma follows from that. □

IV.3. Proof of Theorems 2.8–2.11: Properties of K, A, R, e, and X

We begin by proving that $K(\underline{z})$ is in $C^1(\mathbb{R}_+^k)$ and $C^2(\mathbb{R}_+^k \backslash \underline{0})$.

Lemma 4.20. *(Existence of $K_i(\underline{z})$). Let $K(\underline{z}) \equiv (3/5)\int \phi(\underline{z}, x)^{5/2} dx$. Then*

$$K_i(\underline{z}) \equiv \lim_{\varepsilon \to 0} \varepsilon^{-1}[K(\underline{z} + \varepsilon\underline{e}_i) - K(\underline{z})]$$

exists and is equal to $(3/2)\int \phi(\underline{z}, x)^{3/2}\phi_i(\underline{z}, x)dx$, where $e_i = [\delta_i^i]$ is a unit vector along z_i.

Proof. (i) Consider first the case $\varepsilon \geqq 0$. Then

$$\varepsilon^{-1}[K(\underline{z} + \varepsilon\underline{e}_i) - K(\underline{z})] = (3/5)\int \varepsilon^{-1}[\phi(\underline{z} + \varepsilon\underline{e}_i, x)^{5/2} - \phi(\underline{z}, x)^{5/2}]dx\,. \tag{4.27}$$

Now,

$$\varepsilon^{-1}[\phi(\underline{z} + \varepsilon\underline{e}_i, x)^{5/2} - \phi(\underline{z}, x)^{5/2}] = \varepsilon^{-1}P(\mu)[\phi(\underline{z} + \varepsilon\underline{e}_i, x) - \phi(\underline{z}, x)]\phi(\underline{z} + \varepsilon\underline{e}_i, x)^{3/2}\,, \tag{4.28}$$

where $P(\mu)=(1+\mu^{5/2})^{-1}(1+\mu+\mu^2+\mu^3+\mu^4)$ and $\mu\equiv\phi(\underline{z},x)\phi(\underline{z}+\varepsilon\underline{e}_i,x)^{-1}<1$ by (P-7). Hence $P(\mu)\leqq 5/2$. Moreover, because of (P-9) and Theorem 2.6,

$$\phi(\underline{z}+\varepsilon\underline{e}_i,x)-\phi(\underline{z},x)\leqq\varepsilon\phi_i(\underline{z},x) \tag{4.29}$$

Using (4.29), (4.27), and (4.28) we get:

$$\begin{aligned}\varepsilon^{-1}[\phi(\underline{z}+\varepsilon\underline{e}_i,x)^{5/2}-\phi(\underline{z},x)^{5/2}]&\leqq(5/2)\phi_i(\underline{z},x)\phi(\underline{z}+\varepsilon\underline{e}_i,x)^{3/2}\\&\leqq(5/2)|x-R_i|^{-1}\phi(\underline{z}+\underline{e}_i,x)^{3/2},\end{aligned}$$

where the last inequality follows from (P-7) (assuming $\varepsilon\leqq 1$) and Lemma 4.11(ii). Since $\phi^{3/2}(\underline{z},x)\in L^1\cap L^2$ at least, $|x-R_i|^{-1}\phi^{3/2}(\underline{z}+\underline{e}_i,x)\in L^1$. Hence the lemma follows by Theorem 2.6 and dominated convergence. (ii) In the case $-z_i\leqq\varepsilon\leqq 0$, an analysis similar to the above yields

$$\begin{aligned}\varepsilon^{-1}[\phi(\underline{z}+\varepsilon\underline{e}_i,x)^{5/2}-\phi(\underline{z},x)^{5/2}]&\leqq 5/2\phi_i(\underline{z}+\varepsilon\underline{e}_i,x)\phi(\underline{z},x)^{3/2}\\&\leqq(5/2)|x-R_i|^{-1}\phi(\underline{z},x)^{3/2}\in L^1.\quad\square\end{aligned}$$

Lemma 4.20 assures us that the derivatives of $K(\underline{z})$ along the axis exist. The proof that $K(\underline{z})$ is in fact in $C^1(\mathbb{R}^k_+)$ is provided by the following:

Lemma 4.21. *(Continuity of $K_i(\underline{z})$). $K_i(\underline{z})\equiv(3/2)\int\phi(\underline{z},x)^{3/2}\phi_i(\underline{z},x)dx$ is continuous for all $\underline{z}\in\mathbb{R}^k_+$.*

Proof. Let us prove continuity at $\underline{z}$. Let $\underline{z}^*=2\max_{1\leqq i\leqq k}(z_i)(1,1,\ldots,1)$. Let $A\equiv\{\underline{z}'\in\mathbb{R}^k_+\,|\,\underline{z}^*-\underline{z}'\in\mathbb{R}^k_+\}$. Obviously $\underline{z}\in A$. Consider $z'\in A$, $\underline{z}'\to\underline{z}$. By Corollary 2.7(iii),

$$\phi(\underline{z}',x)^{3/2}\phi_i(\underline{z}',x)\leqq|x-R_i|^{-1}\phi(\underline{z}^*,x)^{3/2}\in L^1.$$

Moreover, as $\underline{z}'\to\underline{z}$, $\phi(\underline{z}',x)^{3/2}\phi_i(\underline{z}',x)\to\phi(\underline{z},x)^{3/2}\phi_i(\underline{z},x)$ everywhere (Theorem 2.6). The lemma follows by dominated convergence. $\square$

The following is a useful alternative expression for $K_i(\underline{z})$.

Lemma 4.22.

$$K_i(\underline{z})=3\lim_{x\to R_i}\left\{\phi(\underline{z},x)-\sum_{j=1}^k z_j\phi_j(\underline{z},x)\right\}. \tag{4.30}$$

Note. Because of (P-3) and Theorem 2.6(iv), the above limit exists.

Proof. From Eqs. (4.1) and (4.2) we have

$$\begin{aligned}F(\underline{z},x)&\equiv\phi(\underline{z},x)-\sum_{j=1}^k z_j\phi_j(\underline{z},x)\\&=(3/2)\sum_{j=1}^k\int|x-y|^{-1}\phi(\underline{z},y)^{1/2}z_j\phi_j(\underline{z},y)dy-\int|x-y|^{-1}\phi(\underline{z},y)^{3/2}dy.\end{aligned} \tag{4.31}$$

The two integrals on the right side of (4.31) are bounded and continuous everywhere, therefore,

$$F(\underline{z},R_i)=(3/2)\sum_{j=1}^{k}\int|y-R_i|^{-1}z_j\phi_j(\underline{z},y)\phi(\underline{z},y)^{1/2}dy-\int|y-R_i|^{-1}\phi(\underline{z},y)^{3/2}dy\,. \tag{4.32}$$

Because of Eq. (4.2) and Theorem 2.6(v) we have

$$\int|y-R_i|^{-1}\phi(\underline{z},y)^{1/2}\phi_j(\underline{z},y)dy=\int|y-R_j|^{-1}\phi(\underline{z},y)^{1/2}\phi_i(\underline{z},y)dy \tag{4.33}$$

for all i,j. From Eqs. (4.32) and (4.33) we get,

$$F(\underline{z},R_i)=(3/2)\int\phi(\underline{z},y)^{1/2}\phi_i(\underline{z},y)\sum_{j=1}^{k}z_j|y-R_j|^{-1}dy-\int|y-R_i|^{-1}\phi(\underline{z},y)^{3/2}dy\,.$$

Combining the last equation and Eqs. (4.1) and (4.2) we finally get

$$F(\underline{z},R_i)=(1/2)\int\phi(\underline{z},y)^{3/2}\phi_i(\underline{z},y)dy=K_i(\underline{z})/3\,,$$

which is Eq. (4.30). Note that to get the first equality we have used

$$\begin{aligned}&\int dy\phi(\underline{z},y)^{1/2}\phi_i(\underline{z},y)\left[\int dw|w-y|^{-1}\phi(\underline{z},w)^{3/2}\right]\\&\quad=\int dy\phi(\underline{z},y)^{3/2}\left[\int dw|w-y|^{-1}\phi(\underline{z},w)^{1/2}\phi_i(\underline{z},w)\right],\end{aligned}$$

which is true by Fubini's theorem, since $\phi(\underline{z},y)^{1/2}\phi_i(\underline{z},y)\in L^1$ (Lemma 4.10) and $\int dw|w-y|^{-1}\phi(\underline{z},w)^{3/2}\in L^\infty$ (Theorem IV.1, [1]). □

The right side of (4.30) can be written in terms of the right sides of the integral Eqs. (4.1) and (4.2). Using the same kind of dominated convergence argument as in the proof of Lemmas 4.20 and 4.21, it is easy to check that K_i is differentiable and

$$\begin{aligned}K_{ij}(\underline{z})&=-3\sum_{l=1}^{k}z_l\phi_{lj}(\underline{z},R_i)\\&=-3\sum_{l=1}^{k}z_l\phi_{ij}(\underline{z},R_l)\,,\end{aligned} \tag{4.34}$$

where the last equality follows by using Theorem 2.6.

Proof of Theorem 2.8. $K\in C^1$, $K\in C^2(\mathbb{R}^k\backslash\underline{0})$ follow from Lemmas 4.20, 4.21, Eq. (4.34) and Theorem 2.6. (i) is proved in Lemma 4.22 and (ii) is Eq. (4.34). □

Proof of Theorem 2.10. (i) Let us start with the convexity of $R(\underline{z})$. Define $\underline{z}\equiv\alpha\underline{z}_1+(1-\alpha)\underline{z}_2$, $\alpha\in[0,1]$. Consider now the following identity:

$$\begin{aligned}&2[R(\underline{z})-\alpha R(\underline{z}_1)-(1-\alpha)R(\underline{z}_2)]\\&\quad=2\int dxdy[\phi(\underline{z},x)^{3/2}-\alpha\phi(\underline{z}_1,x)^{3/2}-(1-\alpha)\phi(\underline{z}_2,x)^{3/2}]\,|x-y|^{-1}\phi(\underline{z},y)^{3/2}\\&\quad\quad-\alpha\int dxdy[\phi(\underline{z},x)^{3/2}-\phi(\underline{z}_1,x)^{3/2}]\,|x-y|^{-1}[\phi(\underline{z},y)^{3/2}-\phi(\underline{z}_1,y)^{3/2}]\\&\quad\quad-(1-\alpha)\int dxdy[\phi(\underline{z},x)^{3/2}-\phi(\underline{z}_2,x)^{3/2}]\,|x-y|^{-1}[\phi(\underline{z},y)^{3/2}-\phi(\underline{z}_2,y)^{3/2}]\,.\end{aligned} \tag{4.35}$$

The last two terms of the left side of (4.35) are negative because $|x|^{-1}$ is a positive kernel. Moreover from Eq. (4.1) we get the following identity

$$\begin{aligned}&\int dxdy|x-y|^{-1}[\phi(\underline{z},x)^{3/2}-\alpha\phi(\underline{z}_1,x)^{3/2}-(1-\alpha)\phi(\underline{z}_2,x)^{3/2}]\phi(\underline{z},y)^{3/2}\\&\quad=-\int dxdy[\phi(\underline{z},x)-\alpha\phi(\underline{z}_1,x)-(1-\alpha)\phi(\underline{z}_2,x)]\phi(z,y)^{3/2}\leqq 0\,,\end{aligned}\tag{4.36}$$

where the last inequality follows from the concavity of $\phi(\underline{z},x)$ (P-9). From (4.35) and (4.36) the convexity follows. The SSA is proved in a similar way. The Virial theorem (Theorem II.23, [1]) yields

$$A(\underline{z})=(5/3)K(\underline{z})+2R(\underline{z})\,,\tag{4.37}$$

and hence the convexity and SSA of A follow from those of K [Corollary 2.9(ii)] and R. (ii) That $e(\underline{z})$ is WSA on $\mathbb{R}^k_+$ is proven in [1], Theorem V.7. See also [3], Theorem 1. (This is in fact Teller's Theorem [5]). □

Proof of Theorem 2.11. Equation (2.4) and Theorem 2.8 imply

$$X_i(\underline{z})=2K_i(\underline{z})-\sum_{j=1}^{k} z_jK_{ji}(\underline{z})\,.\tag{4.38}$$

Using now (2.2), (2.3), and (4.38) we get

$$X_i(\underline{z})=\lim_{x\to R_i} 6N(\underline{z},x)\,,\tag{4.39}$$

with

$$N(\underline{z},x)=\phi(\underline{z},x)-\sum_{j=1}^{k} z_j\phi_j(\underline{z},x)+(\tfrac{1}{2})\sum_{j,l=1}^{k} z_lz_j\phi_{jl}(\underline{z},x)\,.\tag{4.40}$$

Note that in order to obtain (4.39) we have used $\phi\in C^2$ and also Theorem 2.6(v). Note also that the limit in (4.39) in fact exists because of (P-2), Theorem 2.6(iv) and (ii). Let us compute $N(\underline{z}+\underline{\varepsilon},x)-N(\underline{z},x)$ for $\underline{\varepsilon}\in\mathbb{R}^k_+$. Using (P-9) and $\phi\in C^1$ we have,

$$\phi(\underline{z}+\underline{\varepsilon},x)-\phi(\underline{z},x)\geqq\sum_{j=1}^{k}\varepsilon_j\phi_j(\underline{z}+\underline{\varepsilon},x)\,.\tag{4.41}$$

From (4.40) and (4.41),

$$\begin{aligned}N(\underline{z}+\underline{\varepsilon},x)-N(\underline{z},x)\geqq&-\sum_{j=1}^{k} z_j[\phi_j(\underline{z}+\underline{\varepsilon},x)-\phi_j(\underline{z},x)]\\&+(1/2)\sum_{j,l=1}^{k}(\underline{z}+\underline{\varepsilon})_l(\underline{z}+\underline{\varepsilon})_j\phi_{jl}(\underline{z}+\underline{\varepsilon},x)-(1/2)\sum_{l,j=1}^{k} z_lz_j\phi_{jl}(\underline{z},x)\,.\end{aligned}\tag{4.42}$$

The ray-convexity of ϕ_i [Theorem 2.6(iii)] implies:

$$\sum_{j=1}^{k}\varepsilon_j\phi_{ij}(\underline{z},x)\leqq\phi_i(\underline{z}+\underline{\varepsilon},x)-\phi_i(\underline{z},x)\leqq\sum_{j=1}^{k}\varepsilon_j\phi_{ij}(\underline{z}+\underline{\varepsilon},x)\,.$$

We conclude that

$$N(\underline{z}+\underline{\varepsilon},x)-N(\underline{z},x)\geqq(1/2)\sum_{j,l=1}^{k}\varepsilon_l\varepsilon_j\phi_{jl}(\underline{z}+\underline{\varepsilon},x)\,,\tag{4.43}$$

since $\phi_{jl}(\underline{z}+\underline{\varepsilon}, x) \geqq \phi_{jl}(\underline{z}, x)$ [Lemma 4.15(iii)]. Although the right side of (4.43) is negative, it is second order in ε. The following Lemma 4.23 shows that under these conditions $N(\underline{z}, x)$ must in fact be increasing. Hence $X_i(\underline{z})$ is increasing and by Lemma 2.1 X is SSA. □

Lemma 4.23. *Suppose $f:[a,b]\to\mathbb{R}$ is a real-valued function such that for every $x\in[a,b]$ there is a $c(x)$ such that $f(z)-f(x)\geqq c(x)(z-x)^2$ for all $z\in[a,b]$ with $z\geqq x$. Suppose further that $c\in L^1([a,b])$. Then f is increasing, i.e. $z>x\Rightarrow f(z)\geqq f(x)$.*

Proof. Let $N>1$ be an integer and let I_j, for $j=1,\dots,n$ be the interval $I_j=(x+(j-1)(z-x)/n,\ x+j(z-x)/n)$. Then

$$\delta\equiv f(z)-f(x)=\sum_{j=0}^{n}(f(y_{j+1})-f(y_j))$$

with $y_0=x$, $y_{n+1}=z$ and $y_j\in I_j$. Without loss of generality we can assume $c(x)\leqq 0$, all x. Then

$$\delta\geqq\left\{n^{-1}\sum_{j=0}^{n}c(y_j)\right\}\{4(z-x)^2/n\}$$

because $y_{j+1}-y_j\leqq 2(z-x)/n$. Let $d_j=\int_{I_j}c(x)dx$ and $d=\sum_{j=1}^{n}d_j=\int_x^z c\geqq\int_a^b c>-\infty$.

For each j, there exists $y_j\in I_j$ such that $c(y_j)\int_{I_j}1\geqq d_j$, otherwise $\int_{I_j}c<d_j$. Using these y_j we have

$$\delta\geqq\left\{n^{-1}c(x)+n^{-1}\sum_{j=1}^{n}d_j((z-x)/n)^{-1}\right\}\{4(z-x)^2/n\}.$$

Taking $n\to\infty$ proves the lemma. □

Appendix A

Properties of Superadditive and Convex Functions on $\mathbb{R}^k_+$

The definition of superadditive and convex functions on $\mathbb{R}^k_+$, as well as many of their properties, were stated in Section II. Those properties, Lemmas 2.1 to 2.4 and Corollary 2.5, will be proved here.

Proof of Lemma 2.1. (i) $\Rightarrow$ is trivial because $f\in C^2(\mathbb{R}^k_+)$. To prove $\Leftarrow$, define $F(\lambda,\mu)\equiv f(\underline{z}+\lambda\underline{z}_1+\mu\underline{z}_2)-f(\underline{z}+\lambda\underline{z}_1)-f(\underline{z}+\mu\underline{z}_2)+f(\underline{z})$. Then, for $\underline{z}_1,\underline{z}_2\in\mathbb{R}^k_+$,

$$F_\lambda(\lambda,\mu)\equiv(\partial F(\lambda,\mu)/\partial\lambda)=\sum_{i=1}^{k}(z_1)_i\left[f_i(\underline{z}+\lambda\underline{z}_1+\mu\underline{z}_2)-f_i(\underline{z}+\lambda\underline{z}_1)\right],\tag{A.1}$$

and

$$F_{\lambda\mu}(\lambda,\mu)=\sum_{i,j=1}^{k}(z_1)_i(z_2)_j f_{ij}(\underline{z}+\lambda\underline{z}_1+\mu\underline{z}_2)\geqq 0,\tag{A.2}$$

where the last inequality follows from $f_{ij} \geqq 0$ all i,j. But $F_\lambda(\lambda, 0) \geqq 0$, and hence $F_\lambda(\lambda, \mu) \geqq 0$. Also, $F(0, \mu) = 0$, and hence $F(\lambda, \mu) \geqq 0$. (ii) $\Rightarrow$ follows immediately from the definition of an SSA function and the fact that $f \in C^1(\mathbb{R}^k_+)$. To prove $\Leftarrow$ note that if $f \in C^1(\mathbb{R}^k_+)$ and f_i is increasing we get, from (A.1), $F_\lambda(\lambda, \mu) \geqq 0$. But $F(0, \mu) = 0$ and therefore $F(\lambda, \mu) \geqq 0$. □

Proof of Lemma 2.2. This is similar to the previous one, taking into account that $f_{ij} \geqq 0$ all i,j, $i \neq j$, and $\underline{z}_1 \cdot \underline{z}_2 = 0$ imply

$$\sum_{i,j=1}^{k} (z_1)_i (z_2)_j f_{ij} \geqq 0 \quad \square$$

Proof of Lemma 2.3. If f is SSA, taking $\underline{z}_3 = \underline{z}_2$, in definition (iii) we have that $f((1/2)\underline{z}_1 + (1/2)(\underline{z}_1 + 2\underline{z}_2)) \leqq (f(\underline{z}_1)/2) + (f(\underline{z}_1 + 2\underline{z}_2)/2)$. □

Lemma 2.4 is a well known fact for differentiable convex functions. See [8], for example.

Proof of Corollary 2.5.

$$\begin{aligned} f(\underline{z}_1 + \underline{z}_2) - f(\underline{z}_1) &= \int_0^1 \frac{d}{d\lambda}(f(\underline{z}_1 + \lambda \underline{z}_2)) d\lambda \\ &= \int_0^1 \sum_{i=1}^{k} (z_2)_i f_i(\lambda \underline{z}_2 + \underline{z}_1) d\lambda \leqq \sum_{i=1}^{k} (z_2)_i f_i(\underline{z}_2 + \underline{z}_1) \end{aligned}$$

because f_i is increasing, by Lemma 2.4. The other inequality is proved in the same way. □

Appendix B

Positivity of the Pressure Under Separation Relative to a Plane (in the Symmetric Case)

Consider $2k$ nuclei with coordinates $R_1, \ldots, R_k$ and $R_{-1}, \ldots, R_{-k}$ and strictly positive charges $z_1, \ldots, z_k, z_{-1}, \ldots, z_{-k}$ satisfying (for $i = 1, \ldots, k$)

$$z_i = z_{-i}$$

$$R_i^1 = R_{-i}^1, R_i^2 = R_{-i}^2, \; -R_{-i}^3 = R_i^3 > 0 .$$

Let $e(l)$ denote the TF energy for this molecule when R_i^3 is replaced by $R_i^3 + l$, $i > 0$ and by $R_i^3 - l$, $i < 0$. The electron charge λ is immaterial but is fixed at some value $\lambda \leqq 2 \sum_{i=1}^{k} z_i$.

Theorem B.1. *The pressure is strictly positive, i.e.,*

$$e(l) < e(0) \quad \textit{for} \quad l > 0 . \tag{B.1}$$

Proof. The proof consists of showing that if the charge distribution (electron and nuclear) is cut in two parts at the $x^3 = 0$ plane, and then pulled apart by a distance

$2l$, the energy is lowered. Let $\varrho(x)$ be the TF density when $l=0$. Define $\varrho_l(x)$ by

$$\varrho_l(x)=\varrho(x-(0,0,l)),\; x^3\geqq l$$

$$\varrho_l(x)=\varrho(x+(0,0,l)),\; x^3\leqq -l$$

$$\varrho_l(x)=0,\; -l<x^3<l.$$

Clearly $\int\varrho_l=\int\varrho=\lambda$, and we will use ϱ_l as a trial density for the l problem. We will show that $\mathscr{E}_l(\varrho_l)<\mathscr{E}(\varrho)=e(0)$, where $\mathscr{E}_l$ (resp. $\mathscr{E}$) is the energy functional for l (resp. 0). Obviously, $K(\varrho_l)=K(\varrho)$.

Let $D\subset\mathbb{R}^3$ be the domain $\{(x^1,x^2,x^3)|x^3\geqq 0\}$. For any function f: $f:D\to C$, let

$$W_l(f)=\int_D\int_D dxdy\bar{f}(x)f(y)K_l(x,y),$$

$$K_l(x,y)=\{(x^1-y^1)^2+(x^2-y^2)^2+(x^3+y^3+2l)^2\}^{-1/2}. \tag{B.2}$$

In other words, $W_l(f)$ is the Coulomb interaction energy between a charge distribution f, supported on the $x^3\geqq 0$ side of the xy plane, and its (complex conjugate) reflection through the plane $x^3=-l$. It is easy to see that

$$\mathscr{E}_l(\varrho_l)-\mathscr{E}(\varrho)=W_l(\mu)-W_0(\mu),$$

where μ is the charge density for $x^3\geqq 0$ for the $l=0$ problem, namely for $x^3\geqq 0$

$$\mu(x)=-\varrho(x)+\sum_{j=1}^{k} z_j\delta(x-R_j). \tag{B.3}$$

Since $\mu\neq 0$, and $W_0(\mu)=\lim_{l\downarrow 0} W_l(\mu)$, the following Lemma B.2 completes the proof. □

Lemma B.2. *(Reflection Positivity of the Coulomb Potential). Let f be a non-null function with support in $D=\{x|x^3\geqq 0\}$ and with $f\in L^1(D)$. Then, for $l>0$, $W_l(f)>0$, and $W_l(f)$ is a finite, strictly decreasing function of l. Moreover, $W_l(f)$ is a* log *convex function of l, vanishing at $l=\infty$.*

Proof. Using the well-known representation for $|x|^{-1}$, we have that

$$\begin{aligned}K_l(x,y)&=(2\pi^2)^{-1}\int d^3p|p|^{-2}\exp\{i[p^1(x^1-y^1)+p^2(x^2-y^2)\\&\quad+p^3(x^3+y^3+2l)]\}\\&=(2\pi)^{-1}\int d^2p|p|^{-1}\bar{g}_p(x)g_p(y)\exp(-2|p|l)\end{aligned}$$

with $g_p(x)=\exp[ip^1x^1+ip^2x^2-|p|x^3]$.

We have used the fact that

$$\int_{-\infty}^{\infty} dp^3[(p^3)^2+a^2]^{-1}\exp[ip^3(x^3+y^3+2l)]=(\pi/a)\exp[-a(x^3+y^3+2l)]$$

when $x^3+y^3+2l>0$, as it is here. For $p\in\mathbb{R}^2$, let $h(p)=\int_D f(x)g_p(x)d^3x$. Since $f\in L^1(D)$, $|h(p)|\leqq\|f\|_1$, and $h(p)$ is null if and only if f is null. For $l>0$ Fubini's theorem yields

$$W_l(f)=(2\pi)^{-1}\int d^2p|p|^{-1}|h(p)|^2\exp(-2|p|l). \tag{B.4}$$

The representation (B.4) proves the lemma. □

References

1. Lieb, E. H., Simon, B.: Advan. Math. **23**, 22–116 (1977)
2. Lieb, E. H., Simon, B.: Phys. Rev. Letters **31**, 681–683 (1973)
   Lieb, E. H.: Proc. Int. Congress of Math., Vancouver (1974)
   Lieb, E. H.: Rev. Mod. Phys. **48**, 553–569 (1976)
3. Benguria, R., Lieb, E. H.: Ann. Phys. (N.Y.) **110**, 34–45 (1978)
4. Balàsz, N.: Phys. Rev. **156**, 42–47 (1967)
5. Teller, E.: Rev. Mod. Phys. **34**, 627–631 (1962)
6. Reed, M., Simon, B.: Methods of modern mathematical physics. Vol. II, Fourier analysis and self-adjointness. New York: Academic Press 1975
7. Hille, E.: Proc. Nat. Acad. Sci. **62**, 7–10 (1969)
8. Rockafellar, R. T.: Convex analysis. Princeton: University Press 1970
9. Faris, W.: Duke Math. J. **43**, 365–372 (1976)
10. Strichartz, R. S.: J. Math. Mech. **16**, 1031–1060 (1967)
11. Lieb, E. H., Simon, B.: Monotonicity of the electronic contribution to the Born-Oppenheimer energy. J. Phys. B (London) **11**, L537–542 (1979)
12. Brezis, H., Lieb, E. H.: Long range atomic potentials in Thomas-Fermi theory. Commun. math. Phys. (submitted)

Communicated by J. Glimm

Received July 24, 1978

Note Added in Proof

In a recent related work [12], H. Brezis and E. H. Lieb have proved that the interaction among neutral atoms in Thomas-Fermi theory behaves, for large separation l, like Γl^{-7}.

Erratum

The Positivity of the Pressure in Thomas-Fermi Theory

R. Benguria and E. H. Lieb

Commun. Math. Phys. **63**, 193–218 (1978)

The conclusion of Lemma 2.4 should read:

then $\sum_{i=1}^{k} w_i f_i(\underline{z}+\lambda\underline{w})$ is increasing in λ for $\lambda \geqq 0$ and $\underline{z}, \underline{w} \in \mathbb{R}_+^k$.

In the proof of Corollary 2.5 "...f_i is increasing" should be replaced by "...$\Sigma(z_2)_i f_i(\lambda \underline{z}_2 + \underline{z}_1)$ is increasing ...". Lemma 2.4 was used only to prove Corollary 2.5, and the latter was used in the equation following (4.42).

The Thomas–Fermi–von Weizsäcker Theory of Atoms and Molecules*

Rafael Benguria[1], Haim Brezis[2], and Elliott H. Lieb[3]

1. The Rockefeller University, New York, NY 10021, USA, on leave from Universidad de Chile, Santiago, Chile
2. Département de Mathématiques, Université Paris VI, F-75230, Paris Cedex 05, France
3. Departments of Mathematics and Physics, Princeton University, Princeton, NJ 08544, USA

Abstract. We place the Thomas–Fermi–von Weizsäcker model of atoms on a firm mathematical footing. We prove existence and uniqueness of solutions of the Thomas–Fermi–von Weizsäcker equation as well as the fact that they minimize the Thomas–Fermi–von Weizsäcker energy functional. Moreover, we prove the existence of binding for two very dissimilar atoms in the frame of this model.

Introduction

The Thomas–Fermi theory of atoms [1] (TF), attractive because of its simplicity, is not satisfactory because it yields an electron density with incorrect behavior very close and very far from the nucleus. Moreover, it does not allow for the existence of molecules. In order to correct this, von Weizsäcker [2] suggested the addition of an inhomogeneity correction

$$U_w(\rho) = C_w(\nabla\rho)^2/\rho \tag{1}$$

to the kinetic energy density. Here $c_w = h^2/(32\pi^2 m)$, where m is the mass of the electron. This correction has also been obtained as the first order correction to the TF kinetic energy in a semi-classical approximation to the Hartree–Fock theory [3].

The Thomas–Fermi–von Weizsäcker (henceforth TFW) energy functional for nuclei of charges $z_i > 0$ (which need not be integral) located at R_i, $i = 1, \ldots, k$ is defined by

$$\begin{aligned}\xi(\rho) = (3\pi^2)^{-2/3}\int(\nabla\rho^{1/2}(x))^2\,dx + \tfrac{3}{5}\int\rho(x)^{5/3}dx \\ - \int V(x)\rho(x)dx + \tfrac{1}{2}\iint\rho(x)\rho(y)|x-y|^{-1}dxdy,\end{aligned} \tag{2}$$

in units in which $h^2(8m)^{-1}(3/\pi)^{2/3} = 1$ and $|e| = 1$. Here $\rho(x) \geqq 0$ is the electron

* Research supported by U. S. National Science Foundation under Grants MCS78-20455 (R. B.), PHY-7825390 A 01 (H. B. and E. L.), and Army Research Grant DAH 29-78-6-0127 (H. B.)

density, and

$$V(x) = \sum_{i=1}^{k} z_i |x - R_i|^{-1}. \tag{3}$$

While the pure TF problem has been placed on a rigorous mathematical footing [1], no parallel study has been made for the TFW problem. Such a study was undertaken in the Ph.D. thesis of one of us [4]; in this paper some of the results of [4] will be presented together with some newer results.

In this article we will study a rather more general functional, which contains the TFW energy functional (2) as a particular case. In fact, for $\rho(x) \geqq 0$ and $V(x)$ given by (3), let us introduce the functional

$$\xi_p(\rho) = \int |\nabla \rho^{1/2}(x)|^2 dx + \frac{1}{p} \int \rho^p(x) dx - \int V(x)\rho(x) dx + D(\rho, \rho), \tag{4}$$

where

$$D(f, g) \equiv \tfrac{1}{2} \int\int f(x) g(y) |x - y|^{-1} dx dy, \tag{5}$$

for $1 < p < \infty$.

We shall be concerned with the following problem

$$\mathrm{Min}\{\xi_p(\rho) | \rho \in L^1 \cap L^p, \rho(x) \geqq 0, \nabla \rho^{1/2} \in L^2 \text{ and } \int \rho(x) dx = \lambda\}, \tag{I}$$

where λ is a given positive constant, which, physically, is the total electron number.

Our main result is the following:

Theorem 1. *There is a critical value $0 < \lambda_c < \infty$ depending only on p and V such that*

(a) If $\lambda \leqq \lambda_c$, Problem (I) has a unique solution.
(b) If $\lambda > \lambda_c$, Problem (I) has no solution. In addition,
(c) When $p \geqq \frac{4}{3}$, then $\lambda_c \geqq Z \equiv \sum_{i=1}^{k} z_i$.
(d) When $p \geqq \frac{5}{3}$ and $k = 1$ (atomic case), then $\lambda_c > Z$.

Remark 1. Partial results were previously obtained by one of us. Namely in [4] it is proved that for the atomic case ($k = 1$) and $p = \frac{5}{3}$, Problem (I) has a solution if $\lambda \leqq Z$.

Remark 2. Some of the open problems which are raised by our developments are the following:

(i) Suppose $k \geqq 2$ (molecular case) and $p = \frac{5}{3}$. Is $\lambda_c > Z$?
(ii) Find estimates for λ_c.
(iii) Is there binding for atoms?

With respect to the third problem, there is a non-rigorous argument of Balàzs [5] that indicates the possibility of binding for homopolar diatomic molecule in the TFW theory. Also, Gombás [6] applied TFW (including exchange corrections) to study the N_2-molecule (i.e., $z_1 = z_2 = 7$) and found numerically that there is binding. He actually computed the distance between the two centers to be 1.39Å for the configuration of minimum energy. We do not give a proof of binding in the homopolar case, but we will prove that binding occurs for two very dissimilar atoms.

Remark 3. Theorem 1 obviously holds if we replace ξ_p by

$$\xi_p(\rho) = c_1 \int |\nabla \rho^{1/2}(x)|^2 dx + c_2 \int \rho^p(x) dx - \int V(x)\rho(x) dx + D(\rho, \rho),$$

where c_1 and c_2 are positive constants.

The proof of Theorem 1 is divided into several steps. In Sect. 1 we describe some basic properties of $\xi_p(\rho)$. In Sect. 2 we consider the problem

$$\mathrm{Min}\{\xi_p(\rho) \,|\, \rho \in D_p\}$$

where $D_p = \{\rho \,|\, \rho(x) \geqq 0, \rho \in L^3 \cap L^p, \nabla \rho^{1/2} \in L^2, D(\rho, \rho) < \infty\}$, and we prove that the minimum is achieved at a unique ρ_0. Note that D_p contains $\{\rho \,|\, \rho \geqq 0, \rho \in L^3 \cap L^p \cap L^1, \nabla \rho^{1/2} \in L^2\}$. We derive the Euler equation for ρ_0. More precisely we set $\psi = \rho_0^{1/2}$ and we show that

$$-\Delta \psi + \psi^{2p-1} = \varphi \psi, \tag{6}$$

where

$$\varphi(x) = V(x) - \int \psi^2(y) |x - y|^{-1} dy. \tag{7}$$

In Sect. 3 we prove that $\psi \in L^2$ and we obtain some further properties of ψ (ψ is continuous, $\psi(x) \to 0$ as $|x| \to \infty$, etc.). In Sect. 4 we show that if $p \geqq \frac{4}{3}$ then $\int \psi^2(x) dx \geqq Z$. In Sect. 5 we show that if $p \geqq \frac{5}{3}$ and $k = 1$ then $\int \psi^2(x) dx > Z$. In addition $\psi(x) \leqq M e^{-\delta |x|^{1/2}}$ for some appropriate constants M and $\delta > 0$. In Sect. 6 we prove that for every λ

$$E(\lambda) \equiv \mathrm{Inf}\{\xi_p(\rho) \,|\, \int \rho(x) dx = \lambda\} = \mathrm{Inf}\{\xi_p(\rho) \,|\, \int \rho(x) dx \leqq \lambda\}$$

and we conclude the proof of Theorem 1. We also show that $E(\lambda)$ is convex, monotone non-increasing and that $E(\lambda)$ has a finite slope at $\lambda = 0$. This slope is the ground state energy of the corresponding one electron Schrödinger Equation. Using this last fact, binding for dissimilar atoms is proved.

I. Some Basic Properties of ξ_p

In Lemma 2, 3, 4 some properties which are useful in the study of Problem (I) are summarized.

Lemma 2. *For every* $\varepsilon > 0$, *there is a constant* C_ε, *depending on* V *but independent of* ρ, *such that*

$$\int V(x)\rho(x) dx \leqq \varepsilon \|\rho\|_3 + C_\varepsilon D(\rho, \rho)^{1/2}, \tag{8}$$

for every $\rho \geqq 0$.

Proof. Let $\delta > 0$ be a small constant and let $\zeta(x)$ be a smooth function such that

$0 \leqq \zeta \leqq 1$ and

$$\zeta(x) = \begin{cases} 1 & \text{on} \bigcup_{i=1}^{k} B_\delta(R_i) \\ 0 & \text{outside} \bigcup_{i=1}^{k} B_{2\delta}(R_i), \end{cases}$$

where $B_\delta(R_i)$ is the ball of radius δ and centered at R_i. δ is chosen such that all these ball $B_{2\delta}$ are disjoint. Let $V = V\zeta + V(1-\zeta) \equiv V_1 + V_2$. Clearly $V_1 \in L^{3/2}$ and by choosing δ small enough we may assume that $\| V_1 \|_{3/2} < \varepsilon$ Thus

$$\int V_1(x)\rho(x)dx \leqq \varepsilon \| \rho \|_3. \tag{9}$$

On the other hand define the operator B to be

$$(B\rho)(x) = \int \rho(y)|x-y|^{-1}dy, \tag{10}$$

so that (in the sense of distributions) we have

$$-\Delta(B\rho) = 4\pi\rho.$$

Thus,

$$\int |\nabla(B\rho)|^2 dx = 8\pi D(\rho, \rho). \tag{11}$$

We deduce from (11) and Sobolev's inequality that

$$\| B\rho \|_6 \leqq CD(\rho, \rho)^{1/2}. \tag{12}$$

Consequently,

$$\begin{aligned} \int V_2(x)\rho(x)dx &= \frac{1}{4\pi}\int (-\Delta V_2)(x)(B\rho)(x)dx \\ &\leqq C \| \Delta V_2 \|_{6/5} D(\rho, \rho)^{1/2} \end{aligned} \tag{13}$$

(note that $\Delta V_2 \in C_0^\infty$). Combining (9) and (13) we obtain the conclusion. □

Lemma 3. *There exist positive constants α and C such that*

$$\xi_p(\rho) \geqq \alpha(\| \rho \|_3 + \| \rho \|_p^p + \| \nabla\rho^{1/2} \|_2^2 + D(\rho, \rho)) - C$$

Proof. Use Lemma 2 and Sobolev's inequality. □

Lemma 4. *$\xi_p(\rho)$ is strictly convex.*

Proof. The only non-standard fact is that the function $\rho \to \int |\nabla\rho^{1/2}|^2 dx$ is convex (or equivalently subadditive). Indeed let $\rho_1, \rho_2 \in D_p$ and set $\psi_1 = \rho^{1/2}$, $\psi_2 = \rho_2^{1/2}$, $\psi_3 = (\alpha\rho_1 + (1-\alpha)\rho_2)^{1/2}$ with $0 < \alpha < 1$. Thus,

$$\begin{aligned} \psi_3 \nabla\psi_3 &= \alpha\psi_1 \nabla\psi_1 + (1-\alpha)\psi_2 \nabla\psi_2 \\ &= (\alpha^{1/2}\psi_1)(\alpha^{1/2}\nabla\psi_1) + [(1-\alpha)^{1/2}\psi_2][(1-\alpha)^{1/2}\nabla\psi_2] \end{aligned}$$

and by Cauchy–Schwarz inequality

$$\psi_3 \nabla\psi_3 \leqq (\alpha\psi_1^2 + (1-\alpha)\psi_2^2)^{1/2}(\alpha|\nabla\psi_1|^2 + (1-\alpha)|\nabla\psi_2|^2)^{1/2},$$

and therefore,

$$|\nabla\psi_3|^2 \leqq \alpha|\nabla\psi_1|^2 + (1-\alpha)|\nabla\psi_2|^2. \qquad \square$$

II. Minimization of $\xi_p(\rho)$ with $\rho \in D_p$—The Euler Equation

We start with

Lemma 5. Min$\{\xi_p(\rho)|\rho\in D_p\}$ *is achieved at some* $\rho_0 \in D_p$.

Proof. Let $\rho_n \in D_p$ be a minimizing sequence. By Lemma 3 we have

$$\|\rho_n\|_3 \leqq C, \ \|\rho_n\|_p \leqq C, \ \|\nabla\rho_n^{1/2}\|_2 \leqq C, \ D(\rho_n, \rho_n) \leqq C.$$

Therefore, we may extract a subsequence, still denoted by ρ_n, such that

$$\rho_n \to \rho_0 \text{ weakly in } L^3 \text{ and in } L^p, \tag{14}$$

$$\rho_n \to \rho_0 \text{ a.e.,} \tag{15}$$

$$\nabla\rho_n^{1/2} \to \nabla\rho_0^{1/2} \text{ weakly in } L^2. \tag{16}$$

((15) relies on the fact that if Ω is a *bounded* smooth domain then $H^1(\Omega)$ is relatively compact in $L^2(\Omega)$. (14) and (16) implies that $\{\rho_n^{1/2}\}$ is bounded in $H^1(\Omega)$. Hence $\{\rho_n^{1/2}\}$ has a subsequence converging in $L^2(\Omega)$ and a.e.). Hence,

$$\liminf \int |\nabla\rho_n^{1/2}|^2 dx \geqq \int |\nabla\rho_0^{1/2}|^2 dx,$$
$$\liminf \int \rho_n^p(x)dx \geqq \int \rho_0^p(x)dx,$$

$\liminf D(\rho_n, \rho_n) \geqq D(\rho_0, \rho_0)$ (by Fatou's Lemma).
We now prove that

$$\int V(x)\rho_n(x)dx \to \int V(x)\rho_0(x)dx.$$

As in the proof of Lemma 2, we write $V = V_1 + V_2$. Clearly,

$$\int V_1(x)\rho_n(x)dx \to \int V_1(x)\rho_0(x)dx,$$

since $V_1 \in L^{3/2}$. On the other hand

$$\int V_2(x)\rho_n(x)dx \doteq -\frac{1}{4\pi}\int(\Delta V_2)(B\rho_n)dx.$$

It follows from (12) that $B\rho_n \to B\rho_0$ weakly in L^6. Thus,

$$\int V_2(x)\rho_n(x)dx \to \int V_2(x)\rho_0(x)dx.$$

Hence,

$$\xi_p(\rho_0) \leqq \liminf \xi_p(\rho_n) = \mathrm{Inf}\{\xi_p(\rho)|\rho\in D_p\}. \qquad \square$$

We now derive the Euler Equation satisfied by ρ_0. Set $\psi = \rho_0^{1/2}$.

Lemma 6. *The minimizing* ρ_0 *satisfies*

$$-\Delta\psi + \psi^{2p-1} = \varphi\psi, \tag{17}$$

where

$$\varphi = V - B\psi^2 \tag{18}$$

and (17) *holds in the sense of distributions.*

Proof. So far, we know that $\psi \in L^6 \cap L^{2p}$, $\nabla\psi \in L^2$ and $B\psi^2 \in L^6$. Since $V \in L^2_{\text{loc}}$, it follows that $\varphi \in L^2_{\text{loc}}$ and thus $\varphi\psi \in L^1_{\text{loc}}$. On the other hand $\psi^{2p-1} \in L^1_{\text{loc}}$ (since $\psi \in L^{2p}$). Therefore, (17) has a meaning in the sense of distributions. Consider the set $\tilde{D} = \{\zeta \in L^6 \cap L^{2p} | \nabla\zeta \in L^2 \text{ and } D(\zeta, \zeta) < \infty\}$. (Note that we do not assume $\zeta \geqq 0$.) If $\zeta \in \tilde{D}$, then $\rho = \zeta^2 \in D_p$ and

$$\xi_p(\rho) = \int |\nabla\zeta|^2 dx + \frac{1}{p}\int \zeta^{2p} dx - \int V\zeta^2 + D(\zeta, \zeta) \equiv \phi(\zeta).$$

Indeed it suffices to recall that $\nabla\rho^{1/2} = \nabla|\zeta| = \nabla\zeta(sgn\zeta)$ (see [7]). Therefore, we find for every $\zeta \in \tilde{D}$

$$\phi(\psi) \leqq \phi(\zeta)$$

Let $\eta \in C_0^\infty$; using the fact that $d/dt\,\phi(\psi + t\eta)|_{t=0} = 0$ we conclude easily that

$$\int \nabla\psi \cdot \nabla\eta dx + \int \psi^{2p-1}\eta dx = \int \varphi\psi\eta dx. \qquad \square$$

III. Proof that the Minimizing $\psi \in L^2$ and Further Properties of ψ

We first prove that the minimizing ψ is continuous:

Lemma 7. *ψ is continuous on $\mathbb{R}^3$; more precisely $\psi \in C^{0,\alpha}$ for every $\alpha < 1$* (i.e., *for every bounded set $\Omega \subset \mathbb{R}^3$, there is a constant M such that $|\psi(x) - \psi(y)| \leqq M|x - y|^\alpha$ $\forall x, y \in \Omega$*).

Proof. We already know that $B\psi^2 \in L^6$ and (clearly) $V \in L^{3-\delta}_{\text{loc}}(\forall\delta > 0)$. Consequently, $\varphi \in L^{3-\delta}_{\text{loc}}(\forall\delta > 0)$. Since $\psi \in L^6$, it follows that $\varphi\psi \in L^{2-\delta}_{\text{loc}}(\forall\delta > 0)$. Therefore, we have

$$-\Delta\psi \leqq f,$$

where $f = \varphi\psi \in L^{2-\delta}_{\text{loc}}(\forall\delta > 0)$ and in particular $f \in L^q_{\text{loc}}$ for some $q > 3/2$. We may, therefore, apply a result of Stampacchia (see [7], Théorème 5.2) to conclude that $\psi \in L^\infty_{\text{loc}}$. Going back to (17) and using the fact that $\psi \in L^\infty_{\text{loc}}$, we now see that $\Delta\psi \in L^{3-\delta}_{\text{loc}}(\forall\delta > 0)$. The standard elliptic regularity theory [8] implies that $\psi \in C^{0,\alpha}$ for every $\alpha < 1$. $\square$

We now prove an important property of ψ, namely, $\psi \in L^2$. Note that such a fact *cannot* be deduced from the knowledge that $\rho_0 \in D_p$. It is easy to construct a function $\rho \geqq 0$ such that $\rho \in L^3 \cap L^p$, $\nabla\rho^{1/2} \in L^2$, $D(\rho, \rho) < \infty$ and $\int \rho(x)dx = \infty$.

Lemma 8. $\psi \in L^2$.

Proof. Suppose, by contradiction, that $\int \psi^2(x)dx = \infty$. Choose $r_1 > \underset{1 \leqq i \leqq k}{\text{Max}} |R_i|$ such that

$$\int_{|x| < r_1} \psi^2(x)dx \geqq Z + 2\delta,$$

for some $\delta > 0$. We, thus, have

$$(B\psi^2)(x) = \int \psi^2(y)|x-y|^{-1}dy \geqq \int_{|y|<r_1} \psi^2(y)(|x|+|y|)^{-1}dy$$
$$\geqq (Z+2\delta)/(|x|+r_1).$$

Therefore,

$$\varphi(x) = V(x) - (B\psi^2)(x) \leqq \frac{Z}{|x|-r_1} - \frac{Z+2\delta}{|x|+r_1},$$

for $|x| > r_1$. Consequently, there is some $r_2 > r_1$ such that

$$\varphi(x) \leqq -\delta|x|^{-1},$$

for $|x| > r_2$. It follows from (17) that

$$-\Delta\psi + \delta|x|^{-1}\psi \leqq 0, \tag{19}$$

for $|x| > r_2$. We now use a comparison argument. Set

$$\tilde{\psi}(x) = Me^{-2(\delta|x|)^{1/2}},$$

where $M > 0$ is a constant. An easy computation shows that

$$-\Delta\tilde{\psi} + \delta|x|^{-1}\tilde{\psi} \geqq 0, \tag{20}$$

for $x \neq 0$. Hence, by (19) and (20) we have

$$-\Delta(\psi - \tilde{\psi}) + \delta|x|^{-1}(\psi - \tilde{\psi}) \leqq 0, \tag{21}$$

for $|x| > r_2$. We fix M in such a way that

$$\psi(x) \leqq \tilde{\psi}(x), \tag{22}$$

for $|x| = r_2$ (this is possible since $\psi \in L^\infty_{\text{loc}}$). It follows from (21), (22) and the maximum principle that

$$\psi(x) \leqq \tilde{\psi}(x), \tag{23}$$

for $|x| > r_2$. Since we only know that $\psi(x) \to 0$ as $|x| \to \infty$ in a weak sense (namely $\psi \in L^6$), we must justify (23). We use a variant of Stampacchia's method. Fix $\zeta(x) \in C_0^\infty$ with $0 \leqq \zeta \leqq 1$ and

$$\zeta(x) = \begin{cases} 1 & \text{for } |x| < 1 \\ 0 & \text{for } |x| > 2. \end{cases}$$

Set $\zeta_n(x) = \zeta(x/n)$. Multiplying (21) by $\zeta_n(\psi - \tilde{\psi})^+$ (here we set $t^+ = \text{Max}(t, 0)$) and integrating on $[|x| > r_2]$, we find

$$\int_{|x|>r_2} \nabla(\psi - \tilde{\psi})[\nabla\zeta_n(\psi - \tilde{\psi})^+ + \zeta_n\nabla(\psi - \tilde{\psi})^+]dx + \int_{|x|>r_2} \frac{\delta}{|x|}|(\psi - \tilde{\psi})^+|^2\zeta_n dx \leqq 0.$$

In particular it follows that

$$\int_{|x|>r_2} \frac{\delta}{|x|}|(\psi - \tilde{\psi})^+|^2\zeta_n dx \leqq \frac{1}{2}\int_{|x|>r_2} |(\psi - \tilde{\psi})^+|^2\Delta\zeta_n dx. \tag{24}$$

But we have

$$\int_{|x|>r_2} |(\psi - \tilde{\psi})^+ | \Delta\zeta_n dx \leqq \frac{C}{n^2} \int_{n<|x|<2n} \psi^2(x)dx$$
$$\leqq C\left[\int_{n<|x|<2n} \psi^6(x)dx\right]^{1/3},$$

by Hölder's inequality. Since $\psi \in L^6$, we conclude that the right side in (24) tends to zero as $n \to \infty$. Consequently,

$$\int_{|x|>r_2} \delta|x|^{-1}|(\psi - \tilde{\psi})^+|^2 dx = 0$$

and so $\psi \leqq \tilde{\psi}$ for $|x| > r_2$. In particular $\int \psi^2(x)dx < \infty$, a contradiction with the initial assumption. □

We now indicate some further properties of ψ.

Lemma 9. *ψ is bounded on $\mathbb{R}^3$, $\psi(x) \to 0$ as $|x| \to \infty$ and $\psi \in H^2$.*

Proof. By (17) we have,

$$-\Delta\psi \leqq V\psi, \tag{25}$$

and so

$$-\Delta\psi + \psi \leqq (V+1)\psi.$$

Clearly, $(V+1)\psi \in L^2$ and so

$$\psi \leqq (-\Delta + I)^{-1}[(V+1)\psi]. \tag{26}$$

As is well known, the right side in (26) is bound and tends to zero as $|x| \to \infty$. Finally, note that $\psi^{2p-1} \leqq C\psi$ for some constant C and $(B\psi^2)\psi \in L^2$ (since $\psi \in L^3$ and $B\psi^2 \in L^6$). Therefore, we conclude that $\Delta\psi \in L^2$ and so $\psi \in H^2$. □

Lemma 10. *$\psi > 0$ everywhere and ψ is C^∞ except at $x = R_i (1 \leqq i \leqq k)$.*

Proof. From (17) we have,

$$-\Delta\psi + a\psi = 0,$$

where $a \in L^q_{\text{loc}}$ and $q > 3/2$. It follows from Harnack's inequality (see e.g. [9] Corollary 5.3) that either $\psi > 0$ everywhere or $\psi \equiv 0$. We now prove that $\psi \not\equiv 0$ by checking that $\text{Min}\{\xi_p(\rho) | \rho \in D_p\} < 0$. It clearly suffices to consider the case where $V(x) = z_1|x|^{-1}$. Take the trial function $\rho^{1/2}(x) = \gamma \exp[-z_1|x|/4]$. The terms in ρ which are homogeneous of degree one are $-\gamma^2 z_1^2/4$. The remaining terms are proportional to γ^s, $s > 2$. Hence for γ sufficiently small, $\xi_p(\rho) < 0$. Finally, the fact that ψ is C^∞ (except at R_i) follows easily from (17) by a standard bootstrap argument. □

Remark. When $p \geqq 3/2$, there is a simple estimate for ψ, namely

$$\psi^{2(p-1)}(x) \leqq V(x), \tag{27}$$

for every x. Indeed set $u = \psi^{2(p-1)}(x) - V(x)$ so that for $x \neq R_i$ we have

$$\begin{aligned}\Delta u &= 2(p-1)\psi^{2p-3}(\Delta\psi) + (2p-2)(2p-3)\psi^{2p-4}|\nabla\psi|^2 \\ &\geqq 2(p-1)\psi^{2p-3}(\Delta\psi) = 2(p-1)\psi^{2p-3}(\psi^{2p-1} - \varphi\psi).\end{aligned}$$

The function u achieves its maximum at some point $x_0(\neq R_i)$. At x_0 we have $(\Delta u)(x_0) \leqq 0$ and so $\psi^{2(p-1)}(x_0) \leqq \varphi(x_0) \leqq V(x_0)$. Thus $u(x_0) \leqq 0$, and so $u(x) \leqq 0$ everywhere.

IV. Proof that for the Minimizing ψ, $\int \psi^2(x)\,dx \geqq Z$ (when $p \geqq 4/3$)

We start with the following remark:

Lemma 11. *For any* $\zeta \in C_0^\infty$

$$-\int \frac{\Delta\psi}{\psi}\zeta^2\,dx \leqq \int |\nabla\zeta|^2\,dx.$$

Proof. Integrate by parts and use the Cauchy–Schwarz inequality. □

We now prove the main estimate.

Lemma 12. *When* $p \geqq 4/3$, $\int \psi^2(x)\,dx \geqq Z$.

Proof. Let $\zeta_0 \in C_0^\infty$ be a spherically symmetric function such that $\zeta_0 \not\equiv 0$, $\zeta_0(x) = 0$ for $|x| < 1$ and for $|x| > 2$. Set $\zeta_n(x) = \zeta_0(x/n)$. By (17) we have,

$$-\int \frac{\Delta\psi}{\psi}\zeta_n^2\,dx + \int \psi^{2p-2}\zeta_n^2\,dx = \int \varphi\zeta_n^2\,dx. \tag{28}$$

Using Lemma 11, we find

$$-\int \frac{\Delta\psi}{\psi}\zeta_n^2\,dx \leqq \int |\nabla\zeta_n|^2\,dx \leqq Cn. \tag{29}$$

Next we claim that, if $p \geqq 4/3$, then

$$\int \psi^{2p-2}\zeta_n^2\,dx \leqq \varepsilon_n n^2, \tag{30}$$

where $\varepsilon_n \to 0$ as $n \to \infty$. Indeed we have

$$\int \psi^{2p-2}\zeta_n^2\,dx \leqq \int_{n<|x|<2n} \psi^{2p-2}\,dx.$$

If $2p - 2 \geqq 2$, we use the fact that $\psi^{2p-2} \leqq C\psi^2$ in order to obtain (30). If $2p - 2 \leqq 2$, we use Hölder's inequality and we find

$$\int_{n<|x|<2n} \psi^{2p-2}\,dx \leqq C\left[\int_{n<|x|<2n} \psi^2\,dx\right]^{p-1} (n^3)^{2-p}.$$

Assuming $p \geqq 4/3$, we deduce (30). On the other hand, since ζ_n is spherically symmetric, we have

$$\int \varphi\zeta_n^2\,dx = \int [\varphi]\zeta_n^2\,dx,$$

where $[\varphi]$ denotes the spherical average of φ, i.e.,

$$[\varphi](x) = \frac{1}{4\pi|x|^2} \int_{|\Omega|=|x|} \varphi(\Omega)d\Omega = \frac{1}{4\pi} \int_{|\Omega|=1} \varphi(|x|\Omega)d\Omega.$$

By a result of [1] (Eq. (35)) we know that

$$[\varphi](x) \geqq (Z - \lambda_0)/|x|, \text{ for } |x| > \operatorname*{Max}_{1 \leqq i \leqq k} |R_i|, \tag{31}$$

where $\lambda_0 = \int \psi^2(x)dx$. Hence, for large n, we find

$$\int \varphi\zeta_n^2 dx \geqq (Z - \lambda_0) \int \frac{1}{|x|} \zeta_n^2(x)dx = \alpha(Z - \lambda_0)n^2, \tag{32}$$

for some positive constant α. Combining (28), (29), (30), and (32), we find

$$\alpha(Z - \lambda_0)n^2 \leqq Cn + \varepsilon_n n^2.$$

As $n \to \infty$, we conclude that $Z \leqq \lambda_0$. □

Remark. Lemma 12 can also be proved by a direct variational calculation using (4). This is given in Theorem 4.10 of reference [4].

V. $\int\psi^2(x)dx > Z$ when $p \geqq 5/3$ and $k = 1$ (Atomic Case)

We assume now that $V(x) = Z|x|^{-1}$. The main result is the following:

Lemma 13. ([4], *Theorem 4.13*) *Assume* $p \geqq 5/3$, *then* $\int\psi^2(x)dx > Z$. *In addition* $\psi(x) \leqq Me^{-2(\delta|x|)^{1/2}}$ *for some constants* M *and* $0 < 2\delta < \int\psi^2(x)dx - Z$.

Proof. Since the solution of the problem $\mathrm{Min}\{\xi_p(\rho) | \rho \in D_p\}$ is unique, it follows that ρ_0—and therefore ψ—is spherically symmetric. In particular $\varphi = V - B\psi^2$ is also spherically symmetric. On the other hand by (31) we have,

$$\varphi(x) = [\varphi](x) \geqq (Z - \lambda_0)/|x|, \tag{33}$$

for $x \neq 0$. We already know that $\int\psi^2(x)dx \geqq Z$; suppose by contradiction that $\int\psi^2(x)dx = Z$. By (33) we have $\varphi \geqq 0$ and consequently (from (17))

$$-\Delta\psi + \psi^{2p-1} \geqq 0, \tag{34}$$

for $x \neq 0$. We now use a comparison function. Set $\tilde{\psi}(x) = C|x|^{-3/2}$. An easy computation shows that

$$-\Delta\tilde{\psi} + \tilde{\psi}^{2p-1} \leqq 0, \text{ for } |x| > 1, \tag{35}$$

provided $0 < C \leqq C_0$ where $C_0^{2p-2} = 3/4$. We fix the constant $0 < C \leqq C_0$ such that

$$\tilde{\psi}(x) \leqq \psi(x) \text{ for } |x| = 1, \tag{36}$$

(Recall that by Lemma 10, $\psi > 0$). It follows from (34), (35), (36) and the (usual) maximum principle that $\tilde{\psi}(x) \leqq \psi(x)$ for $|x| > 1$. Since $\tilde{\psi} \notin L^2(|x| > 1)$, we obtain a contradiction. Therefore, $\int\psi^2(x)dx > Z$; finally we argue as in the proof of

Lemma 8 and we conclude that for some $M, \psi(x) \leqq M \exp(-2(\delta|x|)^{1/2})$ where $2\delta < \int \psi^2(x)dx - Z$. □

Remark. If the assumptions of Lemma 13 are not satisfied, it may happen that $\int \psi^2 dx = Z$. Consider for example the functional

$$\xi(\rho) = c\int|\nabla\rho^{1/2}(x)|^2 dx + 2/3\int\rho^{3/2}(x)dx - \int V(x)\rho(x)dx + D(\rho,\rho),$$

where $V(x)$ is given by (3). We claim that if $c \leqq 1/16\pi$, then we have $\int\psi^2(x)dx = Z$. Indeed set $u = \psi - 2\varphi$. Recall that $-c\Delta\psi + \psi^2 = \varphi\psi$, and

$$\Delta\varphi = 4\pi\psi^2 \text{ if } x \neq R_i.$$

Thus,

$$\Delta u = \frac{1}{c}(\psi^2 - \varphi\psi) - 8\pi\psi^2 \geqq \frac{1}{c}(\psi^2 - \varphi\psi) - \frac{1}{2c}\psi^2 = \frac{\psi u}{2c}.$$

At a point x_0 where u achieves its maximum we have $(\Delta u)(x_0) \leqq 0$ and so $u(x_0) \leqq 0$ (note that $u(x) \to 0$ as $|x| \to \infty$). Consequently, $u \leqq 0$ everywhere and so $\varphi \geqq 0$ everywhere. Therefore, we must have $Z \geqq \int\psi^2(x)dx$. Since we already know that $Z \leqq \int\psi^2(x)dx$, it follows that $\int\psi^2(x)dx = Z$.

If $c > 1/16\pi$, one can still prove the weaker inequality

$$c\psi \leqq \frac{1}{8\pi}\varphi + (c - 1/16\pi)V, \tag{37}$$

by the same type of arguments. Using (37) we have

$$Z \leqq \int\psi^2 dx \leqq Z(1 + 8\pi(c - 1/16\pi)),$$

because $\varphi + \alpha V \geqq 0$ (with $\alpha > 0$) implies $\int\psi^2(x)dx \leqq Z(1+\alpha)$.

Note that in the one center (atomic) case $\xi_{3/2}$ is scale invariant. In fact, $E_{3/2} = \text{Min}\ \xi_{3/2} = cZ^3$ and $\psi(x) = Z^2\psi(Zx)$.

VI. Proof of Theorem 1 Concluded

We need a final Lemma.

Lemma 14. ([4], *Theorem 4.2) For every* $\lambda > 0$ *we have* $\text{Inf}\{\xi_p(\rho)|\rho\in D_p$ *and* $\int\rho(x)dx = \lambda\} = \text{Inf}\{\xi_p(\rho)|\rho\in D_p$ and $\int\rho(x)dx \leqq \lambda\}$.

Proof. Let $\rho\in D_p$ be such that $\int\rho(x)dx < \lambda$; the Lemma is an obvious consequence of the following claim: There exists a sequence $\rho_n\in D_p$ such that $\int\rho_n(x)dx = \lambda$ and $\liminf \xi_p(\rho_n) \leqq \xi_p(\rho)$. As ρ_n, we choose

$$\rho_n(x) = \rho(x) + \frac{k}{n^3}\zeta_n^2(x)$$

where $\zeta_n(x) = \zeta_0(x/n)$ ($\zeta_0 \in C_0^\infty$ is any function $\zeta_0 \not\equiv 0$) and $k = [\lambda - \int\rho(x)dx]/\int\zeta_0^2(x)dx$, so that $\int\rho_n(x)dx = \lambda$. We now check that $\liminf \xi_p(\rho_n) \leqq \xi_p(\rho)$. Using the subadditivity of the function $\int|\nabla\rho^{1/2}(x)|^2 dx$ and the convexity of ρ^p, we find

$$\xi_p(\rho_n) \leqq \xi_p(\rho) + A_n + B_n + C_n,$$

where

$$A_n = \frac{k}{n^3}\int|\nabla\zeta_n(x)|^2 dx$$

$$B_n = \frac{k}{n^3}\int\rho_n^{p-1}\zeta_n^2 dx$$

$$C_n = \frac{k}{n^3}\int(B\rho)\zeta_n^2 dx + \frac{k^2}{2n^6}\int(B\zeta_n^2)\zeta_n^2 dx$$

We shall prove that A_n, B_n, C_n tend to zero. Indeed we have

$$A_n = \frac{k}{n^2}\int|\nabla\zeta_0(x)|^2 dx \to 0.$$

Next, by Hölder's inequality

$$B_n \leqq \frac{k}{n^3}\left[\int\rho_n^p(x)dx\right]^{p-1/p}\left[\int\zeta_n^{2p}(x)dx\right]^{1/p}.$$

But,

$$\int\rho_n^p(x)dx \leqq C\int[\rho^p(x) + ((k/n^3)\zeta_n^2(x))^p]dx \leqq C$$

and so $B_n \to 0$. Finally, we have

$$C_n \leqq \frac{k}{n^3}\left[\int(B\rho)\rho dx\right]^{1/2}\left[\int(B\zeta_n^2)\zeta_n^2 dx\right]^{1/2} + \frac{k^2}{2n^6}\int(B\zeta_n^2)\zeta_n^2 dx$$

and

$$\frac{1}{n^6}\int(B\zeta_n^2)\zeta_n^2 = \frac{2}{n}D(\zeta_0^2, \zeta_0^2).$$

Therefore, $C_n \to 0$. □

Proof of Theorem 1 concluded: For every $\lambda > 0$ we set $E(\lambda) = \mathrm{Inf}\{\xi_p(\rho)|\rho\in D_p$ and $\int\rho(x)dx \leqq \lambda\}$. It is clear that $E(\lambda)$ is non-increasing and that $E(\lambda)$ is convex. In addition, the same proof as in Lemma 5 shows that there is a unique $\rho_\lambda\in D_p$ such that $\int\rho_\lambda(x) \leqq \lambda$ and $\xi_p(\rho_\lambda) = E(\lambda)$. Set $\lambda_c = \int\psi^2(x)dx$. It is clear that for $\lambda \geqq \lambda_c$ the function $E(\lambda)$ is constant: $E(\lambda) = E(\lambda_c)$; while $E(\lambda)$ is strictly decreasing on the interval $[0, \lambda_c]$. It follows that for $\lambda \leqq \lambda_c$ we have $\int\rho_\lambda(x)dx = \lambda$. Consequently, if $\lambda \leqq \lambda_c$ there is a unique solution for Problem (I). When $\lambda > \lambda_c$, we deduce from Lemma 14 that Problem (I) has no solution. □

By the same methods as used in Lemma 6, the unique minimizing ψ for $E(\lambda)$, $\lambda \leqq \lambda_c$ satisfies

$$-\Delta\psi + U_\psi\psi = -\phi_0\psi, \tag{38}$$

where,

$$U_\psi = \psi^{2p-2} - V + B\psi^2. \tag{39}$$

It is also true (as shown in [1] for the TF problem) that $E(\lambda)$ is differentiable and

$$-\phi_0(\lambda) = \partial E/\partial\lambda. \tag{40}$$

Since ψ satisfies (38) and $\psi(x) > 0$ (by Harnack's inequality as in Lemma 10), we can conclude that ψ and ϕ_0 are, in fact, the lowest eigenfunction and eigenvalue of $-\Delta + U_\psi(x)$.

To summarize what has been proved so far, the function $E(\lambda)$ has many features in common with the $E(\lambda)$ for TF theory [1]. It is convex, non-increasing and has an absolute minimum at some $\lambda_c < \infty$ beyond which $E(\lambda)$ is constant. One important difference is that $\lambda_c > Z$, at least for the atomic case and $p \geqq 5/3$. There is another important difference: in TF theory, $E(\lambda) \sim -c\lambda^{1/3}$ for small λ, i.e., $\partial E/\partial\lambda|_{\lambda=0} = -\infty$. In TFW theory, this is not so as the following shows.

Lemma 15. *Let e_0 be the lowest eigenvalue (ground state energy) of the Schrödinger operator $-\Delta - V(x)$ with V given by (3). Then*

$$\lim_{\lambda \downarrow 0} \lambda^{-1} E(\lambda)/e_0 = 1, \text{ i.e.,}$$
$$-\phi_0(0) = \partial E(\lambda)/\partial\lambda|_{\lambda=0} = e_0.$$

Proof. Let φ be the normalized eigenfunction of $-\Delta - V(x)$ belonging to e_0 and let $\rho_\lambda(x) = \lambda\varphi(x)^2$. Then $\xi_p(\rho_\lambda) \leqq \lambda e_0 + 0(\lambda)$ since the terms in ξ_p of degree higher than the first, while they are positive, are finite and $0(\lambda)$. Conversely,

$$E(\lambda) \geqq \inf_\rho \{ \int |\nabla\rho^{1/2}|^2 - \int V\rho \,|\, \int\rho = \lambda \},$$

but this is precisely the variational problem for the Schrödinger Equation. □

Yet another important distinction with TF theory should be noted. $\psi(x)$ is never zero and therefore $\rho(x) = \psi(x)^2$ does not have compact support. In TF theory, ρ has compact support [1] whenever $\lambda < \lambda_c$.

Binding of Atoms in TFW Theory

Binding does not occur in TF theory [1]; that is Teller's Theorem. Binding can occur in TFW theory as we shall now prove in a special case.

First it is necessary to have a clear definition of what binding means. Given the nuclear coordinates $R_1, \ldots, R_k$, define

$$\tilde{E}(\lambda; \{R_i\}) \equiv E(\lambda; \{R_i\}) + U(\{R_i\}) \tag{41}$$

where

$$U(\{R_i\}) = \sum_{1 \leqq i < j \leqq k} z_i z_j |R_i - R_j|^{-1}. \tag{42}$$

Let us start with two neutral atoms infinitely far apart. The total energy is then $A_{12} = E_1(z_1) + E_2(z_2)$ where E_i is the energy of an atom with nuclear charge z_i. The next step is to distribute the total electron charge so as to minimize the total energy, namely

$$B_{12} = \min_{0 \leqq \lambda \leqq z_1 + z_2} (E_1(z_1 + z_2 - \lambda) + E_2(\lambda)). \tag{43}$$

It may be, and usually is the case in TFW theory, that $B_{12} < A_{12}$, i.e., ions are more stable than atoms. Finally, we bring the atoms together and define

$$C_{12} = \inf_{R_1, R_2} \tilde{E}(z_1 + z_2; R_1, R_2). \tag{44}$$

If $C_{12} < B_{12}$, then binding occurs.

To show that binding is possible, suppose that z_2 is sufficiently small compared to z_1 so that the following is satisfied

(a) $$(\partial E_1/\partial\lambda)\left(\frac{z_1+\lambda_c^1}{2}\right) < -z_2^2/4,$$

(b) $$\lambda_c^1 - z_1 \geqq 2z_2,$$

where λ_c^1 is the critical λ for atom 1; then we claim that

$$B_{12} = E_1(z_1+z_2). \tag{45}$$

This follows from the following observation: $(\partial E_2/\partial\lambda) \geqq (\partial E_2/\partial\lambda)(0) = -z_2^2/4$ by Lemma 15 and convexity. Then the equation $(\partial E_1/\partial\lambda)(z_1+z_2-\lambda) = (\partial E_2/\partial\lambda)(\lambda)$ cannot have a solution for $0 \leqq \lambda \leqq z_1+z_2$.

For two such atoms the lowest total energy occurs when the smaller atom is completely stripped of its electrons which become attached to the larger atom. Now consider the first atom with $R_1 = 0$ and with $\lambda = z_1 + z_2 \leqq \lambda_c^1$. The electric potential φ which is spherically symmetric, will be negative for large R. If the second nucleus z_2 is placed at a point r where $\varphi(r) < 0$, the total energy will be reduced by an amount $z_2\varphi(r)$. Thus,

$$C_{12} < B_{12} + z_2\varphi(r)$$

and binding occurs.

Acknowledgements. One of the authors (H.B.) would like to thank the following for their hospitality during the course of this work: The Princeton University, Courant Institute and The University of Chicago.

References

1. Lieb, E. H., Simon, B.: Adv. Math. **23**, 22–116 (1977)
2. von Weizsäcker, C. F.: Z. Phys. **96**, 431–458 (1935)
3. Kompaneets, A. S., Pavloskii, E. S.: Sov. Phys. JETP **4**, 328–336 (1957)
4. Benguria, R.: The von Weizsäcker and exchange corrections in the Thomas–Fermi theory. Princeton University Thesis: June 1979 (unpublished)
5. Balàzs, N. L.: Phys. Rev. **156**, 42–47 (1967)
6. Gombás, P.: Acta Phys. Hung. **9**, 461–469 (1959)
7. Stampacchia, G.: Equations elliptiques du second ordre à coefficients discontinus. Montreal: Presses de l'Univ. 1965
8. Bers, L., Schechter, M.: Elliptic equations in Partial Differential Equations. New York: Interscience pp. 131–299. 1964
9. Trudinger, N.: Ann. Scuola Norm. Sup. Pisa **27**, 265–308 (1973)

Communicated by A. Jaffe

Received June 9, 1980

Commun. Math. Phys. 85, 15-25 (1982)

Analysis of the Thomas-Fermi-von Weizsäcker Equation for an Infinite Atom Without Electron Repulsion

Elliott H. Lieb*

Departments of Mathematics and Physics, Princeton University, P.O.B. 708, Princeton, NJ 08544, USA

Abstract. The equation

$$\{-\Delta+|\psi(x)|^{2p-2}-|x|^{-1}\}\,\psi(x)=0$$

in three dimensions is investigated. Uniqueness and other properties of the positive solution are proved for $3/2<p<2$. There are two physical interpretations of this equation for $p=5/3$: (i) As the TFW equation for an infinite atom *without* electron repulsion; (ii) The positive solution, ψ, suitably scaled, is asymptotically equal to the solution of the TFW equation for an atom or molecule *with* electron repulsion in the regime where the nuclear charges are large and x is close to one of the nuclei.

I. Introduction

The equation to be analyzed here of primary physical interest is

$$\{-\Delta+|\psi(x)|^{4/3}-|x|^{-1}\}\,\psi(x)=0 \tag{1.1}$$

in three dimensions and with ψ real valued. (1.1) was introduced in [9], wherein it was asserted without proof that (1.1) has a unique, positive solution. The present paper contains that proof. If $z,\gamma,A>0$ and

$$\tilde{\psi}(z,\gamma,A,x)\equiv(z^2/A\gamma)^{3/4}\,\psi(zx/A)\,, \tag{1.2}$$

then

$$\{-A\Delta+\gamma|\tilde{\psi}(x)|^{4/3}-z|x|^{-1}\}\,\tilde{\psi}(x)=0\,. \tag{1.3}$$

and conversely. Thus, (1.3) and (1.1) are equivalent problems.

(1.3) is to be compared with the Thomas-Fermi-von Weizsäcker (TFW) equation for a molecule [2–4, 9, 10, 13, 14]:

$$\{-A\Delta+\gamma|\hat{\psi}(x)|^{4/3}-V(x)+(|x|^{-1}*\hat{\psi}^2)(x)\}\,\hat{\psi}(x)=-\mu\hat{\psi}(x)\,, \tag{1.4}$$

* Work partially supported by U.S. National Science Foundation grant PHY-7825390 A02

with $V(x)=\sum_{j=1}^{k} z_j|x-R_j|^{-1}$ being the potential of k nuclei of charges $z_j \geqq 0$ located at R_j. The term $|x|^{-1} * \hat{\psi}^2$ is called "the electron respulsion." (1.3) has a unique positive solution (denoted by $\tilde{\psi}$) for all $\mu \geqq 0$ [2, 3, 9]. $-\mu$ is the "chemical potential." For all $\mu \geqq 0$, $\psi \in L^2(\mathbb{R}^3)$ [3, 9].

(1.3) is seen to be the TFW equation for an atom ($k=1$, $z_1=z$, $R_1=0$), but with the electron repulsion omitted. It will be shown here that (1.1), which is equivalent to (1.3), also has a unique positive solution, ψ, and $F(\psi)<\infty$ [cf. (1.8)]. The fact that $\mu=0$ in (1.1) and (1.3) means that the electron number, $\lambda=\int \tilde{\psi}^2$, is maximal. We shall see that in this case [for (1.1) and (1.3) but not (1.4)], $\lambda=\infty$, which is physically reasonable since the maximum electron number for the quantum mechanical Bohr atom (without repulsion) is also infinity.

If the foregoing were the only interpretation of (1.3) it would not be especially interesting. However, (1.3) has another interpretation as proved in [9]: Consider the full TFW equation, (1.4), with the scaling $z_j=az_j^0$, $R_j=a^{-1/3}R_j^0$ and $\lambda=a\lambda^0$ with $\lambda^0>0$. If $a\to\infty$ (with A,γ fixed), and if x is close to one of the R_j then $\hat{\psi}(x)\to\tilde{\psi}(x-R_j)$ (with $z\equiv z_j$) in the following sense: Fix $y\in\mathbb{R}^3$ then

$$\lim_{a\to\infty} z_j^{-3/2}\hat{\psi}(R_j+z_j^{-1}y)=\lim_{a\to\infty} z_j^{-3/2}\tilde{\psi}(z_j^{-1}y)=(A\gamma)^{-3/4}\psi(y/A), \tag{1.5}$$

where ψ is the positive solution of (1.1) and with the convergence being pointwise and in L^1_{loc}. [Note that the second expression in (1.5) is, in fact, independent of a. The right side of (1.5) is independent of λ.]

Not only does $\hat{\psi}\to\tilde{\psi}$ as in (1.5) but also the difference between the TFW energy and the TF energy is given, to leading order in a, by ψ:

$$E^{\mathrm{TFW}}-E^{\mathrm{TF}}=D\sum_{j=1}^{k} z_j^2+o(a^2), \tag{1.6}$$

where the constant D is given by [9]

$$D=A^{1/2}\gamma^{-3/2}F(\psi), \tag{1.7}$$

and where the functional F is defined generally for *all real* ψ by

$$F(\psi)=\int|\nabla\psi|^2+\int k(\psi(x),x)\,dx \tag{1.8}$$

with

$$k(\psi,x)\equiv\tfrac{3}{5}|\psi|^{10/3}-|x|^{-1}\psi^2+\tfrac{2}{5}|x|^{-5/2}\geqq 0. \tag{1.9}$$

Note that $k(\psi,x)>0$ and $k(\psi,x)=0$ if and only if $\psi=|x|^{-3/4}$.

The positive solution of (1.1) has been evaluated numerically [8] with the result that

$$\psi(0)=0.9701330$$

$$F(\psi)=17.1676.$$

Generalization. Equation (1.1) has the following obvious generalization

$$\Delta\psi(x)=\{|\psi(x)|^{2p-2}-|x|^{-1}\}\,\psi(x) \tag{1.10}$$

with $p>1$. The physical case, (1.1), corresponds to $p=5/3$. The TFW equation, (1.4) can be generalized in the same way; this was in fact done in [2, 3, 9]. In TF theory, $p=4/3$ and $3/2$ play a special role, while $p=4/3$, $3/2$, and $5/3$ are special in TFW theory [9]. In the analysis here of (1.10), $p=3/2$ and 2 are special (Theorems 3–5 and 7–9). Fortunately, the physical value 5/3 is in the range $3/2<p<2$ in which all theorems are applicable.

The appropriate energy functional is (1.8), but with $k(\psi, x)$ replaced by

$$k_p(\psi, x)=\frac{1}{p}|\psi|^{2p}-|x|^{-1}\psi^2+\frac{p-1}{p}|x|^{-p/(p-1)}\geqq 0\,. \tag{1.11}$$

The corresponding F will be denoted by F_p. However, as will be seen in Sect. III, F_p is useful only when $3/2<p<2$. It is for this range of p that the existence of a positive solution will be proved.

One of the more amusing technical exercises is in Sect. IV where an asymptotic expansion for ψ is established. While the expansion is heuristically obvious, its proof is not, primarily because Δ in (1.10) is essentially a singular perturbation.

II. Properties of Eq. (1.10)

Initially, (1.10) will be interpreted as a distributional equation. Although our main interest is in positive solutions, we shall not restrict ourselves to such and will assume only that ψ is a real valued function. It will turn out that the class of functions such that $F_p(\psi)<\infty$ is the natural class to consider, when $3/2<p<2$, but this will not be assumed initially. The only assumption to be understood in all the theorems is that $p>1$ and

$$\psi\in M=\{\psi\,|\,\nabla\psi\in L^2_{\mathrm{loc}},\ \psi\in L^1_{\mathrm{loc}}\}\,. \tag{2.1}$$

We begin with two "local" theorems.

Theorem 1. *Let B be a bounded open ball in $\mathbb{R}^3$. Let ψ satisfy* (1.10) *in B in the sense of distributions with $\psi\in L^1(B)$ and $\nabla\psi\in L^2(B)$. Then*

(i) *ψ is continuous. More precisely, $\psi\in C^{0,\alpha}(\bar{B})$ for all $0<\alpha<1$. I.e. there is a constant, C, such that $|\psi(x)-\psi(y)|\leqq C|x-y|^\alpha$ for all $x, y\in B$.*

(ii) *If $\psi(x)\geqq 0$, $\forall x\in B$, then either $\psi(x)\equiv 0$ or else $\psi(x)>0$, $\forall x\in B$.*

(iii) *If $0\notin B$ then $\psi\in C^{1,1/2}(\bar{B})$.*

(iv) *If $0\notin B$ and $\psi(x)\geqq 0$, $\forall x\in B$, then ψ is real analytic in B.*

Proof. (i) $\psi\in W^{1,1}$. By the Sobolev imbedding theorem [1],

$$\psi\in W^{0,3/2}=L^{3/2}\Rightarrow\psi\in W^{1,3/2}\Rightarrow\psi\in L^3\Rightarrow\psi\in W^{1,2}\Rightarrow\psi\in L^6$$
$$\Rightarrow f\equiv|x|^{-1}\psi\in L^{2-\varepsilon},\ \forall\varepsilon>0\,;$$

in particular $f\in L^q$ for some $q>3/2$.

If $p<5/2$ then $\psi\in L^6\Rightarrow|\psi|^{2p-1}\in L^{3/2+\varepsilon}$ for some $\varepsilon>0$. Then $\psi\in W^{2,\varepsilon+3/2}$ and thus $\psi\in C^{0,\alpha}$ for some $\alpha>0$. Then the right side of (1.10) is in $L^{3-\varepsilon}$, all $\varepsilon>0$, so $\psi\in W^{2,3-\varepsilon}$. Then $\psi\in C^{0,\alpha}$ for all $\alpha<1$. This proves (i) for $p<5/2$.

If $p \geqq 5/2$ a different argument is needed. The fact that (1.10) holds as a distributional equation means that the right side [call it $h(x)$] must be a distribution and hence must be in L^1_{loc}. Since $|x|^{-1}\psi \in L^1$, $|\psi|^{2p-1} \in L^1_{\mathrm{loc}}$. Kato's inequality [7], $\Delta|\psi| \geqq (\operatorname{sgn}\psi)\Delta\psi$ as distributions (with $\operatorname{sgn}\psi = |\psi|/\psi$ if $\psi \neq 0$ and $\operatorname{sgn}\psi = 0$ if $\psi = 0$), then implies that

$$\Delta|\psi| \geqq (\operatorname{sgn}\psi)\{|\psi|^{2p-2} - |x|^{-1}\}\psi = |\psi|^{2p-1} - |x|^{-1}|\psi| . \tag{2.2}$$

Therefore $-\Delta|\psi| \leqq |x|^{-1}|\psi| \equiv f$. A result of Stampacchia [12, Theoréme 5.2] is that if $f \in L^q_{\mathrm{loc}}$ with $q > 3/2$ then $|\psi| \in L^\infty_{\mathrm{loc}}$. But our f satisfies this condition. Returning to (1.10), $h \in L^{3-\varepsilon}_{\mathrm{loc}}$, $\varepsilon > 0$, and (i) is thus proved as before.

(ii) We have $\Delta\psi = g\psi$ with $g \in L^q$ and $q > 3/2$. By the Harnack inequality (cf. [5]), $\psi = 0$ or $\psi(x) > 0$, $\forall x$.

(iii) $\psi, \Delta\psi \in L^\infty$ and $\nabla\psi \in L^2 \Rightarrow \psi \in W^{2,2} \Rightarrow \nabla\psi \in L^6 \Rightarrow \psi \in W^{2,6} \Rightarrow \psi \in C^{1,1/2}$ (see [1]).

(iv) Assume $\psi(x) > 0$, whence $|\psi|^{2p-1}$ has as many derivatives as ψ has. By a bootstrap argument ψ is C^∞ (see [10, Theorem IV.5]). Since $|x|^{-1}$ is real analytic for $x \neq 0$, by [11, Theorem 5.8.6] ψ is real analytic. □

Theorem 2 (cusp condition [6]). *Assume the hypotheses of Theorem 1 and also that* $0 \in B$. *Then*

$$\psi(0) = -2 \lim_{r \downarrow 0} \int \Omega \cdot \nabla\psi(r\Omega)\, d\Omega , \tag{2.3}$$

where $d\Omega$ *is the normalized invariant measure on the unit sphere. In particular, if* ψ *is spherically symmetric about zero, then*

$$\psi(0) = -2 \lim_{r \downarrow 0} d\psi(r)/dr . \tag{2.4}$$

Proof. Simply integrate (1.10) by parts. □

Theorem 3. *Assume that* $p \geqq 3/2$. *Suppose* ψ *satisfies* (1.10) *in the sense of distributions on all of* $\mathbb{R}^3$. *Then* $|\psi(x)| < |x|^{-1/(2p-2)}$.

Remark. Some condition on p is needed and we believe $p \geqq 3/2$ is the right one. If $p < 3/2$ there cannot be any *positive* ψ satisfying both (1.10) and $\psi(x) \leqq |x|^{-1(2p-2)}$. For then $0 \leqq h = -\Delta\psi \leqq |x|^{(1-2p)/(2p-2)} \equiv f(x)$, where $-h$ is the right side of (1.10). Since $g \equiv |x|^{-1} * f$ is finite for $|x| > 0$ and $g(x) \to 0$ as $|x| \to \infty$, $\psi = 4\pi|x|^{-1} * h$. Since $h \not\equiv 0$, this implies that $\psi(x) \geqq c|x|^{-1}$ for $|x| >$ some R and $c > 0$. This is a contradiction. Thus, $p \geqq 3/2$ is the right condition for positive ψ. It is possible that (1.10) has no positive solution even when $p = 3/2$, for in that case $\psi(x) = |x|^{-1}$ satisfies (1.10) everywhere except at the origin.

Proof. By Theorem 1 we can assume that ψ is continuous. Let $b = -1/(2p-2) \geqq -1$ (it is here that $p \geqq 3/2$ enters). $\Delta|x|^b \leqq 0$. If $g(x) \equiv |\psi(x)| - |x|^b$ we have, by (2.2), that $\Delta g > 0$ on the set $A = \{x \mid g(x) > 0\}$. Since ψ is continuous (Theorem 1) we have that (i) A is open; (ii) $0 \notin A$; (iii) $h(x) \equiv \max[g(x), 0]$ is subharmonic on $\mathbb{R}^3$, i.e. $\Delta h \geqq 0$. If we knew that $h(x) \to 0$ as $|x| \to \infty$ we could then conclude that $h \equiv 0$ and thus that $|\psi(x)| \leqq |x|^b$. But this has to be proved.

We have, in fact, that on A

$$\Delta h \geqq \{|\psi|^{-1/b} - |x|^{-1}\}|\psi| . \tag{2.5}$$

Since $|\psi| = h + |x|^b$ on A, and $(\alpha+\beta)^t > \alpha^t + \beta^t$ for $\alpha, \beta \geqq 0$, $t \geqq 1$,

$$\Delta h \geqq h^{2p-1} \quad \text{on} \quad \mathbb{R}^3 . \tag{2.6}$$

Now let $f(x) = f(|x|)$ be the spherical average of h, i.e. $f(r) = \int h(\Omega r)\, d\Omega$ with $d\Omega$ being the normalized invariant measure on the unit sphere. By averaging (2.6), and using $\mathrm{av}\{h^t\} \geqq \{\mathrm{av}\, h\}^t$ for $t \geqq 1$, we have that $\Delta f \geqq f^{2p-1}$. With $r = |x|$, let $rf(r) \equiv u(r)$, whence

$$u'' \geqq r^{-2p+2} u^{2p-1} \quad \text{for} \quad r > 0 . \tag{2.7}$$

Assume that $u(r) \not\equiv 0$. Since $0 \notin A$, there is a $R_0 > 0$ such that $u(r) = 0$ for $0 \leqq r < R_0$. Since $u'' \geqq 0$ and $u \geqq 0$, u is continuous, convex and non-decreasing in r. Therefore for some $R_1 > 0$, $u(R_1) > 0$ and $u'(R_1) > 0$. Thus $u(r) > br$ for $r >$ some R_2 and with $b > 0$.

Let $w_R(x) \equiv \sigma R(R-r)^{2b}$ be defined on $D_R = \{r | 0 < r < R\}$ with $\sigma^{2p-2} = p(p-1)^{-2}$. On D_R, w_R satisfies $w_R'' \leqq r^{-2p+2} w_R^{2p-1}$. Since $u(r) > br$ for large r, there is some R such that $w_R(R/2) < u(R/2)$. Consider $m(r) \equiv u(r) - w_R(r)$ on D_R and let $E = \{r | m(r) > 0, 0 < r < R\}$. Clearly, (i) m is continuous and $m'' > 0$ on E; (ii) E is non-empty and open; (iii) $\bar{E} \subset D_R$ [since $u(r) = 0$ for $r < R_0$ and $w_R(r) \to \infty$ as $r \to R$]. Therefore m has its maximum on ∂E, but $m = 0$ on ∂E. Therefore E is empty and we have a contradiction.

Thus $u \equiv 0$, which implies that $f \equiv 0 \Rightarrow h \equiv 0 \Rightarrow \psi(x) \leqq |x|^b$. We have to prove that $g(x) < 0$ everywhere. The Harnack inequality argument used in Theorem 1 implies that either $g \equiv 0$ or $g(x) < 0$, all x. Since $g(x) < 0$ for x near zero, the result is proved. □

The following complements Theorem 3.

Theorem 4. *Suppose* $\psi(x) > 0$ *satisfies* (1.10) *on* $\mathbb{R}^3$ *and* ψ *is spherically symmetric. If* $p < 2$ *then* $\psi(x) \geqq (A + |x|)^{-1(2p-2)}$ *with* $A = 1/(p-1)$.

Remark. The hypothesis that ψ is spherically symmetric is not really necessary but it simplifies the proof. It is assumed here because Theorem 5 states that any positive solution is necessarily spherically symmetric when $3/2 \leqq p < 2$.

Proof. Let $f(r) = (A+r)^b$ with $b = -1/(2p-2)$. Then $\Delta f = Wf$ with $W(r) = b(b-1)(A+r)^{-2} + (2b/r)(A+r)^{-1}$. $W(r) > f(r)^{2p-2} - r^{-1}$. Let $B = \{r > 0 | \psi(r) < f(r)\}$, which is open and which is assumed to be non-empty. Let D be a connected component of B. Thus $D = (s, t)$ with $0 \leqq s < t$. Let $h = f - \psi > 0$ on D. On D, $W(r) > \psi(r)^{2p-2} - r^{-1}$, so

$$(\Delta - W)\, h(r) > 0, \quad \text{all} \quad r \in D . \tag{2.8}$$

First, assume t is finite. Multiply (2.8) by f and integrate over the shell $s < |x| < t$. An integration by parts gives

$$K(t) - K(s) > 0 \tag{2.9}$$

Commun. Math. Phys. 85, 15–25 (1982)

with $K(r)=r^2 f(r)^2 (h/f)'(r)$. $(h/f)'(t)\leqq 0$ since $h(t)=0$ and $h(r)>0$ for $s<r<t$. Thus $K(t)\leqq 0$. Likewise, if $s>0$ $K(s)\geqq 0$ and therefore (2.8) is a contradiction. If $s=0$ then $K(s)=0$ since ψ' and f' are bounded near $r=0$.

Now assume $t=\infty$. Integrate (2.9) over $s<|x|<T$. (2.8) holds as before (with T) and $K(s)\geqq 0$. Since $p<2$, $rf(r)^2\to 0$ as $r\to\infty$. $0<(h/f)(r)<1$ on D, so there is a sequence $\{T_n\}$, with $T_n\to\infty$, such that $T_n(h/f)'(T_n)\to 0$. Thus $K(T_n)\to 0$ and (2.8) is again a contradiction. □

Theorem 5. *There is at most one $\psi\not\equiv 0$ that satisfies* (1.10) *on* $\mathbb{R}^3$ *in the sense of distributions with the properties that $\psi(x)\geqq 0$, all x and $\psi\in L^6(\mathbb{R}^3)$. (By Theorem 3, the second property is automatic if $3/2\leqq p<2$.)*

Remark. It is not at all clear whether or not $p\geqq 2$ is special. If it is additionally assumed that $\nabla\psi\in L^2(\mathbb{R}^3)$ then by Theorem 3 and a Sobolev inequality, $\psi\in L^6$ when $p\geqq 3/2$. In Sect. III it will be shown that there is a positive solution to (1.10) for $3/2<p<2$, and this solution indeed has $\nabla\psi\in L^2(\mathbb{R}^3)$. However, the methods of Sect. III are not applicable for $p\geqq 2$.

Proof. Let ψ satisfy (1.10) with the stated properties. By Theorem 1, ψ is continuous and $\psi(x)>0$, all x. If $g\in C_0^\infty$, then $-\int\frac{\Delta\psi}{\psi}|g|^2\leqq\int|\nabla g|^2$. (This fact, which was used in [3], follows by integrating by parts and using the Cauchy-Schwarz inequality.) Thus, for any real $g\in C_0^\infty$,

$$L_\psi(g)=\int|\nabla g|^2+\int W_\psi(x)|g(x)|^2\,dx\geqq 0\,, \tag{2.10}$$

where $W_\psi(x)=-|x|^{-1}+\psi(x)^{2p-2}$. Let f_n be a sequence of spherically symmetric C_0^∞ functions with the properties: (i) $0\leqq f_n(x)\leqq 1$; (ii) $f_n(x)=1$ for $2/n<|x|<n$; (iii) $f_n(x)=0$ for $0\leqq|x|\leqq 1/n$ and $|x|\geqq 2n$; (iv) $\int|\nabla f_n|^3\leqq A$ for some fixed A. Such a sequence is easy to construct. Let $g_n(x)=f_n(x)\,\psi(x)$. $g_n\in C_0^\infty$ by Theorem 1. $\int|\nabla g_n|^2=-\int f_n^2\psi\Delta\psi+\int\psi^2|\nabla f_n|^2$. Thus, $L_\psi(g_n)=\int\psi^2|\nabla f_n|^2\equiv T_n$. Let $\chi_n(x)=0$ if $2/n\leqq|x|\leqq n$ and $\chi_n(x)=1$ otherwise. Then $T_n=\int\chi_n\psi^2|\nabla f_n|^2\leqq\|\chi_n\psi\|_6^2\,\|\nabla f_n\|_3^2$. Since $\chi_n\to 0$ pointwise and $\psi\in L^6$, $T_n\to 0$ as $n\to\infty$.

Now let ψ_1 and ψ_2 be two different solutions to (1.10) with the stated properties. Denote the corresponding two functionals in (2.10) by L_1 and L_2 and the corresponding sequences by $g_n^1=f_n\psi_1$ and $g_n^2=f_n\psi_2$. Then $0\leqq L_1(g_n^2)+L_2(g_n^1)=L_1(g_n^1)+L_2(g_n^2)-\int(\psi_1^{2p-2}-\psi_2^{2p-2})(\psi_1^2-\psi_2^2)f_n^2$. Since $L_1(g_n^1)$ and $L_2(g_n^2)\to 0$, the right side of this inequality is negative for large n if $\psi_1\not\equiv\psi_2$. □

Remark. If $\psi\geqq 0$ is unique then ψ is obviously spherically symmetric. Therefore the following theorem is applicable if $3/2\leqq p<2$.

Theorem 6. *If $\psi\geqq 0$ satisfies* (1.10), *the bound of Theorem 3, and if ψ is spherically symmetric, then $\psi(r)$ is a strictly decreasing function of r.*

Proof. $\Delta\psi\leqq 0$ so ψ is superharmonic. Therefore the minimum of $\psi(x)$ in the set $|x|\leqq R$ must occur at R. If $\psi(r)=\psi(R)$, $r<R$, then $\psi(x)$ is a constant for $r\leqq|x|\leqq R$, but this does not satisfy (1.10). □

III. Existence of a Positive Solution of Eq. (1.10) for $3/2<p<2$

In Sect. II it was shown that any positive distributional solution to (1.10) with $\psi \in M$ has certain nice properties, especially when $p \geqq 3/2$. If $3/2 \leqq p<2$ the solution is unique. In this section it will always be assumed that $3/2<p<2$ (note the inequality $p>3/2$). A positive solution will be shown to exist on $\mathbb{R}^3$ under this condition.

An interesting open question is whether (1.10) has solutions, positive or otherwise, when $p \leqq 3/2$ or $p \geqq 2$. The method given here sheds no light on this question.

Consider the functional F_p given by (1.8), (1.11) and the class of measurable functions

$$G_p=\{\psi \mid \nabla \psi \in L^2, F_p(\psi)<\infty\}. \tag{3.1}$$

G_p is not empty because

$$f_p(x)=(1+|x|^2)^{-1/4(p-1)} \tag{3.2}$$

is in G_p for $3/2<p<2$. We also define

$$W_p(\psi)=\int k_p(\psi(x), x)\, dx \geqq 0, \tag{3.3}$$

$$E_p=\inf\{F_p(\psi) \mid \psi \in G_p\}. \tag{3.4}$$

Remark. The condition $3/2<p<2$ results from the requirement that G_p be non-empty. Note that $k_p(\psi, x)$ has the form $|x|^{-p/(p-1)} h_p(|\psi|\, |x|^{1/(2p-2)})$ and that $h_p(a)=0$ if and only if $a=1$. Suppose $p \geqq 2$. If $k_p(\psi(x), x)$ is to be integrable at infinity then $\psi(x) \sim |x|^{-1/(2p-2)}$. But then $|\nabla \psi|^2$ is not integrable at infinity. If $p \leqq 3/2$ then, for a similar reason, $\psi(x) \sim |x|^{-1/(2p-2)}$ near $x=0$, but then $|\nabla \psi|^2$ is not integrable.

Theorem 7. *Let* $3/2<p<2$. *There exists* $\psi \in G_p$ *such that* (i) $F_p(\psi)=E_p$; (ii) $\psi(x) \geqq 0$; (iii) ψ *satisfies* (1.10) *in the sense of distributions.*

Proof. Let ψ_n be a minimizing sequence for $F_p(\psi)$. $W_p(|\psi|)=W_p(\psi)$ and $\int |\nabla \psi|^2 \geqq \int (\nabla |\psi|)^2$, so we can assume $\psi_n(x) \geqq 0$, all x. Let $b=-(2p-2)^{-1}$. For $\psi \geqq 0$, $k_p(\psi, x) \geqq C_p |x|^{-1}[\psi-|x|^b]^2$ with $C_p>0$. Therefore, if $\psi \in G_p$ then $|x|^{-1/2}\psi - |x|^{b-1/2} \in L^2$. But $|x|^{b-1/2} \in L^2_{\text{loc}}$ so $|x|^{-1/2}\psi \in L^2_{\text{loc}} \Rightarrow \psi \in L^2_{\text{loc}} \Rightarrow \psi \in W^{1,2}(B)$ for any bounded ball, B. Since $F_p(\psi_n) \to E_p$, $\nabla \psi_n$ is bounded in $L^2(\mathbb{R}^3)$ and ψ_n is bounded in $W^{1,2}(B)$. By the Rellich-Kondrachov theorem [1], a bounded set in $W^{1,2}(B)$ is compactly imbedded in $L^2(B)$. By passing to a subsequence we can assume ψ_n has a limit, ψ, in $L^2(B)$. By taking a further subsequence, we can assume $\psi_n \to \psi$ pointwise. This can be done for every B, so we can assume $\psi_n \to \psi$ pointwise in $\mathbb{R}^3$. By passing to a further subsequence we can assume, using the Banach-Alaoglu theorem, that $\nabla \psi_n \to f$ weakly in $L^2(\mathbb{R}^3)$. Clearly, $f=\psi$. Therefore $\liminf \int |\nabla \psi_n|^2 \geqq \int |\nabla \psi|^2$ and, by Fatou's lemma, $\liminf \int k_p(\psi_n(x), x) \geqq \int k_p(\psi(x), x)$. Thus $F_p(\psi) \leqq E_p$, so ψ minimizes.

Now $k_p(\psi(x), x) \in L^1$. By the above $|x|^{-1}\psi^2 \in L^1_{\text{loc}} \Rightarrow |x|^{-1}\psi \in L^1_{\text{loc}}$. Since $k_p(\psi(x), x) \in L^1$, $\psi^{2p} \in L^1_{\text{loc}}$, and thus $\psi^{2p-1} \in L^1_{\text{loc}}$. Hence, the right side of (1.10) is a distribution.

Commun. Math. Phys. 85, 15–25 (1982)

Let $\eta \in C_0^\infty$ and replace ψ by $\psi_t = \psi + t\eta$. Let $E(t) = E_p(\psi_t)$, whence $dE/dt = 0$ at $t=0$. $\int |\nabla \psi_t|^2$ is differentiable with derivative $2\int \nabla\psi \cdot \nabla\eta$ at $t=0$. By dominated convergence, $W_p(\psi_t)$ can be differentiated under the integral sign, whence $0 = \int \nabla\eta \cdot \nabla\psi + \int \eta\{|\psi|^{2p-2} - |x|^{-1}\}\psi$. This is Eq. (1.10) in the sense of distributions. □

The ψ given by Theorem 7 is unique (Theorem 5). ψ satisfies two sum rules, one of which arises from the fact that ψ minimizes F_p. Let us define the following integrals:

$$I_1 = \int |\nabla\psi(x)|^2\, dx\,, \tag{3.5}$$

$$I_2 = \int \{|x|^{-p/(p-1)} - \psi(x)^{2p}\}\, dx\,, \tag{3.6}$$

$$I_3 = \int |x|^{-1}\{|x|^{-1/(p-1)} - \psi(x)^2\}\, dx\,. \tag{3.7}$$

I_1 is finite since $\psi \in G_p$. I_2 and I_3 are finite (and positive) since $|x|^{-1/(2p-2)} > \psi(x) > (A+|x|)^{-1/(2p-2)}$, Theorems 3 and 4, and since $|x|^{-p/(p-1)} \in L^1_{\text{loc}}$. Clearly,

$$F_p(\psi) = I_1 - p^{-1} I_2 + I_3\,. \tag{3.8}$$

The physical interpretation of these integrals is the following: I_1 is the gradient contribution to the kinetic energy. I_2/p is the decrease in the "fermionic" part of the kinetic energy relative to the TF value. I_3 is the increase in the electron-nucleus potential energy relative to the TF value.

Theorem 8. *Let ψ be the minimizing ψ of Theorem 7. Then*

$$I_1 : I_2 : I_3 = 2(2-p)(p-1) : p(3-p) : p^2 - 3p + 4\,.$$

In particular, for $p = 5/3$,

$$I_1 : I_2 : I_3 = 1 : 5 : 4\,.$$

Proof. If (1.10) is multiplied by ψ and integrated, we find $I_2 = I_1 + I_3$. Next, consider $\psi_t(x) \equiv t^{1/(2p-1)}\psi(tx)$ for $t > 0$. Since $g(t) \equiv F_p(\psi_t)$ has its minimum at $t=1$, $dg(t)/dt = 0$ at $t=1$. But $I_1(\psi_t) = t^{-1} I_1$, $I_2(\psi_t) = t^{(4-2p)/(p-1)} I_2$ and $I_3(\psi_t) = t^{(4-2p)/(p-1)} I_3$. The rest is algebra. □

Remark. When $p = 5/3$, Theorem 8 implies the *virial theorem*. The change in kinetic energy is $\delta T = I_1 - p^{-1} I_2$. The change in potential energy is $\delta V = I_3$. Therefore $-2\delta T = \delta V$, as usual.

IV. Asymptotic Expansion for Large r ($3/2 < p < 2$)

We shall be concerned here with the unique, positive solution to (1.10) for $3/2 < p < 2$. By Theorems 3 and 4, $\psi(r) = r^b + O(r^{b-1})$ with $r = |x|$ and $b = -1/(2p-2)$. This suggests an asymptotic expansion of the form

$$\psi(r) = r^b\left\{1 + \sum_{j=1}^{\infty} a_j r^{-j}\right\}. \tag{4.1}$$

The coefficients a_j are determined as follows. If (4.1) is inserted into the right side of (1.10) and then expanded, the coefficient of r^{b-1} is zero and the coefficient of $r^{b-j-1} (j \geqq 1)$ is of the form

$$P_j(A_{j-1}) + (2p-2)\, a_j, \tag{4.2}$$

where $A_j \equiv (a_1, \ldots, a_j)$ and P_j is a polynomial (with $P_1 \equiv 0$). The coefficient of r^{b-j-1} on the left side is zero for $j=0$ and is

$$(b-j+2)(b-j+1)\, a_{j-1} \tag{4.3}$$

for $j \geqq 1$, with $a_0 \equiv 1$. Equating (4.2) and (4.3),

$$a_j = (2p-2)^{-1} [(b-j+2)(b-j+1)\, a_{j-1} - P_j(A_{j-1})]. \tag{4.4}$$

Thus, a_j is determined recursively by $a_1, \ldots, a_{j-1}$. The first three terms of ψ are thus

$$\psi(r) = r^b \{1 - r^{-1}(2p-3)(2p-2)^{-3} - \tfrac{1}{2} r^{-2}(2p-3)(2p-1)^2(2p-2)^{-6} + O(r^{-3})\}. \tag{4.5}$$

The correctness of (4.1), with the rule (4.4), will be proved here. The chief difficulty is that Δ is a singular perturbation: The term $a_j r^{b-j}$ in ψ generates a term r^{b-j-1} on the right side of (1.10), but it generates r^{b-j-2} on the left. While $\Delta\psi$ thus appears to be relatively small, it is not really small because its coefficients (4.3) grow as j^2. For this reason the series (4.1) is probably not convergent. In the proof of Theorem 9, $\Delta\psi$ is controlled by combining it with the leading term in (4.2), namely $(2p-2)r^{-1}\psi$.

Theorem 9. *The asymptotic expansion* (4.1) *and* (4.4) *is correct, i.e. for any* J,

$$\psi(r) = r^b \left\{1 + \sum_{j=1}^{J} a_j r^{-j}\right\} + o(r^{b-J}) \tag{4.6}$$

as $r \to \infty$.

Proof. The proof is by induction. Theorems 3 and 4 assert that (4.6) is true for $J=0$. Assuming (4.6) holds for some $J \geqq 0$, we will prove that (4.6) holds for $J+1$. Write $\psi = \phi + g$ with $\phi(r) = r^b \left\{1 + \sum_{j=1}^{J} a_j r^{-j}\right\}$. For any $\varepsilon > 0$ there is an R_ε such that for all $r > R_\varepsilon$: (i) $\phi(r) > 0$, (ii) $g(r)/\phi(r) < \varepsilon$, (iii) $[\phi(r) + g(r)]^{2p-1} - \phi(r)^{2p-1} \equiv U(r)\, g(r)$ with $|U(r) - (2p-1)r^{-1}| < \varepsilon r^{-1}$.

Equation (1.10) reads

$$[\Delta + W(r)]\, g(r) = h(r), \tag{4.7}$$

with $h = \phi^{2p-1} - r^{-1}\phi - \Delta\phi$ and $W(r) = r^{-1} - U(r)$. Let us examine (4.7) for $r > R_\varepsilon$. As $r \to \infty$, $h(r) = K r^{b-J-2} + o(r^{b-J-2})$ since $a_1, \ldots, a_J$ satisfy (4.4). Moreover, $K = -(2p-2) \cdot$[right side of (4.4) with $j = J+1$]. We also know that $g(r) = o(r^{b-J})$ by the induction hypothesis. Finally, $|W(r) + (2p-2)r^{-1}| < \varepsilon r^{-1}$ by (iii) above. The point of writing (1.10) as (4.7) is that we now regard W as a fixed function that is close to $-(2p-2)r^{-1}$. It is true that W "depends on g", but that information is suppressed.

Commun. Math. Phys. *85*, 15–25 (1982)

Lemma 10 will imply that $g(r)=-K(2p-2)^{-1}r^{b-J-1}+o(r^{b-J-1})$, which is the desired result. To see this, let $z>0$ and define $v_z(r)=\exp[-2(zr)^{1/2}]$. Let $W_z(r)=-z/r$. Then $(\Delta+W_z)v_z=-\frac{3}{2}z^{1/2}r^{-3/2}v_z$. Without loss, assume $K\geqq 0$ [otherwise replace g by $-g$ in (4.7)]. Let $z_0=2p-2$.

Lower Bound for g. Pick $0<\varepsilon<z_0$ and let $z=z_0-\varepsilon$. For $r>R_\varepsilon$, $W(r)\leqq -z/r=W_z(r)$. Pick $\delta>0$ and $A\geqq 0$ and let $f(r)=-Av_z(r)-(K+\delta)z^{-1}r^{b-J-1}<0$. Then

$$(\Delta+W_z)f(r)=\tfrac{3}{2}Az^{1/2}r^{-3/2}v_z(r)+(K+\delta)r^{b-J-2}$$
$$-(K+\delta)z^{-1}(b-J)(b-J-1)r^{b-J-3}.$$

For any fixed $\varepsilon,\delta>0$ we can choose $A\geqq 0$ such that (i) $(\Delta+W_z)f\geqq h=(\Delta+W)g$ for $r>R_\varepsilon$, (ii) $g(R_\varepsilon)\geqq f(R_\varepsilon)$. By Lemma 10 [with $B=\{r|r>R_\varepsilon\}$, $g_1=g$, $g_2=f$, $W_1=W$, $W_2=W_z$, and the facts that $g(r)$ and $f(r)\to 0$ as $r\to\infty$ and $f(r)<0$], $g(r)\geqq f(r)$ for $r>R_\varepsilon$. This implies $\liminf_{r\to\infty} r^{-b+J+1}g(r)\geqq -(K+\delta)/z$. Since δ and ε are arbitrary, $\liminf_{r\to\infty} r^{-b+J+1}g(r)\geqq -K/z_0$.

Upper Bound for g. This is similar to the preceding. For $r>R_\varepsilon$, $-z/r\geqq W(r)\geqq -q/r$ with $q=z_0+\varepsilon$ and $z=z_0-\varepsilon$. Let $G(r)=g(r)-Av_z(r)-\delta r^{b-J-1}$. Then

$$(\Delta+W)G(r)=h(r)+\tfrac{3}{2}Az^{1/2}r^{-3/2}v_z(r)-A(W(r)+zr^{-1})v_z(r)$$
$$-\delta(b-J)(b-J-1)r^{b-J-3}-\delta W(r)r^{b-J-1}.$$

Let $F(r)=-Kq^{-1}r^{b-J-1}\leqq 0$, whence

$$(\Delta+W_q)F(r)<Kr^{b-J-2}.$$

For any fixed $\varepsilon,\delta>0$ we can choose $A\geqq 0$ such that (i) $(\Delta+W)G\geqq(\Delta+W_q)F$ for $r>R_\varepsilon$, (ii) $F(R_\varepsilon)\geqq G(R_\varepsilon)$. By Lemma 10 ($g_1=F$, $g_2=G$, $W_1=W_q$, $W_2=W$), $F(r)\geqq G(r)$. As before this implies $\limsup_{r\to\infty} r^{-b+J+1}g(r)\leqq -K/z_0$.

These two bounds yield the desired result. □

Lemma 10. *Let $B\subset\mathbb{R}^3$ be an open set, let g_i, $i=1,2$, be two functions on $\mathbb{R}^3$ which are continuous on some neighborhood of $\bar{B}$, the closure of B, and let W_i be two functions in $L^1_{\mathrm{loc}}(B)$. Suppose that $[\Delta+W_1]g_1\leqq[\Delta+W_2]g_2$ as distributions on B. In addition suppose that* (i) *$g_1\geqq g_2$ on the boundary of B,* (ii) *if B is unbounded, then for every $\varepsilon>0$ there exists an R_ε such that $g_1(x)-g_2(x)\geqq -\varepsilon$ on $\{x|x\in B, |x|\geqq R_\varepsilon\}$,* (iii) *$0\geqq W_2(x)\geqq W_1(x)$, all $x\in B$,* (iv) *$g_1(x)\leqq 0$ whenever $g_1(x)<g_2(x)$. Then $g_1(x)\geqq g_2(x)$ for all $x\in B$.*

Proof. Let $\phi=g_1-g_2$. Let $D=\{x|x\in B, \phi(x)<0\}$, which is open. Then $\Delta\phi\leqq W_2g_2-W_1g_1$, and thus $\Delta\phi\leqq 0$ on D by (iii) and (iv). Let $U_\varepsilon=\{x||x|<R_\varepsilon\}$. ϕ is superharmonic on $D\cap U_\varepsilon$, but $\phi=0$ on ∂D and $\phi\geqq -\varepsilon$ on ∂U_ε. Thus $\phi\geqq -\varepsilon$ on $D\cap U_\varepsilon$. Since ε is arbitrary and R_ε can be chosen to tend to infinity as $\varepsilon\to 0$, $\phi\geqq 0$ on D, so D is empty. □

Acknowledgement. It is a pleasure to thank Prof. Kenji Yajima for a helpful discussion.

References

1. Adams, R.A.: Sobolev spaces. New York: Academic Press 1975
2. Benguria, R.: The von Weizsäcker and exchange corrections in Thomas-Fermi theory. Ph. D. thesis, Princeton University 1979 (unpublished)
3. Benguria, R., Brezis, H., Lieb, E.H.: The Thomas-Fermi-von Weizsäcker theory of atoms and molecules. Commun. Math. Phys. **79**, 167–180 (1981)
4. Fermi, E.: Un metodo statistico per la determinazione di alcune priorieta dell'atome. Rend. Accad. Naz. Lincei **6**, 602–607 (1927)
5. Gilbarg, D., Trudinger, N.: Elliptic partial differential equations of second order. Berlin, Heidelberg, New York: Springer 1977
6. Kato, T.: On the eigenfunctions of many particle systems in quantum mechanics. Commun. Pure Appl. Math. **10**, 151–171 (1957)
7. Kato, T.: Schrödinger operators with singular potentials. Isr. J. Math. **13**, 135–148 (1973)
8. Liberman, D., Lieb, E.H.: Numerical calculation of the Thomas-Fermi-von Weizsäcker function for an infinite atom without electron repulsion, Los Alamos National Laboratory report (in preparation)
9. Lieb, E.H.: Thomas-Fermi and related theories of atoms and molecules. Rev. Mod. Phys. **53**, 603–641 (1981)
10. Lieb, E.H., Simon, B.: The Thomas-Fermi theory of atoms, molecules, and solids. Adv. Math. **23**, 22–116 (1977)
11. Morrey, C.B., Jr.: Multiple integrals in the calculus of variations. Berlin, Heidelberg, New York: Springer 1966
12. Stampacchia, G.: Equations elliptiques du second ordre à coefficients discontinus. Montréal: Presses de l'Univ. 1965
13. Thomas, L.H.: The calculation of atomic fields. Proc. Camb. Phil. Soc. **23**, 542–548 (1927)
14. von Weizsäcker, C.F.: Zur Theorie der Kernmassen. Z. Phys. **96**, 431–458 (1935)

Communicated by R. Jost

Received November 11, 1981

J. Phys. B: At. Mol. Phys. **18** (1985) 1045–1059. Printed in Great Britain

The most negative ion in the Thomas–Fermi–von Weizsäcker theory of atoms and molecules

Rafael Benguria†§‖ and Elliott H Lieb‡‖

† Departamento de Física, Universidad de Chile, Casilla 5487, Santiago, Chile
‡ Departments of Mathematics and Physics, Princeton University, PO Box 708, Princeton, NJ 08544, USA

Received 4 May 1984

Abstract. Let N_c denote the maximum number of electrons that can be bound to an atom of nuclear charge z, in the Thomas–Fermi–von Weizsäcker theory. It is proved that N_c cannot exceed z by more than one, and thus this theory is in agreement with experimental facts about real atoms. A similar result is proved for molecules, i.e. N_c cannot exceed the total nuclear charge by more than the number of atoms in the molecule.

1. Introduction

The Thomas–Fermi–von Weizsäcker (TFW) theory (von Weizsäcker 1935, Benguria *et al* 1981, Lieb 1981) is defined by the energy functional (see Lieb 1981, § VII) (in units in which $\hbar^2(2m)^{-1} = |e| = 1$, where e and m are the electron charge and mass)

$$\xi(\rho) = A\int (\nabla\rho^{1/2}(x))^2\,dx + \tfrac{3}{5}\gamma\int \rho(x)^{5/3}\,dx - \int V(x)\rho(x)\,dx + D(\rho,\rho) \tag{1}$$

where

$$D(\rho,\rho) = \tfrac{1}{2}\int \rho(x)|x-y|^{-1}\rho(y)\,dx\,dy \tag{2}$$

$$V(x) = \sum_{j=1}^{K} z_j|x-R_j|^{-1}. \tag{3}$$

Here $z_1, z_2, \ldots, z_K \geqslant 0$ are the charges of K fixed nuclei located at $R_1, \ldots, R_K$. The total nuclear charge is denoted by Z, $Z = \Sigma_{j=1}^{K} z_j$. $K = 1$ is the atomic case and here we shall simply write $Z = z_1 = z$. dx is always a three-dimensional integral. $\xi(\rho)$ is defined for electronic densities $\rho(x) \geqslant 0$ such that each of the terms of $\xi(\rho)$ in (1) is finite. In the physical situation, $\gamma = \gamma_{\text{phys}} = (3\pi^2)^{2/3}$ but, for generality, we shall allow γ to be an arbitrary positive constant in what follows. The TFW energy for N (not necessarily an integer) electrons is defined by

$$E(N) = \inf\left(\xi(\rho)\,\middle|\,\int\rho = N\right). \tag{4}$$

§ Work partially supported by Dpto Desarrollo Investigación, Universidad de Chile.
‖ Work partially supported by US National Science Foundation grant PHY-8116101-A02.

On energetic grounds, the value of A should be chosen to reproduce the Scott term in the expression of the atomic or molecular energy $E(N)$ as a function of N and the nuclear charges, (see Lieb 1981, §§ V.B, VII.D). Numerically one finds (Lieb and Liberman 1982, Lieb 1982), $A=0.1859$. However, we should retain A as an arbitrary positive constant. In the original TFW model (von Weizsäcker 1935) the numerical value of A is 1.

It is known (Benguria *et al* 1981, Lieb 1981) that there exists a critical value of N (depending on A, γ and the z_j and R_j), which we denote by N_c, such that for $N \leqslant N_c$ the minimisation problem (4) has a unique solution, whereas for $N > N_c$ there is no solution. In other words, N_c is the maximum number of 'electrons' that can be bound to the atom or molecule. The aim of this paper is to find an upper bound for N_c. The value of N_c is given by $\int \rho$, where $\rho \geqslant 0$ is the unique minimising function of $\xi(\rho)$ without constraints. Let $\psi=\rho^{1/2}$. Then ψ is the *unique* positive solution of the TFW equation (for a saturated system),

$$-A\,\Delta\psi(x)+(\gamma\psi(x)^{4/3}-\phi(x))\psi(x)=0 \tag{5}$$

where

$$\phi(x)=V(x)-|x|^{-1}*\rho \qquad \text{with } \rho=\psi^2. \tag{6}$$

Note that (5) is the Euler equation corresponding to the functional $\xi(\psi^2)$. The only previous rigorous results (Lieb 1981, Benguria *et al* 1981) for N_c were that $Z<N_c<2Z$.

Our main result is the following.

Theorem 1. For a TFW molecule of $K \geqslant 1$ atoms,

$$0<N_c-Z\leqslant 270.74(A/\gamma)^{3/2}K \tag{7}$$

for all choices of $z_1,\ldots,z_K$ and $R_1,\ldots,R_K$. In particular, for the value of A chosen in Lieb (1981) and Lieb and Liberman (1982) to reproduce the Scott term in the energy (i.e. $A=0.1859$) and for the physical $\gamma=(3\pi^2)^{2/3}$,

$$0<N_c-Z<0.7335K. \tag{8}$$

In the TFW model the number of electrons is not generally an integer, but in a real atom N and z are required to be integral. How can theorem 1 be interpreted in the light of this additional requirement? One way is the energetic point of view: since $E(N)$ is strictly decreasing for $N<N_c$ and constant for $N \geqslant N_c$ (Lieb 1981, § VII.A), theorem 1 implies that $E(z)>E(z+1)=E(z+2)$. Thus, the $(z+1)$th electron has a positive binding energy, while the $(z+2)$th does not, and we can say that a singly ionised atom (but not a doubly ionised atom) is stable. This interpretation, however, suffers from the drawback that there is no solution to (5) when $N=z+1$. A second interpretation that leads to the same conclusion about atomic ionisation, but eliminates the problem that (5) has no solution for $N=z+1$, was kindly provided by John Morgan: introduce the Fermi–Amaldi correction (i.e. replace $D(\rho,\rho)$ in (1) by $(1-1/N)D(\rho,\rho)$). This has the effect of replacing z by $z'(N)=Nz/(N-1)$. (It also effectively changes A and γ, but not A/γ.) Theorem 1 now states that a solution to (5) always exists if $N \leqslant z'(N)$ while it never exists if $N-0.74 \geqslant z'(N)$. This implies that a solution exists (with N and z integral) if and only if $N \leqslant z+1$. However it is not clear that $E(z+1)<E(z)$ in this Fermi–Amaldi model.

The best previous upper bound on N_c is, as we said, $N_c<2Z$ (Lieb 1981, theorem 7.23), a result which is valid for both atoms and molecules. It turns out that such a bound also holds for the Hartree (bosonic) atom. More recently one of us (Lieb 1984a, Lieb *et al* 1984) has proved a similar bound for the real Schrödinger equation namely, $N_c < 2z+1$ for an atom and $N_c < 2Z+K$ for a molecule of K atoms. This result (Lieb 1984a, Lieb *et al* 1984) is valid regardless of the statistics of the bound particles. However, if the bound particles are fermions, as is the case for real matter, N_c should presumably not exceed z (for an atom) by more than one or two electrons. This is still a conjecture; however, it has been proved that $N_c/z \to 1$ as $z \to \infty$ for fermions (Lieb *et al* 1984). On the other hand, we know that $N_c - z > 0.2z$ for the Schrödinger equation of an atom with bosonic particles and for large z (Benguria and Lieb 1983, Baumgartner 1983, 1984). See (Lieb 1984b) for a review of the recent literature on the subject.

In the Thomas-Fermi theory, defined by the energy functional (1) with $A=0$, N_c is exactly Z even in the molecular case (Lieb 1981, theorem 3.18, Lieb and Simon 1977). Equation (7) implies that for the TFW atom or molecule $N_c \to Z$ as $A \to 0$. However, we do not expect $N_c(A)$ to be analytic around $A=0$ because the von Weizsäcker correction is a singular perturbation to Thomas-Fermi theory. It is an *open problem* to derive an asymptotic expansion for $N_c(A)$ around $A=0$.

Two other open problems arise from the results of this paper. The first is that while we prove an upper bound for $N_c - Z$, we have no *lower* bound. We conjecture that $N_c - Z \to \text{constant} > 0$ as $Z \to \infty$. The second problem is related to the first: it is highly plausible that $N_c - Z$ is a monotonically increasing function of all the z_j (for fixed $R_1, \ldots, R_K$). Is this true?

This article is organised as follows: in § 2 we give the proof of theorem 1; in § 3 we determine the behaviour of N_c as Z goes to zero. Finally in § 4 we give a bound for the chemical potential of a neutral molecule. Such a bound is independent of the charge of the nuclei.

We should like to emphasise that many of the results herein can be extended in two ways: (i) to spherically symmetric 'smeared out' nuclei; (ii) to the TFW theory in which the exponent $\frac{5}{3}$ in (1) is replaced by some $p \neq \frac{5}{3}$ (cf Lieb 1981). For simplicity and clarity we confine ourselves here to point nuclei and $p = \frac{5}{3}$.

2. Proof of theorem 1

The proof of theorem 1 will be divided into three steps. First, we estimate the excess charge $Q \equiv N_c - Z$ in terms of the electronic density ρ and the TFW potential ϕ evaluated at an arbitrary, but fixed, distance r from all the nuclei. Then we find a local bound for ρ in terms of ϕ. These two estimates do not involve the z_j explicitly. Therefore, if we can prove that at some distance of order one, (i.e. independent of the z_j) the potential ϕ is bounded by a constant independent of the z_j, then the two previous results will imply that Q is less than a constant independent of the z_j, which is basically what theorem 1 says. Proving this last fact about ϕ is our third step. We begin with

Lemma 2. Let $\psi \geq 0$, ϕ be the unique solution pair for the TFW equation (5), (6) with V being the potential (3) for a molecule. Then, the function

$$p(x) = (4\pi A \psi(x)^2 + \phi(x)^2)^{1/2} \tag{9}$$

is subharmonic away from the nuclei, i.e. on $\mathscr{R}^3 \backslash \bigcup_{j=1}^{K} R_j$.

Proof. By direct computation

$$\Delta p = p^{-1}(4\pi A\psi\, \Delta\psi + \phi\, \Delta\phi) + h \tag{10}$$

with

$$h = 4\pi A p^{-3}|\phi\nabla\psi - \psi\nabla\phi|^2 \geqslant 0.$$

By (5), (6) the sum of the first two terms in (10) is (away from the nuclei so that $\Delta V = 0$): $4\pi\gamma\psi^{10/3} \geqslant 0$. Thus, $\Delta p \geqslant 0$, so p is subharmonic. □

Remark. Let

$$W(x) \equiv \gamma\psi(x)^{4/3} - \phi(x) \tag{11}$$

be the 'potential' in the TFW equation (5). Proceeding as in the proof of lemma 2, one can show that

$$(4\pi A\psi(x)^2 + W(x)^2)^{1/2} \tag{12}$$

is subharmonic whenever $W(x) \geqslant 0$.

The next lemma gives a local bound for ψ in terms of ϕ. This bound is independent of the nuclear charges z_j.

Lemma 3. For all $\lambda \in (0, 1)$ and all $x \in \mathscr{R}^3$

$$\gamma\lambda\psi(x)^{4/3} \leqslant \phi(x) + c(\lambda)A^2\gamma^{-3} \tag{13}$$

with

$$c(\lambda) \equiv (9/4)\pi^2\lambda^{-2}(1-\lambda)^{-1}. \tag{14}$$

Proof. Define $u(x) \equiv \psi(x)^{4/3}$. Then, from the TFW equation (5),

$$-\Delta u + (4/3A)(\gamma u - \phi)u + |\nabla u|^2/4u = 0$$

and hence

$$-\Delta u + (4/3A)(\gamma u - \phi)u \leqslant 0. \tag{15}$$

Also, from (6)

$$-\Delta\phi = -4\pi\psi^2 = -4\pi u^{3/2} \qquad x \neq R_j, \text{ all } j. \tag{16}$$

Let $v(x) = \gamma\lambda u(x) - \phi(x) - d$, with d a positive constant. We shall show that $v(x) \leqslant 0$, all x, for appropriate d and λ. From equations (15) and (16),

$$-\Delta v \leqslant -(4\gamma\lambda/3A)(\gamma u - \phi)u + 4\pi u^{3/2}.$$

Let $S = \{x | v(x) > 0\}$. ψ is continuous and goes to zero at infinity; ϕ is continuous away from the R_j and it also goes to zero at infinity (Benguria *et al* 1981, § III). Therefore v is continuous away from all the R_j and goes to $-d$ at infinity. Hence S is open and bounded. Moreover, $R_j \notin S$, all j since $\phi = +\infty$ at the R_j. On S, $\phi < \gamma\lambda u - d$ so

$$\begin{aligned} -\Delta v &\leqslant -(4\gamma\lambda/3A)(\gamma u + d - \gamma\lambda u)u + 4\pi u^{3/2} \\ &\leqslant u[4\pi u^{1/2} - (4/3A)\gamma^2\lambda(1-\lambda)u - (4/3A)\gamma\lambda d] \\ &\leqslant 0 \end{aligned}$$

provided we choose $\lambda \in (0,1)$ and $d = \frac{9}{4}(\pi A/\lambda)^2\gamma^{-3}(1-\lambda)^{-1}$ in order that the quantity in brackets [ ] be non-negative for all possible (unknown) values of $u(x)$. With that choice of λ and d, v is subharmonic on S. On ∂S $v=0$, and therefore S is empty. □

Corollary. For all $x \in \mathcal{R}^3$

$$\phi(x) \geqslant -3^5 2^{-4}\pi^2 A^2\gamma^{-3} \geqslant -150A^2\gamma^{-3}. \tag{17}$$

If x is such that $\phi(x) \leqslant 0$, then $p(x) = (4\pi A\psi(x)^2 + \phi(x)^2)^{1/2}$ satisfies

$$p(x) \leqslant (\tfrac{4}{3})^{3/4} 16\pi^2 A^2\gamma^{-3} \leqslant 196A^2\gamma^{-3}. \tag{18}$$

Proof. By (13) and the fact that $\psi(x)^{4/3} \geqslant 0$, $\phi(x) \geqslant -c(\lambda)A^2\gamma^{-3}$ for all $\lambda \in (0,1)$. Minimising $c(\lambda)$ (at $\lambda = \frac{2}{3}$) gives (17). To prove (18), take $\lambda = \frac{3}{4}$ (which minimises $c(\lambda)/\lambda$), let $a = -\frac{3}{4}cA^2\gamma^{-3}$ and observe from (13) that

$$p(x)^2 \leqslant \max_{a \leqslant \phi \leqslant 0} [4\pi A(3\gamma/4)^{-3/2}(\phi - a)^{3/2} + \phi^2]. \tag{19}$$

The right-hand side of (19) is convex in ϕ, so its maximum occurs either at $\phi = 0$ or $\phi = a$. $\phi = 0$ prevails and gives (18). □

In our next lemma, starting from ρ and ϕ, we introduce a smeared density $\tilde{\rho}$ and potential $\tilde{\phi}$. We find that $\tilde{\rho}$ and $\tilde{\phi}$ satisfy an inequality resembling the Thomas-Fermi equation for smeared nuclei. Then we use a comparison theorem to get an upper bound for the smeared potential $\tilde{\phi}$ in terms of a universal function (independent of the z_j). Finally, noting that ϕ is subharmonic away from the nuclei, we see that essentially the same bound applies to ϕ. In particular, this lemma says that at distances of order one from all the nuclei, in atomic units, ϕ is of order one and, in any case, independent of the z_j. We note, however, that this bound is not satisfactory both very close and very far from the nuclei. Near the nuclei it diverges too fast. On the other hand, the bound is always positive, whereas ϕ is negative at large distances because $Q = N_c - Z$ is strictly positive.

Lemma 4. Let $\psi \geqslant 0$, ϕ be the solution of the TFW equation (5), (6) with V given by (3). Choose any $R > 0$ and define

$$\delta \equiv 25\pi^{-2}\gamma^3 = 2.53\gamma^3 \tag{20}$$

which is independent of the z_j. Suppose $x \in \mathcal{R}^3$ is such that $|x - R_j| > R$ for all $j = 1, 2, \ldots, K$. Then

$$\phi(x) \leqslant A\pi^2 R^{-2} + \delta \sum_{j=1}^{K} (|x - R_j| - R)^{-4}. \tag{21}$$

Proof. Let $W = \gamma\rho^{2/3} - \phi$, $\rho \equiv \psi^2$ and consider the Hamiltonian $H = -A\Delta + W$. H is a non-negative operator, since its ground state, the TFW function ψ, has zero energy (chemical potential). Therefore for any function $b \in L^2$ with $\nabla b \in L^2$ we have

$$A \int |\nabla b(x)|^2 \,\mathrm{d}x + \int W(x) b(x)^2 \,\mathrm{d}x \geqslant 0. \tag{22}$$

We shall choose $b(x)$ to be a translate of the normalised ground state, $e(x)$, of the Laplacian on a ball of radius R with Dirichlet boundary condition. That is, let

$e(x)=(2\pi R)^{-1/2}\sin(\pi|x|R^{-1})/|x|$, for $|x|\leq R$ and $e(x)=0$ otherwise. Clearly, $e(x)$ is spherically symmetric, decreasing and it has compact support. Let $b_x(y)=e(y-x)$ denote the translate of e and define $g(x)=e(x)^2$ and $g_x(y)=g(y-x)$. Let $B=A\int|\nabla b_x(y)|^2\,dy$. Clearly B does not depend on x. With this choice of b, $B=(\pi/R)^2A$. From equation (22) we have,

$$\int W(y)g_x(y)\,dy\geq -B \qquad \text{all } x. \tag{23}$$

Note that $\int W(y)g_x(y)\,dy=(g*W)(x)$, where an asterisk denotes convolution. Define

$$\tilde{\phi}\equiv\phi*g-B. \tag{24}$$

Since $\phi\in L^{3+\varepsilon}+L^{3-\varepsilon}$, $\varepsilon>0$, (Benguria *et al* 1981, proof of lemma 7) and $g\in L^p$, all $p\geq 1$, $\tilde{\phi}$ is continuous and goes to $-B$ at infinity (Lieb 1981, lemma 3.1). Using Hölder's inequality, we have for all x

$$(g*\rho^{2/3})(x)\leq[(g*\rho)(x)]^{2/3}\left(\int g(y)\,dy\right)^{1/3}=[(g*\rho)(x)]^{2/3} \tag{25}$$

where we have used $\int g(y)\,dy=1$. Let us also define

$$\tilde{\rho}\equiv g*\rho. \tag{26}$$

From equations (23)–(26) we obtain for all x

$$B\geq(\phi*g)(x)-\gamma(\rho^{2/3}*g)(x)\geq\tilde{\phi}(x)+B-\gamma\tilde{\rho}(x)^{2/3}.$$

In other words

$$\tilde{\phi}\leq\gamma\tilde{\rho}^{2/3}. \tag{27}$$

Notice that ϕ is subharmonic away from the nuclei and that $\tilde{\phi}=g*\phi-B$ with g being spherically symmetric, positive, of total mass one and having support in a ball of radius R. From this it follows easily that

$$\phi(x)\leq\tilde{\phi}(x)+B \tag{28}$$

for all x such that $|x-R_j|>R$ (for all j). Thus, to prove (21) we need a bound on $\tilde{\phi}$.

From equations (6) and (24), using the bound (27) and the fact that the Laplacian commutes with convolution, we compute

$$-(4\pi)^{-1}\Delta\tilde{\phi}(x)=\tilde{V}(x)-\tilde{\rho}(x)\leq\tilde{V}(x)-\gamma^{-3/2}[\tilde{\phi}_+(x)]^{3/2} \tag{29}$$

with

$$\tilde{V}(x)=\sum_{j=1}^{K}z_jg(x-R_j) \tag{30}$$

and with $\tilde{\phi}_+(x)=\max(\tilde{\phi}(x),0)$.

Note that equation (29) resembles a Thomas–Fermi (TF) equation with smeared nuclei of spherical charge density $z_jg(x-R_j)$. Indeed, let $\hat{\phi}$ be the TF potential for this system (i.e. with equality in (29)):

$$-(4\pi)^{-1}\Delta\hat{\phi}(x)=\tilde{V}(x)-\gamma^{-3/2}\hat{\phi}(x)^{3/2}. \tag{31}$$

It is known from general TF theory that (31) has a unique solution, $\hat{\phi}$, that goes to zero at infinity.

It is easy to see that

$$\tilde{\phi}(x) \leq \hat{\phi}(x) \qquad \text{for all } x \tag{32}$$

by observing that if the set $M = \{x|\hat{\phi}(x) - \tilde{\phi}(x) < 0\}$, then $\hat{\phi} - \tilde{\phi}$ is superharmonic on M and zero on the boundary of M and infinity, so M is empty.

The next step is to bound $\hat{\phi}$. First consider an atom with $V = z/r$, $r = |x|$, and consider the function $f(r) = \delta(r-R)^{-4}$ which satisfies

$$(4\pi)^{-1}\Delta f \leq \gamma^{-3/2} f^{3/2} \qquad \text{for } r > R. \tag{33}$$

Outside the ball of radius R (centred at the origin) $\hat{\phi}$ satisfies

$$(4\pi)^{-1}\Delta\hat{\phi} = \gamma^{-3/2}\hat{\phi}^{3/2}. \tag{34}$$

Again, by a comparison argument (and using the fact that $f(r) - \hat{\phi}(r) = \infty$ when $r = R$)

$$\hat{\phi}(r) \leq f(r) \qquad \text{for } r > R. \tag{35}$$

This, together with (28), proves (21) in the atomic case.

For the molecular case, let $\hat{\phi}_j(x)$ be the solution to (31) for an atom of (smeared) nuclear charge z_j located at R_j. By another comparison argument (Lieb and Simon 1977, theorem V.12 or Lieb 1981, corollary 3.6), $\hat{\phi}(x) \leq \Sigma_{j=1}^{K} \hat{\phi}_j(x)$. This, together with (35) and (28) proves (21). □

We conclude this section with

Proof of Theorem 1: atomic case. Let us start with the atomic case, $V(x) = z/|x|$, in order to expose the ideas most simply. The following facts have been established:

$$p(x) = (4\pi A\psi(x)^2 + \phi(x)^2)^{1/2} \tag{36}$$

is subharmonic for $|x| > 0$.

$$p(x) \leq (4/3)^{3/4} 16\pi^2 A^2 \gamma^{-3} \tag{37}$$

if $\phi(x) \leq 0$.

$$\phi(x) \leq \delta(|x| - R)^{-4} + \pi^2 A R^{-2} \tag{38}$$

for all $|x| > R > 0$, with $\delta = 25\gamma^3\pi^{-2}$ and with arbitrary $R > 0$.

$$\gamma\lambda\psi(x)^{4/3} \leq \phi(x) + c(\lambda)A^2\gamma^{-3} \tag{39}$$

for all $|x|$, $0 < \lambda < 1$ with $c(\lambda) = 9\pi^2[4\lambda^2(1-\lambda)]^{-1}$.

The functions p, ϕ, ψ are functions only of $|x| = r$. As $r \to \infty$, $\psi(r) \to 0$ faster than any power of r (Lieb 1981, theorem 7.24) and $r\phi(r) \to -Q$. Thus,

$$rp(r) \to Q \qquad \text{as} \qquad r \to \infty. \tag{40}$$

The subharmonicity and the fact that $p(r) \to 0$ as $r \to \infty$ imply that $rp(r)$ is monotonically decreasing and convex. (This may be seen from the fact that $\Delta p \geq 0$ is equivalent, in polar coordinates, to $d^2(rp(r))/dr^2 \geq 0$.) Using (40) we conclude that

$$Q \leq rp(r) \qquad \text{for any } r > 0. \tag{41}$$

The same conclusion, (41), can be reached from another viewpoint, which will be important for the molecular case: fix $r > 0$ and consider the domain $D_r = \{x||x| > r\}$. Let P be any *harmonic* function on D_r with $P(x) \to 0$ as $|x| \to \infty$ and $P(x) \geq p(x)$ on the boundary $|x|=r$. Then $P(x) \geqslant p(x)$ for all $x \in D_r$. (Proof: on the set $E = \{x | P(x) < p(x)\} \subset D_r$, $p-P$ is positive, subharmonic and $p-P$ vanishes on the boundary of E, so E is empty.) Now choose $P(x) = rp(r)/|x|$ for $|x|>r$. Then $rp(r) \geqslant |x|p(x)$, for $|x|>r$. However, $|x|p(x) \to Q$ as $|x| \to \infty$, and this establishes (41).

To complete the proof, we make the following specific choices for r, R, λ:

$$r = 0.9086\,(\gamma^3/A)^{1/2} \qquad R = 0.4020\,(\gamma^3/A)^{1/2} \qquad \lambda = 0.7825. \tag{42}$$

If $\phi(r)$ happens to be positive, we use the bound (38), followed by (39) and insert these in (36). (41) then implies that if $\phi(r)>0$ then

$$Q \leqslant 270.74\,(A/\gamma)^{3/2}. \tag{43}$$

(The numbers in (42) were chosen to minimise the coefficient in (43).) On the other hand, if $\phi(r) \leqslant 0$ we can use (37) and (41) to conclude that

$$Q \leqslant 178.03\,(A/\gamma)^{3/2}. \tag{44}$$

Clearly, (43) is the worst case, and this gives theorem 1. Note, however, that if it were to be shown that $\phi(r) \leqslant 0$, then the bound (44) would be valid, and, using the physical values of A and γ, one would obtain $Q \leqslant 0.49$.

Molecular case. Equation (36) is still valid, except that p is subharmonic only on the set $x \neq R_j$ (all $j = 1, \ldots, K$). Equations (37) and (39) are also valid. Equation (38) must be replaced by (21) on the set

$$D_R = \{x \,|\, |x - R_j| > R \text{ for all } j = 1, \ldots, K\}. \tag{45}$$

Now choose r, R and λ as in (42) and consider the smaller domain

$$D_r = \{x \,|\, |x - R_j| > r \text{ for all } j = 1, \ldots, K\}. \tag{46}$$

Consider the following function which is harmonic on D_r:

$$P(x) = Q_1 \sum_{j=1}^{K} |x - R_j|^{-1} \tag{47}$$

where Q_1 is the right-hand side of (43), namely the value of the upper bound for $rp(r)$ computed in the atomic case under the assumption $\phi(r) \geqslant 0$. As explained above, if we can show that $P(x) \geqslant p(x)$ for all x on the boundary of D_r, then $P(x) \geqslant p(x)$ for all $x \in D_r$. Taking the limit $|x| \to \infty$ yields

$$Q = \lim_{|x|\to\infty} |x|\,p(x) \leqslant \lim_{|x|\to\infty} |x|\,P(x) = KQ_1 \tag{48}$$

which is the desired result.

Let x be on the boundary of D_r, so that $|x - R_j| = r$ for some j (say $j = m$). If $\phi(x) \leqslant 0$, the bound (37) is valid and $p(x) \leqslant Q_2/r$, where $Q_2 < Q_1$ is the right-hand side of (44). However, $P(x) \geqslant Q_1 |x - R_m|^{-1} = Q_1/r$, so $P(x) > p(x)$. On the other hand, suppose $\phi(x) \geqslant 0$, in which case we can use (21) and (39). Now use proposition 5 below with the choices $t = \frac{3}{2}$, $s = 2$ and

$$a_j = \delta(|x - R_j| - R)^{-4} \qquad b_j = a_j(4\pi A)^{2/3}/\gamma\lambda \qquad \text{for } j \neq m$$

$$a_m = \delta(|x - R_m| - R)^{-4} + A\pi^2 R^{-2}$$

$$b_m = (a_m + c(\lambda)A^2\gamma^{-3})(4\pi A)^{2/3}/\gamma\lambda.$$

Recalling that $|x-R_m|=r$ we have

$$p(x) \leqslant p_1(r) + \sum_{j\neq m} \hat{p}(x-R_j) \tag{49}$$

where $p_1(r)$ is precisely the number we calculated before in the atomic case and

$$\hat{p}(x-R_j) = (b_j^{3/2}+a_j^2)^{1/2}. \tag{50}$$

By construction, $p_1(r)=Q_1/r$. Thus, $p(x)\leqslant P(x)$ if we can show that $\hat{p}(x-R_j)\leqslant Q_1/|x-R_j|$ for $j\neq m$. Let $|x-R_j|=u\geqslant r$. We require that

$$u^2[4\pi A(\gamma\lambda)^{-3/2}\delta^{3/2}(u-R)^{-6}+\delta^2(u-R)^{-8}]\leqslant Q_1^2. \tag{51}$$

However, the functions $u^2(u-R)^{-6}$ and $u^2(u-R)^{-8}$ are monotonically decreasing in u for $u>R$. Hence, the left-hand side of (51) is less than its value at $u=r$. But this is obviously less than $r^2p_1(r)^2$ which is Q_1^2. □

Proposition 5. Let $0\leqslant s\leqslant 2, 0\leqslant t\leqslant 2$ and let $a_1,\ldots,a_K, b_1,\ldots,b_k$ be $2K$ non-negative numbers. Then

$$\left[\left(\sum_{j=1}^K a_j\right)^s+\left(\sum_{j=1}^K b_j\right)^t\right]^{1/2}\leqslant \sum_{j=1}^K (a_j^s+b_j^t)^{1/2}. \tag{52}$$

Proof. It suffices to prove the proposition for $K=2$ namely, for $a, A, b, B\geqslant 0$,

$$[(a+A)^s+(b+B)^t]^{1/2}\leqslant (a^s+b^t)^{1/2}+(A^s+B^t)^{1/2}. \tag{53}$$

If (53) holds then simply take $a=a_1, A=\Sigma_2^K a_j$ (and similarly for b, B) and use induction. Now $(a+A)^s=(a+A)^2/(a+A)^{2-s}\leqslant(a^2+2aA+A^2)/\max(a^{2-s}, A^{2-s})\leqslant a^s+2a^{s/2}A^{s/2}+A^s$. A similar inequality holds for $(b+B)^t$. Squaring (53) and using these inequalities, it suffices to prove that

$$a^{s/2}A^{s/2}+b^{t/2}B^{t/2}\leqslant(a^s+b^t)^{1/2}(A^s+B^t)^{1/2}.$$

This, however follows from the Cauchy-Schwarz inequality. □

3. Behaviour of N_c for small Z or small γ or large A

Although theorem 1 gives an upper bound for all values of the z_j, it is primarily useful for the large-Z behaviour of Q. In fact, the comparison function f we chose in the proof of lemma 4, (i.e. $f(x)=\delta(|x|-R)^{-4}$ may be too big when we consider small z. Since the atomic $\phi(x)$ is bounded from above by $V(x)=z|x|^{-1}$ and the function g has support on a ball of radius R and total mass 1, $\tilde{\phi}(x)\leqslant z|x|^{-1}$ for $|x|\geqslant R$. In particular $\tilde{\phi}(R)\leqslant z|R|^{-1}$, whereas the comparison function f goes to infinity at $|x|=R$. Therefore it is somewhat better to choose $f(x)=\delta(|x|-\alpha R)^{-4}$, where $\alpha=\alpha(z)=1-(\delta/zR^3)^{1/4}$ is such that $f(R)=zR^{-1}$. Then, proceeding as in the proof of theorem 1, one gets a z_j-dependent bound for Q. Although we do not give any details here, we point out that as z goes to zero, for an atom, this upper bound goes to $3.057A^{3/2}$ with $\gamma=\gamma_{\text{phys}}$. We know, however, that as Z goes to zero, Q vanishes because $Q<Z$ (Lieb 1981, theorem 7.23). Thus, the previous bound is not good for small Z. Getting the behaviour of Q as a function of Z for small Z is the subject of this section. The main result for Q is contained in equation (63) below.

We begin with a normalisation convention. Choose $z_1^0, \ldots, z_K^0 > 0$ such that

$$\sum_{j=1}^{K} z_j^0 = 1 \tag{54}$$

and let $R_1^0, \ldots, R_K^0$ be fixed, distinct points in $\mathscr{R}^3$. Let $Z>0$ be the total nuclear charge in a molecule in which

$$z_j = Z z_j^0 \qquad R_j = (A/Z) R_j^0. \tag{55}$$

In this molecule the length scale is A/Z. In TF theory, by comparison, it is $Z^{-1/3}$. As $Z \to 0$ the atoms move apart. One can also treat the case in which the R_j remain fixed as $Z \to 0$; we do not do so explicitly here, but note that in the limit $Z \to 0$ this is obviously the same as placing all the R_j at one common point.

Let us write the solution to (5) as

$$\psi(x) = Z^2 A^{-3/2} \tilde{\psi}((Z/A)x) \tag{56}$$

whence

$$\int \psi(x)^2 \, dx = Z \int \tilde{\psi}(x)^2 \, dx. \tag{57}$$

The TFW equation (5) then reads

$$(-\Delta + \tilde{\gamma}\tilde{\psi}(x)^{4/3} - \tilde{\phi}(x))\tilde{\psi}(x) = 0 \tag{58}$$

with

$$\tilde{\gamma} = \gamma Z^{2/3}/A \tag{59}$$

$$\tilde{\phi}(x) = \tilde{V}(x) - (|x|^{-1} * \tilde{\psi}^2)(x) \tag{60}$$

$$\tilde{V}(x) = \sum_{j=1}^{K} z_j^0 |x - R_j^0|^{-1}. \tag{61}$$

With this scaling there is only one non-trivial parameter in the problem, $\tilde{\gamma}$. The potential $\tilde{V}$ is that of a molecule with unit total nuclear charge and $A = 1$. Our goal is to elucidate the behaviour of (58) as $\tilde{\gamma} \to 0$. From now on we shall omit the tilde on the various quantities in (58)-(61).

Formally, at least, as $\gamma \to 0$ the solution ψ_γ to (58) approaches the solution ψ_H to the Hartree equation

$$(-\Delta - \phi(x))\psi_H = 0 \tag{62}$$

with ϕ given by (61) and (60) with ψ_H^2. This equation (which was also used in Benguria and Lieb 1983) has a unique positive solution, ψ_H, because the proof in Benguria *et al* (1981), Lieb (1981), that (1) has a unique minimum and that this minimum is the unique solution to (5) only uses the fact that $\gamma \geq 0$. Assuming that $\int \psi^2 \to \int \psi_H^2$, (57) tells us that

$$\lim_{Z \to 0} Q/Z = \int \psi_H^2(x) \, dx - 1. \tag{63}$$

Proving (63) is the goal of this section.

As stated earlier, $1 < \int \psi_{\mathrm{H}}^2 < 2$. For the atomic case $K = 1$, $z_1^0 = 1$, (62) has been solved numerically (Baumgartner 1983) with the result that

$$\int \psi_{\mathrm{H}}(x)^2 \, \mathrm{d}x = 1.21. \tag{64}$$

Thus, as $Z \to 0$, $Q \simeq 0.21Z$ for an atom. The right-hand side of (63) is not known for a molecule, but we conjecture that

$$\int \psi_{\mathrm{H}}(x)^2 \, \mathrm{d}x \leqslant 1 + 0.21 \mathrm{K}. \tag{65}$$

The point $\gamma = 0$ is not special. We shall prove the following general theorem which says that if $\gamma \to \Gamma \geqslant 0$ then the solution $\psi_\gamma \to \psi_\Gamma$ in a very strong sense. In particular there is strong L^2 convergence so that (63) is justified.

Theorem 6. Let ψ_γ (and $\rho_\gamma = \psi_\gamma^2$) denote the unique positive solution to the TFW equation (5) for $\gamma \geqslant 0$, with A fixed and with V in equation (3) fixed. (Note: condition (54) is irrelevant here.) Let $\Gamma \geqslant 0$ be fixed. Then, as $\gamma \to \Gamma$, $\psi_\gamma \to \psi_\Gamma$ in the following senses:

$$\nabla \psi_\gamma \to \nabla \psi_\Gamma \text{ strongly in } L^2. \tag{66}$$

$$\psi_\gamma \to \psi_\Gamma \text{ and } \rho_\gamma \to \rho_\Gamma \text{ strongly in } L^p \tag{67}$$

for all $1 \leqslant p \leqslant \infty$.

$$|x|^{-1} * \psi_\gamma^2 \to |x|^{-1} * \psi_\Gamma^2 \text{ strongly in } L^p \tag{68}$$

for all $3 < p \leqslant \infty$.

$$D(\psi_\gamma^2, \psi_\gamma^2) \to D(\psi_\Gamma^2, \psi_\Gamma^2) \tag{69}$$

(cf (2)).

Proof. Let $\gamma_n \geqslant 0$ be any sequence with $\gamma_n \to \Gamma$ and let $\psi_n \equiv \psi_{\gamma_n}$. Since ψ_Γ is unique, it suffices to show that some subsequence of ψ_n converges to ψ_Γ in the indicated senses. In the appendix it is proved that $\|\psi_\gamma\|_\infty < C_\infty =$ constant, independent of γ. Since $\|\psi_\gamma\|_2^2 < 2$, then for all $2 \leqslant p \leqslant \infty$ we have $\|\psi_\gamma\|_p < C_p =$ constant. With ξ_γ given by (1) and with E_γ being the minimum of ξ_γ we easily find, by considering $\xi_\gamma(\psi_\Gamma^2)$ and $\xi_\Gamma(\psi_\gamma^2)$, that

$$\lim_{\gamma \to \Gamma} E_\gamma = E_\Gamma \tag{70}$$

and also that ψ_n^2 is a minimising sequence for ξ_Γ. By the proof in Lieb (1981) and Benguria *et al* (1981) of the existence of a minimum for ξ_Γ, and the lower semicontinuity of ξ_Γ, we conclude that there is a subsequence (which we continue to denote by ψ_n) such that

$$\nabla \psi_n \to \nabla \psi_\Gamma \text{ strongly in } L^2 \tag{71}$$

$$D(\psi_n^2, \psi_n^2) \to D(\psi_\Gamma^2, \psi_\Gamma^2) \tag{72}$$

$$\psi_n \to \psi_\Gamma \text{ almost everywhere} \tag{73}$$

This proves (66) and (69). ((73) follows from the Rellich-Kondrachov theorem.)

Not only is $\psi_n(x) \leq C_\infty$ for all x, but we also have the bound (with some constants M and α)

$$\psi_n(x) \leq M \exp(-\alpha|x|^{1/2}) \tag{74}$$

for $|\gamma_n - \Gamma|$ small enough and for $|x| > R$ for some fixed R. To prove (74) we note that for some R_1 and some $Q > 0$

$$\int_B \psi_\Gamma(x)^2 \,\mathrm{d}x > Z + Q/2$$

with $B = \{x \,|\, |x| < R_1\}$. Since $H^1(B)$ is compactly imbedded in $L^2(B)$ (Rellich–Kondrachov), there is a further subsequence such that $\psi_n \to \psi_\Gamma$ strongly in $L^2(B)$. Hence the entire sequence (see the remark at the beginning of the proof) converges strongly to ψ_Γ in $L^2(B)$. Thus,

$$\int_B \psi_n(x)^2 \,\mathrm{d}x > Z + Q/4 \tag{75}$$

for n large enough. (74) follows by the proof in Benguria *et al* (1981), lemma 8 or Lieb (1981), lemma 7.24(ii). Consequently, $\psi_n(x) \leq F(x)$ for all x and n large enough, where $F(x) = C_\infty$ for $|x| \leq R$ and $F(x)$ is the right-hand side of (74) for $|x| > R$. Since $F \in L^p$ for $1 \leq p \leq \infty$, (67) follows from (73) for $1 \leq p < \infty$ by dominated convergence.

The upper bound F and (73) also imply (by dominated convergence) that the convergence in (68) is pointwise almost everywhere and that $g_\gamma \equiv |x|^{-1} * \psi_\gamma^2$ is bounded by a function of the form $G = \min(s, t/|x|)$ for suitable s and t. Again, by dominated convergence, we obtain strong convergence in (68) in L^p for $3 < p < \infty$. The L^∞ convergence in (68) follows easily from the L^2 convergence of ρ_γ to ρ_Γ together with the large $|x|$ bound (74).

To prove the L^∞ convergence in (67) note that in view of (74) it suffices to prove L^∞ convergence on bounded sets. But this follows from the fact (Benguria *et al* 1981, lemma 7, or Lieb 1981, theorem 7.9) that for any bounded set S and all $x, y \in S$, $|\psi_n(x) - \psi_n(y)| < M|x-y|^{1/2}$ for some constant M which depends on S but (it is easy to see from the proof) not on n. □

4. Lower bound for the chemical potential of a neutral atom or molecule

Here, we prove a result which is somewhat related to the bound on N_c. We shall show that the chemical potential of a neutral system (K not necessarily one) is bounded from below by a constant independent of the nuclear charge. We *conjecture* that a similar bound from above should hold.

The chemical potential, $-\mu(N) = \mathrm{d}E/\mathrm{d}N$, as a function of N is nonpositive, continuous and monotonically increasing in N (for fixed nuclear charges) since $E(N)$ is convex in N in TFW theory (Lieb 1981, theorem 7.2 and theorem 7.8(iii)). Therefore the binding energy ΔE (or affinity) satisfies $|\Delta E| < \mu_0 Q$, where Q is the added charge ($Q = 1$ for an electron).

The fact that the chemical potential, $-\mu_0 = -\mu(Z)$, for a neutral atom or molecule is bounded independent of the nuclear charges agrees with what is believed to be the case for the Schrödinger equation.

Consider the TFW equation for a neutral system (which is a generalisation of (5)), i.e.

$$-A\,\Delta\psi + (\gamma\psi^{4/3} - \phi)\psi = -\mu_0\psi \tag{76}$$

with ϕ given by (6), (3), or equivalently,

$$-\Delta\phi = -4\pi\psi^2 \qquad x \neq R_i,\ i = 1, 2, \ldots, K. \tag{77}$$

Here $\int \psi^2\,\mathrm{d}x = Z = \Sigma_{i=1}^{K} z_i$. Our bound is the following.

Theorem 7. For a neutral system, i.e. for $\int \psi^2\,\mathrm{d}x = Z$, the chemical potential is bounded from below by,

$$-\mu_0 \geqslant -27\pi^2 A^2 \gamma^{-3} \qquad \text{all } Z \tag{78}$$

in the units chosen in the Introduction. In particular, for the value of A chosen in Lieb (1981) to fit the Scott term in the energy, i.e. $A = 0.1859$, and with $\gamma = \gamma_{\mathrm{phys}}$, $\mu_0 \leqslant 0.0105$.

Remark. Since $N_c > Z$ (Lieb 1981, theorem 7.19), μ_0 is strictly positive.

Proof. First consider the TFW equation with arbitrary $N = \int \psi^2 \leqslant N_c$, in which case the right side of (76) is replaced by $-\mu(N)\psi$. We know that $\mu(N) = -\mathrm{d}E(N)/\mathrm{d}N$ and that $\mu(N)$ is *continuous* and monotonically decreasing (Lieb 1981, theorem 7.8(iii)). Therefore (78) will be proved if we can show that for every $N > Z$, $\mu(N) < 27\pi^2 A^2 \gamma^{-3}$. This, we shall now proceed to do.

For every positive b and for all numbers $\psi \geqslant 0$ we have the algebraic inequality

$$b\psi^2 \leqslant \gamma\psi^{7/3} + d(b)\psi \tag{79}$$

with

$$d(b) = 27 b^4 \gamma^{-3}/256. \tag{80}$$

The TFW equation (with $\int \psi^2 = N > Z$) implies

$$-A\,\Delta\psi + b\psi^2 - \phi\psi \leqslant (d(b) - \mu(N))\psi.$$

Therefore, if $\mu(N) \geqslant d(b)$

$$-A\,\Delta\psi + b\psi^2 - \phi\psi \leqslant 0. \tag{81}$$

Now, as long as b is chosen so that $b > 4(\pi A)^{1/2}$, (77) and (81) imply that $\psi < \beta\phi$, all $x \in \mathcal{R}^3$, where β is the positive root of $b = \beta^{-1} + 4\pi\beta A$. To prove this, let $S = \{x | \psi(x) > \beta\phi(x)\}$. Obviously $R_i \notin S$. Since $\psi - \beta\phi$ is continuous in $\mathcal{R}^3 \backslash \{R_i\}$, S is open. On S,

$$\begin{aligned} -A\,\Delta(\psi - \beta\phi) &\leqslant -b\psi^2 + \phi\psi + 4\pi A\beta\psi^2 \\ &= \beta^{-1}[\beta\phi - \beta(b - 4\pi A\beta)\psi]\psi \\ &= \beta^{-1}(\beta\phi - \psi)\psi \leqslant 0, \end{aligned}$$

where we have used the fact that $b - 4\pi A\beta = \beta^{-1}$. Hence $\psi - \beta\phi$ is subharmonic on S. Moreover $\psi - \beta\phi = 0$ on $\partial S \cup \{\infty\}$. Therefore S is empty and $\beta\phi(x) \geqslant \psi(x)$ for all $x \in \mathcal{R}^3$. Since $\psi \geqslant 0$, ϕ must be non-negative everywhere. On the other hand, $\phi =$

$V-|x|^{-1}*\psi^2$ and $\int\psi^2>Z$; consequently, $\phi(x)<0$ for sufficiently large $|x|$. This is a contradiction, and we conclude that $\mu(N)<d(b)$ whenever $b=\beta^{-1}+4\pi\beta A$ for some $0<\beta<\infty$. Choosing $\beta=(4\pi A)^{-1/2}$ yields the desired result, i.e. $\mu(N)<27\pi^2A^2\gamma^{-3}$. □

Remark. From the asymptotics of the solution of equation (76) we see that $\mu_0^{-1/2}$ somehow measures the range of the electronic density. If our conjecture is true, such a range would be independent of Z.

Acknowledgments

One of us (RB) would like to thank the Physics Department of Princeton University for their hospitality during the course of this work.

Appendix

Here, we give a bound for the L^∞ norm of the solution to the TFW equation. Such a bound is independent of γ, the constant in front of the $\psi^{7/3}$ term. This bound is used in § 3.

Lemma A.1. Let ψ be the positive solution to the TFW equation for a molecule with $V(x)$ given by (3). Then for all $\gamma>0$

$$\|\psi\|_\infty\leq(27/16\pi)^{1/2}(Z/A)^{3/2}\|\psi\|_2$$

with $\|\psi\|_2<(2Z)^{1/2}$ (Lieb 1981, theorem 7.23).

Proof. Because of lemma 9 in Benguria *et al* (1981), $\psi\in L^\infty$. From equation (5), $-A\,\Delta\psi(x)\leq V(x)\psi(x)$. First, consider a single atom with $V(x)=z|x|^{-1}$. In this case, therefore,

$$A\psi(x)\leq(4\pi)^{-1}\int z|y|^{-1}|x-y|^{-1}\psi(y)\,\mathrm{d}y. \tag{A.1}$$

Hence

$$\begin{aligned}8\pi z^{-1}A\psi(x)&\leq\int(|y|^{-2}+|x-y|^{-2})\psi(y)\,\mathrm{d}y\\&=\int|y|^{-2}(\psi(y)+\psi(x-y))\,\mathrm{d}y.\end{aligned} \tag{A.2}$$

We decompose this last integral into two terms. One integral over $\{|y|<r\}$ and the other over $\{|y|>r\}$ for any fixed $r>0$. We have,

$$\int_{|y|<r}|y|^{-2}(\psi(y)+\psi(x-y))\,\mathrm{d}y\leq 8\pi r\|\psi\|_\infty \tag{A.3}$$

$$\int_{|y|>r}|y|^{-2}(\psi(y)+\psi(x-y))\,\mathrm{d}y\leq 2\|\psi\|_2(4\pi/r)^{1/2} \tag{A.4}$$

by Hölder's inequality. Thus, substituting (A.3) and (A.4) into (A.2), we have

$$8\pi z^{-1}A\psi(x)\leq 8\pi r\|\psi\|_\infty+2\|\psi\|_2(4\pi/r)^{1/2} \qquad \text{all } r>0 \tag{A.5}$$

and minimising the right-hand side with respect to r we get

$$8\pi A\psi(x)\leq 6z\|\psi\|_\infty^{1/3}\|\psi\|_2^{2/3}(2\pi)^{2/3} \qquad \text{all } x. \tag{A.6}$$

In the molecular case

$$A\psi(x)\leq(4\pi)^{-1}\sum_{j=1}^{K}\int z_j|y-R_j|^{-1}|x-y|^{-1}\psi(y)\,\mathrm{d}y. \tag{A.7}$$

Using the same analysis (A.2)-(A.6) for each term on the right-hand side of (A.7) (but with $\{|y-R_j|\gtrless r_j\}$ and with r_j depending on j) we have that

$$8\pi A\psi(x)\leq 6Z\|\psi\|_\infty^{1/3}\|\psi\|_2^{2/3}(2\pi)^{2/3}. \tag{A.8}$$

The lemma is proved by taking the supremum over x on the left-hand side of (A.8). □

Remark. By making a similar decomposition, one can show that

$$\|B\psi^2\|_\infty\leq 3(\pi/2)^{1/3}\|\psi\|_2^{4/3}\|\psi\|_\infty^{2/3}\leq\|\psi\|_2^2(Z/A)\times 9\times 2^{-5/3}. \tag{A.9}$$

where

$$(B\psi^2)(x)=\int|x-y|^{-1}\psi(y)^2\,\mathrm{d}y$$

and $\|\psi\|_2^2<2Z$ (Lieb 1981, theorem 7.23). The second inequality in (A.9) comes from lemma A.1. Actually, the sharp constant in the middle term of (A.9) is $3(\pi/6)^{1/3}$, not $3(\pi/2)^{1/3}$. One can show that the maximising ρ for $\|B\rho\|_\infty/(\|\rho\|_1^{2/3}\|\rho\|_\infty^{1/3})$ is $\rho=$ characteristic function of a ball.

References

Baumgartner B 1983 *Lett. Math. Phys.* **7** 439-41
—— 1984 *J. Phys. A: Math. Gen.* **17** 1593-602
Benguria R, Brezis H and Lieb E H 1981 *Commun. Math. Phys.* **79** 167-80
Benguria R and Lieb E H 1983 *Phys. Rev. Lett.* **50** 1771-4
Lieb E H 1981 *Rev. Mod. Phys.* **53** 603-41, Errata 1982 **54** 311
—— 1982 *Commun. Math. Phys.* **85** 15-25
—— 1984a *Phys. Rev. Lett.* **52** 315-7
—— 1984b *Phys. Rev.* A **29** 3018-28
Lieb E H and Liberman D A 1982 *Numerical Calculation of the Thomas-Fermi-von Weizsäcker Function for an Infinite Atom without Electron Repulsion* Los Alamos National Laboratory Report LA 9186-MS
Lieb E H, Sigal I M, Simon B and Thirring W 1984 to be published
Lieb E H and Simon B 1977 *Adv. Math.* **23** 22-116
von Weizsäcker C F 1935 *Z. Phys.* **96** 431-58

Part V

Stability of Matter

With W. Thirring in Phys. Rev. Lett. *35*, 687-689 (1975)

PHYSICAL REVIEW LETTERS

VOLUME 35 | 15 SEPTEMBER 1975 | NUMBER 11

Bound for the Kinetic Energy of Fermions Which Proves the Stability of Matter

Elliott H. Lieb*
Departments of Mathematics and Physics, Princeton University, Princeton, New Jersey 08540

and

Walter E. Thirring
Institut für Theoretische Physik der Universität Wien, A-1090 Wien, Austria
(Received 8 July 1975)

We first prove that $\sum|e(V)|$, the sum of the negative energies of a single particle in a potential V, is bounded above by $(4/15\pi)\int|V|^{5/2}$. This, in turn, implies a lower bound for the kinetic energy of N fermions of the form $\frac{3}{5}(3\pi/4)^{2/3}\int\rho^{5/3}$, where $\rho(x)$ is the one-particle density. From this, using the no-binding theorem of Thomas-Fermi theory, we present a short proof of the stability of matter with a reasonable constant for the bound.

The basis of all theories of bulk matter is the stability theorem of the N-electron Hamiltonian,[1]

$$H_N=\sum_{i=1}^{N}p_i^2-\sum_{i=1}^{N}\sum_{k=1}^{M}Z_k|x_i-R_k|^{-1}+\sum_{i>j}|x_i-x_j|^{-1}+\sum_{k>m}\frac{Z_kZ_m}{|R_k-R_m|}\geqslant AN. \tag{1}$$

Equation (1) has been proved by Dyson and Lenard.[2] Unfortunately, their analysis is complicated and their constant A gigantic, about 10^{14}. Despite subsequent improvements,[3-5] a simple proof yielding a reasonable constant A has not yet been found. In this note we propose to fill this gap.

We start with the observation that if Thomas-Fermi (TF) theory were valid, then the no-binding theorem[6] would yield the desired result because the TF energy is proportional to the number of atoms. Our goal will be to show that the TF energy, with suitably modified constants, is a lower bound to H_N. To show this, we have to demonstrate two things: (i) The TF approximation for the N-fermion kinetic energy, $K\int\rho^{5/3}$, is a lower bound for some $K>0$; (ii) the TF approximation for the electron repulsion, $\iint\rho(x)\rho(y)|x-y|^{-1}d^3x\,d^3y$, can be converted into a bound by a further change of constants. The following is a sketch of our proof. Fine points of rigor, together with some variations of the inequalities given here, will be presented elsewhere.

(i) *Kinetic energy of N fermions.*—Consider the Schrödinger equation for one particle in a potential $V(x)$. Schwinger[7] has derived an upper bound for $N_E(V)$, the number of energy levels with energies $\geqslant E$. For $\alpha>0$, and with $|f(x)|_-\equiv -f(x)$ for $f<0$ and 0 otherwise,

$$\begin{aligned}N_{-\alpha}(V)&\leqslant N_{-\alpha/2}(|V+\alpha/2|_-)\leqslant(4\pi)^{-2}\int d^3x\,d^3y\,|V(x)+\alpha/2|_-|x-y|^{-2}\exp[-(2\alpha)^{1/2}|x-y|]|V(y)+\alpha/2|_-\\&\leqslant(4\pi)^{-1}(2\alpha)^{-1/2}\int d^3x\,|V(x)+\alpha/2|_-^2.\end{aligned} \tag{2}$$

The last inequality is Young's. Consequently, the sum of the negative energy eigenvalues of p^2+V is bounded by

$$\sum_j |e_j(V)| = \int_0^\infty N_{-\alpha}(V)\,d\alpha \leq (8\pi)^{-1}2^{1/2}\int d^3x \int_0^{2|V(x)|_-} d\alpha\, \alpha^{-1/2}[V(x)+\alpha/2]^2 = (4/15\pi)\int d^3x\, |V(x)|_-^{5/2} \tag{3}$$

By comparison, the classical value is

$$(2\pi)^{-3}\int d^3x \int d^3p\, |p^2+V(x)|_- = (15\pi^2)^{-1}\int d^3x\, |V(x)|_-^{5/2}. \tag{4}$$

We conjecture that (4) is actually a bound. In one dimension an analog of (3) holds with $\int |V|_-^{3/2}$, but we have a counterexample that shows that the classical value is not a bound.

Now let ψ be any N-particle normalized antisymmetric wave function of space-spin. Define

$$\rho_\pm(x) = N\int d^3x_2 \cdots d^3x_N \sum_{\sigma_2,\ldots,\sigma_N} |\psi(x,x_2,\ldots,x_N;\pm,\sigma_2,\ldots,\sigma_N)|^2, \tag{5}$$

$$T = \langle\psi| -\sum_{i=1}^N \Delta_i |\psi\rangle, \tag{6}$$

$$K = T\left(\int\rho_+^{5/3} + \int\rho_-^{5/3}\right)^{-1} > 0, \tag{7}$$

and $\pi_\pm$ are projections onto the single-particle spin states. Let $h_i = p_i^2 - (5K/3)[\rho_+^{2/3}(x_i)\pi_{i+} + \rho_-^{2/3}(x_i) \times \pi_{i-}]$ be a single-particle Hamiltonian and

$$H = \sum_{i=1}^N h_i.$$

If E_0 is the fermion ground-state energy of H then $E_0 \leq \langle\psi|H|\psi\rangle = T - (5K/3)(\int\rho_+^{5/3} + \int\rho_-^{5/3})$. On the other hand, E_0 is greater than or equal to the sum of all the negative eigenvalues of $p^2 - (5K/3)\rho_\pm^{2/3}$ together. By (3), $E_0 \geq -(4/15\pi)(5K/3)^{5/2}(\int\rho_+^{5/3} + \int\rho_-^{5/3})$. Combining these two inequalities[8,9] yields

$$K \geq \tfrac{3}{5}(3\pi/2)^{2/3}. \tag{8}$$

With

$$\rho(x) = \rho_+(x) + \rho_-(x), \tag{9}$$

and using the convexity of $\int\rho^{5/3}$, we obtain a weakened version of (8):

$$T \geq \tfrac{3}{5}(3\pi/4)^{2/3}\int\rho(x)^{5/3}d^3x. \tag{10}$$

If (4) were a bound, then the TF constant, $\frac{3}{5}(3\pi^2)^{2/3}$, could be used in (10).

(ii) *Electron repulsion.*—In this paper we shall use TF theory twice; the first use is to derive a theorem about electrostatics. The TF energy functional with $\gamma > 0$ and positive charges Z_k at locations R_k is

$$\mathcal{E}_\gamma(\rho) = (3/5\gamma)\int\rho^{5/3} - \sum_{k=1}^M \int d^3x\,\rho(x)|x-R_k|^{-1}Z_k + \tfrac{1}{2}\int\int\rho(x)\rho(y)|x-y|^{-1}d^3x\,d^3y + \sum_{j<k} Z_jZ_k|R_j-R_k|^{-1}. \tag{11}$$

For any R_j, the minimum of $\mathcal{E}_\gamma(\rho)$ occurs when $\int\rho = \sum_j Z_j$ ($j = 1$ to M), and this in turn has a minimum when the R_j are infinitely separated (no-binding theorem[6]). Thus

$$\mathcal{E}_\gamma(\rho) \geq -3.68\gamma\sum_{j=1}^M Z_j^{7/3}, \tag{12}$$

since a neutral atom of charge Z has an energy $-3.68\gamma Z^{7/3}$ in TF theory.

Consider (11) with R_j being the *electron* coordinates, x_j, $Z_j = 1$, $M = N$, and ρ given by (5) and (9). Multiply (11) by $|\psi|^2$ and integrate, and then use (12). Thus, for all $\gamma > 0$,

$$\langle\psi|\sum_{i<j}|x_i - x_j|^{-1}|\psi\rangle \geq \tfrac{1}{2}\int\int\rho(x)\rho(y)|x-y|^{-1}d^3x\,d^3y - (3/5\gamma)\int d^3x\,\rho(x)^{5/3} - 3.68N. \tag{13}$$

Therefore we have an electrostatic theorem that the TF estimate for the electron repulsion [the first term on the right-hand side of (13)] is a lower bound provided one makes a kinetic energy correction and subtracts an energy proportional to the electron number.

(iii) *Stability of matter.*—Combining (10) and (13), with $\gamma > (4/3\pi)^{2/3}$ and R_k being the *nuclear* coordinates, yields

$$\langle\psi|H_N|\psi\rangle \geq \mathcal{E}_\delta(\rho) - 3.68\,N\gamma, \tag{14}$$

VOLUME 35, NUMBER 11 PHYSICAL REVIEW LETTERS 15 SEPTEMBER 1975

with $1/\delta = (3\pi/4)^{2/3} - 1/\gamma$ and ρ given by (5) and (9). A lower bound is obtained by minimizing $\mathcal{E}_\delta$ over all ρ such that $\int\rho = N$. For simplicity we shall only use the absolute minimum of $\mathcal{E}_\delta$. By (12)

$$\langle\psi|H_N|\psi\rangle \geq -3.68\left(N\gamma + \delta\sum_{j=1}^{M} Z_j^{7/3}\right)^2. \qquad (15)$$

Optimizing (15) with respect to γ yields

$$\langle\psi|H_N|\psi\rangle \geq -2.08\, N\left[1+\left(\sum_{j=1}^{M}\frac{Z_j^{7/3}}{N}\right)^{1/2}\right]^2. \qquad (16)$$

Remarks.—(1) If the fermions are of q species (instead of 2 as in the electron case), then the right-hand side of (10) would acquire a factor $(2/q)^{2/3}$ and the right-hand side of (16) a factor $(q/2)^{2/3}$.

(2) If all $Z_j = Z$, our result (16) gives a $Z^{7/3}$ dependence instead of the known Z^2 bound.[3] If $MZ \leq N$ then $MZ^{7/3}/N \leq Z^{4/3}$, which is an improvement over Ref. 3. If $MZ > N$ then we have to use the $\int\rho = N$ condition in (14). The TF no-binding theorem also holds in the subneutral case. By convexity of the TF energy in $\int\rho$, the minimum occurs for M atoms with equal electron charge N/M. If $MZ \gg N$ the energy per atom is proportional to $(N/M)^{1/3}Z^2$. Then $\langle\psi|H_N|\psi\rangle$ is bounded below by $-aN - bZ^2M^{2/3}N^{1/3}$. While this has the correct Z dependence, it has the wrong M dependence[3]; $M^{2/3}$ should be replaced by $N^{2/3}$. This difficulty is inherent in TF theory. What one needs is a simple proof that if $MZ \gg N$, then one can remove most of the surplus nuclei without affecting the energy. Even for $N=1$ this is not a simple problem. Nevertheless, our present bound is proportional to the *total* particle number, and this is sufficient for proving the existence of the thermodynamic limit.[10]

We thank J. F. Barnes and A. Martin for helpful correspondence and discussions. E. L. thanks the Institut für Theoretische Physik, Universität Wien, for its kind hospitality.

*Work supported by the U. S. National Science Foundation under Grant No. MPS 71-03375-A03 at the Massachusetts Institute of Technology.

[1]The notation is $\hbar = e = 2m = 1$; x_i and p_i are electron variables; R_k and $Z_k > 0$ are nuclear coordinates and charges ($\frac{1}{4} = 1$ Ry).

[2]F. J. Dyson and A. Lenard, J. Math. Phys. (N.Y.) 8, 423 (1967); A. Lenard and F. J. Dyson, J. Math. Phys. (N.Y.) 9, 698 (1968).

[3]A. Lenard, in *Statistical Mechanics and Mathematical Problems*, edited by A. Lenard (Springer, Berlin, 1973).

[4]P. Federbush, J. Math. Phys. (N.Y.) 16, 347, 706 (1975).

[5]J. P. Eckmann, "Sur la Stabilité de Matière" (to be published).

[6]E. Teller, Rev. Mod. Phys. 34, 627 (1962). A rigorous proof of this theorem is given by E. Lieb and B. Simon, "Thomas-Fermi Theory of Atoms, Molecules, and Solids" (to be published). See also E. Lieb and B. Simon, Phys. Rev. Lett. 31, 681 (1973).

[7]J. Schwinger, Proc. Nat. Acad. Sci. 47, 122 (1961).

[8]The connection between $N_E(V)$ and a bound on the kinetic energy was noted by A. Martin (private communication). One can show that the converse holds; i.e., an improvement in (8) implies an improvement in (3).

[9]By numerically solving the three-dimensional variational equation for K, Eq. (7), when $N=1$, J. F. Barnes has shown that (8) holds with the TF constant $\frac{3}{5}(6\pi^2)^{2/3}$ when $N=1$ (private communication).

[10]J. L. Lebowitz and E. H. Lieb, Phys. Rev. Lett. 22, 631 (1969); E. H. Lieb and J. Lebowitz, Adv. Math. 9, 316 (1972).

which is likely to be easily satisfied.

If the decay $D^+ \to \bar{K}^0\pi^+$ is not enhanced, as is to be expected theoretically, the upper limits on all the remaining modes, $K_s\pi^+$, $\pi^+\pi^-$, and K^+K^-, lead to no constraints at all, because these modes are suppressed by $\tan^2\theta_C$ relative to the dominant modes.

Obviously, the nonobservation of charmed hadrons at SPEAR does little to strengthen the case for the hidden-charm interpretation of the newly discovered bosons. How much the case is weakened by the new data is a topic for subjective interpretation of the bounds we have quoted above. In our minds the most damaging result is that two-body decays of D^0 account for less than 10% of its total width. While such a suppression is neither unthinkable nor unprecedented, we find it disturbing not only because it is so small but also because, if 90% of the nonleptonic decays are to three or more particles, it will be difficult to understand the observed charged-particle multiplicity. We disagree with the conclusion of Boyarski *et al.* that their upper limit on $B(D^+ \to \bar{K}^0\pi^+$ or $K^-\pi^+\pi^+)$ violates the expectation of the conventional model by a factor of at least 3.[12] In fact, in the conventional model, with all of its pre-J/ψ baggage of sextet enhancement and $\underline{10}$ suppression, both decays are expected to be absent (i.e., not dominant). An incautious interpretation is that the nonobservation of these modes is good for the model, but we do not wish to go so far. Indeed, it is our feeling that if some of the upper limits, such as those given in Eqs. (2), (9), and (13), were decreased by factors of 2 or 3, the conventional charm scheme[2-6] would require modification.

*Alfred P. Sloan Foundation Fellow; also at Enrico Fermi Institute, University of Chicago, Chicago, Ill. 60637.

†Operated by Universities Research Association Inc. under contract with the U. S. Energy Research and Development Administration.

[1]A. M. Boyarski *et al.*, Phys. Rev. Lett. 35, 196 (1975).

[2]S. L. Glashow, J. Iliopoulos, and L. Maiani, Phys. Rev. D 2, 1285 (1970).

[3]G. Altarelli, N. Cabibbo, and L. Maiani, Nucl. Phys. B88, 285 (1975). The same result was given independently by R. L. Kingsley, S. B. Treiman, F. Wilczek, and A. Zee, Phys. Rev. D 11, 1919 (1975).

[4]Kingsley, Treiman, Wilczek, and Zee, Ref. 3.

[5]M. B. Einhorn and C. Quigg, Phys. Rev. D (to be published).

[6]M. K. Gaillard, B. W. Lee, and J. L. Rosner, Rev. Mod. Phys. 47, 277 (1975).

[7]J.-E. Augustin *et al.*, Phys. Rev. Lett. 34, 764 (1975).

[8]V. Chaloupka *et al.*, Phys. Lett. 50B, 1 (1974).

[9]Y. Dothan and H. Harari, Nuovo Cimento, Suppl. No. 3, 48 (1965). The statement in Ref. 1 that the modes $K^-\pi^+\pi^+$, $\bar{K}^0\bar{K}^0K^+$, $\bar{K}^0\pi^+\eta$, and $\bar{K}^0\pi^+\pi^0$ should occur in the ratios 4:4:3:1 is correct only if they are in the *totally symmetric* $\underline{10}$. It is not true in general.

[10]M. K. Gaillard and B. W. Lee, Phys. Rev. Lett. 33, 108 (1974).

[11]G. Altarelli and L. Maiani, Phys. Lett. 52B, 351 (1974).

[12]The discussion surrounding Table IV of Ref. 6, on which Boyarski *et al.* apparently base their conclusion, clearly warns that these modes may be strongly suppressed.

ERRATUM

BOUND FOR THE KINETIC ENERGY OF FERMIONS WHICH PROVES THE STABILITY OF MATTER. Elliott H. Lieb and Walter E. Thirring [Phys. Rev. Lett. 35, 687 (1975)].

Equation (2), replace $N_{-\alpha/2}(|V+\alpha/2|_-)$ with $N_{-\alpha/2}(-|V+a/2|_-)$.

Equation (13), replace $-3.68N$ with $-3.68N\gamma$.

Equation (15), replace $(\ldots)^2$ with $(\ldots)$.

On page 687, line 15, replace $\geq E$ by $\leq E$.

Stability of Coulomb Systems with Magnetic Fields

I. The One-Electron Atom

Jürg Fröhlich[1,*], Elliott H. Lieb[2,**], and Michael Loss[2,***]

[1] Theoretical Physics, ETH-Hönggerberg, CH-8093 Zürich, Switzerland

[2] Departments of Mathematics and Physics, Princeton University, Jadwin Hall, P.O.B. 708, Princeton, NJ 08544 USA

Abstract. The ground state energy of an atom in the presence of an external magnetic field B (with the electron spin-field interaction included) can be arbitrarily negative when B is arbitrarily large. We inquire whether stability can be restored by adding the self energy of the field, $\int B^2$. For a hydrogenic like atom we prove that there is a critical nuclear charge, z_c, such that the atom is stable for $z<z_c$ and unstable for $z>z_c$.

1. Introduction

The problem of the stability of an atom (i.e. the finiteness of its ground state energy) was solved by the introduction of the Schrödinger equation in 1926. While it is true that Schrödinger mechanics nicely takes care of the $-ze^2/r$ Coulomb singularity at $r=0$ (here $z|e|$ is the nuclear charge), a more subtle problem that has to be considered is the interaction of the atom with an external magnetic field $B(x)$ with vector potential $A(x)$ and $B=\operatorname{curl} A$. In this paper the problem of the one-electron atom in a magnetic field is studied; in a subsequent paper [6] some aspects of the many-electron and many-nucleus problem will be addressed.

Units. Our unit of length will be half the Bohr radius, namely $l=\hbar^2/(2me^2)$. The unit of energy will be 4 Rydbergs, namely $2me^4/\hbar^2=2mc^2\alpha^2$, where α is the fine structure constant $e^2/(\hbar c)$. The magnetic field B is in units of $|e|/(l^2\alpha)$. The vector potential satisfies $B=\operatorname{curl} A$. The magnetic field energy ($\int B^2/8\pi$) is, in these units,

$$\varepsilon\int B^2\,, \qquad 1/\varepsilon=8\pi\alpha^2\,. \tag{1.1}$$

* Work partially supported by U.S. National Science Foundation grant DMS-8405264 during the author's stay at the Institute for Advanced Study, Princeton, NJ, USA

** Work partially supported by U.S. National Science Foundation grant PHY-8116101-A03

*** Work partially supported by U.S. and Swiss National Science Foundation Cooperative Science Program INT-8503858. Current address: Institut f. Mathematik, FU Berlin, Arnimallee 3, D-1000 Berlin 33

The first problem to be considered is one in which the electron spin is neglected. The Hamiltonian in this case is

$$H' = (p-A)^2 - z/|x| \tag{1.2}$$

(with $p = i\nabla$). H' presents no interesting problem as far as stability is concerned because the effect of including A is *always* to raise the ground state energy. (Reason: For any ψ, $(\psi, (p-A)^2\psi) \geqq (|\psi|, p^2|\psi|)$. This is essentially Kato's inequality [4] (see e.g. [13]). On the other hand $(\psi, |x|^{-1}\psi) = (|\psi|, |x|^{-1}|\psi|)$, so we can lower the energy by replacing ψ by $|\psi|$ and setting $A = 0$.)

The problem becomes interesting when the electron spin is included, and this problem is the subject of this paper. The wave function ψ is a two-component (complex valued) spinor:

$$\psi(x) = (\psi_1(x), \psi_2(x)). \tag{1.3}$$

The Hamiltonian is

$$H = (p-A)^2 - \sigma \cdot B(x) - z/|x| \tag{1.4}$$

$$= [\sigma \cdot (p-A)]^2 - z/|x| \tag{1.5}$$

where $\sigma_1, \sigma_2, \sigma_3$ are the Pauli matrices. The first term in (1.5) is the Pauli kinetic energy and is the non-relativistic approximation to the Dirac operator.

The ground state energy $E_0(B,z)$ of H is always finite but depends on B in such a way that $E_0 \to -\infty$ as $B \to \infty$ (for a constant field), roughly as $-(\ln B)^2$, see [1]. What prevents B from spontaneously growing large and driving E_0 towards $-\infty$? (We do not inquire into the source of this B, but simply assume that nature will always contrive to lower the energy, if possible.) The answer, which we shall take as a hypothesis here, is that the price to be paid is the field energy $\int B^2/8\pi$. Thus, we are led to consider (in our units) $H + \varepsilon \int B(x)^2 dx$ and ask whether

$$E(B,z) = E_0(B,z) + \varepsilon \int B^2 \tag{1.6}$$

is bounded below *independent* of B. This problem is important in the analysis of stability in non-relativistic quantum electrodynamics [we have omitted the term $\int E^2$ which makes (1.6) a lower bound and which makes the magnetic field classical]. We define

$$E(z) = \inf_B E(B,z). \tag{1.7}$$

In the remainder of this introduction we shall first outline our results about $E(z)$, then discuss their physical interpretation and finally formulate some preliminary mathematical facts and notation.

We show that there is a critical value of z (called z_c) such that

$$\begin{aligned} E(z) &= -\infty \quad \text{for} \quad z > z_c, \\ E(z) &\text{ is finite} \quad \text{for} \quad z < z_c. \end{aligned} \tag{1.8}$$

The value of z_c is proportional to $1/\alpha^2$ (not $1/\alpha$ as in the case of the Dirac equation or the "relativistic" Schrödinger equation [3]). Section II (in conjunction with [8]) contains the proof that z_c is finite. The fact that $z_c \neq \infty$ is intimately connected with the fact that the equation

$$\sigma \cdot (p - A)\psi = 0 \tag{1.9}$$

has a non-zero solution with $\psi \in H^1$ and $A \in L^6$. When we first worked on this problem we realized this connection and proved (Sect. II) that

$$z_c = \inf \varepsilon \int B^2/(\psi, |x|^{-1}\psi), \tag{1.10}$$

where the infimum is over all solutions to (1.9). (Clearly, any solution to (1.9) has zero kinetic energy so that if z exceeds the right side of (1.10) the total energy can be driven to $-\infty$ by the scaling $\psi(x) \to \lambda^{3/2}\psi(\lambda x)$, $A(x) \to \lambda A(\lambda x)$. The converse is the difficult part of Sect. II.) At first it was unknown whether or not (1.9) has a solution, but now several have been found [8].

Section III gives a lower bound to z_c (which we call z_c^L):

$$z_c > z_c^L = (24.0)/(8\pi\alpha^2) > 17{,}900. \tag{1.11}$$

This is far better than that needed for physics. For all $z < z_c^L$ we also derive a lower bound for $E(z)$:

$$E(z) \geqq -\tfrac{1}{4}z^2 - z^3(32z_c^L)^{-1}(1 - \tfrac{3}{4}z/z_c^L)^{3/2}. \tag{1.12}$$

[Note that $E(z)$ is trivially less than $-\frac{1}{4}z^2$, which is the ground state energy for $B \equiv 0$.]

The B field that causes $E(z)$ to diverge when $z > z_c$ is highly inhomogeneous (both in magnitude and direction) near the nucleus. In astrophysical and other applications [9, 11] one is interested in studying atoms and ions in very strong, external magnetic fields with the property that the *direction of the magnetic field is constant* over distance scales many times the scales of atomic physics, to a very good approximation. Theoretical astrophysicists have carried out large-scale numerical calculations of the spectra of atoms and ions in very strong magnetic fields and have tried to correlate theoretical predictions with experimental data. As a modest contribution to the mathematical foundations of this kind of work, we establish stability of one-electron atoms in *arbitrarily strong* magnetic fields whose direction (but *not* magnitude) is *constant* in a neighborhood of the atom. This is done in Sect. IV, where we prove that $E(z)$ is always finite in this case. An open problem for further investigation is the analysis of $E_0(B, z)$ for magnetic fields that are curl free in a neighborhood of the atom.

Before proceeding to the physical interpretation, we note in passing that the electron g factor was taken to be 2 in (1.4). If we replace the $\sigma \cdot B$ term in (1.4) by $\frac{1}{2}g\sigma \cdot B$ then two cases arise:

$g < 2$. Here we can write the kinetic energy as $\frac{1}{2}g[\sigma \cdot (p - A)]^2 + (1 - \frac{1}{2}g)(p - A)^2$. The first term is nonnegative and the second, when combined with $-z/|x|$ gives a Hamiltonian of type H' in (1.2). This is bounded below by $-\frac{1}{2}z^2/(2 - g)$, and hence $E(z)$ is always finite.

$g > 2$. Here, $E(z) = -\infty$ for all z, *including* $z = 0$. To see this, let B be a field which is constant $= B(0, 0, 1)$ over a large cube of length L, with $A = \frac{1}{2}B(x_2, -x_1, 0)$ inside this cube. Let B drop to zero outside the cube so that $I = \int B^2 < \infty$. Take ψ to be a ground state Landau orbital (cut off in the x_3 direction so that $\psi \in L^2$), i.e.

$$\psi(x) = (\text{const})(1, 0)\exp[-\tfrac{1}{4}B(x_1^2 + x_2^2)]\cos(\pi x_3/L)$$

and $\psi(x)=0$ for $|x_3|>L/2$. With B fixed and with L big enough, we can have $(\psi,[\sigma\cdot(p-A)]^2\psi)\leqq\frac{1}{4}(g/2-1)B$ and $(\psi,\sigma\cdot B\psi)\geqq\frac{1}{2}B$. Also $I\leqq 2B^2L^3$. The total energy (with $z=0$) is less than $-\frac{1}{4}(g/2-1)B+2B^2L^3\varepsilon$. Now, let $\lambda>0$ and replace $\psi(x)$ by $\lambda^{3/2}\psi(\lambda x)$, $A(x)$ by $\lambda A(\lambda x)$ and $B(x)$ by $\lambda^2 B(\lambda x)$. The energy is then less than $-\frac{1}{4}\lambda^2(g/2-1)B+2\lambda B^2L^3\varepsilon$. (This scaling is exact and will be employed frequently in the sequel.) As $\lambda\to\infty$, the energy tends to $-\infty$, so stability never holds.

Since physically $g>2$ because of Quantum Electrodynamics (QED) effects, it is clear that if we try to "improve" (1.4) by replacing $\sigma\cdot B$ by $\frac{1}{2}g\sigma\cdot B$ we shall get an inconsistent theory. The only truly consistent procedure is to include *all* QED effects, and this is outside the scope of this paper.

The foregoing aside about the g-factor leads us to the question of the physical content of the results of this paper, (1.8)–(1.12). There are two ways to view them. The first is to observe that (1.8) and (1.11) show that atomic physics with the Hamiltonian (1.4) contains no seeds of instability for small z (small meaning $z<17{,}900$) and that perturbation theory (in B) can be safely employed for very small B. (Of course one should also analyze the many-electron and many-nucleus problem to be certain about this conclusion. We are unable to do this fully, but in a subsequent paper [6] we do successfully analyze two problems: the one-electron, many-nucleus problem and the one-nucleus, many-electron problem, i.e. the full atom.) The fact that the theory is well behaved for small z is not entirely a trivial matter, especially when the situation is contrasted with that for spin-spin interactions (either electron-electron or electron-nucleus). Here, one adds a two-body term $\sigma^a\cdot\sigma^b|x|^{-3}-3(\sigma^a\cdot x)(\sigma^b\cdot x)|x|^{-5}$, where x is the vector between particles a and b. The $|x|^{-3}$ singularity is not integrable and, in particular it cannot be controlled by the kinetic energy. Thus, a system with this interaction is *always* unstable in our sense. The treatment of the interaction by perturbation theory, is not really a consistent procedure.

Of course, it is always possible to restore stability by cutting off the Coulomb or spin-spin interactions at the Compton wavelength of the electron, but then the theory would depend critically on this wavelength. Stability, in the sense we use it, implies that the Schrödinger equation for electrons and nuclei is independent of the electron's Compton wavelength-in conformity with what is always assumed to be the case.

The second viewpoint is to emphasize the breakdown of (1.4) when $z>z_c$ and to say that magnetic interactions impose an upper bound on $z\alpha^2$. Here we are treading on shaky ground. If we specify ψ and ask what B minimizes $(\psi,H\psi)$ $+\varepsilon\int B^2$, we easily find that Maxwell's equation takes the form

$$2\varepsilon\,\mathrm{curl}\,B(x)=j(x)=2\,\mathrm{Re}\langle\psi,(p-A)\psi\rangle(x)+\mathrm{curl}\langle\psi,\sigma\psi\rangle(x)\,. \tag{1.13}$$

[Notation. $(\psi,H\psi)$ has been used to denote the usual expectation, including the x-integration. $\langle\psi,\sigma\psi\rangle(x)$ denotes the inner product with respect to the spinor indices *only*, and hence it is a function of x. $\langle\psi,\psi\rangle(x)\equiv|\psi(x)|^2=|\psi_1(x)|^2$ $+|\psi_2(x)|^2$.]

The first term in j is the electron current ($p-A$ is the velocity). The second term is the "spin" current; it is conserved. The B field in (1.13) cannot be viewed as external; it is, in fact, generated by the electron as (1.13) shows. It is this B field that causes the breakdown when $z>z_c$. [Technical note. In Sect. II we choose a special pair ψ, A with $\sigma \cdot (p-A)\psi \equiv 0$, so that the right side of (1.13) is zero. For this ψ, the B field we use is not exactly optimal [because (1.13) is not satisfied], but the error becomes inconsequential when we employ the λ-scaling $\psi(x) \to \lambda^{3/2}\psi(\lambda x)$ and $A(x) \to \lambda A(\lambda x)$.]

The instability of (1.6) for $z > z_c$ might indicate a qualitative change in the behaviour of non-relativistic quantum electrodynamics (QED), e.g. some kind of phase transition or an intrinsic instability, as z becomes large. For a compelling argument in this direction we would, however, have to include the term $\int \mathbf{E}^2$ in the Hamiltonian, quantize the electromagnetic field and properly renormalize the theory. Our calculations can be viewed as a quasi-classical approximation to that theory. The fact that this approximation exhibits an instability, for large z, should, by experience, be seen as a warning that the full theory might also exhibit a drastic change in behaviour, for large z.

Physically, our instability result for $z > z_c$ is, of course, quite irrelevant, because $z_c > 17{,}000$. Nuclei with nuclear charge above ~ 100 are not known to exist in nature, and even if nuclei with $z \sim 10{,}000$ existed electrons moving in their field would be highly relativistic particles, so that our use of non-relativistic kinematics is not justified for values of z where the instability occurs. Nevertheless, we feel that it is an interesting mathematical problem to explore the consistency of this model even beyond the domain, where the approximation is justified.

As remarked after (1.12), the interaction given in (1.4) *lowers* the energy. In contrast to this, the Lamb shift, which is obtained from a proper QED calculation (but only in perturbation theory), is a *raising* of the energy. Furthermore the Lamb shift is of order $z^4\alpha^3$ (apart from logarithmic corrections) which contrasts with our lowering (1.12) which is of order $z^3\alpha^2$. Our result is not directly comparable with the Lamb shift since the latter requires a fully quantized theory with renormalization.

Now we turn to the mathematical preliminaries to the rest of this paper. Some notation will be introduced and, more importantly, a careful discussion of the class of functions (A, B, ψ) will be given.

First, consider the B field. In order that (1.1) make sense we obviously require $B \in L^2(\mathbb{R}^3)$. [Notation. For vector fields (A or B)

$$\|A\|_p \equiv \|(A \cdot A)^{1/2}\|_p, \tag{1.14}$$

where $A = (A_1, A_2, A_3)$ and $A \cdot A = \sum |A_i|^2$. For spinors ψ

$$\|\psi\|_p = \|\langle\psi, \psi\rangle^{1/2}\|_p, \tag{1.15}$$

where $\langle\psi, \psi\rangle(x) = |\psi_1(x)|^2 + |\psi_2(x)|^2 = \left\{\sum_i |\langle\psi, \sigma_i\psi\rangle(x)|^2\right\}^{1/2}$. For gradients

$$\|\nabla A\|_2 = \left\|\left(\sum_{i,j} |\partial_i A_j|^2\right)^{1/2}\right\|_2 = \left\{\sum_{i,j} \int |\partial_i A_j|^2\right\}^{1/2} \tag{1.16}$$

with $\partial_i = \partial/\partial x_i$, $i = 1, 2, 3$. A similar formula holds for $\|\nabla\psi\|_2$.] The vector potential, A, satisfies $\operatorname{curl} A = B$, but A is determined only up to a gauge (i.e. $A \to A + \nabla\Phi$). Gauge transformations on ψ (i.e. $\psi \to e^{i\Phi}\psi$) can be nasty ($e^{i\Phi}$ can have very bad differentiability properties). This problem is avoided by fixing a gauge, namely the Coulomb gauge, $\operatorname{div} A = 0$. Additionally, it will be convenient to have the formal identity (when $\operatorname{div} A = 0$)

$$\|B\|_2^2 = \int B^2 = \|\nabla A\|_2^2 . \tag{1.17}$$

The danger is that A might conceivably have bad decay properties at infinity which would prevent the necessary integrations by parts in (1.17). This problem is resolved in Appendix A. Notice that if $\nabla A \in L^2$ and if $A(x) \to 0$ as $|x| \to \infty$ in a weak sense, see [2], then, by the Sobolev inequality

$$\|\nabla A\|_2 \geqq S \|A\|_6 , \tag{1.18}$$

so A is automatically in L^6. Theorem A.1 in the appendix states that when $B \in L^2$ there is a *unique* A satisfying

$$\operatorname{curl} A = B , \quad \operatorname{div} A = 0 \text{ in } \mathscr{D}', \text{ and } \quad A \in L^6 , \tag{1.19}$$

and this A also satisfies (1.17), which implies $\nabla A \subset L^2$. Here, $\mathscr{D}'$ denotes the usual space of distributions. This is the A we shall use (except in Theorems 2.1 and A.2 where only the assumption $A \in L^6$ is used).

Next we turn to the spinor field ψ which obviously must be in L^2. To avoid operator domain questions we shall interpret the first term in (1.5) as a quadratic form $Q = \|\sigma \cdot (p - A)\psi\|_2^2$. Theorem A.2 states that if $\sigma \cdot (p - A)\psi \in L^2$, $\psi \in L^2$, and $A \in L^6$, then automatically $\nabla \psi \in L^2$. This, in turn, implies that $(\psi, |x|^{-1}\psi) < \infty$ by (1.18), or by the well known uncertainty principle for the hydrogen atom,

$$\|\psi\|_2 \|\nabla \psi\|_2 \geqq (\psi, |x|^{-1}\psi) . \tag{1.20}$$

Therefore, we introduce the class of function pairs

$$\mathscr{C} = \{\psi, A | \psi \in H^1(\mathbb{R}^3), \|\psi\|_2 = 1, A \in L^6(\mathbb{R}^3), \operatorname{div} A = 0, \nabla A \in L^2(\mathbb{R}^3)\} . \tag{1.21}$$

[$\psi \in H^1$ means $\psi \in L^2$ and $\nabla \psi \in L^2$. The set of functions f satisfying $f \in L^6(\mathbb{R}^3)$, $\nabla f \in L^2(\mathbb{R}^3)$ is sometimes called $D^{1,2}(\mathbb{R}^3)$; it is the completion of $H^1(\mathbb{R}^3)$, not in the H^1-norm $(\|f\|_2^2 + \|\nabla f\|_2^2)^{1/2}$, but in the norm $\|\nabla f\|_2$.]

For functions in $\mathscr{C}$ the following energy functional is a generalization of $(\psi, H\psi) + \varepsilon \int B^2$.

$$\mathscr{E}(\psi, A) = \|\sigma \cdot (p - A)\psi\|_2^2 + \varepsilon \|B\|_2^2 - z(\psi, |x|^{-1}\psi) , \tag{1.22}$$

and each term in (1.22) is well defined. The ground state energy is

$$E(z) = \inf\{\mathscr{E}(\psi, A) | (\psi, A) \in \mathscr{C}\} . \tag{1.23}$$

Theorem 2.4 states that when $E(z)$ is finite, the infimum in (1.23) is a minimum.

Another class we shall need is

$$\mathscr{F} = \{\psi, A | (\psi, A) \in \mathscr{C} \text{ and } \sigma \cdot (p - A)\psi = 0\} . \tag{1.24}$$

Notice that when $(\psi, A) \in \mathscr{C}$, then each term $p\psi$ and $A\psi$ makes sense as L^2 function. In Sect. II the formula

$$z_c = \varepsilon \inf\{\|B\|_2^2 / (\psi, |x|^{-1}\psi) | (\psi, A) \in \mathscr{F}\} \tag{1.25}$$

will be derived. Theorem 2.5 states that the infimum in (1.25) is actually a minimum.

II. A Basic Theorem and a Formula for z_c

Heuristically, if $\mathscr{E}(\psi, A)$ is unbounded for a certain z, we expect ψ and A to blow up in some sense. The following theorem is essential for understanding this blowup. It is stated in general terms, but its use for our problem will be clarified shortly; it will yield a formula for z_c.

Theorem 2.1. *Let ψ_n be a sequence of spinor valued functions on $\mathbb{R}^3$ and A_n a sequence of vector fields on $\mathbb{R}^3$ satisfying (for some fixed $3<p<\infty$)*

(i) $d_1 \leqq \|\psi_n\|_2 \leqq d_2$ *for some constants* $d_2 \geqq d_1 > 0$.

(ii) $\|\nabla\psi_n\|_2 \to \infty$ *as* $n\to\infty$.

(iii) $\|A_n\|_p \leqq D\|\nabla\psi_n\|_2^s$ *for some* $D>0$, *where* $s = 1-\dfrac{3}{p}$.

(iv) $\|\sigma\cdot(p-A_n)\psi_n\|_2 \leqq C_n\|\nabla\psi_n\|_2$ *for some sequence* $\{C_n\}_{n=1}^{\infty}$ *with* $C_n\to 0$ *as* $n\to\infty$. *Define* $1/\lambda_n = \|\nabla\psi_n\|_2$ *(whence* $\lambda_n\to 0$ *as* $n\to\infty$*),* $\phi_n(x) = \lambda_n^{3/2}\psi_n(\lambda_n x)$ *and* $\alpha_n(x) = \lambda_n A_n(\lambda_n x)$. *Then*

(a) $\liminf_{n\to\infty} \|\alpha_n\|_p \geqq c > 0$.

(b) *There exists a subsequence (which we continue to denote by n) and functions* ϕ *and* α, *and a sequence of points* $x_n \in \mathbb{R}^3$ *such that* $\tilde{\phi}_n(x) \equiv \phi_n(x-x_n) \rightharpoonup \phi(x) \neq 0$ *weakly in* $H^1(\mathbb{R}^3)$, $\tilde{\alpha}_n(x) \equiv \alpha_n(x-x_n) \rightharpoonup \alpha(x) \neq 0$ *weakly in* $L^p(\mathbb{R}^3)$. *Moreover,*

$$\sigma\cdot(p-\alpha)\phi = 0\,. \tag{2.1}$$

(c) *If the original sequence has the property that* ϕ_n *does not converge weakly to zero in* $H^1(\mathbb{R}^3)$, *then the statement in part* (b) *holds with* $x_n \equiv 0$.

Proof. Clearly by (i) and the definition of ϕ_n we have that ϕ_n is uniformly bounded in $H^1(\mathbb{R}^3)$. By (iii)

$$\|\alpha_n\|_p = \lambda_n^{\frac{p-3}{p}}\|A_n\|_p \leqq D\lambda_n^{1-s-3/p} = D\,, \tag{2.2}$$

so α_n is uniformly bounded in $L^p(\mathbb{R}^3)$. By (iv) $\|\sigma\cdot(p-\alpha_n)\phi_n\|_2 \leqq C_n \to 0$ as $n\to\infty$. By the triangle inequality and Hölder's inequality with $q = 2p/(p-2)$ we have

$$C_n \geqq \|\sigma\cdot(p-\alpha_n)\phi_n\|_2 \geqq \|\nabla\phi_n\|_2 - \|\alpha_n\phi_n\|_2 \geqq 1 - \|\alpha_n\|_p\|\phi_n\|_q\,. \tag{2.3}$$

Note that $\|(\sigma\cdot p)\phi\|_2 = \|\nabla\phi\|_2$ and that $\|(\sigma\cdot\alpha)\phi\|_2 = \|\alpha\phi\|_2$. The latter uses the trivial identity $(\sigma\cdot\alpha)^2 = \alpha^2$. The former uses the same identity in Fourier space $(\sigma\cdot p)^2 = p^2$, and this is justified since $\phi\in H^1$. Also note that $2<q<6$. Inequality (2.3) will be used in two ways. Since ϕ_n is uniformly bounded in H^1 we have, by Sobolev's inequality, that $\|\phi_n\|_q \leqq d_q$, and hence $\|\alpha_n\|_p \geqq (1-C_n)/d_q$, which proves (a). On the other hand using (2.2) together with (2.3) we find $\|\phi_n\|_q \geqq (1-C_n)/D$, and hence $\liminf \|\phi_n\|_q \geqq 1/D > 0$. Since $\|\phi_n\|_2 \leqq d_2$ and $\|\phi_n\|_6 \leqq d_6$, Lemma 2.1 and Lemma 2.2 below prove the existence of a sequence $x_n\in\mathbb{R}^3$ and a subsequence ϕ_n such that $\tilde{\phi}_n \rightharpoonup \phi \neq 0$ weakly in $H^1(\mathbb{R}^3)$. By passing, if necessary, to a further subsequence we can assume that $\tilde{\alpha}_n \rightharpoonup \alpha$ weakly in $L^p(\mathbb{R}^3)$. Next we show that $\tilde{\phi}_n\tilde{\alpha}_n \rightharpoonup \phi\alpha$ componentwise weakly in $L^2(\mathbb{R}^3)$. To show that $f_n \rightharpoonup f$ in $L^2(\mathbb{R}^3)$, it suffices to prove that $f_n \rightharpoonup f$ in $L^2(K)$ for every compact $K\subset\mathbb{R}^3$. By the Rellich-Kondrachov theorem, $\tilde{\phi}_n \to \phi$ strongly in $L^q(K)$ (since $2<q<6$), whence $\tilde{\phi}_n\tilde{\alpha} \rightharpoonup \phi\alpha$ in $L^2(K)$. Thus, we have proved that $g_n \equiv \sigma\cdot(p-\tilde{\alpha}_n)\tilde{\phi}_n \rightharpoonup \sigma\cdot(p-\alpha)\phi \equiv g$ weakly in $L^2(\mathbb{R}^3;\mathbb{C}^2)$. But we already noted that $\|g_n\|_2 \to 0$, which implies (by the weak lower semicontinuity of the norm) that $g=0$. Obviously, $\alpha \neq 0$ because otherwise we should have $(\sigma \cdot p)\phi = 0$ with $\phi \neq 0$, which is impossible [recall that $\|(\sigma \cdot p)\phi\|_2 = \|\nabla \phi\|_2$]. This proves (b).

To prove (c) we note a trivial generalization of the Banach-Alaoglu theorem: Since $\phi_n \not\rightharpoonup 0$ we can find a subsequence such that $\phi_n \rightharpoonup \phi$ and $\phi \neq 0$. The rest of the proof is the same as in (b), except that Lemma 2.2 is not needed. □

Lemma 2.1. *Let g be a measurable function on a measure space such that for $p<q<r$ fixed and for some C_p, C_q, C_r all >0,*

(i) $\|g\|_p^p \leqq C_p$,

(ii) $\|g\|_r^r \leqq C_r$,

(iii) $\|g\|_q^q \geqq C_q > 0$.

Then $f(\varepsilon) \equiv \text{meas}\{x \,|\, |g(x)| \geqq \varepsilon\} > C$ for some fixed $\varepsilon, C>0$ depending on p, q, r, C_p, C_q, C_r, but not on g.

Proof. From the fact that $f(\varepsilon)$ is monotone non-increasing and that $\int g^p = p\int_0^\infty f(\varepsilon)\varepsilon^{p-1}d\varepsilon$, we have $C_p \geqq p\int_0^R \varepsilon^{p-1} f(\varepsilon)d\varepsilon \geqq R^p f(R)$ or

$$f(\varepsilon) \leqq \varepsilon^{-p} C_p, \qquad \text{all } \varepsilon > 0. \tag{2.4}$$

Similarly,

$$f(\varepsilon) \leqq \varepsilon^{-r} C_r, \qquad \text{all } \varepsilon > 0. \tag{2.5}$$

Define S and T by

$$qC_p S^{q-p} = \tfrac{1}{4}(q-p)C_q,$$
$$qC_r T^{q-r} = \tfrac{1}{4}(r-q)C_q.$$

From (2.4)

$$q\int_0^S f(\varepsilon)\varepsilon^{q-1}d\varepsilon \leqq qC_p \int_0^S \varepsilon^{q-p-1}d\varepsilon = \tfrac{1}{4}C_q. \tag{2.6}$$

Similarly, from (2.5)

$$q\int_T^\infty f(\varepsilon)\varepsilon^{q-1}d\varepsilon \leqq \tfrac{1}{4}C_q, \tag{2.7}$$

(2.6) and (2.7) imply that $S<T$ and that

$$I \equiv q\int_S^T f(\varepsilon)\varepsilon^{q-1}d\varepsilon \geqq \tfrac{1}{2}C_q.$$

But $I \leqq f(S)|T^q - S^q|$ since f is monotone nonincreasing. This proves the lemma (with $\varepsilon \equiv S$) since S and T are explicitly given independent of f. □

Lemma 2.2 [5]. *Let $1<p<\infty$ and let $\{f_n\}_{n=1}^\infty$, be a uniformly bounded sequence of functions in $W^{1,p}(\mathbb{R}^d)$ with the property that the Lebesgue measure of $\{x \,|\, |f_n(x)| > \varepsilon\} > C$ for some fixed constants C and $\varepsilon > 0$. Then there exists a sequence of translations $\{\tau_n\}_{n=1}^\infty$ of $\mathbb{R}^d$, $\tau_n y = y + x_n$, $F_n(y) \equiv f_n(\tau_n y) = f_n(y + x_n)$, such that, for some subsequence, $F_n \rightharpoonup F$ weakly in $W^{1,p}$ and $F \neq 0$.*

Remark. The proof in [5] was given for real valued functions. It is easy to see that the lemma holds for complex valued functions by considering separately real and imaginary parts. The same argument then carries over to complex spinors. We recall that $W^{1,p}$ consists of all functions in L^p whose first derivatives are in L^p. Note that $W^{1,2}=H^1$.

Let us now apply Theorem 2.1 to the proof of formula (1.10) for z_c. Suppose that z is such that $E=-\infty$. This means there exists a sequence of pairs $(\psi_n, A_n)\in\mathscr{C}$ such that as $n\to\infty$,

$$E_n=\mathscr{E}(\psi_n, A_n)=\|\sigma\cdot(p-A_n)\psi_n\|_2^2-z(\psi_n,|x|^{-1}\psi_n)+\varepsilon\int B_n^2dx \tag{2.8}$$

tends to $-\infty$. We verify the assumptions of Theorem 2.1 for the sequence (ψ_n, A_n). (The usage of ϕ_n, α_n as in Theorem 2.1 will be continued.) (i) is trivial since $\|\psi_n\|_2=1$.

Observe that $-z(\psi_n,|x|^{-1}\psi_n)$ is the only negative term in (2.8) and hence $(\psi_n,|x|^{-1}\psi_n)\to\infty$ as $n\to\infty$. (ii) follows from the inequality $(\psi_n,|x|^{-1}\psi_n)\leqq\|\nabla\psi_n\|_2\|\psi_n\|_2$.

We can choose $E_n<0$ (all n) and we find

$$\|\sigma(p-A_n)\psi_n\|_2^2+\varepsilon\int B_n^2\leqq z(\psi_n,|x|^{-1}\psi_n)\leqq z\|\nabla\psi_n\|_2\,. \tag{2.9}$$

(iv) holds with $C_n=z^{1/2}\|\nabla\psi_n\|_2^{-1/2}$.

From (2.9) we also obtain $\|B_n\|_2^2\leqq(z/\varepsilon)\|\nabla\psi_n\|_2$. On the other hand, Sobolev's inequality gives

$$\|B_n\|_2^2=\sum_i\|\nabla A_{i,n}\|_2^2\geqq S^2\|A_n\|_6^2\,. \tag{2.10}$$

Thus, (iii) holds with $p=6$ and $s=\frac{1}{2}$.

The conclusions (a) and (b) of Theorem 2.1 thus hold for the sequence (ψ_n, A_n). It is easily seen that conclusion (c) also holds, for suppose $\phi_n\rightharpoonup 0$ weakly in $H^1(\mathbb{R}^3,\mathbb{C}^2)$. This would imply that $b_n\equiv(\phi_n,|x|^{-1}\phi_n)\to 0$ as $n\to\infty$. [To prove this, let B_R be the ball of radius R centered at 0 and χ_R its characteristic function. Note that, by Rellich-Kondrachov, $\phi_n\to 0$ strongly in $L^4(B_R)$. Then, writing $b_n=b_n^++b_n^-$, with $b_n^-=(\phi_n,|x|^{-1}\chi_R\phi_n)$, we have that $b_n^-\to 0$. However $b_n^+\leqq R^{-1}$ since $\|\phi_n\|_2=1$. Then let $R\to\infty$.] The energy can be written [using λ_n of Theorem 2.1, $\beta_n(x)\equiv\lambda_n^2B_n(\lambda_n x)$] as

$$E_n=\mathscr{E}(\psi_n, A_n)=\lambda_n^{-1}\{\lambda_n^{-1}\|\sigma\cdot(p-\alpha_n)\phi_n\|_2^2+\varepsilon\|\beta_n\|_2^2-zb_n\}\,. \tag{2.11}$$

If $b_n\to 0$ then, since $E_n<0$, $\beta_n\to 0$ strongly in L^2. By (2.10), $\alpha_n\to 0$ strongly in L^6, which contradicts conclusion (a) of Theorem 2.1.

In the foregoing we did not actually use the fact that $E_n\to-\infty$, but only the facts that $E_n<0$ and that the Coulomb energy diverges. The foregoing analysis was actually the proof of the following

Theorem 2.2. *Let* $(\psi_n, A_n)\in\mathscr{C}$ *be a sequence satisfying*

$$E_n<0 \quad \textit{and} \quad \limsup_{n\to\infty}(\psi_n,|x|^{-1}\psi_n)=\infty\,.$$

Then conclusions (a) *and* (c) *of Theorem 2.1 hold for this sequence. Moreover, for the subsequence given by Theorem 2.1* (c),

$$(\phi_n,|x|^{-1}\phi_n)\to(\phi,|x|^{-1}\phi)\neq 0\,.$$

Remark. For any minimizing sequence (for any z) we can always assume $E_n<0$, since one can always take the pair $\psi=$ ground state hydrogenic function and $A\equiv 0$.

Let us define

$$\hat{z}=\inf_{\mathscr{F}}\varepsilon\|B\|_2^2/(\psi,|x|^{-1}\psi). \tag{2.12}$$

Note that the set $\mathscr{F}$ (defined in Sect. I) is not empty (see [8]).

Theorem 2.3.

$$\hat{z}=z_c. \tag{2.13}$$

Proof. Assume that $E_n\to-\infty$. We shall show that $z\geqq\hat{z}$, which implies $z_c\geqq\hat{z}$. By the remark above (and passing to a subsequence), we can assume that

$$\phi_n\rightharpoonup\phi\neq 0 \text{ weakly in } H^1(\mathbb{R}^3),$$

$$(\phi_n,|x|^{-1}\phi_n)\to(\phi,|x|^{-1}\phi),$$

$$\beta_n\rightharpoonup\beta \text{ weakly in } L^2(\mathbb{R}^3).$$

From this, $\|\beta\|_2\leqq\liminf\|\beta_n\|_2$ and $\|\phi\|_2\leqq\liminf\|\phi_n\|_2=1$. By (2.11)

$$0\geqq\liminf\lambda_nE_n\geqq\varepsilon\|\beta\|_2^2-z(\phi,|x|^{-1}\phi).$$

Since ϕ might not be normalized, define $\hat{\phi}=\phi/\|\phi\|_2$. Then

$$z\geqq\varepsilon\|\beta\|_2^2/(\phi,|x|^{-1}\phi)\geqq\varepsilon\|\beta\|_2^2/(\hat{\phi},|x|^{-1}\hat{\phi})\geqq\hat{z},$$

since $(\hat{\phi},\alpha)\in\mathscr{F}$.

On the other hand, if $z_c>\hat{z}$, then there exists $(\psi,A)\in\mathscr{F}$ such that the ratio on the right side of (2.12) is less than $\bar{z}\equiv\hat{z}+\frac{1}{2}(z_c-\hat{z})$. Define $\psi_n(x)=n^{3/2}\psi(nx)$, $A_n(x)=nA(nx)$. Then for the $\bar{z}$ just defined

$$\mathscr{E}(\psi_n,A_n)=n\{-\bar{z}(\psi,|x|^{-1}\psi)+\varepsilon\|B\|_2^2\},$$

which tends to $-\infty$ as $n\to\infty$. This is a contradiction since $\bar{z}<z_c$. □

Remark. We have repeatedly used the facts that $z<z_c\Rightarrow E>-\infty$ and $z>z_c\Rightarrow E=-\infty$. For the case $z=z_c$ we do not know whether E is finite or $E=-\infty$. This is an open question.

Two natural questions arise. Is there a minimizing $(A,\psi)\in\mathscr{C}$ for $\mathscr{E}$ when $z<z_c$? Is there a minimizing $(A,\psi)\in\mathscr{F}$ for the ratio in (2.12) defining $z_c=\hat{z}$? The answer to both questions is yes.

Theorem 2.4. *When $z<z_c$ there exists a pair $(\psi,A)\in\mathscr{C}$ such that*

$$\mathscr{E}(\psi,A)=E\equiv\inf\{\mathscr{E}(\psi',A')|(\psi',A')\in\mathscr{C}\}.$$

Theorem 2.5. *There exists $(\psi,A)\in\mathscr{F}\equiv\{(\psi',A')\in\mathscr{C}|\sigma\cdot(p-A')\psi'=0\}$ such that $\varepsilon\|B\|_2^2/(\psi,|x|^{-1}\psi)=z_c=\hat{z}$ (given by (2.12)).*

Proof of Theorem 2.4. Let $(\psi_n,A_n)\in\mathscr{C}$ be a minimizing sequence. By Theorem 2.2, $b_n\equiv(\psi_n,|x|^{-1}\psi_n)$ is a bounded sequence (since $z<z_c$). From (2.8) we see that $\|B_n\|_2$ and $\|\sigma\cdot(p-A_n)\psi_n\|_2$ are also bounded sequences (since $E_n<0$). Now,

$$\|\nabla\psi_n\|_2=\|\sigma\cdot p\psi_n\|_2\leqq\|\sigma\cdot(p-A_n)\psi_n\|_2+\|(\sigma\cdot A_n)\psi_n\|_2.$$

However $\|(\sigma \cdot A_n)\psi_n\|_2 = \|A_n\psi_n\|_2 \leqq \|A_n\|_6 \|\psi_n\|_6^{1/2} \|\psi_n\|_2^{1/2}$. By the Sobolev inequality (2.10) applied to A_n and ψ_n, (and with $\|\psi_n\|_2 = 1$),

$$\|\nabla\psi_n\|_2 \leqq \|\sigma \cdot (p - A_n)\psi_n\|_2 + S^{-3/2}\|B_n\|_2 \|\nabla\psi_n\|_2^{1/2}. \tag{2.14}$$

This implies that $\|\nabla\psi_n\|_2$ is also bounded and hence that ψ_n is bounded in H^1.

By passing to a subsequence we have

$$\begin{aligned} \psi_n &\rightharpoonup \psi \quad \text{weakly in } H^1, \\ A_n &\rightharpoonup A \quad \text{weakly in } L^6, \\ B_n &\rightharpoonup B \quad \text{weakly in } L^2, \\ \operatorname{div} A &= 0 \quad \text{and} \quad \operatorname{curl} A = B, \\ (\psi_n, |x|^{-1}\psi_n) &\to (\psi, |x|^{-1}\psi). \end{aligned} \tag{2.15}$$

The proof of the last statement is as in Theorem 2.2. Furthermore, $\psi_n A_n \rightharpoonup \psi A$ in L^2 (as in the proof of Theorem 2.1). By lower semicontinuity of the norms we obtain $E \geqq \mathscr{E}(\psi, A)$. If we knew that $\|\psi\|_2 = 1$ [and hence that $(\psi, A) \in \mathscr{C}$] we would be done. However, $\|\psi\|_2 \leqq 1$ by lower semicontinuity. Suppose that $\gamma = \|\psi\|_2^{-1} > 1$. Define $\hat{\psi} = \gamma\psi$. Then

$$\mathscr{E}(\hat{\psi}, A) = \gamma^2\{\|\sigma \cdot (p - A)\psi\|_2^2 - z(\psi, |x|^{-1}\psi)\} + \varepsilon\|B\|_2^2.$$

The term in { } must be negative [since $\mathscr{E}(\psi, A) \leqq E < 0$]. Therefore, $E \leqq \mathscr{E}(\hat{\psi}, A) < \mathscr{E}(\psi, A) \leqq E$. Hence $\gamma = 1$ and the proof is complete. □

Proof of Theorem 2.5. Let $(\psi_n, A_n) \in \mathscr{F}$ be a minimizing sequence. By scaling

$$\psi_n(x) \to \lambda^{3/2}\psi_n(\lambda x),\ A_n(x) \to \lambda A_n(\lambda x),$$

we can assume that $\|B_n\|_2 = 1$. Also $\|\psi_n\|_2 = 1$ and $\|\nabla\psi_n\|_2 \leqq S^{-3}$ [by (2.14) and $\sigma \cdot (p - A_n)\psi_n = 0$]. Thus ψ_n is bounded in H^1. Again, (2.15) holds for some subsequence. By lower semicontinuity, $\|B\|_2 \leqq \|B_n\|_2$ and $\gamma = \|\psi\|_2^{-1} \geqq 1$. Note that $\psi \neq 0$ by the last line of (2.15). Replacing ψ by $\hat{\psi} = \gamma\psi$ we have that

$$\hat{z}/\varepsilon = \lim\{\|B_n\|_2^2/(\psi_n, |x|^{-1}\psi_n)\} \geqq \|B\|_2^2/(\psi, |x|^{-1}\psi) \geqq \|B\|_2^2/(\hat{\psi}, |x|^{-1}\hat{\psi}).$$

If we can show that $\sigma \cdot (p - A)\psi = 0$, we can conclude from this that $\gamma = 1$ and that (ψ, A) is a minimizing pair. However, $\nabla\psi_n \rightharpoonup \nabla\psi$ and $A_n\psi_n \rightharpoonup A\psi$ weakly in L^2, so $0 = \sigma \cdot (p - A_n)\psi_n \rightharpoonup \sigma \cdot (p - A)\psi$ weakly in L^2. But the weak limit of 0 can only be 0. □

III. A Lower Bound for z_c

In the previous section z_c was shown to be finite (since $\mathscr{F}$ is not empty) and a formula for z_c was given. While we are unable to evaluate that formula, we shall show here that z_c is not too small. The methods of this section are completely different from those of the previous section.

Let $(\psi, A) \in \mathscr{C}$ be given and let

$$T(\psi, A) \equiv \|(p-A)\psi\|_2^2, \tag{3.1}$$

$$Q(\psi) \equiv \frac{1}{4\varepsilon} \int \{\langle\psi, \psi\rangle(x)\}^2 dx, \tag{3.2}$$

$$\mathcal{T}(\psi, A) \equiv \|\sigma \cdot (p-A)\psi\|_2^2 + \varepsilon \|B\|_2^2. \tag{3.3}$$

$$X(\psi, A) \equiv \tfrac{1}{2} T(\psi, A)/Q(\psi). \tag{3.4}$$

Lemma 3.1.

$$\mathcal{T}(\psi, A) \geqq T(\psi, A) \cdot \begin{cases} \frac{1}{2} X(\psi, A) & \text{if } X(\psi, A) \leqq 1 \\ 1 - \{2X(\psi, A)\}^{-1} & \text{if } X(\psi, A) \geqq 1 . \end{cases} \tag{3.5}$$

Proof. Let $0 \leqq t \leqq 1$ and observe that

$$\mathcal{T}(\psi, A) \geqq t \|\sigma \cdot (p-A)\psi\|_2^2 + \varepsilon \|B\|_2^2 .$$

Expanding the first term on the right we get

$$\mathcal{T}(\psi, A) \geqq t T(\psi, A) - t \int B(x) \cdot \langle\psi, \sigma\psi\rangle(x) dx + \varepsilon \|B\|_2^2 . \tag{3.6}$$

Note that in obtaining this result we performed a partial integration in the second term which is easily justified. Minimizing the second and the third term with respect to $B(B(x) = (t/2\varepsilon)\langle\psi, \sigma\psi\rangle(x))$ we find for these two terms the lower bound

$$-t^2 Q(\psi) . \tag{3.7}$$

We have used the identity

$$\langle\psi, \sigma\psi\rangle \cdot \langle\psi, \sigma\psi\rangle = \langle\psi, \psi\rangle^2, \quad \text{all } x . \tag{3.8}$$

The sum of the first term in (3.6) together with (3.7) has its maximum as a function of t at $t_0 = X(\psi, A)$. If $t_0 \leqq 1$ we find

$$\mathcal{T}(\psi, A) \geqq \tfrac{1}{2} T(\psi, A) X(\psi, A) ,$$

and if $t_0 > 1$ we set $t = 1$ and get

$$\mathcal{T}(\psi, A) \geqq T(\psi, A) - Q(\psi) . \quad \square$$

Lemma 3.1 provides us with two alternatives.

Alternative 1. $X(\psi, A) \geqq 1$. In this case $\mathcal{T}(\psi, A) \geqq \frac{1}{2} T(\psi, A)$, and thus

$$\mathscr{E}(\psi, A) \geqq \tfrac{1}{2} T(\psi, A) - z(\psi, |x|^{-1}\psi) \geqq -\tfrac{1}{2} z^2 . \tag{3.9}$$

Here we have used the diamagnetic inequality [4]

$$T(\psi, A) \geqq T(|\psi|, 0) = \|\nabla\phi\|_2^2, \quad \text{with} \quad \phi(x)^2 = \langle\psi, \psi\rangle(x), \tag{3.10}$$

together with the well-known hydrogenic ground state energy. We shall return to this alternative later.

Alternative 2. $X(\psi, A) < 1$. Then

$$\mathscr{E}(\psi, A) \geqq \mathscr{D}(\psi, A) \equiv \tfrac{1}{4} T(\psi, A)^2 / Q(\psi) - z(\psi, |x|^{-1}\psi) . \tag{3.11}$$

The two terms in $\mathcal{D}(\psi, A)$ scale (with x) in the same way, and hence the infimum of $\mathcal{D}$ is either 0 or $-\infty$. Let us define

$$\tilde{z}_c = \sup\{z | \inf \mathcal{D}(\psi, A) = 0\}. \tag{3.12}$$

Clearly

$$\tilde{z}_c \leqq z_c. \tag{3.13}$$

Another expression for $\tilde{z}_c$ [which uses the common x-scaling of the terms in $\mathcal{D}$ and (3.10)] is

$$\tilde{z}_c = \varepsilon \inf_{\phi} \|\phi\|_2^{2} \|\nabla\phi\|_2^{4} \|\phi\|_4^{-4} (\phi, |x|^{-1}\phi)^{-1}. \tag{3.14}$$

Here ϕ is an ordinary real valued function in $H^1(\mathbb{R}^3)$.

As an aside, it is worth mentioning that the fact that $z_c \geqq \tilde{z}_c$ [given by (3.14)] – but not Lemma 3.1 – can be derived directly from the formula (2.12). If $\sigma \cdot (p-A)\psi = 0$, then $0 = \int |\sigma \cdot (p-A)\psi|^2 = \int |(p-A)\psi|^2 - \int B \cdot \langle \psi, \sigma\psi \rangle$. (A justified integration by parts was used in the last term.) Using the Schwarz inequality on the last term, and (3.8), we have

$$T(\psi, A) \leqq \|B\|_2 Q(\psi)^{1/2} (4\varepsilon)^{1/2}. \tag{3.15}$$

Equation (3.14) then follows from (3.15), (3.10) and formula (2.12).

Our next goal is to find a lower bound to the right side of (3.14), which we shall call z_c^L:

$$z_c^L \leqq \tilde{z}_c \leqq z_c. \tag{3.16}$$

(Of course one can try to compute the infimum in (3.14) directly – which leads to an interesting differential equation.) First note that

$$\|\nabla\phi\|_2 \|\phi\|_2 \geqq (\phi, |x|^{-1}\phi), \tag{3.17}$$

which is the uncertainty principle and follows from the hydrogen ground state by scaling in x. Hence

$$\tilde{z}_c \geqq z_c^L \equiv \varepsilon \inf_{\phi} \|\nabla\phi\|_2^{3} \|\phi\|_4^{-4} \|\phi\|_2. \tag{3.18}$$

The minimization problem in (3.18) is equivalent to the following. Let $e<0$ be the ground state energy of $-\Delta - V(x)$. In [7] it is proved that

$$|e|^{1/2} \leqq L^1_{\frac{1}{2},3} \|V\|_2^{2}. \tag{3.19}$$

$L^1_{\frac{1}{2},3}$ is obtained by solving an ordinary differential equation [7] and is found numerically (to 3 significant figures) to be

$$L^1_{\frac{1}{2},3} = 0.0135. \tag{3.20}$$

By choosing $V(x) = C\phi(x)^2$ one deduces from (3.19), that $\|\nabla\phi\|_2^2 \geqq C\|\phi\|_4^4 - C^4 (L^1_{\frac{1}{2},3})^2 \|\phi\|_4^8$ when $\|\phi\|_2 = 1$. Optimizing with respect to C and inserting the result in (3.18) gives

$$z_c^L = \varepsilon (3/4)^{3/2} \{2L^1_{\frac{1}{2},3}\}^{-1} \geqq (24.0)\varepsilon. \tag{3.21}$$

This lower bound to $\tilde{z}_c$ is surprisingly accurate. Inserting the function $\phi(x) = \exp(-|x|)$ in (3.14) yields

$$z_c^L \leqq \tilde{z}_c \leqq 8\pi\varepsilon = (25.1)\varepsilon\,. \tag{3.22}$$

Recalling that $\varepsilon = \{8\pi\alpha^2\}^{-1} = 747.2$, we have that

$$z_c \geqq z_c^L \geqq 17{,}900\,. \tag{3.23}$$

In [8] a solution to $\sigma \cdot (p-A)\psi = 0$ is found which, when inserted into (2.12), yields

$$z_c \leqq 9\pi^3\varepsilon = (279)\varepsilon = 208{,}000\,. \tag{3.24}$$

So far we have shown that if $z \leqq \tilde{z}_c$ then alternative 2 above is irrelevant, for otherwise we should conclude that $E \geqq 0$, which is false. Our next goal is to find a lower bound for E, and we shall do so under the slightly stronger condition that $z \leqq z_c^L$. To this end, we need only consider $(\psi, A) \in \mathscr{C}$ such that $X(\psi, A) \geqq 1$. By Lemma 3.1, a lower bound, $\tilde{E}$, for E is given by

$$E \geqq \tilde{E} = \inf\{T(\psi, A) - Q(\psi) - z(\psi, |x|^{-1}\psi)\} \tag{3.25}$$

under the conditions $(\psi, A) \in \mathscr{C}$ and $T(\psi, A) \geqq 2Q(\psi)$.

The problem posed by (3.25) is too difficult (in particular it is not clear that $A = 0$ is an optimal choice). Therefore we seek a lower bound to the right side of (3.25) as follows. Recall that $T(\psi, A)$ satisfies (for $\|\psi\|_2 = 1$)

$$T(\psi, A) \geqq [4z_c^L Q(\psi)]^{2/3}\,, \tag{3.26}$$

which follows from (3.18) and (3.10),

$$T(\psi, A) \geqq (\psi, |x|^{-1}\psi)^2\,, \tag{3.27}$$

$$T(\psi, A) \geqq 2Q(\psi)\,. \tag{3.28}$$

Define $\tau(\psi)$ by

$$\tau(\psi) = \max\{\text{right sides of (3.26), (3.27), (3.28)}\}\,.$$

Then, if we define E^L by

$$E^L = \inf_{\psi}\{\tau(\psi) - Q(\psi) - z(\psi, |x|^{-1}\psi)\} \tag{3.29}$$

(with $\|\psi\|_2 = 1$), we have that

$$E^L \leqq \tilde{E} \leqq E\,. \tag{3.30}$$

The problem posed by (3.29) is, in fact, algebraic. It is solved in Appendix B with the result that for all $z \leqq z_c^L$

$$E^L = -\tfrac{1}{4}(\tfrac{4}{3})^3 (z_c^L)^2 [3\gamma - 2 + 2(1-\gamma)^{3/2}]\,, \tag{3.31}$$

where $\gamma = \frac{3}{4} z/z_c^L$.

When the right side of (3.31) is Taylor expanded for small z, the leading two terms are

$$\approx -\tfrac{1}{4}z^2 - \frac{1}{32 z_c^L} z^3\,. \tag{3.32}$$

On the other hand using Taylor's theorem with remainder and taking the maximum of d^3E^L/dz^3 in the interval $[0, z]$, we can derive a lower bound to (3.31) for all $z \leqq z_c^L$, which agrees with (3.32) to the first two orders:

$$E \geqq -\tfrac{1}{4}z^2 - (32z_c^L)^{-1}z^3(1-\gamma)^{-3/2}. \tag{3.33}$$

A crude upper bound for E can be obtained with the trial function

$$\begin{gathered} \psi(r) = \begin{pmatrix} \phi(r) \\ 0 \end{pmatrix}, \qquad \phi(r) = (z/2)^{3/2}\pi^{-1/2}e^{-zr/2}, \\ A = \tfrac{1}{2}\alpha^2 z^3 (2/3)^4 e^{-zr/2}(-y, x, 0). \end{gathered} \tag{3.34}$$

(This choice does not satisfy $\operatorname{div} A = 0$, but that does not matter.) A computation with this (ψ, A) gives

$$E \leqq -\tfrac{1}{4}z^2 - \tfrac{1}{2}(\tfrac{2}{3})^8 z^3 \alpha^2 + 2^4 3^{-8} z^4 \alpha^4. \tag{3.35}$$

Remark. We do not know whether E diverges as $z \to z_c$. Of course, E is an upper semicontinuous, monotone decreasing function of z, so $E(z_c) = \lim_{z \to z_c} E(z)$.

IV. A Single Electron Atom in a Magnetic Field of Constant Direction

In the previous two sections we considered a single electron atom in an arbitrary magnetic field and showed that z_c is finite (but huge) and estimated the shift in the ground state energy for $z < z_c^L$. The magnetic field that causes the energy to diverge when $z > z_c$ has to be highly contorted (which is consistent with the example given in [8]). If, on the other hand, certain constraints are placed on B near the nucleus, the divergence will not occur and z_c will be infinite.

In this section we display one such condition-namely that B has a constant direction (but not necessarily constant magnitude) near the nucleus. This is one possible version of the external field problem and is relevant for astrophysics. We shall content ourselves with showing merely that z_c is infinite and will not bother to try to find a good estimate on the energy; in fact we shall obtain $E \geqq -(\text{const})(1 + z^4)$ for all z. The crucial point, of course, is that the bound is independent of B (but it does depend on the size of the region in which the direction of B is constant).

It should be noted that something a bit stronger is actually proved in the following. Namely for *any* B the energy will be bounded below if we replace the troublesome term $\sigma \cdot B$ by $\sigma_3 B_3$. Such a replacement is physically meaningful only when $B_1 = B_2 = 0$.

Let R be a fixed radius and assume that inside the ball K_R of radius R centered at the origin (which is also the location of the nucleus)

$$B(x) = (0, 0, b(x)). \tag{4.1}$$

(The choice of the 3 direction is arbitrary.) $b(x)$ can be anything inside K_R and $B(x)$ can also be anything outside of K_R.

Let $\psi(x)$ be given on $\mathbb{R}^3$, and we want to localize it inside and outside K_R. Define η_1, η_2 both C^∞ such that $\eta_1(x) = 1$ for $x \in K_{R/2}$ and $\eta_1(x) = 0$ for $x \notin K_R$ and

$\eta_1(x)^2+\eta_2(x)^2=1$. Also define $\psi_i(x)=\eta_i(x)\psi(x)$, $i=1,2$. Thus $\psi^2=\psi_1^2+\psi_2^2$. It is easy to see that

$$\|\sigma\cdot(p-A)\psi_1\|_2^2+\|\sigma\cdot(p-A)\psi_2\|_2^2=(\psi,f\psi)+\|\sigma\cdot(p-A)\psi\|_2^2, \tag{4.2}$$

where $f=(\nabla\eta_1)^2+(\nabla\eta_2)^2$. (The cross terms cancel.) We can easily choose η_i such that $f(x)\leqq dR^{-2}$ for some constant, d. Hence we get for $(\psi,A)\in\mathscr{C}$,

$$\mathscr{E}'(\psi,A)\geqq\|\sigma\cdot(p-A)\psi_1\|_2^2-z(\psi_1,|x|^{-1}\psi_1)+\varepsilon\|B\|_2^2-d/R^2-2z/R. \tag{4.3}$$

Here we used the fact that $(\psi_2,|x|^{-1}\psi_2)\leqq 2/R$, since $\psi_2(x)=0$ for $|x|<R/2$ and $\|\psi_2\|_2\leqq 1$.

From now on we drop the subscript 1 and denote ψ_1 by ψ (with $\|\psi\|_2\leqq 1$). Define

$$T_3(\psi,A)=\|(p_3-A_3)\psi\|_2^2\geqq\|p_3|\psi|\,\|_2^2\equiv T_3(\psi), \tag{4.4}$$

$$T_\perp(\psi,A)=\|(p_\perp-A_\perp)\psi\|_2^2\geqq\|p_\perp|\psi|\,\|_2^2=T_\perp(\psi), \tag{4.5}$$

where $p_\perp=(p_1,p_2)$, etc. [The inequality (3.10) holds in any dimensions.] Since $B(x)$ is given by (4.1) on the set where $\psi(x)\neq 0$, we have

$$\|\sigma\cdot(p-A)\psi\|_2^2=T_3(\psi,A)+T_\perp(\psi,A)-\int b\langle\psi,\sigma_3\psi\rangle, \tag{4.6}$$

which can be rewritten as

$$T_3(\psi,A)+U_\perp(\psi,A), \tag{4.7}$$

where

$$U_\perp(\psi,A)\equiv\|\sigma_\perp\cdot(p_\perp-A_\perp)\psi\|_2^2. \tag{4.8}$$

Consider E', which is defined to be the infimum over $(\psi,A)\in\mathscr{C}$ of

$$\mathscr{E}'(\psi,A)=T_3(\psi)+U_\perp(\psi,A)+\varepsilon\|b\|_2^2-z(\psi,|x|_R^{-1}\psi), \tag{4.9}$$

where $|x|_R^{-1}=|x|^{-1}$ if $|x|\leqq R$ and zero otherwise. Here b is defined to be the 3-component of $B=\operatorname{curl}A$, even if B does not point in the 3-direction (note that $\|B\|_2\geqq\|b\|_2$, and that (4.6)–(4.8) is still true, namely $U_\perp=T_\perp-\int b\langle\psi,\sigma_3\psi\rangle$). It is obvious that

$$E\geqq E'-d/R^2-2z/R. \tag{4.10}$$

To analyze E' we observe that each of the four terms in (4.9) involves a 3-dimensional integral, and $\int d^3x=\int dx_3\int dx_\perp$. Think of $\psi, A, B, |x|_R^{-1}$ as functions of $x_\perp$ parameterized by x_3. Then

$$\mathscr{E}'(\psi,A)=T_3(\psi)+\int dx_3\mathscr{E}''(\psi,A), \tag{4.11}$$

$$\mathscr{E}''(\psi,A)=\int dx_\perp|\sigma_\perp\cdot(p_\perp-A_\perp)\psi|^2+\varepsilon\int dx_\perp b^2-z\int_{D(x_3)}dx_\perp\langle\psi,\psi\rangle(x_\perp^2+x_3^2)^{-1/2}, \tag{4.12}$$

where $D(x_3)$ is the domain in $x_\perp$ given by

$$x_\perp^2\leqq R^2-x_3^2. \tag{4.13}$$

To analyze (4.12) we utilize the t trick of Sect. III. For each value of x_3, let $t(x_3)$ be chosen to satisfy $0\leqq t(x_3)\leqq 1$. Replace $|\sigma\cdot(p_\perp-A_\perp)\psi|^2$ in (4.12) by $t(x_3)$ times this quantity and use (3.10) to obtain the lower bound on this first term:

$$t(x_3)T_\perp(\psi)-t(x_3)\int|b|\langle\psi,\psi\rangle dx_\perp\,.$$

[Here $T_\perp$ means $\int dx_\perp(\nabla_\perp|\psi|)^2$.] We used $|\langle\psi,\sigma\psi\rangle|=\langle\psi,\psi\rangle$.

Now minimize with respect to b and then maximize with respect to $t(x_3)$, as in Sect. III. For the first two terms on the right side of (4.12) we obtain the bound

$$\min\{(J_\perp)^2/4\varepsilon,\,\varepsilon(T_\perp/J_\perp)^2\} \tag{4.14}$$

with $(J_\perp)^2=\int dx_\perp\langle\psi,\psi\rangle^2$.

The last term in (4.12) can be bounded below as $-zJ_\perp W(x_3)$, and

$$W(x_3)^2=\int_{D(x_3)}(x_\perp^2+x_3^2)^{-1}dx_\perp=\begin{cases}2\pi\ln(R/|x_3|) & \text{for } |x_3|<R\\ 0 & \text{for } |x_3|\geqq R\,.\end{cases} \tag{4.15}$$

To bound (4.14) below, the Sobolev inequality in $\mathbb{R}^2$ is used:

$$T_\perp\geqq S(J_\perp)^2/g(x_3)^2\,, \tag{4.16}$$

where

$$g(x_3)^2=\int dx_\perp\langle\psi,\psi\rangle\,. \tag{4.17}$$

(The constant S can be found in [7].)

Substituting (4.14)–(4.17) in (4.12),

$$\mathscr{E}''(\psi,A)\geqq J_\perp^2\min\{(4\varepsilon)^{-1},S^2\varepsilon g(x_3)^{-4}\}-zJ_\perp W(x_3)\,. \tag{4.18}$$

Since $J_\perp=J_\perp(x_3)$ is unknown, we simply minimize (4.18) with respect to $J_\perp$ and obtain

$$\mathscr{E}''(\psi,A)\geqq-\tfrac{1}{4}z^2W(x_3)^2\cdot\max\{4\varepsilon,g(x_3)^4/S^2\varepsilon\}\,. \tag{4.19}$$

According to (4.11), (4.19) must be integrated over x_3. Since we do not know which term in the max{ , } in (4.19) holds for any given x_3, we shall simply take the sum of the two. The first yields

$$-\varepsilon z^2\int_{-R}^{R}dx_3W(x_3)^2=-4\pi\varepsilon z^2R\,. \tag{4.20}$$

To control the second possibility we invoke the $T_3(\psi)$ term in (4.11). An application of the Schwarz inequality [12] gives

$$T_3(\psi)\geqq\int dx_3(dg(x_3)/dx_3)^2=\|g'\|_2^2\,. \tag{4.21}$$

It is also a fact that for all x_3 and $g\in L^2(\mathbb{R}^1)$

$$g(x_3)^4\leqq\|g\|_2^2\|g'\|_2^2\leqq T_3(\psi)\,. \tag{4.22}$$

[This follows from $g(x)^2=2\int_{-\infty}^{x}gg'$ and $g(x)^2=-2\int_{x}^{\infty}gg'$. Hence $g(x)^2\leqq\int_{-\infty}^{\infty}|gg'|$. We recall also that $\|g\|_2^2=\int\langle\psi,\psi\rangle d^3x=1$.] Inserting (4.20)–(4.22) in (4.11), we obtain the following lower bound for the second possibility in (4.19).

$$T_3(\psi)\{1-\tfrac{1}{4}z^2S^{-2}\varepsilon^{-1}\int W^2\}=T_3(\psi)\{1-\pi Rz^2/S^2\varepsilon\}\,. \tag{4.23}$$

While the value of $T_3(\psi)$ is unknown, the term (4.23) can be eliminated by the following trick. Call R_0 the original radius inside of which (4.1) holds. A fortiori,

(4.1) holds for any $R<R_0$. If $\{\ \bar{}\ \}$ in (4.23) is nonnegative, use R_0. Otherwise, use

$$R=S^2\varepsilon/\pi z^2\,.$$

Then (4.23)$\geqq 0$ and can be ignored.

Combining all the terms we obtain

$$E\geqq -d/R^2-2z/R-4\pi\varepsilon z^2 R \tag{4.24}$$

with

$$R=\min\{R_0, S^2\varepsilon/\pi z^2\}\,. \tag{4.25}$$

Appendix A

In the following, $\mathscr{D}'$ is the space of distributions.

Theorem A.1. *Let* $B\in L^2(\mathbb{R}^3)$ *be a given vector field and let* $\operatorname{div} B=0$ *in* $\mathscr{D}'$. *Define the vector field*

$$A(x)=\frac{1}{4\pi}\int|x-y|^{-3}(x-y)\times B(y)dy\,. \tag{A.1}$$

Then:

(a) $A\in L^6(\mathbb{R}^3)$ *and* $\operatorname{curl} A=B$, $\operatorname{div} A=0$ *in* $\mathscr{D}'$.

(b) *The distribution* $\partial_i A_j$ *is an* L^2*-function and we have the formula*

$$\sum_i\int|\nabla A_i|^2 d^3x=\int B^2 dx\,.$$

(c) *The* $A(x)$ *given by* (A.1) *is the only vector field having the three properties in* (a) *above.*

Proof. Let us write $A=T(B)$. The kernel in (A.1) is bounded by $|x-y|^{-2}$ and $|x|^{-2}\in L_w^{3/2}$. Let V be a vector field in $L^p(\mathbb{R}^3)$ with $1<p<3$. By the weak Young inequality, $T(V)\in L^r$ where $1/3+1/r=1/p$, and $\|T(V)\|_r\leqq C_p\|V\|_p$ for a suitable constant C_p. By Fubini's theorem

$$(W, T(V))=(T(W), V) \tag{A.2}$$

when $W\in L^q$ and $V\in L^p$, with $q=r'$, $1/q=4/3-1/p$. In (A.2), (W, U) means $\sum_{i=1}^{3}\int\overline{W_i(x)}U_i(x)dx$.

Now we apply (A.2) to $V=B$ and $W=\nabla f$ with $f\in C_0^\infty(\mathbb{R}^3)$,

$$T(W)=-\frac{1}{4\pi}\operatorname{curl}\{|x|^{-1}*\nabla f\}=-\frac{1}{4\pi}\operatorname{curl}\operatorname{grad}\{|x|^{-1}*f\}=0\,.$$

(The first equality, namely exchanging integration and differentiation, follows by dominated convergence.) Then $(\nabla f, A)=0$ for all $f\in C_0^\infty$, and hence $\operatorname{div} A=0$ in $\mathscr{D}'$. A second application of (A.2) is to $V=B$ and $W=\operatorname{curl} G$, with $G\in C_0^\infty$.

$$T(W)=-\frac{1}{4\pi}\operatorname{curl}\operatorname{curl}\{|x|^{-1}*G\}=-G-\nabla\operatorname{div}\frac{1}{4\pi}\{|x|^{-1}*G\}\equiv -G-\nabla g\,.$$

Then $(W, A)=(-G-\nabla g, B)=(-G, B)-(\nabla g, B)$. We claim that $(\nabla g, B)=0$, which will imply that $\operatorname{curl} A=B$ in $\mathscr{D}'$. While $g\in C^\infty$ it does not generally have compact support; otherwise we would have $(\nabla g, B)=0$ since $\operatorname{div} B=0$ in $\mathscr{D}'$.

However $B \in L^2$, and therefore $(\nabla g, B) = \lim_{R \to 0} (\nabla(gf_R), B)$, where $f_R(x) = f(Rx)$ and $f(x)$ is a C_0^∞ function satisfying $f(x) = 1$ for $|x| < 1$, $f(x) = 0$ for $|x| > 2$. Since $(\nabla(gf_R), B) = 0$, we have the desired result, and (a) is proved.

To prove (b) we define $T_j(B) = i\partial_j T(B) = iT(\partial_j B)$ for B smooth and of compact support. It is a standard result about the Riesz transform that $T_j(B)$ has a bounded extension to $L^2(\mathbb{R}^3)$, see [10], so we can assume merely that $B \in L^2$. Furthermore T_j is selfadjoint. Now for any vector field V in $L^2(\mathbb{R}^3)$ we have

$$\sum_j (T_j(B), T_j(V)) = \sum_j (B, T_j^2(V)) = (B, V). \tag{A.3}$$

Indeed, when V is smooth and of compact support $\sum_j T_j^2(V)(x) = V(x) + \frac{1}{4\pi} \nabla \operatorname{div}\{|x|^{-1} * V\}(x)$. Using the previous approximation argument (namely $g \to gf_R$) and the fact that $\operatorname{div} B = 0$ in $\mathscr{D}'$ gives (A.3). Since T_j is bounded, (A.3) is true for all $V \in L^2(\mathbb{R}^3)$. Hence, by setting $V = B$, (b) is also proven.

To prove (c), suppose there were another $\tilde{A}$ with the properties in (a) and let $\alpha = \tilde{A} - A$. Then $\alpha \in L^6$, $\operatorname{curl} \alpha = 0$, $\operatorname{div} \alpha = 0$ in $\mathscr{D}'$. Let $j_\varepsilon(x)$ be a C_0^∞ approximation to the identity and $\alpha_\varepsilon = j_\varepsilon * \alpha$. It is easy to see that $\alpha_\varepsilon \in C^\infty$, $\operatorname{div} \alpha_\varepsilon = 0$, $\operatorname{curl} \alpha_\varepsilon = 0$ and $\alpha_\varepsilon(x) \to 0$ as $|x| \to \infty$. From this, $\Delta \alpha_\varepsilon = -\operatorname{curl} \operatorname{curl} \alpha_\varepsilon + \operatorname{grad} \operatorname{div} \alpha_\varepsilon = 0$. So each component of α_ε is harmonic, but since $\alpha_\varepsilon \to 0$ at ∞, α_ε must be zero for all $\varepsilon > 0$. But as $\varepsilon \to 0$, $\alpha_\varepsilon \to \alpha$ (in L^6 and in $\mathscr{D}'$), so $\alpha = 0$. □

Theorem A.2. *For any $A \in L^6(\mathbb{R}^3)$ and $\psi \in L^2(\mathbb{R}^3)$, $\|\sigma \cdot (p - A)\psi\|_2 < \infty$ implies $\psi \in H^1(\mathbb{R}^3)$.*

Proof. Observe that by assumption

$$\sigma \cdot p\psi = \sigma \cdot A\psi + u, \quad u \in L^2, \quad \text{and} \quad A\psi \in L^{3/2}$$

by Hölder's inequality ($A \in L^6, \psi \in L^2$). Since $(\sigma \cdot p)^{-1} = \sigma \cdot p|p|^{-2}$, we find

$$\psi = \frac{i}{4\pi} \int |x - y|^{-3} \sigma \cdot (x - y) \, [(\sigma \cdot A\psi)(y) + u(y)] dy.$$

Again, by the weak Young inequality, $\psi = v_1 + v_2$, $v_1 \in L^3$, $v_2 \in L^6$ which implies (since $\psi \in L^2$) $\psi \in L^2 \cap L^3$. Hence $A\psi \in L^2$ (again by Hölder's inequality) and thus $\sigma \cdot p\psi \in L^2$.

Appendix B: Proof of Eq. (3.31)

Given ψ, define

$$S(\psi) = (\psi, |x|^{-1}\psi).$$

As ψ ranges over all functions satisfying $\|\psi\|_2 = 1$, $Q(\psi)$ and $S(\psi)$ independently take on all values between 0 and ∞. Therefore we are entitled to think of S and Q simply as an unknown pair of positive numbers.

According to (3.29), then, we have to minimize $e = \tau - Q - zS$ under the conditions $\tau \geqq 2Q$, $\tau \geqq S^2$, $\tau \geqq KQ^{2/3}$ with $K^{3/2} = 4z_c^L$. There are two cases:

case (a): $2Q^{1/3} \geqq K$ or $Q \geqq 2(z_c^L)^2$,

case (b): $2Q^{1/3} < K$.

If case (a) holds, we set $\tau = 2Q$, $S^2 = 2Q$, and then $\tau - Q - zS \geqq 0$, since $z \leqq z_c^L$ and $Q \geqq 2(z_c^L)^2$. If case (b) holds then, similarly,

$$E^L = \min\{KQ^{2/3} - Q - zK^{1/2}Q^{1/3} | Q \leqq 2(z_c^L)^2\}.$$

Change the variable to $Q \equiv 2(z_c^L)^2 x^3$. Then E^L is the minimum of

$$-2(z_c^L)^2[x^3 - 2x^2 + \tfrac{4}{3}\gamma x]$$

subject to $0 \leqq x \leqq 1$ and $\gamma = \frac{3}{4} z/z_c^L \leqq \frac{3}{4}$. The minimum occurs at

$$x = \tfrac{2}{3}[1 - (1-\gamma)^{1/2}]$$

and yields (3.31) for the lower bound.

Acknowledgement. It is a pleasure to acknowledge valuable discussions at the Aspen Center for Physics with David Boulware and Lowell Brown.

References

1. Avron, J., Herbst, I., Simon, B.: Schrödinger operators with magnetic fields: III. Atoms in homogeneous magnetic field. Commun. Math. Phys. **79**, 529–572 (1981)
2. Remark 3 in Brezis, H., Lieb, E.H.: Minimum action solution of some vector field equations. Commun. Math. Phys. **96**, 97–113 (1984)
3. Daubechies, I., Lieb, E.H.: One-electron relativistic molecules with Coulomb interaction. Commun. Math. Phys. **90**, 497–510 (1983)
4. Kato, T.: Schrödinger operators with singular potentials. Israel J. Math. **13**, 135–148 (1972)
5. Lieb, E.H.: On the lowest eigenvalue of the Laplacian for the intersection of two domains. Invent. Math. **74**, 441–448 (1983)
6. Lieb, E.H., Loss, M.: Stability of Coulomb systems with magnetic fields: II. The many-electron atom and the one-electron molecule. Commun. Math. Phys. **104**, 271–282 (1986)
7. Lieb, E.H., Thirring, W.: Inequalities for the moments of the eigenvalues of the Schrödinger Hamiltonian and their relation to Sobolev inequalities. In: Studies in mathematical physics, essays in honor of Valentine Bargmann. Lieb, E.H., Simon, B., Wightman, A.S. (eds.). Princeton, NJ: Princeton University Press 1976
8. Loss, M., Yau, H.T.: Stability of Coulomb systems with magnetic fields: III. Zero energy bound states of the Pauli operator. Commun. Math. Phys. **104**, 283–290 (1986)
9. Michel, F.C.: Theory of pulsar magnetospheres. Rev. Mod. Phys. **54**, 1–66 (1982)
10. Stein, E.M.: Singular integrals and differentiability properties of functions. Princeton, NJ: Princeton University Press 1970
11. Straumann, N.: General relativity and relativistic astrophysics. Berlin, Heidelberg, New York, Tokyo: Springer 1984
12. Hoffmann-Ostenhof, M., Hoffmann-Ostenhof, T.: Schrödinger inequalities and asymptotic behavior of the electron density of atoms and molecules. Phys. Rev. A **16**, 1782–1785 (1977)
13. Avron, J., Herbst, I., Simon, B.: Schrödinger operators with magnetic fields: I. General Interactions. Duke Math. J. **45**, 847–883 (1978)

Communicated by A. Jaffe

Received October 2, 1985; in revised form January 2, 1986

Stability of Coulomb Systems with Magnetic Fields

II. The Many-Electron Atom and the One-Electron Molecule

Elliott H. Lieb* and Michael Loss**

Departments of Mathematics and Physics, Princeton University, Jadwin Hall, P.O. Box 708, Princeton, NJ 08544, USA

Abstract. The analysis of the ground state energy of Coulomb systems interacting with magnetic fields, begun in Part I, is extended here to two cases. Case A: The many electron atom; Case B: One electron with arbitrarily many nuclei. As in Part I we prove that stability occurs if $z\alpha^{12/7} < \text{const}$ (in case A) and $z\alpha^2 < \text{const}$ (in case B), ($z|e| = $ nuclear charge, $\alpha = $ fine structure constant), but a new feature enters in case B. There one *also* requires $\alpha < \text{const}$, regardless of the value of z.

I. Introduction

In the first paper in this series [1] the question of the stability of atoms and molecules in the presence of magnetic fields was raised, and it was answered in the case of the one-electron atom of arbitrary nuclear charge $z|e|$. In the present paper the stability question will be answered in two other cases:

(A) The many electron atom,

(B) The one-electron molecule.

Unfortunately, the stability of the many-electron, many-nucleus system is still an open question.

The reader is referred to the introduction in [1] for the motivation and physical interpretation of this problem. The mathematical essence of the problem is that we want to decide whether or not the energy functional

$$\mathscr{E}(\psi, A, \underline{R}, \underline{z}) \equiv \sum_{j=1}^{N} \int |\sigma_j \cdot (p_j - A(x_j))\psi|^2\, dx + \varepsilon \int B(x)^2\, dx + (\psi, V(X, \underline{R}, \underline{z})\psi) \tag{1.1}$$

* Work partially supported by U.S. National Science Foundation grant PHY-8116101-A03

** Work partially supported by U.S. and Swiss National Science Foundation Cooperative Science Program INT-8503858.

Current address: Institut für Mathematik, FU Berlin, Arnimallee 3, D-1000 Berlin 33

is bounded below by a suitable constant. The three terms in (1.1) are the electronic kinetic energy, the magnetic field energy and the Coulomb energies respectively. The notation is the following:

The energy unit is 4 Rydbergs $=2mc^2\alpha^2$ and $1/\varepsilon=8\pi\alpha^2$, with $\alpha=e^2/\hbar c=1/137$ being the fine structure constant. The charge unit is $|e|$.

$\psi=\psi(x_1,\ldots,x_N,s_1,\ldots,s_N)$ is an arbitrary N particle, antisymmetric (electron) wave function. The particle spatial and spin coordinates are x,s with $s=\pm 1$. X denotes the collection $(x_1,\ldots,x_N)$. The σ_i^j, $j=1,2,3$ denote the Pauli spin matrices. ψ is assumed to be normalized

$$1=\|\psi\|_2^2=(\psi,\psi)=\sum_{s_1\ldots s_N}\int d^{3N}X|\psi(X,s_1,\ldots,s_N)|^2\,. \tag{1.2}$$

$A(x)$ is a vector potential and $B=\operatorname{curl}A$ is the magnetic field which is assumed to be in $L^2(\mathbb{R}^3)$. As explained in [1], for any $B\in L^2$, A exists and is uniquely specified by

$$\operatorname{curl}A=B,\ \operatorname{div}A=0,\ A\in L^6(\mathbb{R}^3)\,. \tag{1.3}$$

The first term in (1.1) is the electron kinetic energy. For particle j it is

$$\|\sigma_j\cdot(p_j-A)\psi\|_2^2=\|(p_j-A)\psi\|_2^2-(\psi,\sigma_j\cdot B\psi)\,. \tag{1.4}$$

The Coulomb term is

$$V(X,\underline{R},\underline{z})=\sum_{1\leq i<j\leq N}|x_i-x_j|^{-1}+\sum_{1\leq i<j\leq K}z^iz^j|R_i-R_j|^{-1}$$
$$-\sum_{i=1}^{N}\sum_{j=1}^{K}z^j|x_i-R_j|^{-1}\,. \tag{1.5}$$

Here we assume that there are K fixed nuclei of charges $z^j|e|$ and distinct locations $R_j\in\mathbb{R}^3$, $j=1,\ldots,K$. The z's and R's will be denoted collectively by $\underline{z}$ and $\underline{R}$. The first term in (1.5) is the electronic repulsion, the second is the nuclear repulsion and the third is the electron-nuclear attraction.

It is useful to have the following notation

$$\tau(\psi,A)\equiv\sum_{j=1}^{N}\|\sigma_j\cdot(p_j-A)\psi\|_2^2+\varepsilon\|B\|_2^2\,, \tag{1.6}$$

$$T(\psi,A)\equiv\sum_{j=1}^{N}\|(p_j-A)\psi\|_2^2\,, \tag{1.7}$$

$$W(\psi,\underline{R},\underline{z})\equiv-(\psi,V(X,\underline{R},\underline{z})\psi)\,. \tag{1.8}$$

We assume $B\in L^2$ and that (1.2) is satisfied. Then, as proved in [1] (with a slight modification to handle the N-coordinate case), in order to make sense of τ and W it is necessary and sufficient to have $\psi\in H^1(\mathbb{R}^{3N})$, i.e. ψ and all its first derivatives are in L^2. The class of all pairs (ψ,A) satisfying the above [and also with ψ normalized as in (1.2)] is denoted by $\mathscr{C}$.

The energy of our system is defined to be

$$E\equiv\inf\{\mathscr{E}(\psi,A,\underline{R},\underline{z})|(\psi,A)\in\mathscr{C},\ \text{all }\underline{R}\}\,. \tag{1.9}$$

This infimum includes an infimum over $\underline{R}$.

From [1] we know that if any single z^j satisfies $z^j > z_c$ (which is evaluated in [1] and which is proportional to α^{-2}), then $E = -\infty$, simply by moving $N-1$ electrons and the other $K-1$ nuclei to infinity. Therefore z_c for the full problem (1.1) is finite. (When $K>1$, z_c is defined to be the largest z such that E is finite whenever all the $z^j < z$.) Our goal here is to show that z_c is not too small for (1.1). Three cases have to be distinguished.

(A) One nucleus (with $R_1 = 0$ and $z^1 \equiv z$) and an arbitrary number, N, of electrons. In Sect. II we find some $\tilde{z}_c$, which is *independent of* N, such that E is finite when $z < \tilde{z}_c$. We also find some $z_c^L < \tilde{z}_c$ for which we can give a lower bound to E (called E^L) when $z < z_c^L$. Both z_c^L and E^L are *independent of* N. The bound on z_c is

$$z_c > z_c^L \geqq -\tfrac{1}{4} + (0.158)\alpha^{-12/7}. \qquad (1.10)$$

Note the exponent 12/7. Is it possible that this can be replaced by 2, as in the one-electron case? We do not know. While our bound on $\tilde{z}_c$ utilizes the electronic Coulomb repulsion in (1.5), we conjecture that the repulsion is not really necessary. This is an interesting open problem.

(B) One electron and an arbitrary number, K, of nuclei. In Sect. III we find, as in case (A), $z_c^L < \tilde{z}_c < z_c$ (with z_c^L and $\tilde{z}_c$ independent of K and proportional to α^{-2}) such that E is finite for $z < z_c$. We also derive a lower bound $E^L < E$ when $z < z_c^L$. However, an important new feature enters here: These results also require that

$$\alpha < \alpha_c \qquad (1.11)$$

for some α_c (which is shown to satisfy $0.32 < \alpha_c < 6.7$). In other words, two conditions are required for stability,

$$z^j\alpha^2 \text{ small (all } j) \text{ and } \alpha \text{ small}. \qquad (1.12)$$

This situation is reminiscent of the relativistic stability problem [2–4], except that there the requirement is $z^j\alpha$ small and α small. It is interesting to note that there are other indications [5, 6] that the stability of field theory requires a bound on the coupling constant (apart from a bound on z). We shall also prove that the requirement (1.11) for stability is real; it is not an artifact of our proof.

(C) Many electrons and many nuclei. We are unable to solve this problem, but the goal would be to prove that E is finite provided $z^j\alpha^2$ is small (all j) and α is small, and that E is then bounded below by -(const)$(N+K)$.

II. Basic Strategy

The following sections are full of technical details, but the common strategy (similar to that used in [1]) is simple. Let us outline it here. Note that the following steps can be carried out even for the full problem, (C), to give an N and K dependent bound on z_c. It is only in cases A and B that we can eliminate this dependence.

The quantities $\tau(\psi, A)$ and $T(\psi, A)$ were defined in (1.6), (1.7); the following quantity Q is also needed. Let $\varrho(x)$ be the one-particle density associated with ψ:

$$\varrho_\psi(x) = \sum_{j=1}^{N} \sum_{s_1,\ldots,s_N} \int |\psi(X, s_1, \ldots, s_N)|^2 \, d^{3N-3}X^j. \qquad (2.1)$$

(X^j means all N variables *except* x_j.) Of course, for fermions we do not have to sum on j. Merely take $j=1$ and then multiply by N. The general expression (2.1) is used because much of the following holds for any statistics (i.e. without symmetry). Then define Q by

$$Q(\psi)=(1/4\varepsilon)\int \varrho_\psi(x)^2\,dx=(1/4\varepsilon)\,\|\varrho_\psi\|_2^2\,. \tag{2.2}$$

Another important quantity is the quantum ground state energy when the $\sigma\cdot B$ and the $\varepsilon\int B^2$ terms are eliminated:

$$E^q(\underline{z})=\inf\{T(\psi,A)-W(\psi,\underline{R},\underline{z})|(\psi,A)\in\mathscr{C},\ \text{all}\ \underline{R}\}\,. \tag{2.3}$$

Of course $E^q<0$. It is well known that E^q is always finite and that the Lieb-Thirring [7] proof of stability carries through for this case [8].

Given ψ, A, and $\underline{R}$, consider the following scaling (with $\lambda>0$):

$$\begin{aligned}\psi(X,\underline{s})&\to\lambda^{3N/2}\,\psi(\lambda X,\underline{s})\,,\\ A(x)&\to\lambda A(\lambda X)\,,\\ B(x)&\to\lambda^2B(\lambda x)\,,\\ \underline{R}&\to(1/\lambda)\,\underline{R}\,.\end{aligned} \tag{2.4}$$

The various quantities scale as

$$\begin{aligned}&W(\psi,\underline{R},\underline{z})\to\lambda W(\psi,\underline{R},\underline{z})\,,\\ &T(\psi,A)\to\lambda^2T(\psi,A),\ \tau(\psi,A)\to\lambda^2\tau(\psi,A)\,,\\ &Q(\psi)\to\lambda^3Q(\psi)\,.\end{aligned} \tag{2.5}$$

If we define

$$W(\psi,\underline{z})=\sup_{\underline{R}}W(\psi,\underline{R},\underline{z})\,, \tag{2.6}$$

then W scales as

$$W(\psi,\underline{z})\to\lambda W(\psi,\underline{z})\,. \tag{2.7}$$

Note that

$$\begin{aligned}E^q(\underline{z})&=\inf_\psi T(\psi,A)-W(\psi,\underline{z})\,,\\ E(\underline{z})&=\inf_\psi \tau(\psi,A)-W(\psi,z)\,.\end{aligned} \tag{2.8}$$

From (2.5)–(2.7) we deduce (as in the case of the one-electron atom) that

$$4|E^q(\underline{z})|T(\psi,A)\geqq W(\psi,\underline{z})^2\geqq W(\psi,\underline{R},\underline{z})^2\,. \tag{2.9}$$

The strategy has 7 steps.

Step 1. In [1, Lemma 3.1] a bound for τ in terms T and Q was derived (which trivially extends to N-particles). There are two cases (depending on ψ and A).

Case 1. $T(\psi, A) \geqq 2Q(\psi)$. Then

$$\tau(\psi, A) \geqq T(\psi, A) - Q(\psi). \tag{2.10}$$

Case 2. $T(\psi, A) \leqq 2Q(\psi)$. Then

$$\tau(\psi, A) \geqq \tfrac{1}{4} T(\psi, A)^2/Q(\psi). \tag{2.11}$$

As will be seen, Case 1 is relevant for determining E^L while Case 2 is relevant for determining $\tilde{z}_c$.

Step 2. (This step is trivial for $K=1$.) Pick some $\underline{z}_0 = (z_0^1, \ldots, z_0^K)$ and consider the rectangle $\underline{z} \prec \underline{z}_0$ (which means $0 \leqq z^j \leqq z_0^j$, all j). For each fixed ψ and $\underline{R}$, the minimum of $W(\psi, \underline{R}, \underline{z})$ in this rectangle occurs at one of the 2^K vertices. This is proved in [2] Lemma 2.3 et. seq. From this it follows that $W(\psi, \underline{z})$, $-E^q(\underline{z})$ and $-E(\underline{z})$ are monotone nondecreasing functions of $\underline{z}$ (with the above order relation). Hence if stability holds for $\underline{\tilde{z}} = (\tilde{z}, \ldots, \tilde{z})$ then it holds when all $z^j \leqq \tilde{z}$.

Step 3 (Definition of $\tilde{z}_c$). Define

$$\delta(\psi, A, \underline{z}) = \tfrac{1}{4} T(\psi, A)^2/Q(\psi) - W(\psi, \underline{z}). \tag{2.12}$$

The two terms of (2.12) scale the same way [see (2.5) and (2.7)], so that the infimum of $\delta(\psi, A, \underline{z})$ (over ψ and A) is either zero or $-\infty$. We define [with $\underline{\tilde{z}} \equiv (\tilde{z}, \ldots, \tilde{z})$]

$$\tilde{z}_c = \sup\{\tilde{z} \,|\, \delta(\psi, A, \underline{\tilde{z}}) \geqq 0 \text{ for all } (\psi, A) \in \mathscr{C}\}. \tag{2.13}$$

Step 4. Suppose that $z^j < \tilde{z}_c$ for all j and let $(\psi, A) \in \mathscr{C}$ be given. If case 1, (2.10), holds then

$$\mathscr{E}(\psi, A, \underline{R}, \underline{z}) \geqq \tfrac{1}{2} T(\psi, A) - W(\psi, \underline{R}, \underline{z}) \geqq 2E^q(\underline{z}), \tag{2.14}$$

by scaling. If case 2 holds then $\delta(\psi, A, \underline{z}) \geqq 0$. In either case $E(\underline{z})$ is finite and thus

$$\tilde{z}_c \leqq z_c. \tag{2.15}$$

Step 5. We want to find a lower bound (which we call z_c^L) to $\tilde{z}_c$. A lower bound on $T(\psi, A)$ is needed and this is provided by the Lieb-Thirring estimate [9]

$$T(\psi, A)^{3/2} \geqq GQ(\psi)/\alpha^2, \tag{2.16}$$

for a universal constant $G = 1.28$, explicated in (3.8). This leads to the bound

$$\delta(\psi, A, \underline{z}) \geqq \tfrac{1}{4}(G\alpha^{-2})\, T(\psi, A)^{1/2} - W(\psi, \underline{z}). \tag{2.17}$$

Combining this with the bound (2.9) [and the trivial fact that we need only consider $W(\psi, \underline{z}) \geqq 0$] we see that $\delta(\psi, A, \underline{z}) \geqq 0$ if

$$|E^q(\underline{z})| \leqq (G/8\alpha^2)^2. \tag{2.18}$$

By (2.13)

$$\tilde{z}_c \geqq z_c^L \equiv \sup\{z \,|\, |E^q(\underline{z})| \leqq (G/8\alpha^2)^2\}, \tag{2.19}$$

[$\underline{z}$ means $(z, \ldots, z)$]. The monotonicity given in Step 2 has been used.

Step 6 (Bound on the energy). Suppose that $z^j \leqq z_c^L$ for all j. Let $(\psi, A) \in \mathscr{C}$ be given. Case 2 is irrelevant since $\delta(\psi, A, \underline{z}) \geqq 0$ by definition. Therefore a lower bound, $E^L(\underline{z})$, to $E(\underline{z})$ can be obtained by the following minimization problem:

$$E^L(\underline{z}) = \min(T - Q - W), \tag{2.20}$$

under the conditions

$$T \geqq 2Q, \; T \geqq (GQ/\alpha^2)^{2/3}, \; T \geqq W^2/4|E^q(\underline{z})|. \tag{2.21}$$

This algebraic problem is solved in Appendix B of [1] and the result is

$$E(\underline{z}) \geqq E^L(\underline{z}) = E^q(\underline{z}) f(\gamma), \tag{2.22}$$

$$f(\gamma) \equiv \tfrac{4}{3}\gamma^{-2}\{3\gamma - 2 + 2(1-\gamma)^{3/2}\}, \tag{2.23}$$

$$\gamma \equiv 6|E^q(\underline{z})|^{1/2}\alpha^2/G. \tag{2.24}$$

Equation (2.22) gives E^L as the exact E^q times a correction factor, f, which depends on γ, where γ is proportional to $|E^q|^{1/2}$. Two things should be noted: By the definition (2.19),

$$\gamma \leqq 3/4, \tag{2.25}$$

when $z^j < z_c^L$ (all j). Second, the function f is monotone increasing in γ on $[0, 1]$.

Step 7. To utilize (2.19) and (2.22) we require a bound on $E^q(\underline{z})$. Let

$$E_L^q(\underline{z}) \leqq E^q(\underline{z}) < 0, \tag{2.26}$$

be any lower bound to E^q. Inserting $E_L^q(\underline{z})$ in (2.19) will give a lower bound to $z_c^L \leqq z_c$. Inserting $E_L^q(\underline{z})$ in (2.24) and then inserting this γ in (2.23) and (2.22) will (assuming that $\gamma \leqq 1$) give a lower bound to E^L. In cases A and B we can get an effective $E_L^q(\underline{z})$ which is independent of N and K. The former uses the Lieb-Thirring technique [7] together with a novel bound on the Coulomb energy. This is done in Sect. III. Case B is controlled by relating it to a relativistic problem solved in [2]; this is done in Sect. IV.

Remark. In case B we deal with only one electron. Given this restriction on N, (2.16) holds with a larger value of G, namely $G = 3.83$. This larger G can be used in Steps 5–7.

III. The Many-Electron Atom

Our first task is to prove the kinetic energy estimate (2.16). Consider the single-particle Schrödinger operator $h = (p - A)^2 - V(x)$, where $V(x) \geqq 0$ and consider also the N-particle operator $H = \sum_j h_j$. The q spin state fermionic ground state energy of H, E, satisfies $E \geqq q \sum_i e_i$, where the e_i are the negative eigenvalues of h. ($q = 2$ in our case.) We have that

$$\sum_i e_i \geqq -|e_1|^{1/2} \sum_i |e_i|^{1/2}, \tag{3.1}$$

where e_1 is the ground state energy. In [1, (3.19)] we quoted a result of [9] that

$$|e_1|^{1/2} \leqq L^1_{\frac{1}{2},3} \|V\|_2^2, \tag{3.2}$$

where $L^1_{1/2,3} = 0.0135$ to three significant figures.

In [9] it is also shown that

$$\sum_i |e_i|^{1/2} \leqq L_{\frac{1}{2},3} \|V\|_2^2. \tag{3.3}$$

Strictly speaking, (3.3) was shown only for $A=0$ in [9] and it is not known whether the (unknown) *sharp* constant L in (3.3) occurs for $A=0$. However, as pointed out in [8, 11], the L actually obtained in [9] holds for all A. The L obtained by using the method of [12] also holds for all A (see [11] for a discussion of the Ito-Nelson integral). The latter method gives a better value for L and the numerical computation is most clearly explained in [10, Eqs. (46)–(51)]. In the notation of [10], we take $a=0.61$ exactly and $b=3.6807$. Then (3.3) holds with

$$L_{\frac{1}{2},3} = b(4\pi)^{-3/2}\Gamma(\tfrac{3}{2})\tfrac{1}{2}a^{-1} = 0.060021, \tag{3.4}$$

to 5 figures. Thus,

$$\sum_i |e_i| \leqq L^1_{\frac{1}{2},3} L_{\frac{1}{2},3} \|V\|_2^4 \leqq (0.000810)\, \|V\|_2^4. \tag{3.5}$$

Now take $V(x) = c\varrho_\psi(x)$, where ϱ_ψ is given by (2.1). Then

$$T(\psi, A) - c\int \varrho_\psi^2 = (\psi, H\psi) \geqq -q\sum_i |e_i|. \tag{3.6}$$

Using (3.5) and (3.6), with $c^{-3} = 4qL^1_{\frac{1}{2},3}L_{\frac{1}{2},3}\int \varrho_\psi^2$, we obtain

$$T(\psi, A) \geqq \tfrac{3}{4}(4qL^1_{\frac{1}{2},3}L_{\frac{1}{2},3})^{-1/3}\{\int \varrho_\psi^2\}^{2/3} \geqq (4.02)\{\int \varrho_\psi^2\}^{2/3}, \tag{3.7}$$

for $q=2$. Thus, (2.16) holds [recalling (2.2)] with

$$G = 8.07/2\pi = 1.28. \tag{3.8}$$

Our second task is to find a lower bound for $E^q(z)$, given by (2.3). Again we use an inequality derived in [7, 9], but with a better constant derived in [10, Eq. (52)]:

$$T(\psi, A) \geqq (2.7709)\int \varrho_\psi(x)^{5/3}\, dx. \tag{3.9}$$

The second term in V, (1.5), is absent since there is only one nucleus, located at $R=0$. The third term contributes the following to W:

$$W_3(\psi, z) = z\int \varrho_\psi(x)\, |x|^{-1}\, dx. \tag{3.10}$$

The first term in V (call its contribution W_1) requires some elaboration. For $x, y \in \mathbb{R}^3$ and $R>0$,

$$|x-y|^{-1} \geqq \{|x|+|y|\}^{-1} \geqq \tfrac{1}{2}Rf(x)f(y), \tag{3.11}$$

$$\begin{aligned} f(x) &= 1/|x| \quad \text{if} \quad |x| \geqq R, \\ &= 0 \quad\quad\; \text{if} \quad |x| < R. \end{aligned} \tag{3.12}$$

Using (3.11) and the positivity of $|\psi|^2$ and $|x_i-x_j|^{-1}$, we have, for any $0 \leqq \sigma \leqq 1$,

$$W_1(\psi, z) \leqq -\tfrac{1}{4} R\sigma \left(\psi, \left\{\left[\sum_{i=1}^{N} f(x_i)\right]^2 - \sum_{i=1}^{N} f(x_i)^2\right\}\psi\right)$$
$$\leqq -\tfrac{1}{4} R\sigma \{[\textstyle\int \varrho_\psi f]^2 - \int \varrho_\psi f^2\}, \tag{3.13}$$

since $\langle(\sum f)^2\rangle \geqq \langle \sum f\rangle^2$.

Combining (3.9), (3.10), (3.13),

$$T(\psi, A) - W(\psi, z) \equiv \mathscr{E}^q(\psi, A) \geqq \mathscr{E}_{R,\sigma}(\varrho_\psi, z)$$
$$\equiv (2.7709) \textstyle\int \varrho_\psi^{5/3} - z \int \varrho_\psi |x|^{-1} + \tfrac{1}{4} R\sigma [\int \varrho_\psi f]^2$$
$$- \tfrac{1}{4} R\sigma \textstyle\int \varrho_\psi f^2 . \tag{3.14}$$

Therefore,

$$E^q(z) \geqq \sup_{0 \leqq \sigma \leqq 1} \sup_{R>0} \inf_{\varrho} \mathscr{E}_{R,\sigma}(\varrho, z) . \tag{3.15}$$

We could, of course, impose the extra condition $\int \varrho = N$ in (3.15) but, as we desire an N-independent bound for E^q, we forego this.

First minimize (3.14) with respect to $\varrho(x)$ for $|x| \leqq R$. Only the first two terms are relevant in this region. Define $\Gamma = (5/3)(2.7709)$. Then $\Gamma \varrho^{2/3}(x) = z/|x|$. The first two terms contribute (for $|x| < R$)

$$-z^{5/2} \Gamma^{-3/2} (16\pi/5) R^{1/2} . \tag{3.16}$$

Next we consider the contributions for $|x| > R$. Here we merely omit the $\varrho^{5/3}$ term and we use $R|x|^{-2} \leqq |x|^{-1}$ in the last term. Let $Y \equiv \int_{|x|>R} \varrho(x)\,|x|^{-1}\,dx$. Then the sum of the last three terms is not less than the minimum (with respect to Y) of $-(z+\tfrac{1}{4}\sigma)Y + \tfrac{1}{4}R\sigma Y^2$. This minimum is

$$-(z+\tfrac{1}{4}\sigma)^2/R\sigma . \tag{3.17}$$

The maximum of this with respect to $\sigma \in [0, 1]$ is

$$-M(z)/R , \tag{3.18}$$

$$M(z) = z \quad \text{if} \quad z \leqq 1/4 ,$$
$$= (z+\tfrac{1}{4})^2 \quad \text{if} \quad z \geqq 1/4 . \tag{3.19}$$

Adding (3.16) and (3.18) and then maximizing with respect to $R > 0$ gives

$$E^q(z) \geqq -3z^{5/3} \Gamma^{-1} (8\pi/5)^{2/3} M(z)^{1/3}$$
$$= -(1.9062) z^{5/3} M(z)^{1/3} . \tag{3.20}$$

As we shall be primarily interested in $z > 1/4$, the little exercise with σ is academic; it was done merely to demonstrate a z^2 (instead of $z^{5/3}$) bound when $z \leqq 1/4$.

With these results we can now bound z_c, see (2.19) and E^L, see (2.22). Since z_c will be large, let us use the bound $z^{5/3} M(z)^{1/3} \leqq (z+\tfrac{1}{4})^{7/3}$ for all $z > 0$. Then, from (2.19)

$$z_c \geqq \tilde{z}_c \geqq z_c^L \geqq -\tfrac{1}{4} + (0.158)\alpha^{-12/7} \geqq 720 . \tag{3.21}$$

This bound (720) is about 25 times smaller than the z_c^L obtained in [1] for the one-electron atom. It is about 290 times less than the *upper* bound on z_c obtained in [13], see also [1, (3.24)]. This upper bound ($z_c \leqq 208{,}000$) also holds, of course, for the full problem with K nuclei and N electrons.

The lower bound (2.22) on the energy is

$$E^L = E^q(z)\, f(\gamma) \tag{3.22}$$

and, using (3.20),

$$\begin{aligned}\gamma &\leqq 6\alpha^2(1.9062)^{1/2}\,(z+\tfrac{1}{4})^{7/6}/G\\ &\leqq (6.47)\,\alpha^2(z+\tfrac{1}{4})^{7/6}\\ &\leqq (0.000345)\,(z+\tfrac{1}{4})^{7/6}.\end{aligned} \tag{3.23}$$

As an illustration, take $z=100$. By (2.23) the fractional change in the energy, $f(\gamma)-1$, is less than 0.013, which is about $1\frac{1}{2}\%$.

IV. The One-Electron Molecule

Our first task is to find a lower bound to E^q in (2.3) with

$$V(x,\underline{R},\underline{z}) = -\sum_{j=1}^{K} z^j|x-R_j|^{-1} + \sum_{i<j} z^i z^j |R_i - R_j|^{-1}. \tag{4.1}$$

Since $N=1$, we can use the diamagnetic inequality (see [1]): $T(\psi, A) \geqq T(|\psi|, 0) \equiv T(\psi) = \|\nabla|\psi|\,\|_2^2$, and hence can assume that ψ is real and positive and $A=0$. Define

$$\bar{V}(x,\underline{R}) = -(2/\pi)\sum_{j=1}^{K} |x-R_j|^{-1} + (12/\pi)\sum_{i<j} |R_i - R_j|^{-1}. \tag{4.2}$$

It is proved in [2, Proposition 2.2] that for all $\psi \in L^2$, $(-\Delta)^{1/4}\psi \in L^2$ and all $\underline{R}$,

$$(\psi, (-\Delta)^{1/2}\psi) \geqq -(\psi, \bar{V}\psi). \tag{4.3}$$

We also have the fact (Schwarz inequality) that

$$\|\nabla\psi\|_2^2 \geqq (\psi, (-\Delta)^{1/2}\psi)^2, \tag{4.4}$$

when $\|\psi\|_2 = 1$.

Given $\underline{z}$, define

$$Z = \max(z^1, \ldots, z^K) \quad \text{and} \quad \underline{Z} = (Z, \ldots, Z). \tag{4.5}$$

As shown in Step 2,

$$E^q(\underline{z}) \geqq E^q(\underline{Z}). \tag{4.6}$$

Suppose that $Z \geqq 6$. Then

$$(\pi Z/2)\,\bar{V}(x,\underline{R}) \leqq V(x,\underline{R},\underline{Z}). \tag{4.7}$$

Combining (4.3), (4.4), (4.7) and with $t=(\psi,(-\Delta)^{1/2}\psi)$

$$E^q(\underset{\sim}{z})\geqq \inf_t\{t^2-(\pi Z/2)t\}=-(\pi Z/4)^2\,. \tag{4.8}$$

[Note: When $K=1$, the exact result is $-(Z/2)^2$.]

By monotonicity (4.6), when $Z<6$

$$E^q(Z)\geqq E^q(6)\geqq -(3\pi/2)^2\,. \tag{4.9}$$

Combining (4.6), (4.8), (4.9) we obtain for *all* $\underset{\sim}{z}$

$$|E^q(\underset{\sim}{z})|^{1/2}\leqq(\pi/4)\max\{6,z^1,\ldots,z^K\}\,. \tag{4.10}$$

Turning now to (2.19) and using (4.10) we have that

$$z_c^L\geqq \sup\left\{z\,\middle|\,\frac{\pi}{4}\max(6,z)\leqq G/8\alpha^2\right\}. \tag{4.11}$$

As remarked at the end of Sect. II, since $N=1$ we are entitled to replace $L_{\frac{1}{2},3}$ by $L^1_{\frac{1}{2},3}$ in (3.7), (3.8), and (2.16). Thus,

$$G=3.83\,, \tag{4.12}$$

in our case.

Suppose that

$$\alpha^2\leqq\alpha_c^2\equiv G/(12\pi)=0.102\,. \tag{4.13}$$

Then, from (4.11)

$$z_c^L\geqq G/(2\pi\alpha^2)=0.609\alpha^{-2}>11{,}400\,. \tag{4.14}$$

(This number, 11,400, compares favorably with 17,900 obtained in [1] for $K=1$.) In the opposite case [(4.13) is violated], the set of z's in (4.13) is empty and our method gives no bound at all on $E(\underset{\sim}{z})$ for $\underset{\sim}{z}\neq 0$. Thus, our method requires *two* conditions for stability

(i) $$\alpha^2 z^j\leqq 0.609 \quad \text{for all } j\,, \tag{4.15}$$

(ii) $$\alpha\leqq\alpha_c=(0.102)^{1/2}=0.319\,. \tag{4.16}$$

One can question whether the condition (4.16) on α is an artifact of our method or whether there really is an α_c (which will, of course, be greater than 0.319 – but finite). The second alternative is correct as we now prove.

Lemma. *Suppose that*

$$\alpha>6.67\,, \tag{4.17}$$

then for every $\underset{\sim}{z}=(z,\ldots,z)$ *with* $z>0$ *there is a* K *such that* $E(\underset{\sim}{z})=-\infty$.

Remark. The right side of (4.17) is not the best bound that can be obtained by the following method.

Proof. In [1] we showed that $E=-\infty$ when $K=1$ if

$$z\alpha^2>\inf\{\textstyle\int B^2\}\{8\pi(\psi,|x|^{-1}\psi)\}^{-1}\equiv P\,, \tag{4.18}$$

where (ψ, A) runs over $\mathscr{F}=\{(\psi, A)\in\mathscr{C}\,|\,\sigma\cdot(p-A)\psi=0\}$. $\mathscr{F}$ is not empty [13]. By taking a particular example, one finds $P\leqq 9\pi^2/8=11.10$. Therefore, if $\alpha^2>P$, we can take $K=1$ and achieve instability for all $z\geqq 1$. Using the above bound, this is also achieved for $z\geqq 1$ if $\alpha>3.34$.

Next, to investigate $z<1$, take any $(\psi, A)\in\mathscr{F}$, whence

$$\mathscr{E}(\psi, A, \underline{R}, \underline{z})=\varepsilon\int B^2+\int \varrho_\psi(x)\, V(x, \underline{R}, \underline{z})\, dx\,, \tag{4.19}$$

with $\varrho_\psi(x)=\langle\psi,\psi\rangle(x)$. We want to show that for suitable α and K, $\mathscr{E}$ is negative for some $\underline{R}$. [If it is negative then, by the scaling (2.4), $\mathscr{E}$ can be made arbitrarily negative.] To show this, it suffices to average $\mathscr{E}$ with some probability density $F(R^1, \ldots, R^K)$, $\int F d^K R=1$, and to show that $\langle\mathscr{E}\rangle\equiv\int \mathscr{E} F d^K R<0$. Take $F=\varrho_\psi(R^1)\ldots\varrho_\psi(R^K)$. The result is

$$\langle\mathscr{E}\rangle=\varepsilon\int B^2-\tfrac{1}{2}zK[2-z(K-1)]\, I(\varrho_\psi)\,, \tag{4.20}$$

$$I(\varrho)=\int\int \varrho(x)\,\varrho(y)\,|x-y|^{-1}\, dx\, dy\,. \tag{4.21}$$

Choose K to be the smallest integer closest to $\frac{1}{2}+1/z$. Then $zK=(z/2)+1+\mu$ with $|\mu|\leqq\frac{1}{2}z$ and $zK[2-z(K-1)]=[1+(z/2)]^2-\mu^2\geqq 1+z>1$. Therefore, if

$$\alpha^2>(4\pi)^{-1}\inf_{\mathscr{F}}\int B^2/I(\varrho)\,, \tag{4.22}$$

instability occurs for all $0<z<1$.

For the particular example in [13] quoted above, one has

$$|B(x)|=12(1+|x|^2)^{-2}, \varrho(x)=[\pi(1+|x|^2)]^{-2}\,,$$

and one computes

$$\int B^2=18\pi^2,\ I(\varrho)=1/\pi\,. \tag{4.23}$$

Therefore, if $\alpha>3\cdot 2^{-1/2}\pi=6.67$, instability also occurs for all $z<1$. □

Acknowledgements. It is a pleasure to thank J. Fröhlich and H.-T. Yau for helpful discussions.

References

1. Fröhlich, J., Lieb, E.H., Loss, M.: Stability of Coulomb systems with magnetic fields. I. The one-electron atom. Commun. Math. Phys. **104**, 251–270 (1986)
2. Daubechies, I., Lieb, E.H.: One electron relativistic molecules with Coulomb interaction. Commun. Math. Phys. **90**, 497–510 (1983)
3. Conlon, J.: The ground state energy of a classical gas. Commun. Math. Phys. **94**, 439–458 (1984)
4. Fefferman, C., de la Llave, R.: Relativistic stability of matter I. Revista Iberoamericana (to appear)
5. Ni, G., Wang, Y.: Vacuum instability and the critical value of the coupling parameter in scalar QED. Phys. Rev. D **27**, 969–975 (1983)
6. Finger, J., Horn, D., Mandula, J.E.: Quark condensation in quantum chromodynamics. Phys. Rev. D **20**, 3253–3272 (1979)
7. Lieb, E.H., Thirring, W.: Bound for the Kinetic energy of fermions which proves the stability of matter. Phys. Rev. Lett. **35**, 687 (1975). Errata **35**, 1116 (1975)
8. Avron, J., Herbst, I., Simon, B.: Schrödinger operators with magnetic fields. I. General interactions. Duke Math. J. **45**, 847–883 (1978)

9. Lieb, E.H., Thirring, W.: Inequalities for the moments of the eigenvalues of the Schrödinger Hamiltonian and their relation to Sobolev inequalities. In: Studies in Mathematical Physics, Essays in honor of Valentine Bargmann, Lieb, E.H., Simon, B., Wightman, A.S. (eds.). Princeton, NJ: Princeton University Press 1976
10. Lieb, E.H.: On characteristic exponents in turbulence. Commun. Math. Phys. **92**, 473–480 (1984)
11. Simon, B.: Functional integration and quantum physics. New York: Academic Press 1979
12. Lieb, E.H.: The number of bound states of one-body Schrödinger operators and the Weyl problem. Proc. Am. Math. Soc. Symp. Pure Math. **36**, 241–252 (1980)
13. Loss, M., Yau, H.-T.: Stability of Coulomb systems with magnetic fields. III. Zero energy bound states of the Pauli operator. Commun. Math. Phys. **104**, 283–290 (1986)

Communicated by A. Jaffe

Received October 10, 1985

PHYSICAL REVIEW
LETTERS

VOLUME 75 | 7 AUGUST 1995 | NUMBER 6

Stability of Matter in Magnetic Fields

Elliott H. Lieb,[1,2] Michael Loss,[3] and Jan Philip Solovej[2]
[1]*Department of Physics, Jadwin Hall, Princeton University, P.O. Box 708, Princeton, New Jersey 08544*
[2]*Department of Mathematics, Fine Hall, Princeton University, Princeton, New Jersey 08544*
[3]*School of Mathematics, Georgia Institute of Technology, Atlanta, Georgia 30332*
(Received 12 April 1995)

In the presence of arbitrarily large magnetic fields, matter composed of electrons and nuclei was known to be unstable if α or Z is too large. Here we prove that matter *is stable* if $\alpha < 0.06$ and $Z\alpha^2 < 0.04$.

PACS numbers: 03.65.–w, 11.10.–z, 12.20.Ds, 31.10.+z

One of the remaining unsolved problems connected with the stability of matter is the inclusion of arbitrary magnetic fields. The model is a caricature of QED, which invites speculations about stability of QED for large fine structure constant α, but that is not our focus here and we refer to [1] for a discussion of these and related matters. The Hamiltonian for N electrons and K fixed nuclei of charge Ze with magnetic field $\boldsymbol{B}(x) = \nabla \times \boldsymbol{A}(x)$, including the field energy, $\varepsilon \int B^2$, is

$$H = \sum_{i=1}^{N} \mathcal{T}_i + V_C + \varepsilon \int B(x)^2 \, d^3x\,, \qquad (1)$$

where $\mathcal{T} \equiv [\boldsymbol{\sigma} \cdot (\boldsymbol{p} + \boldsymbol{A})]^2 = (\boldsymbol{p} + \boldsymbol{A})^2 + \boldsymbol{\sigma} \cdot \boldsymbol{B}$ is the Pauli operator. The Coulomb energy is

$$V_C = -Z \sum_{i=1}^{N} \sum_{j=1}^{K} |x_i - R_j|^{-1} + \sum_{1 \le i < j \le N} |x_i - x_j|^{-1} + Z^2 \sum_{1 \le i < j \le K} |R_i - R_j|^{-1}, \qquad (2)$$

with R_j being the coordinates of the nuclei and x_i the electron coordinates. The energy unit is 4 Ry $= 2mc^2\alpha^2$, $\alpha = e^2/\hbar c$, length unit $= \hbar^2/2me^2$, and $\varepsilon = (8\pi\alpha^2)^{-1}$. Notice that α appears in (1) only through ε.

The negative particles are necessarily spin 1/2 fermions which, for mathematical generality, we assume to exist in $q/2$ flavors (e.g., $q = 6$, corresponding to three leptons). The ground state energy is denoted by E.

Starting with the 1967 pioneering work of Dyson and Lenard we now understand stability for arbitrarily many electrons and nuclei, with $\boldsymbol{B} = 0$, in the context of the nonrelativistic Schrödinger equation. Later it was extended to the "relativistic" Schrödinger equation in which $\boldsymbol{p}^2/2m$ is replaced by $(c^2\boldsymbol{p}^2 + m^2c^4)^{1/2}$ (see [2] for a review). These proofs also hold with the inclusion of a magnetic field coupled to the *orbital motion* of the electrons, i.e., $\boldsymbol{p} \to \boldsymbol{p} + \boldsymbol{A}$, but no Zeeman $\boldsymbol{\sigma} \cdot \boldsymbol{B}$ term.

Stability of matter has two meanings: (i) E is finite for arbitrary N and K; (ii) $E \ge -C_1(N + K)$ for some constant C_1 independent of N, K, and R_j. (ii) obviously implies (i), and it holds in the nonrelativistic case. In the relativistic case, (i) actually implies (ii) (see [3]) but (i) requires two conditions: $Z\alpha \le C_2$ and $\alpha \le C_3$ with C_2 and C_3 being universal constants, the best available values being in [3], Theorems 1 and 2. The inclusion of $\boldsymbol{B}$ changes E, but the point is that while C_1, C_2, C_3 depend on q, they can be chosen to be independent of $\boldsymbol{B}$.

The situation changes dramatically when the magnetic moments of the electrons are allowed to interact with the magnetic field via the $\boldsymbol{\sigma} \cdot \boldsymbol{B}$ term, as in (1). The reason for this is simple: The Pauli operator $\mathcal{T}$ is non-negative, but it is much weaker than $(\boldsymbol{p} + \boldsymbol{A})^2$. Indeed, it can even have square integrable zero modes [4], $\mathcal{T}\psi = 0$, for suitable $A(x)$, which cause instability for large $Z\alpha^2$.

It is known [5] that without the field energy term $\varepsilon \int B^2$ in (1) arbitrarily large $\boldsymbol{B}$ fields can cause arbitrarily negative energies E even for hydrogen. The field energy, hopefully, stabilizes the situation, and our goal is to show that E is finite for (1), *even after minimizing over all possible* $\boldsymbol{B}$ *fields* and all possible R_j.

One of our results on magnetic stability is as follows.

Theorem 1: The ground state energy of H satisfies

$$E \geq -2.6q^{2/3} \max\{Q(Z)^2, Q(5.7q)^2\} N^{1/3} K^{2/3}, \quad (3)$$

with $Q(t) \equiv t + \sqrt{2t} + 2.2$, *provided that*

$$qZ\alpha^2 \leq 0.082 \quad \textit{and} \quad q\alpha \leq 0.12. \quad (4)$$

In (2) all the nuclear charges are set equal to Z. As far as stability is concerned this is no restriction [6], since the energy is concave in each charge Z_j and hence stability holds in the "cube" $\{0 \leq Z_j \leq Z\}_{j=1}^{K}$ if it holds when all $Z_j = Z$. It also follows from this that E is a decreasing function of Z. Moreover, since $\varepsilon \propto \alpha^{-2}$, it follows that E is a decreasing function of α, a fact that will be important later. The form of (3) is the best possible for $Z \geq 1$, as we know from other studies [2].

Our actual condition for stability given after (18) is rather complicated, but very much more general than (4)—which is only representative. The results after (18) show, e.g., that when $\alpha = 1/137$ and $q = 2$, Z can be as large as 1050. The large values of Z and α are important because the comfortable distance of the critical values from the physical values $Z \leq 92$, $\alpha = 1/137$ implies that the effect studied here is merely a small perturbation of the usual $B = 0$ case.

Our proof of Theorem 1 will require a new technique—a running energy-scale renormalization of $\mathcal{T}$. A byproduct of this is a Lieb-Thirring type inequality for $\mathcal{T}$:

Theorem 2: If $\varepsilon_1 \leq \varepsilon_2, \ldots < 0$ *are the negative eigenvalues of* $\mathcal{T} - U$, *for a potential* $-U(x) \leq 0$ *then*

$$\sum |\varepsilon_i| \leq a_\gamma \int U(x)^{5/2} d^3x + b_\gamma \Big(\int B(x)^2 d^3x\Big)^{3/4} \Big(\int U(x)^4 d^3x\Big)^{1/4} \quad (5)$$

for all $0 < \gamma < 1$, *where* $a_\gamma = (2^{3/2}/5)(1-\gamma)^{-1} L_3$ *and* $b_\gamma = 3^{1/4} 2^{-9/4} \pi \gamma^{-3/8} (1-\gamma)^{-5/8} L_3$. *We can take* L_3, *defined below, to be 0.1156.*

More generally, the second term in (5) can be replaced by $(\int B^{3q/2})^{1/q} (\int U^p)^{1/p}$, where $p^{-1} + q^{-1} = 1$.

The investigation of this problem started in [1,7] where type (ii) stability was proved (for suitable Z, α and $q = 2$) for $K = 1$ and arbitrary N [if $(Z + 1/4)\alpha^{12/7} \leq 0.15$] or $N = 1$ and arbitrary K (if $Z\alpha^2 \leq 0.6$ and $\alpha \leq 0.3$). The problem for general N and K was open for nine years, and we present a surprisingly simple solution here.

The bounds in (4) on $Z\alpha^2$ and α are not artifacts. It is shown in [1] and [4] that the zero modes cause $E = -\infty$ when $Z\alpha^2 > 11.11$ for the "hydrogenic" atom, i.e., a single spin 1/2 particle and one nucleus. If the number of nuclei is arbitrary, it is shown in [7] that there is collapse if $\alpha > 6.67$, no matter how small Z is. *Magnetic stability, like relativistic stability, implies a (Z-independent) bound on* α.

Prior to our work a proof of type (ii) stability for (1) with $Z = 1$, $q = 2$, and some sufficiently small α was announced (unpublished) by C. Fefferman and sketched to one of us. Our proof is unrelated to his, considerably simpler and, more importantly, gives physically realistic constants.

We begin our analysis with the observation that length scaling considerations suggest that the key to understanding the stability problem is somehow to replace $\mathcal{T}$, on each energy scale e, by $\mu\mathcal{T}/e$, where μ is a fixed energy but e is variable. On energy scales $e > \mu$ we can use the fact that $\mathcal{T} > 0$ to replace $\mathcal{T}$ by $\mu\mathcal{T}/e$ without spoiling lower bounds. It might seem odd to replace $\mathcal{T}$ by something smaller, but what is really happening is that $\boldsymbol{\sigma} \cdot \boldsymbol{B}$ is being partially controlled by $[1 - \mu e^{-1}](\boldsymbol{p} + \boldsymbol{A})^2$. The idea of replacing $\mathcal{T}$ by a fraction of $\mathcal{T}$ was also used in [1], but no energy dependence was used there.

We shall illustrate this concept by three calculations. The first, (A), will establish magnetic stability by relating it to the stability of relativistic matter (see [3,6,8,9]). The second, (B), will be the proof of Theorem 2. The third, (C), will use essential parts of the second calculation and an electrostatic inequality proved in [3] to prove magnetic stability without resorting to relativistic stability.

(A) Magnetic stability from relativistic stability.—We use stability of relativistic matter in the form proved in [3]. From the corollary of Theorem 1 in [3] with $\beta = 0.5$ we have, for any $0 < q\kappa \leq 0.032$ and $Z\kappa \leq 1/\pi$,

$$\sum_{i=1}^{N} |\boldsymbol{p}_i + \boldsymbol{A}_i| + \kappa V_C \geq 0. \quad (6)$$

(Although Theorem 1 in [3] was stated only for $|\boldsymbol{p}|$, it holds for $|\boldsymbol{p} + \boldsymbol{A}|$ because it relies only on the magnitude of the resolvent, which only gets smaller when $\boldsymbol{A}$ is not zero. That is, $||\boldsymbol{p} + \boldsymbol{A}|^{-s}(x,y)| \leq ||\boldsymbol{p}|^{-s}(x,y)|$ for each $s > 0$ and x, y in $\boldsymbol{R}^3$. This follows at once from a similar bound on the heat kernel $\{\exp[-t(\boldsymbol{p} + \boldsymbol{A})^2]\}(x,y)$ which, in turn, follows from its representation as a path integral. This was pointed out in [5,10]. Only the resolvent powers $|\boldsymbol{p} + \boldsymbol{A}|^{-s}$ enter the proof of Theorem 1 in [3].)

Using (6), H is bounded below by $\overline{H} = \sum_{i=1}^{N} h_i + \varepsilon \int B^2$, where h is the one-body operator $h = \mathcal{T} - \kappa^{-1}|\boldsymbol{p} + \boldsymbol{A}|$. Thus E is bounded below by $\varepsilon \int B^2 + \overline{E}_N$, where $\overline{E}_N = q \sum_{j=1}^{[N/q]} \varepsilon_j$ and $\varepsilon_1 \leq \varepsilon_2 \leq \ldots$ are the eigenvalues of h. For $e > 0$, let $N_{-e}(h)$ be the number of eigenvalues of h less than or equal to $-e$. Choose $\mu > 0$ and note that

$$\overline{E}_N \geq -N\mu - q\int_{\mu}^{\infty} N_{-e}(h)\, de. \quad (7)$$

The crucial step in our proof is noting that the positivity of the operator $\mathcal{T}$ implies that $\mathcal{T} \geq \mu\mathcal{T}/e$ when $e \geq \mu$. Thus $\mathcal{T} \geq \mu e^{-1}\mathcal{T} \geq \mu e^{-1}(\boldsymbol{p} + \boldsymbol{A})^2 - \mu e^{-1}B(x)$

when $e \geq \mu$. By Schwarz's inequality, $\kappa^{-1}|p + A| \leq (1/3)e^{-1}\kappa^{-2}(p + A)^2 + 3e/4$ and hence if we set $\mu = (4/3)\kappa^{-2}$ we obtain

$$h \geq e^{-1}\kappa^{-2}(p + A)^2 - \tfrac{4}{3}e^{-1}\kappa^{-2}B(x) - 3e/4 \equiv h_e .$$

Thus $N_{-e}(h) \leq N_{-e}(h_e)$, and this can be estimated by the Cwikel-Lieb-Rozenblum (CLR) bound [11], i.e., $N_{-e}[(p + A)^2 - U(x)] \leq L_3 \int [U(x) - e]_+^{3/2} d^3x$, where $[a]_+ \equiv \max(a, 0)$ and $L_3 = 0.1156$. In our case

$$N_{-e}(h_e) \leq L_3 \int \left[\frac{4B(x)}{3} - \frac{e^2\kappa^2}{4} \right]_+^{3/2} d^3x . \quad (8)$$

Inserting this bound in (7), a simple calculation yields

$$\overline{E}_N \geq -N\mu - (2\pi/3)q\kappa^{-1}L_3 \int B(x)^2 d^3x .$$

We choose κ so that the field energy terms are nonnegative, i.e., $\kappa \geq (16\pi^2/3)L_3\alpha^2 q = 6.1\alpha^2 q$. We conclude, by (6), that magnetic stability holds if

$$q\alpha \leq 0.071 \quad \text{and} \quad qZ\alpha^2 \leq 0.052 . \quad (9)$$

For $q = 2$, the first condition is $\alpha \leq 1/28$. For $q = 2$ and $\alpha = 1/137$, stability occurs if $Z \leq 490$.

Assuming (9) holds, we then use (6) and choose $\kappa = \min\{0.0315q^{-1}, (\pi Z)^{-1}\}$. Our lower bound on the ground state energy per electron, by this method, is then $-\mu = -(4/3)\kappa^{-2} = -\max\{1345q^2, 13.2Z^2\}$.

Remark: We used the CLR bound in (8). Since the derivation of this bound is not elementary, the reader might wish to use an easier to derive bound—at the cost of worsening the final constants. A useful substitute is

$$N_{-e} \leq 0.1054e^{-1/4} \int [U(x) - e/2]_+^{7/4} d^3x$$

(plus an increased μ), which is in (2.8) of [12] and which can be derived by means originally employed for the Lieb-Thirring inequality. This same remark also applies to our other calculations below.

(B) The Lieb-Thirring inequality.—As before we note that $\sum \varepsilon_i = -\int_0^\infty N_{-e}(\mathcal{T} - U)\,de$. We write $\int_0^\infty = \int_0^\mu + \int_\mu^\infty$. The parameter μ will be optimized below. Noting that $\mathcal{T} \geq (p + A)^2 - B(x)$ and applying the CLR bound in the same fashion as before to $\int_0^\mu$ yields

$$L_3 \int_0^\mu \int [B(x) + U(x) - e]_+^{3/2} d^3x\, de . \quad (10)$$

In $\int_\mu^\infty$ we replace $\mathcal{T}$ by the lower bound $\mu e^{-1}[(p + A)^2 - B(x)]$ and obtain $N_{-e}(\mathcal{T} - U) \leq N_{-e}(\mu e^{-1}[(p + A)^2 - B] - U)$. A further application of the CLR inequality yields the bound on $\int_\mu^\infty$

$$L_e \int_\mu^\infty \int [B(x) + (e/\mu)U(x) - e^2/\mu]_+^{3/2} dx\, de . \quad (11)$$

It is easy to see that for any $0 < \gamma < 1$ the integrand in (10) is bounded above by

$$\sqrt{2}\left([B(x) - \gamma e^2/\mu]_+^{3/2} + [U(x) - (1 - \gamma)e]_+^{3/2}\right).$$

Treating the integrand in (11) in a similar fashion and combining the inequalities we find

$$\sum |\varepsilon_i| \leq \sqrt{2}L_3 \int \left\{ \int_0^\infty [B(x) - \gamma e^2/\mu]_+^{3/2} de + \int_0^\mu [U(x) - (1 - \gamma)e]_+^{3/2} de + \int_\mu^\infty [(e/\mu)U(x) - (1 - \gamma)e^2/\mu]_+^{3/2} de \right\} d^3x .$$

After extending the last two integrals to $\int_0^\infty$, a straightforward computation yields

$$\sum |\varepsilon_i| \leq \sqrt{2}L_3 \int \left\{ \frac{2}{5(1 - \gamma)} U(x)^{5/2} + \frac{3\pi\mu^{1/2}}{16\gamma^{1/2}} B(x)^2 + \frac{3\pi}{128}\mu^{-3/2}(1 - \gamma)^{-5/2}U(x)^4 \right\} d^3x .$$

Optimizing over μ yields (5).

To prove the more general form of (5), replace μe^{-1} by $(\mu e^{-1})^s$, where $s = 2p/3 - 5/3$.

(C) Proof of Theorem 1.—We turn now to our third illustration of the concept of running energy scale and prove the stability directly, not relating it to the relativistic problem. By this method we get the correct dependence of the ground state energy on Z and also somewhat better critical constants than in (9).

Following [3] we first replace the Coulomb potential by a single particle potential in (12) below. We break up R^3 into Voronoi cells defined by the nuclear locations, i.e., $\Gamma_j = \{x : |x - R_j| \leq |x - R_k| \text{ for all } k\}$ is the jth Voronoi cell. Each Γ_j contains a ball centered at R_j with radius $D_j = \min\{|R_j - R_k| : j \neq k\}/2$.

The following bound on V_C is proved in [3]: Choose some $0 < \lambda < 1$. Then

$$V_C \geq -\sum_{i=1}^N W(x_i) + \sum_{j=1}^K \frac{Z^2}{8D_j}, \quad (12)$$

where $W(x) = Z|x - R_j|^{-1} + F_j(x)$ for $x \in \Gamma_j$ with $F_j(x)$ defined by

$$(2D_j)^{-1}(1 - D_j^{-2}|x - R_j|^2)^{-1} \quad \text{for } |x - R_j| \leq \lambda D_j ,$$
$$(\sqrt{2Z} + 1/2)\,|x - R_j|^{-1} \quad \text{for } |x - R_j| > \lambda D_j .$$

The point about this inequality is that the potential W has the same singularity near each nucleus as V_C, and that the rightmost term in (12) is repulsive. This term will be responsible for stabilizing the system.

987

The problem is thus reduced to obtaining a lower bound on $q\sum'\varepsilon_j$, where $\sum'\varepsilon_j$ is the sum of the first $[N/q]$ negative eigenvalues of $\mathcal{T} - W$. Note that Theorem 2 cannot be applied directly to this problem, since W is neither integrable to the power 5/2 nor to the power 4. Instead we have to do the calculations directly.

For $\nu > 0$ (a number that is chosen later) set $W_\nu(x) \equiv [W(x) - \nu]_+$ and note that $W(x) - \nu \le W_\nu(x)$. Then, as in (7), $q\sum'\varepsilon_j \ge -N\nu - q\int_0^\infty N_{-e}(\mathcal{T} - W_\nu)\,de$. Again,

$$\int_0^\infty N_{-e}(\mathcal{T} - W_\nu)\,de \le \int_0^\mu N_{-e}(\mathcal{T} - W_\nu)\,de + \int_\mu^\infty N_{-e}(\mu e^{-1}\mathcal{T} - W)\,de\,, \quad (13)$$

where we have replaced $W_\nu(x)$ by $W(x)$ in the second term. Applying the CLR bound to the first expression on the right side we obtain $L_3\int\int_0^\mu [B(x) + W_\nu(x) - e]_+^{3/2}\,de\,d^3x$, which can be bounded, as in part (B), by

$$\sqrt{2}L_3\int\left\{\int_0^\mu\left[B(x) - \frac{\gamma e^2}{\mu}\right]_+^{3/2} de + \frac{2}{5}(1-\gamma)^{-1}W_\nu(x)^{5/2}\right\}d^3x\,, \quad (14)$$

for any $0 < \gamma < 1$.

The difficulty in dominating the second term in (13) comes from the Coulomb singularity of $W(x)$, which is not fourth power integrable. The singularity can be controlled using the following operator inequality, which follows from the diamagnetic inequality $\int|(p + A)\psi|^2\,d^3x \ge \int|p|\psi||^2\,d^3x$ and Lemma 2a on p. 708 of [13].

$$(p + A)^2 - \frac{Z}{|x|} \ge -\begin{cases} Z^2/4 + \frac{3}{2}ZR^{-1}, & \text{if } |x| \le R\,,\\ Z|x|^{-1}, & \text{if } |x| \ge R\,.\end{cases}$$

Choose $R = \lambda D_j$ and write $(p + A)^2 = \beta(p + A)^2 + (1-\beta)(p + A)^2$ for some $0 < \beta < 1$. Then, by scaling,

$$(\mu/e)\mathcal{T} - W \ge (\mu/e)(1-\beta)(p + A)^2 - (\mu/e)B - \widetilde{W}\,,$$

where $\widetilde{W}(x,e) = \tilde{G}_j(x,e) + F_j(x)$ for $x \in \Gamma_j$ with $\tilde{G}_j(x,e)$ defined by

$$\begin{aligned} &Z^2e/4\beta\mu + 3Z/2\lambda D_j \quad \text{for } |x - R_j| \le \lambda D_j\,,\\ &Z|x - R_j|^{-1} \quad \text{for } |x - R_j| > \lambda D_j\,.\end{aligned}$$

Note that $\widetilde{W}$ depends on e.

Again, as in part (B), we can use the CLR bound on the second term in (13) to obtain (when $1 - \gamma \ge Z^2/4\beta\mu$)

$$\sqrt{2}L_3(1-\beta)^{-3/2}\int\left\{\int_\mu^\infty [B(x) - \gamma e^2/\mu]_+^{3/2}\,de + \mu^{-3/2}\int_0^\infty [e\widetilde{W}(x,e) - (1-\gamma)e^2]_+^{3/2}\,de\right\}d^3x\,. \quad (15)$$

First we compute the last integral in (15), which is

$$\sum_{j=1}^K\int_{\Gamma_j}\int_0^\infty [e\tilde{G}_j(x,e) + eF_j(x) - (1-\gamma)e^2]_+^{3/2}\,de\,d^3x\,.$$

Now split the Γ_j integral into an inner integral $|x - R_j| \le \lambda D_j$ and an outer integral $|x - R_j| > \lambda D_j$. The inner integral yields, using the definitions of $\tilde{G}_j$ and F_j,

$$\frac{3\pi^2}{32}\left(1 - \gamma - \frac{Z^2}{4\beta\mu}\right)^{-5/2}\int_0^\lambda\left[\frac{1}{2(1-r^2)} + \frac{3Z}{2\lambda}\right]^4 r^2\,dr\,D_j^{-1}. \quad (16)$$

To bound the outer integral from above we replace Γ_j by $\boldsymbol{R}^3$ and get

$$(3\pi^2/32)(1-\gamma)^{-5/2}(\sqrt{Z} + \sqrt{1/2})^8(\lambda D_j)^{-1}. \quad (17)$$

Combining (14)–(17) we find that the sum of the negative eigenvalues of $\mathcal{T} - W_\nu$ is bounded below by

$$-a\int W_\nu(x)^{5/2}\,d^3x - b\int B(x)^2\,d^3x - c\sum_{j=1}^{Y} D_j^{-1}. \quad (18)$$

Here $a = q(2\sqrt{2}/5)L_3(1-\gamma)^{-1}$,

$$b = q\frac{3\pi\sqrt{2}}{16}L_3(1-\beta)^{-3/2}(\mu/\gamma)^{1/2},$$

$$c = q\frac{3\pi^2\sqrt{2}}{32}L_3(1-\beta)^{-3/2}\mu^{-3/2} \times\left\{\frac{(\sqrt{Z} + \sqrt{q/4})^8}{\lambda(1-\gamma)^{5/2}} + \left(1 - \gamma - \frac{Z^2}{4\beta\mu}\right)^{-5/2} \times\int_0^\lambda\left[\frac{q}{4(1-r^2)} + \frac{3Z}{2\lambda}\right]^4 r^2\,dr\right\}.$$

To simplify the stability condition we have artificially increased the bounds by recalling that $q \ge 2$ and twice replacing 1/2 by $q/4$ in the definition of c. We choose $\beta = 1/8$, $\gamma = 1/2$, $\lambda = 8/9$, and μ so that $b = (8\pi\alpha^2)^{-1}$. The *stability condition* $c \le Z^2/8$ [see (12)] now depends *only* on the two parameters $X = qZ\alpha^2$ and $Y = q\alpha$. A straightforward but lengthy calculation shows that the stability condition holds if $X = X_0 \equiv 0.082$ and $Y = Y_0 \equiv 0.12$. The condition is monotone in Y, so it holds for $X = X_0, Y \le Y_0$. Although our condition does not hold for *all* $X \le X_0, Y \le Y_0$, we can use the Z monotonicity of E to conclude stability in this range; this proves (4). With the same values of β, γ, and λ and with $q = 2$ the values $Z = 1050$, $\alpha = 1/137$ also give stability.

To derive (3), note that $W(x) \le Q|x - R_j|^{-1}$ for $x \in \Gamma_j$. Using this bound and replacing Γ_j by $\boldsymbol{R}^3$, one easily obtains $-\sqrt{2}\,\pi^2 L_3 q K Q^3 \nu^{-1/2} - N\nu$ as a lower bound on the $-a \int W_\nu^{5/2}$ term in (18). Optimizing over ν yields (3) when $X = X_0, Y \le Y_0$. In this case, $Z \ge Z_0 \equiv 5.7q$. If $X \le X_0$, $Y \le Y_0$ and $Z \ge Z_0$, we get a lower bound on E by increasing α until $X = X_0, Y \le Y_0$; this yields (3) with $Q = Q(Z)$. Otherwise, with $Z < Z_0$, we use the Z monotonicity of E to conclude (3) with $Q = Q(5.7q)$.

This work was partially supported by NSF Grants No. PHY90-19433-A04 (E. H. L.), No. DMS92-07703 (M. L.), and No. DMS92-03829 (J. P. S.).

[1] J. Fröhlich, E. Lieb, and M. Loss, Commun. Math. Phys. **104**, 251 (1986).
[2] E. Lieb, Bull. Am. Math. Soc. **22**, 1 (1990).
[3] E. Lieb and H.-T. Yau, Commun. Math. Phys. **118**, 177 (1988); Phys. Rev. Lett. **61**, 1695 (1988).
[4] M. Loss and H.-T. Yau, Commun. Math. Phys. **104**, 283 (1986).
[5] J. Avron, I. Herbst, and B. Simon, Duke Math. J. **45**, 847 (1978); Commun. Math. Phys. **79**, 529 (1981).
[6] I. Daubechies and E. Lieb, Commun. Math. Phys. **90**, 497 (1983).
[7] E. Lieb and M. Loss, Commun. Math. Phys. **104**, 271 (1986).
[8] J. Conlon, Commun. Math. Phys. **94**, 439 (1984).
[9] C. Fefferman and R. de la Llave, Rev. Math. Iberoamericana **2**, 119 (1986).
[10] J. Combes, R. Schrader, and R. Seiler, Ann. Phys. (N.Y.) **111**, 1 (1978).
[11] E. Lieb, Proc. Am. Math. Soc. Symposia Pure Math. **36**, 241 (1980).
[12] E. Lieb and W. Thirring, in *Studies in Mathematical Physics,* edited by E. H. Lieb, B. Simon, and A. Wightman (Princeton Univ. Press, Princeton, 1976), p. 269.
[13] A. Lenard and F. Dyson, J. Math. Phys. **9**, 698 (1968).

With H.-T. Yau in Commun. Math. Phys. *112*, 147–174 (1987)

The Chandrasekhar Theory of Stellar Collapse as the Limit of Quantum Mechanics

Elliott H. Lieb* and Horng-Tzer Yau**

Departments of Mathematics and Physics, Princeton University, P.O.B. 708, Princeton, NJ 08544, USA

Dedicated to Walter Thirring on his 60th birthday

Abstract. Starting with a "relativistic" Schrödinger Hamiltonian for neutral gravitating particles, we prove that as the particle number $N \to \infty$ and the gravitation constant $G \to 0$ we obtain the well known semiclassical theory for the ground state of stars. For fermions, the correct limit is to fix $GN^{2/3}$ and the Chandrasekhar formula is obtained. For bosons the correct limit is to fix GN and a Hartree type equation is obtained. In the fermion case we also prove that the semiclassical equation has a unique solution – a fact which had not been established previously.

Historical Remarks and Background

There are two principal elementary models of stellar collapse: neutron stars and white dwarfs. In the former there is only one kind of particle which, since it is electrically neutral, interacts only gravitationally. The typical neutron kinetic energy is high, however, so it must be treated relativistically. Unfortunatly, the mass and density are also large enough that general relativistic effects are important. For white dwarfs, on the other hand, there are two kinds of nonneutral particles: electrons and nuclei. Because the density is not too large, it is a reasonable approximation to ignore general relativistic effects (although these effects might be important for stability considerations [29]); the nuclei (because of their large mass) can be treated nonrelativistically but the electrons must be treated relativistically. The Coulomb interaction is usually accounted for by the simple assumption that local neutrality requires the nuclear charge density to be equal to the electron charge density, in which case the problem reduces to calculating the electron density. (There are, in fact, electrostatic exchange and correlation effects [28, 29], but these are small by a factor $\alpha = 1/137$.)

* Work partially supported by U.S. National Science Foundation grant PHY 85-15288-A01
** Work supported by Alfred Sloan Foundation dissertation Fellowship

Under the *assumption* of local neutrality (and no significant exchange and electron-nuclei correlation effects) and neglecting the nuclear kinetic energy, the white dwarf problem is mathematically the same as the neutron star problem – but without general relativistic effects. This problem was formulated by Chandrasekhar in 1931 [2] (and also in [7, 11]) and leads to an equation for the density which we here call the Chandrasekhar equation (1.16, 1.18). The neutron star problem leads to the much more complicated Tolman-Oppenheimer-Volkoff equation which will not concern us. Both are reviewed in [24, 27]. Both equations predict collapse at some critical mass which, in the white dwarf case, is called the Chandrasekhar mass. Clearly, near this mass the elementary theory is not totally adequate.

Quantum mechanics is essential for the stability in both cases. "The black-dwarf material is best likened to a single gigantic molecule in its lowest quantum state" [7]. In all treatments up to now, quantum mechanics enters only through the use of a local equation of state $P(\varrho)$, (P = pressure, ϱ = density) which is that of a degenerate Fermi gas (electrons or neutrons). See [30] for example.

Two years ago Lieb and Thirring [19] decided to investigate whether, starting from the Schrödinger equation for *fermions* one would, indeed, recover the semiclassical Chandrasekhar equation (1.16, 1.18) in the limit N(= particle number) $\to\infty$ and G(= gravitational constant) $\to 0$. More precisely, for fermions the relevant stability parameter should be $GN^{2/3}$, and not GN. Numerically, the critical N is about 10^{57}, so the limit $N\to\infty$ is a very reasonable one to consider. The Chandrasekhar value of the critical mass (with the correct 2/3 exponent) was proved in [19], but only up to a factor of 4. For *bosons*, on the other hand, which have not been considered for astrophysics, Ruffini and Bonazzola [30], Thirring [25], and Messer [21] realized that the relevant parameter should be GN, thus leading to collapse of objects only the size of a mountain. In [19] it was conjectured that, for bosons, (1.18) should be replaced by a Hartree type equation when $N\to\infty$. In a sense this would mean there is no semiclassical limit for bosons (although we shall continue to employ that word) because the Hartree energy involves density gradients, and not just an equation of state. In [19] the Hartree value of the collapse constant was proved to be correct up to a factor of 2.

In this paper we shall prove that the Chandrasekhar (respectively Hartree) equations are exactly correct as $N\to\infty$, $G\to 0$, for *all values* of $GN^{2/3}$ (respectively GN), not just the critical value. In view of Walter Thirring's contributions to, and interest in quantum mechanical stability questions – in particular the stellar collapse problem – it is a great pleasure for us to dedicate this work to him on the occasion of his 60th birthday.

At first it seemed to us that reducing the quantum problem to a semiclassical problem would end the story. But then we realized that a thorough mathematical study of (1.18), e.g. uniqueness of the solution, has not been done. This, it turn out, is in many ways more complicated than the quantum problem, and therefore a large part of this paper is devoted to an analysis of the semiclassical equations.

In Sect. I we state these problems precisely and summarize the main results. Section II contains proofs of the convergence of the quantum energies to the semiclassical energies. The analysis of the semiclassical equations (existence and uniqueness of solutions and qualitative properties) is in Sect. III and IV. The convergence of the quantum density (for fermions) to the semiclassical density is given in Sect. V.

I. Formulation of the Problem and Main Results

Our starting point is the "relativistic" Schrödinger Hamiltonian for N gravitating particles of mass m (in units $\hbar = c = 1$)

$$H_{\kappa N} = \sum_{i=1}^{N} \{(p_i^2 + m^2)^{1/2} - m\} - \kappa \sum_{1 \leq i < j \leq N} |x_i - x_j|^{-1} . \tag{1.1}$$

Here $p^2 = -\Delta$ and $x_i \in \mathbb{R}^3$. $H_{\kappa N}$ can describe a "neutron star" *without* general relativistic effects if we take m = neutron mass and $\kappa = Gm^2$. White dwarfs cannot be described by (1.1) (unless exchange and correlation effects are ignored). A more complicated Hamiltonian is needed in that case and we refer to [19, Sect. 4] for a discussion. Our methods can be extended to the case of several kinds of particles with different masses, but *without* electrostatic interaction. If electrostatic interactions are present, as in white dwarfs, genuinely new ideas are needed. However, if the positive nuclei are *also* fermions, then one can use the inequalities in [19, Sect. 4] to give a lower bound to the energy; unfortunately, this bound will not be the sharp one. It is believed that the semiclassical equation for white dwarfs is nearly the same as for (1.1) as $N \to \infty$, $G \to 0$ provided we take $\kappa = G(m + M/z)^2$ with M = nuclear mass, m = electron mass and z = nuclear charge.

For fermions (e.g. neutrons or electrons) $H_{\kappa N}$ acts on antisymmetric functions of space and spin. For generality we assume q spin states/particle; $q = 2$ in nature, but $q = 1$ would correspond to spin-polarized matter. We also consider $H_{\kappa N}$ without any symmetry restriction. Since the absolute ground state is always symmetric, this is the same as bosons (axion stars?). Technically, $H_{\kappa N}$ is considered as the Friedrich extension of the operator (1.1) with domain $\{\psi \in L^2(\mathbb{R}^{3N}) | \psi$ satisfies fermi statistics (or no statistics in boson case) and $(-\Delta_i)^{1/4}\psi \in L^2(\mathbb{R}^{2N})$ for $i = 1 \ldots N\}$.

The difficulty in going from $H_{\kappa N}$ to the semiclassical Chandrasekhar or Hartree theories as $N \to \infty$ and $G \to 0$ is this: For *one particle*, the operator $h = |p| - Z/|x|$ becomes unbounded below [5, 8–10, 26] when $Z > 2/\pi$. Suppose that, by some fluctuation, $3(\pi\kappa)^{-1}$ particles get very close together. Then they form a trap into which the other particles can fall. Hence we might expect important correlation effects or even collapse for $H_{\kappa N}$ when $N = O(\kappa^{-1})$, in which case the semiclassical point of view wherein the gravitational interaction is treated as a smooth perturbation would be wrong. Something like this *does* happen for bosons and that is why the Hartree equation is the appropriate limiting description. But the interesting (and difficult to prove) fact is that the Pauli principle prevents this from happening for fermions. There is a collapse in that case, but only when $N = O(\kappa^{-3/2})$. The "local equation of state" point of view is valid for fermions.

The quantum energy is defined by

$$E_\kappa^Q(N) = \inf \operatorname{spec} H_{\kappa N} \tag{1.2}$$

in the appropriate space according to the statistics. Later on we shall define the quantum density.

The semiclassical energy functionals, $\mathcal{E}$ from $L^1(\mathbb{R}^3)$ to $\mathbb{R}$ are defined as follows.

Fermions: For $t \in \mathbb{R}^+$, let $\eta = (6\pi^2 t/q)^{1/3}$ and

$$j(t) = q(2\pi^2)^{-1} \int_0^\alpha p^2 \{p^2 + m^2)^{1/2} - m\} dp$$

$$= q(16\pi^2)^{-1} \{\eta(2\eta^2 + m^2)(\eta^2 + m^2)^{1/2} - m^4 \ln [\eta + (\eta^2 + m^2)^{1/2}]\} - tm \ . \tag{1.3}$$

Then

$$\mathscr{E}_\kappa^C(\varrho) \equiv \int j(\varrho(x))dx - \kappa D(\varrho, \varrho) \ , \tag{1.4}$$

where D is the classical gravitational energy

$$D(\varrho, \varrho) = \tfrac{1}{2} \iint \varrho(x)\varrho(y)|x-y|^{-1} dx dy \ . \tag{1.5}$$

$j(\varrho)$ is the ground state kinetic energy density of q-state fermions at density ϱ.

Bosons:

$$\mathscr{E}_\kappa^H(\varrho) = (\varrho^{1/2}, \{(p^2 + m^2)^{1/2} - m\} \varrho^{1/2}) - \kappa D(\varrho, \varrho) \ . \tag{1.6}$$

The superscript C is for Chandrasekhar while H is for Hartree.

Corresponding to these functionals are the minimum energies:

$$E_\kappa^C(N) = \inf \{\mathscr{E}_\kappa^C(\varrho) | \varrho \geqq 0, \ \varrho \in L^{4/3}(\mathbb{R}^3) \ \text{ and } \int \varrho = N\} \ , \tag{1.7}$$

$$E_\kappa^H(N) = \inf \{\mathscr{E}_\kappa^H(\varrho) | \varrho \geqq 0, \ |p|^{1/2} \varrho^{1/2} \in L^2(\mathbb{R}^3) \ \text{ and } \int \varrho = N\} \ . \tag{1.8}$$

Later, we shall omit the subscript κ when it is not necessary.

Recall (see [1, 19] and Lemma 3 for more details) that there is a critical constant $N_f(\kappa)$ which has the properties that $E_\kappa^C(N) = -\infty$ iff $N > N_f(\kappa)$. $N_f(\kappa)$ can be calculated explicitly. Define $\gamma \equiv \frac{3}{4}(6\pi^2/q)^{1/3}$ and $\tau_c \equiv \gamma/\sigma_f$, where $\sigma_f = \sup\{D(\varrho, \varrho)/\int \varrho^{4/3} | \varrho \geqq 0, \ \varrho \in L^{4/3}$ and $\int \varrho = 1\} \approx 1.092$ (see Appendix A). Then

$$N_f(\kappa) = \tau_c^{3/2} \kappa^{-3/2} \approx 4.38\, q^{-1/2} \kappa^{-3/2} \ . \tag{1.9}$$

For bosons, there also exists a critical number $N_b(\kappa)$ which has the properties $E_\kappa^H(N) = -\infty$ iff $N > N_b(\kappa)$ (see [19] and Lemma 4). $N_b(\kappa)$ can be related to $\sigma_b \equiv \sup\{D(\varrho, \varrho)/(\varrho^{1/2}, |p|\varrho^{1/2}) | \varrho \geqq 0, \ |p|^{1/2}\varrho^{1/2} \in L^2$ and $\int \varrho = 1\}$ by the formula

$$N_b(\kappa) = \sigma_b^{-1} \kappa^{-1} \equiv \omega_c \kappa^{-1} \ . \tag{1.10}$$

σ_b is known to satisfy $\pi/4 > \sigma_b > 1/2.7$ (Appendix A).

There are scalings

$$E_\kappa^C(N) = t^{3/2} E_{t\kappa}^C(t^{-3/2} N) \ , \qquad E_\kappa^H(N) = t E_{t\kappa}^H(t^{-1} N) \ , \tag{1.11}$$

which are easy consequences of the transformation $\varrho(x) \to \varrho(t^{-1/2} x)$ (respectively $\varrho(x) \to \varrho(t^{-1/3} x)$). It is convenient to introduce some normalized quantities. For any $\tau > 0$, let

$$\varepsilon_\tau^c(\varrho) = \int j(\varrho) - \tau D(\varrho, \varrho) \ , \tag{1.12}$$

$$e^c(\tau) = \inf \{\varepsilon_\tau^c(\varrho) | \int \varrho = 1 \ , \quad \varrho \geqq 0 \ \text{ and } \int \varrho^{4/3} < \infty\} \ . \tag{1.13}$$

It is easy to see that [with $\tilde{\varrho}(x) \equiv \varrho(N^{1/3} x)$]

$$\mathscr{E}_\kappa^C(\varrho) = N \varepsilon_\tau^c(\tilde{\varrho}) \quad \text{and} \quad E_\kappa^C(N) = N e^c(\tau) \ , \tag{1.14}$$

where $\tau = N^{2/3}\kappa$. Similarly, we have (with $\omega = \kappa N$)

$$\mathscr{E}_\kappa^H(\varrho) = N\varepsilon_\omega^H(\tilde{\varrho}) \quad \text{and} \quad E_\kappa^H(N) = Ne^H(\omega) \ , \tag{1.15}$$

where ε_ω^H and $e^H(\omega)$ are defined analogously to (1.12, 1.13).

Obviously, if we expect to have a nice limit as $N\to\infty$ and $G\to 0$ we should fix the quantities

$$\tau \equiv \kappa N^{2/3} \text{ (fermions)} \ , \qquad \omega \equiv \kappa N \text{ (bosons)} \ .$$

Numerically, $\kappa \approx 10^{-38}$ for neutrons or nuclei and N is about 10^{57} for a neutron star or white dwarf, so this limit is quite justified physically.

Our main theorems can now be stated.

Theorem 1 (fermions). *Fix $\tau = \kappa N^{2/3}$ and q with $\tau < \tau_c$. Then*

$$\lim_{N\to\infty} E_\kappa^Q(N)/E_\kappa^C(N) = 1 \ .$$

If $\tau > \tau_c$ then $\lim_{N\to\infty} E_\kappa^Q(N) = -\infty$.

Theorem 2 (bosons). *Fix $\omega = \kappa N$ with $\omega < \omega_c$. Then*

$$\lim_{N\to\infty} E_\kappa^Q(N)/E_\kappa^H(N) = 1 \ .$$

If $\omega > \omega_c$ then $\lim_{N\to\infty} E_\kappa^Q(N) = -\infty$.

Corollary 1. *Let $N_f^Q(\kappa)$ [respectively $N_b^Q(\kappa)$] be the critical particle number for the stability of* (1.1) *in the fermion (respectively boson) case*, i.e. $N^Q(\kappa) = \sup\{N \mid E_\kappa^Q(N) > -\infty\}$. *Then*

$$1 = \lim_{\kappa\to 0} N_f^Q(\kappa)/N_f(\kappa) = \lim_{\kappa\to 0} N_b^Q(\kappa)/N_b(\kappa)$$

if q is fixed in the fermion case.

Remarks. (a) In fact, the errors between $E_\kappa^Q(N)$ and $E_\kappa^C(N)$ [respectively $E_\kappa^H(N)$] can be estimated (see Sect. II). The difference between the quantum and semiclassical critical particle numbers can be bounded for large N as follows

$$(1+3q^{2/9}N_f(\kappa)^{-2/9})N_f(\kappa) \geqq N_f^Q(\kappa) \geqq (1-20q^{1/9}N_f(\kappa)^{-1/9})N_f(\kappa) \ ,$$

$$(1+2N_b(\kappa)^{-1})N_b(\kappa) \geqq N_b^Q(\kappa) \geqq (1-10N_b(\kappa)^{-1/3})N_b(\kappa) \ .$$

(b) It was proved in [19] that $\lim N_f^Q(\kappa)/N_f(\kappa)$ is between 1 and 1/4 (roughly). Likewise, $\lim N_b^Q(\kappa)/N_b(\kappa)$ is between 1 and 1/2.

Theorems 1 and 2 show that we can study $H_{\kappa N}$ by means of its semiclassical approximations, (1.4) and (1.6), and therefore it behooves us to study the latter. Auchmuty and Beals [1] showed that there is a minimizing ϱ for (1.4) for each $N < N_f$ and that this ϱ has compact support. They did not prove uniqueness. Later, Lions [20, Theorem II.2] proved that any minimizing sequence of ϱ's for (1.4) has (after translation $\varrho(x) \to \varrho(x+y)$) a strongly convergent subsequence in $L^{4/3}(\mathbb{R}^3) \cap L^1(\mathbb{R}^3)$.

The next two theorems summarize some properties of $E_\kappa^C(N)$ and $E_\kappa^H(N)$ which were not previously known, but which are physically important. For them, we need the notion of symmetric decreasing functions and rearrangements. For the convenience of the reader, we collect some basic definitions and facts about this subject in Appendix A. Since the functions we are interested in are all symmetric decreasing, we shall abuse notation by writing, say, $\varrho(r)$, $\dot{\varrho}(r) \equiv d\varrho/dr$, etc. with $r \equiv |x|$ for a function $\varrho: \mathbb{R}^3 \to \mathbb{R}$.

Theorem 3 (fermions). (a) *For each $N < N_f$, there exists a symmetric decreasing minimizer $\varrho_N(x)$ for $E_\kappa^C(N)$. It satisfies the Euler-Lagrange equation for some Lagrange multiplier μ:*

$$j'(\varrho(x)) = [\eta^2(x) + m^2]^{1/2} - m = \{\kappa|x|^{-1} * \varrho - \mu\}_+ , \tag{1.16}$$

where $\{f(x)\}_+ \equiv \max(f(x), 0)$ and $\eta(x) = (6\pi^2 \varrho(x)/q)^{1/3}$.

(b) *Any minimizing ϱ for $E_\kappa^C(N)$ is symmetric decreasing after translation and satisfies* (1.16) *for some μ.*

(c) *There is no minimizing ϱ for $E_\kappa^C(N_f)$ even though $E_\kappa^C(N_f)$ is finite.*

(d) *$E_\kappa^C(N)$ is a strictly concave, monotone decreasing function which is continuous at the end point, N_f, and $E_\kappa^C(N_f) = -mN_f$.*

(e) *Let μ_N be the Lagrange multiplier associated to some minimizer ϱ_N for $N < N_f$. Then the right and left derivatives of $E_\kappa^C(N)$ satisfy $(dE_\kappa^C/dN)_+ \leqq \mu_N \leqq (dE_\kappa^C/dN)_-$.*

(f) *$\mu_N \to \infty$ as $N \to N_f$.*

Remarks. (a) The Euler-Lagrange equation (1.16) is in fact equivalent to the Newtonian limit of the Tolman-Oppenheimer-Volkoff equation ((11.1.13) and (11.3.4) of [27], see also [24]). By differentiating (1.16) with respect to r we have

$$\begin{aligned} \dot{\varrho} j''(\varrho) &= \frac{2\pi^2}{q} [\eta^2(r) + m^2]^{-1/2} \eta(r)^{-1} \dot{\varrho}(r) = -\kappa M(r)/r^2 , \\ M(r) &= 4\pi \int_0^r s^2 \varrho(s) ds . \end{aligned} \tag{1.17}$$

Let $P(r) = \frac{q}{6\pi^2} \int_0^{\alpha(r)} k^4 (k^2 + m^2)^{-1/2} dk$ be the pressure ((11.3.43) of [27]). Then (1.17) can be rewritten as an equation of gravitational-hydrostatic equilibrium:

$$-r^2 \dot{P}(r) = \kappa M(r) \varrho(r) . \tag{1.18}$$

Equation (1.18) is the Newtonian limit of the TOV equation. For historical reasons, we call (1.16), and its equivalent (1.18), the Chandrasekhar equation.

(b) The Euler-Lagrange equation for (1.4) is really $j'(\varrho) - \kappa|x|^{-1} * \varrho - \mu = 0$ when $\varrho(x) > 0$ and $\geqq 0$ when $\varrho(x) = 0$. But, since $j'(0) = 0$, this is equivalent to (1.16).

(c) (1.16) is equivalent to a second order partial differential equation. See (4.7) and Lemma 8.

(d) Theorem 5(b) improves Theorem 3(e).

Theorem 4 (bosons). (a) *For each $N < N_b$, there exists a symmetric decreasing minimizer $\varrho_N(x)$ for $E_\kappa^H(N)$. It satisfies the Euler-Lagrange equation with Lagrange*

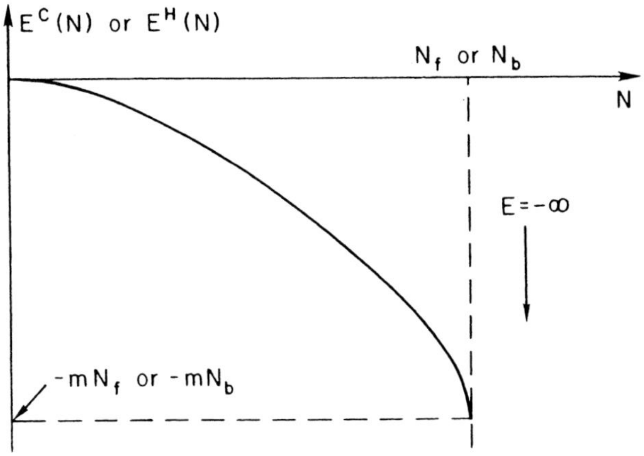

Fig. 1

multiplier v *in the distributional sense (with* $p^2 = -\Delta$*):*

$$[(p^2+m^2)^{1/2}-m]\varrho^{1/2} = (\kappa|x|^{-1} * \varrho - v)\varrho^{1/2} \ . \tag{1.19}$$

(b), (c), (d), (e), (f) *Same as in Theorem 3 mutatis mutandis.*

Figure 1, which is schematic, summarizes parts of Theorems 3 and 4.

The fact that $\mathscr{E}_\kappa^C(\varrho)$ has a local kinetic energy enables us to study $E_\kappa^C(N)$ in more detail. In the next theorem, we show that $E_\kappa^C(N)$ has a unique minimizer up to translation for $N < N_f$. We also show that the central density is strictly increasing to infinity while the radius is strictly decreasing to zero as $N \to N_f$.

The next two theorems are stated only for fermions. While we do not expect that their analogues fail for bosons, to prove them would require a great deal more work. It will be time enough to undertake this work when boson stars are seen in the sky.

Theorem 5 (fermions). (a) *For each* $N < N_f$ *the minimizer* ϱ_N *is unique up to translations* $\varrho(x) \to \varrho(x+y)$ *for* $y \in \mathbb{R}^3$.

(b) $E_\kappa^C(N)$ *is differentiable in* N *and thus* $dE_\kappa^C(N)/dN = -\mu_N$.

(c) *Each* ϱ_N *has compact support. Let* R_N *denote the radius of its support.* $\varrho_N(r)$ *is real analytic for* $r < R_N$. R_N *is a strictly decreasing function of* N. $R_N \to 0$ *as* $N \to N_f$ *and* $R_N \to \infty$ *as* $N \to 0$.

(d) *Let* $\alpha_N \equiv \varrho_N(0)$ *be the central density of* ϱ_N. *Then* α_N *is a strictly increasing function of* N *tending to* ∞ *as* $N \to N_f$.

(e) *Every radial solution of* (1.16) *is a minimizer for* $N = \int \varrho$.

(f) *Any two* ϱ_N*'s intersect at exactly one value of* r.

(g) *If* $N_1 < N_2$ *then* $M_1(r) < M_2(r)$ *for all* $r > 0$ [$M(r)$ *is defined in* (1.17)].

Some of these results are displayed schematically in Fig. 2.

Remarks. (a) We can relate Theorem 5 to the stability theorem given in e.g. [24, 27]. Since the Euler-Lagrange equation is equivalent to the Chandrasekhar equation, Theorem 5 asserts that any solution of the Chandrasekhar equation with given central density is an absolute (global) minimizer for some $E^C(N)$ or, in other words, is stable. However, our result is stronger than the standard result in [24, 27] where only local stability is discussed.

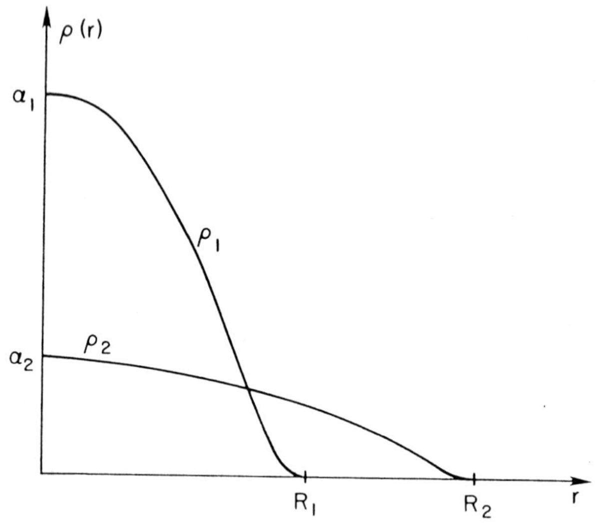

Fig. 2

(b) The normalized semiclassical functional ε_τ^c has properties similar to $\mathscr{E}_\kappa^C$. In particular, $e^c(\tau)$ has a *unique* solution ϱ_τ (say). It is easy to check from the scaling that $\varrho_N(x)=\varrho_\tau(N^{-1/3}x)$.

(c) As will become clear, the uniqueness (up to translations) of the solution amounts to the fact that the radius and central density of the star is a continuous function of the particle number. There are no "phase transitions". While it is clear that the radius is a continuous function of the central density α, it is not obvious that there are some α's which do not correspond to some minimizer. If that were to happen (and we shall show that it does not) then α (and hence the radius) would not be continuous in N.

Equation (1.16) is equivalent to a PDE as shown in (4.7). [Actually, the fact that every solution to (4.7) is a solution to (1.16) is not obvious; in Lemma 8 we prove this for radial solutions.] Theorem 5 translates into statements about the ODE arising from (4.7) in the radial case. This kind of ODE was investigated in an important paper of Ni [23]. He proves that $R(\alpha)$ is a decreasing function of the central density, α, [cf. Theorem 5(c) and (d)] but his uniqueness problem is different from ours since he fixes R while we fix N. It is possible to use Ni's result to prove Theorem 5 [except for (g)]. We originally followed that route and found the proof to be quite complicated. The proof we actually present here is, in our opinion, much more direct. It uses the variational principle in an essential and, we believe, novel way. Admittedly the class of ODE's that can be treated by our method is more restricted than Ni's.

We turn now to the connection between the ground state of $H_{\kappa N}$ and the unique semiclassical density ϱ_N that minimizes $\mathscr{E}_\kappa^C$. Recall that $\varrho_N(x)=\varrho_\tau(N^{-1/3}x)$, where ϱ_τ minimizes ε_τ^c and is independent of N, and $\tau=\kappa N^{2/3}$. Given a normalized N particle function, ψ, of space-spin we define the one-particle density as

$$\varrho_\psi(x)=N\sum_\sigma\int|\psi(x,x_2,\ldots,x_N;\sigma_1,\ldots,\sigma_N)|^2dx_2\ldots dx_N\,. \qquad (1.20)$$

(A similar definition holds for bosons without $\sum_\sigma$.) In some sense, if ψ_N is a "ground state" of $H_{\kappa N}$, and if ϱ_N^Q is the density defined by (1.20) with ψ_N, we expect that $\varrho_N^Q(N^{1/3}x)$ should converge to $\varrho_\tau(x)$ as $N\to\infty$. There are several conceptual difficulties with this expectation as we now explain.

For one thing $H_{\kappa N}$ is translation invariant so it has no L^2 eigenfunction. For this reason, and also because it is physically sensible to consider only functions ψ which are "near" the ground state when the particle number is huge, we first have to introduce the concept of an approximate ground state as in [18].

Definition 1. Fix τ. A *sequence* of normalized wave functions ψ_N is said to be an *approximate ground state* if, as $N\to\infty$,

$$N^{-1}|(\psi_N, H_{\kappa N}\psi_N)-E_N^Q|\to 0 \ . \tag{1.21}$$

Even with this definition there is another problem. In quantum mechanics the fact that ψ_N has a low energy does not imply that the system is localized. To see this, let $g(x)$ be any nonnegative function with $\int g=1$, and define the density matrix Γ by $\Gamma(X,\sigma|X'\sigma')\equiv\int g(y)\psi_N(x_i+y,\sigma)\psi_N(x_i'+y,\sigma')dy$. One finds easily that the energy of Γ, $\mathrm{Tr}\,\Gamma H_{\kappa N}$, equals $(\psi_N, H_{\kappa N}\psi_N)$, but Γ has a one-particle density given by $\varrho_\Gamma=\varrho_N^Q * g$. Thus the single-particle density can be smeared out as much as we please without affecting the energy.

The usual way out of this difficulty is to fix a certain relatively small number of particles and then to discuss the density of the rest. But it is clear that the number that have to be fixed is huge. Thus, this approach is difficult to implement.

What we do instead is to use a localizing potential. We take advantage of the fact that $\varrho_\tau(x)$ has support in a ball of some radius d_τ. Let χ_τ be the characteristic function of this ball and let $\lambda>0$. If we add $-\lambda\int\chi_\tau\varrho$ to ε_τ^c, then the minimum energy would be exactly $e^c(\tau)-\lambda$ and the minimizer would be *uniquely* ϱ_τ [formerly all $\varrho_\tau(x+y)$, $y\in\mathbb{R}$, were minimizers]. Let us do the same to (1.1)

$$H_{\kappa N\lambda}\equiv H_{\kappa N}-\lambda\sum_{i=1}^{N}\chi_\tau(N^{-1/3}x_i) \ . \tag{1.22}$$

Theorem 6. *Fix $\lambda>0$ and $\tau=\kappa N^{2/3}$. Let $\psi_{N\lambda}$ be a sequence of approximate ground states for $H_{\kappa N\lambda}$ as in* (1.21) *and let $\varrho_{N\lambda}^Q(x)$ be the densities as in* (1.20). *Then, for all $\lambda>0$ and $N\to\infty$,*

$$\varrho_{N\lambda}^Q(N^{1/3}x)\to\varrho_\tau(x) \tag{1.23}$$

weakly in $L^{4/3}\cap L^1(\mathbb{R}^3)$.

Remark. The specific choice of χ_τ as a localizing potential is arbitrary. Any other potential for which it is possible to prove the uniqueness of the minimum for $\varepsilon_{\tau\lambda}^c$ would suffice.

II. Convergence of the Quantum Energy to the Semiclassical Energy

The easy part is the upper bound to $E^Q(N)$ and we shall dispose of that first. Our proof is basically the same as that in [19, pp. 503–508] which uses the variational principle with coherent states for fermions and a product state for bosons. Only the main points need be given here.

A.1. Upper Bound to the Quantum Energy (Fermions)

To imitate the proof of [19, Theorem 2] it is only necessary to verify the analogue of [19, Eqs. (45), (46)] for our kinetic energy, namely for all $k, p \in \mathbb{R}^3$

$$[k^2+m^2]^{1/2}-m \leqq |k+p|+|p^2+m^2]^{1/2}-m\ . \tag{2.1}$$

[This follows from the triangle inequality by thinking of (k, m), $(p, -m)$ as two vectors in $\mathbb{R}^4$ with the Euclidean norm.] Then we deduce [19, Eq. (51)] that for every nonnegative $\bar{\varrho}(x)$ with $\int \bar{\varrho}=N$ and every $\xi>0$,

$$E^Q(N) \leqq \mathscr{E}_\kappa^C(\bar{\varrho})+(1.68)\kappa \int \bar{\varrho}^{4/3}+2N\xi^{1/2}+\xi^{-1}\kappa \int \bar{\varrho}^2\ . \tag{2.2}$$

Choosing $\bar{\varrho}$ to be a minimizer, ϱ, for $\mathscr{E}_\kappa^C$ and optimizing ξ we obtain

$$E^Q(N) \leqq E^C(N)+(1.68)\tau N^{1/3} \int f^{4/3}+3N^{7/9}\tau^{1/3}\left(\int f^2\right)^{1/3}\ . \tag{2.3}$$

[Recall $\tau=\kappa N^{2/3}$ and $\varrho(x)=f(N^{-1/3}x)$.] Since $E^C(N)=Ne^c(\tau)$ and $f \in L^2 \cap L^\infty$ for a minimizer (cf. Sect. IV) when $\tau<\tau_c$, we see that (2.3) can be bounded for all $\tau<\tau_c$ as

$$E^Q(N) \leqq E^C(N)\left(1-C_1(\tau, q) N^{-2/9}\right) \tag{2.4}$$

for some function $C_1(\tau, q)$. Theorem 2 of [19] states that $E^Q(N)=-\infty$ when $\tau>\tau_c(1+C_2N^{-2/9})$.

A.2. Upper Bound to the Quantum Energy (Bosons)

Again, we follow [19, p. 505]. Given $\bar{\varrho}(x) \geqq 0$ with $\int \bar{\varrho}=N$, we define the normalized variational function

$$\psi(x_1, \ldots, x_N) \equiv N^{-N/2} \prod_{j=1}^N \bar{\varrho}(x_j)^{1/2}\ . \tag{2.5}$$

Then, adding and subtracting the self interaction, we get

$$(\psi, H_N \psi)=\mathscr{E}^H(\bar{\varrho})+\tfrac{1}{2}\kappa N^{-1} D(\bar{\varrho}, \bar{\varrho})\ . \tag{2.6}$$

Choosing $\bar{\varrho}$ to be a minimizer, ϱ, for $\mathscr{E}^H$ and recalling that $\omega=\kappa N$, $\varrho(x)=f(N^{-1}x)$, we obtain the analogue of (2.4) for all $\omega<\omega_c$:

$$E^Q(N) \leqq E^H(N)\left(1-C_3(\omega) N^{-1}\right)\ . \tag{2.7}$$

When $\omega>[N/(N-1)]\omega_c$, $E^Q(N)=-\infty$.

Now we present the lower bound to $E^Q(N)$ which, apart from the analysis of the semiclassical equation, is the main mathematical point of this paper.

B.1. Lower Bound to the Quantum Energy (Fermions)

As in [19, Eq. (4)] we write H_N as a sum of operators, but here the operators will be more complicated than in [19]. Let P be a partition of $\{1, \ldots, N\}$ into two disjoint sets π_1 and π_2 of sizes L and M respectively, with $L+M=N$. There are $\binom{N}{L}$ such partitions. $X=\{x_1, \ldots, x_N\}$ denotes the N variables in $\mathbb{R}^3$.

$$H_N = \left\{ \frac{N}{L} \binom{N}{L}^{-1} \sum_P h_P \right\} + H^{(2)} \ . \tag{2.8}$$

The h_P and $H^{(2)}$ are given in terms of three positive parameters e, κ', ε with $\varepsilon < 1$ as

$$h_P = (1-\varepsilon) \sum_{i \in \pi_1} \{(p_i^2 + m^2)^{1/2} - m\} - \kappa' e \left[\sum_{i \in \pi_1} \sum_{j \in \pi_2} |x_i - x_j|^{-1} \right.$$

$$\left. - e \sum_{j < k \in \pi_2} |x_j - x_k|^{-1} - \sum_{i \in \pi_1} \delta_i(X) - e \sum_{j \in \pi_2} \delta_j(X) \right] , \tag{2.9}$$

$$H^{(2)} = \varepsilon \sum_{i=1}^{H} \{(p_i^2 + m^2)^{1/2} - m\} - \kappa' e [1 + eM/L] \sum_{i=1}^{N} \delta_i(X) \ . \tag{2.10}$$

In order that (2.8) be an identity we require

$$[2eLM - e^2 M(M-1)]\kappa' = L(N-1)\kappa \ . \tag{2.11}$$

The N functions $\delta_i : \mathbb{R}^{3N} \to \mathbb{R}$ are defined to be

$$\delta_i(X) = \max \{|x_j - x_i|^{-1} | j \neq i\} \ . \tag{2.12}$$

According to Corollary B.2 in the Appendix, $H^{(2)} \geqq -\varepsilon m N$ if the following condition among the parameters, which we shall assume, is satisfied.

$$\varepsilon \geqq (\pi/2) N^{1/3} \kappa' e [1 + eM/L] \ . \tag{2.13}$$

Concerning the h_P we note that they are all unitarily equivalent, so it suffices to study one of them. Call the first L variables $Z = \{z_1, \ldots, z_L\}$ and the last M variables $Y = \{y_1, \ldots, y_M\}$. Since there is no kinetic energy in h_P associated with Y, the y_i can be fixed. Furthermore, for $i = 1, \ldots, L$

$$\delta_i(X) \geqq \delta(z_i | Y) \equiv \max \{|z_i - y_j|^{-1} | 1 \leqq j \leqq M\} \tag{2.14}$$

and, for $i = L + j$ with $1 \leqq j \leqq M$, $\delta_i(X) \geqq \delta_j(Y)$ (since $x_{L+j} = y_j$ by definition). Thus, if we define h^Y on q spin-state fermionic functions of L variables by

$$h^Y \equiv (1-\varepsilon) \sum_{i=1}^{L} \{(p_i^2 + m^2)^{1/2} - m\} - \kappa' e \sum_{i=1}^{L} V^Y(z_i) + U^Y \ , \tag{2.15}$$

$$U^Y \equiv \kappa' e^2 \left[\sum_{1 \leqq i < j \leqq M} |y_i - y_j|^{-1} + \sum_{j=1}^{M} \delta_j(Y) \right] , \tag{2.16}$$

$$V^Y(z) \equiv \sum_{j=1}^{M} |z - y_j|^{-1} - \delta(z|Y) \ , \tag{2.17}$$

we have that for all P

$$h_P \geqq \inf_Y \{\inf \operatorname{spec} (h^Y)\} \ . \tag{2.18}$$

Lemma 1. *If*

$$\tfrac{1}{2} > \varepsilon \geqq 1.7\, q^{1/3} (\kappa')^{2/3} L^{1/3} \ , \tag{2.19}$$

then for all Y, *and with* $\kappa'' \equiv (1-2\varepsilon)^{-1} \kappa'$,

$$h^Y \geqq (1-2\varepsilon) E^C_{\kappa''}(L) - \varepsilon m L \ . \tag{2.20}$$

Proof. Let B_j be the ball centered at y_j of radius $r_j = [2\delta_j(Y)]^{-1}$. These are pairwise disjoint. Let μ_j be the uniform normalized δ-measure on ∂B_j, i.e. $\mu_j(x) = (4\pi r_j^2)^{-1}\delta(|x-y_j|-r_j)$. Let $\mu = \sum_{j=1}^{M} \mu_j$. Let ψ be any normalized fermionic L particle function and $\varrho(x)$ its density [by (1.20) with L in place of N].

Since $D(\varrho - e\mu, \varrho - e\mu) \geqq 0$, one easily derives [using $2D(\mu_j, \mu_j) = r_j^{-1}$, $2D(\mu_j, \mu_k) = |y_j - y_k|^{-1}$ and $(|x|^{-1} * \mu_j)(x) = |x-y_j|^{-1}$ for $|x-y_j| \geqq r_j$]

$$\kappa' D(\varrho, \varrho) \geqq \kappa' e \int \varrho(z) V^Y(z) dz - U^Y . \tag{2.21}$$

Hence, h^Y can be bounded by

$$(\psi, h^Y \psi) \geqq (1-2\varepsilon) K^Q(\psi) - \kappa' D(\varrho, \varrho) + \varepsilon K^Q(\psi) , \tag{2.22}$$

$$K^Q(\psi) \equiv \left(\psi, \left[\sum_{i=1}^{L} (p_i^2 + m^2)^{1/2} - m\right]\psi\right) . \tag{2.23}$$

By Lemma B.3 (Appendix) with $g(x) = \xi^{3/4} \exp(-\pi\xi x^2/2)$ and $(g, |p| g) = 2\xi^{1/2}$, we have

$$\begin{aligned} &(1-2\varepsilon)[K^Q(\psi) + 2\xi^{1/2} L] - \kappa' D(\varrho * g^2, \varrho * g^2) \\ &\geqq (1-2\varepsilon)\mathscr{E}^C_{\kappa''}(\varrho * g^2) \geqq (1-2\varepsilon) E^C_{\kappa''}(L) . \end{aligned} \tag{2.24}$$

The remainder terms which we have to bound below are

$$R = -2(1-2\varepsilon)\xi^{1/2} L + \varepsilon K^Q(\psi) - \tfrac{1}{2}\kappa' \int \varrho(x) w(x-y) \varrho(y) dx dy \tag{2.25}$$

with $w(x) = |x|^{-1} - (g^2 * |x|^{-1} * g^2)(x)$. The integral in (2.25) can be bounded using Young's inequality by $\|w\|_2 \|\varrho\|_{4/3}^2$. Clearly, $\|w\|_2 = C\xi^{-1/4}$ with C a constant; one easily finds $C^2 \leqq 32/(3\pi^{1/2})$. Optimizing the first and last terms in (2.25) with respect to ξ (and replacing $1-2\varepsilon$ by 1) we get

$$R \geqq -(3/2) C^{2/3} (\kappa')^{2/3} L^{1/3} \|\varrho\|_{4/3}^{4/3} + \varepsilon K^Q(\psi) . \tag{2.26}$$

But by (B.10), $K^Q(\psi) \geqq 1.6 q^{-1/3} \|\varrho\|_{4/3}^{4/3} - mL$ since $(p^2+m^2)^{1/2} \geqq |p|$. Thus, condition (2.19) implies $R \geqq -\varepsilon m L$. □

Let us now put our results together to prove Theorem 1. There are five parameters L, M (with $L + M = N$), ε, e, κ'. These must satisfy (2.11), (2.13), (2.19). We set $e = L/M$ and determine κ' from (2.11), whence $\kappa' < \kappa N/L$. Then we take

$$\varepsilon = 1.7 q^{1/3} \kappa^{2/3} N^{2/3} L^{-1/3} = 1.7 q^{1/3} \tau^{2/3} N^{2/9} L^{-1/3} \tag{2.27}$$

so that (2.19) is satisfied (as $N \to \infty$, $L/N \to 1$, so $\varepsilon < 1/2$), The right side of (2.13) is less than $\pi\tau N^{2/3}/M$, so (2.13) will be satisfied [with (2.27)] if we choose

$$M = 1.9 q^{-1/3} \tau^{1/3} N^{7/9} . \tag{2.28}$$

Finally, $L = N - M = N(1 - 0(N^{-2/9}))$ and $\varepsilon = O(N^{-1/9})$.

Our lower bound (2.8, 2.20) is thus

$$E^Q_\kappa(N) \geqq (N/L)(1-2\varepsilon) E^C_{\kappa''}(L) - 2\varepsilon m N \tag{2.29}$$

with $(1-2\varepsilon)\kappa''=\kappa'<\kappa N/L$. Since $N\to E^C(N)$ is concave and $E^C(0)=0$ (Sect. III), $(N/L)E^C(L)\geqq E^C(N)$, and thus

$$E_\kappa^Q(N)\geqq N\{e^c(\tau'')-2\varepsilon m\} \tag{2.30}$$

with $\tau''=(1-2\varepsilon)^{-1}\tau N/L=\tau+O(N^{-1/9})$. This agrees with $Ne^c(\tau)$ to $O(N^{8/9})$ provided $\tau<\tau_c$. □

B.2. Lower Bound to the Quantum Energy (Bosons)

The proof of Theorem 2 closely follows that of Theorem 1 just given, and only the differences will be mentioned below. Everything from (2.8) to (2.18) is the same. Condition (2.19) is not needed and the replacement for Lemma 1 is

Lemma 2. *Let* $\kappa''=(1-\varepsilon)^{-1}\kappa'$. *Acting on* $L^2(\mathbb{R}^{3N})$ *without statistics,* h^Y *satisfies (for all Y)*

$$h^Y\geqq(1-\varepsilon)E_{\kappa''}^H(L)\ . \tag{2.31}$$

The proof is the same as for Lemma 1 up to (2.22), but now we do not split $K^Q(\psi)$ into two pieces. We merely use Lemma B.5 which immediately yields (2.31).

Our only conditions are (2.11) and (2.13). As before, we set $e=L/M$ and determine κ' from (2.11), giving $\kappa'<\kappa N/L$. To satisfy (2.13) we take

$$\varepsilon=\pi N^{4/3}\kappa/M=\pi N^{1/3}\omega/M\ . \tag{2.32}$$

(Recall $N\kappa=\omega$ for besons.) We take $M=N^{2/3}$ so that $\varepsilon=cN^{-1/3}$ and $L/N=1-N^{-1/3}$. Then [again using $(N/L)E^H(L)\geqq E^H(N)$]

$$E_\kappa^Q(N)\geqq N\{e^H(\omega'')-\varepsilon m\} \tag{2.33}$$

with $\omega''=(1-\varepsilon)^{-1}\omega N/L$. This agrees with $Ne^H(\omega)$ to $O(N^{2/3})$ when $\omega<\omega_c$.

III. Properties of Semiclassical Energies

We begin with some a priori bounds on the components of the energy.

Lemma 3 (fermions). *For any* $\varrho\geqq 0$ *with* $\int\varrho=N\leqq N_f$ *we have the following a priori estimates:*

$$\int j(\varrho)dx+mN\leqq\{1-(N/N_f)^{2/3}\}^{-1}[\mathscr{E}^C(\varrho)+mN]\ , \tag{3.1}$$

$$0\leqq E^C(N)+mN\leqq CN[1-(N/N_f)^{2/3}]^{1/2}\ , \tag{3.2}$$

with $C=\frac{3}{2}m\{\int(\varrho_F)^{4/3}\int(\varrho_F)^{2/3}\}^{1/2}\|\varrho_F\|_1^{-1}$, *where* ϱ_F *is any minimizer for* $F(\varrho)$ *in* (A.4). *In particular,* (3.2) *implies that* $E^C(N)$ *is left continuous at* N_f *and* $E^C(N_f)=-mN_f$. *When* $N>N_f$, $E^C(N)=-\infty$.

Proof. Let $j_0(\varrho)=j(\varrho)+m\varrho$. By definition [and recalling $\gamma=\frac{3}{4}(6\pi^2/g)^{1/3}$]

$$mN+\mathscr{E}^C(\varrho)=(N/N_f)^{2/3}[\int j_0(\varrho)-(N_f/N)^{2/3}\kappa D(\varrho,\varrho)]+(1-(N/N_f)^{2/3})\int j_0(\varrho)\ .$$

Since $\int \varrho = N$ and $\kappa = \sigma_f^{-1}\gamma N_f^{-2/3}$ we have from the inequality $\int j_0(\varrho) \geqq \gamma \int \varrho^{4/3}$ that

$$mN + \mathscr{E}^C(\varrho) \geqq (N/N_f)^{2/3}\gamma[\int \varrho^{4/3} - \sigma_f^{-1} N^{-2/3} D(\varrho,\varrho)] + (1-(N/N_f)^{2/3}) \int j_0(\varrho)$$
$$\geqq (1-(N/N_f)^{2/3} \int j_0(\varrho) \ .$$

The last inequality follows from the definition of σ_f in (A.5) and $\int \varrho = N$.

Let $\varrho_\lambda(x) = \lambda^3 \varrho(\lambda x)$, where $\varrho \geqq 0$ is a fixed minimizer for $F(\varrho)$ in (A.4) with $\int \varrho = 1$. By considering $\varrho_\lambda(N^{-1/3}x)$, we have from the variational principle that $mN + E^C(N) \leqq N \int j_0(\varrho_\kappa) - N^{5/3}\kappa D(\varrho_\lambda, \varrho_\lambda)$. Since $(p^2+m^2)^{1/2} \leqq |p| + m^2/2|p|$, we find $j_0(\varrho) \leqq \gamma\varrho^{4/3} + \frac{9}{16} m^2\gamma^{-1}\varrho^{2/3}$. Using this,

$$mN + E^C(N) \leqq N[\gamma\lambda \int \varrho^{4/3} + 9m^2(16\gamma\lambda)^{-1} \int \varrho^{2/3} - \kappa N^{2/3}\lambda D(\varrho,\varrho)] \ .$$

By definition, $D(\varrho,\varrho) = \sigma_f \int \varrho^{4/3}$ and $\kappa = \sigma_f^{-1}\gamma N_f^{-2/3}$, whence

$$mN + E^C(N) \leqq N\{[1-(N/N_f)^{2/3}]\lambda\gamma \int \varrho^{4/3} + 9m^2(16\gamma\lambda)^{-1} \int \varrho^{2/3}\} \ .$$

Optimizing this with respect to λ yields (3.2) and it also yields $E^C(N) = -\infty$ when $N > N_f$. □

Lemma 4 (bosons). *For any $\varrho \geqq 0$ and $\int \varrho = N \leqq N_b$, we have the following a priori estimates*

$$(\varrho^{1/2}, (p^2+m^2)^{1/2}\varrho^{1/2}) \leqq (1-N/N_b)^{-1}(\mathscr{E}^H(\varrho) + mN) \ , \tag{3.3}$$

$$0 \leqq E^H(N) + mN \leqq CN[1-N/N_b]^{1/3} \ , \tag{3.4}$$

where $C = 2m(\varrho_B^{1/2}, |p|\varrho_B^{1/2})^{1/3}(\varrho_B^{1/2}, |p|^{-1/2}\varrho_B^{1/2})^{2/3}\|\varrho_B\|_1^{-1}$ and $\varrho_B^{1/2}$ is any minimizer for $B(\psi)$ in (A.6). Corollary A.2 implies that $C < \infty$. In particular, $E^H(N)$ is left continuous at N_b and $E^H(N_b) = -mN_b$. When $N > N_b$, $E^H(N) = -\infty$.

Proof. Similar to Lemma 3, one uses $(p^2+m^2)^{1/2} \leqq |p| + m^{3/2}|p|^{-1/2}$. □

In order to prove the concavity of $E^C(N)$ and $E^H(N)$, a technical lemma, whose proof is elementary, is needed.

Lemma 5. *Suppose $f: [a,b] \to \mathbb{R}$ is a strictly decreasing concave function. Then for any constant $s > 0$ the functions $g(t) = tf(st^{2/3})$ and $h(t) = tf(st)$ are strictly concave on $[(a/s)^{3/2}, (b/s)^{3/2}]$ and $[a/s, b/s]$ respectively.*

Now we can begin the proofs of Theorems 3 and 4.

Proof of Theorems 3(d) and 4(d). $e^c(\tau)$ in (1.13) is a strictly decreasing concave function because $e^c(\tau) = \inf\{$non-constant linear functions of $\tau\}$. The strict concavity of $E^C(N)$ and $E^H(N)$ follows from the scaling relations (1.14, 1.15) and Lemma 5. The continuity of $E^C(N)$ and $E^H(N)$ at N_f and N_b is in Lemmas 3 and 4. □

Proof of Theorems 3(a), (b), (c), (e). The existence of a minimizer was first proved by Auchmuty and Beals [1]. The fact that a minimizing ϱ is necessarily symmetric decreasing follows from the strong rearrangement inequality in [13] (see Appendix A.1).

Part (b) is a standard result in the calculus of variations.

To prove (c) we use the a-priori estimate (3.2) which reads (under the assumption of a minimizing ϱ when $N=N_f$) $-mN_f=E^C(N_f)=\mathscr{E}^C(\varrho)$. Since $(p^2+m^2)^{1/2}-m>|p|-m$, we have generally (with $\int\varrho=N$)

$$j(\varrho)>\gamma\varrho^{4/3}-mN\ . \tag{3.5}$$

Thus, in our case, $-mN_f>A(\varrho)-mN_f$ with $A(\varrho)=\gamma\int\varrho^{4/3}-\kappa D(\varrho,\varrho)$. But when $N=N_f=\tau_c^{3/2}\kappa^{-3/2}$, $A(\varrho)\geqq 0$ [see (1.9) and (A.5)]. This is a contradiction.

As for (e), the concavity of $E^C(N)$ implies the existence of left-hand and right-hand derivations for all N, and they are equal a.e. Part (e) follows by considering $t\varrho_N$ as a variational function, differentiating $\mathscr{E}^C(t\varrho_N)$ at $t=1+$ and $1-$, and using the Euler-Lagrange equation (1.16). □

Proof of Theorem 4(a), (b), (c), (e). The proof is essentially the same as that for Theorem 3 except for the existence part (a). Here the proof is virtually the same as that for Choquard's equation in [13]. Note that Lemma 4 places an a-priori bound on the kinetic energy which permits us to use a weak compactness argument. The key fact is in (A.2) [and also (A.1)] which permits us to restrict a minimizing sequence to symmetric decreasing functions. Note that the weak compactness argument leads to the existence of a function $\bar\varrho_N$ satisfying $\int\bar\varrho_N\leqq N$ and $\mathscr{E}^H(\bar\varrho_N)\leqq E^H(N)$. Since $E^H(N)$ is strictly monotone decreasing in N, we must have equality in both cases. □

To prove Theorems 3(f) and 4(f), Lemmas 6 and 7 (which are intrinsically different in the two cases) are needed.

Lemma 6 (fermions). *Let ϱ and $\mu>0$ satisfy (1.16) with $\int\varrho=N$. We do not assume ϱ is a minimizer for N, but we do assume ϱ is radial,* i.e. $\varrho(x)=\varrho(r)$, $r=|x|$. *Then*

$$\mathscr{E}^C(\varrho)=\int\{4j(\varrho)-3\varrho j'(\varrho)\}dx\ , \tag{3.6}$$

$$K(\varrho)\equiv\int j(\varrho)dx=\tfrac{5}{2}\,\mathscr{E}^C(\varrho)+\tfrac{3}{2}\,\mu N\ . \tag{3.7}$$

Proof. Multiply (1.17) by $r^3\varrho(r)$ and integrate. Then $-4\pi\int_0^\infty r^3 j''(\varrho)\varrho\dot\varrho dr$ $=4\pi\kappa\int_0^\infty M(r)\varrho(r)rdr$. But the second integrals is $\kappa D(\varrho,\varrho)$ (by Newton's theorem). The first integrals is $-4\pi\int r^3\dfrac{d}{dr}(\varrho j'(\varrho)-j(\varrho))dr$. After integrating by parts [and using $(\varrho j'-j)(0)=0$] it becomes $3\int\{\varrho j'(\varrho)-j(\varrho)\}dx$. This proves (3.6) since $\mathscr{E}^C(\varrho)$ $=K(\varrho)-\kappa D(\varrho,\varrho)$. To prove (3.7), multiply (1.16) by $\varrho(x)$ and integrate. Then $\int\varrho j'(\varrho)dx=2\kappa D(\varrho,\varrho)-\mu N$. Combining this with (3.6) yields (3.7). □

Remark. (3.6) is a *virial theorem.* It can also be proved for minimizing ϱ's by replacing $\varrho(x)$ by $\varrho_\lambda(x)=\lambda^3\varrho(\lambda x)$ and differentiating $\mathscr{E}^C(\varrho_\lambda)$ with respect to λ at $\lambda=1$.

Lemma 7 (bosons). *Suppose $\varrho(x)\geqq 0$, $\varrho^{1/2}\in L^2(\mathbb{R}^3)$, $(\varrho^{1/2},|p|\varrho^{1/2})<\infty$ and ϱ satisfies (1.19) for some v (in the sense of distributions). Let $\int\varrho=N$. Then*

$$K(\varrho)\equiv(\varrho^{1/2},\{(p^2+m^2)^{1/2}-m\}\varrho^{1/2})=2\mathscr{E}^H(\varrho)+vN\ . \tag{3.8}$$

Proof. Since $(\varrho^{1/2}, |p|\varrho^{1/2}) < \infty$ and $\varrho^{1/2} \in L^2$, we have $\int \varrho(|x|^{-1} * \varrho) < (\text{const}) \|\varrho\|^2_{6/5} < \infty$ by Young's and Sobolev's inequalities. Multiply (1.19) by $\varrho^{1/2}$ and integrate. This yields (3.8). Note that although (1.19) holds only in $\mathscr{D}'$, this integration is justified since all the terms are separately finite. □

Proof of Theorems 3 *(f)*, 4 *(f)*. We prove the boson case using Lemma 7. The fermion proof is the same using Lemma 6. Assume, on the contrary, that there is a sequence $N \to N_b$ with minimizers ϱ_N satisfying (1.19) with v_N. Suppose that $v_N \nrightarrow \infty$. Then since $\mathscr{E}^H(\varrho_N) = \mathrm{E}^H(\mathrm{N}) \to \mathrm{E}^H(\mathrm{N}_b) = -\mathrm{m}\mathrm{N}_b$ [Theorem 4(d)], we see from Lemma 7 that $K(\varrho_N)$ is bounded. Recall in the proof of Theorem 4(a) that the proof of the existence of a minimizer for any N needed a bound on $K(\varrho)$ for a minimizing sequence. Formerly we used Lemma 4 to achieve this when $N < N_b$. But now, by our assumption on v_N, we also have uniform boundedness of $K(\varrho_N)$. By the proof of Theorem 4(a) we have a function $\bar{\varrho}$ (weak limit of ϱ_N) with $\int \bar{\varrho} \leqq N_b$ and $\mathscr{E}^H(\bar{\varrho}) = -mN_b$. As in Theorem 4(a), this implies that $\bar{\varrho}$ is a minimizer for $N = N_b$ and this contradicts Theorem 4(c). □

IV. Properties of the Semiclassical Density (Fermions)

Our main goal here is to prove the uniqueness of the minimizer of the semiclassical functional for each $N < N_f$. This will enable us to prove Theorem 5. The main facts about the density ϱ_N are summarized in Fig. 2, which will be explained later. We shall explore all *radial* [i.e. $\varrho(x) = \varrho(|x|)$] solutions to (1.16); this class includes all minimizers by Theorem 3. Henceforth ϱ will be assumed to be radial without further mention. We shall also suppress irrelevant constants [by replacing $\varrho(x)$ by $a\varrho(bx)$] so that (1.16) becomes

$$j'(\varrho(x)) = [(|x|^{-1} * \varrho)(x) - \mu]_+ \tag{4.1}$$

with $j'(t) = (t^{2/3}+1)^{1/2} - 1$. The side condition is $\int \varrho = N < N_f = (3/4\sigma_f)^{3/2}$ in these units [see (A.5) and Lemma 3]. We also have defined

$$M(r) = 4\pi \int_0^r \varrho(s) s^2 ds \tag{4.2}$$

so that

$$V_\varrho \equiv |x|^{-1} * \varrho = r^{-1} M(r) + 4\pi \int_r^\infty t\varrho(t) dt \ . \tag{4.3}$$

Equation (4.1) implies

$$\dot{\varrho} j''(\varrho) = \tfrac{1}{3} (\varrho^{4/3} + \varrho^{2/3})^{-1/2} \dot{\varrho} = -r^{-2} M(r) \ . \tag{4.4}$$

A. Regularity Properties

Since $V_\varrho(r) \approx N/r$ as $r \to \infty$ we see, first of all, that $\mu > 0$, for otherwise for large r, $\varrho \approx (N/r)^{3/2} \notin L^1$. By Newton's theorem, (4.1) implies that ϱ has compact support in a ball of radius

$$R = N/\mu \tag{4.5}$$

and, since $V_\varrho(r)$ is continuous in r and $j'(\varrho)$ is continuous in ϱ with $j'(0)=0$, we have that $\varrho(r)\to 0$ as $r\to R$. Since $dV_\varrho/dr<0$, in the domain $B_R=\{r|r<R\}$ we can write (4.1) as

$$j'(\varrho(x))=V_\varrho(x)-\mu\,,\quad \text{in} \quad B_R \tag{4.6}$$

with $\varrho(R)=0$. We also see [since $j'(\varrho)$ is monotone] that ϱ is *necessarily* monotone decreasing. Since $V_\varrho(r)<N/r$ we can easily iterate (4.6), starting with $j'(\varrho)<N/r-\mu$, to conclude that ϱ is C^∞ in B_R. (Here we have to use the fact that $(j')^{-1}(t)=(t^2+2t)^{3/2}$ is C^∞ for $t>0$).

By applying Δ to (4.6) we get

$$-\Delta\Theta=4\pi(j')^{-1}(\Theta)=4\pi[\Theta^2+2\Theta]^{3/2} \tag{4.7}$$

with $\Theta=V_\varrho-\mu\geqq 0$. Equation (4.7) holds on B_R with the boundary condition $\Theta(R)=0$. Notice that in this version of the problem N is not mentioned. By [22, Theorem 5.8.6] we can go one step further and assert that Θ (and hence ϱ) is real analytic in B_R.

Since ϱ is radial, (4.7) is an ODE. We are looking for a strong solution to (4.7), with $\Theta(0)=\beta<\infty$ and $\dot\Theta(0)=0$. Starting with these initial conditions (with β arbitrary) we can easily (using Picard's method for example) prove that (4.7) has a unique positive solution for each β up to some $R=R(\beta)$ at which $\Theta(R)$ vanishes (which conceivably could be $R=\infty$).

It is clear that every strong solution to (4.7) with the property that $R(\beta)<\infty$ gives a solution to (4.6). Conversely, a solution to (4.6) gives a solution to (4.7) with $\beta=j'(\varrho(0))$. However, it is conceivable that for some $\beta>0$, the solution to (4.7) has $R(\beta)=\infty$. Later on, by an indirect argument, we will show that this does not happen, but it seems worthwhile to give a direct proof now.

Lemma 8. *Let Θ be the radial solution to* (4.7) *obtained by integrating outwards from $r=0$ with the initial conditions $\Theta(0)=\beta$ and $\dot\Theta(0)=0$. Then there is some $0<R(\beta)<\infty$ such that $\Theta(R(\beta))=0$. For $r<R(\beta)$, $\dot\Theta(r)<0$.*

Proof. Integrating (4.7) twice we get, for $r\leqq R(\beta)$,

$$\Theta(r)=\beta-4\pi\int_0^r (t^{-1}-r^{-1})t^2\varrho(t)dt \tag{4.8}$$

with $\varrho=(\Theta^2+2\Theta)^{3/2}$. From (4.8) it is clear that $\dot\Theta(r)<0$. Suppose $R(\beta)=\infty$. We first claim $g(t)\equiv t\varrho(t)\in L^1([0,\infty),dr)$. This follows from $\int_0^r (t^{-1}-r^{-1})tg(t)dt \geqq \int_0^{r/2} (t^{-1}-r^{-1})tg(t)dt\geqq\frac{1}{2}\int_0^{r/2} g(t)dt$; thus, if $g(t)\notin L^1$ we would have from (4.8) that $\Theta(r)<0$ for large r. Next, by Proposition 9 below, $r^{-1}\int_0^r tg(t)\to 0$ as $r\to\infty$. Then, from (4.8), as $r\to\infty$, $\Theta(r)\to\beta-C$ with $C=4\pi\int_0^\infty g(t)dt$. Since $\Theta(r)\geqq 0$, $\beta\geqq C$. Hence $\Theta(r)\geqq 4\pi r^{-1}\int_0^r tg(t)dt\geqq C'/r$ for large r. But, since $\varrho\geqq(2\Theta)^{3/2}$, we have that $\varrho>C''r^{-3/2}$ for large r and hence $\int_0^\infty g(t)dt=\infty$, which is a contradiction. □

Proposition 9. *Suppose* $f:[0,\infty)\to\mathbb{R}^+$ *and* $\int_0^\infty f(t)dt = C < \infty$. *Then* $r^{-1}\int_0^r tf(t)dt \to 0$ *as* $r\to\infty$.

Proof. Let $I(r)=\int_0^r tf(t)dt$. Choose $1>\varepsilon>0$ and write $I(r)=\int_0^{r\varepsilon} tf(t)dt+\int_{r\varepsilon}^r tf(t)dt$. The first integral is bounded by $r\varepsilon\int_0^{r\varepsilon} f(t)dt \leqq r\varepsilon C$. The second is bounded by $rK_\varepsilon(r)$ with $K_\varepsilon(r)=\int_{r\varepsilon}^r f(t)dt$. Since $f\in L^1$, $K_\varepsilon\to 0$ as $r\to\infty$. Thus $\limsup r^{-1}I(r) \leqq \varepsilon C + \lim K_\varepsilon(r) = \varepsilon C$. This holds for all ε, and thus proves the Proposition. □

Let us pause to summarize the situation. For each choice of the central density $\varrho(0)$, which we henceforth call α $(\alpha=(\beta^2+2\beta)^{3/2})$ there is (by Lemma 8) a unique radial solution that satisfies (4.1) for some unique $\mu=\mu(\alpha)>0$. This ϱ is real analytic up to $R=N(\alpha)/\mu(\alpha)$, where $N(\alpha)=\int\varrho$. The qualitative nature of this ϱ is shown in Fig. 2.

Suppose that $N(\alpha)$ were a strictly monotone increasing function of α. Then, since a minimizing ϱ satisfies (4.1) and is radial we would conclude: (i) All radial solutions of (4.1) are minimizers; (ii) for each N the minimizing ϱ is unique. But we do not yet know that $N(\alpha)$ is strictly monotone increasing, and that is the problem we now address. Up to this point the arguments were fairly standard (with the possible exception of Lemma 8) and that is why we were brief.

B. Uniqueness and Comparison Properties of Minimizers

Our strategy will be to first focus on solutions to (4.1) which are minimizers for $\mathscr{E}^C$. Then we will show that all solutions to (4.1) are minimizers.

Lemma 10. *Suppose* ϱ_1 *and* ϱ_2 *are minimizers for* $\mathscr{E}^C$ *with* $\int\varrho_1=N_1$, $\int\varrho_2=N_2$ *respectively. Let* R_1 *and* R_2 *be the radii of their supports and let* $R=\max(R_1,R_2)$. *Suppose that* $\varrho_1(0)>\varrho_2(0)$. [*If* $\varrho_1(0)=\varrho_2(0)$ *then* $\varrho_1\equiv\varrho_2$ *and this is uninteresting.*] *Then for all* $0<r<R$ *we have* $M_1(r)>M_2(r)$ [*see* (4.2)].

Proof. Suppose on the contrary, there is an r_0 such that $M_1(r_0)\leqq M_2(r_0)$. Since $M_1(r)>M_2(r)$ for r sufficiently small, there exists (by continuity) an $\tilde r$, $0<\tilde r\leqq r_0$, such that $M_1(\tilde r)=M_2(\tilde r)\equiv Q$(say). We first note that $\varrho_1(\tilde r)\neq\varrho_2(\tilde r)$. Otherwise from (4.4) we would have $\dot\varrho_1(\tilde r)=\dot\varrho_2(\tilde r)$ which together with $\varrho_1(\tilde r)=\varrho_2(\tilde r)$ would imply $\Theta_1(\tilde r)=\Theta_2(\tilde r)$ and $\dot\Theta_1(\tilde r)=\dot\Theta_2(\tilde r)$. By the uniqueness theorem of ODE, $\Theta_1\equiv\Theta_2$ which is a contradiction.

Now define the set (for $N_j\geqq Q\geqq 0$, $j=1,2$) $\mathscr{A}_Q(N_j)=\left\{\varrho\geqq 0 \mid \int_{|x|\leqq\tilde r}\varrho=Q \text{ and } \int_{|x|\geqq\tilde r}\varrho=N_j-Q\right\}$ which is subset of $\mathscr{A}(N_j)=\{\varrho\geqq 0\mid\int\varrho=N_j\}$. From the variational principle, $E^C(N_1)\leqq \inf_{\varrho\in\mathscr{A}_Q(N_1)}\mathscr{E}^C(\varrho)\equiv E_Q(N_1)$. But since $\varrho_1\in\mathscr{A}_Q(\mathrm{N}_1)$ by assumption, we have in fact $E^C(N_1)=E_Q(N_1)$. Similarly $E^C(N_2)=E_Q(N_2)$. Now define the following sets (with i, o denoting inside and outside)

$$\mathscr{A}^i=\{\varrho^i\geqq 0\mid\varrho^i(x)=0 \quad\text{if}\quad |x|>\tilde r \quad\text{and}\quad \int\varrho^i=Q\}\ ,$$

$$\mathscr{A}_j^o=\{\varrho^o\geqq 0\mid\varrho^o(x)=0 \quad\text{if}\quad |x|\leqq\tilde r \quad\text{and}\quad \int\varrho^o=N_j-Q\}\ .$$

It is easy to check that $S:\mathscr{A}^i\times\mathscr{A}_j^o\to\mathscr{A}_Q(N_j)$ defined by $S(\varrho^i,\varrho^o)=\varrho^i+\varrho^o$ is a bijection. Define a new functional on $\mathscr{A}_j^o$

$$\mathscr{E}_Q^o(\varrho^o)=\int j(\varrho^o)-D(\varrho^o,\varrho^o)-Q\int|x|^{-1}\varrho^o(x)dx\ .$$

Let $\mathscr{E}_Q^i(\varrho^i)\equiv\mathscr{E}^C(\varrho^i)$. Then, for any $\varrho\in\mathscr{A}_Q(N_j)$ it is easy to check that [with $S^{-1}(\varrho)=\varrho^i,\varrho^o]\mathscr{E}^C(\varrho)=\mathscr{E}_Q^i(\varrho^i)+\mathscr{E}_Q^o(\varrho^o)$, where we have used Newton's theorem and the definition of $Q=\int_{|x|\leqq\tilde r}\varrho$.

Let $E_Q^i=\inf\{\mathscr{E}_Q^i(\varrho^i)|\varrho^i\in\mathscr{A}^i\}$ and $E_{Q,j}^o=\inf\{\mathscr{E}_Q^o(\varrho^o)|\varrho^o\in\mathscr{A}_j^o\}$. Then by the variational principle, we have $E_Q(N_j)=E_Q^i+E_{Q,j}^o$. Then, since ϱ_j is a minimizer for $E^C(N_j)$, $\mathscr{E}_Q^i(\varrho_j^i)+\mathscr{E}_Q^o(\varrho_j^o)=\mathscr{E}^C(\varrho_j)=E_Q^i+E_{Q,j}^o$. Since $E_Q^i\leqq\mathscr{E}_Q^i(\varrho_j^i)$ and $E_Q^o(\varrho_j^o)\leqq\mathscr{E}_Q^o(\varrho_j^o)$ by the variational principle, we obtain $E_Q^i=\mathscr{E}_Q^i(\varrho_j^i)$ and $E_{Q,j}^o=\mathscr{E}_Q^o(\varrho_j^o)$. Now let $\tilde\varrho\equiv\varrho_1^i+\varrho_2^o$. It is easy to check that $\tilde\varrho$ is also a minimizer for N_2 by using $\mathscr{E}_Q^i(\varrho_2^i)=\mathscr{E}_Q^i(\varrho_1^i)$. But $\tilde\varrho$ is not continuous at $\tilde r$, which violates the regularity of the minimizer proved above. Another way to reach a contradiction is to note that $\hat\varrho\equiv\varrho_2^i+\varrho_1^o$ is a minimizer for N_1. One of the two functions, $\tilde\varrho$ and $\hat\varrho$ must be *increasing* at $\tilde r$, and this contradicts the symmetric decreasing property of minimizers. □

Remarks. (a) The same method and conclusion apply to minimizers for some other functionals $\mathscr{E}(\varrho)$ which can be written as $\mathscr{E}(\varrho)=\int j(\varrho(x))dx-D(\varrho,\varrho)$.

(b) Lemma 10 does not say $M_1(R)>M_2(R)$. In fact, we shall later see that this is true, but we do not yet know it. If we knew in advance that $M_1(R)>M_2(R)$ the proof of the following Lemma 11 would be trivial.

Lemma 11. *There exist at most one minimizing ϱ for $E^C(N)$ when $N<N_f$.*

Proof. Suppose, on the contrary, that we have two minimizers ϱ_1 and ϱ_2 with $\varrho_1(0)>\varrho_2(0)$. Let R_1 and R_2 be the radii of their supports. By Lemma 10, $M_1(r)>M_2(r)$ for all $0<r<\max(R_1,R_2)$. Then $R_1\leqq R_2$, for otherwise $N=M_2(R_2)<M_1(R_2)<N$. Let $J(z)=4j(z)-3j'(z)z$, whence $J(z)$ is concave since $J'(z)=-1+(z^{2/3}+1)^{-1/2}$ is decreasing. From (3.6), $0=\mathscr{E}^C(\varrho_1)-\mathscr{E}^C(\varrho_2)=\int[J(\varrho_1)-J(\varrho_2)]dx$ $=4\pi\int_0^{R_2}[J(\varrho_1(r))-J(\varrho_2(r))]r^2dr\leqq4\pi\int_0^{R_2}r^2(\varrho_1(r)-\varrho_2(r))J'(\varrho_2(r))dr$. The last inequality is a consequence of the concavity of J. Integrating the last integral by parts and using the definition of $M(r)$, we have $0\leqq-\int_0^{R_2}(M_1(r)-M_2(r))J''(\varrho_2(r))\dot\varrho_2(r)dr$. Since $J''(z)<0$, $\dot\varrho_2(r)\leqq0$ and $M_1(r)>M_2(r)$, this last integral is negative, which is a contradiction. □

Remark. Note that the only property of $j(z)$ used in the above proof is the concavity of $J(z)$. Since the concavity of $J(z)$ is equivalent to the convexity of $z\to j'(z^3)$, Lemma 11 holds for all functionals with $j'(z^3)$ convex.

Lemma 10 says that if $\varrho_1(0)>\varrho_2(0)$ then $N_1\geqq N_2$. But Lemma 11 say that $N_1=N_2$ is impossible if ϱ_1 is not identical to ϱ_2. Therefore we have

Corollary 2. *If ϱ_1 and ϱ_2 are minimizers for N_1 and N_2 respectively and if $\varrho_1(0)>\varrho_2(0)$ then $N_1>N_2$.*

Lemma 12. *Let ϱ^α be the unique bounded nonnegative solution of* (4.1) *with central density $\alpha=\varrho(0)$. Then ϱ is the unique minimizer for $E^C(N)$ with $N=\int\varrho^\alpha$. In other words, all the solutions of* (4.1) *parametrized by their central density are in fact minima of $E^C(N)$ for some N.*

Proof. Let $G=(0,N_f)$ and let $D=\{\alpha|\varrho^\alpha$ is a minimizer for some $N\in G\}$. For each $N\in G$ there is a unique minimizing ϱ_N, and hence a unique central density α_N. We let $\Gamma:G\to D$ denote this map from $N\in G$ to α_N.

(i) Γ is 1:1 by the aforementioned uniqueness of the ODE (4.7) with given initial condition α.

(ii) Γ^{-1} is continuous on D and D is closed in $\mathbb{R}^+$. To prove this we suppose $\alpha_j\to\alpha\in\mathbb{R}^+$ monotonically. By Corollary 2, $N_j=\Gamma^{-1}(\alpha_j)$ is monotone and bounded, so N_j has a limit N. It is clear that $N<N_f$ for otherwise $\mu_j\to\infty$ [Theorem 3(f)] but $V_j(r)$ is uniformly bounded [since $\varrho_j(r)<\alpha_j$ and $\int\varrho_j<N_f$]; these facts would contradict (4.1) when r is small. By continuity of the solution of the ODE with respect to α (this follows e.g. from the contraction mapping principle) $\varrho_j(r)\to\varrho(r)$ pointwise and $\varrho(r)$ satisfies (4.7) with $\varrho(0)=\alpha$. We have to prove that ϱ is a minimizer for N, which will imply $\alpha\in D$ (whence D is closed) and Γ^{-1} is continuous. By (3.1) ϱ_j is uniformly bounded in $L^{4/3}$ so (by passing to a subsequence) $\varrho_j\rightharpoonup\varrho$ weakly in $L^{4/3}$. Then $\mathscr{E}^C(\varrho)\leqq\lim\mathscr{E}^C(\varrho_j)=\lim E^C(N_j)=E^C(N)$. [The lower semicontinuity, namely $\mathscr{E}^C(\varrho)\leqq\lim\mathscr{E}^C(\varrho_j)$, follows as in [13], for example. Clearly $\int j(\varrho)\leqq\lim\int j(\varrho_j)$ since $\varrho_j\rightharpoonup\varrho$. On the other hand $D(\varrho_j,\varrho_j)\to D(\varrho,\varrho)$ since ϱ is symmetric decreasing and $\varrho_j\to\varrho$ pointwise.] Thus, ϱ is a minimizer for N because $\int\varrho\leqq N$ while $E^C(N)$ is strictly decreasing; therefore $\int\varrho=N$.

(iii) Γ^{-1} is 1:1 and continuous from D onto G and also D is closed in $\mathbb{R}^+$. Hence Γ^{-1} is a homeomorphism from D to G. Since G is connected, so is D, and therefore D is a closed interval in $\mathbb{R}^+$. Since the only closed interval in $\mathbb{R}^+$ homeomorphic to $(0,N_f)$ is $\mathbb{R}^+$ itself, we conclude that $D=\mathbb{R}^+$. □

Finally we want to make a qualitative comparison of solutions with different N. First a technical lemma is needed.

Lemma 13. *Suppose Θ_1 and Θ_2 are two nonnegative solutions of* (4.7) *with radii R_1 and R_2 with $R_1<R_2$. Then there exists an $\tilde{r}<R_1$ such that $\Theta_1(\tilde{r})\geqq\Theta_2(\tilde{r})$.*

Proof. This is a standard Sturm comparison argument. Assuming, on the contrary, that $\Theta_2(r)>\Theta_1(r)$ for all $r<R_1$, let $u(r)=\Theta_1(r)/\Theta_2(r)$. Then, from (4.7), u satisfies $\Theta_2\Delta u+2\nabla u\cdot\nabla\Theta_2+(F(\Theta_1)-F(\Theta_2))\Theta_2 u=0$ with $F(t)=4\pi(t^2+2t)^{1/2}(t+2)$. Now $u(R_1)=0$ and $0\leqq u(r)<1$ for $r<R_1$. Hence $g\equiv F(\Theta_1)-F(\Theta_2)<0$. At a maximum, $\Delta u\leqq 0$, $\nabla u=0$, but this contradicts $g<0$. □

Lemma 14. *Let ϱ_1 and ϱ_2 be two solutions to* (4.1) *with central densities $\alpha_1>\alpha_2$. Then $R_1<R_2$ and there is precisely one point r in $(0,R_1)$ at which $\varrho_1(r)=\varrho_2(r)$ (as in Fig. 2).*

Proof. We know $M_1(r)>M_2(r)$ when $0<r\leqq R\equiv\max(R_1,R_2)$. Suppose $R_1\geqq R_2$. Then integrate (4.4) from $r<R_2$ to R_2 for both ϱ_1 and ϱ_2. One has $j'(\varrho_i(r))-j'(\varrho_i(R_2))=-\int_r^{R_2}s^{-2}M_i(s)ds$ for $i=1,2$. Since $j'(\varrho_1(R_2))\geqq 0=j'(\varrho_2(R_2))$ and $M_1(s)>M_2(s)$ for $0<s<R_2$, we easily conclude $j'(\varrho_1(r))>j'(\varrho_2(r))$ and hence

$\varrho_1(r) > \varrho_2(r)$ for $0 \leqq r < R_2$. If $R_1 > R_2$, then this contradicts Lemma 13, so suppose $R_1 = R_2 = R$. Then, similar to Lemma 13, define $u = \Theta_2/\Theta_1$. By Gauss's theorem $\dot{\Theta}_j(R) = N_j R^2$, since $\int \varrho_j = N_j$. Then $u(R) = N_2/N_1 \equiv \delta < 1$. As in Lemma 13 (using $\varrho_1 \geqq \varrho_2$), u can have no maximum for $r < R$. Thus $u \leqq \delta$. However, $N_2 = \int (\Theta_2^2 + 2\Theta_2)^{3/2} \leqq \int (\delta^2 \Theta_1^2 + 2\delta\Theta_1)^{3/2} \leqq \delta^{3/2} \int (\Theta_1^2 + 2\Theta_1)^{3/2} = \delta^{3/2} N_1$, so $\delta \geqq 1$, which is a contradiction. Thus, $R_1 < R_2$.

Since $\alpha_1 > \alpha_2$ the function $f \equiv \varrho_1 - \varrho_2$ which is C^∞ for $t < R_1$ must have at least one zero in $(0, R_1)$. Suppose there is more than one zero. At each zero we have, by (4.4) that $df/dr < 0$ (since $M_1 > M_2$). But it is easy to see that a C^1 function cannot have a negative derivative at *all* its zeros. □

C. *Summary*

Let us summarize the results of A. and B., for this is Theorem 5.

(i) All radial solutions of the Euler-Lagrange equation (4.1) are in L^1 and are minimizers.

(ii) They are real analytic up to the cut off radius R. For $r \approx R$, $\Theta(r) \approx R - r$ so $\varrho(r) \approx (R-r)^{3/2}$ as in Fig. 2.

(iii) They are parametrized by the central density α which goes from 0 to ∞. $\mu(\alpha)$ and $N(\alpha)$ are strictly monotone increasing while $R(\alpha)$ is strictly monotone decreasing. As $\alpha \to \infty$, $\mu(\alpha) \to \infty$, $N(\alpha) \to N_f$, $R(\alpha) \to 0$. As $\alpha \to 0$, $N(\alpha) \to 0$, $R(a) \to \infty$ [to be proved in (vii) below], and hence $\mu(\alpha) = N(\alpha)/R(\alpha) \to 0$. N, μ, R are continuous in α.

(iv) Any two solutions always have exactly one intersection as in Fig. 2.

(v) For each r, the mass $M(r)$ is an increasing function of α.

(vi) *Proof of Theorem 5(b).* In the notation of Theorem 3(e), suppose that $(dE^C/dN)_+ \neq (dE^C/dN)_-$ for some $N < N_f$. In this case it is easy to see that μ_N must be discontinuous at N. However, by (4.5) μ_N is continuous since $N \to \alpha_N$ is continuous (by Lemma 12) and $\alpha \to R(\alpha)$ is continuous [by continuity of the solution of (4.7) with respect to the "initial data" at $r=0$].

(vii) *Proof of Theorem 5(c).* The only fact yet to be proved is that $R_N \to \infty$ as $N \to 0$ [or equivalently, $R(\alpha) \to \infty$ as $\alpha \to 0$]. Suppose on the contrary, $R(\alpha) < R_0$ for all α. Choose a solution Θ_1 of the Lane-Emden equation

$$\Delta\Theta + 4\pi(2\Theta)^{3/2} = 0 \tag{4.9}$$

with $\Theta_1(R_1) = 0$ for some $R_1 > 2R_0$ and $\Theta_1(0) < 1$. Such a Θ_1 always exists since there is a scaling $\Theta(x) \to \lambda^4 \Theta(\lambda x)$. Let $\beta \equiv \frac{1}{8}\Theta_1(R_0)$ and let Θ_β be the solution of (4.7) with $\Theta_\beta(0) = \beta$. Then $\Theta_\beta(r) < \frac{1}{8}\Theta_1(r)$ and $(\Theta_\beta^2 + 2\Theta_\beta)^{1/2}(\Theta_\beta + 2) < (2\Theta_1)^{1/2}$ for $0 \leqq r \leqq R_0$ [since Θ_1 and Θ_β are monotone decreasing and $\Theta_1(R_0) = 8\Theta_\beta(0)$]. But this is impossible as can be seen by the argument given in Lemma 13. □

V. Convergence of the Quantum Density to the Semiclassical Density (Fermions)

Here we prove Theorem 6 for fermions. As explained in Sect. I, we first add a fixed single-particle potential $\lambda\chi_\tau(N^{-1/3}x)$ to $H_{\kappa N}$. (Recall that χ_τ is the characteristic function of the support of ϱ_τ.) Following the method in [18] we next add an additional one parameter single-particle potential $\delta W(N^{-1/3}x)$ and differentiate with respect to δ at $\delta=0$. Two facts have to be established: (i) an extension of Theorem 1 to include the potential $\lambda\chi+\delta W$; (ii) the differentiability of the corresponding semiclassical energy with respect to δ. We shall call (i) Theorem 1′ and (ii) Lemma 15.

For any $W\in C_0^\infty(\mathbb{R}^3)$ and $\delta\in\mathbb{R}$, define [with $H_{\kappa N\lambda}$ defined in (1.22)]

$$H_{\kappa N\lambda\delta}\equiv H_{\kappa N\lambda}+\delta\sum_{i=1}^{N} W(N^{-1/3}x_i)\ ,$$
$$E_{\lambda\delta}^{Q}(N)=\inf\,\mathrm{spec}\,H_{\kappa N\lambda\delta}\ , \tag{5.1}$$

$$\varepsilon_{\tau\lambda\delta}^{c}(\varrho)=\varepsilon_{\tau}^{c}(\varrho)-\lambda\int\chi_\tau\varrho+\delta\int W\varrho\ ,$$
$$e_{\lambda\delta}^{c}(\tau)=\inf\{\varepsilon_{\tau\lambda\delta}^{c}(\varrho)|\varrho\geqq 0,\ \textstyle\int\varrho=1\}\ . \tag{5.2}$$

Theorem 1′ (fermions). *Fix q,λ and $\tau=\kappa N^{2/3}$ with $\tau<\tau_c$. Then $\lim_{N\to\infty} N^{-1}E_{\lambda\delta}^{Q}(N)=e_{\lambda\delta}^{c}(\tau)$.*

Proof. (a) *Upper Bound.* Let $\bar\varrho=\varrho*g^2$ with $g(x)=\xi^{3/4}\exp(-\pi\xi x^2/2)$. Then we have generally $G\equiv|\int(\varrho(x)-\bar\varrho(x))W(N^{-1/3}x)dx|\leqq\int\varrho\,\|W*\bar g^2-W\|_\infty$, where $\bar g(x)=\bar\xi^{3/4}\exp(-\pi\bar\xi x^2/2)$ with $\bar\xi=N^{2/3}\xi$. By the same method as in the proof of Theorem 1, we only have to show that $N^{-1}[1.68\kappa\int\varrho^{4/3}+2N\xi^{1/2}+\xi^{-1}\kappa\int\varrho^2+G]\to 0$ as $N\to\infty$. Note that the minimizing ξ for (2.2) is $\xi=\kappa^{2/3}$, whence $\bar\xi=N^{2/9}\tau^{2/3}\to\infty$. This implies $N^{-1}G=\|W*\bar g^2-W\|_\infty\to 0$ as $N\to\infty$.

(b) *Lower Bound.* We only have to show that [cf. (2.25)] $N^{-1}R'\equiv N^{-1}[R+G]\geqq N^{-1}[-2\xi^{1/2}L+\varepsilon K^{Q}(\psi)-C\kappa'\xi^{-1/4}\|\varrho\|_{4/3}^{2}-G]_-\to 0$ as $N\to\infty$ (with $|f|_-=\min(0,f)$). Let $\xi=\max\{L^{-4/3}\|\varrho\|_{4/3}^{8/3}(\kappa')^{4/3},N^{-1/3}\}$. Then $N^{2/3}\xi\equiv\bar\xi\geqq N^{1/3}$ whence, as in the proof of part (a), $G/N\to 0$ as $N\to\infty$. We take ε and κ' exactly the same as in the proof of Theorem 1. For the first three terms, we can use the same estimate as in Lemma 1 if $\xi=L^{-4/3}\|\varrho\|_{4/3}^{8/3}(\kappa')^{4/3}$. If $\xi=N^{-1/3}$ we have that $\|\varrho\|_{4/3}\leqq(\mathrm{const})N^{17/24}$, and then the result is immediate. □

Lemma 15. *Fix λ and $\tau<\tau_c$. Then $e_{\lambda\delta}^{c}(\tau)$ is differentiable at $\delta=0$ and $de_{\lambda\delta}^{c}(\tau)/d\delta|_{\delta=0}=\int\varrho_\tau W$.*

Proof. For each $\delta>0$ choose a function ϱ_δ satisfying $\delta^{-1}|\varepsilon_{\tau\lambda\delta}^{c}(\varrho_\delta)-e_{\lambda\delta}^{c}(\tau)|\to 0$ as $\delta\to 0$. By the variational principle, $\delta\int\varrho_\tau W\geqq e_{\lambda\delta}^{c}(\tau)-e_{\lambda}^{c}(\tau)\geqq\delta\int\varrho_\delta W+e_{\lambda\delta}^{c}(\tau)-\varepsilon_{\tau\lambda\delta}^{c}(\varrho_\delta)$. To prove the lemma we only have to show that $\int\varrho_\delta W\to\int\varrho_\tau W$ as $\delta\to 0$. Clearly, $e_{\lambda\delta}^{c}(\tau)$ is continuous in δ and therefore, by the assumption about ϱ_δ, we see that ϱ_δ is a minimizing sequence, as $\delta\to 0$, for $e_{\lambda}^{c}(\tau)=e^{c}(\tau)-\lambda$. Since $\varepsilon_{\tau\lambda}^{c}(\varrho_\delta)\geqq\varepsilon_{\tau}^{c}(\varrho_\delta)-\lambda$, we have that ϱ_δ is also a minimizing sequence for $e^{c}(\tau)$ and $\int\varrho_\delta\chi_\tau\to 1$. We also know that the minimizer, ϱ_τ, for $e^{c}(\tau)$ is unique if we center it at the origin (Theorem 5). By Lion's result [20] there exists a sequence $y_\delta\in\mathbb{R}^3$ such that $\varrho_\delta(x+y_\delta)\to\varrho_\tau(x)$ strongly in $L^{4/3}\cap L^1$. But since $\int\chi_\tau\varrho_\delta\to 1$ we must have $y_\delta\to 0$. □

Proof of Theorem 6. For any approximate ground state $\psi_{N\lambda}$ we have, from the variational principle, that

$$\delta\int\varrho_{N\lambda}^{Q}(N^{1/3}x)W(x)dx\geqq N^{-1}\{E_{\lambda\delta}^{Q}(N)-E_{\lambda}^{Q}(N)+E_{\lambda}^{Q}(N)-(\psi_{N\lambda},H_{N\lambda}\psi_{N\lambda})\}\ .$$

By taking the limit $N\to\infty$ and then $\delta\to 0$ (using Theorem 1′ and Lemma 15), $\lim_{N\to\infty}\int \varrho_{N\lambda}^{Q}(N^{1/3}x)W(x)dx=\int \varrho_\tau(x)W(x)$ for any $W\in C_0^\infty(\mathbb{R}^3)$. Hence $\varrho_{N\lambda}^{Q}(N^{1/3}x)\rightharpoonup\varrho_\tau(x)$ in weak $L^{4/3}(\mathbb{R}^3)$. The fact that $\varrho_{N\lambda}^{Q}(N^{1/3}x)\rightharpoonup\varrho_\tau(x)$ also in weak $L^1(\mathbb{R}^3)$ follows form Lemma III.4 of [18]. □

Appendix A: Some Definitions and Basic Facts

A.1. Symmetric Decreasing Rearrangements

Given a function $\psi:\mathbb{R}^3\to\mathbb{C}$, the *symmetric decreasing rearrangement* ψ^* of ψ satisfies $\psi^*:\mathbb{R}^3\to\mathbb{R}^+$, $\psi^*(x)$ depends only on $|x|$ and Lebesgue meas $\{x|\psi^*(x)>a\}$ $=$ Lebesgue meas $\{x|\,|\psi(x)|>a\}$ for all $a>0$. ψ is *symmetric decreasing* if $\psi=\psi^*$. It follows that if $j:\mathbb{R}^+\to\mathbb{R}^+$ and $\varrho(x)\geqq 0$ then $\int j(\varrho^*)dx=\int j(\varrho)dx$. Also $(\psi^*)^2=(\psi^2)^*$. A particular case of the Riesz inequality states that

$$\iint \varrho(x)\varrho(y)|x-y|^{-1}dxdy\leqq\iint \varrho^*(x)\varrho^*(y)|x-y|^{-1}dxdy\ , \tag{A.1}$$

and the strict rearrangement inequality in [13] states that (A.1) is a strict inequality unless $\varrho(x)=\varrho^*(x-y)$ for some $y\in\mathbb{R}^3$.

We also have that

$$(\psi,|p|\psi)\geqq(\psi^*,|p|\psi^*)\ , \tag{A.2}$$

$$(\psi,T\psi)\geqq(\psi^*,T\psi^*)\ , \tag{A.3}$$

with $T=(p^2+m^2)^{1/2}-m=(-\Delta+m^2)^{1/2}-m$. This follows from the proof in [13] and the fact that the kernels $e^{-t|p|}(x,y)$ and $e^{-tT}(x,y)$ are symmetric decreasing functions of $x-y$, as shown in [4].

A.2. Minimizers for the Gravitational Energy

In the fermion case we are concerned with the ratio (for $\varrho\in L^{4/3}\cap L^1$)

$$F(\varrho)=D(\varrho,\varrho)\|\varrho\|_{4/3}^{-4/3}\|\varrho\|_1^{-2/3} \tag{A.4}$$

which is homogeneous in ϱ and dilation invariant. We set

$$\sigma_f\equiv\sup F(\varrho)=1.092\ . \tag{A.5}$$

It is a known fact (proved in [17]) that there is a minimizing $\varrho\equiv\varrho_F$ for $F(\varrho)$. It is unique up to scaling, dilation and translation: $\varrho_F(x)\to a\varrho_F(bx+y)$. It satisfies Emden's equation. The value 1.092 in (A.5) is numerical.

For bosons we consider the ratio

$$B(\psi)=D(|\psi|^2,|\psi|^2)\,(\psi,|p|\psi)^{-1}\|\psi\|_2^{-2}\ , \tag{A.6}$$

$$\sigma_b\equiv\sup\{B(\psi)|\psi\in L^2,(\psi,|p|\psi)<\infty\}\ . \tag{A.7}$$

By (B.10) with $N=1$, $(\psi,|p|\psi)\geqq C\|\psi\|_{8/3}^2$ [actually, there is a Sobolev ineqality $(\psi,|p|\psi)\geqq C\|\psi\|_3^2$]. This, together with the Hardy-Littlewood-Sobolev inequality, $D(\varrho,\varrho)\leqq C\|\varrho\|_{6/5}^2$, shows that $B(\varrho)$ is bounded. In fact, the rearrangement inequalities (A.1), (A.2) permit us to imitate the proof in [17] and show that there is a minimizing ψ for $B(\psi)$ with $\psi \in L^2$ and $(\psi, |p|\psi) < \infty$. See also the proof of Theorem 4(a). We have no precise numerical value for σ_b. However, Theorem 2 states that $\omega_c = 1/\sigma_b$ and, using the results in [19, pp. 503, 504] we have that $\pi/4 > \sigma_b > 1/2.7$.

Another fact we shall need is

Lemma A.1. *Let ψ be any minimizer for $B(\psi)$. Then $\psi \in L^{3/2+\varepsilon}$, all $\varepsilon > 0$.*

Proof. Note that $\psi = \psi^*$ and satisfies

$$|p|\psi = \{W(x) - v\}\psi \tag{A.8}$$

with $W = (\text{const})\, |x|^{-1} * |\psi|^2$ and $v > 0$. Equation (A.8) is just the Euler-Lagrange equation. Since $|\psi|^2 = |\psi^2|^*$ and $|\psi|^2 \in L^1$, $W(x) \to 0$ as $|x| \to \infty$. Thus, writing $f \equiv \{W(x) - v\}\psi$ and using $\psi(x) \geqq 0$, we have that f_+ has compact support in a ball of some radius R. Since

$$\psi = |p|^{-1}\{|p|\psi\} = (2\pi^2)^{-1}|x|^{-2} * f \leqq (2\pi^2)^{-1}|x|^{-2} * f_+$$

we see that $\psi(x) < c|x|^{-2}$ for $|x| > R$. □

Corollary A.2. *If ψ is a minimizer for $B(\psi)$ then $(\psi, |p|^{-1/2}\psi) < \infty$.*

Proof. $|p|^{-1/2}$ is convolution with $|x|^{-5/2}$. By the Hardy-Littlewood-Sobolev inequality $\psi \in L^{12/7} \Rightarrow (\psi, |p|^{-1/2}\psi) < \infty$. □

Appendix B: Some Inequalities

B.1. Domination of the Nearest Neighbor Coulomb Potential by $|p|$

Let $Y = \{y_1, \ldots, y_N\}$ be N arbitrary, but fixed points in $\mathbb{R}^3$ and let $Z = \{z_1, \ldots, z_N\}$ with $z_j \in \mathbb{R}$, $z_j \geqq 0$ be given. Define

$$\bar{z} = \|z\|_3 = \{\sum z_j^3\}^{1/3} . \tag{B.1}$$

For $x \in \mathbb{R}^3$, the function

$$\omega_{YZ}(x) \equiv \max_j |x - y_j|^{-1} z_j \tag{B.2}$$

can be called the *nearest neighbor Coulomb potential.* Let

$$Q = \{f \mid f \in L^2(\mathbb{R}^3),\ \int |\hat{f}(p)|^2 |p| dp < \infty\} , \tag{B.3}$$

where $\hat{f}(p) = \int f(x) \exp(ip \cdot x) dx$ denotes Fourier transform.

Lemma B.1. *For all Y and Z and $f \in Q$*

$$(2\pi)^{-3} \int |\hat{f}(p)|^2 |p| dp \equiv (f, |p| f) \geqq \frac{2}{\pi} (\bar{z})^{-1} (f, \omega_{YZ} f) . \tag{B.4}$$

Proof. If $N = 1$ the lemma is known [8–10, 26] and we shall reduce (B.4) to that case. Let f^* denote the symmetric decreasing rearrangement of f. By (A.2) $(f, |p| f) \geqq (f^*, |p| f^*)$. On the other hand $(f, \omega_{YZ} f) \leqq (f^*, \omega_{YZ}^* f^*)$. We claim that

$\omega_{YZ}^*(x) \leqq \bar{z}|x|^{-1}$, from which (B.4) follows by applying the $N=1$ lemma to f^*. To prove this it is only necessary to note that for all $b>0$ (with μ = Lebesgue measure)

$$\mu\{x|\omega_{YZ}(x)>b\} \leqq \sum_{j=1}^{N} \mu\{x|\ z_j|x-y_j|^{-1}>b\}$$

$$=(4\pi/3)b^{-3}\sum_{j=1}^{N} z_j^3=\mu\{x|\bar{z}|x|^{-1}>b\}. \quad \square \tag{B.5}$$

Corollary B.2. *Let $\psi : \mathbb{R}^{3N} \to \mathbb{C}$ be an N particle function (without any particular statistics) in Q^N. Let $\delta_i : \mathbb{R}^{3N} \to \mathbb{R}$ denote the N functions defined in* (2.12), i.e. $\delta_i(X)=\max\{|x_i-x_j|^{-1}|j \neq i\}$. *Then for each i*

$$(\psi, |p_i|\psi) \geqq (2/\pi)\,(N-1)^{-1/3}(\psi, \delta_i\psi)\ . \tag{B.6}$$

Inequality (B.6) holds without regard to statistics. If, on the other hand, ψ is restricted to be a q-state fermionic function, it is possible to prove that

$$\sum_{i=1}^{N} (\psi, |p_i|\psi) \geqq (\text{const})\, q^{-1/3} \sum_{i=1}^{N} (\psi, \delta_i\psi)\ . \tag{B.7}$$

Since (B.7) is not needed here we defer its proof to a forthcoming paper of ours. Fefferman and de la Llave [6] have proved (B.7) for $q=1$ but their method does not appear to be easily generalizable to $q>1$.

B.2. Semiclassical Lower Bounds to the Kinetic Energy

The single particle kinetic energy operator is $T=(p^2+m^2)^{1/2}-m$ with $p^2=-\Delta$. Let ψ be a normalized wave function for N fermions with q spin states each and let

$$K^Q(\psi)=\left(\psi, \sum_{i=1}^{N} T_i\psi\right) \tag{B.8}$$

be the total kinetic energy. The semiclassical *approximation* to $K^Q(\psi)$ is

$$K^Q(\psi) \approx K(\varrho_\psi) \equiv \int j(\varrho_\psi(x))dx \tag{B.9}$$

with ϱ_ψ being the single particle density defined by (1.20) and with $j(t)$ given by (1.3).

It is conjectured that $K(\varrho_\psi)$ is, in fact, a lower bound to $K^Q(\psi)$ but no one has proved this. Daubechies [4], using the method in [12], has found two lower bounds of the right form for fermions, the first of which we use here.

$$\left(\psi, \left\{\sum_{i=1}^{N} |p_i|\right\}\psi\right) \geqq 1.6q^{-1/3} \int \varrho_\psi(x)^{4/3}dx\ , \tag{B.10}$$

$$K^Q(\psi) \geqq C \int j(C^{-1}\varrho_\psi(x))dx \tag{B.11}$$

with $C=9.6$.

Although (B.10) is important for us, we also need a bound similar to (B.9). This is provided by

Lemma B.3 (fermions). *Let* $g \in Q$ [cf. (B.3)] *with* $\|g\|_2 = 1$. *Then, for all normalized* ψ,

$$K^Q(\psi) \geqq \int j((\varrho_\psi * |g|^2))dx - N(g, |p|g) \; . \tag{B.12}$$

Proof. This is the same as that given in [15, Eqs. (5.14)–(5.22)]. Introduce the coherent states $g_{pq} \in L^2(\mathbb{R}^3)$ by $g_{pq}(x) = g(x-q)\exp(ip\cdot x)$. Here $p, q \in \mathbb{R}^3$. Let π_{pq} be the projection onto g_{pq}. Then for $f \in L^2(\mathbb{R}^3)$

$$(f,f) = (2\pi)^{-3} \iint (f, \pi_{pq} f)dpdq \; , \tag{B.13}$$

$$(f, (V * |g|^2)f) = (2\pi)^{-3} \iint V(q)(f, \pi_{pq} f)dpdq \; , \tag{B.14}$$

$$(f, Tf) \geqq (2\pi)^{-3} \iint \tilde{T}(p)(f, \pi_{pq} f)dpdq - \|f\|_2^2 (g, |p|g) \; . \tag{B.15}$$

Here, $\tilde{T}(p) = (p^2 + m^2)^{1/2} - m$ is a function, not an operator. Equations (B.13) and (B.14) are obvious. Inequality (B.15) is easily proved by writing f as a Fourier integral and then using (2.1) with $k \to p$, $p \to r - p$. Thus, if h is the operator $T - (V * |g|^2)(x)$ and γ is a positive semidefinite operator with $\mathrm{Tr}\gamma = \lambda$ we have (by making an eigenvector expansion of γ) that

$$\mathrm{Tr}\gamma h \geqq (2\pi)^{-3} \iint dpdq [\tilde{T}(p) - V(q)] M(p,q) - \lambda(g, |p|g) \tag{B.16}$$

with $M(p,q) = \mathrm{Tr}\gamma\pi_{pq} \geqq 0$. If, in addition, $0 \leqq \gamma \leqq qI$ then $0 \leqq M(p,q) \leqq q$ and hence

$$\begin{aligned} \mathrm{Tr}\gamma h &\geqq -(2\pi)^{-3} q \iint dpdq [V(q) - \tilde{T}(p)]_+ - \lambda(g, |p|g) \\ &= -(q/6\pi^2) \int (V^2 + 2mV)^{3/2} V + \int j((q/6\pi^2)(V^2 + 2mV)^{3/2}) - \lambda(g, |p|g) \; . \end{aligned} \tag{B.17}$$

Recall that $j(t)$ depends on q.

Let $\varrho(x) \equiv \gamma(x,x)$ and take $(V^2 + 2mV)^{3/2} = (6\pi^2/q)\varrho * |g|^2$. Noting that $\mathrm{Tr}\gamma(V * |g|^2) = \int V(\varrho * |g|^2)$, (B.17) becomes

$$\mathrm{Tr}\gamma T \geqq \int j(\varrho * |g|^2) - \lambda(g, |p|g) \; . \tag{B.18}$$

To apply this to our case, let γ be given by the kernel

$$\begin{aligned} \gamma(x,y) = N \sum_\sigma \int \psi(x, x_2, \ldots, x_N; \sigma_1, \ldots, \sigma_N) \\ \cdot \bar{\psi}(y, x_2, \ldots, x_N, \sigma_1, \ldots, \sigma_N) dx_2 \ldots dx_N \; . \end{aligned} \tag{B.19}$$

Then $\gamma \geqq 0$, $\lambda = \mathrm{Tr}\gamma = N$. The fact that $\gamma \leqq qI$ is standard [16]. Inequality (B.18) becomes (B.12). □

Lemma B.3 can trivially be generalized to operators other than T and to $n \neq 3$ in the following way.

Lemma B.4. *Let* $\tilde{T}: \mathbb{R}^n \to \mathbb{R}^+$ *satisfy* $\tilde{T}(p) \geqq 0$ *and, for all* $p, q \in \mathbb{R}^n$, $\tilde{T}(p) \leqq \tilde{T}(p-q) + \tilde{S}(q)$ *for a suitable nonnegative function* $S: \mathbb{R}^n \to \mathbb{R}^+$. *Let* T *and* S *be the corresponding multiplication operators in momentum space. Let* $g: \mathbb{R}^n \to \mathbb{C}$ *with* $\|g\|_2 = 1$. *Then*

$$(\psi, \sum T_i \psi) \geqq \int J((\varrho_\psi * |g|^2)(x))dx - N(g, Sg) \; . \tag{B.20}$$

Here, $J: \mathbb{R}^+ \to \mathbb{R}^+$ *is defined as follows. Let*

$$\mu(t) = q(2\pi)^{-n} \text{ meas } \{p | \tilde{T}(p) \leqq t\}$$

$$\beta(t) = q(2\pi)^{-n} \int_{\tilde{T}(p) \leqq t} \tilde{T}(p) dp .$$

Then

$$J(t) = \beta(\mu^{-1}(t)) , \tag{B.21}$$

where μ^{-1} *is the inverse function. (It is not always uniquely defined everywhere, but* $\beta \circ \mu^{-1}$ *is.)*

On the other hand, for *bosons* (or, more generally, particles without any statistics) (B.9) is not a good approximation to $K^Q(\psi)$. The following is a good approximation and also a bound.

Lemma B.5 (bosons).

$$K^Q(\psi) \geqq (\varrho_\psi^{1/2}, T\varrho_\psi^{1/2}) . \tag{B.22}$$

The proof of this is that given by Conlon for Lemma 4.2 in [3]. One only has to verify that e^{-tT} has a positive kernel, but this is proved in [4, Remark 1].

Acknowledgements. The authors are grateful to Christopher King and Michael Loss for helpful discussions about this work.

References

1. Auchmuty, J., Beals, R.: Variational solutions of some nonlinear free boundary problems. Arch. Ration. Mech. Anal. **43**, 255–271 (1971). See also Models of rotating stars. Astrophys. J. **165**, L79–L82 (1971)
2. Chandrasekhar, S.: Phil. Mag. **11**, 592 (1931); Astrophys. J. **74**, 81 (1931); Monthly Notices Roy. Astron. Soc. **91**, 456 (1931); Rev. Mod. Phys. **56**, 137 (1984)
3. Conlon, J.: The ground state energy of a classical gas. Commun. Math. Phys. **94**, 439–458 (1984)
4. Daubechies, I.: An uncertainty principle for fermions with generalized kinetic energy. Commun. Math. Phys. **90**, 511–520 (1983)
5. Daubechies, I., Lieb, E.H.: One-electron relativistic molecules with Coulomb interaction. Commun. Math. Phys. **90**, 497–510 (1983)
6. Fefferman, C., de la Llave, R.: Relativistic stability of matter. I. Rev. Math. Iberoamericana **2**, 119–215 (1986)
7. Fowler, R.H.: Monthly Notices Roy. Astron. Soc. **87**, 114 (1926)
8. Herbst, I.: Spectral theory of the operator $(p^2 + m^2)^{1/2} - ze^2/r$. Commun. Math. Phys. **53**, 285–294 (1977); Errata ibid. **55**, 316 (1977)
9. Kato, T.: Perturbation theory for linear operators. Berlin, Heidelberg, New York: Springer 1966. See Remark 5.12, p. 307
10. Kovalenko, V., Perelmuter, M., Semenov, Ya.: Schrödinger operators with $L_w^{1/2}(\mathbb{R}^l)$ potentials. J. Math. Phys. **22**, 1033–1044 (1981)
11. Landau, L.: Phys. Z. Sowjetunion **1**, 285 (1932)
12. Lieb, E.: The number of bound states of one-body Schrödinger operators and the Weyl problem. Proc. Am. Math. Soc. Symp. Pure Math. **36**, 241–252 (1980). See also: Bounds on the eigenvalues of the Laplace and Schrödinger operators. Bull. Am. Math. Soc. **82**, 751–753 (1976)

13. Lieb, E.: Existence and uniqueness of the minimizing solution of Choquard's nonlinear equation. Stud. Appl. Math. **57**, 93–105 (1977)
14. Lieb, E.: Variational principle for many-fermions systems. Phys. Rev. Lett. **46**, 457–459 (1981); Errata ibid. **47**, 69 (1981)
15. Lieb, E.: Thomas-Fermi and related theories of atoms and molecules. Rev. Mod. Phys. **53**, 603–641 (1981); Errata ibid. **54**, 311 (1982)
16. Lieb, E.: Density functionals for Coulomb systems. Int. J. Quant. Chem. **24**, 243–277 (1983)
17. Lieb, E., Oxford, S.: An improved lower bound on the indirect Coulomb energy. Int. J. Quant. Chem. **19**, 427–439 (1981)
18. Lieb, E., Simon, B.: Thomas Fermi theory of atoms, molecules and solids. Adv. Math. **23**, 22–116 (1977)
19. Lieb, E., Thirring, W.: Gravitational collapse in quantum mechanics with relativistic kinetic energy. Ann. Phys. (NY) **155**, 494–512 (1984)
20. Lions, P.-L.: The concentration compactness principle in the calculus of variations; the locally compact case I. Ann. Inst. H. Poincaré Anal. non lineaire **1**, 109–145 (1984)
21. Messer, J.: Lecture Notes in Physics, Vol. 147. Berlin, Heidelberg, New York: Springer 1981
22. Morrey, C.B., Jr.: Multiple integrals in the calculus of variations, Theorem 5.8.6. Berlin, Heidelbereg, New York: Springer 1966
23. Ni, W.M.: Uniqueness of solutions of nonlinear Dirichlet problems. J. Differ. Equations **5**, 289–304 (1983)
24. Straumann, S.: General relativity and relativistic astrophysics. Berlin, Heidelberg, New York: Springer 1984
25. Thirring, W.: Bosonic black holes. Phys. Lett. B**127**, 27 (1983)
26. Weder, R.: Spectral analysis of pseudodifferential operators. J. Funct. Anal. **20**, 319–337 (1975)
27. Weinberg, S.: Gravitation and cosmology. New York: Wiley 1972
28. Hamada, T., Salpeter, E.: Models for zero temperature stars. Astrophys. J. **134**, 683 (1961)
29. Shapiro, S., Teukolsky, S.: Black holes, white dwarfs and neutron stars. New York: Wiley 1983
30. Ruffini, R., Bonazzola, S.: Systems of self-gravitating particles in general relativity and the concept of an equation of state. Phys. Rev. **187**, 1767–1783 (1969)

Communicated by E.H. Lieb

Received March 2, 1987, in revised form March 27, 1987

With I. Daubechies in Commun. Math. Phys. *90*, 497–510 (1983)

One-Electron Relativistic Molecules with Coulomb Interaction

Ingrid Daubechies[†*] and Elliott H. Lieb[†]
Departments of Mathematics and Physics, Princeton University, Princeton, NJ 08544, USA

Abstract. As an approximation to a relativistic one-electron molecule, we study the operator $H=(-\Delta+m^2)^{1/2}-e^2\sum_{j=1}^{K} Z_j|x-R_j|^{-1}$ with $Z_j \geqq 0$, $e^{-2}=137.04$. H is bounded below if and only if $e^2 Z_j \leqq 2/\pi$, all j. Assuming this condition, the system is unstable when $e^2\sum Z_j > 2/\pi$ in the sense that $E_0=\inf\,\mathrm{spec}(H)\to-\infty$ as the $R_j\to 0$, all j. We prove that the nuclear Coulomb repulsion more than restores stability; namely $E_0+0.069e^2\sum_{i<j} Z_iZ_j|R_i-R_j|^{-1}\geqq 0$. We also show that E_0 is an increasing function of the internuclear distances $|R_i-R_j|$.

Introduction

The problem of "stability of matter" consists in proving that a system of charged particles (electrons and nuclei), interacting electromagnetically, does not collapse. In the framework of nonrelativistic Schrödinger quantum mechanics, with Coulomb interactions between the particles, a first proof of this was given by F. Dyson and A. Lenard [1]. A shorter proof, leading to a much better lower bound on the binding energy per electron, was later given by E. Lieb and W. Thirring [2]. The strategy these proofs followed was first to consider the nuclei fixed (i.e. with infinite mass); the general case (nuclei with finite mass) then follows easily. With the K nuclei at fixed positions $R_1,\ldots,R_K$, the problem then consists in proving that 1) the Hamiltonian describing N electrons and K nuclei (including also the repulsion terms between the electrons, and between the nuclei) is bounded below by a constant independent of the R_j; 2) the energy per particle, i.e. the ground state energy of the system of N

[†] Work partially supported by U.S. National Science Foundation grant PHY-8116101-A01
[*] On leave from Vrije Universiteit Brussel, Belgium, and from Interuniversitair Instituut voor Kernwetenschappen, Belgium

electrons and K nuclei, divided by $(N+K)$, is bounded below by some constant.

If electrons were bosons, it is known [3] that matter would not be stable: statement 1) above still holds for bosons, but 2) does not.

As a first approximation to a relativistic approach to the problem of stability of matter, we shall study stability for the Hamiltonian

$$\sum_{j=1}^{N}(-\hbar^2c^2\Delta_j+m^2c^4)^{1/2}-\sum_{j=1}^{N}\sum_{k=1}^{K}Z_ke^2|x_j-R_k|^{-1}+\sum_{\substack{i,j=1\\i<j}}^{N}e^2|x_i-x_j|^{-1}$$
$$+\sum_{\substack{k,l=1\\k<l}}^{K}Z_kZ_le^2|R_k-R_l|^{-1}=H_{N,K}(m,\underline{Z},\underline{R}),\qquad(1.1)$$

where m is the electron mass, and where again the K nuclei are fixed at the distinct positions $R_1,\ldots,R_K$. We shall use the shorthand notation $\underline{Z}$, $\underline{R}$ for the sets $\{Z_j\}_{j=1}^K$, $\{R_j\}_{j=1}^K$.

To simplify our expressions, we rescale the x_j and R_k in units of the Compton wave length of the electron $\hbar/mc$. We also rescale the Hamiltonian by mc^2. In terms of the variables $\tilde{x}_j=(mc/\hbar)x_j$, and with $\tilde{R}_k=(mc/\hbar)R_k$, $\tilde{H}_{N,K}=(mc^2)^{-1}H_{N,K}$ becomes

$$\tilde{H}_{N,K}(\underline{Z},\underline{R})=\sum_{j=1}^{N}(-\tilde{\Delta}_j+1)^{1/2}-\sum_{j=1}^{N}\sum_{k=1}^{K}Z_k\alpha|\tilde{x}_j-\tilde{R}_k|^{-1}$$
$$+\sum_{\substack{i,j=1\\i<j}}^{N}\alpha|\tilde{x}_i-\tilde{x}_j|^{-1}+\sum_{\substack{k,l=1\\k<l}}^{K}Z_kZ_l\alpha|\tilde{R}_k-\tilde{R}_l|^{-1},$$

where $\alpha=e^2/\hbar c\simeq(137.04)^{-1}$.

It is obvious from the definition (1.1) that $H_{N,K}(m,\underline{Z},\underline{R})$ is bounded below by $H_{N,K}(0,\underline{Z},\underline{R})$. This will turn out to be very useful in some of our proofs below. We therefore introduce also a rescaled version of $H_{N,K}(0,\underline{Z},\underline{R})$, with a different scaling, since the Compton wave length is infinity if $m=0$. We rescale $H(0,\underline{Z},\underline{R})$ in units $\hbar c/R_0$, where R_0 is an arbitrary length, and scale x and the R_j by R_0. This yields

$$h_{N,K}(\underline{Z},\underline{R})=\sum_{j=1}^{N}(-\tilde{\Delta}_j)^{1/2}-\sum_{j=1}^{N}\sum_{k=1}^{K}Z_k\alpha|\tilde{x}_j-\tilde{R}_k|^{-1}$$
$$+\sum_{\substack{i,j=1\\i<j}}^{N}\alpha|\tilde{x}_i-\tilde{x}_j|^{-1}+\sum_{\substack{k,l=1\\k<l}}^{K}Z_kZ_l\alpha|\tilde{R}_k-\tilde{R}_l|^{-1},$$

where again the tildes indicate the scaled variables. We shall henceforth always work with these two scaled Hamiltonians, and omit the tildes.

It turns out that, unlike the Schrödinger case, even statement 1) above (i.e. the existence of an R-independent lower bound for the ground state energy of $H_{N,K}$) is not straightforward for $H_{N,K}$ even if we restrict ourselves to the case $N=1$ (1 electron). First of all, there exists only a limited range of the Z_j for which $H_{N,K}$ is bounded below at all. This can already be seen for the simplest case

$K=1$, $N=1$,

$$H_{1,1}(Z)=(-\Delta+1)^{1/2}-Z\alpha|x|^{-1}. \tag{1.2}$$

This operator was studied independently by R. Weder [4] and I. Herbst [5]. A first important fact (which can be found in Kato [6]) is the existence of a critical charge Z_c for the operator $H_{1,1}$, exactly as for the Klein–Gordon or Dirac theories. To be explicit, let

$$H_0=(-\Delta+1)^{1/2}, \quad Z_c=2/(\alpha\pi)\simeq 87.2.$$

Then a) for $Z \leqq Z_c$: $H_{1,1} \geqq 0$ as a form on $Q(H_0)$.

b) for $Z > Z_c$: $H_{1,1}$, is unbounded below as a form on $Q(H_0)$.

Here $Q(H_0)$ is the form domain of H_0, which consists of all the functions f in $L^2(\mathbb{R}^3)$ for which $|p|^{1/2}\hat{f}(p)$ is square integrable, where $\hat{f}$ denotes the Fourier transform of f.

If Z is strictly subcritical (i.e. $Z<Z_c$), more information concerning the spectrum of $H_{1,1}(Z)$ is known [4, 5]. The eigenvalues of $H_{1,1}(Z)$ all lie between 0 and 1; as a matter of fact, they are even separated from 0 by a gap which increases with Z_c-Z.

$$\sigma_{\text{disc}}(H_{1,1})\subset[[1-(Z/Z_c)^2]^{1/2},1). \tag{1.3}$$

Since $(1-\Delta)^{1/2}\leqq 1-\Delta/2$, and since $-\Delta/2-Ze^2|x|^{-1}$ has infinitely many negative eigenvalues, it follows that $H_{1,1}(Z)$ has infinitely many eigenvalues smaller than 1. It turns out, from an argument we give in Sect. 3, that the lowest eigenvalue is nondegenerate, and that the associated ground state is strictly positive.

On the other hand the essential spectrum always starts at 1, and it consists of only an absolutely continuous spectrum: $\sigma_{\text{ess}}(H_{1,1})=\sigma_{\text{abscont}}(H_{1,1})=[1,\infty)$.

The operator $h_{1,1}(Z)=(-\Delta)^{1/2}-Z\alpha|x|^{-1}$ has the same critical Z as $H_{1,1}$: it is bounded below by 0 on $Q(|p|)$ if $Z\leqq Z_c$, and unbounded below if $Z>Z_c$. Its spectrum is much simpler however: $\sigma(h_{1,1})=\sigma_{\text{abscont}}(h_{1,1})=[0,\infty)$.

One can easily extend the proofs in [4, 5] to show that $H_{N,K}$ is bounded below if and only if $Z_j\alpha\leqq 2/\pi$ for all j (we assume that all the R_j are distinct). From now on we shall only consider this case. Even under this condition, however, it is not obvious that an $\underline{R}$-independent lower bound for $H_{N,K}$ exists.

Let us consider the case $N=1$:

$$H_{1,K}=H^0_{1,K}+\sum_{\substack{k,l=1\\k<l}}^{K} Z_kZ_l\alpha|R_k-R_l|^{-1}, \tag{1.4}$$

where

$$H^0_{1,K}=(-\Delta+1)^{1/2}-\sum_{k=1}^{K} Z_k\alpha|x-R_k|^{-1} \tag{1.5}$$

is the 1-electron Hamiltonian without nuclear repulsion. Suppose that $\sum_{k=1}^{K} Z_k\alpha>2/\pi$ (this is possible when $K\geqq 2$). It is a simple consequence of the unboundedness below of $H_{1,1}(Z)$ for $Z\alpha>2/\pi$ that the ground state energy for $H^0_{1,K}$ will tend to $-\infty$ if the internuclear distances shrink to zero:

$$\liminf_{\lambda\to 0}\operatorname{spec} H^0_{1,K}(\underline{Z},\lambda\underline{R})=-\infty, \quad \sum_{k=1}^{K} Z_k\alpha>2/\pi.$$

Under the same conditions however, the nuclear repulsion term in (1.4) tends to $+\infty$, and this may "cancel" the behavior of $E_0 = \inf\operatorname{spec} H^0_{1,K}$. The existence of an $\underline{R}$-independent lower bound for $H_{1,K}(\underline{Z}, \underline{R})$ is therefore only possible if the nuclear repulsion is strong enough to overcome the "collapsing tendency" present in $H^0_{1,K}$. The fact that this is the case is the main content of the following theorem:

Theorem 1. *Let* $h_0 = (-\Delta)^{1/2}$. *Suppose* $z_k \leqq 2/\pi$ *for all* k, *and* $R_k \neq R_l$ *for all* $k \neq l$. *Then* $h_0 - \sum_{k=1}^{K} z_k |x - R_k|^{-1}$, *considered as a form on* $Q(h_0)$, *is bounded below by* $-3\pi \sum_{\substack{k,l=1 \\ k<l}}^{K} z_k z_l |R_k - R_l|^{-1}$.

Since $(-\Delta + 1)^{1/2} \geqq (-\Delta)^{1/2}$, the theorem obviously also holds if we replace h_0 by $H_0 = (-\Delta + 1)^{1/2}$.

In terms of the fine-structure constant α and the nuclear charges $Z_k = z_k/\alpha$, Theorem 1 can be rewritten as

$$E_0 + \sum_{\substack{k,l=1 \\ k<l}}^{K} Z_k Z_l \alpha |R_k - R_l|^{-1} \geqq (1-b) \sum_{\substack{k,l=1 \\ k<l}}^{K} Z_k Z_l \alpha |R_k - R_l|^{-1}, \tag{1.6}$$

where $b = 3\pi\alpha \leqq 6.88 \times 10^{-2}$. The fact that b is strictly smaller than 1 implies that the total energy (including the nuclear repulsion) increases to $+\infty$ if any two nuclei coincide. Thus the nuclear repulsion is not only strong enough to prevent collapse, it even pushes the nuclei apart!

For

$$h_{1,K}(\underline{Z}, \underline{R}) = h^0_{1,K}(\underline{Z}, \underline{R}) + \sum_{\substack{k,l=1 \\ k<l}}^{K} Z_k Z_l \alpha |R_k - R_l|^{-1}, \tag{1.7}$$

with

$$h^0_{1,K}(\underline{Z}, \underline{R}) = (-\Delta)^{1/2} - \sum_{k=1}^{K} Z_k \alpha |x - R_k|^{-1}, \tag{1.8}$$

we can say even more. By scaling one sees that $h_{1,K}(\underline{Z}, \underline{R})$ and $\lambda h_{1,K}(\underline{Z}, \lambda\underline{R})$ are unitarily equivalent. Theorem 1 implies therefore that in this case the total energy is minimal if the nuclei are infinitely far apart.

Incidentally, note that the exact numerical value of α plays a decisive role. This can be understood by the scaling behavior of $H_{1,K}$. By simple scaling one sees that $H_{1,K}(\underline{Z}, \underline{R})$ is unitarily equivalent to $\lambda H_{1,K}(\lambda^{-1}; \underline{Z}, \lambda\underline{R})$, where we define

$$H_{1,K}(\mu; \underline{Z}, \underline{R}) = (-\Delta + \mu^2)^{1/2} - \sum_{k=1}^{K} Z_k \alpha |x - R_k|^{-1} + \sum_{k,l=1}^{K} Z_k Z_l \alpha |R_k - R_l|^{-1}.$$

(The same is true for $H_{N,K}$.) This means essentially that we can scale the mass away: the existence of any $\underline{R}$-independent bound for $H_{1,K}(1; \underline{Z}, \underline{R})$ implies that the zero-mass Hamiltonian $H_{1,K}(0; \underline{Z}, \underline{R}) = h_{1,K}(\underline{Z}, \underline{R})$ is bounded below by zero, independently of $\underline{R}$. (Since $H_{1,K} \geqq h_{1,K}$, the converse statement is trivial.)

This, in turn, implies that the value of α is decisive. Let us illustrate this for the special case $K = 2$, $Z_1\alpha = Z_2\alpha = 2/\pi$. On the one hand, if $H_{1,2}(\underline{Z}, \underline{R})$ is bounded below by an $\underline{R}$-independent constant, then $h_{1,2}(\underline{Z}, \underline{R})$ is bounded below by zero. Thus, for all ψ in $Q[(-\Delta)^{1/2}]$,

$$\langle \psi, h^0_{1,2}(\underline{Z}, \underline{R})\psi \rangle + (4/\pi^2\alpha)|R_1 - R_2|^{-1} \|\psi\|^2 \geqq 0$$

($h^0_{1,2}$ is $h_{1,2}$ without the nuclear repulsion; see (1.8)). On the other hand, by a scaling argument,

$$\inf \operatorname{spec} [h^0_{1,2}(\underline{Z}, \underline{R})] = -C|R_1 - R_2|^{-1}$$

for some positive constant C, which is completely independent of α. It is here that the numerical value of α comes into play: if α is such that $4/(\alpha\pi^2) \geqq C$, then the required bound will hold (for $K = 2$); if $4/(\alpha\pi^2) < C$, the one-electron-two-nucleus system is definitely unstable in this model. By the argument above, this conclusion will hold as well for $H_{1,K}$, i.e. for any positive mass, as for $h_{1,K}$, which has zero mass.

We prove Theorem 1 in Sect. 2.

In Sect. 3 we show that, for one electron, $E^0_{1,K} = \inf \operatorname{spec} H^0_{1,K}(\underline{Z}, \underline{R})$, and $e^0_{1,K} = \inf \operatorname{spec} h^0_{1,K}(\underline{Z}, R)$ are monotone increasing functions of the internuclear distances:

Theorem 2. *Keep the Z_k fixed, where $Z_k\alpha \leqq 2/\pi$ for all k. If for all k, l: $|R_k - R_l| \leqq |R'_k - R'_l|$, then $E^0_{1,K}(\underline{Z}, \underline{R}) \leqq E^0_{1,K}(\underline{Z}, \underline{R}')$ and $e^0_{1,K}(\underline{Z}, \underline{R}) \leqq e^0_{1,K}(\underline{Z}, \underline{R}')$.*

Our proof borrows its basic ideas from [7] and [8], where the analogous statement for the Schrödinger operator $-\Delta - \sum_{k=1}^{K} Z_k e^2 |x - R_k|^{-1}$ was proved (first for dilations, i.e. $R'_k = \lambda R_k$ with $\lambda \geqq 1$, in [7]; this proof was extended to the general situation $|R'_k - R'_l| \geqq |R_k - R_l|$ in [8]).

If all the Z_k are strictly subcritical, i.e. $Z_k\alpha < 2/\pi$ for all k, one can extend a result of [4, 5] and show that $\sigma_{\text{ess}}(H^0_{1,K}) = [1, \infty)$ and analogously $\sigma_{\text{ess}}(h^0_{1,K}) = [0, \infty)$. For any $\underline{Z}$ (provided all the Z_k are not zero) one can find a variational function (e.g. a suitable exponential) such that $\langle \psi, H^0_{1,K}\psi \rangle < 1$. This shows that for $Z_k < Z_c$, all k, $E^0_{1,K}$ is an isolated eigenvalue of $H^0_{1,K}$. (Actually, by the same argument as for $H_{1,1}$, one sees that $H^0_{1,K}$ has infinitely many eigenvalues smaller than 1.) For $h^0_{1,K}$, matters are slightly different. If $\sum_{k=1}^{K} Z_k \leqq Z_c$, one can easily extend arguments of [4, 5] to show that $h^0_{1,K} \geqq 0$, in which case $e^0_{1,K} = 0$, and $h^0_{1,K}$ has no eigenvalues. If $\sum_{k=1}^{K} Z_k > Z_c$, one can again find a variational function ψ for which $\langle \psi, h^0_{1,K}\psi \rangle < 0$, which implies that in that case $e^0_{1,K}$ is also an isolated eigenvalue.

We prove in Sect. 3 that, if $Z_k < Z_c$ for all k, $E^0_{1,K}$ is a nondegenerate eigenvalue of $H^0_{1,K}$, and that the corresponding ground state is strictly positive. The same is true for $e^0_{1,K}$ under the additional assumption $\sum_{k=1}^{K} Z_k > Z_c$.

We have no results in this paper concerning $H_{N,K}$ for $N > 1$. The same scaling argument as for $H_{1,K}$ applies, and the existence of an $\underline{R}$-independent lower bound for $H_{N,K}$ would therefore imply that $h_{N,K}$ (i.e. the same Hamiltonian with $(-\Delta)^{1/2}$ instead of $(-\Delta+1)^{1/2}$) is bounded below by zero. We do not know, however, whether such a lower bound exists for general N, K.

Regardless of the statistics, and neglecting the electron repulsion, we know that if $K=1$ (one nucleus only) $H_{N,1} \geqq 0$ (because $H_{1,1} \geqq 0$). On the other hand, the lower bound (1.6) shows that for $N \leqq 14$ and arbitrary K, we also have $H_{N,K} \geqq 0$ (because $1-14b \geqq 0$, where $b=3\pi\alpha$). For general values of N, K it is obvious that the Fermi statistics and the electron repulsion will have to play a role.

It is clear that the strategy of [2], which used a density functional lower bound on the kinetic energy, and then applied results of Thomas Fermi theory, will not work here. The lower bound on the kinetic energy for the operator $(-\Delta+1)^{1/2}$ has the form [9] $\langle \psi, (-\Delta+1)^{1/2}\psi\rangle \geqq K\int d^3x F(|\psi(x)|^2)$, where $F(a)=\int_0^a du(u^{2/3}+1)^{1/2}$. One sees that for small a, $F(a)$ behaves like $ca^{5/3}+a$, which is similar to the $\rho_\psi^{5/3}$ bound one obtains for $\langle\psi,(-\Delta)\psi\rangle$ [2], and which is caused by the fact that for small p, $(p^2+1)^{1/2}$ behaves like $1+p^2/2$. For large p, however, $(p^2+1)^{1/2}$ behaves linearly in p, which is reflected in the $a^{4/3}$ behavior of $F(a)$ for large a. Hence the lower bound on the kinetic energy is of the form $\int d^3x \rho_\psi(x)^{4/3}$ in the region where ρ_ψ is large; the corresponding Thomas–Fermi functional (including the other contributions to the energy) is not bounded below, and therefore does not lead to a useful lower bound on the ground state energy of $H_{N,K}$.

Heuristically, one can argue that Thomas–Fermi theories are "large Z theories" (i.e. they give the correct asymptotic behavior for $Z\to\infty$; this is rigorously true for the Schrödinger case [10].) Hence it is only natural that the Thomas–Fermi theory corresponding to $H_{N,K}$ is unbounded below, since $H_{N,K}$ itself is unbounded below whenever one Z_j becomes supercritical.

Throughout this paper we consider only the three-dimensional case, i.e. the Hilbert space we use is $L^2(\mathbb{R}^3)$. For functions on $\mathbb{R}^3$, the symbol $\| \quad \|$ will be used to denote the $L^2(\mathbb{R}^3)$-norm only; for operators, this symbol will always mean the norm as a bounded operator from $L^2(\mathbb{R}^3)$ to itself.

2. The Nuclear Repulsion Restores Stability

We shall prove Theorem 1 in two steps: first we shall prove it in the case $Z_k = Z_c$, all k. Then we shall apply a concavity argument to conclude the desired result for all Z_k. In the proof of the first step we shall need the following lemma, which can be considered as a refinement of the result $\||x|^{-1/2}|p|^{-1/2}\| = (\pi/2)^{1/2}$. This norm was evaluated in [5], and by a different method also in [11]; the critical value $Z_c e^2 = 2/\pi$ actually has its origin in this number.

Notation. $B(a,R)=\{x\,|\,|x-a|\leqq R\}$.

Lemma 2.1. *Let ψ be an L^2-function with* supp $\psi \subset B(0,R)$. *We denote by K the bounded operator $K=2\pi^{-1}|x|^{-1/2}|p|^{-1}|x|^{-1/2}$. Then*

$$\langle\psi, K\psi\rangle \leqq \langle\psi,\psi\rangle - \pi^{-3}R^{-2}\left(\int d^3x|x|^{-1/2}|\psi|\right)^2. \tag{2.1}$$

Proof. Note first that since ψ has compact support, the integral $\int d^3x|x|^{-1/2}|\psi(x)|$ converges. By simple scaling, we can obviously assume that $R = 1$, which we shall do henceforth in this proof.

We denote the symmetric decreasing rearrangement of ψ by ψ^*. Since $|x|^{-1/2}$ is symmetric and decreasing, we have $\int \psi^*|x|^{-1/2} \geqq \int |\psi||x|^{-1/2}$. Analogously, using the fact that $|p|^{-1}$ is the same as convolution with $(2\pi^2)^{-1}|x|^{-2}$, one sees that $\langle \psi^*, K\psi^* \rangle \geqq \langle \psi, K\psi \rangle$ (this follows from the generalized rearrangement inequalities proved in [12].) Hence we can restrict our attention to ψ which are symmetric decreasing, i.e. $\psi = \psi^*$.

For any such $\psi(r)$, defined for $0 \leqq r \leqq 1$, we define, for $r \geqq 1, \tilde{\psi}(r) = r^{-3}\psi(r^{-1})$. Consider the function $f(r) = \psi(r)$, $r \leqq 1, f(r) = \tilde{\psi}(r)$, $r \geqq 1$. One easily checks that f is in $L^2(\mathbb{R}^3)$, with $\|f\|^2 = 2\|\psi\|^2$. Straightforward calculation also leads to $\langle f, Kf \rangle = 2\langle \psi, K\psi \rangle + 2\langle \psi, L\psi \rangle$, where L has the integral kernel $L(x,y) = \pi^{-3}|x|^{-1/2}(1 + x^2y^2 - 2x\cdot y)^{-1}|y|^{-1/2}$. Since $\||x|^{-1/2}|p|^{-1/2}\| = (\pi/2)^{1/2}$, we have $\|K\| = 1$, hence $\langle f, Kf \rangle \leqq \langle f, f \rangle$, which implies

$$\langle \psi, K\psi \rangle \leqq \|\psi\|^2 - \langle \psi, L\psi \rangle. \tag{2.2}$$

By computing the spherical average of the kernel $L(x,y)$, we obtain

$$\begin{aligned} \langle \psi, L\psi \rangle &= (8/\pi) \int_0^1 dr \int_0^1 ds\, r^{1/2}\psi(r) s^{1/2}\psi(s) \ln[(1+rs)/(1-rs)] \\ &\geqq (\pi^{-3}/2)[\int d^3x|x|^{-1/2}\psi(x)]^2 \min_{u\in[0,1]} f(u), \end{aligned} \tag{2.3}$$

where $f(u) = u^{-1}\ln[(1+u)/(1-u)]$.

One can easily check that f attains its minimum at $u = 0$, with $f(0) = 2$. Expression (2.1) now follows immediately from (2.2) and (2.3). ■

Remark. The constant π^{-3} is probably not optimal in (2.1). We suspect that the optimal constant is given by functions behaving like $r^{-3/2}$. This is also the typical behavior of functions optimising the expectation value of $|x|^{-1/2}|p|^{-1}|x|^{-1/2}$ (see [11], [13]).

If we define $\psi_n(r) = 0$ for $r < n^{-1}$, $\psi_n(r) = r^{-3/2}$ for $n^{-1} < r \leqq 1$, it turns out that

$$\lim_{n\to\infty} [\langle \psi_n, \psi_n \rangle - \langle \psi_n, K\psi_n \rangle]/[\int d^3x|x|^{-1/2}\psi_n(x)]^2 = a\pi^{-3},$$

with $a = \sum_{k=0}^{\infty} (2k+1)^{-3} = (7/8)\zeta(3) \simeq 1.05$. We think $a\pi^{-3}$ is probably the best constant for (2.1); in any event the sharp constant lies in the interval $[\pi^{-3}, 1.05\pi^{-3}]$.

With the help of Lemma 2.1, we can now prove Theorem 1 for the case where all the Z_j are critical.

Proposition 2.2. *For all* $\psi \in Q(|p|)$, $\|\psi\| = 1$:

$$\langle \psi, |p|\psi \rangle - (2/\pi) \sum_{j=1}^{K} \langle \psi, |x - R_j|^{-1}\psi \rangle \geqq -(12/\pi) \sum_{\substack{j,k=1 \\ j<k}}^{K} |R_j - R_k|^{-1}. \tag{2.4}$$

Proof. Note first that since $|x|^{-1}$ is relatively form-bounded with respect to $|p|$ (see [4–6]), the left hand side of (2.4) makes sense for all ψ in $Q(|p|)$.

Let us rewrite the desired result (2.4) as

$$\forall\psi\in Q(|p|):\langle\psi,|p|\psi\rangle \geqq (2/\pi)\sum_{j=1}^{K}\left\langle\psi,\left[|x-R_j|^{-1}-A\sum_{k\neq j}|R_k-R_j|^{-1}\right]\psi\right\rangle. \tag{2.5}$$

For the time being, we shall not fix A. We shall determine below a value A_0 such that (2.5) holds for all $A\geqq A_0$. It will turn out that $A_0=3$, which then implies (2.4).

For r real, we shall use the notation $r_+=\max(0,r)$. For K arbitrary real numbers $r_1,\ldots,r_K$, we have

$$\sum_{j=1}^{K} r_j \leqq \left(\sum_j r_j\right)_+ \leqq \sum_j (r_j)_+ \leqq \left(\sum_j (r_j)_+^{1/2}\right)^2.$$

Hence (2.5) will certainly be satisfied if

$$\forall\psi\in Q(|p|):\langle\psi,|p|\psi\rangle \geqq \left\langle\psi,\left(\sum_{j=1}^{K} W_j(x)\right)^2\psi\right\rangle, \tag{2.6}$$

where $W_j(x)=[(2/\pi)(|x-R_j|^{-1}-A\sum_{k\neq j}|R_k-R_j|^{-1})_+]^{1/2}$.

For any A, W_j has support in a ball with center R_j and radius t_j, where

$$t_j^{-1}=A\sum_{k\neq j}|R_k-R_j|^{-1}. \tag{2.7}$$

Provided A is large enough, all the balls $B(R_j,t_j)$ are disjoint: by (2.7), we have $t_j \leqq A^{-1}|R_k-R_j|$ for all $k\neq j$. Therefore

$$|R_k-R_j|-(t_j+t_k)\leqq(1-2/A)|R_k-R_j|. \tag{2.8}$$

which shows that $B(R_j,t_j)\cap B(R_k,t_k)=\emptyset$, for all $j\neq k$, if $A>2$. From now on we shall assume $A>2$.

For any j, let f_j be the characteristic function of $B(R_j,t_j)$. Since $W_j=W_jf_j$, and $W_j(x)\leqq(2/\pi)^{1/2}|x-R_j|^{-1/2}$, we have

$$\sum_j W_j(x)=\sum_j W_j(x)f_j(x)\leqq(2/\pi)^{1/2}\sum_j|x-R_j|^{-1/2}f_j(x)\equiv W(x). \tag{2.9}$$

Since $|x-R_j|^{-1/2}f_j\leqq|x-R_j|^{-1/2}$, and $|x|^{-1/2}|p|^{-1/2}$ is bounded, one easily sees that $W|p|^{-1/2}$ is bounded. We shall prove now that as an operator from L^2 to L^2, $\|W|p|^{-1/2}\|\leqq 1$, provided A is larger than some constant which we shall evaluate below.

Let ψ be any function in C_0^∞, so that $W\psi\in L^2$. Then

$$\begin{aligned}\||p|^{-1/2}W\psi\|^2&=(W\psi,|p|^{-1}W\psi)\\&=(2/\pi)\sum_{j=1}^{K}(|x-R_j|^{-1/2}f_j\psi,|p|^{-1}|x-R_j|^{-1/2}f_j\psi)\\&\quad+(2/\pi)\sum_{\substack{j,k=1\\j\neq k}}^{K}(f_j|x-R_j|^{-1/2}\psi,|p|^{-1}f_k|x-R_k|^{-1/2}\psi).\end{aligned} \tag{2.10}$$

The "diagonal" terms in (2.10) can be bounded above by applying Lemma 2.1:

$$(2/\pi)\sum_{j=1}^{K}(f_j\psi,|x-R_j|^{-1/2}|p|^{-1}|x-R_j|^{-1/2}f_j\psi)$$

$$\leqq \sum_{j=1}^{K}\|f_j\psi\|^2-\pi^{-3}\sum_{j=1}^{K}t_j^{-2}m_j^2, \tag{2.11}$$

where

$$m_j=\int d^3x|x-R_j|^{-1/2}f_j|\psi|. \tag{2.12}$$

To find an upper bound for the "non-diagonal" terms in (2.10) we use (2.8) and the fact that $|p|^{-1}$ is the same as convolution with $(2\pi^2)^{-1}|x|^{-2}$. We obtain

$$\begin{aligned}&(2/\pi)(f_j|x-R_j|^{-1/2}\psi,|p|^{-1}f_k|x-R_k|^{-1/2}\psi)\\ &\quad\leqq \pi^{-3}[1-2/A]^{-2}|R_k-R_j|^{-2}m_jm_k,\end{aligned} \tag{2.13}$$

where m_j is defined by (2.12).

Combining (2.11) and (2.13) we obtain

$$\begin{aligned}\||p|^{-1/2}W\psi\|^2 &\leqq \|\psi\|^2-\sum_{j=1}^{K}(\pi^{-3}t_j^{-2}m_j^2)\\ &+\pi^{-3}[1-2/A]^{-2}\sum_{\substack{j,k=1\\ j\neq k}}^{K}m_jm_k|R_j-R_k|^{-2}.\end{aligned}$$

Using $m_jm_k\leqq(m_j^2+m_k^2)/2$ we can rewrite this as

$$\||p|^{-1/2}W\psi\|^2\leqq\|\psi\|^2-\sum_{j=1}^{K}m_j^2b_j,$$

where $b_j=\pi^{-3}t_j^{-2}-\pi^{-3}[1-2/A]^{-2}\sum_{k\neq j}|R_k-R_j|^{-2}$. By the definition (2.7) of t_j, we have

$$b_j\geqq\pi^{-3}[A^2-(1-2/A)^{-2}]\sum_{k\neq j}|R_k-R_j|^{-2}.$$

This shows that all the b_j will be positive, and hence $\||p|^{-1/2}W\psi\|\leqq\|\psi\|$, if $A^2\geqq(1-2/A)^{-2}$, or

$$A\geqq A_0=3. \tag{2.14}$$

Thus, provided (2.14) is satisfied, we have $\||p|^{-1/2}W\psi\|\leqq\|\psi\|$ for all ψ in C_0^∞, hence $\|W|p|^{-1/2}\|=\||p|^{-1/2}W\|=1$. This implies $|p|-W^2=|p|^{1/2}[1-(W|p|^{-1/2})^*(W|p|^{-1/2})]|p|^{1/2}\geqq 0$.

Since $\left(\sum_j W_j\right)^2\leqq W^2$ by (2.9), we have therefore proved (2.6) and *a fortiori* (2.5) for all $A\geqq A_0=3$. ■

Remarks

1. In terms of Z_c and e^2, we can rewrite (2.4) as $\forall\psi\in Q(|p|)$, $\|\psi\|=1$:

$$\left\langle \psi, \left[|p| - Z_c\alpha \sum_{j=1}^{K} |x-R_j|^{-1} \right] \psi \right\rangle \geqq -bZ_c^2\alpha \sum_{\substack{j,k=1\\ j<k}}^{K} |R_j-R_k|^{-1}, \qquad (2.15)$$

where $b=\pi\alpha A_0 \leqq 6.88\times 10^{-2}$.

2. Note that (2.15) implies that $\sum_{j=1}^{K} Z_j\alpha|x-R_j|^{-1}$ is relatively form-bounded with respect to $|p|$, with relative bound $(\max Z_j)/Z_c$. This means that $h_{1,K}$ and $H_{1,K}$ can be defined as form sums, with form domain $Q(|p|)$ if $Z_j<Z_c$ for all j; if $Z_j=Z_c$ for some j, we define $H_{1,K}, h_{1,K}$ to be the Friedrichs extensions of the quadratic forms on $Q(|p|)$.

With Proposition 2.2 ends our first step in the proof of Theorem 1 which is now established if $Z_j=Z_c$ for all j. From this result we shall derive Theorem 1 for all possible values of Z_j by a concavity argument. First we prove another technical lemma:

Lemma 2.3. *Let f be a real function on the n-dimensional unit cube $[0,1]^n$, convex in each variable separately. Let h be the smallest concave function on $[0,1]^n$ (jointly concave in all the variables) agreeing with f on all the vertices of the cube*, i.e.

$$\forall P_k \text{ vertex of } [0,1]^n \quad (k=1,\dots,M, \text{ with } M=2^n): \quad h(P_k)=f(P_k).$$

Then $\forall(x_1,\dots,x_n)\in[0,1]^n$: $h(x_1,\dots,x_n)\geqq f(x_1,\dots,x_n)$.

Remarks

1. We shall apply this lemma to the function $f(x_1,\dots,x_n)=\sum_{i<j}\alpha_{ij}x_ix_j$. If we keep all the variables but one fixed, f is affine in the remaining variable. The requirement that f be convex in each variable separately is therefore obviously satisfied.

2. By changing signs, one immediately sees that the lemma is still true if we exchange "concave" and "convex," if we make h the "largest convex function..." instead of the "smallest concave function...," and if we reverse the inequality sign in the conclusion.

3. Since h is the smallest concave function agreeing with f on the vertices, the conclusion $h\geqq f$ automatically implies that h is the concave envelope of f. The lemma tells us therefore that to construct the concave envelope on a cube of a function which is convex in each variable separately, one only has to consider the values of f at the vertices of the cube.

4. At any point $(x_1,\dots,x_n)$ in the cube, h can be explicitly constructed as follows: for any decomposition of $(x_1,\dots,x_n)$ into a convex combination of the vertices: $x_m=\sum_{j=1}^{M}\lambda_j\,(P_j)_m$, $m=1,\dots,n$, with $\lambda_j\geqq 0$ and $\sum_{j=1}^{M}\lambda_j=1$, we define $F(\underline{\lambda})=\sum_{j=1}^{M}\lambda_j f(P_j)$; h is given by the maximum of $F(\underline{\lambda})$, taken over all possible convex decompositions of $(x_1,\dots,x_n)$.

Proof of Lemma 2.3. Fix $y_2,\ldots,y_n$ so that each is either 0 or 1. Considered as a function of x_1 only, $(h-f)(x_1,y_2,\ldots,y_n)=h(x_1,y_2,\ldots,y_n)-f(x_1,y_2,\ldots,y_n)$ is concave (h is jointly concave, and $-f$ is concave in each variable separately.)

Since h and f agree on the vertices, we have $(h-f)(0,y_2,\ldots,y_n)\geqq 0$, $(h-f)(1,y_2,\ldots,y_n)\geqq 0$. By the concavity of $h-f$ in x_1, this implies $(h-f)(x_1,y_2,\ldots,y_n)\geqq 0$ for all $x_i\in[0,1]$. Now fix x_1, and note that h and f satisfy the stated hypotheses in the variables $y_2,\ldots,y_n$. The complete result follows by iterating the argument n times. ■

With the help of this lemma and Proposition 2.2, we can now prove Theorem 1:

Proof of Theorem 1. For $(z_1,\ldots,z_k)\in[0,1]^K$, we define

$$F(z_1,\ldots,z_K)=\inf_{\psi\in Q(|p|),\|\psi\|=1}\left[\langle\psi,|p|\psi\rangle-(2/\pi)\sum_{k=1}^{K}z_k\langle\psi,|x-R_k|^{-1}\psi\rangle\right].$$

As the infimum of linear functions, F is concave. Define

$$f(z_1,\ldots,z_K)=-(12/\pi)\sum_{\substack{k,l=1\\k<l}}^{K}|R_k-R_l|^{-1}z_kz_l. \tag{2.16}$$

Proposition 2.2 tells us that

$$F(P_k)\geqq f(P_k)\quad k=1,\ldots,2^K,$$

where the P_k, $k=1,\ldots,2^K$ are the vertices of the cube $[0,1]^K$. (When all the components of the vertices are 1, i.e. $P_k=(1,\ldots,1)$, (2.16) follows from Proposition 2.2. When some of the components of P_k are zero, (2.16) still follows from Proposition 2.2, but now with a smaller value for K.) Since F is concave, (2.16) implies that F is larger than h, the smallest concave function agreeing with f at the cornerpoints P_k of the cube. But h is larger than f by Lemma 2.3; by combining these two inequalities we obtain $F\geqq f$, or for all $z_k\in[0,1]$, $k=1,\ldots,K$:

$$\inf_{\psi\in Q(|p|),\|\psi\|=1}\left[\langle\psi,|p|\psi\rangle-(2/\pi)\sum_{k=1}^{K}z_k\langle\psi,|x-R_k|^{-1}\psi\rangle\right]$$

$$\geqq-(12/\pi)\sum_{\substack{k,l=1\\k<l}}^{K}z_kz_l|R_k-R_l|^{-1}.\quad ■ \tag{2.17}$$

Remark. In terms of the fine-structure constant α and of the nuclear charges Z_k, with $Z_k\leqq Z_c=2/(\alpha\pi)$ for all k, (2.17) can be rewritten as

$$|p|-\sum_{k=1}^{K}Z_k\alpha|x-R_k|^{-1}\geqq-b\sum_{\substack{k,l=1\\k<l}}^{K}Z_kZ_l\alpha|R_k-R_l|^{-1},$$

where $b\leqq 6.88\times10^{-2}$.

3. Monotonicity of the Ground State Energy in the Nuclear Coordinates

We prove Theorem 2 by essentially the same method as used by one of us [8] in the proof of the analogous statement for the Schrödinger Hamiltonian

$-\Delta - \sum_{k=1}^{K} Z_k|x - R_k|^{-1}$. The strategy followed there was the following (modulo some epsilons):

1) $E_0(\underline{R}) = \inf \operatorname{spec} H(\underline{R}) = -\lim_{t\to\infty} t^{-1} \ln G(x, x, t, \underline{R})$ (3.1)

for all x, where $G(x, y, t, \underline{R})$ is the kernel $\exp[-tH(\underline{R})](x, y)$.

2) By the Trotter product formula, $G(x, y, t, \underline{R})$ can be approximated by multiple integrals involving $\exp[-(t/n)(-\Delta)](x, y)$ and $\exp[(t/n)Z_k|x - R_k|^{-1}]$. (Here one of the epsilons mentioned above comes into play: in order to ensure convergence of the integrals, one writes $G(x, y,; t, \underline{R})$ as $\lim_{\varepsilon\to 0} G_\varepsilon(x, y; t, \underline{R})$, where G_ε is the kernel of the semigroup corresponding to $-\Delta - \sum_{k=1}^{K} Z_k(|x - R_k|^2 + \varepsilon^2)^{-1/2}$. The whole argument is then carried out for G_ε rather than for G and the limit $\varepsilon \to 0$ is taken at the end.)

3) Since $\exp[as^{-1/2}]$ is completely monotone, it can be written as the Laplace transform of a positive measure; this can be used to write $\exp[(t/n)Z|x - R|^{-1}]$ as an integral, over a positive measure, of $\exp[-s|x - R|^2]$.

4) Since all the kernels in the Trotter product approximation for G have been reduced to positive superpositions of Gaussian kernels (the kernel of $\exp[(t/n)\Delta]$ was already Gaussian in the Schrödinger case [8]), a lemma involving only Gaussian functions (Lemma 4 in [8]) can be applied, leading to the conclusion:

5) If $\underline{R}, \underline{R}'$ are given with $|R_i - R_j| \leqq |R_i' - R_j'| \forall i, j$, and if x, y, x', y' are such that $|x - R_i| \leqq |x' - R_i'|, |y - R_i| \leqq |y' - R_i'|$ for all i, and $|x - y| \leqq |x' - y'|$, then

$$G(x, y, t, \underline{R}) \geqq G(x', y', t, \underline{R}').$$

Combined with 1) this proves the theorem. One uses the fact that given $\underline{R}, \underline{R}'$ and x, one can always find x' satisfying $|x' - R_i'| \geqq |x - R_i|$ by choosing x' far enough away from the R_i'.

The same argument can be used to prove Theorem 2 in our case. Some modifications are needed, however. We shall only describe these in detail, and not repeat the whole proof as given in [8].

The first modification concerns the kernels $\exp[(t/n)(-\Delta + 1)^{1/2}]$ and $\exp[-(t/n)(-\Delta)^{1/2}]$. These are not Gaussian kernels, but they can again be written as positive superpositions of Gaussian kernels. Indeed, $\exp[-bs^{1/2}]$ is again a completely monotone function, and can therefore again, by Bernstein's theorem, be written as a Laplace transform of a positive measure:

$$\exp[-bs^{1/2}] = \int e^{-sv} d\mu(v, b);$$

an explicit formula for μ can be found in any table of Laplace transforms: $d\mu(v, b) = (b\pi^{-1/2}/2)v^{-3/2} \exp[-b^2/4v]dv$. It follows now that

$$\exp[-(t/n)(-\Delta + 1)^{1/2}](x, y) = \int (2v)^{-3/2} \exp[-|x - y|^2/4v] e^{-v} d\mu(v, t/n),$$
$$\exp[-(t/n)(-\Delta)^{1/2}](x, y) = \int (2v)^{-3/2} \exp[-|x - y|^2/4v] d\mu(v, t/n), \quad (3.2)$$

which are indeed positive superpositions of Gaussians.

The second modification is more a set of comments on step 1) than a real modification.

First of all, note that it is enough to prove Theorem 2 for the case $Z_k < Z_c$ for all k. As the infima of decreasing linear functions of the Z_k, $E^0_{1,K}(\underline{Z})$ and $e^0_{1,K}(\underline{Z})$ are decreasing, upper semicontinuous functions of the Z_k, which ensures that

$$E^0_{1,K}(\underline{Z}) = \lim_{\varepsilon\to 0} E^0_{1,K}(\underline{Z-\varepsilon}), e^0_{1,K}(\underline{Z}) = \lim_{\varepsilon\to 0} e^0_{1,K}(\underline{Z-\varepsilon}),$$

where we use the notation $\underline{Z-\varepsilon}$ for the set $\{Z_j - \varepsilon\}$. This implies that if Theorem 2 holds for the case where $Z_k < Z_c$ for all k, it also holds in the case where some of the Z_k are equal to Z_c.

We can therefore restrict ourselves to the case $Z_k < Z_c$, all k. In this case $E^0_{1,K}$ is an isolated eigenvalue (one can easily show $E^0_{1,K} < 1$ by a variational argument, and since $Z_k < Z_c$ for all k, we know that $\sigma_{\mathrm{ess}}(H^0_{1,K}) = [1, \infty)$). For $e^0_{1,K}$, matters are slightly different. If $\sum_{k=1}^{K} Z_k \leqq Z_c$, we know that $e^0_{1,K}(\underline{Z}, \underline{R}) = 0$, independently of $\underline{R}$, in which case the monotonicity in $\underline{R}$ is trivial. We shall therefore not consider this case in the discussion below. Whenever $\sum_{k=1}^{K} Z_k > Z_c$, however, one has again $e^0_{1,K} < 0$ by a variational argument, and since $\sigma_{\mathrm{ess}}(h^0_{1,K}) = [0, \infty)$, this shows that then $e^0_{1,K}$ too is an isolated eigenvalue. We shall therefore assume in the following that $E^0_{1,K}, e^0_{1,K}$ are isolated eigenvalues. We know then that (3.1) will be true, for all x, if there exist ground states for $H^0_{1,K}, h^0_{1,K}$ which are strictly positive. The argument in the next paragraph shows that this is indeed the case: we show that the ground states for $H^0_{1,K}, h^0_{1,K}$ (where $\sum_{k=1}^{K} Z_k > Z_c$ in the last case, and $Z_k < Z_c$ for all k for both Hamiltonians) are nondegenerate and strictly positive, which is an interesting result in its own right.

Since $H^0_{1,K} = H_0 - V(\underline{R}, x), h^0_{1,K} = h_0 - V(\underline{R}, x)$, where $H_0 = (-\Delta + 1)^{1/2}$, $h_0 = (-\Delta)^{1/2}$, and where V is positive, the kernels $\exp[-tH^0_{1,K}](x, y)$, $\exp[-th^0_{1,K}](x, y)$ will be pointwise larger than $\exp[-tH_0](x, y)$ and $\exp[-th_0](x, y)$, respectively. By (3.2) one sees immediately that $\exp[-tH_0](x, y)$, $\exp[-th_0](x, y)$ are strictly positive for all x, y, which implies therefore $\exp[-tH^0_{1,K}](x, y) > 0$, $\exp[-th^0_{1,K}](x, y) > 0$ for all x, y. By the extension of the Perron–Frobenius theorem to operators (see e.g. [14], Theorem XIII.43), this implies that the ground states of $H^0_{1,K}$, $h^0_{1,K}$ (which correspond to the largest eigenvalues of $\exp[-tH^0_{1,K}]$, $\exp[-th^0_{1,K}]$) are nondegenerate, and the corresponding eigenvectors strictly positive.

Taking into account all the remarks made above, the proof given in [8] can now be completely transcribed to our case; this completes our proof of Theorem 2.

References

1. Dyson, F., Lenard, A.: Stability of matter. I. J. Math. Phys. **8**, 423–434 (1967)
Lenard, A., Dyson, F.: Stability of matter. II. J. Math. Phys. **9**, 698–711 (1968)
2. Lieb, E., Thirring, W.: A bound for the kinetic energy of fermions which proves the stability of matter.

Phys. Rev. Lett. **35**, 687–689 (1975); Errata: Phys. Rev. Lett. **35**, 1116 (1975). See also Lieb, E.: The stability of matter. Rev. Mod. Phys. **48**, 553–569 (1976)

3. Dyson, F.: Ground-state energy of a finite system of charged particles. J. Math. Phys. **8**, 1538–1545 (1967)
   Lieb, E.: The $N^{5/3}$ law for bosons. Phys. Lett. **A70**, 71–73 (1979)
4. Weder, R.: Spectral analysis of pseudodifferential operators. J. Funct. Anal. **20**, 319–337 (1975)
5. Herbst, I.: Spectral theory of the operator $(p^2 + m^2)^{1/2} - Ze^2/r$. Commun. Math. Phys. **53**, 285–294 (1977); Errata: Commun. Math. Phys. **55**, 316 (1977)
6. Kato, T.: Perturbation theory for linear operators. Berlin, New York: Springer 1966 (2nd edn. 1976)
7. Lieb, E., Simon, B.: Monotonicity of the electronic contribution to the Born–Oppenheimer energy. J. Phys. **B11**, L537–542 (1978)
8. Lieb, E.: Monotonicity of the molecular electronic energy in the nuclear coordinates. J. Phys. **B15**, L63–L66 (1982)
9. Daubechies, I.: An uncertainty principle for fermions with generalized kinetic energy. Commun. Math. Phys. (1983)
10. Lieb, E., Simon, B.: The Thomas–Fermi theory of atoms, molecules and solids. Adv. Math. **23**, 22–116 (1977)
11. Kovalenko, V., Perelmuter, M., Semenov, Ya.: Schrödinger operators with $L_w^{l/2}(\mathbb{R}^l)$ potentials. J. Math. Phys. **22**, 1033–1044 (1981)
12. Brascamp, H., Lieb, E., Luttinger, M.: A general rearrangement inequality for multiple integrals. J. Funct. Anal. **17**, 227–237 (1974)
13. Lieb, E.: Sharp constants in the Hardy–Littlewood–Sobolev and related inequalities (submitted)
14. Reed, M., Simon, B.: Methods of modern mathematical physics Vol. IV: Analysis of operators. New York: Academic Press 1978

Communicated by J. Fröhlich
Received March 24, 1983

The Stability and Instability of Relativistic Matter

Elliott H. Lieb[1,*] and Horng-Tzer Yau[2,**]

[1] Departments of Mathematics and Physics, Princeton University, P.O. Box 708, Princeton, NJ 08544, USA
[2] School of Mathematics, The Institute for Advanced Study, Princeton, NJ 08540, USA

Abstract. We consider the quantum mechanical many-body problem of electrons and fixed nuclei interacting via Coulomb forces, but with a relativistic form for the kinetic energy, namely $p^2/2m$ is replaced by $(p^2c^2+m^2c^4)^{1/2}-mc^2$. The electrons are allowed to have q spin states ($q=2$ in nature). For one electron and one nucleus instability occurs if $z\alpha>2/\pi$, where z is the nuclear charge and α is the fine structure constant. We prove that stability occurs in the many-body case if $z\alpha\leqq 2/\pi$ and $\alpha<1/(47q)$. For small z, a better bound on α is also given. In the other direction we show that there is a critical α_c (no greater than $128/15\pi$) such that if $\alpha>\alpha_c$ then instability always occurs for *all* positive z (not necessarily integral) when the number of nuclei is large enough. Several other results of a technical nature are also given such as localization estimates and bounds for the relativistic kinetic energy.

I. Introduction

One of the early important successes of quantum mechanics was the interpretation of the stability of the hydrogen atom. The ground state energy of the hydrogen Hamiltonian is finite and thus the hydrogen atom is stable quantum mechanically, even though it is unstable classically. The Coulomb singularity $-ze^2/r$ is controlled by a new feature of Schrödinger mechanics, the uncertainty principle. While the stability of the hydrogen atom is clear and simple, a more subtle question arises when many particles are taken into account. It is convenient to distinguish two notions of stability.

Stability of the first kind: The ground state energy is finite.

Stability of the second kind: The ground state energy is bounded below by a constant times the number of particles.

* Work partially supported by U.S. National Science Foundation grant PHY-85-15288-A02
** The author thanks the Institute for Advanced Study for its hospitality and the U.S. National Science Foundation for support under grant DMS-8601978

The second kind of stability, now commonly known as the stability of matter, was proved in 1967 by Dyson and Lenard [10] – four decades after the invention of Schrödinger mechanics. The Dyson-Lenard analysis clearly showed that the stability of matter depends crucially on the Pauli exclusion principle. The ground state energy (call if E_f) of N fermions interacting with K infinitely massive nuclei via the Coulomb potential is bounded below by a constant time the total particle number, i.e. $E_f \geqq -C_1(N+K)$. On the other hand, if all the particles considered are bosons, Dyson and Lenard [10] showed that the ground state energy (call it E_b) satisfies $E_b \geqq -C_2(N+K)^{5/3}$. Lieb [20] showed that this 5/3 bound is indeed the correct law for infinitely massive nuclei. If the nuclei have finite mass, and are also bosons, Dyson [9] showed by a variational calculation, that the ground state energy of bosons is bounded above by $E_b \leqq -C_3(N+K)^{7/5}$. This clearly shows that bosons are stable in the first sense, but never in the second. Dyson [9] also conjectured a lower bound $E_b \geqq -C_4(N+K)^{7/5}$ and this was finally proved 20 years later by Conlon, Lieb, and Yau [4]. They also proved a related bound for bosonic jellium.

The Dyson-Lenard proof for fermions involved a sequence of inequalities such that the final bound for C_1 is 10^{14} Rydberg. New proofs were given by Federbush [12] and Lieb-Thirring [25] in the seventies. The Lieb-Thirring proof gave a much better bound on C_1 (23 Rydbergs) and related the stability problem to the semiclassical picture of Thomas-Fermi theory. These matters are reviewed in [19].

The aforementioned considerations are all based on the nonrelativistic Schrödinger equation. The kinetic energy operator is the standard $p^2/2m = -\Delta/2m$ (when $\hbar = 1$). One might wonder whether stability still prevails in the relativistic case since the kinetic energy then decreases from $p^2/2m$ to $(p^2+m^2)^{1/2} - m (\hbar = c = 1)$. Historically, Chandrasekhar [2] was one of the first to ask this question, but in the context of gravitational interaction instead of Coulomb interaction. The famous Chandrasekhar model for neutron stars or white dwarfs consists of a semiclassical relativistic kinetic energy and classical gravitational potential energy. This simple model remarkably predicted collapse (i.e. instability of the first kind) and gave a critical mass which is correct, at least approximately. Despite the success of the simple semi-relativistic Chandrasekhar theory, the kinetic energy operator,

$$T = (p^2 + m^2)^{1/2} - m,$$

which it employs is nonlocal and therefore violates a basic physical principle. Nevertheless it is worthwhile studying this operator for several reasons. When $m = 0$, $T = |p|$ and it has the correct inverse length scaling (like the Dirac operator). Unlike the Dirac operator it allows one to formulate a variational principle for the ground state energy and thereby to give a rigorous definition of stability without the necessity of filling the Dirac sea or invoking quantum electrodynamics. In any event, there does not exist a truly relativistic many-body quantum theory at the present time and it is our belief that the study of Schrödinger operators based on T will capture some of the essential features of "the correct theory" when it is eventually formulated.

Let us start with the Hydrogen atom by considering the one particle Hamiltonian $\tilde{H}$ defined by

$$\tilde{H}_1 = (p^2+m^2)^{1/2} - m - \alpha z/|x|\,, \tag{1.1}$$

where $\alpha = e^2$ is the fine structure constant ($\hbar = c = 1$). This operator was studied independently by Weder [29] and Herbst [16]. See also Daubechies' paper [7]. Since the difference between the operator $(p^2+m^2)^{1/2} - m$ and $|p|$ is bounded (more precisely $|p| \geqq (p^2+m^2)^{1/2} - m \geqq |p| - m$), the stability of (1.1) is the same as the stability of

$$H_1 = |p| - \frac{2}{\pi}\beta/|x|\,, \tag{1.2}$$

where

$$\beta = \pi\alpha z/2\,. \tag{1.3}$$

Note that (1.2) is homogeneous under length scaling and therefore $E_1 \equiv \inf \operatorname{spec} H_1$ is either 0 or $-\infty$ by the scaling $\psi(x) \to \lambda^{3/2}\psi(\lambda x)$.

A first important fact about (1.2) is the existence of a critical $\beta_c = 1$, similar to that of the Klein-Gorden or Dirac theories. Kato [17] stated that $\beta_c \geqq 1$ and Herbst [16] showed that $\beta_c = 1$. The ground state energy for the Hamiltonian (1.2) is $E_1 = -\infty$ if $\beta > 1$ and $E_1 = 0$ if $\beta \leqq 1$. (In the Dirac theory $\beta_c = \pi/2$.)

Returning to the many-body case, suppose we have N electrons with coordinates $x_1, \ldots, x_N$ in $\mathbb{R}^3$ and K nuclei with coordinates $R_1, \ldots, R_K$ in $\mathbb{R}^3$ and with positive charges $z_1, \ldots, z_K$. We shall consider the following relativistic Schrödinger Hamiltonian, H_{NK}, for fermions with q spin states ($q = 2$ for real electrons). It is the analogue of (1.2):

$$H_{NK} \equiv \sum_{i=1}^{N} |p_i| + \alpha V_c(x_1, \ldots, x_N; R_1, \ldots, R_K)\,, \tag{1.4}$$

$$\begin{aligned} V_c(x_1, \ldots, x_N; R_1, \ldots, R_K) \equiv & \sum_{1 \leqq i < j \leqq N} |x_1 - x_j|^{-1} - \sum_{i=1}^{N} \sum_{j=1}^{K} z_j |x_i - R_j|^{-1} \\ & + \sum_{1 \leqq i < j \leqq K} z_i z_j |R_i - R_j|^{-1}\,. \end{aligned} \tag{1.5}$$

Note that *charge neutrality is not assumed in* (1.4), *or anywhere else in this paper.*

Mathematically, the Hamiltonian H_{NK} is a quadratic form on the q-state physical subspace $\mathcal{H}^q$ of $L^2(\mathbb{R}^{3N})$. More precisely, $\psi \in \mathcal{H}^q$ if and only if there exists a partition $P = \{\pi_1, \ldots, \pi_q\}$ of $\{1, \ldots, N\}$ such that $\psi(x_1, \ldots, x_N)$ is an antisymmetric function of the variables in each π_j, for all $1 \leqq j \leqq q$. When $q = N$, there is no restriction and the ground state energy for H_{NK} is just the ground state energy for bosons.

Physically, the nuclear kinetic energies should be included in (1.4) since the Born-Oppenheimer approximation (i.e. the neglect of the nuclear kinetic energies) is inadequate in the extreme relativistic regime. For simplicity, we shall confine ourselves to the Born-Oppenheimer approximation.

In reality, our goal is to discuss stability of the second kind for $H_{NK}(m)$, which is given by (1.4) but with $|p|$ replaced by $(p^2+m^2)^{1/2} - m$ there. For this purpose, however, it suffices to study only stability of the first kind for H_{NK} in (1.4). The reason is the following. Let $E_{NK}(R_1, \ldots, R_K)$ denote the ground state energy (= inf spec) of H_{NK} and let E_{NK} be the infimum of $E_{NK}(R_1, \ldots, R_K)$ over all choices of the R's. By simple scaling ($\psi(x_1, \ldots, x_N) \to \lambda^{3N/2}\psi(\lambda x_1, \ldots, \lambda x_N)$ and $R_j \to R_j/\lambda$), we see that E_{NK} is either zero or $-\infty$. On the other hand, if $E_{NK}(m)$ is defined analogously, then, since $|p| - m < (p^2 + m^2)^{1/2} - m < |p|$, we have that $E_{NK} \geqq E_{NK}(m) \geqq E_{NK} - mN$. Thus *stability of the first kind for* H_{NK} (in the sense that E_{NK} is bounded below independent of the R_j) *is equivalent to stability of the second kind for* $H_{NK}(m)$. Our goal then – and that is the purpose of this paper – is to find necessary conditions and sufficient conditions on z and α so that $E_{NK}(R_1, \ldots, R_K) \geqq 0$ for all N and all K and all $R_1, \ldots, R_K$.

If everything is held fixed except for q, then $E_{NK}(R_1, \ldots, R_K)$ is a monotone decreasing function of q. The reason is that specifying q is the same thing as requiring that the admissable wave functions $\psi(x_1, \ldots, x_N)$ are antisymmetric in each of q sets of variables. The number of variables in each set is unimportant, zero being an allowed number. Thus, a valid function for q is trivially a valid function for $q+1$.

A further remark about (1.4) can be made. Using a convexity argument, Daubechies and Lieb [8] proved that the stability of H_{NK} for $z_1 = z_2 = \ldots = z_K = z$ implies the stability of H_{NK} when all the nuclear charges are no greater than z, i.e. $0 \leqq z_j \leqq z$ for all j. With this remark, we shall assume from now on $z_1 = \ldots = z_k = z$.

Let $E_{NK}(\alpha, z)$ denote the dependence of E_{NK} on α and z. We shall use the following terminology: *$H(\alpha, z)$ is stable means that $E_{NK}(\alpha, z) = 0$ for all N and K. Otherwise we say that $H(\alpha, z)$ is unstable.*

The coupling constant of the electrons to the nuclei is $z\alpha = 2\beta/\pi$ and, from the hydrogen atom result, it is clearly necessary to have $\beta \leqq 1$ for stability. It is frequently convenient, therefore, to adopt α and β as the independent variables instead of α and z. When doing so we shall refer to the stability or instability of $H(\alpha, \beta)$ – hopefully without confusion. Indeed α and β are the natural variables from the following point of view. The electron-nuclear coupling is $2\beta/\pi$ while the nuclear-nuclear repulsion constant is $z^2\alpha = (2/\pi)^2\beta^2/\alpha$. Suppose that $K > 1$ and $\beta < 1$, but $K\beta > 1$. Then, if the nuclear-nuclear repulsion is ignored, the K nuclei can come to one common point and the system will collapse – even with only one electron. What discourages this from happening is the repulsion which is proportional to β^2/α. With β fixed, we see that α is required to be small in order that this repulsion prevents collapse. It is a striking fact, and it is the main theme of this paper, that *for every fixed $\beta \leqq 1$ and q there is a critical α (call it $\alpha_c(\beta)$) so that $H(\alpha, \beta)$ is stable when $\alpha < \alpha_c(\beta)$. There is another critical α (call it $\bar{\alpha}_c(\beta)$) so that $H(\alpha, \beta)$ is unstable when $\alpha > \bar{\alpha}_c(\beta)$.* These facts are the reason behind the contention above that α and β are natural. *We do not know whether or not $\alpha_c(\beta) = \bar{\alpha}_c(\beta)$.* Note that by the above monotonicity in z remark, stability for some (α, β_1) implies stability for all (α, β) with $\beta < \beta_1$.

There is an additional piece of information. Suppose that stability occurs for a pair α_1, z. Then stability occurs for a pair α, z if $\alpha \leqq \alpha_1$. The reason for the monotonicity in α is that $\inf\operatorname{spec}(\sum |p_i| + \alpha V_C) \geqq (1 - \alpha/\alpha_1) \inf\operatorname{spec}(\sum |p_i|) + (\alpha/\alpha_1)$ $\inf\operatorname{spec}(\sum |p_1| + \alpha_1 V_c) \geqq 0$.

Before stating our main results in detail, let us review some recent progress with this and related problems that also have the feature of critical coupling constants.

(1) The Chandrasekhar critical mass was established up to a factor of 4 in the framework of the relativistic Schrödinger equation by Lieb-Thirring [26]. Later, Lieb-Yau [27] proved that not only is the Chandrasekhar critical mass exactly correct, but the Chandrasekhar semiclassical equation can be derived rigorously from the relativistic Schrödinger equation in the limit that the gravitational constant $G \to 0$. In particular, in the physically interesting case, the discrepancy between the Chandrasekhar semiclassical critical mass and the quantum mechanical critical mass was shown in [27] to be less than 0.01%.

(2) For the non-relativistic Schrödinger equation, but with magnetic fields present that couple to both the electronic orbital motion and electronic spin, the existence of a critical nuclear charge for the stability of the hydrogen atom was proved by Fröhlich, Lieb, Loss, and Yau [15, 28]. The results were extended to the one-electron molecule and many-electron atom by Lieb and Loss [23]. The stability criteria are very similar to that of the relativistic stability considered in this paper. For stability, one should keep *both* $\alpha^2 z$ and α small. The general case for this model (many electrons and nuclei) remains an interesting open problem.

(3) The relativistic stability of matter itself. For $N=1$ and K arbitrary, Daubechies and Lieb [8] were the first to note the existence of a critical α and β fixed. They proved that H_{1K} is stable in the critical case $\beta = \pi\alpha z/2 = 1$ if $\alpha \leqq 1/3\pi$. The first person to solve a general case for all N and K was Conlon [3], who proved that the Hamiltonian $H(\alpha, z)$ is stable when $z=1$ provided $\alpha \leqq 10^{-200}$ and $q=1$. Using a different method, Fefferman and de la Llave [14] improved Conlon's result for $z=1$ to $\alpha \leqq 1/2.06\pi$, and again $q=1$. The Fefferman-de la Llave proof used computer assisted proofs extensively. Without using a computer, their bound would be worse by a factor 2.5, thereby reducing α to $1/5\pi$. Recently, Fefferman [13] announced a result for the critical case $\beta=1$ provided some numerical computer calculations can be made rigorous. The stability criterion announced in [13] is that stability occurs in the critical case $\beta=1$ if $\alpha \leqq 1/20$ and $q=1$. A complete proof, however, was not available when the present paper was written. Since $H(\alpha, \beta)$ collapses for $\beta > 1$ no matter how small the difference $\beta - 1$ may be, the application of computer assisted proofs to the $\beta=1$ case is delicate and difficult. Fefferman [13] states that "arbitrarily small roundoff errors are apparently fatal."

All the results mentioned above address the situation $q=1$. The methods employed are not, in our opinion, easily generalized to treat arbitrary q, as is done here. The ability to treat arbitrary q without increasing the complexity of the proof as q increases is, in our opinion, one of the main advantages of our method. Another is that we have no intrinsic need to invoke the computer. *The essence of our method is that for all q the many-body problem is reduced to a tractable one-body problem (see e.g. Theorems 6 and 11). This method also makes it possible to prove, for the first time, that stability occurs up to and including the critical value* $\beta=1$.

We should point out that the main tool in proving the nonrelativistic stability of matter, the Thomas-Fermi theory, fails to predict stability in the relativistic case. The semiclassical kinetic energy decreases in the high momentum region from (const) $\int \varrho^{5/3}$ in the nonrelativistic case to (const) $\int \varrho^{4/3}$ in the relativistic case. This semiclassical kinetic energy, $\int \varrho^{4/3}$, cannot control the Coulomb singularity $z\alpha/r$ for *any* $\alpha > 0$. The fact that stability occurs only for some finite $\alpha > 0$ and $z > 0$ is not a trivial matter (see Conlon [3]). A good estimate for α, especially when β is set equal to its critical value 1, is very difficult to achieve and should resolve the following subtle points:

(i) The delicate balance of charge neutrality. If, for example, the attractive term in V_c is changed from $z\alpha\sum\sum|x_i - R_j|^{-1}$ to $z\alpha(1+\varepsilon)\sum\sum|x_i - R_j|^{-1}$ for some $\varepsilon > 0$, then stability will not occur *for any positive* α *and* z. Physically, an attractive gravitational interaction is present and it does alter the Hamiltonian in precisely this manner – collapse does indeed occur. But the gravitational constant is small, and this collapse happens only when N and K are extremely large – the order of a solar mass [26, 27]. Indeed, the problem of determining the critical mass when Coulomb and gravitational interactions are both taken into account is a difficult open problem.

(ii) An improved version of the basic inequality $|p| - \frac{2}{\pi}|x|^{-1} \geqq 0$ is needed. This is apparently crucial since each electron in general feels attractions from more than one nucleus. One may argue that, by virtue of screening, each electron feels only one attraction from its nearest nucleus, but it is difficult to find a simple, precise mathematical statement about screening. Indeed, some corrections (e.g. van der Waals force) are obviously unavoidable and can only be controlled by the kinetic energy.

(iii) The nonlocality of the operator $|p|$. The technical problems caused by this non-locality are serious, especially since the Coulomb potential is long-ranged.

Our main results are the following four theorems about stability and instability.

Theorem 1 (Simple Stability Criterion). *For any* $z > 0$ *and* q, *the Hamiltonian* $H(\alpha, z)$ *is stable if*

$$\alpha \leqq \sup_{z' \geqq z} A_q(z'), \tag{1.6}$$

where

$$A_q(z) = (2/\pi)z^{-1}[1 + q^{1/3}z^{-1/3}C(z)^{-1/3}]^{-1}, \tag{1.7}$$

$$C(z) = 3.0844\{[1.6617 + 1.7258z^{-1} + 0.9533z^{-1/2}]^4 + (4/\pi)^3[1 + (2z)^{-1/2}]^8\}^{-1}. \tag{1.8}$$

Corollary. *Fix* $\beta \equiv z\alpha\pi/2 < 1$. *Then stability occurs if*

$$q\alpha \leqq \begin{cases} 0.062980(1-\beta)^3\beta^{-2} & \text{if} \quad \beta \geqq 0.49910 \\ 0.031774 & \text{if} \quad \beta \leqq 0.49910. \end{cases} \tag{1.9}$$

Remark. There is a number z_1, which is roughly 0.6, such that if $z \geqq z_1$ then the supremum in (1.6) occurs for $z' = z$, while if $z \leqq z_1$ the supremum occurs for $z' = z_1$.

Theorem 2 (Stability criterion for $\beta \leqq 1$). *Fix* $\beta \leqq 1$. *Then the Hamiltonian* $H(\alpha, \beta)$ *is stable if*

$$q\alpha \leqq 1/47.$$

Theorem 3 (Instability for all z and q). *There is a critical value* α_1 *such that if* $\alpha > \alpha_1$ *then* $H(\alpha, z)$ *is unstable for every* $q \geqq 1$ *and every nuclear charge* $z > 0$ *(not necessarily* *integral), no matter how small z may be. This means that if* $\alpha > \alpha_1$, *one can always choose N and K so that* $E_{NK}(\alpha, z) = -\infty$. *In order to achieve this collapse, it is only necessary to use one electron, i.e.* $N=1$. *One can take* $\alpha_1 = 128/15\pi$.

Theorem 4 (Instability dependence on q). *Let* $\beta = \pi\alpha z/2$ *as in* (1.3). *There is a critical value* α_2 *such that if*

$$\alpha > \alpha_2 q^{-1} \beta^{-2}, \tag{1.10}$$

then $H(\alpha, \beta)$ *is always unstable. To achieve this collapse, only* $N = q$ *electrons are needed. One can take* $\alpha_2 = 115, 120$. *Alternatively,* $\alpha > 36 q^{-1/3} z^{2/3}$.

Corollary. *If the electrons are bosons then* $H(\alpha, z)$ *is unstable for all* $\alpha > 0$ *and all fixed* $z > 0$. *The number of electrons necessary to achieve this collapse satisfies* $N \leqq 4\pi^{-2} \alpha_2 z^{-2} \alpha^{-3}$.

Remarks. In view of Theorem 3, the number 115, 120 should not be taken seriously. Its large value merely demonstrates how difficult it is to find simple, rigorous bounds – even variational upper bounds – for the relativistic Coulomb problem.

These theorems, taken together, give a clear picture about the stability of relativistic matter. The relevant parameters for stability are αq (if β is fixed) and $\alpha q^{1/3}$ (if z is fixed). An upper bound for α which is independent of z and q is given in Theorem 3. β is never larger than 1. Theorem 1 clearly fails to predict stability for the critical case $\alpha z = 2/\pi$, but its proof is considerably simpler than that of Theorem 2. It also gives the correct q dependence (when z is fixed), and its bound on α for small z is better than that of Theorem 2.

To gain perspective on how good these bounds are, we specialize our results to the following two cases. First, in the critical case, our upper bound (Theorem 2) and lower bound (Theorem 3) differ by a factor of 128 for $q=1$. Second, for $z=1$ and $q=1$, Theorem 1 predicts stability for $\alpha \leqq 1/3.23\pi$, which is not appreciably worse than the computer assisted proof bound $1/2.06\pi$ in [14]. Our bounds in Theorem 1 and Theorem 2 can certainly be improved, as will become clear in the proofs given below. We refrain from the temptation to optimize our results by complicating the technicalities. Our goal is to give a simple conceptual proof which has the correct q dependence and reasonable estimates.

Our proofs for Theorem 3 and 4 follow the same idea used in [23, 20]. Theorems 1 and 2 are much more difficult. Our basic strategy is first to reduce the Coulomb potential to a one-body potential, W. Then, by localizing the kinetic energy $|p|$, we can control the short distance Coulomb singularity of W, leaving a bounded potential W^* as remainder. The last task is to bound the sum of the negative eigenvalues of $|p| + W^*$, but this is standard and can be done by using semiclassical bounds ([6]).

The following Theorem 5 is a consequence of our localization for $|p|$ and combinatorial ideas in [26]. Theorem 5 was announced in [27, Appendix B], where it was proved for the special case $q = N$. Earlier, Fefferman and de la Llave [14] proved it for $q=1$. This theorem is not needed in the present work, but it is independently interesting. (Note that the definition of δ_i below is the reciprocal of that in [27].)

Theorem 5 (Domination of the nearest neighbor attraction by kinetic energy). *Let $\delta_i = \delta_i(x_1, \ldots, x_N)$ be the nearest neighbor distance for particle i relative to $N-1$ other particles, i.e.*

$$\delta_i \equiv \min\{|x_i - x_j| \,|\, j \neq i\}. \tag{1.11}$$

Let $\psi \in L^2(\mathbb{R}^{3N})$ be an N particle fermionic function of space-spin with q spin states. Then

$$\sum_{i=1}^{N} (\psi, |p_i| \psi) \geqq C_1 q^{-1/3} \sum_{i=1}^{N} (\psi, \delta_i^{-1} \psi), \tag{1.12}$$

$$\sum_{i=1}^{N} (\psi, p_i^2 \psi) \geqq C_2 q^{-2/3} \sum_{i=1}^{N} (\psi, \delta_i^{-2} \psi), \tag{1.13}$$

where

$$C_1 = 0.129, \qquad C_2 = 0.0209. \tag{1.14}$$

The organization of the rest of this paper is as follows:

In Sects. II and III, we prove Theorems 1 and 2 assuming an electrostatic inequality for the Coulomb potential and localization estimates for $|p|$. The theorems used in Sects. II and III are then proved in Sects. IV–VII. The presentation has been broken up this way in order to stress the conceptual underpinnings of Theorems 1 and 2.

Theorem 5 is proved in Sect. V. Some details of our numerical calculations are explained in Sect. VIII. In the final Sect. IX we prove Theorems 3 and 4.

II. Proof of Theorem 1 ($z\alpha < 2/\pi$)

The proofs of Theorems 1 and 2 are conceptually much simpler than the following detailed calculations and technicalities would suggest. There are three main steps for Theorem 1 and five steps for Theorem 2. Step A is the same for both theorems.

Step A. Reduction of the many-body Coulomb potential to a sum of one-body potentials plus a positive constant, namely $-\sum_1^N W(x_i) + C$. This reduces the problem to that of showing that q times the sum of the negative eigenvalues of the operator $|p| - W$ is not less than $-C$.

In the next step we decompose $\mathbb{R}^3$ into regions $B_0, B_1, \ldots, B_K$ where the B_i are disjoint balls centered at the R_i and B_0 is everything else.

Step B. We write $|p| = \beta|p| + (1-\beta)|p|$ with $\beta = z\alpha\pi/2 < 1$. In the balls B_i, $i = 1, \ldots, K$ we use $\beta|p|$ to control the Coulomb singularity of W and prove the operator inequality

$$\beta|p| - \alpha W(x) \geqq -U(x), \tag{2.1}$$

where $U = W$ in B_0 and U is a continuous function inside each ball. Thus $|p| - \alpha W \geqq (1-\beta)|p| - U$.

Step C. The sum of the negative eigenvalues of $(1-\beta)|p| - U$ is bounded by using the semiclassical bound due to Daubechies [6].

Steps B, C, D, and E for Theorem 2 will be explained in Sect. III.

In this section we shall state the basic theorems for steps A and B. These will be proved later in Sects. IV and V. These theorems will be combined here in step C, thus completing the proof of Theorem 1.

Step A. Reduction of the Coulomb Potential to a One-Body Potential

This step has nothing to do with quantum mechanics or the nature of the kinetic energy operator. It has to do with screening in classical potential theory. The total Coulomb potential, V_c, is given in (1.5). There are K nuclei located at distinct points $R_1, \ldots, R_K$ in $\mathbb{R}^3$ and having the same charge, z. There are N electrons.

Introduce the nearest neighbor, or Voronoi, cells $\{\Gamma_j\}_{j=1}^K$ defined by

$$\Gamma_j = \{x \,|\, |x-R_j| \leqq |x-R_k| \text{ for all } k \neq j\}. \tag{2.2}$$

The boundary of Γ_j, $\partial\Gamma_j$, consists of a finite number of planes. Another important quantity is the distance

$$D_j = \mathrm{dist}(R_j, \partial\Gamma_j) = \tfrac{1}{2}\min\{|R_k - R_j| \,|\, j \neq k\}. \tag{2.3}$$

The following theorem will be proved in Sect. IV. Recall (1.5).

Theorem 6 (Reduction of the Coulomb potential). *For any* $0<\lambda<1$

$$V_c(x_1, \ldots, x_N; R_1, \ldots, R_K) \geqq -\sum_{i=1}^{N} W^\lambda(x_i) + \frac{1}{8} z^2 \sum_{j=1}^{K} D_j^{-1} \tag{2.4}$$

and, for x *in the cell* Γ_j, $W^\lambda(x) = W_j^\lambda(x) \equiv G_j(x) + F_j^\lambda(x)$ *with*

$$G_j(x) = z|x-R_j|^{-1} \tag{2.5}$$

$$F_j^\lambda(x) = \begin{cases} \frac{1}{2}D_j^{-1}(1-D_j^{-2}|x-R_j|^2)^{-1} & \textit{for } |x-R_j| \leqq \lambda D_i \\ (\sqrt{2z}+\frac{1}{2})|x-R_j|^{-1} & \textit{for } |x-R_j| > \lambda D_i . \end{cases} \tag{2.6}$$

Theorem 6 says that when the electron-electron and nucleus-nucleus Coulomb repulsion is taken into account, V_c is bounded below by a *positive term* [the last term in (2.4)] consisting of a residue of the nucleus-nucleus repulsion (in fact one quarter of the nearest neighbor repulsion) and an attractive single particle part W^λ. In each cell Γ_j, W_j^λ is essentially the attraction to the nearest nucleus (this is the G_j part of W_j^λ); there is also a small attractive error F_j^λ.

There are two essential points in (2.4). One is that the charge z appearing in G_j is the same as in the original potential V_c. The other is the existence of the positive term. The error term F_j^λ can certainly be improved, especially the long-range part $|x-R_i| > \lambda D_j$; we have not tried to optimize F_j^λ.

It is interesting to compare our Theorem 6 with Baxter's Proposition 1 [1] which says that

$$V_c \geqq -(1+2z)\sum_{j=1}^{N} \delta(x_j)^{-1} \tag{2.7}$$

with

$$\delta(x) = \min\{|x-R_j| \,|\, j=1, \ldots, K\} = |x-R_j| \quad \text{when} \quad x \in \Gamma_j . \tag{2.8}$$

Fefferman and de la Llave [14] later improved this when $z=1$ from $1+2z=3$ to 8/3. Our proof is completely different from both proofs of (2.7), as is Theorem 6 itself. To reiterate the essential points, our bound has the *correct* singularity near the nucleus (namely z and not $1+2z$) and it also has a *positive* repulsive term.

Step B. Control of the Coulomb Singularity in Balls

The following formula is well known. For $f \in L^2$ with Fourier transform $\hat{f}$,

$$(f,|p|f)=(2\pi)^{-3}\int|\hat{f}(p)|^2|p|dp=(2\pi^2)^{-1}\int\int|f(x)-f(y)|^2|x-y|^{-4}dxdy. \quad (2.9)$$

One way to derive this formula is to write

$$(f,|p|f)=\lim_{t\downarrow 0} t^{-1}\{(f,f)-(f,e^{-t|p|}f)\}. \quad (2.10)$$

The convergence is a simple consequence of dominated convergence in Fourier space. The kernel of $\exp(-t|p|)$ can easily be calculated to be

$$e^{-t|p|}(x,y)=\pi^{-2}t[|x-y|^2+t^2]^{-2}. \quad (2.11)$$

Inserting (2.11) in (2.10) yields (2.9).

A formula similar to (2.9) can be derived this way for $(p^2+m^2)^{1/2}$ in place of $|p|$.

$$(f,(p^2+m^2)^{1/2}f)=\tfrac{1}{4}\pi^{-2}m^2\int\int|f(x)-f(y)|^2|x-y|^{-2}K_2(m|x-y|)dxdy, \quad (2.12)$$

where K_2 is a Bessel function. This follows from [11]

$$\exp[-t(p^2+m^2)^{1/2}](x,y)=\tfrac{1}{2}\pi^{-2}m^2t(|x-y|^2+t^2)^{-1}K_2(m(|x-y|^2+t^2)^{1/2}). \quad (2.13)$$

Starting with formula (2.9) we have

Theorem 7 (Kinetic energy in balls). *Let B be a ball of radius D centered at $z\in\mathbb{R}^3$ and let $f\in L^2(B)$. Define*

$$(f,|p|f)_B\equiv\frac{1}{2\pi^2}\int_B\int_B|f(x)-f(y)|^2|x-y|^{-4}dxdy \quad (2.14)$$

and assume this is finite. Then

$$(f,|p|f)_B\geqq D^{-1}\int_B Q(|x-z|/D)|f(x)|^2dx, \quad (2.15)$$

where $Q(r)$ is defined for $0<r\leqq 1$ by

$$\begin{aligned}Q(r)&=2/(\pi r)-Y_1(r),\\ Y_1(r)&=\frac{2}{\pi(1+r)}+\frac{1+3r^2}{\pi(1+r^2)r}\ln(1+r)-\frac{1-r^2}{\pi(1+r^2)r}\ln(1-r)-\frac{4r}{\pi(1+r^2)}\ln r\\ &\leqq 1.56712.\end{aligned} \quad (2.16)$$

The maximum of $Y_1(r)$ occurs at $r\approx 0.225975$ and was computed by S. Knabe. Note that $Y_1(|x|)$ is continuous for all $|x|\leqq 1$.

Using (2.9) we have

Corollary. *If $B_1,\ldots,B_K$ are disjoint balls in $\mathbb{R}^3$ centered at $R_1,\ldots,R_K$ and with radii $D_1,\ldots,D_K$,*

$$|p|\geqq\frac{2}{\pi}\sum_{j=1}^K|x-R_j|^{-1}B_j(x)-\sum_{j=1}^K D_j^{-1}Y_1(|x-R_j|/D_j)B_j(x), \quad (2.17)$$

where $B_j(x)$ is the characteristic function of B_j.

Theorem 7 is proved in Sect. V. Theorem 12, which is the analogue of Theorem 7 with p^2 in place of $|p|$, is stated and proved in Sect. V.

Step C. Semiclassical Bounds and the Conclusion of the Proof of Theorem 1

The problem of showing that $H=\sum|p_i|+\alpha V_c \geqq 0$ has been reduced to the following. In step A we showed that $H \geqq \sum_1^N \tilde{h}_i + C$, where

$$\tilde{h}_i = |p_i| - \alpha W^\lambda(x_i), \tag{2.18}$$

$$C = \frac{1}{8} z^2 \alpha \sum_{j=1}^K D_j^{-1}. \tag{2.19}$$

If we write $|p| = \beta|p| + (1-\beta)|p|$, with $\beta = z\alpha\pi/2$, then step B shows that it suffices to replace $\tilde{h}_i$ in H by h_i where

$$h_i = (1-\beta)|p_i| - U(x_i), \tag{2.20}$$

$$U(x) = \alpha F_j^\lambda(x) + \beta D_j^{-1} Y_1(|x-R_j|/D_j) B_j(x) + z\alpha|x|^{-1}(1-B_j(x)) \quad \text{when} \quad x \in \Gamma_j. \tag{2.21}$$

Proving that $\sum_1^N h_i + C \geqq 0$ for all numbers, N, of q-state fermions amounts to the following inequality in terms of density matrices satisfying $0 \leqq \gamma \leqq q$. [A density matrix is a positive definite trace class operator on $L^2(\mathbb{R}^3)$.]

$$\mathrm{Tr}\,\gamma h \geqq -C \quad \text{for all } \gamma, \tag{2.22}$$

with $h = (1-\beta)|p| - U(x)$. ($\mathrm{Tr}\,\gamma h$ is shorthand for $\sum_k (f_k, hf_k)\gamma_k$, where (f_k, γ_k) are the eigenfunctions and eigenvalues of γ.) For more details see [21].

The tool we shall use to prove (2.22) is Daubechies' extension of the Lieb-Thirring semiclassical bound from p^2 to $|p|$.

Theorem 8 (Daubechies). *Let γ be a density matrix satisfying $0 \leqq \gamma \leqq q$. (q need not be an integer.) Let $U(x)$ be any positive function in $L^4(\mathbb{R}^3)$. Then for $\mu > 0$,*

$$\mathrm{Tr}\,\gamma(\mu|p| - U) \geqq -0.0258 q \mu^{-3} \int U(x)^4 dx. \tag{2.23}$$

To complete the proof we merely insert (2.21) into (2.23). A simple bound is obtained by extending the integral over each Γ_j to an integral over all of $\mathbb{R}^3$. This will give K terms on the right side of (2.23) (each of which scales like D_j^{-1}) to be compared with the K terms in C (2.19). Our condition is then (recalling that $\beta = z\alpha\pi/2$)

$$0.0258 q (1-\beta)^{-3} \left\{ \int_{|x|<1} [\alpha F^\lambda(|x|) + \beta Y_1(|x|)]^4 dx + \int_{|x|>1} [\alpha F^\lambda(|x|) + z\alpha|x|^{-1}]^4 dx \right\} \leqq \tfrac{1}{8} z^2 \alpha \tag{2.24}$$

for some choice of $0 < \lambda < 1$ and where

$$F^\lambda(r) = \begin{cases} \frac{1}{2}(1-r^2)^{-1} & \text{for } 0 \leqq r \leqq \lambda \\ (\sqrt{2z} + \frac{1}{2}) r^{-1} & \text{for } \lambda \leqq r. \end{cases} \tag{2.25}$$

The second integral ($|x|>1$) in (2.24) (call it I_+) is easy to evaluate. It is independent of λ,

$$I_+ = (4/\pi)^3 \beta^4 [1+(2z)^{-1/2}]^8 . \tag{2.26}$$

Next, the integral of Y_1^4 over $|x|<1$ has been done numerically by S. Knabe. The following is actually an upper bound.

$$\int_{|x|<1} Y_1(x)^4 dx = 7.6245 \equiv I_1 . \tag{2.27}$$

We shall take $\lambda = 10/11$. Then

$$\int_{\lambda<|x|<1} F^\lambda(x)^4 dx = (4\pi/10) \left[\frac{1}{2} + (2z)^{1/2}\right]^4 \equiv I_2 , \tag{2.28}$$

$$\int_{|x|<\lambda} F^\lambda(x)^4 dx \leqq (\pi/4)\lambda \int_0^\lambda (1-r^2)^{-4} r dr \equiv I_3$$

$$= (\pi\lambda/24)[(1-\lambda^2)^{-3} - 1] = 22.645 . \tag{2.29}$$

To bound the first integral ($|x|<1$) in (2.24) one can use the triangle inequality $\int(f+g+h)^4 \leqq [(\int f^4)^{1/4} + (\int g^4)^{1/4} + (\int h^4)^{1/4}]^4$. Thus our condition for stability is satisfied if

$$0.0258 q(1-\beta)^{-3}\{[1.6617\beta + 1.0588\alpha(\tfrac{1}{2} + (2z)^{1/2}) + 2.1815\alpha]^4$$
$$+ (4/\pi)^3 \beta^4 [1+(2z)^{-1/2}]^8\} \leqq (2/\pi)^2 \beta^2 / 8\alpha . \tag{2.30}$$

Let us rewrite the stability condition (2.30) as

$$q^{-1} z C(z) \geqq \beta^3 (1-\beta)^{-3} \tag{2.31}$$

with $C(z)$ given by (1.8), namely

$$1/C(z) \equiv (0.0258) 4\pi \{[1.6617 + 1.7258 z^{-1} + 0.9533 z^{-1/2}]^4 + (4/\pi)^3 [1+(2z)^{-1/2}]^8\} . \tag{2.32}$$

By taking the cube root in (2.31) we have that (2.31) is equivalent to the assertion that stability occurs if

$$\alpha \leqq A_q(z) \equiv (2/\pi) z^{-1} \{1 + q^{1/3} z^{-1/3} C(z)^{-1/3}\}^{-1} . \tag{2.33}$$

Using the monotonicity in z for fixed α [8] mentioned in Sect. I, (2.33) can be improved to the statement that stability occurs if

$$\alpha \leqq \sup\{A_q(z') | z' \geqq z\} , \tag{2.34}$$

and this is precisely Theorem 1. □

Next, we address the question of finding a bound on α that depends only on β and not on z. For this purpose return to (2.31) and solve the equation $q^{-1} z C(z) = \beta^3 (1-\beta)^{-3}$. Since $z \to C(z)$ is monotone increasing, this equation has a unique solution. Call it $Z_q(\beta)$. Then stability occurs for any given β if

$$\alpha \leqq \alpha_q^*(\beta) \equiv (2/\pi) \sup\{\beta' / Z_q(\beta') | \beta' \geqq \beta\} . \tag{2.35}$$

Again we have used the aforementioned fact that stability for (α, β') implies stability for (α, β) if $\beta \leqq \beta'$.

Formula (2.35) is correct but lacks transparency. We shall now present a way to find a function $\alpha_q^{**}(\beta)$ which is less than or equal to $\alpha_q^*(\beta)$ but which has the same general features as $\alpha_q^*(\beta)$. It is this function, $\alpha_q^{**}(\beta)$ that is given in the corollary.

Choose an arbitrary z_0. Let $\alpha_0 = A_q(z_0)$ and let $\beta_0 = (2/\pi)\alpha_0 z_0$. Define

$$\alpha_q^{**}(\beta) = \begin{cases} (2/\pi q) C(z_0)(1-\beta)^3 \beta^{-2} & \text{if} \quad \beta \geqq \beta_0 \\ (2/\pi q) C(z_0)(1-\beta_0)^3 \beta_0^{-2} & \text{if} \quad \beta \leqq \beta_0 . \end{cases} \tag{2.36}$$

We claim that $\alpha \leqq \alpha_q^{**}(\beta)$ implies stability. First, suppose that $\beta \geqq \beta_0$. Then we have

$$z \equiv (2/\pi)\alpha^{-1}\beta \geqq (2/\pi)[\alpha_q^{**}(\beta)]^{-1}\beta_0 \geqq (2/\pi)[\alpha_q^{**}(\beta_0)]^{-1}\beta_0 \ (\text{since } \beta \geqq \beta_0) = z_0 .$$

By the monotonicity of C, we have $C(z) \geqq C(z_0)$. Therefore

$$z = (2/\pi)\alpha^{-1}\beta \geqq (2/\pi)[\alpha_q^{**}(\beta)]^{-1}\beta \geqq qC(z)^{-1}\beta^3(1-\beta)^{-3} \ (\text{since } C(z) \geqq C(z_0)) .$$

This is (2.31).

Second, suppose that $\beta \leqq \beta_0$. To prove the stability, we only have to verify (2.35). For this purpose, it suffices to show that $\alpha \leqq \frac{2}{\pi}\beta_0/Z_q(\beta_0)$ with $Z_q(\beta_0)$ solving $q^{-1}zC(z) = \beta_0^3(1-\beta_0)^{-3}$. Since by definition $q^{-1}z_0C(z_0) = \beta_0^3(1-\beta_0)^{-3}$, we have from the uniqueness of the solution of the above equations that $Z_q(\beta_0) = z_0$ and $\alpha_q^{**}(\beta) = (2/\pi q)C(z_0)(1-\beta_0)^3\beta_0^{-2} = \frac{2}{\pi}\beta_0/Z_q(\beta_0)$. Hence $\alpha \leqq \alpha_q^{**}(\beta)$ is the same as $\alpha \leqq \frac{2}{\pi}\beta_0/Z_q(\beta_0)$ and thus stability occurs for (α, β) with $\alpha \leqq \alpha_q^{**}(\beta)$ and $\beta \leqq \beta_0$.

Let us choose $z_0 = 10$. Then $(2/\pi)C(10) = 0.062980$ and $\beta_0 = 0.49910$. This together with (2.36) proves the Corollary of Theorem 1. □

III. Proof of Theorem 2 ($z\alpha \leqq 2/\pi$)

In the proof of Theorem 1 we first reduced the many-body Coulomb potential to a one-body potential in Step A. Then we split the kinetic energy $|p|$ into two pieces. One of them was used to control the Coulomb singularity and the other was used to control the long range part of the potential. If the method of Theorem 1 is used when $z\alpha = 2/\pi$, all of $|p|$ must be used for the singularities and nothing remains to control the long-range potential. In this section both parts of the potential will be controlled without splitting $|p|$, but this requires inventing a suitable localization formula for $|p|$. We shall henceforth take $z\alpha = 2/\pi$; by the monotonicity in z, this case will cover all the cases $z\alpha \leqq 2/\pi$.

There are five steps.

Step A is the same as before. The Coulomb potential V_c is replaced by a one-body potential W^λ plus a positive constant. Henceforth we shall take $\lambda = 0.97$ and omit the superscript on W.

Step B. Here we show that if $\chi_1(x)$ is a C^1 function which is approximately the characteristic function of a ball, and if γ is a density matrix with $0 \leqq \gamma \leqq q$ and if $\chi_2(x)$

is defined by $\chi_1(x)^2+\chi_2(x)^2=1$, then

$$\begin{aligned}\operatorname{Tr}\gamma(|p|-W) \geqq & \operatorname{Tr}\chi_1\gamma\chi_1(|p|-\text{potential energy correction } -W)\\ & +\operatorname{Tr}\chi_2\gamma\chi_2(|p|-\text{potential energy correction } -W)-q\cdot\text{const}. \end{aligned} \tag{3.1}$$

The important aspect of this inequality is this: It might have been thought that since $|p|$ is not a local operator, the potential energy corrections would have to be very long range. In fact they have support only inside a ball which is only slightly larger than the original ball. The long range nature of $|p|$ manifests itself in the term q-constant which depends on $\|\gamma\|$ but not on $N=\operatorname{Tr}\gamma$.

Step C. The ball referred to in step B is taken to be B_1 centered at R_1 (see Sect. II). To control the first term on the right side of (3.1) we have to bound q times the sum of the negative eigenvalues of $|p|$ – potential energy correction – W in a ball, where W is the one-body potential defined in step A.

Step D. For the second term on the right side of (3.1), the localization process in steps B and C are repeated $K-1$ times for nuclei, $2, \ldots, K$. This finally leaves us with a term $\operatorname{Tr}\chi_0\gamma\chi_0(|p|-\text{potential energy corrections})$ where χ_0 is essentially the characteristic function of the complement of the K balls. To estimate this term, Daubechies' semiclassical bound, Theorem 8, is used.

Step E. The above process leads to a lower bound on inf spec(H) in terms of certain integrals which depend on certain parameters that remain to be specified. These numerical facts are presented in this step. The details of the computation are given in Sect. VIII.

Step B. Localization of the Kinetic Energy

By way of comparison we begin by reminding the reader of the IMS localization formula (see [5, Theorem 3.2]) for $p^2=-\Delta$ instead of $|p|$. Let $\chi_0, \chi_1, \ldots, \chi_K$ be real valued functions on $\mathbb{R}^3$ satisfying

$$\sum_{j=0}^{K} \chi_j(x)^2=1 \quad \text{for all } x. \tag{3.2}$$

Then an elementary calculation yields the following operator identity.

$$-\Delta=\sum_{j=0}^{K} \chi_j(x)(-\Delta)\chi_j(x)-\sum_{j=0}^{K}|\nabla\chi_j(x)|^2. \tag{3.3}$$

This is a *localization of* $-\Delta$. If we assume additionally that χ_j has support in some set A_j (which are not pairwise disjoint, of course) then for any $f\in L^2(\mathbb{R}^3)$ and any arbitrary potential V,

$$(f,[-\Delta+V(x)]f)=\sum_{j=0}^{K}(\chi_j f,[-\Delta+V(x)-U(x)]\chi_j f) \tag{3.4}$$

with

$$U(x)=\sum_{j=0}^{K}|\nabla\chi_j(x)|^2. \tag{3.5}$$

The advantage of (3.4) is that in the jth term of (3.4) only $[V(x)-U(x)]1_{A_j}(x)$ appears [where $1_A(x)=1$ if $x\in A$ and $1_A(x)=0$ if $x\notin A$] and one can utilize different bounds on $V-U$ according to the region A_j under consideration. Furthermore, since $\chi_j f$ has support in A_j one can replace $-\Delta$ by the larger operator $-\Delta$ with Dirichlet boundary conditions on ∂A_j. The price one has to pay for all this is the negative potential operator $-U(x)$.

For the operator $|p|$ the following analogue of (3.3) is much more complicated because $|p|$ is not a local operator. We also state its generalization to $(p^2+m^2)^{1/2}$. The proof is immediate starting with (2.9) and (2.12).

Theorem 9 (Localization of kinetic energy-general form). *Let $\chi_0, \ldots, \chi_K$ be Lipschitz continuous functions satisfying* (3.2). *Then for any $f\in L^2(\mathbb{R}^3)$,*

$$(f,|p|f)=\sum_{j=0}^{K}(\chi_j f,|p|\chi_j f)-(f,Lf), \tag{3.6}$$

where L is a bounded operator with the kernel

$$L(x,y)=\frac{1}{2\pi^2}|x-y|^{-4}\sum_{j=0}^{K}[\chi_j(x)-\chi_j(y)]^2. \tag{3.7}$$

More generally,

$$(f,(p^2+m^2)^{1/2}f)=\sum_{j=0}^{K}(\chi_j f,(p^2+m^2)\chi_j f)-(f,L^{(m)}f), \tag{3.8}$$

where $L^{(m)}$ is a bounded operator with the kernel

$$L^{(m)}(x,y)=(2\pi)^{-2}m^2|x-y|^{-2}K_2(m|x-y|)\sum_{j=0}^{K}[\chi_j(x)-\chi_j(y)]^2 \tag{3.9}$$

and K_2 is a Bessel function.

Formula (3.6) was proposed to us by M. Loss, to whom we are grateful.

A simple, but important corollary of Theorem 9 concerns q-state, density matrices. As defined in Sect. II, this is any bounded operator on $L^2(\mathbb{R}^3)$ which satisfies the operator inequality $0\leqq\gamma\leqq q$ and for which $\operatorname{Tr}\gamma<\infty$.

Corollary. *For any density matrix, γ,*

$$\operatorname{Tr}\gamma|p|=\sum_{j=0}^{K}\operatorname{Tr}\gamma_j|p|-\operatorname{Tr}\gamma L, \tag{3.10}$$

where $\gamma_j\equiv\chi_j\gamma\chi_j$, with χ_j being thought of as a multiplication operator.

To exploit (3.10) we now impose a condition on $\chi_0, \ldots, \chi_K$. Let $R_1, \ldots, R_K$ be distinct points in $\mathbb{R}^3$ (namely the nuclear coordinates) and let D_j be given by (2.3). The K disjoint balls $B_j=\{x|x-R_j|<D_j\}$ were defined in Sect. II. Choose some $0<\sigma<1$ and consider the smaller balls

$$B_j^{(\sigma)}=\{x||x-R_j|\leqq(1-\sigma)D_j\}. \tag{3.11}$$

Let $\chi_0, \ldots, \chi_K$ satisfy (3.2) with χ_j supported in $B_j^{(\sigma)}$ for $j=1, \ldots, K$. The explicit choice for χ_j will be made in step D.

First, consider the case $K=1$. We decompose the L of (3.7) into a long-range part, L^0, and a short-range part, L_1^*, with $L=L^0+L_1^*$. Furthermore, $L_1^*(x,y)$ vanishes if x or y is not in B_1 or if $|x-y|>\sigma$, namely

$$L_1^*(x,y)=\begin{cases}\pi^{-2}|x-y|^{-4}[1-\chi_0(x)\chi_0(y)-\chi_1(x)\chi_1(y)]B_1(x)B_1(y) & \text{if } |x-y|\leqq\sigma\\ 0 & \text{if } |x-y|>\sigma,\end{cases} \tag{3.12}$$

where $B_1(x)=1$ if $x\in B_1$ and $B_1(x)=0$ otherwise. Recall that $\chi_0(x)^2+\chi_1(x)^2=1$ in the $K=1$ case. With these conventions, we have the following theorem which will be proved in Sect. VI.

Theorem 10 (Localization of kinetic energy-explicit bound in the one-center case). *For $K=1$, let L_1^* be given by* (3.12) *and $L^0=L-L_1^*$, with L given by* (3.7). *For any positive function, h_1, defined on the ball B_1, let*

$$\theta_1(x)=h_1(x)^{-1}\int_{B_1} L_1^*(x,y)h_1(y)dy. \tag{3.13}$$

Let $\Omega_1=\frac{1}{2}D_1^2\operatorname{Tr}(L^0)^2$, i.e.

$$\Omega_1\equiv\tfrac{1}{2}D_1^2\int[L(x,y)-L_1^*(x,y)]^2dxdy\equiv I^{(1)}+I^{(2)}, \tag{3.14}$$

$$I^{(1)}=\frac{1}{2}\pi^{-4}D_1^2\iint\limits_{\substack{x,y\in B_1^{(\sigma)}\\ ||x|-|y||\geqq\sigma D_1}}|x-y|^{-8}[1-\chi_0(x)\chi_0(y)-\chi_1(x)\chi_1(y)]^2dxdy, \tag{3.15}$$

$$I^{(2)}=\pi^{-4}D_1^2\iint\limits_{\substack{x\in B_1^{(\sigma)}\\ y\notin B_1^{(\sigma)}\\ |y|-|x|\geqq\sigma D_1}}|x-y|^{-8}[1-\chi_0(x)]^2dxdy. \tag{3.16}$$

Then, for any density matrix γ with $\|\gamma\|\leqq q$, and any $\varepsilon>0$,

$$\operatorname{Tr}\gamma|p|\geqq\operatorname{Tr}\chi_1\gamma\chi_1(|p|-U_1^*(x))+\operatorname{Tr}\chi_0\gamma\chi_0(|p|-U_1^*(x))-q(\varepsilon D_1)^{-1}\Omega_1, \tag{3.17}$$

where $U_1^*(x)=0$ for $x\notin B_1$ and, for $x\in B_1$,

$$U_1^*(x)\equiv(\varepsilon/D_1)B_1^{(\sigma)}(x)+\theta_1(x). \tag{3.18}$$

Inequality (3.17) looks complicated, but it is not vastly different from (3.3). The first two terms in (3.17) are the localized kinetic energies (inside and outside the ball B_1). The U_1^* term is a potential energy correction like the U in (3.4), but this potential has support only in the ball B_1. The last term is novel; it involves only the norm of γ and not a trace over γ. One might expect that the non-local nature of $|p|$ would give rise to a long range contribution to U, but these long range effects can be bounded by the norm of γ – as is done in the last term of (3.17).

Step C. Bound on Negative Eigenvalues in a Ball

Our goal is to give a lower bound to $\operatorname{Tr}\chi_1\gamma\chi_1(|p|-W(x)-U_1^*(x))$. The following is our main tool. It will be proved in Sect. VII.

Theorem 11 (Lower bound to the short-range energy in a ball). *Let $C>0$ and $R>0$ and let*

$$H_{CR}=|p|-\frac{2}{\pi}|x|^{-1}-C/R \tag{3.19}$$

be defined on $L^2(\mathbb{R}^3)$ as a quadratic form. Let $0 \leqq \gamma \leqq q$ be a density matrix as before and let χ be any function with support in $B_R = \{x||x| \leqq R\}$. Then

$$\operatorname{Tr} \bar{\chi}\gamma\chi H_{CR} \geqq -4.4827 C^4 R^{-1} q \{(3/4\pi R^3) \int |\chi(x)|^2 dx\}. \tag{3.20}$$

Remark. When $\chi \equiv 1$ in B_R then the factor in braces $\{\ \}$ in (3.20) is 1.

To apply Theorem 11 to our case we take R in Theorem 11 to be $(1-\sigma)D_1$ and we take C to be an upper bound for $(1-\sigma)D_1\{\alpha W(x) + U_1^*(x) - (2/\pi)|x|^{-1}\} = (1-\sigma)D_1\{\alpha F_1(x) + U_1^*(x)\}$ in the ball $|x| \leqq (1-\sigma)D_1$. This computation will be done in Step E.

Step D. The Negative Eigenvalues for the Long Range Potential

Associated with each ball B_j of radius D_j centered at R_j will be a cutoff function χ_j defined by

$$\chi_j(x) = \chi(|x - R_j|/D_j), \tag{3.21}$$

where the universal χ is given by

$$\chi(r) = \begin{cases} 1 & \text{for } r \leqq 1 - 3\sigma \\ \cos[\pi(r - 1 + 3\sigma)/4\sigma] & \text{for } 1 - 3\sigma \leqq r \leqq 1 - \sigma \\ 0 & \text{for } 1 - \sigma \leqq r. \end{cases} \tag{3.22}$$

Here, it is important that $\sigma < 1/3$. We also choose a function $h_j(x)$ for $x \in B_j$,

$$h_j(x) = h(|x - R_j|/D_j), \tag{3.23}$$

$$rh(r) = \begin{cases} 1 & \text{for } r \leqq 1 - 3\sigma \text{ and } 1 - \sigma \leqq r \leqq 1 \\ 2 - \sigma^{-1}|r - 1 + 2\sigma| & \text{for } 1 - 3\sigma \leqq r \leqq 1 - \sigma. \end{cases} \tag{3.24}$$

Starting with Theorem 10, Eq. (3.17), we choose some ε and compute Ω_1, $\theta_1(x)$, $U_1^*(x)$ using (3.13)–(3.16). We also compute some bound

$$C \geqq (1-\sigma)D_1\{\alpha F_1(x) + U_1^*(x)\} \tag{3.25}$$

in $B_1^{(\sigma)}$. By scaling, C does not depend on D_1. Then, using Theorem 11, Eq. (3.20), we have that

$$E = \inf_\gamma \operatorname{Tr}\gamma(|p| - \alpha W) \geqq -qA/D_1 + \inf_\gamma \operatorname{Tr}(1-\chi_1^2)^{1/2}\gamma(1-\chi_1^2)^{1/2}(|p| - \alpha W - U_1^*). \tag{3.26}$$

The first term, qA/D_1, is a sum of two pieces. One is the $q(\varepsilon D_1)^{-1}\Omega_1$ in (3.17); the other is the right side of (3.20) (call it qA_2). The sum is written as qA/D_1 because the various quantities that have been introduced scale in just the right way – so that A really is independent of D_1 and q.

For the second term on the right side of (3.26) we note the identity

$$(1-\chi_1(x)^2)[\alpha W(x) + U_1^*(x)] = (1-\chi_1(x)^2)[\alpha W(x)\beta_1(x) + U_1^*(x)\beta_1(x)], \tag{3.27}$$

where $\beta_1(x) = 1$ if $|x - R_1| \geqq 1 - 3\sigma D_1$ and $\beta_1(x) = 0$ otherwise. Since $(1-\chi_1^2)^{1/2}\gamma(1-\chi_1^2)^{1/2}$ is a q-state density matrix whenever γ is, the last term in (3.26) can be bounded below by

$$\inf_{\gamma} \operatorname{Tr}\gamma(|p|-\alpha W(x)\beta_1(x)-U_1^*(x)\beta_1(x)). \tag{3.28}$$

Now we can apply Theorems 10 and 11 to (3.28), using the ball B_2 in place of B_1. Since $U_1^*(x)=0$ for $x\notin B_1$ we see that $(\alpha W(x)+U_1^*(x))\beta_1(x)=\alpha W(x)$ for $x\notin B_1$. This process can be repeated until all the balls $B_1,\ldots,B_K$ have been used. Our final result (with U_j^* defined as in (3.18) with R_1, D_1 replaced by R_j, D_j) is

$$E\geqq -A\sum_{j=1}^{K} D_j^{-1}+\inf_{\gamma}\operatorname{Tr}\gamma\left[|p|-\left(\alpha W+\sum_{j=1}^{K} U_j^*(x)\right)\prod_{j=1}^{K}\beta_j(x)\right]. \tag{3.29}$$

To bound the last term in (3.29) we use Theorem 8. This will result in a sum of K integrals, one for each cell Γ_j. As in the proof of Theorem 1, a further bound is obtained by pretending that each Γ_j extends to all of $\mathbb{R}^3$. Thus

$$E\geqq -q(A+J)\sum_{j=1}^{K} D_j^{-1}, \tag{3.30}$$

where

$$J=0.0258\int_{|x|>1-3\sigma}[(2/\pi)|x|^{-1}+\alpha F(|x|)+U^*(|x|)]^4\,dx, \tag{3.31}$$

and where $F(r)$ is given in (2.25) with $\lambda=0.97$, and $U^*(x)$ is given by (3.18) with $D_1=1$ there.

From (3.30) and (2.4), stability holds if

$$q(A+J)\leqq \tfrac{1}{8}z^2\alpha=(2\pi^2)^{-1}\alpha^{-1}. \tag{3.32}$$

Step E. Numerical Results

We take $\sigma=0.3$ and $\varepsilon=0.2077$ (recall that λ was previously chosen to be 0.97). Sincc all quantities have the correct length scaling, we shall refer everything to a standard ball of unit radius $D_1=1$. The following are the results of the computations given in Sect. VIII.

Starting with $\chi(r)$ in (3.22) we compute $\Omega_1\equiv\Omega$ in (3.13)–(3.16),

$$I^{(1)}=0.05529,\qquad I^{(2)}=0.06042,$$
$$\Omega=I^{(1)}+I^{(2)}=0.1157,\qquad \varepsilon^{-1}\Omega=0.5571. \tag{3.33}$$

From the definition (3.13) and (3.24) we find that $\theta_1(x)\equiv\theta(|x|)$ satisfies $\theta(r)\leqq\theta^*(r)$ and

$$\theta^*(r)=\begin{cases}(3\pi/32)(2-\sqrt{2})\sigma^{-1}=0.5751 & \text{for } r\leqq 1-\sigma\\ (\pi/64)\sigma^{-5}(1+2\sigma-r)(1-r)^3 & \text{for } 1-\sigma<r\leqq 1.\end{cases} \tag{3.34}$$

Using this we have, from (3.18), that

$$U^*(r)\leqq \varepsilon B^{(\sigma)}(r)+\theta^*(r) \tag{3.35}$$

with $B^{(\sigma)}(r)=1$ for $r<1-\sigma=0.7$ and $B^{(\sigma)}(r)=0$ otherwise.

Next, we want to find some C satisfying (3.25). Since $\lambda=0.97>1-\sigma=0.7$, we need only concern ourselves with the first line of (2.25). Note that α appears in (3.25) in the form $\alpha F_1(x)$ and, since $F_1(x)$ does not depend on z in the region $r<\lambda$, the quantity $\alpha F_1(x)$ is proportional to α when $z\alpha$ is fixed. Our goal is to prove stability when $\alpha<1/47q\leqq 1/47$, and therefore we can replace $\alpha F_1(x)$ by $F_1(x)/47$ in (3.25). Then

$$C=0.7\{0.02086+0.2077+0.5751\}=0.5629 \tag{3.36}$$

satisfies (3.25) for $r<1-\sigma=0.7$.

The right side of (3.20) (with $R=1-\sigma=0.7$) can now be easily calculated. It is

$$qA_2=0.1661q. \tag{3.37}$$

Adding $\varepsilon^{-1}\Omega$ and A_2 we have

$$A=0.7232. \tag{3.38}$$

Finally, the integral in (3.31) must be computed. To bound $\alpha F(r)$ we can use $(1/47)F(r)$ for $r<\lambda$, while for $r>\lambda$ we write $z=2/\pi\alpha$ in (2.25). When $r>\lambda$ this results in two terms in αF, one of which is proportional to $\alpha^{1/2}$ and the other to α. In both terms we can take $\alpha=1/47$. Thus, we bound $\alpha F(r)$ by $0.1753/r$ for $r>\lambda$ and by $(1/94)(1-r^2)^{-1}$ for $r<\lambda$. We then find that

$$\begin{aligned}(0.0258)^{-1}(4\pi)^{-1}J\leqq &\int_{0.1}^{0.7}[2/\pi r+(1/94)(1-r^2)^{-1}+0.2077+0.5751]^4r^2dr\\ &+\int_{0.7}^{0.97}[(2/\pi r+(1/94)(1-r^2)^{-1}+20.20(1.6-r)(1-r)^3]^4r^2dr\\ &+\int_{0.97}^{1}[2/\pi r+0.1753/r+20.20(1.6-r)(1-r)^3]^4r^2dr\\ &+\int_{1}^{\infty}[2/\pi r+0.1753r^{-1}]^4r^2dr.\end{aligned} \tag{3.39}$$

The first integral, J_1, can be bounded by replacing $(1-r^2)^{-1}$ by $(1-(0.7)^2)^{-1}$ and then doing the integral analytically. The second integral, J_2, was done on a computer. In the third integral, J_3, $1.6-r$ was replaced by $1.6-0.97$ and $(1-r^3)$ was replaced by $(1-0.97)^3$; it was then done analytically. The fourth integral, J_4, can be done analytically. We find $J_1\leqq 4.435$, $J_2\leqq 0.17$, $J_3\leqq 0.0135$, and $J_4\leqq 0.435$. Thus

$$J\leqq 1.64 \tag{3.40}$$

and, from (3.32), stability occurs if $\alpha q<1/47$. This completes the proof of Theorem 2. □

IV. An Electrostatic Inequality

Our goal here is to prove Theorem 6 about the Coulomb potential V_c given in (1.5). A similar theorem can be derived for the Yukawa potential $|x|^{-1}\exp(-\mu|x|)$, but we shall not do so here. We recall the definition (2.2) of the K Voronoi cells $\Gamma_1,\ldots,\Gamma_K$ for K nuclei located at $R_1,\ldots,R_K\in\mathbb{R}^3$, and also the radii D_j in (2.3) which is the distance of R_j to $\partial\Gamma_j$. Since Theorem 6 is trivial when $K=1$, we shall assume henceforth that $K>1$. We set

$$V(x)=\sum_{j=1}^{K}|x-R_j|^{-1}, \tag{4.1}$$

which is the potential of K nuclei of unit charge located at the R_j, and

$$\delta(x)=\min\{|x-R_j|\,|\,1\leqq j\leqq K\}, \tag{4.2}$$

which is the distance of a particle at x to the set of K nuclei. We set

$$\Phi(x)=V(x)-\delta(x)^{-1}, \tag{4.3}$$

which is the potential of all the nuclei except for the nucleus in the cell Γ_j in which x is located. Φ is continuous but not differentiable.

Let ν be any Borel measure (possibly signed) on $\mathbb{R}^3$. We say that ν is a bounded measure if $|\nu|(\mathbb{R}^3)<\infty$. In this case $\int\Phi(x)d\nu(x)$ is well defined since Φ is continuous and bounded. We define

$$\mathscr{E}_{\Phi,z}(\nu)=\frac{1}{2}\iint|x-y|^{-1}d\nu(x)d\nu(y)-z\int\Phi(x)d\nu(x)+z^2\sum_{1\leqq i<j\leqq K}|R_i-R_j|^{-1}. \tag{4.4}$$

The first term on the right side of (4.4) is well defined (in the sense that it is either finite or $+\infty$) since $|x-y|^{-1}$ is a positive definite kernel. The following is basic to our analysis.

Lemma 1. *Let ν be any bounded measure, let $z>0$ and let Φ be given by* (4.3). *Then*

$$\mathscr{E}_{\Phi,z}(\nu)\geqq\frac{1}{8}z^2\sum_{j=1}^{K}D_j^{-1}. \tag{4.5}$$

Proof. There is a (positive) measure μ that satisfies the equation

$$|x|^{-1}*\mu=z\Phi \tag{4.6}$$

and μ has support on $\partial\Gamma\equiv\bigcup_{j=1}^{K}\partial\Gamma_j$. In fact, μ can be computed explicitly as

$$\mu=-(z/4\pi)\Delta\Phi. \tag{4.7}$$

More precisely, $\partial\Gamma$ consists of pieces of 2 dimensional planes separating some Γ_i from some Γ_j; on $\partial\Gamma_j$

$$d\mu(x)=-(z/2\pi)\mathbf{n}\cdot\nabla|x-R_j|^{-1}d^2x, \tag{4.8}$$

where d^2x is two-dimensional Lebesgue measure on $\partial\Gamma_j$, and $\mathbf{n}$ is the unit normal pointing *out* of Γ_j. Let

$$A=-\tfrac{1}{2}z\int\delta(x)^{-1}d\mu(x). \tag{4.9}$$

Then

$$\frac{1}{2}\iint|x-y|^{-1}d\mu(x)d\mu(y)=\frac{1}{2}\int\Phi(x)d\mu(x)=\frac{z}{2}\sum_{j=1}^{K}\int|x-R_j|^{-1}d\mu(x)+A$$
$$=\frac{z^2}{2}\sum_{j=1}^{K}\Phi(R_j)+A=z^2\sum_{1\leqq i<j\leqq K}|R_i-R_j|^{-1}+A. \tag{4.10}$$

On the other hand, if each part of $\partial\Gamma$ is counted twice we obtain

$$A=(z^2/8\pi)\sum_{j=1}^{K}\int_{\partial\Gamma_j}|x-R_j|^{-1}\mathbf{n}\cdot\nabla|x-R_j|^{-1}d^2x. \tag{4.11}$$

Let I_j denote the integral in (4.11). The integrand is $\frac{1}{2}\mathbf{n}\cdot\nabla|x-R_j|^{-2}$. With Λ_j denoting the complement of Γ_j in $\mathbb{R}^3$ (so that $\partial\Lambda_j=\partial\Gamma_j$) we have

$$I_j=\frac{1}{2}\int_{\partial\Gamma_j}\mathbf{n}\cdot\nabla|x-R_j|^{-2}d^2x=-\frac{1}{2}\int_{\Lambda_j}\Delta|x-R_j|^{-2}dx=-\int_{\Lambda_j}|x-R_j|^{-4}dx. \tag{4.12}$$

For convenience in evaluating (4.12) we can take $R_j=0$ and assume that Λ_j contains the half-space $\{(x,y,z)|x\geqq D_j\}$; the reason for this is that (assuming $D_j \neq \infty$) there is another nucleus at some R_i such that the midplane between R_j and R_i is given (after rotation of coordinates) by $\{(x,y,z)|x=D_j\}$. Thus

$$I_j\leqq-\int_{-\infty}^{\infty}\int_{-\infty}^{\infty}dydz\int_{D_j}^{\infty}dx(x^2+y^2+z^2)^{-2}=-\pi/D_j, \tag{4.13}$$

and therefore

$$A\leqq-\frac{1}{8}z^2\sum_{j=1}^{K}D_j^{-1}. \tag{4.14}$$

Using (4.6) and (4.10) we have that

$$\mathscr{E}_{\Phi,z}(\nu)=\tfrac{1}{2}\iint|x-y|^{-1}d(\nu-\mu)(x)d(\nu-\mu)(y)-A. \tag{4.15}$$

The integral in (4.15) is nonnegative (since $|x-y|^{-1}$ is positive definite), and the lemma follows from (4.14). □

Proof of Theorem 6. There are N points $x_1,\ldots,x_N$. If x_i is in some cell Γ_j we shall replace the unit point charge at x_i by a unit charge distributed on a sphere S_i but, in general, the center of S_i will not be x^i and the charge distribution on S_i will not be uniform. Also, S_i is not always contained entirely in Γ_j. (If x_i is in more than one Γ_j then an arbitrary choice can be made.) The definition of S_i and the charge distribution ν_i on S_i is the following:

(i) If $|x_i-R_j|\leqq\lambda D_j$, then S_i is the sphere $\partial B_j=\{x||x-R_j|=D_j\}$. The charge ν_i is determined so that its (continuous) potential $V_i\equiv|x|^{-1}*\nu_i$ satisfies

$$V_i(x)=\begin{cases}|x-x_i|^{-1} & \text{for} \quad |x-R_j|\geqq D_j\\ |x-x_i^*|^{-1}|x_i-R_j|^{-1}D_j & \text{for} \quad |x-R_j|\leqq D_j,\end{cases} \tag{4.16}$$

where x_i^* is the image of x_i with respect to S_j, namely

$$x_i^*-R_j=D_j^2|x_i-R_j|^{-2}(x_i-R_j). \tag{4.17}$$

The potential $V_i(x)$ is harmonic inside and outside B_j, and ν_i can be computed from the formula $-\Delta V_i=4\pi\nu_i$, but we shall not need this. It is important to note that ν_i is nonnegative.

(ii) If $|x_i-R_j|>\lambda D_j$ and $x_i\in\Gamma_j$, then S_i is a sphere centered at x_i and of radius t_i given by

$$t_i=|x_i-R_j|(1+\sqrt{2z})^{-1}. \tag{4.18}$$

The charge distribution ν_i on S_i is the uniform one with unit total charge.

Now we apply Lemma 1 with

$$\nu=\sum_{i=1}^{N}\nu_i. \tag{4.19}$$

In order to utilize inequality (4.5) it is necessary to relate $\mathscr{E}_{\Phi,z}(v)$ to V_c. The last term in (4.4) is, of course, exactly the nuclear repulsion. The first term on the right side of (4.4) (call it I) satisfies

$$I = \sum_{1 \leqq i < k \leqq N} \iint |x-y|^{-1} dv_i(x) dv_k(y) + \frac{1}{2} \sum_{i=1}^{N} \iint |x-y|^{-1} dv_i(x) dv_i(y). \quad (4.20)$$

Each $v_i v_k$ integral in (4.20) is less than or equal to $|x_i - x_k|^{-1}$. This is so because, by construction

$$(|x|^{-1} * v_i)(x) \leqq |x - x_i|^{-1}, \quad \text{all } x, \quad (4.21)$$

and hence

$$\int (|x|^{-1} * v_i)(x) dv_k(x) \leqq (|x|^{-1} * v_k)(x_i) \leqq |x_k - x_i|^{-1}. \quad (4.22)$$

The $v_i v_i$ integral in (4.20) is just the self energy of v_i. Call it e_i. There are two cases.

(i) $|x_i - R_j| \leqq \lambda D_j$. Then, from (4.16)

$$\begin{aligned} e_i &= \iint |x-y|^{-1} dv_i(x) dv_i(y) = \int |x - x_i|^{-1} dv_i(x) = V_i(x_i) \\ &= |x_i - x_i^*|^{-1} |x_i - R_j|^{-1} D_j = D_j^{-1}(1 - D_j^{-2}|x_i - R_j|^2)^{-1}. \end{aligned} \quad (4.23)$$

(ii) $|x_i - R_j| > \lambda D_j$ and $x_i \in \Gamma_j$. Here $e_i = 1/t_i$ since v_i is uniformly distributed on a sphere of radius t_i.

To summarize,

$$I \leqq \sum_{1 \leqq i < k \leqq N} |x_i - x_k|^{-1} + \frac{1}{2} \sum_{i=1}^{N} \begin{Bmatrix} \text{Eq. (4.23)} & \text{in case (i)} \\ 1/t_i & \text{in case (ii)} \end{Bmatrix}. \quad (4.24)$$

The second term on the right side of (4.4) is a sum of $z \int \Phi dv_i$. Again, there are two cases.

(i) $|x_i - R_j| \leqq \lambda D_j$. From the definition of W and the fact that $(|x|^{-1} * v_i)(x) = |x - x_i|^{-1}$ when $x \notin \Gamma_j$, we have

$$\int \Phi(x) dv_i(x) = \sum_{k=1}^{K} |x_i - R_k|^{-1} - |x_i - R_j|^{-1}. \quad (4.25)$$

(ii) $|x_i - R_j| > \lambda D_j$ and $x_i \in \Gamma_j$. By the definition of Φ

$$\int \Phi(x) dv_i(x) = \sum_{k=1}^{K} \int |x - R_k|^{-1} dv_i(x) - \int \delta(x)^{-1} dv_i(x), \quad (4.26)$$

where $\delta(x)$ is the distance to the nearest nucleus. Since every R_k (including R_j) is outside S_i, the first term in (4.26) is merely $\sum_{k=1}^{K} |x_i - R_k|^{-1}$. The difficulty in estimating the second term in (4.4) stems from the fact that v_i can have support in several cells – not just Γ_j. We have, however, that for $|x - x_i| = t_i$ and any k,

$$|x - R_k| + t_i = |x - R_k| + |x - x_i| \geqq |R_k - x_i| \geqq |R_j - x_i|. \quad (4.27)$$

Hence $\delta(x) \geqq |R_j - x_i| - t_i$, and therefore in case (ii),

$$\int \Phi(x) dv_i(x) \geqq \sum_{k=1}^{K} |x_i - R_k|^{-1} - (|R_j - x_i| - t_i)^{-1}. \quad (4.28)$$

Using these inequalities and the definition (4.18) we find that

$$\mathscr{E}_{\Phi,z}(v) \leqq V_c + \sum_{i=1}^{N} W^\lambda(x_i), \tag{4.29}$$

with $W^\lambda(x)$ given in (2.5), (2.6). This, together with Lemma 1, proves Theorem 6. □

V. Simple Localization of the Kinetic Energy

Here we shall prove Theorem 7, but before doing so let us motivate Theorem 7 by stating the analogous Theorem 12 below for p^2 instead of $|p|$. This latter theorem is simple to prove, but we have not seen it in the literature.

Theorem 12 (The energy of p^2 in balls). *Let B be a ball of radius R centered at $z \in \mathbb{R}^3$ and let $f \in L^2(B)$ and $\nabla f \in L^2(B)$. Define*

$$(f, p^2 f)_B = \int_B |\nabla f(x)|^2 dx. \tag{5.1}$$

Then

$$(f, p^2 f)_B \geqq R^{-2} \int_B H((x-z)/R) |f(x)|^2 dx, \tag{5.2}$$

where $H(x)$, for $|x|<1$, is any function of the form $H(x) = -h^{-1}(x)\Delta h(x)$ and where h is a smooth, strictly positive function with vanishing normal derivative on the boundary $|x|=1$. In particular, by taking $h(x) = (|x|^2+t)^{-1/4} \exp[\frac{1}{4}|x|^2/(1+t)]$, and then letting $t \to 0$ (using Fatou's lemma) we have that (5.2) *holds with*

$$H(x) = \tfrac{1}{4}|x|^{-2} - Y_2(|x|), \qquad Y_2(r) = 1 + \tfrac{1}{4}r^2. \tag{5.3}$$

Remark. It is important to note that $\frac{1}{4}$, the coefficient of the $|x|^{-2}$ singularity, is precisely the sharp constant for the uncertainty principle in all of $\mathbb{R}^3$, $(f, p^2 f) \geqq \frac{1}{4} \int |f|^2 |x|^{-2} dx$.

Proof. Write $f(x) = g(x)h(x)$ so that $\nabla f = h\nabla g + g\nabla h$. Then

$$\int_B |\nabla f|^2 = \int_B h^2 |\nabla g|^2 + \int_B |g|^2 (\nabla h)^2 + \int_B (\nabla g^2) h \nabla h. \tag{5.4}$$

Integrating the last integral by parts

$$\int_B |\nabla f|^2 \geqq -\int_B g^2 h \Delta h = \int_B f^2 H. \tag{5.5}$$

Equation (5.3) is merely a calculation. □

We turn now to the problem of proving Theorem 7 which is the analogue of Theorem 12 for

$$(f, |p| f)_B = (2\pi^2)^{-1} \int_B \int_B |f(x) - f(y)|^2 |x-y|^{-4} dx dy. \tag{5.6}$$

If B is $\mathbb{R}^3$ then this is just $(f, |p| f)$; see (2.9).

Proof of Theorem 7. Without loss of generality we can take $z=0$ and $R=1$. First, we regularize $|x-y|^{-4}$ to $L_t(x,y) = (|x-y|^2+t)^{-2}$. The theorem will follow by letting $t \to 0$ and using dominated convergence and Fatou's lemma.

With L_t in place of $|x-y|^{-4}$ we have

$$(f,|p|f)_{B,t}=\pi^{-2}\int_B |f(x)|^2 K_t(x)dx-\pi^{-2}\int_B\int_B f(x)\bar{f}(y)L_t(x,y)dxdy\,, \tag{5.7}$$

$$K_t(x)=\int_B L_t(x,y)dy\,. \tag{5.8}$$

The second integral in (5.7) can be bounded above using the Schwarz inequality as follows. Choose a real valued function h with $h(x)>0$ for all $|x|\leqq 1$. Then

$$\int f\bar{f}L_t=\int_B\int_B [f(x)h(y)^{1/2}/h(x)^{1/2}][\bar{f}(y)h(x)^{1/2}/h(y)^{1/2}]L_t(x,y)dxdy$$

$$\leqq \int_B |f(x)|^2\eta_t(x)dx \tag{5.9}$$

with

$$\eta_t(x)=h(x)^{-1}\int_B L_t(x,y)h(y)dy\,. \tag{5.10}$$

We make the choice that h is radial, i.e. $h(x)=h(r)$ with $r=|x|$. To compute K_t and η_t we can do the angular y integration. With $|y|=s$ we have

$$l_t(r,s)\equiv\int L_t(x,y)d\omega_y=[\pi/rs]\{[(r-s)^2+t]^{-1}-[(r+s)^2+t]^{-1}\}\,. \tag{5.11}$$

Combining (5.7)–(5.11) we have that

$$(f,|p|f)_{B,t}\geqq\int_B |f(x)|^2 Q_t(|x|)dx\,, \tag{5.12}$$

with

$$Q_t(r)=\pi^{-2}\int_0^1 l_t(r,s)[1-h(s)/h(r)]s^2ds\,. \tag{5.13}$$

Finally, we choose

$$h(r)=(1+r^2)/r\,. \tag{5.14}$$

(Note that $dh/dr=0$ at $r=1$.) The integrand in (5.13) is then

$$\pi r^{-1}(1+r^2)^{-1}(s-r)(1-rs)\{[(r-s)^2+t]^{-1}-[(r+s)^2+t]^{-1}\}\,. \tag{5.15}$$

At this point we can let $t\to 0$ by recognizing that the integral in (5.13) becomes a principal value integral in the limit, i.e. $Q_t\to Q$ with

$$Q(r)=4\pi^{-1}(1+r^2)^{-1}⨍(s-r)^{-1}(r+s)^{-2}(s-rs^2)ds\,. \tag{5.16}$$

To do this integral (call it I) we set

$$I_1=⨍_0^1(s-r)^{-1}(r+s)^{-2}sds=[2r(1+r)]^{-1}-(4r)^{-1}\ln[(1+r)/(1-r)]\,. \tag{5.17}$$

The remainder of I (namely the rs^2 term) is

$$-\int_0^1 rs(r+s)^{-2}ds-r^2I_1=-r\ln[(1+r)/r]+r(r+1)^{-1}-r^2I_1\,. \tag{5.18}$$

By combining (5.17), (5.18), Eq. (2.16) is derived. The maximum of $Y_1(r)$ was computed numerically by S. Knabe. □

With the help of Theorems 7 and 12, the proof of Theorem 5, which was stated in Sect. 1, can now be given.

Proof of Theorem 5. Fix $0<L<N$ and $M=N-L$ and consider any partition $P=(\pi_1,\pi_2)$ of $\{1,\ldots,N\}$ into two disjoint sets with L integers in π_1 and M integers in π_2. There are $\binom{N}{L}$ such partitions. For each P we define

$$\delta_i(\pi_2)=\min\{|x_i-x_j|\,|\,j\in\pi_2 \quad\text{and}\quad j\neq i \quad\text{if}\quad i\in\pi_2\}. \tag{5.19}$$

First the operator $|p|$ will be considered. Define the N-particle operator

$$h_P=\sum_{i\in\pi_1}|p_i|-\lambda\sum_{i\in\pi_1}\delta_i(\pi_2)^{-1}+\alpha\sum_{i\in\pi_2}\delta_i(\pi_2)^{-1} \tag{5.20}$$

for some λ, $\alpha>0$ to be determined later. Let the N-particle operators H and $\hat{H}$ be given by

$$H=\binom{N}{L}^{-1}\frac{N}{L}\sum_P h_P, \tag{5.21}$$

$$\hat{H}=\sum_{i=1}^N|p_i|-C_1q^{-1/3}\sum_{i=1}^N\delta_i^{-1}. \tag{5.22}$$

If H and $\hat{H}$ are compared we observe that the $|p_i|$ terms are identical. The potential energy terms are more complicated, but we wish to choose λ and α so that $\hat{H}\geqq H$. To this end, fix $x_1,\ldots,x_N$ and let $x_{j(i)}$ be a nearest neighbor of x_i, that is $|x_{j(i)}-x_i|=\min\{|x_k-x_i|\,|k\neq i\}$. It is obvious that $\delta_i(\pi_2)^{-1}\leqq\delta_i^{-1}$, so that the last term in (5.20), when summed on P, is at most $\alpha\tau\sum_1^N\delta_i^{-1}$, where

$$\tau=\binom{N}{L}^{-1}\frac{N}{L}\binom{N-1}{L}=\frac{N-L}{L}. \tag{5.23}$$

To bound the middle, or λ, term in h_P we note that for each $i\in\{1,\ldots,N\}$ there will be $\binom{N-2}{L-1}$ partitions in which $i\in\pi_1$ and $j(i)\in\pi_2$. Therefore this middle sum in h_P, when summed on all partitions, is at least $\lambda v\sum_1^N\delta_i^{-1}$, where

$$v=\binom{N}{L}^{-1}\frac{N}{L}\binom{N-2}{L-1}=\frac{N-L}{N-1}. \tag{5.24}$$

Consequently, $\hat{H}\geqq H$ if

$$C_1q^{-1/3}\leqq(N-L)[\lambda(N-1)^{-1}-\alpha L^{-1}]. \tag{5.25}$$

Assuming (5.25), Theorem 5 will be proved if we show that $(\psi, h_P\psi)\geqq 0$ for every P. Since permutation of the labels in π_1 and π_2 is irrelevant, it suffices to prove this for any one P. To this end we henceforth change notation so that $x_1,\ldots,x_L\in\mathbb{R}^3$ are the variables in the π_1 block and $R_1,\ldots,R_M\in\mathbb{R}^3$ are the variables in the π_2 block. Obviously we can assume that the R_i are fixed and distinct and that ψ is then a function of $x_1,\ldots,x_L$ with q-state Fermi statistics. We shall also drop the subscript P on h_P. Thus, we want to show that $h \geqq 0$ for all choices of the R_i. Since h is a sum of one-body operators, we have to show that for any density matrix γ with $0 \leqq \gamma \leqq q$,

$$\mathrm{Tr}\gamma(|p|-V) \geqq -\alpha \sum_{j=1}^{M} (2D_j)^{-1}, \tag{5.26}$$

where $V(x)$ and D_j are defined by

$$V(x) = -\lambda\delta(x)^{-1}, \tag{5.27}$$

$$2D_j = \min\{|R_j - R_k| \,|\, k=1, \ldots, M \text{ but } k \neq j\}, \tag{5.28}$$

$$\delta(x) = \min\{|x - R_j| \,|\, j=1, \ldots, M\}. \tag{5.29}$$

Under the assumption that $\lambda < 2/\pi$, we write $|p|$ as the sum of two pieces $|p| = (\lambda\pi/2)|p| + (1-\lambda\pi/2)|p|$. We also introduce the Voronoi cells $\Gamma_j = \{x \,|\, |x-R_j| \leqq |x-R_k| \text{ for all } k \neq y\}$ and the balls $B_j \subset \Gamma_j$ defined by $B_j = \{x \in \Gamma_j \,|\, |x-R_j| \leqq D_j\}$. Obviously

$$(f, |p| f) \geqq \sum_{j=1}^{M} (f, |p| f)_{B_j}, \tag{5.30}$$

where the right side is the sum of the kinetic energies in the balls B_j defined in Theorem 7, (2.14). Using Theorem 7, we have that

$$(\lambda\pi/2)(f, |p| f) \geqq (\lambda\pi/2) \sum_{j=1}^{M} D_j^{-1} \int_{B_j} |f(x)|^2 Q(|x-R_j|/D_j)dx, \tag{5.31}$$

with Q given by (2.16). Hence

$$\mathrm{Tr}\gamma(|p|-V) \geqq \mathrm{Tr}\gamma[(1-\lambda\pi/2)|\mathrm{p}| - \lambda W], \tag{5.32}$$

where W is given in each Γ_j by

$$W_j(x) = \begin{cases} |x-R_j|^{-1} & \text{if } |x-R_j| > D_j \\ (\pi/2)D_j^{-1} Y_1(|x-R_j|/D_j) & \text{if } |x-R_j| \leqq D_j \end{cases} \tag{5.33}$$

with Y_1 given in (2.16).

Next, we use the Daubechies bound, Theorem 8,

$$\mathrm{Tr}\gamma[(1-\lambda\pi/2)|p| - \lambda W] \geqq -0.0258q[1-\lambda\pi/2]^{-3}\lambda^4 \int W(x)^4 dx. \tag{5.34}$$

The integral in (5.34) is a sum of integrals over each Γ_j. To obtain a bound we shall merely integrate each $|x-R_j|$ term in W [see (5.33)] over all $|x-R_j| > D_j$ and omit the restriction that $x \in \Gamma_j$. The integral outside each ball B_j is thus

$$\int_{\sim B_j} W_j^4 = 4\pi/D_j. \tag{5.35}$$

The integral inside B_j is (see (2.27))

$$\int_{B_j} W_j^4 = (\pi/2)^4 D_j^{-1} \int_{|x|<1} Y_1(x)^4 dx = 46.418/D_j. \tag{5.36}$$

Combining (5.34)–(5.36) we find that (5.26) is satisfied provided

$$qA\lambda^4(1-\lambda\pi/2)^{-3} \leqq \tfrac{1}{2}\alpha \tag{5.37}$$

with

$$A = 0.0258[4\pi + 46.418] = 1.522 \tag{5.38}$$

and provided $\lambda < 2/\pi$. We shall choose α so that (5.37) is an equality. We shall also write $\lambda = Xq^{-1/3}$. Then (5.25) is satisfied if C_1 satisfies the following for some $0 \leqq X \leqq 2/\pi$ and some $0 < L < N$:

$$C_1 \leqq (N-L)[X(N-1)^{-1} - AX^4(1 - X\pi/2)^{-3}L^{-1}]. \tag{5.39}$$

(Here we have used the bound that $\lambda\pi/2 < X\pi/2$, which holds since $q \geqq 1$.)

Consider the case $N \geqq 3$. To utilize (5.39) we make the following choices

$$X = 1/5 \quad \text{and} \quad L = \{(B/X)^{1/2}N\}, \tag{5.40}$$

where $B \equiv AX^4(1 - X\pi/2)^{-3} = 0.0075486$ and where $\{a\}$ denotes the smallest integer $\geqq a$. Write $L = l + \varepsilon$ with $l = N(B/X)^{1/2}$ and $0 \leqq \varepsilon < 1$. We claim that when $N \geqq 3$,

$$(L-1)X/(N-1) + BN/L \leqq lX/N + BN/l. \tag{5.41}$$

Assuming this for the moment, we would then have that (5.39) is satisfied with

$$C_1 = (X^{1/2} - B^{1/2})^2 \geqq 0.129, \tag{5.42}$$

which proves Theorem 5 when $N \geqq 3$. If $N = 1$ there is nothing to prove. If $N = 2$, Theorem 3 is trivial because it asserts that

$$|p_1| + |p_2| \geqq 0.129 q^{-1/3} |x_1 - x_2|^{-1}, \tag{5.43}$$

but we already have the simple bound $|p_1| \geqq (2/\pi)|x_1 - x_2|^{-1}$ for all x_2.

To prove (5.41), insert $L = l + \varepsilon$ in the left side and multiply by $N(N-1)Ll$ (recalling that $l \equiv N(B/X)^{1/2}$). Then (5.41) is equivalent to

$$Nl - l(l + 2\varepsilon) + N\varepsilon(1 - \varepsilon) \geqq 0. \tag{5.44}$$

Since $l < N/5$, (5.44) holds for $N \geqq 3$.

The proof for p^2 in place of $|p|$ follows the same route, but using Theorem 12 in place of Theorem 7 and using the Lieb-Thirring [25] bound in place of the Daubechies bound. This is

$$\operatorname{Tr}\gamma(\mu p^2 - \lambda W) \geqq -q\sigma\mu^{-3/2}\lambda^{5/2} \int W(x)^{5/2} dx.$$

The best bound for σ is obtained in [22] and is $\sigma = 0.040305$. We split the operator p^2 into $4\lambda p^2 + (1 - 4\lambda)p^2$, and take the μ above to be $(1 - 4\lambda)$. Using Theorem 12, W is given in each cell Γ_j by

$$W(x) = \begin{cases} |x - R_j|^{-2} & \text{if} \quad |x - R_j| > D_j \\ 4D_j^{-2} Y_2(|x - R_j|/D_j) & \text{if} \quad |x - R_j| < D_j. \end{cases} \tag{5.45}$$

The analogue of (5.35), (5.36) using $Y_2(r) = 1 + r^2/4$, is

$$w \equiv D_j^2 \int_{\mathbb{R}^3} W_j(x)^{5/2} dx = 2\pi + 128\pi \int_0^1 (1 + r^2/4)^{5/2} r^2 dr.$$

Using $(1 + r^2/4)^{1/2} \leqq 1 + r^2/8$ in the above integral we find $w < 198.2$.

Setting $\lambda = Xq^{-2/3}$, the analogue of (5.39) is

$$C_2 \leqq (N-L)[X(N-1)^{-1} - AX^{5/2}(1-4X)^{-3/2}L^{-1}] \tag{5.46}$$

with $A = \sigma w = 7.988$. For $N \geqq 3$ we make the following choices:

$$X = 1/20 \quad \text{and} \quad L = \{(B/X)^{1/2}N\}, \tag{5.47}$$

with $B = AX^{5/2}(1-4X)^{-3/2} = 0.006241$. Again, setting $L = l + \varepsilon$ with $l = (B/X)^{1/2}N$, we have to verify (5.41), which is equivalent to (5.44). This inequality is true for $N \geqq 4$ since $l = 0.3533N$. With (5.41) satisfied we have that

$$C_2 \geqq (X^{1/2} - B^{1/2})^2 \geqq 0.0209. \tag{5.48}$$

This proves Theorem 3 for $N \geqq 4$. When $N = 1$ there is nothing to prove, while for $N = 2$ we require

$$p_1^2 + p_2^2 \geqq 0.0209 q^{-2/3} |x_1 - x_2|^{-2}. \tag{5.49}$$

Since $p_1^2 \geqq \frac{1}{4}|x_1 - x_2|^{-2}$ for all x_2, inequality (5.49) is satisfied. For $N = 3$ it suffices to have

$$p_1^2 \geqq 0.0209 q^{-2/3} \{|x_1 - x_2|^{-2} + |x_1 - x_3|^{-2}\}, \tag{5.50}$$

and this is clearly true by the inequality just mentioned. □

Remarks. In the above proof, the inequality for p^2 was proved in a fashion analogous to that for $|p|$ by substituting Theorem 12 for Theorem 7. However, another proof for p^2 can be given by using the IMS localization [see (3.3)] instead of Theorem 12.

VI. Refined Localization of the Kinetic Energy

Proof of Theorem 10 (Sect. III). Starting from the Corollary of Theorem 9, we see from (3.10) that our task is to find an upper bound to $\mathrm{Tr}\gamma L$ with $L = L^0 + L_1^*$ and with

$$L(x,y) = \pi^{-2}|x-y|^4[1 - \chi_0(x)\chi_0(y) - \chi_1(x)\chi_1(y)] \tag{6.1}$$

and

$$L_1^*(x,y) = \begin{cases} L(x,y)B_1(x)B_1(y) & \text{if} \quad |x-y| \leqq \sigma \\ 0 & \text{if} \quad |x-y| > \sigma. \end{cases} \tag{6.2}$$

Recall that B_1 is a ball of radius D_1 centered at the origin. By simple scaling we can, and shall take $D_1 = 1$; we shall also write $B_1 = B$. We have $\chi_1(x) = 0$ unless $|x| \leqq (1-\sigma)$, i.e. unless $x \in B^{(\sigma)}$.

We first bound $\mathrm{Tr}\gamma L^0$. Notice that when $|x| < |y|$, $L^0(x,y) = 0$ unless $|x| \leqq (1-\sigma)$. Using the symmetry of L_0 we can write

$$\mathrm{Tr}\gamma L^0 = 2\mathrm{Re} \iiint_{|x|<|y|} \gamma^{1/2}(x,z)\gamma^{1/2}(z,y)L^0(x,y)B^{(\sigma)}(x)\,dx\,dy\,dz, \tag{6.3}$$

where $\gamma^{1/2}$ is the operator square root of γ. We do the y integration first and then apply Minkowski's inequality to the x integration. For any $\varepsilon>0$,

$$\operatorname{Tr}\gamma L^0 \leqq \varepsilon \int\int |\gamma^{1/2}(x,z)|^2 B^{(\sigma)}(x)dxdz$$
$$+\varepsilon^{-1}\int\int \left| \int_{|y|>|x|} \gamma^{1/2}(z,y)L^0(x,y)dy \right|^2 B^{(\sigma)}(x)dxdz\,. \tag{6.4}$$

The first integral is just

$$\int \gamma(x,x)B^{(\sigma)}(x)dx\,. \tag{6.5}$$

In the second integral we do the z integration before the x integration and obtain

$$\int\int \gamma(y,y')\left(\int_A L^0(x,y)L^0(x,y')dx\right)dydy'\,, \tag{6.6}$$

where A is the region $|x|\leqq \min((1-\sigma),|y|,|y'|)$. The factor in parentheses in (6.6) is the kernel of a positive definite operator, so we can bound (6.6) by

$$\|\gamma\| \int\int_{\Lambda} L^0(x,y)^2 dxdy\,, \tag{6.7}$$

where Λ is the region $|x|\leqq(1-\sigma)$ and $|y|\geqq|x|$. In view of the fact that $L^0(x,y)$ is symmetric and $L^0(x,y)=0$ unless at least one of $|x|$ or $|y|$ is less than $(1-\sigma)$, and given that $\|\gamma\|=q$ by assumption, (6.7) is just $\frac{1}{2}q\operatorname{Tr}(L^0)^2$. Thus,

$$\operatorname{Tr}\gamma L^0 \leqq \varepsilon\int\gamma(x,x)B^{(\sigma)}(x)dx + q\varepsilon^{-1}\Omega_1 \tag{6.8}$$

with $\Omega_1=\frac{1}{2}\operatorname{Tr}(L^0)^2$. The verification of the two integrals for Ω_1 in (3.15), (3.16) is evident if one recognizes that $\chi_0(x)=1$ and $\chi_1(x)=0$ for $|x|\geqq(1-\sigma)$.

Now we turn to $\operatorname{Tr}\gamma L_1^*$. Since γ is a positive operator, its kernel satisfies $|\gamma(x,y)|^2 \leqq\gamma(x,x)\gamma(y,y)$. Hence, since $L_1^*(x,y)>0$ and $h_1(x)>0$,

$$\begin{aligned}\operatorname{Tr}\gamma L_1^* &= \int\int \gamma(x,y)L_1^*(x,y)dxdy\\ &\leqq \int\int [\gamma(x,x)h_1(y)/h_1(x)]^{1/2}[\gamma(y,y)h_1(x)/h_1(y)]^{1/2}L_1^*(x,y)dxdy\\ &\leqq \int\int [\gamma(x,x)h_1(y)/h_1(x)]L_1^*(x,y)dxdy\\ &= \int\gamma(x,x)\theta_1(x)dx\,.\end{aligned} \tag{6.9}$$

The second inequality in (6.9) is the Schwarz inequality, together with the symmetry in x and y. The idea of using the Schwarz inequality in this fashion goes back to Hardy and Littlewood; see [18] for another application.

When inequalities (6.8) and (6.9) are inserted into (3.10), the Corollary of Theorem 9, the result is Theorem 10. □

VII. Estimates of Negative Eigenvalues

Proof of Theorem 11 (Sect. III). It obviously suffices to consider the case $q=1$. Let the kernel of γ be

$$\gamma(x,y)=\sum_\alpha \tau_\alpha f_\alpha(x)\overline{f_\alpha(y)} \tag{7.1}$$

with $0\leqq\tau_\alpha\leqq1$ and $\sum\tau_\alpha<\infty$ and with the f_α being orthonormal. Let $g_\alpha(x) \equiv\chi(x)f_\alpha(x)$. We want to prove that, with $V(x)=2/(\pi|x|)+C/R$,

$$E\equiv\sum_\alpha (g_\alpha,(|p|-V)g_\alpha)\geqq -4.4827(3/4\pi R^3)C^4R^{-1}\|\chi\|_2^2\,. \tag{7.2}$$

By scaling it clearly suffices to prove the theorem for $R=1$, which we assume henceforth.

It is convenient to use Fourier transforms. Let

$$\varrho(p,q)=\iint \bar{\chi}(x)\chi(y)\gamma(x,y)\exp(ip\cdot x-iq\cdot y)dxdy. \tag{7.3}$$

Since $\bar{\chi}\gamma\chi$ is positive semidefinite, so is ϱ, and hence

$$|\varrho(p,q)|\leqq \varrho(p,p)^{1/2}\varrho(q,q)^{1/2}\equiv \mu(p)\mu(q) \tag{7.4}$$

with $\mu(p)=\varrho(p,p)^{1/2}$. From (7.3) and the fact that $0\leqq\gamma\leqq 1$ as an operator,

$$\mu(p)^2=(n_p,\gamma n_p)\leqq(n_p,n_p)=\int|\chi(x)|^2dx\equiv M^2, \tag{7.5}$$

where $n_p(x)=\chi(x)\exp(-ip\cdot x)$ and $M=\|\chi\|_2$. Using the Fourier transform of $|x|^{-1}$, namely

$$4\pi|p|^{-2}=\int|x|^{-1}\exp(ip\cdot x)dx, \tag{7.6}$$

E can be written as

$$E=(2\pi)^{-3}\{\int\varrho(p,p)(|p|-C)dp-\pi^{-3}\iint\varrho(p,q)|p-q|^{-2}dpdq\}. \tag{7.7}$$

Using (7.5) we have that

$$E\geqq(2\pi)^{-3}\inf\{\tilde{E}(\mu)|0\leqq\mu(p)\leqq M \quad \text{for all } p\}, \tag{7.8}$$

where $\tilde{E}(\mu)$ is defined by

$$\tilde{E}(\mu)=\int\mu(p)^2(|p|-C)dp-\pi^{-3}\iint\mu(p)\mu(q)|p-q|^{-2}dpdq. \tag{7.9}$$

To bound the second integral in (7.9), let

$$h(p)=\begin{cases}A^{-2} & \text{if } |p|\leqq A\\ |p|^{-2} & \text{if } |p|>A,\end{cases} \tag{7.10}$$

where A is some constant to be determined later. Employing the same strategy as in (6.9) we have

$$\begin{aligned}&\iint\mu(p)\mu(q)|p-q|^{-2}dpdq\\ &=\int\mu(p)(h(q)/h(p))^{1/2}\mu(q)(h(p)/h(q))^{1/2}|p-q|^{-2}dpdq\leqq\int\mu(p)^2t(p)dp,\end{aligned} \tag{7.11}$$

with

$$\begin{aligned}t(p)&=h(p)^{-1}\int|p-q|^{-2}h(q)dq\\ &=h(p)^{-1}\{\int|p-q|^{-2}q^{-2}dq-s(p)\}=h(p)^{-1}\{\pi^3|p|^{-1}-s(p)\},\end{aligned} \tag{7.12}$$

and with

$$s(p)=\int_{|q|<A}|p-q|^{-2}(q^{-2}-A^{-2})dq. \tag{7.13}$$

To calculate $s(p)$ we use bipolar coordinates, i.e. for any functions f and g

$$\int f(|p-q|)g(|q|)d^3q=(2\pi/|p|)\int_0^\infty\beta f(\beta)\left\{\int_{||p|-\beta|}^{|p|+\beta}\alpha g(\alpha)d\alpha\right\}d\beta. \tag{7.14}$$

Thus,

$$s(p)=(2\pi/|p|)\int_0^A(\beta^{-1}-\beta A^{-2})\left\{\int_{||p|-\beta|}^{|p|+\beta}\alpha^{-1}d\alpha\right\}d\beta$$
$$=(2\pi/|p|)\int_0^{1/\xi}(u^{-1}-u\xi^2)\ln\left(\frac{1+u}{|1-u|}\right)du \tag{7.15}$$

with $\xi=|p|/A$.

We claim that

$$s(p)\geqq\tilde{s}(p)\equiv\begin{cases}(8\pi/3)A & \text{for } |p|\geqq A\\ 4\pi|p|^{-1}[\frac{10}{9}+\frac{\pi^2}{8}-2\xi+\frac{1}{6}\xi^2+\frac{5}{36}\xi^3] & \text{for } |p|\leqq A.\end{cases} \tag{7.16}$$

We shall prove (7.16) later. For now, let us insert (7.16) into (7.12), and then into (7.11) and (7.9),

$$\tilde{E}(\mu)\geqq\int_{|p|>A}\mu(p)^2[8A(3\pi^2)^{-1}-C]dp+\int_{|p|<A}\mu(p)^2[|p|-A^2|p|^{-1}$$
$$+\pi^{-3}A^2\tilde{s}(p)-C]dp. \tag{7.17}$$

We choose

$$A=3\pi^2C/8 \tag{7.18}$$

so that the first integral in (7.17) vanishes. Then, using (7.18) and performing the angular integration,

$$\tilde{E}(\mu)\geqq 4\pi A^4\int_0^1\mu(Aw)^2\left\{w+2w^2/3\pi^2+5w^3/9\pi^2\right.$$
$$\left.-\left[\frac{1}{2}-40/(9\pi^2)\right]w^{-1}-32/(3\pi^2)\right\}w^2dw. \tag{7.19}$$

As is easily seen, the factor { } in (7.19) has its maximum at $w=1$ and it is negative there. Therefore the infimum of the right side of (7.19) over the set $\mu(Aw)\leqq M$ occurs for $\mu(Aw)=M$ for all $0\leqq w\leqq 1$. The right side of (7.19) with $\mu=M$ is

$$-(598/135\pi)A^4M^2. \tag{7.20}$$

Returning to (7.8) and using (7.18) and (7.20) (with 598 replaced by 600) we have that

$$E\geqq-\frac{5}{9}\left(\frac{3\pi}{8}\right)^4C^4M^2. \tag{7.21}$$

Since $M=\|\chi\|_2$, (7.21) is the same as (7.2).

To complete the proof we must bound (7.15) by (7.16). When $u\leqq 1/\xi$, the factor $u^{-1}-u\xi^2\geqq 0$. When $\xi\geqq 1$ (i.e. $|p|\geqq A$), $u\leqq 1$ and we have the bound

$$\ln[(1+u)/(1-u)]\geqq 2u. \tag{7.22}$$

Inserting (7.22) into (7.15) yields the first part of (7.16).

If $|p|\leqq A$, then $\xi<1$. The integral in (7.15) from 0 to 1 can be done explicitly,

$$\int_0^1(u^{-1}-u\xi^2)\ln[(1+u)/(1-u)]du=\pi^2/4-\xi^2. \tag{7.23}$$

To bound the integral from 1 to $1/\xi$, use the fact that for $u>1$,

$$\ln[(1+u)/(u-1)] \geqq 2u^{-1} + \tfrac{2}{3}u^{-3}. \qquad (7.24)$$

Then

$$\int_1^{1/\xi} (u^{-1} - u\xi^2)\ln[(1+u)/(u-1)]\,du$$
$$\geqq \int_1^{1/\xi} (u^{-1} - u\xi^2)\left(2u^{-1} + \frac{2}{3}u^{-3}\right)du = 20/9 - 4\xi + 4\xi^2/3 + 4\xi^3/9. \qquad (7.25)$$

When (7.25) is combined with (7.23) (and the $4\xi^3/9$ term is replaced by the smaller quantity $5\xi^3/18$) the result is the second part of (7.16). □

VIII. Some Numerical Calculations

Our goal here is to derive the bounds (3.33) for Ω and (3.34) for $\theta(r)$.

(A) Evaluation of Ω. Ω is defined as the sum of the two integrals in (3.15), (3.16). Recall that $\sigma = 0.3$ and $\chi_1(x) = \chi(|x|)$ is given in (3.22) while $\chi_0(x)^2 = 1 - \chi_1(x)^2$. We already set $D_1 = 1$.

To evaluate $I^{(1)}$ we use the spherical symmetry of χ and first do the angular integration on x and y. This integral is

$$\int |x-y|^{-8}d\omega_y = 2\pi \int_0^{\pi} (x^2+y^2-2xy\cos\theta)^{-4}\sin\theta d\theta$$
$$= (\pi/3)(|x|\,|y|)^{-1}\{(|x|-|y|)^{-6} - (|x|+|y|^{-6})\}. \qquad (8.1)$$

Thus,

$$I^{(1)} = 4(3\pi^2)^{-1}\int_0^{1-2\sigma} sds \int_{s+\sigma}^{1-\sigma} tdt[(t-s)^{-6}$$
$$-(t+s)^{-6}][1-(1-\chi(s)^2)^{1/2}(1-\chi(t)^2)^{1/2} - \chi(s)\chi(t)]^2. \qquad (8.2)$$

(Note that we integrate over $t>s+\sigma$ and $s,t<1-\sigma$, and then multiply by 2. Since $s<t-\sigma$ and $t<1-\sigma$, we have that $s<1-2\sigma$.) This integral is not elementary, but because it is an integral of a continuous, bounded function over a bounded domain in $\mathbb{R}^2$ it can be confidently evaluated on a computer. The result is (3.33).

To evaluate $I^{(2)}$, the angular integration over y is done first as before, with the result (8.1). Then $I^{(2)}$ is the sum of three integrals according as $|x|<1-3\sigma$, $1-3\sigma \leqq |x| < 1-2\sigma$, $1-2\sigma \leqq |x| \leqq 1-\sigma$. Thus,

$$I^{(2)} = (4/3\pi^2)\int_0^{1-3\sigma} sds \int_{1-\sigma}^{\infty} tdt[(t-s)^{-6} - (t+s)^{-6}]$$
$$+(16/3\pi^2)\int_{1-3\sigma}^{1-2\sigma} sds \int_{1-\sigma}^{\infty} tdt[(t-s)^{-6} - (t+s)^{-6}]\sin^4\left[\frac{\pi}{8\sigma}(1-\sigma-s)\right]$$
$$+(16/3\pi^2)\int_{1-2\sigma}^{1-\sigma} sds \int_{s+\sigma}^{\infty} tdt[(t-s)^{-6} - (t+s)^{-6}]\sin^4\left[\frac{\pi}{8\sigma}(1-\sigma-s)\right]. \qquad (8.3)$$

In each case the t integration can easily be done analytically. This transforms (8.3) into three integrals over the bounded intervals $0 \leqq s < 1-3\sigma$, $1-3\sigma \leqq s < 1-2\sigma$ and $1-2\sigma \leqq s \leqq 1-\sigma$. The integrands are again bounded and continuous so numerical integration can be used. The result is (3.33).

(B) Bound on $\theta(r)$, *Eq.* (3.34). The function $\theta \equiv \theta_1$ is defined in (3.13) with h defined in (3.24). Again we take $D_1 = 1$. The kernel L_1^* is given in (3.12) with $\chi_1 \equiv \chi$ given in (3.22) and $\chi_0^2 \equiv 1-\chi^2$.

We want to compute

$$I(r) = \int L_1^*(x, y) h(|y|) dy \tag{8.4}$$

with $r = |x|$. Since the angular integral of $|x-y|^{-4}$ is less than $\pi(rs)^{-1}(r-s)^{-2}$, with $s = |y|$, we have that

$$I(r) \leqq (1/\pi r) \int_0^1 (r-s)^{-2} h(s) m(r, s) s ds , \tag{8.5}$$

where $m(r, s) = m(s, r)$ and, for $r \leqq s$, $m(r, s)$ is given by

$$m(r,s) = \begin{cases} 1-\cos[\pi(s-\tau)/4\sigma] & \text{for } 0 \leqq r \leqq \tau \leqq s \leqq r+\sigma \\ 1-\cos[\pi(s-r)/4\sigma] & \text{for } \tau \leqq r \leqq s \leqq \min(\tau+2\sigma, r+\sigma) \\ 1-\cos[\pi(2\sigma+\tau-r)/4\sigma] & \text{for } s-\sigma \leqq r \leqq \tau+2\sigma \leqq s \leqq 1 \\ 0 & \text{otherwise}. \end{cases} \tag{8.6}$$

In (8.6), $\tau = 1-3\sigma$.

The arguments of the cosines in (8.6) are all at most $\pi/4$ and one can use the inequality $\cos b \geqq 1 - b^2/2$ for $|b| \leqq \pi/4$. If we use this inequality in (8.6) and then insert the result in (8.5), the integral (8.5) is seen to be elementary but tedious [recall (3.24)]. Finally, $\theta(r) = I(r)/h(r)$.

Let us verify (3.34) when $r \geqq 1-\sigma$. Then $rh(r) = 1$ and thus

$$\theta(r) = (\pi/32\sigma^2) \int_{r-\sigma}^{1-\sigma} sh(s)(r-s)^{-2}(1-\sigma-s)^2 ds . \tag{8.7}$$

The second line of (3.24) is appropriate for this region. In the region $r-\sigma \leqq s \leqq 1-\sigma$ the function $(r-s)^{-2}(1-\sigma-s)^2$ is monotone decreasing in s and so has its maximum at $s = r-\sigma$. Thus,

$$\theta(r) \leqq (\pi/32\sigma^2)\sigma^{-2}(1-r)^2 \int_{r-\sigma}^{1-\sigma} \{2-\sigma(s-1+2\sigma)\} ds , \tag{8.8}$$

and this agrees with (3.34) for $r \geqq 1-\sigma$.

The verification of the $r \leqq 1-\sigma$ case of (3.34) is elementary and we omit the details.

IX. The Occurrence of Collapse for Large α

In the previous sections it was shown that the Hamiltonian H_{NK} (1.4) under consideration is stable if α is small enough. There are two parameters in the problem, $z\alpha$ *and* α. For stability of one electron and one nucleus it is necessary and sufficient that $z\alpha \leqq 2/\pi$, but, assuming this condition, there is stability in the many-body case if $\alpha < \alpha_0/q$ with $\alpha_0 > 1/47$. In this section we shall prove that this stability bound is not just an artifact of our proof but that instability definitely occurs if α is too large. Theorems 3 and 4 will be proved here.

Proof of Theorem 3. The method of proof here is the same as the method employed in [23] to prove the instability of one-electron molecules in a magnetic field. Let $\phi \in L^2(\mathbb{R}^3)$ be real with $\|\phi\|_2 = 1$ and let $\tau = (\phi, |p|\phi)$ which is assumed to be finite. Then

$$E = (\phi, H_{NK}\phi) = \tau - z\alpha \int \phi^2(x) \sum_{j=1}^{K} |x - R_j|^{-1} dx + z^2\alpha \sum_{1 \leqq i < j \leqq K} |R_i - R_j|^{-1} . \tag{9.1}$$

With ϕ fixed let us try to position the R_i so as to minimize the right side of (9.1). This minimum (call it e) is less than any average of E over positions of the R_i. In particular, we use $\psi = \prod_{j=1}^{K} \phi(R_j)^2$ as a probability density for such an average. Then

$$Av(E) = \tau - \sigma[z\alpha K - z^2\alpha K(K-1)/2] = \tau + \tfrac{1}{2}\sigma\{z^2\alpha[K - \tfrac{1}{2} - z^{-1}]^2 - \tfrac{1}{4}z^2\alpha - \alpha - z\alpha\}, \tag{9.2}$$

where

$$\sigma = \int \phi(x)^2 \phi(y)^2 |x-y|^{-1} dx dy . \tag{9.3}$$

Now K can be chosen so that $|K - \frac{1}{2} - z^{-1}| \leqq \frac{1}{2}$. Using this K, we have

$$e \leqq Av(E) \leqq \tau - \tfrac{1}{2}\sigma\alpha . \tag{9.4}$$

If we set $\alpha_1 = 2\tau/\sigma$, then when $\alpha > \alpha_1$, $e < 0$, and we can drive e to $-\infty$ simply by dilation, i.e. $\phi(x) \to \lambda^{3/2}\phi(\lambda x)$ and $R_j \to \lambda R_j/\lambda$ with $\lambda \to \infty$.

To obtain a numerical value for α_1, choose $\phi(x) = \pi^{-1/2}\exp(-r)$ with $r = |x|$. The Fourier transforms of ϕ and ϕ^2 are

$$\hat{\phi}(p) = 8\pi^{1/2}(1+p^2)^{-2}, \qquad \widehat{\phi^2}(p) = 16(4+p^2)^{-2} . \tag{9.5}$$

Then

$$\tau = (2\pi)^{-3} \int \hat{\phi}(p)^2 |p| dp = 8/3\pi , \qquad \sigma = (2\pi)^{-3} \int \widehat{\phi^2}(p)(4\pi/|p|^2) dp = 5/8 , \tag{9.6}$$

and $2\tau/\sigma = 128/15\pi$. □

Proof of Theorem 4. The method of proof here is similar to that used in [20] to prove that the energy of N nonrelativistic bosons interacting with fixed nuclei via Coulomb forces diverges as $-N^{5/3}$. Again, let $\phi \in L^2(\mathbb{R}^3)$ be real with $\|\phi\|_2 = 1$ and $\tau = (\phi, |p|\phi)$. Since there are q spin states, we can put $N = q$ electrons into the state ϕ. The energy is then

$$E = q\tau - z\alpha q \int \phi^2(x) \sum_{j=1}^{K} |x - R_j|^{-1} dx + z^2\alpha \sum_{1 \leqq i < j \leqq K} |R_i - R_j|^{-1} + \frac{1}{2} q(q-1)\sigma \tag{9.7}$$

with σ given in (9.3). Let us first prove the theorem under the condition $q/z \geqq 1$; at the end of the proof we shall show how to handle the case $q/z < 1$.

To construct ϕ we first define $g \in L^2(\mathbb{R}^3)$ by

$$g(x, y, z) = f(x) f(y) f(z), \tag{9.8}$$

where $f \in L^2(\mathbb{R}^1)$ is given by $f(x) = \sqrt{3/2}(1 - |x|)$ for $|x| \leqq 1$ and $f(x) = 0$ for $|x| \geqq 1$. This f has $\|f\|_2 = 1$, and thus $\|g\|_2 = 1$. Let $h \in L^2(\mathbb{R}^3)$ be some other function with compact support and with $(h, |p| h) < \infty$ and $\|h\|_2 = 1$. Define the integers n and K and the positive number λ by

$$\begin{gathered} n = [(q/z)^{1/3}] \geqq 1, \quad K = n^3, \\ \lambda = n^3 z/q = Kz/q, \end{gathered} \tag{9.9}$$

where $[b]$ means integral part of b. Clearly, $1 \geqq \lambda \geqq 1/8$. Finally, we construct a sequence of functions $\phi^{(s)}(x)$, $x \in \mathbb{R}^3$, by

$$\phi^{(s)}(x)^2 = \lambda g(x) + (1 - \lambda) s^{-3} h(x/s + (0, 0, s^2))^2 . \tag{9.10}$$

Now choose some fixed locations $R_1, \ldots, R_K$ of K nuclei. Because of the scaling of h by s^{-1} and translation by $(0, 0, s^2)$, we have that E converges to the following E' as $s \to \infty$:

$$E' = q\lambda\tau - z\alpha\lambda q \int g^2(x) \sum_{j=1}^{K} |x - R_j|^{-1} dx + z^2 \alpha \sum_{1 \leqq i < j \leqq K} |R_i - R_j|^{-1} + \frac{1}{2} \lambda^2 q(q-1) \sigma\alpha, \tag{9.11}$$

where τ now means $(g, |p| g)$ and σ is given in (9.3) with g in place of ϕ.

We claim that it is possible to choose the locations $R_1, \ldots, R_K$ so that

$$\sum_{1 \leqq i < j \leqq K} |R_i - R_j|^{-1} - K \int g^2(x) \sum_{j=1}^{K} |x - R_j|^{-1} dx + \frac{1}{2} K^2 \sigma \leqq -K^{4/3}/6 . \tag{9.12}$$

If (9.12) holds then, recalling (9.9),

$$E' \leqq q\lambda\tau - \tfrac{1}{6} z^2 \alpha (\lambda q/z)^{4/3} . \tag{9.13}$$

Recalling that $\lambda > 1/8$ we have that $E' < 0$ whenever

$$\beta^2 \alpha q \geqq 8(6\tau)^3 (\pi/2)^2 . \tag{9.14}$$

We also have that $\tau = (g, |p| g) \leqq (g, p^2 g)^{1/2} = 3$ (by the Schwarz inequality). Thus, collapse occurs if $\alpha \geqq \alpha_2 q^{-1} \beta^{-2}$ with $\alpha_2 = (\pi/2)^2 8(18)^3 = 115,120$, provided $q/z \geqq 1$.

If, on the other hand, $q/z < 1$ and if $\alpha > \alpha_2 q^{-1} \beta^{-2} = \alpha_2 q^{-1} z^{-2} \alpha^{-2} (2/\pi)^2$, we have that $(\pi/2)^2 (z\alpha)^3 > \alpha_2 z/q > \alpha_2$. Since $\alpha_2 \gg (2/\pi)^5$, we are in the situation that $z\alpha > 2/\pi$, which certainly entails collapse. Therefore, the theorem is proved for all ratios q/z with the α, given above.

There remains to prove (9.12). Choose $n - 1$ numbers $\beta_1, \ldots, \beta_{n-1}$ satisfying $-1 \equiv \beta_0 < \beta_1 < \ldots < \beta_{n-1} < \beta_n \equiv 1$ such that

$$\int_{\beta_j}^{\beta_{j+1}} f(x)^2 dx = 1/n \quad \text{for all } j .$$

Let L_j be the interval $[\beta_{j-1}, \beta_j]$ in $\mathbb{R}^1$ and, with m denoting a triplet (i, j, k), let $\Gamma(m) \subset \mathbb{R}^3$ be the rectangular parallelepiped $L_i \times L_j \times L_k$. Then, for each m,

$$\int_{\Gamma(m)} g^2(x) dx = 1/n^3 = 1/K . \tag{9.15}$$

There are n^3 of these parallelepipeds. To prove (9.12) we shall place one of the R_i's in each $\Gamma(m)$ and average its location with respect to the density $g^2(x)$ restricted to $\Gamma(m)$. If the average satisfies (9.12) then there is surely some choice of the R_i's that satisfies (9.12). Apart from a self energy contribution from each parallelepiped, the average of the left side of (9.12) is zero. Thus the average of the left side is given by the self energy terms

$$W=-\frac{1}{2}n^6\sum_m \int\int_{\Gamma(m)\times\Gamma(m)} g(x)^2 g(y)^2|x-y|^{-1}dxdy\,. \tag{9.16}$$

Each integral is the self energy of a charge density g^2 in $\Gamma(m)$. However $\Gamma(m)$ lies inside a ball $B(m)$ of radius $r(m)=(s^2+t^2+u^2)^{1/2}$, where $2s$, $2t$, and $2u$ are the lengths of $\Gamma(m)$, namely $(\beta_i-\beta_{i-1}),(\beta_j-\beta_{j-1}),(\beta_k-\beta_{k-1})$. The self energy is greater than the *minimum* self energy of a charge $1/K$ distributed in $B(m)$; the minimum occurs for a uniform charge distribution on the boundary of $B(m)$ and is $r(m)^{-1}/K^2$. Thus,

$$W\leqq-\frac{1}{2}\sum_{i=1}^{n}\sum_{j=1}^{n}\sum_{k=1}^{n}(s^2+t^2+u^2)^{-1/2}\,. \tag{9.17}$$

Now $(s^2+t^2+u^2)^{-1/2}>(s+t+u)^{-1}$. Substituting this latter expression in (9.17) and then using the convexity of the function $(s,t,u)\to(s+t+u)^{-1}$ and recalling that $K=n^3$, we have that

$$W\geqq-\tfrac{1}{2}K(a+b+c)^{-1}\,, \tag{9.18}$$

where a, b, and c are the averages of s, t, and u. But $a=b=c=1/n$, and thus (9.12) is proved. □

Acknowledgements. The authors thank Michael Loss for helpful discussions and comments and they thank Stefan Knabe for performing numerical calculations.

References

1. Baxter, J.R.: Inequalities for potentials of particle systems. Ill. J. Math. **24**, 645–652 (1980)
2. Chandrasekhar, S.: Phil. Mag. **11**, 592 (1931); Astro. J. **74**, 81 (1931); Monthly Notices Roy. Astron. Soc. **91**, 456 (1931); Rev. Mod. Phys. **56**, 137 (1984)
3. Conlon, J.G.: The ground state energy of a classical gas. Commun. Math. Phys. **94**, 439–458 (1984)
4. Conlon, J.G., Lieb, E.H., Yau, H.-T.: The $N^{7/5}$ law for charged bosons. Commun. Math. Phys. **116**, 417–448 (1988)
5. Cycon, H.L., Froese, R.G., Kirsch, W., Simon, B.: Schrödinger operators. Berlin, Heidelberg, New York: Springer 1987
6. Daubechies, I.: An uncertainty principle for fermions with generalized kinetic energy. Commun. Math. Phys. **90**, 511–520 (1983)
7. Daubechies, I.: One electron molecules with relativistic kinetic energy: properties of the discrete spectrum. Commun. Math. Phys. **94**, 523–535 (1984)
8. Daubechies, I., Lieb, E.H.: One-electron relativistic molecules with Coulomb interaction. Commun. Math. Phys. **90**, 497–510 (1983)
9. Dyson, F.J.: Ground state energy of a finite system of charged particles. J. Math. Phys. **8**, 1538–1545 (1967)

10. Dyson, F.J., Lenard, A.: Stability of matter I and II. J. Math. Phys. **8**, 423–434 (1967); ibid **9**, 698–711 (1968). See also Lenard's Battelle lecture. In: Lecture Notes in Physics, vol. 23. Berlin, Heidelberg, New York: Springer 1973
11. Erdelyi, A., Magnus, W., Oberhettinger, F., Tricomi, F.G.: Tables of integral transforms, Vol. 1. New York, Toronto, London: McGraw-Hill 1954, p. 75, 2.4 (35)
12. Federbush, P.: A new approach to the stability of matter problem. II. J. Math. Phys. **16**, 706–709 (1975)
13. Fefferman, C.: The N-body problem in quantum mechanics. Commun. Pure Appl. Math. Suppl. **39**, S67–S109 (1986)
14. Fefferman, C., de la Llave, R.: Relativistic stability of matter. I. Rev. Math. Iberoamericana **2**, 119–215 (1986)
15. Fröhlich, J., Lieb, E.H., Loss, M.: Stability of Coulomb systems with magnetic fields. I. The one-electron atom. Commun. Math. Phys. **104**, 251–270 (1986)
16. Herbst, I.: Spectral theory of the operator $(p^2+m^2)^{1/2}-ze^2/r$. Commun. Math. Phys. **53**, 285–294 (1977); Errata ibid **55**, 316 (1977)
17. Kato, T.: Perturbation theory for linear operators. Berlin, Heidelberg, New York: Springer 1966. See remark 5.12, p. 307
18. Kovalenko, V., Perelmuter, M., Semenov, Ya.: Schrödinger operators with $L_w^{l/2}(\mathbb{R}^l)$ potentials. J. Math. Phys. **22**, 1033–1044 (1981)
19. Lieb, E.H.: Stability of matter. Rev. Mod. Phys. **48**, 553–569 (1976)
20. Lieb, E.H.: The $N^{5/3}$ law for bosons. Phys. Lett. **70** A, 71–73 (1979)
21. Lieb, E.H.: Density functionals for Coulomb systems. Int. J. Quant. Chem. **24**, 243–277 (1983)
22. Lieb, E.H.: On characteristic exponents in turbulence. Commun. Math. Phys. **92**, 473–480 (1984)
23. Lieb, E.H., Loss, M.: Stability of Coulomb systems with magnetic fields. II. The many electron atom and the one electron molecule. Commun. Math. Phys. **104**, 271–282 (1986)
24. Lieb, E., Simon, B.: Thomas Fermi theory of atoms, molecules and solids. Adv. Math. **23**, 22–116 (1977)
25. Lieb, E.H., Thirring, W.: Bound for the kinetic energy of fermions which proves the stability of matter. Phys. Rev. Lett. **35**, 687–689 (1975). Errata, ibid **35**, 1116 (1975); see also their article: Inequalities for the moments of the eigenvalues of the Schrödinger Hamiltonian and their relation to Sobolev inequalities. In: Studies in Mathematical Physics, Essays in honor of Valentine Bargmann. Lieb, E.H., Simon, B., Wightman, A.S. (eds.). Princeton, NJ: Princeton University Press 1976
26. Lieb, E.H., Thirring, W.: Gravitational collapse in quantum mechanics with relativistic kinetic energy. Ann. Phys. (NY) **155**, 494–512 (1984)
27. Lieb, E.H., Yau, H.-T.: The Chandrasekhar theory of stellar collapse as the limit of quantum mechanics. Commun. Math. Phys. **112**, 147–174 (1987). See also Lieb, E.H. and Yau, H.-T.: A rigorous examination of the Chandrasekhar theory of stellar collapse. Astro. J. **323**, 140–144 (1987)
28. Loss, M., Yau, H.-T.: Stability of Coulomb systems with magnetic fields. III. Zero energy bound states of the Pauli operator. Commun. Math. Phys. **104**, 283–290 (1986)
29. Weder, R.: Spectral analysis of pseudodifferential operators. J. Funct. Anal. **20**, 319–337 (1975)

Communicated by A. Jaffe

Received May 12, 1988

Stability of Relativistic Matter via Thomas–Fermi Theory

By Elliott H. Lieb[1], Michael Loss[2] and Heinz Siedentop[3]

(1) Departments of Mathematics and Physics, Princeton University, P.O. Box 708, Princeton, NJ 08544-0708, USA
(2) Georgia Institute of Technology, Atlanta, GA 30332-0160, USA
(3) Matematisk institutt, Universitetet i Oslo, Postboks 1053, N-0316 Oslo, Norway

Abstract A Thomas-Fermi-Weizsäcker type theory is constructed, by means of which we are able to give a relatively simple proof of the stability of relativistic matter. Our procedure has the advantage over previous ones in that the lower bound on the critical value of the fine structure constant, α, is raised from 0.016 to 0.77 (the critical value is known to be less than 2.72). When $\alpha = 1/137$, the largest nuclear charge is 59 (compared to the known optimum value 87). Apart from this, our method is simple, for it parallels the original Lieb-Thirring proof of stability of nonrelativistic matter, and it adds another perspective on the subject.

This article is dedicated to our colleagues, teachers, and coauthors Klaus Hepp and Walter Hunziker on the occasion of their sexagesimal birthdays. Their enthusiasm for quantum mechanics as an unending source of interesting physics and mathematics has influenced many.

1 Work partially supported by U.S. National Science Foundation grant PHY95-13072.

2 Work partially supported by U.S. National Science Foundation grant DMS95-00840.

3 Work partially supported by European Union, grant ERBFMRXCT960001.

This paper appeared in Helvetica Physica Acta vol. 69, no. 5/6, 974-984 (1996). Three typographical errors that appeared in the original paper and in the second edition of this 'Selecta' have been corrected here.

1. Introduction

The 'stability of relativistic matter' concerns the N-body Hamiltonian (in units of $\hbar c$)

$$H = \sum_{i=1}^{N} |p_i| + \alpha V_c \, , \tag{1.1}$$

where V_c is the Coulomb potential of K fixed nuclei with nuclear charge Ze, with locations R_j in $\mathbf{R}^3$, and with N electrons. The Coulomb potential is

$$V_c = -V + R + U \, , \tag{1.2}$$

where

$$V := Z \sum_{i=1}^{N} \sum_{j=1}^{K} |x_i - R_j|^{-1} \, , \tag{1.3}$$

$$R := \sum_{1 \le i < j \le N} |x_i - x_j|^{-1} \, , \tag{1.4}$$

$$U := Z^2 \sum_{1 \le i < j \le K} |R_i - R_j|^{-1} \, . \tag{1.5}$$

As usual $p = -i\nabla$ and $|p| = \sqrt{-\Delta}$, and the x_j are the electron coordinates. The electrons are assumed to have q spin states each, $q = 2$ being the physical value. This means that the Hilbert space for the N-electron functions is the N-fold antisymmetric tensor product of $L^2(\mathbf{R}^3; \mathbf{C}^q)$. The constant $\alpha = e^2/\hbar c$ is called the fine structure constant.

We can easily include a magnetic field, which means replacing $|p_i|$ by $|p_i + A(x_i)|$. The vector field, A, is the vector potential (in suitable units) of a magnetic field, $B = \mathrm{curl} A$, and can be arbitrary, as far as the present work is concerned. A mass can be included as well, i.e., $|p_i + A(x_i)|$ can be replaced by $\sqrt{|p_i + A(x_i)|^2 + m^2} - m$. The inclusion of a mass or magnetic field, while it changes the energy, does not affect stability. The reason for this and the requisite changes will be pointed out in the final section. It is for simplicity and clarity that we set $m = 0$ and $A = 0$.

'Stability of matter' means that the operator, H, is bounded below by a universal constant times $N + K$, independent of the R_j and A. In our case, because everything scales as an inverse length, the lower bound for H is either $-\infty$ or 0. Thus, we have to find the conditions under which H is a positive operator.

Many people worked on various aspects of this problem, including J. Conlon (who gave the first proof [C84]), I. Daubechies, C. Fefferman, I. Herbst, T. Kato, E. Lieb, R. de la Llave, R. Weder, and H-T. Yau. A careful, and still current, review of the history is contained in the introduction to [LY88], to which we refer the reader. For present purposes it suffices to note the current state of affairs concerning the best available constants needed for stability, as derived in [LY88]. We can list these in a sequence of remarks as follows:

1. Stability for any given values, α_* and Z_*, implies stability for all $0 \le \alpha < \alpha_*$ and $Z < Z_*$. In fact, we can allow the nuclei to have different charges Z_i, $1 \le i \le K$, provided $Z_i \le Z_*$ for all i. This follows from some simple concavity considerations and has nothing to do with the nature of the proof leading to α_* and Z_*.

2. Theorem 2 of [LY88] has the strongest results, but it is limited to the case of zero magnetic field, $A = 0$. The result is that stability occurs if

$$q\alpha \le 1/47 \quad \textit{and} \quad Z\alpha \le 2/\pi \ . \tag{1.6}$$

It is not clear to us how to incorporate a magnetic field in the proof of Theorem 2, and we leave this as an open problem.

3. Theorem 1 of [LY88] has weaker results, but a simpler proof. That proof generalizes easily to the $A \ne 0$ case, as pointed out in [LLS95]. The result is complicated to state in full generality, but a representative example is that stability holds if

$$q\alpha \le 0.032 \quad \textit{and} \quad Z\alpha \le 1/\pi \ . \tag{1.7}$$

It is possible to let $Z\alpha \to 2/\pi$ at the expense of $q\alpha \to 0$.

4. Instability definitely occurs if $Z\alpha > 2/\pi$, or if $Z_i\alpha > 2/\pi$ for any i. It also occurs if

$$\alpha > 128/(15\pi) \approx 2.72 \tag{1.8}$$

for *any* positive value of Z and any value of q. In other words, if $\alpha > 128/(15\pi)$ and if $Z > 0$ then one can produce collapse with only one electron, $N = 1$, by utilizing sufficiently many nuclei, i.e., by choosing K sufficiently large.

5. Instability also definitely occurs if ([LY88], Theorem 4)

$$\alpha > 36 q^{-1/3} Z^{-2/3} \ , \tag{1.9}$$

which implies that bosonic matter (which can always be thought of as fermionic matter with $q = N$) is *always* unstable. (Note: there is a typographical error in Theorem 4 of [LY88].)

2. Main Results

The proof of the stability of *non*relativistic matter in [LT75] uses a series of inequalities to relate the ground state energy of the Hamiltonian to the Thomas-Fermi energy of the electron density, $\rho(x)$. The chief point is the kinetic energy inequality for an N-electron state Ψ, namely

$$\langle \Psi | \sum_{i=1}^{N} |p_i|^2 \, | \Psi \rangle > \text{const.} \int \rho^{5/3} \ .$$

The same approach will not work in the relativistic case because the corresponding inequality [D83] is, for dimensional reasons,

$$\langle\Psi| \sum_{i=1}^{N} |p_i|\, |\Psi\rangle > \text{const.} \int \rho^{4/3} \, .$$

While $\int \rho^{5/3}$ can control the Coulomb attraction $-Z\alpha \int \rho(x)/|x|$, unfortunately $\int \rho^{4/3}$ cannot do so. For this reason no attempt seems to have been made to imitate the proof in [LT75] of stability in the relativistic case.

However, the Coulomb singularity *can* be controlled by a Weizsäcker type term, namely $(\sqrt{\rho}\,, |p|\, \sqrt{\rho})$. The relativistic kinetic energy can, in turn, be bounded below by a term of this type plus a term of the $\int \rho^{4/3}$ type. This and other essential inequalities will be explained more fully below. With the 'Coulomb tooth' now gone, TF theory with $\int \rho^{4/3}$ can deal adequately with the rest of the Coulomb energy (with the aid of the exchange-correlation energy inequality [LO81], whose remainder term also has the form $\int \rho^{4/3}$).

Before going into details, let us state our main results. First, we define Thomas-Fermi-Weizsäcker (TFW) theory as follows: The class of functions ('densities') to be considered, denoted by $\mathcal{C}$, consists of those nonnegative functions $\rho : \mathbf{R}^3 \to \mathbf{R}^+$ such that $\sqrt{\rho}$ and $\sqrt{|p|\rho}$ have finite $L^2(\mathbf{R}^3)$ norms, i.e.,

$$\mathcal{C} = \left\{ \rho \; : \; \rho(x) \geq 0 \;\; \text{and} \;\; \int_{\mathbf{R}^3} (1 + |p|)\, |\widehat{\sqrt{\rho}}(p)|^2 dp \; < \infty \right\} \, , \tag{2.1}$$

where $\widehat{\sqrt{\rho}}(p) := (2\pi)^{-3/2} \int_{\mathbf{R}^3} \exp[-ip \cdot x] \sqrt{\rho}(x) dx$ denotes the Fourier transform of the function $\sqrt{\rho}(x)$.

Next, we define the functional

$$T(\rho) := \int_{\mathbf{R}^3} |p| |\widehat{\sqrt{\rho}}(p)|^2 dp \equiv (\sqrt{\rho}\,, |p|\, \sqrt{\rho}) \; . \tag{2.2}$$

The TFW functional, with arbitrarily given positive constants β and γ, is then

$$\mathcal{E}(\rho) := \beta T(\rho) + \frac{3}{4}\, \gamma \int_{\mathbf{R}^3} \rho^{4/3}(x) dx - \alpha \int_{\mathbf{R}^3} V(x) \rho(x) dx + \alpha D(\rho\,, \rho) + \alpha U \tag{2.3}$$

with

$$D(\rho\,, \rho) := (1/2) \int_{\mathbf{R}^3} \int_{\mathbf{R}^3} \rho(x) \rho(y) |x - y|^{-1} dx dy \; .$$

The quantity of principal interest is the energy

$$E^{TFW} := \inf \{\mathcal{E}(\rho) \; : \; \rho \in \mathcal{C}\} \; . \tag{2.4}$$

This quantity depends on the parameters α, β and γ and on the nuclear coordinates, R_j. If, however, we try to minimize E over all choices of the nuclear coordinates then the result is either 0 or $-\infty$, as can be easily seen from the fact that all the terms in $\mathcal{E}$ scale, under dilation, as an inverse length.

THEOREM 1. (Stability of TFW theory). *The TFW energy, E^{TFW}, in (2.4) is nonnegative if*

$$\beta \geq \frac{\pi}{2} Z\alpha, \quad \text{and} \quad \gamma \geq 4.8158\ Z^{2/3}\alpha \tag{2.5}$$

On the other hand, if $\beta < (\pi/2)Z\alpha$ then $E = -\infty$ for every choice of the nuclear coordinates.

For the next theorem we have to define the density corresponding to an N-body wave function. If Ψ is an antisymmetric function of N space-spin coordinates, normalized to unity in the usual way, we define

$$\rho_\Psi(x) := N \sum_{1 \leq \sigma_1, \ldots, \sigma_N \leq q} \int_{\mathbf{R}^{3(N-1)}} |\Psi(x, \sigma_1; x_2, \sigma_2; \ldots; x_N, \sigma_N)|^2 dx_2 \cdots dx_N\ . \tag{2.6}$$

THEOREM 2. (TFW theory bounds quantum mechanics). *Let Ψ be any normalized antisymmetric function, with ρ_Ψ defined in (2.6). Choose*

$$\beta = \frac{\pi}{2} Z\alpha \ \text{ and } \ \gamma = \frac{4}{3}\left[1.63 q^{-1/3}\left(1 - \frac{\pi}{2} Z\alpha\right) - 1.68\ \alpha\right]\ . \tag{2.7}$$

Assume that γ is positive. Then, with this definition of the TFW functional (2.3),

$$\langle\Psi|\ H\ |\Psi\rangle \geq \mathcal{E}(\rho_\Psi)\ . \tag{2.8}$$

A corollary of these two theorems is that *our Hamiltonian, H, in (1.1) is stable if*

$$\left(\frac{\pi}{2}\right) Z + 2.2159\ q^{1/3} Z^{2/3} + 1.0307\ q^{1/3} \leq 1/\alpha\ . \tag{2.9}$$

(Cf. (1.9)) In particular, *with $q = 2$ for electrons, relativistic matter is stable if $\alpha < 0.77$ and if Z is not too large. When $\alpha = 1/137$ the allowed Z is 59,* which compares favorably with the best possible value $87 \approx 137(2/\pi)$.

We leave it as a challenge to improve our method so as to achieve the value $137(2/\pi)$ (with a magnetic field present). As noted above, this value has been achieved in [LY88], but without a magnetic field. The most noteworthy point is the large value of the critical fine structure constant: $\alpha_{critical} \geq 0.77$ when $q = 2$.

The bound in (2.9) is, in some respects, similar to Theorem 1 in [LY88], but it is far simpler, clearer and gives the correct q-dependence of α (note that (1.9) gives a similar bound in the other direction). The chief methodological difference is that Theorem 6 is used in [LY88], which bounds the Coulomb potential below by a one-body potential. Here, we use the exchange-correlation inequality (3.9) instead. We repeat that the results above also hold with a magnetic field.

It is to be emphasized that our stability result is really contained in Theorem 2. Theorem 1 only gives a condition for which $\mathcal{E}(\rho) \geq 0$. A better estimate on the TFW functional will, via Theorem 2, yield a better stability bound.

3. Some Essential Inequalities

There are five known inequalities about Coulomb systems that will be needed in our proof of our main theorems. We begin by recalling them.

KINETIC ENERGY LOCALIZATION, [LY88] pp. 186 and 188.

Denote by Γ_j the Voronoi cell in $\mathbf{R}^3$ that contains R_j, i.e., the set

$$\Gamma_j := \{x \in \mathbf{R}^3 \ : \ |x - R_j| \leq |x - R_k| \text{ for all } k\} \ , \tag{3.1}$$

and let D_j be half the distance of the j-th nucleus to its nearest neighbor. These Γ_j are disjoint, except for their boundaries and, being the intersection of half-spaces they are convex sets. The ball centered at R_j with radius D_j is denoted by B_j. Obviously, $B_j \subset \Gamma_j$.

For any function $f \in L^2(\mathbf{R}^3)$ there is the inequality

$$(f, |p|\ f) \geq \sum_{j=1}^{K} \int_{B_j} |f(x)|^2 \left\{ \frac{2}{\pi} |x - R_j|^{-1} - \frac{1}{D_j} Y\left(\frac{|x - R_j|}{D_j}\right) \right\} dx \ . \tag{3.2}$$

The function Y is given, for $0 < r < 1$, by

$$Y(r) = \frac{2}{\pi(1+r)} + \frac{1+3r^2}{\pi r(1+r^2)} \ln(1+r) - \frac{1-r^2}{\pi r(1+r^2)} \ln(1-r) - \frac{4r}{\pi(1+r^2)} \ln r \ . \tag{3.3}$$

Numerically it is found that [LY88] (2.27)

$$4\pi \int_0^1 Y(r)^4 r^2 dr < 7.6245 \ . \tag{3.4}$$

RELATIVISTIC KINETIC ENERGY BOUND FOR FERMIONS, [D83].

Let Ψ and ρ_Ψ be as in (2.6). Then

$$\langle\Psi| \sum_{i=1}^{N} |p_i|\ |\Psi\rangle \geq 1.63 q^{-1/3} \int_{\mathbf{R}^3} \rho_\Psi^{4/3}(x) dx \ . \tag{3.5}$$

A generalization of this, of importance if we wish to include a mass, is

$$\langle\Psi|\sum_{i=1}^{N}[\sqrt{p_i^2+m^2}-m]\,|\Psi\rangle \geq \frac{3}{8}m^4 C\int_{\mathbf{R}^3} g\left((\rho_\Psi(x)/C)^{1/3}m^{-1}\right)dx\,, \tag{3.6}$$

with $C = 0.163q$ (*sic*) and with

$$g(t) := t(1+t^2)^{1/2}(1+2t^2) - \frac{8}{3}t^3 - \ln\left[t+(1+t^2)^{1/2}\right]\,. \tag{3.7}$$

GENERAL KINETIC ENERGY BOUND, [C84], p.454, (and [HO77] for the nonrelativistic case). The following bound follows from a judicious application of Schwarz's inequality.

$$\langle\Psi|\sum_{i=1}^{N}|p_i|\,|\Psi\rangle \geq \left(\sqrt{\rho_\Psi}\,,|p|\,\sqrt{\rho_\Psi}\right)\,. \tag{3.8}$$

This bound holds irrespective of the symmetry type of the wave function.

EXCHANGE AND CORRELATION INEQUALITY, [LO81]. If Ψ is a normalized N-particle wave function there is a lower bound on the interparticle Coulomb repulsion in terms of its density:

$$\langle\Psi|\sum_{1\leq i<j\leq N}|x_i-x_j|^{-1}|\Psi\rangle \geq D(\rho_\Psi,\rho_\Psi) - 1.68\int_{\mathbf{R}^3}\rho_\Psi^{4/3}(x)dx\,. \tag{3.9}$$

(Once again, the antisymmetry of Ψ plays no role in this inequality.)

ELECTROSTATIC INEQUALITY, [LY88], p.196. First, we define a function, Φ on $\mathbf{R}^3$ with the aid of the Voronoi cells mentioned above. In the cell Γ_j, Φ equals the electrostatic potential generated by all the nuclei *except* for the nucleus situated in Γ_j itself, i.e., for x in Γ_j

$$\Phi(x) := Z\sum_{\substack{i=1\\ i\neq j}}^{K}|x-R_i|^{-1}\,. \tag{3.10}$$

If ν is any bounded Borel measure on $\mathbf{R}^3$ (not necessarily positive) then

$$\frac{1}{2}\int_{\mathbf{R}^3}\int_{\mathbf{R}^3}|x-y|^{-1}d\nu(x)d\nu(y) - \int_{\mathbf{R}^3}\Phi(x)d\nu(x) + U \geq \frac{1}{8}Z^2\sum_{j=1}^{K}D_j^{-1}\,. \tag{3.11}$$

4. Proofs of Theorems 1 and 2

To prove Theorem 1 we take $\beta = \pi Z\alpha/2$ (if $\beta > \pi Z\alpha/2$ we simply throw away the excess positive quantity). Using (3.2) with f replaced by $\sqrt{\rho}$, we have that

$$\mathcal{E}(\rho) \geq \mathcal{E}_1(\rho) + \alpha\mathcal{E}_2(\rho) , \tag{4.1}$$

where, by adding and subtracting a term $\int \Phi\rho$, with $\Phi(x)$ as in (3.10),

$$\mathcal{E}_1(\rho) := \frac{3}{4}\gamma \int_{\mathbf{R}^3} \rho^{4/3}(x)dx - \alpha \int_{\mathbf{R}^3} W(x)\rho(x)dx + \alpha \int_{\mathbf{R}^3} \Phi(x)\rho(x)dx \tag{4.2}$$

and

$$\mathcal{E}_2(\rho) := D(\rho\,,\rho) - \int_{\mathbf{R}^3} \Phi(x)\rho(x)dx + U\ . \tag{4.3}$$

The function $W(x)$ is defined as follows: In the Voronoi cell Γ_j it is given by

$$W(x) := \Phi(x) + \begin{cases} Z|x-R_j|^{-1}, & \text{if } |x-R_j| > D_j \\ (\pi Z/2)D_j^{-1}Y\left(|x-R_j|/D_j\right), & \text{if } |x-R_j| \leq D_j\ . \end{cases} \tag{4.4}$$

Note that while the terms $\pm \int \Phi\rho$ that appear in (4.2), (4.3) are merely 'strategic', the presence of the term $\Phi(x)$ in (4.4) is properly part of the potential energy of the electron and is not arbitrary. Actually, this strategic decomposition of $\mathcal{E}(\rho)$ is the one used in the easy part of Fenchel's duality theorem (see [R70], p. 327). This duality principle was used in connection with Thomas-Fermi theory by Firsov [F57] (see also [L81]); the full blown duality theory is not needed for our purposes, so we omit it.

We can now seek lower bounds for $\mathcal{E}_1(\rho)$ and $\mathcal{E}_2(\rho)$ separately. Using Hölder's inequality, for example, one easily concludes that the absolute minimum of $\mathcal{E}_1(\rho)$ is

$$\mathcal{E}_1(\rho) \geq -\frac{\alpha^4}{4\gamma^3}\int_{\mathbf{R}^3} [W(x)-\Phi(x)]_+^4\,dx$$

$$= -\frac{(\alpha Z)^4}{4\gamma^3}\sum_{j=1}^{K}\left(\frac{\pi}{2}\right)^4 \int_{B_j} D_j^{-4}Y\left(|x-R_j|/D_j\right)^4 dx + \int_{\Gamma_j\setminus B_j} |x-R_j|^{-4}dx \tag{4.5}$$

$$\geq -\frac{(\alpha Z)^4}{4\gamma^3}\left\{\left(\frac{\pi}{2}\right)^4 (4\pi)\int_0^1 Y(r)^4 r^2 dr + 3\pi\right\}\sum_{j=1}^{K} D_j^{-1} \tag{4.6}$$

$$> -\frac{(\alpha Z)^4}{4\gamma^3}\left\{7.6245\left(\frac{\pi}{2}\right)^4 + 3\pi\right\}\sum_{j=1}^{K} D_j^{-1}\ . \tag{4.7}$$

The last formula uses (3.4). The second integral in (4.5) is evaluated in (4.6) as $3\pi/D_j$, and the explanation is the following: If we integrate $|x-R_j|^{-4}$ over the exterior of B_j we would obtain $4\pi/D_j$ as the result. However, we know that the Voronoi cell Γ_j lies on one side of the mid-plane defined by the nearest neighbor nucleus. This means that the integral over $\Gamma_j \setminus B_j$ is bounded above by the quantity

$$\frac{4\pi}{D_j} - \frac{1}{D_j}\int_1^\infty dz \int_0^\infty \frac{2\pi r}{(r^2+z^2)^2} dr = \frac{3\pi}{D_j} .$$

The $\mathcal{E}_2$ term can be bounded using (3.11) with $d\nu(x) = \rho(x)dx$. Thus,

$$\mathcal{E}_2(\rho) \geq \frac{Z^2}{8} \sum_{j=1}^{K} \frac{1}{D_j} . \tag{4.8}$$

Combining (4.1), (4.7) and (4.8) we have proved Theorem 1. ■

Theorem 2 is proved by splitting the relativistic kinetic energy $|p|$ into $\beta|p|$ and $(1-\beta)|p|$, with the choice $\beta = \pi Z\alpha/2$. The inequalities (3.5), (3.8) and (3.9) immediately give us Theorem 2. ■

5. Inclusion of Mass and Magnetic Fields

INCLUSION OF MASS. We replace $|p|$ by $\sqrt{p^2+m^2} - m$ and, in the corresponding TFW theory, we replace the right side of (3.5) by the right side of (3.6). It is not easy to carry out the rest of the program in closed form with this more complicated function, however. Moreover, it unfortunately gives a slightly worse constant than before, even when we set $m = 0$; instead of $1.63q^{-1/3}$ in (3.5) we now have $C^{-1/3} \approx 1.37q^{-1/3}$. The new energy will not be positive in the stability regime, as we had before. Instead, it will be a negative constant times N. This new value for the energy is in accord with stability of matter and represents the binding energy of the electron-nuclear system.

Another way to deal with the mass is to observe, simply, that $\sqrt{p^2+m^2} - m > |p| - m$, the effect of which is to add a term $-Nm$ to the energy estimate. This term satisfies the criterion for stability, but it has the defect that is huge in real-world terms, for it equals the rest energy of the electron.

INCLUSION OF MAGNETIC FIELD. Theorem 2, with a magnetic field included, is a consequence of the following two inequalities (proved below) which replace (3.6) and (3.8):

$$\langle\Psi| \sum_{i=1}^{N} [\sqrt{(p_i + A(x_i))^2 + m^2} - m] \, |\Psi\rangle \geq \frac{3}{8} m^4 C \int_{\mathbf{R}^3} g\left((\rho_\Psi(x)/C)^{1/3} m^{-1}\right) dx \, , \tag{5.1}$$

and

$$\langle\Psi| \sum_{i=1}^{N} |p_i + A(x_i)| \, |\Psi\rangle \geq \left(\sqrt{\rho_\Psi} \, , |p| \, \sqrt{\rho_\Psi}\right) . \tag{5.2}$$

As in (3.8), inequality (5.2) holds irrespective of the symmetry type of Ψ.

To define $\sqrt{|p+A|^2+m^2}$, note that if $A \in L^2_{loc}(\mathbf{R}^3;\mathbf{R}^3)$, then $f \mapsto \|(p+A)f\|_2^2$ is a closed quadratic form with $C_0^\infty(\mathbf{R}^3)$ being a form core [K78], [S79-1],[LS81]. Thus it defines a selfadjoint operator and it is then possible to define $\sqrt{|p+A|^2+m^2}$ via the spectral calculus.

The *diamagnetic inequality for the heat kernel* [S79-2] is the pointwise inequality

$$| \exp\left[-t(p+A)^2\right] f(x) \,| \le \exp\left[-tp^2\right] |f|(x) \,. \tag{5.3}$$

Using the formula

$$e^{-|a|} = \frac{1}{\sqrt{\pi}} \int_0^\infty e^{-t-a^2/4t} \frac{dt}{\sqrt{t}} \,, \tag{5.4}$$

which holds for any real number a (and hence for any selfadjoint operator), we obtain the diamagnetic inequality for the 'relativistic heat kernel'

$$\left| \exp\left[-t\sqrt{(p+A)^2+m^2}\right] f(x) \right| \le \exp\left[-t\sqrt{p^2+m^2}\right] |f|(x) \,. \tag{5.5}$$

By using (5.5), and following the proof of (3.6) in [D83] step by step, we obtain (5.1). Likewise, (5.5) and the formula

$$(f,\, \sqrt{(p+A)^2+m^2}\; f) = \lim_{t\to 0} \frac{1}{t} \left\{ (f,f) - (f,\, \exp\left[-t\sqrt{(p+A)^2+m^2}\right] f) \right\} , \tag{5.6}$$

yield

$$(f,\, |p+A|\; f) \ge (|f|,\, |p|\; |f|) \,. \tag{5.7}$$

To prove (5.2) we apply (5.7) to the function $|\Psi|$ and then use (3.8).

References

[C84] Conlon, J.G., *The ground state energy of a classical gas*, Commun. Math. Phys. **94**, 439-458 (1984).

[D83] Daubechies, I., *An uncertainty principle for fermions with generalized kinetic energy*, Commun. Math. Phys. **90**, 511-520 (1983).

[F57] Firsov, O.B., *Calculation of the interaction potential of atoms for small nuclear separations*, Sov. Phys. JETP **5**, 1192-1196 (1957).

[HO77] Hoffmann-Ostenhof, M. and Hoffman-Ostenhof, T., *Schrödinger inequalities and asymptotic behavior of the electronic density of atoms and molecules*, Phys. Rev. A **16**, 1782-1785 (1977).

[K78] Kato, T., *Remarks on Schrödinger operators with vector potentials*, Int. Eq. Operator Theory **1**, 103-113 (1978).

[LS81] Leinfelder, H., Simader, C., *Schrödinger operators with singular magnetic vector potentials*, Math. Z. **176**, 1-19 (1981).

[L81] Lieb, E.H. *Thomas-Fermi and related theories of atoms and molecules*, Rev. Mod. Phys. **53** 603-641 (1981). Errata, ibid **54**, 311 (1982).

[LLS95] Lieb, E.H., Loss, M. and Solovej, J.P., *Stability of Matter in Magnetic Fields*, Phys. Rev. Lett. **75**, 985-989 (1995).

[LO81] Lieb, E.H. and Oxford, S., *Improved lower bound on the indirect Coulomb energy*, Int. J. Quant. Chem. **19**, 427-439 (1981).

[LY88] Lieb, E.H. and Yau, H-T., *The stability and instability of relativistic matter*, Commun. Math. Phys. **118**, 177-213 (1988).

[LT75] Lieb, E.H., and Thirring, W.E., *Bound for the kinetic energy of fermions which proves the stability of matter*, Phys. Rev. Lett. **35**, 687-689 (1975). Erratum, *ibid*, 1116.

[R70] Rockafellar, R.T., *Convex Analysis*, Princeton University Press (1970).

[S79-1] Simon, B., *Maximal and minimal Schrödinger forms*, J. Opt. Theory **1**, 37-47 (1979).

[S79-2] Simon, B., *Kato's inequality and the comparison of semigroups*, J. Funct. Anal. **32**, 97-101 (1979).

Journal of Statistical Physics, Vol. 89, Nos. 1/2, 1997

Stability and Instability of Relativistic Electrons in Classical Electromagnetic Fields

Elliott H. Lieb,[1] Heinz Siedentop,[2] and Jan Philip Solovej[3]

Received October 21, 1996

The stability of matter composed of electrons and static nuclei is investigated for a relativistic dynamics for the electrons given by a suitably projected Dirac operator and with Coulomb interactions. In addition there is an arbitrary classical magnetic field of finite energy. Despite the previously known facts that ordinary nonrelativistic matter with magnetic fields, or relativistic matter without magnetic fields, is already unstable when α, the fine structure constant, is too large, it is noteworthy that the combination of the two is still stable *provided* the projection onto the positive energy states of the Dirac operator, which *defines* the electron, is chosen properly. A good choice is to include the magnetic field in the definition. A bad choice, which always leads to instability, is the usual one in which the positive energy states are defined by the free Dirac operator. Both assertions are proved here.

KEY WORDS: Stability of matter; Schrödinger operators; magnetic fields; relativistic; Dirac operator; instability of matter.

1. INTRODUCTION

The stability of matter concerns the many-electron and many-nucleus quantum mechanical problem and the question whether the ground state energy is finite (stability of the first kind). If so, is it bounded below by a

[1] Department of Mathematics and Physics, Princeton University, Princeton, New Jersey 08544-0708.

[2] Mathematik, Universität Regensburg, D-93040 Regensburg, Germany.

[3] Institut for matematiske fag, Aarhus Universitet, Ny Munkegade, DK-8000 Århus C, Denmark. Current address: Department of Mathematics, University of Copenhagen, Universitetsparken 5, DK-2100 Copenhagen, Denmark.

This paper is dedicated to Bernard Jancovici on the occasion of his 65th birthday.

constant (which is independent of the position of the nuclei) times the number of particles (stability of the second kind)? The linear lower bound is important for thermodynamics, which will not exist in the usual way without it.

The first positive resolution of this problem for the nonrelativistic Schrödinger equation was given by Dyson and Lenard[7,8] and approached differently by Federbush.[10] The constant, i.e., the energy per particle, was considerably improved by Lieb and Thirring in refs. 21 and 22. Following that, the stability of a relativistic version of the Schrödinger equation (in which $\mathbf{p}^2$ is replaced by $\sqrt{\mathbf{p}^2+m^2}$) was proved by Conlon[5] and later improved by Lieb and Yau[23] who showed that matter is stable in this model if and only if the fine structure constant α is small enough and if $Z\alpha \leqslant 2/\pi$. (See ref. 23 for a historical account up to 1995.) A recent result of Lieb, Loss, and Siedentop that we shall use is in ref. 19 and is discussed in Section 3.

In these works the nuclei are fixed in space because they are very massive and because we know that the nuclear motion is largely irrelevant for understanding matter. In other words, if nuclear motion were the only thing that prevented the instability of matter then the world would look very different from what it does. We continue this practice here.

There is, however, a more important quantity that requires some attention, namely magnetic fields. It was noted that the action of such fields on the translational degrees of freedom of the electrons $\mathbf{p} \to \mathbf{p} + e\mathbf{A}$, can lower the energy only by an inconsequential amount. This is a kind of diamagnetic inequality. On the other hand, spin-magnetic field interaction (in which $(\mathbf{p}+e\mathbf{A})^2$ is replaced by the Pauli operator $[\boldsymbol{\sigma}\cdot(\mathbf{p}+e\mathbf{A})]^2 = (\mathbf{p}+e\mathbf{A})^2 + e\boldsymbol{\sigma}\cdot\mathbf{B}$ can cause instability. The energy is then unbounded below if arbitrarily large fields are allowed, but this is so only because the energy of the magnetic field has not been taken into account. Does the field energy, $(8\pi)^{-1}\int B^2$, insure stability? This question was raised for the non-relativistic case in ref. 13 and finally settled in a satisfactory manner in ref. 20 (see also Bugliaro *et al.*[4] and Fefferman[11]). The upshot of this investigation is that stability (of both first and second kinds) requires a bound on both α and on $Z\alpha^2$.

Other related results are the stability of non-relativistic matter with a second quantized, ultra-violet cut-off photon field (Fröhlich *et al.*[12]).

Both the passage to relativistic kinematics (which, in quantum mechanics, means that both the kinetic energy and the Coulomb potential scale with length in the same way, namely like an inverse length) and the introduction of the nonrelativistic Pauli operator require a bound on α and on Z for stability. The combination of the two might be expected to lead to disaster. We find, however, that it does not necessarily do so!

Our main result is that matter is indeed stable with a suitably defined relativistic kinematics. This is shown in Section 3.

The proper way to introduce relativistic kinematics for spin-1/2 particles is via the Dirac operator, but this is unbounded below. A resolution of this problem, due to Dirac, is to permit the electrons to live only in the positive energy subspace of the Dirac operator. This idea was further pursued by Brown and Ravenhall[3] (see also Bethe and Salpeter in their Handbuch article[1]) to give a quantitative description of real atoms.

There are, however, other Dirac operators (which include electromagnetic potentials) whose positive subspace can be used to define the space in which the electrons can live. (To avoid confusion, let us note that the Hamiltonian is formally always the same and includes whatever fields happen to be present. The only point to be resolved is what part of the one-particle Hilbert space is allowed for electrons.) The review articles of Sucher[24, 25, 26] can be consulted here. These choices have also been used in quantum chemistry and other practical calculations, see, e.g., refs. 14 and 15.

All of these choices have in common that there is no creation of electron-positron pairs explaining the name "no-pair Hamiltonian" for the resulting energy operator. (Note that we could also treat positrons or a combination of electrons and positrons interacting by Coulomb forces in a similar way.)

There are three obvious choices to consider. One is the free Dirac operator. This always leads to instability of the first kind when a magnetic field is added unless the particle number is held to some small value (see Section 4). Note also that this choice leads to a non-gauge invariant model: multiplication of a state with the factor $\exp(i\phi(\mathbf{r}))$ for a non-constant gauge is not allowed, since it leads out of the positive spectral subspace.

Remarkably, the Dirac operator that includes the magnetic field always gives stability, if Z and α are not too large, as in the two cases (relativistic without magnetic field and nonrelativistic with magnetic field) mentioned above (see Section 3). This model is gauge invariant.

The third choice which, indeed, is sometimes used, is to include both the one-body attractive electric potential of the nuclei and the magnetic field in the definition of the Dirac operator that defines the positive subspace. If this is done then the question of stability is immediately solved because the remaining terms in the Hamiltonian are positive, and hence the total energy is ipso facto positive. This choice, which is important but trivial in the context of this present inquiry, will not be mentioned further.

Oddly, the instability proof given in Subsection 4.2 is much more complicated than the stability proof (Section 3). This reverses the usual situation.

A preliminary version of this paper appeared in ref. 27; the present version is to be regarded as the original one (as stated in ref. 27) and contains several significant corrections to the preliminary text in ref. 27. In particular, the proof of Theorem 2 and the first half of the proof of Theorem 1 have been corrected and simplified. A summary of this work appears in ref. 28.

2. BASIC DEFINITIONS

The energy of N relativistic electrons in the field of K nuclei with atomic numbers $Z_1,..., Z_K \in \mathbb{R}_+$ located at $\mathbf{R}_1,..., \mathbf{R}_K \in \mathbb{R}^3$ which are pairwise different in a magnetic field $\mathbf{B} = \nabla \times \mathbf{A}$ in the state Ψ is—following the ideas of Brown and Ravenhall[3]—

$$\mathscr{E}[\mathbf{\Psi}, \mathbf{\Psi}] := \left(\mathbf{\Psi}, \left(\sum_{\nu=1}^{N} D_\nu(\mathbf{A}) + \alpha V_c\right) \mathbf{\Psi}\right) + \frac{1}{8\pi} \int_{\mathbb{R}^3} B(\mathbf{r})^2 \, d\mathbf{r}. \tag{1}$$

Here $D_\nu(\mathbf{A}) := \boldsymbol{\alpha} \cdot (-i \nabla_\nu + e \mathbf{A}(\mathbf{r}_\nu)) + m\beta$ is the Dirac operator with vector potential $\mathbf{A}$. Furthermore,

$$V_c := -\sum_{\nu=1}^{N} \sum_{\kappa=1}^{K} \frac{Z_\kappa}{|\mathbf{r}_\nu - \mathbf{R}_\kappa|} + \sum_{\mu, \nu=1, \mu<\nu}^{N} \frac{1}{|\mathbf{r}_\mu - \mathbf{r}_\nu|} + \sum_{\kappa, \lambda=1, \kappa<\lambda}^{K} \frac{Z_\kappa Z_\lambda}{|\mathbf{R}_\kappa - \mathbf{R}_\lambda|} \tag{2}$$

is the Coulomb interaction between the particles, and $B(\mathbf{r}) := |\nabla \times \mathbf{A}(\mathbf{r})|$ is the modulus of the magnetic field. Planck's constant divided by 2π and the velocity of light, are taken to be one in suitable units. The fine structure constant α equals e^2, where $-e$ is the electron charge. Experimentally, α is about 1/137.037. The mass of the electron is denoted by m. The 4×4 matrices α and β are the four Dirac matrices in standard representation, namely

$$\boldsymbol{\alpha} = \begin{pmatrix} 0 & \boldsymbol{\sigma} \\ \boldsymbol{\sigma} & 0 \end{pmatrix}, \quad \sigma_1 = \begin{pmatrix} 0 & 1 \\ 1 & 0 \end{pmatrix}, \quad \sigma_2 = \begin{pmatrix} 0 & -i \\ i & 0 \end{pmatrix}, \quad \sigma_3 = \begin{pmatrix} 1 & 0 \\ 0 & 1 \end{pmatrix},$$

and

$$\beta = \begin{pmatrix} 1 & 0 & 0 & 0 \\ 0 & 1 & 0 & 0 \\ 0 & 0 & -1 & 0 \\ 0 & 0 & 0 & -1 \end{pmatrix}.$$

Finally, the state $\mathbf{\Psi}$ should have finite kinetic energy, i.e., it should be in the Sobolev space $H^{1/2}[(\mathbb{R}^3 \times \{1, 2, 3, 4\})^N]$, and should also be in the electronic Hilbert space of antisymmetric spinors

$$\mathscr{H}_{N, \mathscr{A}} := \bigwedge_{\nu=1}^{N} \mathscr{H}_+, \tag{3}$$

where $\mathscr{H}_+$ is the positive spectral subspace of the Dirac operator $D(\mathscr{A})$ and where $\mathscr{A}$ is some vector potential to be chosen later. The vector potential $\mathscr{A}$ serves to define the positive subspace. The restriction of $\mathscr{E}$ to $\mathscr{H}_{N, \mathscr{A}}$ will be denoted $\mathscr{E}_{\mathscr{A}}$. Two choices will be considered here. One is $\mathscr{A} = \mathbf{0}$, in which case we are talking about the free Dirac operator. This choice, or model, goes back to Brown and Ravenhall.[3] As we shall see in Section 4, the resulting energy functional—apart from being not gauge invariant—is not bounded from below. A natural modification of the model, namely to take $\mathscr{A} := \mathbf{A}$ is not only gauge invariant, but will also turn out to be stable of the second kind (see Section 3).

The quantity of interest is the lowest possible energy

$$E_{N, K} := \inf \mathscr{E}_{\mathscr{A}}$$

where the infimum is taken over all allowed, normalized states $\mathbf{\Psi}$, all allowed vector potentials $\mathbf{A}$, and over all pairwise different nuclear positions $\mathbf{R}_1, \ldots, \mathbf{R}_K$.

In the case of a single nucleus without a magnetic field, the energy form $\mathscr{E}_0$ was shown in ref. 9 to be bounded from below, if and only if $\alpha Z \leqslant \alpha Z_C := 2/(\pi/2 + 2/\pi) > 2/\pi$, which corresponds to $Z_C \approx 124$. We will not be able to reach this value in the general case of many nuclei and when the electron state space is not determined by the free Dirac operator. The reason is that special techniques were used in ref. 9 to handle the one-nucleus case; these techniques took advantage of the weakening of the Coulomb singularities caused by the fact that states in $\mathscr{H}_+$ cannot be localized in space arbitrarily sharply. Unfortunately, we do not know how to implement this observation with magnetic fields and many nuclei.

3. STABILITY WITH THE MODIFIED PROJECTOR

Our proof of the stability of matter when the vector potential $\mathbf{A}$ is included in the definition of the positive energy electron states will depend essentially on three inequalities:

BKS Inequality. For any self-adjoint operator X, the negative (positive) part, $X_{\mp}$ is defined to be $(|X| \mp X)/2$. Given two non-negative self-adjoint operators C and D such that $(C^2 - D^2)_-^{1/2}$ is trace class, we have the trace inequality

$$\operatorname{tr}(C-D)_- \leqslant \operatorname{tr}(C^2 - D^2)_-^{1/2}. \tag{4}$$

This is a special case of a more general inequality of Birman, Koplienko, and Solomyak;[2] in particular, the number 2 in (4) can be replaced by any $p > 1$. A proof for the special case of the inequality needed here is given in Appendix A.

Stability of Relativistic Matter. On $\bigwedge_{\nu=1}^{N} (H^{1/2}(\mathbb{R}^3) \otimes \mathbb{C}^q)$, the fermionic Hilbert space, we have

$$\sum_{\nu=1}^{N} |-i\nabla_\nu - \mathscr{A}| + \tilde{\alpha} V_c \geqslant 0, \tag{5}$$

(where $|\cdots|$ means $\sqrt{(\cdots)^2}$) for all vector fields $\mathscr{A}: \mathbb{R}^3 \to \mathbb{R}^3$ with, e.g., square integrable gradients, if

$$1/\tilde{\alpha} \geqslant 1/\tilde{\alpha}_c := (\pi/2)\, Z + 2.2159 q^{1/3} Z^{2/3} + 1.0307 q^{1/3} \tag{6}$$

and $Z_1, ..., Z_K \leqslant Z$.

We wish to use this inequality for 4-component spinors, i.e., $q = 4$. However, we are interested in the subspace $\mathscr{H}_{N,\mathscr{A}}$ in which the particles are restricted to the positive energy subspace of the Dirac operator, $D(\mathscr{A})$. Although $q = 4$, the "effective" q is really 2, and the analysis in Appendix B is our justification for this. The only thing that really counts in deriving (5) is the bound on the reduced one-body density matrix γ mentioned in Appendix B.

The stability of the relativistic Hamiltonian (5) was first shown by Conlon[5] for $\mathscr{A} = \mathbf{0}$. The best currently available constants with $\mathscr{A} = \mathbf{0}$ are in ref. 23 while (6), which is taken from ref. 19, is the best known result for general $\mathscr{A}$.

Semi-Classical Bound. Given a positive constant μ, a real vector field $\mathscr{A}$ with, e.g., square integrable gradients, and a real-valued function $\varphi \in L^2(\mathbb{R}^3)$ the inequality

$$\operatorname{tr}[(-i\mu\nabla - \mathscr{A})^2 - \varphi]_-^{1/2} \leqslant \frac{L_{1/2,3}}{\mu^3} \int_{\mathbb{R}^3} \varphi_+^2 \tag{7}$$

holds, which is a special case of the Lieb-Thirring inequality (see refs. 22 and 17]). It is known that $L_{1/2,3} \leqslant 0.06003$. The left side of (7) is simply $\sum_j |\lambda_j|^{1/2}$, where the λ_j are the negative eigenvalues of the operator $[\cdots]$.

As an illustration of the usefulness of the trace estimate (4), let us combine it with the Lieb–Thirring inequality (7) (or any other Lieb–Thirring inequality for that matter) to derive some previously known inequalities. The constants obtainable in this way are comparable with the best ones known so far:

Daubechies Inequality. We begin with a "relativistic" inequality that was first proven by Daubechies.[6] By replacing φ by φ^2 in (7), we get using (4)

$$\operatorname{tr}(|-i\nabla-\mathscr{A}|-\varphi)_- \leqslant L_{1/2,3}\int_{\mathbb{R}^3}\varphi_+^4. \tag{8}$$

The constant 0.06003 obtained here should be compared with the number 0.0258 in ref. 6.

Non-Relativistic Magnetic Stability. A non-relativistic analogue of our main problem is to bound the form

$$\tilde{\mathscr{E}}:=\left(\Psi,\left(\sum_{\nu=1}^{N}P_\nu(\mathbf{A})+\alpha V_c\right)\Psi\right)+\frac{1}{8\pi}\int_{\mathbb{R}^3}B(\mathbf{r})^2\,d\mathbf{r} \tag{9}$$

which was treated in ref. 20. Here $P_\nu(\mathbf{A}):=[\boldsymbol{\sigma}\cdot(-i\nabla_\nu+e\mathbf{A}(\mathbf{r}_\nu))]^2$ is the Pauli operator with vector potential $\mathbf{A}$.

First, we note that $x^2\geqslant+\lambda\,|x|-\lambda^2/4$ holds. A constant in the energy form, however, is irrelevant for checking on stability of the second kind. Using (5) it is then enough to show the positivity of

$$-\operatorname{tr}(\lambda P(\mathbf{A})^{1/2}-\kappa\,|-i\nabla+e\mathbf{A}|)_-+\frac{1}{8\pi}\int_{\mathbb{R}^3}B(\mathbf{r})^2\,d\mathbf{r} \tag{10}$$

where we have set $\kappa:=\alpha/\tilde{\alpha}_c$. The trace in this and the next expression are over $L^2(\mathbb{R}^3)\otimes\mathbb{C}^2$. Using the BKS inequality gives the lower bound

$$-\operatorname{tr}[(\lambda^2-\kappa^2)\,|-i\nabla+e\mathbf{A}|^2)-e\lambda^2B(\mathbf{r})]_-^{1/2}+\frac{1}{8\pi}\int_{\mathbb{R}^3}B(\mathbf{r})^2\,d\mathbf{r}.$$

Applying the Lieb–Thirring inequality (7) yields the following sufficient condition for stability (recall that $\alpha=e^2$ and that there are two spin states)

$$2L_{1/2,3}\frac{\lambda^4\alpha}{(\lambda^2-\kappa^2)^{3/2}}\leqslant\frac{1}{8\pi}.$$

Optimizing in λ gives

$$2L_{1/2,3}\frac{16\alpha^2}{3^{3/2}\tilde{\alpha}_c}\leqslant\frac{1}{8\pi},$$

which gives for the physical values $\alpha\approx 1/137.037$ and $q=2$ a range of stability up to $Z\leqslant 1096$, which is to be compared with $Z\leqslant 1050$ in ref. 20.

We turn now to our main result.

Theorem 1. Let $Z_1,..., Z_K\leqslant Z<2/(\pi\alpha)$ and let $\alpha\leqslant\alpha_c$ where α_c is the unique solution of the equation

$$(16\pi L_{1/2,3}\alpha_c)^{2/3}=1-\alpha_c^2/\tilde{\alpha}_c^2,$$

with $\tilde{\alpha}_c:=[(\pi/2)Z+2.2159\cdot 2^{1/3}Z^{2/3}+1.0307\cdot 2^{1/3}]^{-1}$ as in (6). Then $\mathscr{E}_{\mathbf{A}}$ is non-negative.

Numerically, this gives

$$Z\leqslant 56$$

when evaluated with the experimental value $\alpha\approx 1/137.037$ for the fine structure constant. Alternatively, considering hydrogen only, i.e., $Z=1$, we obtain the upper bound

$$\alpha\leqslant 1/8.139$$

for the fine structure constant. It is a challenge to improve this result so that it covers all physical nuclear charges and the physical value of the fine structure constant, as was done for $K=N=1$ and $\mathbf{A}=\mathbf{0}$ in ref. 9.

It is easy to prove (we do not do so here) that, as expected, $\mathscr{E}_{\mathbf{A}}/(\mathbf{\Psi},\mathbf{\Psi})\geqslant m-O(\alpha^2Z^2)$ for small α and Z.

Proof. The first step in our proof is to utilize (5) to replace V_c by the one-body operator $(-1/\tilde{\alpha}_c)\sum_{\nu=1}^N|-i\nabla_\nu+e\mathbf{A}|$, where $\tilde{\alpha}_c$ is given by (6) with $q=2$, as we explained just after (6). (The idea of using the relativistic stability result (5) to bound the Coulomb potential by a one-body operator first appears in ref. 20.) Our energy $\mathscr{E}_{\mathbf{A}}$ is now bounded below by

$$\mathscr{E}'(\mathbf{\Psi}):=\left(\mathbf{\Psi},\sum_{\nu=1}^{N}(D_\nu(\mathbf{A})-\kappa\,|-i\nabla_\nu+e\mathbf{A}(\mathbf{r}_\nu)|)\,\mathbf{\Psi}\right)+\frac{1}{8\pi}\int_{\mathbb{R}^3}B(\mathbf{r})^2\,d\mathbf{r},\qquad(11)$$

where $\kappa:=\alpha/\tilde{\alpha}_c$.

The first term on the right side of (11) is bounded below by the sum of the negative eigenvalues, $-\operatorname{tr}h_-$, of the one-body operator

$$h:=\Lambda^+(D(\mathbf{A})-\kappa\,|-i\nabla+e\mathbf{A}(\mathbf{r})|)\,\Lambda^+,$$

where Λ^+ is the projector onto the positive spectral subspace of $D(\mathbf{A})$.

Let us define

$$S := |D(A)| - \kappa\, |-i\nabla + e\mathbf{A}(\mathbf{r})|,$$

whence $h = \Lambda_+ S \Lambda_+$, because $\Lambda_+ D(\mathbf{A})\, \Lambda_+ = \Lambda_+ |D(\mathbf{A})|\, \Lambda_+$. We note that for any two self-adjoint operators X and ρ with $X \geqslant 0$ and $0 \leqslant \rho \leqslant 1$, $\operatorname{tr} X \geqslant \operatorname{tr} \rho X$. With ρ taken to be the projector onto the negative spectral subspace of h we then have that

$$\begin{aligned} \operatorname{tr} h_- &= -\operatorname{tr} \rho h = -\operatorname{tr} \rho \Lambda_+ S \Lambda_+ \\ &= \operatorname{tr} \rho \Lambda_+ S_- \Lambda_+ - \operatorname{tr} \rho \Lambda_+ S_+ \Lambda_+ \\ &\leqslant \operatorname{tr} \Lambda_+ S_- \Lambda_+ . \end{aligned} \tag{12}$$

We introduce the 4×4 unitary

$$U = \begin{pmatrix} 0 & 1 \\ -1 & 0 \end{pmatrix}$$

and note that $U^{-1} D(\mathbf{A})\, U = -D(\mathbf{A})$. Therefore, $U^{-1} \Lambda_+ U = \Lambda_-$.

It follows from the spectral theorem that for any self-adjoint X, unitary U, and function F

$$F(U^{-1} X U) = U^{-1} F(X)\, U.$$

With $F(t) = |t|$, we then have that $U^{-1}\, |D(\mathbf{A})|\, U = |D(\mathbf{A})|$, and hence $U^{-1} S U = S$.

Therefore, since $U^{-1} \Lambda_+ U = \Lambda_-$, and with $F(t) = \frac{1}{2}(|t| - t) = t_-$, we have that $U^{-1} S_- U = S_-$ and

$$\operatorname{tr} \Lambda_+ S_- \Lambda_+ = \operatorname{tr} \Lambda_- S_- \Lambda_- .$$

Hence, using (12),

$$\operatorname{tr} h_- \leqslant \tfrac{1}{2} \operatorname{tr}(\Lambda_+ S_- \Lambda_+) + \tfrac{1}{2} \operatorname{tr} \Lambda_- S_- \Lambda_- = \tfrac{1}{2} \operatorname{tr} S_- .$$

(Note: much of the preceding discussion was needed only to get the factor 1/2 here. This factor improves our final constants for stability.)

Next, we use the BKS inequality (4) to bound $\operatorname{tr} S_-$ as follows:

$$\operatorname{tr} h_- \leqslant \tfrac{1}{2} \operatorname{tr} S_- \leqslant \tfrac{1}{2} \operatorname{tr}[D(\mathbf{A})^2 - \kappa^2\, |-i\nabla + e\mathbf{A}(\mathbf{r})|^2]_-^{1/2}. \tag{13}$$

However, $D(\mathbf{A})^2 = \begin{pmatrix} Y & 0 \\ 0 & Y \end{pmatrix}$ with $Y = P(\mathbf{A}) + m^2$, and where

$$P(\mathbf{A}) = [\boldsymbol{\sigma} \cdot (-i\nabla + e\mathbf{A})]^2 = |-i\nabla + e\mathbf{A}(\mathbf{r})|^2 + e\boldsymbol{\sigma} \cdot \mathbf{B}(\mathbf{r})$$

is the Pauli operator.

Since $X \mapsto \operatorname{tr} X_-^{1/2}$ is operator monotone decreasing, we see that our lower bound for the energy is monotone increasing in m, and thus it suffices to prove the positivity of $\mathscr{H}_{N,\mathbf{A}}$ in the massless case. The key observation is that our lower bound involves only $\operatorname{tr} S_-$ in the entire one-body space, not the positive energy subspace. The energy would not be obviously monotone in m if we had to restrict functions to the positive subspace, since changing m would also entail changing the space. This problem does not arise in the absence of the positive subspace constraint.

Because of the "diagonal" structure of the operator S, we can drop the factor 1/2 by replacing the trace on $L^2(\mathbb{R}^3) \otimes \mathbb{C}^4$ by the trace on $L^2(\mathbb{R}^3) \otimes \mathbb{C}^2$. This yields

$$\begin{aligned} \mathscr{E}_{\mathbf{A}}[\boldsymbol{\Psi}, \boldsymbol{\Psi}] &\geqslant -\operatorname{tr}[P(\mathbf{A}) - \kappa^2(-i\nabla + e\mathbf{A})^2]_-^{1/2} + \frac{1}{8\pi} \int_{\mathbb{R}^3} B(\mathbf{r})^2 \, d\mathbf{r} \\ &\geqslant -2 \operatorname{tr}[(1-\kappa^2)(-i\nabla + e\mathbf{A})^2 - eB]_-^{1/2} + \frac{1}{8\pi} \int_{\mathbb{R}^3} B(\mathbf{r})^2 \, d\mathbf{r}. \end{aligned} \tag{14}$$

We regard the operator in the second line as acting on functions (of one component only) instead of spinors, which accounts for the factor two (and not one). Finally we apply the Lieb–Thirring inequality (7) to the right hand side yielding (recall that $e^2 = \alpha$)

$$\mathscr{E}_{\mathbf{A}}[\boldsymbol{\Psi}, \boldsymbol{\Psi}] \geqslant [-2L_{1/2,3}\alpha(1 - \alpha^2/\tilde{\alpha}_c^2)^{-3/2} + 1/(8\pi)] \int_{\mathbb{R}^3} B^2(\mathbf{r})^2 \, d\mathbf{r}.$$

Thus we need

$$(16\pi L_{1/2,3}\alpha)^{2/3} \leqslant 1 - \alpha^2/\tilde{\alpha}_c^2. \tag{15}$$

Since the right hand side of this inequality is monotone decreasing in α for positive α, while the left hand side is monotone increasing, there is a unique α_c for which equality holds in (15). Inserting the value (6) with $q = 2$ for $\tilde{\alpha}_c$ yields—together with the second requirement on $Z_1, ..., Z_K$ in the relativistic bound—the claimed stability criterion.

4. INSTABILITY WITH THE FREE DIRAC OPERATOR

In this section we shall discuss the Brown–Ravenhall model.[3] That is to say we consider the energy expression (1) with zero vector potential in the definition of the allowed electronic states (3), i.e., we take only $\mathbf{\Psi} \in \mathscr{H}_{N,\mathbf{0}} = \bigwedge^N \mathscr{H}_+$, where $\mathscr{H}_+$ is the positive spectral subspace of the operator $-\boldsymbol{\alpha} \cdot i\nabla + m\beta$. We shall prove that there is no stability is in this model by showing that for any (sufficiently large) particle number, N, and any $\alpha > 0$ the energy is unbounded below. In other words, "stability of the first kind" is violated. It is nevertheless true, however, that for any choice of particle numbers and nuclear charges there is always a sufficiently small, nonzero α such that the energy is bounded below by zero.

Since the positive spectral subspace $\mathscr{H}_+$ for the free Dirac operator is not invariant under gauge transformations we see that this Brown–Ravenhall model is not gauge invariant. (The previous, modified model discussed in Section 3 is not only stable, it is also gauge invariant.) More precisely, the energy spectrum depends not only on $\mathbf{B}$ but in fact on the full gauge potential $\mathbf{A}$. The Brown–Ravenhall model is therefore physically meaningfully defined only if we make a fixed choice of gauge. The natural choice is the Coulomb gauge (radiation gauge),

$$\nabla \cdot \mathbf{A} = 0,$$

since in quantum electro-dynamics this gauge implies that electrons interact via the usual Coulomb potentials and the coupling to the transverse field is minimal, i.e., derivatives are replaced by covariant derivatives.

The interesting quantity is the lowest energy that the system can have.

Definition 1 (Energy).

$$E_{N,K} := \inf \mathscr{E}_0[\mathbf{\Psi}, \mathbf{\Psi}].$$

where the infimum is taken over all divergence free $\mathbf{A}$ fields, pairwise distinct nuclear locations $\mathbf{R}_1, ..., \mathbf{R}_K$ and normalized, antisymmetric states $\mathbf{\Psi} \in \mathscr{H}_{N,\mathbf{0}}$.

4.1. Stability with Small α and Small Particle Number

Since this result is not a main point of this paper we shall be brief—even sketchy. If a single particle $\mathbf{\Psi}$ is in the positive spectral subspace of $D(\mathbf{0})$ then the action of $D(\mathbf{0})$ on $\mathbf{\Psi}$ is the same as multiplication of each component by $(p^2 + m^2)^{1/2}$ in Fourier space. For such functions we see that $(\mathbf{\Psi}, D(\mathbf{0})\, \mathbf{\Psi})$ exceeds $(\mathbf{\Psi}, |\nabla|\, \mathbf{\Psi})$, so we may as well replace $D(\mathbf{0})$ by $|\nabla|$ and also drop the condition that $\mathbf{\Psi}$ belongs to the positive spectral subspace of $D(\mathbf{0})$.

The next step is to use the lower bound on V_c in (5) so that the energy is now bounded below by a sum of one-body operators, in a manner similar to that in Section 3 with $\tilde{\alpha}_c$ as in Theorem 1:

$$\mathscr{E}_0[\Psi, \Psi] \geqslant \left(\Psi, \sum_{\nu=1}^{N} \left(1 - \frac{\alpha}{\tilde{\alpha}_c}\right) |\nabla_\nu| \, \Psi\right) + \sqrt{\alpha} \int_{\mathbb{R}^3} \mathbf{j} \cdot \mathbf{A} + \frac{1}{8\pi} \int_{\mathbb{R}^3} B^2. \tag{16}$$

(Note again that the "effective spin" q is 2, as can be seen be repeating the above argument.) Here $\mathbf{j}(\mathbf{r})$ is the current in the state Ψ and it is trivially bounded above pointwise by the density $\rho(\mathbf{r})$ in the state Ψ (defined in Appendix B). Therefore the integral involving $\mathbf{A}$ is bounded below by

$$\int_{\mathbb{R}^3} \mathbf{j} \cdot \mathbf{A} \geqslant -\int_{\mathbb{R}^3} \rho A \geqslant -\|A\|_6 \, N^{1/3} \, \|\rho\|_{4/3}^{2/3}.$$

Now $\int B^2 \geqslant \int |\nabla A|^2$ and this is not less than $K_3^{-2} \|A\|_6^2$ by Sobolev's inequality where $K_3 = 4^{1/3}(3\pi)^{-1/2} \pi^{-1/6}$ (see ref. 16, p. 367). Similarly, the kinetic energy $(\Psi, \sum_{\nu=1}^{N} |\nabla_\nu| \, \Psi)$ is bounded below by $1.63 q^{-1/3} \int \rho^{4/3}$, which was proved by Daubechies[6] and which follows from (8). If we use these inequalities and then minimize the energy with respect to the unknown quantity $\|A\|_6$, we easily find that the energy is non-negative as long as

$$1.63(1 - \alpha/\tilde{\alpha}_c) \geqslant 2\pi N^{2/3} K_3^2 q^{1/3} \alpha$$

with $q = 2$.

We shall show in Subsection 4.2 that the condition that $N^{2/3}\alpha$ is small, which—as we just proved—ensures boundedness from below, is in fact also necessary for the energy to be bounded from below.

4.2. Instability for All α and Large Particle Number

The main result of this section is that there is no stability in this model for any fixed, positive α if N and K are allowed to be arbitrary.

Theorem 2 (Instability). There exists a universal number $C > 0$ such that for all values of the parameters $\alpha > 0$, $m \geqslant 0$, $K = 1, 2, 3,...$, and all values of $N = 1, 2, 3,...$, and of $Z_1, Z_2,..., Z_K$ satisfying

$$\sum_{\kappa=1}^{K} Z_\kappa > C \max\{\alpha^{-3/2}, 1\}, \quad N > C \max\{\alpha^{-3/2}, 1\}, \qquad \sum_{\kappa=1}^{K} Z_\kappa^2 > 2$$

we have

$$E_{N,K} = -\infty.$$

Proof. The theorem follows if, for all $E > 0$, we show the existence of three quantities for which $\mathscr{E}_0[\mathbf{\Psi}, \mathbf{\Psi}] \leqslant -E$, with $\mathbf{\Psi} = \psi_1 \wedge \cdots \wedge \psi_N$:

A. A vector potential $\mathbf{A}$ with $\nabla \cdot \mathbf{A} = 0$.

B. Orthonormal spinors $\psi_1, \ldots, \psi_N \in \mathscr{H}_+$.

C. Nuclear coordinates $\mathbf{R}_1, \ldots, \mathbf{R}_K$.

Our construction willed depend on four parameters (to be specified at the very end), $\delta > 0$ a momentum scale, which we shall let tend to infinity, $\theta > 0$, which will be chosen sufficiently small (but independently of N), and P, $A_0 > 0$ which will be chosen as functions of N. Finally we denote by $\mathbf{n}_1, \mathbf{n}_2, \mathbf{n}_3$ the coordinate vectors $(1, 0, 0)$, $(0, 1, 0)$, $(0, 0, 1)$ respectively. We shall use the notation that $\boldsymbol{\omega}_p = \mathbf{p}/|\mathbf{p}|$ is the unit vector in the direction $\mathbf{p} \in \mathbb{R}^3$.

A. The vector potential. We choose the vector potential $\mathbf{A}$ to have Fourier transform

$$\hat{\mathbf{A}}(\mathbf{p}) := A_0 \chi_{B(0,\,5\delta)}(\mathbf{p})(\mathbf{n}_2 \cdot \boldsymbol{\omega}_\mathbf{p})\, \mathbf{n}_3 \times \boldsymbol{\omega}_\mathbf{p},$$

where $\chi_{B(0,\,5\delta)}$ denotes the characteristic function (in $\mathbf{p}$-space) of the ball $B(0, 5\delta)$ centered at 0 with radius 5δ. Note first that $\mathbf{A}$ is real since $\hat{\mathbf{A}}$ is real and $\hat{\mathbf{A}}(\mathbf{p}) = \hat{\mathbf{A}}(-\mathbf{p})$. Moreover, $\mathbf{A}$ is divergence free, i.e., it is in the Coulomb gauge, since $-i\nabla \cdot \hat{\mathbf{A}}(\mathbf{p}) = \mathbf{p} \cdot \hat{\mathbf{A}}(\mathbf{p}) = 0$. We easily estimate the self-energy of the magnetic field $\mathbf{B} = \nabla \times \mathbf{A}$ corresponding to $\mathbf{A}$

$$\frac{1}{8\pi} \int (\nabla \times \mathbf{A})^2 = \frac{1}{8\pi} \int_{\mathbb{R}^3} |\mathbf{p} \times \hat{\mathbf{A}}(\mathbf{p})|^2 \, d\mathbf{p} \leqslant A_0^2 2^{-1} \int_0^{5\delta} p^4 \, dp = 2^{-1} 5^4 A_0^2 \delta^5. \quad (17)$$

Finally, we note for later use that

$$\hat{\mathbf{A}}(\mathbf{p}) \cdot \mathbf{n}_1 = -A_0 \chi_{B(0,\,5\delta)}(\mathbf{p})(\mathbf{n}_2 \cdot \boldsymbol{\omega}_\mathbf{p})^2. \quad (18)$$

B. The orthonormal spinors. For $\mathbf{p}_0 \in \mathbb{R}^3$ define

$$u_{\mathbf{p}_0}(\mathbf{p}) = \sqrt{3/(4\pi)}\, \delta^{-3/2} \begin{pmatrix} \chi_{B(0,\,5\delta)}(\mathbf{p} - \mathbf{p}_0) \\ 0 \end{pmatrix}. \quad (19)$$

We then have a normalized $\psi_{\mathbf{p}_0} \in \mathscr{H}_+$ given by

$$\widehat{\psi_{\mathbf{p}_0}}(\mathbf{p}) = (2E(p)(E(p)+E(0)))^{-1/2} \begin{pmatrix} (E(p)+E(0))\, u_{\mathbf{p}_0}(\mathbf{p}) \\ \mathbf{p}\cdot\boldsymbol{\sigma} u_{\mathbf{p}_0}(\mathbf{p}) \end{pmatrix},$$

where $E(p) = (p^2+m^2)^{1/2}$. Recall that this is the general form of a spinor in the positive spectral subspace $\mathscr{H}_+$ for the free Dirac operator.

For the sake of simplicity we shall henceforth assume that $m=0$. We leave it to the interested reader to check the estimates for the general case $m \neq 0$. We shall indeed consider spinors with momenta p such that we have $p^2(m^2)^{-1} \to \infty$ as $\delta \to \infty$, i.e., $E(p) \approx p$. It is therefore straightforward to estimate the expressions in the general case $m \neq 0$ by the corresponding expressions for $m=0$.

In particular, we have, for $m=0$,

$$\widehat{\psi_{\mathbf{p}_0}}(\mathbf{p}) = 2^{-1/2} \begin{pmatrix} u_{\mathbf{p}_0}(\mathbf{p}) \\ \boldsymbol{\omega}_{\mathbf{p}}\cdot\boldsymbol{\sigma} u_{\mathbf{p}_0}(\mathbf{p}) \end{pmatrix},$$

We shall choose N points $\mathbf{p}_1,..., \mathbf{p}_N \in \mathbb{R}^3$ such that the following conditions are satisfied.

1. $\min_{\nu \neq \mu} |\mathbf{p}_\nu - \mathbf{p}_\mu| > 2\delta$
2. $P \leqslant p_\nu \leqslant 2P$, for all $\nu = 1,..., N$
3. $\boldsymbol{\omega}_{\mathbf{p}_\nu} \cdot \mathbf{n}_1 \geqslant 1-\theta^2$, for all $\nu = 1,..., N$

Condition 1 ensures that the spinors $\psi_{\mathbf{p}_1},..., \psi_{\mathbf{p}_N}$ are orthonormal. The importance of Conditions 2 and 3 will hopefully become clear below.

In order that Conditions 1, 2 and 3 are consistent with having N points (for large N) we must ensure that N balls of radius δ can be packed into the domain defined by Conditions 2 and 3. Since small enough balls can fill at least half the volume of the given region we simply choose P such that

$$2N \leqslant \frac{\mathrm{Vol}(\{\mathbf{p} \mid P \leqslant p \leqslant 2P,\, 1-\theta^2 \leqslant \boldsymbol{\omega}_{\mathbf{p}}\cdot\mathbf{n}_1\})}{(4\pi/3)\,\delta^3} = \frac{7}{2}\theta^2 \frac{P^3}{\delta^3}.$$

Note that the assumption that N is larger than some universal number ensures that the balls are small, i.e., that δ is small enough compared to P. Thus, we have the condition

$$P \geqslant \left(\frac{4N}{7\theta^2}\right)^{1/3} \delta. \tag{20}$$

In particular, since we shall choose θ independently of N we may assume that N is large enough that the above condition implies $P\theta \geqslant 2\delta$. (Since we shall choose $\delta \to \infty$ we see that the momenta of the spinors satisfy $p^2 m^{-2} \to \infty$.)

We are now prepared to calculate $(\boldsymbol{\Psi}, \sum_{\nu=1}^{N} D_\nu(\mathbf{A})\,\boldsymbol{\Psi})$, where $\boldsymbol{\Psi} = \psi_{\mathbf{p}_1} \wedge \cdots \wedge \psi_{\mathbf{p}_N}$. We obtain

$$\left(\boldsymbol{\Psi}, \sum_{\nu=1}^{N} D_\nu(\mathbf{A})\,\boldsymbol{\Psi}\right) = \sum_{\nu=1}^{N} \left(T_\nu + \int e\mathbf{J}_\nu \cdot \mathbf{A}\right), \tag{21}$$

where

$$\mathbf{j}_\nu(\mathbf{r}) := \psi^*_{\mathbf{p}_\nu}(\mathbf{r})\,\boldsymbol{\alpha}\psi_{\mathbf{p}_\nu}(\mathbf{r}), \tag{22}$$

is the current of the ν-th one-electron state $\psi_{\mathbf{p}_\nu}$, and

$$T_\nu := (\psi_{\mathbf{p}_\nu}, (-i\boldsymbol{\alpha} \cdot \nabla + \beta m)\,\psi_{\mathbf{p}_\nu}) = \int E(p)\,|u_{\mathbf{p}_\nu}(\mathbf{p})|^2\,d\mathbf{p} \leqslant p_\nu + \delta, \tag{23}$$

since we have assumed that $m = 0$ and hence $E(p) = p$.

We must evaluate the current integral

$$\begin{aligned}\int \mathbf{j}_\nu \cdot \mathbf{A} &= (2\pi)^{-3/2}\, 2\Re \iint u^*_{\mathbf{p}_\nu}(\mathbf{q}-\mathbf{p})(\hat{\mathbf{A}}(\mathbf{p}) \cdot \boldsymbol{\sigma})(\boldsymbol{\omega}_\mathbf{q} \cdot \boldsymbol{\sigma})\, u_{\mathbf{p}_\nu}(\mathbf{q})\, d\mathbf{p}\, d\mathbf{q} \\ &= (2\pi)^{-3/2}\, 2\Re \iint [\hat{\mathbf{A}}(\mathbf{q}-\mathbf{p}) \cdot \boldsymbol{\omega}_\mathbf{q} u^*_{\mathbf{p}_\nu}(\mathbf{p})\, u_{\mathbf{p}_\nu}(\mathbf{q}) \\ &\qquad + i u^*_{\mathbf{p}_\nu}(\mathbf{p})(\hat{\mathbf{A}}(\mathbf{q}-\mathbf{p}) \times \boldsymbol{\omega}_\mathbf{q}) \cdot \boldsymbol{\sigma} u_{\mathbf{p}_\nu}(\mathbf{q})]\, d\mathbf{p}\, d\mathbf{q}.\end{aligned}$$

We first observe that

$$\begin{aligned}&2\Re \iint [i u^*_{\mathbf{p}_\nu}(\mathbf{p})(\hat{\mathbf{A}}(\mathbf{q}-\mathbf{p}) \times \boldsymbol{\omega}_\mathbf{q}) \cdot \boldsymbol{\sigma} u_{\mathbf{p}_\nu}(\mathbf{q})]\, d\mathbf{p}\, d\mathbf{q} \\ &\qquad = 2\Re \iint [i u^*_{\mathbf{p}_\nu}(\mathbf{p})(\hat{\mathbf{A}}(\mathbf{q}-\mathbf{p}) \times \boldsymbol{\omega}_\mathbf{q}) \cdot \mathbf{n}_2 \sigma_2 u_{\mathbf{p}_\nu}(\mathbf{q})]\, d\mathbf{p}\, d\mathbf{q} = 0.\end{aligned}$$

The terms containing σ_1 and σ_3 vanish as they are clearly imaginary. The term with σ_2 vanishes because of the choice (19) of $u_{\mathbf{p}_\nu}$.

Note that $u^*_{\mathbf{p}_\nu}(\mathbf{p})\, u_{\mathbf{p}_\nu}(\mathbf{q})=0$ unless $|\mathbf{p}-\mathbf{q}|<2\delta$ and $|\mathbf{q}-\mathbf{p}_\nu|\leqslant\delta$. Thus $\boldsymbol{\omega}_{\mathbf{p}_\nu}-\boldsymbol{\omega}_{\mathbf{q}}=p_\nu^{-1}(\mathbf{p}_\nu-\mathbf{q})+\boldsymbol{\omega}_{\mathbf{q}} p_\nu^{-1}(q-pv)$ and we obtain for $|\mathbf{p}_\nu-\mathbf{q}|<2\delta$ that

$$|\boldsymbol{\omega}_{\mathbf{p}_\nu}-\boldsymbol{\omega}_{\mathbf{q}}|\leqslant 2\delta p_\nu^{-1}\leqslant 2\delta P^{-1},$$

where we used that $p_\nu\geqslant P$. Since $\boldsymbol{\omega}_{\mathbf{p}_\nu}\cdot\mathbf{n}_1>1-\theta^2$ we have that $|\boldsymbol{\omega}_{\mathbf{p}_\nu}-\mathbf{n}_1|\leqslant\theta$ and hence

$$|\hat{\mathbf{A}}(\mathbf{q}-\mathbf{p})\cdot(\boldsymbol{\omega}_{\mathbf{q}}-\mathbf{n}_1)|\leqslant(2\delta P^{-1}+\theta)\, A_0\leqslant 2\theta A_0.$$

Hence, since $|\mathbf{p}-\mathbf{q}|<2\delta$ we get from (18) that

$$\hat{\mathbf{A}}(\mathbf{q}-\mathbf{p})\cdot\boldsymbol{\omega}_{\mathbf{q}}\leqslant -A_0[(\boldsymbol{\omega}_{\mathbf{q}-\mathbf{p}}\cdot\mathbf{n}_2)^2-2\theta].$$

Thus,

$$\begin{aligned}\int \mathbf{j}_\nu\cdot\mathbf{A}&=(2\pi)^{-3/2}\, 2\Re\iint \hat{\mathbf{A}}(\mathbf{q}-\mathbf{p})\cdot\boldsymbol{\omega}_{\mathbf{q}} u^*_{\mathbf{p}_\nu}(\mathbf{p})\, u_{\mathbf{p}_\nu}(\mathbf{q})\, d\mathbf{q}\, d\mathbf{p}\\ &\leqslant\frac{3}{(2\pi)^{5/2}} A_0\,\delta^{-3}\iint_{|\mathbf{q}|,\,|\mathbf{p}|<\delta}[(\boldsymbol{\omega}_{\mathbf{q}-\mathbf{p}}\cdot\mathbf{n}_2)^2-2\theta]\, d\mathbf{p}\, d\mathbf{q}\\ &=\frac{3}{(2\pi)^{5/2}} A_0\,\delta^{3}\iint_{|\mathbf{q}|,\,|\mathbf{p}|<1}[(\boldsymbol{\omega}_{\mathbf{q}-\mathbf{p}}\cdot\mathbf{n}_2)^2-2\theta]\, d\mathbf{p}\, d\mathbf{q}.\end{aligned}$$

We now make the choice

$$\theta=\frac{1}{3}\left(\iint_{|\mathbf{q}|,\,|\mathbf{p}|<1} 1\, d\mathbf{p}\, d\mathbf{q}\right)^{-1}\iint_{|\mathbf{q}|,\,|\mathbf{p}|<1}(\boldsymbol{\omega}_{\mathbf{q}-\mathbf{p}}\cdot\mathbf{n}_2)^2\, d\mathbf{p}\, d\mathbf{q}$$

and arrive at

$$\int \mathbf{j}_\nu\cdot\mathbf{A}\leqslant -\tfrac{4}{3}(2\pi)^{-1/2} A_0\theta\delta^3.$$

From (21) and (23) we therefore obtain

$$\left(\boldsymbol{\Psi},\sum_{\nu=1}^{N} D_\nu(\mathbf{A})\,\boldsymbol{\Psi}\right)\leqslant\sum_{\nu=1}^{N}[|\mathbf{p}_\nu|+\delta-\tfrac{4}{3}(2\pi)^{-1/2} A_0 e\delta^3\theta]. \qquad (25)$$

C. The nuclear coordinates. Finally, we show how to choose the nuclear coordinates following an idea in ref. 18. Consider the electronic density of the state Ψ, $\rho(\mathbf{r}) = \sum_{\nu=1}^{N} |\psi_{\mathbf{p}_\nu}(\mathbf{r})|^2$ then

$$(\Psi, V_c\Psi) \leqslant -\sum_{k=1}^{K} \int \frac{Z_\kappa \rho(\mathbf{r})}{|\mathbf{r}-\mathbf{R}_k|} d\mathbf{r} + D(\rho, \rho) + \sum_{\substack{\kappa, \lambda=1 \\ \kappa<\lambda}}^{K} \frac{Z_\kappa Z_\lambda}{|\mathbf{R}_\kappa - \mathbf{R}_\lambda|},$$

were we introduced $D(\rho, \rho) := \frac{1}{2} \iint \rho(\mathbf{r}) \, |\mathbf{r}-\mathbf{r}'|^{-1} \rho(\mathbf{r}') \, d\mathbf{r} \, d\mathbf{r}'$.

Note now that $\int N^{-1}\rho = 1$, i.e., $N^{-1}\rho$ can be considered a probability distribution. We may therefore average $(\Psi, V_c\Psi)$ considered as a function of $\mathbf{R}_1, \ldots, \mathbf{R}_K$ with respect to the probability measure

$$\mathbf{R}_1, \ldots, \mathbf{R}_K \mapsto N^{-1}\rho(\mathbf{R}_1) \cdots N^{-1}\rho(\mathbf{R}_K).$$

We obtain

$$\int (\Psi, V_c\Psi) \, N^{-1}\rho(\mathbf{R}_1) \cdots N^{-1}\rho(\mathbf{R}_K) \, d\mathbf{R}_1 \cdots d\mathbf{R}_K$$
$$= \left[(1-(Z/N))^2 - N^{-2} \sum_{\kappa=1}^{K} Z_\kappa^2 \right] D(\rho, \rho).$$

We shall prove that $[\ldots] < 0$. There are two cases.

(1) $N \geqslant Z$: By moving electrons to infinity we may assume that $Z \leqslant N < Z+1$. Therefore

$$\left[(1-(Z/N))^2 - N^{-2} \sum_{\kappa=1}^{K} Z_\kappa^2 \right] \leqslant Z^{-2} - (Z+1)^{-2} \sum_{\kappa=1}^{K} Z_\kappa^2$$
$$\leqslant Z^{-2} - 2(Z+1)^{-2} < 0.$$

(2) $N < Z$: We may move nuclei to infinity and assume that $N < Z < N + \max_\kappa Z_\kappa$. Therefore

$$\left[(1-(Z/N))^2 - N^{-2} \sum_{\kappa=1}^{K} Z_\kappa^2 \right] \leqslant (\max_\kappa Z_\kappa)^2 N^{-2} - N^{-2} \sum_{\kappa=1}^{K} Z_\kappa^2 \leqslant 0.$$

We can therefore find nuclear positions $\mathbf{R}_1, \ldots, \mathbf{R}_K$ such that $(\Psi, V_c\Psi) \leqslant 0$.

Using these coordinates together with (17) and (25) we get

$$\mathscr{E}_0[\Psi, \Psi] \leqslant \sum_{\nu=1}^{N} [p_\nu + \delta - \tfrac{4}{3}(2\pi)^{-1/2} A_0 e\delta^3\theta] + 2^{-1}5^4 A_0^2 \delta^5.$$

We now choose $P=(4N/7\delta^2)^{1/3}\delta$ in accordance with (20). We than choose A_0 such that $\frac{4}{3}(2\pi)^{-1/2} A_0 e\delta^3\theta = 4P$, i.e.,

$$A_0 = \frac{3}{112^{1/3}}(2\pi)^{1/2}\theta^{-5/3}e^{-1}\delta^{-2}N^{1/3}.$$

Then, using Condition 2 we have $p_\nu - (8\pi/3) A_0 e\delta^3\theta \leqslant -2P$. If we insert these values and estimates above we find

$$\mathscr{E}_0[\Psi, \Psi] \leqslant \delta[-2N^{4/3}(4/(7\theta^2))^{1/3} + N + 3^2 5^4 112^{-2/3}\pi\theta^{-10/3}\alpha^{-1}N^{2/3}].$$

It is now clear that the expression in [ ] will be negative for N sufficiently large. The energy can therefore be made arbitrarily negative by choosing δ large.

APPENDIX A. BKS INEQUALITIES

As a convenience to the reader we give a proof of some cases of the inequalities due to Birman, Koplienko, and Solomyak.(2) The case needed in Section 3 corresponds to $p=2$ below. There we are interested in $(B-A)_-$, but here we treat $(B-A)_+$ to simplify keeping track of signs. The proof is the same. Recall that $X_+ := (|X| + X)/2$.

Theorem 3. Let $p \geqslant 1$ and suppose that A and B are two nonnegative, self-adjoint linear operators on a separable Hilbert space such that $(B^p - A^p)_+^{1/p}$ is trace class. Then $(B-A)_+$ is also trace class and

$$\mathrm{tr}(B-A)_+ \leqslant \mathrm{tr}(B^p - A^p)_+^{1/p}.$$

Proof. Our proof will use essentially only two facts: $X \mapsto X^{-1}$ is operator monotone decreasing on the set of nonnegative self-adjoint operators (i.e., $X \geqslant Y \geqslant 0 \Rightarrow Y^{-1} \geqslant X^{-1}$) and $X \mapsto X^r$ is operator monotone increasing on the set of nonnegative self-adjoint operators for all $0 < r \leqslant 1$. Consequently, $X \mapsto X^{-r}$ is operator monotone decreasing for $0 < r \leqslant 1$.

As a preliminary remark, we can suppose that $B \geqslant A$. To see this, write $B^p = A^p + D$. If we replace B by $[A^p + D_+]^{1/p}$ then $(B^p - A^p)_+ = D_+$ is unchanged, while $X := B - A \mapsto [A^p + D_+]^{1/p} - A$ can only get bigger because $X \mapsto X^{1/p}$ is operator monotone on the set of positive operators. Since the trace is also operator monotone, we can therefore suppose that $D = D_+$, i.e., $B^p - A^p + C^p$ with $A, B, C \geqslant 0$. Our goal is to prove that

$$\mathrm{tr}[(A^p + C^p)^{1/p} - A] \leqslant \mathrm{tr}\, C, \tag{26}$$

under the assumption that C is trace class.

To prove (26) we consider the operator $X := [A^p + C^p]^{1/p} - A$, which is well defined on the domain of A. We assume, at first, that $A^p \geqslant \varepsilon^p$ for some positive number ε. Then, by the functional calculus, and with

$$E := [A^p + C^p]^{(1-p)/p} \qquad \text{and} \qquad P := A^{1-p} - E$$

we have

$$X = E[A^p + C^p] - A^{1-p}A^p = -PA^p + EC^p \tag{27}$$

Clearly, $P \geqslant 0$ and $0 \leqslant P \leqslant \varepsilon^{1-p}$.

Let $Y := EC^p$. We claim that Y is trace class. This follows from $Y^*Y = C^pE^2C^p \leqslant C^pC^{2-2p}C^p = C^2$. Thus, $|Y| \leqslant C$, and hence tr $Y =$ tr $C^{p/2}EC^{p/2} \leqslant$ tr C.

It is also true that P is trace class. To see this, use the integral representation, with suitable $c > 0$, $A^{1-p} = c\int_0^\infty (t + A)^{-1}\, t^{1-p}\, dt$. Use this twice and then use the resolvent formula. In this way we find that

$$P = c\int_0^\infty (A^p + t)^{-1}\, C^p(A^p + C^p + t)^{-1}\, t^{(1-p)/p}\, dt.$$

Since C is trace class, so is C^p, and the integral converges because of our assumed lower bound on A. Thus, P is trace class and hence there is a complete, orthonormal family of vectors $v_1, v_2, \ldots$, each of which is an eigenvector of P.

Since $X \geqslant 0$, the trace of X is well defined by $\sum_{j=1}^\infty (v_j, Xv_j)$ for *any* complete, orthonormal family. The same remark applies to EC^p since it is trace class. Thus, to complete the proof of (26) it suffices to prove that $(v_j, PA^pv_j) \geqslant 0$ for each j. But this number is $\lambda_j(v_j, A^pv_j) \geqslant 0$, where λ_j is the (nonnegative) eigenvalue of P, and the positivity follows from the positivity of A.

We now turn to the case of general $A \geqslant 0$. We can apply the above proof to the operator $A + \varepsilon$ for some positive number ϵ. Thus we have

$$\mathrm{tr}[[(A + \varepsilon)^p + C^p]^{1/p} - (A + \varepsilon)] \leqslant \mathrm{tr}\, C. \tag{28}$$

Let $\varphi_1, \varphi_2, \ldots$ be an orthonormal basis chosen from the domain of A^p. This basis then also belongs to the domain of A and the domain of $[(A + \varepsilon)^p + C^p]^{1/p}$ for all $\varepsilon \geqslant 0$. We then have

$$\mathrm{tr}\, X = \sum_j (\varphi_j, X\varphi_j).$$

Note that a-priori we do not know that the trace is finite, but since the operator is non-negative this definition of the trace is meaningful. Operator monotonicity of $X^{1/p}$ gives

$$(\varphi_j, [[(A+\varepsilon)^p + C^p]^{1/p} - (A+\varepsilon)]\, \varphi_j) \geqslant (\varphi_j, (X-\varepsilon)\, \varphi_j).$$

It therefore follows from (28), followed by Fatou's Lemma applied to sums that

$$\begin{aligned}
\operatorname{tr} C &\geqslant \liminf_{\varepsilon \to 0} \sum_j (\varphi_j, [[(A+\varepsilon)^p + C^p]^{1/p} - (A+\varepsilon)]\, \varphi_j) \\
&\leqslant \sum_j \liminf_{\varepsilon \to 0} (\varphi_j, [[(A+\varepsilon)^p + C^p]^{1/p} - (A+\varepsilon)]\, \varphi_j) \\
&\geqslant \operatorname{tr} X.
\end{aligned}$$

APPENDIX B. COUNTING SPIN STATES

Our goal here is to prove that when $\boldsymbol{\Psi}$ is in $\mathscr{H}_{N,\mathscr{A}}$, the antisymmetric tensor product of the positive energy subspace of the Dirac operator (with or without a magnetic field, $\mathscr{A}$) then the one-body density matrix is bounded by 2 and not merely by 4, as would be the case if there were no restriction to the positive energy subspace. This result will allow us to use 2 instead of 4 in inequalities (6) and (14). We thank Michael Loss for the idea of this proof.

The one-body density matrix is defined in terms of an N-body density matrix (or function) by the partial trace over $N-1$ variables. We illustrate this for functions, but the proof works generally. If $\boldsymbol{\Psi}$ is a function, then

$$\Gamma(\mathbf{r}, \sigma; \mathbf{r}', \sigma') := N \int_{(N-1)} \boldsymbol{\Psi}(\mathbf{r}, \sigma, z_2, z_3,..., z_N)\, \overline{\boldsymbol{\Psi}(\mathbf{r}', \sigma', z_2,..., z_N)}\, dz_2 \cdots dz_N,$$

where z denotes a pair $\mathbf{r}, \sigma$ and dz denotes integration over $\mathbb{R}^3$ and summation over the q "spin" states of σ. We are interested in $q=4$, but that is immaterial for the definition.

The kernel Γ is trace class; in fact its trace is qN. It is also obviously positive definite as an operator. The first remark is that $\Gamma \leqslant 1$ as an operator. To prove this easily, let ψ be any normalized function of one space-spin variable z and define the function of $N+1$ variables $\boldsymbol{\Phi}(z_0,..., z_N) := \psi(z_0)\, \boldsymbol{\Psi}(z_1,..., z_N) + \sum_{j=1}^N (-1)^j\, \psi(z_j)\, \boldsymbol{\Psi}(z_0,..., \hat{z}_j,..., z_N)$, where $\hat{z}_j$ denotes the absence of z_j. This function $\boldsymbol{\Phi}$ is clearly antisymmetric and the integral over all variables of its square is surely nonnegative.

However, this integral is easily computed (using the normalization of ψ and $\mathbf{\Psi}$) to be $(N+1)-(N+1)(\psi, \gamma\psi)$.

The next step is to consider the reduced kernel (without spin) defined by

$$\gamma(\mathbf{r}, \mathbf{r}') := \sum_{\sigma=1}^{q} \Gamma(\mathbf{r}, \sigma; \mathbf{r}', \sigma),$$

which evidently satisfies the operator inequality $0 \leqslant \gamma \leqslant q$, since $\Gamma \leqslant 1$.

The electron density referred to in Section 4.1 is defined by

$$\rho(\mathbf{r}) := \gamma(\mathbf{r}, \mathbf{r}),$$

but it will not be needed in this Appendix. Another quantity of interest is the current, defined by

$$\mathbf{j}(\mathbf{r}) := \sum_{\sigma, \tau} \Gamma(\mathbf{r}, \sigma; \mathbf{r}, \tau)\, \boldsymbol{\alpha}_{\sigma, \tau}.$$

It follows from this that $|\mathbf{j}(\mathbf{r})| \leqslant \rho(\mathbf{r})$ for every $\mathbf{r} \in \mathbb{R}^3$.

Our goal here is to prove the following fact about γ:

If the N-body $\mathbf{\Psi}$ is in $\mathscr{H}_{N, \mathscr{A}}$ then the correspondingly satisfies $0 \leqslant \gamma \leqslant 2$ as an operator.

To prove this we introduce the unitary matrix in spin-space (related to the charge conjugation operator)

$$U = \begin{pmatrix} 0 & 1 \\ -1 & 0 \end{pmatrix}$$

where 1 denotes the unit 2×2 unit matrix. With a slight abuse of notation, we shall also use U to denote the $U \otimes 1$ acting on the full one-particle space, i.e., $(U)(\mathbf{r}, \sigma') = \sum_\sigma U(\sigma', \sigma)\, f(\mathbf{r}, \sigma)$. The important point to note, which is easily verified from the Dirac equation, is that $\psi \in \mathscr{H}_+$ if and only if $U\psi \in \mathscr{H}_-$, the negative spectral subspace of $D(\mathscr{A})$.

Given $f \in L^2(\mathbb{R}^3)$, we define F^τ to be the spinor $F^\tau(\mathbf{r}, \sigma) := f(\mathbf{r})\, \delta_{\sigma, \tau}$. Then evidently $(f, \gamma f) = \sum_\tau (F^\tau, \Gamma F^\tau)$. However, since the matrix U merely permutes the spin indices and possibly changes the sign from $+$ to $-$, we have that $\sum_\tau (F^\tau, \Gamma F^\tau) = \sum_\tau (F^\tau, \Gamma_U F^\tau)$, with $\Gamma_U := U^{-1} \Gamma U$. (Actually, the proof only requires that U be unitary, nothing more.)

We claim that $\Gamma + \Gamma_U \leqslant 1$ in which case we have proved that $(f, \gamma f) \leqslant q/2 = 2$, as claimed. To see this, we note that $\Gamma \leqslant 1$ on $\mathscr{H}_+$ and $\Gamma_U \leqslant 1$ on $\mathscr{H}_-$. Since the two subspaces are orthogonal, $\Gamma + \Gamma_U \leqslant 1$ on the whole spinor space.

ACKNOWLEDGMENTS

The authors thank Michael Loss for valuable discussions, especially with regard to Appendix B. After we had proved the results in this paper, including the inequalities in Appendix A, Huzihiro Araki kindly informed us of the paper by Birman, Koplienko, and Solomyak[2] in which the inequalities of Appendix A were proved 21 years earlier; we are grateful to him for this help. We are also greatful to M. Griesemer for pointing out several errors in the preliminary version of this paper. The authors also thank the following organizations for their support: the Danish Science Foundation, the European Union, TMR grant FMRX-CT 96-0001, the U.S. National Science Foundation, grant PHY95-13072, and NATO, grant CRG96011.

REFERENCES

1. H. A. Bethe and E. E. Salpeter. Quantum mechanics of one- and two-electron atoms. In S. Flügge, editor, *Handbuch der Physik*, XXXV, pages 88–436. Springer, Berlin, 1 edition, 1957.
2. M. S. Birman, L. S. Koplienko, and M. Z. Solomyak. Estimates for the spectrum of the difference between fractional powers of two self-adjoint operators. *Soviet Mathematics* **19**(3):1–6, 1975. Translation of Izvestija vyssich.
3. G. Brown and D. Ravenhall. On the interaction of two electrons. *Proc. Roy. Soc. London A* **208**(A 1095):552–559, September 1951.
4. L. Bugliaro, J. Fröhlich, G. M. Graf, J. Stubbe, and C. Fefferman. A Lieb–Thirring bound for a magnetic Pauli Hamiltonian. *Preprint*, ETH-TH/96-31, 1996.
5. J. G. Conlon. The ground state energy of a classical gas. *Commun. Math. Phys.* **94**(4): 439-458, August 1984.
6. I. Daubechies. An uncertainty principle for Fermions with generalized kinetic energy. *Commun. Math. Phys.* **90**:511–520, September 1983.
7. F. J. Dyson and A. Lenard. Stability of matter I. *J. Math. Phys.* **8**:423–434, 1967.
8. F. J. Dyson and A. Lenard. Stability of matter II. *J. Math. Phys.* **9**:698–711, 1967.
9. W. D. Evans, P. Perry, and H. Siedentop. The spectrum of relativistic one-electron atoms according to Bethe and Salpeter. *Commun. Math. Phys.* **18**:733–746, July 1996.
10. P. Federbush. A new approach to the stability of matter problem. I. *J. Math. Phys.* **16**:347–351, 1975.
11. C. Fefferman. Stability of relativistic matter with magnetic fields. *Proc. Nat. Acad. Sci. USA* **92**:5006–5007, 1995.
12. C. Fefferman, J. Fröhlich, and G. M. Graf. Stability of ultraviolet-cutoff quantum electrodynamics with non-relativistic matter. *Texas Math. Phys. Preprint server* 96–379, 1996.
13. J. Fröhlich, E. H. Lieb, and M. Loss. Stability of Coulomb systems with magnetic fields. I: The one-electron atom. *Commun. Math. Phys.* **104**:251–270, 1986.
14. Y. Ishikawa and K. Koc. Relativistic many-body perturbation theory based on the no-pair Dirac–Coulomb–Breit Hamiltonian: Relativistic correlation energies for the noble-gas sequence through Rn ($Z = 86$), the group-IIB atoms through Hg, and the ions of Ne isoelectronic sequence. *Phys. Rev. A* **50**(6):4733–4742, December 1994.

15. H. J. A. Jensen, K. G. Dyall, T. Saue, and K. Faegri. Jr., Relativistic four-component multiconfigurational self-consistent-field theory for molecules: Formalism. *J. Chem. Physics* **104**(11):4083–4097, March 1996.
16. E. H. Lieb. Sharp constants in the Hardy–Littlewood–Sobolev and related inequalities. *Annals of Mathematics* **118**:349–374, 1983.
17. E. H. Lieb. On characteristic exponents in turbulence. *Commun. Math. Phys.* **92**:473–480, 1984.
18. E. H. Lieb and M. Loss. Stability of Coulomb systems with magnetic fields. II: The many-electron atom and the one-electron molecule. *Commun. Math. Phys.* **104**:271–282, 1986.
19. E. H. Lieb, M. Loss, and H. Siedentop. Stability of relativistic matter via Thomas–Fermi theory. *Helv. Phys. Acta* **69**:974–984, 1996.
20. E. H. Lieb, M. Loss, and J. P. Solovej. Stability of matter in magnetic fields. *Phys. Rev. Lett.* **75**(6):985–989, August 1995.
21. E. H. Lieb and W. E. Thirring. Bound for the kinetic energy of Fermions which proves the stability of matter. *Phys. Rev. Lett.* **35**(11):687–689, September 1975. Erratum: *Phys. Rev. Lett.* **36**(16):11116, October 1975.
22. E. H. Lieb and W. E. Thirring. Inequalities for the moments of the eigenvalues of the Schrödinger Hamiltonian and their relation to Sobolev inequalities. In E. H. Lieb, B. Simon, and A. S. Wightman, editors, *Studies in Mathematical Physics: Essays in Honor of Valentine Bargmann.* Princeton University Press, Princeton, 1976.
23. E. H. Lieb and H.-T. Yau. The stability and instability of relativistic matter. *Common. Math. Phys.* **118**:177-213, 1988.
24. J. Sucher. Foundations of the relativistic theory of many-electron atoms. *Phys. Rev. A* **22**(2):348–362, August 1980.
25. J. Sucher. Foundations of the relativistic theory of many-electron bound states. *International Journal of Quantum Chemistry* **25**:3–21, 1984.
26. J. Sucher. Relativistic many-electron Hamiltonians. *Phys. Scripta* **36**:271–281, 1987.
27. W. Thirring, ed. *The Stability of Matter: From Atoms to Stars, Selecta of Elliott H. Lieb*, Springer-Verlag, Berlin, Heidelberg, New York, 1997.
28. E. H. Lieb, H. Siedentop, and J. P. Solovej. Stability of relativistic matter with magnetic fields. *Phys. Rev. Lett.* **79**:1785, 1997.

Part VI

The Thermodynamic Limit for Real Matter with Coulomb Forces

The stability of matter

Elliott H. Lieb*

Department of Mathematics and Department of Physics, Princeton University, Princeton, New Jersey 08540

A fundamental paradox of classical physics is why matter, which is held together by Coulomb forces, does not collapse. The resolution is given here in three steps. First, the stability of *atom* is demonstrated, in the framework of nonrelativistic quantum mechanics. Next the Pauli principle, together with some facts about Thomas–Fermi theory, is shown, to account for the stability (i.e., saturation) of *bulk* matter. Thomas–Fermi theory is developed in some detail because, as is also pointed out, it is the asymptotically correct picture of heavy atoms and molecules (in the $Z \to \infty$ limit). Finally, a rigorous version of screening is introduced to account for *thermodynamic* stability.

CONTENTS

INTRODUCTION

Some features of the physical world are so commonplace that they hardly seem to deserve comment. One of these is that ordinary matter, either in the form of atoms or in bulk, is held together with Coulomb forces and yet is stable. Nowadays we regard this truly remarkable phenomenon as a consequence of quantum mechanics, but it is far from obvious how the conclusion follows from the premise. It is not necessary to ponder the question very long before realizing that it is a subtle one and that the answer is not to be found in any textbook.

Although the Schrödinger equation is half a century old, it was only in the last few years that the proof of stability was completed. The aim of this paper is to present the full story in a simple and coherent way, highlighting only the main physical and mathematical ideas.

The sense of profound unease about the problem just before the dawn of quantum mechanics is exemplified by this quotation (Jeans, 1915):

> "... there would be a very real difficulty in supposing that the law $1/r^2$ held down to zero values of r. For the force between two charges at zero distance would be infinite; we should have charges of opposite sign continually rushing together and, when once together, no force would be adequate to separate them... Thus the matter in the universe would tend to shrink into nothing or to diminish indefinitely in size... We should however probably be wrong in regarding a molecule as a cluster of electrons and positive charges. A more likely suggestion, put forward by Larmor and others is that the molecule may consist, in part at least, of rings of electrons in rapid orbital motion."

Jeans' words strike a contemporary chord, especially since one aspect of the problem that worried him has not yet been fully resolved. This is that electrons and nuclei have a magnetic dipole–dipole interaction whose energy goes as r^{-3}. Although the angular average of this interaction vanishes, the interaction can cause the collapse that Jeans feared, even with Schrödinger mechanics. A proper quantum electrodynamics is needed to describe the dipolar interaction at very small distances. For that reason spin dependent forces will be ignored in this paper; only nonrelativistic quantum mechanics will be considered.

It is difficult to find a reliable textbook answer even to the question: How does quantum mechanics prevent the collapse of an atom? One possibility is to say that the Schrödinger equation for the hydrogen atom can be solved and the answer seen explicitly. This is hardly satisfactory for the many-electron atom or for the molecule. Another possible answer is the Heisenberg uncertainty principle. This, unfortunately, is a false argument, as shown in Sec. I. There is, however, a much better uncertainty principle, formulated by Sobolev, which does adequately describe the intuitive fact that a particle's kinetic energy increases sufficiently fast, as the wave function is compressed, to prevent collapse. (See Kato, 1951).

The next question to consider is well stated in this quotation from Ehrenfest (in Dyson, 1967):

> "We take a piece of metal. Or a stone. When we think about it, we are astonished that this quantity of matter should occupy so large a volume. Admittedly, the molecules are packed tightly together, and likewise the atoms within each molecule. But why are the atoms themselves so big?... Answer: only the Pauli principle, 'No two electrons in the same state.' That is why atoms are so unnecessarily big, and why metal and stone are so bulky."

Dyson then goes on to say that without the Pauli principle

> "We show that not only individual atoms but matter in bulk would collapse into a condensed high-density phase. The assembly of any two macroscopic objects would release energy comparable to that of an atomic bomb."

Two distinct facts are involved here. One is that matter is stable (or saturates), meaning that the ground

*Work partially supported by U. S. National Science Foundation grant MCS 75–21684.

state energy is bounded below by a constant times the first power, and not a higher power, of the particle number. This was proved for the first time by Dyson and Lenard (Dyson and Lenard, 1967, and Lenard and Dyson, 1968), in a beautiful series of papers. Their method is quite complicated, however, and a simpler proof is given in Sec. IV. In addition, they used sufficiently many inequalities that their estimate (for hydrogen atoms) is about -10^{14} Ry/particle. We will obtain a bound of about -23 Ry/particle. The second fact is that matter would definitely not be stable if electrons were bosons (Dyson, 1967). The energy would increase at least as fast as $-N^{7/5}$.

Therefore, Ehrenfest's surmise that the Pauli principle plays a crucial role in preventing collapse is correct. The problem is to display the essence of the Pauli principle in a clear, succinct and mathematically precise way. Unless this is done the physics of stability will remain unclear.

The key fact is developed in Sec. II: If $\rho(x)$ is the one-particle density of any fermion wave function then the total kinetic energy is bounded below by (constant) $\int\rho(x)^{5/3}dx$.[1] This inequality may be termed the uncertainty principle for fermions. It is simple yet powerful enough to establish stability.

Given this bound, it is then necessary to show how the kinetic energy eventually overcomes the r^{-1} Coulomb singularity. It turns out that Thomas-Fermi (TF) theory is exactly what is needed for this purpose because, as Teller discovered in 1962, atoms do not bind in TF theory. Thus TF theory immediately implies saturation. The necessary facts about TF theory are developed in Sec. III.

There is also another good reason for understanding TF theory in detail. The theory used to be regarded as an uncertain approximation in atomic physics, but it is now known that it is more than that. It happens to be an asymptotically correct theory of atoms and molecules as the nuclear charges tend to infinity. In short, TF theory and the theory of the hydrogen atom constitute two opposite, but rigorous foundations for the many electron problem.

After putting together the results of Sec. II and III in Sec. IV, and thereby proving the stability of bulk matter, we address the third main topic of this paper in Sec. V. Does a sensible thermodynamic limit exist for matter? The problem here centers around the long range r^{-1} nature of the Coulomb potential, not the short range singularity. Put another way, the question is that if matter does not implode, how do we know that it does not explode? Normally systems with potentials that fall off less slowly than $r^{-3-\epsilon}$ for some $\epsilon > 0$ cannot be expected to have a thermodynamic limit. The crucial physical fact was discovered by Newton in 1687: outside an isotropic distribution of charge, all the charge appears to be concentrated at the center. This fact is the basis for screening, but to use it a geometric fact about the packing of balls will be needed. Quantum mechanics as such plays almost no role in Sec. V.

[1] $\int f(x)dx$, or simply $\int f$, always denotes a three-dimensional integral.

The content of this paper can be summarized as follows:

(i) Atoms are stable because of an uncertainty principle,

(ii) Bulk matter is stable because of a stronger uncertainty principle that holds only for fermions;

(iii) Thermodynamics exists because of screening.

My hope is that the necessary mathematics, which is presented as briefly as possible, will not obscure these simple physical ideas.

This paper is based on research carried out over the past few years, and it was my good fortune to have had the benefit of collaboration with J. L. Lebowitz, B. Simon, and W. E. Thirring. Without their insights and stimulation probably none of this would have been carried to fruition. Secs. II and IV come from Lieb and Thirring (1975), Sec. III from Lieb and Simon (1977), and Sec. V from Lieb and Lebowitz (1972).

Lectures given in 1976 at the Centro Internazionale Matematico Estivo in Bressanone were the impetus for writing this paper. The bibliography is not intended to be scholarly, but I believe no theorem or idea has been quoted without proper credit.

I am doubly grateful to S. B. Treiman. He kindly invited me to submit this paper to *Reviews of Modern Physics*, and he also generously devoted much time to reading the manuscript and made many valuable suggestions to improve its clarity.

I. THE STABILITY OF ATOMS

By the phrase "stability of an atom" is meant that the ground state energy of an atom is finite. This is a weaker notion than the concept of H stability of matter, to be discussed in Sec. IV, which means that the ground state energy of a many-body system is not merely bounded below but is also bounded by a constant times the number of particles. This, in turn, is different from thermodynamic stability discussed in Sec. V.

Consider the Hamiltonian for the hydrogenic atom:

$$H = -\Delta - Z|x|^{-1} \tag{1}$$

(using units in which $\hbar^2/2 = 1$, $m = 1$, and $|e| = 1$). H acts on $L^2(\mathbf{R}^3)$, the square integrable functions on 3-space. Why is the ground state energy finite, i.e., why is

$$\langle\psi, H\psi\rangle \geq E_0\langle\psi, \psi\rangle \tag{2}$$

for some $E_0 > -\infty$? The obvious elementary quantum mechanics textbook answer is the *Heisenberg uncertainty principle* (Heisenberg, 1927): If the *kinetic energy* is defined by

$$T_\psi \equiv \int |\nabla\psi(x)|^2 dx, \tag{3}$$

and if

$$\langle |x|^2\rangle_\psi \equiv \int |x|^2 |\psi(x)|^2 dx,$$

then when

$$\langle\psi, \psi\rangle = ||\psi||_2^2 = \int |\psi(x)|^2 dx = 1,$$

$$T_\psi\langle |x|^2\rangle_\psi \geq \tfrac{9}{4}. \tag{4}$$

The intuition behind applying the Heisenberg uncertainty principle (4) to the ground state problem (2) is that if the electron tries to get within a distance R of the nucleus, the kinetic energy T_ψ is at least as large as R^{-2}. Consequently $\langle\psi, H\psi\rangle \geq R^{-2} - Z/R$, and this has a minimum $-Z^2/4$ for $R = 2/Z$.

The above argument is *false*! The Heisenberg uncertainty principle says no such thing, despite the endless invocation of the argument. Consider a ψ consisting of two parts, $\psi = \psi_1 + \psi_2$. ψ_1 is a narrow wave packet of radius R centered at the origin with $\int|\psi_1|^2 = \frac{1}{2}$. ψ_2 is spherically symmetric and has support in a narrow shell of mean radius L and $\int|\psi_2|^2 = \frac{1}{2}$. If L is large then, roughly, $\int|x|^2|\psi(x)|^2\,dx \sim L^2/2$, whereas $\int|x|^{-1}|\psi(x)|^2\,dx \sim 1/2R$. Thus, from (4) we can conclude *only* that $T_\psi > 9/2L^2$ and hence that $\langle\psi, H\psi\rangle \geq 9/2L^2 - Z/2R$. With this wave function, and using *only* the Heisenberg uncertainty principle, we can make E_0 arbitrarily negative by letting $R \to 0$.

A more colorful way to put the situation is this: an electron cannot have both a sharply defined position and momentum. If one is willing to place the electron in two widely separated packets, however, say here and on the moon, then the Heisenberg uncertainty principle *alone* does not preclude each packet from having a sharp position and momentum.

Thus, while Eq. (4) is correct it is a pale reflection of the power of the operator $-\Delta$ to prevent collapse. A better uncertainty principle (i.e., a lower bound for the kinetic energy in terms of some integral of ψ which does not involve derivatives) is needed, one which reflects more accurately the fact that if one tries to compress a wave function *anywhere* then the kinetic energy will increase. This principle was provided by Sobolev (1938) and for some unknown reason his inequality, which is simple and goes directly to the heart of the matter, has not made its way into the quantum mechanics textbooks where it belongs. Sobolev's inequality in three dimensions [unlike (4) its form is dimension dependent] is

$$T_\psi = \int|\nabla\psi(x)|^2\,dx \geq K_s\left\{\int\rho_\psi(x)^3dx\right\}^{1/3} = K_s\|\rho_\psi\|_3\,, \quad (5)$$

where

$$\rho_\psi(x) = |\psi(x)|^2 \quad (6)$$

is the density and

$$K_s = 3(\pi/2)^{4/3} \approx 5.478$$

is known to be the best possible constant. Equation (5) is nonlinear in ρ, but that is unimportant.

A rigorous derivation of (5) would take too long to present but it can be made plausible as follows (Rosen, 1971): K_s is the minimum of

$$K^\psi = \frac{\int|\nabla\psi(x)|^2\,dx}{\{\int|\psi(x)|^6\,dx\}^{1/3}}\,.$$

Let us accept that a minimizing ψ exists (this is the hard part) and that it satisfies the obvious variational equation

$$-(\Delta\psi)(x) - \alpha\psi(x)^5 = 0$$

with $\alpha > 0$. Assuming also that there is a minimizing ψ which is nonnegative and spherically symmetric (this can be proved by a rearrangement inequality), one finds by inspection that

$$\psi(x) = (3/\alpha)^{2/3}(1 + |x|^2)^{-1/2}\,.$$

When this is inserted into the expression for K^ψ the result is $K_s = 3(\pi/2)^{4/3}$. The minimizing ψ is not square integrable, but that is of no concern.

Now let us make a simple calculation to show how good (5) really is. For any ψ

$$\langle\psi, H\psi\rangle \geq K_s\left(\int\rho_\psi(x)^3dx\right)^{1/3} - Z\int|x|^{-1}\rho_\psi(x)dx \equiv h(\rho), \quad (7)$$

and hence when $\langle\psi,\psi\rangle = 1$

$$\langle\psi, H\psi\rangle \geq \min\left\{h(\rho)\colon \rho(x) \geq 0,\ \int\rho = 1\right\}. \quad (8)$$

The latter calculation is trivial (for any potential) since gradients are not involved. One finds that the solution to the variational equation is $\bar\rho(x) = \alpha[|x|^{-1} - R^{-1}]^{1/2}$ for $|x| \leq R$ and $\bar\rho(x) = 0$ for $|x| \geq R$, with $R = K_s\pi^{-4/3}Z^{-1}$. Then

$$h(\bar\rho) = Z^2(\pi/2)^{4/3}/K_s = -\tfrac{4}{3}Z^2\ \mathrm{Ry}.$$

(Recall that one Rydberg = Ry = $\frac{1}{4}$ in these units.) Thus, Eq. (5) leads easily to the conclusion

$$E_0 \geq -\tfrac{4}{3}Z^2\ \mathrm{Ry} \quad (9)$$

and this is an excellent lower bound to the correct $E_0 = -Z^2$ Ry, especially since no differential equation had to be solved.

In anticipation of later developments, a weaker, but also useful, form of Eq. (5) can be derived. By Hölder's inequality[2]

$$\int\rho(x)^{5/3}\,dx \leq \left\{\int\rho(x)^3\,dx\right\}^{1/3}\left\{\int\rho(x)dx\right\}^{2/3} \quad (10)$$

and, since we always take $\int|\psi|^2 = 1$,

$$T_\psi \geq K_s\int\rho_\psi(x)^{5/3}dx\,. \quad (11)$$

Note that there is now an exponent 1 outside the integral. Although K_s is the best constant in (5) it is not the best constant in (11). Call the latter K_1. K_1 is the minimum of

$$\frac{\int|\nabla\psi(x)|^2dx}{\int\rho(x)^{5/3}dx}$$

subject to $\int\rho(x)\,dx = 1$. This leads to a nonlinear Schrödinger equation whose numerical solution yields (J. F. Barnes, private communication)

$$K_1 \approx 9.578\,.$$

In any event

$$K_1 > K^c \equiv (\tfrac{3}{5})(6\pi^2)^{2/3} \approx 9.116,$$

[2] Hölder's inequality states that

$$\left|\int f(x)g(x)\,dx\right| \leq \left\{\int|f(x)|^p\,dx\right\}^{1/p}\left\{\int|g(x)|^q\,dx\right\}^{1/q}$$

when $p^{-1} + q^{-1} = 1$ and $p \geq 1$. To obtain (10) take $f = \rho$, $g = \rho^{2/3}$, $p = 3, q = \frac{3}{2}$.

and hence

$$T_\psi \geq K^c \int \rho_\psi(x)^{5/3} dx \text{ when} \int |\psi(x)|^2 dx = 1 . \qquad (12)$$

K^c is much bigger than K_s; it is the classical value and will be encountered again in Sec. II and in Sec. III, where its significance will be clarified.

We can repeat the minimization calculation analogous to Eq. (8) using the bound (12) and the functional

$$h^c(\rho) = K^c \int \rho(x)^{5/3} dx - Z \int |x|^{-1} \rho(x) dx.$$

(We could, of course, use the better constant K_1.) This time

$$\bar{\rho}(x) = \{\tfrac{3}{5}(Z/K^c)(|x|^{-1} - R^{-1})\}^{3/2} \qquad (13)$$

for $|x| \leq R$. R is determined by $\int \rho = 1$ and one finds that $R = (K^c/Z)(4/\pi^2)^{2/3}$ and

$$E_0 \geq -(9Z^2/5K^c)(\pi^2/4)^{2/3} = -3^{1/3} Z^2 \text{ Ry}. \qquad (14)$$

The quantity $3^{1/3}$ is only 8.2% greater than $\frac{4}{3}$.

The Sobolev inequality (5) or its variant (12) is, for our purposes, a much better uncertainty principle than Heisenberg's—indeed it is also fairly accurate. We now want to extend (12) to the N-particle case in order to establish the stability of bulk matter. The important new fact that will be involved is that the N particles are *fermions*; that is to say the N-particle wave function is an antisymmetric function of the N-space, spin variables.

II. EXTENSION OF THE UNCERTAINTY PRINCIPLE TO MANY FERMIONS

A well known elementary calculation is that of the lowest kinetic energy, T^V of N *fermions* in a cubic box of volume V. For large N one finds that

$$T^V \approx q^{-2/3} K^c V \rho^{5/3}, \qquad (15)$$

where $\rho = N/V$ and q is the number of spin states available to each particle ($q = 2$ for electrons). Equation (15) is obtained by merely adding up the N/q lowest eigenvalues of $-\Delta$ with Dirichlet ($\psi = 0$) boundary conditions on the walls of the box. The important feature of (15) is that it is proportional to $N^{5/3}$ instead of N, as would be the case if the particles were not fermions. The extra factor $N^{2/3}$ is essential for the stability of matter; if electrons were bosons, matter would not be stable.

Equation (15) suggests that Eq. (12), with a factor $q^{-2/3}$ ought to extend to the N-particle case if $\rho(x)$ is interpreted properly. The idea is old, going back to Lenz (1932), who got it from Thomas–Fermi theory. The proof that something like (12) is not only an approximation but is also a lower bound is new.

To say that the N particles are *fermions* with q spin states means that the N-particle wave function $\psi(x_1, \ldots, x_N; \sigma_1, \ldots, \sigma_N)$ defined for $x_i \in R^3$ and $\sigma_i \in \{1, 2, \ldots q\}$ is *antisymmetric* in the pairs (x_i, σ_i). The norm is given by

$$\langle\psi, \psi\rangle = \sum_{\sigma_i = 1}^{q} \int |\psi(x_1, \ldots, x_N; \sigma_1, \ldots, \sigma_N)|^2 dx_1 \ldots dx_N .$$

Define

$$T_\psi = N \sum_{\sigma_i = 1}^{q} \int |\nabla_1 \psi(x_1, \ldots, x_N; \sigma_1, \ldots, \sigma_N)|^2 dx_1 \ldots dx_N \qquad (16)$$

to be the usual kinetic energy of ψ and define

$$\rho_\psi(x) = N \sum_{\sigma_i = 1}^{q} \int |\psi(x, x_2, \ldots, x_N; \sigma_1, \ldots, \sigma_N)|^2 dx_2 \ldots dx_N \qquad (17)$$

to be the single particle density, i.e., the probability of finding a particle at x. The analog of (12) is the following (Lieb and Thirring, 1975):

Theorem 1. If $\langle\psi, \psi\rangle = 1$ then

$$T_\psi \geq (4\pi q)^{-2/3} K^c \int \rho_\psi(x)^{5/3} dx. \qquad (18)$$

Apart from the annoying factor $(4\pi)^{-2/3} \approx 0.185$, (18) says that the intuition behind considering (15) as a lower bound is correct. We believe that $(4\pi)^{-2/3}$ does not belong in (18) and hope to eliminate it someday. Recent work (Lieb, 1976) has improved the constant by a factor $(1.83)^{2/3} = 1.496$, so we are now off from the conjectured constant $q^{-2/3} K^c$ only by the factor 0.277.

The proof of Theorem 1 is not long but it is slightly tricky. It is necessary first to investigate the negative eigenvalues of a one-particle Schrödinger equation when the potential is nonpositive.

Theorem 2. Let $V(x) \leq 0$ be a potential for the one-particle, three dimensional Schrödinger operator $H = -\Delta + V(x)$ on $L^2(R^3)$. For $E < 0$ let $N_E(V)$ be the number of eigenstates of H with energies $\leq E$. Then

$$N_E(V) \leq (4\pi)^{-1}(2|E|)^{-1/2} \int |V(x) - E/2|_-^2 dx, \qquad (19)$$

where $|f(x)|_- = |f(x)|$ if $f(x) \leq 0$ and $|f(x)|_- = 0$ otherwise.

Corollary. If $e_1 \leq e_2 \leq \ldots \leq 0$ are the negative eigenvalues of H (if any) then

$$\sum_{j=1} |e_j| \leq \frac{4\pi}{15\pi^2} \int |V(x)|^{5/2} dx. \qquad (20)$$

Proof. $\sum |e_j| = \int_0^\infty N_{-\alpha}(V) d\alpha$. Insert (19) and do the α integration first and then the x integration. The result is (20). ■

We believe the factor (4π) does not belong in (20).

Proof of Theorem 2. From the Schrödinger equation $H\psi = e\psi$ it is easy to deduce that $N_E(V)$ is equal to the number of eigenvalues which are ≥ 1 of the positive definite Birman–Schwinger operator (Birman, 1961; Schwinger, 1961)

$$B_E(V) = |V|^{1/2}(-\Delta - E)^{-1}|V|^{1/2}. \qquad (21)$$

Essentially Eq. (21) comes from the fact that if $H\psi = e\psi$ then $(-\Delta - e)\psi = |V|\psi$. If one defines $|V|^{1/2}\psi \equiv \phi$, then $B_e \phi = \phi$. Thus B_e has an eigenvalue 1 when e is an eigenvalue. However, B_E is a compact positive semidefinite operator on $L^2(R^3)$ for $E < 0$ and, as an operator, B_E is monotone increasing in E. Thus, if B_E has k eigenvalues ≥ 1, there exist k numbers $e_1 \leq e_2 \leq e_k \leq E$ such that

B_{e_j} has eigenvalue 1.

Consequently $N_E(V) \leqslant \mathrm{Tr} B_E(V)^2$. On the other hand, $N_E(V) \leqslant N_{E/2}(-|V-E/2|_-)$ by the variational principle (draw a graph of $V(x)-E/2$). Thus, since $B_E(V)$ has a kernel, $B_E(x,y)=|V(x)|^{1/2}\exp\{-|E|^{1/2}|x-y|\}[4\pi|x-y|]^{-1} \times |V(y)|^{1/2}$, one has that

$$\begin{aligned} N_E(V) &\leqslant \mathrm{Tr} B_{E/2}(-|V-E/2|_-)^2 \\ &= (4\pi)^{-2}\int\int dxdy\, |V(x)-E/2|_-|V(y)-E/2|_- \\ &\quad \times \exp\{-(2|E|)^{1/2}|x-y|\}|x-y|^{-2}. \end{aligned} \tag{22}$$

Equation (19) results from applying Young's inequality[3] to Eq. (22). Alternatively, one can do the convolution integral by Fourier transforms and note that the Fourier transform of the last factor has a maximum at $p=0$, where it is $4\pi(2|E|)^{-1/2}$. ■

Using (20), which is a statement about the energy levels of a single particle Hamiltonian, we can, surprisingly, prove Theorem 1, which refers to the kinetic energy of N fermions.

Proof of Theorem 1. ψ and hence $\rho_\psi(x)$ are given. Consider the non-positive single particle potential $V(x) \equiv -\alpha\rho_\psi(x)^{2/3}$ where α is given by $(2/3\pi)q\alpha^{3/2}=1$. Next consider the following N-particle Hamiltonian:

$$\tilde{H}_N=\sum_{i=1}^N h_i;\quad h_i=-\Delta_i+V(x_i)$$

on $L^2(\mathbf{R}^3;\mathbf{C}^q)^N$. If E_0 is the *fermion* ground state energy of $\tilde{H}_N$, we have that $E_0 \geqslant q\sum e_j$, where the e_j are the negative eigenvalues of the single particle Hamiltonian h. (We merely fill the lowest negative energy levels q times until the N particles are accounted for; if there are k such levels and if $N<kq$ then $E_0>q\sum e_j$. If $N>kq$, the surplus particles can be placed in wave packets far away from the origin with arbitrarily small kinetic energy.) On the other hand, $E_0 \leqslant \langle\psi,\tilde{H}_N\psi\rangle = T_\psi - \alpha\int\rho_\psi(x)^{5/3}dx$ by the variational principle. If these two inequalities are combined together with (20), which says that $\sum e_j \geqslant -(4/15\pi)\alpha^{5/2}\int\rho_\psi(x)^{5/3}dx$, then (18) is the result. ■

It might not be too much out of place to explain at this point why K^c is called the classical constant. The name does not stem from its antiquity, as in the ideal gas kinetic energy (15), but rather from classical mechanics—more precisely the semiclassical approximation to quantum mechanics. This intuitive idea is valuable.

As the proof of Theorem 1 shows, the constant in (18) for T_ψ is simply related to the constant in (20) for the sum of the eigenvalues. The point is that the semiclassical approximation to this sum is

$$\sum|e_j| \approx (15\pi^2)^{-1}\int|V(x)|^{5/2}dx,$$

and this, in turn, would yield (18) without the $(4\pi)^{-2/3}$ factor. The semiclassical approximation is obtained by saying that a region of volume $(2\pi)^3$ in the six-dimensional phase space (p,x) can accommodate one eigenstate. Hence, integrating over the set $\theta(H)$, in which $H(p,x)=p^2+V(x)$ is negative,

$$\begin{aligned} \sum e_j &\approx (2\pi)^{-3}\int\int_{\theta(H)} dx\,dp\,\{p^2+V(x)\} \\ &= (2\pi)^{-3}\int dx\, 4\pi\int_0^{|V(x)|^{1/2}} p^2\,dp\,\{p^2+V(x)\} \\ &= -(15\pi^2)^{-1}\int|V(x)|^{5/2}dx. \end{aligned}$$

If a coupling constant g is introduced, and if V is replaced by gV, then it is a theorem that the semiclassical approximation is asymptotically exact as $g\to\infty$ for any V in $L^{5/2}(\mathbf{R}^3)$.

Theorem 1 gives a lower bound to the kinetic energy of fermions which is crucial for the H stability of matter as developed in Sec. IV. To appreciate the significance of Theorem 1 it should be compared with the one-particle Sobolev bound (12). Suppose that $\rho(x)=0$ outside some fixed domain Ω of volume V. Then since

$$\int_\Omega \rho(x)^{5/3}dx \geqslant \left\{\int_\Omega\rho(x)\,dx\right\}^{5/3}\left\{\int_\Omega 1\right\}^{-2/3}=N^{5/3}V^{-2/3}$$

by Hölder's inequality, one sees that T_ψ grows at least as fast as $N^{5/3}$. Using Eq. (12) alone, one would only be able to conclude that T_ψ grows as N. This distinction stems from the Pauli principle, i.e., the antisymmetric nature of the N-particle wave function. As we shall see, this $N^{5/3}$ growth is essential for the stability of matter because without it the ground state energy of N particles with Coulomb forces would grow at least as fast as $-N^{7/5}$ instead of $-N$.

The Fermi pressure is needed to prevent a collapse, but to learn how to exploit it we must first turn to another chapter in the theory of Coulomb systems, namely Thomas–Fermi theory.

III. THOMAS–FERMI THEORY

The statistical theory of atoms and molecules was invented independently by Thomas and Fermi (Thomas, 1927; Fermi, 1927). For many years the TF theory was regarded as an uncertain approximation to the N-particle Schrödinger equation and much effort was devoted to trying to determine its validity (e.g., Gombás, 1949). It was eventually noticed numerically (Sheldon, 1955) that molecules did not appear to bind in this theory, and then Teller (1962) proved this to be a general theorem.

It is now understood that TF theory is really a large Z theory (Lieb and Simon, 1977); to be precise it is exact in the limit $Z\to\infty$. For finite Z, TF theory is qualita-

---

[3]Young's inequality states that

$$\left|\int\int f(x)g(x-y)h(y)\,dxdy\right| \leq \left\{\int|f(x)|^p dx\right\}^{1/p} \times\left\{\int|g(x)|^q\right\}^{1/q} \times\left\{\int|h(x)|^r dx\right\}^{1/r}$$

when $p^{-1}+q^{-1}+r^{-1}=2$ and $p,q,r\geq 1$. For (22) take $p=r=2$ and $q=1$.

Rev. Mod. Phys. *48*, 553-569 (1976)

tively correct in that it adequately describes the bulk of an atom or molecule. It is not precise enough to give binding. Indeed, it should not do so because binding in TF theory would imply that the cores of atoms bind, and this does not happen. Atomic binding is a fine quantum effect. Nevertheless, TF theory deserves to be well understood because it is exact in a limit; the TF theory is to the many-electron system as the hydrogen atom is to the few-electron system. For this reason the main features of the theory are presented here, mostly without proof.

A second reason for our interest in TF theory is this: in the next section the problem of the H stability of matter will be reduced to a TF problem. The knowledge that TF theory is H stable (this is a corollary of the no binding theorem) will enable us to conclude that the true quantum system is H stable.

The Hamiltonian for N electrons with k static nuclei of charges $z_i > 0$ and locations R_i is

$$H_N = \sum_{i=1}^{N} -\Delta_i - V(x_i) + \sum_{1\le i<j\le N} |x_i - x_j|^{-1} + U(\{z_j, R_j\}_{j=1}^k), \tag{23}$$

where

$$V(x) = \sum_{j=1}^{k} z_j |x - R_j|^{-1}, \tag{24a}$$

and

$$U(\{z_j, R_j\}_{j=1}^k) = \sum_{1\le i<j\le k} z_i z_j |R_i - R_j|^{-1}. \tag{24b}$$

The nuclear–nuclear repulsion U is, of course, a constant term in H_N but it is included for two reasons:

(i) We wish to consider the dependence on the R_i of

$$E_N^Q(\{z_j, R_j\}_{j=1}^k) \equiv \text{the ground state energy of } H_N. \tag{25}$$

(ii) Without U the energy will not be bounded by N.

The nuclear kinetic energy is not included in H_N. For the H-stability problem we are only interested in finding a lower bound to E_N^Q, and the nuclear kinetic energy adds a positive term. In other words,

$$\inf_{\{R_i\}} E_N^Q(\{z_j, R_j\}_{j=1}^k)$$

is smaller than the ground state energy of the true Hamiltonian [defined in Eq. (58)] in which the nuclear kinetic energy is included. Later on when we do the proper thermodynamics of the whole system we shall have to include the nuclear kinetic energy.

The problem of estimating E_N^Q is as old as the Schrödinger equation. The TF theory, as interpreted by Lenz (1932), reads as follows: For fermions having q spin states ($q = 2$ for electrons) define the *TF energy functional*:

$$\mathcal{E}(\rho) = q^{-2/3} K^c \int \rho(x)^{5/3} - \int V(x)\rho(x) + \tfrac{1}{2}\int\int \rho(x)\rho(y)|x-y|^{-1}dxdy + U(\{z_j, R_j\}_{j=1}^k) \tag{26}$$

for *non-negative* functions $\rho(x)$. Then for $\lambda \ge 0$

$$E_\lambda^{TF} \equiv \inf\left\{\mathcal{E}(\rho): \int \rho(x)dx = \lambda\right\} \tag{27}$$

is the TF energy for λ electrons (λ need not be an integer, of course). When $\lambda = N$, the minimizing ρ is supposed to approximate the ρ_ψ given by (17), wherein ψ is the true ground state wave function, and E_N^{TF} is supposed to approximate E_N^Q.

The intuitive idea behind TF theory is this: If ψ is any fermion wave function and T_ψ and ρ_ψ are given by Eqs. (16) and (17), then the first term in (26) is supposed to approximate T_ψ. This is based on the box kinetic energy (15). The last three terms in (26) represent, respectively, the electron–nuclear, electron–electron, and nuclear-nuclear Coulomb energy. E_λ^{TF} in (27) is then the "ground state energy" of (26).

The second and fourth terms on the right side of (26) are exact but the first and third are not. The first is to some extent justified by the kinetic energy inequality, Theorem 1; the third term will be discussed later. In any event, Eqs. (26) and (27) *define* TF theory.

It would be too much to try to reproduce here the details of our analysis of TF theory. A short summary of some of the main theorems will have to suffice.

The first question is whether or not E_λ^{TF} (which, by simple estimates using Young's and Hölder's inequalities, can be shown to be finite for all λ) is a minimum as distinct from merely an infimum. The distinction is crucial because the *TF equation* [the Euler–Lagrange equation for (26) and (27)]

$$\tfrac{5}{3}K^c q^{-2/3}\rho^{2/3}(x) = \max\{\phi(x) - \mu, 0\} \tag{28}$$

with

$$\phi(x) = V(x) - \int \rho(y)|x-y|^{-1}dy \tag{29}$$

has a solution with $\int \rho = \lambda$ if and only if there is a minimizing ρ for E_λ^{TF}. The basic theorem is as follows.

Theorem 3. If $\lambda \le Z \equiv \sum_{j=1}^k z_j$ then

(i) $\mathcal{E}(\rho)$ has a minimum on the set $\int \rho(x)dx = \lambda$.

(ii) This minimizing ρ (call it ρ_λ^{TF}) is unique and satisfies (28) and (29). μ is non-negative, and $-\mu$ is the chemical potential, i.e.,

$$-\mu = dE_\lambda^{TF}/d\lambda. \tag{30}$$

(iii) There is no other solution to (28) and (29) (for any μ) with $\int \rho = \lambda$ other than ρ_λ^{TF}.

(iv) When $\lambda = Z$, $\mu = 0$. Otherwise $\mu > 0$, i.e., E_λ^{TF} is strictly decreasing in λ.

(v) As λ varies from 0 to Z, μ varies continuously from $+\infty$ to 0.

(vi) μ is a convex, decreasing function of λ.

(vii) $\phi_\lambda^{TF}(x) > 0$ for all x and λ. Hence when $\lambda = Z$

$$\tfrac{5}{3}K^c q^{-2/3}\rho_Z^{TF}(x)^{2/3} = \phi_Z^{TF}(x).$$

If $\lambda > Z$ then $E^{TF}(\lambda)$ is not a minimum and (28) and (29) have no solution with $\int \rho = \lambda$. Negative ions do not exist in TF theory. Nevertheless, E_λ^{TF} exists and $E_\lambda^{TF} = E_Z^{TF}$ for $\lambda \ge Z$.

The proof of Theorem 3 is an exercise in functional

analysis. Basically, one first shows that $\mathcal{E}(\rho)$ is bounded below so that $E_\lambda^{\rm TF}$ exists. The Banach–Alaoglu theorem is used to find an $L^{5/3}$ weakly convergent sequence of ρ's such that $\mathcal{E}(\rho)$ converges to $E_\lambda^{\rm TF}$. Then one notes that $\mathcal{E}(\rho)$ is weakly lower semicontinuous so that a minimizing ρ exists under the subsidiary condition that $\int\rho\le\lambda$. The uniqueness comes from an important property of $\mathcal{E}(\rho)$, namely that it is *convex*. This also implies that the minimizing ρ satisfies $\int\rho=\lambda$.

A major point to notice is that a solution of the TF *equation* is obtained as a by-product of minimizing $\mathcal{E}(\rho)$; a direct proof that the TF equation has a solution would be very complicated.

Only in the case $\lambda\le Z$ is $\rho_\lambda^{\rm TF}(x)$ positive for all x, when $\lambda<Z$, $\mu>0$ and, since $\phi_\lambda^{\rm TF}(x)$ goes to zero as $|x|$ goes to infinity, Eq. (28) implies that $\rho_\lambda^{\rm TF}(x)$ vanishes outside some bounded set.

Apart from the details presented in Theorem 3, the main point is that TF theory is well defined. In particular the density $\rho^{\rm TF}$ is unique—a state of affairs in marked contrast to that of Hartree–Fock theory (Hartree, 1927-28; Fock, 1930; Slater, 1930; Lieb and Simon, 1973).

The TF density $\rho_\lambda^{\rm TF}$ has the following properties:

Theorem 4. If $\lambda\le Z$ then

(i) $(\frac{5}{3})K^c q^{-2/3}\rho_\lambda^{\rm TF}(x)^{2/3}\sim z_i|x-R_i|^{-1}$ (31)

near each R_i.

(ii) In the neutral case, $\lambda=Z=\sum_{j=1}^k z_j$,

$$|x|^6\rho_Z^{\rm TF}(x)\to(3/\pi)^3[\tfrac{5}{3}K^c q^{-2/3}]^3 \tag{32}$$

as $|x|\to\infty$, irrespective of the distribution of the nuclei.

(iii) $\phi_\lambda^{\rm TF}(x)$ and $\rho_\lambda^{\rm TF}(x)$ are real analytic in x away from all the R_i, on all of 3-space in the neutral case and on $\{x:\phi_\lambda^{\rm TF}(x)>\mu\}$ in the positive ionic case.

Equation (32) is especially remarkable: at large distances one loses all knowledge of the nuclear charges and configuration. Property (i) recalls the singularity found in the minimization of $h^c(\rho)$ [see Eq. (13)].

Equation (31) can be seen from (28) and (29) by inspection. Equation (32) is more subtle but it is consistent with the observation that (28) and (29) can be rewritten (when $\mu=0$) as

$$-(4\pi)^{-1}\Delta\,\phi_Z^{\rm TF}(x)=-\{(\tfrac{3}{5})q^{2/3}\phi_Z^{\rm TF}(x)/K^c\}^{3/2}$$

away from the R_i. If it is *assumed* that $\phi_Z^{\rm TF}(x)$ goes to zero as a power of $|x|$ then (32) follows. This observation was first made by Sommerfeld (1932). The proof that a power law falloff actually occurs is somewhat subtle and involves potential theoretic ideas such as that used in the proof of Lemma 8.

As pointed out earlier, the connection between TF theory and the Schrödinger equation is best seen in the limit $Z\to\infty$. Let the number, k, of nuclei be held fixed, but let $N\to\infty$ and $z_i\to\infty$ in such a way that the degree of ionization N/Z is constant, where

$$Z=\sum_{j=1}^k z_j\,.$$

To this end we make the following definition: Fix $\{z_j,R_j\}_{j=1}^k$ and λ. It is not necessary to assume that $\lambda\le Z$. For each $N=1,2,\ldots$ define a_N by $\lambda a_N=N$. In H_N (23) replace z_j by z_ja_N and R_j by $R_ja_N^{-1/3}$. This means that the nuclei come together as $N\to\infty$. If they stay at fixed positions then that is equivalent, in the limit, to isolated atoms, i.e., it is equivalent to starting with all the nuclei infinitely far from each other. Finally, for the nuclear configuration $\{a_Nz_j,a_N^{-1/3}R_j\}_{j=1}^k$ let ψ_N be the ground state wave function, $E_N^{\rm Q}$ the ground state energy, and $\rho_N^{\rm Q}(x)$ be the single particle density as defined by Eq. (17).[4]

It is important to note that there is a simple and obvious scaling relation for TF theory, namely

$$E_{\lambda a}^{\rm TF}(\{az_j,a^{-1/3}R_j\}_{j=1}^k)=a^{7/3}E_\lambda^{\rm TF}(\{z_j,R_j\}_{j=1}^k) \tag{33}$$

and the densities for the two systems are related by

$$\rho_{\lambda a}^{\rm TF}(a^{-1/3}x)=a^2\rho_\lambda^{\rm TF}(x) \tag{34}$$

for any $a\ge0$. Hence, for the above sequence of systems parametrized by a_N,

$$a_N^{-7/3}E_N^{\rm TF}(\{a_Nz_j,a_N^{-1/3}R_j\}_{j=1}^k)=E_\lambda^{\rm TF}(\{z_j,R_j\}_{j=1}^k)\,, \tag{35}$$

$$a_N^{-2}\rho_N^{\rm TF}(a_N^{-1/3}x)=\rho_\lambda^{\rm TF}(x) \tag{36}$$

for all N.

If, on the other hand, the nuclei are held fixed then one can prove that

$$\lim_{N\to\infty}a_N^{-7/3}E_N^{\rm TF}(\{a_Nz_j,R_j\})=\sum_{j=1}^k E_{\lambda_j}^{\rm TF}(z_j)\,, \tag{37}$$

where $E_{\lambda_j}^{\rm TF}(z)$ is the energy of an isolated atom of nuclear charge z. The λ_j are determined by the condition that $\sum_{j=1}^k\lambda_j=\lambda$ if $\lambda\le Z$ (otherwise, $\sum_{j=1}^k\lambda_j=Z$) and that the chemical potentials of the k atoms are all the same. Another way to say this is that the λ_j minimize the right side of Eq. (37). With the nuclei *fixed*, the analog of (36) is

$$\lim_{N\to\infty}a_N^{-2}\rho_N^{\rm TF}(a^{-1/3}(x-R_j))=\rho_{\lambda_j}^{\rm TF}(x)\,. \tag{38}$$

The right side of Eq. (38) is the ρ for a single atom of nuclear charge z and electron charge λ_j. Equations (37) and (38) are a precise statement of the fact that isolated atoms result from fixing the R_j.

The TF energy for an isolated, *neutral atom* of nuclear charge Z is found numerically to be

$$E_Z^{\rm TF}=-(2.21)q^{2/3}(K^c)^{-1}Z^{7/3}\,. \tag{39}$$

For future use, note that $E_Z^{\rm TF}$ is proportional to $1/K^c$. Thus, if one considers a TF theory with K^c replaced by some other constant $\alpha>0$, as will be necessary in Sec. IV, then Eq. (39) is correct if K^c is replaced by α.

Theorem 5. With $a_N=N/\lambda$ and $\{z_j,R_j\}_{j=1}^k$ fixed

(i) $a_N^{-7/3}E_N^{\rm Q}(\{a_Nz_j,a_N^{-1/3}R_j\}_{j=1}^k)$ has a limit as $N\to\infty$.

(ii) This limit is $E_\lambda^{\rm TF}(\{z_j,R_j\}_{j=1}^k)$.

(iii) $a_N^{-7/3}E_N^{\rm Q}(\{a_Nz_j,R_j\}_{j=1}^k)$ has a limit as $N\to\infty$. This limit is the right side of (37).

(iv) $a_N^{-2}\rho_N^{\rm Q}(a_N^{-1/3}x;\{a_Nz_j,a_N^{-1/3}R_j\}_{j=1}^k)$ also has a limit as

[4] If $E_N^{\rm Q}$ is degenerate, ψ_N can be *any* ground state wave function as far as Theorem 5 is concerned. If $E_N^{\rm Q}$ is not an eigenvalue, but merely inf spec H_N, then it is possible to define an approximating sequence ψ_N, with $\rho_N^{\rm Q}$ still given by Eq. (17), in such a way that Theorem 5 holds. We omit the details of this construction here.

$N \to \infty$. If $\lambda \le Z = \sum_{j=1}^{k} z_j$, this limit is $\rho_\lambda^{TF}(x)$ and the convergence is in weak $L^1(\mathbf{R}^3)$. If $\lambda > Z$, the limit is $\rho_Z^{TF}(x)$ in weak $L^1_{loc}(\mathbf{R}^3)$.

(v) For fixed nuclei, $a_N^{-2}\rho_N^Q(a_N^{-1/3}(x - R_j); \{a_N z_j, R_j\}_{j=1}^k)$ has a limit [in the same sense as (iv)] which is the right side of (38).

The proof of Theorem 5 does not use anything introduced so far. It is complicated, but elementary. One partitions 3-space into boxes with sides of order $Z^{-1/3}$. In each box the potential is replaced by its maximum (respectively, mimimum) and one obtains an upper (respectively, lower) bound to E_N^Q by imposing Dirichlet ($\psi = 0$) (respectively, Neumann ($\nabla\psi = 0$)) boundary conditions on the boxes. The upper bound is essentially a Hartree-Fock calculation. The $-r^{-1}$ singularity near the nuclei poses a problem for the lower bound, and it is tamed by exploiting the concept of angular momentum barrier.

What Theorem 5 says, first of all, is that the true quantum energy has a limit on the order of $Z^{7/3}$ when the ratio of electron to nuclear charge is held fixed. Second, this limit is given correctly by TF theory as is shown in Eq. (35). The requirement that the nuclei move together as $Z^{-1/3}$ should be regarded as a refinement rather than as a drawback, for if the nuclei are fixed a limit also exists but it is an uninteresting one of isolated atoms.

Theorem 5 also says that the density ρ_N^Q is proportional to Z^2 and has a scale length proportional to $Z^{-1/3}$. If $\lambda > Z$, Theorem 5 states that the surplus charge moves off to infinity and the result is a neutral molecule. This means that large atoms or molecules cannot have a *negative* ionization proportional to the total nuclear charge; at best they can have a negative ionization which is a vanishingly small fraction of the total charge. This result is physically obvious for electrostatic reasons, but it is nice to have a proof of it.

Theorem 5 also resolves certain "anomalies" of TF theory:

(a) In real atoms or molecules the electron density falls off exponentially, while in TF theory (Theorem 4) the density falls off as $|x|^{-6}$.

(b) The TF atom shrinks in size as $Z^{-1/3}$ [cf. Eq. (36)] while real large atoms have roughly constant size.

(c) In TF theory there is no molecular binding, as we shall show next, but there is binding for real molecules.

(d) In real moleucles the electron density is finite at the nuclei, but in TF theory it goes to infinity as $z_j |x - R_j|^{-3/2}$ (Theorem 4).

As Theorem 5 shows, TF theory is really a theory of heavy atoms or molecules. A large atom looks like a stellar galaxy, poetically speaking. It has a core which shrinks as $Z^{-1/3}$ and which contains most of the electrons. The density (on a scale of Z^2) is not finite at the nucleus because, as the simplest Bohr theory shows, the S-wave electrons have a density proportional to Z^3 which is infinite on a scale of Z^2. Outside the core is a mantle in which the density is proportional to (cf. Theorem 4) $(3/\pi)^3[(\frac{5}{3})K^c 2^{-2/3}]^3 Z^2/(Z^{1/3}|x|)^6$, which is *independent* of Z! This density is correct to infinite distances on a length scale $Z^{-1/3}$. The core and the mantle contain 100% of the electrons as $Z \to \infty$. The third region is a transition region to the outer shell, and while it may contain many electrons, it contains only a vanishingly small fraction of them. The fourth region is the outer shell in which chemistry and binding takes place. TF theory has nothing to say about this region. The fifth region is the one in which the density drops off exponentially.

Thus, TF theory deals only with the core and the mantle in which the bulk of the energy and the electrons reside. There ought not to be binding in TF theory, and indeed there is none, because TF energies are proportional to $Z^{7/3}$ and binding energies are of order one. The binding occurs in the fourth layer.

An important question is what is the next term in the energy beyond the $Z^{7/3}$ term of TF theory. Several corrections have been proposed: (e.g., Dirac, 1930; Von Weizsäcker, 1935; Kirzhnits, 1957; Kompaneets and Pavlovskii, 1956; Scott, 1952). With the exception of the last, all these corrections are of order $Z^{5/3}$. Scott (as late as 1952!) said there should be a $Z^{6/3}$ correction because TF theory is not able to treat correctly the innermost core electrons. Let us give a heuristic argument. Recall that in Bohr theory each inner electron alone has an energy proportional to Z^2. As these inner electrons are unscreeened, their energies should be independent of the presence or absence of the electron-electron repulsion. In other words, the Z^2 correction for a molecule should be precisely a sum of corrections, one for each atom. The atomic correction should be the difference between the Bohr energy and the $Z^{7/3}$ TF energy for an atom in which the electron-electron repulsion is *neglected*. We already calculated the TF energy for such an "atom" in Eq. (14) (put $Z = 1$ there and then use scaling; also replace K^c by $q^{-2/3}K^c$). Thus, for a *neutral* atom *without* electron-electron repulsion

$$\tilde{E}_Z^{TF} = -(3^{1/3}/4)q^{2/3}Z^{7/3}. \tag{40}$$

For the Bohr atom, each shell of energy $-Z^2/4n^2$ has n^2 states, so

$$\frac{Z}{q} = \frac{N}{q} = \sum_{n=1}^{L} n^2 + (L+1)^2\phi = \frac{L^3}{3} + \frac{L^2}{2} + \frac{L}{6} + (L+1)^2\phi$$

with $0 \le \phi \le 1$ being the fraction of the $(L+1)$th shell that is filled. One finds $L \approx (3Z/q)^{1/3} - \frac{1}{2} - \phi + o(1)$ and

$$E_Z^{Bohr} = -\frac{Z^2}{4} q\left\{\phi + \sum_{n=1}^{L} 1\right\} \approx \tilde{E}_Z^{TF} + \frac{q}{8}Z^2.$$

Thus, to the next order, the energy should be

$$E_N^Q(\{z_j, R_j\}_{j=1}^k) = E_N^{TF}(\{z_j, R_j\}_{j=1}^k) + \tfrac{1}{4}\sum_{j=1}^{k} z_j^2 + \text{lower order}, \tag{41}$$

since $q = 2$ for electrons. Note that $E_N^{TF} \sim q^{2/3}$ while the Scott correction is proportional to q.

It is remarkable that Eq. (41) gives a *precise conjecture* about the next correction. It is simple to understand physically, yet we do not have the means to prove it.

The third main fact about TF theory is that there is no binding. This was proved by Teller in 1962. Considering the effort that went into the study of TF theory since its inception in 1927, it is remarkable that the no binding phenomenon was not seriously noticed until the computer study of Sheldon in 1955. Teller's original proof involved some questionable manipulation with δ functions and for that reason his result was questioned. His ideas were basically right, however, and we have made them rigorous.

Theorem 6 (no binding). If there are at least two nuclei, write the nuclear attraction $V(x)=\sum_{j=1}^{k} z_j|x-R_j|^{-1}$ as the sum of two pieces, $V=V^1+V^2$ where $V^1(x)=\sum_{j=1}^{m} z_j|x-R_j|^{-1}$ and $1\leq m<k$. Let $E_\lambda^{\mathrm{TF},1}$ be the TF energy for the nuclei $1,\ldots,m$ (with $U=\sum_{1\leq i<j\leq m} z_i z_j|R_i-R_j|^{-1}$, of course) and let $E_\lambda^{\mathrm{TF},2}$ be the same for the nuclei $m+1,\ldots,k$. Given λ, let $\lambda_1\geq 0$ and $\lambda_2=\lambda-\lambda_1\geq 0$ be chosen to minimize the sum of the energies of the separate molecules, i.e., $E_{\lambda_1}^{\mathrm{TF},1}+E_{\lambda_2}^{\mathrm{TF},2}$. (If $\lambda=Z=\sum_{j=1}^{k} z_j$ then by Theorem 3, $\lambda_1=\sum_{j=1}^{m} z_j$.) Then

$$E_\lambda^{\mathrm{TF}}\geq E_{\lambda_1}^{\mathrm{TF},1}+E_{\lambda_2}^{\mathrm{TF},2}\,. \tag{42}$$

Since the right side of Eq. (42) is the energy of two widely separated molecules, with the relative nuclear positions unchanged within each molecule, Theorem 6 says that the TF energy is unstable under *every* decomposition of the big molecule into smaller molecules. In particular, a molecule is unstable under decomposition into isolated atoms, and Theorem 9 is a simple consequence of this fact. One would suppose that if λ and the z_j are fixed, but the R_j are replaced by αR_j then

$$E_\lambda^{\mathrm{TF}}(\{z_j,\alpha R_j\}_{j=1}^{k}) \text{ is monotone decreasing in } \alpha\,.$$

In other words, the "pressure" is always positive. This is an unproved *conjecture*, but it has been proved (Balàzs, 1967) in the case $k=2$ and $z_1=z_2$.

An interesting side remark is the following.

Theorem 7. If the TF energy (26), (27) is redefined by excluding the repulsion term U in (26), then the inequality in (42) is reversed.

Thus, the nuclear repulsion is essential for the no binding theorem 6.

Another useful fact for some further developments of the theory, especially the TF theory of solids and the TF theory of screening (Lieb and Simon, 1977) is the following lemma (also attributed to Teller), which is used to prove the main no binding theorem 6.

Lemma 8. Fix $\{R_j\}_{j=1}^{k}$ and fix $\mu\geq 0$ in the TF equation (28) but not $\{z_j\}_{j=1}^{k}$. (This means that as the z_j's are varied λ will vary, but always $0\leq\lambda\leq Z=\sum z_j$. If $\mu=0$ then $\lambda=Z$ always.) If $\{z_j^1\}_{j=1}^{k}$ and $\{z_j^2\}_{j=1}^{k}$ are two sets of z's such that

$$z_j^1\leq z_j^2 \text{ all } j, \text{ and } z_1^1<z_1^2$$

and if λ_1 and λ_2 are the corresponding λ's for the two sets, then for all x

$$\phi_{\lambda_1}^{\mathrm{TF}}(x)\leq\phi_{\lambda_2}^{\mathrm{TF}}(x)$$

and hence

$$\rho_{\lambda_1}^{\mathrm{TF}}(x)\leq\rho_{\lambda_2}^{\mathrm{TF}}(x)\,.$$

There is strict inequality when $\mu=0$. In short, increasing some z_j increases the density everywhere, not just on the average.

The proof of Lemma 8 involves a beautifully simple potential theoretic argument which we cannot resist giving.

Proof of Lemma 8. We want to prove $\phi_1^{\mathrm{TF}}(x)\leqslant\phi_2^{\mathrm{TF}}(x)$ for all x and will content ourselves here with proving only $\leqslant$ when $\mu=0$. Let $B=\{x:\phi_1^{\mathrm{TF}}(x)>\phi_2^{\mathrm{TF}}(x)\}$. B is an open set and B does not contain any R_i for which $z_i^1<z_i^2$ by the TF equation (29). Let $\psi(x)=\phi_1^{\mathrm{TF}}(x)-\phi_2^{\mathrm{TF}}(x)$. If $x\in B$ then $\psi(x)>0$ and, by (28), $\rho_1^{\mathrm{TF}}(x)\geqslant\rho_2^{\mathrm{TF}}(x)$. For $x\in B$, $-(4\pi)^{-1}\Delta\psi(x)=\rho_2^{\mathrm{TF}}(x)-\rho_1^{\mathrm{TF}}(x)\leqslant 0$, so ψ is *subharmonic* on B [i.e., $\psi(x)\leqslant$ the average of ψ on any sphere contained in B and centered at x]. Hence ψ has its maximum on the boundary of B or at ∞, at all of which points $\psi=0$. Therefore B is the empty set. ■

In the $\mu=0$ case it is easy to show how Theorem 6 follows from Lemma 8.

Proof of Theorem 6 when $\lambda=\sum_{j=1}^{k} z_j$. The proof when $\lambda<\sum z_j$ uses the same ideas but is more complicated. Since $\lambda=\sum z_j$ then $\lambda_1=\sum_{j=1}^{m} z_j$, $\lambda_2=\sum_{j=m+1}^{k} z_j$ and $\mu=0$ for all three systems. For $\alpha>0$ let $f(\alpha)=E^{\mathrm{TF}}(\alpha z_1,\ldots,\alpha z_m,z_{m+1},\ldots,z_k;R_1,\ldots,R_k)-E^{\mathrm{TF}}(\alpha z_1,\ldots,\alpha z_m;R_1,\ldots,R_m)-E^{\mathrm{TF}}(z_{m+1},\ldots,z_k;R_{m+1},\ldots,R_k)$, where the three E^{TF} are defined for neutral systems (i.e., $\mu=0$ for all α). The goal is to show that $f(1)\geqslant 0$. Since $f(0)=0$, it is enough to show that $df(\alpha)/d\alpha\geqslant 0$. From (26) and (27) it is true, and almost obvious, that

$$\frac{\partial E^{\mathrm{TF}}}{\partial z_i}=-\int\rho^{\mathrm{TF}}(y)|y-R_i|^{-1}dy+\sum_{j\neq i} z_j|R_i-R_j|^{-1}$$

$$=\lim_{x\to R_i}\phi^{\mathrm{TF}}(x)-z_i|x-R_i|^{-1}.$$

This is the TF version of the Feynman–Hellmann theorem; notice how the nuclear–nuclear repulsion comes in here. Thus,

$$\frac{df(\alpha)}{d\alpha}=\sum_{i=1}^{m}\lim_{x\to R_i} z_i\eta_\alpha(x)\,,$$

where $\eta_\alpha(x)=\phi_1^{\mathrm{TF}}(x)-\phi_2^{\mathrm{TF}}(x)$ and ϕ_1^{TF} is the potential for $\{\alpha z_1,\ldots,\alpha z_m,z_{m+1},\ldots,z_k;R_1,\ldots,R_k\}$ and ϕ_2^{TF} is the potential for $\{\alpha z_1,\ldots,\alpha z_m;R_1,\ldots,R_m\}$. $\phi_1^{\mathrm{TF}}(x)\geqslant\phi_2^{\mathrm{TF}}(x)$ for all x by Lemma 8, and hence $\eta_\alpha(x)\geqslant 0$. ■

Theorem 6 has a natural application to the stability of matter problem. As will be shown in the next section, the TF energy (27) is, with suitably modified constants, a lower bound to the true quantum energy E_N^Q for *all Z*. By Theorem 3 (iv) and Theorem 6 we have the following theorem.

Theorem 9. Fix $\{z_j,R_j\}_{j=1}^{k}$ and let $Z=\sum_{j=1}^{k} z_j$. Then for all $\lambda\geqslant 0$

$$E_\lambda^{\mathrm{TF}}\geqslant E_Z^{\mathrm{TF}}\geqslant-(2.21)q^{2/3}(K^c)^{-1}\sum_{j=1}^{k} z_j^{7/3}. \tag{43}$$

The latter constant, 2.21, is obtained by numerically solving the TF equation for a single, neutral atom (J. F. Barnes, private communication). By scaling, Eq. (43)

holds for an choice of K^c in the definition (26) of $\mathcal{E}(\rho)$.

Theorem 9 is what will be needed for the H stability of matter because it says that the TF system is H stable, i.e., the energy is bounded below by a constant times the nuclear particle number (assuming that the z_j are bounded, of course).

Another application of Theorem 6 that will be needed is the following strange *inversion of the role of electrons and nuclei* in TF theory. It will enable us to give a lower bound to the true quantum-mechanical *electron-electron* repulsion. This theorem has nothing to do with quantum mechanics per se; it is really a theorem purely about electrostatics even though it is derived from the TF no binding theorem.

Theorem 10. Suppose that $x_1, \ldots, x_N$ are any N distinct points in 3-space and define

$$V_X(y) = \sum_{j=1}^{N} |y - x_j|^{-1}. \tag{44}$$

Let $\gamma > 0$ and let $\rho(x)$ be any non-negative function such that $\int \rho(x)\,dx < \infty$ and $\int \rho(x)^{5/3} dx < \infty$. Then

$$\sum_{1\le i<j\le N} |x_i - x_j|^{-1} \ge -\frac{1}{2}\int\int \rho(x)|x-y|^{-1}\rho(y)\,dx\,dy + \int \rho(y)V_X(y)\,dy - (2.21)N/\gamma - \gamma\int \rho(y)^{5/3} dy. \tag{45}$$

Proof. Consider $\mathcal{E}(\rho)$ (26) with $q=1$, $k=N$, K^c replaced by γ, $z_j \equiv 1$ and $R_j \equiv x_j$, $j=1,\ldots,N$. Let $\lambda = \int \rho(x)\,dx$. Then $\mathcal{E}(\rho) \ge E_\lambda^{\mathrm{TF}}$ (by definition) and $E_\lambda^{\mathrm{TF}} \ge -(2.21)N/\gamma$ by Theorem 9. The difference of the two sides in Eq. (45) is just $\mathcal{E}(\rho) + (2.21)N/\gamma$. ■

IV. THE STABILITY OF BULK MATTER

The various results of the last two sections can now be assembled to prove that the ground state energy (or infimum of the spectrum, if this not an eigenvalue) of H_N is bounded below by an extensive quantity, namely the total number of particles, independent of the nuclear locations $\{R_j\}$. This is called the H stability of matter to distinguish it from thermodynamic stability introduced in the next section. As explained before, the inclusion of the nuclear kinetic energy, as will be done in the next section, can only raise the energy.

The first proof of the N boundedness of the energy was given by Dyson and Lenard (Dyson and Lenard, 1967, Lenard and Dyson, 1968). Their proof is a remarkable analytic tour de force, but a chain of sufficiently many inequalities was used that they ended up with an estimate of something like -10^{14} Ry/particle. Using the results of the previous sections we will end up with -23 Ry/particle [see Eq. (55)].

We have in mind, of course, that the nuclear charges z_j, if they are not all the same, are bounded above by some fixed charge z.

Take any fermion $\psi(x_1, \ldots, x_N; \sigma_1, \ldots, \sigma_N)$ which is normalized and antisymmetric in the (x_i, σ_i). Define the kinetic energy T_ψ and the single particle density ρ_ψ as in (16) and (17). We wish to compute a lower bound to

$$E_\psi^Q \equiv \langle\psi, H_N\psi\rangle \tag{46}$$

with H_N being the N-particle Hamiltonian given in (23) and $\langle\psi,\psi\rangle = 1$.

For the third term on the right side of (23) Theorem 10 can be used with ρ taken to be ρ_ψ. Then, for any $\gamma > 0$

$$\left\langle\psi, \sum_{1\le i<j\le N} |x_i - x_j|^{-1}\psi\right\rangle \ge \frac{1}{2}\int\int \rho_\psi(x)|x-y|^{-1}\rho_\psi(y)\,dx\,dy - (2.21)N\gamma^{-1} - \gamma\int \rho_\psi(y)^{5/3} dy. \tag{47}$$

Notice how the first and second terms on the right side of (45) combine to give $+\frac{1}{2}$ since

$$\left\langle\psi, \left\{\int \rho_\psi(y)V_X(y)\,dy\right\}\psi\right\rangle = \int\int \rho_\psi(x)|x-y|^{-1}\rho_\psi(y)\,dx\,dy. \tag{48}$$

To control the kinetic energy in (23) Theorem 1 is used; the total result is then

$$E_\psi^Q \ge \alpha\int \rho_\psi(x)^{5/3} dx - \int V(x)\rho_\psi(x)\,dx + \frac{1}{2}\int\int \rho_\psi(x)|x-y|^{-1}\rho_\psi(y)\,dx\,dy + U(\{z_j, R_j\}_{j=1}^{k}) - (2.21)N\gamma^{-1} \tag{49}$$

with

$$\alpha = (4\pi q)^{-2/3}K^c - \gamma. \tag{50}$$

Restrict γ, which was arbitrary, so that $\alpha > 0$. Then, apart from the constant term $-(2.21)N\gamma^{-1}$, Eq. (49) is just $\mathcal{E}_\alpha(\rho_\psi)$, the Thomas–Fermi energy functional $\mathcal{E}$ applied to ρ_ψ, but with $q^{-2/3}K^c$ replaced by α. Since $\mathcal{E}_\alpha(\rho_\psi) \ge E_{\alpha,N}^{\mathrm{TF}} \equiv \inf\{\mathcal{E}_\alpha(\rho): \int\rho = N\}$ (by definition), and since the neutral case always has the lowest TF energy, as shown in Theorem 9, we have that

$$\mathcal{E}_\alpha(\rho_\psi) \ge -(2.21)\alpha^{-1}\sum_{j=1}^{k} z_j^{7/3}. \tag{51}$$

Thus we have proved the following:

Theorem 11. If ψ is a normalized, antisymmetric function of space and spin of N variables, and if there are q spin states associated with each particle then, for any $\gamma > 0$ such that α defined by Eq. (50) is positive,

$$\langle\psi, H_N\psi\rangle \ge -(2.21)\left\{N\gamma^{-1} + \alpha^{-1}\sum_{j=1}^{k} z_j^{7/3}\right\}. \tag{52}$$

The optimum choice for γ is

$$\gamma = (4\pi q)^{-2/3}K^c\left[\left(\sum_{j=1}^{k}\frac{z_j^{7/3}}{N}\right)^{1/2} + 1\right]^{-1}$$

in which case

$$E_N^Q \ge -(2.21)\frac{(4\pi q)^{2/3}N}{K^c}\left\{1 + \left[\sum_{j=1}^{k}\frac{z_j^{7/3}}{N}\right]^{1/2}\right\}^2. \tag{53}$$

This is the desired result, but some additional remarks are in order.

(1) Since $[1 + a^{1/2}]^2 \le 2 + 2a$,

$$E_N^Q \geq -(4.42)(4\pi q)^{2/3}(K_c)^{-1}\left\{N + \sum_{j=1}^{k} z_j^{7/3}\right\}. \quad (54)$$

Thus, provided the nuclear charges z_j are bounded above by some fixed z, E_N^Q is indeed bounded below by a constant times the *total* particle number $N+k$.

(2) Theorem 11 does not presuppose neutrality.

(3) For electrons, $q=2$ and the prefactor in Eq. (53) is $-(2.08)N$. As remarked after Theorem 1, the unwanted constant $(4\pi)^{2/3}$ has been improved to $[4\pi/(1.83)]^{2/3}$. Using this, the prefactor becomes $-(1.39)N$. If $z_j = 1$ (hydrogen atoms) and $N=k$ (neutrality) then

$$E_N^Q \geq -(5.56)N = -(22.24)N \text{ Ry}. \quad (55)$$

(4) The power law $z^{7/3}$ cannot be improved upon for large z because Theorem 5 asserts that the energy of an atom is indeed proportional to $z^{7/3}$ for large z.

(5) It is also possible to show that matter is indeed bulky. This will be proved for any ψ and any nuclear configuration (not just the minimum energy configuration) for which $E_\psi^Q \leq 0$. The minimizing nuclear configuration is, of course, included in this hypothesis. Then

$$0 \geq E_\psi^Q = \tfrac{1}{2} T_\psi + \langle \psi, H'_N \psi \rangle,$$

where H'_N is Eq. (23) but with a factor $\frac{1}{2}$ multiplying $\sum_{i=1}^N \Delta_i$. By Theorem 11, $\langle \psi, H'_N \psi \rangle \geq 2E_N$, where E_N is the right side of Eq. (53) (replace K^c by $K^c/2$ there). Therefore, the first important fact is that

$$T_\psi \leq 4|E_N|,$$

and this is bounded above by the total particle number.

Next, for any $p \geq 0$, it is easy to check that there is a $C_p > 0$ such that for any non-negative $\rho(x)$,

$$\left\{\int \rho(x)^{5/3} dx\right\}^{p/2} \int |x|^p \rho(x)\, dx \geq C_p \left\{\int \rho(x)\, dx\right\}^{1+5p/6}.$$

It is easy to find a minimizing ρ for this and to calculate C_p: $\rho(x)^{2/3} = 1 - |x|^p$ for $|x| \leq 1$; $\rho(x)=0$, otherwise.

Since T_ψ satisfies Eq. (18) we have that

$$\left\langle \psi, \sum_{i=1}^{N} |x_i|^p \psi \right\rangle = \int |x|^p \rho_\psi(x)\, dx \geq C'_p N (N^{5/3}/|E_N|)^{p/2},$$

with $C'_p = C_p (K^c/4)^{p/2} (4\pi q)^{-p/3}$.

If it is assumed that $\sum z_j^{7/3}/N$ is bounded, and hence that $(N^{5/3}/|E_N|)^{p/2} > A N^{p/3}$ for some A, we reach the conclusion that the radius of the system is at least of the order $N^{1/3}$, as it should be.

The above analysis did not use any specific property of the Coulomb potential, such as the virial theorem. It is also applicable to the more general Hamiltonian $H_{n,k}$ in Eq. (58).

(6) The q dependence was purposely retained in Eq. (53) in order to say something about *bosons*. If $q=N$, then it is easy to see that the requirement of antisymmetry in ψ is *no restriction* at all. In this case then, one has simply

$$E_N^Q = \inf \text{spec } H_N$$

over all of $L^2(\mathbf{R}^3)^N$. Therefore

$$E_N^Q(\text{bosons}) \geq \frac{-(2.21)(4\pi)^{2/3}}{K^c} N^{5/3} \left\{1 + \left[\sum_{j=1}^{k} \frac{z_j^{7/3}}{N}\right]^{1/2}\right\}^2. \quad (56)$$

It was shown by Dyson and Lenard (Dyson and Lenard, 1967) that

$$E_N^Q(\text{bosons}) \geq -(\text{const})N^{5/3},$$

and by Dyson (Dyson, 1967) that

$$E_N^Q(\text{bosons}) \leq -(\text{const})N^{7/5}. \quad (57)$$

Proving Eq. (57) was not easy. Dyson had to construct a rather complicated variational function related to the type used in the BCS theory of superconductivity. Therefore *bosons are not stable* under the action of Coulomb forces, but the exact power law is not yet known. Dyson has conjectured that it is $\frac{7}{5}$.

In any event, the essential point has been made that Fermi statistics is essential for the stability of matter. The uncertainty principle for one particle, even in the strong form (5), together with intuitive notions that the electrostatic energy ought not to be very great, are insufficient for stability. The additional physical fact that is needed is that the kinetic energy increases as the $\frac{5}{3}$ power of the fermion density.

V. THE THERMODYNAMIC LIMIT

Having established that E_N^Q is bounded below by the total particle number, the next question to consider is whether, under appropriate conditions, E_N^Q/N has a limit as $N \to \infty$, as expected. More generally, the same question can be asked about the free energy per particle when the temperature is not zero and the particles are confined to a box.

It should be appreciated that the difficulty in obtaining the lower bound to E_N^Q came almost entirely from the r^{-1} *short range* singularity of the Coulomb potential. Other potentials, such as the Yukawa potential, with the same singularity would present the same difficulty which would be resolved in the same way. The singularity was tamed by the $\rho^{5/3}$ behavior of the fermion kinetic energy.

The difficulty for the thermodynamic limit is different. It is caused by the *long range* r^{-1} behavior of the Coulomb potential. In other words, we are faced with the problem of explosion rather than implosion. Normally, a potential that falls off with distance more slowly than $r^{-3-\epsilon}$ for some $\epsilon > 0$ does *not* have a thermodynamic limit. Because the charges have different signs, however, there is hope that a cancellation at large distances may occur.

An additional physical hypothesis will be needed, namely *neutrality*. To appreciate the importance of neutrality consider the case that the electrons have positive, instead of negative charge. Then $E_N^Q > 0$ because every term in Eq. (23) would be positive. While the H-stability question is trivial in this case, the thermodynamic limit is not. If the particles are constrained to be in a domain Ω whose volume $|\Omega|$ is proportional to N, the particles will repel each other so strongly that they will all go to the boundary of Ω in order to minimize the electrostatic energy. The minimum electrostatic energy will be of the order $+N^2|\Omega|^{-1/3}\sim +N^{5/3}$. Hence no thermodynamic limit will exist.

When the system is neutral, however, the energy can be expected to be extensive, i.e., $O(N)$. For this to be so, different parts of the system far from each other must be approximately independent, despite the long range nature of the Coulomb force. The fundamental physical, or rather electrostatic, fact that underlies this is *screening;* the distribution of the particles must be sufficiently neutral and isotropic locally so that according to Newton's theorem (13 below) the electric potential far away will be zero. The problem is to express this idea in precise mathematical form.

We begin by defining the Hamiltonian for the *entire* system consisting of k nuclei, each of charge z and mass M, and n electrons ($\hbar^2/2=1$, $m=1$, $|e|=1$):

$$H_{n,k} = -\sum_{j=1}^{n}\Delta_j - \frac{1}{M}\sum_{j=n+1}^{n+k}\Delta_j - z\sum_{i=1}^{n}\sum_{j=n+1}^{n+k}|x_i-y_j|^{-1} + \sum_{1\le i<j\le n}|x_i-x_j|^{-1} + z^2\sum_{n+1\le i<j\le n+k}|y_i-y_j|^{-1}. \tag{58}$$

The first and second terms in Eq. (58) are, respectively, the kinetic energies of the electrons and the nuclei. The last three terms are, respectively, the electron–nuclear, electron–electron, and nuclear–nuclear Coulomb interactions. The electron coordinates are x_i and the nuclear coordinates are y_i. The electrons are fermions with spin $\frac{1}{2}$; the nuclei may be either bosons or fermions.

The basic neutrality hypotheses is that n and k are related by

$$n = kz\,. \tag{59}$$

It is assumed that z is rational.

The thermodynamic limit to be discussed here can be proved under more general assumptions, i.e., we can have several kinds of negative particles (but they must all be fermions in order that the basic stability estimate of Sec. IV holds) and several kinds of nuclei with different statistics, charges, and masses. Neutrality must always hold, however. Short range forces and hard cores, in addition to the Coulomb forces, can also be included with a considerable sacrifice in simplicity of the proof. See (Lieb and Lebowitz, 1972).

$H_{n,k}$ acts on square integrable functions of $n+k$ variables (and spin as well). To complete the definition of $H_{n,k}$ we must specify boundary conditions: choose a domain Ω (an open set, which need not be connected) and require that $\psi=0$ if x_i or y_i are on the boundary of Ω.

For each non-negative integer j, choose an n_j and a corresponding k_j determined by Eq. (59), and choose a domain Ω_j. The symbol N_j will henceforth stand for the *pair* (n_j, k_j) and

$$|N_j| \equiv n_j + k_j\,.$$

We require that the densities

$$\rho_j \equiv |N_j|\,|\Omega_j|^{-1} \tag{60}$$

be such that

$$\lim_{j\to\infty}\rho_j = \rho\,. \tag{61}$$

ρ is then the density in the thermodynamic limit. Here we shall choose the Ω_j to be a sequence of *balls* of radii R_j and shall denote them by B_j.

It can be shown tha the *same* thermodynamic limit for the energy and free enrgy holds for any sequence N_j, Ω_j and depends *only* on the limiting ρ and β, and not on the "shape" of the Ω_j, provided the Ω_j go to infinity in some reasonable way.

The basic quantity of interest is the *canonical partition function*

$$Z(N,\Omega,\beta) = \mathrm{Tr}\exp(-\beta H_{n,k})\,, \tag{62}$$

where the trace is on $L^2(\Omega)^{|N|}$ and $\beta = 1/T$, T being the temperature in units in which Boltzmann's constant is unity.

The *free energy per unit volume* is

$$F(N,\Omega,\beta) = -\beta^{-1}\ln Z(N,\Omega,\beta)/|\Omega| \tag{63}$$

and the problem is to show that with

$$F_j = F(N_j, \Omega_j, \beta)\,, \tag{64}$$

then

$$\lim_{j\to\infty} F_j \equiv F(\rho,\beta) \tag{65}$$

exists. A similar problem is to show that

$$E(N,\Omega) \equiv |\Omega|^{-1}\inf_{\psi}\langle\psi, H_{n,k}\psi\rangle/\langle\psi,\psi\rangle\,, \tag{66}$$

the *ground state energy per unit volume,* has a limit

$$e(\rho) = \lim_{j\to\infty} E_j\,, \tag{67}$$

where

$$E_j = E(N_j, \Omega_j)\,.$$

The proof we will give for the limit $F(\rho,\beta)$ will hold equally well for $e(\rho)$ because E_j can be substituted for F_j in all statements.

The basic strategy consists of two parts. The easiest part is to show that F_j is bounded below. We already know this for E_j by the results of Sec. IV. The second step is to show that in some sense the sequence F_j is decreasing. This will then imply the existence of a limit.

Theorem 12. Given N, Ω, and β there exists a constant C depending only on $\rho = |N|/|\Omega|$ and β such that

$$F(N,\Omega,\beta) \ge C\,. \tag{68}$$

Proof. Write $H_{n,k} = H_A + H_B$, where

$$H_A = -\frac{1}{2}\left\{\sum_{i=1}^{n}\Delta_i + \frac{1}{M}\sum_{j=n+1}^{n+k}\Delta_j\right\}$$

is half the kinetic energy. Then $H_B \ge b|N|$, with b depending only on z, by the results of Sec. IV (increasing the mass by a factor of 2 in H_B only changes the constant b). Hence $Z(N,\Omega,\beta) \le \exp(-\beta b|N|)\,\mathrm{Tr}\exp(-\beta H_A)$. However, $\mathrm{Tr}\exp(-\beta H_A)$ is the partition function of an ideal gas and it is known by explicit computation that it is bounded above by $e^{\beta d|N|}$ with d depending only on $\rho = |N|/|\Omega|$ and β. Thus

$$F(N,\Omega,\beta) \ge (b-d)\rho\,. \qquad \blacksquare$$

For the second step, two elementary but basic inequalities used in the general theory of the thermodynamic limit are needed and they will be described next.

A. Domain partition inequality: Given a domain Ω and the particle numbers $N=(n,k)$, let π be a partition of Ω into l disjoint domains $\Omega^1,\ldots,\Omega^l$. Likewise N is partitioned into l integral parts (some of which may be zero):

$$N=N^1+\cdots+N^l.$$

Then for any such partition, π, of Ω and N

$$Z(N,\Omega,\beta)=\mathrm{Tr}\exp(-\beta H_{n,k})\geq \mathrm{Tr}^{\pi}\exp(-\beta H_N^{\pi}). \tag{69}$$

Here Tr^{π} means trace over

$$\mathcal{H}^{\pi}\equiv L^2(\Omega^1)^{|N^1|}\otimes\cdots\otimes L^2(\Omega^l)^{|N^l|},$$

and H_N^{π} is defined as in (58) but with Dirichlet ($\psi=0$) boundary conditions for the N^i particles on the boundary of Ω^i (for $i=1,\ldots,l$).

Simply stated, the first N^1 particles are confined to Ω^1, the second N^2 to Ω^2, etc. The interaction among the particles in different domains is still present in H_N^{π}. Equation (69) can be proved by the Peierls–Bogoliubov variational principle for $\mathrm{Tr}e^X$. Alternatively, (69) can be viewed simply as the statement that the insertion of a hard wall, infinite potential on the boundaries of the Ω^i only decreases Z; the further restriction of a definite particle number to each Ω^i further reduces Z because it means that the trace is then over only the H_N^{π}-invariant subspace, $\mathcal{H}^{\pi}$, of the full Hilbert space.

B. Inequality for the interdomain interaction: The second inequality is another consequence of the convexity of $A\to \mathrm{Tr}e^A$ (Peierls–Bogoliubov inequality):

$$\mathrm{Tr}e^{A+B}\geq \mathrm{Tr}e^A\exp\langle B\rangle, \tag{70}$$

where

$$\langle B\rangle\equiv \mathrm{Tr}Be^A/\mathrm{Tr}e^A. \tag{71}$$

Some technical conditions are needed here, but Eqs. (70) and (71) will hold in our application.

To exploit (70), first make the same partition π as in inequality A and then write

$$H_N^{\pi}=H_0+W(X), \tag{72}$$

$$H_0=H^1+\cdots+H^l, \tag{73}$$

with H^i being that part of the total Hamiltonian (58) involving only the N^i particles in Ω^i, and H^i is defined with the stated Dirichlet boundary conditions on the boundary of Ω^i. $W(X)$, with X standing for all the coordinates, is the interdomain Coulomb interaction. In other words, $W(X)$ is that part of the last three terms on the right side of (58) which involves coordinates in different blocks of the partition π. Technically, W is a small perturbation of H_0.

With

$$A=-\beta H_0 \text{ and } B=-\beta W \tag{74}$$

in (70), we must calculate $\langle W\rangle$. Since $e^A=e^{-\beta H_0}$ is a simple tensor product of operators on each $L^2(\Omega^i)^{|N^i|}$, W is merely the *average* interdomain Coulomb energy in a canonical ensemble in which the Coulomb interaction is present in each subdomain but the l domains are independent of each other. In other words, let $q^i(x)$, $x\in\Omega^i$ denote the average *charge* density in Ω^i for this ensemble of independent domains, namely

$$q^i(x)=\sum_{j=1}^{|N^i|}q_j\int_{|\Omega^i|^{|N^i|-1}}\exp(-\beta H^i)(X^i,X^i)d\hat{x}_j/Z(N^i,\Omega^i,\beta) \tag{75}$$

with the following notation: X^i stands for the coordinates of the $|N^i|$ particles in Ω^i, $d\hat{x}_j$ means integration over all these coordinates (in Ω^i) with the exception of x_j, and x_j is set equal to x; q_j is the charge (-1 or $+z$) of the jth particle; $\exp(-\beta H^i)(X^i,Y^i)$ is a kernel (x-space representation) for $\exp(-\beta H^i)$. $q^i(x)$ vanishes if $x\notin\Omega^i$.

With the definitions (75) one has that

$$\langle W\rangle=\sum_{i<j}\int_{\Omega_i}\int_{\Omega_j}q^i(x)q^j(y)|x-y|^{-1}dxdy. \tag{76}$$

Equation (70), together with (76) and (74), is the desired inequality for the interdomain interaction. It is quite general in that an analogous inequality holds for arbitrary two-body potentials. Neither specific properties of the Coulomb potential nor neutrality was used.

Now we come to the crucial point at which screening is brought in. The following venerable result from the *Principia Mathematica* is essential.

Theorem 13 (Newton). Let $\rho(x)$ be an integrable function on 3-space such that $\rho(x)=\rho(y)$ if $|x|=|y|$ (isotropy) and $\rho(x)=0$ if $|x|>R$ for some $R>0$. Let

$$\phi(x)=\int\rho(y)|x-y|^{-1}dy \tag{77}$$

be the Coulomb potential generated by ρ. Then if $|x|\geq R$

$$\phi(x)=|x|^{-1}\int\rho(y)dy. \tag{78}$$

The important point is that an isotropic, *neutral* charge distribution generates zero potential outside its support, irrespective of how the charge is distributed radially.

Suppose that N^i is *neutral*, i.e., the electron number $=z$ times the nucleon number for each subdomain in Ω. Suppose also that the subdomain Ω^i is a *ball* of radius R^i centered at a^i. Then since H^i is rotation invariant, $q^i(x)=q^i(y)$ if $|x-a^i|=|y-a^i|$, $\int q^i(x)dx=0$ (by neutrality) and $q^i(x)=0$ if $|x-a^i|>R^i$. Then, by Theorem 13, *every* term in Eq. (76) involving q^i vanishes, because when $j\neq i$, $q^j(y)=0$ if $|y-a^i|<R^i$ since Ω^j is disjoint from Ω^i. Consequently the average interdomain interaction, $\langle W\rangle$, *vanishes*.

In the decomposition, π, of Ω into $\Omega^1,\ldots,\Omega^l$ and N into $N^1,\ldots,N^l$ we will arrange matters such that

(i) $\Omega^1,\ldots,\Omega^{l-1}$ are balls,

(ii) $N^1,\ldots,N^{l-1}$ are neutral,

(iii) $N^l=0$.

Then $\langle W\rangle=0$ and, using Eqs. (69) and (70)

$$Z(N,\Omega,\beta)\geq \mathrm{Tr}^{\pi}\exp(-\beta H_N^{\pi})\geq\prod_{i=1}^{l-1}Z(N^i,\Omega^i,\beta)e^{-\beta\langle W\rangle}=\prod_{i=1}^{l-1}Z(N^i,\Omega^i,\beta). \tag{79}$$

In addition to (i), (ii), (iii) it will also be necessary to arrange matters such that when Ω is a ball B_K in the chosen sequence of domains, then the subdomains $\Omega^1, \ldots, \Omega^{l-1}$ in the partition of B_k are *also* smaller balls in the same sequence. With these requirements in mind the *standard sequence*, which depends on the limiting density ρ, is defined as follows:

(1) Choose $\rho > 0$.

(2) Choose any N_0 satisfying the neutrality condition (59).

(3) Choose R_0 such that

$$28(4\pi/3)\rho R_0^3 = |N_0| . \tag{80}$$

(4) For $j \geq 1$ let

$$R_j = (28)^j R_0 , \quad N_j = (28)^{3j-1} N_0 \tag{81}$$

be the radius of the ball B_j and the particle number in that ball.

It will be noted that the density in all the balls except the first is

$$\rho_j = \rho, \; j \geq 1 , \tag{82}$$

while the density in the smallest ball is much bigger:

$$\rho_0 = 28\rho . \tag{83}$$

This has been done so that when a ball $B_K, K \geq 1$ is packed with smaller balls in the manner to be described below, the density in each ball will come out right; the higher density in B_0 compensates for the portion of B_K not covered by smaller balls. The radii increase geometrically, namely by a factor of 28.

The number 28 may be surprising until it is realized that the objective is to be able to pack B_K with balls of type B_{K-1}, B_{K-2}, etc., in such a way that as much as possible of B_K is covered and *also* that very little of B_K is covered by very small balls. If the ratio of radii were too close to unity, then the packing of B_K would be inefficient from this point of view. In short, if the number 28 is replaced by a much smaller number the analog of the following basic geometric theorem will not be true.

Theorem 14 (Cheese theorem). For j a positive integer define the integer $m_j \equiv (27)^{j-1}(28)^{2j}$. Then for each positive integer $K \geq 1$ it is possible to pack the ball B_K of radius R_K (given by 81) with

$$\bigcup_{j=0}^{K-1} (m_{K-j} \text{ balls of radius } R_j).$$

"Pack" means that all the balls in the union are disjoint.

We will not give a proof of Theorem 14 here, but note that it entails showing that m_1 balls of radius R_{K-1} can be packed in B_K in a cubic array, then that m_2 balls of radius R_{K-2} can be packed in a cubic array in the interstitial region, etc.

Theorem 14 states that B_K can be packed with $(28)^2$ balls of type B_{K-1}, $(27)(28)^4$ balls of type B_{K-2}, etc. If f_{K-j} is the fraction of the volume of B_K occupied by all the balls of radius R_j in the packing, then

$$f_j = m_j(R_{K-j}/R_K)^3 = \tfrac{1}{27}\gamma^j \tag{84}$$

with

$$\gamma = \tfrac{27}{28} < 1 . \tag{85}$$

The packing is *asymptotically complete* in the sense that

$$\lim_{K\to\infty} \sum_{j=0}^{K-1} f_{K-j} = \tfrac{1}{27} \sum_{j=1}^{\infty} \gamma^j = 1 . \tag{86}$$

It is also "geometrically rapid" because the fraction of $|B_K|$ that is uncovered is

$$\sum_{j=K+1}^{\infty} f_j = \gamma^K . \tag{87}$$

The necessary ingredients having been assembled, we can now prove the following theorem.

Theorem 15. Given ρ and $\beta > 0$, the thermodynamic limits $F(\rho, \beta)$ and $e(\rho)$ (65, 67) exist for the sequence of balls and particle numbers specified by (80) and (81).

Proof. Let F_K given by Eq. (64) be the free energy per unit volume for the ball B_K with N_K particles in it. For $K \geq 1$, partition B_K into disjoint domains $\Omega^1, \ldots, \Omega^l$, where the Ω^i for $i = 1, \ldots, l-1$ designate the smaller balls referred to in Theorem 14, and Ω^l (which is the "cheese" after the holes have been removed) is the remainder of B_K. The smaller balls are copies of B_j, $0 \leq j \leq K-1$; in each of these place N_j particles according to (81). $N^l = 0$. The total particle number in B_K is then

$$\sum_{j=0}^{K-1} N_j m_{K-j} = N_0\{(27)^{K-1}(28)^{2K} + \sum_{j=1}^{K-1} (28)^{3j-1}(27)^{K-j-1}(28)^{2K-2j}\}$$

$$= N_0(28)^{3K-1} = N_K$$

as it should be.

Use the basic inequality (79); $\langle W \rangle = 0$ since all the smaller balls are neutral and Ω^l contains no particles. Thus, taking logarithms and dividing by $|B_K|$, we have for $K \geq 1$ that

$$F_K \leq \sum_{j=0}^{K-1} F_j f_{K-j} \tag{88}$$

with $f_j = \gamma^j/27$ and $\gamma = \frac{27}{28}$. This inequality can be rewritten as

$$F_K = \sum_{j=0}^{K-1} \frac{F_j \gamma^{K-j}}{27} - d_K \tag{89}$$

with $d_K \geq 0$. Equation (89) is a *renewal equation* which can be solved explicitly by inspection:

$$F_K = -\gamma d_K - \sum_{j=1}^{K} \frac{d_j}{28} + \frac{F_0}{28} . \tag{90}$$

We now use the first step, Theorem 13, on the boundedness of F_K. Since $F_K \geq C$, $\sum_{j=1}^{\infty} d_j$ must be finite, for otherwise (90) would say that $F_K \to -\infty$. The convergence of the sum implies that $d_K \to 0$ as $K \to \infty$. Hence the limit exists; specifically

$$F = \lim_{K\to\infty} F_K = -\sum_{j=1}^{\infty} \frac{d_j}{28} + \frac{F_0}{28} . \quad \blacksquare \tag{91}$$

Theorem 15 is the desired goal, namely the existence of the thermodynamic limit for the free energy (or ground state energy) per unit volume. There are, however, some additional points that deserve comment.

(A) For each given limiting density ρ, a particular sequence of domains, namely balls, and particle numbers was used. It can be shown that the same limit is reached for general domains, with some mild conditions on their shape including, of course, balls of different radii than that used here. The argument involves packing the given domains with balls of the standard sequence and vice versa. The proof is tedious, but standard, and can be found in (Lieb and Lebowitz, 1972).

(B) Here we have considered the thermodynamic limit for real matter, in which all the particles are mobile. There are, however, other models of some physical interest. One is *jellium* in which the positive nuclei are replaced by a *fixed, uniform background* of positive charge. With the aid of an additional trick the thermodynamic limit can also be proved for this model (Lieb and Narnhofer, 1975). Another, more important model is one in which the nuclei are *fixed point charges* arranged periodically in a lattice. This is the model of solid state physics. Unfortunately, local rotation invariance is lost and Newton's Theorem 13 cannot be used. This problem is still open and its solution will require a deeper insight into screening.

(C) An absolute physical requirement for $\beta F(\rho, \beta)$, as a function of $\beta = 1/T$, is that it be *concave*. This is equivalent to the fact that the specific heat is non-negative since (specific heat) $= -\beta^2 \partial^2 \beta F(\rho, \beta)/\partial\beta^2$. Fortunately it is true. From the definitions (57), (58) we see that $\ln Z(N, \Omega, \beta)$ is convex in β for *every* finite system and hence $\beta F(N, \Omega, \beta)$ is concave. Since the limit of a sequence of concave functions is always concave, the limit $\beta F(\rho, \beta)$ is concave in β.

(D) Another absolute requirement is that $F(\rho, \beta)$ be *convex* as a function of ρ. This is called *thermodynamic stability* as distinct from the lower bound H stability of the previous sections. It is equivalent to the fact that the compressibility is non-negative, since (compressibility)$^{-1} = \partial P/\partial\rho = \rho\partial^2 F(\rho, \beta)/\partial\rho^2$. Frequently, in approximate theories (e.g., van der Waals' theory of the vapor-liquid transition, some field theories, or some theories of magnetic systems in which the magnetization per unit volume plays the role of ρ), one introduces an F with a double bump. Such an F is nonphysical and never should arise in an exact theory.

For a finite system, F is defined only for integral N, and hence not for all real ρ. It can be defined for all ρ by linear interpolation, for example, but even so it can neither be expected to be, nor is it generally, convex, except in the limit. The idea behind the following proof is standard.

Theorem 16. The limit function $F(\rho, \beta)$ is a convex function of ρ for each fixed β. $E(\rho)$ is also a convex function of ρ.

Proof: This means that for $\rho = \lambda\rho^1 + (1-\lambda)\rho^2, 0 \le \lambda \le 1$,

$$F(\rho, \beta) \le \lambda F(\rho^1, \beta) + (1-\lambda) F(\rho^2, \beta) \tag{92}$$

and similarly for $E(\rho)$. As F is bounded *above* on bounded ρ intervals (this can be proved by a simple variational calculation), it is sufficient to prove (92) when $\lambda = \frac{1}{2}$. To avoid technicalities (which can be supplied) and concentrate on the main idea, we shall here prove (92) when ρ^2 and ρ^1 are rationally related: $a\rho^1 = b\rho^2$, a and b positive integers. Choose any neutral particle number M and define a sequence of balls B_j with radii as given in (81) and with $28(4\pi/3)\rho R_0^3 = (a+b)|M|$. For the ρ system take $N_0 = (a+b)M, N_j = (28)^{3j-1}N_0$, $j \ge 1$. For the ρ^1 (respectively, ρ^2) system take $N_0^1 = 2bM, N_j^1 = (28)^{3j-1}N_0^1$ [respectively, $N_0^2 = 2aM, N_j^2 = (28)^{3j-1}N_0^2$]. Consider the ρ system. In the canonical partition π of B_K into smaller balls (Theorem 14) note that the number of balls B_j is m_{K-j} and this number is *even*. In *half* of these balls place N_j^1 particles and in the other half place N_j^2 particles, $0 \le j \le K-1$. Then in place of (88) we get

$$F_K(\rho) \le \frac{1}{2}\sum_{j=0}^{K-1} f_{K-j}[F_j(\rho^1) + F_j(\rho^2)] \tag{93}$$

in an obvious notation. Inserting (89) on the right side of (93),

$$F_K(\rho) \le \tfrac{1}{2}[F_K(\rho^1) + F_K(\rho^2)] + \tfrac{1}{2}(d_K^1 + d_K^2). \tag{94}$$

Since $\lim_{K\to\infty} d_K^{1,2} = 0$, we can take the limit $K \to \infty$ in Eq. (94) and obtain (92). ■

(E) The convexity in ρ^1 and concavity in β of $F(\rho, \beta)$ has another important consequence. Since F is bounded below (Theorem 13) and bounded above (by a simple variational argument) on bounded sets in the (ρ, β) plane, the convexity/concavity implies that it is *jointly continuous* in (ρ, β). This, together with the monotonicity in K of $F_K + \gamma d_K$ (see (90)), implies by a standard argument using Dini's theorem that *the thermodynamic limit is uniform* on bounded (ρ, β) sets. This uniformity is sometimes overlooked as a basic desideratum of the thermodynamic limit. Without it one would have to fix ρ and β *precisely* in taking the limit—an impossible task experimentally. With it, it is sufficient to have merely an increasing sequence of systems such that $\rho_j \to \rho$ and $\beta_j \to \beta$. The same result holds for $e(\rho)$.

(F) An application of the uniformity of the limit for $e(\rho)$ is the following. Instead of confining the particles to a box (Dirichlet boundary condition for $H_{n,k}$) one could consider $H_{n,k}$ defined on all of $L^2(\mathbf{R}^3)|N|$, i.e., no confinement at all. In this case

$$E_N^Q \equiv \inf_\psi \langle\psi, H_{n,k}\psi\rangle/\langle\psi, \psi\rangle$$

is just the ground state energy of a neutral molecule and it is expected that $E_N^Q/|N|$ has a limit. Indeed, this limit exists and it is simply

$$\lim_{n\to\infty} E_N^Q/|N| = \lim_{\rho\to 0} \rho^{-1}e(\rho).$$

There is no analog of this for $F(\rho, \beta)$ because removing the box would cause the partition function to be infinite even for a finite system.

(G) The ensemble used here is the *canonical* ensemble. It is possible to define and prove the existence of the thermodynamic limit for the *microcanonical* and *grand canonical* ensembles and to show that all three ensembles are equivalent (i.e., that they yield the same values for all thermodynamic quantities, such as the pressure). (See Lieb and Lebowitz, 1972.)

(H) Charge neutrality was essentially for taming the long range Coulomb force. What happens if the system is *not neutral*? To answer this let N_j, Ω_j be a sequence of pairs of particle numbers and domains, but without (59) being satisfied. Let $Q_j = zk_j - n_j$ be the net charge, $\rho_j = |N_j|/|\Omega_j|$ as before, and $\rho_j \to \rho$. One expects that if

(i) $Q_j|\Omega_j|^{-2/3} \to 0$ then the *same* limit $F(\rho, \beta)$ is achieved as if $Q_j = 0$.

On the other hand, if

(ii) $Q_j|\Omega_j|^{-2/3} \to \infty$ then there is no limit for $F(N_j, \Omega_j, \beta)$. More precisely $F(N_j, \Omega_j, \beta) \to \infty$ because the minimum electrostatic energy is too great. Both of these expectations can be proved to be correct.

The interesting case is if

(iii) $\lim_{j\to\infty} Q_j|\Omega_j|^{-2/3} = \sigma$ exists. Then one expects a *shape dependent limit* to exist as follows. Assume that the Ω_j are geometrically similar, i.e., $\Omega_j = \lambda_j \Omega_0$ with $|\Omega_0| = 1$ and $|N_j|\lambda_j^{-3} = \rho_j$ with $\rho_j \to \rho$. Let C be the *electrostatic capacity* of Ω_0; it depends upon the shape of Ω_0. The capacity of Ω_j is then $C_j = C\lambda_j$. From elementary electrostatics theory the expectation is that

$$\lim_{j\to\infty} F(N_j, \Omega_j, \beta) = F(\rho, \beta) + \sigma^2/2C\,. \qquad (95)$$

Note that $(Q_j^2/2C_j)|\Omega_j|^{-1} \to \sigma^2/2C$.

Equation (95) *can be proved* for ellipsoids and balls. The proof is as complicated as the result is simple. With work, the proof could probably be pushed through for other domains Ω_0 with smooth boundaries.

The result (95) is amazing and shows how special the Coulomb force is. It says that the surplus charge Q_j goes to a thin layer near the surface. There, only its electrostatic energy, which overwhelms its kinetic energy, is significant. The bulk of Ω_j is neutral and uninfluenced by the surface layer because the latter generates a constant potential inside the bulk. It is seldom that one has two strongly interacting subsystems and that the final result has no cross terms, as in Eq. (95).

(I) There might be a temptation, which should be avoided, to suppose that the thermodynamic limit describes a single phase system of uniform density. The temptation arises from the construction in the proof of Theorem 15 in which a large domain B_K is partitioned into smaller domains having essentially constant density. Several phases can be present inside a large domain. Indeed, if β is very large a solid is expected to form, and if the average density, ρ, is smaller than the equilibrium density, ρ_s, of the solid a dilute gas phase will also be present. The location of the solid inside the larger domain will be indeterminate.

From this point of view, there is an amusing, although expected, aspect to the theorem given in Eq. (95). Suppose that β is very large and that $\rho < \rho_s$. Suppose, also, that a surplus charge $Q = \sigma V^{2/3}$ is present, where V is the volume of the container. In *equilibrium*, the surplus charge will never be bound to the surface of the solid, for that would give rise to a larger free energy than in (95).

(J) The inequality (53) of Sec. IV, together with known facts about the ideal gas, permit one to derive upper and lower bounds to the free energy and pressure for any neutral mixture of electrons and various nuclei. These bounds are absolutely rigorous and involve no approximation whatsoever (beyond the assumption of nonrelativistic Schrödinger mechanics with purely Coulomb forces).

If one has bounds on the free energy per unit volume

$$F^L(\rho, \beta) \le F(\rho, \beta) \le F^U(\rho, \beta)\,, \qquad (96)$$

then since the pressure P is equal to $-F + \rho\partial F/\partial\rho$, and since F is convex in ρ, one has that

$$P \le -F + \rho \min_{\epsilon>0} \epsilon^{-1}\{F(\rho+\epsilon, \beta) - F(\rho, \beta)\}\,,$$
$$P \ge -F + \rho \max_{\epsilon>0} \epsilon^{-1}\{F(\rho, \beta) - F(\rho-\epsilon, \beta)\}\,. \qquad (97)$$

Inserting (96) into (97) yields bounds on P.

Equation (96) comes from bounds on Z [see Eq. (63)]. Using (70) and $\rho = \rho_{\rm nuc} + \rho_{\rm el}$

$$F^U(\rho, \beta) = F^0_{\rm el}(\rho_{\rm el}, \beta) + F^0_{\rm nuc}(\rho_{\rm nuc}, \beta) + \langle W\rangle/|\Omega|\,, \qquad (98)$$

where F^0 is the ideal gas free energy, and $\langle W\rangle/|\Omega|$ is the average *total* Coulomb energy per unit volume in the ideal gas state. This can easily be computed in terms of exchange integrals. To obtain F^L, choose $0<\gamma<1$ and write $H_{n,k} = (1-\gamma)T_{\rm el} + T_{\rm nuc} + h(\gamma)$, where T is the kinetic energy operator, and $h(\gamma) = \gamma T_{\rm el} + W$. $h(\gamma)$ is bounded below by $A/\gamma \equiv$[right side of Eq. (53)]$/\gamma$. Thus

$$Z \le \exp[-\beta h(\gamma)]\,\mathrm{Tr}\exp[-\beta((1-\gamma)T_{\rm el} + T_{\rm nuc})]$$

and

$$F^L = F^0_{\rm nuc}(\rho_{\rm nuc}, \beta) + \max_{0<\gamma<1}\{(1-\gamma)F^0_{\rm el}(\rho_{\rm el}, (1-\gamma)\beta) + \gamma^{-1}A/|\Omega|\}\,. \qquad (99)$$

A numerical evaluation of these bounds will be presented elsewhere.

As a final remark, the existence of the thermodynamic limit (and hence the existence of intensive thermodynamic variables such as the pressure) does not establish the existance of a unique *thermodynamic state*. In other words, it has not been shown that correlation functions, which always exist for finite systems, have unique limits as the volume goes to infinity. Indeed, unique limits might not exist if several phases are present. For well behaved potentials there are techniques available for proving that a state exists when the density is small, but these techniques do not work for the long range Coulomb potential. Probably the next chapter to be written in this subject will consist of a proof that correlation functions are well defined in the thermodynamic limit when ρ or β is small.

REFERENCES

Balàzs, N., 1967, "Formation of stable molecules within the statistical theory of atoms," Phys. Rev. **156**, 42–47.

Barnes, J. F., 1975, private communication.

Birman, M. S., 1961, Mat. Sb. **55** (97), 125–174 ["The spectrum of singular boundary value problems," Am. Math. Soc. Transl. Ser. 2 **53**, 23–80 (1966)].

Dirac, P. A. M., 1930, "Note on exchange phenomena in the Thomas atom," Proc. Camb. Phil. Soc. **26**, 376–385.

Dyson, F. J., 1967, "Ground-state energy of a finite system of charged particles," J. Math. Phys. **8**, 1538–1545.

Dyson, F. J., and A. Lenard, 1967, "Stability of matter. I." J. Math. Phys. **8**, 423–434.

Fermi, E., 1927, "Un metodo statistico per la determinazione di alcune prioretà dell'atome," Rend. Acad. Naz. Lincei **6**,

602–607.

Fock, V., 1930, "Näherungsmethode zur Lösung des quantenmechanischen Mehrkörperproblems," Z. Phys. **61**, 126–148; see also V. Fock, 1930, "Selfconsistent field" mit austausch für Natrium," Phys. **62**, 795–805.

Gombás, P., 1949, *Die statistischen Theorie des Atomes und ihre Anwendungen* (Springer Verlag, Berlin).

Hartree, D. R., 1927–1928, "The wave mechanics of an atom with a non-Coulomb central field. Part I. Theory and methods," Proc. Camb. Phil. Soc. **24**, 89–110.

Heisenberg, W., 1927, "Über den anschaulichen Inhalt der quanten-theoretischen Kinematik und Mechanik," Z. Phys. **43**, 172–198.

Jeans, J. H., 1915, *The Mathematical Theory of Electricity and Magnetism* (Cambridge University, Cambridge) 3rd ed., p. 168.

Kato, T., "Fundamental properties of Hamiltonian operators of Schrödinger type," Trans. Am. Math. Soc. **70**, 195–211.

Kirzhnits, D. A., 1957, J. Exptl. Theoret. Phys. (USSR) **32**, 115–123 [Engl. transl. "Quantum corrections to the Thomas-Fermi equation," Sov. Phys. JETP **5**, 64–71 (1957)].

Kompaneets, A. S., and E. S. Pavlovskii, 1956, J. Exptl. Theoret. Phys. (USSR) **31**, 427–438 [Engl. transl. "The self-consistent field equations in an atom," Sov. Phys. JETP **4**, 328–336 (1957)].

Lenard, A., and F. J. Dyson, 1968, "Stability of matter. II," J. Math. Phys. **9**, 698–711.

Lenz, W., 1932, "Über die Anwendbarkeit der statistischen Methode auf Ionengitter," Z. Phys. **77**, 713–721.

Lieb, E. H., 1976, "Bounds on the eigenvalues of the Laplace and Schrödinger operators," Bull. Am. Math. Soc., in press.

Lieb, E. H., and J. L. Lebowitz, 1972, "The constitution of matter: existence of thermodynamics for systems composed of electrons and nuclei," Adv. Math. **9**, 316–398. See also J. L. Lebowitz and E. H. Lieb, 1969, "Existence of thermodynamics for real matter with Coulomb forces," Phys. Rev. Lett. **22**, 631–634.

Lieb, E. H., and H. Narnhofer, 1975, "The thermodynamic limit for jellium," J. Stat. Phys. **12**, 291–310; Erratum: J. Stat. Phys. **14**, No. 5 (1976).

Lieb, E. H., and B. Simon, "On solutions to the Hartree–Fock problem for atoms and molecules," J. Chem. Phys. **61**, 735–736. Also, a longer paper in preparation.

Lieb, E. H., and B. Simon, 1977, "The Thomas–Fermi theory of atoms, molecules and solids," Adv. Math., in press. See also E. H. Lieb and B. Simon, 1973, "Thomas–Fermi theory revisited," Phys. Rev. Lett. **33**, 681–683.

Lieb, E. H., and W. E. Thirring, 1975, "A bound for the kinetic energy of fermions which proves the stability of matter," Phys. Rev. Lett. **35**, 687–689; Errata: Phys. Rev. Lett. **35**, 1116. For more details on kinetic energy inequalities and their application, see also E. H. Lieb and W. E. Thirring, 1976, "Inequalities for the moments of the Eigenvalues of the Schrödinger Hamiltonian and their relation to Sobolev inequalities," in *Studies in Mathematical Physics: Essays in Honor of Valentine Bargmann,* edited by E. H. Lieb, B. Simon, and A. S. Wightman (Princeton University, Princeton).

Rosen, G., 1971, "Minimum value for c in the Sobolev inequality $\|\phi\|_3 \leq c\|\nabla\phi\|^3$," SIAM J. Appl. Math. **21**, 30–32.

Schwinger, J., 1961, "On the bound states of a given potential," Proc. Nat. Acad. Sci. (U.S.) **47**, 122–129.

Scott, J. M. C., 1952, "The binding energy of the Thomas Fermi atom," Phil. Mag. **43**, 859–867.

Sheldon, J. W., 1955, "Use of the statistical field approximation in molecular physics," Phys. Rev. **99**, 1291–1301.

Slater, J. C., 1930, "The theory of complex spectra," Phys. Rev. **34**, 1293–1322.

Sobolev, S. L., 1938, Mat. Sb. **46**, 471. See also S. L. Sobolev, 1950, "Applications of functional analysis in mathematical physics," Leningrad; Am. Math. Soc. Transl. Monographs **7** (1963).

Sommerfeld, A., 1932, "Asymptotische Integration der Differential-gleichung des Thomas–Fermischen Atoms," Z. Phys. **78**, 283–308.

Teller, E., 1962, "On the stability of molecules in the Thomas–Fermi theory," Rev. Mod. Phys. **34**, 627–631.

Thomas, L. H., 1927, "The calculation of atomic fields," Proc. Camb. Phil. Soc. **23**, 542–548.

Von Weizsäcker, C. F., 1935, "Zur Theorie der Kernmassen," Z. Phys. **96**, 431–458.

VOLUME 22, NUMBER 13 PHYSICAL REVIEW LETTERS 31 MARCH 1969

EXISTENCE OF THERMODYNAMICS FOR REAL MATTER WITH COULOMB FORCES

J. L. Lebowitz*
Belfer Graduate School of Science, Yeshiva University, New York, New York 10033

and

Elliott H. Lieb†
Department of Mathematics, Massachusetts Institute of Technology, Cambridge, Massachusetts 02139
(Received 3 February 1969)

It is shown that a system made up of nuclei and electrons, the constituents of ordinary matter, has a well-defined statistical-mechanically computed free energy per unit volume in the thermodynamic (bulk) limit. This proves that statistical mechanics, as developed by Gibbs, really leads to a proper thermodynamics for macroscopic systems.

In this note we wish to report the solution to a classic problem lying at the foundations of statistical mechanics.

Ever since the daring hypothesis of Gibbs and others that the equilibrium properties of matter could be completely described in terms of a phase-space average, or partition function, $Z = \mathrm{Tr}e^{-\beta H}$, it was realized that there were grave difficulties in justifying this assumption in terms of basic microscopic dynamics and that such delicate matters as the ergodic conjecture stood in the way. These questions have still not been satisfactorily resolved, but more recently still another problem about Z began to receive attention: Assuming the validity of the partition function, is it true that the resulting properties of matter will be extensive and otherwise the same as those postulated in the science of thermodynamics? In particular, does the thermodynamic, or bulk, limit exist for the free energy derived from the partition function, and if so, does it have the appropriate convexity, i.e., stability properties?

To be precise, if N_j are an unbounded, increasing sequence of particle numbers, and Ω_j a sequence of reasonable domains (or boxes) of volume V_j such that $N_j/V_j \to \text{constant} = \rho$, does the free energy per unit volume

$$f_j = -kT(V_j)^{-1} \ln Z(\beta, N_j, \Omega_j) \tag{1}$$

approach a limit [called $f(\beta,\rho)$] as $j \to \infty$, and is this limit independent of the particular sequence and shape of the domains? If so, is f convex in the density ρ and concave in the temperature β^{-1}? Convexity is the same as thermodynamic stability (non-negative compressibility and specific heat).

Various authors have evolved a technique for proving the above,[1,2] but always with one severe drawback. It had to be assumed that the interparticle potentials were short range (in a manner to be described precisely later), thereby excluding the Coulomb potential which is the true potential relevant for real matter. In this note we will indicate the lines along which a proof for Coulomb forces can be and has been constructed. The proof itself, which is quite long, will be given elsewhere.[3] We will also list here some additional results for charged systems that go beyond the existence and convexity of the limiting free energy.

To begin with, a sine qua non for thermodynamics is the stability criterion on the N-body Hamiltonian $H=E_K+V$. It is that there exists a constant $B \geqslant 0$ such that for all N,

$$V(r_1, \cdots, r_N) > -BN \quad \text{(classical mechanics)}, \tag{2}$$

$$E_0 > -BN \quad \text{(quantum mechanics)}, \tag{3}$$

where E_0 is the ground-state energy in infinite space. (Classical stability implies quantum-mechanical stability, but not conversely.) Heuristically, stability insures against collapse. From the mathematical point of view, it provides a lower bound to f_j in (1). We wish to emphasize that stability of the Hamiltonian (H stability), while necessary, is insufficient for assuring the existence of thermodynamics. For example, it is trivial to prove H stability for charged particles all of one sign, and it is equally obvious that the thermodynamic limit does not exist in this case.

It is not too difficult to prove classical and thus also quantum-mechanical H stability for a wide variety of short-range potentials or for charged particles having a hard core.[2,4] But real charged particles require quantum mechanics and the recent proof of H stability by Dyson and Lenard[5] is as difficult as it is elegant. They show that stability will hold for any set of charges and masses provided that the negative particles and/or the positive ones are fermions.

The second requirement in the canonical proofs[1] is that the potential be tempered, which is to say that there exist a fixed r_0 and constants $C \geqslant 0$ and $\epsilon > 0$ such that if two groups of N_a and N_b particles are separated by a distance $r > r_0$, their interparticle energy is bounded by

$$V(N_a \oplus N_b) - V(N_a) - V(N_b) \leqslant C r^{-(3+\epsilon)} N_a N_b. \tag{4}$$

Tempering is roughly the antithesis of stability because the requirements that the forces are not too repulsive at infinity insures against "explosion." Coulomb forces are obviously not tempered and for this reason the canonical proofs have to be altered. Our proof, however, is valid for a mixture of Coulomb and tempered potentials and this will always be understood in the theorems below. It is not altogether useless to include tempered potentials along with the true Coulomb potentials because one might wish to consider model systems in which ionized molecules are the elementary particles.

Prior to explaining how to overcome the lack of tempering we list the main theorems we are able to prove. These are true classically as well as quantum mechanically. But first three definitions are needed:

(D1) We consider s species of particles with charges e_i, particle numbers $N^{(i)}$, and densities $\rho^{(i)}$. In the following N and ρ are a shorthand notation for s-fold multiplets of numbers. The conditions for H stability (see above) are assumed to hold.

(D2) A neutral system is one for which $\sum_1^s N^{(i)} \times e_i = 0$, alternatively $\sum_1^s \rho^{(i)} e_i = 0$.

(D3) The ordinary s-species grand canonical partition function is

$$\sum_{N^{(s)}=0}^{\infty} \cdots \sum_{N^{(1)}=0}^{\infty} \prod_1^s z_i^{N^{(i)}} Z(N, \Omega). \tag{5}$$

The neutral grand canonical partition function is the same as (5) except that only neutral systems enter the sum.

The theorems are the following:

(T1) The canonical, thermodynamic limiting free energy per unit volume $f(\beta, \rho)$ exists for a neutral system and is independent of the shape of the domain for reasonable domains. Furthermore, $f(\beta, \rho^{(1)}, \rho^{(2)}, \cdots)$ is concave in β^{-1} and jointly convex in the s variables $(\rho^{(1)}, \cdots, \rho^{(s)})$.

(T2) The thermodynamic limiting microcanonical[6] entropy per unit volume exists for a neutral system and is a concave function of the energy per unit volume. It is also independent of domain shape for reasonable shapes and it is equal to the entropy computed from the canonical free energy.

(T3) The thermodynamic limiting free energy per unit volume exists for both the ordinary and the neutral grand canonical ensembles and are independent of domain shape for reasonable domains. Moreover, they are equal to each other and to the neutral canonical free energy per unit volume.

Theorem 3 states that systems which are not charge neutral make a vanishingly small contribution to the grand canonical free energy. While this is quite reasonable physically, it does raise an interesting point about nonuniform convergence because the ordinary and neutral partition functions are definitely not equal if we switch off the charge before passing to the thermodynamic limit, whereas they are equal if the limits are taken in the reverse order.

An interesting question is how much can charge neutrality be nonconserved before the free energy per unit volume deviates appreciably from its neutral value? The answer is in theorem 4.

(T4) Consider the canonical free energy with a surplus (i.e., imbalance) of charge Q and take the thermodynamic limit in either of three ways: (a) $QV^{-2/3}\to 0$; (b) $QV^{-2/3}\to\infty$; (c) $QV^{-2/3}\to$ const. In case (a) the limit is the same as for the neutral system while in case (b) the limit does not exist, i.e., $f\to\infty$. In case (c) the free energy approaches a limit equal to the neutral-system free energy plus the energy of a surface layer of charge Q as given by elementary electrostatics.

We turn now to a sketch of the method of proof and will restrict ourselves here to the neutral canonical ensemble. As usual, one first proves the existence of the limit for a standard sequence of domains. The limit for an arbitrary domain is then easily arrived at by packing that domain with the standard ones. The basic inequality that is needed is that if a domain Ω containing N particles is partitioned into D domains $\Omega_1, \Omega_2, \cdots, \Omega_D$ containing $N_1, N_2, \cdots, N_D$ particles, respectively, and if the interdomain interaction be neglected, then

$$Z(N,\Omega)\geq\prod_1^D Z(N_i,\Omega_i). \tag{6}$$

If Ω is partitioned into subdomains, as above, plus "corridors" of thickness $>r_0$ which are devoid of particles, one can use (4) to obtain a useful bound on the tempered part of the omitted interdomain interaction energy. We will refer to these energies as surface terms.

The normal choice[1] for the standard domains are cubes C_j containing N_j particles, with C_{j+1} being composed of eight copies of C_j together with corridors, and with $N_{j+1}=8N_j$. Neglecting surface terms one would have from (6) and (1)

$$f_{j+1}\leq f_j. \tag{7}$$

Since f_j is bounded below by H stability, (7) implies the existence of a limit. To justify neglect of the surface terms one makes the corridors increase in thickness with increasing j; although V_j^C, the corridor volume, approaches ∞ one makes $V_j^C/V_j\to 0$ in order that the limiting density not vanish. The positive ϵ of (4) allows one to accomplish these desiderata.

Obviously, such a strategy will fail with Coulomb forces, but fortunately there is another way to bound the interdomain energy. The essential point is that it is not necessary to bound this energy for all possible states of the systems in the subdomains; it is only necessary to bound the "average" interaction between domains, which is much easier. This is expressed mathematically by using the Peierls-Bogoliubov inequality[7] to show that

$$Z(N,\Omega)\geq e^{-\beta U}\prod_1^D Z(N_i,\Omega_i), \tag{8}$$

where U is the average interdomain energy in an ensemble where each domain is independent. U consists of a Coulomb part, U_C, and a tempered part, U_t, which can be readily bounded.[1]

We now make the observation, which is one of the crucial steps in our proof, that independently of charge symmetry U_C will vanish if the subdomains are spheres and are overall neutral. The rotation invariance of the Hamiltonian will produce a spherically symmetric charge distribution in each sphere and, as Newton[8] observed, two such spheres would then interact as though their total charges (which are zero) were concentrated at their centers.

With this in mind we choose spheres for our standard domains. Sphere S_j will have radius $R_j=p^j$ with p an integer. The price we pay for using spheres instead of cubes is that a given one, S_k, cannot be packed arbitrarily full with spheres S_{k-1} only. We prove, however, that it can be packed arbitrarily closely (as $k\to\infty$) if we use all the previous spheres $S_{k-1}, S_{k-2}, \cdots S_0$. Indeed for the sequence of integers $n_1, n_2, \cdots, n_j=(p-1)^{j-1}p^{2j}$ we can show that we can simultaneously pack n_j spheres S_{k-j} into S_k for $1\leq j\leq k$. The fractional volume of S_k occupied by the S_{k-j} spheres is $\varphi_j=p^{-3j}n_j$, and from (8) we then have

$$f_k\leq\varphi_1 f_{k-1}+\varphi_2 f_{k-2}+\cdots+\varphi_k f_0, \tag{9}$$

and

$$\sum_1^\infty\varphi_j=1. \tag{10}$$

VOLUME 22, NUMBER 13 PHYSICAL REVIEW LETTERS 31 MARCH 1969

[Note that the inequality (6) is correct as it stands for pure Coulomb forces because U_C in (8) is identically zero. If short-range potentials are included there will also be surface terms, as in the cube construction, but these present only a technical complication that can be handled in the same manner as before.[1]] While Eq. (9) is more complicated than (7), it is readily proven explicitly that f_k approaches a limit as $k \to \infty$. [Indeed, it follows from the theory of the renewal equation[9] that (9) will have a limit if $\sum_1^\infty j\varphi_j < \infty$.]

The possibility of packing spheres this way is provided by the following geometrical theorem which plays the key role in our analysis. We state it without proof, but we do so in d dimensions generally and use the following notation: σ_d = volume of a unit d-dimensional sphere $= \frac{4}{3}\pi$ in three dimensions and $\alpha_d = (2^d - 1)2d^{\frac{1}{2}}$.

(T5) Let $p \geq \alpha_d + 2^d\sigma_d^{-1}$ be a positive integer. For all positive integers j, define radii $r_j = p^{-j}$ and integers $n_j = (p-1)^{j-1}p^{j(d-1)}$. Then it is possible to place simultaneously $\bigcup_j(n_j$ spheres of radius $r_j)$ into a unit d-dimensional sphere so that none of them overlap.

The minimum value of p required by the theorem in three dimensions is 27.

Many of the ideas presented here had their genesis at the Symposium on Exact Results in Statistical Mechanics at Irvine, California, in 1968, and we should like to thank our colleagues for their encouragement and stimulation: M. E. Fisher, R. Griffiths, O. Lanford, M. Mayer, D. Ruelle, and especially A. Lenard.

---

*Work supported by Air Force Office of Scientific Research, U. S. Air Force under Grant No. AFOSR 68-1416.

†Work supported by National Science Foundation Grant No. GP-9414.

[1]These developments are clearly expounded in M. E. Fisher, Arch. Ratl. Mech. Anal. 17, 377 (1964); D. Ruelle, Statistical Mechanics (W. A. Benjamin, Inc., New York, 1969). For a synopsis, see also J. L. Lebowitz, Ann. Rev. Phys. Chem. 19, 389 (1968).

[2]R. B. Griffiths, Phys. Rev. 176, 655 (1968), and footnote 6a in A. Lenard and F. J. Dyson [J. Math. Phys. 9, 698 (1968)]; O. Penrose, in Statistical Mechanics, Foundations and Applications, edited by T. Bak (W. A. Benjamin, Inc., New York, 1967), p. 98.

[3]E. H. Lieb and J. L. Lebowitz, "The Constitution of Matter," to be published.

[4]L. Onsager, J. Phys. Chem. 43, 189 (1939); M. E. Fisher and D. Ruelle, J. Math. Phys. 7, 260 (1966).

[5]F. J. Dyson and A. Lenard, J. Math. Phys. 8, 423 (1967); A. Lenard and F. J. Dyson, J. Math. Phys. 9, 698 (1968); F. J. Dyson, J. Math. Phys. 8, 1538 (1967).

[6]R. B. Griffiths, J. Math. Phys. 6, 1447 (1965).

[7]K. Symanzik, J. Math. Phys. 6, 1155 (1965).

[8]I. Newton, in Mathematical Principles, translated by A. Motte, revised by F. Cajori (University of California Press, Berkeley, Calif., 1934), Book 1, p. 193, propositions 71, 76.

[9]W. Feller, An Introduction to Probability Theory and Its Applications (J. Wiley & Sons, New York, 1957), 2nd ed. Vol. 1, p. 290.

Note: This paper is the announcement of the existence of the thermodynamic limit for particles interacting via Coulomb forces. The full version of this work appears in item 58 of the list of publications:

E.H. Lieb and J.L. Lebowitz, *The Constitution of Matter: Existence of Thermodynamics for Systems Composed of Electrons and Nuclei*, Adv. in Math. **9**, 316–398 (1972).

An abridged version (item 65 in the list of publications) appeared in the first and second editions of this *Selecta* but is omitted in the third edition for space reasons. A different abridged version also appears in Section V of the paper, *The Stability of Matter*, Rev. Mod. Phys. **48**, 553–569 (1976), which is included as item VI.1 in this edition.

With H. Narnhofer in J. Stat. Phys. *12*, 291–310 (1975)

The Thermodynamic Limit for Jellium

Elliott H. Lieb[1,2] and Heide Narnhofer[3,4]

Received December 27, 1974

The thermodynamic limit of the free energy, energy, pressure, and entropy is established for a neutral system of charged particles interacting with a fixed, uniformly charged background (jellium).

KEY WORDS: Thermodynamic limit; jellium; charged particles; uniform background; neutral system; free energy density; quantum mechanics; equilibrium statistical mechanics.

1. INTRODUCTION

In 1938 Wigner[1] introduced a model for matter which is now called jellium. One supposes that the electrons in a solid provide a uniform, constant charge background in which the heavier nuclei move. The Hamiltonian for the system consisting of N particles with coordinates $\mathbf{X} = \{\mathbf{x}_1, ..., \mathbf{x}_N\}$ in a three-dimensional domain Λ is

$$H = (2m)^{-1} \sum_{i=1}^{N} p_i^2 + e^2 U(\mathbf{X})$$

$$U(\mathbf{X}) = \sum_{i<j}^{N} |\mathbf{x}_i - \mathbf{x}_j|^{-1} - \rho \sum_{i=1}^{N} \varphi(\mathbf{x}_i) + \tfrac{1}{2}\rho^2 \int_{\Lambda} \varphi(\mathbf{x})\, d\mathbf{x} \tag{1}$$

Work partially supported by the National Science Foundation, grant GP31674X.

[1] On leave from Departments of Mathematics and Physics, MIT, Cambridge, Massachusetts.
[2] Departments of Mathematics and Physics, Princeton University, Princeton, New Jersey.
[3] Bell Laboratories, Murray Hill, New Jersey.
[4] On leave from Institute for Theoretical Physics, University of Vienna, Austria.

and where

$$\rho\varphi(\mathbf{x}) = \rho \int_\Lambda |\mathbf{x} - \mathbf{y}|^{-1}\, d\mathbf{y}$$

is the Coulomb potential produced by the background of charge density ρ.

Throughout the following we shall set $m = e^2 = 1$, and $\hbar = 1$ in the quantum case. Thus the Bohr radius is equal to unity and the energy unit, the Rydberg (Ry), is equal to one-half. The dimensionless length r_s is equal to $[3/(4\pi\rho)]^{1/3}$. Whenever the distinction is necessary, we shall assume $\rho > 0$ and that the particles are negative.

We shall show that for neutral systems, i.e., $\rho|\Lambda| = N$, the thermodynamic functions per unit volume (free energy, energy, entropy, pressure) exist as $\Lambda \to \infty$.

It is also possible to consider the one- and two-dimensional versions of this problem, where the Coulomb potential $|\mathbf{x}|^{-1}$ is replaced by $-|x|$ and $-\ln|\mathbf{x}|$, respectively. In the one-dimensional, classical case, Baxter[(2)] calculated the partition function exactly. For that case, Kunz[(3)] showed that the one-particle distribution function exists and that it has crystalline ordering, i.e., the Wigner lattice exists for all temperatures. Brascamp and Lieb[(4)] showed the same to be true in the quantum mechanical case for one-component fermions when β is large enough. Although we do not deal with the one-dimensional problem here, our methods would apply in that case. In two dimensions there are difficulties connected with the long-range nature of the $-\ln|\mathbf{x}|$ potential, and we shall not discuss this here.

The problem of jellium is closely related to the same problem for real matter treated by Lebowitz and Lieb[(5,6),5] and their methods will be employed here. The difficulty with jellium is that the background is held rigid by definition and one cannot freely constrain the particles to lie in balls without at the same time imparting an enormous electrostatic energy to the system. On the other hand, the fixed background considerably simplifies the H-stability question. (Cf. Dyson and Lenard.[(8)]) The connection between the jellium and the real matter problems is discussed by Narnhofer and Thirring.[(9)]

In Section 2 we use H-stability to get an upper bound on the partition function Z. The H-stability itself is proved in the appendix.

Section 3 deals with the classical case. We first treat a distinguished sequence of domains, which are balls, and then we treat general domains. The usual results are obtained, except that since the free energy is not a convex function of the density for jellium, the compressibility can be negative and the grand canonical ensemble is not equivalent to the canonical ensemble.

[5] See also Penrose and Smith.[(7)]

As in Ref. 5, we show that a system with an excess charge $Q \sim |\Lambda|^{2/3}$ has an excess free energy $-(2\beta)^{-1}Q^2/C$, where C is the capacity of Λ.

In Section 4 we outline the proof when weakly tempered potentials are also present. Although the thermodynamic limit exists in this case, we lose continuity in ρ—at least by our methods. This is an open question. The inclusion of hard cores is also not covered by our method and this, too, is an open question.

In Section 5 we explain the additional techniques needed for the quantum case. An open question here is to show the equivalence of different boundary conditions; we use Dirichlet conditions. A related problem is to show that the particle density and the electrostatic potential stay suitably bounded as $N \to \infty$.

2. *H*-STABILITY

The condition of H-stability is that the Hamiltonian is bounded below by a constant times N. It is sufficient to require that the potential energy alone has this property, since the kinetic energy operator is positive. For real matter one is obliged to consider the total Hamiltonian because the interaction energy of a positive and a negative particle has no lower bound. The proof of H-stability in this latter case is very difficult and was given by Dyson and Lenard[8,10] and recently a new proof was given by Federbush.[11] It is essential here that the electrons be fermions, thereby excluding classical particles.

For jellium, on the other hand, one can easily find a lower bound on U, by using an idea due to Onsager.[12] This is given in the appendix. A different proof and a different bound are also given in Ref. 10. Our bound is

$$U > -0.9N/r_s \tag{2}$$

and we emphasize that this result holds for all N and all domains, connected or not, and requires only that the background have charge density $(3/4\pi)r_s^{-3}$ or zero everywhere. This lower bound is surprisingly accurate. In Ref. 13 a numerical evaluation for the body-centered cubic lattice of particles in a uniform background gives

$$U_{\min} \leqslant -0.896N/r_s \tag{3}$$

when the system is neutral.

The significance of the lower bound, and the only place it will be used here, is to establish an upper bound for the partition function Z, i.e.,

$$Z \leqslant Z_{\text{ideal}}e^{\xi N} \tag{4}$$

where ξ is some constant and Z_{ideal} is the partition function of ideal, non-interacting particles. Thus, defining

$$g = V^{-1} \ln Z \tag{5}$$

for a domain of volume V, one has that g is bounded above.

3. CLASSICAL PARTICLES WITH PURELY COULOMB FORCES

3.1. Canonical Ensemble (Spherical Domains)

Fix the density ρ. Let $\{B_k\}_{k=0}^{\infty}$ be a sequence of balls of radii $R_k = R_0(1 + p)^k$, where $p = 26$ and the volume of $B_0 \equiv |B_0|$ is ρ^{-1}. Let $N_k = (1 + p)^{3k}$ be the number of particles in B_k, whence $\rho_k = N_k/|B_k| = \rho$. Let $n_j = p^{j-1}(1 + p)^{2j}$. According to Ref. 5, Section III, one can pack B_K with $\cup_{j=0}^{K-1}$ (n_{K-j} balls B_j) so that they do not overlap, and

$$\lim_{K\to\infty} |B_K|^{-1} \sum_{j=0}^{K-1} n_{K-j}|B_j| = 1 \tag{6}$$

The part of B_K not covered by the above packing will be called D_K.

At this point the principal difference between the proof for the jellium model and the proof for a system of positive and negative particles appears. In the latter, the N_K particles are constrained to be in the balls B_j, $j < K$, and the domain D_K is left empty. For jellium this cannot be done because the domain D_K would then not be neutral and the electrostatic energy of the system would be too large. Even though $|D_K|/|B_K| \to 0$ as $K \to \infty$, N_K^{-1} (the electrostatic energy of D_K) would go to infinity.

We proceed as follows: Let Z_k, $k = 0, 1, 2,\ldots$, be the configurational partition function of the ball B_k with N_k particles and with a uniform background of density ρ:

$$Z_k = (N_k!)^{-1} \int_{(B_k)^{N_k}} \exp[-\beta U(\mathbf{x}_1,\ldots, \mathbf{x}_{N_k})]\, d\mathbf{x}_1 \ldots d\mathbf{x}_{N_k} \tag{7}$$

Let $Z_K{}^D$ be the configurational partition function of D_K with M_K particles, where

$$M_K = N_K - \sum_{j=0}^{K=1} n_{K-j}N_j = N_K p^K(1 + p)^{-K} \tag{8}$$

D_K is understood to have a uniform background of density ρ. Clearly, $\rho D_k = M_k$ and $M_k/N_k \to 0$ exponentially fast.

The fundamental inequality, to be found in Ref. 5, Section IIE, is that

$$\ln Z_K \geqslant \sum_{j=0}^{K=1} n_{K-j} \ln Z_j + \ln Z_K{}^D \tag{9}$$

This inequality exploits Newton's electrostatic theorem and the fact that all the subdomains, except D_K, are both spherical and neutral; therefore the average interdomain interaction is zero.

The next step is to estimate $Z_K{}^D$. Using Jensen's inequality,

$$\ln Z_K{}^D \geqslant M_K \ln|D_K| - \ln(M_K!) - \beta\langle U\rangle_{D_K}$$

where

$$\begin{aligned}\langle U\rangle_{D_K} = {} & \tfrac{1}{2}M_K(M_K - 1)|D_K|^{-2}\int\int_{D_K} |\mathbf{x} - \mathbf{y}|^{-1}\, d\mathbf{x}\, d\mathbf{y} \\ & + \tfrac{1}{2}\rho^2\int\int_{D_K} |\mathbf{x} - \mathbf{y}|^{-1}\, d\mathbf{x}\, d\mathbf{y} \\ & - \rho M_K|D_K|^{-1}\int\int_{D_K} |\mathbf{x} - \mathbf{y}|^{-1}\, d\mathbf{x}\, d\mathbf{y}\end{aligned} \tag{10}$$

Since $M_K = \rho\int_{D_K} d\mathbf{x}$,

$$\langle U\rangle_{D_K} = -\tfrac{1}{2}\rho|D_K|^{-1}\int\int_{D_K} |\mathbf{x} - \mathbf{y}|^{-1}\, d\mathbf{x}\, d\mathbf{y} < 0 \tag{11}$$

Thus, defining

$$g_K = |B_K|^{-1} \ln Z_K \tag{12}$$

and

$$\gamma = p(1 + p)^{-1} < 1 \tag{13}$$

we have, for large K,

$$g_K \geqslant p^{-1}\sum_{j=0}^{K-1} \gamma^{K-j}g_j + \gamma^K\rho(1 - \ln\rho) \tag{14}$$

where Stirling's formula for $M_K!$ has been used. As shown in Ref. 5, Section IVD, this inequality implies that g_K *has a limit*, $g(\beta, \rho)$, *for this special, ρ-dependent sequence of domains* B_K.

3.2. Canonical Ensemble (General Domains)

Let ρ be fixed. We take a regular sequence of domains $\{\Lambda_j\}_{j=1}^{\infty}$ tending to infinity which satisfies conditions A (Van Hove limit) and B (ball condition) given in Ref. 5, Section V, and which *also* satisfies the condition that $\rho|\Lambda_j| = j$. To get a lower bound on $Z(\Lambda_j)$, we pack Λ_j with balls B_k of the standard sequence appropriate to ρ given above and distribute the j particles with constant density in the balls and in D_j, which is the complement of the B_k in Λ_j. As above, we have

$$|\Lambda_j|g_j \equiv \ln Z(\Lambda_j) \geqslant \sum_k m_{jk} \ln Z_k + \ln Z_j{}^D \tag{15}$$

where m_{jk} is the number of balls B_k in the packing of Λ_j. Met M_j be the number of particles in D_j, i.e.,

$$M_j = j - \sum_k m_{jk}(1 + p)^{3k}$$

Then, as before,

$$\ln Z_j{}^D \geqslant M_j \ln|D_j| - \ln(M_j!) - \beta\langle U\rangle_{D_j} \tag{16}$$

and $\langle U\rangle_{D_j} \leqslant 0$. Following the proof in Ref. 5, Section V,

$$\liminf_{j\to\infty} g_j \geqslant g(\beta, \rho) \tag{17}$$

where $g(\beta, \rho)$ is the limit for the standard balls.

An upper bound to $Z(\Lambda_j)$ can be found by embedding Λ_j in a minimum standard ball $B_{K(j)}$ and packing $B_{K(j)}\backslash\Lambda_j$ with balls B_k. Let D_j be as before, i.e., $B_{K(j)}\backslash(\Lambda_j \cup B_k)$ and $M_j = \rho|D_j|$. Then

$$\begin{aligned} \ln Z_{K(j)} \geqslant \ln Z(\Lambda_j) + \sum m'_{jk} \ln Z_k + M_j \ln|D_j| - \ln(M_j!) \\ - \beta\langle U(D_j, D_j)\rangle - \beta\langle U(D_j, \Lambda_j)\rangle \end{aligned} \tag{18}$$

In the last four terms we use Jensen's inequality for the integration over the coordinates of the particles in D_j: $\langle U(D_j, D_j)\rangle$ is the average Coulomb energy in D_j in an ensemble in which the particles are free; $\langle U(D_j, \Lambda_j)\rangle$ is the average interdomain interaction between D_j and Λ_j when the particles in D_j are free and the particles in Λ_j are fully interacting. The last term is zero because the average total charge distribution in D_j is zero. The term $\langle U(D_j, D_j)\rangle$ is negative as before. Thus we can use the argument of Ref. 5, Section V, to conclude that

$$\limsup_{j\to\infty} g_j \leqslant g(\beta, \rho) \tag{19}$$

The result of these inequalities is that for any regular sequence of domains $\{\Lambda_j\}$ and particle numbers $N_j = j$ such that $N_j = \rho|\Lambda_j|$,

$$\lim_{j\to\infty} g_j = g(\beta, \rho) \tag{20}$$

While this establishes the existence and shape independence of the thermodynamic limit for each fixed ρ, we do not yet know anything about the dependence of $g(\beta, \rho)$ on ρ or whether the limit is uniform in ρ. We next discuss how such a relationship can be obtained.

3.3. Scaling Relations

Let $\{\Lambda_j\}_{j=1}^{\infty}$ be a regular sequence of domains for a given ρ, i.e., $\rho|\Lambda_j| = j$. Let $\eta > 0$ be fixed and define the following:

$$\rho' = \rho\eta^3, \qquad \beta' = \beta\eta^{-1}, \qquad \Lambda_j' = \eta^{-1}\Lambda_j = \{\eta^{-1}\mathbf{x}|\mathbf{x} \in \Lambda_j\} \tag{21}$$

Thus $\rho'|\Lambda_j'| = j$.

If one considers the integral defining $Z(\beta, \rho; \Lambda_j)$ and changes integration variables $\mathbf{x}$ to $\mathbf{y} = \eta^{-1}\mathbf{x}$, then one derives

$$|\Lambda_j|g(\beta, \rho; \Lambda_j) = |\Lambda_j'|g(\beta\eta^{-1}, \rho\eta^3; \Lambda_j') + 3j \ln \eta \tag{22}$$

Since the thermodynamic limit is independent of the sequence of domains, one has that

$$g(\beta, \rho) = \eta^{-3}g(\beta\eta^{-1}, \rho\eta^3) + 3\rho \ln \eta \tag{23}$$

for all $\eta > 0$. Now let $\eta = \rho^{-1/3}$, whence

$$g(\beta, \rho) = \rho g(\beta\rho^{1/3}, 1) - \rho \ln \rho = \rho\bar{g}(\beta\rho^{1/3}) + \rho(1 - \ln \rho) \tag{24}$$

where $\bar{g}(\cdot) \equiv g(\cdot, 1) - 1$.

From the basic definition of $g(\beta, \rho; \Lambda_j)$ one has that these functions, and hence their limits also, are convex functions of β. Therefore the function $t \to \bar{g}(t)$ is convex in t.

For finite j, let $\{\Lambda_j'\}$ be a regular sequence of domains with $|\Lambda_j'| = j$, and define

$$\bar{g}(\beta) = g(\beta, 1; \Lambda_j') - 1 \tag{25}$$

Then

$$\lim_{j\to\infty} \bar{g}_j(\beta) = \bar{g}(\beta) \tag{26}$$

3.4. Properties of the Thermodynamic Limit

3.4.1. Uniformity of the Limit. Since $\bar{g}(t)$ is bounded on finite t intervals, its convexity implies that it is continuous. Furthermore, each $\bar{g}_j(\cdot)$ has the same properties from (25). Thus the sequence of functions $\bar{g}_j(\beta\rho^{1/3})$ is continuous in ρ and has a continuous limit $\bar{g}(\beta\rho^{1/3})$ and the limit is essentially monotone as one sees from (14). Hence Dini's theorem tells us that the limit is uniform on compact ρ intervals.

3.4.2. Pressure and Compressibility. For a normal thermodynamic system, $g(\beta, \rho)$ is *concave* in ρ. This implies positive compressibility and, since the pressure is zero at zero density, it implies positive pressure. For jellium this is unfortunately not true. Using (24), and assuming differentiability, we obtain

$$\beta P/\rho = 1 - \tfrac{1}{3}\beta\rho^{1/3}\dot{\bar{g}}(\beta\rho^{1/3}) \tag{27}$$

where the dot denotes derivative, and

$$\beta\kappa^{-1} = \beta\frac{dP}{d\rho} = 1 - \frac{4}{9}\beta\rho^{1/3}\dot{\bar{g}}(\beta\rho^{1/3}) - \frac{\beta^2}{9}\rho^{2/3}\ddot{\bar{g}}(\beta\rho^{1/3}) \tag{28}$$

Note that $\dddot{g} > 0$. This implies that for $t > 0$, $\ddot{g}(t) \geqslant \ddot{g}(0) = 0$. These formulas show that P and κ can have either sign. In fact, for fixed β, they are both negative for sufficiently high density since, from (3), one sees that the potential energy will go as $\rho^{4/3}$ for large ρ, i.e., $g(t) \sim t$ for large t. Since $\ddot{g}(t)$ is monotone, $t\ddot{g}(t)$ is also monotone. This implies that there is always exactly one value, $(\beta\rho^{1/3})_c$, of $\beta\rho^{1/3}$ at which the pressure is zero. Without any constraint on the volume, classical jellium would collapse to a density $\rho^{1/3} = (\beta\rho^{1/3})_c\beta^{-1}$. This fact is not unrelated to the absence of H-stability for real matter without Fermi statistics.

3.5. Systems That Are Not Neutral

We wish to consider a sequence of systems with fixed background density ρ, but where $N \neq \rho|\Lambda|$. Define $Q_j \equiv -N_j + \rho|\Lambda_j|$ to be the net charge in Λ_j, and consider a sequence of domains Λ_j of *fixed shape* of capacitance $C_j = c|\Lambda_j|^{1/3}$. If $Q_j|\Lambda_j|^{-2/3} \to \sigma$, the result to be proved is that

$$g_j(\beta, \rho) \to g(\beta, \rho) - \tfrac{1}{2}\sigma^2 c \tag{29}$$

Note that σ can have either sign. If $|\sigma| = \infty$, then $g_j(\beta, \rho) \to -\infty$. This last statement is easily proved by noting that $|\Lambda_j|^{-1}\min\{U(\mathbf{x})|\mathbf{x}_i \in \Lambda_j\} \to +\infty$ when $|Q_j|\,|\Lambda_j|^{-2/3} \to +\infty$.

In order to simplify matters we shall prove the theorem only for balls, in which case $c = (4\pi/3)^{-1/3}$.

Let B be a ball of radius R and let B' be a concentric ball of radius $R' > R$. Note that a uniform charge density τ placed in $\Sigma \equiv B'\backslash B$ produces a constant potential $\tau\Phi(\Sigma)$ inside B. This same charge density in Σ has a self-energy $\tau^2 S(\Sigma)$. If $R \to \infty$ and $R'/R \to 1$, then

$$\Phi(\Sigma)|\Sigma|/S(\Sigma) \to 2 \tag{30}$$

Let $Z(N, B')$ be the partition function for N particles in B' with background density ρ. A lower bound to $Z(N, B')$ can be obtained as follows:

1. Restrict the configurations to N_1 particles in B and $N_2 = N - N_1$ particles in Σ.
2. Let $U_1(\mathbf{X}_1)$ [resp. $U_2(\mathbf{X}_2)$] be the potential energy of the particles and background in B [resp. Σ] and let $U_{12}(\mathbf{X}_1, \mathbf{X}_2)$ be the interdomain energy, where $\mathbf{X}_1$ and $\mathbf{X}_2$ are the particle coordinates. Then

$$Z(N, B') \geqslant (N_1!\, N_2!)^{-1} \int_{B^{N_1}} \exp[-\beta U_1(\mathbf{X}_1)] \times \int_{\Sigma^{N_2}} \exp\{-\beta[U_2(X_2) + U_{12}(\mathbf{X}_1, \mathbf{X}_2)]\} \tag{31}$$

3. Use Jensen's inequality on the second integral together with the aforementioned constancy of the potential $\Phi(\Sigma)$. Thus,

$$\begin{aligned}\ln Z(N, B') \geqslant \ln Z(N_1, B) &+ \ln\{|\Sigma|^{N_2}/N_2!\} \\ &- \beta S(\Sigma)\left[\tfrac{1}{2}\rho^2 + \binom{N_2}{2}|\Sigma|^{-2} - N_2\rho|\Sigma|^{-1}\right] \\ &- \beta\Phi(\Sigma)[\rho - N_2|\Sigma|^{-1}][\rho|B'| - N_1]\end{aligned} \tag{32}$$

Now we consider a sequence of balls B_j of radii R_j with background density ρ and particle numbers $N_j = j$, $j = 1, 2, \ldots$. For $Q_j \equiv -j + \rho|B_j|$ negative we first use (32) with $N = j$, $B' = B_j$, $R = R_j - 1$, and $N_1 = \rho|B|$. Then we use (32) with $N = N_1 = j$, $B = B_j$, and $|B'| = j/\rho$. When $Q_j > 0$, we first use (32) with $N = N_1 = j$, $B' = B_j$, and $|B| = j/\rho$. Then we use (32) with $N_1 = j$, $B = B_j$, $R' = R_j + 1$, and $N = |B'|\rho$. Using the fact that $Q_j|B_j|^{-2/3} \to \sigma$ and (30), we obtain the desired result (29).

3.6. Microcanonical Ensemble

The existence of the thermodynamic limit for the microcanonical ensemble can be demonstrated using the methods of Ref. 5, Section VIII. There, the energy as a function of entropy was given for the quantum case. The corresponding classical equation is as follows: Let $\Gamma(N, \Lambda) = (\Lambda \times \mathbf{R}^3)^N$ be the phase space (including momentum). For σ real, let

$$\Delta(\sigma, N, \Lambda) = \{A \subset \Gamma(N, \Lambda) | \mu(A) = e^{\sigma|\Lambda|}\} \tag{33}$$

where μ is Lebesgue measure. Let

$$\epsilon(A, N, \Lambda) = |\Lambda|^{-1}\int_A H(\mathbf{X}, \mathbf{P})e^{-\sigma|\Lambda|} \tag{34}$$

where

$$H(\mathbf{X}, \mathbf{P}) = U(\mathbf{X}) + \sum \mathbf{p}_i^2/2m$$

Then we define

$$\epsilon(\sigma, N, \Lambda) = \inf\{\epsilon(A, N, \Lambda) | A \in \Delta(\sigma, N, \Lambda)\} \tag{35}$$

to be the energy per unit volume as a function of the entropy per unit volume, σ.

Obviously, when Λ_1 and Λ_2 are disjoint,

$$\Delta(\sigma_1 + \sigma_2, N_1 + N_2, \Lambda_1 \cup \Lambda_2) \supset \Delta(\sigma_1, N_1, \Lambda_1) \times \Delta(\sigma_2, N_2, \Lambda_2) \tag{36}$$

Hence

$$\begin{aligned}&|\Lambda|\epsilon(\sigma_1 + \sigma_2, N_1 + N_2, \Lambda_1 \cup \Lambda_2) \\ &\qquad \leqslant |\Lambda_1|\epsilon(\sigma_1, N_1, \Lambda_1) + |\Lambda_2|\epsilon(\sigma_2, N_2, \Lambda_2) + \langle U(\Lambda_1, \Lambda_2)\rangle\end{aligned} \tag{37}$$

where $\langle U \rangle$ is the average over A_1 and A_2 of the interaction energy between Λ_1 and Λ_2. [This may require passing to a subsequence in (35) for Λ_1 and Λ_2.]

Now we are in the same position as in (16); the existence and appropriate convexity properties of $\epsilon(\sigma)$ and $\sigma(\epsilon)$ follow. See Ref. 5, Section VIII for details.

The one essential difference from the systems studied in Ref. 5, Section VIII is that for jellium we do not obtain convexity of ϵ as a function of ρ, but only as a function of σ. This lack does not alter the equivalence of the canonical and microcanonical ensembles.

3.7. The Grand Canonical Ensemble

If one considers the grand canonical ensemble (GCE) for fixed Λ, fixed background density ρ, and fixed chemical potential μ, then the GCE partition function Ξ will exist. From the results of Section 3.5 the thermodynamic limit of $\pi = |\Lambda|^{-1} \ln \Xi$ will exist and $\pi = \rho\mu + g(\rho)$ as in Theorem 7.1 of Ref. 5, Section VII. If, on the other hand, one defines Ξ for neutral jellium by requiring that $\rho = N/|\Lambda|$ for each N, then Ξ will diverge, even for finite Λ. This is a consequence of (3) that $g(N, \Lambda) \sim N^{4/3}$ for large N. In the quantum case with fermions, this divergence will not occur since the kinetic energy is proportional to $N^{5/3}$. Although the thermodynamic limit of π for neutral jellium would then exist, it would not be equivalent to the canonical partition ensemble because of the lack of convexity of the free energy in ρ.

4. ADDITIONAL POTENTIALS

As was shown in Ref. 5, additional short-range forces among the particles can be included without any conceptual difficulty, but with a great deal of technical difficulty, provided they are tempered and provided that these forces are integrable. This means that hard cores are excluded. We do not say that the thermodynamic limit does not exist when hard cores are present—it probably does—but only that our method is not adequate. The difficulty arises in (11), where $\ln Z_K{}^D$ is estimated by Jensen's inequality in terms of $\langle U \rangle$. We made this estimate in order to show that the energy of the particles in D_K was not too large. If some other method could be found to show this, then perhaps hard cores could be included, but in our estimate, $\langle U \rangle = +\infty$ and $\ln Z_K{}^D \geqslant -\infty$ when hard cores are present.

There is another serious difficulty when additional potentials, even nice ones, are present. The scaling relation of Section 3.3 does not hold, and hence the continuity with respect to ρ that was used in Section 3.4.1 cannot be established that way.

5. QUANTUM MECHANICAL PARTICLES

We first remark that it is immaterial for our purposes whether the particles are bosons or fermions. In contrast to the situation for real particles, H-stability (2) holds in the classical sense and therefore Fermi statistics is not required. We shall construct the proof for fermions and it will obviously be valid for bosons as well. Dirichlet boundary conditions will be employed, i.e., $\psi = 0$ on the boundary of Λ.

5.1. Canonical Ensemble (Spherical Domains)

By well-known arguments (see Ref. 5, Section II), a lower bound to Z_K can be obtained by constraining the particles to lie in various subdomains. In this way we arrive at precisely the same inequality (9) as for the classical case. The problem is to show that $\ln Z_K{}^D$ is not too small. For this purpose it would be sufficient to find one wave function ψ for the M_K particles in D_K such that $\langle H_K \rangle \equiv \langle \psi, H_K \psi \rangle <$ (positive constant)M_K, where H_K is the total Hamiltonian of the M_K particles in D_K, including the background self-energy. Then, by the Peierls–Bogoliubov inequality,

$$\ln Z_K{}^D \geqslant -\beta \langle H_K \rangle \tag{38}$$

A natural suggestion would be to take a determinantal wave function that vanishes on the boundary of D_K, but this will not work for the reason that the single-particle density will not be a constant and consequently the estimate $\langle U \rangle_{D_K} < 0$ [Eq. (11)] will not hold. On the other hand, suppose one could find M_K points $\mathbf{Y} = \{\mathbf{y}_1, \ldots, \mathbf{y}_{M_K}\}$ in D_K such that:

(a) $U(\mathbf{Y}) <$ (positive constant)M_K.
(b) $|\mathbf{y}_i - \mathbf{y}_j| > 2h$ for some fixed $h > 0$.
(c) The distance of $\mathbf{y}_i$ to the boundary of D_K is $> h$ for all i.

Then one could construct a product wave function in which the single-particle wave functions have support in balls of radius h centered at the $\mathbf{y}_i$ and which are spherically symmetric about the $\mathbf{y}_i$. The kinetic energy would be proportional to h^{-2}. Due to the peculiar shape of D_K, we are unable to find such a $\mathbf{Y}$. What Eq. (11) shows is that there certainly exists a $\mathbf{Y}$ satisfying condition (a) but we do not know if it satisfies (b) and (c).

It is in fact possible to find a $\mathbf{Y}$ such that condition (a) is satisfied and condition (b) is *effectively* satisfied. To do this, define

$$U'(\mathbf{Y}) = U(\mathbf{Y}) + \sum_{i<j}^{M_K} L(\mathbf{y}_i - \mathbf{y}_j) \tag{39}$$

where $U(\mathbf{Y})$ is the Coulomb potential as before and

$$\begin{aligned} L(\mathbf{y}) &= 2(\pi/|\mathbf{y}|)^2 \qquad \text{for} \quad |\mathbf{y}| \leqslant 1 \\ &= 0 \qquad\qquad\quad \text{for} \quad |\mathbf{y}| > 1 \end{aligned} \tag{40}$$

Using (11), we obtain

$$\langle U' \rangle_{D_K} = \langle U \rangle_{D_K} + \binom{M_K}{2} |D_K|^{-2} \int_{D_K} \int L(\mathbf{x} - \mathbf{y})\, d\mathbf{x}\, d\mathbf{y}$$

$$\leqslant \binom{M_K}{2} |D_K|^{-1} 8\pi^3 \leqslant 4\pi^3 M_K \rho \qquad (41)$$

Therefore there exists a $\mathbf{Y}$ such that

$$U'(\mathbf{Y}) \leqslant 4\pi^3 M_K \rho \qquad (42)$$

Let $d_i = \frac{1}{2} \min\{1, \min_{j \neq i} |\mathbf{y}_i - \mathbf{y}_j|\}$. Construct a product trial function ψ using single-particle functions $\{\varphi_i\}_{i=1}^{M_K}$ centered at $\mathbf{y}_i$ and having support in a ball of radius d_i of the form

$$\varphi_i(\mathbf{x}) = (2\pi d_i)^{-1/2} |\mathbf{x} - \mathbf{y}_i|^{-1} \sin[\pi |\mathbf{x} - \mathbf{y}_i| / d_i] \qquad (43)$$

The kinetic energy of φ_i is $(\pi/d_i)^2/2$. The potential energy of ψ can be evaluated as follows: The particle–particle energy is the same as if the particles were located at $\mathbf{y}_i$, by Newton's theorem. The interaction of a smeared-out particle with the background is changed by the amount

$$\rho \int_{|\mathbf{x}|<d_i} d\mathbf{x} \int_{|\mathbf{y}|<d_i} d\mathbf{y}\, |\mathbf{x} - \mathbf{y}|^{-1} [\varphi_i(\mathbf{x} + \mathbf{y}_i)^2 - \delta(\mathbf{x})]$$

$$= \xi \rho\, d_i^2 \leqslant \xi \rho \qquad (44)$$

where ξ is a constant, assuming that the ball of radius d_i lies entirely in D_K. Thus the total energy $\langle \psi, H_K \psi \rangle$ is less than

$$U'(Y) + \xi \rho M_K + 2\pi^2 M_K \leqslant (2\pi^2) M_K [1 + 2\pi\rho + \xi\rho(2\pi^2)^{-1}] \qquad (45)$$

This result is exactly what conditions (a) and (b) would give.

Condition (c) is more difficult, for it requires that the coordinates in $\mathbf{Y}$ are not too close to the boundary. If one tries to introduce

$$U''(\mathbf{Y}) = U'(\mathbf{Y}) + \sum_i d(\mathbf{y}_i)^{-2}$$

where $d(\mathbf{y}_i)$ is the distance to the boundary, one will find that $\langle U'' \rangle_{D_K} = \infty$ since $d(\mathbf{x})^{-2}$ is not integrable.

Since we are unable to deal with this problem directly, we shall modify our basic construction for the ball packing in such a way that the balls have a minimum spacing of some length independent of K.

Let $\{B_k\}_{k=0}^{\infty}$ be a sequence of balls of radii $R_k = R_0'(1 + p)^k (1 - \frac{1}{2}\theta^k)$ with $\theta = (1 + p)^{-1}$, $p = 26$, and R_0' chosen so that $\rho |B_0| = 1$. Let $N_k = \rho |B_k|$, whence N_k is an integer. As shown in Ref. 5, Section IV, it is possible to pack B_K with n_{K-j} balls B_j so that the distance of every ball to the boundary is not less than $4h$ and the distance between balls is not less than $8h$, where

$h = R_0'(1 - \theta)/8$. As in Section 3.1, the part of B_K not covered by the packing will be called D_K.

Let us label the individual balls in the packing of B_K by a superscript i, namely B^i, and let R^i be its radius. Around B^i we construct two concentric, spherical shells S^i and T^i of radii $(R^i, R^i + h)$ and $(R^i + h, R^i + 2h)$, respectively. *Inside* B_K we also construct two concentric, spherical shells S_K and T_K of radii $(R_K - h, R_K)$ and $(R_K - 2h, R_K - h)$, respectively. All these shells are disjoint and lie in D_K and we denote by $D_K' \subset D_K$ the complement of the shells in D_K, and define $D_K'' = D_K' \cup T^i \cup T_K$.

We wish to find a $\mathbf{Y} = \{\mathbf{y}_1, \ldots, \mathbf{y}_{M_K}\}$ with $\mathbf{y}_i \in D_K''$, and a corresponding product wave function ψ such that $\langle H_K \rangle$ is not too large. To this end, let

$$\begin{aligned} f(\mathbf{y}) &= 1, && \mathbf{y} \in D_K' \\ &= 0, && \mathbf{y} \notin D_K'' \\ &= f^i, && \mathbf{y} \in T^i \\ &= f_K, && \mathbf{y} \in T_K \end{aligned} \tag{46}$$

where

$$\begin{aligned} f^i &= 1 + [(R^i + h)^3 - (R^i)^3][(R^i + 2h)^3 - (R^i + h)^3]^{-1} < 2 \\ f_K &= 1 + [(R_K - h)^3 - (R_K - 2h)^3]^{-1}[R_K{}^3 - (R_K - h)^3] < 3 \end{aligned} \tag{47}$$

whence

$$\int_{T^i} (f - 1) = \int_{S^i} 1$$

and similarly for T_K, S_K, and $\int f = |D_K|$.

Let

$$F(\mathbf{Y}) = \prod_{i=1}^{M_K} f(\mathbf{y}_i)$$

and let

$$\langle U' \rangle_F = \int F(\mathbf{Y}) U'(\mathbf{Y}) \Big/ \int F(\mathbf{Y}) \tag{48}$$

The part involving L is [using (41)]

$$\binom{M_K}{2} |D_K|^{-2} \int \int f(\mathbf{x}) f(\mathbf{y}) L(\mathbf{x} - \mathbf{y}) \leqslant 9 \cdot 4\pi^3 M_K \rho$$

since $f(\mathbf{x}) \leqslant 3$. The part involving $U(\mathbf{Y})$ is

$$\begin{aligned} &\tfrac{1}{2}\rho^2 \int_{D_K} d\mathbf{x} \int_{D_K} d\mathbf{y} |\mathbf{x} - \mathbf{y}|^{-1} [1 - f(\mathbf{x})][1 - f(\mathbf{y})] \\ &\quad - \tfrac{1}{2}\rho |D_K|^{-1} \int_{D_K} d\mathbf{x} \int_{D_K} d\mathbf{y}\, |\mathbf{x} - \mathbf{y}|^{-1} f(\mathbf{x}) f(\mathbf{y}) \end{aligned} \tag{49}$$

The second term is negative. The first is the Coulomb energy of double shells, each pair of which is neutral and spherically symmetric. By Newton's theorem, this is just the sum of the self-energies of each pair. Let E_k be the self-energy of the two shells S and T surrounding a ball B_k in the packing and let W_K be the self-energy of the shells S_K and T_K. Then

$$\langle U' \rangle_F \leqslant 36\pi^3 M_K \rho + \sum_{j=0}^{K-1} n_{K-j} E_j + W_K \leqslant \text{const} \times M_K \tag{50}$$

The latter inequality comes from an elementary calculation of E_k and W_K.

The conclusion is that there exists a $\mathbf{Y}$ with $\mathbf{y}_i \in D''_K$ such that $U'(\mathbf{Y})$ is bounded by a constant times $|D_K|$. Now we construct a trial function ψ as before with φ_i given by (43) except that

$$d_i = \tfrac{1}{2} \min\{2h, 1, \min_{j \neq i} |\mathbf{y}_i - \mathbf{y}_j|\}$$

Then

$$\begin{aligned} |B_K|^{-1} \ln Z_K{}^D &\geqslant -\beta |B_K|^{-1} \langle \psi, H_K \psi \rangle \\ &\geqslant -\beta \times \text{const} \times |D_K| \, |B_K|^{-1} \end{aligned} \tag{51}$$

and this goes to zero as $K \to \infty$ like γ^K [Eq. (13)]. Thus the thermodynamic limit is established as in the classical case.

5.2. Canonical Ensemble (General Domains)

Let ρ be fixed. We take a regular sequence of domains $\{\Lambda_j\}_{j=1}^{\infty}$ tending to infinity which satisfies conditions A and B of Ref. 5, Section V. Also, $|\Lambda_j|\rho = j$. In addition, we require some conditions on the sequence which are not required in the classical case. These are the following:

(i) Let $h > 0$ and let $\Lambda_j{}^h$ and Λ_j^{2h} be the domains

$$\begin{aligned} \Lambda_j{}^h &= \{\mathbf{x} \in R^3 | \mathbf{x} \notin \Lambda_j, d(\mathbf{x}; \partial\Lambda_j) \leqslant h\} \\ \Lambda_j^{2h} &= \{\mathbf{x} \in R^3 | \mathbf{x} \notin \Lambda_j{}^h \cup \Lambda_j, d(\mathbf{x}; \partial\Lambda_j{}^h) \leqslant h\} \end{aligned} \tag{52}$$

where $d(\cdot\,;\,\cdot)$ is the Euclidean distance. We require that $|\Lambda_j{}^h|/|\Lambda_j^{2h}|$ be bounded in j for each fixed h.

(ii) Consider the charge density

$$\begin{aligned} \sigma_j{}^h(\mathbf{x}) &= 1, & \mathbf{x} &\in \Lambda_j{}^h \\ &= -|\Lambda_j{}^h|/|\Lambda_j^{2h}|, & \mathbf{x} &\in \Lambda_j^{2h} \end{aligned} \tag{53}$$

Thus $\sigma_j{}^h$ is neutral. Let $\varphi_j{}^h(\mathbf{x})$ be the Coulomb potential of σ_j. We require that there exists a function $C(h) < \infty$ such that for all $\mathbf{x} \in \Lambda_j^{2h} \cup \Lambda_j{}^h \cup \Lambda_j$

$$|\varphi_j{}^h(\mathbf{x})| < C(h) \tag{54}$$

and that

$$\lim_{h\to 0} C(h) = 0$$

(iii) Let E_j^h be the Coulomb self-energy of the double layer σ_j^h. We require that

$$\lim_{j=\infty} |\Lambda_j|^{-1} E_j^h = 0 \qquad (55)$$

Conditions (i) and (ii) obviously imply (iii) since

$$E_j^h \leqslant \tfrac{1}{2}[\sup_{\mathbf{x}}|\sigma_j^h(\mathbf{x})|\,|\varphi_j^h(\mathbf{x})|][|\Lambda_j^{2h}| + |\Lambda_j^h|]$$

and

$$[|\Lambda_j^{2h}| + |\Lambda_j^h|]/|\Lambda_j| \to 0$$

by the Van Hove limit.

(iv) Define $\tilde{\Lambda}_j^h$, $\tilde{\Lambda}_j^{2h}$, $\tilde{\sigma}_j^h(x)$, and $\tilde{E}_j^h$ similarly to the above except $\mathbf{x} \notin \Lambda_j$ (resp. $\mathbf{x} \notin \Lambda_j^h \cup \Lambda_j$) is replaced by $\mathbf{x} \in \Lambda_j$ (resp. $\mathbf{x} \in \Lambda_j \backslash \Lambda_j^h$). That is, the double layer $\tilde{\sigma}_j^h$ is now inside Λ_j. We require that

$$\lim_{j\to\infty} |\Lambda_j|^{-1} \tilde{E}_j^h \to 0 \qquad (56)$$

We do not require that the analogs of (i) and (ii) hold.

It is clear that for any reasonable sequence of domains, such as cubes or ellipsoids, these conditions will be satisfied. We shall not attempt to determine geometric conditions on the Λ_j so that (i)–(iv) hold.

Let Λ_j contain j particles. As in the classical case, we derive a lower bound for $Z(\Lambda_j)$. The kinetic energy for D_j can be handled as in Section 5.1. The only essential difference from inequality (15) is that we have to add the self-energy of the double layer $\tilde{\sigma}_j^h$ inside Λ_j and that of the double layers of the balls in the packing of Λ_j. Call this latter quantity $W(j)$. Thus, on the right side of the inequality we must add $-\beta \tilde{E}_j^h - \beta W(j)$. Using condition (iv), we have that

$$\liminf g_j \geqslant g(\beta, \rho) \qquad (57)$$

An upper bound for $Z(\Lambda_j)$ is also obtained as in the classical case (18). For the domain D_j we choose a vector state ψ and have to compute

$$E(\psi) = \langle H_K \rangle + \langle U(D_j, \Lambda_j) \rangle \qquad (58)$$

The kinetic energy part of $\langle H_K \rangle$ can be handled as in Section 5.1. If M_j is the number of particles in D_j, we have to find M_j points in $D_j'' = D_j -$

$\{\bigcup S^i \cup \Lambda_j^h\}$ in such a way that $E(\psi)$ is not too large. The novel feature is that $U(D_j, \Lambda_j)$ involves the additional term

$$\sum_{i=1}^{M_j} w(\mathbf{x}_i) \qquad \text{where} \qquad w(\mathbf{x}) = \int_{\Lambda_j} \rho_j(\mathbf{y})|\mathbf{x} - \mathbf{y}|^{-1}\, d\mathbf{y}$$

and $\rho_j(\mathbf{y})$ is the average charge distribution (including the background) in Λ_j in the canonical distribution. Although Λ_j is neutral, $w \neq 0$ because Λ_j is not spherical. To find these M_j points we again average over all allowed configuration in D_j''.

The self-energy of the double layers S^i and T^i is small for the same reason as in Section 5.1. The problem then reduces to computing E_j^h as defined in condition (iii), together with the energy of the charge distribution σ_j^h in the potential w. Condition (iii) states that $|\Lambda_j|^{-1}E_j^h \to 0$, so we can ignore it. The latter contribution, Δ_j, can be bounded as follows:

$$\begin{aligned}
|\Delta_j| &= \left| \int w(\mathbf{x})\rho\sigma_j^h(\mathbf{x}) \right| \\
&= \left| \int_{\Lambda_j} d\mathbf{x} \int d\mathbf{y}\ \rho_j(x)|\mathbf{x} - \mathbf{y}|^{-1}\rho\sigma_j^h(\mathbf{x}) \right| \\
&= \rho\left| \int_{\Lambda_j} d\mathbf{x}\ \rho_j(\mathbf{x})\varphi_j^h(\mathbf{x}) \right| \leqslant \rho C(h) \int_{\Lambda_j} d\mathbf{x}\ |\rho_j(\mathbf{x})| \\
&\leqslant \rho C(h) \int_{\Lambda_j} |\rho_+(\mathbf{x})| + |\rho_-(\mathbf{x})| \leqslant 2j\rho C(h)
\end{aligned} \tag{59}$$

where $\rho_+ = \rho$ is the background charge and $\rho_-(\mathbf{x})$ is the average particle charge distribution.

Now we divide by $|\Lambda_j| = j$ and let $j \to \infty$. For each fixed h we obtain

$$\limsup_{j\to\infty} g_j \leqslant g(\beta, \rho) + \beta\rho C(h) \tag{60}$$

Since h is arbitrary, we can now let $h \to 0$ and, recalling condition (ii), obtain

$$\lim_{j\to\infty} g_j = g(\beta, \rho) \tag{61}$$

which is the desired result.

5.3. Scaling Relations

In Section 3.3 we showed that $g(\beta, \rho) = \rho(1 - \ln \rho) + \rho\bar{g}(\beta\rho^{1/3})$. Such a simple relation will not hold quantum mechanically. To obtain a similar result quantum mechanically, we have to add another parameter; the simplest is α, the square of the electric charge. Thus $H = K + U \rightarrow K + \alpha U$, where K is the kinetic energy operator.

We make a scale change which now involves α:

$$\rho' = \rho\eta^3, \qquad \beta' = \beta\eta^{-2}, \qquad \alpha' = \alpha\eta, \qquad \Lambda_j' = \eta^{-1}\Lambda_j \tag{62}$$

Then, as in Section 3.3 [Eq. (22)],

$$|\Lambda_j| g(\beta, \rho, \alpha; \Lambda_j) = |\Lambda_j'| g(\beta\eta^{-2}, \rho\eta^3, \alpha\eta; \Lambda_j') \tag{63}$$

Again, choosing $\eta = \rho^{-1/3}$, and taking the limit $j \rightarrow \infty$, we obtain

$$g(\beta, \rho, \alpha) = \rho g(\beta\rho^{2/3}, 1, \alpha\rho^{-1/3}) \tag{64}$$

This equation tells us nothing that we did not know before, i.e., α is an inessential parameter. But it does tell us something important about the continuity with respect to ρ. Define $\gamma = \beta\alpha$. Then

$$\ln Z = \ln \operatorname{Tr} \exp(-\beta K - \gamma U) \tag{65}$$

is a *jointly* convex function of (β, γ) for $\beta > 0$. Thus, when $\gamma > 0$, the thermodynamic limit $g(\beta, \rho, \gamma\beta^{-1})$ is convex, and hence continuous, in (β, γ). Hence the function $g(x, 1, y)$ is continuous in (x, y) when $x, y > 0$. Therefore $g(\beta, \rho, \alpha)$ is continuous in ρ for $\rho > 0$.

5.4. Properties of the Thermodynamic Limit and Related Questions

The results given for the classical case in Sections 3.4–3.6 and 4 hold for the quantum case. The conclusions of Section 3.7 have to be modified. In summary one has:

(i) Uniformity and continuity of the limit.

(ii) Unusual behavior of the pressure and compressibility.

(iii) Equivalence of canonical and microcanonical ensembles. The existence of the thermodynamic limit of the microcanonical ensemble includes as a special case the existence of the limiting ground state energy per unit volume. This is also true classically.

(iv) Existence of the grand canonical pressure even for strictly neutral systems because for large ρ, the quantum kinetic energy, which behaves like $\rho^{5/3}$, will dominate the electrostatic $-\rho^{4/3}$ term. We shall not prove this statement since the lack of convexity in ρ prevents the grand canonical ensemble from being equivalent to the canonical ensemble.

(v) The possibility of adding tempered potentials, with the same caveat as in Section 4.

APPENDIX. A LOWER BOUND FOR THE CLASSICAL AND QUANTUM MECHANICAL GROUND-STATE ENERGY

Consider a bounded, measurable set Λ with a uniform charge density ρ and N point particles of charge -1. We do not assume that Λ is spherical, that the points are constrained to lie in Λ, or that the total charge is zero.

To find a lower bound for the total electrostatic energy we use an idea of Onsager[12] to replace point charge distributions by charges smeared around the initial points. In fact one can show, by taking functional derivatives, that the best smearing is a uniform charge distribution inside a ball of radius a.

We define

U_{BB} the self-energy of the background;
U_i the interaction energy of the particle i at position $\mathbf{x}_i$ with the background;
U_{ij} the interaction of two particles at positions $\mathbf{x}_i$ and $\mathbf{x}_j$;
$\hat{U}_{ij}(a)$ the interaction (or twice the self-energy when $i = j$) of balls of total charge -1 with centers $\mathbf{x}_i$ and $\mathbf{x}_j$;
$\hat{U}_i(a)$ the interaction of such a ball with the background.

Then, with $\mathbf{X} = \{\mathbf{x}_1,..., \mathbf{x}_N\}$

$$U(\mathbf{X}) = U_{BB} + \sum_{i=1}^{N} U_i + \sum_{i<j} U_{ij}$$

$$\begin{aligned} U(X) = U_{BB} + \sum_{i=1}^{N} \hat{U}_i + \frac{1}{2}\sum_{i,j}^{N} \hat{U}_{ij} \qquad &(\alpha)\\ + \sum_i (U_i - \hat{U}_i) \qquad &(\beta)\\ - \frac{1}{2}\sum_i \hat{U}_{ii} \qquad &(\gamma)\\ + \sum_{i<j} (U_{ij} - \hat{U}_{ij}) \qquad &(\delta)\end{aligned}$$

Let us consider these terms individually:

The (α) terms are evidently positive, being the total electrostatic energy of the background charge and the charged balls.

In (δ), a term $U_{ij} - \hat{U}_{ij}$ is zero if the two balls do not overlap by Newton's theorem. For overlapping balls, a simple calculation shows that $U_{ij} - \hat{U}_{ij} > 0$. Thus ($\delta$) is positive.

For (β) we calculate

$$U_i - \hat{U}_i \geqslant -(2\pi/5)\rho\alpha^2 \tag{A.1}$$

The above is an equality whenever the ball lies completely in Λ.

For (γ)

$$\hat{U}_{ii} = 6/(5a) \tag{A.2}$$

Our lower bound is (β) + (γ), and the best bound is obtained when a is

$$a_{\max} = \rho^{-1/3}(3/4\pi)^{1/3} = r_s \tag{A.3}$$

With this value we obtain

$$U(\mathbf{X}) \geqslant -0.9Nr_s^{-1} = -0.9N\rho^{1/3}(4\pi/3)^{1/3} \tag{A.4}$$

for all $\mathbf{X}$.

For the quantum mechanical case, when the particles are spin-$\frac{1}{2}$ fermions, we consider a sequence of domains $\{\Lambda_j\}$ which tend to infinity in the sense of Van Hove and we constrain the particles to lie in Λ_j. We also suppose that $\rho|\Lambda_j| = j$, the number of particles, although this neutrality restriction is not essential in what follows.

Let

$$E_j = \inf_{\psi}\langle\psi|H|\psi\rangle|\Lambda_j|^{-1} \geqslant \inf_{\psi}|\Lambda_j|^{-1}\langle\psi|K|\psi\rangle + \inf_{\psi}|\Lambda_j|^{-1}\langle\psi|U|\psi\rangle$$

For $m = 1$ and j large

$$\begin{aligned}\inf\langle\psi, K\psi\rangle \approx j\rho^{2/3}(6/5)(3\pi^2)^{2/3} &= jr_s^{-2}(6/5)(9\pi/4)^{2/3} \\ &= 2.21jr_s^{-2}\ \text{Ry}\end{aligned} \tag{A.5}$$

Therefore, to leading order in j,

$$j^{-1}H \geqslant r_s^{-2}(2.21 - 0.45r_s)\ \text{Ry} \tag{A.6}$$

It makes sense here, unlike the situation for the classical or the Bose problem, to ask for the r_s that gives the lowest ground-state energy per particle. We obtain

$$r_s = 9.82 \tag{A.7}$$

and

$$j^{-1}H \geqslant -0.0229\ \text{Ry} \tag{A.8}$$

ACKNOWLEDGMENT

We would like to thank Prof. W. Thirring for stimulating discussions and for his hospitality at the Institute for Theoretical Physics, University of Vienna, where this work was started.

REFERENCES

1. E. P. Wigner, *Trans. Faraday Soc.* (*London*) **34**:678 (1938).
2. R. J. Baxter, *Proc. Camb. Phil. Soc.* **59**:779 (1963).
3. H. Kunz, *Ann. Phys.* (*N.Y.*) **85**:303 (1974).

4. H. J. Brascamp and E. H. Lieb, Some inequalities for Gaussian measures and the long-range order of the one-dimensional plasma, in *Proceedings of the Conference on Functional Integration*, London (1974), to be published.
5. E. H. Lieb and J. L. Lebowitz, *Adv. in Math.* **9**:316 (1972).
6. J. L. Lebowitz and E. H. Lieb, *Phys. Rev. Lett.* **22**:631 (1969).
7. O. Penrose and E. R. Smith, *Commun. Math. Phys.* **26**:53 (1972).
8. F. J. Dyson and A. Lenard, *J. Math. Phys.* **8**:423 (1967).
9. H. Narnhofer and W. Thirring, Convexity properties of Coulomb systems, *Acta Phys. Austr.* (1975), to be published.
10. A. Lenard and F. J. Dyson, *J. Math. Phys.* **9**:698 (1968).
11. P. Federbush, *J. Math. Phys.* **16**:706 (1975).
12. L. Onsager, *J. Phys. Chem.* **43**:189 (1939).
13. R. A. Goldwell-Horstall and A. A. Maradudin, *J. Math. Phys.* **1**:395 (1960).

Erratum

The Thermodynamic Limit for Jellium[1]

Elliott H. Lieb[2] and Heide Narnhofer[3]

Received September 22, 1975

On p. 309, Eq. (A.5) should read

$$\inf\langle\psi, K\psi\rangle \approx j\rho^{2/3}(3/5)(3\pi^2)^{2/3} = jr_s^{-2}(3/5)(9\pi/4)^{2/3} = 2.21 jr_s^{-2}\ \mathrm{Ry} \tag{A.5}$$

Accordingly, the resulting expressions should read as follows:

$$j^{-1}H \geqslant r_s^{-2}(2.21 - 1.80r_s)\ \mathrm{Ry} \tag{A.6}$$

$$r_s = 2.46 \tag{A.7}$$

$$j^{-1}H \geqslant -0.367\ \mathrm{Ry} \tag{A.8}$$

---

[1] This paper appeared in *J. Stat. Phys.* **12**(4):291 (1975).

[2] Department of Mathematics and Physics, Princeton University, Princeton, New Jersey.
[3] Institute for Theoretical Physics, University of Vienna, Vienna, Austria.

Part VII

Quantum Electrodynamics

Self-Energy of Electrons in Non-Perturbative QED

Elliott H. Lieb*
Departments of Mathematics and Physics
Princeton University, Princeton, New Jersey 08544-0708

Michael Loss†
School of Mathematics
Georgia Institute of Technology, Atlanta, Georgia 30332-0160 ‡§

Abstract

Various models of charged particles interacting with a quantized, ultraviolet cutoff radiation field (but not with each other) are investigated. Upper and lower bounds are found for the self- or ground state-energies without mass renormalization. For N fermions the bounds are proportional to N, but for bosons they are sublinear, which implies 'binding', and hence that 'free' bosons are never free. Both 'relativistic' and non-relativistic kinematics are considered. Our bounds are non-perturbative and differ significantly from the predictions of perturbation theory.

1 Introduction

Quantum electrodynamics (QED), the theory of electrons interacting with photons (at least for small energies) is one of the great successes of physics. Among its major achievements is the explanation of the Lamb shift and the anomalous magnetic moment of the electron. Nevertheless, its computations, which are entirely based on perturbation theory, created some uneasiness among the practitioners. The occurrence of

*Supported in part by NSF Grant PHY-98 20650.

†Supported in part by NSF Grant DMS-95 00840.

§This work was presented by E.H.L. at the University of Alabama, Birmingham - Georgia Tech International Conference on Differential Equations and Mathematical Physics, Birmingham, March 15-19, 1999 and first appears in the proceedings of that conference published by the American Mathematical Society and International Press, "Studies in Advanced Mathematics", vol 16, pp. 279-293 (2000). A second version, with some typos corrected, appears in the proceedings of 'Conference Moshe Flato' in Dijon in September, 1999, published by Kluwer, "Mathematical Physics Studies", vol.21, pp. 327-344 (2000). The present version, containing additional corrections, was prepared for the third edition of the *Selecta*.

infinities was and is especially vexing. Moreover, a truly nontrivial, 3+1-dimensional example of a relativistically invariant field theory has not yet been achieved.

There are, however, unresolved issues at a much earlier stage of QED that hark back to black-body radiation, the simplest and historically first problem involving the interaction of matter with radiation. The conceptual problems stemming from black-body radiation were partly resolved by quantum mechanics, i.e., by the the non-relativistic Schrödinger equation, which is, undoubtedly, one of the most successful of theories, for it describes matter at low energies almost completely. It is mathematically consistent and there are techniques available to compute relevant quantities. Moreover, it allows us to explain certain facts about bulk matter such as its stability, it extensivity, and the existence of thermodynamic functions. What has not been as successful, so far, is the incorporation of radiation phenomena, the very problem quantum mechanics set out to explain.

It ought to be possible to find a mathematically consistent theory, free of infinities, that describes the interaction of non-relativistic matter with radiation at moderate energies, such as atomic binding energies. It should not be necessary, as some physicists believe, to view QED as a low energy part of a consistent high energy theory.

From such a theory one could learn a number of things that have not been explained rigorously. i) The decay of excited states in atoms. This problem has been investigated in some ultraviolet cutoff models in [BFS] and in a massive photon model in [OY]. See also the review of Hogreve [H]. ii) Non-relativistic QED could be a playground for truly non-perturbative calculations and it could shed light on renormalization procedures. In fact, this was the route historically taken by Kramers that led to the renormalization program of Dyson, Feynman, Schwinger and Tomonaga. iii) Last but not least, one could formulate and answer the problems of stability of bulk matter interacting with the radiation field.

It has been proved in [F],[LLS] that stability of non-relativistic matter (with the Pauli Hamiltonian) interacting with classical magnetic fields holds provided that the fine-structure constant, $\alpha = e^2/\hbar c$, is small enough. It is certain, that the intricacies and difficulties of this classical field model will persist and presumably magnify in QED.

The same may be expected from a relativistic QED since replacing the Pauli Hamiltonian by a Dirac operator leads to a similar requirement on α [LSS]. Indeed, stability of matter in this model (the Brown-Ravenhall model) requires that the electron (positron) be defined in terms of the positive (negative) spectral subspace of the Dirac operator *with* the magnetic vector potential $A(x)$, instead of the free Dirac operator without $A(x)$. This observation, that perturbation theory, if there is one, must start from the dressed electrons rather than the electrons unclothed by its magnetic field, might ultimately be important in a non-perturbative QED.

The first, humble, step is to understand electrons that interact with the radiation field but which are free otherwise. In order for this model to make sense an ultraviolet cutoff has to be imposed that limits the energy of photon modes. The simplest question, which is the one we address in this paper, is the behavior of the self-energy of the electron as the cutoff tends to infinity (with the bare mass of the electron fixed). The self-energy of the electron diverges as the cutoff tends to infinity and it has to be subtracted for each electron in any interacting theory. The total energy will still depend strongly on the cutoff because of the interactions. This dependence will, hopefully, enter through an effective mass which will be set equal to the physical mass (mass renormalization). The resulting theory should be essentially Schrödinger's mechanics, but slightly modified by so-called radiative corrections.

Lest the reader think that the self-energy problem is just a mathematical exercise, consideration of the many-body problem will provide a counterexample. Imagine N charged bosons interacting with the radiation field, but neglect any interaction among them such as the Coulomb repulsion. We say that these particles bind if the energy of the combined particles is less than the energy of infinitely separated particles. As we shall show, charged bosons indeed bind and they do it in such a massive way that it will be very likely that this cannot be overcome by the Coulomb repulsion. In particular, the energy of a charged many-boson system is *not* extensive, and from this perspective it is fortunate that stable, charged bosons do not exist in nature.

The situation is very different for fermions. We are not able to show that they do not bind, but we can show — and this is one of the main results of our paper — that the self-energy is extensive, i.e., bounded above and below by a constant times N.

We thus have strong evidence that there is *no* consistent description of a system of stable charged relativistic or non-relativistic bosons interacting with the radiation field, while the Pauli exclusion principle, on the other hand, is able to prevent the above mentioned pathology.

In the remainder of the section we explain our notation and state the results. In the subsequent sections we sketch the proof of some of them but for details we refer the reader to [LL].

We measure the energy in units of mc^2 where m is the bare mass of the electron, the length in units of the Compton wave length $\ell_C = \hbar/mc$ of the bare electron. We further choose $\ell_C^{-1}\sqrt{\hbar c}$ as the unit for the vector potential A and $\ell_C^{-2}\sqrt{\hbar c}$ as the unit for the magnetic field B. As usual, $\alpha = e^2/\hbar c \approx 1/137.04$ is the fine structure constant.

In the expression below, $A(x)$ denotes an ultraviolet cutoff radiation field localized in a box $L \times L \times L$ with volume $V = L^3$,

$$A(x) = \frac{1}{\sqrt{2V}} \sum_{|k|<\Lambda} \sum_{\lambda=1,2} \frac{1}{\sqrt{|k|}} \varepsilon_\lambda(k) \left[a_\lambda(k) e^{ix\cdot k} + a_\lambda^*(k) e^{-ix\cdot k} \right] . \tag{1.1}$$

The index $k = 2\pi n/L$ where $n \in \mathbb{Z}^3$, and the word cutoff refers to the restriction to all values of k with $|k| < \Lambda$.

The vectors $\varepsilon_\lambda(k)$ are the polarization vectors and are normalized in such a way that

$$\varepsilon_\lambda(k) \cdot \varepsilon_{\lambda'}(k) = \delta_{\lambda,\lambda'} , \qquad \varepsilon_\lambda(k) \cdot k = 0 . \tag{1.2}$$

The operators $a_\lambda(k)$ and $a_\lambda^*(k)$ satisfy the commutation relation

$$\left[a_\lambda(k), a_{\lambda'}^*(k') \right] = \delta_{\lambda,\lambda'} \delta(k, k') , \tag{1.3}$$

while all others commute with each other.

The energy of the radiation field can now be conveniently written as

$$H_f = \sum_{|k|<\Lambda} \sum_{\lambda=1,2} |k| a_\lambda^*(k) a_\lambda(k) . \tag{1.4}$$

These operators act on the Hilbert space generated by the polynomials in $a_\lambda^*(k)$ acting on the vacuum $|0\rangle$.

The self energy of (one or more) particles is the **ground state energy** of the Hamiltonian

$$H = \text{kinetic energy } + H_f \ , \tag{1.5}$$

where, as usual, the ground state energy of H is defined to be

$$E_0 = \inf_\Psi \frac{\langle \Psi, H\ \Psi\rangle}{\langle \Psi, \Psi\rangle} \ . \tag{1.6}$$

Typically, in the inquiry about the self–energy problem, i.e., the problem of computing the self–energy for fixed, albeit small, α and for large Λ, one proceeds via perturbation theory. First order perturbation theory will predict an energy of the order of $\alpha\Lambda^2$, and a higher order power counting argument confirms the asymptotically large Λ dependence of that calculation. Our theorems below show that the predictions of perturbation theory for the self–energy problem are wrong, if one is interested in the large Λ asymptotics of the energy. If perturbation theory works at all, then it works only for a range of α that vanishes as Λ increases. In fact we deduce from the upper bound in Theorem 1.1 that the size of this range shrinks at least as $\Lambda^{-2/5}$.

All the theorems below are asymptotic statements for *large* Λ and for *fixed* α, and all the constants are *independent of the volume.* For actual bounds we refer the reader to [LL]. The first result concerns the self energy of a nonrelativistic electron interacting with the radiation field. The Hamiltonian is given by

$$H = \frac{1}{2}(p + \sqrt{\alpha}A(x))^2 + H_f \ , \tag{1.7}$$

where $p = -i\nabla$ and acts on $L^2(\mathbb{R}^3) \otimes \mathcal{F}$, where $\mathcal{F}$ denotes the photon Fock space.

Theorem 1.1 *The ground state energy, E_0, of the operator (1.7) satisfies the asymptotic bounds (with positive constants C_1, C_2)*

$$C_1\alpha^{1/2}\Lambda^{3/2} \ < \ E_0 \ < \ C_2\alpha^{2/7}\Lambda^{12/7} \tag{1.8}$$

We do not know how to get upper and lower bounds that are of the same order in Λ, but we suspect that $\Lambda^{12/7}$ is the right exponent. This is supported by the following theorem in which the $p \cdot A$ term is omitted.

Theorem 1.2 *The ground state energy E_0 of the operator*

$$\frac{1}{2}\left[p^2 + \alpha A(x)^2\right] + H_f \tag{1.9}$$

satisfies the asymptotic bounds (with positive constants C_1, C_2)

$$C_1\alpha^{2/7}\Lambda^{12/7} \le E_0 \le C_2\alpha^{2/7}\Lambda^{12/7} \ . \tag{1.10}$$

While these results are not of direct physical relevance (since E_0 is not observable), the many-body problem is of importance since it reveals a dramatic difference between bosons and fermions.

Theorem 1.3 *The ground state energy of N bosons, $E_0^{boson}(N)$, with Hamiltonian*

$$H(N) = \sum_{j=1}^{N} \frac{1}{2}(p_j + \sqrt{\alpha}A(x_j))^2 + H_f \tag{1.11}$$

satisfies the asymptotic bounds (with positive constants C_1, C_2)

$$C_1\sqrt{N}\sqrt{\alpha}\Lambda^{3/2} \le E_0^{boson}(N) \le C_2 N^{5/7}\alpha^{2/7}\Lambda^{12/7} . \tag{1.12}$$

Thus, the energy $E_0^{boson}(N)$ is not extensive, i.e., it costs a huge energy to separate charged bosons (in the absence of their Coulomb repulsion). This has to be contrasted with the next theorem about fermions. The Hamiltonian is the same as before but it acts on the Hilbert space

$$\mathcal{F} \otimes \wedge_{j=1}^{N} L^2(\mathbb{R}^3; \mathbb{C}^2) , \tag{1.13}$$

where the wedge product indicates that the antisymmetric tensor product is taken.

Theorem 1.4 *The ground state energy, $E_0^{fermion}(N)$, of N charged fermions interacting with the radiation field satisfies the asymptotic bounds (with positive constants C_1, C_2)*

$$C_1\alpha^{1/2}\Lambda^{3/2}N \;\le\; E_0^{fermion}(N) \;\le\; C_2\alpha^{2/7}\Lambda^{12/7}N . \tag{1.14}$$

The **"relativistic" kinetic energy** for an electron is

$$T^{\rm rel} = |p + \sqrt{\alpha}A(x)| = \sqrt{[p + \sqrt{\alpha}A(x)]^2} \tag{1.15}$$

with $p = -i\nabla$. (Really, we should take $\sqrt{[p + \sqrt{\alpha}A(x)]^2 + 1}$, but since $x < \sqrt{x^2+1} < x + 1$, the difference is bounded by N.)

Consider, first, the $N = 1$ body problem with the Hamiltonian

$$H = T^{\rm rel} + H_f . \tag{1.16}$$

By simple length scaling (with a simultaneous scaling of the volume V) we easily see that $E_0 = \inf \operatorname{spec}(H) = C\Lambda$. Our goal here is to show that the constant, C, is strictly positive and to give an effective lower bound for it. But we would like to do more, namely investigate the dependence of this constant on α. We also want to show, later on, that for N fermions the energy is bounded below by a positive constant times $N\Lambda$. Our proof will contain some novel — even bizarre — features.

Theorem 1.5 *For the Hamiltonian in 1.16 there are positive constants, C, C', C'' and α_0 , independent of α and N, and such that*

$$\begin{aligned} E_0 &\leq C\sqrt{\alpha}\Lambda \\ E_0 &\geq C'\sqrt{\alpha}\Lambda \text{ for } \alpha \leq \alpha_0 \\ E_0 &\geq C''\Lambda \text{ for } \alpha \geq \alpha_0 \ . \end{aligned}$$

The generalization of this to N fermions is similar to the nonrelativistic generalization, except that the power of Λ is the same on both sides of the inequalities.

Theorem 1.6 *For N fermions with Hamiltonian*

$$H_N = \sum_{i=1}^{N} T^{\text{rel}}(x_i) + H_f$$

there are positive constants C, C', C'', and α_0, independent of α and N, such that

$$\begin{aligned} E_0 &\leq CN\sqrt{\alpha}\Lambda \\ E_0 &\geq C'N\sqrt{\alpha}\Lambda \quad \text{for } \alpha \leq \alpha_0 \\ E_0 &\geq C''N\Lambda \quad \text{for } \alpha \geq \alpha_0 \ . \end{aligned} \tag{1.17}$$

We close this introduction by mentioning one last result about the Pauli–operator. The kinetic energy expression is given by

$$T^{\text{Pauli}} = [\sigma \cdot (p + \sqrt{\alpha}A(x))]^2 = (p + \sqrt{\alpha}A(x))^2 + \sqrt{\alpha}\, \sigma \cdot B(x) \ . \tag{1.18}$$

where σ denotes the vector consisting of the Pauli matrices. Observe that this term automatically accounts for the spin–field interaction. Our result for the self energy of a Pauli electron is the following.

Theorem 1.7 *For the Hamiltonian with Pauli kinetic energy,*

$$\frac{1}{2}[\sigma \cdot (p + \sqrt{\alpha}A(x))]^2 + H_f, \tag{1.19}$$

there are positive constants C, C', C'', and α_0, such that the ground state energy satisfies the asymptotic bounds

$$\begin{aligned} E_0 &\leq C_3\sqrt{\alpha}\Lambda^{3/2} \\ E_0 &\geq C_1\alpha\Lambda \quad \text{for } \alpha \leq \alpha_0 \\ E_0 &\geq C_2\alpha^{1/3}\Lambda \quad \text{for } \alpha \geq \alpha_0 \ . \end{aligned} \tag{1.20}$$

For N fermions, the bounds above are multiplied by N (and the constants are changed).

For the details of the proof, we refer the reader to [LL]. We believe that the upper bound is closer to the truth since the main contributions to the self energy should come from the fluctuations of the A^2 term.

Theorem 1.7 has the following consequence for stability of matter interacting with quantized fields. It was shown in [LLS] that a system of electrons and nuclei interacting with Coulomb forces, with the Pauli kinetic energy for the electrons and with a *classical* magnetic field energy is stable (i.e., the ground state energy is bounded below by N) if and only if α is small enough. In [BFG],[FFG] this result was extended to *quantized*, ultraviolet cutoff magnetic fields (as here). Among other things, it was shown in [FFG] that the ground state energy, E_0, of the electrons and nuclei problem is bounded below by $-\alpha^2\Lambda N$ for small α. Theorem 1.7 implies, as a corollary, that for small α the *total energy* (including Coulomb energies) is bounded below by $+\alpha\Lambda N$. In other words, among the three components of energy (kinetic, field and Coulomb), the first two overwhelm the third — for small α, at least.

All of these statements are true without mass renormalization and the situation could conceivably be more dramatic when the mass is renormalized. In any case, the true physical questions concern energy differences, and this question remains to be addressed.

2 Non-Relativistic Energy Bounds

Theorem 1.1: We sketch a proof of Theorem 1.1. It is clear by taking the state $V^{-1/2} \otimes |0\rangle$ (where $|0\rangle$ denotes the photon vacuum) that the ground state energy is bounded above by (const)$\alpha\Lambda^2$, which is the same result one gets from perturbation theory. Since the field energy in this state vanishes, such a computation ignores the tradeoff between the kinetic energy of the electron and the field energy. Thus, it is important to quantify this tradeoff. The main idea is to estimate the field energy in terms of selected modes.

Consider the operators (field modes), parametrized by $y \in \mathbb{R}^3$,

$$L(y) = \frac{1}{\sqrt{2V}} \sum_{|k|<\Lambda,\lambda} \sqrt{|k|}a_\lambda(k)\overline{v}_\lambda(k)e^{ik\cdot y} , \tag{2.1}$$

with some arbitrary complex coefficients $v_\lambda(k)$. The following lemma is elementary

Lemma 2.1

$$H_f \geq \int w(y)L^*(y)L(y)\mathrm{d}y \tag{2.2}$$

provided that the functions $v_\lambda(k)$ and w are chosen such that, as matrices,

$$|k|\delta_{\lambda',\lambda}\delta(k',k) \geq \frac{1}{2V}\sqrt{|k|}\overline{v}_\lambda(k)\widehat{w}(k-k')v_{\lambda'}(k')\sqrt{|k'|} , \tag{2.3}$$

or equivalently, that

$$\sum_{|k|<\Lambda,\lambda} \frac{|f_\lambda(k)|^2}{|v_\lambda(k)|^2} \geq \sum_{|k|,|k'|<\Lambda,\lambda,\lambda'} \frac{1}{2V}\overline{f_\lambda(k)}f_{\lambda'}(k')\widehat{w}(k-k') \tag{2.4}$$

for all $f_\lambda(k)$, where $\widehat{w}(k) = \int e^{ik\cdot x}w(x)\mathrm{d}x$.

For the proof, one simply notes that both sides of (2.2) are quadratic forms in the creation and annihilation operators, and hence (2.3) and (2.4) are necessary and sufficient conditions for (2.2) to be true. ■

Corollary 2.2 *Assuming (2.2), the following two inequalities hold.*

$$H_f \geq -\frac{1}{4V} \sum_{|k|<\Lambda,\lambda} |k||v_\lambda(k)|^2 \int |w(y)| \mathrm{d}y + \frac{1}{4} \begin{cases} \int w(y)(L(y)+L^*(y))^2 \mathrm{d}y \\ -\int w(y)(L(y)-L^*(y))^2 \mathrm{d}y \ . \end{cases} \tag{2.5}$$

To prove this, note that

$$L^*L = LL^* - \frac{1}{2V} \sum_{|k|<\Lambda,\lambda} |k||v_\lambda(k)|^2 \ , \tag{2.6}$$

and, quite generally for operators,

$$\pm(LL + L^*L^*) \leq L^*L + LL^* \ . \tag{2.7}$$

■

Returning to the proof of Theorem 1.1 we start with the lower bound. Denote by

$$\Pi(x) = \frac{-i}{\sqrt{2V}} \sum_{|k|<\Lambda,\lambda} \sqrt{|k|}\varepsilon_\lambda(k) \left(a_\lambda(k)e^{ik\cdot x} - a_\lambda^*(k)e^{-ik\cdot x}\right) \ . \tag{2.8}$$

This operator is canonically conjugate to $A(x)$ in the sense that we have the commutation relations

$$i[\Pi_i(x), A_j(x)] = \delta_{i,j} \frac{1}{(3\pi)^2} \Lambda^3 \ . \tag{2.9}$$

For our calculation, it is important to note that

$$\mathrm{div}\ \Pi(x) = 0 \ . \tag{2.10}$$

Hence from (2.9) and (2.10) we get that

$$\sum_{j=1}^{3} [p_j + \sqrt{\alpha}A_j(x), \Pi_j(x)] = \sqrt{\alpha}\frac{i}{3\pi^2}\Lambda^3 \ . \tag{2.11}$$

The inequality

$$\frac{1}{2}(p+\sqrt{\alpha}A(x))^2 + 2a^2\Pi(x)^2 \geq -ai\sum_{j=1}^{3}[p_j + \sqrt{\alpha}A_j(x), \Pi_j(x)] \ , \tag{2.12}$$

valid for all positive numbers a, yields

$$\frac{1}{2}(p+\sqrt{\alpha}A(x))^2 + H_f \geq a\sqrt{\alpha}\frac{1}{3\pi^2}\Lambda^3 + H_f - 2a^2\Pi(x)^2 \,. \tag{2.13}$$

We choose

$$v_\lambda(k) = 3\pi\Lambda^{-3/2}\varepsilon_{\lambda,j}(k) \tag{2.14}$$

(where $j = 1, 2, 3$ indexes the polarization vector component) and

$$w(y) = 12\beta\,\delta(x-y)\,. \tag{2.15}$$

Eqn. (2.3) is satisfied if and only if $\beta \leq 3\pi^2\Lambda^{-3}/2$, in which case Corollary 2.2 yields an inequality which, after summing on j, is

$$H_f \geq \beta\,\Pi(x)^2 - \frac{\beta}{(2\pi)^2}\Lambda^4\,. \tag{2.16}$$

In eqn. (2.13) we choose $2a^2 = \beta = 3\pi^2\Lambda^{-3}/2$, i.e., $a = (\sqrt{3}\pi)/(2\Lambda^{3/2})$. This yields the *lower bound*

$$H \geq \frac{1}{2\pi}\sqrt{\frac{\alpha}{3}}\,\Lambda^{3/2} - \frac{3}{8}\Lambda\,. \tag{2.17}$$

The idea of using a commutator, as in (2.12), (2.13) to estimate the ground state energy, goes back to the study of the polaron [LY].

For the upper bound we take a simple trial function of the form

$$\phi(x)\otimes\Psi \tag{2.18}$$

where $\Psi \in \mathcal{F}$ is normalized and $\phi(x)$ is a *real* function normalized in $L^2(\mathbb{R}^3)$. An upper bound to the energy is thus given by

$$\frac{1}{2}\int|\nabla\phi(x)|^2\mathrm{d}x + \frac{\alpha}{2}\int\phi(x)^2\left(\Psi, A(x)^2\Psi\right)\mathrm{d}x + (\Psi, H_f\Psi)\,. \tag{2.19}$$

It is not very difficult to see that the last two terms can be concatenated into the following expression.

$$\frac{1}{2}\int\left(\Psi, \left[\Pi(x)^2 + \alpha A(x)(-\Delta+\phi(x)^2)A(x)\right]\Psi\right)\mathrm{d}x \;-\frac{1}{2}\mathrm{Tr}\sqrt{P-\Delta P}\,. \tag{2.20}$$

Here, P is the projection onto the divergence free vector fields with ultraviolet cutoff Λ. This can be deduced by writing the field energy in terms of $\Pi(x)$ and $A(x)$. The first term in (2.20) is a sum of harmonic oscillators whose zero point energy is given by

$$\frac{1}{2}\mathrm{Tr}\sqrt{P\,(-\Delta+\alpha\phi(x)^2)\,P} \tag{2.21}$$

and hence

$$\frac{1}{2}\mathrm{Tr}\sqrt{P\,(-\Delta+\alpha\phi(x)^2)\,P}-\frac{1}{2}\mathrm{Tr}\sqrt{P(-\Delta)P}\ , \tag{2.22}$$

is an exact expression for the ground state energy of (2.20) Using the operator monotonicity of the square root we get as an upper bound on (2.20)

$$\frac{1}{2}\int|\nabla\phi(x)|^2\mathrm{d}x+\frac{1}{2}\sqrt{\alpha}\ \mathrm{Tr}\sqrt{P\phi(x)^2P}\ . \tag{2.23}$$

As a trial function we use

$$\phi(x)=\mathrm{const.}K^{-3/2}\int\left(1-\frac{|k|}{K}\right)_+^3 e^{ik\cdot x}\ \mathrm{d}k\ . \tag{2.24}$$

Optimizing the resulting expression over K yields the stated result. For details we refer the reader to [LL]. ■

It is natural to ask, how good this upper bound is. If we neglect the cross terms in $(p+A)^2$, i.e., we replace the kinetic energy by $p^2+\alpha A(x)^2$, then we have Theorem 1.2, which we prove next.

Theorem 1.2: The upper bound was already given in Theorem 1.1 because $<p\cdot A>=0$ in the state (2.18). Loosely speaking equation (2.9) expresses the Heisenberg uncertainty principle for the field operators. An uncertainty principle that is quite a bit more useful is the following.

Lemma 2.3 *The following inequality holds in the sense of quadratic forms*

$$\Pi(x)^2\geq\frac{1}{4}\frac{1}{(3\pi)^4}\Lambda^6\frac{1}{A(x)^2}\ . \tag{2.25}$$

For the proof note that $[A_j(x),A_k(y)]=0$ and compute

$$i[\Pi(x)_j,\frac{A_j(x)}{A(x)^2}]=\frac{1}{(3\pi)^2}\Lambda^3\left[\frac{1}{A(x)^2}-2\left(\frac{A_j(x)}{A(x)^2}\right)^2\right]\ , \tag{2.26}$$

and summing over j we obtain that

$$i\sum_{j=1}^{3}[\Pi(x)_j,\frac{A_j(x)}{A(x)^2}]=\frac{1}{(3\pi)^2}\Lambda^3\frac{1}{A(x)^2} \tag{2.27}$$

Our statement follows from the Schwarz inequality. ■

To prove Theorem 1.2 we return to Lemma 2.1 and choose $v_\lambda(k)=\varepsilon_\lambda(k)$ and $w(x)$ any function ≤ 1. Corollary 2.2 applied to each of the 3 components of $\Pi(x)$ then yields

$$H_f\geq\frac{1}{4}\int w(x-y)\Pi(y)^2\mathrm{d}y-\Lambda^4\frac{3}{8\pi^2}\int w(y)\mathrm{d}y\ , \tag{2.28}$$

for every $x \in \mathbb{R}^3$. By Lemma 2.1 the right side is bounded below by

$$\Lambda^6 \int w(x-y)\frac{1}{A(y)^2}\mathrm{d}y - \Lambda^4 \int w(y)\mathrm{d}y \ , \tag{2.29}$$

and hence

$$\begin{aligned} \langle \Psi, H\Psi \rangle \ &\geq \ \frac{1}{2}\int \langle \nabla\Psi(x), \nabla\Psi(x)\rangle \mathrm{d}x + \frac{\alpha}{2}\int \langle \Psi(x), A(x)^2\Psi(x)\rangle \mathrm{d}x \\ &+ \ \Lambda^6 \int w(x-y)\langle \Psi(y), \frac{1}{A(x)^2}\Psi(y)\rangle \mathrm{d}y\mathrm{d}x \\ &- \ \Lambda^4 \int w(y)\mathrm{d}y \int \langle \Psi(x), \Psi(x)\rangle \mathrm{d}x \ . \end{aligned} \tag{2.30}$$

By Schwarz's inequality the second and third term together are bounded below by

$$\sqrt{\frac{\alpha}{2}}\Lambda^3 \int \langle \Psi(x), \Psi(y)\rangle \frac{w(x-y)}{\sqrt{\int w(z)\mathrm{d}z}}\mathrm{d}x\mathrm{d}y \ . \tag{2.31}$$

If we restate our bound in terms of Fourier space variables we get

$$\int \left[\frac{|p|^2}{2} + \sqrt{\frac{\alpha}{2}}\,\Lambda^3 \frac{\widehat{w}(p)}{\sqrt{\widehat{w}(0)}}\right] \langle \widehat{\Psi}(p), \widehat{\Psi}(p)\rangle \mathrm{d}p - \Lambda^4\widehat{w}(0)\int \langle \widehat{\Psi}(p), \widehat{\Psi}(p)\rangle \mathrm{d}p \ . \tag{2.32}$$

Choosing the function $\widehat{w}(p)$ to be $(2\pi)^3\Lambda^{-18/7}$ times the characteristic function of the ball of radius $\Lambda^{6/7}$, we have that $w(x) \leq 1$ and it remains to optimize (2.32) over all normalized states $\widehat{\Psi}(p)$. This is easily achieved by noting that the function

$$\frac{1}{2}|p|^2 + \sqrt{\frac{\alpha}{2}}\Lambda^3 \frac{\widehat{w}(p)}{\sqrt{\widehat{w}(0)}} \tag{2.33}$$

is everywhere larger than $\Lambda^{12/7}$. Strictly speaking, the function $w(x)$ should be positive in order for the argument that led to (2.31) to be valid. This can be achieved with a different choice of $w(x)$, like the one in (2.24), that is more complicated but does not change the argument in an essential way.

3 Non-Relativistic Many-Body Energies

A problem that has to be addressed is the energy of N particles (bosons or fermions) interacting with the radiation field. If $E_0 = E_0(1)$ is the energy of one particle (which we estimated in the preceding section) then, ideally, the energy, $E_0(N)$, of N particles (which trivially satisfies $E_N \leq NE$, since the N particles can be placed infinitely far apart) ought to be, *exactly*,

$$E_0(N) = NE_0 \tag{3.1}$$

in a correct QED. In other words, in the absence of nuclei and Coulomb potentials, there should be *no binding* caused by the field energy H_f. This is what we seem to observe experimentally, but this important topic does not seem to have been discussed in the QED literature.

Normally, one should expect binding, for the following mathematical reason: The first particle generates a field, $A(x)$, and energy E_0. The second particle can either try to generate a field $A(x+y)$, located very far away at y or the second particle can try to take advantage of the field $A(x)$, already generated by the first particle, and achieve an insertion energy lower than E_0.

Indeed, this second phenomenon happens for bosons — as expected. For fermions, however, the Paul principle plays a crucial role (even in in the absence of Coulomb attractions). We show that $E_0(N) \geq CNE_0$ for fermions, but we are unable to show that the universal constant $C = 1$. Even if $C < 1$, the situation could still be saved by mass renormalization, which drives the bare mass to zero as Λ increases, thereby pushing particles apart.

3.1 Bosons

Theorem 1.3; This theorem concerns the ground state energy of N charged bosons. the Hamiltonian is given by 1.11 acting on the Hilbert space of symmetric functions tensored with the photon Fock space $\mathcal{F}$. It states, basically, that $C_1\sqrt{N}\sqrt{\alpha}\Lambda^{3/2} \leq E_0^{boson}(N) \leq C_2 N^{5/7}\alpha^{2/7}\Lambda^{12/7}$.

The proof follows essentially that of the one particle case. The interesting fact is that it implies *binding* of *charged bosons* (in the absence of the Coulomb repulsion). The binding energy is defined by

$$\Delta E(N) = E_0(N) - NE_0(1)$$

and satisfies the bounds

$$\begin{aligned} \Delta E(N) &\geq C_1\sqrt{N}\sqrt{\alpha}\,\Lambda^{3/2} - C_2 N\alpha^{2/7}\Lambda^{12/7} \\ \Delta E(N) &\leq C_2 N^{2/7}\alpha^{2/7}\Lambda^{12/7} - C_1 N\sqrt{\alpha}\,\Lambda^{3/2} \end{aligned} \tag{3.2}$$

which can be made negative for appropriately chosen N and Λ. There will be binding for all large enough N, irrespective of the cutoff Λ. It also has to be remarked that the Coulomb repulsion will, in all likelihood, not alter this result since it has an effect on energy scales of the order of Λ and not $\Lambda^{12/7}$ or $\Lambda^{3/2}$.

3.2 Fermions

The real issue for physics is what happens with fermions. We cannot show that there is no binding but we can show that the energy is extensive as in Theorem 1.4. The Hamiltonian is the same as (1.11) but it acts on antisymmetric functions tensored with $\mathcal{F}$. (Spin can be ignored for present purposes.)

Rough sketch of the proof of Theorem 1.4.

The difficulty in proving this theorem stems from the fact that the field energy is not extensive in any obvious way.

Define $\underline{X} = (x_1, \cdots, x_N)$ and define the function

$$n_j(\underline{X}, R) = \#\{x_i \neq x_j \ : \ |x_i - x_j| < R\} \, .$$

This function counts the number of electrons that are within a distance R of the jth electron. Note that this function is not smooth, so that all the following computations have to be modified. (See [LL].) We save half of the kinetic energy and write

$$H = \frac{1}{4}\sum_{j=1}^{N}(p_j + \sqrt{\alpha}A(x_j))^2 + H' .$$

We apply the commutator estimate (2.11) and (2.13) to the pair

$$i[p_j + \sqrt{\alpha}A(x_j), \frac{1}{\sqrt{N_j(\underline{X}, R)+1}}\Pi(x_j)]$$

and obtain the bound (with the caveat mentioned above), for all $a > 0$,

$$H' \geq a\sqrt{\alpha}\Lambda^3 \sum_{j=1}^{N} \frac{1}{\sqrt{N_j(\underline{X}, R)+1}} - a^2 \sum_{j=1}^{N} \frac{1}{N_j(\underline{X}, R)+1}\Pi(x_j)^2 + H_f \tag{3.3}$$

The next two steps are somewhat nontrivial and we refer the reader to [LL]. First one notes that the modes $F(x_i)$ and $F(x_j)$ are essentially orthogonal (i.e., they commute) if $|x_i - x_j| > \Lambda^{-1}$. Ignoring the technical details of how this is implemented, the key observation is that the last two terms in (3.3) can be estimated from below by $-N\Lambda$ provided $a = \Lambda^{-3/2}$.

The next ingredient is a new Lieb-Thirring type estimate involving the function $N_j(\underline{X}, R)$. It is here and only here that the Pauli exclusion principle is invoked.

Theorem 3.1 *On the space $\wedge_{j=1}^{N} L^2\mathbb{R}^3; \mathbb{C}^q)$ of antisymmetric functions*

$$\sum_{j=1}^{N}(p_j + \sqrt{\alpha}A(x_j))^2 \geq \frac{C}{q^{2/3}}\frac{1}{R^2}\sum_{j=1}^{N} N_j(\underline{X}, R)^{2/3} \tag{3.4}$$

with $C \geq 0.00127$. An analogous inequality holds for the relativistic case as well:

$$\sum_{j=1}^{N} |p_j + \sqrt{\alpha}A(x_j)| \geq \frac{C}{q^{1/3}}\frac{1}{R}\sum_{j=1}^{N} N_j(\underline{X}, R)^{1/3} \tag{3.5}$$

By using the kinetic energy previously saved together with (3.3) and the previous discussion, we get

$$H \geq \sum_{j=1}^{N} \left\{ N_j(\underline{X}, R)^{2/3} + \sqrt{\alpha}\Lambda^{3/2} \frac{1}{\sqrt{N_j(\underline{X}, R)+1}} \right\} - N\Lambda .$$

By minimizing over N_j the desired lower bound in Theorem (1.4) is obtained. The upper bound is fairly elementary and is omitted. ■

4 Relativistic Energy Bounds

Theorem 1.5: Sketch of Proof.

An upper bound for E_0 is easy to obtain, but it is indirect. Note that

$$|p + \sqrt{\alpha}A(x)| \leq \varepsilon[p + \sqrt{\alpha}A(x)]^2 + (4\varepsilon)^{-1} \tag{4.1}$$

for any $\varepsilon > 0$. Take $\Psi = f(x) \otimes |0\rangle$ with $|0\rangle$ being the Fock space vacuum. Using (4.1)

$$\begin{aligned} (\Psi, H\Psi) &\leq \varepsilon \int_{\mathbb{R}^3} \{\alpha\langle 0|A(x)^2|0\rangle |f(x)|^2 + |\nabla f(x)|^2\} dx + \varepsilon^{-1} \\ &= \frac{\varepsilon\alpha\Lambda^2}{4\pi} + \int |\nabla f|^2 + \frac{1}{4\varepsilon}, \end{aligned} \tag{4.2}$$

since $\langle 0|A(x)^2|0\rangle = (2\pi)^{-3}\int_{|k|<\Lambda} |k|^{-1}dk = \Lambda^2/4\pi^2$. We can now let $f(x) \to V^{-\frac{1}{2}}$ and take $\varepsilon = (\pi/\alpha)^{1/2}\Lambda^{-1}$, whence

$$E_0 \leq (\alpha/4\pi)^{1/2}\Lambda . \tag{4.3}$$

Now we turn to the lower bound for H.

Step 1: Since $x \to \sqrt{x}$ is operator monotone,

$$T > T_1 = |p_1 + \sqrt{\alpha}A_1(x)| , \tag{4.4}$$

where the subscript 1 denotes the 1 component (i.e., the x-component) of a vector. By replacing T by T_1, we are now in a position to remove A_1 by a gauge transformation - but it has to be an *operator-valued gauge transformation.* The use of such a gauge transformation is a novelty, as far as we are aware, in QED.

To effect the gauge transformation, set

$$\varphi(x) = \frac{1}{\sqrt{2V}} \sum_{h,\lambda} \frac{\varepsilon^\lambda(k)_1}{\sqrt{(k)}} [a_\lambda(k) + a^*_\lambda(-k)] \frac{e^{ik_1x_1} - 1}{ik_1} e^{ik_\perp x_\perp} \tag{4.5}$$

with $x_\perp = (x_2, x_3)$. Then $[A_j(x), \varphi(x)] = 0, \quad j = 1,2,3$ and $-i\frac{\partial}{\partial x_1}\exp\,[i\varphi(x)] = -A_1(x)$. The unitary $U = \exp\,[i\varphi(x)]$ is a gauge transformation, but it is *operator-valued.* We have

$$\begin{aligned} U^{-1}|p_1 + A_1(x)|U &= |p_1| \\ U^{-1}a_\lambda(k)U &= a_\lambda(k) + f_\lambda(k,x) \\ U^{-1}a^*_\lambda(k)U &= a^*_\lambda(k) + \bar{f}_\lambda(k,x) \\ \widetilde{H}_f = U^{-1}H_fU &= \sum_{k,\lambda} |k|[a^*_\lambda(k) + \bar{f}_\lambda(k,x)][a_\lambda(k) + f_\lambda(k,x)] \end{aligned} \tag{4.6}$$

with

$$f_\lambda(k,x) = \sqrt{\frac{\alpha}{2V}} \sum_{k,\lambda} \frac{\varepsilon_\lambda(k)_1}{|k|} \frac{e^{-ik_1x_1} - 1}{k_1} e^{-ik_\perp x_\perp} . \tag{4.7}$$

Since $p_\perp$ does not appear in our new Hamiltonian,

$$\widetilde{H} = U^{-1}HU \ = \ |p_1| + \ \widetilde{H}_f \ , \tag{4.8}$$

the variable $x_\perp$ appears only as a parameter, and thus we can set $x_\perp$ = constant = (0,0), by translation invariance, and replace $\mathbb{R}^3$ by $\mathbb{R}^1 = \mathbb{R}$.

From now on $x_1 = x$ and, $p_1 = p = -i\ d/dx$.

Step 2: The dependence on x now appears in $\widetilde{H}_f$ instead of in the kinetic energy, $|p|$. For each x we can try to put $\widetilde{H}_f$ into its ground state, which is that of a displaced harmonic oscillator. But, since this state depends on x, to do so will require a great deal of kinetic energy, $|p_1|$.

Let Ψ be a normalized wave-function, i.e., a function on $L^2(\mathbb{R}) \otimes \mathcal{F}$. We write it as ψ_x where $\psi_x \in \mathcal{F}$. Thus, with $\langle \cdot\ ,\ \cdot \rangle$ denoting the inner product on $\mathcal{F}$, $\int_{\mathbb{R}} \langle \psi_x, \psi_x \rangle dx = 1$.

Decompose $\mathbb{R}$ as the disjoint union of intervals of length ℓ/Λ, where ℓ is a parameter to be determined later. Denote these intervals by I_j, $j = 1, 2, \ldots$. A simple Poincaré type inequality gives, for $g\ :\ L^2(\mathbb{R}) \to \mathbb{C}$,

$$(g, |p|g) \geq C_1 \frac{\Lambda}{\ell} \sum_j \int_{I_j} \{|g(x)|^2 - |\bar{g}_j|^2\} dx\ ,$$

where $\bar{g}_j = \frac{\Lambda}{\ell} \int_{I_j} g(x)dx$ is the average of g in I_j. Then

$$(\Psi, |p|\Psi)\ \geq C_1 \frac{\Lambda}{\ell} \sum_j \int_{I_j} \{\langle \psi_x, \psi_x \rangle\ -\ \langle \bar{\psi}_j, \bar{\psi}_j \rangle\ \} dx\ . \tag{4.9}$$

Step 3: Next, we analyze $\widetilde{H}_f$. We think of this as an operator on $\mathcal{F}$, parameterized by $x \in \mathbb{R}$. We would like $\widetilde{H}_f$ to have a gap so we define

$$H_x \ = \ \frac{\Lambda}{2} \sum_{\varepsilon\Lambda \leq |k| \leq \Lambda} \ \sum_\lambda \ [a^+_\lambda(k) + \bar{f}_\lambda(k,x)] \ \cdot\ [\ h.c.\ ] \tag{4.10}$$

Clearly, $\widetilde{H}_f \geq H_x$ and

$$(\Psi, \widetilde{H}\ \Psi) \geq \frac{\Lambda}{\ell} \sum_j \int_{I_j} \langle \psi_x, \psi_x \rangle - \langle \bar{\psi}_j, \bar{\psi}_j \rangle + \langle \psi_x, H_x \psi_x \rangle dx\ . \tag{4.11}$$

For each interval I_j we can minimize (4.11) subject to $\int_{I_j} \langle \psi_x, \psi_x \rangle dx$ fixed. This leads to

$$(h_j\ \psi)_x = e\ \psi_x \tag{4.12}$$

with

$$(h_j\ \psi)_x = \frac{\Lambda}{\ell}\ \psi_x - \frac{\Lambda}{\ell}\ \bar{\psi}_j + H_x\ \psi_x \tag{4.13}$$

Obviously, this eigenvalue problem (4.12, 4.13) is the same for all intervals I_j, so we shall drop the subscript j and try to find the minimum e.

A lower bound to h_j (and hence to e) can be found by replacing H_x by

$$\widehat{H}_x = \frac{\Lambda}{2}(1 - \Pi_x) ,$$

where $\Pi_x = |g_x\rangle\langle g_x|$ is the projector onto the ground state, $|g_x\rangle$, of H_x.

If we substitute $\widehat{H}_x$ into (4.13) the corresponding eigenvalue equation (4.12) becomes soluble. Multiply (4.12) on the left by $\langle g_x|$, whence

$$\left(\frac{\Lambda}{\ell} - e\right)\langle g_x, \psi_x\rangle = \frac{\Lambda}{\ell}\langle g_x, \bar{\psi}\rangle \tag{4.14}$$

Then, substitute (4.14) into (4.13) and integrate $\int_I dx$ to find

$$\frac{1}{2}\Lambda^3\ell^{-2}\left(\int \Pi_x dx\right)\bar{\psi} = \left(\frac{\Lambda}{\ell} - e\right)\left(\frac{\Lambda}{2} - e\right)\bar{\psi} . \tag{4.15}$$

We know that $e < \Lambda/2$ because we could take ψ_x = constant as a trial function, and then use $\widetilde{H}_x \leq \Lambda/2$. Also, $e < \Lambda/\ell$, because we could take $\Psi = \delta_{x0}|g_{x0}\rangle$.

Step 4: Eq. (4.15) will give us a lower bound to e if we can find an upper bound to $Y = (\Lambda/\ell)\int_I \Pi_x dx$. To do this note that

$$\begin{aligned} Y^2 &\leq \text{Trace } Y^2 = \left(\frac{\Lambda}{\ell}\right)^2 \int_I\int_I |\langle g_x, g_y\rangle|^2 dxdy \\ &= \left(\frac{\Lambda}{\ell}\right)^2 \int_I\int_I \exp\{-\frac{\alpha}{2V}\sum_{\varepsilon\Lambda\leq|k|\leq\Lambda}\sum_{\lambda}|f_\lambda(k,x) - f_\lambda(k,y)|^2 dxdy\} \end{aligned} \tag{4.16}$$

Noting that $\sum_{\lambda=1}^2 e_\lambda(k)_1^2 = 1 - k_1^2/k^2$, the quantity $\{\quad\}$ in (4.16) becomes (as $V \to \infty$)

$$\{\ \ \} = -\frac{2\alpha}{(2\pi)^3}\int_{\Lambda/2<|k|<\Lambda}\frac{1}{|k|^3k_1^2}(k_\perp^2)[\sin\frac{k_1}{2}(x-y)] \tag{4.17}$$

After some algebra we find that

$$\left(\frac{1}{\ell} - \frac{e}{\Lambda}\right)\left(\frac{1}{2} - \frac{e}{\Lambda}\right) \leq \frac{1}{2\ell}(\text{Trace } Y^2)^{1/2} \leq \frac{1}{2\ell}\sqrt{K_\ell(\alpha)} \tag{4.18}$$

where

$$\begin{aligned} K_\ell(\alpha) &= \int_0^1\int_0^1 \exp\left[-\alpha\frac{\ell}{\pi^2}|x-y|\int_0^{|x-y|\ell/4}\left(\frac{\sin t}{t}\right)^2 dt\right] dxdy . \\ &\leq \int_{-1/2}^{1/2}\exp[-\alpha x^2\ell^2/8\pi]\, dx . \end{aligned} \tag{4.19}$$

We find that

$$K_\ell(\alpha) \sim 1 - \alpha\ell^2/96\pi, \quad \ell^2\alpha \text{ small} \qquad (4.20)$$
$$\sim \sqrt{2}\pi(\alpha\ell^2)^{1/2}, \quad \ell^2\alpha \text{ large}$$

If α is small we take $\ell \sim \alpha^{-1/2}$. If α is large we take $\ell = 2$. This establishes our theorem for $N = 1$. ■

Theorem 1.6: Sketch of Proof.

For $N > 1$ we can decompose $\mathbb{R}^3$ into cubic boxes $B_j, j = 1, 2, 3, \ldots$ of size $\ell\Lambda$ and "borrow" $\frac{1}{2}|p + A(x)|^2$ kinetic energy from each particle. That is, $H_N = H_N^{1/2} + \frac{1}{2}T_N$ with $T_N = \sum_{i=1}^N T(x_i)$. The Pauli principle will then yield an energy for $\frac{1}{2}T_N$ that is bounded below by (const.) $(n_{j-1})^{4/3}$, where n_j is the particle number in box B_j.

References

[BFG] L. Bugliaro, J. Fröhlich and G.M. Graf, *Stability of quantum electrodynamics with nonrelativistic matter*, Phys.Rev. Lett. **77** (1996), 3494–3497.

[BFS] V. Bach, J. Fröhlich and I.M. Sigal, *Renormalization group analysis of spectral problems in quantum field theory*, Adv. Math. **137** (1998), 205–298; *Quantum electrodynamics of confined nonrelativistic particles*, *ibid* 299–395.

[F] C. Fefferman, *Stability of Coulomb systems in a magnetic field*, Proc. Natl. Acad. Sci. USA, **92** (1995), 5006–5007.

[FFG] C. Fefferman, J. Fröhlich and G.M. Graf, *Stabilty of nonrelativistic quantum mechanical matter coupled to the (ultraviolet cutoff) radiation field*, Proc. Natl. Acad. Sci. USA **93** (1996), 15009–15011; *Stability of ultraviolet cutoff quantum electrodynamics with non-relativistic matter*, Commun. Math. Phys. **190** (1997), 309–330.

[H] H. Hogreve, Math. Reviews **99e:81051a, b** Amer. Math. Soc. (1999).

[LL] E.H. Lieb and M. Loss, *Remarks about the ultraviolet problem in QED*, (in preparation).

[LLS] E.H. Lieb, M. Loss and J. P. Solovej, *Stability of Matter in Magnetic Fields*, Phys. Rev. Lett. *75*, 985-989 (1995).

[LSS] E.H. Lieb, H. Siedentop and J.P. Solovej, *Stability and Instability of Relativistic Electrons in Magnetic Fields*, J. Stat. Phys. **89** (1997), 37-59.

[LY] E.H.Lieb and K. Yamazaki, *Ground State Energy and Effective Mass of the Polaron*, Phys. Rev. **111** (1958), 728-733.

[OY] T. Okamoto and K. Yajima, *Complex scaling technique in nonrelativistic massive QED*, Ann. Inst. H. Poincaré Phys. Théor. **42** (1985), 311-327.

Commun. Math. Phys. 213, 673 – 683 (2000)

Communications in
Mathematical
Physics

Renormalization of the Regularized Relativistic Electron-Positron Field*

Elliott H. Lieb[1], **Heinz Siedentop**[2,**]

[1] Departments of Mathematics and Physics, Jadwin Hall, Princeton University, P.O.B. 708, Princeton, NJ 08544-0708, USA. E-mail: lieb@princeton.edu
[2] Mathematik, Universität Regensburg, 93040 Regensburg, Germany

Received: 8 March 2000 / Accepted: 7 July 2000

Abstract: We consider the relativistic electron-positron field interacting with itself via the Coulomb potential defined with the physically motivated, positive, density-density quartic interaction. The more usual normal-ordered Hamiltonian differs from the bare Hamiltonian by a quadratic term and, by choosing the normal ordering in a suitable, self-consistent manner, the quadratic term can be seen to be equivalent to a renormalization of the Dirac operator. Formally, this amounts to a Bogolubov-Valatin transformation, but in reality it is non-perturbative, for it leads to an inequivalent, fine-structure dependent representation of the canonical anticommutation relations. This non-perturbative redefinition of the electron/positron states can be interpreted as a mass, wave-function and charge renormalization, among other possibilities, but the main point is that a non-perturbative definition of normal ordering might be a useful starting point for developing a consistent quantum electrodynamics.

1. Introduction

In relativistic quantum electrodynamics (QED) the quantized electron-positron field $\Psi(x)$, which is an operator-valued spinor, is written formally as

$$\Psi(x) := a(x) + b^*(x), \tag{1}$$

where $a(x)$ annihilates an electron at x and $b^*(x)$ creates a positron at x. (We use the notation that x denotes a space-spin point, namely $x = (\mathbf{x}, \sigma) \in \Gamma = \mathbb{R}^3 \times \{1, 2, 3, 4\}$ and $\int d^3x$ denotes integration over $\mathbb{R}^3$ and a summation over the spin index.) More precisely, we take the Hilbert space, $\mathfrak{H}$ of $L^2(\mathbb{R}^3)$ spinors, i.e., $\mathfrak{H} = L^2(\mathbb{R}^3) \otimes \mathbb{C}^4$. To specify the one-electron space we choose a subspace, $\mathfrak{H}_+$ of $\mathfrak{H}$ and denote the orthogonal

** *Current address:* Mathematik, Universität München, Theresienstraße 39, 80333 München, Germany. E-mail: hkh@rz.mathematik.uni-muenchen.de

projector onto this subspace by P_+. The one-positron space is the (anti-unitary) charge conjugate, $C\mathfrak{H}_-$, of the orthogonal complement, $\mathfrak{H}_-$ with projector P_-. In position space, $(C(\psi))(\mathbf{x}) := i\beta\alpha_2\overline{\psi(\mathbf{x})}$, whereas in momentum space $(C(\hat{\psi}))(\mathbf{p}) := i\beta\alpha_2\overline{\hat{\psi}(-\mathbf{p})}$. For later purposes we note explicitly that the Hilbert space $\mathfrak{H}$ can be written as the orthogonal sum

$$\mathfrak{H} = \mathfrak{H}_+ \oplus \mathfrak{H}_-. \tag{2}$$

If f_ν is an orthonormal basis for $\mathfrak{H}_+$ with $\nu \geq 0$ and an orthonormal basis for $\mathfrak{H}_-$ with $\nu < 0$ then $a(x) := \sum_{\nu=0}^{\infty} a(f_\nu) f_\nu(x)$, and $b(x) := \sum_{\nu=-\infty}^{-1} b(f_\nu)\overline{f_\nu(x)}$, where $a^*(f)$ creates an electron in the state $P_+ f$, etc. (For further details about the notation see Thaller [7], and also Helffer and Siedentop [4] and Bach et al [2, 1].

For free electrons and positrons, $\mathfrak{H}_+$ and $\mathfrak{H}_-$ are, respectively, the positive and negative energy solutions of the free Dirac operator

$$D_0 = \boldsymbol{\alpha} \cdot \mathbf{p} + m_0\beta, \tag{3}$$

in which $\boldsymbol{\alpha}$ and β denote the four 4×4 Dirac matrices and $\mathbf{p} = -i\nabla$. The number m_0 is the *bare mass* of the electron/positron. Perturbation theory is defined in terms of this splitting of $\mathfrak{H}$ into $\mathfrak{H}_+$ and $\mathfrak{H}_-$. The electron/positron lines in Feynman diagrams are the resolvents of D_0 split in this way. As we shall see, this splitting may not be the best choice, ultimately, and another choice (which will require a fine-structure dependent, inequivalent representation of the canonical anticommutation relations) might be more useful.

The Hamiltonian for the free electron-positron field is

$$\mathbb{H}_0 = \int d^3x : \Psi(x)^* D_0 \Psi(x) :$$

with this choice of $\mathfrak{H}_+$. The symbol : : denotes normal ordering, i.e., anti-commuting all a^*, b^* operators to the left (but ignoring the anti-commutators).

In this paper we investigate the effect of the Coulomb interaction among the particles, which is a quartic form in the $a^\#$, $b^\#$ operators. (# denotes either a star or no star.) The normal ordering will give rise to extra constants and certain quadratic terms. Our main point will be that by making an appropriate choice of $\mathfrak{H}_+$ that is different from the usual one mentioned above, we can absorb the additional quadratic terms into a mass, wave function, and charge renormalization. This choice of $\mathfrak{H}_+$ has to be made self consistently and, as we show, can be successfully carried out if some combination of the fine structure constant α (=1/137 in nature) and the ultraviolet cutoff Λ is small enough (but "small" is actually "huge" since the condition is, roughly, $\alpha \log \Lambda < 1$).

The physical significance of our construction is not immediate, but it does show that certain annoying terms, when treated *non-perturbatively*, can be incorporated into the renormalization program. It is not at all clear that what we have called wave function renormalization, for example, really corresponds to a correct interpretation of that renormalization, or even what the true meaning of wave function renormalization is. It could also be interpreted as a renormalization of Planck's constant. Likewise, the charge renormalization we find is not mandatory. But our main point is that the effect is mathematically real and is best dealt with by a change of the meaning of the one-particle electron/positron states. It should be taken into account in the non-perturbative QED that is yet to be born. Of course, there are other renormalization effects in QED that we do not consider. In particular, the magnetic field has not been included and so we do not have a Ward identity to help us. Our result here is only a small part of the bigger picture of a proper, non-perturbative QED. Some other QED renormalizations involving the magnetic field are discussed in [6].

The last section of this paper contains a brief discussion of some possible interpretations of our findings and the interested reader is urged to look at that section.

Our starting point is the unrenormalized ("bare") Hamiltonian for a quantized electron/positron field interacting with itself via Coulomb's law, namely

$$\mathbb{H}^{\text{bare}} := \int d^3x : \Psi^*(x) D_0 \Psi(x) : \\ + \frac{\alpha}{2} \int d^3x \int d^3y W(\mathbf{x}, \mathbf{y}) : \Psi^*(x)\Psi(x) :: \Psi^*(y)\Psi(y) :, \quad (4)$$

where W is a symmetric ($W(x, y) = W(y, x)$) interaction potential. The case of interest here is the Coulomb potential $W(x, y) = \delta_{\sigma,\tau} |\mathbf{x}-\mathbf{y}|^{-1}$ and regularized versions thereof. We will refer to the first term on the right side of (4) as the kinetic energy $\mathbb{T}$ and to the second term as the interaction energy $\mathbb{W}^{\text{bare}}$.

Our choice of the product of the two normal ordered factors in the interaction, namely $: \rho(x) : : \rho(y) :$, is taken from the book of Bjorken and Drell [3] (Eq. (15.28) without magnetic field). It is quite natural, we agree, to start with the closest possible analog to the classical energy of a field. Of course, this term is only partly normal ordered in the usual sense and therefore it does not have zero expectation value in the unperturbed vacuum or one-particle states. It is positive, however, as a Coulomb potential ought to be.

The fully normal ordered interaction (which is not positive) is defined to be

$$\mathbb{W}_\alpha^{\text{ren}} = \frac{\alpha}{2} \int d^3x \int d^3y W(\mathbf{x}, \mathbf{y}) : \Psi^*(x)\Psi^*(y)\Psi(y)\Psi(x) : . \quad (5)$$

Bear in mind that this definition entails a determination of the splitting (2) of $\mathfrak{H}$.

We introduce a normal ordered ("dressed") Hamiltonian

$$\mathbb{H}_\alpha^{\text{ren}} := \int d^3x : \Psi^*(x) D_{Z,m} \Psi(x) : + \mathbb{W}_\alpha^{\text{ren}}. \quad (6)$$

(Again, this depends on the splitting of $\mathfrak{H}$.) The operator $D_{Z,m}$ differs from D_0 in two respects:

$$D_{Z,m} = Z^{-1}(\boldsymbol{\alpha} \cdot \mathbf{p} + m\beta), \quad (7)$$

where $Z < 1$ and $m > m_0$. We call m_0 the bare mass and m the "physical" mass – up to further renormalizations, which we do not address in this paper. The factor Z is called the "wave function renormalization" because the "wave function renormalization" in the standard theory is defined by the condition that the electron-positron propagator equals Z times a non-interacting propagator, which is the inverse of a Dirac operator. To the extent that $\mathbb{W}_\alpha^{\text{ren}}$ can be neglected, the propagator with just the first term on the right side of (6) would, indeed, be Z times a Dirac operator with mass m. One could also interpret Z in other ways, e.g., as a renormalization of the speed of light, but such a choice would be considerably more radical. Some other interpretations are discussed in the last section.

Our goal is to show that by a suitable choice of Z, m, and the electron space $\mathfrak{H}_+$, we have – up to physically unimportant infinite additive constants – asymptotic equality between $\mathbb{H}^{\text{bare}}$ and $\mathbb{H}_\alpha^{\text{ren}}$ for momenta that are much smaller than the cutoff Λ

$$\mathbb{H}^{\text{bare}} = \mathbb{H}_\alpha^{\text{ren}} \qquad \text{(for small momenta).} \tag{8}$$

The additional normal ordering in (6) will introduce some quadratic terms and it is these terms to which we turn our attention and which we would like to identify as renormalization terms. Whether our picture is really significant physically, or whether it is just a mathematical convenience remains to be seen. What it does do is indicate a need for starting with a non-conventional, non-perturbative choice of the free particle states, namely the choice of $\mathfrak{H}_+$.

2. Calculation of the New Quadratic Terms

The canonical anti-commutation relations are

$$\begin{aligned}\{a(f), a(g)\} &= \{a^*(f), a^*(g)\} = \{a(f), b(g)\} = \{a^*(f), b^*(g)\}\\ &= \{a^*(f), b(g)\} = \{a(f), b^*(g)\} = 0,\end{aligned} \tag{9}$$

and

$$\{a(f), a^*(g)\} = (f, P_+g), \ \{b^*(f), b(g)\} = (f, P_-g). \tag{10}$$

Formally, this is equivalent to

$$\begin{aligned}\{a(x), a(y)\} &= \{a^*(x), a^*(y)\} = \{a(x), b(y)\} = \{a^*(x), b^*(y)\}\\ &= \{a^*(x), b(y)\} = \{a(x), b^*(y)\} = 0,\end{aligned} \tag{11}$$

and

$$\{a(x), a^*(y)\} = P_+(x, y), \ \{b^*(x), b(y)\} = P_-(x, y), \tag{12}$$

where $P_\pm(x, y)$ is the integral kernel of the projector $P_\pm$.

In order to compare the bare Hamiltonian $\mathbb{H}^{\text{bare}}$ with the renormalized one $\mathbb{H}_\alpha^{\text{ren}}$ we must face a small problem; the momentum cutoff Λ, which is necessary in order to get finite results, will spoil the theory at momenta comparable to this cutoff. Thus, our identification of the difference as a wave function and mass renormalization will be exact only for momenta that are small compared to Λ. We will also require (although it is not clear if this is truly necessary or whether it is an artifact of our method) that $\alpha \ln(\Lambda/m_0)$ is not too large.

The potential energy $\mathbb{W}^{\text{bare}}$ clearly has positive singularities in it if the unregularized Coulomb interaction $W(x, y) = |\mathbf{x} - \mathbf{y}|^{-1}\delta_{\sigma,\tau}$ is taken. We therefore cut off the field operators for high momenta above Λ. (Another choice would be to cut off the Fourier transform of $|\mathbf{x} - \mathbf{y}|^{-1}$ at Λ and the result would be essentially the same; our choice is motivated partly by computational convenience.) Our cutoff procedure is to require that

$$\Psi(x) := (2\pi)^{-3/2} \int_{\{\mathbf{p}\in\mathbb{R}^3 \mid |\mathbf{p}|<\Lambda\}} \widehat{\Psi}(\mathbf{p}, \sigma) e^{-i\mathbf{p}\cdot\mathbf{x}} d\mathbf{p} \tag{13}$$

holds, with $\Lambda > 0$. We could also take a smooth cutoff without any significant change. This regularization keeps the positivity of $\mathbb{W}^{\text{bare}}$. To make all terms finite we also need a volume cutoff for one of the difference terms. However, the volume singularity only occurs as an additive constant, which we drop. The renormalized Hamiltonian is free of infrared divergent energies and so does not need a volume cutoff.

Next, we calculate the difference

$$\begin{aligned}\mathbb{R} := \mathbb{W}^{\text{bare}} - \mathbb{W}_\alpha^{\text{ren}} = \frac{\alpha}{2}\int d^3x \int d^3y\, W(\mathbf{x},\mathbf{y})[a^*(x)a(y)P_+(x,y)\\ + a^*(x)b^*(y)P_+(x,y) + b(x)a(y)P_+(x,y) - b^*(y)b(x)P_+(x,y)\\ - a^*(y)a(x)P_-(x,y) + P_-(x,y)P_+(x,y) + a(x)b(y)P_-(x,y)\\ + b^*(x)a^*(y)P_-(x,y) + b(x)^*b(y)P_-(x,y)]\,. \qquad (14)\end{aligned}$$

Thus, (4) can be rewritten as

$$\begin{aligned}\mathbb{H}^{\text{bare}} &= \int d^3x : \Psi^*(x)D_0\Psi(x) :\\ &+ \frac{\alpha}{2}\int d^3x \int d^3y \frac{P_+(x,y) - P_-(x,y)}{|\mathbf{x}-\mathbf{y}|} : \Psi^*(x)\Psi(y) : + \mathbb{W}_\alpha^{\text{ren}}\\ &+ \frac{\alpha}{2}\int d^3x \int d^3y \frac{P_+(x,y)P_-(x,y)}{|\mathbf{x}-\mathbf{y}|}. \qquad (15)\end{aligned}$$

The first two terms on the right side are one-particle operators (i.e., quadratic in the field operators). Let us call their sum $\mathbb{A}$, which we write (formally, since A is a differential operator) as

$$\mathbb{A} = \int d^3x \int d^3y A(x,y) : \Psi^*(x)\Psi(y) : . \qquad (16)$$

The last term is a cutoff dependent constant, which happens to be infinite. (One can show that this term is positive and, if we put in a momentum cutoff Λ and volume (infrared) cutoff V, this integral is proportional to $V\Lambda^4$.)

We can make $\mathbb{A}$ positive by choosing the normal ordering appropriately. That is, we choose P_+ to be the projector onto the positive spectral subspace of the operator A. Thus, we are led to the nonlinear equation for the unknown A,

$$A = D_0 + \frac{\alpha}{2}R, \qquad (17)$$

where the operator R has the integral kernel

$$R(x,y) := \frac{P_+(x,y) - P_-(x,y)}{|\mathbf{x}-\mathbf{y}|} = \frac{(\operatorname{sgn} A)(x,y)}{|\mathbf{x}-\mathbf{y}|}, \qquad (18)$$

with $\operatorname{sgn} t$ being the sign of t.

For physical reasons and to simplify matters we restrict the search for a solution to (17) to translationally invariant operators, i.e., 4×4 matrix-valued Fourier multipliers. Moreover, we make the ansatz

$$A(\mathbf{p}) := \boldsymbol{\alpha} \cdot \omega_{\mathbf{p}} g_1(|\mathbf{p}|) + \beta g_0(|\mathbf{p}|) \qquad (19)$$

with real functions g_1 and g_0, where $\omega_{\mathbf{p}}$ is the unit vector in the $\mathbf{p}$ direction. In other words, we try to make A look as much as possible like a Dirac operator. With this ansatz we have $\operatorname{sgn} A(\mathbf{p}) = A(\mathbf{p})/(g_1(|\mathbf{p}|)^2 + g_0(|\mathbf{p}|)^2)^{1/2}$. Thus, recalling (13), (17) can be fulfilled, if

$$g_0(|\mathbf{p}|) := m_0 + \frac{\alpha}{4\pi^2}\int_{|\mathbf{q}|<\Lambda} d\mathbf{q}\frac{1}{|\mathbf{p}-\mathbf{q}|^2}\frac{g_0(|\mathbf{q}|)}{(g_1(|\mathbf{q}|)^2 + g_0(|\mathbf{q}|)^2)^{1/2}}, \tag{20}$$

$$g_1(|\mathbf{p}|) := |\mathbf{p}| + \frac{\alpha}{4\pi^2}\int_{|\mathbf{q}|<\Lambda} d\mathbf{q}\frac{\omega_\mathbf{p}\cdot\omega_\mathbf{q}}{|\mathbf{p}-\mathbf{q}|^2}\frac{g_1(|\mathbf{q}|)}{(g_1(|\mathbf{q}|)^2 + g_0(|\mathbf{q}|)^2)^{1/2}} \tag{21}$$

is solvable. Note that the bare mass m_0 appears in (20).

3. Determination of the Dressed Electron

We will find a solution of the system (20), (21) by a fixed point argument. To this end we first integrate out the angle on the right-hand side. Setting $u := |\mathbf{p}|$ and $v := |\mathbf{q}|$ we get

$$g_0(u) = m_0 + \frac{\alpha}{2\pi}\int_0^\Lambda dv\frac{v}{u}Q_0\left(\frac{1}{2}\left(\frac{u}{v}+\frac{v}{u}\right)\right)\frac{g_0(v)}{(g_1(v)^2 + g_0(v)^2)^{1/2}}, \tag{22}$$

$$g_1(u) = u + \frac{\alpha}{2\pi}\int_0^\Lambda dv\frac{v}{u}Q_1\left(\frac{1}{2}\left(\frac{u}{v}+\frac{v}{u}\right)\right)\frac{g_1(v)}{(g_1(v)^2 + g_0(v)^2)^{1/2}}, \tag{23}$$

where, for $z > 1$,

$$Q_0(z) = \frac{1}{2}\log\frac{z+1}{z-1} \tag{24}$$

and

$$Q_1(z) = \frac{z}{2}\log\frac{z+1}{z-1} - 1. \tag{25}$$

The case of zero bare mass is particularly easy. We can choose $g_0 = 0$, in which case g_1 is obtained by integration. We get

$$\begin{aligned} g_1(u) &= u + \frac{\alpha}{2\pi}\int_0^\Lambda dv\frac{v}{u}Q_1\left(\frac{1}{2}\left(\frac{u}{v}+\frac{v}{u}\right)\right) \\ &= u + \frac{\alpha}{2\pi}u\int_0^{\Lambda/u} dv\, v\, Q_1\left((1/v+v)/2\right) \\ &= u + \frac{\alpha}{2\pi}u\left[\frac{2}{3}\log|(\Lambda/u)^2 - 1| - \frac{1}{3}(\Lambda/u)^2\right. \\ &\quad \left.+\frac{\Lambda}{6u}\log\left|\frac{(\Lambda/u)+1}{(\Lambda/u)-1}\right|\left(3+(\Lambda/u)^2\right)\right]. \end{aligned} \tag{26}$$

The function g_1 behaves asymptotically for small u or large Λ as $g_1(u) = \frac{\alpha}{2\pi}\frac{4}{3}u\log(\Lambda/u)$. Unfortunately, because of the $\log u$, the operator A can never look like the renormalized Dirac operator in (7). This asymptotic expansion can be seen either from (26) or else by noting that the integrand is a continuous function, except for $v = 1$, that decays at infinity as $4/(3v)$. This follows from the large v expansion

$$vQ_1\left((1/v+v)/2\right) = \frac{4}{3}v^{-1} + \frac{8}{15}v^{-3} + O(v^{-5}). \tag{27}$$

For positive bare masses we solve the system (22) and (23) by a fixed point argument. To this end we define the following set of pairs of functions, for $\epsilon, \delta > 0$,

$$S_{\epsilon,\delta} := \{\mathbf{g} = (g_0, g_1) \mid \forall_{u \in [0,\Lambda]} \mathbf{g}(u) \in [m_0, (1+\delta)m_0] \times [u, (1+\epsilon)u]\}. \tag{28}$$

Note that with the metric generated by the sup norm $S_{\epsilon,\delta}$ is a complete metric space. Next define $T : S_{\epsilon,\delta} \to S_{\epsilon,\delta}$ by the right-hand side of (23) and (22), i.e.,

$$T_0(\mathbf{g})(u) = m_0 + \frac{\alpha}{2\pi} \int_0^\Lambda dv \frac{v}{u} Q_0\left(\frac{1}{2}\left(\frac{u}{v} + \frac{v}{u}\right)\right) \frac{g_0(v)}{(g_1(v)^2 + g_0(v)^2)^{1/2}}, \tag{29}$$

$$T_1(\mathbf{g})(u) = u + \frac{\alpha}{2\pi} \int_0^\Lambda dv \frac{v}{u} Q_1\left(\frac{1}{2}\left(\frac{u}{v} + \frac{v}{u}\right)\right) \frac{g_1(v)}{(g_1(v)^2 + g_0(v)^2)^{1/2}}. \tag{30}$$

Lemma 1. *If Y, defined by*

$$Y := \alpha \operatorname{arsinh}(\Lambda/m_0)/\pi, \tag{31}$$

satisfies $Y < 9/50$, if $\epsilon \geq 50Y/(9-50Y)$, and if $\delta \geq Y/(1-Y)$, then T maps $S_{\epsilon,\delta}$ into $S_{\epsilon,\delta}$.

Note that $\operatorname{arsinh}(x) = \log(x + \sqrt{x^2+1})$, i.e., Y grows logarithmically in Λ/m_0.

Proof. Obviously, $T_0(\mathbf{g})(u) \geq m_0$ and $T_1(\mathbf{g})(u) \geq u$.

To bound g_0 we return to (20) and note the bound (which holds in $S_{\epsilon,\delta}$)

$$\begin{aligned} g_0(|\mathbf{p}|) &\leq m_0 + \frac{\alpha}{4\pi^2} \int_{|\mathbf{q}|<\Lambda} d\mathbf{q} \frac{1}{|\mathbf{p}-\mathbf{q}|^2} \frac{(1+\delta)m_0}{(|\mathbf{q}|^2 + m_0^2)^{1/2}} \\ &\leq m_0 + \frac{\alpha}{4\pi^2} \int_{|\mathbf{q}|<\Lambda} d\mathbf{q} \frac{1}{\mathbf{q}^2} \frac{(1+\delta)m_0}{(|\mathbf{q}|^2 + m_0^2)^{1/2}} \\ &= m_0[1 + \frac{\alpha}{\pi}(1+\delta) \operatorname{arsinh}(\Lambda/m_0)]. \end{aligned} \tag{32}$$

The second inequality holds because the convolution of two symmetric decreasing functions is symmetric decreasing.

Next we turn to g_1, which is a bit more complicated. We split the integration region in (21) into $A = \{\mathbf{q} \mid |\mathbf{q}| < 2|\mathbf{p}| \text{ and } |\mathbf{q}| < \Lambda\}$ and $B = \{\mathbf{q} \mid |\mathbf{q}| \geq 2|\mathbf{p}| \text{ and } |\mathbf{q}| < \Lambda\}$. In region A we use $\omega_\mathbf{p} \cdot \omega_\mathbf{q} \leq 1$ and hence the contribution to g_1 from this region is bounded above by

$$\frac{\alpha}{4\pi^2} \int_{|\mathbf{q}|<\Lambda} d\mathbf{q} \frac{1}{|\mathbf{p}-\mathbf{q}|^2} \frac{(1+\epsilon)2|\mathbf{p}|}{(|\mathbf{q}|^2 + m_0^2)^{1/2}} \leq 2|\mathbf{p}| \frac{\alpha}{\pi}(1+\epsilon) \operatorname{arsinh}(\Lambda/m_0)], \tag{33}$$

for the same reason as in (32).

In region B we use that $(|\mathbf{p}|^2 - |\mathbf{q}|^2)^2 \geq 9|\mathbf{q}|^4/16$. We also note, for the integration over B, that we can take the mean of the integrand for $\mathbf{q}$ and $-\mathbf{q}$. In other words, we can bound this term as follows:

$$
\begin{aligned}
&\frac{\alpha}{4\pi^2}\left|\int_B d\mathbf{q}\frac{\omega_\mathbf{p}\cdot\omega_\mathbf{q}}{|\mathbf{p}-\mathbf{q}|^2}\frac{g_1(|\mathbf{q}|)}{(g_1(|\mathbf{q}|)^2+g_0(|\mathbf{q}|)^2)^{1/2}}\right|\\
&\quad\le\frac{\alpha}{8\pi^2}\left|\int_B d\mathbf{q}\left(\frac{\omega_\mathbf{p}\cdot\omega_\mathbf{q}}{|\mathbf{p}-\mathbf{q}|^2}-\frac{\omega_\mathbf{p}\cdot\omega_\mathbf{q}}{|\mathbf{p}+\mathbf{q}|^2}\right)\frac{g_1(|\mathbf{q}|)}{(g_1(|\mathbf{q}|)^2+g_0(|\mathbf{q}|)^2)^{1/2}}\right|\\
&\quad\le\frac{\alpha}{8\pi^2}\int_B d\mathbf{q}\frac{4|\mathbf{p}||\mathbf{q}|}{(|\mathbf{p}|^2-|\mathbf{q}|^2)^2}\frac{(1+\epsilon)|\mathbf{q}|}{(|\mathbf{q}|^2+m_0^2)^{1/2}}\\
&\quad\le\frac{8\alpha(1+\epsilon)}{9\pi^2}|\mathbf{p}|\int_B d\mathbf{q}\frac{1}{|\mathbf{q}|^2}\frac{1}{(|\mathbf{q}|^2+m_0^2)^{1/2}}\\
&\quad\le\frac{32\alpha(1+\epsilon)}{9\pi}|\mathbf{p}|\operatorname{arsinh}(\Lambda/m_0).
\end{aligned}
\tag{34}
$$

Thus, we obtain the bound

$$
g_1(|\mathbf{p}|)\le|\mathbf{p}|\{1+\frac{50\alpha}{9\pi}(1+\epsilon)\operatorname{arsinh}(\Lambda/m_0)\}.
\tag{35}
$$

Theorem 1. *If* $Y:=\alpha\operatorname{arsinh}(\Lambda/m_0)/\pi<9/50$, *if* $(2+\epsilon+\delta)Y<1$ *and if* ϵ,δ *satisfy the conditions of Lemma 1 then* T *is a contraction.*

Proof. Thanks to Lemma 1 we only need to establish the contraction property. We first note that for positive real numbers $x, y, \tilde{x}, \tilde{y}$,

$$
\left|\frac{x}{(x^2+y^2)^{1/2}}-\frac{\tilde{x}}{(\tilde{x}^2+\tilde{y}^2)^{1/2}}\right|\le\frac{|\eta|}{\xi^2+\eta^2}((x-\tilde{x})^2+(y-\tilde{y})^2)^{1/2},
\tag{36}
$$

where (ξ,η) is some point on the line between (x,y) and $(\tilde{x},\tilde{y})$. Thus we get

$$
\begin{aligned}
&|[T_0(g)-T_0(\tilde{g})](u)|+|[T_1(g)-T_1(\tilde{g})](u)|\\
&\le\frac{\alpha}{2\pi}\int_0^\Lambda dv\frac{v}{u}\left[Q_0\left(\frac{1}{2}\left(\frac{u}{v}+\frac{v}{u}\right)\right)\left|\frac{g_0(v)}{\sqrt{g_0(v)^2+g_1(v)^2}}-\frac{\tilde{g}_0(v)}{\sqrt{\tilde{g}_0(v)^2+\tilde{g}_1(v)^2}}\right|\right.\\
&\qquad\left.+Q_1\left(\frac{1}{2}\left(\frac{u}{v}+\frac{v}{u}\right)\right)\left|\frac{g_1(v)}{\sqrt{g_0(v)^2+g_1(v)^2}}-\frac{\tilde{g}_1(v)}{\sqrt{\tilde{g}_0(v)^2+\tilde{g}_1(v)^2}}\right|\right]\\
&\le\frac{\alpha}{2\pi}\int_0^\Lambda dv\frac{v}{u}\left[Q_0\left(\frac{1}{2}\left(\frac{u}{v}+\frac{v}{u}\right)\right)\frac{(1+\epsilon)v}{v^2+m_0^2}+Q_1\left(\frac{1}{2}\left(\frac{u}{v}+\frac{v}{u}\right)\right)\frac{(1+\delta)m_0}{v^2+m_0^2}\right]\\
&\qquad\times\left[(g_0(v)-\tilde{g}_0(v))^2+(g_1(v)-\tilde{g}_1(v))^2\right]^{1/2}\\
&\le\frac{\alpha}{2\pi}\int_0^\Lambda dv\frac{v}{u}\left[Q_0\left(\frac{1}{2}\left(\frac{u}{v}+\frac{v}{u}\right)\right)\frac{(1+\epsilon)v}{v^2+m_0^2}\right.\\
&\qquad\left.+Q_1\left(\frac{1}{2}\left(\frac{u}{v}+\frac{v}{u}\right)\right)\frac{(1+\delta)m_0}{v^2+m_0^2}\right]\|\mathbf{g}-\tilde{\mathbf{g}}\|\\
&\le\frac{\alpha}{2\pi}\int_0^\Lambda dv\frac{v}{u}Q_0\left(\frac{1}{2}\left(\frac{u}{v}+\frac{v}{u}\right)\right)\frac{(1+\epsilon)v+(1+\delta)m_0}{v^2+m_0^2}\|\mathbf{g}-\tilde{\mathbf{g}}\|\\
&\le(2+\epsilon+\delta)Y\|\mathbf{g}-\tilde{\mathbf{g}}\|,
\end{aligned}
\tag{37}
$$

Invent. math. (2001)
Digital Object Identifier (DOI) 10.1007/s002220100159

Inventiones mathematicae

Ground states in non-relativistic quantum electrodynamics

Marcel Griesemer[1,*], **Elliott H. Lieb**[2,**], **Michael Loss**[3,***]

[1] Department of Mathematics, University of Alabama at Birmingham, Birmingham, AL 35294, USA (e-mail: marcel@math.uab.edu)
[2] Departments of Physics and Mathematics, Jadwin Hall, Princeton University, P. O. Box 708, Princeton, NJ 08544, USA (e-mail: lieb@princeton.edu)
[3] School of Mathematics, Georgia Tech, Atlanta, GA 30332, USA (e-mail: loss@math.gatech.edu)

Oblatum 21-IX-2000 & 8-IV-2001
Published online: 18 June 2001

Abstract. The excited states of a charged particle interacting with the quantized electromagnetic field and an external potential all decay, but such a particle should have a true ground state – one that minimizes the energy and satisfies the Schrödinger equation. We prove quite generally that this state exists for *all values* of the fine-structure constant and the ultraviolet cutoff. We also show the same thing for a many-particle system under physically natural conditions.

1 Introduction

An established picture of an atom or molecule is that even in the presence of a quantized radiation field there is a ground state. The excited states that exist in the absence of coupling to the field are expected to melt into resonances, which means that they eventually decay with time into the ground state plus free photons. This picture has been established by Bach, Fröhlich and Sigal in [8] for sufficiently small values of the various parameters that define the theory. Here we show that a ground state exists for all values of the parameters (including a variable g-factor) in the one particle case and, under a physically appropriate assumption, in the many-particle case.

* Work partially supported by the Faculty Development Program of UAB.
** Work partially supported by U.S. National Science Foundation grant PHY 98-20650-A01.
*** Work partially supported by U.S. National Science Foundation grant DMS 00-70589.

We know that the Hamiltonian for the system is bounded below, so a ground state energy always exists in the sense of being the infimum of the spectrum, but the existence of a genuine normalizable solution to the eigenvalue equation is a more delicate matter that has received a great deal of attention, especially in recent years. A physically simple example where no ground state exists (as far as we believe now) is the free particle (i.e., particle plus field). In the presence of an external potential, however, like the Coulomb potential of a nucleus, a ground state should exist.

The difficulty in establishing this ground state comes from the fact that the bottom of the spectrum lies in the continuum (i.e., essential spectrum), not below it, as is the case for the usual Schrödinger equation. We denote the bottom of spectrum of the free-particle Hamiltonian for N particles with appropriate statistics by $E^0(N)$. The "free-particle" Hamiltonian includes the interparticle interaction (e.g., the Coulomb repulsion of electrons) but it does not include the interaction with a fixed external potential, e.g., the interaction with nuclei. When the latter is included we denote the bottom of the spectrum by $E^V(N)$. It is not hard to see in many cases that $E^V(N) < E^0(N)$, but despite this inequality $E^V(N)$ is, nevertheless, the bottom of the essential spectrum. The reason is that we can always add arbitrarily many, arbitrarily 'soft' photons that add arbitrarily little energy. It is the soft photon problem that is our primary concern here.

The main point of this paper is to show how to overcome this infrared problem and to show, quite generally for a one-particle system, that a ground state exists for all values of the particle mass, the coupling to the field (fine-structure constant $\alpha = e^2/\hbar c$), the magnetic g-factor, and the ultraviolet cutoff Λ of the electromagnetic field frequencies, provided a bound state exists when the field is turned off. This result implies, in particular, that for a fixed ultraviolet cutoff renormalization of the various physical quantities will not affect the existence of a ground state. Of course, nothing can be said about the limit as the cutoff tends to infinity. We also include a large class of interactions much more general than the usual Coulomb interaction.

The model we discuss has been used quite frequently in field theory. In its classical version it was investigated by Kramers [12] who seems to have been the first to point out the possibility of renormalization. The quantized version was investigated by Pauli and Fierz [24] in connection with scattering theory. Most importantly, it was used by Bethe [9] to obtain a suprisingly good value for the Lamb shift.

Various restricted versions of the problem have been attacked successfully. In the early seventies Fröhlich investigated the infrared problem in translation invariant models of scalar electrons coupled to scalar bosons [13]. In [14] he proved that for an electron coupled to a massive field a unique ground state exists for fixed total momentum. External potentials were not considered in these papers.

The first rigorous result on the bound state problem, to our knowledge, is due to Arai and concerns one particle confined by an x^2-potential, the interaction with the photons being subject to the dipole-approximation.

For this model, which is explicitly solvable, Arai proved existence and uniqueness of the ground state [2]. Later, Spohn showed by perturbation theory that this result extends to perturbed x^2-potentials [26]. No bounds on the parameters were needed to obtain these results but the methods oviously do not admit extensions to more realistic models.

Bach, Fröhlich and Sigal [6,7] initiated the study of the full nonrelativistic QED model (the same model as considered in the present paper) under various simplifying assumptions. In [8] the existence of a ground state in this model for particles subject to an external binding potential was proved for $\alpha\Lambda$ small enough. The main achievement of this paper is the elimination of the earlier simplifying assumptions, especially the infrared regularization. This is the first, and up to now the only paper where a 'first principles' QED model was successfully analyzed, but with a restricted parameter range. A weaker result for the same system but with simplifications such as infrared regularization were independently obtained by Hiroshima by entirely different methods; he also showed uniqueness of the ground state, assuming its existence [17].

In a parallel development Arai and Hirokawa [3,4], Spohn [28], and Gerard [15] investigated the ground-state problem for systems similar to the one of Bach et al in various degrees of mathematical generality. Arai and Hirokawa proved existence in what they called a "generalized spin-boson model". If specialized to the case of a non-relativistic N-particle system interacting with bosons, their result proves the existence of a ground state for confined particles and small α. This result was extended in a recent preprint [4] to account for non-confined particles and systems with the true infrared singularity of QED. Concrete results in the infrared-singular case concern only special models, however, such as the Wigner-Weisskopf Hamiltonian which describes a two-level atom. Hirokawa continued this work in a recent preprint [16]. Spohn and Gerard both proved existence of a ground state for a confined particle and arbitrary coupling constant, the result of Gerard being somewhat more general [28,15].

For the existence of a ground state in the case of massive bosons, which is a typical intermediate result in the cited works, a short and elegant proof was given by Derezinski and Gerard [11,15] for the case of linear coupling, in which the A^2 term is omitted, and a confining potential. Some of their ideas are used in our paper.

The Hamiltonian for N-particles has four parts which are described precisely in the next section,

$$H^V = T + V + I + H_f \,. \tag{1}$$

The dependence of H^V on N is not noted explicitly. The first term, $T = \sum_{j=1}^N T_j$, is the kinetic energy with "minimal coupling" in the Coulomb gauge (i.e., p is replaced by $p + \sqrt{\alpha}\, A$, where A is the magnetic vector potential satisfying $\mathrm{div} A = 0$), $V(X)$, with $X = (x_1, x_2, ..., x_N)$ is the external potential, typically a Coulomb attraction to one or more nuclei. In any case, we assume that V is a sum of one-body potentials, i.e.,

$$V(X) = \sum_{j=1}^{N} v(x_j) .$$

The particle-particle interaction, I, has the important feature that it is translation invariant and, of course, symmetric in the particle labels. Both I and V could be spin dependent, but we shall not burden the notation with this latter possibility. Typically I is a Coulomb repulsion, but we do not have to assume that I is merely a sum of two-body potentials. The only requirements are: (1) the negative part of v vanishes at infinity; (2) the negative part of I satisfies clustering, i.e., the intercluster part of I_- tends to 0 when the spacing between any two clusters tends to infinity; (3) $V_- + I_-$ are dominated by the kinetic energy as in (5). Another assumption we make (in the N-particle case) is that there is binding, as described below.

The natural choice for the kinetic energy is the "Pauli operator" $T = (p + \sqrt{\alpha}\, A(x))^2 + \sqrt{\alpha}\, \sigma \cdot B$, but we can generalize this to include the case of the usual kinetic energy $T = (p + \sqrt{\alpha}\, A(x))^2$ by introducing a 'g-factor', $g \in \mathbb{R}$. Thus, we take

$$T = (p + \sqrt{\alpha}\, A(x))^2 + \frac{g}{2}\sqrt{\alpha}\, \sigma \cdot B . \tag{2}$$

Note that T is a positive operator if $0 \leq g \leq 2$ and may not be otherwise. Nevertheless, the Hamiltonian is always bounded below because of the ultraviolet cutoff we shall impose on the A field, which implies that $(g/2)\sqrt{\alpha}\, \sigma \cdot B + H_f$ is always bounded below.

We believe that the "relativistic" operator $T = |p + \sqrt{\alpha} A(x)|$, presents no real difficulty either, but we do not want to overburden this paper with a lengthy proof. This problem is currently under investigation.

A model that is frequently discussed is the "Pauli-Fierz" model, but it is not entirely clear how this is defined since several variants appear in [24]. One version uses $T = (p + \sqrt{\alpha} A(x))^2 + \sqrt{\alpha}\, \sigma \cdot B$, which is one of the models under consideration. Another variant uses a linearized version of this operator, $T = p^2 + 2\sqrt{\alpha} p \cdot A(x) + \sqrt{\alpha}\, \sigma \cdot B$ or $T = p^2 + 2\sqrt{\alpha} p \cdot A(x)$. These variants are *not gauge invariant* and, therefore, depend on the choice of gauge for A. Our method is applicable to these linearized models in some gauges, but not in others. We omit further discussion of this point since these variants are not the most relevant ones for quantum electrodynamics.

There is one important point that as far as we know, has not been mentioned in the QED context. This is the binding condition. Our proof of the existence of a ground state uses, as input, the assumption that

$$E^V(N) < E^V(N') + E^0(N - N') \qquad \text{for all } N' < N. \tag{3}$$

Our work can be generalized (but we shall not do so here) to the case in which the external potential is that of attractive nuclei and these nuclei are also dynamical particles. Then, of course, it is necessary to work in the center of mass system and then (3) must be replaced by the condition that $E^V(N)$ is less than the lowest two-cluster threshold. While this condition, or (3) in the static case, are physically necessary for the existence of a ground state, the validity of these conditions cannot be taken for granted.

We prove inequality (3) for one particle ($N = 1$) quite generally, using only the assumption that the ordinary Schrödinger operator $p^2 + v$ has a negative energy ground state. This certainly holds for the Coulomb potential. Indeed, one could expect, on physical grounds, that there could be binding even if $p^2 + v$ has no negative energy bound state, because the interaction with the field increases the effective mass of the particle – and hence the binding energy. The same argument shows that when there are N particles at least one them is necessarily bound, i.e., $E^V(N) < E^0(N)$. When we consider more than one particle, we are not able to show (3) for all $N' > 1$, even if $\sum p_j^2 + I + V$ has a ground state.

In the Coulomb case, it is possible to show (but we shall not do so here) that condition (3) is satisfied if the nuclear charge Z is large enough. The basic idea is that if breakup into two groups, one of them with N' particles close to the nucleus and the second consisting of $N - N'$ particles far away occurs, then there will be an attractive Coulomb tail acting on the separated particles at a distance R away with net attractive potential $(Z - N')/R$. However, to localize one of these particles within a distance R of the nucleus will require a field energy localization error of the order of C/R, by dimensional analysis arguments. If $(Z - N') > C$ then the energy can be decreased by bringing one of the unbound particles close to the nucleus.

Section 2 introduces the precise definition of our problem and the main result Theorem 2.1.

In Sect. 3, we show how to prove that $E^V(1) < E^0(1)$. More generally, $E^V(N) < E^0(N)$ if $v \le 0$.

Our strategy to establish a ground state is the usual one of showing that a minimizing sequence of trial vectors for the energy actually has a weak limit that, in fact, is a minimizer. The problem here is that one can easily construct minimizing sequences that converge weakly to zero by choosing vectors with too many soft photons. To avoid this we take a special sequence.

To define this sequence we first consider an artificial model in which the photons have a mass, i.e., $\omega(k) = \sqrt{k^2 + m^2}$. Here there is no soft photon problem and we show in Sect. 4 that this model has a ground state Φ_m.

In Sect. 5 we show that as $m \to 0$ the Φ_m sequence is minimizing. Then in Sect. 6 we use the Schrödinger equation for Φ_m to deduce certain properties of Φ_m which we call infrared bounds. One of these was proved in [8] but we need one more, which is new.

With these bounds we can show in Sect. 7 that Φ_m has a strong limit as $m \to 0$, which is a minimizer for H^V.

Acknowledgement. We thank Professor Fumio Hiroshima for a useful correspondence concerning equation (53) and for sending us his preprint [18].

2 Definitions and main theorem

The Hamiltonian for N particles interacting with the quantized radiation field and with a given external potential $V(X)$, with $X = (x_1, x_2, ..., x_N)$ and $x_j \in \mathbb{R}^3$, is

$$H^V = \sum_{j=1}^{N} \left\{ \left(p_j + \sqrt{\alpha} A(x_j)\right)^2 + \frac{g}{2} \sqrt{\alpha}\, \sigma_j \cdot B(x_j) \right\} + V(X) + I(X) + H_f . \tag{4}$$

The unit of energy is $Mc^2/2$, where M is the particle mass, the unit of length is $\ell_c = 2\hbar/Mc$, twice the Compton wavelength of the particle, and $\alpha = e^2/\hbar c$ is the dimensionless "fine structure constant" ($= 1/137$ in nature). The electric charge of the particle is e. The unit of time is the time it takes a light wave to travel a Compton wavelength, i. e., the speed of light is $c = 1$.

The operator $p = -i\nabla$, while A is the (ultraviolet cutoff) magnetic vector potential (we use the Coulomb, or radiation gauge). The unit of $A^2(x)$ is $Mc^2/2\ell_c$. The magnetic field is $B = \mathrm{curl} A$ and the unit of B is $\alpha^{3/2}$ times the quantity $M^2 e^3 c/4\hbar^3$, which is the value of B for which the magnetic length $(\hbar c/eB)^{1/2}$ equals the Bohr radius $2\hbar^2/Me^2$.

The reader might wonder why we use these units, which seem to be more appropriate for a relativistic theory than for the nonrelativistic theory we are considering. Why not use the Bohr radius as the unit of length, for example? Our reason is that we want to isolate the electric charge, which is the quantity that defines the interaction of matter with the electromagnetic field, in precisely one place, namely α. "Atomic units" have the charge built into the length, etc. and we find this difficult to disentangle.

The Hilbert space is an appropriate subspace of

$$\mathcal{H} = \otimes^N L^2(\mathbb{R}^3; \mathbb{C}^2) \otimes \mathcal{F},$$

where $\mathcal{F}$ is the Fock space for the photon field. We have in mind Fermi statistics (the antisymmetric subspace of $\otimes^N L^2(\mathbb{R}^3; \mathbb{C}^2)$) and the $\mathbb{C}^2$ is to accomodate the electron spin. We can also deal with "Boltzmann statistics", in which case we would set $g = 0$ and use $\otimes^N L^2(\mathbb{R}^3)$, or bose statistics, in which case we would set $g = 0$ and use the symmetric subspace of $\otimes^N L^2(\mathbb{R}^3)$. These generalizations are mathematically trivial and we do not discuss them further.

For our purposes we assume that for every $\varepsilon > 0$ there exists a constant $a(\varepsilon)$ such that the negative part of the potentials, $V_-(X)$ and $I_-(X)$, satisfy

$$V_- + I_- \leq \varepsilon \sum_{j=1}^{N} p_j^2 + a(\varepsilon) \tag{5}$$

as quadratic forms on $\mathcal{H}$.

The vector potential is

$$A(x) = \sum_{\lambda=1,2} \int_{|k|<\Lambda} \frac{1}{\sqrt{|k|}} \left[\varepsilon_\lambda(k) a_\lambda(k) + \varepsilon_\lambda(-k) a_\lambda^*(-k)\right] e^{ik\cdot x} d^3k \tag{6}$$

where the operators a_λ, a_λ^* satisfy the usual commutation relations

$$[a_\lambda(k), a_\nu^*(q)] = \delta(k-q)\delta_{\lambda,\nu}\,, \quad [a_\lambda(k), a_\nu(q)] = 0, \quad \text{etc} \tag{7}$$

and the vectors $\varepsilon_\lambda(k)$ are the two possible orthonormal polarization vectors perpendicular to k. They are chosen for convenience in (59,60).

The number Λ is the ultraviolet cutoff on the wavenumbers k. Our results hold for all finite Λ. The details of the cutoff in (6) are quite unimportant, except for the requirement that rotation symmetry in k-space is maintained. E.g., a gaussian cutoff can be used instead of our sharp cutoff. We avoid unnecessary generalisations.

The field energy H_f, sometimes called $d\Gamma(\omega)$ is given by

$$H_f = \sum_{\lambda=1,2} \int_{\mathbb{R}^3} \omega(k) a_\lambda^*(k) a_\lambda(k) d^3k\,. \tag{8}$$

The energy of a photon is $\omega(k)$ and the physical value of interest to us is

$$\omega(k) = |k|, \tag{9}$$

in our units. Indeed, any continuous function that is bounded below by const.$|k|$ for small $|k|$ is acceptable. In the process of proving the existence of a ground state for H we will first study the unphysical "massive photon" case, in which

$$\omega_m(k) = \sqrt{k^2 + m^2} \tag{10}$$

for some $m > 0$, called the 'photon mass'.

In the remainder of this paper, unless otherwise stated, we shall always assume that there is no restriction on α, Λ and g and that $\omega(k)$ can be either as in (9) or as in (10).

By Lemma A.5 we see easily that H^V is bounded below for all values of the parameters, including $m = 0$. Thus, H^V defines a closable quadratic form and hence a selfadjoint operator, the Friedrich's extension. We denote this extension again by H^V.

Our main theorem is

2.1. Theorem (Existence of a ground state). *Assume that the binding condition (3) and the condition (5) hold. Then there is a vector Φ in the N-particle Hilbert space $\mathcal{H}$ such that*

$$H^V \Phi = E^V(N)\Phi \, . \tag{11}$$

3 Upper bound

We shall prove the binding condition (3) for one particle and, with an additional assumption, for the N particle case as well. This is that if N particles are present then at least one of them binds.

As we mentioned before, our requirement that the system without the radiation field has a bound state is somewhat unnaturally restrictive, since one expects that the radiation field *enhances binding*, at least in the single particle case; this has been shown to be true in the "dipole", or Kramers approximation [19]. We are able to show in the *one-particle* case, that the photon field cannot decrease the binding energy. It is quite possible that there could be binding even when the operator $p^2 + v$ does not have a negative energy state, but we cannot shed any light on that question.

For the one-particle case the situation is less delicate than the N-particle case.

3.1. Theorem (Binding of at least one particle). *Assume that the one-particle Hamiltonian $p^2 + v(x)$ has a negative energy bound state with eigenfunction $\phi(x)$ and energy $-e_0$. Then,*

$$E^V(1) \le E^0(1) - e_0 \, , \tag{12}$$

i.e., binding continues to exist when the field is turned on.

For the N-particle case we make the additional assumption that $v(x) \le 0$ for all x. Then,

$$E^V(N) \le E^0(N) - e_0 \, , \tag{13}$$

i.e., at least one particle is bound.

Proof. It suffices to prove that $E^V(N) \le E^0(N) + \varepsilon - e_0$ for all $\varepsilon > 0$. There is a normalized vector $F \in \mathcal{H}$ such that $\left(F, H^0 F\right) < E^0(N) + \varepsilon$. ($F$ is antisymmetric according to the Pauli principle.) We use the notation $\langle \cdot, \cdot \rangle$ to denote the inner product in Fock space and spin space. Then we can write $\left(F, H^0 F\right) = \int G(X) d^{3N} X$ with

$$\begin{aligned} G(X) = \\ \sum_{j=1}^{N} \big\{ &\langle (-i\nabla_j + \sqrt{\alpha} A(x_j))F, (-i\nabla_j + \sqrt{\alpha} A(x_j))F \rangle (X) \\ &+ \sqrt{\alpha}\,(g/2) \langle F, \sigma_j \cdot B(x_j) F \rangle (X) \big\} + \langle F, (I + H_f) F \rangle (X) \, . \end{aligned} \tag{14}$$

As a (unnormalized) variational trial vector we take the vector $\psi = \left[\sum_{j=1}^N \phi(x_j)^2\right]^{1/2} F$. Recall that $\phi(x) \geq 0$ since ϕ is the ground state of $p^2 + v$. We also recall the Schwarz inequality

$$\left| \frac{\sum_{j=1}^N \phi(x_j)\nabla\phi(x_j)}{\left[\sum_{j=1}^N \phi(x_j)^2\right]^{1/2}} \right|^2 \leq \sum_{j=1}^N \left|\nabla\phi(x_j)\right|^2 . \tag{15}$$

Using (15), integration by parts, and the fact that ϕ satisfies the Schrödinger equation $(p^2 + v)\phi = -e_0\phi$, we easily find that

$$\begin{aligned} &\left(\psi, \left[H^V - (E^0(N) + \varepsilon - e_0)\right]\psi\right) \\ &\leq \int \left\{G(X) - (E^0(N) + \varepsilon)\langle F, F\rangle(X)\right\} \sum_{j=1}^N \phi(x_j)^2 \, d^{3N}X \\ &+ \int \sum_{j \neq k} v(x_k)\phi(x_j)^2 \langle F, F\rangle(X) d^{3N}X . \end{aligned} \tag{16}$$

When $N = 1$ the last term in (16) is not present so no assumption about the potential v is needed. When $N > 1$ we can omit the last term because it is negative by assumption.

Now, by the $\mathbb{R}^3$-translation invariance of H^0, for every $y \in \mathbb{R}^3$ there is a "translated" vector F_y so that $G(X) \to G(X+(y, ..., y))$ and $\langle F_y, F_y\rangle(X) = \langle F, F\rangle(X + (y, ..., y))$. (This is accomplished by the unitary operator on $\mathcal{H}$ that takes $x_j \to x_j + y$ for every j and $a_\lambda(k) \to exp(ik \cdot y)a_\lambda(k)$.) Thus, if we denote the quantity in {} in (16) by $W(X)$, and if we define ψ_y by replacing F by F_y in the definition of ψ, we have

$$\begin{aligned} &\Omega(y) = \left(\psi_y, H^V - (E^0 + \varepsilon - e_0)\, \psi_y\right) \\ &\leq \int W(X + (y, ..., y)) \sum_{j=1}^N \phi(x_j)^2 d^{3N}X = \int W(X) \sum_{j=1}^N \phi(x_j - y)^2 d^{3N}X . \end{aligned} \tag{17}$$

Note that $\int \Omega(y)dy \leq N \int W(X)d^{3N}X$. But $\int W(X)d^{3N}X = (F, (H^0 - E^0(N) - \varepsilon)\, F)$ and this is strictly negative by assumption. Hence, for some $y \in \mathbb{R}^3$ we have that $\Omega(y) < 0$ and thus $\psi_y \neq 0$, which proves the theorem. □

Remark: [Alternative theorem]
It may be useful to note, briefly, a different proof of Theorem 3.1, for *long range* potentials $v(x)$, such as the attractive Coulomb potential $-Z/|x|$, which shows that the bottom of the spectrum of H^V lies strictly below E^0. Unfortunately, this proof does not show that the difference is at least e_0. We

sketch it for the one-body case. Using the notation of the proof above, the first step is to replace F by $F_R = u(x_1/R)F$ where u is a smooth function with support in a ball of radius 1. One easily finds that $\left(F_R,\, H^0\, F_R\right) / (F_R,\, F_R) = E^0 + \varepsilon + c/R^2$, where c is a constant that depends only on u and not on ε and R. On the other hand $(F_R,\, V\, F_R) / (F_R,\, F_R) \le -Z/R$, to use the Coulomb potential as an example. To complete the argument, choose $R = 2c/Z$ and then choose $\varepsilon = c/R^2$. What we have used here is the fact that localization 'costs' a kinetic energy R^{-2}, while the potential energy falls off slower than this, e.g., R^{-1}.

4 Ground state with massive photons

As we emphasized in the introduction, not every minimizing sequence converges to the minimizer for our $m = 0$ problem, i.e., with $\omega(k) = |k|$. The situation is much easier for the massive case (10). The Hamiltonian in this case is given by (4) and H_f is given by (8) with (10). To emphasize the dependence on m we denote this Hamiltonian and field energy by H_m^V and $H_f(m)$, respectively. Likewise, $E^V(m, N)$ and $E^0(m, N)$ denote the mass dependent energies, as defined before.

We emphasize that the vector potential is still given by (6), but we could, if we wished, easily replace $|k|^{-1/2}$ in (6) by $(k^2 + m^2)^{-1/4}$.

It will be shown in this section that H_m^V has a ground state. More precisely we prove

4.1. Theorem (Existence of ground state). *Assume that for some fixed value of the ultraviolet cutoff Λ there is binding for the Hamiltonian H_m^V, that is, $E^V(m, N) < \Sigma^V(m, N)$ where $\Sigma^V(m, N) = min\{E^V(m, N') + E^0(m, N - N') : \text{all } N' < N\}$ is the "lowest two-cluster threshold". Then $E^V(m, N)$ is an eigenvalue, i.e., there exists a state Φ_m in $\mathcal{H}$ such that $H_m^V \Phi_m = E^V(m, N)\Phi_m$.*

Proof. Let us first show that it suffices to prove that for any normalized sequence Ψ^j, $j = 1, 2, \ldots$, (not necessarily minimizing) tending weakly to zero

$$\liminf_{j\to\infty} \left(\Psi^j,\, H_m^V \Psi^j\right) > E^V(m, N)\,. \tag{18}$$

To prove this let Φ^j be some *minimizing* sequence, i.e., assume that

$$\|\Phi^j\| = 1\,, \tag{19}$$

and that

$$\left(\Phi^j,\, H_m^V \Phi^j\right) \to E^V(m, N)\,. \tag{20}$$

By the Banach Alaoglu Theorem we can assume that this sequence, as well as the sequence $H_m^V \Phi^j$ converge weakly in the sense that for any $\Psi \in \mathcal{H}$ with $(\Psi, H_m^V \Psi) < \infty$ we have that

$$\left(\Psi, H_m^V \Phi^j\right) \to \left(\Psi, H_m^V \Phi_m\right), \tag{21}$$

where Φ_m is the weak limit of Φ^j. Our goal is to show that $(\Phi_m, H_m^V \Phi_m) = E^V(m, N)$ and that $\|\Phi_m\| = 1$.

Write $\Phi^j = \Phi_m + \Psi^j$. Obviously Ψ^j as well as $H_m^V \Psi^j$ go weakly to zero. Thus

$$\begin{aligned}
0 &= \lim_{j\to\infty} \left(\Phi^j, \left(H_m^V - E^V(m, N)\right)\Phi^j\right) \\
&= \lim_{j\to\infty} \left((\Phi_m + \Psi^j), \left(H_m^V - E^V(m, N)\right)\left(\Phi_m + \Psi^j\right)\right) \\
&= \lim_{j\to\infty} \left(\Psi^j, \left(H_m^V - E^V(m, N)\right)\Psi^j\right) + \left(\Phi_m, \left(H_m^V - E^V(m, N)\right)\Phi_m\right)
\end{aligned}$$

where we used that the cross terms vanish. Since $H_m^V - E^V(m, N) \geq 0$ this shows that Φ_m minimizes the energy, and, furthermore, that

$$0 \geq \lim_{j\to\infty} \left(\Psi^j, \left(H_m^V - E^V(m, N)\right)\Psi^j\right) \geq \delta \liminf_{j\to\infty} \|\Psi^j\|^2$$

for some positive constant δ. The second inequality is trivial if $\liminf_{j\to\infty} \|\Psi^j\|^2 = 0$ and otherwise follows from our assumption (18). This proves that Ψ^j converges strongly to zero along a subsequence, which implies that $\|\Phi_m\| = 1$. Hence Φ_m is a normalized ground state. Thus, it suffices to prove (18).

The steps that lead to a proof of (18) are quite standard. The only difficulty is that one has to localize in Fock space, which we describe first. We follow [11] with some necessary modifications and some simplifications.

Recall that, when the $a_\lambda^\#$ operators are viewed in x-space

$$a_\lambda(f) : \mathcal{F} \to \mathcal{F}\ ,\quad a_\lambda^*(g) : \mathcal{F} \to \mathcal{F}\ , \tag{22}$$

they obey the commutation relations

$$\left[a_\lambda(f), a_\lambda^*(g)\right] = \int_{\mathbb{R}^3} \overline{f}(x) g(x) d^3x =: (f, g)\ . \tag{23}$$

Consider now two smooth localization functions j_1 and j_2 that satisfy $j_1^2 + j_2^2 = 1$ and j_1 is supported in a ball of radius P. The first derivatives of j_1 and j_2 are of order $1/P$.

The operators

$$\begin{aligned}
c_\lambda(f) &= a_\lambda(j_1 f) \otimes \mathcal{I} + \mathcal{I} \otimes a_\lambda(j_2 f)\ , \\
c_\lambda^*(g) &= a_\lambda^*(j_1 g) \otimes \mathcal{I} + \mathcal{I} \otimes a_\lambda^*(j_2 g)
\end{aligned} \tag{24}$$

act both on the space $\mathcal{F} \otimes \mathcal{F}$. Note that

$$\left[c_\lambda(f), c_\lambda^*(g)\right] = (f, g) . \tag{25}$$

Thus, these new creation and anihilation operators create another Fock space $\mathcal{F}^l$ that is a subspace of $\mathcal{F} \otimes \mathcal{F}$ and is isomorphic to the old Fock space $\mathcal{F}$. Hence, there exists a map

$$U : \mathcal{F} \to \mathcal{F}^l \tag{26}$$

that is an invertible isometry between Fock spaces. It is uniquely specified by the properties

$$a^\# = U^* c^\# U , \tag{27}$$

and the vacuum in $\mathcal{F}$ is mapped to the vacuum in $\mathcal{F} \otimes \mathcal{F}$.

The map U^* is defined on $\mathcal{F}^l$ only, but we can extend it to all of $\mathcal{F} \otimes \mathcal{F}$ by setting $U^* F = 0$ whenever $F \in \mathcal{F} \otimes \mathcal{F}$ is perpendicular to $\mathcal{F}^l$. In other words U^* is a partial isometry between Fock spaces where $U^* U = \mathcal{I}$ on $\mathcal{F}$, and where UU^* is the orthogonal projection onto $\mathcal{F}^l$. We continue to denote the extended map by U^*.

Let ϕ and $\overline{\phi}$ be smooth nonnegative functions, with $\phi^2 + \overline{\phi}^2 = 1$, ϕ identically one on the unit ball, and vanishing outside the ball of radius 2. Set $\phi_R(X) = \phi(X/R)$. It is a standard calculation to show that for any Ψ with finite energy

$$\begin{aligned}\left(\Psi, H_m^V \Psi\right) =& \left(\phi_R \Psi, H_m^V \phi_R \Psi\right) + \left(\overline{\phi_R} \Psi, H_m^V \overline{\phi_R} \Psi\right) \\ &- \left(\Psi, (\nabla \phi_R)^2 \Psi\right) - \left(\Psi, (\nabla \overline{\phi_R})^2 \Psi\right) .\end{aligned} \tag{28}$$

The last two terms in (28) are bounded by $const./R^2$.

One goal will be to show that for any Ψ with finite energy

$$\begin{aligned}&\left(\Psi, \phi_R H_m^V \phi_R \Psi\right) = \\ &\left(\Psi, \phi_R U^* \left\{H_m^V \otimes \mathcal{I} + \mathcal{I} \otimes H_f\right\} U \phi_R \Psi\right) + o(1) .\end{aligned} \tag{29}$$

The error term $o(1)$ vanishes as both R and P go to infinity and depends otherwise only on the energy of Ψ. Notice that the invertible map U depends on the cutoff parameter P as well. (29) will be proved in Lemma A.1 in the appendix. The intuition behind the estimate (29) is that localized electrons interact only weakly with far away photons. Those photons are described solely by their own field energy.

An immediate consequence of (29) is the estimate

$$\begin{aligned}\left(\Psi, \phi_R H_m^V \phi_R \Psi\right) \geq& (E^V(m, N) + m)\|\phi_R \Psi\|^2 \\ &- m(\phi_R \Psi, U^* \mathcal{I} \otimes P_2 U \phi_R \Psi) + o(1) ,\end{aligned} \tag{30}$$

To bound the norm of the first term we need to estimate

$$
\begin{aligned}
&\Big\| \sum_{j=1}^{N} R(\omega_m(k))\varepsilon_\lambda(k)\cdot(p_j+\sqrt{\alpha}\,\widetilde{A}(x_j))(1-e^{-ik\cdot x_j})\widetilde{\Phi}_m \Big\| \\
&= \sup_{\|\eta\|\le 1} \Big| \sum_{j=1}^{N} \big(\varepsilon_\lambda(k)\cdot(p_j+\sqrt{\alpha}\,\widetilde{A}(x_j))R(\omega_m(k))\eta,\,(1-e^{-ik\cdot x_j})\widetilde{\Phi}_m\big) \Big| \\
&\le \sup_{\|\eta\|\le 1} \Big[\sum_{j=1}^{N} \|(p_j+\sqrt{\alpha}\,\widetilde{A}(x_j))R(\omega_m(k))\eta\|^2\Big]^{1/2} \Big[\sum_{j=1}^{N} \|(1-e^{-ik\cdot x_j})\widetilde{\Phi}_m\|^2\Big]^{1/2}
\end{aligned}
\tag{56}
$$

Next estimate the square of the first factor to get

$$
\begin{aligned}
&\Big(\eta,\, R(\omega_m(k))\Big[\sum_{j=1}^{N} (p_j+\sqrt{\alpha}\,\widetilde{A}(x_j))^2\Big] R(\omega_m(k))\eta\Big) \\
&\le a\,\big(\eta,\, R(\omega_m(k))H_m^V R(\omega_m(k))\eta\big) + b \\
&\le a\,(\eta,\, R(\omega_m(k))\eta) + (aE^V(m,N)+b)\,\big(\eta,\, R(\omega_m(k))^2\eta\big) \\
&\le C\frac{(\Lambda+1)}{|k|^2} \qquad \text{for } |k|\le\Lambda
\end{aligned}
\tag{57}
$$

where a and b are independent of m. Since $\sup_{m<1} E^V(m,N) < \infty$ the constant C is also independent of m. Finally the second factor in (56) is bounded by $|k|\,\||X|\widetilde{\Phi}_m\|$. The term containing the Pauli matrices in (54) is estimated similarly. In conjuction with (54), (52), (56) and (57) this shows that

$$
\|a_\lambda(k)\Phi_m\| \le C\sqrt{\alpha}(\Lambda+1)^{1/2}\frac{\chi_\Lambda(k)}{|k|^{1/2}}\||X|\widetilde{\Phi}_m\| \tag{58}
$$

This, together with Lemma 6.2, proves the theorem. □

Next we differentiate (54) with respect to k. There is a slight problem with this calculation since the polarization vectors cannot be defined in a smooth fashion globally. We make the following choice for the polarization vectors.

$$
\varepsilon_1(k) = \frac{(k_2,-k_1,0)}{\sqrt{k_1^2+k_2^2}} \tag{59}
$$

and

$$
\varepsilon_2(k) = \frac{k}{|k|}\wedge\varepsilon_1(k)\,. \tag{60}
$$

6.3. Theorem (Photon derivative bound). *Assume that there is binding, i.e.,* $\Sigma^V(m,N) - E^V(m,N) > 0$. *Assume that* Φ_m *is a normalized ground state for the many-body Hamiltonian* H_m^V, $m \geq 0$. *Then for* $|k| < \Lambda$ *and* $(k_1, k_2) \neq (0,0)$

$$\|\nabla_k a_\lambda(k)\Phi_m\| < \frac{Q\sqrt{\alpha}(1+|g|)}{|k|^{1/2}\sqrt{k_1^2+k_2^2}}, \tag{61}$$

where Q *is a finite constant independent of* Φ_m, g, α, Λ, *and depends on* m *only through the binding energy* $\Sigma^V(m,N) - E^V(m,N) > 0$.

Proof. We differentiate (54) with respect to k and obtain

$$\begin{aligned}
&\nabla_k(a_\lambda(k))\widetilde{\Phi}_m = \\
&2\sqrt{\alpha}R(\omega_m(k))^2\frac{k}{|k|}\varepsilon_\lambda(k)\cdot\sum_{j=1}^N(p_j+\sqrt{\alpha}\,\widetilde{A}(x_j))\frac{(1-e^{-ik\cdot x_j})}{|k|^{1/2}}\widetilde{\Phi}_m + \\
&2\sqrt{\alpha}R(\omega_m(k))\nabla_k(\varepsilon_\lambda(k))\cdot\sum_{j=1}^N(p_j+\sqrt{\alpha}\,\widetilde{A}(x_j))\frac{(1-e^{-ik\cdot x_j})}{|k|^{1/2}}\widetilde{\Phi}_m + \\
&2\sqrt{\alpha}R(\omega_m(k))\varepsilon_\lambda(k)\cdot\sum_{j=1}^N(p_j+\sqrt{\alpha}\,\widetilde{A}(x_j))\nabla_k\left(\frac{(1-e^{-ik\cdot x_j})}{|k|^{1/2}}\right)\widetilde{\Phi}_m \\
&+\frac{g}{2}\sqrt{\alpha}\,\nabla_k\left(iR(\omega_m(k))\frac{k\wedge\varepsilon_\lambda(k)}{\sqrt{|k|}}\cdot\sum_{j=1}^N\sigma_j e^{-ik\cdot x_j}\widetilde{\Phi}_m\right) \\
&-\nabla_k\left(R(\omega_m(k))\omega_m(k)iw_\lambda(k,X)\widetilde{\Phi}_m\right).
\end{aligned}$$

The norms of the first and third terms can estimated precisely the same way as in (56) and (57), and yields a bound of the form

$$\frac{C\sqrt{\alpha}}{|k|^{3/2}}\|\,(1+|X|)\,\widetilde{\Phi}_m\|\,. \tag{62}$$

For the second term, a straightforward calculation shows that

$$|\nabla_k\varepsilon_i(k)| \leq \frac{\text{const.}}{\sqrt{k_1^2+k_2^2}} \quad \text{for } i = 1,2. \tag{63}$$

The last term is dealt with in a similar fashion as the previous ones. Using the steps in (56) and (57), this leads to the bound

$$\|\nabla_k(a_\lambda(k))\widetilde{\Phi}_m\| \leq \frac{C\sqrt{\alpha}}{|k|^{1/2}\sqrt{k_1^2+k_2^2}}\|\,(1+|X|)\,\widetilde{\Phi}_m\|\,. \tag{64}$$

The fourth term can be estimated in the same fashion to yield a similar result.

Differentiating (51) leads to the same estimate with $\widetilde{\Phi}_m$ replaced by Φ_m. This, together with Lemma 6.2, proves the theorem. □

As for the proof of Theorem 6.1, our somewhat formal calculations above are rigorously justified in Appendix B.

7 Proof of Theorem 2.1

The proof will be done in two steps.

Proof. **Step 1.** The Hamiltonian H_m^V has a normalized ground state Φ_m, by Theorem 4.1. Pick a sequence $m_1 > m_2 > \dots$ tending to zero and denote the corresponding eigenvectors by Φ_j. This sequence is a minimizing sequence for H_0^V by Theorem 5.3. Since $\|\Phi_j\|$ is bounded there is a subsequence (call it again Φ_j) which has a weak limit Φ. Since $H_0^V - E^V(0, N) \geq 0$ and by the lower semi-continuity of non-negative quadratic forms (in our case, $H_0^V - E^V(0, N)$)

$$0 \leq \left(\Phi, \left(H_0^V - E^V(0, N)\right)\Phi\right) \leq \liminf_{j\to\infty} \left(\Phi_j, \left(H_0^V - E^V(0, N)\right)\Phi_j\right) = 0\,.$$

Hence Φ will be a (normalized) ground state if we show that $\|\Phi\| = 1$ (i.e. $\Phi_j \to \Phi$ strongly). It is important to note, however that if we write $\Phi_j = \{\Phi_{j,0}, \Phi_{j,1}, \dots, \Phi_{j,n}, \dots\}$, where $\Phi_{j,n}$ is the n-photon component of Φ_j then it suffices to prove the L^2 norm-convergence of each $\Phi_{j,n}$. The reason is the uniform bound on the total average photon number; see the remark after Theorem 6.1 which implies

$$\sum_{n\geq N} \|\Phi_{j,n}\|^2 \leq \text{const}\, N^{-1}\,.$$

Likewise, it suffices to prove the strong L^2 convergence in the bounded domain in which $|X| < R$ for each finite R. The reason for this is the exponential decay given in Lemma 6.2, which is uniform by Lemma 5.4. Finally, by Theorem 6.1, $\Phi_{j,n}(X, k_1, \dots, k_n)$ vanishes if $|k_i| > \Lambda$ for some i. So it suffices to show L^2 convergence for $\Phi_{j,n}$ restricted to

$$\Omega = \{(X, k_1, \dots, k_n) : |X| < R;\ |k_i| < \Lambda,\ i = 1, \dots, n\} \subset \mathbb{R}^{3(N+n)}$$

for each $R > 0$.

Step 2. For each $p < 2$ and $R > 0$ we show that $\Phi_{j,n}$ restricted to Ω is a bounded sequence in $W^{1,p}(\Omega)$. The key to this bound is (61) and

$$(a_\lambda(k)\Phi_j)_{n-1}(X, k_1, \dots, k_{n-1}) = \sqrt{n}\Phi_{j,n}(X, k, k_1, \dots, k_{n-1}) \qquad (65)$$

where the arguments $\lambda, \lambda_1, \dots, \lambda_{n-1}$ and the spin indices have been supressed. By the symmetry of $\Phi_{j,n}$, (65), Hölder's inequality and (61)

$$\int_{B_R} dX \int_{|k_1|,\dots,|k_n|<\Lambda} dk_1 \dots dk_n \sum_{i=1}^{n} |\nabla_{k_i} \Phi_{j,n}(X, k_1, \dots, k_n)|^p$$

$$= n^{1-p/2} \int_{B_R} dX \int_{|k_1|,\dots,|k_n|<\Lambda} dk_1 \dots dk_n |\nabla_{k_1} (a_\lambda(k_1)\Phi_j)_{n-1}(X, k_2, \dots, k_n)|^p$$

$$\leq C \int_{|k_1|<\Lambda} dk_1 \Bigg(\int_{B_R} dX \int_{|k_2|,\dots,|k_n|<\Lambda} dk_2 \dots dk_n$$

$$\times |\nabla_{k_1} (a_\lambda(k_1)\Phi_j)_{n-1}(X, k_2, \dots, k_n)|^2 \Bigg)^{p/2}$$

$$\leq C \int_{|k_1|<\Lambda} dk_1 \|\nabla_{k_1} a_\lambda(k_1)\Phi_j\|^p \leq \text{const} \tag{66}$$

independent of j. The constant C depends on all the parameters, but is finite because $|k_i| \leq \Lambda$ in the integration. Similarly, by Hölder's inequality

$$\|\chi(|X| < R)\nabla_X \Phi_{j,n}\|_p^p \leq C\|\chi(|X| < R)\nabla_X \Phi_{j,n}\|_2^p$$

$$\leq C(\Phi_j, \sum_{i=1}^{N} p_i^2 \Phi_j)^{p/2}$$

which is uniformly bounded by Lemma (A.5).

Since the classical derivative of $a_\lambda(k)\Phi_m$ is not defined in all of Ω one has to check that the weak derivative coincides with the classical derivative a.e.. Because of our definitions (59) and (60), the classical derivative is not defined along the 3-axis.

One has to show that

$$\int_\Omega \partial_i \psi \Phi_{j,n} = \lim_{\varepsilon \to 0} \int_{\Omega_\varepsilon} \partial_i \psi \Phi_{j,n} = -\lim_{\varepsilon \to 0} \int_{\Omega_\varepsilon} \psi \partial_i \Phi_{j,n} \quad i = 1, \dots, 3(N+n).$$

for any test function $\psi \in C_c^\infty(\Omega)$. Here Ω_ε is Ω with an ε cylinder around the 3-axis removed in each k-ball. The first equality is trivial; it is the second equality that has to be checked. This amounts to showing that the boundary term, coming from the integration by parts vanishes in the limit as ε tends to zero. But this follows immediately from Theorem 6.1.

This shows that $\Phi_{j,n}$ as a function of all its $3(N+n)$ variables, is in the Sobolev space $W^{1,p}(\Omega)$ and that $\sup_j \|\Phi_{j,n}\|_{W^{1,p}(\Omega)} < \infty$. Since $\Phi_{n,j}$ converges weakly in $L^2(\Omega)$ it converges weakly in $L^p(\Omega)$ and since the sequence is bounded in $W^{1,p}(\Omega)$, $\nabla\Phi_{n,j}$ converges weakly to $\nabla\Phi_n$.

The Rellich-Kondrachov theorem (see [22] Theorem 8.9) states that such a sequence converges *strongly* in $L^q(\Omega)$ if $1 \leq q < [3p(N+n)/3(N+n) - p]$. The boundedness of Ω is crucial here. For our purposes we need $q = 2$, and hence we have to pick p such that

$$2 > p > \frac{2 \cdot 3(N+n)}{2 + 3(N+n)} \tag{67}$$

which is possible for each N and n. We conclude that $\Phi_{j,n} \to \Phi_n$ strongly in $L^2(\Omega)$ as $j \to \infty$, for each n and R. This proves the theorem. □

Remark: Theorem 6.3 essentially says that the derivative is almost, but not quite in L^2. For high dimensions, $3(N+n)$, the required p is as close as we please to 2 if we require $q = 2$, but $p = 2$ is not allowed. The way out of the difficulty was to prove a uniform bound on the number operator and use this to say that that it suffices to prove strong convergence for each n separately. With n then fixed, it is possible to find a $p < 2$ that yields $q = 2$. It is, therefore, crucial to have the derivative in every L^p space with $p < 2$. The resolution of the problem of the infrared singularity is thus seen to be a delicate matter.

A Appendix: Localization estimates

In this appendix we collect a few facts which we use several times in this paper. Generally we worry about localizations of Hamiltonians in configuration space. While this is standard for Schrödinger operators it is somewhat more complicated in the presence of the radiation field. This is chiefly due to the problem of localization of photons.

We begin by stating a few well known facts about partitions of unity. Let β denote one of the 2^N subsets of the set of integers $1, 2, ..., N$. Its complement is denoted by β^c. As shown in [20] there exists a family of smooth functions j_β having the following four properties.

(i)

$$\sum_\beta j_\beta^2 = 1 \,. \tag{68}$$

(ii) For $\beta \neq \{1, \ldots, N\}$ (including the empty set) the j_β's are homogeneous of degree 0 and live outside the ball of radius R centered at the origin and

$$\text{supp } j_\beta \subset \{X : \min_{i \in \beta, j \in \beta^c} (|x_i - x_j|, |x_j|) \geq c|X|\} \,, \tag{69}$$

where C is some positive constant.

(iii) In the case where $\beta = \{1, \ldots, N\}$, j_β is compactly supported.

Corresponding to these electron localizations we define photon localizations.

For given $\beta \neq \{1, \ldots, N\}$ consider the function

$$g_1(y; \beta, X) = \Pi_{j \in \beta^c} \left(1 - \chi\left(\frac{y - x_j}{P}\right)\right) \tag{70}$$

where χ is a smoothed characteristic function of the unit ball. Define $g_2(y; \beta, X) = 1 - g_1(y, \beta, X)$. In the variable y, the function g_1 is supported away from the particles in β^c while g_2 lives close to the particles in β^c. Next define, for $i = 1, 2$,

$$j_i(y;\beta,X) = \frac{g_i(y;\beta,X)}{\sqrt{g_1(y;\beta,X)^2 + g_2(y;\beta,X)^2}} \,. \tag{71}$$

Certainly $j_1^2 + j_2^2 = 1$ and a simple computation shows that

$$|\nabla j_i| \leq \frac{\text{const.}}{P} \,. \tag{72}$$

In the case where $\beta = \{1,\ldots,N\}$ the construction of j_1 and j_2 is similar to the above one except that the function g_1 depends on y, is equal to one in a neighborhood of the origin and is compactly supported.

With the help of j_1 and j_2 the photons can now be localized as was done in Sect. 4. Let $U_\beta(X)\colon \mathcal{F} \to \mathcal{F} \otimes \mathcal{F}$ be the corresponding isometric transformation, i.e., the one that is defined via the relation

$$U_\beta(X)a^\#(h)U_\beta^*(X) = a^\#(j_1h) \otimes \mathcal{I} + \mathcal{I} \otimes a^\#(j_2h) \,. \tag{73}$$

The tensor product indicated is a tensor product between Fock spaces.

We denote by H_β the Hamiltonian of the form (4) with photon mass, but only for the particles in the set β. More precisely this operator acts on $L^2(\mathbb{R}^{3|\beta|}) \otimes \mathcal{F}$. By H^{β^c} we denote the Hamiltonian of the form (4) with photon mass, but only for the particles in the set β^c where the interaction with the nuclei has been dropped. This operator acts on $L^2(\mathbb{R}^{3|\beta^c|}) \otimes \mathcal{F}$. In particular we keep the interaction among those particles. In the case where $\beta = \{1,\ldots,N\}$ the Hamiltonian $H^{\beta^c} = H_f(m)$.

A.1. Lemma (Localization of the Hamiltonian). *For every β*

$$j_\beta H j_\beta = U_\beta^*(X) j_\beta \left[H_\beta \otimes \mathcal{I} + \mathcal{I} \otimes H^{\beta^c}\right] j_\beta U_\beta(X) + o(1) \,. \tag{74}$$

For $\beta \neq \{1,\ldots,N\}$, $o(1) \to 0$ as first $R \to \infty$ and then $P \to \infty$. If $\beta = \{1,\ldots,N\}$ then $o(1) \to 0$ as $P \to \infty$ for every fixed $R > 0$.

Proof. Our immediate aim is to compare the field energy H_f with the localized field energy $U_\beta^*(X)[H_f \otimes \mathcal{I} + \mathcal{I} \otimes H_f]U_\beta(X)$. For simplicity the various indices are supressed and $U_\beta(X)$ is replaced by U_β. The variable X plays no role here. Pick an orthonormal basis $\{g_j\}_{j=1}^\infty$ of $L^2(\mathbb{R}^3)$ in $H^{1/2}(\mathbb{R}^3)$. States of the form

$$\zeta = const.a_{\lambda_{i_1}}^*(g_{i_1})\cdots a_{\lambda_{i_k}}^*(g_{i_k})|0> \tag{75}$$

where k is finite, form an orthonormal basis in the Fock space. The field energy acts on such states as

$$H_f\zeta = \sum_{j=1}^k a_{\lambda_{i_1}}^*(g_{i_1})\cdots a_{\lambda_{i_j}}^*(\omega g_{i_j})\cdots a_{\lambda_{i_k}}^*(g_{i_k})|0> \,. \tag{76}$$

Thus, we have that

$$H_f\zeta = U_\beta^* \sum_{j=1}^{k} c_{\lambda_{i_1}}^*(g_{i_1})\cdots c_{\lambda_{i_j}}^*(\omega g_{i_j})\cdots c_{\lambda_{i_k}}^*(g_{i_k})U_\beta|0> \qquad (77)$$

and

$$H_f = U_\beta^*\left[H_f\otimes\mathcal{I}+\mathcal{I}\otimes H_f\right]U_\beta + E_f \qquad (78)$$

where the error E_f is given by

$$E_f\zeta = U_\beta^* \sum_{j=1}^{k} c_{\lambda_{i_1}}^*(g_{i_1})\cdots(a_{\lambda_{i_j}}^*([j_1,\omega]g_{i_j})\otimes\mathcal{I}+\mathcal{I}\otimes a_{\lambda_{i_j}}^*([j_2,\omega]g_{i_j}))$$
$$\cdots c_{\lambda_{i_k}}^*(g_{i_k})U_\beta|0> \ . \qquad (79)$$

Thus E_f is given by the operator (the λ's are omitted)

$$E_f = U_\beta^* \sum_k \left[a^*([j_1,\omega]g_k)\otimes\mathcal{I}+\mathcal{I}\otimes a^*([j_2,\omega]g_k)\right]$$
$$\times\left[a\ (j_1 g_k)\otimes\mathcal{I}+\mathcal{I}\otimes a\ (j_2 g_k)\right]U_\beta\ . \qquad (80)$$

The expression for the operator E_f does not look hermitian but it is, remembering that U_β^* is a partial isometry. Standard estimates lead to

$$|(\Psi, E_f\Psi)| \le (\|[\omega, j_1]\| + \|[\omega, j_2]\|)\,(\Psi, [\mathcal{N}+1]\,\Psi)\ . \qquad (81)$$

where $\mathcal{N}$ is the number operator.

Here $\|[j_1,\omega]\|$ denotes the operator norm associated with the kernel $[j_1,\omega]$. This norm can be estimated using the formula

$$[j_1,\omega] = \left[j_1,\omega^2\right]\frac{1}{\omega} + \omega^2\left[j_1,\frac{1}{\omega}\right]. \qquad (82)$$

Recalling the definition of j_1, the operator norm of the first term is easily seen to be bounded by a *const.*$/P$. Likewise, the second term, using the formula

$$\frac{1}{\sqrt{p^2+m^2}} = \frac{1}{\pi}\int_0^\infty \frac{1}{t+p^2+m^2}\frac{dt}{\sqrt{t}}\,, \qquad (83)$$

can be estimated by *const.*$/P$. The term $\|[j_2,\omega]\|$ is estimated in a similar fashion. The estimate (81) immediately shows that for a general state Φ we have that

$$\left|(\Phi, \left[H_f - U_\beta^*[H_f\otimes\mathcal{I}+\mathcal{I}\otimes H_f]U_\beta\right]\Phi)\right| \le \frac{const.}{P}(\Phi,\mathcal{N}\Phi)\ . \qquad (84)$$

Since the photons have a mass we can estimate the number operator in terms of the field energy. The field energy is relatively bounded with respect to the Hamiltonian, i.e., $H_f \le aH_m^V + b$ for some positive constants a and b, and thus we obtain

$$\left|(\Phi, \left[H_f - U_\beta^*[H_f \otimes \mathcal{I} + \mathcal{I} \otimes H_f]U_\beta\right]\Phi)\right| \le \frac{const.}{Pm}\left(\Phi, \left[aH_m^V + b\right]\Phi\right). \tag{85}$$

Note that this estimate had nothing to do with the electron, in particular the x–space cutoff is not present in the calculation.

Next we have to compare $\sum_{j=1}^N (p_j + \sqrt{\alpha}\, A(x_j))^2$ with

$$U_\beta^*(X) j_\beta\left[\sum_{i\in\beta}(p_i+\sqrt{\alpha}\, A(x_i))^2 \otimes \mathcal{I} + \mathcal{I} \otimes \sum_{j\in\beta^c}(p_j+\sqrt{\alpha}\, A(x_j))^2\right] j_\beta U_\beta(X)\,.$$

This time the X–space cutoff is important. We would like to estimate the difference

$$\begin{aligned} & j_\beta \sum_{i\in\beta}\left[(p_i + \sqrt{\alpha}\, A(x_i))^2 - U_\beta^*(X)(p_i + \sqrt{\alpha}\, A(x_i))^2 \otimes \mathcal{I}\, U_\beta(X)\right] j_\beta \\ & + j_\beta \sum_{i\in\beta^c}\left[(p_i + \sqrt{\alpha}\, A(x_i))^2 - U_\beta^*(X)\mathcal{I} \otimes (p_i + \sqrt{\alpha}\, A(x_i))^2 U_\beta(X)\right] j_\beta\,. \end{aligned} \tag{86}$$

It suffices to treat the first term, the other is similar. It can be easily expressed as

$$j_\beta\left[\sum_{i\in\beta}(p_i + \sqrt{\alpha}\, A(x_i))Q_i + Q_i(p_i + \sqrt{\alpha}\, A(x_i)) - Q_i^2\right] j_\beta \tag{87}$$

where

$$Q_i = p_i + \sqrt{\alpha}\, A(x_i) - U_\beta^*(X)(p_i + \sqrt{\alpha}\, A(x_i)) \otimes \mathcal{I}\, U_\beta(X)\,. \tag{88}$$

Using the form boundedness of the kinetic energy with respect to the full Hamiltonian, we have

$$\left(\Psi, \sum_{j=1}^N (p_j + \sqrt{\alpha}\, A(x_j))^2 \Psi\right) \le a\big(\Psi, H_m^V \Psi\big) + b(\Psi, \Psi) \tag{89}$$

for positive constants a and b. Thus, using Schwarz' inequality it suffices to show that

$$\|Q_i j_\beta \Psi\| = o(1) \quad \text{for} \quad i \in \beta\,, \tag{90}$$

as R (the localization radius for the electrons) tends to infinity. Denote by

$$h_{i,x}^\lambda(y) = (2\pi)^{-3/2}\int_{|k|<\Lambda} \frac{1}{\omega(k)}\varepsilon_i^\lambda(k) e^{ik\cdot(y-x)} d^3k\,. \tag{91}$$

Explicitly, Q_i is given by

$$\begin{aligned}
&p_i - U_\beta^*(X) p_i \otimes \mathcal{I} U_\beta(X) \\
&+ U_\beta^*(X)\Big[\sum_\lambda a_\lambda\big([j_1 - 1]h_x^\lambda\big) \otimes \mathcal{I} + \mathcal{I} \otimes a_\lambda\big(j_2 h_x^\lambda\big)\Big] U_\beta(X) \\
&+ U_\beta^*(X)\Big[\sum_\lambda a_\lambda^*\big([j_1 - 1]h_x^\lambda\big) \otimes \mathcal{I} + \mathcal{I} \otimes a_\lambda^*\big(j_2 h_x^\lambda\big)\Big] U_\beta(X)\,, \qquad (92)
\end{aligned}$$

and it suffices to estimate each of these terms separately. Each of the last two terms can be brought into the form

$$\left\| U_\beta^*(X) a^\#(f) \otimes \mathcal{I} U_\beta(X) j_\beta \Psi \right\| \qquad (93)$$

where f is one of the functions

$$[j_1(y,\beta,X) - 1]h_{1,x_j}^\lambda(y) \ \text{ or } \ j_2(y,\beta,X) h_{1,x_j}^\lambda(y) \qquad j \in \beta\,. \qquad (94)$$

The terms (93) are estimated by

$$\sup_X \left\{ j_\beta(X) \left\| [j_1 - 1]h_{i,x_j}^\lambda \right\|_2 \right\} \sqrt{(\Psi, (\mathcal{N}+1)\Psi)} \qquad (95)$$

respectively

$$\sup_X \left\{ j_\beta(X) \left\| j_2 h_{1,x_j}^\lambda \right\|_2 \right\} \sqrt{(\Psi, (\mathcal{N}+1)\Psi)}\,. \qquad (96)$$

In both formulas the index j is in β. The function j_β lives in the region where $|x_i - x_j| \geq cR$ for $i \in \beta$ and $j \in \beta^c$. The function $j_1 - 1$ (and likewise j_2) is not zero only if $|y - x_j| \leq P$ for some $j \in \beta^c$. Thus, $j_\beta(X)(j_1 - 1)(y)$ and $j_\beta(X) j_2(y)$ are nonzero only if $|y - x_i| \geq cR - P$. As $cR - P$ gets large only the tail of the function h^λ contributes to the integral which can be made as small as we please. The number operator is bounded by the field energy times $1/m$ which in turn is bounded by the full energy.

To estimate the first term in (92) we calculate

$$\begin{aligned}
&p_i - U_\beta^*(X) p_i \otimes \mathcal{I} U_\beta(X) = \\
&U_\beta^*(X) \sum_k \left[a^*([p_i, j_1]g_k) \otimes \mathcal{I} + \mathcal{I} \otimes a^*([p_i, j_2]g_k)\right] \times \qquad (97) \\
&\left[a\ (j_1 g_k) \otimes \mathcal{I} + \mathcal{I} \otimes a\ (j_2 g_k)\right] U_\beta(X)\,.
\end{aligned}$$

Note that the tensor product in the first line is different from the second. In the first the identity acts on $L^2(\mathbb{R}^{3|\beta^c|}) \otimes \mathcal{F} \otimes \mathcal{F}$ while in the second $\otimes$ indicates the tensor product of the Fock spaces only. The functions g_k

indicates a basis of $L^2(\mathbb{R}^3)$. The operators $U^*_\beta(X)$ and $U_\beta(X)$ have unit norm. Thus

$$\begin{aligned}&\Big\| U^*_\beta(X) \sum_k \big[a^*([p_i, j_1]g_k) \otimes \mathcal{I} + \mathcal{I} \otimes a^*([p_i, j_2]g_k)\big] \times \\ &\big[a\ (j_1 g_k) \otimes \mathcal{I} + \mathcal{I} \otimes a\ (j_2 g_k)\big] U_\beta(X)\Psi \Big\| \\ &\leq (\|[p_i, j_1]\| + \|[p_i, j_2]\|)\ \|\sqrt{\mathcal{N}+1}\Psi\| ,\end{aligned} \tag{98}$$

where $\|\cdot\|$ indicates that the operator norm has been taken. The norms of the commutators are of the order $1/P$ and hence vanish as $P \to \infty$. Since the photons have a mass we can estimate the number operator in terms of the field energy.

Similar consideration apply to the β^c term in (86). The only difference is that instead of (95) and (96) we have

$$\sup_X \big\{ j_\beta(X) \big\| j_1 h^\lambda_{x_j} \big\|_2 \big\} \sqrt{(\Psi, (\mathcal{N}+1)\Psi)} \tag{99}$$

respectively

$$\sup_X \big\{ j_\beta(X) \big\| [j_2 - 1] h^\lambda_{x_j} \big\|_2 \big\} \sqrt{(\Psi, (\mathcal{N}+1)\Psi)} , \tag{100}$$

with $j \in \beta^c$. Again this terms tend to zero as $P \to \infty$. The proof for the case where $\beta = \{1, \dots, N\}$ is similar but simpler since the operator U_β does not depend on X.

Finally, we have to compare the $\sigma \cdot B$ term with its localized counterparts. The estimates are similar to, but much easier than the estimates for $(p + \sqrt{\alpha}\, A(x))^2$ and are omitted for the convenience of the reader and authors who, by now, are exhausted. □

A simple consequence of Lemma A.1 is the following.

A.2. Corollary. *Let ϕ be a smooth function on $\mathbb{R}^{3N}$ such that $j_\beta \phi \equiv 0$ for $\beta = \{1, \dots, N\}$. Thus, ϕ depends on R. Then, as operators,*

$$\phi H \phi \geq \big(\Sigma^V(m, N) + o(1)\big)\, \phi^2 . \tag{101}$$

Here, $\Sigma^V(m, N) = \min_{1 \leq N' < N}(E^V(N') + E^0(N - N'))$ and $o(1)$ vanishes as $R \to \infty$.

Proof. By the IMS localization formula we have that

$$\phi H \phi = \sum_\beta \phi j_\beta H j_\beta \phi - \phi^2 \sum_\beta |\nabla j_\beta|^2 , \tag{102}$$

where the second term goes to 0 as $R \to \infty$. With our assumption on ϕ only the sets β with $\beta^c \neq \emptyset$ contribute. From Lemma A.1 we get that

$$\phi H\phi = \sum_{\beta} U_\beta^*(X)\phi j_\beta \left[H_\beta \otimes \mathcal{I} + \mathcal{I} \otimes H^{\beta^c}\right] j_\beta \phi U_\beta(X) + o(1) \quad (103)$$

as first $R \to \infty$ then $P \to \infty$. Certainly $H_\beta \geq E^V(m, |\beta|)$ and $H^{\beta^c} \geq E^0(m, |\beta^c|)$ from which the statement immediately follows.

A.3. Lemma. *Let Ψ_n be a normalized sequence in $\mathcal{H}$ whose energy is uniformly bounded and such that for any $\Phi \in \mathcal{H}$ with finite energy,*

$$(\Psi_n, \Phi) \to 0\,, \text{ and } (\Psi_n, H\Phi) \to 0\,. \quad (104)$$

Then

$$(\phi_R\Psi_n, U^*\mathcal{I} \otimes P_2 U\phi_R\Psi_n) \to 0\,. \quad (105)$$

Here U is the Fock space localization U_β that corresponds to $\beta = \{1, ..., N\}$.

Proof. Since the energy of Ψ_n is uniformly bounded we also know that

$$\left(\Psi_n, H^0(m)\Psi_n\right) \leq C\,. \quad (106)$$

is uniformly bounded

Let us describe the operator $\mathcal{I} \otimes P_2 U$ in more detail. Recall that

$$Ua^*(h_{i_1}) \cdots a^*(h_{i_k})|0> = c^*(h_{i_1}) \cdots c^*(h_{i_k})U|0> \quad (107)$$

where $|0>$ denotes the vacuum vector in Fock space and $U|0> = |0> \otimes |0>$. Hence, using the definition of $c^*(h)$, we find that

$$\mathcal{I} \otimes P_2 Ua^*(h_{i_1}) \cdots a^*(h_{i_k})|0> = a^*(h_{i_1} j_1) \cdots a^*(h_{i_k} j_1)|0> \otimes |0>\,. \quad (108)$$

The projection P_2 annihilates the photons in the second factor. In other words, the operator $\mathcal{I} \otimes P_2 U$ when acting on a state

$$\Psi = \left\{\Psi^0, \Psi^1(y_1), \Psi^2(y_1, y_2), \cdots\right\} \quad (109)$$

produces the localized state $\Gamma(j_1)\Psi \otimes |0>$ where

$$\Gamma(j_1)\Psi = \left\{\Psi^0, j_1(y_1)\Psi^1(y_1), j_1(y_1)j_1(y_2)\Psi^2(y_1, y_2), \cdots\right\}. \quad (110)$$

It follows that

$$\begin{aligned}\left(\phi_R\Psi_n, U^*\mathcal{I} \otimes P_2 U\phi_R\Psi_n\right) &= \|\mathcal{I} \otimes P_2 U\phi_R\Psi_n\|^2 \\ &= \|\Gamma(j_1)\phi_R\Psi_n\|^2\end{aligned}$$

Next, we show that (106) implies that

$$\Gamma(j_1)\phi_R\Psi_n \to 0\,. \quad (111)$$

To achieve that we note first that on account of the positive mass we have that $(\Psi_n, \mathcal{N}\Psi_n)$ is uniformly bounded. Since Ψ_n is of the form

$$\left\{\Psi_n^0, \Psi_n^1(X, y_1), \Psi_n^2(X, y_1, y_2), \cdots\right\}$$

we know that $\sum_{k\geq M}\left(\Psi_n^k, \Psi_n^k\right) \leq \text{const}/M$. It is therefore sufficient to prove (111) for each function

$$\Psi_n^M(X, y_1, \cdots, y_k)\ .$$

From the lemma below we learn that

$$\sum_{j=1}\left(\Psi_n, p_j^2\Psi_n\right) \tag{112}$$

is uniformly bounded. Thus, we can write (111) as

$$\Gamma(j_1)\phi_R\left(1+\sum_{j=1}p_j^2+H_f\right)^{-1/2}\left(1+\sum_{j=1}p_j^2+H_f\right)^{1/2}\Psi_n \tag{113}$$

which vanishes as $n\to\infty$ since $\|(1+\sum_{j=1}p_j^2+H_f)^{1/2}\Psi_n\|$ is uniformly bounded and since

$$\Gamma(j_1)\phi_R\left(1+\sum_{j=1}p_j^2+H_f\right)^{-1/2}$$

is compact on every finite particle subspace. Compactness follows from the fact that for continuous functions f and g vanishing at infinity the operator $f(i\nabla)g(x)$ is compact.

A.4. Lemma (Bound on $A(x)^2$). *For each $x\in\mathbb{R}^3$ and ultraviolet cutoff Λ write $A(x)=D(x)+D^*(x)$ where D contains the annihilation operators in $A(x)$ and D^* the creation operators. Similarly, write $B(x)=E(x)+E^*(x)$. As operator bounds*

$$\begin{aligned}
H_f &\geq \frac{1}{8\pi\Lambda}D^*(x)D(x)\\
H_f+\frac{\Lambda}{2} &\geq \frac{1}{8\pi\Lambda}D(x)D^*(x)\\
H_f+\frac{\Lambda}{8} &\geq \frac{1}{32\pi\Lambda}A(x)^2\\
H_f &\geq \frac{3}{8\pi\Lambda^3}E^*(x)E(x)\\
H_f+\frac{3\Lambda}{4} &\geq \frac{3}{8\pi\Lambda^3}E(x)E^*(x)\\
H_f+\frac{3\Lambda}{16} &\geq \frac{3}{32\pi\Lambda^3}B(x)^2\ .
\end{aligned} \tag{114}$$

Proof. We write $A(x) = D(x) + D^*(x)$ with

$$D(x) = \sum_\lambda \int_{|k|<\Lambda} |k|^{-1/2} \varepsilon_\lambda(k) \exp[ik \cdot x] a_\lambda(k) d^3k\,.$$

There are thus four terms in $A(x)^2$. Using the Schwarz inequality, the (DD) term can be bounded above by $(D^*D)/2 + (DD^*)/2$. On the other hand, $(DD^*) = (D^*D) + \Gamma$, where Γ is the commutator $\int 2/|k| = 4\pi\Lambda^2$; the factor 2 comes from the two polarizations $\lambda = 1, 2$. Altogether, we obtain

$$A(x)^2 \leq 4D^*(x)D(x) + 4\pi\Lambda^2\,.$$

Finally, we use the Schwarz inequality again to obtain

$$\begin{aligned} &\sum_\lambda \int \overline{h_\lambda(k)} a^*_\lambda(k) d^3k \sum_\lambda \int h_\lambda(k) a_\lambda(k) d^3k \\ \leq &\sum_\lambda \int |h_\lambda(k)|^2/|k| d^3k \sum_\lambda \int |k| a^*_\lambda(k) a_\lambda(k) d^3k. \end{aligned}$$

In our case, $h_\lambda(k) = \varepsilon_\lambda(k) \exp[ik \cdot x]/\sqrt{|k|}$, so $\sum_\lambda \int |h_\lambda(k)|^2/|k| d^3k = 8\pi\Lambda$.
For $B = \mathrm{curl} A$, replace Γ by $2\pi\Lambda^4$ and replace $|h_\lambda(k)|$ by $\sqrt{|k|}$. □

As a corollary of Lemma A.4 we have the following.

A.5. Lemma (Bound on $(p + A(x))^2$**).** *For any* $\varepsilon > 0$ *there are constants* $\delta(\varepsilon) > 0$ *and* $C(\varepsilon) < \infty$ *such that*

$$\sum_{j=1}^N \left\{ (p_j + \sqrt{\alpha}\, A(x_j))^2 + \frac{g}{2} \sqrt{\alpha}\, \sigma_j \cdot B(x_j) \right\} + \varepsilon H_f \geq \delta(\varepsilon) \sum_{j=1}^N p_j^2 - C(\varepsilon)\,. \tag{115}$$

The constants $\delta(\varepsilon)$, $C(\varepsilon)$ *depend on* α, g, Λ, N.

Proof. In addition to Lemma A.4, use the facts that for any $0 < \mu, \nu < 1$, $(p_j + \sqrt{\alpha}\, A(x_j))^2 \geq (1 - \mu)p^2 + (1 - 1/\mu)\alpha A(x_j)^2$ and $2\sigma_j \cdot B(x_j) \geq -\nu B(x_j)^2 - 1/\nu$. □

B Appendix: Verification of infrared bounds

The proofs of the infrared bounds in Sect. 6 are somewhat formal. In particular, we carried out the calculations tacitly assuming that $a_\lambda(k)\widetilde{\Phi}_m$ (which is itself only defined for almost every k) is in the domain of the Operator $\widetilde{H}_m$. One can actually prove this when $\widetilde{H}_m$ is self-adjointly realized in terms of the Friedrichs' extension and thereby make all the formal computations in Sect. 6 rigorous. Instead of doing so, we give here alternative proofs of the

Theorems in Sect. 6 which avoid any reference to a domain of $\widetilde{H}_m$. All the arguments can be carried out on the level of quadratic forms.

We recall that $\widetilde{H}_m$ is the Hamiltonian H_m^V after an "operator-valued gauge transformation". Our remarks here about quadratic forms in relation to $\widetilde{H}_m$ could just as well be applied to H_m^V itself.

In order to keep the notation simple, we give the proof of the infrared bounds for the case of a single charged particle ($N = 1$) with no magnetic moment, i.e., $g = 0$. There is no difficulty in deriving these bounds for the general case.

Denote by $\mathcal{S}$ the set of all finite linear combinations of vectors that are products of $C_c^\infty(\mathbb{R}^3)$-functions and states in $\mathcal{F}$ that have only a finite number of photons. It is well known that this set is dense in $\mathcal{H}$, and that the quadratic form $(\Psi, \Psi)_+ := (\Psi, (\widetilde{H}_m - E^V(m, 1) + 1)\Psi)$ is defined for all Ψ in $\mathcal{S}$ and is bounded below by $\|\Psi\|^2$. Hence this quadratic form is closable and the closure of $\mathcal{S}$ in this inner product is a Hilbert space $Q(\widetilde{H}_m)$ with inner product $(\cdot, \cdot)_+$ and norm $\|\Psi\|_+ = \sqrt{(\Psi, \Psi)_+}$.

An eigenfunction $\widetilde{\Phi}_m$ of $\widetilde{H}_m$ in the weak sense is a vector in $Q(\widetilde{H}_m)$ such that

$$(\Psi, \widetilde{\Phi}_m)_+ = e(\Psi, \widetilde{\Phi}_m) \tag{116}$$

for some real number e and for all $\Psi \in Q(\widetilde{H}_m)$. It is in this sense that we proved in Sect. 4 that a ground state exists for the model with massive photons. (This implies that $\widetilde{\Phi}_m$ is in an eigenstate of the Friedrichs' extension of $\widetilde{H}_m$).

Define the smeared operators

$$a(f) = \sum_\lambda \int a_\lambda(k)\overline{f(k, \lambda)}dk\ , \tag{117}$$

where $f(k, \lambda)$ is any function in $L^2(\mathbb{R}^3; \mathbb{C}^2)$. It is not difficult to show that $a(f)\widetilde{\Phi}_m$ is in the form domain of $\widetilde{H}_m$. To this end define $a_R(f) = R[\mathcal{N} + R]^{-1}a(f)$. Here R is some large real number (which we eventually take towards infinity) and $\mathcal{N}$ is the number operator.

It is straightforward to see that $a_R(f)$ and $a_R^*(f)$ are bounded operators on $Q(\widetilde{H}_m)$ for every $R > 0$, i.e.,

$$\|a_R(f)\Psi\|_+ \le C(R)\|\Psi\|_+\ , \tag{118}$$

and similarly for $a_R^*(f)$.

Generally, the constant $C(R)$ tends to ∞ as R tends to ∞. For an eigenfunction of $\widetilde{H}_m$, however, this is not the case. Simple but tedious commutator estimates reveal that for any eigenfunction $\widetilde{\Phi}_m$ there exists a constant C independent of R such that

$$\left(a_R(f)\widetilde{\Phi}_m, a_R(f)\widetilde{\Phi}_m\right)_+ \le C\left(a_R(f)\widetilde{\Phi}_m, a_R(f)\widetilde{\Phi}_m\right)\ . \tag{119}$$

The point is that $(a_R(f)\widetilde{\Phi}_m, a_R(f)\widetilde{\Phi}_m)_+ = (a_R^*(f)a_R(f)\widetilde{\Phi}_m, \widetilde{\Phi}_m)_+$ plus terms that are uniformly bounded in R. By the previous statement we know that $a_R^*(f)a_R(f)\widetilde{\Phi}_m$ is in $Q(\widetilde{H}_m)$ and hence

$$\big(a_R^*(f)a_R(f)\widetilde{\Phi}_m, \widetilde{\Phi}_m\big)_+ = e\big(a_R^*(f)a_R(f)\widetilde{\Phi}_m, \widetilde{\Phi}_m\big) = e\big(a_R(f)\widetilde{\Phi}_m, a_R(f)\widetilde{\Phi}_m\big) . \tag{120}$$

The last expression, however, is bounded uniformly in R, since the condition $\widetilde{\Phi}_m \in Q(\widetilde{H}_m)$ implies that the expectation value of the field energy in $\widetilde{\Phi}_m$ is finite which in turn bounds the last expression in (120). Here we use the fact that the photons have a mass.

From this it follows easily that for a subsequence of R's tending to infinity, $a_R(f)\widetilde{\Phi}_m$ has a weak limit in $Q(\widetilde{H}_m)$. Since $a_R(f)\widetilde{\Phi}_m \to a(f)\widetilde{\Phi}_m$ strongly this shows that $a(f)\widetilde{\Phi}_m \in Q(\widetilde{H}_m)$.

Proof of Theorem 6.1. We shall use the abreviation

$$\sum_\lambda \int \ldots \mathrm{d}k = \sum\!\!\!\!\!\!\int \ldots \mathrm{d}k . \tag{121}$$

For our special choice of gauge

$$\widetilde{A}^i(x) = a(G^i) + a^*(G^i),\ i = 1, 2, 3 , \tag{122}$$

where we set

$$G_\lambda^i(k, x) = \varepsilon_\lambda^i(k)|k|^{-1/2}(e^{ik\cdot x} - 1)\chi_\Lambda(k) . \tag{123}$$

Next, pick any Ψ in $\mathcal{S}$ and calculate (recalling the definition of w in Sect. 6 equation (48))

$$\big(\Psi, \big(\widetilde{H}_m - E^V(m, 1)\big)a(f)\widetilde{\Phi}_m\big) = -2\big(\Psi, (f, G)(p + \widetilde{A})\widetilde{\Phi}_m\big) - \big(\Psi, a(\omega f)\widetilde{\Phi}_m\big) + i\big(\Psi, (f, \omega w)\widetilde{\Phi}_m\big) \tag{124}$$

with $\omega(k) = \sqrt{k^2 + m^2}$. This extends, using an approximation argument, to all $\Psi \in Q(\widetilde{H}_m)$ and, in particular, to $a(f)\widetilde{\Phi}_m$. Here we note that, on account of Lemma A.4 and the assumption on the potential, $\Psi \in Q(\widetilde{H}_m)$ implies that $(p + \widetilde{A})\Psi \in \mathcal{H}$. Hence

$$\begin{aligned} 0 &\le \big(a(f)\widetilde{\Phi}_m, \big(\widetilde{H}_m - E^V(m, 1)\big)a(f)\widetilde{\Phi}_m\big) \\ &= -2\big(a(f)\widetilde{\Phi}_m, (f, G)(p+\widetilde{A})\widetilde{\Phi}_m\big) \\ &\quad - \big(a(f)\widetilde{\Phi}_m, a(\omega f)\widetilde{\Phi}_m\big) + i\big(a(f)\widetilde{\Phi}_m, (f, \omega w)\widetilde{\Phi}_m\big) , \end{aligned} \tag{125}$$

which yields the inequality

$$\big(a(f)\widetilde{\Phi}_m, a(\omega f)\widetilde{\Phi}_m\big) \le -2\big(a(f)\widetilde{\Phi}_m, (f, G)(p + \widetilde{A})\widetilde{\Phi}_m\big) + i\big(a(f)\widetilde{\Phi}_m, (f, \omega w)\widetilde{\Phi}_m\big) \tag{126}$$

for all f in $L^2(\mathbb{R}^3;\mathbb{C}^2)$. Pick f of the form $\omega(k)^{-1/2}q(k,\lambda)g_i(k,\lambda)$ where g_i is an orthonormal basis of $L^2(\mathbb{R}^3;\mathbb{C}^2)$ and $q(k,\lambda)$ a bounded function. Summing over this basis, we get on the left side of (126)

$$\sum_i \big(a(\omega^{-1/2}qg_i)\widetilde{\Phi}_m, a(\omega^{1/2}qg_i)\widetilde{\Phi}_m\big) = \sum\!\!\!\!\!\!\int |q(k,\lambda)|^2 \big\|a_\lambda(k)\widetilde{\Phi}_m\big\|^2 \mathrm{d}k\,, \tag{127}$$

and on the right side

$$-2\big(a(\omega^{-1}|q|^2G)\widetilde{\Phi}_m, (p+\widetilde{A})\widetilde{\Phi}_m\big) + i\big(a(|q|^2w)\widetilde{\Phi}_m, \widetilde{\Phi}_m\big)\,. \tag{128}$$

Hence

$$\begin{aligned}\sum\!\!\!\!\!\!\int |q(k,\lambda)|^2 \big\|a_\lambda(k)\widetilde{\Phi}_m\big\|^2 \mathrm{d}k \le &-2\big(a(\omega^{-1}|q|^2G)\widetilde{\Phi}_m, (p+\widetilde{A})\widetilde{\Phi}_m\big)\\ &+ i\big(a(|q|^2w)\widetilde{\Phi}_m, \widetilde{\Phi}_m\big)\,.\end{aligned} \tag{129}$$

The right side can be written as

$$\begin{aligned}-2\sum\!\!\!\!\!\!\int \frac{|q(k,\lambda)|^2}{\omega(k)}\big(a_\lambda(k)\widetilde{\Phi}_m, G_\lambda(k)(p+\widetilde{A})\widetilde{\Phi}_m\big)\mathrm{d}k&\\ + i\sum\!\!\!\!\!\!\int |q(k,\lambda)|^2\big(a_\lambda(k)\widetilde{\Phi}_m, w_\lambda\widetilde{\Phi}_m\big)\mathrm{d}k\,,&\end{aligned} \tag{130}$$

and, applying Schwarz's inequality, this is bounded above by

$$\begin{aligned}&2\Big[\sum\!\!\!\!\!\!\int |q(k,\lambda)|^2\|a_\lambda(k)\widetilde{\Phi}_m\|^2\mathrm{d}k\Big]^{1/2}\times\\ &\quad\Big[\Big[\sum\!\!\!\!\!\!\int_{|k|\le\Lambda} \omega(k)^{-2}|q(k,\lambda)|^2\|G_\lambda(p+\widetilde{A})\widetilde{\Phi}_m\|^2\mathrm{d}k\Big]^{1/2}\\ &\qquad + \Big[\sum\!\!\!\!\!\!\int_{|k|\le\Lambda} |q(k,\lambda)|^2\|w_\lambda\widetilde{\Phi}_m\|^2\mathrm{d}k\Big]^{1/2}\Big]\,.\end{aligned} \tag{131}$$

Hence we obtain the bound

$$\begin{aligned}&\sum\!\!\!\!\!\!\int |q(k,\lambda)|^2\|a(k)\widetilde{\Phi}_m\|^2\mathrm{d}k \le\\ &\quad 8\Big[\sum\!\!\!\!\!\!\int_{|k|\le\Lambda} \omega(k)^{-2}|q(k,\lambda)|^2\|G_\lambda(k)\cdot(p+\widetilde{A})\widetilde{\Phi}_m\|^2\mathrm{d}k\\ &\qquad + \sum\!\!\!\!\!\!\int_{|k|\le\Lambda} |q(k,\lambda)|^2\|w_\lambda\widetilde{\Phi}_m\|^2\mathrm{d}k\Big]\,.\end{aligned} \tag{132}$$

Since, $\operatorname{div}_x G_\lambda = 0$ we have that $G_\lambda \cdot (p + \widetilde{A})\widetilde{\Phi}_m = (p + \widetilde{A}) \cdot G_\lambda \widetilde{\Phi}_m$. Moreover, $(p + \widetilde{A})^2$ is relatively form bounded with respect to $\widetilde{H}_m$. But, as in the proof of exponential decay (Lemma 6.2), we have for each $i = 1, 2, 3$

$$\left(G_\lambda^i \widetilde{\Phi}_m, \left(\widetilde{H}_m - E^V(m,1)\right) G_\lambda^i \widetilde{\Phi}_m\right) = \left(\widetilde{\Phi}_m, \left|\nabla_x G_\lambda^i\right|^2 \widetilde{\Phi}_m\right) \tag{133}$$

and we arrive at the bound

$$\begin{aligned} &\sum\!\!\!\!\!\!\int |q(k,\lambda)|^2 \left\|a_\lambda(k)\widetilde{\Phi}_m\right\|^2 \mathrm{d}k \le \\ &C \sum\!\!\!\!\!\!\int_{|k|\le\Lambda} \frac{|q(k,\lambda)|^2}{\omega(k)^2} \left[\left\|G_\lambda \widetilde{\Phi}_m\right\|^2 + \left\||\nabla_x G_\lambda|\widetilde{\Phi}_m\right\|^2 + \omega(k)^2 \left\|w_\lambda \widetilde{\Phi}_m\right\|^2\right] \mathrm{d}k\,, \end{aligned} \tag{134}$$

where C is some constant independent of m. Since $q(k, \lambda)$ is arbitrary we obtain for almost every k and each λ that

$$\begin{aligned} &\left\|a_\lambda(k)\widetilde{\Phi}_m\right\|^2 \le \\ &C\omega(k)^{-2}\left[\left\|G_\lambda \widetilde{\Phi}_m\right\|^2 + \left\||\nabla_x G_\lambda|\widetilde{\Phi}_m\right\|^2 + \omega(k)^2 \left\|w_\lambda \widetilde{\Phi}_m\right\|^2\right] \chi_\Lambda(k)\,. \end{aligned} \tag{135}$$

The right side is bounded by

$$\frac{C}{|k|} \left\||x|\widetilde{\Phi}_m\right\|^2 \chi_\Lambda(k)\,, \tag{136}$$

which is finite on account of the exponential decay of $\widetilde{\Phi}_m$. □

Proof of Theorem 6.3. First some notation: For any function $f(k)$ define

$$(\Delta_h f)(k) = f(k+h) - f(k), \tag{137}$$

and

$$\Delta_{-h} a(f) = a(\Delta_h f) \tag{138}$$

Returning to (126) with f replaced by $\Delta_h f$ we have

$$\begin{aligned} &\left(\Delta_{-h} a(f)\widetilde{\Phi}_m, a(\omega \Delta_h f)\widetilde{\Phi}_m\right) \le \\ &-2\left(\Delta_{-h} a(f)\widetilde{\Phi}_m, (\Delta_h f, G)(p+\widetilde{A})\widetilde{\Phi}_m\right) + i\left(\Delta_{-h} a(f)\widetilde{\Phi}_m, (\Delta_h f, \omega w)\widetilde{\Phi}_m\right) \end{aligned} \tag{139}$$

which can be rewritten as

$$\begin{aligned} \left(\Delta_{-h} a(f)\widetilde{\Phi}_m, \Delta_{-h} a(\omega f)\widetilde{\Phi}_m\right) &\le \left(\Delta_{-h} a(f)\widetilde{\Phi}_m, a((\Delta_h \omega) f(\cdot + h))\widetilde{\Phi}_m\right) \\ &\quad - 2\left(\Delta_{-h} a(f)\widetilde{\Phi}_m, (f, \Delta_{-h} G)(p + \widetilde{A})\widetilde{\Phi}_m\right) \\ &\quad + i\left(\Delta_{-h} a(f)\widetilde{\Phi}_m, (f, \Delta_{-h}(\omega w))\widetilde{\Phi}_m\right)\,. \end{aligned} \tag{140}$$

Notice that without the first term on the right of the inequality sign, the structure of this inequality is the same as (126), except that, of course, $\Delta_{-h}a(f)$ plays the role of $a(f)$, $\Delta_{-h}G$ plays the role of G and $\Delta_{-h}(\omega w)$ plays the role of ωw. Thus, without this term we would obtain immediately the estimate analogous to (134),

$$
\begin{aligned}
&\not\!\!\!\!\sum\!\!\!\!\!\!\int |q(k,\lambda)|^2 \left\| (\Delta_{-h}a_\lambda)(k)\widetilde{\Phi}_m \right\|^2 \mathrm{d}k \\
&\le C \not\!\!\!\!\sum\!\!\!\!\!\!\int \frac{|q(k,\lambda)|^2}{\omega(k)^2} \Big[\left\| \Delta_{-h}G_\lambda \widetilde{\Phi}_m \right\|^2 \\
&\quad + \left\| |\nabla_x \Delta_{-h} G_\lambda| \widetilde{\Phi}_m \right\|^2 + \omega^2 \left\| \Delta_{-h}(\omega w_\lambda)\widetilde{\Phi}_m \right\|^2 \Big] \mathrm{d}k \,.
\end{aligned}
\tag{141}
$$

The remaining term in equation (140), after summing over the functions $qg_i/\sqrt{\omega}$, turns into

$$
\not\!\!\!\!\sum\!\!\!\!\!\!\int \left((\Delta_{-h}a_\lambda)(k)\widetilde{\Phi}_m , a_\lambda(k-h)\widetilde{\Phi}_m \right) \frac{|q(k,\lambda)|^2}{\omega(k)} (\Delta_h \omega)(k-h) \mathrm{d}k \tag{142}
$$

which, by Schwarz's inequality, is bounded above by

$$
\begin{aligned}
&\left[\not\!\!\!\!\sum\!\!\!\!\!\!\int |q(k,\lambda)|^2 \left\| (\Delta_{-h}a_\lambda)(k)\widetilde{\Phi}_m \right\|^2 \mathrm{d}k \right]^{1/2} \\
&\times \left[\not\!\!\!\!\sum\!\!\!\!\!\!\int \left\| a_\lambda(k-h)\widetilde{\Phi}_m \right\|^2 \frac{|q(k,\lambda)|^2}{\omega(k)^2} |\Delta_h \omega(k-h)|^2 \mathrm{d}k \right]^{1/2} .
\end{aligned}
\tag{143}
$$

This, together with (141), yields

$$
\begin{aligned}
&\not\!\!\!\!\sum\!\!\!\!\!\!\int |q(k,\lambda)|^2 \left\| (\Delta_{-h}a_\lambda)(k)\widetilde{\Phi}_m \right\|^2 \mathrm{d}k \\
&\le C \not\!\!\!\!\sum\!\!\!\!\!\!\int \frac{|q(k,\lambda)|^2}{\omega(k)^2} \Big[\left\| \Delta_{-h}G_\lambda \widetilde{\Phi}_m \right\|^2 + \left\| |\nabla_x \Delta_{-h} G_\lambda| \widetilde{\Phi}_m \right\|^2 \\
&\quad + \omega(k)^2 \left\| \Delta_{-h}(\omega w_\lambda)\widetilde{\Phi}_m \right\|^2 \Big] \mathrm{d}k \\
&\quad + C \not\!\!\!\!\sum\!\!\!\!\!\!\int \frac{|q(k,\lambda)|^2}{\omega(k)^2} \left\| a_\lambda(k-h)\widetilde{\Phi}_m \right\|^2 |\Delta_h \omega(k-h)|^2 \mathrm{d}k \,.
\end{aligned}
\tag{144}
$$

Again, since q is arbitrary we obtain for every fixed λ

$$
\begin{aligned}
\left\| (\Delta_{-h}a_\lambda)(k)\widetilde{\Phi}_m \right\|^2 &\le \frac{C}{\omega(k)^2} \Big[\left\| \Delta_{-h}G_\lambda \widetilde{\Phi}_m \right\|^2 + \left\| |\nabla_x \Delta_{-h} G_\lambda| \widetilde{\Phi}_m \right\|^2 \\
&\quad + \omega(k)^2 \left\| \Delta_{-h}(\omega w_\lambda)\widetilde{\Phi}_m \right\|^2 + \left\| a_\lambda(k-h)\widetilde{\Phi}_m \right\|^2 |\Delta_h \omega(k-h)|^2 \Big] .
\end{aligned}
\tag{145}
$$

Combining this with (136) we get

$$\begin{aligned}&\left\|(\Delta_{-h}a_\lambda)(k)\widetilde{\Phi}_m\right\|^2\\ &\le \frac{C}{\omega(k)^2}\left[\left\|\Delta_{-h}G_\lambda\widetilde{\Phi}_m\right\|^2+\left\||\nabla_x\Delta_{-h}G_\lambda|\widetilde{\Phi}_m\right\|^2+\omega(k)^2\left\|\Delta_{-h}(\omega w_\lambda)\widetilde{\Phi}_m\right\|^2\right]\\ &\quad+\frac{C}{\omega(k)^2|k-h|}\left\||x|\widetilde{\Phi}_m\right\|^2\chi_\Lambda(k-h)|\Delta_h\omega(k-h)|^2 .\end{aligned} \tag{146}$$

The polarization vectors defined in (59), (60), are differentiable away from the 3-axis. The same straightforward estimates as in Sect. 6 lead to

$$\begin{aligned}&\left\|(\Delta_{-h}a_\lambda)(k)\widetilde{\Phi}_m\right\|^2\\ &\le C\left[\frac{1}{|k|\left(k_1^2+k_2^2\right)}+\frac{1}{|k-h|\left((k_1-h_1)^2+(k_2-h_2)^2\right)}\right]|h|^2\||x|\widetilde{\Phi}_m\|^2\end{aligned} \tag{147}$$

which hold for all $|k| < \Lambda$ and small $|h|$ with a constant C that is independent of m.

Next, we observe that for $k \neq 0$ fixed, there exist a sequence of h values, say h_l, tending to zero. so that $h^{-1}(\Delta_{-he_j}a_\lambda)(k)\widetilde{\Phi}_m$ converges weakly to some element $v_j(k)$ which satisfies the estimate

$$\|v_j(k)\|^2 \le C\frac{1}{|k|\left(k_1^2+k_2^2\right)}\left\||x|\widetilde{\Phi}_m\right\|^2 . \tag{148}$$

Here e_j is the j-th canonical basis vector. Next we identify $v_j(k)$ as the weak derivative of $a_\lambda(k)\widetilde{\Phi}_m$. This weak derivative, by definition, can be computed via the expression

$$-\left(\Psi, a(\partial_j\phi)\widetilde{\Phi}_m\right) \tag{149}$$

where Ψ is any state in $\mathcal{H}$ and ϕ is any test function in $C_c^\infty(\mathbb{R}^3)$. Clearly the above expression equals

$$\lim_{h\to 0}\sum\!\!\!\!\!\!\int \left(\Psi, (\Delta_{-he_j}a_\lambda)(k)\widetilde{\Phi}_m\right)\phi(k)\mathrm{d}k . \tag{150}$$

But along the sequence h_l

$$\lim_{l\to\infty}\sum\!\!\!\!\!\!\int \left(\Psi, \Delta_{-h_le_j}a_\lambda(k)\widetilde{\Phi}_m\right)\phi(k)\mathrm{d}k = \int(\Psi, v_j(k))\phi(k)\mathrm{d}k , \tag{151}$$

which identifies $v_j(k)$ as the (negative) weak derivative of $a_\lambda(k)\widetilde{\Phi}_m$. □

References

1. S. Agmon, Lectures on exponential decay of solutions of second order elliptic equations: Bounds on eigenfunctions of N-body Schrödinger operators, Mathematical Notes 29, Princeton University Press (1982)
2. A. Arai, Rigorous theory of spectra and radiation for a model in quantum electrodynamics, J. Math. Phys. **24**, 1896–1910 (1983)
3. A. Arai, M. Hirokawa, On the existence and uniqueness of ground states of a generalized spin-boson model, J. Funct. Anal. **151**, 455–503 (1997)
4. A. Arai, M. Hirokawa, Ground states of a general class of quantum field Hamiltonians, Rev. Math. Phys. **12**, 1085–1135 (2000), mp_arc 99–179 (1999)
5. A. Arai, M. Hirokawa, F. Hiroshima, On the absence of eigenvectors of Hamiltonians in a class of massless quantum field models without infrared cutoff, J. Funct. Anal. **168**, 470–497 (1999)
6. V. Bach, J. Fröhlich, I.M. Sigal, Mathematical theory of nonrelativistic matter and radiation, Lett. Math. Phys. **34**, 183–201 (1995)
7. V. Bach, J. Fröhlich, I.M. Sigal, Quantum electrodynamics of confined non-relativistic particles, Adv. Math. **137**, 299–395 (1998)
8. V. Bach, J. Fröhlich, I.M. Sigal, Spectral analysis for systems of atoms and molecules coupled to the quantized radiation field, Commun. Math. Phys. **207**, 249–290 (1999)
9. H. Bethe, The electromagnetic shift of energy levels, Phys. Rev. **72**, 339–342 (1947)
10. J. Combes, L. Thomas, Asymptotic behavior of eigenfunctions for multiparticle Schrödinger operators, Commun. Math. Phys. **34**, 251–270 (1973)
11. J. Dereziński, C. Gérard, Asymptotic completeness in quantum field theory. Massive Pauli-Fierz Hamiltonians, Rev. Math. Phys. **11**, 383–450 (1999)
12. M. Dresden, H.A. Kramers, Between tradition and revolution, Springer Verlag (1987)
13. J. Fröhlich, On the infrared problem in a model of scalar electrons and masselss scalar bosons, Ann. Inst. H. Poincaré **19**, 1–103 (1973)
14. J. Fröhlich, Existence of dressed one-electron states in a class of persistent models, Fortschritte Phys. **22**, 159–198 (1974)
15. C. Gérard, On the existence of ground states for massless Pauli-Fierz Hamiltonians, Ann. Henri Poincaré **1**, 443–459 (2000), mp_arc 99–158 (1999)
16. M. Hirokawa, Remarks on the ground state energy of the spin-boson model, An application of the Wigner-Weisskopf model, Rev. Math. Phys. **13**, 221–251 (2001), mp_arc 00–239 (2000)
17. F. Hiroshima, Ground states of a model in nonrelativistic quantum electrodynamics I and II, J. Math. Phys. **40**, 6209–6222 (1999), **41**, 661–674 (2000)
18. F. Hiroshima, The self-adjointness and relative bound of the Pauli-Fierz Hamiltonian in quantum electrodynamics for arbitrary coupling constants, preprint (October, 2000)
19. F. Hiroshima, H. Spohn, Enhanced binding through coupling to a quantum field, Mathematical Physics Preprint Archive, mp_arc 01–39 (2001)
20. W. Hunziker, I.M. Sigal, The general theory of N-body quantum systems, in: Mathematical quantum theory. II. Schrödinger operators (Vancouver, BC, 1993), 35–72, CRM Proc. Lecture Notes, 8, Amer. Math. Soc., Providence, RI, 1995
21. E.H. Lieb, M. Loss, Self-Energy of Electrons in Non-perturbative QED, in: Differential Equations and Mathematical Physics, University of Alabama, Birmingham, 1999, R. Weikard, G. Weinstein, eds., 255–269, Internat. Press (1999). arXiv math-ph/9908020, mp_arc 99–305
22. E.H. Lieb, M. Loss, Analysis, Graduate Studies in Mathematics, American Mathematical Society, 1997
23. A. O'Connor, Exponential decay of bound-state wave functions, Commun. Math. Phys. **32**, 319–340 (1973)
24. W. Pauli, M. Fierz, Zur Theorie der Emission langwelliger Lichtquanten, Nuovo Cimento **15**, 167–188 (1938)

25. M. Reed, B. Simon, Methods of modern mathematical physics, vol 4, Theorem XIII.39, Academic Press (1978)
26. H. Spohn, Asymptotic completeness for Rayleigh scattering, J. Math. Phys. **38**, 2281–2296 (1997)
27. H. Spohn, Ground state(s) of the spin-boson Hamiltonian, Commun. Math. Phys. **123**, 277–304 (1989)
28. H. Spohn, Ground state of a quantum particle coupled to a scalar Bose field, Lett. Math. Phys. **44**, 9–16 (1998)

Part VIII

Bosonic Systems

Bosonic Systems

This section contains papers on the ground state energies of several kinds of many-boson systems.

Item VIII.2 *Ground state energy of the low density Bose gas*, with J. Yngvason, proves a conjecture that goes back to Bogolubov's 1947 work: At low density ϱ the ground state energy per particle, $e_0(\varrho)$, for a system of (3-dimensional) bosons with a short-range, two-body potential v of scattering length a is $e_0(\varrho) \sim 4\pi(\hbar^2/2m)\varrho a$. Actually, the proof in VIII.2 is for nonnegative v (such as a hard core potential) and the question of what happens otherwise is still open. The new conjecture made in the paper is that the formula holds for any short-range v with the property that there are no many-body bound states.

This work was extended in two directions. One direction is to 2-dimensions. The relevant formula was proposed only as late as 1971, by Schick: $e_0(\varrho) \sim 4\pi(\hbar^2/2m)\varrho/|\ln(\varrho a^2)|$. This is proved in *The Ground State Energy of a Dilute Two-dimensional Bose Gas*, with J. Yngvason, item 258 in the publication list.

The second direction concerns bosons in a trap, which models the experimental situation. A one-body potential $V(x)$ is added to the Hamiltonian. For low average density, the ground state energy is supposedly described by the Gross-Pitaevskii energy. This is proved in two papers (with J. Yngvason and R. Seiringer): Item VIII.3 *Bosons in a Trap: A Rigorous Derivation of the Gross-Pitaevskii Energy Functional* and *A Rigorous Derivation of the Gross-Pitaevskii Energy Functional for a Two-dimensional Bose Gas* (item 263 in the publication list).

When one turns to long-range potentials the situation is different. For charged bosons in a neutralizing background ("jellium"), the "weak-coupling" situation corresponds to *high* density. The asymptotic formula was proposed by L. Foldy in 1961: $e_0(\varrho) \sim C\varrho^{1/4}$, with C being a definite constant. This was finally proved (with J. P. Solovej) in *Ground State Energy of the One-Component Charged Bose Gas*, item 265, which is reproduced here as item VIII.7.

All these extensions are briefly summarized in a review paper: Item VIII.4, *The Bose Gas: A Subtle Many-Body Problem*.

Two other papers, on charged bosons, which were in section V of earlier editions, are reproduced in this section. A system of negatively charged bosons with fixed nuclei, as in the stability of matter problem, significantly fails the stability condition. The energy is proportional to $-N^{5/3}$ instead of $-N$ (the $N^{5/3}$ law). If on the other hand, the positive particles have a finite mass and are therefore dynamic, the situation improves somewhat, but not enough. The energy is proportional to $-N^{7/5}$. F. Dyson proved an upper bound of this kind in 1967 and the lower bound was done in 1988 with J. Conlon and H-T. Yau.

Earlier work on many-boson systems is in publication list items 5, 12–16. These results along with a survey of the state of the field, is contained in the review article *The Bose fluid*, item 18. Items 12 and 13 *Exact Analysis of an interacting Bose gas* (with W. Liniger) contain an exact solution for one-dimensional bosons with a repulsive delta-function interaction. This is the only soluble model with short-range, two-body potentials, and it verifies Bogolubov's conjecture for the ground state energy in the weak-coupling limit. Items 14 and 15 (with W. Liniger and A. Sakakura) give a non-rigorous method for calculating the energy for the jellium and short-range models mentioned above, the significant point being that the calculations are done in configuration-space instead of momentum-space, as was, and continues to be, customary.

VOLUME 80, NUMBER 12 PHYSICAL REVIEW LETTERS 23 MARCH 1998

Ground State Energy of the Low Density Bose Gas

Elliott H. Lieb[1] and Jakob Yngvason[2]

[1]*Department of Physics, Jadwin Hall, Princeton University, P.O. Box 708, Princeton, New Jersey 08544*

[2]*Institut für Theoretische Physik, Universität Wien, Boltzmanngasse 5, A 1090 Vienna, Austria*

(Received 27 October 1997)

Now that the properties of low temperature Bose gases at low density, ρ, can be examined experimentally it is appropriate to revisit some of the formulas deduced by many authors four to five decades ago. One of these is that the leading term in the energy/particle is $2\pi\hbar^2\rho a/m$, where a is the scattering length. Owing to the delicate and peculiar nature of bosonic correlations, four decades of research have failed to establish this plausible formula rigorously. The only known lower bound for the energy was found by Dyson in 1957, but it was 14 times too small. The correct bound is proved here. [S0031-9007(98)05619-1]

PACS numbers: 03.75.Fi, 05.30.Jp, 67.40.–w

With the renewed experimental interest in low density, low temperature Bose gases, some of the formulas posited four and five decades ago have been dusted off and reexamined. One of these is the leading term in the ground state energy. In the limit of small particle density, ρ,

$$e_0(\rho) \approx \mu 4\pi\rho a\,, \tag{1}$$

where $e_0(\rho)$ is the ground state energy (g.s.e.) per particle in the thermodynamic limit, a is the scattering length (assumed positive) of the two-body potential v for bosons of mass m, and $\mu \equiv \hbar^2/2m$.

Is Eq. (1) correct? In particular, is it true for the hard-sphere gas? While there have been many attempts at a rigorous proof of (1) in the past 40 years, none has been found so far. Our aim here is to supply that proof for finite range, positive potentials. As remarked below, (1) cannot hold unrestrictedly; more than $a > 0$ is needed.

An upper bound for $e_0(\rho)$ agreeing with (1) is not easy to derive, but it was achieved for hard spheres by a variational calculation [1], which can be extended to include general, positive potentials of finite range. What remained unknown was a good lower bound. The only one available is Dyson's [1], and that is about *fourteen times smaller* than (1). In this paper we shall provide a lower bound of the desired form, and thus prove (1). We can also give explicit error bounds for small enough values of the dimensionless parameter $Y \equiv 4\pi\rho a^3/3$:

$$e_0(\rho) \geq \mu 4\pi\rho a(1 - CY^{1/17}) \tag{2}$$

for some fixed C (which is not evaluated explicitly because C and the exponent $1/17$ are only of academic interest). The bound (2) holds for *all non-negative, finite range, spherical, two-body potentials.* A typical experimental value [2] is $Y \approx 10^{-5}$. Dyson's upper bound is $\mu 4\pi\rho a(1 + 2Y^{1/3})(1 - Y^{1/3})^{-2}$.

We conjecture that (1) requires only a positive scattering length *and* the absence of any many-body, negative energy bound state. If there are such bound states then (1) is certainly wrong, but this obvious caveat does not seem to have been clearly emphasized before. There is a "nice" potential with positive scattering length, no two-body bound state, but with a three-body bound state [3].

Our method also obviously applies to the positive temperature free energy [because Neumann boundary conditions give an upper bound to the solution to the heat (or Bloch) equation].

We also give some explicit bounds for *finite* systems, which might be useful for experiments with traps, but we concentrate here on the thermodynamic limit for simplicity. For traps with slowly varying confining potentials, V_{ext}, our method will prove that the leading term in the energy is given by the well known local density approximation [4], which minimizes the gaseous energy (1) plus the confining energy, with respect to $\rho(\mathbf{x})$, namely,

$$\mathcal{E}(\rho) \equiv \int [V_{\text{ext}}(\mathbf{x})\rho(\mathbf{x}) + \mu 4\pi a\rho(\mathbf{x})^2]\, d^3\mathbf{x}$$

is minimized subject to $\int \rho = N =$ number of particles.

The fact that Dyson's lower bound was not improved for four decades, despite many attempts, attests to the fact that bosons are subtle quantum mechanical objects which can have peculiar correlations unknown to fermions. For example, there is the nonthermodynamic $N^{7/5}$ law for the charged Bose gas that was discovered by Dyson [5], confirmed only 20 years later [6], and which defies any simple physical interpretation.

The first understanding of (1) goes back to Bogoliubov [7], who also introduced the notion of "pairing" in Helium (which resurfaced in the BCS theory for fermions). Later, there were several derivations of (1) (and higher order) [8,9]. The method of the pseudopotential, which is an old idea of Fermi's, was closest to the Bogoliubov analysis. The "exact" pseudopotential was constructed in [10], but it did not help to make this appealing idea more rigorous. Most of the derivations were in momentum space, the exception being [9], which works directly in physical space and which can handle both long and short range potentials. See [11] for a review. All these methods rely on special assumptions about the ground state (e.g., selecting special terms in a perturbation expansion, which

 0031-9007/98/80(12)/2504(4)$15.00

likely diverges) and it is important to derive a fundamental result like (1) without extra assumptions.

In all of this earlier work one key fact was not understood, or at least not clearly stated in connection with the derivation of (1). It is that there are two different regimes, even at low density, with very different physics, even though the simple formula (1) seems to depend only on the scattering length. Recall that the (two-body) scattering length is defined, for a spherically symmetric potential, v, by

$$-\mu u_0''(r) + \frac{1}{2} v(r) u_0(r) = 0, \tag{3}$$

with $u_0(0) = 0$, $u_0(r) > 0$ (which is equivalent to the absence of negative energy bound states, and which is true for non-negative v). As $r \to \infty$, $u(r) \approx r - a$. [Note the $v/2$ and not v in (3) because of the reduced mass.] Thus, a depends on m in a nontrivial way, and there are two extremes:

Potential energy dominated region.—The hard sphere [$v(r) = \infty$ for $r < a$], is the extreme case here; the scattering length is independent of m, and the energy is mostly (entirely) *kinetic*. We see this from (1) because $-m\partial e_0/\partial m$ is the kinetic energy (Hellmann-Feynman theorem). In this regime the potential is so dominant that it forces the energy to be mostly kinetic. The ground-state (g.s.) wave function is highly correlated.

Kinetic energy dominated region.—The typical case is a very "soft" potential. Then $a \approx (m/\hbar^2) \int_0^\infty v(r) r^2\, dr$, which implies, from (1), that e_0 hardly depends on m. Thus, the energy is almost all *potential.* The g.s. wave function is essentially the noninteracting one in this limit.

In other words, "scattering length" is not a property of v alone, and the low density gas, viewed from the perspective of the bosons, looks quite different in the two regimes. Nevertheless, as (1) says, the energy cannot distinguish the two cases. Whether Bose-Einstein condensation itself can notice the difference remains to be seen. Condensation will not be touched upon here, except to note that so far *the only case with two-body interactions in which Bose-Einstein condensation has been rigorously established is hard core lattice bosons, but only at half filling* [12].

Dyson [1] effectively converted region 1 into region 2. We shall make use of his important idea, which substitutes a very soft potential for the original one (even a hard core) at the price of sacrificing the kinetic energy.

We assume that the N particles are in a $L \times L \times L$ cubic box, Ω. The particle density is then $\rho = NL^{-3}$. It is well known that the energy per particle in the thermodynamic limit, $e_0(\rho)$, does not depend on the details of (reasonable) Ω, so we are free to use a cube and take $N \to \infty$ through any sequence we please, as far as $e_0(\rho)$ is concerned. We set $N = kM$ with k an integer and M the cube of an integer, because we shall want to divide up Ω into M smaller cubes (called *cells*) of length $\ell = (k/\rho)^{1/3}$. We will take $M \to \infty$ with ℓ and $k = \rho\ell^3$ fixed, but large.

The N-body Schrödinger operator is

$$H = -\mu \sum_{i=1}^{N} \Delta_i + \sum_{1 \le i < j \le N} v(\mathbf{x}_i - \mathbf{x}_j). \tag{4}$$

For boundary conditions we impose Neumann (zero derivative) boundary conditions on Ω. It is well known that Neumann boundary conditions give the lowest possible g.s.e. for H, and hence its use is appropriate for a discussion of a *lower* bound for the g.s.e. Denote this Neumann g.s.e. by $E_0(N, L)$.

Now divide Ω into M cells and impose Neumann conditions on each cell, which, as stated before, lowers the energy further. We also neglect the interaction between particles in different cells; this, too, can only lower the energy because $v \ge 0$.

A lower bound for $E_0(N, L)$ is obtained by distributing the N particles in the M cells and then finding a lower bound for the energy in these cells, which are now independent. We then add these M energies. Finally, we minimize the total energy over *all choices* of the particle number in each cell (subject to the total number being N). Despite the independence of the cells, the latter problem is not easy. In particular, something has to be invoked to make sure that we do not end up with some cells having too large a number of particles and some cells having too few.

With L, N and $M = N/(\rho\ell^3)$ fixed, let Mc_n, for $n = 0, 1, 2, \ldots$ denote the number of cells containing exactly n particles. Then the particle number and cell number constraints are

$$\sum_{n \ge 0} n c_n = k = \rho\ell^3, \qquad \sum_{n \ge 0} c_n = 1, \tag{5}$$

and our energy bound is

$$E_0(N, L) \ge M \min \sum_{n \ge 0} c_n E_0(n, \ell), \tag{6}$$

where the minimum is over all $c_n \ge 0$ satisfying (5).

The minimization would be easy if we knew that $E_0(n, \ell)$ (or a good lower bound for it) is convex in n, for then the optimum would be $c_n = \delta_{n,k}$. This convexity is very plausible, but we cannot prove it (except in the thermodynamic limit, where it amounts to thermodynamic stability). What we do know instead is *superadditivity:*

$$E_0(n + n', \ell) \ge E_0(n, \ell) + E_0(n', \ell) \tag{7}$$

for all n, n', and this turns out to be an adequate substitute for controlling the large n terms in (6).

Equation (7) is an immediate consequence of the positivity of the potential and it is used as follows. Suppose, provisionally, that we have a lower bound of the form

$$E_0(n, \ell) \ge K(\ell) n(n - 1), \quad \text{for } 0 \le n \le 4k, \tag{8}$$

VOLUME 80, NUMBER 12 PHYSICAL REVIEW LETTERS 23 MARCH 1998

with $K(\ell)$ independent of n for $0 \leq n \leq 4k$. In fact, we shall later prove that for small enough ρ (and hence small enough k) and suitable ℓ, (8) holds with

$$K(\ell) \geq \mu 4\pi a \ell^{-3}(1 - C' Y^{1/17}), \qquad (9)$$

with C' some constant. [However, the analysis we give now, leading to (12), does not depend on this particular form of $K(\ell)$.]

Split the sum in (6) into two pieces: $0 \leq n < 4k$ and $4k \leq n$. Let $t \equiv \sum_{n<4k} n c_n \leq k$, so that $k - t = \sum_{n \geq 4k} n c_n$. From (8) and Cauchy's inequality (and $\sum_{n<4k} c_n \leq 1$)

$$\sum_{n<4k} c_n E_0(n, \ell) \geq K(\ell) t(t-1). \qquad (10)$$

On the other hand, if $n \geq 4k$ then, by (7), $E_0(n,\ell) \geq (n/8k) E_0(4k, \ell)$, so

$$\sum_{n \geq 4k} c_n E_0(n, \ell) \geq \frac{k - t}{2} K(\ell)(4k - 1).$$

Upon adding (10) and (11) the factor $t(t-1) + (k - t)(2k - 1/2)$ is obtained. Although the number t is unknown, we note that this factor is monotone decreasing in t in the interval $0 \leq t \leq k$ [which is where t lies, by (5)]. Thus, we can set $t = k$ and obtain the same bound as if we had convexity, i.e.,

$$E_0(N, L) \geq N K(\ell)(\rho \ell^3 - 1). \qquad (12)$$

In summary, if we can show (8) for a box of a *fixed* size ℓ, for all particle numbers up to $n = 4\rho\ell^3$, then we will have obtained our goal, (2), in the thermodynamic limit *provided* we can show that the K in (8) satisfies (9) with the constant C' when ℓ is large compared to the mean particle spacing, i.e., $\rho\ell^3 > C'' Y^{-1/17}$. Then the C in (2) equals $C' + C''$.

We now focus on a single cell and denote the n coordinates $(\mathbf{x}_1, ..., \mathbf{x}_n)$ collectively by $\mathbf{X}$. The first step in proving (9) is to replace the total potential, $\sum_{i<j} v(\mathbf{x}_i - \mathbf{x}_j)$, by a lesser quantity, $\mathcal{W}_v(\mathbf{X})$, the *nearest-neighbor potential* defined by

$$\mathcal{W}_v(\mathbf{X}) \equiv \frac{1}{2} \sum_{i=1}^{n} v(\mathbf{x}_i - \mathbf{x}_{j(i)}), \qquad (13)$$

where $j(i)$ is the nearest-neighbor to particle i in the configuration $\mathbf{X}$; i.e., particle i "feels" only its nearest neighbor. Hence, we replace H by the smaller operator

$$\widetilde{H} \equiv \mathcal{T} + \mathcal{W}_v \leq H, \qquad (14)$$

where $\mathcal{T} = -\mu \sum \Delta_i$ is the kinetic energy in (4). Since $v \geq 0$, the g.s.e. of $\widetilde{H}$ satisfies $\widetilde{E}_0(n, \ell) \leq E_0(n, \ell)$.

To get into the kinetic energy dominated region, we wish to replace v in (13) by a gentler potential U. To this end we generalize Lemma 1 of [1] and simplify its proof.

LEMMA 1.—Let $v(r) \geq 0$ and $v(r) = 0$ for $r > R_0$. Let $U(r) \geq 0$ be any function satisfying $\int U(r) r^2 dr \leq 1$ and $U(r) = 0$ for $r < R_0$. Let $\mathcal{B} \subset \mathbf{R}^3$ be star-shaped (convex suffices) with respect to 0. Then, for all functions ϕ,

$$\int_{\mathcal{B}} \mu |\nabla\phi(\mathbf{x})|^2 + \left[\frac{1}{2} v(r) - \mu a U(r)\right] |\phi(\mathbf{x})|^2 d^3\mathbf{x} \geq 0. \qquad (15)$$

Proof.—Actually, (15) holds with $\mu|\nabla\phi(\mathbf{x})|^2$ replaced by the (smaller) *radial kinetic energy*, $\mu|\partial\phi(\mathbf{x})/\partial r|^2$, and thus it suffices to prove the analog of (15) for the integral along each radial line, and to assume that $\phi(\mathbf{x}) = u(r)/r$ along this line, with $u(0) = 0$. Let us first prove (15) when U is a delta function at some radius $R \geq R_0$, i.e., $U(r) = R^{-2}\delta(r - R)$. Then, it is enough to show, for all u, that

$$\int_0^R \mu |u'(r) - u(r)/r|^2 + \frac{1}{2} v(r)|u(r)|^2 dr \geq \mu a |u(R)|^2 R^{-2}. \qquad (16)$$

If the length of the radial line is less than R then (16) is trivial. Otherwise, normalize u by $u(R) = R - a$, and ask for the minimum of the left side of (16) under the condition that $u(0) = 0$, $u(R) = R - a$. This is a simple problem in the calculus of variations and leads to the scattering length equation (3). If we substitute the solution into (16), integrate by parts, and note that $u_0(r) = r - a$ for $r > R_0$, we find that (16) is true if $a \leq R$, which is true since $u_0 \geq 0$. Finally, by linearity and the fact that $U(r) = \int r^{-2}\delta(r - s) U(s) s^2 ds$, the δ-function case implies the general case. Q.E.D.

We select our U by picking some $R \gg R_0$ and setting

$$U(r) = 3(R^3 - R_0^3)^{-1} \quad \text{for } R_0 < r < R \qquad (17)$$

and $U(r) = 0$ otherwise. Later on we shall choose R, and we shall take

$$R_0 \ll R \ll \rho^{-1/3} \ll \ell. \qquad (18)$$

By further decomposing a cube into Voronoi cells (which are always convex), Dyson [1] deduces from Lemma 1 that $\widetilde{H}$ is bounded below by a nearest-neighbor potential, as in (13), i.e.,

$$H > \widetilde{H} > \mu a \mathcal{W}_U(\mathbf{X}), \qquad (19)$$

where $\mathcal{W}_U$ is as in (13), with v replaced by U. For the hard core case, Dyson estimates the *minimum* (over all $\mathbf{X}$) of $\mathcal{W}_U(\mathbf{X})$, for a U similar to (17), and gets a lower bound for all ρ, but 14 times smaller than (1). We follow another route. An important quantity for us will be the *average* value of $\mathcal{W}_U(\mathbf{X})$ in a cell, denoted by $\langle \mathcal{W}_U \rangle$.

To compute $\langle U(\mathbf{x}_1 - \mathbf{x}_{j(1)}) \rangle$, for example, it is easiest to do the $\mathbf{x}_2, \ldots, \mathbf{x}_n$ integrations over the cell first and then the $\mathbf{x}_1$ integration. Provided $\mathbf{x}_1$ is in the smaller cube which is a distance R from the cell boundary [whose volume is $(\ell - 2R)^3$], the probability that $R_0 < |\mathbf{x}_j - \mathbf{x}_1| < R$ is $4\pi(R^3 - R_0^3)/3\ell^3$. Thus, performing the $\mathbf{x}_1$ integration over the smaller cube, and then adding similar contributions from $U(\mathbf{x}_2 - \mathbf{x}_{j(2)})$, etc., and using (17), we get

$$\langle \mathcal{W}_U \rangle \geq \frac{3n(\ell - 2R)^3}{(R^3 - R_0^3)\ell^3}[1 - (1 - Q)^{n-1}] \tag{20}$$

$$\geq \frac{4\pi}{\ell^3} n(n-1)\left(1 - \frac{2R}{\ell}\right)^3 \frac{1}{1 + Q(n-1)}. \tag{21}$$

with

$$Q = 4\pi(R^3 - R_0^3)/3\ell^3 \ll 1. \tag{22}$$

In (21) we used $[1 - x]^{n-1} \leq [1 + (n-1)x]^{-1}$ for $0 \leq x \leq 1$. Note that (21) is of the form (8).

By similar reasoning, we obtain the upper bound

$$\langle \mathcal{W}_U \rangle \leq \frac{3n}{R^3 - R_0^3}[1 - (1 - Q)^{n-1}] \tag{23}$$

$$\leq \frac{4\pi}{\ell^3} n(n-1). \tag{24}$$

Since $U(r)^2 = 4\pi(Q\ell^3)^{-1}U(r)$, we also obtain

$$\langle \mathcal{W}_U^2 \rangle \leq 4\pi n(Q\ell^3)^{-1}\langle \mathcal{W}_U \rangle. \tag{25}$$

We can now use Lemma 1 and these averages to obtain (8) and (9). Instead of using (19) alone, we pick some $0 < \varepsilon \ll 1$ and, borrowing a bit of kinetic energy, define

$$\widehat{H} \equiv \varepsilon \mathcal{T} + (1 - \varepsilon)\mu a \mathcal{W}_U(\mathbf{X}). \tag{26}$$

By Lemma 1 and $v \geq 0$, we have

$$H > \widetilde{H} > \widehat{H}. \tag{27}$$

We shall derive (8) and (9) from a lower bound to $\widehat{H}$.

Although ε is small, we regard $H_0 \equiv \varepsilon\mathcal{T}$ as our unperturbed Hamiltonian and $V \equiv (1 - \varepsilon)\mu a \mathcal{W}_U(\mathbf{X})$ as a perturbation of H_0. The ground state wave function for H_0 is $\Psi_0(\mathbf{X}) = \ell^{-3n/2}$ and $H_0\Psi_0 = \lambda_0\Psi_0 = 0$ (Neumann conditions). The second eigenvalue of H_0 is $\lambda_1 = \varepsilon\mu\pi/\ell^2$. Note that the ground state expectation, $\langle\Psi_0|\mathcal{W}_U|\Psi_0\rangle$, is precisely the average $\langle \mathcal{W}_U \rangle$ mentioned in (20)–(25).

Temple's inequality [13] states that when a perturbation V is non-negative (as here) and when $\lambda_1 - \lambda_0 \geq \langle\Psi_0|V|\Psi_0\rangle$ then the g.s.e., E_0, of the perturbed Hamiltonian $H = H_0 + V$ satisfies

$$E_0 \geq \lambda_0 + \langle\Psi_0|V|\Psi_0\rangle - \frac{\langle\Psi_0|V^2|\Psi_0\rangle - \langle\Psi_0|V|\Psi_0\rangle^2}{\lambda_1 - \lambda_0 - \langle\Psi_0|V|\Psi_0\rangle}. \tag{28}$$

We apply this to our case with $\lambda_1 - \lambda_0 = \varepsilon\mu\pi/\ell^2$ and $V = (1 - \varepsilon)\mu a \mathcal{W}_U$. We neglect the (positive) term $\langle\Psi_0|V|\Psi_0\rangle^2$ in (28) and we use (21), (24), and (25). We also use $1 - \varepsilon < 1$ in two appropriate places and find

$$\frac{E_0(n,\ell)}{\mu a\langle \mathcal{W}_U \rangle} \geq (1 - \varepsilon)\left(1 - \frac{4\pi a n}{Q\ell} \frac{1}{\varepsilon\pi - a\ell^2\langle \mathcal{W}_U \rangle}\right). \tag{29}$$

Apart from some higher order errors, (29) is just what we need in (8) and (9). Let us denote the order of the main error by Y^α, and we would like to show that $\alpha = 1/17$ suffices. The errors are the following:

From the $(1 - \varepsilon)$ factor, we need $\varepsilon \leq O(Y^\alpha)$.

From the $Q(n + 1)$ error in (21) we need $Q\rho\ell^3 \leq O(Y^\alpha)$.

From the R/ℓ error in (21) we need $R/\ell \leq O(Y^\alpha)$.

From (29) we need $a\ell^2\langle \mathcal{W}_U \rangle/\varepsilon \leq O(Y^\alpha)$ and $\rho\ell^5 a/R^3\varepsilon \leq O(Y^\alpha)$.

All these desiderata can be met with $\varepsilon = Y^\alpha$, $R/\ell = Y^\alpha$, $Q = O(Y^\alpha)$, $\rho R^3 = Y^{2\alpha}$, and $\alpha = 1/17$—as claimed.

The partial support of U.S. National Science Foundation Grant No. PHY95-13072A01 (EHL) and the Adalsteinn Kristjansson Foundation of the University of Iceland (JY) is gratefully acknowledged.

[1] F. J. Dyson, Phys. Rev. **106**, 20 (1957).
[2] W. Ketterle and N. J. van Druten, in *Advances in Atomic, Molecular and Optical Physics*, edited by B. Bederson and H. Walther (Academic Press, New York, 1996), Vol. 37, p. 181.
[3] B. Baumgartner, J. Phys. A **30**, L741 (1997).
[4] J. Oliva, Phys. Rev. B **39**, 4197 (1989).
[5] F. J. Dyson, J. Math. Phys. (N.Y.) **8**, 1538 (1967).
[6] J. G. Conlon, E. H. Lieb, and H-T. Yau, Commun. Math. Phys. **116**, 417 (1988).
[7] N. N. Bogoliubov, J. Phys. (U.S.S.R.) **11**, 23 (1947); N. N. Bogoliubov and D. N. Zubarev, Sov. Phys. JETP **1**, 83 (1955).
[8] K. Huang and C. N. Yang, Phys. Rev. **105**, 767–775 (1957); T. D. Lee, K. Huang, and C. N. Yang, Phys. Rev. **106**, 1135–1145 (1957); K. A. Brueckner and K. Sawada, Phys. Rev. **106**, 1117–1127 (1957); **106**, 1128–1135 (1957); S. T. Beliaev, Sov. Phys. JETP **7**, 299–307 (1958); T. T. Wu, Phys. Rev. **115**, 1390 (1959); N. Hugenholtz and D. Pines, Phys. Rev. **116**, 489 (1959); M. Girardeau and R. Arnowitt, Phys. Rev. **113**, 755 (1959); T. D. Lee and C.N. Yang, Phys. Rev. **117**, 12 (1960).
[9] E. H. Lieb, Phys. Rev. **130**, 2518 (1963); E. H. Lieb and A. Y. Sakakura, Phys. Rev. **133**, A899 (1964); E. H. Lieb and W. Liniger, Phys. Rev. **134**, A312 (1964).
[10] E. H. Lieb, Proc. Nat. Acad. Sci. U.S.A. **46**, 1000 (1960).
[11] E. H. Lieb, *The Bose Fluid*, in Lecture Notes in Theoretical Physics VIIC, edited by W. E. Brittin (University of Colorado Press, Boulder, Colorado, 1964), p. 175.
[12] E. H. Lieb, T. Kennedy, and S. Shastry, J. Stat. Phys. **53**, 1019 (1988).
[13] G. Temple, Proc. R. Soc. London A **119**, 276 (1928).

PHYSICAL REVIEW A, VOLUME 61, 043602

Bosons in a trap: A rigorous derivation of the Gross-Pitaevskii energy functional

Elliott H. Lieb

Department of Physics and Department of Mathematics, Jadwin Hall, Princeton University, P.O. Box 708, Princeton, New Jersey 08544

Robert Seiringer and Jakob Yngvason

Institut für Theoretische Physik, Universität Wien, Boltzmanngasse 5, A 1090 Vienna, Austria

(Received 31 August 1999; published 6 March 2000)

The ground-state properties of interacting Bose gases in external potentials, as considered in recent experiments, are usually described by means of the Gross-Pitaevskii energy functional. We present here a rigorous proof of the asymptotic exactness of this approximation for the ground-state energy and particle density of a dilute Bose gas with a positive interaction.

PACS number(s): 03.75.Fi, 05.30.Jp, 67.40.Db, 71.35.Lk

I. INTRODUCTION

Recent experimental breakthroughs in the treatment of dilute Bose gases have renewed interest in formulas for the ground state and its energy derived many decades ago. One of these is the Gross-Pitaevskii (GP) formula for the energy in a trap [1–3], such as is used in the actual experiments. We refer to [4] for an up-to-date review of this approximation and its applications. One of the inputs needed for its justification is the ground-state energy per unit volume of a dilute, thermodynamically infinite, homogeneous gas. This latter quantity has been known for many years, but it was only very recently that it was derived rigorously [5] for suitable interparticle potentials. Consequently, it is appropriate now to use this result to go one step further and derive the GP formula rigorously.

The starting point for our investigation is the Hamiltonian for N identical bosons

$$H^{(N)}=\sum_{i=1}^{N} [-\nabla_i^2+V(\mathbf{x}_i)]+\sum_{i<j} v(|\mathbf{x}_i-\mathbf{x}_j|) \quad (1.1)$$

acting on totally *symmetric*, square integrable wave functions of $(\mathbf{x}_1,\ldots,\mathbf{x}_N)$ with $\mathbf{x}_i\in\mathbb{R}^3$. Units have here been chosen so that $\hbar=2m=1$, where m is the mass. We consider external potentials V that are measurable and locally bounded and tend to infinity for $|\mathbf{x}|\to\infty$ in the sense that $\inf_{|\mathbf{x}|\geq R}V(\mathbf{x})\to\infty$ for $R\to\infty$. The potential is then bounded below and for convenience we assume that its minimum value is zero. The ground state of $-\nabla^2+V(\mathbf{x})$ provides a natural energy unit, $\hbar\omega$, and the corresponding length unit, $\sqrt{\hbar/m\omega}$, describes the extension of the potential. We shall measure all energies and lengths in these units. In the available experiments V is typically $\sim|\mathbf{x}|^2$ and $\sqrt{\hbar/m\omega}$ of the order 10^{-6} m.

The particle interaction v is assumed to be positive, spherically symmetric, and decay faster than $|\mathbf{x}|^{-3}$ at infinity. In particular, the scattering length, denoted by a, should be finite. We recall that the (two-body) scattering length is defined by means of the solution $u(r)$ of the zero-energy scattering equation

$$-u''(r)+\tfrac{1}{2}v(r)u(r)=0 \quad (1.2)$$

with $u(0)=0$; by definition, $a=\lim_{r\to\infty}[r-u(r)/u'(r)]$. Let $v_1(r)$ be a fixed potential with scattering length a_1. Then $v(r)=(a_1/a)^2v_1(a_1r/a)$ has scattering length a. In the following we regard v_1 as fixed, but vary a (in fact, we shall take $a=a_1/N$). The ground-state energy $E^{\rm QM}$ of Eq. (1.1) depends on the potentials V and v, besides N, but with V fixed and $v(r)=(a_1/a)^2v_1(a_1r/a)$, the notation $E^{\rm QM}(N,a)$ is justified. The corresponding eigenfunction will be denoted $\Psi_0^{(N)}$. It is unique up to a phase that can be chosen such that the wave function is strictly positive where the interaction is finite [6]. The particle density is defined by

$$\rho_{N,a}^{\rm QM}(\mathbf{x})=N\int_{\mathbb{R}^{3(N-1)}}|\Psi_0^{(N)}(\mathbf{x},\mathbf{x}_2,\ldots,\mathbf{x}_N)|^2d\mathbf{x}_2\cdots d\mathbf{x}_N. \quad (1.3)$$

The *Gross-Pitaevskii (GP) energy functional* is defined as

$$\mathcal{E}^{\rm GP}[\Phi]=\int [|\nabla\Phi(\mathbf{x})|^2+V(\mathbf{x})|\Phi(\mathbf{x})|^2+4\pi a|\Phi(\mathbf{x})|^4]d\mathbf{x} \quad (1.4)$$

where Φ is a function on $\mathbb{R}^3$. For a given N the corresponding GP energy, denoted $E^{\rm GP}(N,a)$, is defined as the infimum of $\mathcal{E}[\Phi]$ under the normalization condition

$$\int_{\mathbb{R}^3}|\Phi(\mathbf{x})|^2d\mathbf{x}=N. \quad (1.5)$$

It has the simple scaling property

$$E^{\rm GP}(N,a)=NE^{\rm GP}(1,Na). \quad (1.6)$$

What Eq. (1.6) shows is that the GP functional (1.4) together with the normalization condition (1.5) has one characteristic parameter, namely, Na. (Recall that lengths are measured in the unit $\sqrt{\hbar/m\omega}$ associated with V so a is dimensionless.) Thus, if we want to investigate the nontrivial aspects of GP theory we have to consider a limit in which $N\to\infty$ with Na fixed. This explains the seemingly peculiar limit in Theorems I.1 and I.2. As $Na\to\infty$ the GP energy functional simplifies, since the gradient term becomes small compared to the other terms, and the so called Thomas-Fermi limit described in Theorem II.2 results. In some typical experiments a is about

1050-2947/2000/61(4)/043602(13)/$15.00

10^{-3}, while N varies from 10^3 to 10^7. Thus a_1 in Theorems I.1 and I.2 varies from 1 to about 10^4.

In the next section it will be shown that the infimum of the energy functional (1.4), under the subsidiary condition (1.5), is obtained for a unique, strictly positive function, denoted $\Phi^{\rm GP}$. The GP density is, by definition,

$$\rho_{N,a}^{\rm GP}(\mathbf{x})=\Phi^{\rm GP}(\mathbf{x})^2. \tag{1.7}$$

It satisfies

$$\rho_{N,a}^{\rm GP}(\mathbf{x})=N\rho_{1,Na}^{\rm GP}(\mathbf{x}). \tag{1.8}$$

The main result of this paper concerns the behavior of the quantum mechanical ground-state energy $E^{\rm QM}(N,a)$ when N is large, but a is small, so that Na is $O(1)$. It is important to note that although the density tends to infinity for $N\to\infty$ [by Eq. (1.8)] we are still concerned with *dilute* systems in the sense that $a^3\bar{\rho}\ll 1$, where

$$\bar{\rho}=\frac{1}{N}\int \rho_{N,a}^{\rm GP}(\mathbf{x})^2 d\mathbf{x} \tag{1.9}$$

is the *mean GP density*. [Note the exponent 2 in Eq. (1.9).] In fact, since $a\sim N^{-1}$, $a^3\bar{\rho}\sim N^{-2}$.

The precise statement of the limit theorem for the energy is as follows.

Theorem I.1 (The GP energy is the dilute limit of the QM energy). For every fixed a_1

$$\lim_{N\to\infty}\frac{E^{\rm QM}(N,a_1/N)}{N}=E^{\rm GP}(1,a_1) \tag{1.10}$$

and the convergence is uniform on bounded intervals of a_1.

While we do not prove anything about Bose-Einstein condensation, which necessarily involves the full one-body density matrix $\rho^{(1)}(\mathbf{x},\mathbf{x}')$, we can make an assertion about the diagonal part of the density matrix, $\rho^{\rm QM}(\mathbf{x})=\rho^{(1)}(\mathbf{x},\mathbf{x})$:

Theorem I.2 (The GP density is the dilute limit of the QM density). For every fixed a_1

$$\lim_{N\to\infty}\frac{1}{N}\rho_{N,a_1/N}^{\rm QM}(\mathbf{x})=\rho_{1,a_1}^{\rm GP}(\mathbf{x}) \tag{1.11}$$

in the sense of weak convergence in L^1.

For the proof of Theorem I.1 we establish upper and lower bounds on $E^{\rm QM}(N,a)$ in terms of $E^{\rm GP}(N,a)$ with controlled errors. Theorem I.2 follows from Theorem I.1 by variation of the external potential. The upper bound is obtained in Sec. III by a variational calculation that generalizes the upper bound of Dyson [7] for a homogeneous gas of hard spheres. We also derive an upper bound on the chemical potential, i.e., the energy increase when one particle is added to the system. This upper bound is used in the proof of the lower bound of the energy in Sec. IV. The main ingredient for the lower bound, however, is the bound fpr the homogeneous case established in [5]. In addition, some basic properties of the minimizer of the GP functional are used in the proof and we consider them next.

II. THE GROSS-PITAEVSKII ENERGY FUNCTIONAL

The GP functional is defined by Eq. (1.4) for $\Phi\in\mathcal{D}$ with

$$\mathcal{D}=\{\Phi:\nabla\Phi\in L^2(\mathbb{R}^3),V|\Phi|^2\in L^1(\mathbb{R}^3),$$
$$\Phi\in L^4(\mathbb{R}^3)\cap L^2(\mathbb{R}^3)\}, \tag{2.1}$$

where $f\in L^p(\mathbb{R}^n)$ means $\int_{\mathbb{R}^n}|f(\mathbf{x})|^p d\mathbf{x}<\infty$. The corresponding GP energy is given by

$$E^{\rm GP}(N,a)=\inf\{\mathcal{E}^{\rm GP}[\Phi]:\quad \Phi\in\mathcal{D}_N\} \tag{2.2}$$

with

$$\mathcal{D}_N=\mathcal{D}\cap\left\{\Phi:\int|\Phi(\mathbf{x})|^2d\mathbf{x}=N\right\}. \tag{2.3}$$

The basic facts about the GP functional are summarized in the following theorem.

Theorem II.1 (Existence and properties of a minimizer). The infimum in Eq. (2.2) is a minimum, i.e., there is a $\Phi^{\rm GP}\in\mathcal{D}_N$ such that $E^{\rm GP}(N,a)=\mathcal{E}^{\rm GP}[\Phi^{\rm GP}]$. This $\Phi^{\rm GP}$ is unique up to a phase factor, which can be chosen so that $\Phi^{\rm GP}$ is strictly positive. $\Phi^{\rm GP}$ is at least once continuously differentiable, and if V is C^∞ then also $\Phi^{\rm GP}$ is C^∞. The energy $E^{\rm GP}(N,a)$ is continuously differentiable in a and hence [by Eq. (1.6)] also in N. The minimizer $\Phi^{\rm GP}$ solves the Gross-Pitaevskii equation

$$-\nabla^2\Phi+V\Phi+8\pi a|\Phi|^2\Phi=\mu\Phi \tag{2.4}$$

(in the sense of distributions) with

$$\mu=dE^{\rm GP}(N,a)/dN=E^{\rm GP}(N,a)/N+4\pi a\bar{\rho}. \tag{2.5}$$

Here $\bar{\rho}$ is the mean density (1.9).

The GP energy functional is mathematically quite similar to the energy functional of Thomas–Fermi–von Weizsäcker theory and Theorem II.1 can be proved by the methods of Sec. VII in [8]. For completeness, the proof is given in Appendix A. With additional properties of V one can draw further conclusions about $\Phi^{\rm GP}$:

Proposition II.1 (Symmetry and monotonicity). If V is spherically symmetric and monotone increasing, then $\Phi^{\rm GP}$ is spherically symmetric and monotone decreasing.

Proof. Let Φ^* be the symmetric-decreasing rearrangement of $\Phi^{\rm GP}$ (see [9]). Then $\mathcal{E}^{\rm GP}[\Phi^*]\leq\mathcal{E}^{\rm GP}[\Phi^{\rm GP}]$. □

Proposition II.2 (Log concavity). If V is convex, then $\Phi^{\rm GP}$ is log concave, i.e., $\Phi^{\rm GP}(\mathbf{x})^\lambda\Phi^{\rm GP}(\mathbf{y})^{(1-\lambda)}\leq\Phi^{\rm GP}(\lambda\mathbf{x}+(1-\lambda)\mathbf{y})$, for all $\mathbf{x},\mathbf{y}\in\mathbb{R}^3$, $\lambda\in(0,1)$.

Proof. Using the Trotter product formula it suffices to show that the solutions $u(t,\mathbf{x})$ of the equations

$$\frac{\partial u}{\partial t}-\nabla^2u=0,\quad \frac{\partial u}{\partial t}+Vu=0,\quad \frac{\partial u}{\partial t}+8\pi au^3=\mu u$$

are log concave, if $u(0,\mathbf{x})$ is log concave. The first follows from the fact that the convolution of two log concave functions is log concave, the second follows easily from convexity of V, and the third is shown in [10]. □

The GP theory has a well defined limit if $Na \to \infty$. It is sometimes referred to as the "Thomas-Fermi limit" of GP theory because the gradient term vanishes in this limit. For simplicity we restrict ourselves to *homogeneous external potentials* V, i.e.,

$$V(\lambda \mathbf{x}) = \lambda^s V(\mathbf{x}) \tag{2.6}$$

for some $s>0$.

Theorem II.2 (Large Na limit). Let V be homogeneous of order s and let $\mathcal{F}$ be the functional

$$\mathcal{F}[\rho] = \int_{\mathbb{R}^3} [V(\mathbf{x})\rho(\mathbf{x}) + 4\pi a \rho(\mathbf{x})^2] d\mathbf{x} \tag{2.7}$$

with $\rho(\mathbf{x}) \geq 0$, $\mathbf{x} \in \mathbb{R}^3$. Let $F(N,a)$ be the infimum of $\mathcal{F}$ under the condition $\int \rho = N$. By scaling, $F(N,a) = NF(1,Na)$ and $F(1,Na) = (Na)^{s/(s+3)} F(1,1)$. In the limit $Na \to \infty$ we have

$$\lim_{Na \to \infty} \frac{E^{\mathrm{GP}}(1,Na)}{(Na)^{s/(s+3)}} = F(1,1). \tag{2.8}$$

The minimizing density of $\mathcal{F}$ under the condition $\int \rho = 1$ and with $a=1$ is given by

$$\rho_{1,1}^{\mathrm{F}}(\mathbf{x}) = (8\pi)^{-1} [\tilde{\mu} - V(\mathbf{x})]_+ \tag{2.9}$$

with $\tilde{\mu} = F(1,1) + 4\pi \int (\rho_{1,1}^{\mathrm{F}})^2$, and $[t]_+ = t$ for $t>0$ and 0 otherwise. Moreover,

$$\lim_{Na \to \infty} \rho_{1,Na}^{\mathrm{GP}}(\mathbf{x}) = \rho_{1,1}^{\mathrm{F}}(\mathbf{x}) \tag{2.10}$$

strongly in $L^2(\mathbb{R}^3)$.

Proof. Since $\mathcal{E}^{\mathrm{GP}}[\sqrt{\rho}] \geq \mathcal{F}[\rho]$ it is clear that $E^{\mathrm{GP}}(1,Na) \geq F(1,Na)$. For the converse we write ρ in the form $\rho(\mathbf{x}) = (Na)^{-3/(s+3)} \tilde{\rho}((Na)^{-1/(s+3)}\mathbf{x})$ and obtain

$$\mathcal{E}^{\mathrm{GP}}[\sqrt{\rho}] = (Na)^{s/(s+3)} \int [(Na)^{-(s+2)/(s+3)} |\nabla \sqrt{\tilde{\rho}}|^2 + V\tilde{\rho} + 4\pi \tilde{\rho}^2] d\mathbf{x},$$

$$\mathcal{F}[\rho] = (Na)^{s/(s+3)} \int (V\tilde{\rho} + 4\pi\tilde{\rho}^2) d\mathbf{x}.$$

In particular, $F(1,Na) = (Na)^{s/(s+3)} F(1,1)$, and with $\tilde{\rho} = \rho_{1,1}^{\mathrm{F}}$ we obtain

$$E^{\mathrm{GP}}(1,Na) \leq F(1,Na) + (Na)^{-2/(s+3)} \int |\nabla \sqrt{\rho_{1,1}^{\mathrm{F}}}|^2.$$

(After a slight regularization, we may assume that $\int |\nabla \sqrt{\rho_{1,1}^{\mathrm{F}}}|^2 < \infty$.) In the limit $Na \to \infty$ the gradient term vanishes, and thus the limit of the energies is proved. Now

$$\lim_{Na \to \infty} \frac{\mathcal{E}^{\mathrm{GP}}[\sqrt{\rho_{1,Na}^{\mathrm{GP}}}]}{(Na)^{s/(s+3)}} = \lim_{Na \to \infty} \left(\frac{\mathcal{F}[\rho_{1,Na}^{\mathrm{GP}}]}{(Na)^{s/(s+3)}} + (Na)^{-(s+2)/(s+3)} \int |\nabla \sqrt{\rho_{1,Na}^{\mathrm{GP}}}|^2 \right) = F(1,1).$$

Since $F(1,1)$ is the minimum of $\mathcal{F}/(Na)^{s/(s+3)}$, the second term vanishes for $Na \to \infty$, and it follows that $\rho_{1,Na}^{\mathrm{GP}}$ is a minimizing sequence for $\int (V\rho + 4\pi\rho^2)$. Since both terms in the functional are non-negative, they must converge individually; in particular, $\|\rho_{1,Na}^{\mathrm{GP}}\|_2$ converges to $\|\rho_{1,1}^{\mathrm{F}}\|_2$. On the other hand $\rho_{1,Na}^{\mathrm{GP}}$ converges weakly to $\rho_{1,1}^{\mathrm{F}}$ by uniqueness of the minimizer. Together with the convergence of the norms this implies strong convergence.

The solution of the variational equation for $\rho_{1,1}^{\mathrm{F}}$ is simply $\rho_{1,1}^{\mathrm{F}} = (8\pi)^{-1}[\tilde{\mu} - V]_+$ with $\tilde{\mu}$ given by $\tilde{\mu} = F(1,1) + 4\pi \int (\rho_{1,1}^{\mathrm{F}})^2$. □

Lemma II.1 (Virial theorem). When V is homogeneous of order s, as in Eq. (2.6), the minimizer of the GP functional satisfies

$$\frac{2}{3} \int |\nabla \Phi^{\mathrm{GP}}(\mathbf{x})|^2 d\mathbf{x} - \frac{s}{3} \int \Phi^{\mathrm{GP}}(\mathbf{x})^2 V(\mathbf{x}) d\mathbf{x} + 4\pi a \int \Phi^{\mathrm{GP}}(\mathbf{x})^4 d\mathbf{x} = 0. \tag{2.11}$$

Proof. Define Φ_k by

$$\Phi_k(\mathbf{x}) = k^{1/2} \Phi^{\mathrm{GP}}(k^{1/3}\mathbf{x}).$$

Because Φ^{GP} is the minimizer of $\mathcal{E}^{\mathrm{GP}}[\Phi]$, it must be true that

$$\left. \frac{\partial}{\partial k} \mathcal{E}^{\mathrm{GP}}[\Phi_k] \right|_{k=1} = 0.$$

This leads to the virial theorem (2.11). □

In the proof of the lower bound we shall also consider the GP energy functional in a finite box. For $R>0$ we denote by Λ_R a cube centered at the origin, with side length $2R$. The energy functional $\mathcal{E}_R^{\mathrm{GP}}$ in the box is simply Eq. (1.4) with the integration reduced to Λ_R. The corresponding minimizer, denoted by Φ_R^{GP}, satisfies Neumann conditions at the boundary of Λ_R, and is strictly positive. Analogously to Eqs. (1.7), (1.9), and (2.5) we define ρ_R^{GP}, $\bar{\rho}_R$, and μ_R. The corresponding energy will be denoted by $E_R^{\mathrm{GP}}(N,a)$. Then we have the following:

Lemma II.2 (GP energy in a box).

$$\lim_{R \to \infty} E_R^{\mathrm{GP}}(N,a) = E^{\mathrm{GP}}(N,a). \tag{2.12}$$

Proof. Using $N^{1/2}\Phi^{\mathrm{GP}}\chi_R / \|\Phi^{\mathrm{GP}}\chi_R\|_2$ as a test function for $\mathcal{E}_R^{\mathrm{GP}}$, where χ_R denotes the characteristic function of Λ_R, we immediately get

$$\lim_{R \to \infty} E_R^{\mathrm{GP}}(N,a) \leq E^{\mathrm{GP}}(N,a). \tag{2.13}$$

Let Θ_R be a C^∞ function on $\mathbb{R}^3$, with $\Theta_R=0$ outside Λ_R, and $\Theta_R=1$ inside Λ_{R-1}. We use $N^{1/2}\Phi_R^{\rm GP}\Theta_R/\|\Phi_R^{\rm GP}\Theta_R\|_2$ as a test function for $\mathcal{E}^{\rm GP}$. Since $\nabla\Theta_R$ is bounded and

$$\lim_{R\to\infty}\int_{\Lambda_R\setminus\Lambda_{R-1}}|\Phi_R^{\rm GP}|^2=0$$

[because V tends to infinity and $\mathcal{E}_R^{\rm GP}[\Phi_R^{\rm GP}]$ is bounded by Eq. (2.13)], we have

$$E^{\rm GP}(N,a)\leq\liminf_{R\to\infty}E_R^{\rm GP}(N,a).$$

This completes the proof of Eq. (2.12). □

III. UPPER BOUNDS

A. Upper bound for the quantum mechanical energy

It will now be shown that for all N and small values of $a\bar\rho^{1/3}$ [with $\bar\rho=\int\rho^{\rm GP}(\mathbf{x})^2d\mathbf{x}/N$, cf. Eq. (1.9)]

$$E^{\rm QM}(N,a)\leq E^{\rm GP}(N,a)[1+O(a\bar\rho^{1/3})]. \tag{3.1}$$

This upper bound, which holds for all positive, spherically symmetric v with finite scattering length, is derived by means of the variational principle. We generalize a method of Dyson [7], who proved an upper bound for the homogeneous Bose gas with hard-sphere interaction. Consider as a trial function

$$\Psi=F(\mathbf{x}_1,\ldots,\mathbf{x}_N)G(\mathbf{x}_1,\ldots,\mathbf{x}_N) \tag{3.2}$$

with

$$F(\mathbf{x}_1,\ldots,\mathbf{x}_N)=\prod_{i=1}^N F_i(\mathbf{x}_1,\ldots,\mathbf{x}_i),$$

$$G(\mathbf{x}_1,\ldots,\mathbf{x}_N)=\prod_{i=1}^N g(\mathbf{x}_i) \tag{3.3}$$

where

$$F_i(\mathbf{x}_1,\ldots,\mathbf{x}_i)=f(t_i),$$

$$t_i=\min\{|\mathbf{x}_i-\mathbf{x}_j|,j=1,\ldots,i-1\}, \tag{3.4}$$

with a function f satisfying

$$0\leq f\leq 1,\ f'\geq 0, \tag{3.5}$$

and

$$g(\mathbf{x})=\Phi^{\rm GP}(\mathbf{x})/\|\Phi^{\rm GP}\|_\infty. \tag{3.6}$$

The function f will be specified later. This trial function is not symmetric in the particle coordinates, but the expectation value $\langle\Psi|H^{(N)}\Psi\rangle/\langle\Psi|\Psi\rangle$ is still an upper bound to the bosonic ground-state energy because the Hamiltonian is symmetric and its ground-state wave function is positive. Hence the bosonic ground-state energy is equal to the *absolute* ground-state energy [7,11].

The physical meaning of the trial function can be understood as follows: The G part describes independent particles, each with the GP wave function. The F part means that the particles are inserted into the system one at a time, taking into account the particles previously inserted, but without adjusting their wave function (cf. [7]). Although a wave function of this form cannot describe all correlations present in the true ground state, it captures the leading term in the energy for dilute systems.

For the computation of the kinetic energy we use

$$\int_{\mathbb{R}^{3N}}\Psi\nabla_k^2\Psi=\int_{\mathbb{R}^{3N}}(G\nabla_k^2G)F^2-\int_{\mathbb{R}^{3N}}G^2|\nabla_kF|^2, \tag{3.7}$$

where ∇_k denotes the gradient with respect to $\mathbf{x}_k$, $k=1,\ldots,N$. We write

$$\epsilon_{ik}=\begin{cases}1 & \text{for } i=k\\ -1 & \text{for } t_i=|\mathbf{x}_i-\mathbf{x}_k|\\ 0 & \text{otherwise.}\end{cases} \tag{3.8}$$

Let n_i be the unit vector in the direction of $\mathbf{x}_i-\mathbf{x}_{j(i)}$, when $\mathbf{x}_{j(i)}$ is the nearest to $\mathbf{x}_i$ of the points $(\mathbf{x}_1,\ldots,\mathbf{x}_{i-1})$. [Note that $j(i)$ really depends on all the points $\mathbf{x}_1,\ldots,\mathbf{x}_i$ and not just on the index i. Except for a set of zero measure, $j(i)$ is unique.] Then

$$G\nabla_kF=\sum_i\Psi F_i^{-1}\epsilon_{ik}n_if'(t_i), \tag{3.9}$$

and after summation over k

$$\begin{aligned}\sum_k G^2|\nabla_kF|^2&=|\Psi|^2\sum_{i,j,k}\epsilon_{ik}\epsilon_{jk}(n_i\cdot n_j)F_i^{-1}F_j^{-1}f'(t_i)f'(t_j)\\&\leq 2|\Psi|^2\sum_i F_i^{-2}f'(t_i)^2+2|\Psi|^2\\&\quad\times\sum_{k\leq i<j}|\epsilon_{ik}\epsilon_{jk}|F_i^{-1}F_j^{-1}f'(t_i)f'(t_j).\end{aligned} \tag{3.10}$$

The expectation value can thus be bounded as follows:

$$\begin{aligned}\frac{\langle\Psi|H^{(N)}\Psi\rangle}{\langle\Psi|\Psi\rangle}\leq{}&2\sum_{i=1}^N\frac{\int|\Psi|^2F_i^{-2}f'(t_i)^2}{\int|\Psi|^2}\\&+\sum_{j<i}\frac{\int|\Psi|^2v(|\mathbf{x}_i-\mathbf{x}_j|)}{\int|\Psi|^2}\\&+2\sum_{k\leq i<j}\frac{\int|\Psi|^2|\epsilon_{ik}\epsilon_{jk}|F_i^{-1}F_j^{-1}f'(t_i)f'(t_j)}{\int|\Psi|^2}\\&+\sum_{i=1}^N\frac{\int|\Psi|^2[-g(\mathbf{x}_i)^{-1}\nabla_i^2g(\mathbf{x}_i)+V(\mathbf{x}_i)]}{\int|\Psi|^2}.\end{aligned} \tag{3.11}$$

For $i<p$, let $F_{p,i}$ be the value that F_p would take if the point $\mathbf{x}_i$ were omitted, i.e.,

$$F_{p,i}=f(|\mathbf{x}_p-\mathbf{x}_{k(p)}|), \qquad (3.12)$$

where $\mathbf{x}_{k(p)}$ is the nearest to $\mathbf{x}_p$ of the points $(\mathbf{x}_1,\ldots,\mathbf{x}_{i-1},\mathbf{x}_{i+1},\ldots,\mathbf{x}_{p-1})$. The reason for introducing these functions is that one wants to decouple the integration over $\mathbf{x}_i$ from the integrations over the other variables. (Note that $F_{p,i}$ is independent of $\mathbf{x}_i$.) Analogously, one defines $F_{p,ij}$ by omitting $\mathbf{x}_i$ and $\mathbf{x}_j$. This simultaneously decouples $\mathbf{x}_i$ and $\mathbf{x}_j$ from the other variables. The functions F_i occur both in the numerator and the denominator so one needs estimates from below and above. Since

$$F_p=\min\{F_{p,ij},f(|\mathbf{x}_p-\mathbf{x}_j|),f(|\mathbf{x}_p-\mathbf{x}_i|)\}, \qquad (3.13)$$

one has (recall that $f\leq 1$)

$$F_{p,ij}^2 f(|\mathbf{x}_p-\mathbf{x}_i|)^2 f(|\mathbf{x}_p-\mathbf{x}_j|)^2\leq F_p^2\leq F_{p,ij}^2. \qquad (3.14)$$

Hence, with $j<i$,

$$F_{j+1}^2\cdots F_{i-1}^2F_{i+1}^2\cdots F_N^2\leq F_{j+1,j}^2\cdots F_{i-1,j}^2F_{i+1,ij}^2\cdots F_{N,ij}^2 \qquad (3.15)$$

and

$$F_j^2\cdots F_N^2\geq F_{j+1,j}^2\cdots F_{i-1,j}^2F_{i+1,ij}^2\cdots F_{N,ij}^2 \times\left(1-\sum_{k=1,k\neq i,j}^N[1-f(|\mathbf{x}_j-\mathbf{x}_k|)^2]\right) \times\left(1-\sum_{k=1,k\neq i}^N[1-f(|\mathbf{x}_i-\mathbf{x}_k|)^2]\right). \qquad (3.16)$$

We now consider the first two terms in Eq. (3.11). In the numerator of the first term for each fixed i we use the estimate

$$f'(t_i)^2\leq\sum_{j=1}^{i-1}f'(|\mathbf{x}_i-\mathbf{x}_j|)^2, \qquad (3.17)$$

and in the second term we use $F_i\leq f(|\mathbf{x}_i-\mathbf{x}_j|)$. For fixed i and j one eliminates $\mathbf{x}_i$ and $\mathbf{x}_j$ from the rest of the integrand by using Eq. (3.15) and $F_j\leq 1$ in the numerator and Eq. (3.16) in the denominator to do the $\mathbf{x}_i$ and $\mathbf{x}_j$ integrations. With the transformation $\boldsymbol{\eta}=\mathbf{x}_i-\mathbf{x}_j$, $\boldsymbol{\chi}=(\mathbf{x}_i+\mathbf{x}_j)/2$ one gets

$$\int[2f'(|\mathbf{x}_i-\mathbf{x}_j|)^2+v(|\mathbf{x}_i-\mathbf{x}_j|)f(|\mathbf{x}_i-\mathbf{x}_j|)^2]\times g(\mathbf{x}_i)^2g(\mathbf{x}_j)^2d\mathbf{x}_id\mathbf{x}_j =\int[2f'(|\boldsymbol{\eta}|)^2+v(|\boldsymbol{\eta}|)f(|\boldsymbol{\eta}|)^2]\times g\left(\boldsymbol{\chi}+\frac{1}{2}\boldsymbol{\eta}\right)^2g\left(\boldsymbol{\chi}-\frac{1}{2}\boldsymbol{\eta}\right)^2d\boldsymbol{\eta}d\boldsymbol{\chi}. \qquad (3.18)$$

By the Cauchy-Schwarz inequality

$$\int g\left(\boldsymbol{\chi}+\frac{1}{2}\boldsymbol{\eta}\right)^2g\left(\boldsymbol{\chi}-\frac{1}{2}\boldsymbol{\eta}\right)^2d\boldsymbol{\chi}\leq\int g(\boldsymbol{\chi})^4d\boldsymbol{\chi}, \qquad (3.19)$$

and one obtains

$$\int[2f'(|\mathbf{x}_i-\mathbf{x}_j|)^2+v(|\mathbf{x}_i-\mathbf{x}_j|)f(|\mathbf{x}_i-\mathbf{x}_j|)^2]\times g(\mathbf{x}_i)^2g(\mathbf{x}_j)^2d\mathbf{x}_id\mathbf{x}_j \leq\int g(\boldsymbol{\chi})^4d\boldsymbol{\chi}\int[2f'(|\boldsymbol{\eta}|)^2+v(|\boldsymbol{\eta}|)f(|\boldsymbol{\eta}|)^2]d\boldsymbol{\eta} \equiv 2\int g(\boldsymbol{\chi})^4d\boldsymbol{\chi}J. \qquad (3.20)$$

In the denominator one gets, using that $0\leq g\leq 1$,

$$\int\left(1-\sum_{p=1,p\neq i}^N[1-f(|\mathbf{x}_p-\mathbf{x}_i|)^2]\right)g(\mathbf{x}_i)^2d\mathbf{x}_i \geq\int g(\mathbf{x})^2d\mathbf{x}-N\int[1-f(|\mathbf{x}_p-\mathbf{x}_i|)^2] \equiv\int g(\mathbf{x})^2d\mathbf{x}-NI. \qquad (3.21)$$

The same factor comes from the $\mathbf{x}_j$ integration, the remaining factors are identical in numerator and denominator, and so finally the first and second term are bounded by

$$\sum_{i=1}^N(i-1)\frac{2\int g(\mathbf{x})^4d\mathbf{x}J}{\left[\int g(\mathbf{x})^2d\mathbf{x}-NI\right]^2}\leq N^2\frac{\int g(\mathbf{x})^4d\mathbf{x}J}{\left[\int g(\mathbf{x})^2d\mathbf{x}-NI\right]^2}. \qquad (3.22)$$

A similar argument is now applied to the third term of Eq. (3.11). Note that the contributions from $k=i$ and $k<i$ are the same. Therefore

$$\sum_{k=1}^i\int|\epsilon_{ik}|f(t_i)f'(t_i)|\epsilon_{jk}|f(t_j)f'(t_j)g(\mathbf{x}_i)^2g(\mathbf{x}_j)^2d\mathbf{x}_id\mathbf{x}_j \leq 2\sum_{k=i}^{i-1}\int f(|\mathbf{x}_i-\mathbf{x}_k|)f'(|\mathbf{x}_i-\mathbf{x}_k|)f(|\mathbf{x}_j-\mathbf{x}_k|) \times f'(|\mathbf{x}_j-\mathbf{x}_k|)g(\mathbf{x}_i)^2g(\mathbf{x}_j)^2d\mathbf{x}_id\mathbf{x}_j. \qquad (3.23)$$

With $g\leq 1$ one gets

$$2\sum_{k=1}^{i-1}\int f(|\mathbf{x}_i|)f'(|\mathbf{x}_i|)f(|\mathbf{x}_j|)f'(|\mathbf{x}_j|)d\mathbf{x}_id\mathbf{x}_j =2(i-1)\left(\int f(|\mathbf{x}|)f'(|\mathbf{x}|)d\mathbf{x}\right)^2 \equiv 2(i-1)K^2. \qquad (3.24)$$

The summation over i and j gives

$$\sum_{j=1}^{N}\sum_{i=1}^{j-1}(i-1)=\frac{1}{6}N(N-1)(N-2). \tag{3.25}$$

In the denominator one gets the same factors as above, and so the third term is bounded by

$$\frac{2}{3}N^3\frac{K^2}{\left[\int g(\mathbf{x})^2 d\mathbf{x}-NI\right]^2}. \tag{3.26}$$

Next consider the last term of Eq. (3.11). Define $\tilde{e}$ by

$$\int [-g(\mathbf{x})\nabla^2 g(\mathbf{x})+V(\mathbf{x})g(\mathbf{x})^2]d\mathbf{x}\equiv\tilde{e}\int g(\mathbf{x})^2 d\mathbf{x}. \tag{3.27}$$

After eliminating $\mathbf{x}_i$ from the integrands in the numerator and the denominator and using $F_i\leq 1$ one sees that the term is bounded above by

$$N\frac{\tilde{e}\int g(\mathbf{x})^2 d\mathbf{x}}{\int g(\mathbf{x})^2 d\mathbf{x}-NI}. \tag{3.28}$$

Putting all terms together we obtain as an upper bound for the ground-state energy:

$$\frac{E^{\rm QM}}{N}\leq\frac{\tilde{e}\int g(\mathbf{x})^2 d\mathbf{x}}{\int g(\mathbf{x})^2 d\mathbf{x}-NI}+N\frac{\int g(\mathbf{x})^4 d\mathbf{x}J}{\left[\int g(\mathbf{x})^2 d\mathbf{x}-NI\right]^2}+\frac{2}{3}N^2\frac{K^2}{\left[\int g(\mathbf{x})^2 d\mathbf{x}-NI\right]^2} \tag{3.29}$$

with I, J, K, and $\tilde{e}$ defined by Eqs. (3.21), (3.20), (3.24), and (3.27). It remains to choose f. We take for $b>a$ (we shall soon fix b)

$$f(r)=\begin{cases}(1+\epsilon)u(r)/r & \text{for } r\leq b\\ 1 & \text{for } r>b\end{cases} \tag{3.30}$$

where $u(r)$ is the solution of the scattering equation

$$-u''(r)+\frac{1}{2}v(r)u(r)=0 \tag{3.31}$$

with

$$u(0)=0,\quad \lim_{r\to\infty}u'(r)=1, \tag{3.32}$$

and ϵ is determined by requiring f to be continuous. Convexity of u gives

$$r\geq u(r)\geq\begin{cases}0 & \text{for } r\leq a\\ r-a & \text{for } r>a,\end{cases}$$

$$1\geq u'(r)\geq\begin{cases}0 & \text{for } r\leq a\\ 1-\dfrac{a}{r} & \text{for } r>a.\end{cases} \tag{3.33}$$

These estimates imply

$$J\leq(1+\epsilon)^2 4\pi a, \tag{3.34}$$

$$I\leq 4\pi\left(\frac{a^3}{3}+ab(b-a)\right), \tag{3.35}$$

$$K\leq 4\pi(1+\epsilon)a\left(b-\frac{a}{2}\right), \tag{3.36}$$

$$0\leq\epsilon\leq\frac{a}{b-a}. \tag{3.37}$$

For Eq. (3.34) we used partial integration. By definition [Eq. (1.9)]

$$\bar{\rho}=\frac{1}{N}\int(\rho^{\rm GP})^2=N\frac{\int g^4}{\left(\int g^2\right)^2}, \tag{3.38}$$

and we choose b such that

$$\frac{4\pi}{3}\bar{\rho}b^3=\frac{\int g^4}{\int g^2}=\frac{\bar{\rho}}{\|\rho^{\rm GP}\|_\infty}\equiv c. \tag{3.39}$$

With this choice the factor in the denominators in Eq. (3.29) is bounded by

$$\int g^2-NI\geq\int g^2\left(1-\frac{a}{b}\right)^3. \tag{3.40}$$

Note that $c\leq 1$, and $a<b$ holds provided

$$\frac{a^3}{b^3}=\frac{4\pi}{3}a^3\|\rho^{\rm GP}\|_\infty<1. \tag{3.41}$$

Collecting the estimates Eqs. (3.34)–(3.37), we finally obtain the following theorem.

Theorem III.1 (Upper bound for the QM energy).

$$E^{\mathrm{QM}} \leq \frac{\int [|\nabla \Phi^{\mathrm{GP}}|^2 + V(\Phi^{\mathrm{GP}})^2]}{\left(1-\frac{a}{b}\right)^3} + 4\pi a \int (\Phi^{\mathrm{GP}})^4 \frac{1+\frac{2}{c}\left(\frac{a}{b}\right)-\frac{2}{c}\left(\frac{a}{b}\right)^2+\frac{1}{2c}\left(\frac{a}{b}\right)^3}{\left(1-\frac{a}{b}\right)^8} \tag{3.42}$$

with b and c defined by Eq. (3.39).

Remark III.1 (Negative potentials with hard core). Equation (3.1) can be extended to include partially negative potentials of the form

$$v(r) = \begin{cases} \infty & \text{for } 0 \leq r \leq d \\ -|w(r)| & \text{for } d < r \leq R_0 \\ 0 & \text{for } r > R_0, \end{cases} \tag{3.43}$$

as long as $f'(r)^2 + \frac{1}{2}v(r)f(r)^2 \geq 0$ for all r. With f from Eq. (3.30), this is the case for sufficiently shallow potentials. The potential energy is then negative, and the estimates used for Eq. (3.22) are no longer valid. But

$$\sum_{j<i} v(|\mathbf{x}_i - \mathbf{x}_j|)F_i^2 \leq \sum_i v(t_i)F_i^2, \tag{3.44}$$

and because of $2f'(r)^2 + v(r)f(r)^2 \geq 0$ we get

$$2f'(t_i)^2 + v(t_i)f(t_i)^2 \leq \sum_{j=1}^{i-1} 2f'(|\mathbf{x}_i - \mathbf{x}_j|)^2 + v(|\mathbf{x}_i - \mathbf{x}_j|)f(|\mathbf{x}_i - \mathbf{x}_j|)^2, \tag{3.45}$$

which replaces Eq. (3.17). Note that for potentials as in Eq. (3.43) f satisfies Eq. (3.5), as long as $a > 0$.

Remark III.2 (Homogeneous gas). For the special case of a homogeneous Bose gas (i.e., $V=0$) in a box of volume $\mathcal{V}$, the GP density is simply

$$\rho^{\mathrm{GP}}(\mathbf{x}) = N/\mathcal{V} = \bar{\rho}, \tag{3.46}$$

for all $\mathbf{x}$ in the box, and the GP energy is given by

$$E^{\mathrm{GP}}(N,a) = 4\pi a \frac{N^2}{\mathcal{V}}. \tag{3.47}$$

Our method also applies here, if we impose periodic boundary conditions on the box. Therefore our upper bound is a generalization of a result by Dyson [7], who proved an analogous bound for the special case of a homogeneous Bose gas with hard-sphere interaction.

B. Upper bound for the chemical potential

By the same method as in the previous subsection one can derive a bound on the increase of the energy when one particle is added to the system. This bound will be needed for the derivation of the lower bound to the energy.

Theorem III.2 (Upper bound for the chemical potential). Let $E^*(N,a)$ denote the infimum of the functional

$$\mathcal{E}^*[\Phi] = \int [|\nabla \Phi(\mathbf{x})|^2 + V(\mathbf{x})|\Phi(\mathbf{x})|^2 + 8\pi a \|\Phi\|_\infty^2 |\Phi(\mathbf{x})|^2] d\mathbf{x} \tag{3.48}$$

with $\int |\Phi|^2 = N$. Let Φ^* be the positive minimizer of $\mathcal{E}^*$ (its existence is guaranteed by the same arguments as for the GP functional itself), and $\bar{\rho}^* = \int \Phi^{*4}/N$. Then

$$E^{\mathrm{QM}}(N+1,a) \leq E^{\mathrm{QM}}(N,a) + E^*(1,Na)[1 + O(a\bar{\rho}^{*1/3})]. \tag{3.49}$$

Proof. Let $\Psi_0^{(N)}$ be the ground-state wave function of $H^{(N)}$. As test wave function for $H^{(N+1)}$ we take

$$\Psi(\mathbf{x}_1, \ldots, \mathbf{x}_{N+1}) = \Psi_0^{(N)}(\mathbf{x}_1, \ldots, \mathbf{x}_N)\Phi^*(\mathbf{x}_{N+1})f(t_{N+1}), \tag{3.50}$$

where f and t_{N+1} are defined as in Eqs. (3.30) and (3.4), i.e., $rf(r)$ is essentially the zero-energy scattering solution and t_{N+1} is the distance of $\mathbf{x}_{N+1}$ from its nearest neighbor. We have

$$\langle \Psi | H^{(N+1)} \Psi \rangle = \int f^2 \Phi^{*2} \left(\sum_{i=1}^N (-\overline{\Psi_0^{(N)}} \nabla_i^2 \Psi_0^{(N)} + V(\mathbf{x}_i)|\Psi_0^{(N)}|^2) + \sum_{i<j}^N v(|\mathbf{x}_i - \mathbf{x}_j|)|\Psi_0^{(N)}|^2 \right) + \int |\Psi_0^{(N)}|^2 \Phi^{*2} \left(\sum_{i=1}^N |\nabla_i f|^2 \right)$$
$$+ \int |\Psi_0^{(N)}|^2 f^2 [-\Phi^* \nabla_{N+1}^2 \Phi^* + V(\mathbf{x}_{N+1})\Phi^{*2}] + \int |\Psi_0^{(N)}|^2 \Phi^{*2} \left(|\nabla_{N+1} f|^2 + \sum_{i=1}^N v(|\mathbf{x}_{N+1} - \mathbf{x}_i|) f^2 \right). \tag{3.51}$$

For f one uses the estimates

$$f \leq 1, \quad f(t_{N+1})^2 \geq 1 - \sum_{i=1}^N [1 - f(|\mathbf{x}_{N+1} - \mathbf{x}_i|)^2], \tag{3.52}$$

and for the derivatives one has

$$|\nabla_{N+1} f|^2 = f'(t_{N+1})^2 = \sum_{i=1}^{N} |\nabla_i f|^2 \tag{3.53}$$

and

$$f'(t_{N+1})^2 \leq \sum_{i=1}^{N} f'(|\mathbf{x}_{N+1} - \mathbf{x}_i|)^2. \tag{3.54}$$

After division by the norm of Ψ Eq. (3.51) becomes

$$E^{\mathrm{QM}}(N+1,a) \leq E^{\mathrm{QM}}(N,a) + \frac{\int |\Psi_0^{(N)}|^2(-\Phi^* \nabla_{N+1}^2 \Phi^* + V\Phi^{*2})}{\int |\Psi_0^{(N)}|^2 \left[N - N\Phi^{*2} \int (1-f^2) \right]}$$
$$+ \frac{\int |\Psi_0^{(N)}|^2 \|\Phi^*\|_\infty^2 \sum_{i=1}^{N} [2f'(|\mathbf{x}_{N+1}-\mathbf{x}_i|)^2 + v(|\mathbf{x}_{N+1}-\mathbf{x}_i|) f(|\mathbf{x}_{N+1}-\mathbf{x}_i|)^2]}{\int |\Psi_0^{(N)}|^2 \left[N - N\|\Phi^*\|_\infty^2 \int (1-f^2) \right]}. \tag{3.55}$$

$\Psi_0^{(N)}$ does not depend on $\mathbf{x}_{N+1}$. One integrates first over $\mathbf{x}_{N+1}$ and then over the remaining variables. In analogy with the estimates (3.34) and (3.40) one gets

$$\int [2f'(|\mathbf{x}|)^2 + v(|\mathbf{x}|) f(|\mathbf{x}|)^2] d\mathbf{x} \leq 8\pi a[1 + O(a\bar{\rho}^{*1/3})] \tag{3.56}$$

and

$$\|\Phi^*\|_\infty^2 \int [1 - f(|\mathbf{x}|)^2] d\mathbf{x} \leq O(a\bar{\rho}^{*1/3}). \tag{3.57}$$

This implies

$$E^{\mathrm{QM}}(N+1,a) \leq E^{\mathrm{QM}}(N,a) + \frac{E^*(N,a)}{N}[1 + O(a\bar{\rho}^{*1/3})]. \tag{3.58}$$

By scaling, $E^*(N,a) = NE^*(1,Na)$ and Eq. (3.49) follows. □

Remark III.3 (Box with Neumann conditions). Equation (3.49) also holds for a box with Neumann boundary conditions. To see this we note that Neumann conditions give the lowest energy for the quadratic form $\langle \Psi | H^{(N+1)} \Psi \rangle$; therefore it is possible to use Eq. (3.50) as a test function, even if f does not fulfill Neumann conditions. If $\Psi_0^{(N)}$ and Φ^* do, the calculation above is still valid, since for

$$\int |\nabla(gf)|^2 = -\int f^2 g \nabla^2 g + \int g^2 |\nabla f|^2 \tag{3.59}$$

only boundary conditions for g are needed.

Note also that in the homogeneous case, i.e., $V=0$ in the box, $E^*(N,a) = 2E^{\mathrm{GP}}(N,a)$.

IV. LOWER BOUNDS

A. The homogeneous case

In [5] the following lower bound was established for the ground-state energy, E^{hom}, of a Bose gas in a box of side length L with Neumann boundary conditions and v of finite range:

$$E^{\mathrm{hom}}(N,L) \geq 4\pi a \frac{N^2}{L^3}(1 - CY^{1/17}) \tag{4.1}$$

with $Y = a^3 N/L^3$ and C a constant. The estimate holds for all Y small enough and $L/a \gg Y^{-6/17}$ (note that this implies $N \gg Y^{-1/17}$). In the thermodynamic limit the constant C can be taken to be $C = 8.9$, but this value is only of academic interest, because the error term $-CY^{1/17}$ is not believed to reflect the true state of affairs. Presumably, it does not even have the correct sign.

The restriction of a finite range can be relaxed. In fact, Eq. (4.1) holds (with a different constant C) for all positive, spherically symmetric v with

$$v(r) \leq \mathrm{const} \times r^{-(3+\epsilon+1/5)} \quad \text{for } r \text{ large enough}, \quad \epsilon > 0. \tag{4.2}$$

More generally, if

$$v(r) \leq \mathrm{const} \times r^{-(3+\epsilon)} \quad \text{for } r \text{ large enough}, \quad \epsilon > 0, \tag{4.3}$$

then Eq. (4.1) holds at least with the exponent 1/17 replaced by an exponent $O(\epsilon)$. We prove these assertions in Appendix B.

In the next section we shall stick to the estimate (4.1) for simplicity, but in the limit $N\to\infty$ the explicit form of the error term is not significant so a decrease of the potential as in Eq. (4.3) is sufficient for the limit Theorems I.1 and I.2.

B. The lower bound in the inhomogeneous case

Our generalization of Eq. (4.1) to the inhomogeneous case is as follows:

Theorem IV.1 (Lower bound for the QM energy). Let v be positive, spherically symmetric, and decrease at infinity like Eq. (4.2). Its scattering length is $a=a_1/N$ with a_1 fixed, as explained in the Introduction. Then as $N\to\infty$

$$E^{\rm QM}(N,a)\geq E_R^{\rm GP}(N,a)(1-{\rm const}\times N^{-1/10}) \tag{4.4}$$

for all R large enough, where $E_R^{\rm GP}$ is the GP energy in a cube with side length $2R$, center at the origin, and Neumann boundary conditions; the constant in Eq. (4.4) depends only on a_1 and R.

Proof. As in [5] the lower bound will be obtained by dividing space into cubic boxes with Neumann conditions at the boundary, which only lowers the energy. Moreover, interactions among particles in different boxes are dropped. Since $v\geq 0$, this, too, lowers the energy. For the lower bound one has to estimate the energy for a definite particle number in each box and then optimize over all distributions of the N particles among the boxes.

Step 1 (Finite box): The first step is to show that all the particles can be assumed to be in some large but finite box. Since

$$K(R)=\inf_{|\mathbf{x}|>R} V(\mathbf{x}) \tag{4.5}$$

tends monotonically to ∞ with R, one knows that the energy of a particle outside a cube Λ_R of side length $2R$ and center at the origin is at least $K(R)$. Hence

$$E^{\rm QM}(N,a)\geq \inf_{0\leq n\leq N}\{E_R^{\rm QM}(N-n,a)+nK(R)\}, \tag{4.6}$$

where $E_R^{\rm QM}(N-n,a)$ denotes the energy of $N-n$ particles in Λ_R, with Neumann conditions at the boundary. We now apply Theorem III.2 (which holds also in a cube with Neumann conditions). Applying the theorem n times and noting that $E^*(1,Na)$ is monotone in N we have

$$E_R^{\rm QM}(N-n,a)\geq E_R^{\rm QM}(N,a)-nE^*(1,Na)[1+O(a\bar\rho^{*1/3})]. \tag{4.7}$$

Hence there is a constant K' (that depends only on Na), such that $K(R)>K'$ implies that the infimum is obtained at $n=0$. This is fulfilled for all sufficiently large R, independently of N if Na is fixed. So we can restrict ourselves to estimating the energy in Λ_R with Neumann boundary conditions.

Step 2 (Trading V for $-\rho_R^{\rm GP}$): We shall now use the GP equation to eliminate V from the problem, effectively replacing it by $-8\pi a\rho_R^{\rm GP}$. We write the wave function in Λ_R^N as

$$\Psi(\mathbf{x}_1,\dots,\mathbf{x}_N)=f(\mathbf{x}_1,\dots,\mathbf{x}_N)\prod_{i=1}^N \Phi_R^{\rm GP}(\mathbf{x}_i), \tag{4.8}$$

where $\Phi_R^{\rm GP}$ denotes the the minimizer of the GP functional in Λ_R; since it is strictly positive, every wave function can be written in this form. Note also that $\Phi_R^{\rm GP}$ and f obey Neumann conditions. We have

$$\begin{aligned}\langle\Psi|H\Psi\rangle=&\sum_{i=1}^N\int|\Psi|^2\Phi_R^{\rm GP}(\mathbf{x}_i)^{-1}\\&\times[-\nabla_i^2+V(\mathbf{x}_i)]\Phi_R^{\rm GP}(\mathbf{x}_i)d^N\mathbf{x}\\&+\sum_{i=1}^N\int\prod_{k=1}^N\Phi_R^{\rm GP}(\mathbf{x}_k)^2|\nabla_i f|^2d^N\mathbf{x}\\&+\sum_{i<j}^N\int|\Psi|^2v(|\mathbf{x}_i-\mathbf{x}_j|)d^N\mathbf{x},\end{aligned} \tag{4.9}$$

where the integrals are over Λ_R^N. Using the GP equation (2.4) this becomes

$$\begin{aligned}\langle\Psi|H\Psi\rangle=\sum_{i=1}^N\int\prod_{k=1}^N\rho_R^{\rm GP}(\mathbf{x}_k)\Bigg[&|\nabla_i f|^2\\&+\left(\mu_R-8\pi a\rho_R^{\rm GP}(\mathbf{x}_i)+\sum_{j=1}^{i-1}v(|\mathbf{x}_i-\mathbf{x}_j|)\right)|f|^2\Bigg].\end{aligned} \tag{4.10}$$

Inserting the value (2.5) for μ_R gives

$$\frac{\langle\Psi|H\Psi\rangle}{\langle\Psi|\Psi\rangle}-E_R^{\rm GP}=4\pi a\bar\rho_R N+Q(f) \tag{4.11}$$

with

$$Q(f)=\sum_{i=1}^N\frac{\int_{\Lambda_R^N}\prod_{k=1}^N\rho_R^{\rm GP}(\mathbf{x}_k)\left(|\nabla_i f|^2+\sum_{j=1}^{i-1}v(|\mathbf{x}_i-\mathbf{x}_j|)|f|^2-8\pi a\rho_R^{\rm GP}(\mathbf{x}_i)|f|^2\right)}{\int_{\Lambda_R^N}\prod_{k=1}^N\rho_R^{\rm GP}(\mathbf{x}_k)|f|^2}. \tag{4.12}$$

Step 3 (division into boxes): $Q(f)$ is a normalized quadratic form on the weighted L^2 space $L^2(\Lambda_R^N,\Pi_{k=1}^N\rho_R^{\rm GP}(\mathbf{x}_k)d\mathbf{x}_k)$, and can be minimized by dividing the cube Λ_R into smaller cubes with side length L, labeled by α, distributing the N particles among the boxes, and optimizing over all distributions. We therefore have

$$\inf_f Q(f) \geq \inf_{\{n_\alpha\}} \sum_\alpha \inf_{f_\alpha} Q_\alpha(f_\alpha), \tag{4.13}$$

where the infimum is taken over all distributions of the particles with $\Sigma n_\alpha = N$, and $Q_\alpha(f)$ is given by

$$Q_\alpha(f) = \sum_{i=1}^{n_\alpha} \frac{\int_\alpha \prod_{k=1}^{n_\alpha} \rho_R^{\mathrm{GP}}(\mathbf{x}_k)\left(|\nabla_i f|^2 + \sum_{j=1}^{i-1} v(|\mathbf{x}_i - \mathbf{x}_j|)|f|^2 - 8\pi a \rho_R^{\mathrm{GP}}(\mathbf{x}_i)|f|^2\right)}{\int_\alpha \prod_{k=1}^{n_\alpha} \rho_R^{\mathrm{GP}}(\mathbf{x}_k)|f|^2}, \tag{4.14}$$

where the integrals are over $\mathbf{x}_k$ in the box α, $k = 1, \ldots, n_\alpha$. Note that here $f = f(\mathbf{x}_1, \ldots, \mathbf{x}_{n_\alpha})$, and Eq. (4.14) is the same as Eq. (4.12) with N replaced by n_α and Λ_R with the box α.

We now want to use Eq. (4.1) and therefore we must approximate ρ_R^{GP} by constants in each box. Let $\rho_{\alpha,\max}$ and $\rho_{\alpha,\min}$, respectively, denote the maximal and minimal values of ρ_R^{GP} in box α. With

$$\Phi^{(i)}(\mathbf{x}_1, \ldots, \mathbf{x}_{n_\alpha}) = f(\mathbf{x}_1, \ldots, \mathbf{x}_{n_\alpha}) \prod_{k=1, k\neq i}^{n_\alpha} \Phi_R^{\mathrm{GP}}(\mathbf{x}_k), \tag{4.15}$$

one has

$$\frac{\int \prod_k \rho_R^{\mathrm{GP}}(\mathbf{x}_k)\left(|\nabla_i f|^2 + \sum_{j=1}^{i-1} v(|\mathbf{x}_i - \mathbf{x}_j|)|f|^2\right)}{\int \prod_k \rho_R^{\mathrm{GP}}(\mathbf{x}_k)|f|^2} \geq \frac{\rho_{\alpha,\min}}{\rho_{\alpha,\max}} \frac{\int |\nabla_i \Phi^{(i)}|^2 + \sum_{j=1}^{i-1} v(|\mathbf{x}_i - \mathbf{x}_j|)|\Phi^{(i)}|^2}{\int |\Phi^{(i)}|^2}. \tag{4.16}$$

This holds for all i, and if we use $\rho_R^{\mathrm{GP}}(\mathbf{x}_i) \leq \rho_{\alpha,\max}$ in Eq. (4.14), we get

$$Q_\alpha(f) \geq \frac{\rho_{\alpha,\min}}{\rho_{\alpha,\max}} E^{\mathrm{hom}}(n_\alpha, L) - 8\pi a \rho_{\alpha,\max} n_\alpha, \tag{4.17}$$

where E^{hom} is the energy in a box without an external potential.

Remark: If we had not taken Step 2 and used instead the division into boxes directly on the original Hamiltonian (1.1) we would be considering the minimization of

$$\sum_\alpha E^{\mathrm{hom}}(n_\alpha, L) + V_{\alpha,\min} n_\alpha. \tag{4.18}$$

Such a procedure, however, would not lead to the GP energy. To see this, consider the special case of no interaction, i.e., $v = 0$ and hence also $E^{\mathrm{hom}}(n_\alpha, L) = 0$. The minimum of Eq. (4.18) is then simply $N \min_{\mathbf{x}} V(\mathbf{x})$, whereas the GP energy is in this case N times the ground-state energy of $-\nabla^2 + V$.

Step 4 (Minimizing in each box): Dropping the subsidiary condition $\Sigma n_\alpha = N$ can only lower the infimum. Hence it is sufficient to minimize each Q_α separately. To justify the use of Eq. (4.1), we have to ensure that n_α is large enough. But if the minimum is taken for some $\bar{n}_\alpha$, we have

$$\frac{\rho_{\alpha,\min}}{\rho_{\alpha,\max}}[E^{\mathrm{hom}}(\bar{n}_\alpha + 1, L) - E^{\mathrm{hom}}(\bar{n}_\alpha, L)] \geq 8\pi a \rho_{\alpha,\max}, \tag{4.19}$$

and using Theorem III.2, which states that

$$E^{\mathrm{hom}}(\bar{n}_\alpha + 1, L) - E^{\mathrm{hom}}(\bar{n}_\alpha, L) \leq 8\pi a \frac{\bar{n}_\alpha}{L^3}[1 + O(\bar{n}_\alpha a^3/L^3)], \tag{4.20}$$

we see that $\bar{n}_\alpha$ is at least $\sim \rho_{\alpha,\max} L^3$. We shall later choose $L \sim N^{-1/10}$, so the conditions needed for Eq. (4.1) are fulfilled for N large enough, since $\rho_{\alpha,\max} \sim N$ and hence $\bar{n}_\alpha \sim N^{7/10}$, $L/a \sim N^{9/10}$, and $Y_\alpha \sim N^{-2}$. Thus we have (for large enough N)

$$Q_\alpha(f) \geq 4\pi a \left(\frac{\rho_{\alpha,\min}}{\rho_{\alpha,\max}} \frac{n_\alpha^2}{L^3} (1 - C Y_\alpha^{1/17}) - 2 n_\alpha \rho_{\alpha,\max} \right). \tag{4.21}$$

We now use $Y_\alpha = a^3 n_\alpha / L^3 \leq a^3 N / L^3 \equiv Y$, and drop the requirement that n_α has to be an integer. The minimum of Eq. (4.21) is obtained for

$$n_\alpha = \frac{\rho_{\alpha,\max}^2}{\rho_{\alpha,\min}} \frac{L^3}{(1 - C Y^{1/17})}. \tag{4.22}$$

This gives for Eq. (4.11)

$$E^{\rm QM}(N,a)-E_R^{\rm GP}(N,a)$$
$$\geq 4\pi a\bar{\rho}_R N-4\pi a\sum_\alpha \rho_{\alpha,\min}^2 L^3\left(\frac{\rho_{\alpha,\max}^3}{\rho_{\alpha,\min}^3}\frac{1}{(1-CY^{1/17})}\right). \tag{4.23}$$

Now $\rho_R^{\rm GP}$ is differentiable by Lemma A.6, and strictly positive. Since all the boxes are in the fixed cube Λ_R there are constants $C'<\infty$, $C''>0$, such that

$$\rho_{\alpha,\max}-\rho_{\alpha,\min}\leq NC'L,\quad \rho_{\alpha,\min}\geq NC''. \tag{4.24}$$

Therefore we have, for Y and L small,

$$\frac{\rho_{\alpha,\max}^3}{\rho_{\alpha,\min}^3}\frac{1}{(1-CY^{1/17})}\leq 1+DY^{1/17}+D'L \tag{4.25}$$

with suitable constants D and D'. Also,

$$4\pi a\sum_\alpha \rho_{\alpha,\min}^2 L^3\leq 4\pi a\int(\rho_R^{\rm GP})^2\leq E_R^{\rm GP}(N,a), \tag{4.26}$$

and hence

$$E^{\rm QM}(N,a)\geq E_R^{\rm GP}(N,a)(1-DY^{1/17}-D'L). \tag{4.27}$$

As a last step it remains to optimize the length L. Recall that $Y=a^3N/L^3$ and Na is fixed. The exponents of N in both error terms in Eq. (4.27) are the same for

$$L\sim aN^{9/10}\sim N^{-1/10}. \tag{4.28}$$

The final result, therefore, is

$$E^{\rm QM}(N,a)\geq E_R^{\rm GP}(N,a)(1-D''N^{-1/10}). \tag{4.29}$$

□

C. The limit theorems

By Theorems III.1 and IV.1 we have (with $a=a_1/N$)

$$E^{\rm GP}(N,a)[1+O(N^{-2/3})]\geq E^{\rm QM}(N,a)$$
$$\geq E_R^{\rm GP}(N,a)[1-O(N^{-1/10})]. \tag{4.30}$$

Dividing by N and taking the limit $N\to\infty$ we have

$$E^{\rm GP}(1,a_1)\geq \lim_{N\to\infty}\frac{E^{\rm QM}(N,a_1/N)}{N}\geq E_R^{\rm GP}(1,a_1) \tag{4.31}$$

for all R large enough. Using Lemma II.2 and taking the limit $R\to\infty$ we finally prove Theorem I.1.

The convergence of the energy implies also the convergence of the densities: We replace V by $V+\delta W$ with $W\in L^\infty$ and $\delta\in\mathbb{R}$, and denote the corresponding energies by $E_\delta(N,a)$. It is no restriction to assume that $V+\delta W\geq 0$ for small $|\delta|$. $E_\delta^{\rm QM}(N,a_1/N)/N$ is concave in δ (it is an infimum over linear functions), and converges for each δ to $E_\delta^{\rm GP}(1,a_1)$ as $N\to\infty$. This implies convergence of the derivatives and we have (Feynman-Hellmann principle)

$$\frac{\partial}{\partial\delta}E_\delta^{\rm QM}(N,a)\bigg|_{\delta=0}=\int_{\mathbb{R}^3}W\rho_{N,a}^{\rm QM},$$
$$\frac{\partial}{\partial\delta}E_\delta^{\rm GP}(N,a)\bigg|_{\delta=0}=\int_{\mathbb{R}^3}W\rho_{N,a}^{\rm GP} \tag{4.32}$$

with $\rho_{N,a}^{\rm QM}$ given by Eq. (1.3). In the weak L^1 sense we thus have

$$\lim_{N\to\infty}\frac{1}{N}\rho_{N,a_1/N}^{\rm QM}(\mathbf{x})=\rho_{1,a_1}^{\rm GP}(\mathbf{x}), \tag{4.33}$$

which proves Theorem I.2.

V. CONCLUSIONS

We have proved that the GP energy functional correctly describes the energy and particle density of a Bose gas in a trap to leading order in the small parameter $\bar{\rho}a^3$ (where $\bar{\rho}$ is the mean density and a is the scattering length) in the limit where the particle number N tends to infinity, but a tends to zero with Na fixed.

ACKNOWLEDGMENT

The work was partially supported by the National Science Foundation, Grant No. PHY 98-20650.

APPENDIX A

In this appendix we prove Theorem II.1. The proof is split into several lemmas.

Lemma A.1 (Strict convexity). For $\rho\geq 0$, $\sqrt{\rho}\in\mathcal{D}$, $\mathcal{E}^{\rm GP}[\sqrt{\rho}]$ *is strictly convex in* ρ.

Proof. The second term in Eq. (1.4) is linear, the third quadratic in ρ. So it suffices to show that the first term is convex. Let ρ_1 and ρ_2 be given, with $\Phi_1=\rho_1^{1/2}$ and $\Phi_2=\rho_2^{1/2}$ in $\mathcal{D}_N$. Then also $\Phi=[\alpha\rho_1+(1-\alpha)\rho_2]^{1/2}\in\mathcal{D}_N$ for all $0<\alpha<1$. We have

$$\begin{aligned}\Phi\nabla\Phi&=\alpha\Phi_1\nabla\Phi_1+(1-\alpha)\Phi_2\nabla\Phi_2\\&=(\alpha^{1/2}\Phi_1)(\alpha^{1/2}\nabla\Phi_1)+[(1-\alpha)^{1/2}\Phi_2]\\&\quad\times[(1-\alpha)^{1/2}\nabla\Phi_2]\\&\leq[\alpha\Phi_1^2+(1-\alpha)\Phi_2^2]^{1/2}[\alpha|\nabla\Phi_1|^2\\&\quad+(1-\alpha)|\nabla\Phi_2|^2]^{1/2}\\&=\Phi(\alpha|\nabla\Phi_1|^2+(1-\alpha)|\nabla\Phi_2|^2)^{1/2}.\end{aligned}$$

Hence

$$|\nabla\Phi|^2\leq\alpha|\nabla\Phi_1|^2+(1-\alpha)|\nabla\Phi_2|^2.$$

□

Remark. Because $V\geq 0$, $\mathcal{E}^{\rm GP}[\Phi]$ is also convex in $\Phi\in\mathcal{D}$. But since the domain $\mathcal{D}_N$ is not convex, it is necessary to consider $\rho\mapsto\mathcal{E}^{\rm GP}[\sqrt{\rho}]$.

Lemma A.2 (Minimizer). For all N there exists a minimizing $\Phi_\infty \in \mathcal{D}_N$, with $\mathcal{E}^{\rm GP}[\Phi_\infty]=E^{\rm GP}(N,a)$. Moreover, $|\Phi_\infty|^2$ is unique.

Proof. Let Φ_n be a minimizing sequence in $\mathcal{D}_N$, i.e., $\lim_{n\to\infty}\mathcal{E}^{\rm GP}[\Phi_n]=E^{\rm GP}$. It is clear that there exists a constant C, such that $\|\nabla\Phi_n\|_2<C$, $\|\Phi_n\|_4<C$ and $\int|\Phi_n|^2V<C$ for all n (recall that V is non-negative). Hence the sequence belongs to a weakly compact set in L^4, as well as in the Sobolev space $H^1=\{\Phi:\|\Phi\|_2^2+\|\nabla\Phi\|_2^2<\infty\}$, and in the space L_V^2, defined by the L^2 norm $\|\Phi\|_V=[\int|\Phi(\mathbf{x})|^2V(\mathbf{x})d\mathbf{x}]^{1/2}$. Thus, there exists a $\Phi_\infty\in\mathcal{D}$ and a weakly convergent subsequence, again denoted by Φ_n, such that

$$\Phi_n \rightharpoonup \Phi_\infty \quad \text{in } L^2\cap L^4\cap L_V^2$$

$$\nabla\Phi_n \rightharpoonup \nabla\Phi_\infty \quad \text{in } L^2.$$

Because the L^4 norm, the Sobolev norm, and the L_V^2 norm are all weakly lower semicontinuous, we have

$$\liminf_{n\to\infty}\mathcal{E}^{\rm GP}[\Phi_n]\geq\mathcal{E}^{\rm GP}[\Phi_\infty],$$

and it remains only to show that $\Phi_\infty\in\mathcal{D}_N$. Since $|\Phi_n|^2$ converges to $|\Phi_\infty|^2$ in $L^1_{\rm loc}$ it is clear that $\|\Phi_\infty\|_2^2\leq N$. Moreover,

$$\int_B|\Phi_n|^2\xrightarrow{n\to\infty}\int_B|\Phi_\infty|^2\leq\|\Phi_\infty\|_2^2$$

for all bounded regions B. If $\|\Phi_\infty\|_2^2=N-\epsilon$ with $\epsilon>0$, then there exists a constant M_B for all B, such that

$$\int_{\mathbb{R}^3\setminus B}|\Phi_n|^2\geq\epsilon$$

for all $n>M_B$. Since $\lim_{|\mathbf{x}|\to\infty}V(\mathbf{x})=\infty$, this would imply $\int V|\Phi_n|^2\to\infty$, which is impossible because Φ_n is a minimizing sequence for the functional $\mathcal{E}^{\rm GP}$. Hence $\|\Phi_\infty\|_2^2=N$.

The uniqueness of $|\Phi_\infty|^2$ follows immediately from strict convexity, Lemma A.1. □

Lemma A.3 (GP equation). Every minimizing Φ_∞ satisfies the Gross-Pitaevskii equation (2.4). Conversely, every solution to Eq. (2.4), with μ given by Eq. (2.5), is a minimizer for $\mathcal{E}^{\rm GP}$.

Proof. Pick a function $f\in C_0^\infty$. The stationarity of $\mathcal{E}^{\rm GP}$ at Φ_∞ implies

$$\frac{\partial}{\partial\epsilon}(\mathcal{E}^{\rm GP}[\Phi_\infty+\epsilon f]+\mu\|\Phi_\infty+\epsilon f\|_2^2)|_{\epsilon=0}=0$$

with a Lagrange parameter μ to take account of the subsidiary condition $\|\Phi\|_2^2=N$. With f real valued one obtains

$$-\nabla^2\,\mathrm{Re}\,\Phi_\infty+V\,\mathrm{Re}\,\Phi_\infty+8\pi a|\Phi_\infty|^2\,\mathrm{Re}\,\Phi_\infty=\mu\,\mathrm{Re}\,\Phi_\infty$$

and an analogous equation for $\mathrm{Im}\,\Phi_\infty$ with f purely imaginary. The value of μ is obtained by multiplying the GP equation with Φ_∞ and integrating. By the same argument, every solution Φ to the GP equation satisfies $\mathcal{E}^{\rm GP}[\Phi]=E^{\rm GP}$ and is thus a minimizer. □

Lemma A.4 (Uniqueness). The minimizing Φ_∞ is unique up to a constant phase factor. This factor can be chosen so that Φ_∞ is strictly positive.

Proof. Since $\mathcal{E}^{\rm GP}[|\Phi|]\leq\mathcal{E}^{\rm GP}[\Phi]$ (by an analogous computation as in the proof of Lemma A.1), we know that $|\Phi_\infty|$ is a minimizer and hence a solution to the GP equation. It is thus an eigenstate of the Hamiltonian $H=-\nabla^2+W$ with $W=V+8\pi a|\Phi_\infty|^2$ (recall that $|\Phi_\infty|^2$ is unique), and since it is non-negative, it must be a ground state. Since Φ_∞ solves the same equation it is also a ground state. Now $W\in L^2_{\rm loc}$ and $\lim_{|\mathbf{x}|\to\infty}W(\mathbf{x})=\infty$, so the ground state of H is unique up to a phase and without zeros (see [6], Sec. XIII.47). □

The unique strictly positive minimizer is denoted by $\Phi^{\rm GP}$.

Lemma A.5 (Exponential falloff). For all $t>0$ there exists an M_t, such that $\Phi^{\rm GP}(\mathbf{x})\leq M_te^{-t|\mathbf{x}|}$. In particular, $\Phi^{\rm GP}\in L^\infty$.

Proof. Put $W=V+8\pi a(\Phi^{\rm GP})^2$ and let $t>0$. The GP equation implies

$$(-\nabla^2+t^2)\Phi^{\rm GP}=-(W-\mu-t^2)\Phi^{\rm GP}.$$

Using the Yukawa potential $Y_t(\mathbf{x})=(4\pi|\mathbf{x}|)^{-1}\exp(-t|\mathbf{x}|)$ we can rewrite this as

$$\Phi^{\rm GP}(\mathbf{x})=-\int Y_t(\mathbf{x}-\mathbf{y})[W(\mathbf{y})-\mu-t^2]\Phi^{\rm GP}(\mathbf{y})d\mathbf{y}.$$

Since $\Phi^{\rm GP}>0$, and $W(\mathbf{y})-\mu-t^2>0$ for $|\mathbf{y}|>R$ with R large enough, we also have

$$\Phi^{\rm GP}(\mathbf{x})\leq-\int_{|\mathbf{y}|<R}Y_t(\mathbf{x}-\mathbf{y})[W(\mathbf{y})-\mu-t^2]\Phi^{\rm GP}(\mathbf{y})d\mathbf{y}.$$

Now $W\Phi^{\rm GP}\in L^2_{\rm loc}$, and hence

$$M_t=\sup_{\mathbf{x}}\int_{|\mathbf{y}|<R}\frac{\exp\{t(|\mathbf{x}|-|\mathbf{x}-\mathbf{y}|)\}}{4\pi|\mathbf{x}-\mathbf{y}|}\times[W(\mathbf{y})-\mu-t^2]\Phi^{\rm GP}(\mathbf{y})d\mathbf{y}<\infty.$$

□

Lemma A.6 (Regularity). $\Phi^{\rm GP}(\mathbf{x})$ is once continuously differentiable in $\mathbf{x}\in\mathbb{R}^3$, and $\nabla\Phi^{\rm GP}$ is Hölder continuous of order 1. If $V\in C^\infty$, then $\Phi^{\rm GP}\in C^\infty$. Moreover, $E^{\rm GP}(N,a)$ is continuously differentiable in a and hence in N [by Eq. (1.6)], and $dE^{\rm GP}(N,a)/dN$ satisfies Eq. (2.5).

Proof. The last lemma and the GP equation imply $\nabla^2\Phi^{\rm GP}\in L^\infty_{\rm loc}$. Thus $\nabla\Phi^{\rm GP}$ exists and is Hölder continuous (see [9], Sec. 10.2). The C^∞ property follows by a bootstrap argument. The differentiability with respect to the parameter a may be shown by a Feynman-Hellmann type argument like analogous statements (e.g., with respect to differentiability nuclear charges) in Thomas-Fermi theory [8]. Equation (2.5) follows immediately from Eq. (1.6) and $E^{\rm GP}(1,Na)=\mathcal{E}^{\rm GP}[\Phi^{\rm GP}_{1,Na}]$. □

Lemmas A.1–A.6 complete the proof of Theorem II.1.

APPENDIX B

In this appendix we show that Eq. (4.1) holds for nonnegative potentials satisfying Eq. (4.2), and that a similar estimate with 1/17 replaced by $O(\varepsilon)$ holds under the condition (4.3).

We cut the potential at a finite radius $\tilde{R}$ which, because of $v \geqslant 0$, can only decrease the energy. We thus define

$$\tilde{v}(r) = v(r)\Theta(\tilde{R}-r) \tag{B1}$$

and denote the corresponding scattering length by $\tilde{a} \leqslant a$. Let u be the zero-energy scattering solution for the potential v [cf. Eq. (1.2)] and put

$$h(r) = r - \frac{u(r)}{u'(r)}. \tag{B2}$$

The difference $a-\tilde{a}$ can be estimated as follows. Since $v(r)$ and $\tilde{v}(r)$ agree for $r \leqslant \tilde{R}$, the same holds for the corresponding scattering solutions. Moreover, $\tilde{a} = h(\tilde{R})$. Hence

$$a-\tilde{a} = \int_{\tilde{R}}^{\infty} h'(r)dr = \int_{\tilde{R}}^{\infty} \frac{u(r)u''(r)}{u'(r)^2}dr$$

$$\leqslant \int_{\tilde{R}}^{\infty} \frac{u''(r)}{u(r)} r^2 dr = \frac{1}{2}\int_{\tilde{R}}^{\infty} v(r) r^2 dr, \tag{B3}$$

where convexity of u has been used to derive the inequality. We remark that for $\tilde{R} \to 0$ this simple estimate gives the *Spruch-Rosenberg* inequality [12]

$$a \leqslant \frac{1}{2}\int_0^{\infty} v(r) r^2 dr. \tag{B4}$$

Assuming Eq. (4.2) one obtains

$$\tilde{a} \geqslant a\left(1 - \text{const}\left(\frac{a}{\tilde{R}}\right)^{1/5+\epsilon}\right). \tag{B5}$$

Equation (4.1) holds in any case with a replaced by $\tilde{a}$, and if we we take $\tilde{R} \propto a Y^{-5/17+\epsilon'}$ with $\epsilon' > 0$ then the error in Eq. (B5) is of higher order than the leading error term in Eq. (4.1). We have thus established Eq. (4.1) under the condition (4.2). If only the weaker condition (4.3) holds, then the additional error term may be $O(Y^{(5/17-\epsilon')\epsilon})$.

To see the significance of condition (4.3) we also estimate $a-\tilde{a}$ from below:

$$a-\tilde{a} \geqslant \int_{\tilde{R}}^{\infty} u(r)u''(r) \geqslant \frac{1}{2}\int_{\max(\tilde{R},a)}^{\infty} v(r)(r-a)^2 dr. \tag{B6}$$

In order that a be finite the last integral must converge, i.e., a slower decrease than $1/r^3$ is not allowed.

[1] E.P. Gross, Nuovo Cimento **20**, 454 (1961).
[2] L.P. Pitaevskii, Zh. Eksp. Teor. Fiz. **40**, 646 (1961) [Sov. Phys. JETP **13**, 451 (1961)].
[3] E.P. Gross, J. Math. Phys. **4**, 195 (1963).
[4] F. Dalfovo, S. Giorgini, L.P. Pitaevskii, and S. Stringari, Rev. Mod. Phys. **71**, 463 (1999).
[5] E.H. Lieb and J. Yngvason, Phys. Rev. Lett. **80**, 2504 (1998).
[6] M. Reed and B. Simon, *Methods of Modern Mathematical Physics IV* (Academic Press, New York, 1978).
[7] F.J. Dyson, Phys. Rev. **106**, 20 (1957).
[8] E.H. Lieb, Rev. Mod. Phys. **53**, 603 (1981).
[9] E. H. Lieb and M. Loss, *Analysis* (Amererican Mathematical Society, Providence, RI, 1997).
[10] P.L. Lions, Appl. Anal. **12**, 267 (1981).
[11] E.H. Lieb, Phys. Rev. **130**, 2518 (1963); see also E.H. Lieb and A.Y. Sakakura, Phys. Rev. A **133**, A899 (1964); E.H. Lieb and W. Liniger, *ibid.* **134**, A312 (1964).
[12] L. Spruch and L. Rosenberg, Phys. Rev. **116**, 1034 (1959).

Proceedings of the XIIIth International Congress on Mathematical Physics, London, July 18-24, 2000. A. Fokas, A. Grigoryan, T. Kibble, B. Zegarlinski editors, International Press, 2001.

The Bose Gas: A Subtle Many-Body Problem

Elliott H. Lieb

ABSTRACT. Now that the properties of the ground state of quantum-mechanical many-body systems (bosons) at low density, ρ, can be examined experimentally it is appropriate to revisit some of the formulas deduced by many authors 4-5 decades ago. One of these is that the leading term in the energy/particle is $4\pi a\rho$ where a is the scattering length of the 2-body potential. Owing to the delicate and peculiar nature of bosonic correlations (such as the strange $N^{7/5}$ law for charged bosons), four decades of research failed to establish this plausible formula rigorously. The only previous lower bound for the energy was found by Dyson in 1957, but it was 14 times too small. The correct asymptotic formula has recently been obtained jointly with J. Yngvason and this work will be presented. The reason behind the mathematical difficulties will be emphasized. A different formula, postulated as late as 1971 by Schick, holds in two-dimensions and this, too, will be shown to be correct. Another problem of great interest is the existence of Bose-Einstein condensation, and what little is known about this rigorously will also be discussed. With the aid of the methodology developed to prove the lower bound for the homogeneous gas, two other problems have been successfully addressed. One is the proof (with Yngvason and Seiringer) that the Gross-Pitaevskii equation correctly describes the ground state in the 'traps' actually used in the experiments. The other is a very recent proof (with Solovej) that Foldy's 1961 theory of a high density gas of charged particles correctly describes its ground state energy.

1. Introduction

Schrödinger's equation of 1926 defined a new mechanics whose Hamiltonian is based on classical mechanics, but whose consequences are sometimes non-intuitive from the classical point of view. One of the most extreme cases is the behavior of the ground (= lowest energy) state of a many-body system of particles. Since the ground state function $\Psi(x_1, ..., x_N)$ is automatically symmetric in the coordinates $x_j \in R^3$ of the N particles, we are dealing necessarily with *'bosons'*. If we imposed the Pauli exclusion principle (antisymmetry) instead, appropriate for electrons, the outcome would look much more natural and, oddly, more classical. Indeed, the Pauli principle is *essential* for understanding the stability of the ordinary matter that surrounds us.

1991 *Mathematics Subject Classification.* 81V70, 35Q55, 46N50.

The author was supported in part by U.S. National Science Foundation grant PHY 98-20650 A01.

Recent experiments have confirmed some of the bizarre properties of bosons close to their ground state, but the theoretical ideas go back to the 1940's - 1960's. The first sophisticated analysis of a gas or liquid of *interacting* bosons is due to Bogolubov in 1947. His approximate theory as amplified by others, is supposed to be exact in certain limiting cases, and some of those cases have now been verified rigorously (for the ground state energy) – 3 or 4 decades after they were proposed.

The discussion will center around three main topics.

1. The dilute, repulsive Bose gas at low density (2D and 3D) (with Jakob Yngvason).

2. Repulsive bosons in a trap (as used in recent experiments) and the 'Gross-Pitaevskii equation (with Robert Seiringer and Jakob Yngvason).

3. Foldy's 'jellium' model of charged particles in a neutralizing background (with Jan Philip Solovej)

The discussion below of topic 1 is taken from **[1]** (in 3-dimensions) and **[2]** (in 2-dimensions). That for topic 2 is taken from **[3]** (in 3-dimensions) and **[4]** (in 2-dimensions). Topic 3 is from **[5]**. See also **[6, 7]**.

Topic 1 (3-dimensions) was the starting point and contains essential ideas. It is explained here in some detail and is taken, with minor modifications (and corrections), from **[6]**. In terms of technical complexity, however, the third topic is the most involved and can be treated here only very briefly.

2. The Dilute Bose Gas in 3D

We consider the Hamiltonian for N bosons of mass m enclosed in a cubic box Λ of side length L and interacting by a spherically symmetric pair potential $v(|\vec{x}_i - \vec{x}_j|)$:

$$H_N = -\mu \sum_{i=1}^{N} \Delta_i + \sum_{1 \le i < j \le N} v(|\vec{x}_i - \vec{x}_j|). \tag{2.1}$$

Here $\vec{x}_i \in \mathbb{R}^3$, $i = 1, \ldots, N$ are the positions of the particles, Δ_i the Laplacian with respect to $\vec{x}_i$, and we have denoted $\hbar^2/2m$ by μ for short. (By choosing suitable units μ could, of course, be eliminated, but we want to keep track of the dependence of the energy on Planck's constant and the mass.) The Hamiltonian (2.1) operates on *symmetric* wave functions in $L^2(\Lambda^N, d\vec{x}_1 \cdots d\vec{x}_N)$ as is appropriate for bosons. The interaction potential will be assumed to be *nonnegative* and to decrease faster than $1/r^3$ at infinity.

We are interested in the ground state energy $E_0(N, L)$ of (2.1) in the *thermodynamic limit* when N and L tend to infinity with the density $\rho = N/L^3$ fixed. The energy per particle in this limit

$$e_0(\rho) = \lim_{L \to \infty} E_0(\rho L^3, L)/(\rho L^3). \tag{2.2}$$

Our results about $e_0(\rho)$ are based on estimates on $E_0(N, L)$ for finite N and L, which are important, e.g., for the considerations of inhomogeneous systems in **[3]**. To define $E_0(N, L)$ precisely one must specify the boundary conditions. These should not matter for the thermodynamic limit. To be on the safe side we use Neumann boundary conditions for the lower bound, and Dirichlet boundary conditions for the upper bound since these lead, respectively, to the lowest and the highest energies.

For experiments with dilute gases the *low density asymptotics* of $e_0(\rho)$ is of importance. Low density means here that the mean interparticle distance, $\rho^{-1/3}$ is much larger than the *scattering length* a of the potential, which is defined as follows. The zero energy scattering Schroedinger equation

$$-2\mu\Delta\psi + v(r)\psi = 0 \tag{2.3}$$

has a solution of the form, asymptotically as $r \to \infty$ (or for all $r > R_0$ if $v(r) = 0$ for $r > R_0$),

$$\psi(x) = 1 - a/|x| \tag{2.4}$$

This is the same as

$$a = \lim_{r\to\infty} r - \frac{u_0(r)}{u_0'(r)}, \tag{2.5}$$

where u_0 solves the zero energy scattering equation,

$$-2\mu u_0''(r) + v(r)u_0(r) = 0 \tag{2.6}$$

with $u_0(0) = 0$. (The factor 2 in (2.6) comes from the reduced mass of the two particle problem.)

An important special case is the hard core potential $v(r) = \infty$ if $r < a$ and $v(r) = 0$ otherwise. Then the scattering length a and the radius a are the same.

Our main result is a rigorous proof of the formula

$$e_0(\rho) \approx 4\pi\mu\rho a \tag{2.7}$$

for $\rho a^3 \ll 1$, more precisely of

THEOREM 2.1 (Low density limit of the ground state energy).

$$\lim_{\rho a^3\to 0} \frac{e_0(\rho)}{4\pi\mu\rho a} = 1. \tag{2.8}$$

This formula is independent of the boundary conditions used for the definition of $e_0(\rho)$.

The genesis of an understanding of $e_0(\rho)$ was the pioneering work [**11**] of Bogolubov, and in the 50's and early 60's several derivations of (2.8) were presented [**12**], [**13**], even including higher order terms:

$$\frac{e_0(\rho)}{4\pi\mu\rho a} = 1 + \tfrac{128}{15\sqrt{\pi}}(\rho a^3)^{1/2} + 8\left(\tfrac{4\pi}{3} - \sqrt{3}\right)(\rho a^3)\log(\rho a^3) + O(\rho a^3) \tag{2.9}$$

These early developments are reviewed in [**14**]. They all rely on some special assumptions about the ground state that have never been proved, or on the selection of special terms from a perturbation series which likely diverges. The only rigorous estimates of this period were established by Dyson, who derived the following bounds in 1957 for a gas of hard spheres [**15**]:

$$\frac{1}{10\sqrt{2}} \leq \frac{e_0(\rho)}{4\pi\mu\rho a} \leq \frac{1+2Y^{1/3}}{(1-Y^{1/3})^2} \tag{2.10}$$

with $Y = 4\pi\rho a^3/3$. While the upper bound has the asymptotically correct form, the lower bound is off the mark by a factor of about 1/14. But for about 40 years this was the best lower bound available!

Under the assumption that (2.8) is a correct asymptotic formula for the energy, we see at once that understanding it physically, much less proving it, is not a simple matter. Initially, the problem presents us with two lengths, $a \ll \rho^{-1/3}$ at low density. However, (2.8) presents us with another length generated by the solution to the problem. This length is the de Broglie wavelength, or 'uncertainty principle' length

$$\ell_c = (\rho a)^{-1/2}.$$

The reason for saying that ℓ_c is the de Broglie wavelength is that in the hard core case all the energy is kinetic (the hard core just imposes a $\psi = 0$ boundary condition whenever the distance between two particles is less than a). By the uncertainty principle, the kinetic energy is proportional to an inverse length squared, namely ℓ_c. We then have the relation (since ρa^3 is small)

$$a \ll \rho^{-1/3} \ll \ell_c$$

which implies, physically, that it is impossible to localize the particles relative to each other (even though ρ is small) bosons in their ground state are therefore 'smeared out' over distances large compared to the mean particle distance and their individuality is entirely lost. They cannot be localized with respect to each other without changing the kinetic energy enormously.

Fermions, on the other hand, prefer to sit in 'private rooms', i.e., ℓ_c is never bigger than $\rho^{-1/3}$ by a fixed factor. In this respect the quantum nature of bosons is much more pronounced than for fermions.

Since (2.8) is a basic result about the Bose gas it is clearly important to derive it rigorously and in reasonable generality, in particular for more general cases than hard spheres. The question immediately arises for which interaction potentials one may expect it to be true. A notable fact is that it *not true for all* v with $a > 0$, since there are two body potentials with positive scattering length that allow many body bound states [**16**]. Our proof, presented in the sequel, works for nonnegative v, but we conjecture that (2.8) holds if $a > 0$ and v has no N-body bound states for any N. The lower bound is, of course, the hardest part, but the upper bound is not altogether trivial either.

Before we start with the estimates a simple computation and some heuristics may be helpful to make (2.8) plausible and motivate the formal proofs.

With u_0 the scattering solution and $f_0(r) = u_0(r)/r$, partial integration gives

$$\begin{aligned}\int_{|\vec{x}|\leq R}\{2\mu|\nabla f_0|^2 + v|f_0|^2\}d\vec{x} &= 4\pi\int_0^R\{2\mu[u_0'(r) - (u_0(r)/r)]^2 + v(r)[u_0(r)]^2\}dr \\ &= 8\pi\mu a|u_0(R)|^2/R^2 \to 8\pi\mu a \quad \text{for } R\to\infty,\end{aligned} \tag{2.11}$$

if u_0 is normalized so that $f_0(R) \to 1$ as $R \to \infty$. Moreover, for positive interaction potentials the scattering solution minimizes the quadratic form in (2.11) for each R with $u_0(0) = 0$ and $u_0(R)$ fixed as boundary conditions. Hence the energy $E_0(2, L)$ of two particles in a large box, i.e., $L \gg a$, is approximately $8\pi\mu a/L^3$. If the gas is sufficiently dilute it is not unreasonable to expect that the energy is essentially a sum of all such two particle contributions. Since there are $N(N-1)/2$ pairs, we are thus lead to $E_0(N, L) \approx 4\pi\mu a N(N-1)/L^3$, which gives (2.8) in the thermodynamic limit.

This simple heuristics is far from a rigorous proof, however, especially for the lower bound. In fact, it is rather remarkable that the same asymptotic formula holds both for 'soft' interaction potentials, where perturbation theory can be expected to be a good approximation, and potentials like hard spheres where this is not so. In the former case the ground state is approximately the constant function and the energy is *mostly potential*: According to perturbation theory $E_0(N,L) \approx N(N-1)/(2L^3)\int v(|\vec{x}|)d\vec{x}$. In particular it is *independent of* μ, i.e. of Planck's constant and mass. Since, however, $\int v(|\vec{x}|)d\vec{x}$ is the first Born approximation to $8\pi\mu a$ (note that a depends on μ!), this is not in conflict with (2.8). For 'hard' potentials on the other hand, the ground state is *highly correlated*, i.e., it is far from being a product of single particle states. The energy is here *mostly kinetic*, because the wave function is very small where the potential is large. These two quite different regimes, the potential energy dominated one and the kinetic energy dominated one, cannot be distinguished by the low density asymptotics of the energy. Whether they behave differently with respect to other phenomena, e.g., Bose-Einstein condensation, is not known at present.

Bogolubov's analysis [**11**] presupposes the existence of Bose-Einstein condensation. Nevertheless, it is correct (for the energy) for the one-dimensional delta-function Bose gas [**17**], despite the fact that there is (presumably) no condensation in that case. It turns out that BE condensation is not really needed in order to understand the energy. As we shall see, 'global' condensation can be replaced by a 'local' condensation on boxes whose size is independent of L. It is this crucial understanding that enables us to prove Theorem 1.1 without having to decide about BE condensation.

An important idea of Dyson was to transform the hard sphere potential into a soft potential at the cost of sacrificing the kinetic energy, i.e., effectively to move from one regime to the other. We shall make use of this idea in our proof of the lower bound below. But first we discuss the simpler upper bound, which relies on other ideas from Dyson's beautiful paper [**15**].

2.1. Upper Bound. The following generalization of Dyson's upper bound holds [**3**], [**10**]:

THEOREM 2.2 (Upper bound). *Define* $\rho_1 = (N-1)/L^3$ *and* $b = (4\pi\rho_1/3)^{-1/3}$. *For nonnegative potentials* v, *and* $b > a$ *the ground state energy of (2.1) with periodic boundary conditions satisfies*

$$E_0(N,L)/N \leq 4\pi\mu\rho_1 a\frac{1-\frac{a}{b}+\left(\frac{a}{b}\right)^2+\frac{1}{2}\left(\frac{a}{b}\right)^3}{\left(1-\frac{a}{b}\right)^8}. \tag{2.12}$$

For Dirichlet boundary conditions the estimate holds with (const.)$/L^2$ *added to the right side. Thus in the thermodynamic limit and for all boundary conditions*

$$\frac{e_0(\rho)}{4\pi\mu\rho a} \leq \frac{1-Y^{1/3}+Y^{2/3}-\frac{1}{2}Y}{(1-Y^{1/3})^8}. \tag{2.13}$$

provided $Y = 4\pi\rho a^3/3 < 1$.

Remark. The bound (2.12) holds for potentials with infinite range, provided $b > a$. For potentials of finite range R_0 it can be improved for $b > R_0$ to

$$E_0(N,L)/N \leq 4\pi\mu\rho_1 a\frac{1-\left(\frac{a}{b}\right)^2+\frac{1}{2}\left(\frac{a}{b}\right)^3}{\left(1-\frac{a}{b}\right)^4}. \tag{2.14}$$

Proof. We first remark that the expectation value of (2.1) with any trial wave function gives an upper bound to the bosonic ground state energy, even if the trial function is not symmetric under permutations of the variables. The reason is that an absolute ground state of the elliptic differential operator (2.1) (i.e., a ground state without symmetry requirement) is a nonnegative function which can be symmetrized without changing the energy because (2.1) is symmetric under permutations. In other words, the absolute ground state energy is the same as the bosonic ground state energy.

Following [**15**] we choose a trial function of the following form

$$\Psi(x_1,\ldots,x_N) = F_1(x_1)\cdot F_2(x_1,x_2)\cdots F_N(x_1,\ldots,x_N). \tag{2.15}$$

More specifically, $F_1 \equiv 1$ and F_i depends only on the distance of x_i to its nearest neighbor among the points $x_1,\ldots,x_{i-1}$ (taking the periodic boundary into account):

$$F_i(x_1,\ldots,x_i) = f(t_i), \quad t_i = \min\left(|x_i - x_j|, j = 1,\ldots,i-1\right), \tag{2.16}$$

with a function f satisfying

$$0 \le f \le 1, \quad f' \ge 0. \tag{2.17}$$

The intuition behind the ansatz (2.15) is that the particles are inserted into the system one at the time, taking into account the particles previously inserted. While such a wave function cannot reproduce all correlations present in the true ground state, it turns out to capture the leading term in the energy for dilute gases. The form (2.16) is computationally easier to handle than an ansatz of the type $\prod_{i<j} f(|x_i - x_j|)$, which might appear more natural in view of the heuristic remarks at the end of the last subsection.

The function f is chosen to be

$$f(r) = \begin{cases} f_0(r)/f_0(b) & \text{for } 0 \le r \le b, \\ 1 & \text{for } r > b, \end{cases} \tag{2.18}$$

with $f_0(r) = u_0(r)/r$. The estimates (2.12) and (2.14) are obtained by somewhat lengthy computations similar as in [**15**], but making use of (2.11). For details we refer to [**3**] and [**10**].

A test wave function with Dirichlet boundary condition may be obtained by localizing the wave function (2.15) on the length scale L. The energy cost per particle for this is (const.)$/L^2$. □

2.2. Lower Bound. In the beginning it was explained why the lower bound for the bosonic ground state energy of (2.1) is not easy to obtain. The three different length scales for bosons will play a role in the proof below.

- The scattering length a.
- The mean particle distance $\rho^{-1/3}$.
- The 'uncertainty principle length' ℓ_c, defined by $\mu\ell_c^{-2} = e_0(\rho)$, i.e., $\ell_c \sim (\rho a)^{-1/2}$.

Our lower bound for $e_0(\rho)$ is as follows.

THEOREM 2.3 (Lower bound in the thermodynamic limit). *For a positive potential v with finite range and Y small enough*

$$\frac{e_0(\rho)}{4\pi\mu\rho a} \ge (1 - CY^{1/17}) \tag{2.19}$$

with C a constant. If v does not have finite range, but decreases at least as fast as $1/r^{3+\varepsilon}$ at infinity with some $\varepsilon > 0$, then an analogous bound to (2.19) holds, but with C replaced by another constant and 1/17 by another exponent, both of which may depend on ε.

It should be noted right away that the error term $-CY^{1/17}$ in (2.19) is of no fundamental significance and is not believed to reflect the true state of affairs. Presumably, it does not even have the right sign. We mention in passing that C can be taken to be 8.9 [**10**].

As mentioned in the Introduction a lower bound on $E_0(N,L)$ for finite N and L is of importance for applications to inhomogeneous gases, and in fact we derive (2.19) from such a bound. We state it in the following way:

THEOREM 2.4 (Lower bound in a finite box). *For a positive potential v with finite range there is a $\delta > 0$ such that the ground state energy of (2.1) with Neumann conditions satisfies*

$$E_0(N,L)/N \geq 4\pi\mu\rho a\left(1 - CY^{1/17}\right) \tag{2.20}$$

for all N and L with $Y < \delta$ and $L/a > C'Y^{-6/17}$. Here C and C' are constants, independent of N and L. (Note that the condition on L/a requires in particular that N must be large enough, $N >$ (const.)$Y^{-1/17}$.) As in Theorem 2.3 such a bound, but possibly with other constants and another exponent for Y, holds also for potentials v of infinite range decreasing faster than $1/r^3$ at infinity.

The first step in the proof of Theorem 2.4 is a generalization of a lemma of Dyson, which allows us to replace v by a 'soft' potential, at the cost of sacrificing kinetic energy and increasing the effective range.

LEMMA 2.5. *Let $v(r) \geq 0$ with finite range R_0. Let $U(r) \geq 0$ be any function satisfying $\int U(r)r^2dr \leq 1$ and $U(r) = 0$ for $r < R_0$. Let $\mathcal{B} \subset \mathbf{R}^3$ be star shaped with respect to 0 (e.g. convex with $0 \in \mathcal{B}$). Then for all differentiable functions ψ*

$$\int_{\mathcal{B}} \left[\mu|\nabla\psi|^2 + \tfrac{1}{2}v|\psi|^2\right] \geq \mu a \int_{\mathcal{B}} U|\psi|^2. \tag{2.21}$$

Proof. Actually, (2.21) holds with $\mu|\nabla\phi(\vec{x})|^2$ replaced by the (smaller) radial kinetic energy, $\mu|\partial\phi(\vec{x})/\partial r|^2$, and it suffices to prove the analog of (2.21) for the integral along each radial line with fixed angular variables. Along such a line we write $\phi(\vec{x}) = u(r)/r$ with $u(0) = 0$. We consider first the special case when U is a delta-function at some radius $R \geq R_0$, i.e.,

$$U(r) = \frac{1}{R^2}\delta(r - R). \tag{2.22}$$

For such U the analog of (2.21) along the radial line is

(2.23)
$$\int_0^{R_1} \{\mu[u'(r) - (u(r)/r)]^2 + \tfrac{1}{2}v(r)[u(r)]^2\}dr \geq \begin{cases} 0 & \text{if } R_1 < R \\ \mu a|u(R)|^2/R^2 & \text{if } R \leq R_1 \end{cases}$$

where R_1 is the length of the radial line segment in $\mathcal{B}$. The case $R_1 < R$ is trivial, because $\mu|\partial\psi/\partial r|^2 + \frac{1}{2}v|\psi|^2 \geq 0$. (Note that positivity of v is used here.) If $R \leq R_1$ we consider the integral on the left side of (2.23) from 0 to R instead of R_1 and minimize it under the boundary condition that $u(0) = 0$ and $u(R)$ is a fixed constant. Since everything is homogeneous in u we may normalize this value to $u(R) = R - a$. This minimization problem leads to the zero energy scattering equation (2.6). Since v is positive, the solution is a true minimum and not just a stationary point.

Because $v(r) = 0$ for $r > R_0$ the solution, u_0, satisfies $u_0(r) = r - a$ for $r > R_0$. By partial integration,

(2.24)

$$\int_0^R \{\mu[u_0'(r) - (u_0(r)/r)]^2 + \tfrac{1}{2}v(r)[u_0(r)]^2\}dr = \mu a|R-a|/R \geq \mu a|R-a|^2/R^2.$$

equation But $|R-a|^2/R^2$ is precisely the right side of (2.23) if u satisfies the normalization condition.

This derivation of (2.21) for the special case (2.22) implies the general case, because every U can be written as a superposition of δ-functions, $U(r) = \int R^{-2}\delta(r - R)\,U(R)R^2dR$, and $\int U(R)R^2dR \leq 1$ by assumption. □

By dividing Λ for given points $\vec{x}_1, \ldots, \vec{x}_N$ into Voronoi cells $\mathcal{B}_i$ that contain all points closer to $\vec{x}_i$ than to $\vec{x}_j$ with $j \neq i$ (these cells are star shaped w.r.t. $\vec{x}_i$, indeed convex), the following corollary of Lemma 2.5 can be derived in the same way as the corresponding Eq. (28) in [**15**].

COROLLARY 2.6. *For any U as in Lemma 2.5*

$$H_N \geq \mu a W \tag{2.25}$$

with

$$W(\vec{x}_1, \ldots, \vec{x}_N) = \sum_{i=1}^N U(t_i), \tag{2.26}$$

where t_i is the distance of $\vec{x}_i$ to its nearest neighbor among the other points $\vec{x}_j$, $j = 1, \ldots, N$, i.e.,

$$t_i(\vec{x}_1, \ldots, \vec{x}_N) = \min_{j,\, j\neq i} |\vec{x}_i - \vec{x}_j|. \tag{2.27}$$

(Note that t_i has here a slightly different meaning than in (2.16), where it denoted the distance to the nearest neighbor among the $\vec{x}_j$ with $j \leq i-1$.)

Dyson considers in [**15**] a one parameter family of U's that is essentially the same as the following choice, which is convenient for the present purpose:

$$U_R(r) = \begin{cases} 3(R^3 - R_0^3)^{-1} & \text{for } R_0 < r < R \\ 0 & \text{otherwise.} \end{cases} \tag{2.28}$$

We denote the corresponding interaction (2.26) by W_R. For the hard core gas one obtains

$$E(N,L) \geq \sup_R \inf_{(\vec{x}_1,\ldots,\vec{x}_N)} \mu a W_R(\vec{x}_1, \ldots, \vec{x}_N) \tag{2.29}$$

where the infimum is over $(\vec{x}_1, \ldots, x_N) \in \Lambda^N$ with $|\vec{x}_i - \vec{x}_j| \geq R_0 = a$, because of the hard core. At fixed R simple geometry gives

$$\inf_{(\vec{x}_1,\ldots,\vec{x}_N)} W_R(\vec{x}_1, \ldots, \vec{x}_N) \geq \left(\frac{A}{R^3} - \frac{B}{\rho R^6}\right) \tag{2.30}$$

with certain constants A and B. An evaluation of these constants gives Dyson's bound

$$E(N,L)/N \geq \frac{1}{10\sqrt{2}} 4\pi\mu\rho a. \tag{2.31}$$

The main reason this method does not give a better bound is that R must be chosen quite big, namely of the order of the mean particle distance $\rho^{-1/3}$, in order to guarantee that the spheres of radius R around the N points overlap. Otherwise the infimum of W_R will be zero. But large R means that W_R is small. It should also be noted that this method does not work for potentials other than hard spheres: If $|\vec{x}_i - \vec{x}_j|$ is allowed to be less than R_0, then the right side of (2.29) is zero because $U(r) = 0$ for $r < R_0$.

For these reasons we take another route. We still use Lemma 2.21 to get into the soft potential regime, but we do *not* sacrifice *all* the kinetic energy as in (2.25). Instead we write, for $\varepsilon > 0$

$$H_N = \varepsilon H_N + (1-\varepsilon)H_N \geq \varepsilon T_N + (1-\varepsilon)H_N \tag{2.32}$$

with $T_N = -\sum_i \Delta_i$ and use (2.25) only for the part $(1-\varepsilon)H_N$. This gives

$$H_N \geq \varepsilon T_N + (1-\varepsilon)\mu a W_R. \tag{2.33}$$

We consider the operator on the right side from the viewpoint of first order perturbation theory, with εT_N as the unperturbed part, denoted H_0.

The ground state of H_0 in a box of side length L is $\Psi_0(\vec{x}_1,\dots,\vec{x}_N) \equiv L^{-3N/2}$ and we denote expectation values in this state by $\langle\cdot\rangle_0$. A computation, cf. Eq. (21) in [**1**], gives

$$\begin{aligned} 4\pi\rho\left(1-\tfrac{1}{N}\right) &\geq \langle W_R\rangle_0/N \\ &\geq 4\pi\rho\left(1-\tfrac{1}{N}\right)\left(1-\tfrac{2R}{L}\right)^3\left(1+4\pi\rho(1-\tfrac{1}{N})(R^3-R_0^3)/3\right)^{-1}. \end{aligned} \tag{2.34}$$

The rationale behind the various factors is as follows: $(1-\frac{1}{N})$ comes from the fact that the number of pairs is $N(N-1)/2$ and not $N^2/2$, $(1-2R/L)^3$ takes into account the fact that the particles do not interact beyond the boundary of Λ, and the last factor measures the probability to find another particle within the interaction range of the potential U_R for a given particle.

The first order result (2.34) looks at first sight quite promising, for if we let $L\to\infty$, $N\to\infty$ with $\rho = N/L^3$ fixed, and subsequently take $R\to\infty$, then $\langle W_R\rangle_0/N$ converges to $4\pi\rho$, which is just what is desired. But the first order result (2.34) is not a rigorous bound on $E_0(N,L)$, we need *error estimates*, and these will depend on ε, R and L.

We now recall *Temple's inequality* [**18**] for the expectations values of an operator $H = H_0 + V$ in the ground state $\langle\cdot\rangle_0$ of H_0. It is a simple consequence of the operator inequality

$$(H-E_0)(H-E_1) \geq 0 \tag{2.35}$$

for the two lowest eigenvalues, $E_0 < E_1$, of H and reads

$$E_0 \geq \langle H\rangle_0 - \frac{\langle H^2\rangle_0 - \langle H\rangle_0^2}{E_1 - \langle H\rangle_0} \tag{2.36}$$

provided $E_1 - \langle H\rangle_0 > 0$. Furthermore, if $V \geq 0$ we may use $E_1 \geq E_1^{(0)}$= second lowest eigenvalue of H_0 and replace E_1 in (2.36) by $E_1^{(0)}$.

From (2.34) and (2.36) we get the estimate

$$\frac{E_0(N,L)}{N} \geq 4\pi\mu a\rho\,(1-\mathcal{E}(\rho,L,R,\varepsilon)) \tag{2.37}$$

with

$$\begin{aligned} 1-\mathcal{E}(\rho,L,R,\varepsilon) &= (1-\varepsilon)\left(1-\tfrac{1}{\rho L^3}\right)\left(1-\tfrac{2R}{L}\right)^3\left(1+\tfrac{4\pi}{3}\rho(1-\tfrac{1}{N})(R^3-R_0^3)\right)^{-1} \\ &\times \left(1-\frac{\mu a(\langle W_R^2\rangle_0-\langle W_R\rangle_0^2)}{\langle W_R\rangle_0(E_1^{(0)}-\mu a\langle W_R\rangle_0)}\right). \end{aligned} \tag{2.38}$$

To evaluate this further one may use the estimates (2.34) and the bound

$$\langle W_R^2\rangle_0 \leq 3\frac{N}{R^3-R_0^3}\langle W_R\rangle_0 \tag{2.39}$$

which follows from $U_R^2 = 3(R^3-R_0^3)^{-1}U_R$ together with the Cauchy-Schwarz inequality. A glance at the form of the error term reveals, however, that it is *not* possible here to take the thermodynamic limit $L\to\infty$ with ρ fixed: We have $E_1^{(0)} = \varepsilon\pi\mu/L^2$ (this is the kinetic energy of a *single* particle in the first excited state in the box), and the factor $E_1^{(0)} - \mu a\langle W_R\rangle_0$ in the denominator in (2.38) is, up to unimportant constants and lower order terms, $\sim(\varepsilon L^{-2} - a\rho^2L^3)$. Hence the denominator eventually becomes negative and Temple's inequality looses its validity if L is large enough.

As a way out of this dilemma we divide the big box Λ into cubic *cells* of side length ℓ that is kept *fixed* as $L\to\infty$. The number of cells, L^3/ℓ^3, on the other hand, increases with L. The N particles are distributed among these cells, and we use (2.38), with L replaced by ℓ, N by the particle number, n, in a cell and ρ by n/ℓ^3, to estimate the energy in each cell with *Neumann* conditions on the boundary. This boundary condition leads to lower energy than any other boundary condition. For each distribution of the particles we add the contributions from the cells, neglecting interactions across boundaries. Since $v\geq 0$ by assumption, this can only lower the energy. Finally, we minimize over all possible choices of the particle numbers for the various cells adding up to N. The energy obtained in this way is a lower bound to $E_0(N,L)$, because we are effectively allowing discontinuous test functions for the quadratic form given by H_N.

In mathematical terms, the cell method leads to

$$E_0(N,L)/N \geq (\rho\ell^3)^{-1}\inf\sum_{n\geq 0} c_n E_0(n,\ell) \tag{2.40}$$

where the infimum is over all choices of coefficients $c_n\geq 0$ (relative number of cells containing exactly n particles), satisfying the constraints

$$\sum_{n\geq 0} c_n = 1, \qquad \sum_{n\geq 0} c_n n = \rho\ell^3. \tag{2.41}$$

The minimization problem for the distributions of the particles among the cells would be easy if we knew that the ground state energy $E_0(n,\ell)$ (or a good lower bound to it) were convex in n. Then we could immediately conclude that it is best to have the particles as evenly distributed among the boxes as possible, i.e., c_n would be zero except for the n equal to the integer closest to $\rho\ell^3$. This would give

$$\frac{E_0(N,L)}{N} \geq 4\pi\mu a\rho\,(1-\mathcal{E}(\rho,\ell,R,\varepsilon)) \tag{2.42}$$

i.e., replacement of L in (2.37) by ℓ, which is independent of L. The blow up of $\mathcal{E}$ for $L \to \infty$ would thus be avoided.

Since convexity of $E_0(n,\ell)$ is not known (except in the thermodynamic limit) we must resort to other means to show that $n = O(\rho\ell^3)$ in all boxes. The rescue comes from *superadditivity* of $E_0(n,\ell)$, i.e., the property

$$E_0(n+n',\ell) \geq E_0(n,\ell) + E_0(n',\ell) \tag{2.43}$$

which follows immediately from $v \geq 0$ by dropping the interactions between the n particles and the n' particles. The bound (2.43) implies in particular that for any $n, p \in \mathbb{N}$ with $n \geq p$

$$E(n,\ell) \geq [n/p]\, E(p,\ell) \geq \frac{n}{2p} E(p,\ell) \tag{2.44}$$

since the largest integer $[n/p]$ smaller than n/p is in any case $\geq n/(2p)$.

The way (2.44) is used is as follows: Replacing L by ℓ, N by n and ρ by n/ℓ^3 in (2.37) we have for fixed R and ε

$$E_0(n,\ell) \geq \frac{4\pi\mu a}{\ell^3} n(n-1) K(n,\ell) \tag{2.45}$$

with a certain function $K(n,\ell)$ determined by (2.38). We shall see that K is monotonously decreasing in n, so that if $p \in \mathbb{N}$ and $n \leq p$ then

$$E_0(n,\ell) \geq \frac{4\pi\mu a}{\ell^3} n(n-1) K(p,\ell). \tag{2.46}$$

We now split the sum in (2.40) into two parts. For $n < p$ we use (2.46), and for $n \geq p$ we use (2.44) together with (2.46) for $n = p$. The task is thus to minimize

$$\sum_{n<p} c_n n(n-1) + \tfrac{1}{2} \sum_{n \geq p} c_n n(p-1) \tag{2.47}$$

subject to the constraints (2.41). Putting

$$k := \rho\ell^3 \quad \text{and} \quad t := \sum_{n<p} c_n n \leq k \tag{2.48}$$

we have $\sum_{n \geq p} c_n n = k - t$, and since $n(n-1)$ is convex in n, and $\sum_{n<p} c_n \leq 1$ the expression (2.47) is

$$\geq t(t-1) + \tfrac{1}{2}(k-t)(p-1). \tag{2.49}$$

We have to minimize this for $1 \leq t \leq k$. If $p \geq 4k$ the minimum is taken at $t = k$ and is equal to $k(k-1)$. Altogether we have thus shown that

$$\frac{E_0(N,L)}{N} \geq 4\pi\mu a\rho \left(1 - \frac{1}{\rho\ell^3}\right) K(4\rho\ell^3, \ell). \tag{2.50}$$

What remains is to take a closer look at $K(4\rho\ell^3, \ell)$, which depends on the parameters ε and R besides ℓ, and choose the parameters in an optimal way. ¿From (2.38) and (2.39) we obtain

$$\begin{aligned} K(n,\ell) &= (1-\varepsilon)\left(1 - \tfrac{2R}{\ell}\right)^3 \left(1 + \tfrac{4\pi}{3}\rho(1 - \tfrac{1}{n})(R^3 - R_0^3)\right)^{-1} \\ &\quad \times \left(1 - \frac{3}{\pi} \frac{an}{(R^3 - R_0^3)(\varepsilon\ell^{-2} - 4a\ell^{-3} n(n-1))}\right). \end{aligned} \tag{2.51}$$

The estimate (2.45) with this K is valid as long as the denominator in the last factor in (2.51) is ≥ 0, and in order to have a formula for all n we can take 0 as a trivial lower bound in other cases or when (2.45) is negative. As required for (2.46), K is monotonously decreasing in n. We now insert $n = 4\rho\ell^3$ and obtain

$$\begin{aligned} K(4\rho\ell^3,\ell) &\geq (1-\varepsilon)\left(1-\tfrac{2R}{\ell}\right)^3\left(1+(\text{const.})Y(\ell/a)^3(R^3-R_0^3)/\ell^3\right)^{-1} \\ &\times \left(1-\frac{\ell^3}{(R^3-R_0^3)}\frac{(\text{const.})Y}{(\varepsilon(a/\ell)^2-(\text{const.})Y^2(\ell/a)^3)}\right) \end{aligned} \tag{2.52}$$

with $Y = 4\pi\rho a^3/3$ as before. Also, the factor

$$\left(1-\frac{1}{\rho\ell^3}\right) = (1-(\text{const.})Y^{-1}(a/\ell)^3) \tag{2.53}$$

in (2.50) (which is the ratio between $n(n-1)$ and n^2) must not be forgotten. We now make the ansatz

$$\varepsilon \sim Y^\alpha, \quad a/\ell \sim Y^\beta, \quad (R^3-R_0^3)/\ell^3 \sim Y^\gamma \tag{2.54}$$

with exponents α, β and γ that we choose in an optimal way. The conditions to be met are as follows:

- $\varepsilon(a/\ell)^2 - (\text{const.})Y^2(\ell/a)^3 > 0$. This holds for all small enough Y, provided $\alpha + 5\beta < 2$ which follows from the conditions below.
- $\alpha > 0$ in order that $\varepsilon \to 0$ for $Y \to 0$.
- $3\beta - 1 > 0$ in order that $Y^{-1}(a/\ell)^3 \to 0$ for $Y \to 0$.
- $1 - 3\beta + \gamma > 0$ in order that $Y(\ell/a)^3(R^3 - R_0^3)/\ell^3 \to 0$ for $Y \to 0$.
- $1 - \alpha - 2\beta - \gamma > 0$ to control the last factor in (2.52).

Taking

$$\alpha = 1/17, \quad \beta = 6/17, \quad \gamma = 3/17 \tag{2.55}$$

all these conditions are satisfied, and

$$\alpha = 3\beta - 1 = 1 - 3\beta + \gamma = 1 - \alpha - 2\beta - \gamma = 1/17. \tag{2.56}$$

It is also clear that $2R/\ell \sim Y^{\gamma/3} = Y^{1/17}$, up to higher order terms. This completes the proof of Theorems 3.1 and 3.2, for the case of potentials with finite range. By optimizing the proportionality constants in (2.54) one can show that $C = 8.9$ is possible in Theorem 1.1 [**10**]. The extension to potentials of infinite range decreasing faster than $1/r^3$ at infinity is obtained by approximation by finite range potentials, controlling the change of the scattering length as the cut-off is removed. See Appendix B in [**3**] for details. A slower decrease than $1/r^3$ implies infinite scattering length. □

The exponents (2.55) mean in particular that

$$a \ll R \ll \rho^{-1/3} \ll \ell \ll (\rho a)^{-1/2}, \tag{2.57}$$

whereas Dyson's method required $R \sim \rho^{-1/3}$ as already explained. The condition $\rho^{-1/3} \ll \ell$ is required in order to have many particles in each box and thus $n(n-1) \approx n^2$. The condition $\ell \ll (\rho a)^{-1/2}$ is necessary for a spectral gap $\gg e_0(\rho)$ in Temple's inequality. It is also clear that this choice of ℓ would lead to a far too big energy and no bound for $e_0(\rho)$ if we had chosen Dirichlet instead of Neumann boundary conditions for the cells. But with the latter the method works!

3. The Dilute Bose Gas in 2D

The two-dimensional theory, in contrast, began to receive attention only relatively late. The first derivation of the asymptotic formula was, to our knowledge, done by Schick [**19**], as late as 1971! He found

$$e_0(\rho) \simeq \frac{4\pi\mu\rho}{|\ln(\rho a^2)|}. \tag{3.1}$$

The scattering length a in (3.1) is defined using the zero energy scattering equation (2.3) but instead of $\psi(r) \approx 1-a/r$ we now impose the asymptotic condition $\psi(r) \approx \ln(r/a)$. This is explained in the appendix to [**2**].

Note that the answer could not possibly be $e_0(\rho) \simeq 4\pi\mu\rho a$ because that would be dimensionally wrong. But $e_0(\rho)$ must essentially be proportional to ρ, which leaves no room for an a dependence — which is ridiculous! It turns out that this dependence comes about in the $\ln(\rho a^2)$ factor.

One of the intriguing facts about (3.1) is that the energy for N particles is *not equal* to $N(N-1)/2$ times the energy for two particles in the low density limit — as is the case in three dimensions. The latter quantity, $E_0(2, L)$, is, asymptotically, for large L, equal to $8\pi\mu L^{-2}\left[\ln(L^2/a^2)\right]^{-1}$. Thus, if the $N(N-1)/2$ rule were to apply in 2D, (3.1) would have to be replaced by the much smaller quantity $4\pi\mu\rho\left[\ln(L^2/a^2)\right]^{-1}$. In other words, L, which tends to ∞ in the thermodynamic limit, has to be replaced by the mean particle separation, $\rho^{-1/2}$ in the logarithmic factor. Various poetic formulations of this curious fact have been given, but it remains true that the non-linearity is something that does not occur in more than two-dimensions and its precise nature is hardly obvious, physically. This anomaly is the main reason that the 2D result is not a trivial extension of [**1**].

The proof of (3.1) is in [**2**]. The (relative) error terms to (3.1) given in [**2**] are $|\ln(\rho a^2)|^{-1}$ for the upper bound and $|\ln(\rho a^2)|^{-1/5}$

To prove (3.1) the essential new step is to modify Dyson's lemma for 2D. The rest of the proof parallels that for 3D. The 2D version of Lemma 2.5 is [**2**]:

LEMMA 3.1. *Let $v(r) \geq 0$ with finite range R_0. Let $U(r) \geq 0$ be any function satisfying $\int U(r)\ln(r/a)rdr \leq 1$ and $U(r) = 0$ for $r < R_0$. Let $\mathcal{B} \subset \mathbf{R}^3$ be star shaped with respect to 0 (e.g. convex with $0 \in \mathcal{B}$). Then for all differentiable functions ψ*

$$\int_{\mathcal{B}} \left[\mu|\nabla\psi|^2 + \tfrac{1}{2}v|\psi|^2\right] \geq \mu \int_{\mathcal{B}} U|\psi|^2. \tag{3.2}$$

4. Bose-Einstein Condensation

Let us comment very briefly on the notion of Bose-Einstein condensation (BEC). Given the normalized ground state wave function $\Psi_0(\vec{x}_1, \dots, \vec{x}_N)$ we can form the one-body density matrix which is an operator on $L^2(\mathbf{R}^n)$ ($n = 2$ or 3) given by the kernel

$$\gamma(\vec{x}, \vec{y}) = N \int_{\text{BOX}^{N-1}} \Psi_0(\vec{x}, \vec{x}_2, \dots, \vec{x}_N)\Psi_0(\vec{y}, \vec{x}_2, \dots, \vec{x}_N) d\vec{x}_2 \cdots d\vec{x}_N.$$

Then $\int \gamma(\vec{x}, \vec{x})d\vec{x} = \text{Trace}(\gamma) = N$. BEC is the assertion that this operator has an eigenvalue of order N. Since γ is a positive kernel and, hopefully, translation invariant in the thermodynamic limit, the eigenfunction belonging to the largest eigenvalue must be the constant function (volume)$^{-1/2}$. Therefore, another way to say that BEC exists is that

$$\int\int \gamma(\vec{x}, \vec{y}) d\vec{x} d\vec{y} = O(N).$$

Unfortunately, this is something that is frequently invoked but never proved — except for one special case: hard core bosons on a lattice at half-filling (i.e., $N =$ half the number of lattice sites). The proof is in [**20**].

The problem remains open after about 70 years. It is not at all clear that BEC is essential for superfluidity, as frequently claimed. Our construction in section 2 shows that BEC exists on a length scale of order $\rho^{-1/3}Y^{-1/17}$ which, unfortunately, is not a 'thermodynamic' length like volume$^{1/3}$.

5. Gross-Pitaevskii Equation for Trapped Bosons

In the recent experiments on Bose condensation, the particles have to be confined in a cold 'trap' instead of a 'box' and we are certainly not at the 'thermodynamic limit'. For a 'trap' we add a *slowly* varying confining potential V, with $V(\vec{x}) \to \infty$ as $|\vec{x}| \to \infty$. The Hamiltonian becomes

$$H = \sum_{i=1}^{N} -\mu\Delta_i + V(\vec{x}_i) + \sum_{1\leq i<j\leq N} v(|\vec{x}_i - \vec{x}_j|) \tag{5.1}$$

If $v = 0$, then

$$\Psi_0(\vec{x}_1, \ldots, \vec{x}_N) = \prod_{i=1}^{N} \Phi_0(\vec{x}_i)$$

with Φ_0= normalized ground state of $-\Delta + V(\vec{x})$ with eigenvalue $= \lambda$.

The idea is to use the information about the thermodynamic limiting energy of the dilute Bose gas in a box to find the ground state energy of (5.1). This has been done with Robert Seiringer and Jakob Yngvason [**3, 4**].

As we saw in Sections 2 and 3 there is a difference in the ρ dependence between two and three dimensions, so we can expect a related difference now. We discuss 3D first.

5.1. Three Dimensions. Associated with the quantum mechanical ground state energy problem is the GP energy functional

$$\mathcal{E}^{\rm GP}[\Phi] = \int_{R^3} \left(\mu|\nabla\Phi|^2 + V|\Phi|^2 + 4\pi\mu a|\Phi|^4\right) d^3\vec{x} \tag{5.2}$$

with the subsidiary condition

$$\int_{R^3} |\Phi|^2 = N$$

and corresponding energy

$$E^{\rm GP}(N, a) = \inf\{\mathcal{E}^{\rm GP}[\Phi] \ : \ \int |\Phi|^2 = N\} = \mathcal{E}^{\rm GP}[\Phi^{\rm GP}]. \tag{5.3}$$

As before, a is the scattering length of v.

It is not hard to prove that for every choice of the real number N there is a unique minimizer for $\Phi^{\rm GP}$ for $\mathcal{E}^{\rm GP}$.

Relation of $\mathcal{E}^{\rm GP}$ and E_0: If $v = 0$ then clearly $\Phi^{\rm GP} = \sqrt{N}\,\Phi_0$, and then $\mathcal{E}^{\rm GP} = N\lambda = E_0$. In the other extreme, if $V(\vec{x}) = 0$ for $\vec{x}$ inside a large box of volume L^3 and $V(\vec{x}) = \infty$ otherwise, then we take $\Phi^{\rm GP} \approx \sqrt{N/L^3}$ and we get $E^{\rm GP}(N) = 4\pi\mu a N^2/L^3$ = previous, homogeneous E_0 in the low density regime. (In this case, the gradient term in $\mathcal{E}^{\rm GP}[\Phi]$ plays no role.)

In general, we expect that $E^{\rm GP} = E_0$ in a suitable limit. This limit has to be chosen so that *all three* terms in $\mathcal{E}^{\rm GP}[\Phi]$ make a contribution. It turns out that fixing Na is the right thing to do (and this can be quite good experimentally since Na can range from about 1 to 1000).

THEOREM 5.1. *If $a_1 = Na$ is fixed,*

$$\lim_{N\to\infty} \frac{E_0(N,a)}{E^{\rm GP}(N,a)} = 1$$

Moreover, the GP density is a limit of the QM density (convergence in weak L_1-sense):

$$\lim_{N\to\infty} \frac{1}{N}\rho^{\rm QM}_{N,a}(\vec{x}) = \left|\Phi^{\rm GP}_{1,Na}(\vec{x})\right|^2 .$$

We could imagine, instead, an $N \to \infty$ limit in which $a \gg N^{-1}$, i.e. $a_1 \to \infty$, but still $\bar{\rho}a^3 \to 0$, where $\bar{\rho}$ is the average density, given by

$$\bar{\rho} = \frac{1}{N}\int (\Phi^{\rm GP})^4. \tag{5.4}$$

In this case we simply omit the $\int|\nabla\Phi|^2$ term in $\mathcal{E}^{\rm GP}$; this theory is usually called 'Thomas-Fermi theory', but it has nothing to do with the fermionic theory invented by Thomas and Fermi in 1927. It is appropriate for the case in which Na is much bigger than 1, e.g., $Na \approx 1000$.

PROOF. **Outline:** Getting an upper bound for E_0 in terms of $E^{\rm GP}$ is relatively easy, as before. The problem is the lower bound.

One might suppose that one decomposes R^3 into small boxes as before, with Neumann b.c. on each box, and in each box one approximates $V(\vec{x})$ by a constant.

This will NOT work, even if $v = 0$, because all the particles will then want to be in the box with the smallest value of V. The gradient term will vanish and we will not get $E^{GP} = N\lambda$.

The trick is to write the quantum Ψ_0 as

$$\Psi_0 = \textstyle\prod_{i=1}^N \Phi^{\rm GP}(\vec{x}_i)F(\vec{x}_1,\ldots,\vec{x}_N).$$

This leads to a variational problem for F instead of Ψ_0. Partial integration and the variational equation for $\Phi^{\rm GP}$ lead to the replacements:

$$\text{External potential: } V(\vec{x}) \quad\to\quad -8\pi a|\Phi^{\rm GP}(\vec{x})|^2$$

and

$$\text{Measure: } \textstyle\prod_{i=1}^N d\vec{x}_i \quad\to\quad \prod_{i=1}^N |\Phi^{\rm GP}(\vec{x}_i)|^2 d\vec{x}_i.$$

We have to show that, with these replacements, the energy is bounded below by $-4\pi\mu a\int|\Phi^{\rm GP}|^4$ (up to small errors). *Now* we can effectively use the Neumann box method on *this* F problem.

There are still plenty of technical difficulties, but the back of the problem has been broken by recasting it in terms of F instead of Ψ_0. □

5.2. Two Dimensions. In view of the two-dimensional result (3.1) we might suppose that we must replace $4\pi\mu a\Phi^4$ in (5.2) by $4\pi\mu\Phi^4/|\ln(\Phi^2a^2)|$, as suggested by (3.1). This is indeed correct, and a limit theorem for the energy and density is obtained that is similar to Theorem 5.1. There is a difference, however. This time one has to fix the ratio $N/\ln a$ instead of Na.

Since the logarithm is such a slowly varying function it turns out that one can still get the correct limit and yet simplify the problem by replacing $|\ln(\Phi^2a^2)|^{-1}$ in the functional by the constant $|\ln(\bar{\rho}a^2)|^{-1}$, where $\bar{\rho}$ is the average density in (5.4), but for 2D, of course.

6. The Charged Bose Gas

The setting now changes abruptly. Instead of particles interacting with a short-range potential $v(\vec{x}_i - \vec{x}_j)$ they interact via the Coulomb potential

$$v(|\vec{x}_i - \vec{x}_j|) = |\vec{x}_i - \vec{x}_j|^{-1}$$

(in 3 dimensions). There are N particles in a large box of volume L^3 as before, with $\rho = N/L^3$. We know from the work in [**21**] that a nice thermodynamic limit exists for $e_0 = E_0/N$.

To offset the huge Coulomb repulsion (which would drive the particles to the walls of the box) we add a uniform negative background of precisely the same charge, namely density ρ. Our Hamiltonian is thus

$$H = -\mu \sum_{i=1}^{N} \Delta_i - V(\vec{x}_i) + \sum_{1\le i<j\le N} v(\vec{x}_i - \vec{x}_j) + C \tag{6.1}$$

with

$$V(\vec{x}) = \rho \int_{\text{BOX}} |\vec{x} - \vec{y}|^{-1} d^3y \qquad \text{and} \qquad C = \frac{1}{2}\rho \int_{\text{BOX}} V(\vec{x}) d^3x \, .$$

Despite the fact that the Coulomb potential is positive definite, each particle interacts only with others and not with itself. Thus, E_0 can be (and is) negative (just take Ψ =const). This time, *large* ρ is the 'weakly interacting' regime.

Another way in which this problem is different from the previous one is that *perturbation theory is correct to leading order.* If one computes $(\Psi, H\Psi)$ with Ψ =const, one gets the right first order answer, namely 0. It is the next order in $1/\rho$ that is interesting, and this is *entirely* due to correlations. In 1961 Foldy [**22**] calculated this correlation energy according to the prescription of Bogolubov's 1947 theory. That theory was not exact for the dilute Bose gas, as we have seen, even to first order . We are now looking at *second* order, which should be even worse. Nevertheless, there was good physical intuition that this calculation should be asymptotically *exact.* *It is!* and this was recently proved with Jan Philip Solovej [**5**].

The Bogolubov theory states that the main contribution to the energy comes from pairing of particles into momenta $\vec{k}, -\vec{k}$ and is the bosonic analogue of the BCS theory of superconductivity which came a decade later. I.e., Ψ_0 is a sum of products of germs of the form $\exp\{i\vec{k}\cdot(\vec{x}_i - \vec{x}_j)\}$.

Foldy's energy, based on Boglubov's ansatz, has now been proved. His calculation yields an upper bound. The lower bound is the hard part, and Solovej and I do this using the decomposition into 'Neumann boxes' as above. But unlike the short range case, many complicated gymnastics are needed to control the long range $|\vec{x}|^{-1}$ Coulomb potential. The two bounds agree to leading order in ρ, namely $\rho^{1/4}$ (although there is an unimportant technical point that different boundary conditions are used for the two bounds).

THEOREM 6.1. *For large* ρ

$$e_0(\rho) \approx -0.402\, r_s^{-3/4} \frac{me^4}{\hbar^2} \tag{6.2}$$

where $r_s = (3/4\pi\rho)^{1/3} e^2 m/\hbar^2$.

This is the *first example* (in more than 1 dimension) in which Bogolubov's pairing theory has been rigorously validated. It has to be emphasized, however, that Foldy and Bogolubov rely on the existence of Bose-Einstein condensation. We neither make such a hypothesis nor does our result for the energy imply the existence of such condensation. As we said earlier, it is sufficient to prove condensation in small boxes of fixed size.

Incidentally, the one-dimensional example for which Bogolubov's theory is asymptotically exact to the first two orders (high density) is the repulsive delta-function Bose gas [**17**]

To appreciate the $-\rho^{1/4}$ nature of (6.2), it is useful to compare it with what one would get if the bosons had infinite mass, i.e., the first term in (6.1) is dropped. Then the energy would be proportional to $-\rho^{1/3}$ as shown in [**21**]. Thus, the effect of quantum mechanics is to lower $\frac{1}{3}$ to $\frac{1}{4}$.

It is supposedly true that there is a critical mass above which the ground state should show crystalline ordering (Wigner crystal), but this has never been proved and it remains an intriguing open problem, even for the infinite mass case. Since the relevant parameter is r_s, large mass is the same as small ρ, and is outside the region where a Bogolubov approximation can be expected to hold.

Another important remark about the $-\rho^{1/4}$ law is its relation to the $-N^{7/5}$ law for a *two*-component charged Bose gas. Dyson [**23**] proved that the ground state energy for such a gas was at least as negative as $-(\text{const})N^{7/5}$ as $N \to \infty$. Thus, thermodynamic stability (i.e., a linear lower bound) fails for this gas. Years later, a lower bound of this $-N^{7/5}$ form was finally established in [**24**], thereby proving that this law is correct. The connection of this $-N^{7/5}$ law with the jellium $-\rho^{1/4}$ law (for which a corresponding lower bound was also given in [**24**]) was pointed out by Dyson [**23**] in the following way. Assuming the correctness of the $-\rho^{1/4}$ law, one can treat the 2-component gas by treating each component as a background for the other. What should the density be? If the gas has a radius L and if it has N bosons then $\rho = NL^{-3}$. However, the extra kinetic energy needed to compress the gas to this radius is NL^{-2}. The total energy is then $NL^{-2} - N\rho^{1/4}$, and minimizing this with respect to L leads to the $-N^{7/5}$ law. A proof going in the other direction is in [**24**].

A problem somewhat related to bosonic jellium is *fermionic* jellium. Graf and Solovej [**25**] have proved that the first two terms are what one would expect, namely

$$e(\rho) = C_{TF}\rho^{5/3} - C_D\rho^{4/3} + o(\rho^{4/3}),$$

where C_{TF} is the usual Thomas-Fermi constant and C_D is the usual Dirac exchange constant.

Let us conclude with a few more details about some of the technicalities.

6.1. Foldy's Calculation, Pairing Theory and Ideas in the Rigorous Proof. Foldy uses periodic boundary conditions for $-\Delta$, so the problem is on a torus. In order to make the Coulomb potential periodic and to take care of the background he replaces $|x-y|^{-1}$ by

$$\sum_{p\neq 0} L^{-3}|p|^{-2}\exp(ip(x-y)),$$

where the sum is over 'periodic momenta', p. (Note the $p\neq 0$ condition, so the spatial average of this potential is 0).

Foldy's Hamiltonian is

$$H' = \sum_{i=1}^{N} -\tfrac{1}{2}\Delta_i + \sum_{i<j}\sum_{p\neq 0} L^{-3}|p|^{-2}\exp(ip(x_i-x_j)),$$

in which the $p\neq 0$ condition is supposed to make up for the background that is not explicitly included in H'. It is 'physically clear' that this device works, but a rigorous proof of it is not easy.

The next step is to use the second quantization formalism, which is the one used by Bogolubov and which is a very convenient bookkeeping device (but it has to be noted that it is no more than a convenient device and it does not introduce any new physics or mathematics).

$$H' = \sum_{p} |p|^2 a_p^* a_p + \sum_{p\neq 0} L^{-3}|p|^{-2}\sum_{k,q} a_{k+p}^* a_{q-p}^* a_q a_k,$$

where the a_p operators satisfy the usual bosonic commutation relations. An important observation is that since $p\neq 0$ there are no terms with 3 or 4 $a_0^\sharp$. This is different from the situation with the usual short range potential treated in Section 2, and it means that the leading term in perturbation theory vanishes. Everything now comes from the terms with two $a_0^\sharp$.

6.1.1. *The Bogolubov approximation:* The motivation is Bose condensation: Almost all particles are in the state of momentum $p=0$ created by a_0^*. Thus:

Step 1 in the Bogolubov approximation: Keep only quartic terms with precisely two $a_0^\sharp$ (and ignore terms with one or no $a_0^\sharp$). We obtain the reduced Hamiltonian

$$\begin{aligned} H'' &= \sum_{p} |p|^2 a_p^* a_p + \sum_{p\neq 0} L^{-3}|p|^{-2}\Big[a_p^* a_0^* a_p a_0 \\ &\quad + a_0^* a_{-p}^* a_0 a_{-p} + a_p^* a_{-p}^* a_0 a_0 + a_0^* a_0^* a_p a_{-p}\Big] \end{aligned}$$

Step 2 in the Bogolubov approximation: Replace the *operators* $a_0^\sharp$ by the number $\sqrt{N}$. We then obtain:

$$\begin{aligned} H''' &= \sum_{p\neq 0} |p|^2 a_p^* a_p + \rho|p|^{-2}\Big[a_p^* a_p + a_{-p}^* a_{-p} \\ &\quad + a_p^* a_{-p}^* + a_p a_{-p}\Big] \end{aligned}$$

This is a quadratic Hamiltonian and can be diagonalized by completing the square:

$$\begin{aligned} H''' &= \sum_p A_p(a_p^* + \beta_p a_{-p})(a_p + \beta_p a_{-p}^*) \\ &\quad + A_p(a_{-p}^* + \beta_p a_p)(a_{-p} + \beta_p a_p^*) \\ &\quad - 2\sum_{p\neq 0} A_p \beta_p^2 \end{aligned}$$

The last term comes from the commutator $[a_p, a_q^*] = \delta_{pq}$.

$$\begin{aligned} A_p(1+\beta_p^2) &= \frac{1}{2}|p|^2 + \rho|p|^{-2} \\ 2A_p\beta_p &= \rho|p|^{-2} \end{aligned}$$

The ground state energy is given by the last term above.

$$e = \lim_{L\to\infty} -\frac{2}{L^3}\sum_{p\neq 0} A_p\beta_p^2 = \int A_p\beta_p^2 = C_F\rho^{5/4}.$$

In this approximation the ground state wave function ψ satisfies

$$(a_p + \beta_p a_{-p}^*)\psi = 0,$$

for all $p \neq 0$.

In the original language (in which a_0 an operator) this corresponds to a function of the form

$$\begin{aligned} \psi &= 1 + \sum_{i<j} f(x_i - x_j) \\ &\quad + c \sum_{\substack{i,j,l,k \\ \text{different}}} f(x_i - x_j) f(x_l - x_k) + \ldots \end{aligned}$$

where $\hat{f}(p) = \beta_p$. In fact, $\hat{f}(p) = G(|p|^4/\rho)$, with G independent of ρ. Thus f varies on a length scale $\rho^{-1/4}$ (which is the typical inter*pair* distance).

6.1.2. *Ideas in the rigorous proof:* As in the short range case in Section 2, there is no need to prove Bose condensation globally. It is enough to do so on a short scale.

- Localize particles, by means of Neumann bracketing, in "small" boxes of size ℓ. The constant function (i.e., the 'condensate') is not affected by this localization since the constant function in a small box satisfies Neumann boundary conditions. The function f discussed above is not affected very much if $\ell \gg \rho^{-1/4}$. We choose ℓ close to $\rho^{-1/4}$.
- We control the Coulomb interaction between boxes by using an averaging method in [**24**]. The error in neglecting intercell interactions can be shown to be dominated by $N/\ell \ll N\rho^{1/4}$.
- We establish condensation on the scale ℓ by noting that the first non-zero Neumann eigenvalue is $\sim \ell^{-2}$. If N_+ is the expected number N_+ of particles not in the condensate in the "small box", their energy is bounded from below by $N_+\ell^{-2} \sim N_+\rho^{1/2}$. If this is to be consistent with the total energy $-N\rho^{1/4}$ we should expect to have $N_+ \ll N\rho^{-1/4}$, i.e., local condensation exists.
- This is established by means of a bootstrap procedure.

- Having established local condensation one starts the hard work of proving that the Bogolubov 'quadratic' approximation really gives the leading term in the energy. There is however a new difficulty that arises from the finite size of the small boxes in which we are working. Neumann, and not periodic boundary conditions must be used. Since we no longer have 'momentum conservation' in the small boxes, diagonalization of the Hamiltonian by 'completing the square' is not the simple algebraic problem it was before.
- Nevertheless, it can all be carried to a successful conclusion

References

[1] E.H. Lieb, J. Yngvason, *Ground State Energy of the low density Bose Gas*, Phys. Rev. Lett. **80**, 2504–2507 (1998). arXiv math-ph/9712138, mp_arc 97-631

[2] E.H. Lieb, J. Yngvason, *The Ground State Energy of a Dilute Two-dimensional Bose Gas*, J. Stat. Phys. (in press) arXiv math-ph/0002014, mp_arc 00-63.

[3] E.H. Lieb, R. Seiringer, and J. Yngvason, *Bosons in a Trap: A Rigorous Derivation of the Gross-Pitaevskii Energy Functional*, Phys. Rev A **61**, 043602-1 – 043602-13 (2000)mp_arc 99-312, arXiv math-ph/9908027 (1999).

[4] E.H. Lieb, R. Seiringer and J. Yngvason, *A Rigorous Derivation of the Gross-Pitaevskii Energy Functional for a Two-dimensional Bose Gas*, Commun. Math. Phys. (in press). arXiv cond-mat/0005026, mp_arc 00-203.

[5] E.H. Lieb, J.P. Solovej, *Ground State Energy of the One-Component Charged Bose Gas*, Commun. Math. Phys. bf 217, 127–163 (2001). arXiv cond-mat/0007425, mp_arc 00-303.

[6] E.H. Lieb, J. Yngvason, *The Ground State Energy of a Dilute Bose Gas*, in *Differential Equations and Mathematical Physics, University of Alabama, Birmingham, 1999*, R. Weikard and G. Weinstein, eds., 271-282 Amer. Math. Soc./Internat. Press (2000). arXiv math-ph/9910033, mp_arc 99-401.

[7] E.H. Lieb, R. Seiringer, J. Yngvason, *The Ground State Energy and Density of Interacting Bosons in a Trap*, in *Quantum Theory and Symmetries*, Goslar, 1999, H.-D. Doebner, V.K. Dobrev, J.-D. Hennig and W. Luecke, eds., pp. 101-110, World Scientific (2000). arXiv math-ph/9911026, mp_arc 99-439.

[8] W. Ketterle, N. J. van Druten, *Evaporative Cooling of Trapped Atoms*, in B. Bederson, H. Walther, eds., Advances in Atomic, Molecular and Optical Physics, **37**, 181–236, Academic Press (1996).

[9] F. Dalfovo, S. Giorgini, L.P. Pitaevskii, and S. Stringari, *Theory of Bose-Einstein condensation in trapped gases*, Rev. Mod. Phys. **71**, 463–512 (1999).

[10] R. Seiringer, Diplom thesis, University of Vienna, (1999).

[11] N.N. Bogolubov, J. Phys. (U.S.S.R.) **11**, 23 (1947); N.N. Bogolubov and D.N. Zubarev, Sov. Phys.-JETP **1**, 83 (1955).

[12] K. Huang, C.N. Yang, Phys. Rev. **105**, 767-775 (1957); T.D. Lee, K. Huang, and C.N. Yang, Phys. Rev. **106**, 1135-1145 (1957); K.A. Brueckner, K. Sawada, Phys. Rev. **106**, 1117-1127, 1128-1135 (1957).; S.T. Beliaev, Sov. Phys.-JETP **7**, 299-307 (1958); T.T. Wu, Phys. Rev. **115**, 1390 (1959); N. Hugenholtz, D. Pines, Phys. Rev. **116**, 489 (1959); M. Girardeau, R. Arnowitt, Phys. Rev. **113**, 755 (1959); T.D. Lee, C.N. Yang, Phys. Rev. **117**, 12 (1960).

[13] E.H. Lieb, *Simplified Approach to the Ground State Energy of an Imperfect Bose Gas*, Phys. Rev. **130** (1963), 2518–2528. See also Phys. Rev. **133** (1964), A899–A906 (with A.Y. Sakakura) and Phys. Rev. **134** (1964), A312–A315 (with W. Liniger).

[14] E.H. Lieb, *The Bose fluid*, in W.E. Brittin, ed., Lecture Notes in Theoretical Physics VIIC, Univ. of Colorado Press, pp. 175–224 (1964).

[15] F.J. Dyson, *Ground-State Energy of a Hard-Sphere Gas*, Phys. Rev. **106**, 20–24 (1957).

[16] B. Baumgartner, *The Existence of Many-particle Bound States Despite a Pair Interaction with Positive Scattering Length*, J. Phys. A **30** (1997), L741–L747.

[17] E.H. Lieb, W. Liniger, *Exact Analysis of an Interacting Bose Gas. I. The General Solution and the Ground State*, Phys. Rev. **130**, 1605-1616 (1963); E.H. Lieb, *Exact Analysis of an Interacting Bose Gas. II. The Excitation Spectrum*, Phys. Rev. **130**, 1616-1624 (1963).

[18] G. Temple, *The theory of Rayleigh's Principle as Applied to Continuous Systems*, Proc. Roy. Soc. London A **119** (1928), 276–293.

[19] M. Schick, *Two-Dimensional System of Hard Core Bosons*, Phys. Rev. A **3**, 1067-1073 (1971).

[20] T. Kennedy, E.H. Lieb, S. Shastry, *The XY Model has Long-Range Order for all Spins and all Dimensions Greater than One*, Phys. Rev. Lett. textbf 61, 2582-2584 (1988).

[21] E.H. Lieb, H. Narnhofer, *The Thermodynamic Limit for Jellium*, J. Stat. Phys. **12**, 291-310 (1975). Errata J. Stat. Phys. **14**, 465 (1976).

[22] L.L. Foldy, *Charged Boson Gas*, Phys. Rev. **124**, 649-651 (1961); Errata *ibid* **125**, 2208 (1962).

[23] F.J. Dyson, *Ground State Energy of a Finite System of Charged Particles*, J. Math. Phys. **8**, 1538-1545 (1967).

[24] J. Conlon, E.H. Lieb, H-T. Yau *The $N^{7/5}$ Law for Charged Bosons*, Commun. Math. Phys. **116**, 417-448 (1988).

[25] G.M. Graf, J.P. Solovej, *A correlation estimate with applications to quantum systems with Coulomb interactions*, Rev. Math. Phys., 6 (No. 5a, Special Issue) 977-997 (1994). mp_arc 93-60.

DEPARTMENTS OF PHYSICS AND MATHEMATICS, PRINCETON UNIVERSITY, PRINCETON, NEW JERSEY 08544-0708

E-mail address: lieb@princeton.edu

THE $N^{5/3}$ LAW FOR BOSONS

Elliott H. LIEB [1]
Departments of Mathematics and Physics, Princeton University, Princeton, NJ 08540, USA

Received 13 December 1978

Non-relativistic negative bosons interacting with infinite mass positive particles via Coulomb forces are shown to be unstable in the sense that $E_0 \leqslant -CN^{5/3}$. This agrees with the previously known lower bound $E_0 \geqslant -AN^{5/3}$.

In a celebrated series of papers [1–3], Dyson and Lenard proved that matter is quantum-mechanically stable under the action of Coulomb forces provided all the particles of at least one sign of charge (say negative) are fermions. In other words, the ground state energy E_0 satisfies

$$E_0 \geqslant -A_f N \quad \text{(negative fermions)}\,, \tag{1}$$

where N is the number of negative particles. It is not necessary to assume neutrality or that the positive particles have finite mass. It is necessary to assume that all the charges are bounded however. (1) was subsequently rederived by Federbush [4] and by Lieb and Thirring [5]. The best current value [6] is

$$A_f \leqslant -22.24 \text{ Ry}\,, \tag{2}$$

for electrons and protons, and with the assumption of neutrality.

If all the particles are bosons or, what is the same thing, are not subject to any statistics, the best available lower bound [1–7] is

$$E_0 \geqslant -A_b N^{5/3} \quad \text{(all bosons)}\,, \tag{3}$$

with [6]

$$A_b \leqslant 14.01 \text{ Ry}\,, \tag{4}$$

when the positive and negative particles have charges $\pm e$ and the negative particle mass is m_e. The positive particles can have infinite mass.

[1] Work partially supported by U.S. National Science Foundation grants MCS 75-21684 A02 and INT 78-01160.

Dyson [8] then proved, by a complicated variational calculation, that in the boson case

$$E_0 \leqslant -BN^{7/5}\,, \tag{5}$$

if all particles have finite mass. It was conjectured [1,8] that $N^{7/5}$ is the correct law for bosons and not $N^{5/3}$. While this question might have only moderate practical importance (it would be relevant for π^- mesons and ^{4}He nuclei, for example), it has great theoretical importance and it is to be hoped that its solution will soon be forthcoming. It is interesting because at this point there is no simple, compelling physical argument, as distinguished from a computational argument [9], why the $N^{7/5}$ law is correct. Subtle correlation effects are yet to be understood fully and rigorously.

The purpose of this paper is to add a minor commentary on the problem. By means of a simple variational calculation, it will be shown that the $N^{5/3}$ law is indeed correct if the positive particles have *infinite mass* and charge $z|e| > 0$, i.e.

$$E_0 \leqslant -CN^{5/3}\,. \tag{6a}$$

If the negative particles have mass m_e and charge $-|e|$, and if the system is neutral, then

$$C \geqslant (1/108) z^{4/3} \text{ Ry}\,. \tag{6b}$$

Thus, the limits $N \to \infty$ and the mass of the positive particles $\to \infty$ are not interchangeable if the $N^{7/5}$ conjecture is correct.

We note in passing that (1) alone does not imply that $e_0 = \lim_{N\to\infty} E_0/N$ exists. However, the method developed to prove the existence of the thermodynamic limit [10] also proves that this limit exists. Likewise (3) and (6a) do not imply that $\lim_{N\to\infty} E_0/N^{5/3}$ exists. The aforementioned method [10] is not suitable for this task and we do not know an adequate substitute.

For simplicity of exposition we assume there is one kind each of positive and negative particle. The N negative particles have a mass $=1/2$ and charge -1. If $\hbar^2 = 1$, then one Rydberg $= 1/4$. The K infinite mass positive particles have charge $z > 0$ and we assume $N = Kz$ (neutrality) with $K = 8n^3$, n an integer. Other cases can easily be handled by this method but we omit details for simplicity. The hamiltonian for the negative particles is

$$H_{N,R} = -\sum_{i=1}^{N} \{\Delta_i + V_R(r_i)\} + \sum_{1\leqslant i<j\leqslant N} |r_i - r_j|^{-1} + U(R)\,, \tag{7}$$

where $R = \{R_1, \ldots, R_K\}$ is the collection of fixed coordinates of the positive particles and

$$V_R(r) = z\sum_{j=1}^{K} |r - R_j|^{-1}\,,$$

$$U(R) = z^2 \sum_{1\leqslant i<j\leqslant k} |R_i - R_j|^{-1}\,. \tag{8}$$

We want to find a normalized $\psi(r_1, \ldots, r_N)$ and R (depending on N) such that $\langle\psi|H_{N,R}|\psi\rangle \leqslant -CN^{5/3}$.

ψ is chosen to be a simple product of identical functions:

$$\psi(r_1, \ldots, r_N) = \prod_{i=1}^{N} \phi_\lambda(r_i)\,. \tag{9}$$

ϕ_λ depends on the parameter λ (which will turn out to be proportional to $N^{1/3}$) as follows:

$$\phi_\lambda(r) = \lambda^{3/2} g(\lambda r)\,, \tag{10}$$

where $g(r)$ is the fixed, normalized function

$$g(r) = f(x)f(y)f(z)\,, \quad r = (x, y, z)\,, \tag{11}$$

and

$$f(x) = \sqrt{3/2}\,[1 - |x|]\,, \quad |x| \leqslant 1\,, \qquad = 0\,, \quad |x| \geqslant 1\,. \tag{12}$$

The explicit choice in (12) is neither important nor optimal.

With

$$T = \int |\nabla g(r)|^2\, d^3r = 9\,, \tag{13}$$

$$\langle\psi|H_{N,R}|\psi\rangle \leqslant \lambda^2 NT + \lambda W(N, R)\,, \tag{14}$$

$$W(N,R) = \tfrac{1}{2}N^2 \iint g^2(r) g^2(r') |r - r'|^{-1}\, d^3r\, d^3r' - N\int g^2(r) V_R(r)\, d^3r + U(R)\,. \tag{15}$$

There is $\leqslant$ in (14) because we should have $N(N-1)$ instead of N^2 in (15). We claim R can be chosen so that

$$W(N,R) \leqslant -(12)^{-1/2} z^{2/3} N^{4/3}\,. \tag{16}$$

If so, (6) is proved by minimizing (14) with respect to λ.

To prove (16), let $0 = a(0) < a(1) < \ldots < a(n) = 1$ (recall $K = 8n^3$) and define the real intervals $L(j) = [a(j), a(j+1)]$ for $0 \leqslant j \leqslant n-1$ and $L(j) = [-a(-j), -a(-j-1)]$ for $-n \leqslant j \leqslant -1$. Choose the $a(j)$ such that

$$\int_{L(j)} f(x)^2\, dx = (2n)^{-1}\,, \quad \text{for all } j\,. \tag{17}$$

The rectangles $\Gamma(i, j, k) = L(i) \times L(j) \times L(k)$, $-n \leqslant i, j, k < n$ satisfy

$$\int_{\Gamma(i,j,k)} g(r)^2\, d^3r = 1/K\,. \tag{18}$$

Returning to (15), place *one* R_i in each of the K rectangles and then average $W(N,R)$ over the positions, R_i, of the positive particles within the rectangles with a *relative* weight $g(R_1)^2 \ldots g(R_K)^2$. This average is given by

$$W(N) = -\tfrac{1}{2}N^2 \sum_m \iint_{\Gamma(m)} g(r)^2 g(r')^2 \times |r - r'|^{-1}\, d^3r\, d^3r'\,, \tag{19}$$

with $m = (i, j, k)$. Since the weight is nonnegative, there is at least one choice of R (with one particle per rectangle) such that $W(N, R) \leqslant W(N)$. Thus, we are done if $W(N) \leqslant$ right side of (16).

Each integral in (19) is the self-energy of a charge $1/K$ confined to a rectangle. This, in turn, is greater than the *minimum* self-energy of a charge $1/K$ confined to lie only in a circumscribed sphere. If $\Gamma(m)$ has sides of length (s, t, u), this sphere has radius $\rho(m) = \frac{1}{2}[s^2 + t^2 + u^2]^{1/2}$. It is well known that the minimum self-energy occurs when the charge is distributed uniformly on the surface of the sphere and is $\rho(m)^{-1} \times (1/K)^2$. Thus

$$W(N) \leqslant -\tfrac{1}{2}z^2 \sum_m \rho(m)^{-1} \leqslant -z^2K[\sigma^2 + \tau^2 + \mu^2]^{-1/2} \quad \text{(by convexity)}, \tag{20}$$

where σ, τ and μ are the mean lengths of the sides of the rectangles. However, $\sigma = \tau = \mu = 1/n$, and (16) is proved.

The author would like to thank the Research Institute for Mathematical Sciences, Kyoto University, in particular Professors H. Araki and K. Itô, for their generous hospitality.

References

[1] F.J. Dyson and A. Lenard, J. Math. Phys. 8 (1967) 423; A. Lenard and F.J. Dyson, J. Math. Phys. 9 (1968) 698.

[2] F.J. Dyson, in: Brandeis University Summer Institute in Theoretical physics (1966) eds. M. Chretien, E.P. Gross and S. Deser (Gordon and Breach, New York 1968), Vol. 1, p. 179.

[3] A. Lenard, in: Statistical mechanics and mathematical problems, ed. A. Lenard, Lecture Notes in Physics (Springer, New York, 1973), Vol. 20, p. 114.

[4] P. Federbush, J. Math. Phys. 16 (1975) 347, 706.

[5] E.H. Lieb and W.E. Thirring, Phys. Rev. Lett. 35 (1975) 687.

[6] E.H. Lieb, Rev. Mod. Phys. 48 (1976) 553.

[7] D. Brydges and P. Federbush, J. Math. Phys. 17 (1976) 2133.

[8] F.J. Dyson, J. Math. Phys. 8 (1967) 1538.

[9] L.L. Foldy, Phys. Rev. 124 (1961) 649;
M. Girardeau and R. Arnowitt, Phys. Rev. 113 (1959) 755;
M. Girardeau, Phys. Rev. 127 (1962) 1809;
W.H. Bassichis and L.L. Foldy, Phys. Rev. 133 (1964) A935;
W.H. Bassichis, Phys. Rev. 134 (1964) A543;
J.M. Stephen, Proc. Phys. Soc. (London) 79 (1962) 994;
D.K. Lee and E. Feenberg, Phys. Rev. 137 (1965) A731;
D. Wright, Phys. Rev. 143 (1966) 91;
E.H. Lieb, Phys. Rev. 130 (1963) 2518;
E.H. Lieb and A.Y. Sakakura, Phys. Rev. 133 (1964) A899.

[10] E.H. Lieb and J.L. Lebowitz, Adv. Math. 9 (1972) 316.

The $N^{7/5}$ Law for Charged Bosons

Joseph G. Conlon[1]*, Elliott H. Lieb[2]** and Horng–Tzer Yau[2]***
[1] Department of Mathematics, University of Missouri, Columbia, MO 65211, USA
[2] Departments of Mathematics and Physics, Princeton University, P.O.B. 708, Princeton, NJ 08544, USA

Abstract. Non-relativistic bosons interacting with Coulomb forces are unstable, as Dyson showed 20 years ago, in the sense that the ground state energy satisfies $E_0 \leqq -AN^{7/5}$. We prove that 7/5 is the correct power by proving that $E_0 \geqq -BN^{7/5}$. For the non-relativistic bosonic, one-component jellium problem, Foldy and Girardeau showed that $E_0 \leqq -CN\rho^{1/4}$. This 1/4 law is also validated here by showing that $E_0 \geqq -DN\rho^{1/4}$. These bounds prove that the Bogoliubov type paired wave function correctly predicts the order of magnitude of the energy.

I. Introduction and Background

Twenty years ago Dyson and Lenard [5] proved the stability of ordinary non-relativistic matter with Coulomb forces, namely that the ground state energy, E_0, of an N-particle system satisfies $E_0 \geqq -A_1 N$ for some universal constant A_1. In ordinary matter, the negative particles (electrons) are fermions. At the same time, Dyson [4] proved that bosonic matter is definitely not stable; if all the particles (positive as well as negative) are bosons then $E_0 \leqq -A_2 N^{7/5}$ for some $A_2 > 0$. Dyson and Lenard [5] did prove, however, that $E_0 \geqq -A_3 N^{5/3}$ in the boson case, and thus the open problem was whether the correct exponent for bosons is 5/3 or 7/5 or something in between.

In this paper we prove that the $N^{7/5}$ law is correct for bosons by obtaining a lower bound $E_0 \geqq -A_4 N^{7/5}$. As is well known, the bosonic energy is the absolute lowest energy when no symmetry restriction is imposed.

It may appear that the difference between 5/3 and 7/5 is insignificant, especially since bosonic matter does not exist experimentally, but that impression would fail to take into account the essential difference between the ground states implied by

* Work partially supported by U.S. National Science Foundation grant DMS 8600748
** Work partially supported by U.S. National Science Foundation grant PHY85-15288-A01
***Work supported by Alfred Sloan Foundation dissertation fellowship

the two laws. The $-A_3N^{5/3}$ lower bound can be derived by using a semiclassical estimate which leads to a Thomas–Fermi type theory. This estimate is the same as that used by Lieb and Thirring [15] to give a simple proof of the stability of matter in the fermion case. Correlations are unimportant in this estimate. The $N^{7/5}$ law, on the other hand, is much more subtle. To get the upper bound, $-A_2N^{7/5}$, Dyson had to use an extremely complicated variational function which contains delicate correlations. It is the same kind of function proposed by Bogoliubov [1] (see also [10] for a review) in his theory of the many-boson system and in which particles of equal and opposite momenta are paired. It is also very similar to the Bardeen–Cooper–Schrieffer pair function of superconductivity. Since this kind of wave function plays such an important role in physics, it is important to know whether it is correct, and in proving the $N^{7/5}$ law for the energy we are, in a certain sense, validating this function.

The Hamiltonian to be considered is

$$H_N = -\sum_{i=1}^{N} \Delta_i + \sum_{1\leqq i<j\leqq N} e_i e_j |x_i - x_j|^{-1}, \tag{1.1}$$

which is relevant for N charged particles with coordinates labeled $x_1, \dots, x_N \in \mathbf{R}^3$. The charges satisfy $e_i = \pm 1$, all i and $\hbar^2/2m = 1$. The neutral case is $\sum e_i = 0$. In Sect. II we shall prove the $N^{7/5}$ lower bound for H_N which is stated precisely in Theorem 1.2 below. The neutrality condition is not imposed in this theorem. If, however, the system is very non-neutral, with N_- negative and N_+ positive particles with $N_- + N_+ = N$ and $N_+ \gg N_-$, we expect that the bounds (1.7) and (1.8) are not optimal. One should have $E_0 \geqq -A_5 N_-^{7/5}$ instead; this is indeed true but, for simplicity of exposition, this generalization is deferred to Sect. V, Theorem 5.1.

A closely related system that we shall consider in Sect. III is jellium. In this case there is a domain Λ, in which there is a fixed constant density, ρ_B, of positive charge called the background. There are also N negative particles of charge -1 and the jellium Hamiltonian is

$$H^J_{N\Lambda} = -\sum_{i=1}^{N} \{\Delta_i + V(x_i)\} + \sum_{1\leqq i<j\leqq N} |x_i - x_j|^{-1} + \tfrac{1}{2}\rho_B \int_\Lambda V(x)dx, \tag{1.2}$$

where $V(x) = \rho_B \int_\Lambda |x-y|^{-1} dy$ is the potential generated by the background. We do *not* restrict ourselves to the neutral case, $N = \rho_B L^3$, in Sect. III. As boundary conditions we can take either $\psi \in L^2(\mathbf{R}^{3N})$ or else $\psi \in L^2(\Lambda^N)$ with Dirichlet Boundary conditions. Clearly E_0 for the former is less than E_0 for the latter. In the physics literature one usually imposes neutrality and takes Λ to be a cube, $\psi \in L^2(\Lambda^N)$ with periodic boundary conditions and, in addition, the potential is replaced by an interaction solely among the negative particles in which the $k=0$ Fourier component of the $1/r$ potential is omitted. It is not a trivial matter to show rigorously that this periodic problem is the same, in the thermodynamic limit, as the more physical problem (1.2) which we consider here—even in the neutral case. Here, again, the bosonic energy is the absolute lowest.

Let us briefly review what is known rigorously about these two problems.

A. Jellium. Foldy [7] was the first to apply Bogoliubov's method to the neutral bosonic jellium problem (with the periodic boundary conditions mentioned above) and obtained, for large ρ_B and in the thermodynamic limit,

$$E_0 \approx -1.933 N \rho_B^{1/4}. \tag{1.3}$$

A proof of (1.3) was, and is lacking, but later, Girardeau [8] proved that (1.3) is an upper bound to E_0 (for large ρ_B and with the same conditions). Another non-rigorous derivation of (1.3) that does not use Bogoliubov's method was given by Lieb and Sakakura [13]. In Sect. III we shall derive the following lower bound for the real problem (1.2).

Theorem 1.1. *With $H^J_{N\Lambda}$ given by (1.2) on $L^2(\mathbf{R}^{3N})$, the ground state energy satisfies, for all N and Λ*

$$E_0 \geqq -A_6 N \rho_B^{1/4}, \tag{1.4}$$

for some universal constant A_6. A bound for A_6 is given in (3.19); In the limit $\rho_B \to \infty$ we can take $A_6 = 8.57$.

Theorem 1.1 is generalized in Theorem 3.1. Note that our lower bound (1.4) is close to the upper bound (1.3) (with a factor about 4.5).

The existence of the thermodynamic limit for jellium was proved by Lieb and Narnhofer [12]. This limit will not concern us here, but a useful result in the appendix of [12] contains a lower bound to the potential energy terms in (1.2), and hence to the ground state energy of (1.2) for all N. This bound is

$$E_0 \geqq -(0.9)(4\pi/3)^{1/3} N \rho_B^{1/3}. \tag{1.5}$$

A result similar to (1.5) is given in [3].

It is not easy to give a heuristic derivation of the $\rho_B^{1/4}$ law. Dyson [4] gives one, but we prefer the following point of view. The reason that $E_0 < 0$ is that the negative particles stay away from each other. If λ is the correlation length (i.e. the radius of a ball surrounding any one particle in which there is, on the average, an absence of one particle) then the potential energy, P, is roughly $P \approx -N/\lambda$. On the other hand, let us study the kinetic energy, K. Most of the particles will be in the zero momentum state. A correlation length λ can be achieved by decomposing Λ into $n \equiv (L/\lambda)^3$ boxes of size λ. If there is one single particle wave function in each box, with Dirichlet boundary conditions, its kinetic energy will be λ^{-2} and thus $K = n\lambda^{-2} = L^3\lambda^{-5}$. Minimization of $K + P = -N\lambda^{-1} + L^3\lambda^{-5}$ with respect to λ (recalling $N = \rho_B L^3$ for neutrality) yields $\lambda^4 = 5/\rho_B$ and $E_0 = -\frac{4}{5}N\lambda^{-1} = -\frac{4}{5}N(\rho_B/5)^{1/4}$. In addition we learn that $K/P = -1/5$, which is very different from the usual virial theorem value $-1/2$.

The difficulty with the above argument is its apparent inconsistency. If we put n particles into boxes, as stated above, then K will be $n\lambda^{-2}$ but also P will be $-n\lambda^{-1}$, not $N\lambda^{-1}$. Nevertheless, it is true that the Bogoliubov pair wave function has the properties $K \approx n\lambda^{-2}$ and $P = -N\lambda^{-1}$ mentioned above. How it achieves this is not easy to understand; one *must*, apparently, study the problem in momentum space.

If the kinetic energy were $|p|^\alpha$ with $1 \leqq \alpha < 2$, instead of p^2, we would predict, by the same argument, that E_0 would then be of the order $-N\rho_B^{1/(2+\alpha)}$ and $\lambda \approx \rho_B^{-1/(2+\alpha)}$. This conclusion does indeed agree with what is obtained from an appropriately modified Bogoliubov function. When $\alpha = 1$ (the relativistic case) we get $-N\rho_B^{1/3}$ which agrees with the lower bound (1.5).

B. The Two-Component System. For simplicity let us consider the neutral system with N bosons of each charge. In [5] and [15] it is proved that $E_0 \geqq -A_3 N^{5/3}$. Indeed, if one kind of particle is infinitely massive then the $N^{5/3}$ law is correct—as proved by Lieb [9]. Moreover, the $N^{5/3}$ upper bound in [9] is very simple and semiclassical—correlations are unnecessary.

The $N^{7/5}$ result for particles all of finite mass is subtle. For (1.1) Dyson obtained (for large N and $\sum e_i = 0$)

$$E_0 \leqq -5.001 \times 10^{-7} N^{7/5}. \tag{1.6}$$

Surely, the coefficient in (1.6) is too small. Our lower bound for the energy, proved in Sect. II, is the following:

Theorem 1.2. *Let H_N be given by (1.1) with $e_i = \pm 1$. Neutrality is not assumed. Then, on $L^2(\mathbf{R}^{3N})$*

$$H_N \geqq -0.30 N^{7/5} \tag{1.7}$$

for sufficiently large N.

Generalizations of Theorem 1.2 are given in Theorems 2.1 and 5.1. The former is a generalization to the Yukawa potential while the latter treats the nonneutral case $N_- \ll N_+$,

$$E_0 \geqq -A_7 N^{7/5} \tag{1.8}$$

for some constant A_7.

Let us recall Dyson's heuristic derivation [4] of (1.7) from (1.4). There are two parts to the energy: (i) a local kinetic energy and electrostatic correlation energy and (ii) a global kinetic energy needed to localize the system in a region of radius R. The latter is approximately $K_{\text{global}} \approx N/R^2$. The former is approximately $E_{\text{local}} \approx -A_5 N\rho^{1/4}$ with $\rho = N/R^3$. Here we have taken over the one-component jellium result (1.4) even though we are considering a two-component system; the reason is that the electrostatic correlation energy comes primarily from the fact, as we said, that particles of like charge stay away from each other and therefore the energy in the two-component and one-component systems are comparable. If we now minimize $E = E_{\text{global}} + E_{\text{local}}$ with respect to R we find $R \approx N^{-1/5}$ and $E \approx -N^{7/5}$. A check on the consistency of this is that the correlation length satisfies $\lambda \approx \rho^{-1/4} = (N/R^3)^{-1/4} = N^{-2/5} \ll R$.

In the present paper we begin with the $N^{7/5}$ problem and prove (1.7) in Sect. II. Our analysis is based on Conlon's paper [2] in which the following was proved about the two-component system in a box. A symmetric wave function connotes a function that is separately symmetric in the positive and negative charge spatial variables, i.e. a bosonic function.

Theorem 1.3. *Let Λ be a cube in $\mathbf{R}^3$ and suppose that $\psi(x_1, \ldots, x_N)$ is any symmetric, infinitely differentiable, $L^2(\mathbf{R}^{3N})$ normalized function with support in Λ^N. Let*

$$K(\psi) \equiv \sum_{i=1}^{N} \langle \psi, -\Delta_i \psi \rangle \equiv \langle \psi, T\psi \rangle \tag{1.9}$$

be the kinetic energy and define γ_ψ *by*

$$K(\psi) = N\gamma_\psi^2/L^2, \tag{1.10}$$

where L is the length of Λ. *Let* H_N^ν *be the two-component Hamiltonian analogous to (1.1) but with the Coulomb potential replaced by the Yukawa potential* $Y_\nu(x) = |x|^{-1}\exp(-\nu|x|)$, *namely*

$$H_N^\nu = -\sum_{i=1}^{N} \Delta_i + \sum_{1 \leq i < j \leq N} e_i e_j Y_\nu(x_i - x_j).$$

The $e_i = \pm 1$ *as before and neutrality is not assumed. Then, if* $\gamma_\psi \leqq N^{1/3}$,

$$\langle \psi, H_N^\nu \psi \rangle \geqq -A_8 N^{7/5} \tag{1.11}$$

for some constant A_8, *which is independent of* ν, *N and L.*

Theorem 1.3 is proved in [2] by a succession of inequalities that turn the Bogoliubov ansatz [1] into a rigorous bound (with a different constant, of course). It concerns the local energy and, being intrinsically quantum-mechanical, has no classical, analogue. The reason that Theorem 1.3 does not imply the $N^{7/5}$ law, Theorem 1.2, is the condition that $\gamma_\psi \leqq N^{1/3}$ (alternatively, $K(\psi) \leqq N^{5/3}/L^2$). We do not know in advance what the radius, R, is for an energy minimizer. If, for example, $K(\psi) = N^{5/3}$ and $R \gg 1$, we could not use Theorem 1.3. Thus, we are faced with what might be called an infrared problem which our analysis in Sect. II solves.

To get the constant in (1.7) we need a good value for A_8 in (1.11). A value can be deduced from [2], but the constant there is not optimum. It turns out that restricting $\gamma_\psi \leqq N^{1/3-\delta}$ for some $\delta > 0$ is sufficient for the analysis in Sect. II. Under this condition the following improvement of Theorem 1.3 is possible, and is proved in Sect. IV.

Theorem 1.4. *Assume the hypotheses of Theorem 1.3 except that* $\gamma_\psi \leqq N^{1/3}$ *is replaced by* $\gamma_\psi \leqq N^{1/3-\delta}$ *for some fixed* $\delta > 0$. *The parameters* ν *and L can depend on N, but we assume that* $N^{-1/5}\nu L$ *stays bounded as* $N \to \infty$. *Then, for sufficiently large N,*

$$\langle \psi, H_N^\nu \psi \rangle \geqq -0.30 N^{7/5}. \tag{1.12}$$

The analysis in Sect. III of the jellium problem, leading to (1.4), uses the $N^{7/5}$ result of Sect. II. This may seem a bit odd in view of Dyson's heuristic discussion in which one uses the jellium result to understand the $N^{7/5}$ theorem. Our procedure is to bound the jellium energy in arbitrarily large boxes in terms of the energy in a box of size $l = \rho_B^{-1/8}$. In such a box the particle number (with neutrality) is $n = \rho_B l^3 = \rho_B^{5/8}$. But then, by the $N^{7/5}$ theorem (with the background being thought of as N particles in a simple, smeared out state), $E_{\text{box}} \geqq -An^{7/5} = -A\rho_B^{7/8} = -An\rho_B^{1/4}$. By adding up the boxes we obtain $E_0 \geqq -AN\rho_B^{1/4}$.

Our work here leads to many questions, of which the following are a few.

Open Problems

(1) Find the correct coefficient in (1.4) for large ρ_B in the jellium problem. Is Foldy's constant in (1.3) correct?

(2) Find the correct coefficient A_7 in (1.8) as $N \to \infty$ for the two-component

problem. The bound in Theorem 1.4 is within a factor of 11 of what one would get heuristically from a calculation using the Bogoliubov function. This is discussed in Sect. IV. This bound translates into the bound (1.7) of Theorem 1.2. We should emphasize, however, that Bogoliubov's method does not predict an exact value for the asymptotic constant in Theorem 1.2. The reason for this is that in the Bogoliubov method one is forced to work in cubes and, in the Bogoliubov function, most particles are in the lowest momentum mode of the cube. The size of the cube can be taken to be the size of the system, namely $N^{-1/5}$. Thus the energy depends critically on the lowest eigenvalue of the Laplacian in a cube and this depends on boundary conditions. The lowest eigenvalue will be uncertain because of the boundary conditions and will be of order $N^{2/5}$. The uncertainty in the energy will be of order $N^{7/5}$.

(3) What can be said about the correlation functions at high density? Is Bogoliubov's ansatz really correct or does it merely give a good account of the energy?

(4) As shown in Sect. II, the statement $E_0 \geqq -AN^{7/5}$ for the two-component system is equivalent, via the virial theorem, to $-P(\psi) \leqq 2A^{1/2}N^{7/10}K(\psi)^{1/2}$ for all ψ. Here $K(\psi)$ is the kinetic energy and $P(\psi)$ the potential energy of ψ. Now let us replace p^2 by $|p|^\alpha$ in the kinetic energy. In the heuristic discussion above we surmised that the jellium energy should be $-C_\alpha \rho_B^{1/(2+\alpha)}$. Then, by the uncertainty principle argument relating the jellium energy to the two-component energy given before, we would have (with $K_{\text{global}} \approx NR^{-\alpha}$) $R \approx N^{-1/(\alpha-1)(\alpha+3)}$ and $E_0 = -A_\alpha N^{(\alpha^2+3\alpha-3)/(\alpha^2+2\alpha-3)}$. This statement about E_0 is equivalent, via the virial theorem, to

$$-P(\psi) \leqq \alpha[A_\alpha/(\alpha-1)]^{1-1/\alpha} N^{(\alpha^2+3\alpha-3)/\alpha(\alpha+3)} K(\psi)^{1/\alpha}. \tag{1.13}$$

We conjecture that these formal calculations are correct as $N \to \infty$. If so, it is interesting to look at the $\alpha = 1$ case (relativistic bosons). In this case, $E_0 = -\infty$ for large enough N, which is correct, but (1.13) continues to make sense. Namely, for $\alpha = 1$,

$$-P(\psi) \leqq CN^{1/4}K(\psi). \tag{1.14}$$

We conjecture that (1.14) is true for large N and we remark that in [3] it is proved that (1.14) holds with $N^{1/4}$ replaced by $N^{1/3}$. Since the bosonic energy is the absolute lowest, (1.13) and (1.14) are independent of statistics.

II. The $N^{7/5}$ Theorem

Our strategy to prove Theorem 1.2 is to decompose $\mathbf{R}^3$ into cubic boxes of size $l = N^{-\varepsilon}$ with ε some small number less than 1/5. This l is large compared to the expected size of the system, $N^{-1/5}$, but we do not know this fact in advance. It will be necessary to localize H_N in these boxes and to control the interaction between boxes.

The main difficulty in localizing the Hamiltonian (1.1) comes from the localization of the Coulomb potential. The effects of localization on the kinetic energy can be easily computed to be of order Nl^{-2}, where l is the cutoff length. For the potential energy, however, even a small amount of net charge will produce enormous potential energies, and therefore charge neutrality must be preserved very carefully. Our basic strategy is first to replace the Coulomb potential by a Yukawa cutoff and then, by averaging over all possible box locations, the errors can be controlled.

For $\mu > 0$, let

$$Y_\mu(r) \equiv r^{-1} e^{-\mu r} \tag{2.1}$$

be the Yukawa potential with range μ^{-1}. For $\kappa > 0$ and N a positive integer, let

$$H_{\kappa N}^{\mu l} \equiv -\sum_{i=1}^{N} \Delta_i + \kappa \sum_{1 \leqq i < j \leqq N} e_i e_j Y_\mu(x_i - x_j) \tag{2.2}$$

be the Hamiltonian of N charged bosons interacting pairwise by the Yukawa potential with coupling constant κ. As before, $e_i = \pm 1$ but neutrality is not assumed. $H_{\kappa N}^{\mu l}$ is defined as a quadratic form on $L^2([0,l]^{3N})$ with Dirichlet boundary conditions. We shall drop μ or l or κ whenever they are equal to $0, \infty$ or 1 respectively. Since the Hamiltonian (2.2) is symmetric under permutations separately on positive or negative particles, the ground state automatically obeys Bose statistics, and we shall assume this henceforth. Let

$$E_{\kappa N}^{\mu l} = \inf \operatorname{spec} H_{\kappa N}^{\mu l} \tag{2.3}$$

be the ground state energy. Then a trivial scaling yields, for the Coulomb Hamiltonian in all of $\mathbf{R}^3$,

$$E_{\kappa N} = \kappa^2 E_N. \tag{2.4}$$

To fix a partition, let g be a piecewise C^1 function on $\mathbf{R}$ defined by

$$g(t) = \cos(\pi t/2), \quad -1 \leqq t \leqq 1 \tag{2.5}$$

and zero otherwise. Then $\sum_{j \in \mathbf{Z}} g^2(t+j) = 1$ for all $t \in \mathbf{R}$. Let $\chi(x) \equiv g(x^1) g(x^2) g(x^3)$, with $x = (x^1, x^2, x^3)$ and let $\chi_{u\lambda}(x) \equiv \chi(x + u + \lambda)$. Here $\lambda \in \mathbf{Z}^3$ and $u \in [0,1]^3 \equiv \Gamma$. Then

$$\sum_{\lambda \in \mathbf{Z}^3} \chi_{u\lambda}^2(x) = 1 \quad \forall x \in \mathbf{R}^3, \quad u \in \Gamma. \tag{2.6}$$

A function h which is of central importance in our localization is defined by

$$h(x,y) \equiv \int_\Gamma du \sum_{\lambda \in \mathbf{Z}^3} \chi_{u\lambda}^2(x) \chi_{u\lambda}^2(y). \tag{2.7}$$

Then h depends only on the difference $z \equiv x - y$ and

$$\begin{aligned} h(z) \equiv h(x-y) &= \int_\Gamma du \sum_{\lambda \in \mathbf{Z}^3} \chi^2(x+u+\lambda) \chi^2(y+u+\lambda) \\ &= \int_{\mathbf{R}^3} du\, \chi^2(x+u) \chi^2(y+u) = (\chi^2 * \chi^2)(z). \end{aligned} \tag{2.8}$$

An easy computation shows that

$$\begin{aligned} (g^2 * g^2)(t) &= \frac{1}{8}\left[4 - 2|t| + \frac{3}{\pi} \sin \pi |t| + (2 - |t|) \cos \pi t\right], \quad |t| \leqq 1 \\ &\sim \tfrac{3}{4} - \tfrac{1}{8}\pi^2 t^2 + O(t^4) \end{aligned} \tag{2.9}$$

and zero otherwise. Hence $h(z)$ is a C^4 function and

$$|h(z) - a_0 - a_1|z|^2| \leqq a_2|z|^4 \tag{2.10}$$

with $a_0 = (3/4)^3, a_1 = -(\frac{3}{4})^2(\pi^2/8)$ and a_2 some constant of order 1.

Let $h_l(x) \equiv h(x/l)$.

We now define localized kinetic and potential energies. Let $\alpha = (u, \alpha_1, \ldots, \alpha_N) \in \Gamma \times \mathbf{Z}^{3N}$ be a multi-index and let $\int d\alpha \equiv \int_\Gamma du \sum_{\alpha_1 \in \mathbf{Z}^3} \cdots \sum_{\alpha_N \in \mathbf{Z}^3}$. If $\beta = (v, \beta_1, \ldots, \beta_N)$ is another multi-index, denote $\delta(u-v) \prod_{i=1}^N \delta_{\alpha_i \beta_i}$ by $\delta_{\alpha\beta}$. (Here, $\delta(u-v)$ is the Dirac δ-function and $\delta_{\alpha_i \beta_i}$ is the Kronecker delta.) For any $l > 0$, let

$$\psi_\alpha^l(x_1, \ldots, x_N) \equiv \prod_{k=1}^N \chi_{u\alpha_k}(x_k/l)\psi(x_1, \ldots, x_N), \tag{2.11}$$

$$V_\alpha^\mu \equiv \sum_{1 \leqq i < j \leqq N} e_i e_j Y_\mu(x_i - x_j)\delta_{\alpha_i, \alpha_j}. \tag{2.12}$$

Then by using (2.6) and (2.7) one has the identity

$$\begin{aligned}\int d\alpha \langle \psi_\alpha^l, V_\alpha^\mu \psi_\alpha^l \rangle &= \sum_{1 \leqq i < j \leqq N} e_i e_j \int dX \int d\alpha \prod_{k=1}^N \chi_{u\alpha_k}^2(x_k/l)\psi^2(x_1, \ldots, x_N) Y_\mu(x_i - x_j)\delta_{\alpha_i \alpha_j} \\ &= \left\langle \psi, \sum_{1 \leqq i < j \leqq N} e_i e_j Y_\mu(x_i - x_j) h_l(x_i - x_j)\psi \right\rangle. \end{aligned} \tag{2.13}$$

Similarly, since for any $f \in C_0^\infty(\mathbf{R}^3)$

$$\langle f\chi, -\Delta(f\chi)\rangle = \langle f, -\Delta(\chi^2 f)\rangle + \langle f, |\nabla\chi|^2 f\rangle,$$

one has the following estimate for the kinetic energy with $C_0 = \sup_x |\nabla\chi|^2(x) < 3(\pi/2)^2$ (and recalling (1.9)):

$$\int d\alpha \langle \psi_\alpha^l, T\psi_\alpha^l \rangle \leqq \langle \psi, T\psi \rangle + C_0 N l^{-2}. \tag{2.14}$$

We emphasize that, definition (2.12), particles in different boxes *do not interact*. Hence

$$\begin{aligned} &\left\langle \psi, \left[T + \sum_{1 \leqq i < j \leqq N} e_i e_j Y_\mu(x_i - x_j) h_l(x_i - x_j) \right] \psi \right\rangle + C_0 N l^{-2} \\ &\geqq \int d\alpha \langle \psi_\alpha^l, (T + V_\alpha^\mu)\psi_\alpha^l \rangle \geqq \left(\int d\alpha \|\psi_\alpha^l\|_2^2\right) \inf_{\sum n_\sigma = N} \sum_{\sigma \in \mathbf{Z}^3} E_{n_\sigma}^{\mu l} \\ &= \inf_{\sum n_\sigma = N} \sum_{\sigma \in \mathbf{Z}^3} E_{n_\sigma}^{\mu l}. \end{aligned} \tag{2.15}$$

Here $E_{n_\sigma}^{\mu l}$ is the ground state energy of a n_σ-particle system with Yukawa cutoff μ in a box of size l (see (2.2)). The sub-systems need to be neutral.

To complete the localization, one has to relate the potential $Y_\mu(z)h_l(z)$ to the Coulomb potential. Let

$$f_{\mu l}(z) \equiv a_0|z|^{-1} - Y_\mu(z)h_l(z). \tag{2.16}$$

The coefficient a_0 in front of r^{-1} is added for the purpose of normalization. Clearly, $f_{\mu l}(0) = a_0 \mu$. It will be shown in Lemma 2.1 below that $f_{\mu l}$ has a positive Fourier transform if $\mu l \geqq C_3$ for some fixed constant C_3. Hence by (2.15), (2.4) and (2.20)

$$E_N = a_0^{-2} E_{a_0,N} \geqq a_0^{-2}\left[-C_0 N l^{-2} - \tfrac{1}{2} a_0 \mu N + \inf_{\sum n_\sigma = N} \sum_{\sigma \in Z^3} E_{n_\sigma}^{\mu l} \right]. \tag{2.17}$$

Equation (2.17) is the localization estimate which we need to prove Theorem 1.2. Note that the correction terms are remarkably simple.

Let $l = N^{-\varepsilon}$ with ε some small number ($\varepsilon < 1/5$) and $\mu = C_3 N^{\varepsilon}$. Our goal is to apply Theorem 1.4 in each box.

Let ϕ be a n-particle wave function satisfying $\langle \phi, H_n^{\mu l} \phi \rangle \leqq 0$. Then one has the trivial estimate (recall definition (2.2))

$$\tfrac{1}{2}\langle \phi, T\phi \rangle \leqq -\tfrac{1}{2}\langle \phi, H_{2,n}^{\mu l}\phi \rangle \leqq -\tfrac{1}{2} \inf \operatorname{spec} H_{2,n}^{\mu l}. \tag{2.18}$$

But $H_{2,n}^{\mu l}$ can be bounded below by $-C_5 n^{5/3}$ [see (A.23) which is the stability of matter bound with Yukawa cutoff derived in the Appendix]. Hence the hypothesis of Theorem 1.4 is satisfied for each box with $l = N^{-\varepsilon}$ and $\gamma_\psi = N^{1/3-\varepsilon}$. We also have that $\sum_{\sigma \in \mathbf{Z}^3} (n_\sigma)^{7/5} \leqq \left(\sum_{\sigma \in \mathbf{Z}^3} n_\sigma \right)^{7/5} = N^{7/5}$.

Let us now combine Theorem 1.4 with (2.17), temporarily ignoring the possibility that the particle number in some boxes may not be large. This yields

$$E_N \geqq -0.30 a_0^{-2} N^{7/5} - a_0^{-2} C_0 N^{1+2\varepsilon} - C_3 a_0^{-1} N^{1+\varepsilon}. \tag{2.19}$$

To eliminate the last two terms as $N \to \infty$ we simply take $\varepsilon < 1/5$.

Despite the aforementioned problem about the particle number in each box, (2.19) is correct as $N \to \infty$. To prove this, note that in any box we can use $E_\sigma \geqq -C_5 n_\sigma^{5/3}$. However, $\sup\{\sum n_\sigma^{5/3} | \sum n_\sigma \leqq N \text{ and } n_\sigma < N^p\} \leqq N^{2p/3} \sum n_\sigma = N^{1+2p/3}$. If we take $0 < p < 3/5$, this shows that boxes with small particle number can be neglected.

Finally, to complete the proof of Theorem 1.2 we have to eliminate the a_0^{-2} factor in the first term in (2.19). This can be done as follows. Note that $a_0 = h(0)^3 = (\int g^4)^3$ with g given in (2.5), and g satisfies $\int g^2 = 1$. If g is replaced by $\eta(t) = 1$ for $|t| \leqq \frac{1}{2}$ and $\eta(t) = 0$ otherwise, we would have $a_0 = 1$. But we cannot do this because $\int |\nabla g|^2$, the coefficient of the l^{-2} term in (2.14), would be infinite. Since $N l^{-2}$ is small on a scale of $N^{7/5}$, the remedy is to take $g \approx \eta$ and $\int |\nabla g|^2$ finite, but large. As $N \to \infty, g \to \eta$ and $a_0 \to 1$. Note that Lemma 2.1 does not depend on the special choice (2.5) we made for g.

This concludes the proof of Theorem 1.2 and we turn to Lemma 2.1.

Lemma 2.1. *Let $K: \mathbf{R}^3 \to \mathbf{R}$ be given by*

$$K(z) = r^{-1}\{e^{-\nu r} - e^{-\omega r} h(z)\}$$

with $r = |z|$ and $\omega > \nu \geqq 0$. Let h satisfy (i) *h is a C^4 function of compact support;* (ii) *$h(z) = 1 + ar^2 + O(r^3)$ near $z = 0$. Let $h(z) = h(-z)$, so that K has a real Fourier transform. Then there is a constant C_3 (depending on h) such that if $\omega - \nu \geqq C_3$ then*

K has a positive Fourier transform and, moreover,

$$\sum_{1\le i<j\le N} e_i e_j K(x_i - x_j) \geqq \tfrac{1}{2}(\nu - \omega)N \tag{2.20}$$

for all $x_1,\ldots,x_N \in \mathbf{R}^3$ *and* $e_i = \pm 1$.

Proof. Let $F(z) = [h(z) - 1 - ar^2]r^{-1}(1+r^5)^{-1}$. $K(z)$ can thus be decomposed as

$$K(z) = Y_\nu(z) - Y_\omega(z) - are^{-\omega r} - (1+r^5)e^{-\omega r}F(z).$$

The Fourier transforms of the first three terms are $4\pi/(p^2+\nu^2)$, $-4\pi/(p^2+\omega^2)$ and $-8\pi a(3\omega^2 - p^2)(p^2+\omega^2)^{-3}$ respectively. For the last term note that $F(z)$ is of order r^2 and r^{-4} near the origin and near infinity, respectively and $\Delta^2 F(z)$ is of order r^{-2} and r^{-8} near the origin and near infinity, respectively. Therefore, ΔF and $\Delta^2 F \in L^1(\mathbf{R}^3)$ and hence (with $\hat{}$ denoting Fourier transform)

$$|(1+p^2)^2 \hat{F}(p)| \leqq 4\pi C_1$$

for some constant C_1. But the Fourier transform of $(1+r^5)e^{-\omega r}$ can be shown to satisfy $|((1+r^5)e^{-\omega r})\hat{}\,| \leqq 16\pi\omega(\omega^2+p^2)^{-2}$ if $\omega \geqq C_2$ for some constant C_2. Hence

$$\begin{aligned}|[(1+r^5)e^{-\omega r}F(z)]\hat{}\,| &= C_1[8\pi\omega(\omega^2+p^2)^{-2}]*[8\pi(1+p^2)^{-2}]\\ &= C_1[e^{-\omega r}\cdot e^{-r}]\hat{} = 8\pi C_1(\omega+1)[(\omega+1)^2+p^2]^{-2}.\end{aligned}$$

We can now put all these Fourier transform together to yield the estimate

$$\hat{K}(p) \geqq 4\pi[(p^2+\nu^2)^{-1} - (p^2+\omega^2)^{-1} - 6|a|(p^2+\omega^2)^{-2} - 2C_1(\omega+1)(\omega^2+p^2)^{-2}].$$

Hence $\hat{K}(p) \geqq 0$ for all p if $\omega - \nu$ is large enough. To conclude the proof of Lemma 2.1, one only has to note the identity

$$\sum_{1\le i<j\le N} e_i e_j K(x_i - x_j) = \tfrac{1}{2}\int \hat{K}(p)\left[\left|\sum_{j=1}^N e_j e^{ipx_j}\right|^2 - N\right]dp$$

which implies (2.20) since $\int \hat{K} = K(0) = \omega - \nu$. □

Lemma 2.1 is applied to (2.16) with $\nu = 0$ and the requirement is that $\mu l \geqq C_3$. However, our energy bound does not depend on the fact that we started with a Coulomb potential in (1.1). By the foregoing construction and Lemma 2.1 we have the following generalization of Theorem 1.2.

Theorem 2.1. *Let* $e_i = \pm 1$ *and let*

$$H_N^\nu = -\sum_{i=1}^N \Delta_i + \sum_{1\le i<j\le N} e_i e_j Y_\nu(x_i - x_j) \tag{2.21}$$

be defined on $L^2(\mathbf{R}^{3N})$ *with* $Y_\nu(x) = |x|^{-1}\exp(-\nu|x|)$. ν *can depend on N, but suppose that as* $N \to \infty$, $N^{-2/5}\nu \to 0$. *Then, for sufficiently large N,*

$$H_N^\nu \geqq -0.30\,N^{7/5}. \tag{2.22}$$

Proof. As in (2.16), we write $f_{\mu l} = a_0 Y_\nu - Y_\mu h_l$. In order to apply our foregoing construction, the assumptions of Lemma 2.1 and Theorem 1.4 must be satisfied, namely $l(\mu - \nu) \geqq C_3$, $l^{-1} \geqq N^\varepsilon$ and $l\mu N^{-1/5} < \infty$ as $N \to \infty$. On the other hand, the correction terms resulting from the localization (cf. (2.17)) should be of lower order. Hence we must have $l^{-1} \leqq o(N^{1/5})$ and $(\mu - \nu) \leqq o(N^{2/5})$. It is easy to check that $\mu = \max(N^{1/5+\varepsilon}, 2\nu)$ and $l = N^{1/5}\mu^{-1}$ satisfy all the requirements. □

Returning to the Coulomb case, (1.1), we note the following *virial type theorem.*

Theorem 2.2 *Let the hypotheses be as in Theorem 1.2 and let* ψ *be any normalized (not necessarily symmetric) function in* $L^2(\mathbf{R}^{3N})$. *Let* $K(\psi)$ *and* $P(\psi)$ *denote the kinetic and potential energies of* ψ *(see (1.9) and* $P(\psi) = \langle \psi, \sum e_i e_j |x_i - x_j|^{-1} \psi \rangle$). *Then*

$$-P(\psi) \leqq 2A^{1/2} N^{7/10} K(\psi)^{1/2}, \tag{2.23}$$

where $A = -N^{-7/5} \inf \operatorname{spec}(H_N)$.

Proof. Replacing $\psi(x_i)$ by $\lambda^{3N/2}\psi(\lambda x_i)$ we find that $\lambda^2 K(\psi) + \lambda P(\psi) \geqq -AN^{7/5}$. Then $-P(\psi) \geqq \lambda^{-1} AN^{7/5} + \lambda K(\psi)$. Optimizing this with respect to λ yields (2.23). □

III. The $\rho^{1/4}$ Law for Jellium

We shall prove Theorem 1.1 in this section by localizing the jellium Hamiltonian to a box of size $l = \rho_B^{-1/8}$. The localized Hamiltonian can thus be estimated by relating it to Theorem 2.1. In localizing the jellium Hamiltonian (1.2), one should be cautious about the fact that, after averaging over all translations, the coupling constant in the two-particle Coulomb interactions changes from 1 to a_0 [see (2.9)–(2.16)], while that of the particle-background remains unchanged. A straightforward localization as in Sect. II will fail to preserve the charge neutrality. We shall solve this difficulty by replacing the uniform background charge density, ρ_B, in each small box by a *non-uniform* background charge density which depends on the cutoff functions.

Let χ_Λ be the characteristic function of the big domain, Λ. For $\tau \in \mathbf{Z}^3$, $l > 0$, $\mu > 0$ and $\alpha = (u, \alpha_1, \ldots, \alpha_N)$, and recalling (2.6), (2.7), *et.seq.*, let

$$\rho_{B\tau}^{lu}(y) \equiv \chi_\Lambda(y)\chi_{u\tau}^2(y/l)\rho_B, \tag{3.1}$$

$$V_{B\tau\mu}^{lu}(x) \equiv \int Y_\mu(x-y)\rho_{B\tau}^{lu}(y)dy, \tag{3.2}$$

$$V_{B\mu}^{l}(x) \equiv \rho_B \int_\Lambda Y_\mu(x-y)h_l(x-y)dy. \tag{3.3}$$

Then using (2.11) one has the following definitions of $\mathscr{L}(\psi)$ and $\mathscr{R}(\psi)$ and localization estimate:

$$\begin{aligned}\mathscr{L}(\psi) \equiv \int d\alpha \Big\langle \psi_\alpha^l, \Big[-\sum_{i=1}^N \Big\{ \Delta_i + \sum_{\tau\in\mathbf{Z}^3} V_{B\tau\mu}^{lu}(x_i)\delta_{\alpha_i,\tau} \Big\} \\ + \sum_{1\leqq i<j\leqq N} Y_\mu(x_i - x_j) + \sum_{\tau\in\mathbf{Z}^3} \tfrac{1}{2}\int \rho_{B\tau}^{lu}(y) V_{B\tau\mu}^{lu}(y)dy \Big] \psi_\alpha^l \Big\rangle \\ \leqq \Big\langle \psi, \Big[-\sum_{i=1}^N \{\Delta_i + V_{B\mu}^l(x_i)\} + \sum_{1\leqq i<j\leqq N} Y_\mu(x_i-x_j)h_l(x_i-x_j) \\ + \tfrac{1}{2}\rho_B \int_\Lambda V_{B\mu}^l(y)dy \Big] \psi \Big\rangle + C_0 N l^{-2} \equiv \mathscr{R}(\psi).\end{aligned} \tag{3.4}$$

Equation (3.4) may appear to be complicated, but the proof is just a reordering of indices. Recall Eq. (2.6),

$$\int d\alpha \left\langle \psi^l_{\alpha}, \sum_{j=1}^{N} \sum_{\tau\in Z^3} V^{lu}_{B\tau\mu}(x_j)\delta_{\alpha_j,\tau}\psi^l_{\alpha}\right\rangle$$

$$= \sum_{j=1}^{N} \left\langle \psi, \left[\rho_B \int_{\Lambda} dy \int d\alpha \sum_{\tau\in Z^3}\delta_{\alpha_j,\tau} Y_\mu(x_j-y)\chi^2_{u\tau}(y/l)\prod_{i=1}^{N} \chi^2_{u\alpha_i}(x_i/l)\right]\psi\right\rangle$$

$$= \sum_{j=1}^{N} \left\langle \psi, \rho_B\int_{\Lambda} dy\, Y_\mu(x_j-y)\left[\int_{\Gamma} du \sum_{\alpha_j}\chi^2_{u\alpha_j}(y/l)\chi^2_{u\alpha_j}(x_j/l)\right]\psi\right\rangle$$

$$= \sum_{j=1}^{N} \langle\psi, V^l_{B\mu}(x_j)\psi\rangle,$$

$$\int d\alpha\left\langle \psi^l_{\alpha}, \sum_{j=1}^{N}\sum_{\tau\in Z^3}\iint \rho^{lu}_{B\tau}(y)\rho^{lu}_{B\tau}(y')Y_\mu(y-y')dydy'\psi^l_{\alpha}\right\rangle$$

$$=\left\langle\psi, \rho_B^2\int_{\Lambda} dy\int_{\Lambda} dy'\left\{\int du\sum_{\tau\in Z^3}\chi^2_{u\tau}(y/l)\chi^2_{u\tau}(y'/l)\right\}\sum_{\alpha_1,\dots,\alpha_N\in Z^3}\prod_{j=1}^{N}\chi^2_{u\alpha_j}(x_j/l)\psi\right\rangle$$

$$=\rho_B\int_{\Lambda} V^l_{B\mu}(y)dy.$$

For the other terms in (3.4) one can use (2.13) and (2.14).

As in Sect. II, one can use the positive definiteness of $f_{\mu l}$ (Lemma 2.1) to yield the bound

$$\mathscr{R}(\psi)\leqq\langle\psi, H^J_{N\Lambda}(a_0,\rho_B)\psi\rangle + C_0Nl^{-2}+N\mu a_0. \tag{3.5}$$

In (3.5) $H^J_{N\Lambda}(a_0,\rho_B)$ is the jellium Hamiltonian (1.2) but with all the potential energy terms multiplied by a_0. To utilize (3.5) we have to relate the energy of $H^J_{N\Lambda}(a_0,\rho_B)$ to that of $H^J_{N\Lambda}$. By simple scaling this is given by

$$\inf\operatorname{spec} H^J_{N,a_0\Lambda}(1,\rho_B) = a_0^{-2}\inf\operatorname{spec} H^J_{N\Lambda}(a_0,\rho_B a_0^3). \tag{3.6}$$

Let $l=\rho_B^{-1/8}$ and $\mu=C_6\rho_B^{1/8}$. Then the last two terms in (3.5) are of order at most $N\rho_B^{1/4}$. To complete the proof of Theorem 1.1, one only has to show that $\mathscr{L}(\psi)\geqq -C_7N\rho_B^{1/4}$.

For each fixed τ and multi-index α, consider the localized Hamiltonian

$$-\sum_{j=1}^{N}[\Delta_j\delta_{\tau\alpha_j}+V^{lu}_{B\tau\mu}(x_j)\delta_{\tau\alpha_j}] + \sum_{i<j}\delta_{\tau\alpha_i}\delta_{\tau\alpha_j}Y_\mu(x_i-x_j)+\tfrac{1}{2}\int\rho^{lu}_{B\tau}(y)V^{lu}_{B\tau\mu}(y)dy. \tag{3.7}$$

Our goal is to estimate the ground state of (3.7). Suppose $\alpha_1=\alpha_2=\dots=\alpha_M=\tau$ and $\alpha_j\neq\tau$ for $j>M$. Let $\rho^{lu}_{B\tau}(y)\equiv\tilde\rho_B(y)$ and $\tilde V^\mu_B\equiv Y_\mu*\tilde\rho_B$. Then (3.7) becomes

$$H^\mu_{BM}\equiv -\sum_{j=1}^{M}\{\Delta_j+\tilde V^\mu_B(x_j)\}+\sum_{i<j}^{M} Y_\mu(x_i-x_j)+\tfrac{1}{2}\int\tilde\rho_B(y)\tilde V^\mu_B(y)dy. \tag{3.8}$$

Note that, by definition (3.1),

$$n_B\equiv\int\tilde\rho_B(y)dy\leqq l^3\rho_B. \tag{3.9}$$

Recall that the density function ρ_ϕ for an M-particle normalized wave function,

ϕ, is defined by

$$\rho_\phi(x) = M\int |\phi(x, x_2, \ldots, x_M)|^2 dx_2 \cdots dx_M.$$

Therefore, if one defines

$$D_\mu(f) \equiv \tfrac{1}{2}\int\int f(x) f(y) Y_\mu(x-y) dx dy, \tag{3.10}$$

and

$$\Omega(\phi) \equiv \left\langle \phi, \left[-\sum_{j=1}^{M} \Delta_j + \sum_{1 \leq i < j \leq M} Y_\mu(x_i - x_j) \right] \phi \right\rangle - D_\mu(\rho_\phi),$$

an easy calculation yields

$$\langle \phi, H^\mu_{BM} \phi \rangle = \Omega(\phi) + D_\mu(\rho_\phi - \tilde{\rho}_B). \tag{3.11}$$

Let $Q \equiv \int \rho_\phi - \int \tilde{\rho}_B$ be the value of the total charge in the small box. The following lemma is needed to bound the last term in (3.11).

Lemma 3.1 *Let $U = \{x \,|\, |x| \leq d\}$ be a ball of radius d and let $f: U \to \mathbf{R}$ be a (not necessarily positive) density satisfying $\int_U f = Q$. Then*

$$D_\mu(f) \geq \tfrac{1}{2}(Q^2/d)[1 + \mu d + \mu^2 d^2/3]^{-1}. \tag{3.12}$$

Proof. $D_\mu(f)$ can be written as

$$D_\mu(f) = \sup_h \int_U fh - \frac{1}{8\pi} \int_{\mathbf{R}^3} [|\nabla h|^2 + \mu^2 h^2].$$

To prove (3,10) we merely take (with $r = |x|$) $h(x) = \alpha$ for $r \leq d$ and $h(x) = \alpha d e^{-\mu(r-d)}/r$ for $r \geq d$. Then $\int fh = \alpha Q$. The $r < d$ part of the second integral is $\alpha^2 \mu^2 d^3/6$. The $r > d$ part can be calculated by integrating by parts, using $(-\Delta + \mu^2)h = 0$, and $d^2 hh'|_{r=d} = -\alpha^2(\mu d^2 + d)$. This $r > d$ part is $\frac{1}{2}\alpha^2 d(1 + d\mu)$. Maximizing with respect to α yields (3.12). □

Remark. Equation (3.12) is sharp when $\mu = 0$ or $\mu \to \infty$ with fixed d.

Returning to (3.11), recall that $l = \rho_B^{-1/8}$ and $\mu = C_6 \rho_B^{1/8}$. The $l \times l \times l$ cube fits into a ball of radius $d = 3^{1/2} l/2$. Applying (3.12) with $\mu d = \sqrt{3} C_6/2 \equiv C_7$ we find that, with $C_8 = 3^{-1/2}[1 + C_7 + C_7^2/3]^{-1}$,

$$D_\mu(\rho_\phi - \tilde{\rho}_B) \geq C_8 (M - n_B)^2 \rho_B^{1/8}. \tag{3.13}$$

Finally, we have to estimate $\Omega(\phi)$. For this purpose we introduce a "duplication of variables" trick. Consider the Hamiltonian on $L^2(\mathbf{R}^{3M})$.

$$H^\mu_{2M} = -\sum_{j=1}^{2M} \Delta_i + \sum_{1 \leq i < j \leq 2M} e_i e_j Y_\mu(x_i - x_j), \tag{3.14}$$

where $e_i = 1$ for $i \leq M$ and $e_i = -1$ for $i > M$. Let Φ be a normalized trial function defined by

$$\Phi(x_1, \ldots, x_{2M}) = \phi(x_1, \ldots, x_M)\phi(x_{M+1}, \ldots, x_{2M}).$$

A simple calculation yields

$$\Omega(\phi) = \tfrac{1}{2}\langle \Phi, H^\mu_{2M} \Phi \rangle. \tag{3.15}$$

By (A.23) in the appendix,

$$\Omega(\phi) \geqq -\tfrac{1}{2}(4.016)(2M)^{5/3}. \tag{3.16}$$

Let us divide the possible values of M into two cases.

(a) $M \leqq \rho_B^{11/32}$. Here we use (3.11), (3.16) and $D_\mu(\rho_\phi - \tilde{\rho}_B) \geqq 0$ to conclude that

$$\langle \phi, H_{BM}^{\mu} \phi \rangle \geqq -(6.375)M^{5/3} \geqq -(6.375)M\rho_B^{11/48}. \tag{3.17}$$

(b) $M > \rho_B^{11/32}$. Here we use Theorem 2.1, (3.11), (3.13) and (3.15) to conclude that for large enough ρ_B

$$\langle \phi, H_{BM}^{\mu} \phi \rangle \geqq -C_9(\rho_B)M^{7/5} + C_8(M - n_B)^2 \rho_B^{1/8}, \tag{3.18}$$

where $C_9(\rho_B) \to (0.30)2^{2/5}$ as $\rho_B \to \infty$. The statement "large enough ρ_B" comes from the condition in Theorem 2.1 that $\mu M^{-2/5} \to 0$ as $M \to \infty$. By our assumption $\mu M^{-2/5} \leqq C_6 \rho_B^{-1/80}$, and this goes to zero as $\rho_B \to \infty$. If, in (3.18), we recall that $n_B \leqq \rho_B l^3 = \rho_B^{5/8}$ and $M > \rho_B^{11/32}$, it is easy to see that the right side of (3.18) is bounded below by $-C_{10}(\rho_B)M\rho_B^{1/4}$ and that $C_{10}(\rho_B) \to (0.30)2^{2/5}$ as $\rho_B \to \infty$.

Using these results (a) and (b), and summing over boxes, and recalling (3.6), we conclude that

$$E_0 \geqq -[C_{11}(\rho_B)a_0^{-5/4} + C_0]N\rho_B^{1/4} \tag{3.19}$$

with $C_{11}(\rho_B) \to (0.30)2^{2/5}$ as $\rho_B \to \infty$. Recall from Sect. II that $a_0 = (3/4)^3$ and $C_0 < 3(\pi/2)^2$. Note that (3.19) holds for all N; we did not take the limit $N \to \infty$ in deriving (3.19). With the bounds just given, the factor [ ] in (3.19) is 8.57 when $\rho_B \to \infty$.

This completes the proof of Theorem 1.1. This theorem can be generalized to the case of Yukawa potentials as in Theorem 2.1. It can also be generalized in another direction as follows.

Theorem 3.1. *Consider the modified jellium Hamiltonian with variable background charge density*

$$H_N^J = -\sum_{i=1}^{N} \{\Delta_i + V(x_i)\} + \sum_{1 \leqq i < j \leqq N} |x_i - x_j|^{-1} + \tfrac{1}{2}\int \rho(x)\rho(y)|x-y|^{-1}dxdy, \tag{3.20}$$

with $V(x) = \int \rho(y)|x-y|^{-1}dy$. *The density* ρ *satisfies* $-\infty < \rho(x) \leqq \rho_B$ *with* $\rho_B \geqq 0$. *Then the ground state energy satisfies*

$$E_0 \geqq -A_9 N\rho_B^{1/4} \tag{3.21}$$

and A_9 *satisfies the same bound as* A_6, *given in Theorem 1.1.*

The proof is an easy generalization of the one for constant $\rho(x) = \rho_B$ in Λ just given.

IV. Computation of Constants

Our main goal in this section is to obtain the constant 0.30 in the inequality (1.12). The calculation will consist of optimizing the methodology of [2]. We shall first make a heuristic calculation of the ground state energy E_0 of the Hamiltonian H_N

in (1.1) by using a modification of Bogoliubov's method, and will return to a proper proof of Theorem 1.4 later after Eq. (4.23). We find heuristically that $E_0 \sim -0.028N^{7/5}$ but we are not conjecturing that this constant is sharp. In the following we shall closely follow the notation of [2].

Heuristic Calculations. We consider the case $v = 2\pi/L$ (essentially the Coulomb case) and introduce periodic boundary conditions on Λ for the Hamiltonian H_N. Thus we have

$$\langle\psi, H_N^v\psi\rangle = \langle\psi, T\psi\rangle + \tfrac{1}{2}\sum_{k\in\mathbf{Z}^3} v(k)[\langle\psi, A_k^* A_k\psi\rangle - N]. \tag{4.1}$$

The operator T is the kinetic energy operator, which we write in the second quantized form (see [10] or [7], for example) as

$$T = L^{-2}4\pi^2 \sum_{k\in\mathbf{Z}^3} k^2[a_k^* a_k + b_k^* b_k]. \tag{4.2}$$

Here L is the length of a side of Λ. The operators a_k, b_k are annihilation operators corresponding to the two species of bosons and $k\in\mathbf{Z}^3$. The charge density operator A_k is given by

$$A_k = \sum_{n\in\mathbf{Z}^3} a_{n+k}^* a_n - b_{n+k}^* b_n. \tag{4.3}$$

The $v(k)$ is just the Fourier transform of the Yukawa potential Y_v (with $v = 2\pi/L$) divided by the volume of Λ. We take this value of $v > 0$ to avoid the singularity at $k = 0$. Thus $v(k)$ is given by

$$v(k) = [\pi L(|k|^2 + 1)]^{-1}. \tag{4.4}$$

In Bogoliubov's approximation one makes the ansatz

$$A_k \simeq \sum_{|m|\leqq D\gamma} [S_{k,m}^* + T_{k,m}]. \tag{4.5}$$

Here, D and γ are constants which will be defined later in (4.14) and (4.23). The operators $S_{k,m}, T_{k,m}$ are defined as in (2.8) of [2] by

$$S_{k,m} = \begin{cases} a_n^* a_{n+k} & \text{if } m = (n, 1) \\ -b_n^* b_{n+k} & \text{if } m = (n, -1) \end{cases},$$
$$T_{k,m} = \begin{cases} a_n^* a_{n-k} & \text{if } m = (n, 1) \\ -b_n^* b_{n-k} & \text{if } m = (n, -1) \end{cases}. \tag{4.6}$$

In (4.6) $n\in\mathbf{Z}^3$ and ± 1 indicates the charge species; $|m|$ is defined to be $|n|$. The operators $a_n^\#, b_n^\#$ with $|n| \leqq D\gamma$ are to be thought of as scalars subject to the constraint

$$\sum_{|n|\leqq D\gamma} a_n^* a_n = \sum_{|n|\leqq D\gamma} b_n^* b_n = \frac{N}{2}. \tag{4.7}$$

Hence if $|m| \leqq D\gamma$ and $|k| > 2D\gamma$ the $S_{k,m}$ and $T_{k,m}$ are just annihilation operators. The expression (4.1) then becomes quadratic in creation and annihilation operators. One can compute its ground state energy exactly in the case when $D\gamma = 0$ but also to a good approximation when $D\gamma > 0$. We do this by writing (4.1) as

$$\langle \psi, H_N^v \psi \rangle = 4\pi^2 L^{-2} \sum_{|k| \leqq D\gamma} k^2 \langle \psi, [a_k^* a_k + b_k^* b_k] \psi \rangle$$

$$+ \tfrac{1}{2} \sum_{|k| > D\gamma N^\delta} v(k) I_k(\varepsilon_k, \psi) + \text{lower order terms}, \tag{4.8}$$

where $\delta > 0$ is a small positive number. The expression $I_k(\varepsilon, \psi)$ is given, for general ε, by

$$I_k(\varepsilon, \psi) = \left\langle \psi, \left[\varepsilon \sum_{|m| \leqq D\gamma} [S_{k,m}^* S_{k,m} + T_{k,m}^* T_{k,m}] \right.\right.$$

$$\left.\left. + \left(\sum_{|m| \leqq D\gamma} S_{k,m}^* + T_{k,m} \right)^* \left(\sum_{|m| \leqq D\gamma} S_{k,m}^* + T_{k,m} \right) - N \right] \psi \right\rangle. \tag{4.9}$$

The number ε_k is given by the formula

$$\varepsilon_k = 8\pi^2 k^2 [N L^2 v(k)]^{-1}. \tag{4.10}$$

One can compute exactly the ground state energy of $I_k(\varepsilon, \psi)$. It is given in [2] as

$$I_k(\varepsilon, \psi) \geqq N \left\{ \left[\left(\frac{\varepsilon}{n_0} \right)^2 + 2 \left(\frac{\varepsilon}{n_0} \right) \right]^{1/2} - 1 - \frac{\varepsilon}{n_0} \right\}, \tag{4.11}$$

where

$$n_0 = \sum_{|m| \leqq D\gamma} 1. \tag{4.12}$$

The right-hand side of (4.11) can be achieved if the numbers a_n, b_n satisfy

$$a_n^* a_n = b_n^* b_n = \frac{N}{n_0}, \quad |n| \leqq D\gamma. \tag{4.13}$$

We shall take γ to be large, $\gamma \geqq N^\delta$, and fix D to be the finite number

$$D = \pi^{-1} (5/12)^{1/2} = 0.645/\pi. \tag{4.14}$$

Then, to leading order of magnitude the ground state energy of the second sum in (4.8) is given by

$$\frac{-2N}{L} \left(\frac{N L n_0}{8\pi^3} \right)^{1/4} I, \tag{4.15}$$

where I is the integral

$$I = \int_0^\infty [1 + \xi^4 - (\xi^8 + 2\xi^4)^{1/2}] d\xi. \tag{4.16}$$

Observe that

$$0 < I < J = \int_0^\infty \frac{d\xi}{1 + 2\xi^4}. \tag{4.17}$$

Numerical values for I, J are given by

$$I = .806, \quad J = \pi 2^{1/4}/4 = 0.934. \tag{4.18}$$

The integral I can be expressed exactly in terms of elliptic integrals [7]. Since γ is large n_0 is given to leading order as

$$n_0 = 2 \int_{|x| \leqq D\gamma} dx = \frac{8\pi}{3}(D\gamma)^3. \tag{4.19}$$

Hence (4.15) is given by

$$-B_0 \frac{N}{L}(NL\gamma^3)^{1/4}, \tag{4.20}$$

where

$$B_0 = 2(3\pi^2)^{-1/4} D^{3/4} I. \tag{4.21}$$

The formula (4.20) gives the second sum in (4.8) to leading order of magnitude. Next we need to calculate the first sum in (4.8) which is the macroscopic kinetic energy of the low lying occupied states subject to (4.13). This is clearly given to leading order of magnitude by

$$\begin{aligned}\frac{2N}{n_0} \sum_{|k| \leqq D\gamma} \frac{4\pi^2}{L^2}|k|^2 &= \frac{2N}{n_0}\frac{4\pi^2}{L^2} \int_{|x| \leqq D\gamma} |x|^2 dx \\ &= \left(\frac{12D^2\pi^2}{5}\right)\frac{N\gamma^2}{L^2} = \frac{N\gamma^2}{L^2}.\end{aligned} \tag{4.22}$$

The total energy of the system then, according to this calculation, is

$$\frac{12\pi^2}{5} N\left(\frac{D\gamma}{L}\right)^2 - 2(3\pi^2)^{-1/4} I N^{5/4}\left(\frac{D\gamma}{L}\right)^{3/4}. \tag{4.23}$$

If we minimize this expression with respect to $D\gamma/L$ we obtain the value $-0.028N^{7/5}$.

Proof of Theorem 1.4. In the following calculation we shall ignore all terms of lower order then $N^{7/5}$ since we are only concerned with proving the inequality (1.12) for $N \to \infty$. Let γ be the γ_ψ defined by (1.10). We can assume without loss of generality (by changing Λ to be a sufficiently large cube) that the δ in Theorem 1.4 is less than $\frac{1}{6}$ and that $\gamma \geqq N^\delta$. Define μ by $\mu L = \max(N^{1/10}, \nu L)$ so that $\mu L \to \infty$ and $N^{-1/5}\mu L < \infty$ as $N \to \infty$. If $\nu L < N^{1/10}$, let us write $Y_\nu = (Y_\nu - Y_\mu) + Y_\mu$ and write

$$\langle \psi, H_N^\nu \psi \rangle = N^{-1/10} K(\psi) + P_\nu(\psi) - P_\mu(\psi) + (1 - N^{-1/10}) K(\psi) + P_\mu(\psi). \tag{4.24}$$

Here, $P_\nu(\psi)$ denotes the potential energy terms in H_N^ν with the Yukawa potential Y_ν. Since $Y_\nu - Y_\mu$ is positive definite, $P_\nu(\psi) - P_\mu(\psi) \geqq -\frac{1}{2}(\mu - \nu)N \geqq -\frac{1}{2}\mu N \geqq -\frac{1}{2}N^{11/10}/L$. By the uncertainty principle in a box, $K(\psi) \geqq CN/L^2$, whence the first three terms on the right side of (4.24) are at least $CN^{9/10}/L^2 - \frac{1}{2}N^{11/10}/L$. Minimizing this with respect to L, we find that these terms are bounded below by $-CN^{13/10} \gg -N^{7/5}$. For the last two terms on the right side of (4.24) we can clearly replace $N^{-1/10}$ by zero in the limit $N \to \infty$. Thus we need prove Theorem 1.4 only under the condition $\nu L \geqq N^{1/10}$ and $N^{-1/5}\nu L$ bounded.

By taking Λ to be four times as big, we can suppose that ψ is supported in Q^N, where Q is a smaller cube of size $L/4$. We can then replace $Y_\nu(x)$ in H_N^ν by its periodic extension

$$Y_\nu^L(x) = \sum_{n\in\mathbf{Z}^3} Y_\nu(x+nL), \tag{4.25}$$

because the difference in the two potential energies is at most $W(N)=N^2L^{-1}e^{-\nu L/2}$ (with the factor N^2 coming from the number of interaction terms). Since $\nu L \geqq N^{1/10}$, $W(N)<N^\varepsilon/L$ as $N\to\infty$ for every $\varepsilon>0$. As before, we can borrow $N^{-1/10}K(\psi)\geqq N^{9/10}/L^2$ to control $W(N)$.

Using Y_ν^L in H_N^ν, we then have that (4.1) is an identity provided that H_N^ν is now understood to contain Y_ν^L and provided (4.4) is replaced by

$$v(k) = [\pi L\{k^2 + (\nu L/2\pi)^2\}]^{-1}. \tag{4.26}$$

Clearly, $v(k)\geqq[\pi L(|k|^2+1)]^{-1}$. Now we are ready to bound the various terms in (4.1).

First we bound the potential energy terms for $|k|<N^\delta D\gamma$ from below by

$$-\tfrac{1}{2}N \sum_{|k|\leqq N^\delta D\gamma} v(k) \geqq -CN^{1+\delta}D\gamma/L. \tag{4.27}$$

If, as before, we combine a small portion of the kinetic energy with (4.27) we obtain a lower bound which is lower order than $N^{7/5}$.

Next consider terms in the potential energy which have $|k|>N^\delta D\gamma$. We define $S_{k,m}$, and $T_{k,m}$ again as in (4.6) but this time for all m with $|m|\leqq|k|/N^\delta$. Let us assume for the moment that the system is neutral so that the number of negative particles is $N/2$. We shall return to the nonneutral case after Eq. (4.68). Since $\gamma\leqq N^{1/3-\delta}$, Lemma 2.2 of [2] becomes

$$\langle\psi, T\psi\rangle \geqq \frac{4\pi^2}{NL^2} \sum_{|k|>N^\delta D\gamma} [1-CN^{-\delta}]k^2 C_k(\psi), \tag{4.28}$$

and $C_k(\psi)$ is given by

$$C_k(\psi) = \sum_{r=0}^{\infty} 2^{-4r} \sum_{|m|}^{r} \langle\psi|S_{k,m}^* S_{k,m} + T_{k,m}^* T_{k,m}|\psi\rangle, \tag{4.29}$$

where $\sum_{|m|}^{r}$ is a sum over $(2^r-1)N^{1/3-\delta/2}\leqq|m|<(2^{r+1}-1)N^{1/3-\delta/2}$. Note that the constant $4\pi^2/NL^2$ in (4.28) is better than that in [2]. This is due to the improved summation procedure in (4.29). Hence we have

$$\langle\psi, H_N^\nu\psi\rangle \geqq \tfrac{1}{2} \sum_{|k|>N^\delta D\gamma} v(k) I_k(\varepsilon_k,\psi), \tag{4.30}$$

where

$$I_k(\varepsilon,\psi) = \varepsilon C_k(\psi) + \langle\psi|A_k^* A_k - N|\psi\rangle, \tag{4.31}$$

$$\varepsilon_k = 8\pi^2 k^2 [1-CN^{-\delta}][NL^2 v(k)]^{-1}. \tag{4.32}$$

Since the term $CN^{-\delta}$ in (4.32) is lower order, we shall ignore it in future computations.

Now let $a > 1$ be a positive number which we shall fix later. We define

$$n_0 = \#\{m : |m| \leqq D\gamma\},$$
$$n_r = \#\{m : a^{r-1} D\gamma < |m| \leqq a^r D\gamma\}, \quad r = 1, 2, 3, \ldots . \tag{4.33}$$

Evidently we have, to leading order,

$$n_0 = \frac{8\pi}{3}(D\gamma)^3,$$

$$n_r = \frac{8\pi}{3}(D\gamma)^3 a^{3(r-1)}[a^3 - 1]. \tag{4.34}$$

We define N_r, $r = 1, 2, \ldots$, to be the maximum possible number of particles, consistent with the given $K(\psi)$, such that $|k| > a^{r-1} D\gamma$. Thus we have

$$4\pi^2 N_r (a^{r-1} D\gamma)^2 L^{-2} = N\gamma^2 L^{-2}, \tag{4.35}$$

which yields

$$N_r = N[4\pi^2 (a^{r-1} D)^2]^{-1}. \tag{4.36}$$

We define $N_0 = N$.

The key inequality in [2] is

$$I_k(\varepsilon, \psi) \geqq \sum_{r=0}^{\infty} [\alpha_r - (1 + \eta_r) N_r] + E_k. \tag{4.37}$$

The term E_k is a constant times the number of particles with momentum n satisfying $|n| \geqq C|k|/N^\delta$. It follows that the expression

$$\sum_k v(k) E_k \tag{4.38}$$

can be combined with a small portion of the kinetic energy to yield a lower order term. We shall therefore concentrate on the sum on the right-hand side of (4.37). The η_r are defined as

$$\eta_r = \varepsilon / [n_r p_r], \tag{4.39}$$

and

$$p_r = 2^{4t} \quad \text{if} \quad (2^t - 1) N^{1/3 - \delta} \leqq a^r D\gamma < (2^{t+1} - 1) N^{1/3 - \delta/2}. \tag{4.40}$$

The α_r are the positive roots of the polynomial equation (in μ)

$$1 + \sum_{r=0}^{\infty} N_r [\eta_r N_r - \mu]^{-1} + N_r [\eta_r N_r + \mu]^{-1} = 0. \tag{4.41}$$

We order the roots α_r in the following manner: Let α_0 be the unique root of (4.41) which has $\alpha_0 > \eta_0 N_0$. The roots $\alpha_r, r = 1, 2, \ldots$, are the unique roots of (4.41) which satisfy $\eta_{r-1} N_{r-1} > \alpha_r > \eta_r N_r$. We define $\beta_r(k)$ by

$$-\beta_r(k) = \alpha_r - (1 + \eta_r) N_r, \tag{4.42}$$

where α_r is determined from (4.41) after setting $\varepsilon = \varepsilon_k$ in the definition, (4.39), of η_r.

Let us define

$$\tfrac{1}{2} \sum_{|k| > N^\delta D\gamma} v(k) \beta_r(k) = B_r N (N L \gamma^3)^{1/4} / L. \tag{4.43}$$

Note that B_r is a constant plus correction terms which tend to zero as $NL\gamma^3 \to \infty$. In the following computation (cf. (4.66)), it will be found that $NL\gamma^3$ indeed tends to infinity, and thus we are able to neglect these correction terms.

If we define B by

$$B = \sum_{r=0}^{\infty} B_r, \tag{4.44}$$

we have that

$$\langle \psi, H_N^{\nu} \psi \rangle \geqq -BN(NL\gamma^3)^{1/4}/L. \tag{4.45}$$

We need then to estimate β_r and B_r, for $r = 0, 1, 2, \ldots$. We first consider the case $r = 0$. The root α_0 of (4.41) is clearly bounded below by the unique positive root of the equation

$$N_0[\eta_0 N_0 - \mu]^{-1} + N_0[\eta_0 N_0 + \mu]^{-1} + 1 = 0. \tag{4.46}$$

Hence we obtain

$$\beta_0 \leqq N_0\{1 + \eta_0 - [\eta_0^2 + 2\eta_0]^{1/2}\}. \tag{4.47}$$

Now substituting the values for $\beta_0(k)$ and performing the sum in (4.43) we obtain

$$B_0 = 2(3\pi^2)^{-1/4} D^{3/4} I. \tag{4.48}$$

In the calculation for (4.48) we have used the fact that $p_0 = 1$ in (4.39). In fact $p_r = 1$ provided $r \leqq C \log N$, since $\gamma \leqq N^{1/3-\delta}$. Note that (4.48) and (4.21) are identical.

Next, we wish to estimate β_r and B_r when $r = 1, 2, \ldots$. Now α_r is bounded below by the unique root, μ, of the equation

$$1 + \sum_{j=0}^{r} N_j[\eta_j N_j - \mu]^{-1} + N_j[\eta_j N_j + \eta_r N_r]^{-1} = 0, \tag{4.49}$$

which lies in the interval $\eta_{r-1} N_{r-1} > \mu > \eta_r N_r$. Let $\alpha_{r,1}$ be the root of the polynomial equation which is the same as (4.49) except that the terms $N_j/(\eta_j N_j - \mu)$, $j = 0, \ldots, r-1$ are replaced by $N_j/(\eta_j N_j - \eta_r N_r)$, $j = 0, \ldots, r-1$. Thus $\alpha_{r,1}$ is *larger* than the corresponding root of (4.49). Next, let $\alpha_{r,2}$ be the root of the polynomial equation which is the same as (4.49) except that the terms $N_j/(\eta_j N_j - \mu)$, $j = 0, \ldots, r-1$ are replaced by $N_j/(\eta_j N_j - \alpha_{r,1})$. It is clear that $\alpha_{r,2}$ is *smaller* than the corresponding root of (4.49). We can define the quantities $\beta_{r,1}, B_{r,1}, B^1$ and $\beta_{r,2}, B_{r,2}, B^2$ to correspond to the roots $\alpha_{r,1}, \alpha_{r,2}$ respectively in exactly the some manner as β_r, B_r, B correspond to α_r.

We calculate $\alpha_{r,1}$. To do this we write the corresponding polynomial equation in the form

$$N_r[\eta_r N_r - \mu]^{-1} + 1 + h_{r,1}/(2\eta_r) = 0, \tag{4.50}$$

where $h_{r,1}$ is given by the equation

$$h_{r,1} = 1 + 2\sum_{j=0}^{r-1} [\eta_j/\eta_r + N_r/N_j]^{-1} + [\eta_j/\eta_r - N_r/N_j]^{-1}. \tag{4.51}$$

From (4.50) it follows that

$$\beta_{r,1} = (1 + \eta_r)N_r - \alpha_{r,1} = N_r[1 + 2\eta_r/h_{r,1}]^{-1}. \tag{4.52}$$

We wish now to fix the values of a and D in an optimal way. We do this by making the approximation $h_{r,1} \simeq 1$ and optimizing the value of B^1 based on this. With this approximation we have, then, an approximate value for $B_{r,1}$ obtained by summing (4.52),

$$B_{r,1} \simeq \tfrac{1}{2} J[\pi(a^{r-1}D)]^{-2}\left[\frac{1}{3\pi^2} D^3 a^{3(r-1)}(a^3-1)\right]^{1/4}. \tag{4.53}$$

Summing (4.53) from $r = 1, \ldots, \infty$ we have

$$\sum_{r=1}^{\infty} B_{r,1} \simeq \tfrac{1}{2}(3\pi^2)^{-1/4}\pi^{-2} J D^{-5/4} g(a), \tag{4.54}$$

where $g(a)$ is the function of a given by

$$g(a) = (a^3-1)^{1/4} a^{5/4}[a^{5/4}-1]^{-1}. \tag{4.55}$$

We shall take $a = 2$, which is close to the minimum for g, and the corresponding value for g is $g(2) = 2.81$. From (4.48) and (4.54) we then have

$$B_0 + \sum_{r=1}^{\infty} B_{r,1} \simeq \tfrac{1}{2}(3\pi^2)^{-1/4}[4ID^{3/4} + 2.81\pi^{-2} J D^{-5/4}]. \tag{4.56}$$

The value of D is chosen to minimize the right side of (4.56). This yields the value

$$D = (1/\pi)[14.05J/12I]^{1/2} = 1.16/\pi. \tag{4.57}$$

It is of some interest to compare this value of D with the value of D given in (4.14), namely $D = .645/\pi$, which was used in the previous heuristic calculation. With D chosen as in (4.57), Eq. (4.56) yields

$$B_0 + \sum_{r=1}^{\infty} B_{r,1} \simeq 0.53. \tag{4.58}$$

Having fixed a and D we obtain an upper bound for B. The expression $h_{r,1}$ is given from (4.51) and (4.34), (4.36) as

$$h_{r,1} = 1 + 2\sum_{j=1}^{r-1}\{[a^{3(r-j)} + a^{2(j-r)}]^{-1} + [a^{3(r-j)} - a^{2(j-r)}]^{-1}\}$$
$$+ 2\left[a^{3(r-1)}(a^3-1) + \frac{a^{2(1-r)}}{4\pi^2 D^2}\right]^{-1} + 2\left[a^{3(r-1)}(a^3-1) - \frac{a^{2(1-r)}}{4\pi^2 D^2}\right]^{-1}. \tag{4.59}$$

It is easy to see from (4.59) that

$$1 < h_{r,1} < 5/3. \tag{4.60}$$

If we use the lower bound in (4.60) we obtain from (4.50) an upper bound on $\alpha_{r,1}$,

$$\alpha_{r,1} < \eta_r N_r + (1 + 1/2\eta_r)^{-1} N_r = \eta_r N_r[1 + 2(1 + 2\eta_r)^{-1}] \leqq 3\eta_r N_r. \tag{4.61}$$

We may now use the upper bound (4.61) to obtain a *lower* bound on $\alpha_{r,2}$. In view of (4.61), $\alpha_{r,2}$ is bounded below by the root of the equation

$$N_r[\eta_r N_r - \mu]^{-1} + 1 + h_{r,2}/(2\eta_r) = 0, \tag{4.62}$$

where $h_{r,2}$ is given by the equation

$$h_{r,2} = 1 + 2\sum_{j=0}^{r-1} [\eta_j/\eta_r + N_r/N_j]^{-1} + [\eta_j/\eta_r - 3N_r/N_j]^{-1}. \tag{4.63}$$

If we express $h_{r,2}$ in a similar fashion to (4.59) it is easy to conclude that $h_{r,2} < 5/3$. We conclude then that

$$\beta_{r,2} \leqq N_r\left[1 + \frac{6\eta_r}{5}\right]^{-1}. \tag{4.64}$$

Hence from (4.58) we have

$$B \leqq 0.53(5/3)^{1/4} = 0.60. \tag{4.65}$$

Thus (4.45) and (4.65) yield a lower bound on the energy, $\langle \psi, H_N^v \psi \rangle$, of the wave function, ψ, in terms of $\gamma = \gamma_\psi$ and L.

To obtain a lower bound on the energy in terms of N alone, we have to use the fact that γ, L and N are not really independent when the energy is negative and when the hypotheses of Theorem 1.4 are satisfied. To see this let us divide the kinetic energy into two parts. One part is estimated by using the definition of γ as $N\gamma^2/L^2$. The other part is put together with the potential energy and use is made of (4.45). In all, then, we have for any $\lambda, 0 < \lambda < 1$, the inequality

$$\langle \psi, H_N^v \psi \rangle \geqq \frac{\lambda N \gamma^2}{L^2} - \frac{B}{(1-\lambda)^{1/4}} N^{5/4} \left(\frac{\gamma}{L}\right)^{3/4}. \tag{4.66}$$

The factor $(1-\lambda)^{-1/4}$ in (4.66) is obtained by applying scaling to (4.45). Minimizing (4.66) with respect to γ/L yields

$$\langle \psi, H_N^v \psi \rangle \geqq -(5/8)(3/8)^{3/5} B^{8/5} N^{7/5} \lambda^{-3/5} (1-\lambda)^{-2/5}. \tag{4.67}$$

The maximum value of $h(\lambda) = \lambda^{3/5}(1-\lambda)^{2/5}$ for $0 < \lambda < 1$ is obtained at $\lambda = 3/5$ with $h(3/5) = .510$. Hence (4.67) yields

$$\langle \psi, H_N^v \psi \rangle \geqq -0.30 N^{7/5}. \tag{4.68}$$

We have proven (4.68) under the assumption that H_N^v is the Hamiltonian of a *neutral* system. However for the argument of Sect. II to be valid we need to know that (4.68) holds even for nonneutral systems. The neutrality assumption entered in our calculations only in the inequality (4.28) and it did so in the following way. The estimate in Lemma 2.2 of [2] leads to a denominator 2 $\max(N_+, N_-)$ instead of N in (4.28). It is only when $N_+ = N_- = N/2$ that we get (4.28). If the system is not neutral and the ratio of negative particles to the total number of particles is given by

$$\frac{N_-}{N} = \frac{(1-\xi)}{2}, \quad 0 \leqq \xi \leqq 1, \tag{4.69}$$

then the coefficient $4\pi^2/NL^2$ of the sum in (4.28) must be decreased to $4\pi^2/(1+\xi)NL^2$, which in turn leads to the inequality

$$\langle \psi, H_N^v \psi \rangle \geqq -0.30(1+\xi)^{2/5} N^{7/5}. \tag{4.70}$$

The inequality (4.70) gives an $N^{7/5}$ lower bound for a nonneutral system which has a slightly larger constant than the constant 0.30 for the neutral case. We wish to show that the constant 0.30 still holds for the nonneutral case in the situation where we apply this inequality in Sect. II. The Hamiltonian H_N^ν can be written as

$$H_N^\nu = W_N^\nu + H_N^\omega, \tag{4.71}$$

where W_N^ν is the N-body potential energy obtained from the function

$$r^{-1}(e^{-\nu r} - e^{-\omega r}) = \int_\nu^\omega e^{-ur} du. \tag{4.72}$$

We choose $\omega = \nu + N^{1/5}$. The inequality (4.70) applies to H_N^ω. In fact, the inequality becomes better since $\omega > \nu$, which implies that $v(k)$ becomes smaller. Our bound (4.70) is monotone in $v(k)$.

To bound W_N^ν from below, let us suppose the particles are fixed at points $x_1, \ldots, x_N$ with the negative particles being at $x_i, i = 1, \ldots, N_-$. We define a density $\rho(x)$, by

$$\rho(x) = \sum_{i=1}^N e_i \delta(x - x_i). \tag{4.73}$$

It is clear that

$$W_N^\nu = \frac{1}{2}\int_\nu^\omega du \int e^{-u|x-y|} \rho(x)\rho(y) dx dy - \tfrac{1}{2}N^{6/5}. \tag{4.74}$$

The following lemma and proof is due to Federbush [6, 3]. It can also be proved by the method of Lemma 3.1.

Lemma 4.1. *Let $\Lambda \subset \mathbf{R}^3$ be a cube of side length L. Let $f: \Lambda \to \mathbf{R}$ be a (not necessarily positive) density with $Q = \int_\Lambda f dx$. Let $\mu \geqq 0$. Then there is a constant C_{14} independent of μ, f, L such that*

$$D_\mu(f) \equiv \tfrac{1}{2}\int f(x) f(y) \exp[-\mu|x-y|] dx dy \geqq C_{14} Q^2 \mu L (1 + \mu^2 L^2)^{-2}.$$

Proof. Assume $f \in L^2(\Lambda)$ and write

$$\begin{aligned} D_\mu(f) &= \tfrac{1}{2}\langle f, e^{-\mu|x-y|} f\rangle = 4\pi\mu \langle f, (-\Delta + \mu^2)^{-2} f\rangle \\ &= 4\pi\mu \|(-\Delta + \mu^2)^{-1} f\|_2^2 \geqq 4\pi\mu |\langle g, (-\Delta + \mu^2)^{-1} f\rangle|^2 / \|g\|_2^2 \\ &\geqq 4\pi\mu |\langle h, f\rangle|^2 / \|(-\Delta + \mu^2) h\|_2^2, \end{aligned}$$

where g is any function in $L^2(\mathbf{R}^3)$ and where $g = (-\Delta + \mu^2)h$. Let H be a C^∞ function with $H(x) = 1$ for $|x| \leqq 2$ and $H(x) = 0$ for $|x| \geqq 3$. Finally, take $h(x) = H(x/L)$. □

From (4.70) to (4.74) and Lemma 4.1 we conclude that

$$\langle \psi, H_N^\nu \psi\rangle \geqq C_{14} \nu L (1 + \nu^2 L^2)^{-2} \xi^2 N^{11/5} - \tfrac{1}{2} N^{6/5} - 0.30(1 + \xi)^{2/5} N^{7/5}. \tag{4.75}$$

The inequality (4.75) shows that (4.68) holds for large N, even in the nonneutral case, provide $\nu L N^{-1/5} < \infty$ as $N \to \infty$. (Recall that, as stated in the beginning, it is only necessary to prove Theorem 1.4 when $\nu L > N^{1/10}$.) This concludes the proof of Theorem 1.4.

V. The Nonneutral Case

Consider the Hamiltonian H_N of (1.1) and its generalization H_N^ν of Theorem 2.1 or (2.21) acting on N_- negative particles and N_+ positive particles with $N_- \leqq N_+, N_- + N_+ = N$. Our goal here is to generalize Theorems 1.2 and 2.1 as follows.

Theorem 5.1. *Let H_N^ν be as in Theorem 2.1, and let there be N_- negative and N_+ positive particles with $N_- \leqq N_+$. The parameter ν can depend on N_- and N_+ but we suppose that $N_-^{-2/15}\nu \to 0$ as $N_- \to \infty$. (Note the difference from Theorem 2.1.) Then*

$$H_N^\nu \geqq -A_5 N_-^{7/5} \tag{5.1}$$

for some constant, A_5.

The proof follows the same lines as in Sect. II. One must modify it in two respects, however. First, it is necessary to prove that the interaction energy depends only on the number of negative particles. Second, we need to localize the kinetic energy in a somewhat different way than in Sect. II. Basically we only want to localize the kinetic energy of a positive particle if it lies in a box containing a negative particle. If we were to localize the kinetic energy of all positive particles, the cost in energy would be proportional to the number of positive particles and this of course could be much larger than $N_-^{7/5}$.

We solve the problem of the interaction energy in Lemma 5.3 below, but first we require the two preliminary Lemmas 5.1 and 5.2. The first is independently interesting.

Lemma 5.1. *Suppose that K and $L: \mathbf{R}^3 \to \mathbf{R}^+$ are two nonnegative functions (not necessarily symmetric) that satisfy the following (5.2), for some fixed, positive integer s,*

$$sL(x) \geqq K(y) \quad \text{whenever} \quad |x| \leqq |y|. \tag{5.2}$$

Let $x_1, \ldots, x_{N_-}$ and $y_1, \ldots, y_{N_+}$ be points in $\mathbf{R}^3$ that satisfy

$$\sum_{i=1}^{N_-} K(y_j - x_i) - \sum_{1 \leqq k \leqq N_+, k \neq j} L(y_j - y_k) > 0 \tag{5.3}$$

for each $j = 1, \ldots, N_+$. Then $N_+ \leqq CsN_-$, where C is some universal geometric constant (60 will suffice).

Proof. We shall use the following geometric fact. There exists a finite set of closed, solid, circular cones in $\mathbf{R}^3$, each with apex at the origin and each with solid angle $\pi/3$ such that their union is all of $\mathbf{R}^3$. The minimum number of cones required for this is some integer C, and it is easy to see that $C \leqq 60$. Denote these cones by $B_1, \ldots, B_C$. Let Y denote the set of y_i points.

Now, without loss of generality, assume $x_1 = 0$. Let $Y_1 = \{y_i | y_i \in B_1\}$ be the points in B_1, and let Z_1 be those s points in Y_1 which are closest to x_1. (If there is an ambiguity, make an arbitrary choice; if Y_1 has fewer than s points then $Z_1 \equiv Y_1$.) Next, apply this process to the remainder $Y \setminus Y_1$ and thereby obtain Z_2 with respect to B_2. Continuing in this way we obtain $Z_1, \ldots, Z_C$ and $Y_1, \ldots, Y_C$. Let $Z = \bigcup_{i=1}^{C} Z_i$, whence Z has at most sC points.

Take $y_j \notin Z$ and consider the contribution to the left side of (5.3) coming from x_1 and Z. This contribution is

$$A_j = K(y_j) - \sum_{y_k \in Z} L(y_j - y_k).$$

We claim that $A_j \leqq 0$. If $y_j \in B_\alpha$ then the second sum in A_j is not less than $\sum_{y_k \in Z_\alpha} L(y_j - y_k)$. But $|y_j - y_k|^2 = |y_j|^2 + |y_k|^2 - 2y_j \cdot y_k \leqq |y_j|^2 + |y_k|^2 - |y_j||y_k| \leqq |y_j|^2$, since $|y_k| \leqq |y_j|$. Thus, $|y_j - y_k| \leqq |y_j|$ and thus $sL(y_j - y_k) \geqq K(y_j)$. Given that $y_j \in B_\alpha$, Z_α has s points and thus $A_j \leqq 0$.

If we now remove x_1 and Z from the system we obtain a reduced system with a new $N_- = N_- - 1$ and with a new $N_+ \geqq N_+ - sC$, and that satisfies (5.3) for all y_j in the new system. The construction can now be repeated with x_2 and then x_3 and so on until we obtain a final system with $N_- = 0$ and a final $\tilde{N}_+ \geqq N_+ - sCN_-$. This clearly cannot satisfy (5.3) if $\tilde{N}_+ > 0$. □

Lemma 5.2. *Let $K: \mathbf{R}^3 \to \mathbf{R}^+$ be given as in Lemma 2.1 by*

$$K(x) = r^{-1}\{e^{-\nu r} - e^{-\omega r}h(x)\}$$

with $r = |x|$ and $\omega > \nu \geqq 0$. Here we assume only that $h: \mathbf{R}^3 \to \mathbf{R}$ satisfies (i) $-H \leqq h(x) \leqq 1$ *for all x and some finite $H \geqq 0$;* (ii) *h is continuous in some neighborhood of $x = 0$. Then there is a positive integer s such that*

$$sK(x) \geqq K(y) \quad \textit{whenever} \quad |x| \leqq |y|. \tag{5.4}$$

The integer s depends only on $\omega - \nu \equiv \rho$ and on h. For fixed h, s is a nonincreasing function of ρ.

Proof. For (5.4) we can restrict our attention to the case $\nu = 0, \omega = \rho$ because multiplication of this K by $e^{-\nu r}$ only makes inequality (5.4) stronger. There is an $R > 0$ such that h is continuous in $B_R = \{x \mid |x| \leqq R\}$. Since $K(x) \geqq r^{-1}\{1 - e^{-\rho r}\}$, which is decreasing in r and since $K(x) \leqq r^{-1}\{1 + He^{-\rho r}\}$, we have that $K(y)/K(x) \leqq (1 + He^{-\rho r})/(1 - e^{-\rho r})$ with $r = |y|$. The maximum of this ratio for $r \geqq R$ occurs at $r = R$ and is $s_1 = (1 + He^{-\rho R})/(1 - e^{-\rho R})$. Thus $s_1 K(y) \geqq K(x)$ when $|x| \leqq |y|$ and $|y| \geqq R$. On the other hand, when $|x| \leqq |y| \leqq R$, consider the function $F_R(x, y) \equiv K(y)/K(x)$ defined in the closed set $T = \{(x, y): |x| \leqq |y| \leqq R\}$. There are 2 cases. Case (i): $h(0) = 1$. Then K is continuous on B_R with $K(0) = \rho$. Moreover, $K(x) \geqq r^{-1}(1 - e^{-\rho r})$ on B_R. Thus $F_R(x, y)$ is continuous and so has a maximum on T. Case (ii): $h(0) < 1$. Then $|x|K(x)$ is a continuous function on B_R and it is never zero, so $|x|K(x) \geqq t$ for some $t > 0$. Hence, $F_R(x, y) \leqq t^{-1}|x|K(y) \leqq t^{-1}(1 + H)$. Thus, (5.4) is satisfied for any integer $s \geqq \max\{s_1, \max_T F_R(x, y)\}$. To prove the monotonicity of s, consider $K(x)$ with $\nu \geqq 0$ and $\omega = \nu + \rho$. Let $F(x, y) = K(y)/K(x)$. Since $s \geqq 1$, we only consider x, y such that $F(x, y) \geqq 1$. Hence $(\partial F/\partial \rho)(x, y) = [(\partial K/\partial \rho)(y) - F(\partial K/\partial \rho)(x)]/K(x) = -[|y|K(y) - e^{-\nu|y|} - |x|K(y) + Fe^{-\nu|x|}]/K(x)$. One concludes that $(\partial F/\partial \rho)(x, y) \leqq 0$ if $|x| \leqq |y|$ and if $F(x, y) \geqq 1$. Hence s is monotone in ρ. □

Lemma 5.3. *Suppose $\nu, \rho, l \geqq 0$ and let*

$$f(x) = a_0 Y_\nu(x) - Y_{\rho+\nu}(x) h_l(x)$$

with $h_l(0) = a_0, h_l(x) = h(x/l)$ and with $h(x)$ given by (2.8). This f is a generalization of $f_{\mu l}$ given in (2.16). Given $x_1, \ldots, x_N \in \mathbf{R}^3$ and $e_i = \pm 1$, let N_- (respectively N_+) be the number of e_i which are -1 (respectively $+1$). Assume that $N_- \leqq N_+$. Finally, suppose that $\rho l \geqq C_3$ (which is defined in Lemma 2.1 and which depends only on h), so that $\hat{f} \geqq 0$. Then there is a constant C_{13} depending only on h and not on v, ρ, l, such that

$$\sum_{i<j} e_i e_j f(x_i - x_j) \geqq -C_{13} \rho N_- . \tag{5.5}$$

Proof. Let W denote the left side of (5.5). Combining Lemmas 5.1 and 5.2, there is an s (which depends on h and on ρl (by scaling)) so that whenever $N_+ \geqq CsN_-$ we can eliminate $N_+ - CsN_-$ positive particles without increasing W. Thus we can assume $N_+ \leqq CsN_-$. Furthermore, this s can only decrease when ρl increases, so we can take s to be the value it has when $\rho l = C_3$ (which depends only on h). Thus s depends only on h. Now since $\hat{f} \geqq 0$, we have $W \geqq -\frac{1}{2}(N_+ + N_-) f(0) = -\frac{1}{2}(N_+ + N_-)\rho a_0 \geqq -\frac{1}{2}(Cs+1)\rho a_0 N_-$. □

We return next to the problem of localizing the kinetic energy similarly to Sect. II. For any $\alpha = (u, \alpha_1, \ldots, \alpha_N) \in \Gamma \times \mathbf{Z}^{3n}$ we define ψ_α^l as in (2.11). We adopt the convention that the negative particles are labelled $1, \ldots, N_-$ and the positive particles are labelled $N_- + 1, \ldots, N$. Let S_α be the α_i which correspond to the negative particles,

$$S_\alpha = \{\alpha_1, \ldots, \alpha_{N_-}\}. \tag{5.6}$$

We denote by $\bar{S}_\alpha$ the set of nearest neighbors in $\mathbf{Z}^3$ of S_α, so

$$\bar{S}_\alpha = \{m \in \mathbf{Z}^3 \,|\, |m - \alpha_i| \leqq \sqrt{3} \text{ for some } \alpha_i \in S_\alpha\}. \tag{5.7}$$

Let N_α be "the number of positive particles which lie in a box occupied by a negative particle" and $\bar{N}_\alpha$ "the number of positive particles which lie in the same box as a negative particle or a nearest neighbor of such box." By this is meant

$$\begin{aligned} N_\alpha &= \#\{j > N_- \,|\, \alpha_j \in S_\alpha\}, \\ \bar{N}_\alpha &= \#\{j > N_- \,|\, \alpha_j \in \bar{S}_\alpha\}. \end{aligned} \tag{5.8}$$

The definition of $S_\alpha, \bar{S}_\alpha, N_\alpha, \bar{N}_\alpha$ depend only on $\alpha \in \mathbf{Z}^{3N}$. Finally we define the kinetic energy operator T_α (which also depends only on α) to be the kinetic energy of the negative particles plus the kinetic energy of "the positive particles which lie in a box occupied by a negative particle," namely

$$T_\alpha = \sum_{i=1}^{N_-} -\Delta_i + \sum_{\{j > N_- | \alpha_j \in S_\alpha\}} -\Delta_j . \tag{5.9}$$

We then have the following lemma:

Lemma 5.4. *Let C_0 be the constant in (2.14). The kinetic energy is bounded below (recalling the definition of $\int d\alpha$ before (2.11)) as*

$$\langle \psi, T\psi \rangle \geqq \tfrac{1}{2} \int d\alpha \langle \psi_\alpha^l, T_\alpha \psi_\alpha^l \rangle - C_0 l^{-2} [N_- + 27 \int \bar{N}_\alpha \| \psi_\alpha^l \|^2 d\alpha]. \tag{5.10}$$

Proof. We use (2.14) to bound $\langle \psi, -\Delta_i \psi \rangle$ below for $i \leqq N_-$, namely

$$\left\langle \psi, \sum_{i=1}^{N_-} -\Delta_i \psi \right\rangle \geqq \int d\alpha \left\langle \psi_\alpha^l, \sum_{i=1}^{N_-} -\Delta_i \psi_\alpha^l \right\rangle - C_0 l^{-2} N_- .$$

Now suppose $i > N_-$ and consider a fixed α. Then we have the inequality

$$\tfrac{1}{2}\int dx_i |\nabla_i(\chi_{u\alpha_i}(x_i/l)\psi)|^2 \leqq \int dx_i |\nabla_i \chi_{u\alpha_i}(x_i/l)|^2 |\psi|^2 + \int dx_i \chi^2_{u\alpha_i}(x_i/l)|\nabla_i \psi|^2. \quad (5.11)$$

Now use the fact that

$$|\nabla_i \chi_{u\alpha_i}(x_i/l)|^2 \leqq C_0 l^{-2} \sum_{\lambda\in\mathbf{Z}^3} g(\lambda-\alpha_i)\chi^2_{u\alpha_i}(x_i/l), \quad (5.12)$$

where $g(z)$ is the function $g(z)=1$ if $|z| \leqq \sqrt{3}$, $g(z)=0$ if $|z| > \sqrt{3}$. Hence we have for all $i > N_-$,

$$\tfrac{1}{2}\langle \psi^l_\alpha, -\Delta_i \psi^l_\alpha\rangle \leqq \int \prod_{j=1}^{N} \chi^2_{u\alpha_j}(x_j/l)|\nabla_i\psi|^2 dx + C_0 l^{-2} \sum_{\lambda\in\mathbf{Z}^3} g(\lambda-\alpha_i)\|\psi^l_{\alpha_1,\ldots,\lambda,\ldots,\alpha_N}\|^2, \quad (5.13)$$

with λ being in the ith position in the last sum. For $i > N_-$ let T^i_α be the ith term in the kinetic energy T_α in (5.9), namely $T^i_\alpha = -\Delta_i$ if $\alpha_i \in S_\alpha$, and $T^i_\alpha = 0$ otherwise. We have then from (5.13),

$$\tfrac{1}{2}\sum_{\alpha_i\in\mathbf{Z}^3}\langle \psi^l_\alpha, T^i_\alpha \psi^l_\alpha\rangle \leqq \sum_{\alpha_i\in\mathbf{Z}^3}\int \prod_{j=1}^{N}\chi^2_{u\alpha_j}(x_j/l)|\nabla_i\psi|^2 dx + 27C_0 l^{-2}\sum_{\alpha_i\in\bar{S}_\alpha}\|\psi^l_\alpha\|^2. \quad (5.14)$$

The number 27 is the number of nearest neighbors of a point in $\mathbf{Z}^3$. If we sum (5.14) with respect to all α_j for $j \neq i$, and then sum over i, and then integrate over $u \in \Gamma$, we obtain the inequality (5.10). □

The following lemma is also needed for the proof of Theorem 5.1.

Lemma 5.5. *Let ψ^l_α be the localized wave function (2.11). Let V^μ_α be given by (2.12) and T_α by (5.9). Assume that $1 \leqq \mu \leqq N^{2/15}_-$. Then there is a constant $C = C(\mu l)$, depending only on μl such that, with the notation of (5.8), there is the estimate*

$$\tfrac{1}{2}\langle \psi^l_\alpha, T_\alpha \psi^l_\alpha\rangle + \langle \psi^l_\alpha, V^\mu_\alpha \psi^l_\alpha\rangle - 27C_0 l^{-2}\bar{N}_\alpha \|\psi^l_\alpha\|^2 \geqq -C(\mu l)N^{7/5}_- \|\psi^l_\alpha\|^2. \quad (5.15)$$

Proof. We analyze the left side of (5.15) similarly to (2.15). Since there is no interaction between boxes, the left-hand side of (5.15) is bounded below by

$$\left[\sum_{\sigma\in\mathbf{Z}^3} E_\sigma\right]\|\psi^l_\alpha\|^2, \quad (5.16)$$

where E_σ is the ground state energy of the following Hamiltonian, H_σ, depending on σ. There are three cases: $\sigma\in S_\alpha$, $\sigma\in\bar{S}_\alpha\backslash S_\alpha$ and $\sigma\notin\bar{S}_\alpha$. If $\sigma\in S_\alpha$ and n_σ of the i, $1\leqq i\leqq N$, have $\alpha_i=\sigma$ with n^-_σ of these satisfying $i\leqq N_-$, then H_σ is the Hamiltonian

$$H_\sigma = \tfrac{1}{2}T + V^\mu - 27C_0 l^{-2} n^+_\sigma \quad (5.17)$$

acting on n_σ particles in a box of size l, n^-_σ of which are negative, n^+_σ positive. Here, $V^\mu = \sum_{i<j} e_i e_j Y_\mu(x_i - x_j)$. If $\sigma\in\bar{S}_\alpha\backslash S_\alpha$, then $n^-_\sigma = 0$ and H_σ is the Hamiltonian

$$H_\sigma = V^\mu - 27C_0 l^{-2} n_\sigma \quad (5.18)$$

acting on n_σ positive particles in a box of size l. If $\sigma\notin\bar{S}_\alpha$ then H_σ is

$$H_\sigma = V^\mu \quad (5.19)$$

acting on n_σ positive particles.

We estimate the ground state energies E_σ. In the case of (5.19) we clearly have $E_\sigma \geqq 0$. In the case of (5.17) we use (4.75). Taking into account the factor $\frac{1}{2}$ in (5.17), which gives a factor $2^{1/4}$ in the $N^{7/5}$ law (cf. (4.66)), and with $\tau \equiv \mu l$, (4.75) reads (with $v = \mu$ and $\omega = 2\mu$ instead of $\omega = \mu + N^{1/5}$)

$$\begin{aligned} H_\sigma = {} & C_{14}\mu\tau(1+\tau^2)^{-2}(n_\sigma^+ - n_\sigma^-)^2 - \tfrac{1}{2}\mu n_\sigma \\ & - (0.30)2^{1/4} n_\sigma[\max(2n_\sigma^+, 2n_\sigma^-)]^{2/5} - 27C_0\mu^2\tau^{-2}n_\sigma^+ . \end{aligned} \tag{5.20}$$

The quantities in (5.20) satisfy $n_\sigma^- \leqq N_-$, τ is fixed, $1 \leqq \mu \leqq N_-^{2/15}$ and n_σ^+ is arbitrary. We minimize the left side of (5.20) with respect to n_σ^+. One can show that

$$H_\sigma \geqq -A(\tau)[\mu^2 n_\sigma^- + \mu^3 + (n_\sigma^-)^{7/5}] \tag{5.21}$$

for some A depending only on τ.

To bound (5.18), one simply notes that in this case

$$H_\sigma \geqq \tfrac{1}{2}(\sqrt{3}l)^{-1} n_\sigma(n_\sigma - 1)\exp(-\sqrt{3}\tau) - 27C_0 l^{-2} n_\sigma \geqq -B(\tau)l^{-3}$$

for some $B(\tau)$ independent of n_σ. By using the fact that $1 \leqq \mu \leqq N_-^{2/15}$, one has $H_\sigma \geqq -D(\tau)N_-^{2/5}$.

Now, putting together the bounds for (5.17), (5.18) and (5.19), we conclude that for some $F(\tau)$

$$\sum_{\sigma\in\mathbf{Z}^3} E_\sigma \geqq -F(\tau)\sum_\sigma^{(1)}\{\mu^2 n_\sigma^- + \mu^3 + n_\sigma^{-\,7/5}\} - F(\tau)\sum_\sigma^{(2)} N_-^{2/5},$$

where the first sum is over S_α and the second is over $\bar{S}_\alpha \backslash S_\alpha$. The number of points in S_α is N_- while the number of points in $\bar{S}_\alpha$ is at most 27 N_-. Using the facts that $\mu \leqq N_-^{2/15}$, $\sum^{(1)} n_\sigma^- = N_-$, and the convexity of $n \to n^{7/5}$, the lemma is proved. □

Proof of Theorem 5.1. Step 1. Starting with v, we define $\mu = N_-^{2/15}$, and $l = C_3 N_-^{-2/15}$, where C_3 is given in Lemma 2.1. As in Sect. II we write $a_0 Y_v = f + Y_\mu h_l$, with $f = a_0 Y_v - Y_\mu h_l$ as in Lemma 5.3. By Lemma 5.3, the contribution to the potential energy from f is bounded below by $-C_{13}(\mu - v)N_- \geqq -C_{13}N_-^{17/15}$ for large N. This can be neglected compared to $N_-^{7/5}$.

Step 2. Lemma 5.4 is used to localize the kinetic energy. The term $-C_0 l^{-2} N_-$ in (5.10) can be neglected since $l^{-2} = (C_3)^{-2} N_-^{4/15}$.

Step 3. The first and third terms on the right side of (5.10) is combined with the $Y_\mu h_l$ part of the potential energy. We localize this potential energy as in (2.13). The first and third terms of (5.10) plus the localized potential energy is just the left side of (5.15). To prove the theorem we merely have to sum the right side of (5.15) over α, but this is exactly $-C(C_3)N_-^{7/5}$ by the normalization condition on ψ. □

Appendix: Thomas–Fermi Theory and the Stability of Matter with Yukawa Potentials

Our main goal here is to establish a lower bound to the energy and an upper bound to the kinetic energy for quantum mechanical particles interacting with

Yukawa, instead of Coulomb potentials. We consider N movable particles with charge -1 and coordinates $x_1, \ldots, x_N \in \mathbf{R}^3$ and K fixed particles with coordinates $R_1, \ldots, R_K \in \mathbf{R}^3$ and charges $z_1, \ldots, z_K \geqq 0$. The movable particles will be considered to be fermions with q spin states, so that $q = N$ corresponds to the boson case. The Hamiltonian is

$$H = -\sum_{i=1}^{N} \{\Delta_i + V(x_i)\} + \sum_{1 \leqq i < j \leqq N} Y_\mu(x_i - x_j) + U, \tag{A.1}$$

with

$$V(x) = \sum_{j=1}^{K} z_j Y_\mu(x - R_j),$$

$$U = \sum_{1 \leqq i < j \leqq K} z_i z_j Y_\mu(R_i - R_j). \tag{A.2}$$

$Y_\mu(x) = |x|^{-1} \exp\{-\mu|x|\}$ is the Yukawa potential. It is positive definite and satisfies

$$(-\Delta + \mu^2) Y_\mu = 4\pi\delta. \tag{A.3}$$

The energy is

$$E = \inf\{(\psi, H_n \psi) |\, \|\psi\|_2 = 1 \text{ and all } R_1, \ldots, R_K\}. \tag{A.4}$$

The method of [15] will be used, which means that we first have to examine the Thomas–Fermi (TF) functional

$$\mathscr{E}(\rho) = \tfrac{3}{5} q^{-2/3} \gamma \int \rho^{5/3}(x) dx - \int V(x)\rho(x) dx + \tfrac{1}{2} \int\int \rho(x)\rho(y) Y_\mu(x-y) dx dy + U \tag{A.5}$$

and corresponding energy

$$E^{\mathrm{TF}} = \inf\{\mathscr{E}(\rho) | \rho \in L^{5/3} \cap L^1\}. \tag{A.6}$$

Notice that in (A.6) we do not impose $\int \rho = N$. This constraint could easily be dealt with, but it is not needed in this paper.

One of our results will be that $E^{\mathrm{TF}} - U$ is a monotone decreasing function of μ.

A. The Thomas–Fermi Problem. By the methods of [14], a minimizer exists for (A.6) and satisfies $\gamma q^{-2/3} \rho(x)^{2/3} = \max(\phi(x), 0)$ with

$$\phi(x) = V(x) - (Y_\mu * \rho)(x). \tag{A.7}$$

Lemma A.1. *$\phi(x) \geqq 0$, all x, and therefore the* TF *equation becomes*

$$\gamma q^{-2/3} \rho(x)^{2/3} = \phi(x). \tag{A.8}$$

Proof. Let $B = \{x | \phi(x) < 0\}$. On $B, \rho(x) = 0$ and $R_i \notin B$, all i (because $\phi(R_i) = \infty$). Therefore $-\Delta\phi = -\mu^2 \phi \geqq 0$ on B, so ϕ is superharmonic on B. Since $\phi = 0$ on $\partial B, \phi \geqq 0$ on B which implies that B is empty. □

Lemma A.2. *Let $z_1, \ldots, z_K \geqq 0$ and $\tilde{z}_1, z_2, \ldots, z_K > 0$ be two sets of charges with $\tilde{z}_1 \geqq z_1$. Then, for all x, $\tilde{\phi}(x) \geqq \phi(x)$.*

Proof. Let $\psi = \tilde{\phi} - \phi$ and $B = \{x | \psi < 0\}$. Clearly, $R_1 \notin B$. On $B, \tilde{\rho} \leqq \rho$ so $(-\Delta + \mu^2)\psi = 4\pi(\rho - \tilde{\rho}) \geqq 0$. Thus ψ is superharmonic on B and again B is empty. □

Lemma A.3. *Let $z_1, \ldots, z_M > 0$ and $z_{M+1}, \ldots, z_K > 0$ be two sets of charges located* *at* $R_1, \ldots, R_K$. *Then*

$$E(z_1, \ldots, z_K) > E(z_1, \ldots, z_M) + E(z_{M+1}, \ldots, z_K). \tag{A.9}$$

Proof. This is Teller's theorem for the Yukawa potential and is proved as in [14] using Lemma A.2. □

Lemma A.3 is given in [16, p. 237].

Next, we turn to the question of monotonicity with respect to μ.

Lemma A.4. *Suppose* $\mu_1 > \mu_2$, *with given fixed charges* $z_i > 0$ *and locations* R_i. *Then* $\phi_2(x) \geqq \phi_1(x)$, *for all* x.

Proof. Let $\psi = \phi_2 - \phi_1$ and $B = \{x | \psi(x) < 0\}$. Then $(-\Delta + \mu_i^2)\phi_i(x) = \sum z_j \delta(x - R_j) - \rho_i(x)$. By subtracting these two equations, and using the fact that $\rho_1 > \rho_2$ on B, we find that $-\Delta\psi > \mu_1^2\phi_1 - \mu_2^2\phi_2 > 0$. Again, B is empty. □

Let us define

$$N^c = \int \rho, \tag{A.10}$$

where ρ is the solution to (A.8). N^c is the maximum negative charge for the TF system (A.5).

Lemma A.5. *If* $\mu_1 > \mu_2$, *with fixed* z_i *and* R_i, *then*

$$N_1^c \leqq N_2^c \quad \text{and} \quad E_1^{\mathrm{TF}} - U_1 \geqq E_2^{\mathrm{TF}} - U_2. \tag{A.11}$$

Proof. $N_1^c \leqq N_2^c$ is a trivial consequence of Lemma A.4 and (A.8). By multiplying (A.8) by ρ and integrating, we have that

$$E - U = -\tfrac{2}{5}\int V\rho - \tfrac{1}{10}\int\int \rho(x)\rho(y)\, Y_\mu(x-y)\,dx\,dy. \tag{A.12}$$

Since $\mu_1 > \mu_2$, $\rho_1(x) \leqq \rho_2(x)$ and $Y_{\mu_1}(x) < Y_{\mu_2}(x)$, for all x. This, together with (A.12), proves the lemma. □

Let us now compare the Yukawa TF problem with the Coulomb TF problem, which corresponds to $\mu = 0$. For the Coulomb problem $N^c = Z \equiv \sum_1^K z_j$ [14]. By Lemmas A.3 and A.5 we have that

$$E^{\mathrm{TF}} \geqq \sum_{j=1}^{K} E^{\mathrm{TF,atom}}(z_j) \geqq \sum_{J=1}^{K} E^{\mathrm{TF,atom}}_{\mathrm{Coulomb}}(z_j). \tag{A.13}$$

The latter inequality follows from the fact that $U = 0$ for an atom. For the TF Coulomb atom [14], $E^{\mathrm{TF}}(z) = -(3.679)\gamma^{-1} q^{2/3} z^{7/3}$. Thus, for the Yukawa problem,

$$E^{\mathrm{TF}} \geqq -(3.679)\gamma^{-1} q^{2/3} \sum_{j=1}^{K} z_j^{7/3}. \tag{A.14}$$

Another lower bound for $E^{\mathrm{TF,atom}}(z)$ can be obtained by dropping the $\rho\rho Y_\mu$ term in (A.5). The resulting minimization problem is trivial: $q^{-2/3}\gamma\rho(x)^{2/3} = V(x) = zY_\mu(x)$ for an atom. Since $\int Y_\mu^{5/2} = 4\pi(2\pi/5\mu)^{1/2}$, (A.13) implies

$$E^{\mathrm{TF}} \geqq -4q\mu^{-1/2}\gamma^{-3/2}(2\pi/5)^{3/2} \sum_{j=1}^{K} z_j^{5/2}. \tag{A.15}$$

B. The Quantum-Mechanical Problem. Returning to the Hamiltonian in (A.1), we want to find a lower bound to $\langle\psi, H\psi\rangle$ for any normalized N-particle function, ψ. The one-particle density of ψ is defined by

$$\rho_\psi(x) = N \int |\psi(x, x_2, \ldots, x_N)|^2 dx_2 \cdots dx_N, \tag{A.16}$$

and $\langle\psi, H\psi\rangle$ will be bounded in terms of ρ_ψ.

To bound the particle-particle energy we use the trick in [15]. Consider (A.5) with $q = 1, K = N, \gamma = \delta$ (arbitrary), $R_i = x_i$ and $z_i = 1$ for $i = 1, \ldots, N$. Then, inserting ρ_ψ in (A.1) and using (A.14),

$$\sum_{1 \leqq i < j \leqq N} Y_\mu(x_i - x_j) \geqq \tfrac{1}{2} \iint \rho_\psi(x)\rho_\psi(y) Y_\mu(x-y)dxdy - \tfrac{3}{5}\delta \int \rho_\psi^{5/3} - 3.679N/\delta. \tag{A.17}$$

To bound the kinetic energy, we use the bound in [15] (recall that $q = N$ for bosons):

$$K(\psi) = \left\langle \psi, -\sum_{i=1}^{N} \Delta_i \psi \right\rangle \geqq K_3 N^{-2/3} \int \rho_\psi(x)^{5/3} dx. \tag{A.18}$$

In [15], the constant K_3 is given as $\frac{3}{5}(3\pi/2)^{2/3} = 1.69$, but this constant was subsequently improved. The best bound at present is in [11] where it is shown that we can take $K_3 = 2.7709$.

Combining (A.17), (A.18) we have the following bound for any normalized ψ

$$\langle\psi, H\psi\rangle \geqq \mathscr{E}(\rho_\psi) - (3.679)N/\delta, \tag{A.19}$$

with $q = 1$ and $\gamma = \frac{5}{3}K_3 N^{-2/3} - \delta$ in (A.5). We choose

$$\delta = \tfrac{5}{3}K_3 N^{-1/6}\left[N^{1/2} + \left(\sum_{j=1}^{K} z_j^{7/3}\right)^{1/2}\right]^{-1}, \tag{A.20}$$

which implies that $\gamma > 0$. Using the bound (A.14) we obtain

Theorem A.1. *With H given by (A.1), the following holds for all normalized ψ:*

$$\langle\psi, H\psi\rangle \geqq -\tfrac{3}{5}(3.679)K_3^{-1}N^{2/3}\left[N^{1/2} + \left(\sum_{j=1}^{K} z_j^{7/3}\right)^{1/2}\right]^2 \tag{A.21}$$

with $K_3 = 2.7709$.

The final task is to apply Theorem A.1 to H_N in (1.1). Suppose that K particles have $e_i = +1$ and M particles have $e_i = -1$ with $K + M = N$. By ignoring the positive kinetic energy of the positive particles, (A.21) can be used with $(N, K) \to (M, K)$. Alternatively, the roles of positive and negative particles can be interchanged, so we can also replace (N, K) in (A.21) by (K, M). The two bounds can then be averaged and an expression of the form $\frac{1}{2}(K^{2/3} + M^{2/3})(K^{1/2} + M^{1/2})^2$ is obtained. However, given that $K + M = N$, $K^{2/3} + M^{2/3}$ has its maximum at $K = M = N/2$. So does $K^{1/2} + M^{1/2}$. Thus we have

Theorem A.2. *With H_N given by (1.1), the following holds for all normalized ψ.*

$$\langle\psi, H_N\psi\rangle \geqq -1.004N^{5/3}. \tag{A.22}$$

A virial type theorem, analogous to Theorem 2.2, can be obtained from (A.22). Another application is the following.

Theorem A.3. *Suppose ψ is normalized and $\langle\psi, H_N\psi\rangle \leqq 0$. Then*

$$K(\psi) \leqq 4.016 N^{5/3}. \tag{A.23}$$

Proof. $0 \geqq K(\psi) + P(\psi) = \frac{1}{2}K(\psi) + \langle\psi, H_{N,1/2}\psi\rangle$ where $H_{N,1/2}$ is given by (1.1) but with Δ_i replaced by $\frac{1}{2}\Delta_i$. By scaling, the analogue of (A.22) is $\langle\psi, H_{N,1/2}\psi\rangle \geqq -2(1.004)N^{5/3}$. □

Acknowledgements. We are grateful to Michael Loss for many helpful discussions.

References

1. Bogoliubov, N. N.: On the theory of superfluidity. J. Phys. (USSR) **11**, 23–32 (1947)
2. Conlon, J. G.: The ground state energy of a Bose gas with Coulomb interaction II. Commun. Math. Phys. **108**, 363–374 (1987). See also part I, ibid **100**, 355–397 (1985)
3. Conlon, J. G.: The ground state energy of a classical gas. Commun. Math. Phys. **94**, 439–458 (1984)
4. Dyson, F. J.: Ground-state energy of a finite system of charged particles. J. Math. Phys. **8**, 1538–1545 (1967)
5. Dyson, F. J., Lenard, A.: Stability of matter I and II. J. Math. Phys. **8**, 423–434 (1967); ibid **9**, 698–711 (1968)
6. Federbush, P.: A new approach to the stability of matter problem II. J. Math. Phys. **16**, 706–709 (1975)
7. Foldy, L. L.: Charged boson gas. Phys. Rev. **124**, 649–651 (1961); Errata ibid **125**, 2208 (1962)
8. Girardeau, M.: Ground state of the charged Bose gas. Phys. Rev. **127**, 1809–1818 (1962)
9. Lieb, E. H.: The $N^{5/3}$ law for bosons. Phys. Lett. **70A**, 71–73 (1979)
10. Lieb, E. H.: The Bose fluid. In: Lectures in Theoretical Physics, Vol. VII C, pp. 175–224. Brittin, W. E. (ed.). Boulder: University of Colorado Press 1965
11. Lieb, E. H.: On characteristic exponents in turbulence. Commun. Math. Phys. **92**, 473–480 (1984)
12. Lieb, E. H., Narnhofer, H.: The thermodynamic limit for jellium. J. Stat. Phys. **12**, 291–310 (1975). Errata, ibid **14**, 465 (1976)
13. Lieb, E. H., Sakakura, A. Y.: Simplified approach to the ground-state energy of an imperfect Bose gas II. Charged Bose gas at high density. Phys. Rev. **133**, A899–A906 (1964)
14. Lieb, E. H., Simon, B.: The Thomas–Fermi theory of atoms, molecules and solids. Adv. Math. **23**, 22–116 (1977). See also, Lieb, E. H.: Thomas–Fermi and related theories of atoms and molecules. Rev. Mod. Phys. **53**, 603–641 (1981); Errata ibid **54**, 311 (1982)
15. Lieb, E. H., Thirring, W. E.: Bound for the kinetic energy of fermions which proves the stability of matter. Phys. Rev. Lett. **35**, 687–689 (1975). Errata, ibid **35**, 1116 (1975).
16. Thirring, W. E.: A course in mathematical physics, Vol. 4. Quantum mechanics of large systems. New York, Wien: Springer 1983

Communicated by A. Jaffe

Received October 21, 1987

Commun. Math. Phys. 217, 127 – 163 (2001)

Communications in **Mathematical Physics**

Ground State Energy of the One-Component Charged Bose Gas*

Elliott H. Lieb[1,], Jan Philip Solovej[2,***]**

[1] Departments of Physics and Mathematics, Jadwin Hall, Princeton University, PO Box 708, Princeton, NJ 08544-0708, USA. E-mail: lieb@princeton.edu

[2] Department of Mathematics, University of Copenhagen, Universitetsparken 5, 2100 Copenhagen, Denmark. E-mail: solovej@math.ku.dk

Received: 23 August 2000 / Accepted: 5 October 2000

Dedicated to Leslie L. Foldy on the occasion of his 80th birthday

Abstract: The model considered here is the "jellium" model in which there is a uniform, fixed background with charge density $-e\rho$ in a large volume V and in which $N = \rho V$ particles of electric charge $+e$ and mass m move – the whole system being neutral. In 1961 Foldy used Bogolubov's 1947 method to investigate the ground state energy of this system for bosonic particles in the large ρ limit. He found that the energy per particle is $-0.402\, r_s^{-3/4} me^4/\hbar^2$ in this limit, where $r_s = (3/4\pi\rho)^{1/3}e^2m/\hbar^2$. Here we prove that this formula is correct, thereby validating, for the first time, at least one aspect of Bogolubov's pairing theory of the Bose gas.

1. Introduction

Bogolubov's 1947 pairing theory [B] for a Bose fluid was used by Foldy [F] in 1961 to calculate the ground state energy of the one-component plasma (also known as "jellium") in the high density regime – which is the regime where the Bogolubov method was thought to be exact for this problem. Foldy's result will be verified rigorously in this paper; to our knowledge, this is the first example of such a verification of Bogolubov's theory in a three-dimensional system of bosonic particles.

Bogolubov proposed his approximate theory of the Bose fluid [B] in an attempt to explain the properties of liquid Helium. His main contribution was the concept of pairing of particles with momenta k and $-k$; these pairs are supposed to be the basic constituents of the ground state (apart from the macroscopic fraction of particles in the "condensate", or $k = 0$ state) and they are the basic unit of the elementary excitations of the system. The pairing concept was later generalized to fermions, in which case the pairing was between

** Work partially supported by U.S. National Science Foundation grant PHY98 20650-A01.

*** Work partially supported by EU TMR grant, by the Danish Research Foundation Center MaPhySto, and by a grant from the Danish Research Council.

particles having opposite momenta and, at the same time, opposite spin. Unfortunately, this appealing concept about the boson ground state has neither been verified rigorously in a 3-dimensional example, nor has it been conclusively verified experimentally (but pairing has been verified experimentally for superconducing electrons).

The simplest question that can be asked is the correctness of the prediction for the ground state energy (GSE). This, of course, can only be exact in a certain limit – the "weak coupling" limit. In the case of the charged Bose gas, interacting via Coulomb forces, this corresponds to the *high density* limit. In gases with short range forces the weak coupling limit corresponds to low density instead.

Our system has N bosonic particles with unit positive charge and coordinates x_j, and a uniformly negatively charged "background" in a large domain Ω of volume V. We are interested in the thermodynamic limit. A physical realization of this model is supposed to be a uniform electron sea in a solid, which forms the background, while the moveable "particles" are bosonic atomic nuclei. The particle number density is then $\rho = N/V$ and this number is also the charge density of the background, thus ensuring charge neutrality.

The Hamiltonian of the one-component plasma is

$$H = \frac{1}{2}\sum_{j=1}^{N} p_j^2 + U_{pp} + U_{pb} + U_{bb}, \tag{1}$$

where $p = -i\nabla$ is the momentum operator, $p^2 = -\Delta$, and the three potential energies, particle-particle, particle-background and background-background, are given by

$$U_{pp} = \sum_{1\le i<j\le N} |x_i - x_j|^{-1}, \tag{2}$$

$$U_{pb} = -\rho \sum_{j=1}^{N} \int_{\Omega} |x_j - y|^{-1}\, d^3y, \tag{3}$$

$$U_{bb} = \tfrac{1}{2}\rho^2 \int_{\Omega}\int_{\Omega} |x - y|^{-1}\, d^3x d^3y. \tag{4}$$

In our units $\hbar^2/m = 1$ and the charge is $e = 1$. The "natural" energy unit we use is two Rydbergs, $2Ry = me^4/\hbar^2$. It is customary to introduce the dimensionless quantity $r_s = (3/4\pi\rho)^{1/3}e^2m/\hbar^2$. High density is small r_s.

The Coulomb potential is infinitely long-ranged and great care has to be taken because the finiteness of the energy per particle in the thermodynamic limit depends, ultimately, on delicate cancellations. The existence of the thermodynamic limit for a system of positive and negative particles, with the negative ones being fermions, was shown only in 1972 [LLe] (for the free energy, but the same proof works for the ground state energy). Oddly, the jellium case is technically a bit harder, and this was done in 1976 [LN] (for both bosons and fermions). One conclusion from this work is that neutrality (in the thermodynamic limit) will come about automatically – even if one does not assume it – provided one allows any excess charge to escape to infinity. In other words, given the background charge, the choice of a neutral number of particles has the lowest energy in the thermodynamic limit. A second point, as shown in [LN], is that e_0 is independent of the shape of the domain Ω provided the boundary is not too wild. For Coulomb systems this is not trivial and for real magnetic systems it is not even generally true. We take advantage of this liberty and assume that our domain is a cube $[0, L] \times [0, L] \times [0, L]$ with $L^3 = V$.

We note the well-known fact that the lowest energy of H in (1) without any restriction about "statistics" (i.e., on the whole of $\otimes^N L^2(\mathbb{R}^3)$) is the same as for bosons, i.e., on the symmetric subspace of $\otimes^N L^2(\mathbb{R}^3)$. The fact that bosons have the lowest energy comes from the Perron–Frobenius Theorem applied to $-\Delta$.

Foldy's calculation leads to the following theorem about the asymptotics of the energy for small r_s, which we call Foldy's law.

Theorem 1.1 (Foldy's Law). *Let E_0 denote the ground state energy, i.e., the bottom of the spectrum, of the Hamiltonian H acting in the Hilbert space $\otimes^N L^2(\mathbb{R}^3)$. We assume that $\Omega = [0, L] \times [0, L] \times [0, L]$. The ground state energy per particle, $e_0 = E_0/N$, in the thermodynamic limit $N, L \to \infty$ with $N/V = \rho$ fixed, in units of $me^4/\hbar^2$, is*

$$\begin{aligned}\lim_{V\to\infty} E_0/N = e_0 &= -0.40154\, r_s^{-3/4} + o(\rho^{1/4}) \\ &= -0.40154\left(\frac{4\pi}{3}\right)^{1/4}\rho^{1/4} + o(\rho^{1/4}),\end{aligned} \tag{5}$$

where the number -0.40154 is, in fact, the integral

$$A = \frac{1}{\pi}6^{1/4}\int_0^\infty \left\{p^2(p^4+2)^{1/2} - p^4 - 1\right\} dp = -\frac{3^{1/4}4\,\Gamma(3/4)}{5\sqrt{\pi}\,\Gamma(5/4)} \approx -0.40154. \tag{6}$$

Actually, our proof gives a result that is more general than Theorem 1.1. We allow the particle number N to be totally arbitrary, i.e., we do not require $N = \rho V$. Our lower bound is still given by (5), where now ρ refers to the background charge density.

In [F] 0.40154 is replaced by 0.80307 since the energy unit there is 1 Ry. The main result of our paper is to prove (5) by obtaining a lower bound on E_0 that agrees with the right side of (5) An upper bound to E_0 that agrees with (5) (to leading order) was given in 1962 by Girardeau [GM], using the variational method of himself and Arnowitt [GA]. Therefore, to verify (5) to leading order it is only necessary to construct a rigorous lower bound of this form and this will be done here. It has to be admitted, as explained below, that the problem that Foldy and Girardeau treat is slightly different from ours because of different boundary conditions and a concommitant different treatment of the background. We regard this difference as a technicality that should be cleared up one day, and do not hesitate to refer to the statement of 1.1 as a theorem.

Before giving our proof, let us remark on a few historical and conceptual points. Some of the early history about the Bose gas, can be found in the lecture notes [L].

Bogolubov's analysis starts by assuming periodic boundary condition on the big box Ω and writing everything in momentum (i.e., Fourier) space. The values of the momentum, k are then discrete: $k = (2\pi/L)(m_1, m_2, m_3)$ with m_i an integer. A convenient tool for taking care of various $n!$ factors is to introduce second quantized operators $a_k^\#$ (where $a^\#$ denotes a or a^*), but it has to be understood that this is only a bookkeeping device. Almost all authors worked in momentum space, but this is neither necessary nor necessarily the most convenient representation (given that the calculations are not rigorous). Indeed, Foldy's result was reproduced by a calculation entirely in x-space [LS]. Periodic boundary conditions are not physical, but that was always chosen for convenience in momentum space.

We shall instead let the particle move in the whole space, i.e., the operator H acts in the Hilbert space $L^2(\mathbb{R}^{3N})$, or rather, since we consider bosons, in the the subspace consisting of the N-fold fully symmetric tensor product of $L^2(\mathbb{R}^3)$. The background potential defined in (2) is however still localized in the cube Ω. We could also have confined the particles to Ω with Dirichlet boundary conditions. This would only raise the ground state energy and thus, for the lower bound, our setup is more general.

There is, however, a technical point that has to be considered when dealing with Coulomb forces. The background never appears in Foldy's calculation; he simply removes the $k = 0$ mode from the Fourier transform, ν of the Coulomb potential (which is $\nu(k) = 4\pi|k|^{-2}$, but with k taking the discrete values mentioned above, so that we are thus dealing with a "periodized" Coulomb potential). The $k = 0$ elimination means that we set $\nu(0) = 0$, and this amounts to a subtraction of the average value of the potential – which is supposed to be a substitute for the effect of a neutralizing background. It does not seem to be a trivial matter to prove that this is equivalent to having a background, but it surely can be done. Since we do not wish to overload this paper, we leave this demonstration to another day. In any case the answers agree (in the sense that our rigorous lower bound agrees with Foldy's answer), as we prove here. If one accepts the idea that setting $\nu(0) = 0$ is equivalent to having a neutralizing background, then the ground state energy problem is finished because Girardeau shows [GM] that Foldy's result is a true upper bound within the context of the $\nu(0) = 0$ problem.

The potential energy is quartic in the operators $a_k^\#$. In Bogolubov's analysis only terms in which there are four or two $a_0^\#$ operators are retained. The operator a_0^* creates, and a_0 destroys particles with momentum 0 and such particles are the constituents of the "condensate". In general there are no terms with three $a_0^\#$ operators (by momentum conservation) and in Foldy's case there is also no four $a_0^\#$ term (because of the subtraction just mentioned).

For the usual short range potential there is a four $a_0^\#$ term and this is supposed to give the leading term in the energy, namely $e_0 = 4\pi\rho a$, where a is the "scattering length" of the two-body potential. Contrary to what would seem reasonable, this number, $4\pi\rho a$ is *not* the coefficient of the four $a_0^\#$ term, and to to prove that $4\pi\rho a$ is, indeed, correct took some time. It was done in 1998 [LY] and the method employed in [LY] will play an essential role here. But it is important to be clear about the fact that the four $a_0^\#$, or "mean field" term is absent in the jellium case by virtue of charge neutrality. The leading term in this case presumably comes from the two $a_0^\#$ terms, and this is what we have to prove. For the short range case, on the other hand, it is already difficult enough to obtain the $4\pi\rho a$ energy that going beyond this to the two $a_0^\#$ terms is beyond the reach of rigorous analysis at the moment.

The Bogolubov ansatz presupposes the existence of Bose–Einstein condensation (BEC). That is, most of the particles are in the $k = 0$ mode and the few that are not come in pairs with momenta k and $-k$. Two things must be said about this. One is that the only case (known to us) in which one can verify the correctness of the Bogolubov picture at weak coupling is the *one-dimensional* delta-function gas [LLi] – in which case there is presumably *no* BEC (because of the low dimensionality). Nevertheless the Bogolubov picture remains correct at low density and the explanation of this seeming contradiction lies in the fact that BEC is not needed; what is really needed is a kind of condensation on a length scale that is long compared to relevant parameters, but which is fixed and need not be as large as the box length L. This was realized in [LY] and the main idea there was to decompose Ω into fixed-size boxes of appropriate length and use Neumann boundary conditions on these boxes (which can only lower the energy, and which is fine since we want a lower bound). We shall make a similar decomposition here, but, unlike the case in [LY] where the potential is purely repulsive, we must deal here with the Coulomb potential and work hard to achieve the necessary cancellation.

The only case in which BEC has been proved to exist is in the hard core lattice gas at half-filling (equivalent to the spin-1/2 XY model) [KLS].

Weak coupling is sometimes said to be a "perturbation theory" regime, but this is not really so. In the one-dimensional case [LLi] the asymptotics near $\rho = 0$ is extremely difficult to deduce from the exact solution because the "perturbation" is singular. Nevertheless, the Bogolubov calculation gives it effortlessly, and this remains a mystery.

One way to get an excessively negative lower bound to e_0 for jellium is to ignore the kinetic energy. One can then show easily (by an argument due to Onsager) that the potential energy alone is bounded below by $e_0 \sim -\rho^{1/3}$. See [LN]. Thus, our goal is to show that the kinetic energy raises the energy to $-\rho^{1/4}$. This was done, in fact, in [CLY], but without achieving the correct coefficient $-0.803(4\pi/3)^{1/4}$. Oddly, the $-\rho^{1/4}$ law was proved in [CLY] by first showing that the *non-thermodynamic* $N^{7/5}$ law for a *two-component* bosonic plasma, as conjectured by Dyson [D], is correct.

The [CLY] paper contains an important innovation that will play a key role here. There, too, it was necessary to decompose $\mathbf{R}^3$ into boxes, but a way had to be found to eliminate the Coulomb interaction *between* different boxes. This was accomplished by *not* fixing the location of the boxes but rather averaging over all possible locations of the boxes. This "sliding localization" will play a key role here, too. This idea was expanded upon in [GG]. Thus, we shall have to consider only one finite box with the particles and the background charge in it independent of the rest of the system. However, a price will have to be paid for this luxury, namely it will not be entirely obvious that the number of particles we want to place in each box is the same for all boxes, i.e., $\rho\ell^3$, where ℓ is the length of box. Local neutrality, in other words, cannot be taken for granted. The analogous problem in [LY] is easier because no attractive potentials are present there. We solve this problem by choosing the number, n, in each box to be the number that gives the lowest energy in the box. This turns out to be close to $n = \rho\ell^3$, as we show and as we know from [LN] must be the case as $\ell \to \infty$.

Finally, let us remark on one bit of dimensional analysis that the reader should keep in mind. One should not conclude from (5) that a typical particle has energy $\rho^{1/4}$ and hence momentum $\rho^{1/8}$ or de Broglie wavelength $\rho^{-1/8}$. This is *not* the correct picture. Rather, a glance at the Bogolubov–Foldy calculation shows that the momenta of importance are of order $\rho^{-1/4}$, and the seeming paradox is resolved by noting that the number of excited particles (i.e., those not in the $k = 0$ condensate) is of order $N\rho^{-1/4}$. This means that we can, hopefully, localize particles to lengths as small as $\rho^{-1/4+\epsilon}$, and cut off the Coulomb potential at similar lengths, without damage, provided we do not disturb the condensate particles. It is this clear separation of scales that enables our asymptotic analysis to succeed.

2. Outline of the Proof

The proof of our Main Theorem 1.1 is rather complicated and somewhat hard to penetrate, so we present the following outline to guide the reader.

2.1. Section 3. Here we localize the system whose size is L into small boxes of size ℓ independent of L, but dependent on the intensive quantity ρ. Neumann boundary conditions for the Laplacian are used in order to ensure a lower bound to the energy. We always think of operators in terms of quadratic forms and the Neumann Laplacian in a box Q is defined for all functions in $\psi \in L^2(Q)$ by the quadratic form

$$(\psi, -\Delta_{\text{Neumann}}\psi) = \int_Q |\nabla\psi(x)|^2\, dx.$$

The lowest eigenfunction of the Neumann Laplacian is the constant function and this plays the role of the condensate state. This state not only minimizes the, kinetic energy, but it is also consistent with neutralizing the background and thereby minimizing the Coulomb energy. The particles not in the condensate will be called "excited" particles.

To avoid localization errors we take $\ell \gg \rho^{-1/4}$, which is the relevant scale as we mentioned in the Introduction. The interaction among the boxes is controlled by using the sliding method of [CLY]. The result is that we have to consider only interactions among the particles and the background in each little box separately.

The N particles have to be distributed among the boxes in a way that minimizes the total energy. We can therefore not assume that each box is neutral. Instead of dealing with this distribution problem we do a simpler thing which is to choose the particle number in each little box so as to achieve the absolute minimum of the energy in that box. Since all boxes are equivalent this means that we take a common value n as the particle number in each box. The total particle number which is n times the number of boxes will not necessarily equal N, but this is of no consequence for a lower bound. We shall show later, however, that it equality is nearly achieved, i.e., the the energy minimizing number n in each box is close to the value needed for neutrality.

2.2. Section 4. It will be important for us to replace the Coulomb potential by a cutoff Coulomb potential. There will be a short distance cutoff of the singularity at a distance r and a large distance cutoff of the tail at a distance R, with $r \leq R \ll \ell$. One of the unusual features of our proof is that r are R are not fixed once and for all, but are readjusted each time new information is gained about the error bounds.

In fact, already in Sect. 4 we give a simple preliminary bound on n by choosing $R \sim \rho^{-1/3}$, which is much smaller than the relevant scale $\rho^{-1/4}$, although the choice of R that we shall use at the end of the proof is of course much larger than $\rho^{-1/4}$, but less than ℓ.

2.3. Section 5. There are several terms in the Hamiltonian. There is the kinetic energy, which is non-zero only for the excited particles. The potential energy, which is a quartic term in the language of second quantization, has various terms according to the number of times the constant function appears. Since we do not have periodic boundary conditions we will not have the usual simplification caused by conservation of momentum, and the potential energy will be correspondingly more complicated than the usual expression found in textbooks.

In this section we give bounds on the different terms in the Hamiltonian and use these to get a first control on the condensation, i.e., a control on the number of particles $\widehat{n}_+$ in each little box that are not in the condensate state.

The difficult point is that $\widehat{n}_+$ is an operator that does not commute with the Hamiltonian and so it does not have a sharp value in the ground state. We give a simple preliminary bound on its average $\langle\widehat{n}_+\rangle$ in the ground state by again choosing $R \sim \rho^{-1/3}$. In order to control the condensation to an appropriate accuracy we shall eventually need not only a bound on the average, $\langle\widehat{n}_+\rangle$, but also on the fluctuation, i.e, on $\langle\widehat{n}_+^2\rangle$. This will be done in Sect. 8 using a novel method developed in Appendix A for localizing off-diagonal matrices.

2.4. Section 6. The part of the potential energy that is most important is the part that is quadratic in the condensate operators $a_0^\#$ and quadratic in the excited variables $a_p^\#$ with $p \neq 0$. This, together with the kinetic energy, which is also quadratic in the $a_p^\#$, is the part of the Hamiltonian that leads to Foldy's law. Although we have not yet managed to eliminate the non-quadratic part up to this point we study the main "quadratic" part of the Hamiltonian. It is in this section that we essentially do Foldy's calculation.

It is not trivial to diagonalize the quadratic form and thereby reproduce Foldy's answer because there is no momentum conservation. In particular there is no simple relation between the resolvent of the Neumann Laplacian and the Coulomb kernel. The former is defined relative to the box and the latter is defined relative to the whole of $\mathbb{R}^3$. It is therefore necessary for us to localize the wavefunction in the little box away from the boundary. On such functions the boundary condition is of no importance and we can identify the kinetic energy with the Laplacian in all of $\mathbb{R}^3$. This allows us to have a simple relation between the Coulomb term and the kinetic energy term since the Coulomb kernel is in fact the resolvent of the Laplacian in all of $\mathbb{R}^3$.

When we cut off the wavefunction near the boundary we have to be very careful because we must not cut off the part corresponding to the particles in the condensate. To do so would give too large a localization energy. Rather, we cut off only functions with sufficiently large kinetic energy so that the localization energy is relatively small compared to the kinetic energy. The technical lemma needed for this is a double commutator inequality given in Appendix B.

2.5. Section 7. At this point we have bounds available for the quadratic part (from Sect. 6) and the annoying non-quadratic part (from Sect. 5) of the Hamiltonian. These depend on r, R, n, $\langle\widehat{n}_+\rangle$, and $\langle\widehat{n}_+^2\rangle$. We avail ourselves of the bounds previously obtained for n and $\langle\widehat{n}_+\rangle$ and now use our freedom to choose different values for r and R to bootstrap to the desired bounds on n and $\langle\widehat{n}_+\rangle$, i.e., we prove that there is almost neutrality and almost condensation in each little box.

2.6. Section 8. In order to control $\langle\widehat{n}_+^2\rangle$ we utilize, for the first time, the new method for localizing large matrices given in Appendix A. This method allows us to restrict to states with small fluctuations in $\widehat{n}_+$, and thereby bound $\langle\widehat{n}_+^2\rangle$, provided we know that the terms that do not commute with $\widehat{n}_+$ have suffciently small expectation values. We then give bounds on these $\widehat{n}_+$ "off-diagonal" terms. Unfortunately, these bounds are in terms of positive quantities coming from the Coulomb repulsion, but for which we actually do not have independent a-priori bounds. Normally, when proving a lower bound to a Hamiltonian, we can sometimes control error terms by absorbing them into positive terms in the Hamiltonian, which are then ignored. This may be done even when we do not have an a-priori bound on these positive terms. If we want to use Theorem A.1 in Appendix A, we will need an absolute bound on the "off-diagonal" terms and we can therefore not use the technique of absorbing them into the positive terms. The decision when to use the theorem in Appendix A or use the technique of absorption into positive terms is resolved in Sect. 9.

2.7. Section 9. Since we do not have an a-priori bound on the positive Coulomb terms as described above we are faced with a dichotomy. If the positive terms are, indeed, so large that enough terms can be controlled by them we do not need to use the localization technique of Appendix A to finish the proof of Foldy's law. The second possibility is that the positive terms are bounded in which case we can use this fact to control the terms that do commute with $\widehat{n}_+$ and this allows us to use the localization technique in Appendix A to finish the proof of Foldy's law. Thus, the actual magnitude of the positive repulsion terms is unimportant for the derivation of Foldy's law.

3. Reduction to a Small Box

As described in the previous sections we shall localize the problem into smaller cubes of size $\ell \ll L$. We shall in fact choose ℓ as a function of ρ in such a way that $\rho^{1/4}\ell \to \infty$ as $\rho \to \infty$.

We shall localize the kinetic energy by using Neumann boundary conditions on the smaller boxes.

We shall first, however, describe how we may control the electostatic interaction between the smaller boxes using the sliding technique of [CLY].

Let t, with $0 < t < 1/2$, be a parameter which we shall choose later to depend on ρ in such a way that $t \to 0$ as $\rho \to \infty$.

The choice of ℓ and t as functions of ρ will be made at the end of Sect. 9 when we complete the proof of Foldy's law.

Let $\chi \in C_0^\infty(\mathbb{R}^3)$ satisfy supp $\chi \subset [(-1+t)/2, (1-t)/2]^3$, $0 \leq \chi \leq 1$, $\chi(x) = 1$ for x in the smaller box $[(-1+2t)/2, (1-2t)/2]^3$, and $\chi(x) = \chi(-x)$. Assume that all m-th order derivatives of χ are bounded by $C_m t^{-m}$, where the constants C_m depend only on m and are, in particular, independent of t. Let $\chi_\ell(x) = \chi(x/\ell)$. Let $\eta = \sqrt{1-\chi}$. We shall assume that χ is defined such that η is also C^1. Let $\eta_\ell(x) = \eta(x/\ell)$. Using χ we define the constant γ by $\gamma^{-1} = \int \chi(y)^2\,dy$, and note that $1 \leq \gamma \leq (1-2t)^{-3}$. We also introduce the Yukawa potential $Y_\nu(x) = |x|^{-1}e^{-\nu|x|}$ for $\nu > 0$.

As a preliminary to the following Lemma 3.1 we quote Lemma 2.1 in [CLY].

Lemma. *Let* $K : \mathbb{R}^3 \to \mathbb{R}$ *be given by*

$$K(z) = r^{-1}\left\{e^{-\nu r} - e^{-\omega r}h(z)\right\}$$

with $r = |z|$ *and* $\omega > \nu \geq 0$. *Let* h *satisfy* (i) *h is a* C^4 *function of compact support;* (ii) $h(z) = 1 + ar^2 + O(r^3)$ *near* $z = 0$. *Let* $h(z) = h(-z)$, *so that* K *has a real Fourier transform. Then there is a constant,* C_3 *(depending on h) such that if* $\omega - \nu \geq C_3$ *then* K *has a positive Fourier transform and, moreover,*

$$\sum_{1\leq i<j\leq N} e_i e_j K(x_i - x_j) \geq \frac{1}{2}(\nu - \omega)N$$

for all $x_1, \ldots x_N \in \mathbb{R}^3$ *and all* $e_i = \pm 1$.

Lemma 3.1 (Electrostatic decoupling of boxes using sliding). *There exists a function of the form* $\omega(t) = Ct^{-4}$ *(we assume that* $\omega(t) \geq 1$ *for* $t < 1/2$*) and a constant* γ *with* $1 \leq \gamma \leq (1-2t)^{-3}$ *such that if we set*

$$w(x,y) = \chi_\ell(x)Y_{\omega(t)/\ell}(x-y)\chi_\ell(y) \tag{7}$$

then the potential energy satisfies

$$
\begin{aligned}
&U_{pp} + U_{pb} + U_{bb} \\
&\quad \geq \gamma \sum_{\lambda \in \mathbb{Z}^3} \int_{\mu \in [-\frac{1}{2}, \frac{1}{2}]^3} d\mu \Big\{ \sum_{1 \leq i < j \leq N} w\left(x_i + (\mu + \lambda)\ell, x_j + (\mu + \lambda)\ell\right) \\
&\qquad - \rho \sum_{j=1}^{N} \int_{\Omega} w\left(x_j + (\mu + \lambda)\ell, y + (\mu + \lambda)\ell\right) dy \\
&\qquad + \tfrac{1}{2}\rho^2 \iint_{\Omega \times \Omega} w\left(x + (\mu + \lambda)\ell, y + (\mu + \lambda)\ell\right) dx\, dy \Big\} - \frac{\omega(t) N}{2\ell}.
\end{aligned}
$$

Proof. We calculate

$$
\begin{aligned}
&\sum_{\lambda \in \mathbb{Z}^3} \int_{\mu \in [-1/2, 1/2]^3} d\mu\, \gamma \chi(x + (\mu + \lambda)) Y_\omega(x - y) \chi(y + (\mu + \lambda)) \\
&\qquad = \int \gamma \chi(x + z) Y_\omega(x - y) \chi(y + z)\, dz = h(x - y) Y_\omega(x - y),
\end{aligned}
$$

where we have set $h = \gamma \chi * \chi$. Note that $h(0) = 1$ and that h satisfies all the assumptions in Lemma 2.1 in [CLY]. We then conclude from Lemma 2.1 in [CLY] that the Fourier transform of the function $F(x) = |x|^{-1} - h(x) Y_{\omega(t)}(x)$ is non-negative, where ω is a function such that $\omega(t) \to \infty$ as $t \to 0$. [The detailed bounds from [CLY] show that we may in fact choose $\omega(t) = Ct^{-4}$, since $\omega(t)$ has to control the 4th derivative of h.] Note, moreover, that $\lim_{x \to 0} F(x) = \omega(t)$. Hence

$$
\begin{aligned}
&\sum_{1 \leq i < j \leq N} F(y_i - y_j) - \rho \sum_{j=1}^{N} \int_{\ell^{-1}\Omega} F(y_j - y)\, dy \\
&\qquad + \tfrac{1}{2}\rho^2 \iint_{\ell^{-1}\Omega \times \ell^{-1}\Omega} F(x - y)\, dx\, dy \geq -\frac{N \omega(t)}{2}.
\end{aligned}
$$

The lemma follows by writing $|x|^{-1} = F(x) + h(y) Y_{\omega(t)}(x)$ and by rescaling from boxes of size 1 to boxes of size ℓ. □

As explained above we shall choose the parameters t and ℓ as functions of ρ at the very end of the proof. We shall choose them in such a way that $t \to 0$ and $\rho^{1/4}\ell \to \infty$ as $\rho \to \infty$. Moreover, we will have conditions of the form

$$
\rho^{-\tau}(\rho^{1/4}\ell) \to 0, \quad \text{and } t^{\nu}(\rho^{1/4}\ell) \to \infty
$$

as $\rho \to \infty$, where τ, ν are universal constants.

Consider now the n-particle Hamiltonian

$$
H^n_{\mu,\lambda} = -\tfrac{1}{2} \sum_{j=1}^{n} \Delta^{(j)}_{Q_{\mu,\lambda}} + \gamma W_{\mu,\lambda}, \tag{8}
$$

where we have introduced the Neumann Laplacian $\Delta^{(j)}_{Q_{\mu,\lambda}}$ of the cube $Q_{\mu,\lambda} = (\mu + \lambda)\ell + \left[-\frac{1}{2}\ell, \frac{1}{2}\ell\right]^3$ and the potential

$$\begin{aligned} W_{\mu,\lambda}(x_1, \dots, x_n) = &\sum_{1\le i<j\le n} w\left(x_i + (\mu+\lambda)\ell, x_j + (\mu+\lambda)\ell\right) \\ &- \rho \sum_{j=1}^{n} \int_\Omega w\left(x_j + (\mu+\lambda)\ell, y + (\mu+\lambda)\ell\right) dy \\ &+ \tfrac{1}{2}\rho^2 \iint_{\Omega\times\Omega} w\left(x + (\mu+\lambda)\ell, y + (\mu+\lambda)\ell\right) dx\, dy. \end{aligned}$$

Lemma 3.2 (Decoupling of boxes). *Let $E^n_{\mu,\lambda}$ be the ground state energy of the Hamiltonian $H^n_{\mu,\lambda}$ given in (8) considered as a bosonic Hamiltonian. The ground state energy E_0 of the Hamiltonian H in (1) is then bounded below as*

$$E_0 \ge \sum_{\lambda\in\mathbb{Z}^3} \int_{\mu\in[-\frac{1}{2},\frac{1}{2}]^3} \inf_{1\le n\le N} E^n_{\mu,\lambda}\, d\mu - \frac{\omega(t)N}{2\ell}.$$

Proof. If $\Psi(x_1, \dots, x_N) \in L^2(\mathbb{R}^{3N})$ is a symmetric function. Then

$$(\Psi, H\Psi) \ge \sum_{\lambda\in\mathbb{Z}^3} \int_{\mu\in[-\frac{1}{2},\frac{1}{2}]^3} (\Psi, \widetilde{H}_{\mu,\lambda}\Psi)\, d\mu - \frac{\omega(t)N}{2\ell},$$

where

$$\begin{aligned} (\Psi, \widetilde{H}_{\mu,\lambda}\Psi) = &\sum_{j=1}^{N} \int_{x_j\in Q_{\mu,\lambda}} |\nabla_j \Psi(x_1, \dots, x_N)|^2\, dx_1 \dots dx_N \\ &+ \gamma \int W_{\mu,\lambda}(x_1, \dots, x_N) |\Psi(x_1, \dots, x_N)|^2\, dx_1 \dots dx_N. \end{aligned}$$

The lemma follows since it is clear that $(\Psi, \widetilde{H}_{\mu,\lambda}\Psi) \ge \inf_{1\le n\le N} E^n_{\mu,\lambda}$. □

For given μ the Hamiltonians $H^n_{\mu,\lambda}$ fall in three groups depending on λ. The first kind for which $Q_{\lambda,\mu} \cap \Omega = \emptyset$. They describe boxes with no background. The optimal energy for these boxes are clearly achieved for $n = 0$. The second kind for which $Q_{\lambda,\mu} \subset \Omega$. These Hamiltonians are all unitarily equivalent to γH^n_ℓ, where

$$\begin{aligned} H^n_\ell = &\sum_{j=1}^{n} \left(-\tfrac{1}{2}\gamma^{-1}\Delta_{\ell,j} - \rho \int w(x_j, y)\, dy\right) \\ &+ \sum_{1\le i<j\le n} w(x_i, x_j) + \tfrac{1}{2}\rho^2 \iint w(x, y)\, dx\, dy, \end{aligned} \tag{9}$$

where $-\Delta_\ell$ is the Neumann Laplacian for the cube $[-\ell/2, \ell/2]^3$. Finally, there are operators of the third kind for which $Q_{\mu,\lambda}$ intersects both Ω and its complement. In this case the particles only see part of the background. If we artificially add the missing background only the last term in the potential $W_{\mu,\lambda}$ increases. (The first term does not change and the second can only decrease.) In fact it will increase by no more than

$$\tfrac{1}{2}\rho^2 \iint w(x,y)\,dx\,dy \le \tfrac{1}{2}\rho^2 \iint_{\substack{x\in[-\ell/2,\ell/2]^3\\ y\in[-\ell/2,\ell/2]^3}} |x-y|^{-1}\,dx\,dy \le C\rho^2\ell^5.$$

Thus the operator $H^n_{\mu,\lambda}$ of the third kind are bounded below by an operator which is unitarily equivalent to $\gamma H^n_\ell - C\rho^2\ell^5$.

We now note that the number of boxes of the third kind is bounded above by $C(L/\ell)^2$. The total number of boxes of the second or third kind is bounded above by $(L+\ell)^3/\ell^3 = (1+L/\ell)^3$.

We have therefore proved the following result.

Lemma 3.3 (Reduction to one small box). *The ground state energy E_0 of the Hamiltonian H in (1) is bounded below as*

$$E_0 \ge (1+L/\ell)^3\gamma \inf_{1\le n\le N} \inf \operatorname{Spec} H^n_\ell - C(L/\ell)^2\rho^2\ell^5 - \frac{\omega(t)N}{2\ell},$$

where H^n_ℓ is the Hamiltonian defined in (9).

In the rest of the paper we shall study the Hamiltonian (9).

4. Long and Short Distance Cutoffs in the Potential

The potential in the Hamiltonian (9) is w given in (7). Our aim in this section to replace w by a function that has long and short distance cutoffs.

We shall replace the function w by

$$w_{r,R}(x,y) = \chi_\ell(x)V_{r,R}(x-y)\chi_\ell(y), \tag{10}$$

where

$$V_{r,R}(x) = Y_{R^{-1}}(x) - Y_{r^{-1}}(x) = \frac{e^{-|x|/R} - e^{-|x|/r}}{|x|}. \tag{11}$$

Here $0 < r \le R \le \omega(t)^{-1}\ell$. Note that for $x \ll r$ then $V_{r,R}(x) \approx r^{-1} - R^{-1}$ and for $|x| \gg R$ then $V_{r,R}(x) \approx |x|^{-1}e^{-|x|/R}$.

In this section we shall bound the effect of replacing w by $w_{r,R}$. We shall not fix the cutoffs r and R, but rather choose them differently at different stages in the later arguments.

We first introduce the cutoff R alone, i.e., we bound the effect of replacing w by $w_R(x,y) = \chi_\ell(x)V_R(x-y)\chi_\ell(y)$, where $V_R(x) = |x|^{-1}e^{-|x|/R} = Y_{R^{-1}}(x)$. Thus, since $R \le \omega(t)^{-1}\ell$, the Fourier transforms satisfy

$$\widehat{Y}_{\omega/\ell}(k) - \widehat{V}_R(k) = 4\pi\left(\frac{1}{k^2+(\omega(t)/\ell)^2} - \frac{1}{k^2+R^{-2}}\right) \ge 0.$$

(We use the convention that $\hat{f}(k) = \int f(x)e^{-ikx}\,dx$.) Hence $w(x,y) - w_R(x,y) = \chi_\ell(x)\left(Y_{\omega/\ell} - V_R\right)(x-y)\chi_\ell(y)$ defines a positive semi-definite kernel. Note, moreover, that $\left(Y_{\omega/\ell} - V_R\right)(0) = R^{-1} - \omega/\ell \leq R^{-1}$ Thus,

$$\sum_{1\leq i<j\leq n} w(x_i,x_j) - \rho\sum_{j=1}^{n}\int w(x_j,y)\,dy + \tfrac{1}{2}\rho^2\iint w(x,y)\,dx\,dy$$
$$-\left(\sum_{1\leq i<j\leq n} w_R(x_i,x_j) - \rho\sum_{j=1}^{n}\int w_R(x_j,y)\,dy + \tfrac{1}{2}\rho^2\iint w_R(x,y)\,dx\,dy\right)$$
$$=\frac{1}{2}\iint\left[\sum_i^n \delta(x-x_i) - \rho\right](w-w_r)(x,y)\left[\sum_i^n \delta(y-x_i) - \rho\right]dx\,dy$$
$$-\frac{1}{2}\sum_i^n \chi_\ell(x_i)^2\left(Y_{\omega/\ell} - V_R\right)(0) \geq -\tfrac{1}{2}n\left(Y_{\omega/\ell} - V_R\right)(0) = -\tfrac{1}{2}nR^{-1}. \tag{12}$$

We now bound the effect of replacing w_R by $w_{r,R}$. I.e., we are replacing $V_R(x) = |x|^{-1}e^{-|x|/R}$ by $|x|^{-1}\left(e^{-|x|/R} - e^{-|x|/r}\right)$. This will lower the repulsive terms and for the attractive term we get

$$-\rho\sum_{j=1}^{n}\int w_R(x_j,y)\,dy \geq -\rho\sum_{j=1}^{n}\int w_{r,R}(x_j,y)\,dy$$
$$-n\rho\sup_x\int \chi_\ell(x)\frac{e^{-|x-y|/r}}{|x-y|}\chi_\ell(y)\,dy \tag{13}$$
$$\geq -\rho\sum_{j=1}^{n}\int w_{r,R}(x_j,y)\,dy - Cn\rho r^2.$$

If we combine the bounds (12) and (13) we have the following result.

Lemma 4.1 (Long and short distance potential cutoffs). *Consider the Hamiltonian*

$$H_{\ell,r,R}^n = \sum_{j=1}^{n}\left(-\tfrac{1}{2}\gamma^{-1}\Delta_{\ell,j} - \rho\int w_{r,R}(x_j,y)\,dy\right) + \sum_{1\leq i<j\leq n} w_{r,R}(x_i,x_j)$$
$$+\tfrac{1}{2}\rho^2\iint w_{r,R}(x,y)\,dx\,dy, \tag{14}$$

where $w_{r,R}$ is given in (10) and (11) with $0 < r \leq R \leq \omega(t)^{-1}\ell$ and $-\Delta_\ell$ as before is the Neumann Laplacian for the cube $[-\ell/2,\ell/2]^3$. Then the Hamiltonian H_ℓ^n defined in (9) obeys the lower bound

$$H_\ell^n \geq H_{\ell,r,R}^n - \tfrac{1}{2}nR^{-1} - C_1 n\rho r^2.$$

A similar argument gives the following result.

Lemma 4.2. *With the same notation as above we have for $0 < r' \leq r \leq R \leq R' \leq \omega(t)^{-1}\ell$ that*

$$H_{\ell,r',R'}^n \geq H_{\ell,r,R}^n - \tfrac{1}{2}nR^{-1} - C_1 n\rho r^2.$$

Proof. Simply note that $V_{r',R'}(x) - V_{r,R}(x) = Y_{R'^{-1}}(x) - Y_{R^{-1}}(x) + Y_{r^{-1}}(x) - Y_{r'^{-1}}(x)$ and now use the same arguments as before. □

Corollary 4.3 (The particle number n cannot be too small). *There exists a constant $C > 0$ such that if $\omega(t)^{-1}\rho^{1/3}\ell > C$ then $H_\ell^n \geq 0$ if $n \leq C\rho\ell^3$.*

Proof. Choose $R = \rho^{-1/3}$ and $r = \frac{1}{2}R$. Then we may assume that $R \leq \omega(t)^{-1}\ell$ since $\omega(t)^{-1}\rho^{1/3}\ell$ is large. From Lemma 4.1 we see immediately that

$$H_\ell^n \geq -\sum_{j=1}^n \rho \int w_{r,R}(x_j, y)\,dy + \tfrac{1}{2}\rho^2 \iint w_{r,R}(x,y)\,dx\,dy - Cn\rho R^2$$

$$\geq -n \sup_x \rho \int w_{r,R}(x,y)\,dy + \tfrac{1}{2}\rho^2 \iint w_{r,R}(x,y)\,dx\,dy - Cn\rho R^2.$$

The corollary follows since $\sup_x \int w_{r,R}(x,y)\,dy \leq 4\pi R^2$ and with the given choice of R and r it is easy to see that $\frac{1}{2}\iint w_{r,R}(x,y)\,dx\,dy \geq cR^2\ell^3$. □

5. Bound on the Unimportant Part of the Hamiltonian

In this section we shall bound the Hamiltonian $H_{\ell,r,R}^n$ given in (14). We emphasize that we do not necessarily have neutrality in the cube, i.e., n and $\rho\ell^3$ may be different. We are simply looking for a lower bound to $H_{\ell,r,R}^n$, that holds for all n. The goal is to find a lower bound that will allow us to conclude that the optimal n, i.e., the value for which the energy of the Hamiltonian is smallest, is indeed close to the neutral value.

We shall express the Hamiltonian in second quantized language. This is purely for convenience. We stress that we are not in any way changing the model by doing this and the treatment is entirely rigorous and could have been done without the use of second quantization.

Let u_p, $\ell p/\pi \in (\mathbb{N} \cup \{0\})^3$ be an orthonormal basis of eigenfunctions of the Neumann Laplacian $-\Delta_\ell$ such that $-\Delta_\ell u_p = |p|^2 u_p$. I.e.,

$$u_p(x_1, x_2, x_3) = c_p \ell^{-3/2} \prod_{j=1}^3 \cos\left(\frac{p_j \pi (x_j + \ell/2)}{\ell}\right),$$

where the normalization satisfies $c_0 = 1$ and in general $1 \leq c_p \leq \sqrt{8}$. The function $u_0 = \ell^{-3/2}$ is the constant eigenfunction with eigenvalue 0. We note that for $p \neq 0$ we have

$$(u_p, -\Delta_\ell u_p) \geq \pi^2 \ell^{-2}. \tag{15}$$

We now express the Hamiltonian $H_{\ell,r,R}^n$ in terms of the creation and annihilation operators $a_p = a(u_p)$ and $a_p^* = a(u_p)^*$.

Define

$$\widehat{w}_{pq,\mu\nu} = \iint w_{r,R}(x,y) u_p(x) u_q(y) u_\mu(x) u_\nu(y)\,dx\,dy.$$

We may then express the two-body repulsive potential as

$$\sum_{1\leq i<j\leq n} w_{r,R}(x_i, x_j) = \tfrac{1}{2} \sum_{pq,\mu\nu} \widehat{w}_{pq,\mu\nu} a_p^* a_q^* a_\nu a_\mu,$$

where the right-hand side is considered restricted to the n-particle subspace. Likewise the background potential can be written

$$-\rho \sum_{j=1}^{n} w_{r,R}(x_j, y)\,dy = -\rho \ell^3 \sum_{pq} \widehat{w}_{0p,0q} a_p^* a_q$$

and the background-background energy

$$\tfrac{1}{2}\rho^2 \iint w_{r,R}(x, y)\,dx\,dy = \tfrac{1}{2}\rho^2 \ell^6 \widehat{w}_{00,00}.$$

We may therefore write the Hamiltonian as

$$\begin{aligned} H^n_{\ell,r,R} = {} & \tfrac{1}{2}\gamma^{-1} \sum_p |p|^2 a_p^* a_p + \tfrac{1}{2} \sum_{pq,\mu\nu} \widehat{w}_{pq,\mu\nu} a_p^* a_q^* a_\nu a_\mu \\ & - \rho\ell^3 \sum_{pq} \widehat{w}_{0p,0q} a_p^* a_q + \tfrac{1}{2}\rho^2\ell^6 \widehat{w}_{00,00}. \end{aligned} \tag{16}$$

We also introduce the operators $\widehat{n}_0 = a_0^* a_0$ and $\widehat{n}_+ = \sum_{p\neq 0}$. These operators represent the number of particles in the condensate state created by a_0^* and the number of particle *not* in the condensate. Note that on the subspace where the total particle number is n, both of these operators are non-negative and $\widehat{n}_+ = n - \widehat{n}_0$.

Using the bounds on the long and short distance cutoffs in Lemma 4.1 we may immediately prove a simple bound on the expectation value of $\widehat{n}_+$.

Lemma 5.1 (Simple bound on the number of excited particles). *There is a constant $C > 0$ such that if $\omega(t)^{-1}\rho^{1/3}\ell > C$ then for any state such that the expectation $\langle H^n_\ell \rangle \leq 0$, the expectation of the number of excited particles satisfies $\langle \widehat{n}_+ \rangle \leq C n \rho^{-1/6}\left(\rho^{1/4}\ell\right)^2$.*

Proof. We simply choose $r = R = \rho^{-1/3}$ in Lemma 4.1. This is allowed since $R \leq \omega(t)^{-1}\ell$ is ensured from the assumption that $\omega(t)^{-1}\rho^{1/3}\ell$ is large. We then obtain

$$H^n_\ell \geq \sum_{j=1}^{n} -\tfrac{1}{2}\gamma^{-1}\Delta_{\ell,j} - \tfrac{1}{2} n R^{-1} - C n \rho r^2 \geq \sum_{j=1}^{n} -\tfrac{1}{2}\gamma^{-1}\Delta_{\ell,j} - C n \rho^{1/3}.$$

The bound on $\langle \widehat{n}_+ \rangle$ follows since the bound on the gap (15) implies that $\langle \sum_{j=1}^{n} -\Delta_{\ell,j} \rangle \geq \langle \widehat{n}_+ \rangle \pi^2 \ell^{-2}$. □

Motivated by Foldy's use of the Bogolubov approximation it is our goal to reduce the Hamiltonian $H^n_{\ell,r,R}$ so that it has only what we call quadratic terms, i.e., terms which contain precisely two $a_p^\#$ with $p \neq 0$. More precisely, we want to be able to ignore all terms containing the coefficients

- $\widehat{w}_{00,00}$.
- $\widehat{w}_{p0,q0} = \widehat{w}_{0p,0q}$, where $p, q \neq 0$. These terms are in fact quadratic, but do not appear in the Foldy Hamiltonian. We shall prove that they can also be ignored.
- $\widehat{w}_{p0,00} = \widehat{w}_{0p,00} = \widehat{w}_{00,p0} = \widehat{w}_{00,0p}$, where $p \neq 0$.
- $\widehat{w}_{pq,\mu 0} = \widehat{w}_{\mu 0,pq} = \widehat{w}_{qp,0\mu} = \widehat{w}_{0\mu,qp}$, where $p, q, \mu \neq 0$.

- $\widehat{w}_{pq,\mu\nu}$, where $p, q, \mu, \nu \neq 0$. The sum of all these terms form a non-negative contribution to the Hamiltonian and can, when proving a lower bound, either be ignored or used to control error terms.

We shall consider these cases one at a time.

Lemma 5.2 (Control of terms with $\widehat{w}_{00,00}$). *The sum of the terms in $H^n_{\ell,r,R}$ containing $\widehat{w}_{00,00}$ is equal to*

$$\tfrac{1}{2}\widehat{w}_{00,00}\left[\left(\widehat{n}_0 - \rho\ell^3\right)^2 - \widehat{n}_0\right]$$

$$= \tfrac{1}{2}\widehat{w}_{00,00}\left[\left(n - \rho\ell^3\right)^2 + (\widehat{n}_+)^2 - 2\left(n - \rho\ell^3\right)\widehat{n}_+ - \widehat{n}_0\right].$$

Proof. The terms containing $\widehat{w}_{00,00}$ are

$$\tfrac{1}{2}\widehat{w}_{00,00}\left(a_0^* a_0^* a_0 a_0 - 2\rho\ell^3 a_0^* a_0 + \rho^2\ell^6\right) = \tfrac{1}{2}\widehat{w}_{00,00}\left(a_0^* a_0 - \rho\ell^3\right)^2 - \tfrac{1}{2}\widehat{w}_{00,00} a_0^* a_0$$

using the 0commutation relation $[a_p, a_q^*] = \delta_{p,q}$. □

Lemma 5.3 (Control of terms with $\widehat{w}_{p0,q0}$). *The sum of the terms in $H^n_{\ell,r,R}$ containing $\widehat{w}_{p0,q0}$ or $\widehat{w}_{0p,0q}$ with $p, q \neq 0$ is bounded below by*

$$-4\pi[\rho - n\ell^{-3}]_+ \widehat{n}_+ R^2 - 4\pi\widehat{n}_+^2\ell^{-3}R^2,$$

where $[t]_+ = \max\{t, 0\}$.

Proof. The terms containing $\widehat{w}_{p0,q0}$ or $\widehat{w}_{0p,0q}$ are

$$\sum_{\substack{p\neq 0\\ q\neq 0}}\left(\tfrac{1}{2}\widehat{w}_{p0,q0} a_p^* a_0^* a_0 a_q + \tfrac{1}{2}\widehat{w}_{0p,0q} a_0^* a_p^* a_q a_0 - \rho\ell^3\widehat{w}_{0p,0q} a_p^* a_q\right)$$

$$= (\widehat{n}_0 - \rho\ell^3)\sum_{\substack{p\neq 0\\ q\neq 0}}\widehat{w}_{p0,q0} a_p^* a_q.$$

Note that $\widehat{n}_0$ commutes with $\sum_{\substack{p\neq 0\\ q\neq 0}}\widehat{w}_{p0,q0} a_p^* a_q$.

We have that

$$\widehat{w}_{p0,q0} = \ell^{-3}\int\int w_{r,R}(x,y)\,dy\, u_p(x) u_q(x)\,dx.$$

Hence

$$\sum_{\substack{p\neq 0\\ q\neq 0}}\widehat{w}_{p0,q0} a_p^* a_q = \ell^{-3}\int\int w_{r,R}(x,y)\,dy\left(\sum_{p\neq 0} u_p(x) a_p^*\right)\left(\sum_{p\neq 0} u_p(x) a_p^*\right)^* dx.$$

$$\leq \ell^{-3}\sup_{x'}\int w_{r,R}(x',y)\,dy\int\left(\sum_{p\neq 0} u_p(x) a_p^*\right)\left(\sum_{p\neq 0} u_p(x) a_p^*\right)^* dx.$$

$$= \ell^{-3}\sup_{x'}\int w_{r,R}(x',y)\,dy\sum_{p\neq 0} a_p^* a_p = \ell^{-3}\sup_{x'}\int w_{r,R}(x',y)\,dy\,\widehat{n}_+.$$

Since

$$\sup_x \int w_{r,R}(x,y)\,dy \le \int V_{r,R}(y)\,dy \le 4\pi R^2$$

we obtain the operator inequality

$$0 \le \sum_{\substack{p\neq 0\\ q\neq 0}} \widehat{w}_{p0,q0} a_p^* a_q \le 4\pi \ell^{-3} R^2 \widehat{n}_+,$$

and the lemma follows.

Before treating the last two types of terms we shall need the following result on the structure of the coefficients $\widehat{w}_{pq,\mu\nu}$.

Lemma 5.4. *For all $p', q' \in (\pi/\ell)(\mathbb{N}\cup\{0\})^3$ and $\alpha \in \mathbb{N}$ there exists $J^\alpha_{p'q'} \in \mathbb{R}$ with $J^\alpha_{p'q'} = J^\alpha_{q'p'}$ such that for all $p, q, \mu, \nu \in (\pi/\ell)(\mathbb{N}\cup\{0\})^3$ we have*

$$\widehat{w}_{pq,\mu\nu} = \sum_\alpha J^\alpha_{p\mu} J^\alpha_{q\nu}. \tag{17}$$

Moreover we have the operator inequalities

$$0 \le \sum_{p,p'\neq 0} \widehat{w}_{pp',00} a_p^* a_{p'} = \sum_{p,p'\neq 0} \widehat{w}_{p0,0p'} a_p^* a_{p'} \le 4\pi \ell^{-3} R^2 \widehat{n}_+ \tag{18}$$

and

$$0 \le \sum_{p,p',m\neq 0} \widehat{w}_{pm,mp'} a_p^* a_{p'} \le r^{-1} \widehat{n}_+.$$

Proof. The operator $\mathcal{A}$ with integral kernel $w_{r,R}(x,y)$ is a non-negative Hilbert–Schmidt operator on $L^2(\mathbb{R}^3)$ with norm less than $\sup_k \hat{V}_{r,R}(k) \le 4\pi R^2$. Denote the eigenvalues of $\mathcal{A}$ by λ_α, $\alpha = 1, 2, \ldots$ and corresponding orthonormal eigenfunctions by φ_α. We may assume that these functions are real. The eigenvalues satisfy $0 \le \lambda_\alpha \le 4\pi R^2$. We then have

$$\widehat{w}_{pq,\mu\nu} = \sum_\alpha \lambda_\alpha \int u_p(x) u_\mu(x) \varphi_\alpha(x)\,dx \int u_q(y) u_\nu(y) \varphi_\alpha(y)\,dy.$$

The identity (17) thus follows with $J^\alpha_{p\mu} = \lambda_\alpha^{1/2} \int u_p(x) u_\mu(x) \varphi_\alpha(x)\,dx$.

If P denotes the projection onto the constant functions we may also consider the operator $(I-P)\mathcal{A}(I-P)$. Denote its eigenvalues and eigenfunctions by λ'_α and φ'_α. Then again $0 \le \lambda'_\alpha \le 4\pi R^2$. Hence we may write

$$\widehat{w}_{p0,0p'} = \ell^{-3} \sum_\alpha \lambda'_\alpha \int u_p(x) \varphi'_\alpha(x)\,dx \int u_{p'}(y) \varphi'_\alpha(y)\,dy.$$

Thus, since all φ'_α are orthogonal to constants we have

$$\sum_{p,p'\neq 0} \widehat{w}_{p0,0p'} a_p^* a_{p'}$$
$$= \ell^{-3} \sum_\alpha \lambda'_\alpha \left(\sum_{p\neq 0} \int u_p(x)\varphi'_\alpha(x)\, dx\, a_p^* \right) \left(\sum_{p\neq 0} \int u_p(x)\varphi'_\alpha(x)\, dx\, a_p^* \right)^*$$
$$= \ell^{-3} \sum_\alpha \lambda'_\alpha a^*\left(\varphi'_\alpha\right) a\left(\varphi'_\alpha\right).$$

The inequalities (18) follow immediately from this.

The fact that $\sum_{p,p',m\neq 0} \widehat{w}_{pm,mp'} a_p^* a_{p'} \geq 0$ follows from the representation (17). Moreover, since the kernel $w_{R,r}(x,y)$ is a continuous function we have that $w_{r,R}(x,x) = \sum_\alpha \lambda_\alpha \varphi_\alpha(x)^2$ for almost all x and hence

$$\sum_{m\neq 0} \widehat{w}_{pm,mp'} = \int u_p(x) u_{p'}(x) w_{r,R}(x,x)\, dx - \widehat{w}_{p0,0p'}.$$

We therefore have

$$\sum_{p,p',m\neq 0} \widehat{w}_{pm,mp'} a_p^* a_{p'} \leq \sum_{p,p'\neq 0} \int u_p(x) u_{p'}(x) W_{r,R}(x,x)\, dx\, a_p^* a_{p'}$$
$$= \int w_{r,R}(x,x) \left(\sum_{p\neq 0} u_p(x) a_p^* \right) \left(\sum_{p\neq 0} u_p(x) a_p^* \right)^* dx$$
$$\leq \sup_{x'} w_{r,R}(x',x') \int \left(\sum_{p\neq 0} u_p(x) a_p^* \right) \left(\sum_{p\neq 0} u_p(x) a_p^* \right)^* dx$$
$$= \sup_{x'} w_{r,R}(x',x') \widehat{n}_+$$

and the lemma follows since $\sup_{x'} w_{r,R}(x',x') \leq r^{-1}$. □

Lemma 5.5 (Control of terms with $\widehat{w}_{p0,00}$). *The sum of the terms in $H^n_{\ell,r,R}$ containing $\widehat{w}_{p0,00}$, $\widehat{w}_{0p,00}$, $\widehat{w}_{00,p0}$, or $\widehat{w}_{00,0p}$, with $p \neq 0$ is, for all $\varepsilon > 0$, bounded below by*

$$-\varepsilon^{-1} 4\pi \ell^{-3} R^2 \widehat{n}_0 \widehat{n}_+ - \varepsilon \widehat{w}_{00,00} (\widehat{n}_0 + 1 - \rho\ell^3)^2, \tag{19}$$

and by

$$\sum_{p\neq 0} \widehat{w}_{p0,00} \left((n-\rho\ell^3) a_p^* a_0 + a_0^* a_p (n - \rho\ell^3) \right)$$
$$- \varepsilon^{-1} 4\pi \ell^{-3} R^2 \widehat{n}_0 \widehat{n}_+ - \varepsilon \widehat{w}_{00,00} (\widehat{n}_+ - 1)^2. \tag{20}$$

Proof. The terms containing $\widehat{w}_{p0,00}$, $\widehat{w}_{0p,00}$,$\widehat{w}_{00,p0}$, or $\widehat{w}_{00,0p}$ are

$$\begin{aligned}
&\sum_{p\neq 0}\tfrac{1}{2}\widehat{w}_{p0,00}\left(2a_p^*a_0^*a_0a_0+2a_0^*a_0^*a_0a_p-2\rho\ell^3a_0^*a_p-2\rho\ell^3a_p^*a_0\right)\\
&\quad=\sum_{p\neq 0}\widehat{w}_{p0,00}\left((\widehat{n}_0-\rho\ell^3)a_p^*a_0+a_0^*a_p(\widehat{n}_0-\rho\ell^3)\right)\\
&\quad=\sum_\alpha\sum_{p\neq 0}J_{p0}^\alpha J_{00}^\alpha\left(a_p^*a_0(\widehat{n}_0+1-\rho\ell^3)+(\widehat{n}_0+1-\rho\ell^3)a_0^*a_p\right).
\end{aligned}$$

In the last term we have used the representation (17) and the commutation relation $[\widehat{n}_0, a_0] = a_0$. For all $\varepsilon > 0$ we get that the above expression is bounded below by

$$\begin{aligned}
&\varepsilon^{-1}\sum_\alpha\sum_{p,p'\neq 0}J_{p0}^\alpha J_{p'0}^\alpha\widehat{n}_0a_p^*a_{p'}-\varepsilon\sum_\alpha\left(J_{00}^\alpha\right)^2(\widehat{n}_0+1-\rho\ell^3)^2\\
&\qquad\qquad=-\varepsilon^{-1}\sum_{p,p'\neq 0}\widehat{w}_{p0,0p'}\widehat{n}_0a_p^*a_{p'}-\varepsilon\widehat{w}_{00,00}(\widehat{n}_0+1-\rho\ell^3)^2.
\end{aligned}$$

The bound (19) follows from (18).

The second bound (20) follows in the same way if we notice that the terms containing $\widehat{w}_{p0,00}$, $\widehat{w}_{0p,00}$,$\widehat{w}_{00,p0}$, or $\widehat{w}_{00,0p}$ may be written as

$$\begin{aligned}
&\sum_{p\neq 0}\widehat{w}_{p0,00}\left((n-\rho\ell^3)a_p^*a_0+a_0^*a_p(n-\rho\ell^3)\right)\\
&\qquad\qquad+\sum_\alpha\sum_{p\neq 0}J_{p0}^\alpha J_{00}^\alpha\left(a_p^*a_0(1-\widehat{n}_+)+(1-\widehat{n}_+)a_0^*a_p\right).\qquad\square
\end{aligned}$$

Lemma 5.6 (Control of terms with $\widehat{w}_{pq,m0}$). *The sum of the terms in $H^n_{\ell,r,R}$ containing $\widehat{w}_{pq,m0}$, $\widehat{w}_{pq,0m}$, $\widehat{w}_{p0,qm}$, or $\widehat{w}_{0p,qm}$, with $p,q,m\neq 0$ is bounded below by*

$$-\varepsilon^{-1}4\pi\ell^{-3}R^2\widehat{n}_0\widehat{n}_+-\varepsilon\widehat{n}_+r^{-1}-\varepsilon\sum_{p,m,p',m'\neq 0}\widehat{w}_{mp',pm'}a_m^*a_{p'}^*a_{m'}a_p,$$

for all $\varepsilon > 0$.

Proof. The terms containing $\widehat{w}_{pq,m0}$, $\widehat{w}_{pq,0m}$, $\widehat{w}_{p0,qm}$, or $\widehat{w}_{0p,qm}$ are

$$\begin{aligned}
\sum_{pqm\neq 0} \widehat{w}_{pqm0}\left(a_p^* a_q^* a_m a_0 + a_0^* a_m^* a_q a_p\right) &= \sum_\alpha \left(\left(\sum_{q\neq 0} J_{q0}^\alpha a_q^* a_0\right)\left(\sum_{pm\neq 0} J_{pm}^\alpha a_p^* a_m\right)\right. \\
&\quad \left. + \left(\sum_{pm\neq 0} J_{pm}^\alpha a_p^* a_m\right)^* \left(\sum_{q\neq 0} J_{q0}^\alpha a_q^* a_0\right)^*\right) \\
&\geq -\sum_\alpha \left(\varepsilon^{-1}\left(\sum_{q\neq 0} J_{q0}^\alpha a_q^* a_0\right)\left(\sum_{q\neq 0} J_{q0}^\alpha a_0^* a_q\right)\right. \\
&\quad \left. +\varepsilon\left(\sum_{pm\neq 0} J_{pm}^\alpha a_m^* a_p\right)\left(\sum_{pm\neq 0} J_{pm}^\alpha a_p^* a_m\right)\right).
\end{aligned}$$

Using that $J_{pm}^\alpha = J_{mp}^\alpha$ we may write this as

$$\begin{aligned}
&-\varepsilon^{-1}\sum_{qq'\neq 0} \widehat{w}_{q0,0q'} a_q^* a_{q'} a_0 a_0^* - \varepsilon \sum_{p,m,p',m'\neq 0} \widehat{w}_{mp',pm'} a_m^* a_p a_{p'}^* a_{m'} \\
&= -\varepsilon^{-1}\sum_{qq'\neq 0} \widehat{w}_{q0,0q'} a_q^* a_{q'} a_0 a_0^* - \varepsilon \sum_{p,m,p',m'\neq 0} \widehat{w}_{mp',pm'} a_m^* a_{p'}^* a_{m'} a_p \\
&\quad - \varepsilon \sum_{p,m,m'\neq 0} \widehat{w}_{mp,pm'} a_m^* a_{m'}.
\end{aligned}$$

The lemma now follows from Lemma 5.4. □

6. Analyzing the Quadratic Hamiltonian

In this section we consider the main part of the Hamiltonian. This is the "quadratic" Hamiltonian considered by Foldy. It consists of the kinetic energy and all the terms with the coefficients $\widehat{w}_{pq,00}$, $\widehat{w}_{00,pq}$ $\widehat{w}_{p0,0q}$, and $\widehat{w}_{0p,q0}$with $p, q \neq 0$, i.e.,

$$\begin{aligned}
H_{\text{Foldy}} &= \tfrac{1}{2}\gamma^{-1}\sum_p |p|^2 a_p^* a_p \\
&\quad + \tfrac{1}{2}\sum_{pq\neq 0} \widehat{w}_{pq,00}\left(a_p^* a_0^* a_0 a_q + a_0^* a_p^* a_q a_0 + a_p^* a_q^* a_0 a_0 + a_0^* a_0^* a_p a_q\right) \quad (21) \\
&= \tfrac{1}{2}\gamma^{-1}\sum_p |p|^2 a_p^* a_p + \sum_{pq\neq 0} \widehat{w}_{pq,00}\left(a_p^* a_q a_0^* a_0 + \tfrac{1}{2} a_p^* a_q^* a_0 a_0 + \tfrac{1}{2} a_0^* a_0^* a_p a_q\right).
\end{aligned}$$

In order to compute all the bounds we found it necessary to include the first term in (20) into the "quadratic" Hamiltonian. We therefore define

$$\begin{aligned} H_Q = \tfrac{1}{2}\gamma^{-1}\sum_p |p|^2 a_p^* a_p + \sum_{p\neq 0} \widehat{w}_{p0,00}\left((n-\rho\ell^3)a_p^* a_0 + a_0^* a_p (n-\rho\ell^3)\right) \\ + \sum_{pq\neq 0} \widehat{w}_{pq,00}\left(a_p^* a_q a_0^* a_0 + \tfrac{1}{2} a_p^* a_q^* a_0 a_0 + \tfrac{1}{2} a_0^* a_0^* a_p a_q\right). \end{aligned} \tag{22}$$

Note that $H_{\text{Foldy}} = H_Q$ in the neutral case $n = \rho\ell^3$. Our goal is to give a lower bound on the ground state energy of the Hamiltonian H_Q.

For the sake of convenience we first enlarge the one-particle Hilbert space $L^2\left([-\ell/2, \ell/2]^3\right)$. In fact, instead of considering the symmetric Fock space over $L^2\left([-\ell/2, \ell/2]^3\right)$ we now consider the symmetric Fock space over the one-particle Hilbert space $L^2\left([-\ell/2, \ell/2]^3\right) \oplus \mathbb{C}$. Note that the larger Fock space of course contains the original Fock space as a subspace. On the larger space we have a new pair of creation and annihilation operators that we denote $\widetilde{a}_0^*$ and $\widetilde{a}_0$. These operators merely create vectors in the $\mathbb{C}$ component of $L^2\left([-\ell/2, \ell/2]^3\right) \oplus \mathbb{C}$, and so commute with all other operators.

We shall now write

$$\widetilde{a}_p = \begin{cases} a_p, & \text{if } p \neq 0 \\ \widetilde{a}_0, & \text{if } p = 0 \end{cases} \quad \text{and} \quad \widetilde{a}_p^* = \begin{cases} a_p^*, & \text{if } p \neq 0 \\ \widetilde{a}_0^*, & \text{if } p = 0 \end{cases}. \tag{23}$$

We now define the Hamiltonian

$$\begin{aligned} \widetilde{H}_Q = \tfrac{1}{2}\gamma^{-1}\sum_p |p|^2 \widetilde{a}_p^* \widetilde{a}_p + \sum_{p} \widehat{w}_{p0,00}\left((n-\rho\ell^3)\widetilde{a}_p^* a_0 + a_0^* \widetilde{a}_p (n-\rho\ell^3)\right) \\ + \sum_{pq} \widehat{w}_{pq,00}\left(\widetilde{a}_p^* \widetilde{a}_q a_0^* a_0 + \tfrac{1}{2} \widetilde{a}_p^* \widetilde{a}_q^* a_0 a_0 + \tfrac{1}{2} a_0^* a_0^* \widetilde{a}_p \widetilde{a}_q\right), \end{aligned} \tag{24}$$

where we no longer restrict p, q to be different from 0. Note that for all states on the larger Fock space for which $\langle \widetilde{a}_0^* \widetilde{a}_0 \rangle = 0$ we have $\langle \widetilde{H}_Q \rangle = \langle H_Q \rangle$.

For any function $\varphi \in L^2\left([-\ell/2, \ell/2]^3\right)$ we introduce the creation operator

$$\widetilde{a}^*(\varphi) = \sum_p (u_p, \varphi)\widetilde{a}_p^*.$$

Note that the sum includes $p = 0$. the difference from $a^*(\varphi)$ is given by $\widetilde{a}^*(\varphi) - a^*(\varphi) = (u_0, \varphi)\left(\widetilde{a}_0^* - a_0^*\right)$.

Then $[\widetilde{a}(\varphi), \widetilde{a}^*(\psi)] = (\varphi, \psi)$. We have introduced the "dummy" operator $\widetilde{a}_0^*$ in order for this relation to hold. One could just as well have stayed in the old space, but then the relation above would hold only for functions orthogonal to constants.

For any $k \in \mathbb{R}^3$ denote $\chi_{\ell,k}(x) = e^{ikx}\chi_\ell(x)$ and define the operators

$$b_k^* = \widetilde{a}^*(\chi_{\ell,k})a_0 \quad \text{and} \quad b_k = \widetilde{a}(\chi_{\ell,k})a_0^*$$

They satisfy the commutation relations

$$\begin{aligned} [b_k, b_{k'}^*] &= a_0^* a_0 \left(\chi_{\ell,k}, \chi_{\ell,k'}\right) - \widetilde{a}(\chi_{\ell,k})\widetilde{a}^*(\chi_{\ell,k'}) \\ &= a_0^* a_0 \widehat{\chi_\ell^2}(k'-k) - \widetilde{a}(\chi_{\ell,k})\widetilde{a}^*(\chi_{\ell,k'}) \end{aligned} \tag{25}$$

We first consider the kinetic energy part of the Hamiltonian. We shall bound it using the double commutator bound in Appendix B. First we need a well known comparisson between the Neumann Laplacian and the Laplacian in the whole space.

Lemma 6.1 (Neumann resolvent is bigger than free resolvent). *Let P_ℓ denote the projection in $L^2(\mathbb{R}^3)$ that projects onto $L^2([-\ell/2, \ell/2]^3)$ (identified as a subspace). Then if $-\Delta$ denotes the Laplacian on all of $\mathbb{R}^3$ and $-\Delta_\ell$ is the Neumann Laplacian on $[-\ell/2, \ell/2]^3$ we have the operator inequality*

$$(-\Delta_\ell + a)^{-1} \geq P_\ell(-\Delta + a)^{-1} P_\ell,$$

for all $a > 0$.

Proof. It is clear that for all $f \in L^2(\mathbb{R}^3)$

$$\| P_\ell(-\Delta_\ell + a)^{1/2} P_\ell(-\Delta + a)^{-1/2} f \|^2 \leq \| f \|^2,$$

and hence

$$\| (-\Delta + a)^{-1/2} P_\ell(-\Delta_\ell + a)^{1/2} P_\ell f \|^2 \leq \| f \|^2.$$

Now simply use this with $f = (-\Delta_\ell + a)^{-1/2} u$. □

Lemma 6.2 (The kinetic energy bound). *There exists a constant $C' > 0$ such that if $C't < 1$, where t is the parameter used in the definition of χ_ℓ in Sect. 3, we have*

$$\left\langle \sum_p |p|^2 \widetilde{a}_p^* \widetilde{a}_p \right\rangle \geq (2\pi)^{-3}(1 - C't)^2 n^{-1} \int_{\mathbb{R}^3} \frac{|k|^4}{|k|^2 + (\ell t^3)^{-2}} \langle b_k^* b_k \rangle \, dk$$

for all states with $\langle \widetilde{a}_0^* \widetilde{a}_0 \rangle = 0$ and particle number equal to n, i.e., $\left\langle \sum_p a_p^* a_p \right\rangle^2 = \left\langle \left(\sum_p a_p^* a_p \right)^2 \right\rangle = n^2$.

Proof. Let s, with $0 < s \leq t$, be a parameter to be chosen below. Recall that t is the parameter used in the definition of χ_ℓ in Section 3. Then since $\chi_\ell^2 + \eta_\ell^2 = 1$ we have

$$\begin{aligned}
-\Delta_\ell &\geq \frac{(-\Delta_\ell)^2}{-\Delta_\ell + ((\ell s)^{-2}} = \tfrac{1}{2}(\chi_\ell^2 + \eta_\ell^2)\frac{(-\Delta_\ell)^2}{-\Delta_\ell + (\ell s)^{-2}} + \tfrac{1}{2}\frac{(-\Delta_\ell)^2}{-\Delta_\ell + (\ell s)^{-2}}(\chi_\ell^2 + \eta_\ell^2) \\
&= \chi_\ell \frac{(-\Delta_\ell)^2}{-\Delta_\ell + (\ell s)^{-2}} \chi_\ell + \eta_\ell \frac{(-\Delta_\ell)^2}{-\Delta_\ell + (\ell s)^{-2}} \eta_\ell \\
&\quad + \left[\left[\frac{(-\Delta_\ell)^2}{-\Delta_\ell + (\ell s)^{-2}}, \chi_\ell\right], \chi_\ell\right] + \left[\left[\frac{(-\Delta_\ell)^2}{-\Delta_\ell + (\ell s)^{-2}}, \eta_\ell\right], \eta_\ell\right] \\
&\geq \chi_\ell \frac{(-\Delta_\ell)^2}{-\Delta_\ell + (\ell s)^{-2}} \chi_\ell + \eta_\ell \frac{(-\Delta_\ell)^2}{-\Delta_\ell + (\ell s)^{-2}} \eta_\ell \\
&\quad - C(\ell t)^{-2} \frac{-\Delta_\ell}{-\Delta_\ell + (\ell s)^{-2}} - C\ell^{-2} s^2 t^{-4},
\end{aligned}$$

where the last inequality follows from Lemma B.1 in Appendix B. We can now repeat this calculation to get

$$-\Delta_\ell \geq \chi_\ell \left(\frac{(-\Delta_\ell)^2}{-\Delta_\ell + (\ell s)^{-2}} - C(\ell t)^{-2} \frac{-\Delta_\ell}{-\Delta_\ell + (\ell s)^{-2}} \right) \chi_\ell$$
$$+ \eta_\ell \left(\frac{(-\Delta_\ell)^2}{-\Delta_\ell + (\ell s)^{-2}} - C(\ell t)^{-2} \frac{-\Delta_\ell}{-\Delta_\ell + (\ell s)^{-2}} \right) \eta_\ell - C\ell^{-2} s^2 t^{-4}$$
$$- C(\ell t)^{-2} \left(\left[\left[\frac{-\Delta_\ell}{-\Delta_\ell + (\ell s)^{-2}}, \chi_\ell \right], \chi_\ell \right] + \left[\left[\frac{-\Delta_\ell}{-\Delta_\ell + (\ell s)^{-2}}, \eta_\ell \right], \eta_\ell \right] \right).$$

If we therefore use (53) in Lemma B.1 and recall that $s \leq t$ we arrive at

$$-\Delta_\ell \geq \chi_\ell \left(\frac{(-\Delta_\ell)^2}{-\Delta_\ell + (\ell s)^{-2}} - C(\ell t)^{-2} \frac{-\Delta_\ell}{-\Delta_\ell + (\ell s)^{-2}} \right) \chi_\ell$$
$$+ \eta_\ell \left(\frac{(-\Delta_\ell)^2}{-\Delta_\ell + (\ell s)^{-2}} - C(\ell t)^{-2} \frac{-\Delta_\ell}{-\Delta_\ell + (\ell s)^{-2}} \right) \eta_\ell - C\ell^{-2} s^2 t^{-4}.$$

Note that for $\alpha > 0$ we have

$$\alpha \frac{(-\Delta_\ell)^2}{-\Delta_\ell + (\ell s)^{-2}} - C(\ell t)^{-2} \frac{-\Delta_\ell}{-\Delta_\ell + (\ell s)^{-2}} \geq -C\alpha^{-1} s^2 t^{-4} \ell^{-2}.$$

Thus if we also assume that $\alpha < 1$ we have

$$-\Delta_\ell \geq (1-\alpha) \chi_\ell \frac{(-\Delta_\ell)^2}{-\Delta_\ell + (\ell s)^{-2}} \chi_\ell - C\alpha^{-1} s^2 t^{-4} \ell^{-2}.$$

Thus if u is a normalized function on $L^2(\mathbb{R}^3)$ which is orthogonal to constants we have according to the bound on the gap (15) that for all $0 < \delta < 1$

$$(u, -\Delta_\ell u) \geq (1-\delta)(1-\alpha) \left(u, \chi_\ell \frac{(-\Delta_\ell)^2}{-\Delta_\ell + (\ell s)^{-2}} \chi_\ell u \right)$$
$$- C(1-\delta)\alpha^{-1} s^2 t^{-4} \ell^{-2} + \delta \pi^2 \ell^{-2}.$$

We choose $\alpha = \delta = C' s t^{-2}$ for an appropriately large constant $C' > 0$ and assume that s and t are such that δ is less than 1. Then

$$(u, -\Delta_\ell u) \geq (1 - C' s t^{-2})^2 \left(u, \chi_\ell \frac{(-\Delta_\ell)^2}{-\Delta_\ell + (\ell s)^{-2}} \chi_\ell u \right).$$

If we now use Lemma 6.1 we may write this as

$$(u, -\Delta_\ell u) \geq (1 - C' s t^{-2})^2 \left(u, \chi_\ell \Delta_\ell \frac{1}{-\Delta + (\ell s)^{-2}} \Delta_\ell \chi_\ell u \right)$$
$$= (1 - C' s t^{-2})^2 \left(u, \chi_\ell \frac{(-\Delta)^2}{-\Delta + (\ell s)^{-2}} \chi_\ell u \right),$$

where in the last inequality we have used that $\Delta \chi = \Delta_\ell \chi$ and $\chi \Delta = \chi \Delta_\ell$.

We now choose $s = t^3$ and we may then write this inequality in second quantized form as

$$\left\langle \sum_p |p|^2 \widetilde{a}_p^* \widetilde{a}_p \right\rangle \geq (2\pi)^{-3}(1 - C't)^2 \int_{\mathbb{R}^3} \frac{|k|^4}{|k|^2 + (\ell t^3)^{-2}} \langle \widetilde{a}^*(\chi_{\ell,k})\widetilde{a}(\chi_{\ell,k})\rangle \, dk$$

using that $\langle \widetilde{a}_0^* \widetilde{a}_0 \rangle = 0$. Since we consider only states with particle number n the inequality still holds if we insert $n^{-1} a_0 a_0^*$ as in the statement of the lemma. □

With the same notation as in the above lemma we may write

$$w_{r,R}(x, y) = (2\pi)^{-3} \int \hat{V}_{r,R}(k) \chi_{\ell,k}(x) \overline{\chi_{\ell,k}(y)} \, dk.$$

The last two sums in the Hamiltonian (24) can therefore be written as

$$(2\pi\ell)^{-3} \int \hat{V}_{r,R}(k) \Big[(n - \rho\ell^3)\ell^{-3/2} \left(\widehat{\chi}_\ell(k) b_k^* + \overline{\widehat{\chi}_\ell(k)} b_k \right) + \tfrac{1}{2} \left(b_k^* b_k + b_{-k}^* b_{-k} + b_k^* b_{-k}^* + b_k b_{-k} \right) \Big] dk - \sum_{pq} \widehat{w}_{pq,00} \widetilde{a}_p^* \widetilde{a}_q.$$

Note that it is important here that the potential $w_{r,R}$ contains the localization function χ_ℓ.

Thus, since $\hat{V}_{r,R}(k) = \hat{V}_{r,R}(-k)$ and $\overline{\widehat{\chi}_\ell(k)} = \widehat{\chi}_\ell(-k)$ we have for states with $\langle \widetilde{a}_0^* \widetilde{a}_0 \rangle = 0$ that

$$\langle \widetilde{H}_Q \rangle \geq \int_{\mathbb{R}^3} \langle h_Q(k) \rangle \, dk - \sum_{pq} \widehat{w}_{pq,00} \left\langle \widetilde{a}_p^* \widetilde{a}_q \right\rangle, \tag{26}$$

where

$$\begin{aligned} h_Q(k) = {} & \frac{(1 - C't)^2}{4(2\pi)^3 \gamma n} \frac{|k|^4}{|k|^2 + (\ell t^3)^{-2}} \left(b_k^* b_k + b_{-k}^* b_{-k} \right) \\ & + \frac{\hat{V}_{r,R}(k)}{2(2\pi\ell)^3} \Big[(n - \rho\ell^3)\ell^{-3/2} \left(\widehat{\chi}_\ell(k)(b_k^* + b_{-k}) + \overline{\widehat{\chi}_\ell(k)}(b_k + b_{-k}^*) \right) \\ & + \left(b_k^* b_k + b_{-k}^* b_{-k} + b_k^* b_{-k}^* + b_k b_{-k} \right) \Big]. \end{aligned} \tag{27}$$

Theorem 6.3 (Simple case of Bogolubov's method). *For arbitrary constants $\mathcal{A} \geq \mathcal{B} > 0$ and $\kappa \in \mathbb{C}$ we have the inequality*

$$\begin{aligned} & \mathcal{A}(b_k^* b_k + b_{-k}^* b_{-k}) + \mathcal{B}(b_k^* b_{-k}^* + b_k b_{-k}) + \kappa(b_k^* + b_{-k}) + \overline{\kappa}(b_k + b_{-k}^*) \\ & \geq -\tfrac{1}{2}(\mathcal{A} - \sqrt{\mathcal{A}^2 - \mathcal{B}^2})([b_k, b_k^*] + [b_{-k}, b_{-k}^*]) - \frac{2|\kappa|^2}{\mathcal{A} + \mathcal{B}}. \end{aligned}$$

Proof. We may complete the square

$$\begin{aligned}&\mathcal{A}(b_k^*b_k + b_{-k}^*b_{-k}) + \mathcal{B}(b_k^*b_{-k}^* + b_kb_{-k}) + \kappa(b_k^* + b_{-k}) + \overline{\kappa}(b_k + b_{-k}^*)\\ &= D(b_k^* + \alpha b_{-k} + a)(b_k + \alpha b_{-k}^* + \overline{a}) + D(b_{-k}^* + \alpha b_k + \overline{a})(b_{-k} + \alpha b_{-k}^* + a)\\ &\qquad - D\alpha^2([b_k, b_k^*] + [b_{-k}, b_{-k}^*]) - 2D|a|^2,\end{aligned}$$

if

$$D(1+\alpha^2) = \mathcal{A}, \quad 2D\alpha = \mathcal{B}, \quad aD(1+\alpha) = \kappa.$$

We choose the solution $\alpha = \mathcal{A}/\mathcal{B} - \sqrt{\mathcal{A}^2/\mathcal{B}^2 - 1}$. Hence

$$D\alpha^2 = \mathcal{B}\alpha/2 = \tfrac{1}{2}(\mathcal{A} - \sqrt{\mathcal{A}^2 - \mathcal{B}^2}), \quad D|a|^2 = \frac{|\kappa|^2}{D(1+\alpha^2+2\alpha)} = \frac{|\kappa|^2}{\mathcal{A}+\mathcal{B}}. \qquad \square$$

Usually when applying Bogolubov's method the commutator $[b_k, b_k^*]$ is a positive constant. In this case the lower bound in the theorem is actually the bottom of the spectrum of the operator. If moreover, $\mathcal{A} > \mathcal{B}$ the bottom is actually an eigenvalue. In our case the commutator $[b_k, b_k^*]$ is not a constant, but according to (25) we have

$$[b_k, b_k^*] \le \int \chi_\ell(x)^2\,dx a_0^* a_0 \le \ell^3 a_0^* a_0. \tag{28}$$

From this and the above theorem we easily conclude the following bound.

Lemma 6.4 (Lower bound on quadratic Hamiltonian). *On the subspace with n particles we have*

$$H_Q \ge -In^{5/4}\ell^{-3/4} - \tfrac{1}{2}\left(n - \rho\ell^3\right)^2 \widehat{w}_{00,00} - 4\pi n^{5/4}\ell^{-3/4}(n\ell)^{-1/4},$$

where $I = \frac{1}{2}(2\pi)^{-3}\int_{\mathbb{R}^3} f(k) - (f(k)^2 - g(k)^2)^{1/2}\,dk$ *with*

$$g(k) = 4\pi\frac{1}{k^2 + (n^{1/4}\ell^{-3/4}R)^{-2}} - 4\pi\frac{1}{k^2 + (n^{1/4}\ell^{-3/4}r)^{-2}}$$

and

$$f(k) = g(k) + \tfrac{1}{2}\gamma^{-1}(1 - C't)^2\frac{|k|^4}{|k|^2 + (n^{1/4}\ell^{1/4}t^3)^{-2}}.$$

Proof. We consider a state with $\langle \widetilde{a}_0^*\widetilde{a}_0\rangle = 0$. Then $\langle H_Q\rangle = \langle \widetilde{H}_Q\rangle$. We shall use (26). Note first that

$$\left\langle \sum_{pq} \widehat{w}_{pq,00}\widetilde{a}_p^*\widetilde{a}_q \right\rangle = \left\langle \sum_{p,q\neq 0} \widehat{w}_{p0,0q} a_p^* a_q \right\rangle \le 4\pi\ell^{-3}R^2\widehat{n}_+ \le 4\pi\ell^{-1}n$$

by (18) and the fact that $R \le \ell$. We may of course rewrite $\ell^{-1}n = n^{5/4}\ell^{-3/4}(n\ell)^{-1/4}$.

By Theorem 6.3, (27) and (28) we have

$$h_Q(k) \ge -(\mathcal{A}_k - \sqrt{\mathcal{A}_k^2 - \mathcal{B}_k^2})n\ell^3 - \frac{\hat{V}_{r,R}(k)^2(n-\rho\ell^3)^2}{2(2\pi)^6\ell^9(\mathcal{A}_k + \mathcal{B}_k)}\left|\widehat{\chi}_\ell(k)\right|^2,$$

where

$$\mathcal{B}_k = \frac{\hat{V}_{r,R}(k)}{2(2\pi\ell)^3}, \quad \mathcal{A}_k = \mathcal{B}_k + \frac{(1-C't)^2}{4(2\pi)^3\gamma n}\frac{|k|^4}{|k|^2+(\ell t^3)^{-2}}.$$

Since $\mathcal{A}_k > \mathcal{B}_k$ we have that

$$h_Q(k) \geq -(\mathcal{A}_k - \sqrt{\mathcal{A}_k^2 - \mathcal{B}_k^2})n\ell^3 - \frac{\hat{V}_{r,R}(k)(n-\rho\ell^3)^2}{2(2\pi)^3\ell^6}\left|\widehat{\chi}_\ell(k)\right|^2.$$

Note that

$$\begin{aligned}\int &\frac{\hat{V}_{r,R}(k)(n-\rho\ell^3)^2}{2(2\pi)^3\ell^6}\left|\widehat{\chi}_\ell(k)\right|^2 dk \\ &= \tfrac{1}{2}\left(\frac{n}{\ell^3}-\rho\right)^2 \iint \chi_\ell(x)V_{r,R}(x-y)\chi_\ell(y)\,dx\,dy = \tfrac{1}{2}\left(n-\rho\ell^3\right)^2 \widehat{w}_{00,00}.\end{aligned}$$

The lemma now follows from (26) by a simple change of variables in the k integral. □

As a consequence we get the following bound for the Foldy Hamiltonian.

Corollary 6.5 (Lower bound on the Foldy Hamiltonian). *The Foldy Hamiltonian in (21) satisfies*

$$H_{\text{Foldy}} \geq -In^{5/4}\ell^{-3/4} - 4\pi n^{5/4}\ell^{-3/4}(n\ell)^{-1/4}. \tag{29}$$

There is constant $C > 0$ such that if $\rho^{1/4}R > C$, $\rho^{1/4}\ell t^3 > C$, and $t < C^{-1}$ then the Foldy Hamiltonian satisfies the bound

$$H_{\text{Foldy}} \geq \tfrac{1}{4}\sum_p |p|^2 a_p^* a_p - Cn^{5/4}\ell^{-3/4}. \tag{30}$$

Proof. Lemma 6.4 holds for all ρ hence also if we had replaced ρ by n/ℓ^3 in this case we get (29).

The integral I satisfies the bound

$$I \leq \tfrac{1}{2}(2\pi)^{-3}\int_{\mathbb{R}^3} \max\left\{g(k),\, \tfrac{1}{2}g(k)^2(f(k)-g(k))^{-1}\right\} dk.$$

By Corollary 4.3 we may assume that $n \geq c\rho\ell^3$. Hence I is bounded by a constant as long as $\rho^{1/4}R$ and $\rho^{1/4}\ell t^3$ are sufficiently large and t is sufficiently small (which also ensures that γ is close to 1). Note that we do not have to make any assumptions on r. Moreover, if this is true we also have that $n\ell \geq c\rho\ell^4$ is large and hence $(n\ell)^{-1}$ is small. This would give the bound in the corollary except for the first positive term. The above argument, however, also holds (with different constants) if we replace the kinetic energy in the Foldy Hamiltonian by $\frac{1}{2}\left(\gamma^{-1}-\frac{1}{2}\right)\sum_p |p|^2 a_p^* a_p$ (assuming that $\gamma < 2$). This proves the corollary. □

Note that if

$$n^{1/4}\ell^{-3/4}R \to \infty,\ n^{1/4}\ell^{-3/4}r \to 0,\ n^{1/4}\ell^{1/4}t^3 \to \infty,\ \text{and } t \to 0 \qquad (31)$$

it follows by dominated convergence that I converges to

$$\begin{aligned}\tfrac{1}{2}(2\pi)^{-3}\int_{\mathbb{R}^3} 4\pi|k|^{-2} + \tfrac{1}{2}|k|^2 - \left((4\pi|k|^{-2} + \tfrac{1}{2}|k|^2)^2 - (4\pi|k|^{-2})^2\right)^{1/2} dk \\ = (2/\pi)^{3/4}\int_0^\infty 1 + x^4 - x^2\left(x^4+2\right)^{1/2} dx = -\left(\frac{4\pi}{3}\right)^{1/4} A,\end{aligned}$$

where A was given in (6). Thus if we can show that $n \sim \rho\ell^3$ we see that the term $-In^{5/4}\ell^{-3/4} \sim -I\rho^{1/4}n$ agrees with Foldy's calculation (5) for the little box of size ℓ.

Our task is now to show that indeed $n \sim \rho\ell^3$, i.e., that we have approximate neutrality in each little box and that the term above containing the integral I is indeed the leading term.

7. Simple Bounds on n and $\widehat{n}_+$

The Lemmas 4.1, 5.2, 5.3, 5.5, and 5.6 together with Lemma 6.4 or Corollary 6.5 control all terms in the Hamiltonian H_ℓ^n except the positive term

$$\tfrac{1}{2}\sum_{p,m,p',m'\neq 0} \widehat{w}_{mp',pm'} a_m^* a_{p'}^* a_{m'} a_p.$$

If we use (30) in Corollary 6.5 together with the other bounds we obtain the following bound if $\rho^{1/4}R$ and $\rho^{1/4}\ell t^3$ are sufficiently large and t is sufficiently small

$$\begin{aligned}H_\ell^n \geq{}& \tfrac{1}{4}\sum_p |p|^2 a_p^* a_p - Cn^{5/4}\ell^{-3/4} - \tfrac{1}{2}nR^{-1} - Cn\rho r^2 \\ &+ \tfrac{1}{2}\widehat{w}_{00,00}\left[\left(\widehat{n}_0 - \rho\ell^3\right)^2 - \widehat{n}_0\right] \\ &- 4\pi[\rho - n\ell^{-3}]_+\widehat{n}_+R^2 - 4\pi\widehat{n}_+^2\ell^{-3}R^2 \\ &- \varepsilon^{-1}8\pi\ell^{-3}R^2\widehat{n}_0\widehat{n}_+ - \varepsilon\widehat{w}_{00,00}(\widehat{n}_0 + 1 - \rho\ell^3)^2 \\ &- \varepsilon\widehat{n}_+ r^{-1} + (\tfrac{1}{2} - \varepsilon)\sum_{p,m,p',m'\neq 0} \widehat{w}_{mp',pm'} a_m^* a_{p'}^* a_{m'} a_p.\end{aligned}$$

The assumptions on $\rho^{1/4}R$, $\rho^{1/4}\ell t^3$, and t are needed in order to bound the integral I above by a constant. If we choose $\varepsilon = 1/4$, use $\widehat{w}_{00,00} \leq 4\pi R^2\ell^{-3}$ and ignore the last positive term in the bound above we arrive at

$$\begin{aligned} H_\ell^n \geq & \tfrac{1}{4}\sum_p |p|^2 a_p^* a_p - C n^{5/4}\ell^{-3/4} - \tfrac{1}{2} n R^{-1} - C n \rho r^2 + \tfrac{1}{4}\widehat{w}_{00,00}\left(\widehat{n}_0 - \rho\ell^3\right)^2 \\ & - 4\pi[\rho - n\ell^{-3}]_+ \widehat{n}_+ R^2 - 4\pi \widehat{n}_+^2 \ell^{-3} R^2 \\ & - 32\pi\ell^{-3} R^2 \widehat{n}_0 \widehat{n}_+ - 4\pi R^2 \ell^{-3}\left(\widehat{n}_0 - \tfrac{1}{2}\rho\ell^3 + \tfrac{1}{4}\right) - \tfrac{1}{4}\widehat{n}_+ r^{-1} \\ \geq & \tfrac{1}{4}\sum_p |p|^2 a_p^* a_p - C n^{5/4}\ell^{-3/4} - \tfrac{1}{2} n R^{-1} - C n \rho r^2 + \tfrac{1}{4}\widehat{w}_{00,00}\left(\widehat{n}_0 - \rho\ell^3\right)^2 \\ & - 48\pi\ell^{-3} R^2 n \widehat{n}_+ - 4\pi R^2 \ell^{-3}\left(\widehat{n}_0 + \tfrac{1}{4}\right) - \tfrac{1}{4}\widehat{n}_+ r^{-1}, \end{aligned} \tag{32}$$

where in the last inequality we have used that $\rho\ell^3 \leq 2n$, $\widehat{n}_0 \leq n$ and $\widehat{n}_+ \leq n$.

Lemma 7.1 (Simple bound on n). *Let $\omega(t)$ be the function described in Lemma 3.1. There is a constant $C > 0$ such that if $(\rho^{1/4}\ell)t^3 > C$ and $(\rho^{1/4}\ell)\rho^{-1/12}$, t, and $\omega(t)(\rho^{1/4}\ell)^{-1}$ are smaller than C^{-1} then for any state with $\langle H_\ell^n\rangle \leq 0$ we have $C^{-1}\rho\ell^3 \leq n \leq C\rho\ell^3$.*

Proof. The lower bound follows from Corollary 4.3. To prove the upper bound on n we choose $R = \omega(t)^{-1}\ell$ (the maximally allowed value) and $r = b\omega(t)^{-1}\ell$, where we shall choose b sufficiently small, in particular $b < 1/2$. We then have that $\rho^{1/4}R = \omega(t)^{-1}\rho^{1/4}\ell$ is large. Moreover $\widehat{w}_{00,00} \geq CR^2\ell^{-3} = C\omega(t)^{-2}\ell^{-1}$ for some constant $C > 0$ and we get from (32) and Lemma 5.1 that

$$\begin{aligned} \langle H_\ell^n\rangle \geq \ell^{-1}[& -C n^{5/4}\ell^{1/4} - \tfrac{1}{2} n\omega(t) - C b^2 \omega(t)^{-2} n^2 + C\omega(t)^{-2}\left(\langle\widehat{n}_0\rangle - \rho\ell^3\right)^2 \\ & - 48\pi\omega(t)^{-2}\rho^{-1/6}(\ell\rho^{1/4})^2 n^2 - 4\pi\omega(t)^{-2}\left(n + \tfrac{1}{4}\right) - \tfrac{1}{4} n b^{-1}\omega(t)], \end{aligned}$$

where we have again used that $c\rho\ell^3 \leq n$, $\widehat{n}_0 \leq n$ and $\widehat{n}_+ \leq n$. Note that

$$n^{5/4}\ell^{1/4} \leq C\omega(t)^{-2} n^2 (\rho^{1/4}\ell)^{-2}\rho^{-1/4}\omega(t)^2$$

and $n\omega(t) \leq C\omega(t)^{-2} n^2 \rho^{-1}\omega(t)^3$. From Lemma 5.1 we know that $\langle\widehat{n}_0\rangle \geq n(1 - C\rho^{-1/6}(\ell\rho^{1/4})^2)$. By choosing b small enough we see immediately that $n \leq C\rho\ell^3$. □

Using this result as an input in (32) we can get a better bound on n than above and a better bound on $\langle\widehat{n}_+\rangle$ than given in Lemma 5.1. In particular, the next lemma in fact implies that we have near neutrality, i.e., that n is nearly $\rho\ell^3$.

Lemma 7.2 (Improved bounds on n and $\langle\widehat{n}_+\rangle$). *There exists a constant $C > 0$ such that if $(\rho^{1/4}\ell)t^3 > C$ and $(\rho^{1/4}\ell)\rho^{-1/12}$, t, and $\omega(t)(\rho^{1/4}\ell)^{-1}$ are smaller than C^{-1} then for any state with $\langle H_\ell^n\rangle \leq 0$ we have $\langle\sum_p |p|^2 a_p^* a_p\rangle \leq C\rho^{5/4}\ell^3(\rho^{1/4}\ell)$ and*

$$\langle\widehat{n}_+\rangle \leq C n \rho^{-1/4}(\rho^{1/4}\ell)^3 \quad \textit{and} \quad \left(\frac{n - \rho\ell^3}{\rho\ell^3}\right)^2 \leq C\rho^{-1/4}(\rho^{1/4}\ell)^3.$$

For any other state with $\langle H_{\ell,r',R'}^n\rangle' \leq 0$ we have the same bound on $\langle\widehat{n}_+\rangle'$ if $r' \leq \rho^{-3/8}(\rho^{1/4}\ell)^{1/2}$ and $R' \geq a(\rho^{1/4}\ell)^{-2}\ell$ where $a > 0$ is an appropriate constant.

Proof. Inserting the bound $n \leq C\rho\ell^3$ into (32) gives

$$H_\ell^n \geq \tfrac{1}{4}\sum_p |p|^2 a_p^* a_p - C\rho^{5/4}\ell^3 - \tfrac{1}{2}\rho\ell^3 R^{-1} - C\rho^2\ell^3 r^2 + \tfrac{1}{4}\widehat{w}_{00,00}\left(\widehat{n}_0 - \rho\ell^3\right)^2$$
$$- CR^2\rho\widehat{n}_+ - CR^2\left(\rho + \tfrac{1}{4}\ell^{-3}\right) - \tfrac{1}{4}\widehat{n}_+ r^{-1}.$$

We now choose $r = \rho^{-3/8}(\rho^{1/4}\ell)^{1/2}$ and $R = a(\rho^{1/4}\ell)^{-2}\ell$, where we shall choose a below, independently of ρ, $\rho^{1/4}\ell$, and t. Note that since $\omega(t)(\rho^{1/4}\ell)^{-2}$ is small we may assume that $R \leq \omega(t)^{-1}\ell$ as required and since $(\rho^{1/4}\ell)\rho^{-1/12}$ is small we may assume that $r \leq R$. Moreover $r^{-1} = \rho^{-1/8}(\rho^{1/4}\ell)^{3/2}\ell^{-2}$ and $R^2\rho = a^2(\rho^{1/4}\ell)^{-4}\ell^2\rho = a^2\ell^{-2}$. Hence, since $\sum_p |p|^2 a_p^* a_p \geq \pi^2\ell^{-2}\widehat{n}_+$ (see 15), we have

$$H_\ell^n \geq \tfrac{1}{8}\sum_p |p|^2 a_p^* a_p + \left(\tfrac{\pi^2}{8} - a^2 - \tfrac{1}{4}\rho^{-1/8}(\rho^{1/4}\ell)^{3/2}\right)\ell^{-2}\widehat{n}_+$$
$$+ \tfrac{1}{4}\widehat{w}_{00,00}\left(\widehat{n}_0 - \rho\ell^3\right)^2$$
$$- (\tfrac{1}{2a} + C)\rho^{5/4}\ell^3(\rho^{1/4}\ell) - Ca^2\rho^{5/4}\ell^3(\rho^{1/4}\ell)^{-5}(1 + (\rho^{1/4}\ell)^{-3}\rho^{-1/4}).$$

By choosing a appropriately (independently of ρ, $\rho^{1/4}\ell$, and t) we immediately get the bound on $\langle\sum_p |p|^2 a_p^* a_p\rangle$ and the bound $\ell^{-2}\langle\widehat{n}_+\rangle \leq C\rho^{5/4}\ell^3(\rho^{1/4}\ell)$, which implies the stated bound on $\langle\widehat{n}_+\rangle$. The bound on $(n - \rho\ell^3)^2(\rho\ell^3)^{-2}$ follows since we also have $\widehat{w}_{00,00}\langle(\widehat{n}_0 - \rho\ell^3)^2\rangle \leq C\rho^{5/4}\ell^3(\rho^{1/4}\ell)$ and

$$\widehat{w}_{00,00}\langle\left(\widehat{n}_0 - \rho\ell^3\right)^2\rangle \geq CR^2\ell^{-3}\left(\langle\widehat{n}_0\rangle - \rho\ell^3\right)^2$$
$$\geq Ca^2(\rho^{1/4}\ell)^{-4}\ell^2\left(n - \rho\ell^3 - nC\rho^{-1/4}(\ell\rho^{1/4})^3\right)^2,$$

where we have used the bound on $\langle\widehat{n}_+\rangle$ which we have just proved.

The case when $\langle H_{\ell,r',R'}^n\rangle' \leq 0$ follows in the same way because we may everywhere replace H_ℓ^n by $H_{\ell,r',R'}^n$ and use Lemma 4.2 instead of Lemma 4.1. Note that in this case we already know the bound on n since we still assume the existence of the state such that $\langle H_\ell^n\rangle \leq 0$. □

8. Localization of $\widehat{n}_+$

Note that Lemma 7.2 may be interpreted as saying that we have neutrality and condensation, in the sense that $\langle\widehat{n}_+\rangle$ is a small fraction of n, in each little box. Although this bound on $\langle\widehat{n}_+\rangle$ is sufficient for our purposes we still need to know that $\langle\widehat{n}_+^2\rangle \sim \langle\widehat{n}_+\rangle^2$. We shall however not prove this for a general state with negative energy. Instead we shall show that we may change the ground state, without changing its energy expectation significantly, in such a way that the possible $\widehat{n}_+$ values are bounded by $Cn\rho^{-1/4}(\rho^{1/4}\ell)^3$. To do this we shall use the method of localizing large matrices in Lemma A.1 of Appendix A.

We begin with any normalized n-particle wavefunction Ψ of the operator H_ℓ^n. Since Ψ is an n-particle wave function we may write $\Psi = \sum_{m=0}^n c_m\Psi_m$, where for all $m = 1, 2, \ldots, n$, Ψ_m, is a normalized eigenfunctions of $\widehat{n}_+$ with eigenvalue m. We may now consider the $(n+1)\times(n+1)$ Hermitean matrix $\mathcal{A}$ with matrix elements $\mathcal{A}_{mm'} = \left(\Psi_m, H^n_{\ell,r,R}\psi'_m\right)$.

We shall use Lemma A.1 for this matrix and the vector $\psi = (c_0, \dots, c_n)$. We shall choose M in Lemma A.1 to be of the order of the upper bound on $\langle \widehat{n}_+ \rangle$ derived in Lemma 7.2, e.g., M is the integer part of $n\rho^{-1/4}(\rho^{1/4}\ell)^3$. Recall that with the assumption in Lemma 7.2 we have $M \gg 1$. With the notation in Lemma A.1 we have $\lambda = (\psi, \mathcal{A}\psi) = (\Psi, H^n_{\ell,r,R}\Psi)$. Note also that because of the structure of $H^n_{\ell,r,R}$ we have, again with the notation in Lemma A.1, that $d_k = 0$ if $k > 3$. We conclude from Lemma A.1 that there exists a normalized wavefunction $\widetilde{\Psi}$ with the property that the corresponding $\widehat{n}_+$ values belong to an interval of length M and such that

$$\left(\Psi, H^n_{\ell,r,R}\Psi\right) \geq \left(\widetilde{\Psi}, H^n_{\ell,r,R}\widetilde{\Psi}\right) - CM^{-2}(|d_1| + |d_2|).$$

We shall discuss d_1, d_2, which depend on Ψ, in detail below, but first we give the result on the localization of $\widehat{n}_+$ that we shall use.

Lemma 8.1 (Localization of $\widehat{n}_+$). *There is a constant $C > 0$ with the following property. If $(\rho^{1/4}\ell)t^3 > C$ and $(\rho^{1/4}\ell)\rho^{-1/12}$, t, and $\omega(t)(\rho^{1/4}\ell)^{-1}$ are less than C^{-1} and $r \leq \rho^{3/8}(\rho^{1/4}\ell)^{1/2}$, $R \geq C(\rho^{1/4}\ell)^{-2}\ell$, and Ψ is a normalized wavefunction such that*

$$\left(\Psi, H^n_{\ell,r,R}\Psi\right) \leq 0 \quad \text{and} \quad \left(\Psi, H^n_{\ell,r,R}\Psi\right) \leq -C(n\rho^{-1/4}(\rho^{1/4}\ell)^3)^{-2}(|d_1| + |d_2|) \quad (33)$$

then there exists a normalized wave function $\widetilde{\Psi}$, which is a linear combination of eigenfunctions of $\widehat{n}_+$ with eigenvalues less than $Cn\rho^{-1/4}(\rho^{1/4}\ell)^3$ only, such that

$$\left(\Psi, H^n_{\ell,r,R}\Psi\right) \geq \left(\widetilde{\Psi}, H^n_{\ell,r,R}\widetilde{\Psi}\right) - C(n\rho^{-1/4}(\rho^{1/4}\ell)^3)^{-2}(|d_1| + |d_2|). \quad (34)$$

Here d_1 and d_2, depending on Ψ, are given as explained in Lemma A.1.

Proof. As explained above we choose M to be of order $n\rho^{-1/4}(\rho^{1/4}\ell)^3$. We then choose $\widetilde{\Psi}$ as explained above. Then (34) holds. We also know that the possible $\widehat{n}_+$ values of $\widetilde{\Psi}$ range in an interval of length M. We do not know however, where this interval is located. The assumption (33) will allow us to say more about the location of the interval.

In fact, it follows from (33), (34) that $\left(\widetilde{\Psi}, H^n_{\ell,r,R}\widetilde{\Psi}\right) \leq 0$. It is then a consequence of Lemma 7.2 that $\left(\widetilde{\Psi}, \widehat{n}_+\Psi\right) \leq Cn\rho^{-1/4}(\rho^{1/4}\ell)^3$. This of course establishes that the allowed $\widehat{n}_+$ values are less than $C'n\rho^{-1/4}(\rho^{1/4}\ell)^3$ for some constant $C' > 0$. □

Our final task in this section is to bound d_1 and d_2. We have that $d_1 = (\Psi, H^n_{\ell,r,R}(1)\psi)$, where $H^n_{\ell,r,R}(1)$ is the part of the Hamiltonian $H^n_{\ell,r,R}$ containing all the terms with the coefficents $\widehat{w}_{pq,\mu\nu}$ for which precisely one or three indices are 0. These are the terms bounded in Lemmas 5.5 and 5.6. These lemmas are stated as one-sided bounds. It is clear from the proof that they could have been stated as two sided bounds. Alternatively we may observe that $H^n_{\ell,r,R}(1)$ is unitarily equivalent to $-H^n_{\ell,r,R}(1)$. This follows by applying the unitary transform which maps all operators a^*_p and a_p with $p \neq 0$ to $-a^*_p$ and $-a_p$. From Lemmas 5.5 and 5.6 we therefore immediately get the following bound on d_1.

Lemma 8.2 (Control of d_1). *With the notation above we have for all $\varepsilon > 0$*

$$|d_1| \le \varepsilon^{-1} 8\pi \ell^{-3} R^2 (\Psi, \widehat{n}_0 \widehat{n}_+ \Psi) + \varepsilon \left(\Psi, \left(\widehat{n}_+ r^{-1} + \widehat{w}_{00,00}(\widehat{n}_0 + 1 - \rho\ell^3)^2\right)\Psi\right)$$
$$+ \varepsilon\left(\Psi, \sum_{p,m,p',m'\neq 0} \widehat{w}_{mp',pm'} a_m^* a_{p'}^* a_{m'} a_p \Psi\right).$$

Likewise, we have that $d_2 = (\Psi, H^n_{\ell,r,R}(2)\psi)$, where $H^n_{\ell,r,R}(2)$ is the part of the Hamiltonian $H^n_{\ell,r,R}$ containing all the terms with precisely two a_0 or two a_0^*. i.e., these are the terms in the Foldy Hamiltonian, which do not commute with $\widehat{n}_+$.

Lemma 8.3 (Control of d_2). *There exists a constant $C > 0$ such that if $(\rho^{1/4}\ell)t^3 > C$ and $(\rho^{1/4}\ell)\rho^{-1/12}$, t, and $\omega(t)(\rho^{1/4}\ell)^{-1}$ are less than C^{-1} and Ψ is a wave function with $(\Psi, H^n_\ell \Psi) \le 0$ then with the notation above we have*

$$|d_2| \le C\rho^{5/4}\ell^3(\rho^{1/4}\ell) + 4\pi\ell^{-3}R^2 (\Psi, \widehat{n}_+\widehat{n}_0\Psi).$$

Proof. If we replace all the operators a_p^* and a_p with $p \neq 0$ in the Foldy Hamiltonian by $-ia_p^*$ and ia_p we get a unitarily equivalent operator. This operator however differs from the Hamiltonian H_{Foldy} only by a change of sign on the part that we denoted $H^n_{\ell,r,R}(2)$. Since both operators satisfy the bound in Corollary 6.5 we conclude that

$$|d_2| \le \left(\Psi, \left[\tfrac{1}{2}\gamma^{-1}\sum_p |p|^2 a_p^* a_p + \tfrac{1}{2}\sum_{pq\neq 0} \widehat{w}_{pq,00}\left(a_p^* a_0^* a_0 a_q + a_0^* a_p^* a_q a_0\right)\right]\Psi\right)$$
$$+ Cn^{5/4}\ell^{-3/4}.$$

Note that both sums above define positive operators. This is trivial for the first sum. For the second it follows from (18) in Lemma 5.4 since $a_0^* a_0$ commutes with all a_p^* and a_p with $p \neq 0$. The lemma now follows from (18) and from Lemma 7.2. □

9. Proof of Foldy's Law

We first prove Foldy's law in a small cube. Let Ψ be a normalized n-particle wave function. We shall prove that with an appropriate choice of ℓ

$$(\Psi, H^n_\ell \Psi) \ge \left(\tfrac{4\pi}{3}\right)^{1/3} A\rho\ell^3 \left(\rho^{1/4} + o\left(\rho^{1/4}\right)\right), \tag{35}$$

where A is given in (6). Note that $A < 0$. It then follows from Lemma 3.3 that

$$E_0 \ge (1 + L/\ell)^3 \gamma \left(\tfrac{4\pi}{3}\right)^{1/3} A\rho\ell^3 \left(\rho^{1/4} + o\left(\rho^{1/4}\right)\right) - C(L/\ell)^2\rho^2\ell^5 - \frac{\omega(t)N}{2\ell}.$$

Thus, since $N = \rho L^3$ we have

$$\lim_{L\to\infty} \frac{E_0}{N} \ge \gamma \left(\tfrac{4\pi}{3}\right)^{1/3} A \left(\rho^{1/4} + o\left(\rho^{1/4}\right)\right) - C\rho^{1/4}\omega(t)\left(\rho^{1/4}\ell\right)^{-1}.$$

Foldy's law (5) follows since we shall choose (see below) t and ℓ in such a way that as $\rho \to \infty$ we have $t \to 0$ and hence $\gamma \to 1$ and $\omega(t)(\rho^{1/4}\ell)^{-1} \to 0$ (see condition (41) below).

It remains to prove (35). First we fix the long and short distance potential cutoffs

$$R = \omega(t)^{-1}\ell, \quad \text{and} \quad r = \rho^{-3/8}(\rho^{1/4}\ell)^{-1/2}. \tag{36}$$

We may of course assume that $\left(\Psi, H_\ell^n \Psi\right) \leq 0$. Thus n satisfies the bound in Lemma 7.2. We proceed in two steps. In Lemma 9.1 Foldy's law in the small boxes is proved under the restrictive assumption given in (37) below. Finally, in Theorem 9.2 Foldy's law in the small boxes is proved by considering the alternative case that (37) fails. Let us note that, logically speaking, this could have been done in the reverse order. I.e., we could, instead, have begun with the case that (37) fails. At the end of the section we combine Theorem 9.2 with Lemma 3.3 to show that Foldy's law in the small box implies Foldy's law Theorem 1.1.

At the end of this section we show how to choose ℓ and t so that Theorem 9.2 implies (35) and hence Theorem 1.1, as explained above.

Lemma 9.1 (Foldy's law for H_ℓ^n: restricted version). *Let R and r be given by (36). There exists a constant $C > 0$ such that if $(\rho^{1/4}\ell)t^3 > C$ and $(\rho^{1/4}\ell)\rho^{-1/12}$, t, and $\omega(t)(\rho^{1/4}\ell)^{-1}$ are less than C^{-1} then, whenever*

$$\begin{aligned} & n\ell^{-3}R^2\,(\Psi, \widehat{n}_+\Psi) \\ & \leq C^{-1}\left(\Psi, \left(\widehat{w}_{00,00}(\widehat{n}_0 - \rho\ell^3)^2 + \sum_{p,m,p',m'\neq 0} \widehat{w}_{mp',pm'} a_m^* a_{p'}^* a_{m'} a_p\right)\Psi\right), \end{aligned} \tag{37}$$

we have that

$$\begin{aligned} \left(\Psi, H_\ell^n \Psi\right) \geq -In^{5/4}\ell^{-3/4} - C\rho^{5/4}\ell^3 \Big(&\omega(t)(\rho^{1/4}\ell)^{-1} + \omega(t)^{-2}\rho^{-1/8}(\rho^{1/4}\ell)^{13/2} \\ &+ +\rho^{-1/8}(\rho^{1/4}\ell)^{7/2}\Big), \end{aligned}$$

with I as in Lemma 6.4.

Proof. We assume $\left(\Psi, H_\ell^n \Psi\right) \leq 0$. We proceed as in the beginning of Sect. 7, but we now use (29) of Corollary 6.5 instead of (30). We then get

$$\begin{aligned} H_\ell^n \geq & -In^{5/4}\ell^{-3/4} - 4\pi n^{5/4}\ell^{-3/4}(n\ell)^{-1/4} - \tfrac{1}{2}nR^{-1} - Cn\rho r^2 \\ & + \tfrac{1}{2}\widehat{w}_{00,00}\left[\left(\widehat{n}_0 - \rho\ell^3\right)^2 - \widehat{n}_0\right] \\ & - 4\pi[\rho - n\ell^{-3}]_+\widehat{n}_+R^2 - 4\pi\widehat{n}_+^2\ell^{-3}R^2 \\ & - \varepsilon^{-1}8\pi\ell^{-3}R^2\widehat{n}_0\widehat{n}_+ - \varepsilon\widehat{w}_{00,00}(\widehat{n}_0 + 1 - \rho\ell^3)^2 \\ & - \varepsilon\widehat{n}_+r^{-1} + (\tfrac{1}{2} - \varepsilon)\sum_{p,m,p',m'\neq 0} \widehat{w}_{mp',pm'} a_m^* a_{p'}^* a_{m'} a_p. \end{aligned}$$

If we now use the assumption (37) and the facts that $\widehat{n}_+ \leq n$, $\widehat{n}_0 \leq n$, and $\widehat{w}_{00,00} \leq 4\pi R^2\ell^{-3}$ we see with appropriate choices of ε and C that

$$\begin{aligned} H_\ell^n \geq & -In^{5/4}\ell^{-3/4} - 4\pi n^{5/4}\ell^{-3/4}(n\ell)^{-1/4} - \tfrac{1}{2}nR^{-1} - Cn\rho r^2 - CR^2\ell^{-3}(n+1) \\ & - CR^2\ell^{-3}|n - \rho\ell^3|(\widehat{n}_+ + 1) - C\widehat{n}_+r^{-1}. \end{aligned}$$

If we finally insert the choices of R and r and use Lemma 7.2 we arrive at the bound in the lemma. □

Theorem 9.2 (Foldy's law for H_ℓ^n). *There exists a $C > 0$ such that if $(\rho^{1/4}\ell)t^3 > C$ and $(\rho^{1/4}\ell)\rho^{-1/12}$, t, and $\omega(t)(\rho^{1/4}\ell)^{-1}$ are less than C^{-1} then for any normalized n-particle wave function Ψ we have*

$$(\Psi, H_\ell^n\Psi) \geq -In^{5/4}\ell^{-3/4} - C\rho^{5/4}\ell^3\Big(\omega(t)(\rho^{1/4}\ell)^{-1} + \omega(t)^{-1}\rho^{-1/16}(\rho^{1/4}\ell)^{29/4} + \rho^{-1/8}(\rho^{1/4}\ell)^{7/2}\Big), \quad (38)$$

where I is defined in Lemma 6.4 with r and R as in (36).

Proof. According to Lemma 9.1 we may assume that

$$n\ell^{-3}R^2(\Psi, \widehat{n}_+\Psi) \geq C^{-1}\Big(\Psi, \Big(\widehat{w}_{00,00}(\widehat{n}_0 - \rho\ell^3)^2 + \sum_{p,m,p',m'\neq 0} \widehat{w}_{mp',pm'}a_m^*a_{p'}^*a_{m'}a_p\Big)\Psi\Big), \quad (39)$$

where C is at least as big as the constant in Lemma 9.1. We still assume that $(\Psi, H_\ell^n\Psi) \leq 0$.

We begin by bounding d_1 and d_2 using Lemmas 8.2 and 8.3. We have from Lemmas 7.2 and 8.3 that

$$\begin{aligned}|d_2| &\leq C\rho^{5/4}\ell^3(\rho^{1/4}\ell) + C\ell^{-1}\omega(t)^{-2}n^2\rho^{-1/4}(\rho^{1/4}\ell)^3\\ &\leq C[n\rho^{-1/4}(\rho^{1/4}\ell)^3]^2\rho^{5/4}\ell^3\left((\rho^{1/4}\ell)^{-11} + \omega(t)^{-2}(\rho^{1/4}\ell)^{-7}\right)\\ &\leq C[n\rho^{-1/4}(\rho^{1/4}\ell)^3]^2\rho^{5/4}\ell^3\omega(t)^{-2}(\rho^{1/4}\ell)^{-7}.\end{aligned}$$

In order to bound d_1 we shall use (39). Together with Lemma 8.2 this gives (choosing $\varepsilon = 1/2$ say)

$$|d_1| \leq C\ell^{-3}R^2n(\Psi, \widehat{n}_+\Psi) + \tfrac{1}{2}\left(\Psi, \left(\widehat{n}_+r^{-1} + \widehat{w}_{00,00}(n - \rho\ell^3 + 1)\right)\Psi\right).$$

Inserting the choices for r and R and using Lemma 7.2 gives

$$|d_1| \leq C[n\rho^{-1/4}(\rho^{1/4}\ell)^3]^2\rho^{5/4}\ell^3\left(\omega(t)^{-2}(\rho^{1/4}\ell)^{-7} + \rho^{-1/8}(\rho^{1/4}\ell)^{-17/2}\right),$$

where we have also used that we may assume that $\rho^{-1/8}(\rho^{1/4}\ell)^{-9/2}$ is small. The assumption (33) now reads

$$(\Psi, H_{\ell,r,R}^n\Psi) \leq -C\rho^{5/4}\ell^3\left(\omega(t)^{-2}(\rho^{1/4}\ell)^{-7} + \rho^{-1/8}(\rho^{1/4}\ell)^{-17/2}\right).$$

If this is not satisfied we see immediately that the bound (38) holds.

Thus from Lemma 8.1 it follows that we can find a normalized n-particle wavefunction $\widetilde{\Psi}$ with

$$(\widetilde{\Psi}, \widehat{n}_+\widetilde{\Psi}) \leq Cn\rho^{-1/4}(\rho^{1/4}\ell)^3 \quad \text{and} \quad \left(\widetilde{\Psi}, \widehat{n}_+^2\widetilde{\Psi}\right) \leq Cn^2\rho^{-1/2}(\rho^{1/4}\ell)^6 \quad (40)$$

such that

$$(\Psi, H_{\ell,r,R}^n\Psi) \geq (\widetilde{\Psi}, H_{\ell,r,R}^n\widetilde{\Psi}) - C\rho^{5/4}\ell^3\left(\omega(t)^{-2}(\rho^{1/4}\ell)^{-7} + \rho^{-1/8}(\rho^{1/4}\ell)^{-17/2}\right).$$

In order to analyze $\left(\widetilde{\Psi}, H^n_{\ell,r,R}\widetilde{\Psi}\right)$ we proceed as in the beginning of Sect. 7. This time we use Lemmas 4.1, 5.2, 5.3, 5.5, and 5.6 together with Lemma 6.4 instead of Corollary 6.5. We obtain

$$\begin{aligned}
H^n_{\ell,r,R} \geq \tfrac{1}{2}\widehat{w}_{00,00}&\left[\left(n-\rho\ell^3\right)^2 + (\widehat{n}_+)^2 - 2\left(n-\rho\ell^3\right)\widehat{n}_+ - \widehat{n}_0\right]\\
&- 4\pi[\rho - n\ell^{-3}]_+\widehat{n}_+R^2 - 4\pi\widehat{n}_+^2\ell^{-3}R^2 - \varepsilon\widehat{n}_+r^{-1} - \varepsilon^{-1}8\pi\ell^{-3}R^2\widehat{n}_0\widehat{n}_+\\
&- \varepsilon\widehat{w}_{00,00}(\widehat{n}_+ - 1)^2 + (\tfrac{1}{2} - \varepsilon)\sum_{p,m,p',m'\neq 0}\widehat{w}_{mp',pm'}a^*_m a^*_{p'} a_{m'} a_p\\
&- \tfrac{1}{2}\left(n-\rho\ell^3\right)^2\widehat{w}_{00,00} - 4\pi n^{5/4}\ell^{-3/4}(n\ell)^{-1/4} - In^{5/4}\ell^{-3/4}.
\end{aligned}$$

This time we shall however not choose ε small, but rather big. Note that since $w_{r,R}(x,y) \leq r^{-1}$ we have $\sum_{p,m,p',m'\neq 0}\widehat{w}_{mp',pm'}a^*_m a^*_{p'} a_{m'} a_p \leq r^{-1}\widehat{n}_+(\widehat{n}_+ - 1)$, which follows immediately from

$$\begin{aligned}
&\sum_{p,m,p',m'\neq 0}\widehat{w}_{mp',pm'}a^*_m a^*_{p'} a_{m'} a_p\\
&\quad = \iint w_{r,R}(x,y)\Big(\sum_{p,m\neq 0}u_m(x)u_p(y)a_m a_p\Big)^*\sum_{p,m\neq 0}u_m(x)u_p(y)a_m a_p\,dx\,dy.
\end{aligned}$$

We therefore have

$$\begin{aligned}
H^n_{\ell,r,R} \geq &- In^{5/4}\ell^{-3/4} - 4\pi n^{5/4}\ell^{-3/4}(n\ell)^{-1/4} - CR^2\ell^{-3}\widehat{n}_0\\
&- C\ell^{-3}R^2|\rho\ell^3 - n|\widehat{n}_+ - 4\pi\widehat{n}_+^2\ell^{-3}R^2 - \varepsilon\widehat{n}_+r^{-1} - \varepsilon^{-1}8\pi\ell^{-3}R^2\widehat{n}_0\widehat{n}_+\\
&- \varepsilon CR^2\ell^{-3}\widehat{n}_+^2 - \varepsilon\widehat{n}_+^2r^{-1}.
\end{aligned}$$

If we now insert the choices of r and R, take the expectation in the state given by $\widetilde{\Psi}$, and use (40) and the bound on n from Lemma 7.2 we arrive at

$$\begin{aligned}
(\widetilde{\Psi}, H^n_{\ell,r,R}\widetilde{\Psi}) \geq &- In^{5/4}\ell^{-3/4} - C\rho^{5/4}\ell^3\Big[(\rho^{1/4}\ell)^{-1} + \omega(t)^{-2}(\rho^{1/4}\ell)^{-1}\\
&+ \omega(t)^{-2}\rho^{-1/8}(\rho^{1/4}\ell)^{11/2} + \omega(t)^{-2}\rho^{-1/4}(\rho^{1/4}\ell)^8 + \varepsilon\rho^{-1/8}(\rho^{1/4}\ell)^{7/2}\\
&+ \varepsilon^{-1}\omega(t)^{-2}(\rho^{1/4}\ell)^5 + \varepsilon\omega(t)^{-2}\rho^{-1/4}(\rho^{1/4}\ell)^8 + \varepsilon\rho^{-1/8}(\rho^{1/4}\ell)^{19/2}\Big].
\end{aligned}$$

If we now choose $\varepsilon = \omega(t)^{-1}\rho^{1/16}(\rho^{1/4}\ell)^{-9/4}$ we arrive at (38). □

Completion of the proof of Foldy's law, Theorem 1.1. We have accumulated various errors and we want to show that they can all be made small. There are basically two parameters that can be adjusted, ℓ and t. Instead of ℓ it is convenient to use $X = \rho^{1/4}\ell$. We shall choose X as a function of ρ such that $X \to \infty$ as $\rho \to \infty$. From Lemma 7.1 we know that for some fixed $C > 0$ $C^{-1}\rho\ell^3 \leq n \leq C\rho\ell^3$. Hence according to (31) with r and R

given in (36) we have that $I \to -\left(\frac{4\pi}{3}\right)^{1/3} A$ as $\rho \to \infty$ if

$$\omega(t)^{-1}X \to \infty, \tag{41}$$

$$\rho^{1/4}X \to \infty, \tag{42}$$

$$t^3 X \to \infty, \tag{43}$$

$$t \to 0. \tag{44}$$

The hypotheses of Theorem 9.2 are valid if (41), (43), (44), and

$$\rho^{-1/12}X \to 0 \tag{45}$$

hold. From Lemma 7.2, for which the hypotheses are now automatically satisfied, we have that $n = \rho\ell^3(1 + O(\rho^{-1/8}X^{3/2})$ and from (45) we see that n is $\rho\ell^3$ to leading order.

With these conditions we find that the first term on the right side of (38) is, in the limit $\rho \to \infty$, exactly Foldy's law. The conditions that the other terms in (38) are of lower order are

$$(X/\omega(t))^{4/25}\rho^{-1/100}X \to 0, \tag{46}$$

$$\rho^{-1/28}X \to 0 \tag{47}$$

together with (41).

It remains to show that we can satisfy the conditions (41–47). Condition (42) is trivially satisfied since both ρ and X tend to infinity. Since $\omega(t) \sim t^{-4}$ for small t we see that (43) is implied by (41). Condition (45) is implied by (47), which is in turn implied by (41) and (46). The remaining two conditions (41) and (46) are easily satisfied by an approriate choice of X and t as functions for ρ with $X \to \infty$ and $t \to 0$ as $\rho \to \infty$. In fact, we simply need $\rho^{1/116}t^{-16/29} \gg X \gg t^{-4}$.

The bound (35) has now been established. Hence Foldy's law Theorem 1.1 follows as discussed in the beginning of the section.

Appendix

A. Localization of Large Matrices

The following theorem allows us to reduce a big Hermitean matrix, $\mathcal{A}$, to a smaller principal submatrix without changing the lowest eigenvalue very much. (The kth supra- (resp. infra-) diagonal of a matrix $\mathcal{A}$ is the submatrix consisting of all elements $a_{i,i+k}$ (resp. $a_{i+k,i}$).)

Theorem A.1 (Localization of large matrices). *Suppose that $\mathcal{A}$ is an $N \times N$ Hermitean matrix and let $\mathcal{A}^k$, with $k = 0, 1, \ldots, N-1$, denote the matrix consisting of the kth supra- and infra-diagonal of $\mathcal{A}$. Let $\psi \in \mathbf{C}^N$ be a normalized vector and set $d_k = (\psi, \mathcal{A}^k\psi)$ and $\lambda = (\psi, \mathcal{A}\psi) = \sum_{k=0}^{N-1} d_k$. ($\psi$ need not be an eigenvector of $\mathcal{A}$.)*

Choose some positive integer $M \le N$. Then, with M fixed, there is some $n \in [0, N - M]$ and some normalized vector $\phi \in \mathbf{C}^N$ with the property that $\phi_j = 0$ unless $n + 1 \le j \le n + M$ (i.e., ϕ has length M) and such that

$$(\phi, \mathcal{A}\phi) \le \lambda + \frac{C}{M^2}\sum_{k=1}^{M-1} k^2|d_k| + C\sum_{k=M}^{N-1}|d_k|, \tag{48}$$

where $C > 0$ is a universal constant. (Note that the first sum starts with $k = 1$.)

Proof. It is convenient to extend the matrix $\mathcal{A}_{i,j}$ to all $-\infty < i, j < +\infty$ by defining $\mathcal{A}_{i,j} = 0$ unless $1 \leq i, j \leq N$. Similarly, we extend the vector ψ and we define the numbers d_k and the matrix $\mathcal{A}^k$ to be zero when $k \notin [0, N-1]$. We shall give the construction for M odd, the M even case being similar.

For $s \in \mathbf{Z}$ set $f(s) = A_M[M + 1 - 2|s|]$ if $2|s| < M$ and $f(s) = 0$ otherwise. Thus, $f(s) \neq 0$ for precisely M values of s. Also, $f(s) = f(-s)$. A_M is chosen so that $\sum_s f(s)^2 = 1$.

For each $m \in \mathbf{Z}$ define the vector $\phi^{(m)}$ by $\phi_j^{(m)} = f(j-m)\psi_j$. We then define $K^{(m)} = (\phi^{(m)}, \mathcal{A}\phi^{(m)}) - (\lambda + \sigma)(\phi^{(m)}, \phi^{(m)})$. (The number σ will be chosen later.) After this, we define $K = \sum_m K^{(m)}$. Using the fact that $\sum_s f(s)^2 = 1$, we have that

$$\sum_m (\phi^{(m)}, \mathcal{A}\phi^{(m)}) = \sum_m \sum_{k=0} (\phi^{(m)}, \mathcal{A}^k\phi^{(m)}) = \sum_s \sum_k f(s)f(k+s)(\psi, \mathcal{A}^k\psi)$$
$$= \sum_s \sum_{k=0} f(s)f(k+s)d_k$$

and

$$\lambda = \lambda \sum_m (\phi^{(m)}, \phi^{(m)}) = \sum_s \sum_{k=0} f(s)^2(\psi, \mathcal{A}^k\psi) = \sum_s \sum_k f(s)^2 d_k \tag{49}$$

Hence

$$K = \sum_m K^{(m)} = -\sigma - \sum_{k=1}^{N-1} d_k\gamma_k \tag{50}$$

with

$$\gamma_k = \frac{1}{2}\sum_s [f(s) - f(s+k)]^2 . \tag{51}$$

Let us choose $\sigma = -\sum_{k=1}^{N-1} d_k\gamma_k$. Then, $\sum_m K^{(m)} = 0$. Recalling that not all of the $\phi^{(m)}$ equal zero, we conclude that there is at least one value of m such that (i) $\phi^{(m)} \neq 0$ and (ii) $(\phi^{(m)}, \mathcal{A}\phi^{(m)}) \leq (\lambda + \sigma)(\phi^{(m)}, \phi^{(m)})$.

This concludes the proof of (48) except for showing that $\gamma_k \leq C\frac{k^2}{k^2+M^2}$ for all M and k. This is evident from the easily computable large M asymptotics in (51). □

B. A Double Commutator Bound

Lemma B.1. *Let $-\Delta_N$ be the Neumann Laplacian of some bounded open set $\mathcal{O}$. Given $\theta \in C^\infty(\overline{\mathcal{O}})$ with* supp $|\nabla\theta| \subset \mathcal{O}$ *satisfying $\|\partial_i\theta\| \leq Ct^{-1}$, $\|\partial_i\partial_j\theta\| \leq Ct^{-2}$, $\|\partial_i\partial_j\partial_k\theta\| \leq Ct^{-3}$, for some $0 < t$ and all $i, j, k = 1, 2, 3$. Then for all $s > 0$ we have the operator inequality*

$$\left[\left[\frac{(-\Delta_N)^2}{-\Delta_N + s^{-2}}, \theta\right], \theta\right] \geq -Ct^{-2}\frac{-\Delta_N}{-\Delta_N + s^{-2}} - Cs^2t^{-4}. \tag{52}$$

We also have the norm bound

$$\left\|\left[\left[\frac{-\Delta_N}{-\Delta_N + s^{-2}}, \theta\right], \theta\right]\right\| \leq C(s^2t^{-2} + s^4t^{-4}). \tag{53}$$

Proof. We calculate the commutator

$$\left[\frac{(-\Delta_N)^2}{-\Delta_N+s^{-2}},\theta\right]=s^{-2}\frac{1}{-\Delta_N+s^{-2}}[-\Delta_N,\theta]\frac{1}{-\Delta_N+s^{-2}}(-\Delta_N)$$
$$+\frac{-\Delta_N}{-\Delta_N+s^{-2}}[-\Delta_N,\theta].$$

Likewise we calculate the double commutator

$$\left[\left[\frac{(-\Delta_N)^2}{-\Delta_N+s^{-2}},\theta\right],\theta\right]=-\frac{-\Delta_N}{-\Delta_N+s^{-2}}[[-\Delta_N,\theta]\theta]\frac{-\Delta_N}{-\Delta_N+s^{-2}}$$
$$+[[-\Delta_N,\theta]\theta]\frac{-\Delta_N}{-\Delta_N+s^{-2}}+\frac{-\Delta_N}{-\Delta_N+s^{-2}}[[-\Delta_N,\theta]\theta] \tag{54}$$
$$-2s^{-4}\frac{1}{-\Delta_N+s^{-2}}[-\Delta_N,\theta]\frac{1}{-\Delta_N+s^{-2}}[\theta,-\Delta_N]\frac{1}{-\Delta_N+s^{-2}}.$$

Note that $[[-\Delta_N,\theta]\theta]=-2(\nabla\theta)^2$ and thus the first term above is positive.

We claim that

$$[-\Delta_N,\theta][\theta,-\Delta_N]\leq -Ct^{-2}\Delta_N+Ct^{-4}. \tag{55}$$

To see this we simply calculate

$$[-\Delta_N,\theta][\theta,-\Delta_N]=-\sum_{i,j}^{3}\left(4\partial_i(\partial_i\theta)(\partial_j\theta)\partial_j+(\partial_i^2\theta)(\partial_j^2\theta)+2(\partial_i\theta)(\partial_i\partial_j^2\theta)\right)$$

The last two terms are bounded by Ct^{-4}. For the first term we have by the Cauchy-Schwarz inequality for operators, $BA^*+AB^*\leq\varepsilon^{-1}AA^*+\varepsilon BB^*$, for all $\varepsilon>0$, that

$$-\sum_{i,j}^{3}\partial_i(\partial_i\theta)(\partial_j\theta)\partial_j=\sum_{i,j}^{3}(\partial_i(\partial_i\theta))\left(\partial_j(\partial_j\theta)\right)^*\leq -3\sum_{i}^{3}\partial_i(\partial_i\theta)(\partial_i\theta)\partial_i$$

and this is bounded above by $-3t^{-2}\Delta_N$ and we get (55). Inserting (55) into (54), recalling that the first term is positive, we obtain

$$\left[\left[\frac{(-\Delta_N)^2}{-\Delta_N+s^{-2}},\theta\right],\theta\right]\geq -2(\nabla\theta)^2\frac{-\Delta_N}{-\Delta_N+s^{-2}}-2\frac{-\Delta_N}{-\Delta_N+s^{-2}}(\nabla\theta)^2$$
$$-Ct^{-2}\frac{-\Delta_N}{-\Delta_N+s^{-2}}-Cs^2t^{-4}.$$

Again using the Cauchy–Schwarz inequality, we have

$$2(\nabla\theta)^2\frac{-\Delta_N}{-\Delta_N+s^{-2}}+2\frac{-\Delta_N}{-\Delta_N+s^{-2}}(\nabla\theta)^2$$
$$\leq 2t^{-2}\left(\frac{-\Delta_N}{-\Delta_N+s^{-2}}\right)^{1/2}(\nabla\theta)^4\left(\frac{-\Delta_N}{-\Delta_N+s^{-2}}\right)^{1/2}+2t^{-2}\left(\frac{-\Delta_N}{-\Delta_N+s^{-2}}\right)$$
$$\leq Ct^{-2}\frac{-\Delta_N}{-\Delta_N+s^{-2}},$$

and (52) follows.

The bound (53) is proved in the same way. Indeed,

$$\left[\left[\frac{-\Delta_N}{-\Delta_N + s^{-2}}, \theta\right], \theta\right] = -s^{-2} \frac{1}{-\Delta_N + s^{-2}} [[-\Delta_N, \theta], \theta] \frac{1}{-\Delta_N + s^{-2}}$$
$$+ 2s^{-2} \frac{1}{-\Delta_N + s^{-2}} [-\Delta_N, \theta] \frac{1}{-\Delta_N + s^{-2}} [\theta - \Delta_N] \frac{1}{-\Delta_N + s^{-2}},$$

and (53) follows from $[[-\Delta_N, \theta]\theta] = -2(\nabla\theta)^2$ and (55). □

References

[B] Bogolubov, N.N.: J. Phys. (U.S.S.R.) **11**, 23 (1947); Bogolubov, N.N. and Zubarev, D.N.: Sov. Phys. JETP **1**, 83 (1955)

[CLY] Conlon, J.G.. Lieb, E.H. and Yau, H-T.: The $N^{7/5}$ law for charged bosons. Commun. Math. Phys. **116**, 417–448 (1988)

[D] Dyson, F.J.: Ground-state energy of a finite system of charged particles. J. Math. Phys. **8**, 1538–1545 (1967)

[F] Foldy, L.L.: Charged boson gas. Phys. Rev. **124**, 649–651 (1961); Errata. ibid **125**, 2208 (1962)

[GM] Girardeau, M.: Ground state of the charged Bose gas. Phys. Rev. **127**, 1809–1818 (1962)

[GA] Girardeau, M. and Arnowitt, R.: Theory of many-boson systems: Pair theory. Phys.Rev. **113**, 755–761 (1959)

[GG] Graf, G.M.: Stability of matter through an electrostatic inequality. Helv. Phys. Acta **70**, 72–79 (1997)

[KLS] Kennedy, T., Lieb, E.H. and Shastry, S.: The XY model has long-range order for all spins and all dimensions greater than one. Phys. Rev. Lett. **61**, 2582–2585 (1988)

[L] Lieb, E.H.: The Bose fluid. In: *Lecture Notes in Theoretical Physics VIIC*, edited by W.E. Brittin. Univ. of Colorado Press, 1964, pp. 175–224

[LS] Lieb, E.H. and Sakakura, A.Y.: Simplified approach to the ground state energy of an imperfect Bose gas II. Charged Bose gas at high density. Phys. Rev. A **133**, 899–906 (1964)

[LLe] Lieb, E.H. and Lebowitz, J.L.: The constitution of matter: Existence of thermodynamics for systems composed of electrons and nuclei. Adv. in Math. **9**, 316–398 (1972)

[LLi] Lieb, E.H. and Liniger, W.: Exact analysis of an interacting Bose gas I. The general solution and the ground state. Phys. Rev. **130**, 1605–1616 (1963). See Fig. 3

[LN] Lieb, E.H. and Narnhofer, H.: The thermodynamic limit for jellium. J. Stat. Phys. **12**, 295–310 (1975); Errata. **14**, 465 (1976)

[LY] Lieb, E.H. and Yngvason, J.: Ground state energy of the low density Bose gas. Phys. Rev. Lett. **80**, 2504–2507 (1998)

Communicated by M. Aizenman

Correction to "Ground State Energy of the One-Component Charged Bose Gas"

Elliott H. Lieb and Jan Philip Solovej

The proof of Lemma B.1 of [1] contains an unjustified operator inequality. In the last estimate on p. 162 the Cauchy-Schwarz inequality was used incorrectly. The lemma is however still correct as stated. We shall show this below. The operator inequality to be proven is that

$$(\nabla\theta)^2\frac{-\Delta_N}{-\Delta_N+s^{-2}}+\frac{-\Delta_N}{-\Delta_N+s^{-2}}(\nabla\theta)^2\leq Ct^{-2}\frac{-\Delta_N}{-\Delta_N+s^{-2}}+Cs^2t^{-4}, \tag{1}$$

where $-\Delta_N$ is the Neumann Laplacian of some bounded open set $\mathcal{O}\subset\mathbb{R}^n$, $s>0$, and $\theta\in C^\infty(\overline{\mathcal{O}})$ is constant near the boundary of $\mathcal{O}$ and satisfies the estimates $\|\partial^\alpha\theta\|_\infty\leq Ct^{-|\alpha|}$, for some $t>0$ and all multi-indices α with $|\alpha|\leq 3$.

The proof of (1) is a little technical. For the application in the paper the following estimate, in which the Cauchy-Schwarz inequality has been used correctly, would have sufficed.

$$\begin{aligned}(\nabla\theta)^2\frac{-\Delta_N}{-\Delta_N+s^{-2}}+\frac{-\Delta_N}{-\Delta_N+s^{-2}}(\nabla\theta)^2 &\leq st(\nabla\theta)^4+(st)^{-1}\left(\frac{-\Delta_N}{-\Delta_N+s^{-2}}\right)^2\\ &\leq Cst^{-3}+C(st)^{-1}\frac{-\Delta_N}{-\Delta_N+s^{-2}}.\end{aligned}$$

In order to prove (1) we shall use the two operator inequalities

$$[-\Delta_N,f][f,-\Delta_N]\leq C\|\nabla f\|_\infty^2(-\Delta_N)+C\left(\sum_i\|\partial_i^2 f\|_\infty\right)^2 \tag{2}$$

and

$$\begin{aligned}f(-\Delta_N)f &= -\sum_i\partial_i f^2\partial_i+\sum_i[\partial_i f,f\partial_i]\leq -C\sum_i\partial_i f^2\partial_i+C\sum_i(\partial_i f)^2\\ &\leq C\|f\|_\infty^2(-\Delta_N)+C\|\nabla f\|^2,\end{aligned} \tag{3}$$

where f is a smooth function with compact support in $\mathcal{O}$, which we identify as a multiplication operator.

We begin by rewriting the left side of (1).

$$\begin{aligned}&(\nabla\theta)^2\frac{-\Delta_N}{-\Delta_N+s^{-2}}+\frac{-\Delta_N}{-\Delta_N+s^{-2}}(\nabla\theta)^2\\ &=\int_0^\infty(\nabla\theta)^2\frac{-\Delta_N}{(-\Delta_N+s^{-2}+u)^2}+\frac{-\Delta_N}{(-\Delta_N+s^{-2}+u)^2}(\nabla\theta)^2du\\ &=\int_0^\infty\Big(\frac{1}{-\Delta_N+s^{-2}+u}(\nabla\theta)^2\frac{-\Delta_N}{-\Delta_N+s^{-2}+u}+\frac{-\Delta_N}{-\Delta_N+s^{-2}+u}(\nabla\theta)^2\frac{1}{-\Delta_N+s^{-2}+u}\Big)du\\ &\quad+\int_0^\infty\Big(\frac{1}{-\Delta_N+s^{-2}+u}[-\Delta_N,(\nabla\theta)^2]\frac{-\Delta_N}{(-\Delta_N+s^{-2}+u)^2}\\ &\quad+\frac{-\Delta_N}{(-\Delta_N+s^{-2}+u)^2}[(\nabla\theta)^2,-\Delta_N]\frac{1}{-\Delta_N+s^{-2}+u}\Big)du.\end{aligned} \tag{4}$$

The first integral we estimate using a Cauchy-Schwarz inequality

$$\begin{aligned}
&\int_0^\infty \frac{1}{-\Delta_N + s^{-2} + u}(\nabla\theta)^2\frac{-\Delta_N}{-\Delta_N + s^{-2} + u} + \frac{-\Delta_N}{-\Delta_N + s^{-2} + u}(\nabla\theta)^2\frac{1}{-\Delta_N + s^{-2} + u}du \\
&\le t^2\int_0^\infty \frac{1}{-\Delta_N + s^{-2} + u}(\nabla\theta)^2(-\Delta_N)(\nabla\theta)^2\frac{1}{-\Delta_N + s^{-2} + u} + t^{-2}\frac{-\Delta_N}{(-\Delta_N + s^{-2} + u)^2}du \\
&\le Ct^{-2}\frac{-\Delta_N}{-\Delta_N + s^{-2}} + Cs^2t^{-4},
\end{aligned}$$

where in the last estimate we have used (3) with $f = (\nabla\theta)^2$.

The last integral in (4) we estimate again using a Cauchy-Schwarz inequality this time together with (2) with $f = (\nabla\theta)^2$

$$\begin{aligned}
&\int_0^\infty \Big(\frac{1}{-\Delta_N + s^{-2} + u}[-\Delta_N, (\nabla\theta)^2]\frac{-\Delta_N}{(-\Delta_N + s^{-2} + u)^2} \\
&\qquad \frac{-\Delta_N}{(-\Delta_N + s^{-2} + u)^2}[(\nabla\theta)^2, -\Delta_N]\frac{1}{-\Delta_N + s^{-2} + u}\Big)du \\
&\le t^4\int_0^\infty \frac{1}{-\Delta_N + s^{-2} + u}[-\Delta_N, (\nabla\theta)^2][(\nabla\theta)^2, -\Delta_N]\frac{1}{-\Delta_N + s^{-2} + u}du \\
&\quad + t^{-4}\int_0^\infty \frac{(-\Delta_N)^2}{(-\Delta_N + s^{-2} + u)^4}du \\
&\le t^4\int_0^\infty \frac{1}{-\Delta_N + s^{-2} + u}(Ct^{-6}(-\Delta_N) + Ct^{-8})\frac{1}{-\Delta_N + s^{-2} + u}du + s^2t^{-4} \\
&\le Ct^{-2}\frac{-\Delta_N}{-\Delta_N + s^{-2}} + Cs^2t^{-4}.
\end{aligned}$$

This proves (1).

References

[1] Elliott H. Lieb and Jan Philip Solovej, Ground State Energy of the One-Component Charged Bose Gas, *Commun. Math. Phys.* **217**, 127–163 (2001).

Publications of Elliot H. Lieb

1. Second Order Radiative Corrections to the Magnetic Moment of a Bound Electron, Phil. Mag. Vol. **46**, 311–316 (1955).
2. A Non-Perturbation Method for Non-Linear Field Theories, Proc. Roy. Soc. **241A**, 339–363 (1957).
3. (with K. Yamazaki) Ground State Energy and Effective Mass of the Polaron, Phys. Rev. **111**, 728–733 (1958).
4. (with H. Koppe) Mathematical Analysis of a Simple Model Related to the Stripping Reaction, Phys. Rev. **116**, 367–371 (1959).
5. Hard Sphere Bose Gas – An Exact Momentum Space Formulation, Proc. U.S. Nat. Acad. Sci. **46**, 1000–1002 (1960).
6. Operator Formalism in Statistical Mechanics, J. Math. Phys. **2**, 341–343 (1961).
7. (with D.C. Mattis) Exact Wave Functions in Superconductivity, J. Math. Phys. **2**, 602–609 (1961).
8. (with T.D. Schultz and D.C. Mattis) Two Soluble Models of an Antiferromagnetic Chain, Annals of Phys. (N.Y.) **16**, 407–466 (1961).
9. (with D.C. Mattis) Theory of Ferromagnetism and the Ordering of Electronic Energy Levels, Phys. Rev. **125**, 164–172 (1962).
10. (with D.C. Mattis) Ordering Energy Levels of Interacting Spin Systems, J. Math. Phys. **3**, 749–751 (1962).
11. New Method in the Theory of Imperfect Gases and Liquids, J. Math. Phys. **4**, 671–678 (1963).
12. (with W. Liniger) Exact Analysis of an Interacting Bose Gas. I. The General Solution and the Ground State, Phys. Rev. **130**, 1605–1616 (1963).
13. Exact Analysis of an Interacting Bose Gas. II. The Excitation Spectrum, Phys. Rev. **130**, 1616–1624 (1963).
14. Simplified Approach to the Ground State Energy of an Imperfect Bose Gas, Phys. Rev. **130**, 2518–2528 (1963).
15. (with A. Sakakura) Simplified Approach to the Ground State Energy of an Imperfect Bose Gase. II. The Charged Bose Gas at High Density, Phys. Rev. **133**, A899–A906 (1964).
16. (with W. Liniger) Simplified Approach to the Ground State Energy of an Imperfect Bose Gas. III. Application to the One-Dimensional Model, Phys. Rev. **134**, A312–A315 (1964).

---

* means the paper appeared in the first and/or second editions.
† means the paper appears in the third edition.

17. (with T.D. Schultz and D.C. Mattis) Two-Dimensional Ising Model as a Soluble Problem of Many Fermions, Rev. Mod. Phys. **36**, 856–871 (1964).
18. The Bose Fluid, *Lectures in Theoretical Physics, Vol. VIIC,* (Boulder summer school), University of Colorado Press, 175–224 (1965).
19. (with D.C. Mattis) Exact Solution of a Many-Fermion System and its Associated Boson Field, J. Math. Phys. **6**, 304–312 (1965).
20. (with S.Y. Larsen, J.E. Kilpatrick and H.F. Jordan) Suppression at High Temperature of Effects Due to Statistics in the Second Virial Coefficient of a Real Gas, Phys. Rev. **140**, A129–A130 (1965).
21. (with D.C. Mattis) Book *Mathematical Physics in One Dimension*, Academic Press, New York (1966).
22. Proofs of Some Conjectures on Permanents, J. of Math. and Mech. **16**, 127–139 (1966).
23. Quantum Mechanical Extension of the Lebowitz-Penrose Theorem on the van der Waals Theory, J. Math. Phys. **7**, 1016–1024 (1966).
24. (with D.C. Mattis) Theory of Paramagnetic Impurities in Semiconductors, J. Math. Phys. **7**, 2045–2052 (1966).
25. (with T. Burke and J.L. Lebowitz) Phase Transition in a Model Quantum System: Quantum Corrections to the Location of the Critical Point, Phys. Rev. **149**, 118–122 (1966).
26. Some Comments on the One-Dimensional Many-Body Problem, unpublished Proceedings of Eastern Theoretical Physics Conference, New York (1966).
27. Calculation of Exchange Second Virial Coefficient of a Hard Sphere Gas by Path Integrals, J. Math. Phys. **8**, 43–52 (1967).
28. (with Z. Rieder and J.L. Lebowitz) Properties of a Harmonic Crystal in a Stationary Nonequilibrium State, J. Math. Phys. **8**, 1073–1078 (1967).
29. Exact Solution of the Problem of the Entropy of Two-Dimensional Ice, Phys. Rev. Lett. **18**, 692–694 (1967).
30. Exact Solution of the F Model of an Antiferroelectric, Phys. Rev. Lett. **18**, 1046–1048 (1967).
31. Exact Solution of the Two-Dimensional Slater KDP Model of a Ferroelectric, Phys. Rev. Lett. **19**, 108–110 (1967).
32. The Residual Entropy of Square Ice, Phys. Rev. **162**, 162–172 (1967).
33. Ice, Ferro- and Antiferroelectrics, in *Methods and Problems in Theoretical Physics, in honour of R.E. Peierls*, Proceedings of the 1967 Birmingham conference, North-Holland, 21–28 (1970).
34. Exactly Soluble Models, in *Mathematical Methods in Solid State and Superfluid Theory*, Proceedings of the 1967 Scottish Universities' Summer School of Physics, Oliver and Boyd, Edinburgh 286–306 (1969).
35. The Solution of the Dimer Problems by the Transfer Matrix Method, J. Math. Phys. **8**, 2339–2341 (1967).
36. (with M. Flicker) Delta Function Fermi Gas with Two Spin Deviates, Phys. Rev. **161**, 179–188 (1967).
37. Concavity Properties and a Generating Function for Stirling Numbers, J. Combinatorial Theory **5**, 203–206 (1968).

38. A Theorem on Pfaffians, J. Combinatorial Theory **5**, 313–319 (1968).
39. (with F.Y. Wu) Absence of Mott Transition in an Exact Solution of the Short-Range One-Band Model in One Dimension, Phys. Rev. Lett. **20**, 1445–1448 (1968).
40. Two Dimensional Ferroelectric Models, J. Phys. Soc. (Japan) **26** (supplement), 94–95 (1969).
41. (with W.A. Beyer) Clusters on a Thin Quadratic Lattice, Studies in Appl. Math. **48**, 77–90 (1969).
42. (with C.J. Thompson) Phase Transition in Zero Dimensions: A Remark on the Spherical Model, J. Math. Phys. **10**, 1403–1406 (1969).
†*43. (with J.L. Lebowitz) The Existence of Thermodynamics for Real Matter with Coulomb Forces, Phys. Rev. Lett. **22**, 631–634 (1969).
44. Two Dimensional Ice and Ferroelectric Models, in *Lectures in Theoretical Physics, XI D*, (Boulder summer school) Gordon and Breach, 329–354 (1969).
45. Survey of the One Dimensional Many Body Problem and Two Dimensional Ferroelectric Models, in *Contemporary Physics: Trieste Symposium 1968*, International Atomic Energy Agency, Vienna, vol. 1, 163–176 (1969).
46. Models, in *Phase Transitions*, Proceedings of the 14th Solvay Chemistry Conference, May 1969, Interscience, 45–56 (1971).
47. (with H. Araki) Entropy Inequalities, Commun. Math. Phys. **18**, 160–170 (1970).
48. (with O.J. Heilmann) Violation of the Non-Crossing Rule: The Hubbard Hamiltonian for Benzene, Trans. N.Y. Acad. Sci. **33**, 116–149 (1970). Also in Annals N.Y. Acad. Sci. **172**, 583–617 (1971). (Awarded the 1970 Boris Pregel award for research in chemical physics.)
49. (with O.J. Heilmann) Monomers and Dimers, Phys. Rev. Lett. **24**, 1412–1414 (1970).
50. Book Review of "Statistical Mechanics" by David Ruelle, Bull. Amer. Math. Soc. **76**, 683–687 (1970).
51. (with J.L. Lebowitz) Thermodynamic Limit for Coulomb Systems, in *Systèmes a un Nombre Infini de Degrés de Liberté*, Colloques Internationaux de Centre National de la Recherche Scientifique **181**, 155–162 (1970).
52. (with D.B. Abraham, T. Oguchi and T. Yamamoto) On the Anomolous Specific Heat of Sodium Trihydrogen Selenite, Progr. Theor. Phys. (Kyoto) **44**, 1114–1115 (1970).
53. (with D.B. Abraham) Anomalous Specific Heat of Sodium Trihydrogen Selenite – An Associated Combinatorial Problem, J. Chem. Phys. **54**, 1446–1450 (1971).
54. (with O.J. Heilmann, D. Kleitman and S. Sherman) Some Positive Definite Functions on Sets and Their Application to the Ising Model, Discrete Math. **1**, 19–27 (1971).
55. (with Th. Niemeijer and G. Vertogen) Models in Statistical Mechanics, in *Statistical Mechanics and Quantum Field Theory*, Proceedings of 1970

Ecole d'Eté de Physique Théorique (Les Houches), Gordon and Breach, 281–326 (1971).
56. (with H.N.V. Temperley) Relations between the 'Percolation' and 'Colouring' Problem and Other Graph-Theoretical Problems Associated with Regular Planar Lattices: Some Exact Results for the 'Percolation' Problem, Proc. Roy. Soc. **A322**, 251–280 (1971).
57. (with M. de Llano) Some Exact Results in the Hartree-Fock Theory of a Many-Fermion System at High Densities, Phys. Letts. **37B**, 47–49 (1971).
58. (with J.L. Lebowitz) The Constitution of Matter: Existence of Thermodynamics for Systems Composed of Electrons and Nuclei, Adv. in Math. **9**, 316–398 (1972).
59. (with F.Y. Wu) Two Dimensional Ferroelectric Models, in *Phase Transitions and Critical Phenomena*, C. Domb and M. Green eds., vol. 1, Academic Press 331–490 (1972).
60. (with D. Ruelle) A Property of Zeros of the Partition Function for Ising Spin Systems, J. Math. Phys. **13**, 781–784 (1972).
61. (with O.J. Heilmann) Theory of Monomer-Dimer Systems, Commun. Math. Phys. **25**, 190–232 (1972). Errata **27**, 166 (1972).
62. (with M.L. Glasser and D.B. Abraham) Analytic Properties of the Free Energy for the "Ice" Models, J. Math. Phys. **13**, 887–900 (1972).
63. (with D.W. Robinson) The Finite Group Velocity of Quantum Spin Systems, Commun. Math. Phys. **28**, 251–257 (1972).
64. (with J.L. Lebowitz) Phase Transition in a Continuum Classical System with Finite Interactions, Phys. Lett. **39A**, 98–100 (1972).
*65. (with J.L. Lebowitz) Lectures on the Thermodynamic Limit for Coulomb Systems, in *Statistical Mechanics and Mathematical Problems*, Battelle 1971 Recontres, Springer Lecture Notes in Physics **20**, 136–161 (1973).
66. (with J.L. Lebowitz) Lectures on the Thermodynamic Limit for Coulomb Systems, in *Lectures in Theoretical Physics XIV B*, (Boulder summer school), Colorado Associated University Press, 423–460 (1973).
67. Convex Trace Functions and the Wigner-Yanase-Dyson Conjecture, Adv. in Math. **11**, 267–288 (1973).
68. (with M.B. Ruskai) A Fundamental Property of Quantum Mechanical Entropy, Phys. Rev. Lett. **30**, 434–436 (1973).
69. (with M.B. Ruskai) Proof of the Strong Subadditivity of Quantum-Mechanical Entropy, J. Math. Phys. **14**, 1938–1941 (1973).
70. (with K. Hepp) On the Superradiant Phase Transition for Molecules in a Quantized Radiation Field: The Dicke Maser Model, Annals of Phys. (N.Y.) **76**, 360–404 (1973).
71. (with K. Hepp) Phase Transition in Reservoir Driven Open Systems with Applications to Lasers and Superconductors, Helv. Phys. Acta **46**, 573–602 (1973).
72. (with K. Hepp) The Equilibrium Statistical Mechanics of Matter Interacting with the Quantized Radiation Field, Phys. Rev. **A8**, 2517–2525 (1973).
73. (with K. Hepp) Constructive Macroscopic Quantum Electrodynamics, in *Constructive Quantum Field Theory*, Proceedings of the 1973 Erice Sum-

mer School, G. Velo and A. Wightman, eds., Springer Lecture Notes in Physics **25**, 298–316 (1973).
74. The Classical Limit of Quantum Spin Systems, Commun. Math. Phys. **31**, 327–340 (1973).
75. (with B. Simon) Thomas–Fermi Theory Revisited, Phys. Rev. Lett. **31**, 681–683 (1973).
76. (with M.B. Ruskai) Some Operator Inequalities of the Schwarz Type, Adv. in Math. **12**, 269–273 (1974).
77. Exactly Soluble Models in Statistical Mechanics, lecture given at the 1973 I.U.P.A.P. van der Waals Centennial Conference on Statistical Mechanics, Physica **73**, 226–236 (1974).
78. (with B. Simon) On Solutions to the Hartree–Fock Problem for Atoms and Molecules, J. Chem. Physics **61**, 735–736 (1974).
79. Thomas–Fermi and Hartree–Fock Theory, lecture at 1974 International Congress of Mathematicians, Vancouver. Proceedings, Vol. 2, 383–386 (1975).
80. Some Convexity and Subadditivity Properties of Entropy, Bull. Amer. Math. Soc. **81**, 1–13 (1975).
81. (with H.J. Brascamp and J.M. Luttinger) A General Rearrangement Inequality for Multiple Integrals, Jour. Funct. Anal. **17**, 227–237 (1975).
82. (with H.J. Brascamp) Some Inequalities for Gaussian Measures and the Long-Range Order of the One-Dimensional Plasma, lecture at Conference on Functional Integration, Cumberland Lodge, England. *Functional Integration and its Applications*, A.M. Arthurs ed., Clarendon Press, 1–14 (1975).
83. (with K. Hepp) The Laser: A Reversible Quantum Dynamical System with Irreversible Classical Macroscopic Motion, in *Dynamical Systems*, Battelle 1974 Rencontres, Springer Lecture Notes in Physics **38**, 178–208 (1975). Also appears in *Melting, Localization and Chaos*, Proc. 9th Midwest Solid State Theory Symposium, 1981, R. Kalia and P. Vashishta eds., North-Holland, 153–177 (1982).
†*84. (with P. Hertel and W. Thirring) Lower Bound to the Energy of Complex Atoms, J. Chem. Phys. **62**, 3355–3356 (1975).
†*85. (with W. Thirring) Bound for the Kinetic Energy of Fermions which Proves the Stability of Matter, Phys. Rev. Lett. **35**, 687–689 (1975). Errata **35**, 1116 (1975).
86. (with H.J. Brascamp and J.L. Lebowitz) The Statistical Mechanics of Anharmonic Lattices, in the proceedings of the 40th session of the International Statistics Institute, Warsaw, **9**, 1–11 (1975).
87. (with H.J. Brascamp) Best Constants in Young's Inequality, Its Converse and Its Generalization to More Than Three Functions, Adv. in Math. **20**, 151–172 (1976).
88. (with H.J. Brascamp) On Extensions of the Brunn–Minkowski and Prékopa–Leindler Theorems, Including Inequalities for Log Concave Functions and with an Application to the Diffusion Equation, J. Funct. Anal. **22**, 366–389 (1976).

89. (with J.F. Barnes and H.J. Brascamp) Lower Bounds for the Ground State Energy of the Schroedinger Equation Using the Sharp Form of Young's Inequality, in *Studies in Mathematical Physics*, Lieb, Simon, Wightman eds., Princeton Press, 83–90 (1976).
90. Inequalities for Some Operator and Matrix Functions, Adv. in Math. **20**, 174–178 (1976).
†*91. (with H. Narnhofer) The Thermodynamic Limit for Jellium, J. Stat. Phys. **12**, 291–310 (1975). Errata J. Stat. Phys. **14**, 465 (1976).
†*92. The Stability of Matter, Rev. Mod. Phys. **48**, 553–569 (1976).
93. Bounds on the Eigenvalues of the Laplace and Schroedinger Operators, Bull. Amer. Math. Soc. **82**, 751–753 (1976).
94. (with F.J. Dyson and B. Simon) Phase Transitions in the Quantum Heisenberg Model, Phys. Rev. Lett. **37**, 120–123 (1976). (See no. 104.)
†*95. (with W. Thirring) Inequalities for the Moments of the Eigenvalues of the Schrödinger Hamiltonian and Their Relation to Sobolev Inequalities, in *Studies in Mathematical Physics*, E. Lieb, B. Simon, A. Wightman eds., Princeton University Press, 269–303 (1976).
96. (with B. Simon and A. Wightman) Book *Studies in Mathematical Physics: Essays in Honor of Valentine Bargmann*, Princeton University Press (1976).
97. (with B. Simon) Thomas–Fermi Theory of Atoms, Molecules and Solids, Adv. in Math. **23**, 22–116 (1977).
98. (with O. Lanford and J. Lebowitz) Time Evolution of Infinite Anharmonic Oscillators, J. Stat. Phys. **16**, 453–461 (1977).
99. The Stability of Matter, Proceedings of the Conference on the Fiftieth Anniversary of the Schroedinger equation, Acta Physica Austriaca Suppl. XVII, 181–207 (1977).
100. Existence and Uniqueness of the Minimizing Solution of Choquard's Non-Linear Equation, Studies in Appl. Math. **57**, 93–105 (1977).
101. (with J. Fröhlich) Existence of Phase Transitions for Anisotropic Heisenberg Models, Phys. Rev. Lett. **38**, 440–442 (1977).
†*102. (with B. Simon) The Hartree–Fock Theory for Coulomb Systems, Commun. Math. Phys. **53**, 185–194 (1977).
103. (with W. Thirring) A Lower Bound for Level Spacings, Annals of Phys. (N.Y.) **103**, 88–96 (1977).
104. (with F. Dyson and B. Simon) Phase Transitions in Quantum Spin Systems with Isotropic and Non-Isotropic Interactions, J. Stat. Phys. bft 18, 335–383 (1978).
105. Many Particle Coulomb Systems, lectures given at the 1976 session on statistical mechanics of the International Mathematics Summer Center (C.I.M.E.). In *Statistical Mechanics*, C.I.M.E. 1 Ciclo 1976, G. Gallavotti, ed., Liguore Editore, Naples, 101–166 (1978).
†*106. (with R. Benguria) Many-Body Atomic Potentials in Thomas–Fermi Theory, Annals of Phys. (N.Y.) **110**, 34–45 (1978).
†*107. (with R. Benguria) The Positivity of the Pressure in Thomas–Fermi Theory, Commun. Math. Phys. **63**, 193–218 (1978). Errata **71**, 94 (1980).

108. (with M. de Llano) Solitons and the Delta Function Fermion Gas in Hartree–Fock Theory, J. Math. Phys. **19**, 860–868 (1978).
109. (with J. Fröhlich) Phase Transitions in Anisotropic Lattice Spin Systems, Commun. Math. Phys. **60**, 233–267 (1978).
110. (with J. Fröhlich, R. Israel and B. Simon) Phase Transitions and Reflection Positivity. I. General Theory and Long Range Lattice Models, Commun. Math. Phys. **62**, 1–34 (1978). (See no. 124.)
111. (with M. Aizenman and E.B. Davies) Positive Linear Maps Which are Order Bounded on C* Subalgebras, Adv. in Math. **28**, 84–86 (1978).
†*112. (with M. Aizenman) On Semi-Classical Bounds for Eigenvalues of Schrödinger Operators, Phys. Lett. **66A**, 427–429 (1978).
113. New Proofs of Long Range Order, in *Proceedings of the International Conference on Mathematical Problems in Theoretical Physics* (June 1977), Springer Lecture Notes in Physics, **80**, 59–67 (1978).
114. Proof of an Entropy Conjecture of Wehrl, Commun. Math. Phys. **62**, 35–41 (1978).
115. (with B. Simon) Monotonicity of the Electronic Contribution to the Born–Oppenheimer Energy, J. Phys. B. **11**, L537–L542 (1978).
116. (with O. Heilmann) Lattice Models for Liquid Crystals, J. Stat. Phys. **20**, 679–693 (1979).
117. (with H. Brezis) Long Range Atomic Potentials in Thomas–Fermi Theory, Commun. Math. Phys. **65**, 231–246 (1979).
†*118. The $N^{5/3}$ Law for Bosons, Phys. Lett. **70A**, 71–73 (1979).
119. A Lower Bound for Coulomb Energies, Phys. Lett. **70A**, 444–446 (1979).
120. Why Matter is Stable, Kagaku **49**, 301–307 and 385–388 (1979). (In Japanese).
121. The Number of Bound States of One-Body Schrödinger Operators and the Weyl Problem, Symposium of the Research Inst. of Math. Sci., Kyoto University, (1979).
122. Some Open Problems About Coulomb Systems, in *Proceedings of the Lausanne 1979 Conference of the International Association of Mathematical Physics*, Springer Lecture Notes in Physics, **116**, 91–102 (1980).
†*123. The Number of Bound States of One-Body Schrödinger Operators and the Weyl Problem, *Proceedings of the Amer. Math. Soc. Symposia in Pure Math.*, **36**, 241–252 (1980).
124. (with J. Fröhlich, R. Israel and B. Simon) Phase Transitions and Reflection Positivity. II. Lattice Systems with Short-Range and Coulomb Interactions. J. Stat. Phys. **22**, 297–347 (1980). (See no. 110.)
125. Why Matter is Stable, Chinese Jour. Phys. **17**, 49–62 (1980). (English version of no. 120).
126. A Refinement of Simon's Correlation Inequality, Commun. Math. Phys. **77**, 127–135 (1980).
127. (with B. Simon) Pointwise Bounds on Eigenfunctions and Wave Packets in N-Body Quantum Systems. VI. Asymptotics in the Two-Cluster Region, Adv. in Appl. Math. **1**, 324–343 (1980).
128. The Uncertainty Principle, article in *Encyclopedia of Physics*, R. Lerner and G. Trigg eds., Addison Wesley, 1078–1079 (1981).

†*129. (with S. Oxford) An Improved Lower Bound on the Indirect Coulomb Energy, Int. J. Quant. Chem. **19**, 427–439 (1981).
†*130. (with R. Benguria and H. Brezis) The Thomas–Fermi–von Weizsaecker Theory of Atoms and Molecules, Commun. Math. Phys. **79**, 167–180 (1981).
131. (with M. Aizenman) The Third Law of Thermodynamics and the Degeneracy of the Ground State for Lattice Systems, J. Stat. Phys. **24**, 279–297 (1981).
132. (with J. Bricmont, J. Fontaine, J. Lebowitz and T. Spencer) Lattice Systems with a Continuous Symmetry III. Low Temperature Asymptotic Expansion for the Plane Rotator Model, Commun. Math. Phys. **78**, 545–566 (1981).
133. (with A. Sokal) A General Lee–Yang Theorem for One-Component and Multi-component Ferromagnets, Commun. Math. Phys. **80**, 153–179 (1981).
†*134. Variational Principle for Many-Fermion Systems, Phys. Rev. Lett. **46**, 457–459 (1981). Errata **47**, 69 (1981).
135. Thomas–Fermi and Related Theories of Atoms and Molecules, in *Rigorous Atomic and Molecular Physics*, G. Velo and A. Wightman, eds., Plenum Press, 213–308 (1981).
†*136. Thomas–Fermi and Related Theories of Atoms and Molecules, Rev. Mod. Phys. **53**, 603–641 (1981). Errata **54**, 311 (1982). (Revised version of no. 135.)
137. Statistical Theories of Large Atoms and Molecules, in *Proceedings of the 1981 Oaxtepec conference on Recent Progress in Many-Body Theories,* Springer Lecture Notes in Physics, **142**, 336–343 (1982).
138. Statistical Theories of Large Atoms and Molecules, Comments Atomic and Mol. Phys. **11**, 147–155 (1982).
†*139. Analysis of the Thomas–Fermi–von Weizsäcker Equation for an Infinite Atom without Electron Repulsion, Commun. Math. Phys. **85**, 15–25 (1982).
140. (with D.A. Liberman) Numerical Calculation of the Thomas–Fermi–von Weizsäcker Function for an Infinite Atom without Electron Repulsion, Los Alamos National Laboratory Report, LA-9186-MS (1982).
†*141. Monotonicity of the Molecular Electronic Energy in the Nuclear Coordinates, J. Phys. B.: At. Mol. Phys. **15**, L63–L66 (1982).
142. Comment on "Approach to Equilibrium of a Boltzmann Equation Solution", Phys. Rev. Lett. **48**, 1057 (1982).
143. Density Functionals for Coulomb Systems, in *Physics as Natural Philosophy: Essays in honor of Laszlo Tisza on his 75th Birthday*, A. Shimony and H. Feshbach eds., M.I.T. Press, 111–149 (1982).
144. An L^p Bound for the Riesz and Bessel Potentials of Orthonormal Functions, J. Funct. Anal. **51**, 159–165 (1983).
145. (with H. Brezis) A Relation Between Pointwise Convergence of Functions and Convergence of Functionals, Proc. Amer. Math. Soc. **88**, 486–490 (1983).

†*146. (with R. Benguria) A Proof of the Stability of Highly Negative Ions in the Absence of the Pauli Principle, Phys. Rev. Lett. **50**, 1771–1774 (1983).
147. Sharp Constants in the Hardy–Littlewood–Sobolev and Related Inequalities, Annals of Math. **118**, 349–374 (1983).
148. Density Functionals for Coulomb Systems (a revised version of no. 143), Int. Jour. Quant. Chem. **24**, 243–277 (1983). An expanded version appears in *Density Functional Methods in Physics*, R. Dreizler and J. da Providencia eds., Plenum Nato ASI Series **123**, 31–80 (1985).
149. The Significance of the Schrödinger Equation for Atoms, Molecules and Stars, lecture given at the Schrödinger Symposium, Dublin Institute of Advanced Studies, October 1983, unpublished Proceedings.
†*150. (with I. Daubechies) One Electron Relativistic Molecules with Coulomb Interaction, Commun. Math. Phys. **90**, 497–510 (1983).
151. (with I. Daubechies) Relativistic Molecules with Coulomb Interaction, in *Differential Equations, Proc. of the Conference held at the University of Alabama in Birmingham, 1983*, I. Knowles and R. Lewis eds., Math. Studies Series, **92**, 143–148, North-Holland (1984).
152. Some Vector Field Equations, in *Differential Equations, Proc. of the Conference held at the University of Alabama in Birmingham, 1983*, I. Knowles and R. Lewis eds., Math. Studies Series **92**, 403–412 North-Holland (1984).
153. On the Lowest Eigenvalue of the Laplacian for the Intersection of Two Domains, Inventiones Math. **74**, 441–448 (1983).
154. (with J. Chayes and L. Chayes) The Inverse Problem in Classical Statistical Mechanics, Commun. Math. Phys. **93**, 57–121 (1984).
155. On Characteristic Exponents in Turbulence, Commun. Math. Phys. **92**, 473–480 (1984).
†*156. Atomic and Molecular Negative Ions, Phys. Rev. Lett. **52**, 315–317 (1984).
†*157. Bound on the Maximum Negative Ionization of Atoms and Molecules, Phys. Rev. **29A**, 3018–3028 (1984).
158. (with W. Thirring) Gravitational Collapse in Quantum Mechanics with Relativistic Kinetic Energy, Annals of Phys. (N.Y.) **155**, 494–512 (1984).
*159. (with I.M. Sigal, B. Simon and W. Thirring) Asymptotic Neutrality of Large-Z Ions, Phys. Rev. Lett. **52**, 994–996 (1984). (See no. 185.)
†*160. (with R. Benguria) The Most Negative Ion in the Thomas–Fermi–von Weizsäcker Theory of Atoms and Molecules, J. Phys. B: At. Mol. Phys. **18**, 1045–1059 (1985).
161. (with H. Brezis) Minimum Action Solutions of Some Vector Field Equations, Commun. Math. Phys. **96**, 97–113 (1984).
162. (with H. Brezis) Sobolev Inequalities with Remainder Terms, J. Funct. Anal. **62**, 73–86 (1985).
163. Baryon Mass Inequalities in Quark Models, Phys. Rev. Lett. **54**, 1987–1990 (1985).
†*164. (with J. Fröhlich and M. Loss) Stability of Coulomb Systems with Magnetic Fields I. The One-Electron Atom, Commun. Math. Phys. **104**, 251–270 (1986).

†*165. (with M. Loss) Stability of Coulomb Systems with Magnetic Fields II. The Many-Electron Atom and the One-Electron Molecule, Commun. Math. Phys. **104**, 271–282 (1986).
†*166. (with W. Thirring) Universal Nature of van der Waals Forces for Coulomb Systems, Phys. Rev. A **34**, 40–46 (1986).
167. Some Ginzburg–Landau Type Vector-Field Equations, in *Nonlinear systems of Partial Differential Equations in Applied Mathematics*, B. Nicolaenko, D. Holm and J. Hyman eds., Amer. Math. Soc. Lectures in Appl. Math. **23**, Part 2, 105–107 (1986).
168. (with I. Affleck) A Proof of Part of Haldane's Conjecture on Spin Chains, Lett. Math. Phys. **12**, 57–69 (1986).
169. (with H. Brezis and J-M. Coron) Estimations d'Energie pour des Applications de $\mathbf{R}^3$ a Valeurs dans S^2, C.R. Acad. Sci. Paris **303** Ser. 1, 207–210 (1986).
170. (with H. Brezis and J-M. Coron) Harmonic Maps with Defects, Commun. Math. Phys. **107**, 649–705 (1986).
171. Some Fundamental Properties of the Ground States of Atoms and Molecules, in *Fundamental Aspects of Quantum Theory*, V. Gorini and A. Frigerio eds., Nato ASI Series B, Vol. 144, Plenum Press (1986).
172. (with T. Kennedy) A Model for Crystallization: A Variation on the Hubbard Model, in *Statistical Mechanics and Field Theory: Mathematical Aspects*, Springer Lecture Notes in Physics **257**, 1–9 (1986).
173. (with T. Kennedy) An Itinerant Electron Model with Crystalline or Magnetic Long Range Order, Physica **138A**, 320–358 (1986).
174. (with T. Kennedy) A Model for Crystallization: A Variation on the Hubbard Model, Physica **140A**, 240–250 (1986) (Proceedings of IUPAP Statphys 16, Boston).
175. (with T. Kennedy) Proof of the Peierls Instability in One Dimension, Phys. Rev. Lett. **59**, 1309–1312 (1987).
176. (with I. Affleck, T. Kennedy and H. Tasaki) Rigorous Results on Valence-Bond Ground States in Antiferromagnets, Phys. Rev. Lett. **59**, 799–802 (1987).
†*177. (with H.-T. Yau) The Chandrasekhar Theory of Stellar Collapse as the Limit of Quantum Mechanics, Commun. Math. Phys. **112**, 147–174 (1987).
*178. (with H.-T. Yau) A Rigorous Examination of the Chandrasekhar Theory of Stellar Collapse, Astrophys. Jour. **323**, 140–144 (1987).
179. (with F. Almgren) Singularities of Energy Minimizing Maps from the Ball to the Sphere, Bull. Amer. Math. Soc. **17**, 304–306 (1987). (See no. 190.)
180. Bounds on Schrödinger Operators and Generalized Sobolev Type Inequalities, Proceedings of the International Conference on Inequalities, University of Birmingham, England, 1987, Marcel Dekker Lecture Notes in Pure and Appl. Math., W.N. Everitt ed., volume 129, pages 123–133 (1991).
181. (with I. Affleck, T. Kennedy and H. Tasaki) Valence Bond Ground States in Isotropic Quantum Antiferromagnets, Commun. Math. Phys. **115**, 477–528 (1988).

182. (with T. Kennedy and H. Tasaki) A Two Dimensional Isotropic Quantum Antiferromagnet with Unique Disordered Ground State, J. Stat. Phys. **53**, 383–416 (1988).
183. (with T. Kennedy and S. Shastry) Existence of Néel Order in Some Spin 1/2 Heisenberg Antiferromagnets, J. Stat. Phys. **53**, 1019–1030 (1988).
184. (with T. Kennedy and S. Shastry) The XY Model has Long-Range Order for all Spins and all Dimensions Greater than One, Phys. Rev. Lett. **61**, 2582–2584 (1988).
†*185. (with I.M. Sigal, B. Simon and W. Thirring) Approximate Neutrality of Large-Z Ions, Commun. Math. Phys. **116**, 635–644 (1988). (See no. 159.)
†*186. (with H.-T. Yau) The Stability and Instability of Relativistic Matter, Commun. Math. Phys. **118**, 177–213 (1988).
*187. (with H.-T. Yau) Many-Body Stability Implies a Bound on the Fine Structure Constant, Phys. Rev. Lett. **61**, 1695–1697 (1988).
†*188. (with J. Conlon and H.-T. Yau) The $N^{7/5}$ Law for Charged Bosons, Commun. Math. Phys. **116**, 417–448 (1988).
189. (with F. Almgren and W. Browder) Co-area, Liquid Crystals, and Minimal Surfaces, in *Partial Differential Equations*, S.S. Chern ed., Springer Lecture Notes in Math. **1306**, 1–22 (1988).
190. (with F. Almgren) Singularities of Energy Minimizing Maps from the Ball to the Sphere: Examples, Counterexamples and Bounds, Ann. of Math. **128**, 483–530 (1988).
191. (with F. Almgren) Counting Singularities in Liquid Crystals, in *IXth International Congress on Mathematical Physics*, B. Simon, A. Truman, I.M. Davies eds., Hilger, 396–409 (1989). This also appears in: *Symposia Mathematica, vol. XXX,* Ist. Naz. Alta Matem. Francesco Severi Roma, 103–118, Academic Press (1989); *Variational Methods,* H. Berestycki, J-M. Coron, I. Ekeland eds., Birkhäuser, 17–36 (1990); How many singularities can there be in an energy minimizing map from the ball to the sphere?, in *Ideas and Methods in Mathematical Analysis, Stochastics, and Applications,* S. Albeverio, J.E. Fenstad, H. Holden, T. Lindstrom eds., Cambridge Univ. Press, vol. 1, 394–408 (1992).
192. (with F. Almgren) Symmetric Decreasing Rearrangement can be Discontinuous, Bull. Amer. Math. Soc. **20**, 177–180 (1989).
193. (with F. Almgren) Symmetric Decreasing Rearrangement is Sometimes Continuous, Jour. Amer. Math. Soc. **2**, 683–773 (1989). A summary of this work (using 'rectifiable currents') appears as The (Non)continuity of Symmetric Decreasing Rearrangement in *Symposia Mathematica, vol. XXX,* Ist. Naz. Alta Matem. Francesco Severi Roma, 89–102, Academic Press (1989) and in *Variational Methods*, H. Berestycki, J-M. Coron, I. Ekeland eds., Birkhäuser, 3–16 (1990).
194. Two Theorems on the Hubbard Model, Phys. Rev. Lett. **62**, 1201–1204 (1989). Errata **62**, 1927 (1989).
195. (with J. Conlon and H.-T. Yau) The Coulomb gas at Low Temperature and Low Density, Commun. Math. Phys. **125**, 153–180 (1989).
196. Gaussian Kernels have only Gaussian Maximizers, Invent. Math. **102**, 179–208 (1990).

†*197. Kinetic Energy Bounds and their Application to the Stability of Matter, in *Schrödinger Operators*, Proceedings Sønderborg Denmark 1988, H. Holden and A. Jensen eds., Springer Lecture Notes in Physics **345**, 371–382 (1989). Expanded version of no. 180.

†*198. The Stability of Matter: From Atoms to Stars, 1989 Gibbs Lecture, Bull. Amer. Math. Soc. **22**, 1–49 (1990).

199. Integral Bounds for Radar Ambiguity Functions and Wigner Distributions, J. Math. Phys. **31**, 594–599 (1990).

200. On the Spectral Radius of the Product of Matrix Exponentials, Linear Alg. and Appl.**141**, 271–273 (1990).

201. (with M. Aizenman) Magnetic Properties of Some Itinerant-Electron Systems at $T > 0$, Phys. Rev. Lett. **65**, 1470–1473 (1990).

202. (with H. Siedentop) Convexity and Concavity of Eigenvalue Sums, J. Stat. Phys. **63**, 811–816 (1991).

203. (with J.P. Solovej) Quantum Coherent Operators: A Generalization of Coherent States, Lett. Math. Phys. **22**, 145–154 (1991).

204. The Flux-Phase Problem on Planar Lattices, Helv. Phys. Acta **65**, 247–255 (1992). Proceedings of the conference "Physics in Two Dimensions", Neuchatel, August 1991.

205. Atome in starken Magnetfeldern, Physikalische Blätter **48**, 549–552 (1992). Translation by H. Siedentop of the Max-Planck medal lecture (1 April 1992) "Atoms in strong magnetic fields".

206. Absence of Ferromagnetism for One-Dimensional Itinerant Electrons, in *Probabilistic Methods in Mathematical Physics*, Proceedings of the international workshop Siena, May 1991, F. Guerra, M. Loffredo and C. Marchioro eds., World Scientific pp. 290–294 (1992). A shorter version appears in *Rigorous Results in Quantum Dynamics*, J. Dittrich and P. Exner eds., World Scientific, pp. 243–245 (1991).

207. (with J.P. Solovej and J. Yngvason) Heavy Atoms in the Magnetic Field of a Neutron Star, Phys. Rev. Lett. **69**, 749–752 (1992).

†*208. (with J.P. Solovej) Atoms in the Magnetic Field of a Neutron Star, in *Differential Equations with Applications to Mathematical Physics*, W.F. Ames, J.V. Herod and E.M. Harrell II eds., Academic Press, pages 221–237 (1993). Also in *Spectral Theory and Scattering Theory and Applications*, K. Yajima, ed., Advanced Studies in Pure Math. **23**, 259–274, Math. Soc. of Japan, Kinokuniya (1994). This is a summary of nos. 215, 216. Earlier summaries also appear in: (a) *Méthodes Semi-Classiques, Colloque internatinal (Nantes 1991)*, Asterisque **210**, 237–246 (1991); (b) *Some New Trends on Fluid Dynamics and Theoretical Physics*, C.C. Lin and N. Hu eds., 149–157, Peking University Press (1993); (c) *Proceedings of the International Symposium on Advanced Topics of Quantum Physics, Shanxi*, J.Q. Lang, M.L. Wang, S.N. Qiao and D.C. Su eds., 5–13, Science Press, Beijing (1993).

209. (with M. Loss and R. McCann) Uniform Density Theorem for the Hubbard Model, J. Math. Phys. **34**, 891–898 (1993).

210. Remarks on the Skyrme Model, in *Proceedings of the Amer. Math. Soc. Symposia in Pure Math. 54*, part 2, 379–384 (1993). (Proceedings of summer research institute on differential geometry at UCLA, July 8–28, 1990.)
211. (with E. Carlen) Optimal Hypercontractivity for Fermi Fields and Related Non-Commutative Integration Inequalities, Commun. Math. Phys. **155**, 27–46 (1993).
212. (with E. Carlen) Optimal Two-Uniform Convexity and Fermion Hypercontractivity, in *Quantum and Non-Commutative Analysis*, Proceedings of June, 1992 Kyoto Conference, H. Araki et al. eds., Kluwer (1993), pp. 93–111. (Condensed version of no. 211.)
213. (with M. Loss) Fluxes, Laplacians and Kasteleyn's Theorem, Duke Math. Journal **71**, 337–363 (1993).
214. (with V. Bach, R. Lewis and H. Siedentop) On the Number of Bound States of a Bosonic N-Particle Coulomb System, Zeits. f. Math. **214**, 441–460 (1993).
215. (with J.P. Solovej and J. Yngvason) Asymptotics of Heavy Atoms in High Magnetic Fields: I. Lowest Landau Band Region, Commun. Pure Appl. Math. **47**, 513–591 (1994).
216. (with J.P. Solovej and J. Yngvason) Asymptotics of Heavy Atoms in High Magnetic Fields: II. Semiclassical Regions, Commun. Math. Phys. **161**, 77–124 (1994).
†*217. (with V. Bach, M. Loss and J.P. Solovej) There are No Unfilled Shells in Unrestricted Hartree–Fock Theory. Phys. Rev. Lett. **72**, 2981–2983 (1994).
218. (with K. Ball and E. Carlen) Sharp Uniform Convexity and Smoothness Inequalities for Trace Norms, Invent. Math. **115**, 463–482 (1994).
219. Coherent States as a Tool for Obtaining Rigorous Bounds, *Proceedings of the Symposium on Coherent States, past, present and future*, Oak Ridge, D.H. Feng, J. Klauder and M.R. Strayer eds., World Scientific (1994), pages 267–278.
220. The Hubbard model – Some Rigorous Results and Open Problems, in Proceedings of 1993 conference in honor of G.F. Dell'Antonio, *Advances in Dynamical Systems and Quantum Physics*, S. Albeverio et al. eds., pp. 173–193, World Scientific (1995). A revised version appears in Proceedings of 1993 NATO ASI *The Hubbard Model*, D. Baeriswyl et al. eds., pp. 1–19, Plenum Press (1995). A further revision appears in *Proceedings of the XIth International Congress of Mathematical Physics*, Paris, 1994, D. Iagolnitzer ed., pp. 392–412, International Press (1995).
221. (with V. Bach and J.P. Solovej) Generalized Hartree–Fock Theory of the Hubbard Model, J. Stat. Phys. **76**, 3–90 (1994).
222. The Flux Phase of the Half-Filled Band, Phys. Rev. Lett. **73**, 2158–2161 (1994).
223. (with M. Loss) Symmetry of the Ginzburg–Landau Minimizer in a Disc, Math. Res. Lett. **1**, 701–715 (1994).
224. (with J.P. Solovej and J. Yngvason) Quantum Dots, in *Proceedings of the Conference on Partial Differential Equations and Mathematical Physics*,

University of Alabama, Birmingham, 1994, I. Knowles, ed., International Press (1995), pages 157–172.

†*225. (with J.P. Solovej and J. Yngvason) Ground States of Large Quantum Dots in Magnetic Fields, Phys. Rev. B **51**, 10646–10665 (1995).

226. (with J. Freericks) The Ground State of a General Electron–Phonon Hamiltonian is a Spin Singlet, Phys. Rev. B **51**, 2812–2821 (1995).

227. (with B. Nachtergaele) The Stability of the Peierls Instability for Ring Shaped Molecules, Phys. Rev. B **51**, 4777–4791 (1995).

228. (with B. Nachtergaele) Dimerization in Ring-Shaped Molecules: The Stability of the Peierls Instability in *Proceedings of the XIth International Congress of Mathematical Physics*, Paris, 1994, D. Iagolnitzer ed., pp. 423–431, International Press (1995).

229. (with B. Nachtergaele) Bond Alternation in Ring-Shaped Molecules: The Stability of the Peierls Instability. In Proceedings of the conference *The Chemical Bond*, Copenhagen 1994, Int. J. Quant. Chem. Chem. **58**, 699–706 (1996).

230. Fluxes and Dimers in the Hubbard Model, in Proceedings of the International Congress of Mathematicians, Zürich, 1994, S.D. Chatterji ed., vol. 2, pp. 1279–1280, Birkhäuser (1995).

†*231. (with M. Loss and J. P. Solovej) Stability of Matter in Magnetic Fields, Phys. Rev. Lett. **75**, 985–989 (1995).

†*232. (with O. J. Heilmann) Electron Density near the Nucleus of a large Atom, Phys. Rev A **52**, 3628–3643 (1995). (Reprinted without appendices.)

†*233. (with A. Iantchenko and H. Siedentop) Proof of a Conjecture about Atomic and Molecular Cores Related to Scott's Correction, J. reine u. ang. Math. **472**, 177–195 (1996).

234. (with L. Thomas) Exact Ground State Energy of the Strong-Coupling Polaron, Commun. Math. Phys. **183**, 511–519 (1997). Errata **188**, 499–500 (1997).

235. (with L. Cafarelli and D. Jerison) On the Case of Equality in the Brunn–Minkowski Inequality for Capacity, Adv. in Math. **117**, 193–207 (1996).

†*236. (with M. Loss and H. Siedentop) Stability of Relativistic Matter via Thomas–Fermi Theory, Helv. Phys. Acta **69**, 974–984 (1996).

237. Some of the Early History of Exactly Soluble Models, in *Proceedings of the 1996 Northeastern University conference on Exactly Soluble Models,* Int. Jour. Mod. Phys. B **11**, 3–10 (1997).

†*238. (with H. Siedentop and J. P. Solovej) Stability and Instability of Relativistic Electrons in Magnetic Fields, J. Stat. Phys. **89**, 37–59 (1997).

239. (with H. Siedentop and J-P. Solovej) Stability of Relativistic Matter with Magnetic Fields, Phys. Rev. Lett. **79**, 1785–1788 (1997).

240. Stability of Matter in Magnetic Fields, in *Proceedings of the Conference on Unconventional Quantum Liquids, Evora, Portugal, 1996* Zeits. f. Phys. B **933**, 271–274 (1997).

241. Birmingham in the Good Old Days, in *Proceedings of the Conference on Unconventional Quantum Liquids, Evora, Portugal, 1996* Zeits. f. Phys. B **933**, 125–126 (1997).

242. (with M. Loss) book *Analysis*, American Mathematical Society (1997).
243. Doing Math with Fred, in *In Memoriam Frederick J. Almgren Jr., 1937–1997*, Experimental Math. **6**, 2–3 (1997).
†244. (with J.P. Solovej and J. Yngvason) Asymptotics of Natural and Artificial Atoms in Strong Magnetic Fields, in *The Stability of Matter: From Atoms to Stars, Selecta of E. H. Lieb*, W. Thirring ed., second edition, Springer Verlag, pp. 145–167 (1997). This is a summary of nos. 207, 208, 215, 216, 224, 225.
245. Stability and Instability of Relativistic Electrons in Classical Electromagnetic Fields, in Proceedings of Conference on Partial Differential Eqations and Mathematical Physics, Georgia Inst. of Tech., March, 1997, Amer. Math. Soc. Contemporary Math. series. E. Carlen, E. Harrell, M. Loss eds., **217**, 99–108 (1998),
†246. (with J. Yngvason) Ground State Energy of the Low Density Bose Gas, Phys. Rev. Lett. **80**, 2504–2507 (1998). arXiv math-ph/9712138, mp_arc 97-631.
247. (with J. Yngvason) A guide to Entropy and the Second Law of Thermodynamics, Notices of the Amer. Math. Soc. **45**, 571–581 (1998). arXiv math-ph/9805005, mp_arc 98-339. http://www.ams.org/notices/199805/lieb.pdf. See no. 266.
248. (with D. Hundertmark and L.E. Thomas) A Sharp Bound for an Eigenvalue Moment of the One-Dimensional Schroedinger Operator, Adv. Theor. Math. Phys. **2**, 719–731 (1998). arXiv math-ph/9806012, mp_arc 98-753.
249. (with E. Carlen) A Minkowski Type Trace Inequality and Strong Subadditivity of Quantum Entropy, in Amer. Math. Soc. Transl. (2), **189**, 59–69 (1999).
250. (with J. Yngvason) The Physics and Mathematics of the Second Law of Thermodynamics, Physics Reports **310**, 1–96 (1999). arXiv cond-mat/9708200, mp_arc 97-457.
251. Some Problems in Statistical Mechanics that I would like to see Solved, 1998 IUPAP Boltzmann prize lecture, Physica A **263**, 491–499 (1999).
252. (with P. Schupp) Ground State Properties of a Fully Frustrated Quantum Spin System, Phys. Rev. Lett. **83**, 5362–5365 (1999). arXiv math-ph/9908019, mp_arc 99-304.
253. (with P. Schupp) Singlets and Reflection Symmetric Spin Systems, Physica A **279**, 378–385 (2000). arXiv math-ph/9910037, mp_arc 99-404.
†254. (with R. Seiringer and J.Yngvason) Bosons in a Trap: A Rigorous Derivation of the Gross–Pitaevskii Energy Functional, Phys. Rev. A **61**, 043602-1 – 043602-13 (2000). arXiv math-ph/9908027, mp_arc 99-312.
255. (with J. Yngvason) The Ground State Energy of a Dilute Bose Gas, in *Differential Equations and Mathematical Physics, University of Alabama, Birmingham, 1999*, R. Weikard and G. Weinstein, eds., 271–282, Internat. Press (2000). arXiv math-ph/9910033, mp_arc 99-401.
†256. (with M. Loss) Self-Energy of Electrons in Non-perturbative QED, in *Differential Equations and Mathematical Physics, University of Alabama,*

Birmingham, 1999, R. Weikard and G. Weinstein, eds. 255–269, Amer. Math. Soc./Internat. Press (2000). arXiv math-ph/9908020, mp_arc 99-305.

257. (with R. Seiringer and J. Yngvason) The Ground State Energy and Density of Interacting Bosons in a Trap, in *Quantum Theory and Symmetries, Goslar, 1999*, H.-D. Doebner, V.K. Dobrev, J.-D. Hennig and W. Luecke, eds., pp. 101–110, World Scientific (2000). arXiv math-ph/9911026, mp_arc 99-439.
258. (with J. Yngvason) The Ground State Energy of a Dilute Two-dimensional Bose Gas, J. Stat. Phys. **103**, 509–526 (2001). arXiv math-ph/0002014, mp_arc 00-63.
259. (with J. Yngvason) A Fresh Look at Entropy and the Second Law of Thermodynamics, Physics Today, **53**, 32–37 (April 2000). arXiv math-ph/0003028, mp_arc 00-123. See also **53**, 11–14, 106 (October 2000).
260. Lieb–Thirring Inequalities, in *Encyclopaedia of Mathematics, Supplement vol. 2,* pp. 311–313, Kluwer (2000). arXiv math-ph/0003039, mp_arc 00-132.
261. Thomas–Fermi Theory, in *Encyclopaedia of Mathematics, Supplement vol. 2,* pp. 455–457, Kluwer (2000). arXiv math-ph/0003040, mp_arc 00-131.

†262. (with H. Siedentop) Renormalization of the Regularized Relativistic Electron–Positron Field, Commun. Math. Phys. **213**, 673–684 (2000). arXiv math-ph/0003001 mp_arc 00-98.

263. (with R. Seiringer and J. Yngvason) A Rigorous Derivation of the Gross–Pitaevskii Energy Functional for a Two-dimensional Bose Gas, Commun. Math. Phys. (in press). arXiv cond-mat/0005026, mp_arc 00-203.

†264. (with M. Griesemer and M. Loss) Ground States in Non-relativistic Quantum Electrodynamics, Invent. Math. (in press). arXiv math-ph/0007014, mp_arc 00-313.

†265. (with J.P. Solovej) Ground State Energy of the One-Component Charged Bose Gas, Commun. Math. Phys. **217**, 127–163 (2001). arXiv cond-mat/0007425, mp_arc 00-303.

266. (with J. Yngvason) The Mathematics of the Second Law of Thermodynamics, in "Visions in Mathematics, Towards 2000", A. Alon, J. Bourgain, A. Connes, M. Gromov and V. Milman, eds., GAFA 2000, no. 1, Birkhäuser, p. 334–358 (2000). See no. 247. mp_arc 00-332.

†267. The Bose Gas: A Subtle Many-Body Problem, in *Proceedings of the XIIIth International Congress on Mathematical Physics,* London, A. Fokas, A. Grigoryan, T. Kibble, B. Zegarlinski eds., Internat. Press (2001). arXiv math-ph/0009009, mp_arc 00-351.

268. (with J. Freericks and D. Ueltschi) Segregation in the Falicov-Kimball Model. arXiv math-ph/0107003, mp_arc 01-243.
269. (with G.K. Pedersen) Convex Multivariable Trace Functions. arXiv math.OA/0107062.

Druck: Strauss Offsetdruck, Mörlenbach
Verarbeitung: Schäffer, Grünstadt